D1195766

GREENE'S PROTECTIVE GROUPS IN ORGANIC SYNTHESIS

BICENTENNIAL
1807
WILEY
2007
BICENTENNIAL

The Wiley Bicentennial—Knowledge for Generations

Each generation has its unique needs and aspirations. When Charles Wiley first opened his small printing shop in lower Manhattan in 1807, it was a generation of boundless potential searching for an identity. And we were there, helping to define a new American literary tradition. Over half a century later, in the midst of the Second Industrial Revolution, it was a generation focused on building the future. Once again, we were there, supplying the critical scientific, technical, and engineering knowledge that helped frame the world. Throughout the 20th Century, and into the new millennium, nations began to reach out beyond their own borders and a new international community was born. Wiley was there, expanding its operations around the world to enable a global exchange of ideas, opinions, and know-how.

For 200 years, Wiley has been an integral part of each generation's journey, enabling the flow of information and understanding necessary to meet their needs and fulfill their aspirations. Today, bold new technologies are changing the way we live and learn. Wiley will be there, providing you the must-have knowledge you need to imagine new worlds, new possibilities, and new opportunities.

Generations come and go, but you can always count on Wiley to provide you the knowledge you need, when and where you need it!

WILLIAM J. PESCE
PRESIDENT AND CHIEF EXECUTIVE OFFICER

PETER BOOTH WILEY
CHAIRMAN OF THE BOARD

GREENE'S PROTECTIVE GROUPS IN ORGANIC SYNTHESIS

Fourth Edition

PETER G. M. WUTS
Kalexsyn, Inc

and

THEODORA W. GREENE
The Rowland Institute for Science

WILEY-INTERSCIENCE
A John Wiley & Sons, Inc., Publication

Published by John Wiley & Sons, Inc., Hoboken, New Jersey
Published simultaneously in Canada

Limit of Liability/Disclaimer of Warranty: While the publisher and author have used their best efforts
in preparing this book, they make no representations or warranties with respect to the accuracy
or completeness of the contents of this book and specifically disclaim any implied warranties of
merchantability or fitness for a particular purpose. No warranty may be created or extended by sales
representatives or written sales materials. The advice and strategies contained herein may not be
suitable for your situation. You should consult with a professional where appropriate. Neither the
publisher nor author shall be liable for any loss of profit or any other commercial damages, including
but not limited to special, incidental, consequential, or other damages.

For general information on our other products and services or for technical support, please contact our
Customer Care Department within the United States at (800) 762-2974, outside the United States at
(317) 572-3993 or fax (317) 572-4002.

Wiley also publishes its books in a variety of electronic formats. Some content that appears in print
may not be available in electronic formats. For more information about Wiley products, visit our
web site at www.wiley.com.

Library of Congress Cataloging-in-Publication Data:

Wuts, Peter G. M.
  Greene's protective groups in organic synthesis. – 4th ed. / Peter G. M. Wuts,
Theodora W. Greene
       p. cm.
  Greene's name appears first on the earlier edition.
  Includes index.
  ISBN-13: 978-0-471-69754-1 (cloth)
  ISBN-10: 0-471-69754-0 (cloth)
1. Organic compounds–Synthesis.   2. Protective groups (Chemistry) I.
Greene, Theodora W., 1931-Protective groups in organic synthesis.   II. Title.

  QD262.G665 2006
  547.2–dc22   2006016601

Printed in the United States of America

10 9 8 7 6 5

CONTENTS

PREFACE TO THE FOURTH EDITION

After completing the mammoth third edition, I never imagined that a fourth edition would eventuate because of the sheer volume of literature that must be examined to cover the subject comprehensively. Nonetheless, I took on the task with the encouragement and help of my wife, Lizzie, who agreed to assist me with this one, since Theo was not able to. As with the last edition, the searches were primarily done by hand because databases such as Scifinder fail to be selective and have such a prodigious output that no one can be expected to filter all that material in a reasonable amount of time. Nevertheless, Scifinder was used to locate material in journals which were not readily accessible. In recent years, in both corporate and academic America, there has also been a trend to do away with physical libraries, which makes doing a literature search extremely difficult, especially if you like reading the literature at home in a comfortable chair. Reading journals on a computer screen may be easy for Spock, but I find it difficult and stressful. With limited access to hard copies of some of the literature, I may have missed some things. For this I apologize and will not be offended if the author sends me the material for inclusion in a possible future edition. The literature search is complete through the end of 2005.

With that said, the fourth edition contains over 3100 new references compared to the 2349 new citations in the third edition. In keeping with the tradition of the past, I tried to include material covering new methods for existing protective groups along with new groups that have been developed. When the authors disclosed the information, I also provided the rationale for the choice of a given protective group. In that synthetic chemistry is still not sufficiently developed to do away with protective groups altogether, I have included many examples that highlight selective protection and deprotection, especially when the selectivity might not be totally obvious or expected. Issues of unexpected reactivity are also included, since these cases should help in choosing a group during the development of a synthetic plan. On the whole, this is a book of options for the synthetic chemist, since no one method is suitable for all occasions. Also, many of the published methods have not been tested in complex situations; thus it is impossible to determine which method of a particular set might be the best, and, as such, no attempt was made to try and order the various methods that appear in a section. The issue of functional group compatibility is often not addressed in papers describing new methods, and this further complicates the evaluation process. Comparative studies for either protection or deprotection are rarely done and as a result, trial and error and chemical intuition must be used to define the most suitable method in a given situation.

All sections of the book have seen some expansion, especially the chapters on alcohol and amine protection. I had considered adding a section that covered areas such as diene protection as metal complexes and Diels–Alder adducts, but the use of these is rather limited. The Reactivity Charts of Chapter 10 have not been altered, but a new chart covering selectivity in silyl group deprotection has been added. The overall format of the book has been retained and in some of the larger sections, similar methods have been grouped together. A new area has emerged since the last edition, and this is the use of fluorous protective groups. These have been included and placed in the appropriate sections rather than having collected them together.

The completion of this project was aided by a number of people. First of all this work would not have been started without the encouragement and dedication of my wife, Lizzie, who looked up and downloaded many of the references and then typed every new reference into an Endnote™ database. She double-checked the entire set in order to prevent errors. She also read through the entire manuscript to check it for punctuation, grammar, and consistency. She has a degree in Near Eastern Medieval History, thus I take full responsibility for any chemical errors. I must also thank her for not complaining about becoming a book widow while I spent countless hours on this project over a period of ~3 years. A special note of thanks must be extended to Peter Green, the Pfizer Michigan site head, who approved giving Lizzie access to the company library system even though she was not an employee. I would also like to thank Jake Szmuszkovicz, Raymond Conrow, and Martin Lang for providing me with references to be included in the fourth edition, and finally I wish to thank Joseph Muchowski for bringing an error in the third edition, now corrected, to my attention.

PETER G. M. WUTS

July 2006

PREFACE TO THE THIRD EDITION

Organic synthesis has not yet matured to the point where protective groups are not needed for the synthesis of natural and unnatural products; thus, the development of new methods for functional group protection and deprotection continues. The new methods added to this edition come from both electronic searches and a manual examination of all the primary journals through the end of 1997. We have found that electronic searches of *Chemical Abstracts* fail to find many new methods that are developed during the course of a synthesis, and issues of selectivity are often not addressed. As with the second edition, we have attempted to highlight unusual and potentially useful examples of selectivity for both protection and deprotection. In some areas the methods listed may seem rather redundant, such as the numerous methods for THP protection and deprotection, but we have included them in an effort to be exhaustive in coverage. For comparison, the first edition of this book contains about 1500 references and 500 protective groups, the second edition introduces an additional 1500 references and 206 new protective groups, and the third edition adds 2349 new citations and 348 new protective groups.

Two new sections on the protection of phosphates and the alkyne-CH are included. All other sections of the book have been expanded, some more than others. The section on the protection of alcohols has increased substantially, reflecting the trend of the nineties to synthesize acetate- and propionate-derived natural products. An effort was made to include many more enzymatic methods of protection and deprotection. Most of these are associated with the protection of alcohols as esters and the protection of carboxylic acids. Here we have not attempted to be exhaustive, but hopefully, a sufficient number of cases are provided that illustrate the true power of this technology, so that the reader will examine some of the excellent monographs and review articles cited in the references. The Reactivity Charts in Chapter 10 are identical to those in the first edition. The chart number appears beside the name of each protective group when it is first introduced. No attempt was made to update these Charts, not only because of the sheer magnitude of the task, but because it is nearly impossible in a two-dimensional table to address adequately the effect that electronic and steric controlling elements have on a particular instance of protection or deprotection. The concept of fuzzy sets as outlined by Lofti Zadeh would be ideally suited for such a task.

The completion of this project was aided by the contributions of a number of people. I am grateful to Rein Virkhaus and Gary Callen, who for many years forwarded me references when they found them, to Jed Fisher for the information he contributed on phosphate protection, and to Todd Nelson for providing me a preprint of his excellent review article on the deprotection of silyl ethers. I heartily thank Theo Greene for checking and rechecking the manuscript—all 15 cm of it—for spelling and consistency and for the arduous task of checking all the references for accuracy. I thank Fred Greene for reading the manuscript, for his contribution to Chapter 1 on the use of protective groups in the synthesis of himastatin, and for his contribution to the introduction to Chapter 9, on phosphates. I thank my wife, Lizzie, for encouraging me to undertake the third edition, for the hours she spent in the library looking up and photocopying hundreds of references, and for her understanding while I sat in front of the computer night after night and numerous weekends over a two-year period. She is the greatest!

PETER G. M. WUTS

Kalamazoo, Michigan
June 1998

PREFACE TO THE SECOND EDITION

Since publication of the first edition of this book in 1981, many new protective groups and many new methods of introduction or removal of known protective groups have been developed: 206 new groups and approximately 1500 new references have been added. Most of the information from the first edition has been retained. To conserve space, generic structures used to describe Formation/Cleavage reactions have been replaced by a single line of conditions, sometimes with explanatory comments, especially about selectivity. Some of the new information has been obtained from on-line searches of *Chemical Abstracts*, which have limitations. For example, *Chemical Abstracts* indexes a review article about protective groups only if that word appears in the title of the article. References are complete through 1989. Some references, from more widely circulating journals, are included for 1990.

Two new sections on the protection for indoles, imidazoles, and pyrroles and protection for the amide –NH are included. They are separated from the regular amines because their chemical properties are sufficiently different to affect the chemistry of protection and deprotection. The Reactivity Charts in Chapter 8 are identical to those in the first edition. The chart number appears beside the name of each protective group when it is first discussed.

A number of people must be thanked for their contributions and help in completing this project. I am grateful to Gordon Bundy, who loaned me his card file, which provided many references that the computer failed to find, and to Bob Williams, Spencer Knapp, and Tohru Fukuyama for many references on amine and amide protection. I thank Theo Greene who checked and rechecked the manuscript for spelling and consistency and for the herculean task of checking all the references to make sure that my 3's and 8's and 7's and 9's were not interchanged—all done without a single complaint. I thank Fred Greene who read the manuscript and provided valuable suggestions for its improvement. My wife Lizzie was a major contributor to getting this project finished, by looking up and photocopying references, by turning on the computer in an evening ritual, and by typing many sections of the original book, which made the changes and additions much easier. Without her understanding and encouragement, the volume probably would never have been completed.

PETER G. M. WUTS

Kalamazoo, Michigan
May 1990

PREFACE TO THE FIRST EDITION

The selection of a protective group is an important step in synthetic methodology, and reports of new protective groups appear regularly. This book presents information on the synthetically useful protective groups (~500) for five major functional groups: $-OH$, $-NH$, $-SH$, $-COOH$, and $>C=O$. References through 1979, the best method(s) of formation and cleavage, and some information on the scope and limitations of each protective group are given. The protective groups that are used most frequently and that should be considered first are listed in Reactivity Charts, which give an indication of the reactivity of a protected functionality to 108 prototype reagents.

The first chapter discusses some aspects of protective group chemistry: the properties of a protective group, the development of new protective groups, how to select a protective group from those described in this book, and an illustrative example of the use of protective groups in a synthesis of brefeldin. The book is organized by functional group to be protected. At the beginning of each chapter are listed the possible protective groups. Within each chapter protective groups are arranged in order of increasing complexity of structure (e.g., methyl, ethyl, t-butyl, ..., benzyl). The most efficient methods of formation or cleavage are described first. Emphasis has been placed on providing recent references, since the original method may have been improved. Consequently, the original reference may not be cited; my apologies to those whose contributions are not acknowledged. Chapter 8 explains the relationship between reactivities, reagents, and the Reactivity Charts that have been prepared for each class of protective groups.

This work has been carried out in association with Professor Elias J. Corey, who suggested the study of protective groups for use in computer-assisted synthetic analysis. I appreciate his continued help and encouragement. I am grateful to Dr. J. F. W.

McOmie (Ed., *Protective Groups in Organic Chemistry,* Plenum Press, New York and London, 1973) for his interest in the project and for several exchanges of correspondence, and to Mrs. Mary Fieser, Professor Frederick D. Greene, and Professor James A. Moore for reading the manuscript. Special thanks are also due to Halina and Piotr Starewicz for drawing the structures, and to Kim Chen, Ruth Emery, Janice Smith, and Ann Wicker for typing the manuscript.

THEODORA W. GREENE

Harvard University
September 1980

ABBREVIATIONS

PROTECTIVE GROUPS

In some cases, several abbreviations are used for the same protective group. We have listed the abbreviations as used by an author in his original paper, including capital and lowercase letters. Occasionally, the same abbreviation has been used for two different protective groups. This information is also included.

ABO	2,7,8-trioxabicyclo[3.2.1]octyl
Ac	acetyl
ACBZ	4-azidobenzyloxycarbonyl
ACE	*O*-bis(2-Acetoxyethoxy)methyl
AcHmb	2-acetoxy-4-methoxybenzyl
Acm	acetamidomethyl
Ad	1-adamantyl
ADMB	4-acetoxy-2,2-dimethylbutanoate
Adoc	1-adamantyloxycarbonyl
Adpoc	1-(1-adamantyl)-1-methylethoxycarbonyl
Alloc or AOC	allyloxycarbonyl
AOC or Alloc	allyloxycarbonyl
Allocam	allyloxycarbonylaminomethyl
Als	allylsulfonyl
AMB	2-(acetoxymethyl)benzoyl
AMPA	(2-azidomethyl)phenylacetate
AN	4-methoxyphenyl or anisyl
Ans	anisylsulfonyl

Anpe	2-(4-acetyl-2-nitrophenyl)ethyl
p-AOM	p-anisyloxymethyl or (4-methoxyphenoxy)methyl
APAC	2-allyloxyphenylacetate
Aqmoc	anthraquinone-2-ylmethoxycarbonyl
Azb	p-azidobenzyl
Azm	azidomethyl
AZMB	2-(azidomethyl)benzoate
Bam	benzamidomethyl
BBA	butane-2,3-bisacetal
Bbc	but-2-ynylbisoxycaronyl
BDIPS	biphenyldiisopropylsilyl
BDMS	biphenyldimethylsilyl
	benzyldimethylsilyl
Bdt	1,3-benzodithiolan-2-yl
Betsyl or Bts	benzothiazole-2-sulfonyl
Bhcmoc	6-bromo-7-hydroxycoumarin-4-ylmethoxycarbonyl
BHQ	8-bromo-7-hydroxyquinoline-2-ylmethyl
BHT	2,6-di-t-butyl-4-methylphenyl
Bic	5-benzisoxazolylmethoxycarbonyl
Bim	5-benzisoazolylmethylene
Bimoc	benz[f]inden-3-ylmethoxycarbonyl
BIPSOP	N-2,5-bis(triisopropylsiloxy)pyrrolyl
BMB	o-(benzoyloxymethyl)benzoyl
Bmpc	2,4-dimethylthiophenoxycarbonyl
Bmpm	bis(4-methoxyphenyl)-1′-pyrenylmethyl
Bn	benzyl
Bnf	fluorousbenzyl
Bnpeoc	2,2-bis(4′-nitrophenyl)ethoxycarbonyl
Bns	benzylsulfonate
BOB	benzyloxybutyrate
BOC	t-butoxycarbonyl
Bocdene	2-(t-butylcarbonyl)ethylidene
BOM	benzyloxymethyl
Bpoc	1-methyl-1-(4-biphenyl)ethoxycarbonyl
BSB	benzoSTABASE
Bsmoc	1,1-dioxobenzo[b]thiophene-2-ylmethoxycarbonyl
BTM	t-butylthiomethyl
Bts or Betsyl	benzothiazole-2-sulfonyl
B^{t}SE	2-t-butylsulfonylethyl
Bts-Fmoc	2,7-bis(trimethylsilyl)fluorenylmethoxycarbonyl
Bum	t-butoxymethyl
Bus	t-butylsulfonyl
t-Bumeoc	1-(3,5-di-t-butylphenyl)-1-methylethoxycarbonyl
Bz	benzoyl
CAEB	2-[(2-chloroacetoxy)ethyl]benzoyl

Cam	carboxamidomethyl
CAMB	2-(chloroacetoxymethyl)benzoyl
Cbz or Z	benzyloxycarbonyl
CEM	2-cyanoethoxymethyl
CDA	cyclohexane-1,2-diacetal
CDM	2-cyano-1,1-dimethylethyl
CE or Cne	2-cyanoethyl
Cee	1-(2-chloroethoxy)ethyl
CEE	1-(2-cyanoethoxy)ethyl
Ceof	cyclic ethyl orthoformate
cHex	cyclohexyl
Chx	cyclohexyl
Cin	cinnamyl
ClAzab	4-azido-3-chlorobenzyl
Climoc	2-chloro-3-indenylmethoxycarbonyl
Cms	carboxymethylsulfenyl
CNAP	2-naphthylmethoxycarbonyl
Cne or CE	2-cyanoethyl
Coc	cinnamyloxycarbonyl
CPC	*p*-chlorophenylcarbonyl
CPDMS	(3-cyanopropyl)dimethylsilyl
Cpeoc	2-(cyano-1-phenyl)ethoxycarbonyl
Cpep	1-(4-chlorophenyl)-4-methoxypiperidin-4-yl
CPTr	4,4′,4″-tris(4,5-dichlorophthalimido)-triphenylmethyl
CTFB	4-trifluoromethylbenzyloxycarbonyl
CTMP	1-[(2-chloro-4-methyl)phenyl]-4-methoxypiperidin-4-yl
Cyclo-SEM	5-trimethylsilyl-1,3-dioxane
Cys	cysteine
DAM	di-*p*-anisylmethyl or bis(4-methoxyphenyl)methyl
DATE	1,1-di-*p*-anisyl-2,2,2-trichloroethyl
DB-*t*-BOC	1,1-dimethyl-2,2-dibromoethoxycarbonyl
DBD-Tmoc	2,7-di-*t*-butyl[9-(10,10-dioxo-10,10,10,10-tetra=hydrothioxanthyl)]methoxycarbonyl
DBS	dibenzosuberyl
DCP	dichlorophthalimide
Dcpm	dicyclopropylmethyl
Ddm or Dmbh	bis(4-methoxyphenyl)methyl
Dde	2-(4,4-dimethyl-2,6-dioxocyclohexylidene)ethyl
Ddz	1-methyl-1-(3,5-dimethoxyphenyl)ethoxycarbonyl
DEM	diethoxymethyl
DEIPS	diethylisopropylsilyl
Desyl	2-oxo-1,2-diphenylethyl
Dim	1,3-dithianyl-2-methyl

Dmab	4-{*N*-[1-(4,4-dimethyl-2,6-dioxocyclohexylidene)-3-methylbutyl]amino}benzyl
DMB	"3',5'-dimethoxybenzoin"
Dmb	2,4-dimethoxybenzyl
DMBM	[(3,4-dimethoxybenzyl)oxy]methyl
DMIPS	dimethylisopropylsilyl
DMN	2,3-dimethylmaleimide
Dmoc	dithianylmethoxycarbonyl
Dmp	2,4-dimethyl-3-pentyl
Dmp	dimethylphosphinyl
DMP	dimethoxyphenyl
	dimethylphenacyl
DMPM	3,4-dimethoxybenzyl
DMTC	dimethylthiocarbamate
DMT or DMTr	di(*p*-methoxyphenyl)phenylmethyl or dimethoxytrityl
DMTr or DMT	di(*p*-methoxyphenyl)phenylmethyl or dimethoxytrityl
DNAP	2-(dimethylamino)-5-nitrophenyl
DNB	*p,p'*-dinitrobenzhydryl
DNMBS	4-(4',8'-dimethoxynaphthylmethyl)benzenesulfonyl
DNP	2,4-dinitrophenyl
Dnpe	2-(2,4-dinitrophenyl)ethyl
Dnpeoc	2-(2,4-dinitrophenyl)ethoxycarbonyl
DNs	2,4-dinitrobenzenesulfonyl
DNse	2-(2,4-dinitrophenylsulfonyl)ethoxycarbonyl
Dnseoc	2-dansylethoxycarbonyl
Dobz	*p*-(dihydroxyboryl)benzyloxycarbonyl
Doc	2,4-dimethylpent-3-yloxycarbonyl
Dod	bis(4-methoxyphenyl)methyl
DOPS	dimethyl[1,1-dimethyl-3-(tetrahydro-2*H*-pyran-2-yloxy)propyl]silyl
DPA	diphenylacetyl
DPIPS	diphenylisopropylsilyl
DPM or Dpm	diphenylmethyl
DPMS	diphenylmethylsilyl
Dpp	diphenylphosphinyl
Dppe	2-(diphenylphosphino)ethyl
Dppm	(diphenyl-4-pyridyl)methyl
DPSE	2-(methyldiphenylsilyl)ethyl
DPSide	diphenylsilyldiethylene
Dpt	diphenylphosphinothioyl
DPTBOS	*t*-Butoxydiphenylsilyl
DPTBS	diphenyl-*t*-butoxysilyl or diphenyl-*t*-butylsilyl
Dtb-Fmoc	2,6-di-*t*-butyl-9-fluorenylmethoxycarbonyl
DTBMS	di-*t*-butylmethylsilyl

DTBS	di-*t*-butylsilylene
DTE	2-(hydroxyethyl)dithioethyl or "dithiodiethanol"
Dts	dithiasuccinimidyl
E-DMT	1,2-ethylene-3,3-bis(4′4″-dimethoxytrityl)
EE	1-ethoxyethyl
EOM	ethoxymethyl
^FCbz	fluorous benzyloxycarbonyl
Fcm	ferrocenylmethyl
Flu	fluorenyl
Fm	9-fluorenylmethyl
Fmoc	9-fluorenylmethoxycarbonyl
Fpmp	1-(2-fluorophenyl)-4-methoxypiperidiny-4-yl
GUM	guaiacolmethyl
HAPE	1-[2-(2-hydroxyalkyl)phenyl]ethanone
HBn	2-hydroxybenzyl
Hdoc	hexadienyloxycarbonyl
HFB	hexafluoro-2-butyl
HIP	1,1,1,3,3,3-hexafluoro-2-phenylisopropyl
Hoc	cyclohexyloxycarbonyl
HSDIS	(hydroxystyryl)diisopropylsilyl
HSDMS	(hydroxystyryl)dimethylsilyl
hZ or homo Z	homobenzyloxycarbonyl
IDTr	3-(imidazol-1-ylmethyl)-4′,4″-dimethoxytriphenylmethyl
IETr	4,4′-dimethoxy-3″-[*N*-(imidazolylethyl)carbamoyl]trityl
iMds	2,6-dimethoxy-4-methylbenzenesulfonyl
Ipaoc	1-isopropylallyloxycarbonyl
Ipc	isopinocampheyl
IPDMS	isopropyldimethylsilyl
Lev	levulinoyl
LevS	4,4-(ethylenedithio)pentanoyl
LevS	levulinoyldithioacetal ester
LMM*o*(*p*)NBz	6-(levulinyloxymethyl)-3-methoxy-2-nitrobenzoate
MAB	2-{{[(4-methoxytrityl)thio]methylamino}methyl}benzoate
MAQ	2-(9,10-anthraquinonyl)methyl or 2-methyleneanthraquinone
MBE	1-methyl-1-benzyloxyethyl
Mbh	bis(4-methylphenyl)methyl
MBF	2,3,3a,4,5,6,7,7a-octahydro-7,8,8-trimethyl-4,7-methanobenzofuran-2-yl
MBS or Mbs	*p*-methoxybenzenesulfonyl
MCPM	1-methyl-1′-cyclopropylmethyl
Mds	2,6-dimethyl-4-methoxybenzenesulfonyl

MDPS	methylene-bis-(diisopropylsilanoxanylidene
Me	methyl
ME	methoxyethyl
MEC	α-methylcinnamyl
Mee	methoxyethoxyethyl
MeOAc	methoxyacetate
MEM	2-methoxyethoxymethyl
Menpoc	α-methylnitropiperonyloxycarbonyl
MeOZ or Moz	p-methoxybenzyloxycarbonyl
Mes	mesityl or 2,4,6-trimethylphenyl
MIP	methoxyisopropyl or 1-methyl-1-methoxyethyl
MM	menthoxymethyl
MMT or MMTr	p-methoxyphenyldiphenylmethyl
MMTr or MMT	p-methoxyphenyldiphenylmethyl
MMPPOC	2-(3,4-methylenedioxy-6-nitrophenypropyloxycarbonyl
MOB	2-{[(4-methoxytritylthio)oxy]methyl}benzoate
Mocdene	2-(methoxycarbonyl)ethylidene
MoEt	2-N-(morpholino)ethyl
MOM	methoxymethyl
MOMO	methoxymethoxy
Moz or MeOZ	p-methoxybenzyloxycarbonyl
MP	p-methoxyphenyl
Mpe	3-methyl-3-pentyl
MPM or PMB	p-methoxyphenylmethyl or p-methoxybenzyl
Mps	p-methoxyphenylsulfonyl
Mpt	dimethylphosphinothioyl
Ms	methanesulfonyl or mesyl
MSE	2-(methylsulfonyl)ethyl
Msib	4-(methylsulfinyl)benzyl
Mspoc	2-methylsulfonyl-3-phenyl-1-prop-2-enyloxy
Msz	4-methylsulfinylbenzyloxycarbonyl
MTAD	4-methyl-1,2,4-triazoline-3,5-dione
Mtb	2,4,6-trimethoxybenzenesulfonyl
Mte	2,3,5,6-tetramethyl-4-methoxybenzenesulfonyl
MTHP	4-methoxytetrahydropyranyl
MTM	methylthiomethyl
MTMB	4-(methylthiomethoxy)butyryl
MTMECO	2-(methylthiomethoxy)ethoxycarbonyl
MTMT	2-(methylthiomethoxymethyl)benzoyl
Mtpc	4-(methylthio)phenoxycarbonyl
Mtr	2,3,6-trimethyl-4-methoxybenzenesulfonyl
Mts	2,4,6-trimethylbenzenesulfonyl or mesitylenesulfonyl
Mtt	4-methoxytrityl or 4-methyltrityl
Nap	2-napthylmethyl

NBOM	nitrobenzyloxymethyl
NBM	nitrobenzyloxymethyl
NDMS	2-norbornyldiemethylsilyl
Ne	2-nitroethyl
Noc	4-nitrocinnamyloxycarbonyl
Nosyl or Ns	2- or 4-nitrobenzenesulfonyl
Npe or npe	2-(nitrophenyl)ethyl
Npeoc	2-(4-nitrophenyl)ethoxycarbonyl
Npeom	[1-(2-nitrophenyl)ethoxy]methyl
Npes	2-(4-nitrophenyl)ethylsulfonyl
NPPOC	2-(2-nitrophenyl)propyloxycarbonyl
NPS or Nps	2-nitrophenylsulfenyl
NpSSPeoc	2-[(2-nitrophenyl)dithio]-1-phenylethoxycarbonyl
Npys	3-nitro-2-pyridinesulfenyl
Ns or Nosyl	2- or 4-nitrobenzenesulfonyl
Nse	2-(4-nitrophenylsulfonyl)ethoxycarbonyl
NVOC or Nvoc	3,4-dimethoxy-6-nitrobenzyloxycarbonyl or 6-nitroveratryloxycarbonyl
OBO	2,6,7-trioxabicyclo[2.2.2]octyl
O-DMT	3,3′-oxybis(dimethoxytrityl)
ONB	*o*-nitrobenzyl
PAB	*p*-acylaminobenzyl
PAB	acetoxybenzyl
PAC_H	2-[2-(benzyloxy)ethyl]benzoyl
PAC_M	2-[2-(4-methoxybenzyloxy)ethyl]benzoyl
Paloc	3-(3-pyridyl)allyloxycarbonyl or 3-(3-pyridyl)prop-2-enyloxycarbonyl
Pbf	2,2,4,6,7-pentamethyldihydrobenzofuran-5-sulfonyl
PeNB	pentadienylnitrobenzyl
PeNP	pentadienylnitropiperonyl
Peoc	2-phosphonioethoxycarbonyl
Peoc	2-(triphenylphosphonio)ethoxycarbonyl
Pet	2-(2′-pyridyl)ethyl
Pf	9-phenylfluorenyl
Pfp	pentafluoropenyl
Phamc	phenylacetamidomethyl
PhAc	4-phenylacetoxybenzyloxycarbonyl
Phenoc	4-methoxyphenacyloxycarbonyl
Pic	picolinate
Pim	phthalimidomethyl
Pixyl or Px	9-(9-phenyl)xanthenyl
PMB or MPM	*p*-methoxybenzyl or *p*-methoxyphenylmethyl
PMBM	*p*-methoxybenzyloxymethyl
Pmc	2,2,5,7,8-pentamethylchroman-6-sulfonyl
Pme	pentamethylbenzenesulfonyl

PMP	*p*-methoxyphenyl
PMS	*p*-methylbenzylsulfonyl
Pms	2-[phenyl(methyl)sulfonio]ethoxycarbonyl
PNB	*p*-nitrobenzyl or *p*-nitrobenzoate
*p*NBZ	*p*-nitrobenzoate
PNP	*p*-nitrophenyl
PNPE	2-(4-nitrophenyl)ethyl
PNZ	*p*-nitrobenzylcarbonyl
POC	propargyloxycarbonyl
POM	4-pentenyloxymethyl
POM	pivaloyloxymethyl
POM	[(*p*-phenylphenyl)oxy]methyl
POMB	2-(prenyloxy)methylbenzoate
Ppoc	2-triphenylphosphonioisopropoxycarbonyl
Pp	2-phenyl-2-propyl
Ppt	diphenylthiophosphinyl
Pre	prenyl
Preoc	prenyloxycarbonyl
Proc or Poc	propargyloxycarbonyl
PSB	*p*-siletanylbenzyl
PSE	2-(phenylsulfonyl)ethyl
Psoc	(2-phenyl-2-trimethylsilyl)ethoxycarbonyl
Psec	2-(phenylsulfonyl)ethoxycarbonyl
PTE	2-(4-nitrophenyl)thioethyl
PTM	phenylthiomethyl
PTMSE	(2-phenyl-2-trimethylsilyl)ethyl
Pv	pivaloyl
Px or pixyl	9-(9-phenyl)xanthenyl
Pyet	1-(α-pyridyl)ethyl
Pyoc	2-(2′- or 4′-pyridyl)ethoxycarbonyl
Qn	2-quinolinylmethyl
Qm	2-quinolinylmethyl
QUI	4-quinolinylmethyl
SATE	*S*-acetylthioethyl
Scm	*S*-carboxymethylsulfenyl
SEE	1-[2-(trimethylsilyl)ethoxy]ethyl
SEM	2-(trimethylsilyl)ethoxymethyl
SES	2-(trimethylsilyl)ethanesulfonyl
SIBA	1,1,4,4-tetraphenyl-1,4-disilanylidene
Sisyl	tris(trimethylsilyl)silyl
SMOM	(phenyldimethylsilyl)methoxymethyl
Snm	*S*-(*N*′-methyl-*N*′-phenylcarbamoyl)sulfenyl
SOB	4-trialkylsilyloxybutyrate
STABASE	1,1,4,4-tetramethyldisilylazacyclopentane
TAB	2-{[[(methyl(tritylthio)amino]methyl}benzoate

Tacm	trimethylacetamidomethyl
TBDMS or TBS	*t*-butyldimethylsilyl
TBDPS	*t*-butyldiphenylsilyl
Tbf-DMTr	4-(17-tetrabenzo[*a*,*c*,*g*,*i*]fluorenylmethyl-4′,4″-dimethoxytrityl
Tbfmoc	17-tetrabenzo[*a*,*c*,*g*,*i*]fluorenylmethoxycarbonyl
TBDPSE	*t*-butyldiphenylsilylethyl
TBDS	tetra-*t*-butoxydisiloxane-1,3-diylidene
TBMPS	*t*-butylmethoxyphenylsilyl
TBS or TBDMS	*t*-butyldimethylsilyl
TBTr	4,4′,4″-tris(benzyloxy)triphenylmethyl
TCB	2,2,2-trichloro-1,1-dimethylethyl
TcBOC	1,1-dimethyl-2,2,2-trichloroethoxycarbonyl
TCP	*N*-tetrachlorophthalimido
Tcroc	2-(trifluoromethyl)-6-chromonylmethyleneoxycarbonyl
Tcrom	2-(trifluoromethyl)-6-chromonylmethylene
TDE	(2,2,2-trifluoro-1,1-diphenyl)ethyl
TDG	thiodiglycoloyl
TDS	thexyldimethylsilyl or tris(2,6-diphenylbenzyl)silyl
Teoc	2-(trimethylsilyl)ethoxycarbonyl
TES	triethylsilyl
Tf	trifluoromethanesulfonyl
TFA	trifluoroacetyl
Tfav	4,4,4-trifluoro-3-oxo-1-butenyl
Thexyl	2,3-dimethyl-2-butyl
THF	tetrahydrofuranyl
THP	tetrahydropyranyl
TIBS	triisobutylsilyl
TIPDS	1,3-(1,1,3,3-tetraisopropyldisiloxanylidene)
TIPS	triisopropylsilyl
TIX	trimethylsilylxylyl
TLTr	4,4′,4″-tris(levulinoyloxy)triphenylmethyl
Tmb	2,4,6-trimethylbenzyl
Tmob	trimethoxybenzyl
TMPM	trimethoxyphenylmethyl
TMS	trimethylsilyl
Tms	(2-methyl-2-trimethylsilyl)ethyl
TMSE or TSE	2-(trimethylsilyl)ethyl
TMSEC	2-(trimethylsilyl)ethoxycarbonyl
TMSP	2-trimethylsilylprop-2-enyl
TMTr	tris(*p*-methoxyphenyl)methyl
TOB	2-{[(tritylthio)oxy]methyl}benzoate
Tos or Ts	*p*-toluenesulfonyl
Tom	triisopropylsilyloxymethyl

TPS	triphenylsilyl
TPTE	2-(4-triphenylmethylthio)ethyl
Tr	triphenylmethyl or trityl
TrtF$_7$	2,3,4,4′,4″,5,6-heptafluorotriphenylmethyl
Tritylone	9-(9-phenyl-10-oxo)anthryl
Troc	2,2,2-trichloroethoxycarbonyl
Ts or Tos	p-toluenesulfonyl
Tsc	2-(4-trifluoromethylphenylsulfonyl)ethoxycarbonyl
TSE or TMSE	2-(trimethylsilyl)ethyl
Tse	2-(p-toluenesulfonyl)ethyl
Tsoc	triisopropylsiloxycarbonyl
Tsv	p-toluenesulfonylvinyl
Voc	vinyloxycarbonyl
Xan	xanthenyl
Z or Cbz	benzyloxycarbonyl

REAGENTS

9-BBN	9-borabicyclo[3.3.1]nonane
bipy	2,2′-bipyridine
BOP Reagent	benzotriazol-1-yloxytris(dimethylamino) phosphonium hexafluorophosphate
BOP-Cl	bis(2-oxo-3-oxazolidinyl)phosphinic chloride
BroP	bromotris(dimethylamino)phosphonium hexafluorophosphate
Bt	benzotriazol-1-yl or 1-benzotriazolyl
BTEAC	benzyltriethylammonium chloride
CAL	Candida antarctica lipase
CAN	ceric ammonium nitrate
CMPI	2-chloro-1-methylpyridinium iodide
cod	cyclooctadiene
cot	cyclooctatetraene
CSA	camphorsulfonic acid
DABCO	1,4-diazabicyclo[2.2.2]octane
DBN	1,5-diazabicyclo[4.3.0]non-5-ene
DBAD	di-t-butyl azodicarboxylate
DBU	1,8-diazabicyclo[5.4.0]undec-7-ene
DCC	dicyclohexylcarbodiimide
DDQ	2,3-dichloro-5,6-dicyano-1,4-benzoquinone
DEAD	diethyl azodicarboxylate
DIAD	diisopropyl azodicarboxylate
DIBAL-H	diisobutylaluminum hydride
DIPEA	diisopropylethylamine
DMAC	N,N-dimethylacetamide

DMAP	4-*N*,*N*-dimethylaminopyridine
DMB	2,4-dimethoxybenzyl
DMDO	2,2-dimethyldioxirane
DME	1,2-dimethoxyethane
DMF	*N*,*N*-dimethylformamide
DMPU	1,3-dimethyl-3,4,5,6-tetrahydro-2(1*H*)-pyrimidinone
DMS	dimethyl sulfide
DMSO	dimethyl sulfoxide
dppb	1,4-bis(diphenylphosphino)butane
dppe	1,2-bis(diphenylphosphino)ethane
DTE	dithioerythritol
DTT	dithiothreitol
EDC or EDCI	1-ethyl-3-(3-dimethylaminopropyl)carbodiimide (or 1-[3-(dimethylamino)propyl]-3-ethylcarbodimide) hydrochloride
EDCI or EDC	1-ethyl-3-(3-(dimethylaminopropyl)carbodiimide
EDTA	ethylenediaminetetraacetic acid
HATU	*N*-[(dimethylamino)(3*H*-1,2,3-triazolo(4,5-*b*)pyridin-3-yloxy)methylene]-*N*-methylmethanaminium hexafluorophosphate, previously known as *O*-(7-azabenzotriazol-1-yl)-1,1,3,3-tetramethyluronium hexafluorophosphate.
HMDS	1,1,1,3,3,3-hexamethyldisilazane
HMPA	hexamethylphosphoramide
HMPT	hexamethylphosphorous triamide
HOAt	7-aza-1-hydroxybenzotriazole
HOBT	1-hydroxybenzotriazole
Im	imidazol-1-yl or 1-imidazolyl
IPA	isopropyl alcohol
IPCF (=IPCC)	isopropenyl chloroformate (isopropenyl chlorocarbonate)
KHMDS	potassium hexamethyldisilazide
LAH	lithium aluminum hydride
LDBB	lithium 4,4′-di-*t*-butylbiphenylide
MAD	methylaluminumbis(2,6-di-*t*-butyl-4-methylphenoxide
MCPBA	*m*-chloroperoxybenzoic acid
MoOPH	oxodiperoxymolybdenum(pyridine)hexamethylphosphoramide
ms	molecular sieves
MSA	methanesulfonic acid
MTB	methylthiobenzene
MTBE	*t*-butyl methyl ether
NBS	*N*-bromosuccinimide
Ni(acac)$_2$	nickel acetylacetonate

NMM	*N*-methylmorpholine
NMO	*N*-methylmorpholine *N*-oxide
NMP	*N*-methylpyrrolidinone
P	polymer support
Pc	phthalocyanine
PCC	pyridinium chlorochromate
PdCl$_2$(tpp)$_2$	dichlorobis[tris(2-methylphenyl)phosphine] palladium
Pd$_2$(dba)$_3$	tris(dibenzylideneacetone)dipalladium
PG	protective group
PhI(OH)OTs	[hydroxy(tosyloxy)iodo]benzene
PPL	porcine pancreatic lipase
PPTS	pyridinium *p*-toluenesulfonate
proton sponge	1,8-bis(dimethylamino)naphthalene
Pyr	pyridine
Rh$_2$(pfb)$_4$	rhodium perfluorobutyrate
ScmCl	methoxycarbonylsulfenyl chloride
SMEAH	sodium bis(2-methoxyethoxy)aluminum hydride
Su	succinimidyl
TAS-F	tris(dimethylamino)sulfonium difluorotrimethylsilicate
TBAF	tetrabutylammonium fluoride
TEA	triethylamine
TEBA or TEBAC	triethylbenzylammonium chloride
TEBAC or TEBA	triethylbenzylammonium chloride
TESH	triethylsilane
Tf	trifluoromethanesulfonyl
TFA	trifluoroacetic acid
TFAA	trifluoroacetic anhydride
TFMSA or TfOH	trifluoromethanesulfonic acid
TfOH or TFMSA	trifluoromethanesulfonic acid
THF	tetrahydrofuran
THP	tetrahydropyran
TMEDA	*N,N,N'',N''*-tetramethylethylenediamine
TMOF	trimethyl orthoformate
TPAP	tetrapropylammonium perruthenate
TPP	tetraphenylporphyrin
TPPTS	sulfonated triphenylphosphine
TPS	triisopropylbenzensulfonyl chloride
Tr$^+$BF$_4^-$ or Ph$_3$C$^+$BF$_4^-$	triphenylcarbenium tetrafluoroborate
TrS$^-$Bu$_4$N$^+$	tetrabutylammonium triphenylmethanethiolate
Ts	toluenesulfonyl

1

THE ROLE OF PROTECTIVE GROUPS IN ORGANIC SYNTHESIS

PROPERTIES OF A PROTECTIVE GROUP

When a chemical reaction is to be carried out selectively at one reactive site in a multifunctional compound, other reactive sites must be temporarily blocked. Many protective groups have been, and are being, developed for this purpose. A protective group must fulfill a number of requirements. It must react selectively in good yield to give a protected substrate that is stable to the projected reactions. The protective group must be selectively removed in good yield by readily available, preferably nontoxic reagents that do not attack the regenerated functional group. The protective group should form a derivative (without the generation of new stereogenic centers) that can easily be separated from side products associated with its formation or cleavage. The protective group should have a minimum of additional functionality to avoid further sites of reaction. All things considered, no protective group is the best protective group. Currently, the science and art of organic synthesis, contrary to the opinions of some, has a long way to go before we can call it a finished and well-defined discipline, as is amply illustrated by the extensive use of protective groups during the synthesis of multifunctional molecules. Greater control over the chemistry used in the building of nature's architecturally beautiful and diverse molecular frameworks, as well as unnatural structures, is needed when one considers the number of protection and deprotection steps often used to synthesize a molecule.

Greene's Protective Groups in Organic Synthesis, Fourth Edition, by Peter G. M. Wuts and Theodora W. Greene
Copyright © 2007 John Wiley & Sons, Inc.

HISTORICAL DEVELOPMENT

Since a few protective groups cannot satisfy all these criteria for elaborate substrates, a large number of mutually complementary protective groups are needed and, indeed, are available. In early syntheses the chemist chose a standard derivative known to be stable to the subsequent reactions. In a synthesis of callistephin chloride the phenolic $-OH$ group in **1** was selectively protected as an acetate.[1] In the presence of silver ion the aliphatic hydroxyl group in **2** displaced the bromide ion in a bromoglucoside. In a final step the acetate group was removed by basic hydrolysis.

Other classical methods of cleavage include acidic hydrolysis (eq. 1), reduction (eq. 2), and oxidation (eq. 3):

(1) $ArO-R \rightarrow ArOH$

(2) $RO-CH_2Ph \rightarrow ROH$

(3) $RNH-CHO \rightarrow [RNHCOOH] \rightarrow RNH_3^+$

Some of the original work in the carbohydrate area in particular reveals extensive protection of carbonyl and hydroxyl groups. For example, a cyclic diacetonide of glucose was selectively cleaved to the monoacetonide.[2] A summary[3] describes the selective protection of primary and secondary hydroxyl groups in a synthesis of gentiobiose, carried out in the 1870s, as triphenylmethyl ethers.

DEVELOPMENT OF NEW PROTECTIVE GROUPS

As chemists proceeded to synthesize more complicated structures, they developed more satisfactory protective groups and more effective methods for the formation and cleavage of protected compounds. At first a tetrahydropyranyl acetal was prepared,[4] by an acid-catalyzed reaction with dihydropyran, to protect a hydroxyl group. The acetal is readily cleaved by mild acid hydrolysis, but formation of this acetal introduces a new stereogenic center. Formation of the 4-methoxytetrahydropyranyl ketal[5] eliminates this problem.

Catalytic hydrogenolysis of an O-benzyl protective group is a mild, selective method introduced by Bergmann and Zervas[6] to cleave a benzyl carbamate ($>NCO-OCH_2C_6H_5 \rightarrow >NH$) prepared to protect an amino group during peptide syntheses. The method also has been used to cleave alkyl benzyl ethers, stable compounds prepared to protect alkyl alcohols; benzyl esters are cleaved by catalytic hydrogenolysis under neutral conditions.

Three selective methods to remove protective groups have received attention: "assisted," electrolytic, and photolytic removal. Four examples illustrate "assisted removal" of a protective group. A stable allyl group can be converted to a labile vinyl ether group (eq. 4)[7]; a β-haloethoxy (eq. 5)[8] or a β-silylethoxy (eq. 6)[9] derivative is cleaved by attack at the β-substituent; and a stable o-nitrophenyl derivative can be reduced to the o-amino compound, which undergoes cleavage by nucleophilic displacement (eq. 7)[10]:

$$(4) \quad ROCH_2CH=CH_2 \quad \xrightarrow{t\text{-BuO}^-} \quad [ROCH=CHCH_3] \quad \xrightarrow{H_3O^+} \quad ROH$$

$$(5) \quad RO-CH_2-CCl_3 + Zn \quad \longrightarrow \quad RO^- + CH_2=CCl_2$$

$$(6) \quad RO-CH_2-CH_2-SiMe_3 \quad \xrightarrow{F^-} \quad RO^- + CH_2=CH_2 + FSiMe_3$$

R = alkyl, aryl, R′CO−, or R′NHCO−

(7)

The design of new protective groups that are cleaved by "assisted removal" is a challenging and rewarding undertaking.

Removal of a protective group by electrolytic oxidation or reduction is useful in some cases. An advantage is that the use and subsequent removal of chemical oxidants or reductants (e.g., Cr or Pb salts; Pt– or Pd–C) are eliminated. Reductive cleavages have been carried out in high yield at −1 to −3 V (vs. SCE), depending on the group; oxidative cleavages in good yield have been realized at 1.5–2 V (vs. SCE). For systems possessing two or more electrochemically labile protective groups, selective cleavage is possible when the half-wave potentials, $E_{1/2}$, are sufficiently different; excellent selectivity can be obtained with potential differences on the order of 0.25 V. Protective groups that have been removed by electrolytic oxidation or reduction are described at the appropriate places in this book; a review article by Mairanovsky[11] discusses electrochemical removal of protective groups.[12]

Photolytic cleavage reactions (e.g., of o-nitrobenzyl, phenacyl, and nitrophenyl-sulfenyl derivatives) take place in high yield on irradiation of the protected compound for a few hours at 254–350 nm. For example, the o-nitrobenzyl group, used to protect alcohols,[13] amines,[14] and carboxylic acids,[15] has been removed by irradiation. Protective groups that have been removed by photolysis are described at the appropriate places in this book; in addition, the reader may wish to consult five review articles.[16–20]

One widely used method involving protected compounds is solid-phase synthesis[21–24] (polymer-supported reagents). This method has the advantage of simple workup by filtration and automated syntheses, especially of polypeptides, oligonucleotides, and oligosaccharides.

Internal protection, used by van Tamelen in a synthesis of colchicine, may be appropriate[25]:

SELECTION OF A PROTECTIVE GROUP FROM THIS BOOK

To select a specific protective group, the chemist must consider in detail all the re-actants, reaction conditions, and functionalities involved in the proposed synthetic scheme. First he or she must evaluate all functional groups in the reactant to deter-mine those that will be unstable to the desired reaction conditions and require pro-tection. The chemist should then examine reactivities of possible protective groups, listed in the Reactivity Charts, to determine compatibility of protective group and reaction conditions. A guide to these considerations is found in Chapter 10. (The protective groups listed in the Reactivity Charts in that chapter were the most widely used groups at the time the charts were prepared in 1979 in a collaborative effort with other members of Professor Corey's research group.) He or she should consult the complete list of protective groups in the relevant chapter and consider their proper-ties. It will frequently be advisable to examine the use of one protective group for sev-eral functional groups (i.e., a 2,2,2-trichloroethyl group to protect a hydroxyl group as an ether, a carboxylic acid as an ester, and an amino group as a carbamate). When several protective groups are to be removed simultaneously, it may be advantageous to use the same protective group to protect different functional groups (e.g., a ben-zyl group, removed by hydrogenolysis, to protect an alcohol and a carboxylic acid). When selective removal is required, different classes of protection must be used (e.g., a benzyl ether cleaved by hydrogenolysis but stable to basic hydrolysis, to protect an alcohol, and an alkyl ester cleaved by basic hydrolysis but stable to hydrogenolysis, to protect a carboxylic acid). One often overlooked issue in choosing a protective group is that the electronic and steric environments of a given functional group will greatly influence the rates of formation and cleavage. For an obvious example, a tertiary ac-etate is much more difficult to form or cleave than a primary acetate.

If a satisfactory protective group has not been located, the chemist has a number of alternatives: Rearrange the order of some of the steps in the synthetic scheme so that a functional group no longer requires protection or a protective group that was reactive in the original scheme is now stable; redesign the synthesis, possibly making use of latent functionality[26] (i.e., a functional group in a precursor form; e.g., anisole as a precursor of cyclohexanone). Or, it may be necessary to include the synthesis of a new protective group in the overall plan or better yet, design new chemistry that avoids the use of a protective group.

Several books and chapters are associated with protective group chemistry. Some of these cover the area[27, 28]; others deal with more limited aspects. Protective groups continue to be of great importance in the synthesis of three major classes of naturally occuring substances—peptides,[22] carbohydrates,[24] and oligonucleotides[23]—and significant advances have been made in solid-phase synthesis,[22–24] including automated procedures. The use of enzymes in the protection and deprotection of functional groups has been reviewed.[29] Special attention is also called to a review on selective deprotection of silyl ethers.[30]

SYNTHESIS OF COMPLEX SUBSTANCES. TWO EXAMPLES (AS USED IN THE SYNTHESIS OF HIMASTATIN AND PALYTOXIN) OF THE SELECTION, INTRODUCTION, AND REMOVAL OF PROTECTIVE GROUPS

Synthesis of Himastatin

Himastatin, isolated from an actinomycete strain (ATCC) from the Himachal Pradesh State in India and active against gram-positive microorganisms and a variety of tumor probe systems, is a $C_{72}H_{104}N_{14}O_{20}$ compound, **1**.[31] It has a novel bisindolyl structure in which the two halves of the molecule are identical. Each half contains a cyclic peptidal ester containing an L-tryptophanyl unit, D-threonine, L-leucine, D-[(R)-5-hydroxy]piperazic acid, (S)-2-hydroxyisovaleric acid, and D-valine. Its synthesis[32] illustrates several important aspects of protective group usage.

Synthesis of himastatin involved the preparation of the pyrroloindoline moiety **A**, its conversion to the bisindolyl unit **A'$_2$**, synthesis of the peptidal ester moiety **B**, the subsequent joining of these units (**A'$_2$** and two **B** units), and cyclization leading to himastatin. The following brief account focuses on the protective group aspects of the synthesis.

Unit A (Scheme 1)

The first objective was the conversion of L-tryptophan into a derivative that could be converted to pyrroloindoline **3**, possessing a *cis* ring fusion and a *syn* relationship of the carboxyl and hydroxyl groups. This was achieved by the conversions shown in Scheme 1. A critical step was **e**. Of many variants tried, the use of the trityl group on the NH$_2$ of tryptophan and the *t*-butyl group on the carboxyl resulted in stereospecific oxidative cyclization to afford **3** of the desired *cis–syn* stereochemistry in good yield.

Himastatin

1

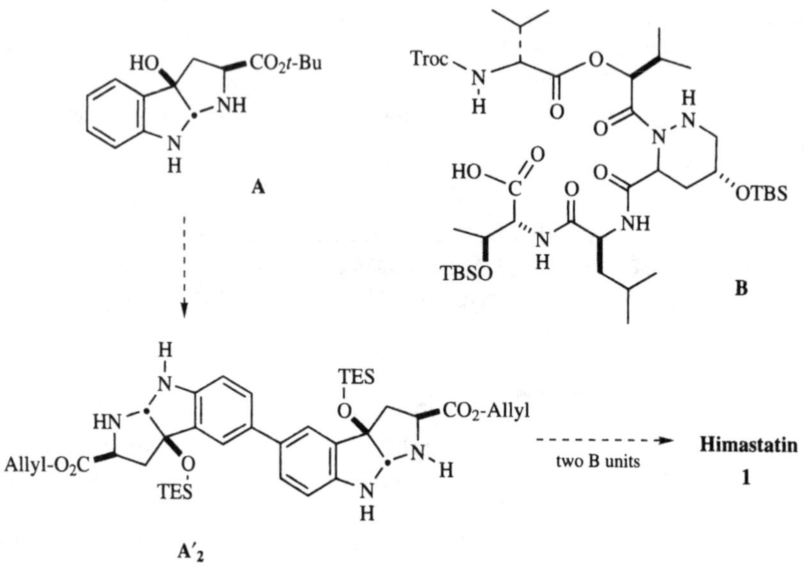

A
B
A′2
two B units
Himastatin
1

Bisindolyl Unit A′$_2$ (Schemes 2 and 3)

The conversion of **3** to **8** is summarized in Scheme 2. The trityl group (too large and too acid-sensitive for the ensuing steps) was removed from N and both N's were protected by Cbz (benzyloxycarbonyl) groups. Protection of the tertiary OH specifically as the robust TBS (*t*-butyldimethylsilyl) group was found to be necessary for the sequence involving the electrophilic aromatic substitution step, **5** to **6**, and the Stille coupling steps (**6** + **7** → **8**).

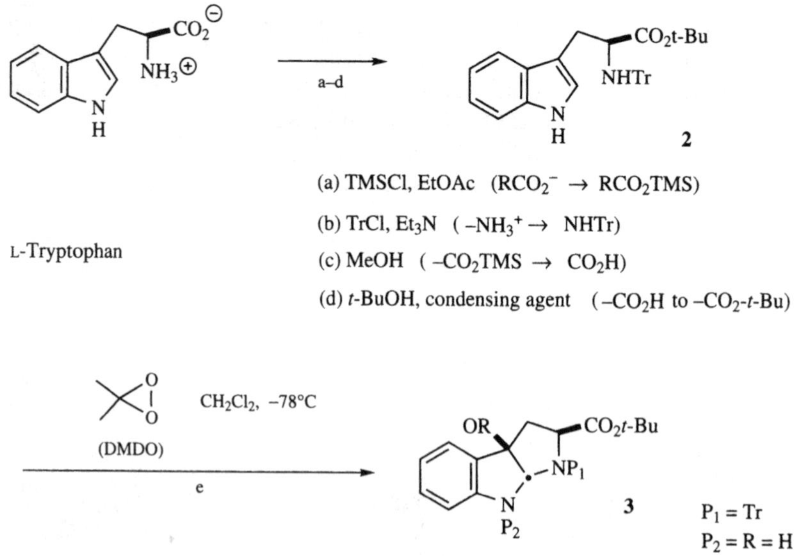

a–d
2
L-Tryptophan
(a) TMSCl, EtOAc (RCO2− → RCO2TMS)
(b) TrCl, Et3N (–NH3+ → NHTr)
(c) MeOH (–CO2TMS → CO2H)
(d) t-BuOH, condensing agent (–CO2H to –CO2-t-Bu)
(DMDO)
CH2Cl2, –78°C
e
3
P1 = Tr
P2 = R = H

Scheme 1

3 P_1 = Tr; P_2 = R = H

4 P_1 = P_2 = R = H

5 P_1 = P_2 = Cbz; R = TBS

a

b

c

6 R = TBS; P_1 = P_2 = Cbz; X = I

$\downarrow$ d

7 R = TBS; P_1 = P_2 = Cbz; X = SnMe$_3$

6 $\downarrow$ e

8 R = TBS; P_1 = P_2 = Cbz; X = [dimer]

(a) HOAc, MeOH, CH$_2$Cl$_2$ (N-Trityl → NH)
(b) (i) CbzCl, pyridine, CH$_2$Cl$_2$ (both NH's → N-Cbz)
　　(ii) TBSCl, DBU, CH$_3$CN (29% from 2) (–OH → OTBS)
(c) ICl, 2,6-di-t-butylpyridine, CH$_2$Cl$_2$ (75%) (X = H → X = I)
(d) Me$_6$Sn$_2$, Pd(Ph$_3$P)$_4$, THF (86%) (X = I → X = SnMe$_3$)
(e) 6, Pd$_2$dba$_3$, Ph$_3$As, DMF, 45°C, (79%) (6 + 7 → 8)

Scheme 2

The TBS group then had to be replaced (two steps, Scheme 3: a and b) by the more easily removable TES (triethylsilyl) group to permit deblocking at the last step in the synthesis of himastatin. Before combination of the bisindolyl unit with the peptidal ester unit, several additional changes in the state of protection at the two nitrogens

8 → a

9 P = H; P′ = P″ = Cbz

$\downarrow$ b

10 P = TES; P′ = P″ = Cbz

$\downarrow$ c

11 P = TES; P′ = P″ = H

$\downarrow$ d

12 P = TES; P′ = FMOC; P″ = H

13 P = FMOC; R = H

$\downarrow$ f

14 P = FMOC; R = allyl

$\downarrow$ g

15 P = H; R = allyl

e

(a) TBAF, THF, (91%) (TBSO– → HO–)
(b) TESCl, DBU, DMF (92%) (HO– → TESO–)
(c) H$_2$, Pd/C, EtOAc (100%) (both NCbz's → NH)
(d) FMOC-HOSU, pyridine, CH$_2$Cl$_2$ (95%) (NH → NFMOC)
(e) TESOTf, lutidine, CH$_2$Cl$_2$ (–CO$_2$-t-Bu → –CO$_2$H)
(f) allyl alcohol, DBAD, Ph$_3$P, CH$_2$Cl$_2$ (90% from 12) (–CO$_2$H → –CO$_2$–allyl)
(g) piperidine, CH$_3$CN (74%) (NFMOC → NH)

Scheme 3

and the carboxyl of **8** were needed (Schemes 2 and 3). The Cbz protective groups were removed from both N's, and the more reactive pyrrolidine N was protected as the FMOC (fluorenylmethoxycarbonyl) group. At the carboxyl, the *t*-butyl group was replaced by the allyl group. [The smaller allyl group was needed for the later condensation of the adjacent pyrrolidine nitrogen of **15** with the threonine carboxyl of **24** (Scheme 5); also, the allyl group can be cleaved by the Pd(Ph$_3$P)$_4$–PhSiH$_3$ method, conditions under which many protective groups (including, of course, the other protective groups in **25**; see Scheme 6) are stable.] Returning to Scheme 3, the FMOC groups on the two equivalent pyrrolidine N's were then removed, affording **15**.

Peptidal Ester Unit B (Schemes 4 and 5)

Several of these steps are common ones in peptide synthesis and involve standard protective groups. Attention is called to the 5-hydroxypiperazic acid. Its synthesis (Scheme 4) has the interesting feature of the introduction of the two nitrogens in protected form as BOC (*t*-butoxycarbonyl) groups in the same step. Removal of the BOC groups and selective conversion of the nitrogen furthest from the carboxyl group into the *N*-Teoc (2-trimethylsilylethoxycarbonyl) group, followed by hydrolysis of the lactone and TBS protection of the hydroxyl, afforded the piperazic acid entity **16** in a suitable form for combination with dipeptide **18** (Scheme 5). Because of the greater reactivity of the leucyl −NH$_2$ group of **18** in comparison to the piperazyl −N$_\alpha$H group in **16**, it was not necessary to protect this piperazyl NH in the condensation of **18** and **16** to form **19**. In the following step (**19** + **20** → **21**), this somewhat hindered piperazyl NH **is** condensed with the acid chloride **20**. Note that the hydroxyl in **20** is protected by the FMOC group—not commonly used in

(a) TFA   (both –NBOC's → NH)

(b) TeocCl, pyridine   (–NH → *N*-Teoc)

(c) LiOH   (lactone → –CO$_2^-$ + HO–)

(d) TBSOTf, lutidine   (–OH → –OTBS)

Scheme 4

FMOC-L-Leucine

17

several steps

D-Threonine

(a) EDCI, DMAP, CH$_2$Cl$_2$
(b) piperidine, CH$_3$CN (NHFMOC to –NH$_2$)
(76%)

18

piperazic acid 16 (from Scheme 4)
HATU, HOAt, collidine, CH$_2$Cl$_2$
(95%)

19

20

collidine, CH$_2$Cl$_2$

21

a, b

23 P = Teoc
   R = Allyl

c–e

24 P = R = H

(a) piperidine, CH$_3$CN (96%) (–OFMOC → –OH)
(b) Troc-D-val (22), IPCC, Et$_3$N, DMAP, CH$_2$Cl$_2$
(c) ZnCl$_2$, CH$_3$NO$_2$ (–NTeoc → –NH)
(d) TBSOTf, lutidine, CH$_2$Cl$_2$ (reprotection of any OH's inadvertently
    deblocked in step c)
(e) Pd(Ph$_3$P)$_4$, PhSiH$_3$, THF (–CO$_2$–allyl → –CO$_2$H)
    (b → e: 72% yield)

Scheme 5

hydroxyl protection. A requirement for the protective group on this hydroxyl was
that it be removable (for the next condensation: **21** + Troc-D-valine **22** → **23**) under
conditions that would leave unaltered the −COO−allyl, the N-Teoc, and the OTBS
groups. The FMOC group (cleavage by piperidine) met this requirement. Choice of
the Troc (2,2,2-trichloroethoxycarbonyl) group for N-protection of valine was based
on the requirements of removability, without affecting OTBS and OTES groups, and
stability to the conditions of removal of allyl from −COO−allyl [easily met by use
of Pd(Ph$_3$P)$_4$ for this deblocking].

25 R = allyl
R′ = Troc

15

+

24

(a) HATU, HOAt, collidine, CH$_2$Cl$_2$, −10°C → rt (65%)
(b) Pd(Ph$_3$P)$_4$, PhSiH$_3$, THF (−CO$_2$-allyl → −CO$_2$H)
(c) Pb/Cd, NH$_4$OAc, THF (N-Troc → NH)

26 R = R′= H
(56%)

(d) HATU, HOAt,
i-Pr$_2$NEt$_2$, DMF

(e) TBAF, THF, HOAc
(−OTBS and −OTES
→ −OH)
(35% from **26**)

27 P = TES
P′ = TBS

1 HIMASTATIN
P = P′ = H

Scheme 6

Himastatin 1 (Scheme 6)

Of special importance to the synthesis was the choice of condensing agents and conditions.[33] HATU-HOAt[34] was of particular value in these final stages. Condensation of the threonine carboxyl of **24** (from Scheme 5) with the pyrrolidine N's of the bisindolyl compound **15** (from Scheme 3) afforded **25**. Removal of the allyl groups from the tryptophanyl carboxyls and the Troc groups from the valine amino nitrogens, followed by condensation (macrolactamization), gave **27**. Removal of the six silyl groups (the two quite hindered TES groups and the four, more accessible, TBS groups) by fluoride ion afforded himastatin.

Synthesis of Palytoxin Carboxylic Acid

Palytoxin carboxylic acid, C$_{123}$H$_{213}$NO$_{53}$, Figure 1 (R^1–R^8 = H), derived from palytoxin, C$_{129}$H$_{223}$N$_3$O$_{54}$, contains 41 hydroxyl groups, one amino group, one ketal, one hemiketal, and one carboxylic acid, in addition to some double bonds and ether linkages.

The total synthesis[35] was achieved through the synthesis of eight different segments, each requiring extensive use of protective group methodology, followed by the appropriate coupling of the various segments in their protected forms.

The choice of what protective groups to use in the synthesis of each segment was based on three aspects: (a) the specific steps chosen to achieve the synthesis of each

1: R^1 = OMe, R^2 = Ac, R^3 = $(t\text{-Bu})Me_2Si$, R^4 = 4-MeOC$_6$H$_4$CH$_2$, R^5 = Bz, R^6 = Me, R^7 = acetonide, R^8 = Me$_3$SiCH$_2$CH$_2$OCO
2: Palytoxin carboxylic acid: R^1 = OH, R^2-R^8 = H

Figure 1. Palytoxin carboxylic acid.

segment; (b) the methods to be used in coupling the various segments, and (c) the conditions needed to deprotect the 42 blocked groups in order to liberate palytoxin carboxylic acid in its unprotected form. (These conditions must be such that the functional groups already deprotected are stable to the successive deblocking conditions.) Kishi's synthesis employed only eight different protective groups for the 42 functional groups present in the fully protected form of palytoxin carboxylic acid (Figure 1, **1**). A few additional protective groups were used for "end group" protection in the synthesis and sequential coupling of the eight different segments. The synthesis was completed by removal of all of the groups by a series of five different methods. The selection, formation, and cleavage of these groups are described below.

For the synthesis of the C.1–C.7 segment, the C.1 carboxylic acid was protected as a methyl ester. The C.5 hydroxyl group was protected as the *t*-butyldimethylsilyl (TBS) ether. This particular silyl group was chosen because it improved the chemical yield and stereochemistry of the Ni(II)/Cr(II)-mediated coupling reaction of segment C.1–C.7 with segment C.8–C.51. Nine hydroxyl groups were protected as *p*-methoxyphenylmethyl (MPM) ethers, a group that was stable to the conditions used in the synthesis of the C.8–C.22 segment. These MPM groups were eventually cleaved oxidatively by treatment with 2,3-dichloro-5,6-dicyano-1,4-benzoquinone (DDQ).

The C.2 hydroxyl group was protected as an acetate, since cleavage of a *p*-methoxyphenylmethyl (MPM) ether at C.2 proved to be very slow. An acetyl group was also used to protect the C.73 hydroxyl group during synthesis of the right-hand half of the molecule (C.52–C.115). Neither a p-methoxyphenyl-methyl (MPM) nor a t-butyldimethylsilyl (TBS) ether was satisfactory at C.73: Dichlorodicyanobenzoquinone (DDQ) cleavage of a p-methoxyphenylmethyl (MPM) ether at C.73 resulted in oxidation of the cis–trans dienol at C.78–C.73 to a cis–trans dienone. When C.73 was protected as a t-butyldimethylsilyl (TBS) ether, Suzucki coupling of segment C.53–C.75 (in which C.75 was a vinyl iodide) to segment C.76–C.115 was too slow. In the synthesis of segment C.38–C.51, the C.49 hydroxyl group was also protected at one stage as an acetate, to prevent benzoate migration from C.46. The C.8 and C.53 hydroxyl groups were protected as acetates for experimental convenience. A benzoate ester, more electron-withdrawing than an acetate ester, was used to protect the C.46 hydroxyl group to prevent spiroke-talization of the C.43 and C.51 hydroxyl groups during synthesis of the C.38–C.51 segment. Benzoate protection of the C.46 hydroxyl group also increased the stabil-ity of the C.47 methoxy group (part of a ketal) under acidic cleavage conditions. Benzoates rather than acetates were used during the synthesis of the C.38–C.51 segment since they were more stable and better chromophores in purification and characterization.

Several additional protective groups were used in the coupling of the eight dif-ferent segments. A tetrahydropyranyl (THP) group was used to protect the hydroxyl group at C.8 in segment C.8–C.22, and a t-butyldiphenylsilyl (TBDPS) group was used for the hydroxyl group at C.37 in segment C.23–C.37. The TBDPS group at C.37 was later removed by $Bu_4N^+F^-$/THF in the presence of nine p-methoxyphenyl-methyl (MPM) groups. After the coupling of segment C.8–C.37 with segment C.38–C.51, the C.8 THP ether was hydrolyzed with pyridinium p-toluenesulfonate (PPTS) in methanol-ether, 42°, in the presence of the bicyclic ketal at C.28–C.33 and the cyclic ketal at C.43–C.47. (As noted above, the resistance of this ketal to these acidic conditions was due to the electron-withdrawing effect of the benzoate at C.46.) A cyclic acetonide (a 1,3-dioxane) at C.49–C.51 was also removed by this step and had to be reformed (acetone/PPTS) prior to the coupling of segment C.8–C.51 with segment C.1–C.7. After coupling of these segments to form segment C.1–C.51, the new hydroxyl group at C.8 was protected as an acetate, and the acetonide at C.49–C.51 was, again, removed without alteration of the bicyclic ketal at C.28–C.33 or the cyclic ketal at C.43–C.47, still stabilized by the benzoate at C.46.

The synthesis of segment C.77–C.115 from segments C.77–C.84 and C.85–C.115 involved the liberation of an aldehyde at C.85 from its protected form as a dithioac-etal, $RCH(SEt)_2$, by mild oxidative deblocking (I_2/NaHCO$_3$, acetone, water) and the use of the p-methoxyphenyldiphenylmethyl (MMTr) group to protect the hydroxyl group at C.77. The C.77 MMTr ether was subsequently converted to a primary alcohol (PPTS/MeOH-CH$_2$Cl$_2$, rt) without affecting the 19 t-butyldimethylsilyl (TBS) ethers or the cyclic acetonide at C.100–C.101.

The C.100–C.101 diol group, protected as an acetonide, was stable to (a) the Wittig reaction used to form the cis double bond at C.98–C.99 and (b) all of the conditions used in the buildup of segment C.99–C.115 to fully protected palytoxin carboxylic acid (Figure 1, 1).

The C.115 amino group was protected as a trimethylsilylethyl carbamate ($Me_3SiCH_2CH_2OCONHR$), a group that was stable to the synthesis conditions and cleaved by the conditions used to remove the t-butyldimethylsilyl (TBS) ethers.

Thus the 42 functional groups in palytoxin carboxylic acid (39 hydroxyl groups, one diol, one amino group, and one carboxylic acid) were protected by eight different groups:

1 methyl ester	$-COOH$
5 acetate esters	$-OH$
20 t-butyldimethylsilyl (TBS) ethers	$-OH$
9 p-methoxyphenylmethyl (MPM) ethers	$-OH$
4 benzoate esters	$-OH$
1 methyl "ether"	$-OH$ of a hemiketal
1 acetonide	1,2-diol
1 $Me_3SiCH_2CH_2OCO$	$-NH_2$

The protective groups were then removed in the following order by the five methods listed below:

(1) To cleave p-methoxyphenylmethyl (MPM) ethers: DDQ (dichlorodicyanobenzoquinone)/t-BuOH–CH_2Cl_2–phosphate buffer (pH 7.0), 4.5 h.

(2) To cleave the acetonide: 1.18 N $HClO_4$–THF, 25°C, 8 days.

(3) To hydrolyze the acetates and benzoates: 0.08 N $LiOH/H_2O$–MeOH–THF, 25°C, 20 h.

(4) To remove t-butyldimethylsilyl (TBS) ethers and the carbamoyl ester ($Me_3SiCH_2CH_2OCONHR$): $Bu_4N^+F^-$, THF, 22°C, 18 h $\rightarrow$ THF–DMF, 22°C, 72 h.

(5) To hydrolyze the methyl ketal at C.47, no longer stabilized by the C.46 benzoate: HOAc–H_2O, 22°C, 36 h.

This order was chosen so that DDQ (dichlorodicyanobenzoquinone) treatment would not oxidize a deprotected allylic alcohol at C.73 and so that the C.47 hemiketal would still be protected (as the ketal) during basic hydrolysis (Step 3).

And so the skillful selection, introduction, and removal of a total of 12 different protective groups has played a major role in the successful total synthesis of palytoxin carboxylic acid (Figure 1, **2**).

1. A. Robertson and R. Robinson, *J. Chem. Soc.*, 1460 (1928).

2. E. Fischer, *Ber.*, **28**, 1145 (1895); see p. 1165.

3. B. Helferich, *Angew. Chem.*, **41**, 871 (1928).

4. W. E. Parham and E. L. Anderson, *J. Am. Chem. Soc.*, **70**, 4187 (1948).

5. C. B. Reese, R. Saffhill, and J. E. Sulston, *J. Am. Chem. Soc.*, **89**, 3366 (1967).

6. M. Bergmann and L. Zervas, *Chem. Ber.*, **65**, 1192 (1932).

7. J. Cunningham, R. Gigg, and C. D. Warren, *Tetrahedron Lett.*, 1191, (1964).

8. R. B. Woodward, K. Heusler, J. Gosteli, P. Naegeli, W. Oppolzer, R. Ramage, S. Ranganathan, and H. Vorbruggen, J. Am. Chem. Soc., **88**, 852 (1966).

9. P. Sieber, *Helv. Chim. Acta*, **60**, 2711 (1977).

10. I. D. Entwistle, *Tetrahedron Lett.*, **555**, (1979).

11. V. G. Mairanovsky, *Angew. Chem., Int. Ed. Engl.*, **15**, 281 (1976).

12. See also M. F. Semmelhack and G. E. Heinsohn, *J. Am. Chem. Soc.*, **94**, 5139 (1972).

13. S. Uesugi, S. Tanaka, E. Ohtsuka, and M. Ikehara, *Chem. Pharm. Bull.*, **26**, 2396 (1978).

14. S. M. Kalbag and R. W. Roeske, *J. Am. Chem. Soc.*, **97**, 440 (1975).

15. L. D. Cama and B. G. Christensen, *J. Am. Chem. Soc.*, **100**, 8006 (1978).

16. V. N. R. Pillai, Synthesis, 1, (1980).

17. P. G. Sammes, *Q. Rev., Chem. Soc.*, **24**, 37 (1970); see pp. 66–68.

18. B. Amit, U. Zehavi, and A. Patchornik, *Isr. J. Chem.*, **12**, 103 (1974).

19. V. N. R. Pillai, "Photolytic Deprotection and Activation of Functional Groups," *Org. Photochem.*, **9**, 225 (1987).

20. V. Zehavi, "Applications of Photosensitive Protecting Groups in Carbohydrate Chemistry," *Adv. Carbohydr. Chem. Biochem.*, **46**, 179 (1988).

21. (a) R. B. Merrifield, *J. Am. Chem. Soc.*, **85**, 2149 (1963); (b) P. Hodge, *Chem. Ind. (London)*, 624 (1979); (c) C. C. Leznoff, *Acc. Chem. Res.*, **11**, 327 (1978); (d) *Solid Phase Synthesis*, E. C. Blossey and D. C. Neckers, Eds., Halsted, New York, 1975; *Polymer-Supported Reactions in Organic Synthesis*, P. Hodge and D. C. Sherrington, Eds., Wiley-Interscience, New York, 1980. A comprehensive review of polymeric protective groups by J. M. J. Fréchet is included in this book. (e) D. C. Sherrington and P. Hodge, *Synthesis and Separations Using Functional Polymers*, Wiley-Interscience, New York (1988).

22. **Peptides:** (a) *Methods in Molecular Biology*, Vol. **35**: *Peptide Synthesis Protocols*, M. W. Pennington and B. M. Dunn, Eds., Humana Press, Totowa, NJ, 1994, pp. 91–169; (b) *Synthetic Peptides*, G. Grant, Ed., W. H. Freeman & Co.; New York, 1992; (c) *Novabiochem 97/98, Catalog*, Technical Section S1-S85 (this section contains many useful details on the use and manipulation of protective groups in the amino acid–peptide area); (d) J. M. Stewart and J. D. Young, *Solid Phase Peptide Synthesis*, 2nd ed., Pierce Chemical Company, Rockford, IL (1984); (e) E. Atherton and R. C. Sheppard, *Solid Phase Peptide Synthesis. A Practical Approach*, Oxford-IRL Press, New York (1989); (f) *Innovation and Perspectives in Solid Phase Synthesisa: Peptides, Polypeptides and Oligonucleotides: Collected Papers, First International Symposium: Macro-Organic Reagents and Catalysts*, R. Epton, Ed., SPCC, UK, 1990; *Collected Papers, Second International Symposium*, R. Epton, Ed., Intercept Ltd, Andover, UK, 1992; (g) V. J. Hruby and J-P. Meyer, "The Chemical Synthesis of Peptides," in *Bioorganic Chemistry: Peptides and Proteins*, S. M. Hecht, Ed., Oxford University Press, New York, 1998, Chapter 2, pp. 27–64.

23. **Oligonucleotides:** (a) S. L. Beaucage and R. P. Iyer, Tetrahedron, **48**, 2223 (1992); **49**, 1925 (1993); **49**, 6123 (1993); **49**, 10441 (1993); (b) J. W. Engels and E. Uhlmann, *Angew. Chem., Int. Ed. Engl.*, **28**, 716 (1989); (c) S. L. Beaucage and M. H. Caruthers, "The Chemical Synthesis of DNA/RNA," in *Bioorganic Chemistry: Nucleic Acids*, S. M. Hecht, Ed., Oxford University Press, New York, 1996, Chapter 2, pp. 36–74.

24. **Oligosaccharides:** (a) S. J. Danishefsky and M. T. Bilodeau, "Glycals in Organic Synthesis: The Evolution of Comprehensive Strategies for the Assembly of Oligosaccharides and Glycoconjugates of Biological Consequence, "*Angew. Chem., Int. Ed. Engl.*, **35**, 1380 (1996); (b) P. H. Seeberger and S. J. Danishefsky, *Acc. Chem. Res.*, **31**, 685 (1998); (c) P. H. Seeberger, M. T. Bilodeau and S. J. Danishefsky, *Aldrichchimica Acta*, **30**, 75

(1997); (d) J. Y. Roberge, X. Beebe, and S. J. Danishefsky, "Solid Phase Convergent Synthesis of N-Linked Glycopeptides on a Solid Support," *J. Am. Chem. Soc.*, **120**, 3915 (1998); (e) B. O. Fraser-Reid et al., "The Chemical Synthesis of Oligosaccharides," in *Bioorganic Chemistry: Oligosaccharides*, S. M. Hecht, Ed., Oxford University Press, New York, 1999, Chapter 3, pp. 89–133; (f) K. C. Nicolaou et al., "The Chemical Synthesis of Complex Carbohydrates," *Ibid.*, Chapter 4, pp. 134–173.

25. E. E. van Tamelen, T. A. Spencer, Jr., D. S. Allen, Jr. and R. L. Orvis, *Tetrahedron*, **14**, 8 (1961).

26. D. Lednicer, *Adv. Org. Chem.*, **8**, 179 (1972).

27. (a) H. Kunz and H. Waldmann, "Protecting Groups," in *Comprehensive Organic Synthesis*, B. M. Trost, Ed., Pergamon Press, Oxford, United Kingdom, 1991, Vol. 6, pp. 631–701; (b) P. J. Kocienski, *Protecting Groups*, Georg Theime Verlag, Stuttgart and New York, 1994; (c) *Protective Groups in Organic Chemistry*, J. F. W. McOmie, Ed., Plenum, New York and London, 1973.

28. *Organic Syntheses*, Wiley-Interscience, New York, *Collect. Vols. I–IX*, 1941–1998, **75**, 1997; W. Theilheimer, Ed., *Synthetic Methods of Organic Chemistry*, S. Karger, Basel, Vols. 1–52, 1946–1997; E. Müller, Ed., *Methoden der Organischen Chemie* (Houben-Weyl), G. Thieme Verlag, Stuttgart, Vols. 1–21f, 1958–1995; *Spec. Period. Rep.: General and Synthetic Methods*, Royal Society of Chemistry **1–16** (1978–1994).; S. Patai, Ed., *The Chemistry of Functional Groups*, Wiley-Interscience, New York, Vols. 1–51, 1964–1997.

29. (a) H. Waldmann and D. Sebastian, "Enzymatic Protecting Group Techniques," *Chem. Rev.*, **94**, 911 (1994); (b) *Enzyme Catalysis in Organic Synthesis: A Comprehensive Handbook*, K. Drauz and H. Waldmann, Eds., VCH, Weinheim, 1995, Vol. 2, pp. 851–889.

30. T. D. Nelson and R. D. Crouch, "Selective Deprotection of Silyl Ethers," *Synthesis*, **1031** (1996).

31. (a) K.-S. Lam, G. A. Hesler, J. M. Mattel, S. W. Mamber and S. Forenza, *J. Antibiot.* **43**, 956 (1990); (b) J.E. Loet, D. R. Schroeder, B.S. Krishnan, and J. A. Matson, *ibid.*, **43**, 961 (1990); (c) J. E. Leet, D. R. Schroeder, J. Golik, J. A. Matson, T. W. Doyle, K. S. Lam, S. E. Hill, M. S. Lee, J. L. Whitney, and B. S. Krishnan, *ibid.*, **49**, 299 (1996); (d) T. M. Kamenecka and S. J. Danishefsky, "Studies in the Total Synthesis of Himastatin: A Revision of the Stereochemical Assignment," *J. Angew. Chem. Int. Ed.*, 37, 2993 (1998).

32. T. M. Kamenecka and S.J. Danishefsky, "The Total Synthesis of Himastatin: Confirmation of its Stereostructure," *Angew. Chem. Int. Ed.*, 37 2995 (1998). We thank Professor Danishefsky for providing us with preprints of the himastatin communications (refs. 31d and 32).

33. (a) J. M. Humphrey and A. R. Chamberlin, *Chem. Rev.* **97**, 2241 (1997); (b) A. Ehrlich, H.-U. Heyne, R. Winter, M. Beyermann, H. Haber, L. A. Carpino, and M. Bienert, *J. Org. Chem.*, **61**, 8831 (1996).

34. HOAt, 7-aza-1-hydroxybenzotriazole; HATU (CAS Registry No. 148893-10-1), *N*-[(dimethylamino)(3*H*-1,2,3-triazolo(4,5-*b*)pyridin-3-yloxy)methylene]-*N*-methyl-methanaminium hexafluorophosphate, previously known as O-(7-azabenzotriazol-1-yl)-1,1,3,3-tetramethyluronium hexafluorophosphate. [Note Assignment of structure to HATU as a guanidinium species rather than as a uronium species—that is, attachment of the (Me$_2$NC=NMe$_2$)$^+$ unit to N3 of 7-azabenzotriazole 1-*N*-oxide instead of to the *O*—is based on X-ray analysis (ref. 33b).]

35. R. W. Armstrong, J.-M. Beau, S. H. Cheon, W. J. Christ, H. Fujioka, W.-H. Ham, L. D. Hawkins, H. Jin, S. H. Kang, Y. Kishi, M. J. Martinelli, W. W. McWhorter, Jr., M. Mizuno, M. Nakata, A. E. Stutz, F. X. Talamas, M. Taniguchi, J. A. Tino, K. Ueda, J.-i. Uenishi, J. B. White, and M. Yonaga, *J. Am. Chem. Soc.*, **111**, 7530–7533 (1989). See also *idem., ibid.*, **111**, 7525 (1989).

2

PROTECTION FOR THE HYDROXYL GROUP, INCLUDING 1,2- AND 1,3-DIOLS

Greene's Protective Groups in Organic Synthesis, Fourth Edition, by Peter G. M. Wuts and
Theodora W. Greene
Copyright © 2007 John Wiley & Sons, Inc.

Substituted Ethyl Ethers   74

Methoxy-Substituted Benzyl Ethers   120

Chiral Ketones   345

Cyclic Ortho Esters   346

Silyl Derivatives   353

ETHERS

Hydroxyl groups are present in a number of compounds of biological and synthetic interest, including nucleosides, carbohydrates, steroids, macrolides, polyethers, and the side chain of some amino acids.[1a] During oxidation, acylation, halogenation with phosphorus or hydrogen halides, or dehydration reactions of these compounds, a hydroxyl group must be protected. In polyfunctional molecules, selective protection becomes an issue that has been addressed by the development of a number of new methods. Ethers are among the most used protective groups in organic synthesis. They vary from the simplest, most stable, methyl ether to the more elaborate, substituted, trityl ethers developed for use in nucleotide synthesis. They are formed and removed under a wide variety of conditions. Some of the ethers that have been used extensively to protect alcohols are included in Reactivity Chart 1.[1a,b]

1. (a) See ref. 23 (oligonucleotides) and 24 (oliogsaccharides) in Chapter 10; (b) see also C. B. Reese, "Protection of Alcoholic Hydroxyl Groups and Glycol Systems," in *Protective Groups in Organic Chemistry*, J. F. W. McOmie, Ed., Plenum, New York and London, 1973, pp. 95–143; H. M. Flowers, "Protection of the Hydroxyl Group," in *The Chemistry of the Hydroxyl Group*, S. Patai, Ed., Wiley-Interscience, New York, 1971, Vol. 10/2, pp. 1001–1044; C. B. Reese, *Tetrahedron*, **34**, 3143 (1978), see pp. 3145–3150; V. Amarnath and A. D. Broom, *Chem. Rev.*, **77**, 183 (1977), see pp. 184–194; M. Lalonde and T. H. Chan, "Use of Organosilicon Reagents as Protective Groups in Organic Synthesis," *Synthesis*, 817 (1985); P. Kocienski, *Protecting Groups*, 3rd Ed., Thieme Medical Publishers, New York, 2004, p. 184; B. C. Ranu and S. Bhar, "Dealkylation of Ethers. A Review," *Org. Prep. Proced. Int.*, **28**, 371 (1996). S. A. Weissman and D. Zewge, "Recent Advances in Ether Dealkylation," *Tetrahedron*, **61**, 7833 (2005). F. Guibe, "Allylic Protecting Groups and Their Use in Complex Environment. Part I: Allylic Protection of Alcohols," *Tetrahedron*, **53**, 13509 (1997); F. Guibe, "Allylic Protecting Groups and Their Use in Complex Environment. Part II: Allylic Protecting Groups and Their Removal Through Catalytic Palladiium π-Allyl Methodology," *Tetrahedron*, **54**, 2969 (1998).

Methyl Ether: ROMe (Chart 1)

Formation

1. Me_2SO_4, NaOH, Bu_4NI, organic solvent, 60–90% yield.[1] This is an excellent and general method that can easily be scaled up.

2. MeI or Me_2SO_4,[2] NaH or KH, THF. This is the standard method for introducing the methyl ether function onto hindered and unhindered alcohols.

3. Me_2SO_4, DMSO, DMF, $Ba(OH)_2$, BaO, rt, 18 h, 88% yield.[3]

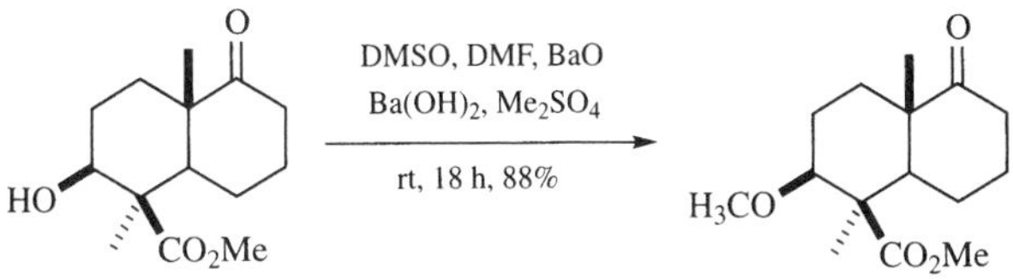

4. MeI, CsOH, DMF, TBAI, 4-Å molecular sieves (ms), CH_3CN, 23°C, 1 h, 88% yield.[4]

5. MeI, solid KOH, DMSO, 20°C, 5–30 min, 85–90% yield.[5]

6. $TMSCHN_2$, 40% HBF_4, CH_2Cl_2, 0°C, 79% yield. This is a safe alternative to the use of diazomethane (74–93% yield).[6,7]

7. CH_2N_2, silica gel, 0–10°C, 100% yield.[8]

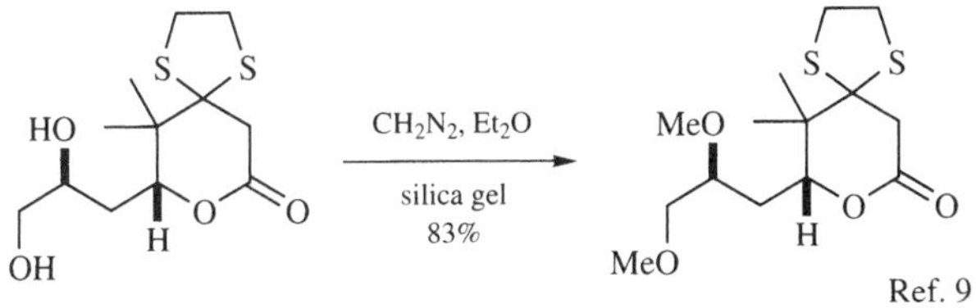

Ref. 9

8. CH_2N_2, HBF_4, CH_2Cl_2, Et_3N, 25°C, 1 h, 95% yield.[10,11] Hydroxyl amines will *O*-alkylate without the acid catalyst.[12]

9. CH_2N_2, $SnCl_2$, CH_3CN, Et_2O, 75% yield.[13]

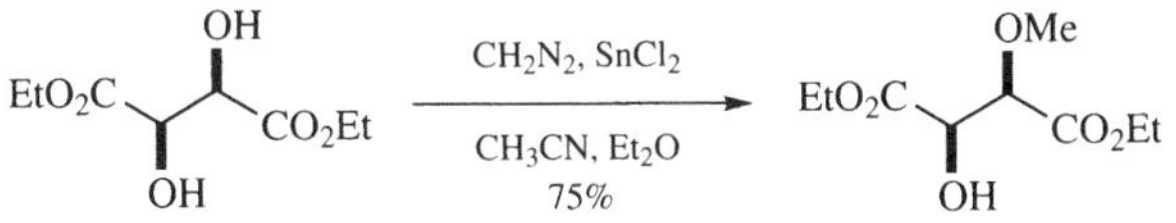

10. $(MeO)_2POH$, cat. TsOH, 90–100°C, 12 h, 60% yield.[14]

11. Me_3OBF_4, 3 days, 55% yield.[15] A simple large-scale preparation of this reagent has been described.[16] This reagent was used in conjunction with Proton-Sponge in CH_2Cl_2 (3 h, 0°C, 90% yield) to give a methyl ether without acyl migration. It should be noted that the use of MeOTf (*highly toxic*) in this case failed to give satisfactory results.[17] This method can also be used on aldols without reversion.[18]

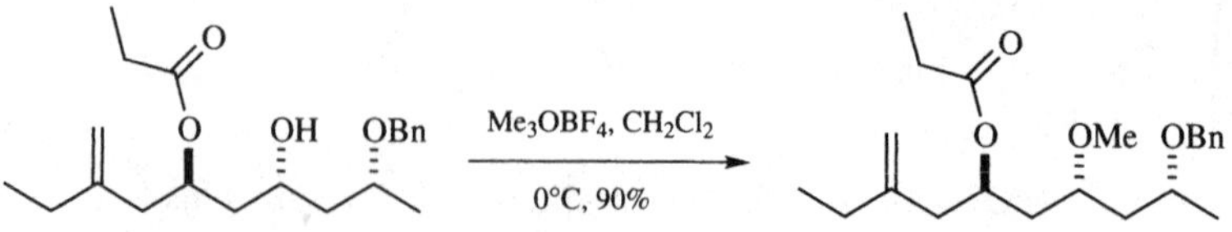

12. CF_3SO_3Me, CH_2Cl_2, Pyr, 80°C, 2.5 h, 85–90% yield.[19,20] The use of 2,6-di-t-butyl-4-methylpyridine as a base is also very effective.[21]

13. CF_3SO_3Me, LHMDS, THF, HMPA, 89% yield.[22]

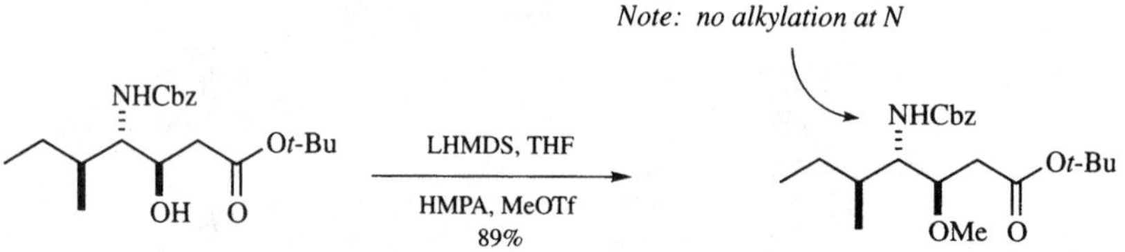

Note: no alkylation at N

14. Because of the increased acidity and reduced steric requirement of the carbo-hydrate hydroxyl, t-BuOK can be used as a base to achieve ether formation.[23]

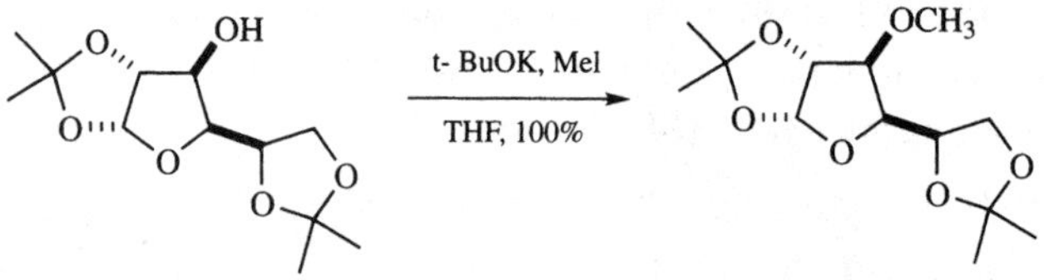

15. MeI, Ag_2O, 93% yield.[24]

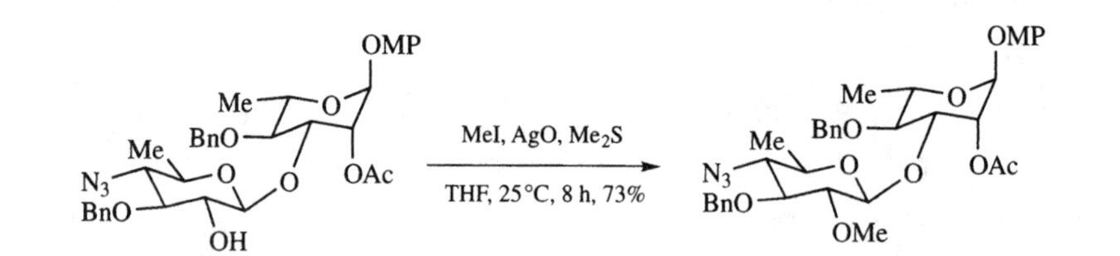

This method, when modified with a catalytic amount of dimethyl sulfide, was the only method found satisfactory for the methylation of the glycoside in the following scheme.[25]

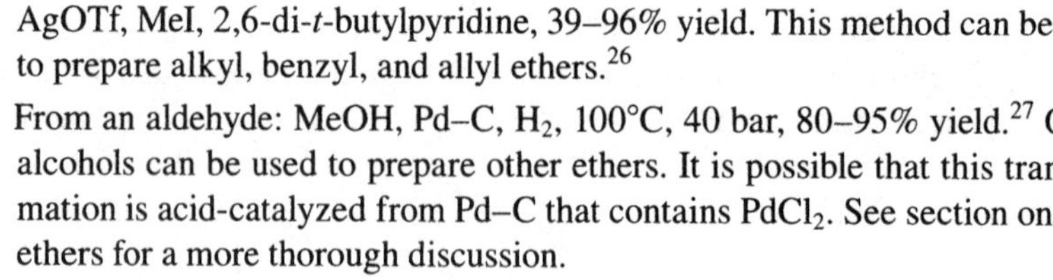

16. AgOTf, MeI, 2,6-di-t-butylpyridine, 39–96% yield. This method can be used to prepare alkyl, benzyl, and allyl ethers.[26]

17. From an aldehyde: MeOH, Pd–C, H_2, 100°C, 40 bar, 80–95% yield.[27] Other alcohols can be used to prepare other ethers. It is possible that this transfor-mation is acid-catalyzed from Pd–C that contains $PdCl_2$. See section on TES ethers for a more thorough discussion.

18. From a MOM ether: $Zn(BH_4)_2$, TMSCl, 87% yield.[28]

19.

Ref. 29

Cleavage[30]

1. Me_3SiI, $CHCl_3$, 25°C, 6 h, 95% yield.[31] A number of methods have been reported in the literature for the *in situ* formation of Me_3SiI[32] since Me_3SiI is somewhat sensitive to handle. This reagent also cleaves many other ether type protective groups, but selectivity can be maintained by control of the reaction conditions and the inherent rate differences between functional groups.

2. BBr_3, NaI, 15-crown-5.[33] Methyl esters are not cleaved under these conditions.[34]

3. BBr_3, EtOAc, 1 h, 95% yield.[35]

4. BBr_3, CH_2Cl_2, high yields.[36]

This method is probably the most commonly used method for the cleavage of methyl ethers because it generally gives excellent yields with a variety of structural types. The solid complex $BBr_3 \cdot Me_2S$ that is more easily handled can also be used.[37] BBr_3 will cleave ketals.

5. $BF_3 \cdot Et_2O$, $HSCH_2CH_2SH$, HCl, 15 h, 82% yield.[38,39]

6. $MeSSiMe_3$ or $PhSSiMe_3$, ZnI_2, Bu_4NI.[40] In this case the 6-O-methyl ether was cleaved selectively from permethylated glucose.

7. $SiCl_4$, NaI, CH_2Cl_2, CH_3CN, 80–100% yield.[41]

8. AlX_3 (X = Br, Cl), EtSH, 25°C, 0.5–3 h, 95–98% yield.[42]

9. t-BuCOCl or AcCl, NaI, CH_3CN, 37 h, rt, 84% yield.[43] In this case the methyl ether is replaced by a pivalate or acetate group that can be hydrolyzed with base.

10. Ac_2O, $FeCl_3$, 80°C, 24 h.[44] In this case the methyl ether is converted to an acetate. The reaction proceeds with complete racemization. Benzyl and allyl ethers are also cleaved.

11. AcCl, NaI, CH_3CN.[45]

12. Me_2BBr, CH_2Cl_2, 0–25°C, 3–18 h, 75–93% yield. Tertiary methyl ethers give the tertiary bromide.[46]

13. $BI_3 \cdot Et_2NPh$, benzene, rt, 3–4 h, 94% yield.[47]

14. TMSCl, cat. H_2SO_4, Ac_2O, 71–89% yield.[48]

15. $AlCl_3$, Bu_4NI, CH_3CN, 83% yield.[49,50]

16. The following method works well for methyl ethers that have a hydroxyl within 2.3–2.8 Å.[51]

77% if done in 1 pot

17. Treatment of a methyl ether with $RuCl_3$, $NaIO_4$ converts it into a ketone.[52]

1. A. Merz, *Angew. Chem., Int. Ed. Engl.*, **12**, 846 (1973).
2. M. E. Jung and S. M. Kaas, *Tetrahedron Lett.*, **30**, 641 (1989).
3. J. T. A. Reuvers and A. de Groot, *J. Org. Chem.*, **51**, 4594 (1986).
4. E. E. Dueno, F. Chu, S.-I. Kim and K. W. Jung, *Tetrahedron Lett.*, **40**, 1843 (1999).
5. R. A. W. Johnstone and M. E. Rose, *Tetrahedron*, **35**, 2169 (1979).
6. T. Aoyama and T. Shioiri, *Tetrahedron Lett.*, **31**, 5507 (1990).
7. Y. Lin and G. B. Jones, *Org. Lett.*, **7**, 71 (2005).
8. K. Ohno, H. Nishiyama, and H. Nagase, *Tetrahedron Lett.*, **20**, 4405 (1979).
9. T. Nakata, S. Nagao, N. Mori, and T. Oishi, *Tetrahedron Lett.*, **26**, 6461 (1985).
10. M. Neeman and W. S. Johnson, *Org. Synth., Collect. Vol. V*, 245 (1973).
11. A. B. Smith, III, K. J. Hale, L. M. Laakso, K. Chen, and A. Riera, *Tetrahedron Lett.*, **30**, 6963 (1989).
12. M. Somei and T. Kawasaki, *Heterocycles.*, **29**, 1251 (1989).
13. A. Gateau-Olesker, J. Cléophax, and S. D. Géro, *Tetrahedron Lett.*, **27**, 41 (1986).
14. Y. Kashman, *J. Org. Chem.*, **37**, 912 (1972).
15. H. Meerwein, G. Hinz, P. Hofmann, E. Kroning, and E. Pfeil, *J. Prakt. Chem.*, **147**, 257 (1937).

16. M. J. Earle, R. A. Fairhurst, R. G. Giles, and H. Heaney, *Synlett*, 728 (1991).

17. I. Paterson and M. M. Coster, *Tetrahedron Lett.*, **43**, 3285 (2002).

18. G. R. Pettit and M. P. Grealish, *J. Org. Chem.*, **66**, 8640 (2001); P. L. DeRoy and A. B. Charette, *Org. Lett.*, **5**, 4163 (2003).

19. J. Arnarp and J. Lönngren, *Acta Chem. Scand. Ser. B*, **32**, 465 (1978).

20. R. E. Ireland, J. L. Gleason, L. D. Gegnas, and T. K. Highsmith, *J. Org. Chem.*, **61**, 6856 (1996).

21. J. A. Marshall and S. Xie, *J. Org. Chem.*, **60**, 7230 (1995).

22. K. Tomioka, M. Kanai, and K. Koga, *Tetrahedron Lett.*, **32**, 2395 (1991).

23. P. G. M. Wuts and S. R. Putt, unpublished results.

24. A. E. Greene, C. L. Drian, and P. Crabbe, *J. Am. Chem. Soc.*, **102**, 7583 (1980).

25. D. B. Werz and P. H. Seeberger, *Angew. Chem. Int. Ed.*, **44**, 6315 (2005).

26. R. M. Burk, T. S. Gac, and M. B. Roof, *Tetrahedron Lett.*, **35**, 8111 (1994).

27. V. Bethmont, F. Fache, and M. Lemaive, *Tetrahedron Lett.*, **36**, 4235 (1995).

28. H. Kotsuki, Y. Ushio, N. Yoshimura, and M. Ochi, *J. Org. Chem.*, **52**, 2594 (1987).

29. H. Dvorakova, D. Dvorak, J. Srogl, and P. Kocovsky, *Tetrahedron Lett.*, **36**, 6351 (1995).

30. For a review of alkyl ether cleavage, see B. C. Ranu and S. Bhar, *Org. Prep. Proced. Int.*, **28**, 371 (1996).

31. M. E. Jung and M. A. Lyster, *J. Org. Chem.*, **42**, 3761 (1977).

32. M. E. Jung and T. A. Blumenkopf, *Tetrahedron Lett.*, **19**, 3657 (1978); G. A. Olah, A. Husain, B. G. B. Gupta, and S. C. Narang, *Angew. Chem., Int. Ed. Engl.*, **20**, 690 (1981); T.-L. Ho and G. Olah, *Synthesis* 417 (1977). For a review on the uses of Me₃SiI, see A. H. Schmidt, *Aldrichimica Acta*, **14**, 31 (1981).

33. H. Niwa, T. Hida, and K. Yamada, *Tetrahedron Lett.*, **22**, 4239 (1981).

34. M. E. Kuehne and J. B. Pitner, *J. Org. Chem.*, **54**, 4553 (1989).

35. H. Shimomura, J. Katsuba, and M. Matsui, *Agric. Biol. Chem.*, **42**, 131 (1978).

36. M. Demuynck, P. De Clercq, and M. Vandewalle, *J. Org. Chem.*, **44**, 4863 (1979); P. A. Grieco, M. Nishizawa, T. Oguri, S. D. Burke, and N. Marinovic, *J. Am. Chem. Soc.*, **99**, 5773 (1977).

37. P. G. Williard and C. B. Fryhle, *Tetrahedron Lett.*, **21**, 3731 (1980).

38. G. Vidari, S. Ferrino, and P. A. Grieco, *J. Am. Chem. Soc.*, **106**, 3539 (1984).

39. M. Node, H. Hori, and E. Fujita, *J. Chem. Soc., Perkin Trans. 1*, 2237 (1976).

40. S. Hanessian and Y. Guindon, *Tetrahedron Lett.*, **21**, 2305 (1980); R. S. Glass, *J. Organomet. Chem.*, **61** 83 (1973); I. Ojima, M. Nihonyangi, and Y. Nagai, *J. Organmet. Chem.*, **50**, C26 (1973).

41. M. V. Bhatt and S. S. El-Morey, *Synthesis*, 1048 (1982).

42. M. Node, K. Nishide, M. Sai, K. Ichikawa, K. Fuji, and E. Fujita, *Chem. Lett.*, **8**, 97 (1979).

43. A. Oku, T. Harada, and K. Kita, *Tetrahedron Lett.*, **23**, 681 (1982).

44. B. Ganem and V. R. Small, Jr., *J. Org. Chem.*, **39**, 3728 (1974).

45. T. Tsunoda, M. Amaike, U. S. F. Tambunan, Y. Fujise, S. Ito, and M. Kodama, *Tetrahedron Lett.*, **28**, 2537 (1987).

46. Y. Guindon, C. Yoakim, and H. E. Morton, *Tetrahedron Lett.*, **24**, 2969 (1983).

47. C. Narayana, S. Padmanabhan, and G. W. Kabalka, *Tetrahedron Lett.*, **31**, 6977 (1990).

48. J. C. Sarma, M. Borbaruah, D. N. Sarma, N. C. Barua, and R. P. Sharma, *Tetrahedron*, **42**, 3999 (1986).

49. T. Akiyama, H. Shima, and S. Ozaki, *Tetrahedron Lett.*, **32**, 5593 (1991).

50. E. D. Moher, J. L. Collins, and P. A. Grieco, *J. Am. Chem. Soc.*, **114**, 2764 (1992).

51. A. Boto, D. Hernandez, R. Hernandez, and E. Suarez, *Org. Lett.*, **6**, 3785 (2004).

52. L. E. Overman, D. J. Ricca, and V. D. Tran, *J. Am. Chem. Soc.*, **119**, 12031 (1997).

Substituted Methyl Ethers

Methoxymethyl Ether (MOM Ether): CH_3OCH_2OR (Chart 1)

Formation

1. CH_3OCH_2Cl, *i*-Pr_2NEt, 0°C, 1 h → 25°C, 8 h, 86% yield.[1] This is the most commonly employed procedure for introduction of the MOM group. The reagent **chloromethylmethyl ether is reported to be carcinogenic, and dichloromethylmethyl ether, a by-product in its preparation, is considered even more toxic.** A preparation that does not produce any of the dichloro ether has been reported.[2]

2. CH_3OCH_2Cl,[3] NaH, THF, 80% yield.[4]

3. MOMBr, DIPEA, CH_2Cl_2, 0°C, 6 h, 72% yield.[1,5]

4. NaI increases the reactivity of MOMCl by the *in situ* preparation of MOMI, which facilitates the protection of tertiary alcohols.[6]

5. For the selective protection of diols: Bu_2SnO, benzene, reflux; MOMCl, Bu_4NI, rt, 87% yield.[7]

6. Selective formation of MOM ethers has been achieved in a diol system.[8]

7. Mono MOM derivatives of diols can be prepared from the ortho esters by diisobutylaluminum hydride reduction (46–98% yield). In general, the most hindered alcohol is protected.[9]

In the case of allylic or propargylic diols, the nonallylic (propargylic) alcohol is protected.[10]

8. MOMCl, Al_2O_3, ultrasound, 68–92% yield.[11]

9. MOMCl, CH_2Cl_2, Na–Y Zeolite, reflux, 70–91% yield.[12]

10. The Avermectin derivative was protected under the illustrated mild and nearly neutral conditions.[13] The reagent is easily prepared from the thiol and $CH_2(OMe)_2$ with $BF_3 \cdot Et_2O$ activation.

11. $CH_2(OMe)_2$, Nafion H.[14]

12. $CH_2(OMe)_2$, SAC-13 (commercially available), 72–96% yield. This method very efficiently produces the i-PrOCH$_2$OR derivative (82–100% yield) from the isopropyl acetal.

13. $CH_2(OMe)_2$, CH_2Cl_2, TfOH, 4 h, 25°C, 65% yield.[15] This method is suitable for the formation of primary, secondary, allylic, and propargylic MOM ethers. Tertiary alcohols fail to give complete reaction. 1,3-Diols give methylene acetals (89% yield).

14. $CH_2(OMe)_2$, CH_2=CHCH$_2$SiMe$_3$, Me$_3$SiOTf, P_2O_5, 93–99% yield.[16] This method was used to protect the 2′-OH of ribonucleosides and deoxyribonucleosides as well as the hydroxyl groups of several other carbohydrates bearing functionality such as esters, amides, and acetonides.

15. $CH_2(OEt)_2$, Montmorillonite clay (H^+), 72–80% for nonallylic alcohols, 56% for a propargylic alcohol.[17] Amberlyst 15 has been used as a catalyst.[18]

16. $CH_2(OMe)_2$, $MoO_2(acac)_2$, CHCl$_3$, reflux, 63–95% yield.[19]

17. $CH_2(OCH_3)_2$, anhydrous FeCl$_3$–MS (3 Å), 1–3 h, 70–99% yield.[20]

18. $CH_2(OMe)_2$, TsOH, LiBr, 9 h, rt, 71–100% yield.[21]

19. $CH_2(OMe)_2$, cat. P_2O_5, CHCl$_3$, 25°C, 30 min, 95% yield.[22]

20. $CH_2(OMe)_2$, Me$_3$SiI or CH_2=CHCH$_2$SiMe$_3$, I$_2$, 76–95% yield.[23]

21. $CH_2(OMe)_2$, TsOH, LiBr, 9 h, rt, 71–100% yield.[21]

22. $CH_2(OMe)_2$, ZrCl$_4$, rt, 93–98% yield. TBDMS and THP ethers are converted to MOM ethers directly by this method.[24]

23. $CH_2(OMe)_2$, $Sc(OTf)_3$, $CHCl_3$, reflux, 77–98% yield.[25]

24. From a stannylmethyl ether: electrolysis, MeOH, 90% yield.[26]

25. From a trimethylsilyl glycoside: TMSOTf or TFA or $BF_3 \cdot Et_2O$, $CH_3OCH_2OCH_3$, 54–66% yield.[27]

26. From a PMB ether: $CH_2(OMe)_2$, MOMBr, $SnBr_2$, $ClCH_2CH_2Cl$, rt, 57–81% yield. Phenolic PMB ethers were not converted efficiently. A BOM ether was prepared using this method.[28]

27. The following reaction works best for secondary alcohols. Primary and tertiary alcohols give yields in the 50–60% range, whereas secondary alcohols give yields from 84–98% .[29]

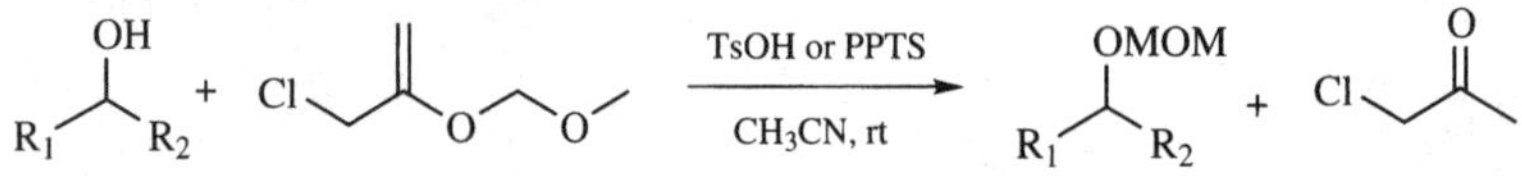

Cleavage

1. Trace concd. HCl, MeOH, 62°C, 15 min.[30]

2. 6 M HCl, aq. THF, 50°C, 6–8 h, 95% yield.[84] An attempt to cleave the MOM group with acid in the presence of a dimethyl acetal resulted in the cleavage of both groups, probably by intramolecular assistance.[32]

3. Concd. HCl, isopropyl alcohol (IPA), 65% yield.[33]

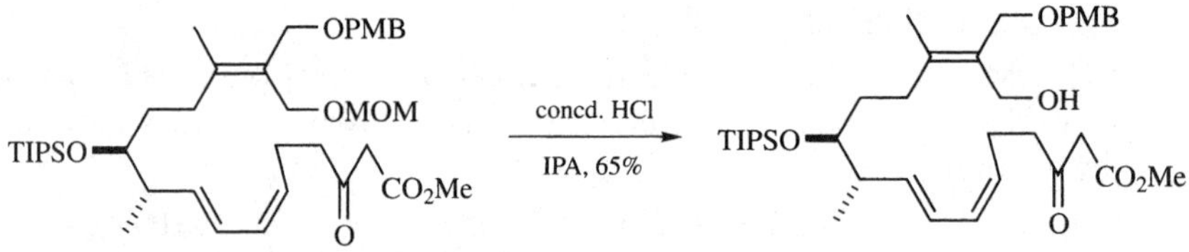

Other methods attempted for the cleavage of this MOM group were unsuccessful.

4. Pyridinium p-toluenesulfonate, t-BuOH or 2-butanone, reflux, 80–99% yield.[34] This method is useful for allylic alcohols. MEM ethers are also cleaved under these conditions. PPTS (t-BuOH, 84°C, 8 h, 45% yield) has been used to cleave a MOM in the presence of a PMB group, which is somewhat acid-sensitive.[35]

5. AcCl, MeOH, 0°C, 4 d, 93% yield.[36] This is a method of generating HCl *in situ*. Note that the acid labile PMB group was retained.

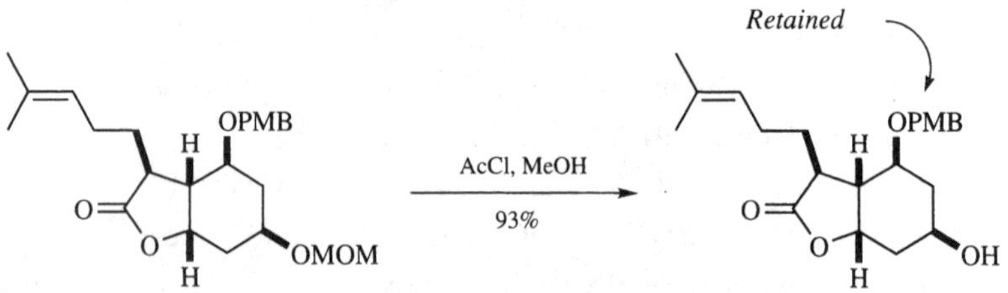

6. CF_3COOH, CH_2Cl_2, >85% yield.[37]

An attempt to deprotect a MOM ether in a synthesis of Pamamycin 621A resulted in participation of the PMB ether and the formation of the formaldehyde acetal which is very difficult to cleave.[38] Since PMB ethers can be cleaved with TFA, the formaldehyde acetal probably forms after loss of the PMB group rather than by participation through an oxacarbenium ion.

A similar problem was encountered when BBr$_3$ was used in an attempt to remove a MOM group with a proximal PMB ether.[39] The methylene acetal was also formed from a MOM ether during an attempt to remove a TBDPS ether with HF/pyridine.[40]

7. Dowex-50W-X2, aq. MeOH, 42–97% yield.[41]

Other methods resulted in skeletal rearrangement. This study also showed that the rate of acid-catalyzed MOM cleavage increases in the order: primary (30 h) < secondary (8 h) < tertiary (0.5–2 h). MOM ethers of tertiary alcohols are cleaved in excellent yield (94–97% yield).

8. 50% AcOH, cat. H$_2$SO$_4$, reflux, 10–15 min, 80% yield.[42]

9. MOM ethers can be converted directly to an acetate (FeCl$_3$, Ac$_2$O, 2–9 h, 20–95% yield), which is easily hydrolyzed to the alcohol.[43] InI$_3$/Ac$_2$O converts MOM and THP ethers to acetates.[44]

10. Ac_2O, $BF_3 \cdot Et_2O$, 4°C, 89% yield[45] This reagent combination converts the MOM ether to the $AcOCH_2OR$ ether which is cleavable with base.

11. PhSH, $BF_3 \cdot Et_2O$, 98% yield.[46] With dimethylsulfide as the cation scavenger an adjacent PMB ether is stable.[47]

12. 1,3-dithiane, $BF_3 \cdot Et_2O$, 84% yield.

13. Ph_3CBF_4, CH_2Cl_2, 25°C.[48]

Ref. 49

14. Catechol boron halides, particularly the bromide,

are effective reagents for the cleavage of MOM ethers. The bromide also cleaves the following groups in the order: MOMOR ≈ MEMOR > t-BOC > Cbz ≈ t-BuOR > BnOR > allylOR > t-BuO₂CR ≈ 2° alkylOR > BnO₂CR > 1° alkylOR ≫ alkylO₂CR. The t-butyldimethylsilyl (TBDMS), t-butyldiphenylsilyl (TBDPS) and the PMB groups are stable to this reagent.[50] The chloride is less reactive and thus may be more useful for achieving selectivity in multifunctional substrates. Yields are generally > 83%.[51] If the reaction is run in AcOH, formyl acetals are not formed in cases having a 1,3-disposed alcohol.[52] It appears that the reagent should be used in >1 equivalent because a methylene-bridged dimer was formed during a synthesis of Epoxydictymene with <1 eq.[53]

15. $(i\text{-PrS})_2BBr$, MeOH, 94% yield.[54] This method has the advantage that 1,2- and 1,3-diols do not give formyl acetals as is sometimes the case in cleaving MOM groups with neighboring hydroxyl groups.[55] The reagent also cleaves MEM groups and, under basic conditions, affords the $i\text{-PrSCH}_2OR$ derivatives.

16. Me$_2$SiCl$_2$, TBAB, 4-Å sieves, CH$_2$Cl$_2$, 0°C, 6 h, 47% yield.[56]

17. Me$_2$BBr, CH$_2$Cl$_2$, −78°C, then NaHCO$_3$/H$_2$O, 87–95% yield.[57] This reagent also cleaves the MEM, MTM, and acetal groups. An ester, a BOC, and a TIPS group were unaffected by this reagent in a synthesis of the Didemnins.[58]

18. Me$_3$SiBr, CH$_2$Cl$_2$, 0–78°C, 10 min, −10°C, 4 h, 93% yield.[59] Since the reagent is unstable and fumes in air, a method for generating TMSBr *in situ* from TMSCl and TBAB has been used to advantage.[60] A BOC and a benzyl ether were unaffected. This reagent also cleaves the acetonide, THP, trityl and *t*-BuMe$_2$Si groups. Esters, methyl and benzyl ethers, *t*-butyldiphenylsilyl ethers and amides are reported to be stable.[61]

19. LiBF$_4$, CH$_3$CN, H$_2$O, 72°C, 100% yield.[62] Note that the SEM group is also removed. LiBF$_4$ disproportionates to LiF and BF$_3$ upon heating, which no doubt has its mechanistic implications.

20. MgBr$_2$, ether, BuSH, rt, 40–97% yield. Tertiary and allylic MOM derivatives seem to give low yields, but this is not always the case as with the example below. MTM and SEM ethers are also cleaved, but MEM ethers are stable.[63,64]

21. ZrCl$_4$, IPA, reflux, 93–97% yield.[24]

22. Sc(OTf)$_3$, CH$_3$CN, HOCH$_2$CH$_2$CH$_2$OH, reflux, 1–4 h, 79–98% yield. THP ethers are similarly cleaved.[65]

23. Bi(OTf)$_3$, THF, H$_2$O, rt, 15–60 min, 86–95% yield. Both phenolic and alkanolic MOM ethers are readily removed.[66] TBS ether is unstable to these conditions.

24. AlCl$_3$, NaI, CH$_3$CN, CH$_2$Cl$_2$, 0°C, 25 min, >70% yield.[67]

25. The thermolysis of MOM, MEM, and THP ether in ethylene or propylene glycol at 120–160°C releases the alcohol under neutral conditions, but tertiary derivatives give some by-products that are consistent with a carbenium ion intermediate.[68]

26. There are times when the MOM group is not such an innocent bystander and participates in some unexpected and surprising reactions.[69]

27. The following was an attempt to prepare the silyl enol ether, but the reaction gave the unexpected silyl acetal.[70]

1. G. Stork and T. Takahashi, *J. Am. Chem. Soc.*, **99**, 1275 (1977).

2. R. J. Linderman, M. Jaber, and B. D. Griedel *J. Org. Chem.*, **59**, 6499 (1994); J. M. Chong and L. Shen, *Synth. Commum.*, **28**, 2801 (1998); M. Reggelin and S. Doerr, *Synlett*, 1117 (2004).

3. For a review of α-monohalo ethers in organic synthesis, see T. Benneche, *Synthesis*, 1 (1995).

4. A. F. Kluge, K. G. Untch, and J. H. Fried, *J. Am. Chem. Soc.*, **94**, 7827 (1972).

5. D. Askin, R. P. Volante, R. A. Reamer, K. M. Ryan, and I. Shinkai, *Tetrahedron Lett.*, **29**, 277 (1988).

6. K. Narasaka, T. Sakakura, T. Uchimaru, and D. Guédin-Vuong, *J. Am. Chem. Soc.*, **106**, 2954 (1984).

7. S. David, A. Thieffry, and A. Veyrières, *J. Chem. Soc., Perkin Trans. 1*, 1796 (1981).

8. M. Ihara, M. Suzuki, K. Fukumoto, T. Kametani, and C. Kabuto, *J. Am. Chem. Soc.*, **110**, 1963 (1988).

9. M. Takasu, Y. Naruse, and H. Yamamoto, *Tetrahedron Lett.*, **29**, 1947 (1988).

10. R. W. Friesen and C. Vanderwal, *J. Org. Chem.*, **61**, 9103 (1996).

11. B. C. Ranu, A. Majee and A. R. Das, *Synth. Commun.*, **25**, 363 (1995).

12. P. Kumar, S. V. N. Raju, R. S. Reddy, and B. Pandey, *Tetrahedron Lett.*, **35**, 1289 (1994).

13. B. F. Marcune, S. Karady, U.-H Dolling, and T. J. Novak, *J. Org. Chem.*, **64**, 2446 (1999).

14. G. A. Olah, A. Husain, B. G. B. Gupta, and S. C. Narang, *Synthesis*, 471 (1981).

15. M. P. Groziak and A. Koohang, *J. Org. Chem.*, **57**, 940 (1992).

16. S. Nishino and Y. Ishido, *J. Carbohydr. Chem.*, **5**, 313 (1986).

17. U. A. Schaper, *Synthesis*, 794 (1981).

18. N. W. Boaz and B. Venepalli, *Org. Proc. Res. Dev.*, **5**, 127 (2001).

19. M. L. Kantam and P. L. Santhi, *Synlett*, 429 (1993).

20. H. K. Patney, *Synlett*, 567 (1992).

21. J.-L. Gras, Y.-Y. K. W. Chang, and A. Guerin, *Synthesis*, 74 (1985).

22. K. Fuji, S. Nakano, and E. Fujita, *Synthesis*, 276 (1975).

23. G. A. Olah, A. Husain, and S. C. Narang, *Synthesis*, 896 (1983).

24. G. V. M. Sharma, K. L. Reddy, P. S. Lakshmi, and P. R. Krishna, *Tetrahedron Lett.*, **45**, 9229 (2004).

25. B. Karimi and L. Ma'mani, *Tetrahedron Lett.*, **44**, 6051 (2003).

26. J.-i. Yoshida, Y. Ishichi, K. Nishiwaki, S. Shiozawa, and S. Isoe, *Tetrahedron Lett.*, **33**, 2599 (1992).

27. K. Jansson and G. Magnusson, *Tetrahedron*, **46**, 59 (1990).

28. T. Oriyama, M. Kimura, and G. Koga, *Bull. Chem. Soc. Jpn.*, **67**, 885 (1994).

29. Y. Watanabe and T. Ikemoto, *Tetrahedron Lett.*, **45**, 5795 (2004).

30. J. Auerbach and S. M. Weinreb, *J. Chem. Soc., Chem. Commun.*, 298 (1974).

31. A. I. Meyers, J. L. Durandetta, and R. Munavu, *J. Org. Chem.*, **40**, 2025 (1975).

32. M. L. Bremmer, N. A. Khatri, and S. M. Weinreb, *J. Org. Chem.*, **48**, 3661 (1983).

33. D. G. Hall and P. Deslogchamps, *J. Org. Chem.*, **60**, 7796 (1995).

34. H. Monti, G. Léandri, M. Klos-Ringquet, and C. Corriol, *Synth. Commun.*, **13**, 1021 (1983).

35. A. K. Ghosh, Y. Wang, and J. T. Kim, *J. Org. Chem.*, **66**, 8973 (2001).

36. S. Amano, N. Takemura, M. Ohtusuka, S. Ogawa, and N. Chida, *Tetrahedron Lett.*, **55**, 3855 (1999).

37. R. B. Woodward and 48 co-workers, *J. Am. Chem. Soc.*, **103**, 3210 (1981).

38. M. A. Calter and F. C. Bi, *Org. Lett.*, **2**, 1529 (2000).

39. X. Wang and J. A. Porco, Jr., *J. Am. Chem. Soc.*, **125**, 6040 (2003).

40. C. Aïssa, R. Riveiros, J. Ragot, and A. Furstner, *J. Am. Chem. Soc.*, **125**, 15512 (2003).

41. H. Seto and L. N. Mander, *Synth. Commun.*, **22**, 2823 (1992).

42. F. B. Laforge, *J. Am. Chem. Soc.*, **55**, 3040 (1933).

43. M. P. Bosch, I. Petschen, and A. Guerrero, *Synthesis*, 300 (2000).

44. B. C. Ranu and A. Hajra, *J. Chem. Soc. Perkin Trans. 1*, 355 (2001).

45. D. F. Ewing, V. Glacon, G. Mackenzie, and C. Len, *Tetrahedron Lett.*, **43**, 989 (2002).

46. G. R. Kieczykowski and R. H. Schlessinger, *J. Am. Chem. Soc.*, **100**, 1938 (1978).

47. J. Cossy, F. Pradaux, and S. BouzBouz, *Org. Lett.*, **3**, 2233 (2001).

48. T. Nakata, G. Schmid, B. Vranesic, M. Okigawa, T. Smith-Palmer, and Y. Kishi, *J. Am. Chem. Soc.*, **100**, 2933 (1978).

49. J. M. Schkeryantz and S. J. Danishefsky, *J. Am. Chem. Soc.*, **117**, 4722 (1995).

50. L. A. Paquette, Z. Gao, Z. Ni, and G. F. Smith, *Tetrahedron Lett.*, **38**, 1271 (1997); L. A. Paquette, Z. Gao, Z. Ni, and G. F. Smith, *J. Am. Chem. Soc.*, **120**, 2543 (1998).

51. R. K. Boeckman, Jr., and J. C. Potenza, *Tetrahedron Lett.*, **26**, 1411 (1985).

52. C. Yu, B. Liu, and L. Hu, *Tetrahedron Lett.*, **41**, 819 (2000).

53. L. A. Paquette, L.-Q. Sun, D. Friedrich, and P. B. Savage, *J. Am. Chem. Soc.*, **119**, 8438 (1997).

54. E. J. Corey, D. H. Hua, and S. P. Seitz, *Tetrahedron Lett.*, **25**, 3 (1984).

55. B. C. Barot and H. W. Pinnick, *J. Org. Chem.*, **46**, 2981 (1981).

56. A. K. Ghosh and W. Liu, *J. Org. Chem.*, **62**, 7908 (1997).

57. Y. Guindon, H. E. Morton, and C. Yoakim, *Tetrahedron Lett.*, **24**, 3969 (1983).

58. A. J. Pfizenmayer, J. M. Ramanjulu, M. D. Vera, X. B. Ding, D. Xiao, W. C. Chen, and M. M. Joullié, *Tetrahedron*, **55**, 313 (1999).

59. T. Hosoya, E. Takashiro, T. Matsumoto, and K. Suzuki, *J. Am. Chem. Soc.*, **116**, 1004 (1994).

60. A. K. Ghosh, W. M. Liu, Y. B. Xu, and Z. D. Chen, *Angew, Chem. Int. Ed Engl.* **35**, 74 (1996).

61. S. Hanessian, D. Delorme, and Y. Dufresne, *Tetrahedron Lett.*, **25**, 2515 (1984). For *in situ* prepared TMSBr, see R. B. Woodward and 48 co-workers, *J. Am. Chem. Soc.*, **103**, 3213 (note 2) (1981).

62. R. E. Ireland and M. D. Varney, *J. Org. Chem.*, **51**, 635 (1986).

63. S. Kim, I. S. Kee, Y. H. Park, and J. H. Park, *Synlett*, 183 (1991).

64. L. N. Mander and R. J. Thomson, *J. Org. Chem.*, **70**, 1654 (2005).

65. H. M. I. Osborn and N. A. O. Williams, *Org. Lett.*, **6**, 3111 (2004).

66. S. V. Reddy, R. J. Rao, U. S. Kumar, and J. M. Rao, *Chem. Lett.*, **32**, 1038 (2003).

67. E. D. Moher, P. A. Grieco, and J. L. Collins, *J. Org. Chem.*, **58**, 3789 (1993).

68. H. Miyake, T. Tsumura, and M. Sasaki, *Tetrahedron Lett.*, **45**, 7213 (2004).

69. N. A. Powell and W. R. Roush, *Org. Lett.*, **3**, 453 (2001).

70. T. L. Graybill, E. G. Casillas, K. Pal, and C. A. Townsend, *J. Am. Chem. Soc.*, **121**, 7729 (1999).

Methylthiomethyl Ether (MTM Ether): CH_3SCH_2OR (Chart 1)

Methylthiomethyl ethers are quite stable to acidic conditions. Most ethers and 1,3-dithianes are stable to the neutral mercuric chloride used to remove the MTM

group. One problem with the MTM group is that it is sometimes difficult to introduce.

Formation

1. NaH, DME, CH_3SCH_2Cl, NaI, 0°C, 1 h to 25°C, 1.5 h, >86% yield.[1]
2. CH_3SCH_2I, DMSO, Ac_2O, 20°C, 12 h, 80–90% yield.[2]
3. DMSO, Ac_2O, AcOH, 20°C, 1–2 days, 80%.[3]
4. CH_3SCH_2Cl, $AgNO_3$, Et_3N, benzene, 22–80°C, 4–24 h, 60–80% yield.[4]
5. DMSO, molybdenum peroxide, benzene, reflux, 7–20 h, ≈60% yield.[5] This method was used to monoprotect 1,2-diols. The method is not general because oxidation to α-hydroxy ketones and diketones occurs with some substrates. Based on the mechanism and on the results, it would appear that overoxidation has a strong conformational dependence.
6. MTM ethers can be prepared from MEM and MOM ethers by treatment with Me_2BBr to form the bromomethyl ether, which is trapped with MeSH and (*i*-Pr)$_2$NEt, 87–91% yield. This method may have some advantage since the preparation of MTM ethers directly is not always simple. Acetals in the presence of thiols are converted *O,S*-acetals.[6]
7. CH_3SCH_3, CH_3CN, $(PhCOO)_2$, 0°C, 2 h, 75–95% yield.[7,8] Acetonides, THP ethers, alkenes, ketones, The Fmoc group[9] and epoxides all survive these conditions.
8. $(COCl)_2$, DMSO, −78°C to −50°C; Et_3N, −78°C to −15°C.[10]

Cleavage

1. $HgCl_2$, CH_3CN, H_2O, 25°C, 1–2 h, 88–95% yield.[1] If 2-methoxyethanol is substituted for water, the MTM ether is converted to a MEM ether. Similarly, substitution with methanol affords a MOM ether.[11] If the MTM ether has an adjacent hydroxyl, it is possible to form the formylidene acetal as a by-product of cleavage.[12]
2. $HgCl_2$, $CaCO_3$, MeCN, H_2O.[1] The calcium carbonate is used as an acid scavenger for acid sensitive substrates.

3. MeI, acetone, H_2O, $NaHCO_3$, heat a few hours, 80–95% yield.[3]
4. Electrolysis: applied voltage = 10 V, AcONa, AcOH; K_2CO_3, MeOH, H_2O, 80–95% yield.[13]
5. MgI_2, ether, Ac_2O, rt, 90–100% yield. Cleavage occurs to give a mixture of acetate and an acetoxymethyl ether that is reported to be very acid- and base- sensitive.[14]
6. Me_3SiCl, Ac_2O, 90% .[15] Treatment of the resulting acetoxymethyl ether with acid or base readily affords the free alcohol.

7. Ph_3CBF_4, CH_2Cl_2, 5–30 min, 80–95% yield.[16] The mechanism of this cleavage has been determined to involve complex formation by the trityl cation with the sulfur, followed by hydrolysis, rather than by hydride abstraction.[17]

In this case the use of $HgCl_2$, $AgNO_3$, and MeI gave extensive decomposition.

8. $Hg(OTf)_2$, CH_2Cl_2, H_2O, Na_2HPO_4.[18]

9. $AgNO_3$, THF, H_2O, 2,6-lutidine, 25°C, 45 min, 88–95% yield.[1] These conditions can be used to cleave an MTM ether in the presence of a dithiane.[19]

10. $MgBr_2$, n-BuSH, Et_2O, rt, 0.5–3 h, 83–85% yield.[20]

1. E. J. Corey and M. G. Bock, *Tetrahedron Lett.*, **16**, 3269 (1975).

2. K. Yamada, K. Kato, H. Nagase, and Y. Hirata, *Tetrahedron Lett.*, **17**, 65 (1976).

3. P. M. Pojer and S. J. Angyal, *Aust. J. Chem.*, **31**, 1031 (1978).

4. K. Suzuki, J. Inanaga, and M. Yamaguchi, *Chem. Lett.*, **8**, 1277 (1979).

5. Y. Masuyama, M. Usukura, and Y. Kurusu, *Chem. Lett.*, **11**, 1951 (1982).

6. H. E. Morton and Y. Guindon, *J. Org. Chem.*, **50**, 5379 (1985). Y. Guindon, M. A. Bernstein, and P. C. Anderson, *Tetrahedron Lett.*, **28**, 2225 (1987).

7. J. C. Medina, M. Salomon, and K. S. Kyler, *Tetrahedron Lett.*, **29**, 3773 (1988).

8. P. Garner and J. U. Yoo, *Tetrahedron Lett.*, **34**, 1275 (1993).

9. P. Garner, S. Dey, Y. Huang, and X. Zhang, *Org. Lett.*, **1**, 403 (1999).

10. D. R. Williams, F. D. Klinger, and V. Dabral, *Tetrahedron Lett.*, **29**, 3415 (1988).

11. P. K. Chowdhury, D. N. Sarma, and R. P. Sharma, *Chem. Ind, (London)*, 803 (1984).

12. M. P. Wachter and R. E. Adams, *Synth. Commun.*, **10**, 111 (1980).

13. T. Mandai, H. Yasunaga, M. Kawada, and J. Otera, *Chem. Lett.*, **13**, 715 (1984).

14. P. K. Chowdhury, *J. Chem. Res., Synop.*, 68 (1992).

15. D. N. Sarma, N. C. Barua, and R. P. Sharma, *Chem. Ind. (London)* 223 (1984); N. C. Barur, R. P. Sharma and J. N. Baruah, *Tetrahedron Lett.*, **24**, 1189 (1983).

16. P. K. Chowdhury, R. P. Sharma, and J. N. Baruah, *Tetrahedron Lett.*, **24**, 4485 (1983).

17. H. Niwa and Y. Miyachi, *Bull. Chem. Soc. Jpn.*, **64**, 716 (1991).

18. G. E. Keck, E. P. Boden, and M. R. Wiley, *J. Org. Chem.*, **54**, 896 (1989).

19. E. J. Corey, D. H. Hua, B.-C. Pan, and S. P. Seitz, *J. Am. Chem. Soc.*, **104**, 6818 (1982).

20. S. Kim, I. S. Kee, Y. H. Park, and J. H. Park, *Synlett*, 183 (1992).

(Phenyldimethylsilyl)methoxymethyl Ether (SMOM−OR):
$C_6H_5(CH_3)_2SiCH_2OCH_2OR$

Formation

SMOMCl, *i*-PrEt$_2$N, CH$_3$CN, 3 h, 40°C, 87–91% yield.[1] Diols are selectively protected using the stannylene methodology.

Cleavage

AcOOH, KBr, AcOH, NaOAc, 1.5 h, 20°C, 82–92% yield.[1] The SMOM group is stable to Bu$_4$NF; NaOMe/MeOH; 4 *N* NaOH/dioxane/methanol; *N*-iodosuccinimide, cat. trifluoromethanesulfonic acid.

1. G. J. P. H. Boons, C. J. J. Elie, G. A. van der Marel, and J. H. van Boom, *Tetrahedron Lett.*, **31**, 2197 (1990).

Benzyloxymethyl Ether (BOM−OR): PhCH$_2$OCH$_2$OR

Formation

1. PhCH$_2$OCH$_2$Cl, (*i*-Pr)$_2$NEt, 10–20°C, 12 h, 95% yield.[1,2] Bu$_4$NI can be added to increase the reactivity for protection of more hindered alcohols.

2. PhCH$_2$OCH$_2$Cl, NaI, proton sponge [1,8-bis(dimethylamino)naphthalene], 84% yield.[3] BOMBr can also be used.[4]

Cleavage

1. Na, NH$_3$, EtOH.[1,5] A trisubstituted epoxide was stable to these conditions.

2. Li, NH$_3$.[6,7] As expected benzyl groups are also cleaved.

3. LiDBB, THF, −78°C, 88% yield.[8] Contrary to expectation hydrogenolysis with Pd(OH)$_2$ failed to remove the BOM group without also reducing the olefin. See #5 below.

4. PhSH, BF$_3$·Et$_2$O, CH$_2$Cl$_2$, −78°C, 95% yield.[9,10]

5. H_2, 1 atm, Pd–C, EtOAc–hexane, 68% yield.[11] This method is compatible with an N–O bond and an aziridine.[12]

6. H_2, 1 atm, 10% Pd–C, 0.01 N HClO$_4$, in 80% THF/H$_2$O, 25°C.[13] Without the acid, the overall deprotection was sluggish.

7. Transfer hydrogenation: 1-methyl-1,4-cyclohexadiene, Pd/C, CaCO$_3$, EtOH, 100% yield.[13] This method was compatible with a disubstituted olefin. Benzyl groups are also cleaved.

8. Transfer hydrogenation: HCO$_2$H, MeOH, Pd black, rt, 1.5 h. These conditions also remove a Cbz group from an amine.[14] Ammonium formate can also be used as the hydrogen source.[15]

9. Bromocatecholborane, 71% yield. A 3° TMS and 1° TBS ether were retained.[16]

10. HCl, MeOH, 56% yield.[17]

11. MeOH, Dowex 50W-X8, rt, 5–6 days, 90% yield.[18]

12. AlH$_2$Cl, AlHCl$_2$, or BH$_3$ in toluene or THF. See the section on SEM ethers for a selectivity study of these reagents with the SEM, MTM, EOM (ethoxymethyl), and p-AOM groups.[19]

13. HCl, NaI, 97% yield based on 67% conversion.[20]

14. LiBF$_4$, CH$_3$CN, H$_2$O, reflux, 62% yield.[21] LiBF$_4$ upon heating dissociates into LiF and BF$_3$.

Note that the methyl ketal is also cleaved

1. G. Stork and M. Isobe, *J. Am. Chem. Soc.*, **97**, 6260 (1975).

2. D. A. Evans, S. L. Bender, and J. Morris, *J. Am. Chem. Soc.*, **110**, 2506 (1988).

3. S. F. Martin, W.-C. Lee, G. J. Pacofsky, R. P. Gist, and T. A. Mulhern, *J. Am. Chem. Soc.*, **116**, 4674 (1994).

4. D. A. Evans, S. L. Bender, and J. Morris, *J. Am. Chem. Soc.*, **110**, 2506 (1988).

5. W. C. Still and D. Mobilio, *J. Org. Chem.*, **48**, 4785 (1983).

6. H. Nagaoka, W. Rutsch, G. Schmid, H. Iio, M. R. Johnson, and Y. Kishi, *J. Am. Chem. Soc.*, **102**, 7962 (1980).

7. J. A. Marshall and G. P. Luke, *J. Org. Chem.*, **58**, 6229 (1993).

8. C.-H. Tan and A. B. Holmes, *Chem. Eur. J.*, **7**, 1845 (2001).

9. K. Suzuki, K. Tomooka, E. Katayama, T. Matsumoto, and G.-P. C. Tsuchihashi, *J. Am. Chem. Soc.*, **108**, 5221 (1986).

10. K. C. Nicolaou, C.-K. Hwang, M. E. Duggan, D. A. Nugiel, Y. Abe, K. B. Reddy, S. A. DeFrees, D. R. Reddy, R. A. Awartani, S. R. Conley, F. P. J. T. Rutjes, and E. A. Theodorakis, *J. Am. Chem. Soc.*, **117**, 10227 (1995).

11. D. Tanner and P. Somfai, *Tetrahedron*, **43**, 4395 (1987).

12. T. Katoh, E. Itoh, T. Yoshino, and S. Terashima, *Tetrahedron*, **53**, 10229 (1997).

13. A. B. Smith, III, V. A. Doughty, Q. Lin, L. Zhuang, M. D. McBriar, A. M. Boldi, W. H. Moser, N. Murase, K. Nakayama, and M. Sobukawa, *Angew. Chem. Int. Ed.*, **40**, 191 (2001).

14. A. G. Myer, D. Y. Gin, and D. H. Rogers, *J. Am. Chem. Soc.*, **116**, 4697 (1994).

15. M. Izumi, K. Wada, H. Yuasa, and H. Hashimoto, *J. Org. Chem.*, **70**, 8817 (2005).

16. K. Lee and J. K. Cha, *J. Am. Chem. Soc.*, **123**, 5590 (2001).

17. R. S. Coleman and E. B. Grant, *J. Am. Chem. Soc.*, **116**, 8795 (1994).

18. W. R. Roush, M. R. Michaelides, D. F. Tai, and W. K. M. Chong, *J. Am. Chem. Soc.*, **109**, 7575 (1987).

19. I. Bajza, Z. Varga, and A. Liptak, *Tetrahedron Lett.*, **34**, 1991 (1993).

20. P. A. Wender, N. F. Badham, S. P. Conway, P. E. Floreancig, T. E. Glass, J. B. Houze, N. E. Krauss, D. Lee, D. G. Marquess, P. L. McGrane, W. Meng, M. G. Natchus, A. J. Shuker, J. C. Sutton, and R. E. Taylor, *J. Am. Chem. Soc.*, **119**, 2757 (1997).

21. G. E. Keck and A. P. Truong, *Org. Lett.*, **7**, 2153 (2005).

p-Methoxybenzyloxymethyl Ether (PMBM−OR):
$MeOC_6H_4CH_2OCH_2OR$ and

[(3,4-Dimethoxybenzyl)oxy]methyl Ether (DMBM−OR):
$(MeO)_2C_6H_3CH_2OCH_2OR$

The [(3,4-dimethoxybenzyl)oxy]methyl group has been used similarly to the PMBM group except, that as expected, it is more easily cleaved (DDQ, CH_2Cl_2, *t*-BuOH, phosphate buffer, pH 6.0, 23°C, 110 min, 88% yield). In fact, it was successfully removed where a PMBM ether could not be cleaved.[1]

Formation

1. *p*-$MeOC_6H_4CH_2OCH_2Cl$, (*i*-Pr)$_2$NEt (DIPEA), CH_2Cl_2, 78–100% yield.[2,3]

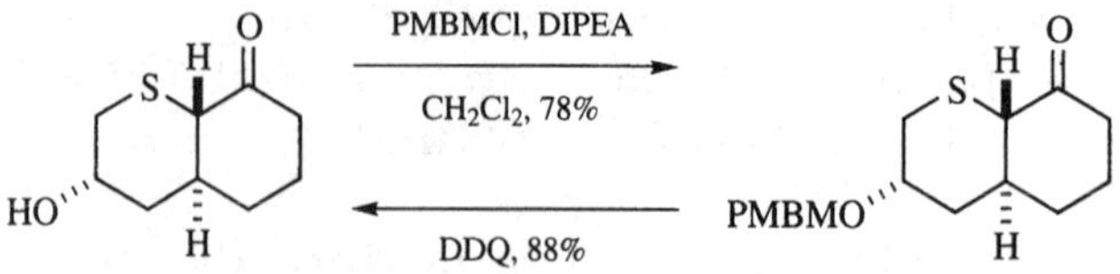

2. Lithium alkoxides react with PMBMCl to form the ethers.[4]

3. p-MeOC$_6$H$_4$CH$_2$OCH$_2$SCH$_3$, CuBr$_2$, TBAB, MS4A, CH$_2$Cl$_2$, 58–95% yield.[5]

Cleavage

1. DDQ, H$_2$O, rt, 1–10 h, 63–96% yield.[3]
2. 3:1 THF-6 M HCl, 50°C, 6 h.[1]

1. H. Kigoshi, K. Suenaga, T. Mutou, T. Ishigaki, T. Atsumi, H. Ishiwata, A. Sakakura, T. Ogawa, M. Ojika, and K. Yamada, *J. Org. Chem.*, **61**, 5326 (1996).
2. G. Guanti, L. Banfi, E. Narisano, and S. Thea, *Tetrahedron Lett.*, **32**, 6943 (1991).
3. A. P. Kozikowski and J.-P. Wu, *Tetrahedron Lett.*, **28**, 5125 (1987).
4. J. A. Marshall and W. Y. Gung, *Tetrahedron Lett.*, **30**, 7349 (1989).
5. D. Sawada and Y. Ito, *Tetrahedron Lett.*, **42**, 2501 (2001).

p-Nitrobenzyloxymethyl Ether: NO$_2$C$_6$H$_4$CH$_2$OCH$_2$OR

Formation

1. NO$_2$C$_6$H$_4$CH$_2$OCH$_2$-Py$^+$ Cl$^-$, TBAB, DMF, 75°C.[1]
2. NO$_2$C$_6$H$_4$CH$_2$OCH$_2$SCH$_3$, CuBr$_2$, TBAB, MS4A, CH$_2$Cl$_2$, rt, 70–77% yield.[2]

Cleavage

1. TBAF, THF, 25°C, 24 h.[1]
2. The section on the cleavage of 4-nitrobenzyl ethers should be consulted since those methods are expected to be applicable in this case as well.

1. G. R. Gough, T. J. Miller, and N. A. Mantick, *Tetrahedron Lett.*, **37**, 981 (1996).
2. D. Sawada and Y. Ito, *Tetrahedron Lett.*, **42**, 2501 (2001).

o-Nitrobenzyloxymethyl Ether (NBM−OR): 2-NO$_2$C$_6$H$_4$CH$_2$OCH$_2$OR

Formation

1. This group was developed for 2′-protection in ribonucleotide synthesis.[1,2]

2. From a diol: Bu_2SnO, then $2\text{-}NO_2C_6H_4CH_2OCH_2Cl$.[3]

Cleavage

1. t-BuOH, H_2O, pH 3.7, long-wave UV for 4.5 h.[3]
2. Photolysis: $h\nu$, Pyrex filtered, 0.1 M sodium citrate buffer, pH 3.5, t-BuOH, 25°C, 2 h.[1,2]
3. Hydrogenolysis or nitro group reduction should cleave this group. See the section on the nitrobenzyl ether.

[(R)-1-(2-Nitrophenyl)ethoxy]methyl Ether ((R)-npeom−OR)

This group was developed for 2′-OH protection in ribonucleotide synthesis.[4,5] Its advantage is that the reduced steric hindrance of this and related groups improves coupling yields.[6] It is introduced on a diol using the stannylene method and the chloride. It is cleaved by photolysis (10 mM $MgCl_2$, 50 mM Tris·HCl, pH 8, H_2O, 25°C).

1. A. Stutz and S. Pitsch, *Synlett*, 930 (1999).
2. S. Pitsch, *Helv. Chim. Acta*, **80**, 2286 (1997).
3. M. E. Schwartz, R. R. Breaker, G. T. Asteriadis, J. S. deBear, and G. R. Gough, *Biorg. Med. Chem. Lett.*, **2**, 1019 (1992).
4. A. Stutz, C. Hobartner, and S. Pitsch, *Helv. Chim. Acta*, **83**, 2477 (2000).
5. S. Pitsch, P. A. Weiss, X. Wu, D. Ackermann, and T. Honegger, *Helv. Chim. Acta*, **82**, 1753 (1999).
6. S. Pitsch, *Chimia*, **55**, 320 (2001).

(4-Methoxyphenoxy)methyl Ether (p-AOM−OR), (p-Anisyloxymethyl Ether): $ROCH_2OC_6H_4$−4-OCH_3

Formation[1]

1. p-AOMCl, $PhCH_2N^+Et_3Cl^-$, CH_3CN, 50% NaOH, rt, 46–91% yield.

2. p-AOMCl, $(i$-Pr$)_2$NEt, CH$_2$Cl$_2$, reflux.
3. p-AOMCl, DMF, 18-crown-6, K$_2$CO$_3$, rt.

Cleavage

1. CAN, CH$_3$CN, H$_2$O, 0°C, 0.5 h, 60–98% yield.[1] In some cases the addition of pyridine improves the yields.[2]
2. CAN, CH$_3$CN, H$_2$O, 2,6-pyridinedicarboxylic acid N-oxide (PDNO), 0°C, 20 min, 77% yield. the N-oxide was essential for this cleavage to work.[3]

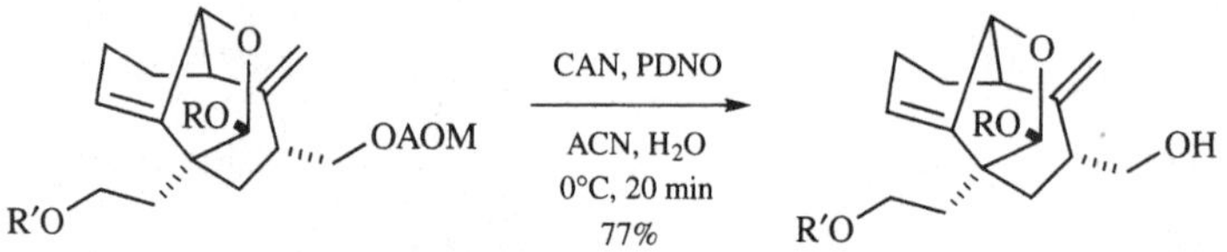

3. BH$_3$, toluene converts the p-AOM ether into a methyl ether. For a stability comparison of this group with MTM, SEM, BOM and EOM to various hydride reagents see the section on SEM ethers.[4]

1. Y. Masaki, I. Iwata, I. Mukai, H. Oda, and H. Nagashima, *Chem. Lett.*, **18**, 659 (1989).
2. D. L. Clive, Y. X. Bo, Y. Tao, S. Daigneault, and Y. J. Meignam, *J. Am. Chem. Soc.*, **120**, 10332 (1998).
3. D. L. J. Clive and S. Sun, *Tetrahedron Lett.*, **42**, 6267 (2001).
4. I. Bajza, Z. Varga, and A. Liptak, *Tetrahedron Lett.*, **34**, 1991 (1993).

Guaiacolmethyl Ether (GUM−OR): 2-MeOC$_6$H$_4$OCH$_2$OR

Formation/Cleavage

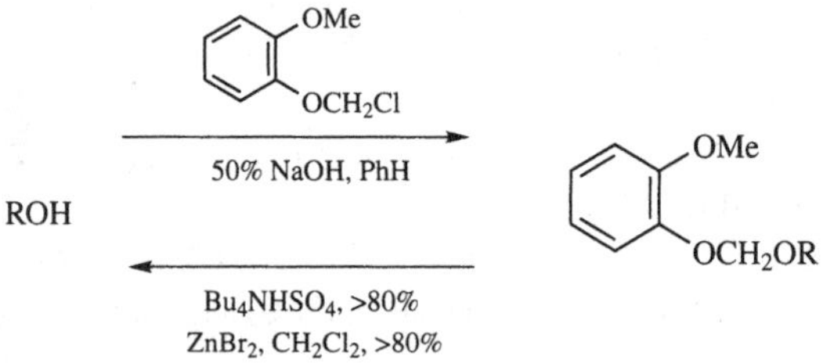

It is possible to introduce this group selectively onto a primary alcohol in the presence of a secondary alcohol. The derivative is stable to KMnO$_4$, m-chloroperoxybenzoic acid, LiAlH$_4$, and CrO$_3$·Pyr. Since this derivative is similar to the p-methoxyphenyl ether it should also be possible to remove it oxidatively. The GUM ethers are less stable than the MEM ethers in acid but have comparable stability to the SEM ethers. It is possible to remove the GUM ether in the presence of a MEM ether.[1]

1. B. Loubinoux, G. Coudert, and G. Guillaumet, *Tetrahedron Lett.*, **22**, 1973 (1981).

[(*p*-Phenylphenyl)oxy]methyl Ether (POM−OR): 4-C$_6$H$_5$C$_6$H$_4$OCH$_2$OR

This group was developed to impart crystallinity to an intermediate in a synthesis of PNU-140690. The derivative is formed from POMCl (from a 3° alcohol: toluene, DIPEA, reflux, 5 h, 76% yield) and can be cleaved with H$_2$SO$_4$ (THF, MeOH, rt, 84% yield).[1]

1. K. S. Fors, J. R. Gage, R. F. Heier, R. C. Kelly, W. R. Perrault, and N. Wicienski, *J. Org. Chem.*, **63**, 7348 (1998).

t-Butoxymethyl Ether: *t*-BuOCH$_2$OR

The advantage of this ether is that it can be introduced under relatively neutral conditions whereas the *t*-Bu group is introduced under acidic conditions, but can be cleaved by typical conditions used to cleave the *t*-Bu ether.

Formation

1. *t*-BuOCH$_2$Cl,[1] Et$_3$N, −20°C → 20°C, 3 h, 54–80% yield.[2]
2. *t*-BuOCH$_2$SO$_2$Ph, LiBr, TEA, toluene, 2–4 days, 70–92% yield.[3]
3. *t*-BuOCH$_2$SCH$_2$CH$_3$, CuBr$_2$, TBAB, MS4, CH$_2$Cl$_2$, rt, 4 h, 69–91% yield.[4]

Cleavage

CF$_3$COOH, H$_2$O, 20°C, 48 h, 85–90% yield.[2] The *t*-butoxymethyl ether is stable to hot glacial acetic acid; aqueous acetic acid, 20°C; and anhydrous trifluoroacetic acid.

1. For an improved preparation of this reagent, see J. H. Jones, D. W. Thomas, R. M. Thomas, and M. E. Wood, *Synth. Commun.*, **16**, 1607 (1986).
2. H. W. Pinnick and N. H. Lajis, *J. Org. Chem.*, **43**, 3964 (1978).
3. M. Julia, D. Uguen, and D. Zhang, *Synlett*, 503 (1991).
4. D. Sawada and Y. Ito, *Tetrahedron Lett.*, **42**, 2501 (2001).

4-Pentenyloxymethyl Ether (POM−OR)[1]: CH$_2$=CHCH$_2$CH$_2$CH$_2$OCH$_2$OR

Formation

POMCl, (*i*-Pr)$_2$NEt, CH$_2$Cl$_2$.[2] The related pentenyl glycosides, prepared by the usual methods, were used to protect the anomeric center.[2]

Cleavage

NBS, CH$_3$CN, H$_2$O, 62–90% yield.[2–4] The POM group has been selectively removed in the presence of an ethoxyethyl ether, TBDMS ether, benzyl ether, *p*-methoxybenzyl ether, an acetate, and an allyl group. Because the hydrolysis of a pentenyl 2-acetoxyglycoside was so much slower than a pentenyl 2-benzyloxyg-lycoside, the 2-benzyl derivative could be cleaved selectively in the presence of the 2-acetoxy derivative.[5] The POM group is stable to 75% AcOH but is cleaved by 5% HCl.

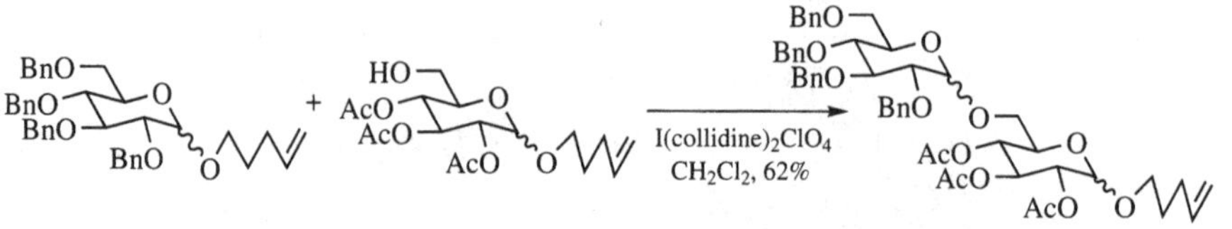

Cleavage of the POM group in the presence of neighboring hydroxyls can result in the formation of methylene acetals.[2]

1. The chemistry of the 4-pentenyloxy group has been reviewed by B. Fraser-Reid, U. E. Udodong, Z. Wu, H. Ottosson, J. R. Merritt, C. S. Rao, C. Roberts, and R. Madsen, *Synlett*, 927 (1992).

2. Z. Wu, D. R. Mootoo, and B. Fraser-Reid, *Tetrahedron Lett.*, **29**, 6549 (1988).

3. D. R. Mootoo, V. Date, and B. Fraser-Reid, *J. Am. Chem. Soc.*, **110**, 2662 (1988).

4. For a discussion of the factors that influence the rate of NBS induced n-pentenylglycoside hydrolysis, see C. W. Andrews, R. Rodebaugh, and B. Fraser-Reid, *J. Org. Chem.*, **61**, 5280 (1996).

5. D. R. Mootoo, P. Konradsson, U. Udodong, and B. Fraser-Reid, *J. Am. Chem. Soc.*, **110**, 5583 (1988).

Siloxymethyl Ether: RR'_2SiOCH_2OR'', $R' = Me$, $R = t$-Bu; $R = $ Thexyl, $R' = Me$; $R = t$-Bu, $R' = Ph$, $R = R' = i$-Pr (tom$-$OR)

These groups are sterically less demanding than the corresponding silyl ethers, but are cleaved by the same conditions as the silyl ethers.

Formation

1. RR'_2SiOCH_2Cl, $(i$-Pr$)_2NEt$, CH_2Cl_2, 73–92% yield.[1,2]

2. $(i$-Pr$)_3SiOCH_2SCH_3$, $CuBr_2$, TBAB, MS4A, CH_2Cl_2, 90–100% yield. Phenols can also be protected with this method.[3]

3. The stannylene method can be used to monoprotect a diol.[4,5]

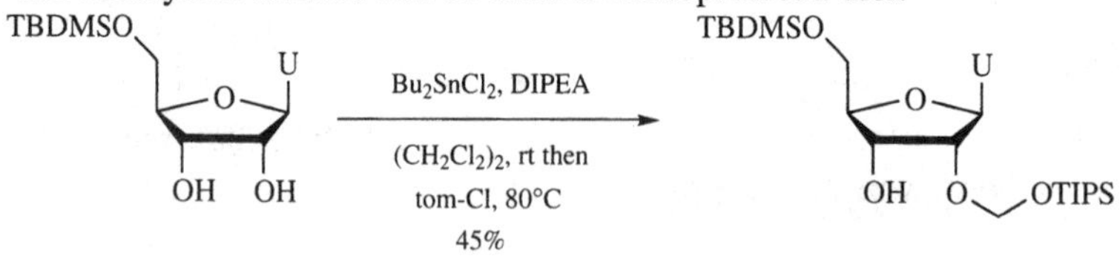

Cleavage

1. Bu_4NF, THF, 70–80% yield.[1] TBAF buffered with AcOH has also been used.[6]

2. Et_4NF, CH_3CN, rt, 64–75% yield.[1]

3. AcOH, H_2O.[1]

1. L. L. Gundersen, T. Benneche, and K. Undheim, *Acta Chem. Scand.*, **43**, 706 (1989).

2. E. Vedejs and J. D. Little, *J. Org. Chem.*, **69**, 1788 (2004).

3. D. Sawada and Y. Ito, *Tetrahedron Lett.*, **42**, 2501 (2001).

4. A. Stutz, C. Hobartner, and S. Pitsch, *Helv. Chim. Acta*, **83**, 2477 (2000).

5. D. A. Berry, K.-Y. Jung, D. S. Wise, A. D. Sercel, W. H. Pearson, H. Mackie, J. B. Randolph, and R. L. Somers, *Tetrahedron Lett.*, **45**, 2457 (2004); X. Wu and S. Pitsch, *Nucleic Acids Res.*, **26**, 4315 (1998).

6. C. Höbartner, R. Rieder, C. Kreutz, B. Puffer, K. Lang, A. Polonskaia, A. Serganov, and R. Micura, *J. Am. Chem. Soc.*, **127**, 12035 (2005).

2-Methoxyethoxymethyl Ether (MEM−OR): $CH_3OCH_2CH_2OCH_2OR$ (Chart 1)

MEM ethers are similar to the MOM and SEM ethers in their stability to protic acids but are more sensitive to Lewis acids because the additional ether improves its ability to coordinate a Lewis acid more strongly than the MOM or SEM ether.

Formation

1. NaH or KH, MEMCl, THF or DME, 0°C, 10–60 min, >95% yield.[1]

2. $MEMN^+Et_3Cl^-$, CH_3CN, reflux, 30 min, >90% yield.[1]

3. MEMCl, $(i\text{-}Pr)_2NEt$ (DIPEA), CH_2Cl_2, 25°C, 3 h, quant.[1]

4. The MEM group has been introduced on one of two sterically similar but electronically different alcohols in a 1,2-diol.[2]

Cleavage

1. $ZnBr_2$, CH_2Cl_2, 25°C, 2–10 h, 90% yield.[1] When a MEM protected diol was cleaved using $ZnBr_2$ in EtOAc, 1,3-dioxolane formation occurred,[3] but this can be prevented by the use of *in situ* prepared TMSI.[4]

2. $TiCl_4$, CH_2Cl_2, 0°C, 20 min, 95% yield.[1,5]

3. Me_2BBr, CH_2Cl_2, −78°C; $NaHCO_3$, H_2O, 87–95% yield.[6] This method also cleaves MTM and MOM ethers and ketals.

4. $(i\text{-}PrS)_2BBr$, DMAP; K_2CO_3, H_2O.[7] In this case the MEM ether is converted into the $i\text{-}PrSCH_2$-ether that can be cleaved using the same conditions used to cleave

the MTM ether. In one case where the related 2-chloro-1,3,2-dithioborolane was used for MEM ether cleavage, a thiol ($-OCH_2SCH_2CH_2SH$) was isolated as a by-product in 29% yield.[8]

5. Pyridinium p-toluenesulfonate, t-BuOH or 2-butanone, heat, 80–99% yield.[9] This method also cleaves the MOM ether and has the advantage that it cleanly cleaves allylic ethers that could not be cleaved by Corey's original procedure.

6. TFA, CH_2Cl_2, 90% yield.

7. HCO_2H, MeOH, 65°C, 4 h, 97% yield.[10]

8. Me_3SiCl, NaI, CH_3CN, −20°C, 79% .[11] Allylic and benzylic ethers tend to form some iodide as a by-product, but less iodide is formed than when Me_3SiI is used directly.

9. BBr 2 eq. CH_2Cl_2, −78°C.[12] Benzyl, allyl, methyl, THP, TBDMS and

TBDPS ethers are all stable to these conditions. A primary MEM group could be selectively removed in the presence of a hindered secondary MEM group.

10.

11. HBF_4, CH_2Cl_2, 0°C, 3 h, 50–60% yield.[14]

12. $CeCl_3$, CH_3CN, reflux.[15]

13. CBr_4, IPA, reflux, 94% yield.[16] MOM groups are also cleaved (87–97% yield).

14. In a study of the deprotection of the MEM ethers of hydroxyproline and serine derivatives, it was found that the MEM group was stable to conditions that normally cleave the t-butyl and BOC groups [CF_3COOH, CH_2Cl_2, 1:1 (v:v)]. The MEM group was also stable to 0.2 N HCl but not stable to 2.0 N HCl or HBr-AcOH.[17]

Removal Time in TFA/CH$_2$Cl$_2$(v/v)

	1:4	1:1	1:0
Z-Hyp(t-Bu)−ONb	45 min	15 min	5 min
Z-Hyp(MEM)OMe	10 h	6 h	2 h

Hyp = hydroxyproline, Nb = 4-nitrobenzoate

15. (a) *n*-BuLi, THF; (b) Hg(OAc)$_2$, H$_2$O, THF, 81% yield.[18] In this case, conventional methods to remove the MEM group were unsuccessful.

16. BBr For a further discussion of this reagent refer to the section on MOM ethers.[19] TBS ethers are stable to these conditions but a BOC group was not.[20]

17. Ph$_2$BBr, CH$_2$Cl$_2$, −78°C, 71% yield.[21]

18. MgBr$_2$, Et$_2$O, 77–95% yield.[22] MOM, SEM, and MTM ethers are also cleaved with this reagent.

19. Aq. HBr, THF, rt, 72 h, 74% yield.[19]

20. FeCl$_3$, Ac$_2$O, −45°C; K$_2$CO$_3$, MeOH, 90% yield.[24] A TBDMS group and an acetonide were not affected by these conditions.

21. CAN, Ac_2O, rt, 24 h, 80–98% yield. These conditions result in the formation of $ROCH_2OAc$, which would then be hydrolyzed to the alcohol.[25]

22. $H_2ZnCl_2Br_2$, THF, rt, 1 h, 84% or Li_2ZnBr_4, THF, rt, 48 h, 94% yield.[26] *t*-Butyl esters ethers are stable and TBS ethers are cleaved very slowly.

1. E. J. Corey, J.-L. Gras, and P. Ulrich, *Tetrahedron Lett.*, **17**, 809 (1976).

2. G. H. Posner, A. Haces, W. Harrison, and C. M. Kinter, *J. Org. Chem.*, **52**, 4836 (1987).

3. J. A. Boynton and J. R. Hanson, *J. Chem. Res., Synop.*, 378 (1992).

4. K. C. Nicolaou, E. W. Yue, S. La Greca, A Nadin, Z. Yang, J. E. Leresche, T. Tsuri, Y. Naniwa, and F. De Riccardis, *Chem. Eur. J.*, **1**, 467 (1995).

5. O. Miyata, T. Shinada, I. Ninomiya, and T. Naito, *Tetrahedron Lett.*, **32**, 3519 (1991).

6. Y. Quindon, H. E. Morton, and C. Yoakim, *Tetrahedron Lett.*, **24**, 3969 (1983).

7. E. J. Corey, D. H. Hua, and S. P. Seitz, *Tetrahedron Lett.*, **25**, 3 (1984).

8. M. Bénéchie and F. Khuong-Huu, *J. Org. Chem.*, **61**, 7133 (1996).

9. H. Monti, G. Léandri, M. Klos-Ringuet, and C. Corriol, *Synth. Commun.*, **13**, 1021 (1983).

10. P. A. Procopiou, B. Cox, B. E. Kirk, M. G. Lester, A. D. McCarthy, M. Sareen, P. J. Sharratt, M. A. Snowden, S. J. Spooner, N. S. Watson, and J. Widdowson, *J. Med. Chem.*, **39**, 1413 (1996).

11. J. H. Rigby and J. Z. Wilson, *Tetrahedron Lett.*, **25**, 1429 (1984).

12. D. R. Williams and S. Sakdarat, *Tetrahedron Lett.*, **24**, 3965 (1983).

13. L. A. Paquette, J. Chang, and Z. Liu, *J. Org. Chem.*, **69**, 6441 (2004).

14. N. Ikota and B. Ganem, *J. Chem. Soc., Chem. Commun.*, 869 (1978).

15. L. A. Paquette, J. Chang, and Z. Liu, *J. Org. Chem.*, **69**, 6441 (2004); G. Sabitha, R. S. Babu, M. Rajkumar, R. Srividya, and J. S. Yadav, *Org. Lett.*, **3**, 1149 (2001).

16. A. S.-Y. Lee, Y.-J. Hu, and S.-F. Chu, *Tetrahedron*, **57**, 2121 (2001).

17. D. Vadolas, H. P. Germann, S. Thakur, W. Keller, and E. Heidemann, *Int. J. Pept. Protein Res.*, **25**, 554 (1985).

18. R. E. Ireland, P. G. M. Wuts, and B. Ernst, *J. Am. Chem. Soc.*, **103**, 3205 (1981).

19. R. K. Boeckman, Jr., and J. C. Potenza, *Tetrahedron Lett.*, **26**, 1411 (1985).

20. D. L. Boger, S. Miyazaki, S. H. Kim, J. H. Wu, S. L. Castle, O. Loiseleur, and Q. Jin, *J. Am. Chem. Soc.*, **121**, 10004 (1999).

21. M. Shibasaki, Y. Ishida, and N. Okabe, *Tetrahedron Lett.*, **26**, 2217 (1985).

22. S. Kim, Y. H. Park, and I. S. Kee, *Tetrahedron Lett.*, **32**, 3099 (1991).

23. D. R. Williams, P. A. Jass, H.-L. A. Tse, and R. D. Gaston, *J. Am. Chem. Soc.*, **112**, 4552 (1990).

24. R. A. Holton, R. R. Juo, H. B. Kim, A. D. Williams, S. Harusawa, R. E. Lowenthal, and S. Yogai, *J. Am. Chem. Soc.*, **110**, 6558 (1988).

25. K. Tanemura, T. Suzuki, Y. Nishida, K. Satsumabayashi, and T. Horaguchi, *Chem. Lett.*, 1012 (2001).

26. J. M. Herbert, J. G. Knight, and B. Sexton, *Tetrahedron*, **52**, 15257 (1996).

2-Cyanoethoxymethyl Ether (CEM): $ROCH_2OCH_2CH_2CN$

The CEM group was developed as a 2'-hydroxy protective group for oligoribonucleotide synthesis. It is introduced with poor selectivity using the stannylene method. It is not completely stable to $MeNH_2$ but is stable to NH_3, which allows for cyanoethyl cleavage on the phosphate residue with retention of the CEM group. It is cleaved with TBAF.[1]

1. T. Ohgi, Y. Masutomi, K. Ishiyama, H. Kitagawa, Y. Shiba, and J. Yano, *Org. Lett.*, **7**, 3477 (2005).

Bis(2-chloroethoxy)methyl Ether: $ROCH(OCH_2CH_2Cl)_2$ (Chart 1)

The mixed ortho ester formed from tri(2-chloroethyl) orthoformate (100°C, 10 min to 2 h, 76% yield) is more stable to acid than the unsubstituted derivative, but can be cleaved with 80% AcOH (20°C, 1 h).[1]

1. T. Hata and J. Azizian, *Tetrahedron Lett.*, **10**, 4443 (1969).

2,2,2-Trichloroethoxymethyl Ether: $Cl_3CCH_2OCH_2OR$

Formation

1. $Cl_3CCH_2OCH_2Cl$, NaH or KH, LiI, THF, 5 h, 70–90% yield.[1]
2. $Cl_3CCH_2OCH_2Cl$, (*i*-Pr)$_2$NEt, CH_2Cl_2, 30–60% yield.[1]
3. $Cl_3CCH_2OCH_2Br$, 1,8-bis(dimethylamino)naphthalene (proton sponge), CH_3CN, 0–25°C, >87% yield.[2]

Cleavage

1. Zn–Cu or Zn–Ag, MeOH, reflux, 97% .[1]
2. Zn, MeOH, Et$_3$N, AcOH, reflux 4 h, 90–100% .[1]
3. Li, NH_3.[1]
4. SmI_2, THF, 25°C, 71% yield.[2]
5. 6% Na(Hg), MeOH, THF, >66% yield.[2]

1. R. M. Jacobson and J. W. Clader, *Synth. Commun.*, **9**, 57 (1979).
2. D. A. Evans, S. W. Kaldor, T. K. Jones, J. Clardy, and T. J. Stout, *J. Am. Chem. Soc.*, **112**, 7001 (1990).

2-(Trimethylsilyl)ethoxymethyl Ether (SEM−OR): $Me_3SiCH_2CH_2OCH_2OR$

SEM ethers are stable to the acidic conditions (AcOH, H_2O, THF, 45°C, 7 h) that are used to cleave tetrahydropyranyl and *t*-butyldimethylsilyl ethers. Overall, this is a very robust protective group that is often difficult to remove.[1]

Formation

1. $Me_3SiCH_2CH_2OCH_2Cl$, (*i*-Pr)$_2$NEt (DIPEA), CH_2Cl_2, 35–40°C, 1–5 h, 86–100% yield.[2]

2. $Me_3SiCH_2CH_2OCH_2Cl$, 2,6-di-*tert*-butylpyridine, 48 h, 56% yield. Other bases resulted in much lower selectivity and the formation of considerable bis-SEM ethers.[3]

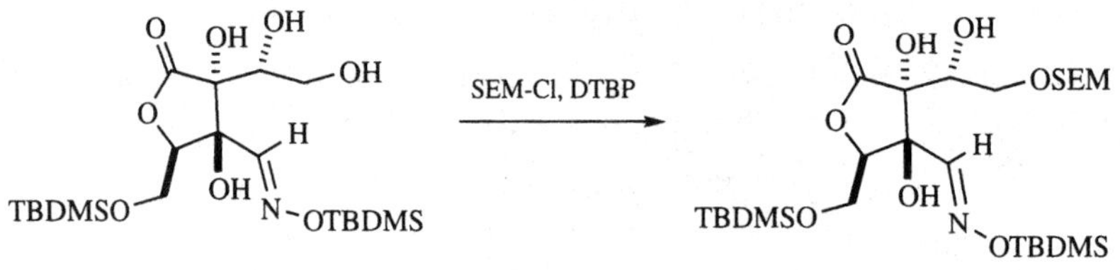

3. The above conditions failed in this example unless Bu_4NI was added to prepare SEMI *in situ*.[4]

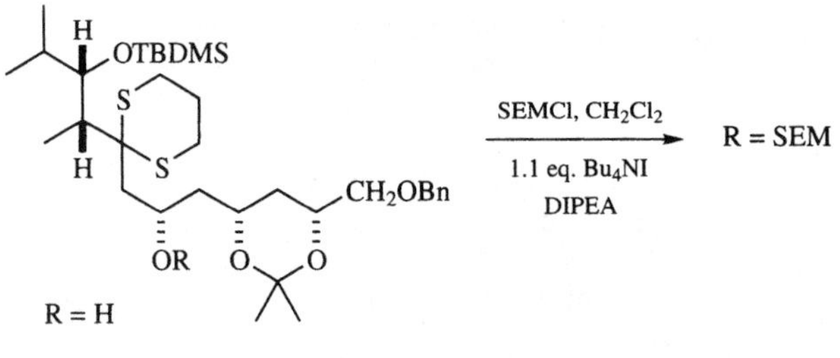

4. SEMCl, KH, THF, 0°C → rt, 1 h, 87% yield.[5]

5. *t*-BuMgCl, THF, rt, 5 min, then Bu_4NI, SEMCl, rt, 20–30 h, 78–84% yield. These conditions prevent alkylation of the nitrogen in the nucleoside bases.[6]

Cleavage

1. Bu_4NF, THF, or HMPA, 45°C, 8–12 h, 85–95% yield.[2,7] The cleavage of 2-(trimethylsilyl)ethyl glycosides is included here because they are functionally equivalent to the SEM group. They can be prepared by oxymercuration of a glycal with $Hg(OAc)_2$ and $TMSCH_2CH_2OH$, by the reaction of a glycosyl halide using Koenig–Knorr conditions, by a Fischer glycosylation, and by a glycal rearrangement.[4] *N,N*-Dimethylpropyleneurea can be used to replace the carcinogenic HMPA (45–80% yield).[8] An improved isolation procedure utilizing the insolubility of Bu_4NClO_4 in water has been developed for isolations where tetrabutylammonium fluoride is used.[9]

2. Bu$_4$NF, DMPU, 4-Å molecular sieves, 45–80°C, 80–95% yield.[8] These conditions were especially effective in cleaving tertiary SEM derivatives and avoid the use of the toxic HMPA.

3. CsF, DMF, 130°C, >89% yield.[10] HMPA has also been used as a solvent.[11] DMPU can be used as a HMPA replacement.

4. TFA, CH$_2$Cl$_2$ (2:1, v:v), 0°C, 30 min, 93% yield.[12]

The 4,6-*O*-benzylidene group is also cleaved under these conditions, but the anomeric linkage between sugars is not affected. Anomeric trimethylsilylethyl groups are also cleaved with BF$_3$·Et$_2$O[13] or Ac$_2$O/FeCl$_3$ (this reagent also cleaves the BOM group).[14] The anomeric trimethylsilylethyl group is hydrolyzed much faster than the other alkyl glycosides.[15]

5. LiBF$_4$, CH$_3$CN, 70°C, 3–8 h, 81–90% yield.[16] This system of reagents also cleaves benzylidene acetals. This reagent was used when conventional reagents failed to cleave the glycosidic TMSEt group. It is interesting to note that the β-anomers are cleaved more rapidly than the α-anomers and that the furanoside derivatives are not cleaved. TBS, MOM, and BOM ethers are also cleaved under these conditions.

6. MgBr$_2$, *n*-BuSH, Et$_2$O, rt, 3–24 h, 49–97% yield. MOM and MTM ethers are also cleaved, but MEM and TBDMS ethers are stable. These conditions have resulted in the formation of an ethyl thioether.[17]

7. MgBr$_2$, Et$_2$O, CH$_3$NO$_2$, 1–6 h, rt, 64–99% yield. The addition of nitromethane greatly improves the reaction which is now compatible with silylated cyanohydrins, TBS and TBDPS ethers, acetonides and a Troc group.[18] In the presence of HSCH$_2$CH$_2$CH$_2$SH, aldehydes are converted to dithianes.[19]

8. ZnCl$_2$·Et$_2$O, 99% yield[20] or BF$_3$·Et$_2$O, CH$_2$Cl$_2$, 0–25°C, 2 h.[21] In these examples a simple trimethylsilylethyl ether was cleaved, but the method is also applicable to SEM deprotection.[19,22]

9. BCl$_3$, toluene, 2,6-di-*tert*-butyl-4-methylpyridine, −78°C, 87% yield. In the synthesis of ditriptophenaline, an FMOC group did not survive the basic TBAF or BF$_3$·Et$_2$O.[23]

10. 1.5% Methanolic HCl, 16 h, 80–94% yield. These conditions do not cleave the MEM group.[24] 1% Sulfuric acid in methanol has also been used.[25]

11. Concd. HF, CH$_3$CN, >76% yield.[26] Note that a trimethylsilylethyl ester was not cleaved under these conditions. A dithiane can also be cleaved because of internal participation of the released alcohol during a SEM deprotection.[27]

12. I$_2$, sunlamp, 92% yield.[28]

13. Pyridine·HF, THF, 2.5 h, 0–25°C, 79% yield.[29]

14. CBr$_4$, MeOH, reflux, 10–18 h, 88–98% yield. These conditions produce HBr *in situ*. The TES, TBS, TBDPS, and TIPS ethers are also cleaved, but when IPA is used as the solvent TIPS and TBDPS ethers are stable.[30]

15. A study of the reductive cleavage of a series of alkoxymethyl ethers using the glucose backbone shows that, depending on the reagent, excellent selectivity can be obtained for deprotection vs. methyl ether formation for most of the common protective groups.[31]

1. R = H
2. R = Me

Relative Cleavage Rates for Selected Ethers of a Primary Alcohol

Ether R′ =	AlH$_2$Cl		AlHCl$_2$		BH$_3$/THF		BH$_3$/Toluene	
	Percent 1	Percent 2	Percent 1	Percent 2	Percent 1	Percent 2	Percent 1	Percent 2
MTM	100	0	100	0	85	15	100	0
SEM	0	0	100	0	100	0	100	0
BOM	0	0	89	11	98	2	100	0
pAOM	45	55	32	68	12	86	0	100
EOM	0	0	100	0	0	0	100	0

For secondary derivatives, the selectivity and reactivity varies somewhat. To what extent this is a function of the highly functionalized glucose derivative has not been determined. The table below gives the cleavage selectivity for the following reaction.

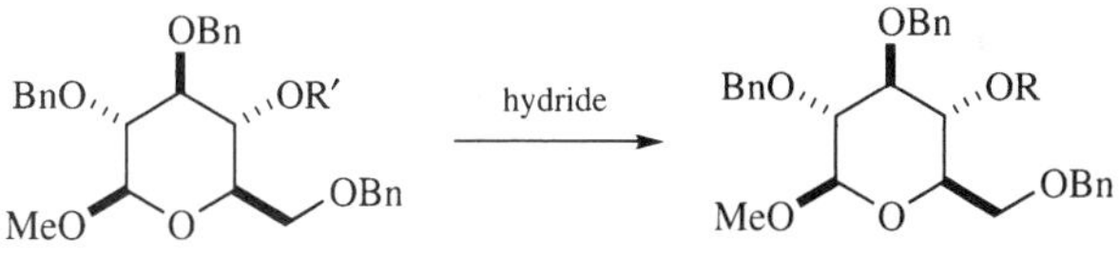

1. R = H
2. R = Me

Relative Cleavage Rates of Various Ethers of a Secondary Alcohol

Ether R' =	AlH₂Cl		AlHCl₂		BH₃/THF		BH₃/Toluene	
	Percent 1	Percent 2	Percent 1	Percent 2	Percent 1	Percent 2	Percent 1	Percent 2
SEM	100	0	82	18	0	0	100	0
BOM	100	0	90	10	0	0	100	0
pAOM	0	0	0	0	0	0	0	100
EOM	100	0	100	0	0	0	100	0

1. See, for example, Q. Zeng, S. Bailey, T.-Z Wang, and L. A. Paquette, *J. Org. Chem.*, **63**, 137 (1998).
2. B. H. Lipshutz and J. J. Pegram, *Tetrahedron Lett.*, **21**, 3343 (1980).
3. K. D. Freeman-Cook and R. L. Halcomb, *J. Org. Chem.*, **65**, 6153 (2000).
4. B. H. Lipshutz, R. Moretti, and R. Crow, *Tetrahedron Lett.*, **30**, 15 (1989).
5. D. R. Williams, P. A. Jass, H.-L. A. Tse, and R. D. Gaston, *J. Am. Chem. Soc.*, **112**, 4552 (1990).
6. T. Wada, M. Tobe, T. Nagayama, K. Furusawa, and M. Sekine, *Tetrahedron Lett.*, **36**, 1683 (1995).
7. T. Kan, M. Hashimoto, M. Yanagiya, and H. Shirahama, *Tetrahedron Lett.*, **29**, 5417 (1988).
8. B. H. Lipschutz and T. A. Miller, *Tetrahedron Lett.*, **30**, 7149 (1989).
9. J. C. Craig and E. T. Everhart, *Synth. Commun.*, **20**, 2147 (1990).
10. K. Suzuki, T. Matsumoto, K. Tomooka, K. Matsumoto, and G.-I. Tsuchihashi, *Chem. Lett.*, **16**, 113 (1987).
11. R. E. Ireland, R. S. Meissner, and M. A. Rizzacasa, *J. Am. Chem. Soc.*, **115**, 7166 (1993).
12. K. Jansson, T. Frejd, J. Kihlberg and G. Magnusson, *Tetrahedron Lett.*, **29**, 361 (1988). For an other case, see R. H. Schlessinger, M. A. Poss, and S. Richardson, *J. Am. Chem. Soc.*, **108**, 3112 (1986).
13. A. Hasegawa, Y. Ito, H. Ishida, and M. Kiso, *J. Carbohydr. Chem.*, **8**, 125 (1989); K. Jansson, T. Frejd, J. Kihlberg, and G. Magnusson, *Tetrahedron Lett.*, **27**, 753 (1986).
14. K. P. R. Kartha, M. Kiso, and A. Hasegawa, *J. Carbohydr. Chem.*, **8**, 675 (1989).
15. K. Jansson, G. Noori, and G. Magnusson, *J. Org. Chem.*, **55**, 3181 (1990).

16. B. H. Lipshutz, J. J. Pegram, and M. C. Morey, *Tetrahedron Lett.*, **22**, 4603 (1981).

17. S. Kim, I. S. Kee, Y. H. Park, and J. H. Park, *Synlett*, 183 (1991); S. Bailey, A. Teerawutgulrag, and E. J. Thomas, *J. Chem. Soc., Chem. Commun.*, 2521 (1995).

18. J.-C. Jung, R. Kache, K. K. Vines, Y.-S. Zheng, P. Bijoy, M. Valluri, and M. A. Avery, *J. Org. Chem.*, **69**, 9269 (2004).

19. A. Vakalopoulos and H. M. R. Hoffmann, *Org. Lett.*, **2**, 1447 (2000).

20. H. C. Kolb and H. M. R. Hoffman, *Tetrahedron: Asymmetry.*, **1**, 237 (1990).

21. S. D. Burke and G. J. Pacofsky, *Tetrahedron Lett.*, **27**, 445 (1986).

22. F. E. Wincott and N. Usman, *Tetrahedron Lett.*, **35**, 6827 (1994).

23. L. E. Overman and D. V. Paone, *J. Am. Chem. Soc.*, **123**, 9465 (2001).

24. B. M. Pinto, M. M. W. Buiting, and K. B. Reimer, *J. Org. Chem.*, **55**, 2177 (1990).

25. A. A. Kandil and K. N. Sellsor, *J. Org. Chem.*, **50**, 5649 (1985).

26. J. D. White and M. Kawasaki, *J. Am. Chem. Soc.*, **112**, 4991 (1990).

27. P. G. Steel and E. J. Thomas, *J. Chem. Soc., Perkin Trans. I*, 371 (1997).

28. S. Karim, E. R. Parmee, and E. J. Thomas, *Tetrahedron Lett.*, **32**, 2269 (1991).

29. K. Sugita, K. Shigeno, C. F. Neville, H. Sasai, and M. Shibasaki, *Synlett*, 325 (1994).

30. M.-Y. Chen and A. S.-Y. Lee, *J. Org. Chem.*, **67**, 1384 (2002).

31. I. Bajza, Z. Varga, and A. Liptak, *Tetrahedron Lett.*, **34**, 1991 (1993).

Menthoxymethyl Ether (MM−OR)

This protective group was developed to determine the enantiomeric excess of chiral alcohols. It is anticipated that many of the methods used to cleave the MOM group would be effective for the MM group as well.

Formation

Menthoxymethyl chloride, DIPEA, CH_2Cl_2, rt, overnight, 77–95% yield.[1]

Cleavage

1. $ZnBr_2$, CH_2Cl_2.[1]
2. TMSOTf, TMSOMe, $ClCH_2CH_2Cl$, 0°C to rt, 98% yield. The MM ether is converted to a simple MOM ether. When the TMSOMe was left out of the reaction, neighboring group participation occurred to give a 1,3-dioxane.[2]

1. D. Dawkins and P. R. Jenkins, *Tetrahedron: Asymmetry*, **3**, 833 (1992).
2. R. J. Linderman, K. P. Cusack, and M. R. Jaber, *Tetrahedron Lett.*, **37**, 6649 (1996).

O-Bis(2-acetoxyethoxy)methyl (ACE) orthoester, $(CH_3CO_2CH_2CH_2O)_2CH-OR$

This orthoester was developed for RNA synthesis. It is cleaved by hydrolysis of the acetates to produce *O*-bis(2-hydroxyethoxy)methyl orthoester during the general deprotection of the bases followed by treatment with acid at pH 3 for 10 min at 55°C. The *O*-bis(2-hydroxyethoxy)methyl orthoester is 10 times more labile to acid than is the acetylated derivative. It is formed by orthoester exchange with the alcohol using PPTS as a catalyst at 55°C for 3 h under high vacuum. This group greatly improves RNA synthesis over existing methods.[1]

1. S. A. Scaringe, F. E. Wincott, and M. H. Caruthers, *J. Am. Chem. Soc.*, **120**, 11820 (1998), R. Micura, *Angew. Chem. Int. Ed.*, **41**, 2265 (2002).

Tetrahydropyranyl Ether (THP−OR): (Chart 1)

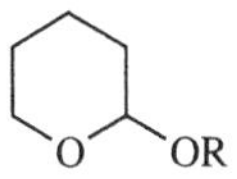

The introduction of a THP ether onto a chiral molecule results in the formation of diastereomers because of the additional stereogenic center present in the tetrahydropyran ring. This can make the interpretation of NMR spectra somewhat troublesome at times. Even so, this was an extensively used protective group in chemical synthesis because of its low cost, ease of installation, general stability to most nonacidic reagents, and the ease with which it can be removed. *Generally, almost any acidic reagent or reagent that generates an acid in situ can be used to introduce the THP group.* Although still used, it has largely been replaced with the TBS ether since it does not introduce an additional chiral center. Its relative stability compared to some other acetals discussed in following sections is illustrated below.[1]

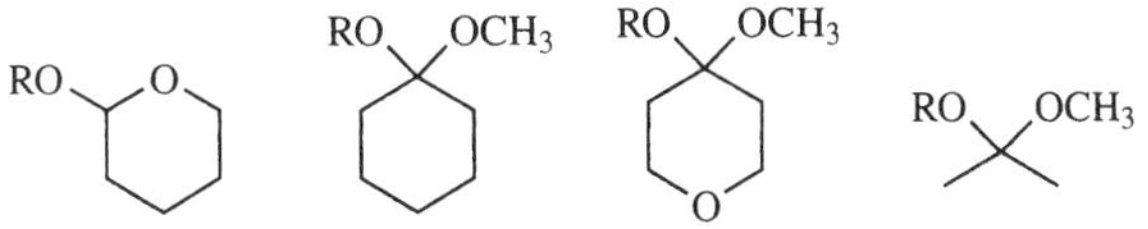

| Relative stability = 1 | 3 | 8 | 20 |

Formation

1. Dihydropyran, TsOH, CH_2Cl_2, 20°C, 1.5 h, 100% yield.[2]
2. The following method proceeds under nonacidic conditions: 2-hydroxytetrahydropyran, Ph$_3$P, DEAD, THF, 52–86% yield. The method is also effective for phenolic THP derivatives.[3]

3. Pyridinium p-toluenesulfonate (PPTS), dihydropyran, CH_2Cl_2, 20°C, 4 h, 94–100% yield.[4] The lower acidity of PPTS makes this a very mild method that has excellent compatibility with most functional groups. This is probably one of the simplest methods.

4. Reillex 425·HCl, dihydropyran, 86°C, 1.5 h, 84–98% yield.[5] The Reillex resin is a macroreticular polyvinylpyridine resin and is thus an insoluble form of the PPTS catalyst.

5. Amberlyst H-15 (SO_3H ion exchange resin), dihydropyran, hexane, 1–2 h, 95% yield.[6]

6. Dihydropyran, Dowex-50wx2, toluene, 10–355 min, 78–95% yield. These conditions were developed to monoprotect symmetrical 1,ω-diols.[7] Aqueous $NaHSO_4$[8] and HCl[9] as a catalyst shows good selectivity for the monoprotection of 1,ω-diols.

7. Dihydropyran, sulfonated charcoal, 3-Å ms, CH_2Cl_2, 67–98% yield.[10] Sulfated zirconia has also been used as a catalyst with similar effectiveness.[11]

8. Dihydropyran, $Zr(O_3PCH_3)_{1.2}(O_3PC_6H_4SO_3H)_{0.8}$, CH_2Cl_2, 70–94% yield. Phenols are similarly protected.[12]

9. Dihydropyran, K-10 clay, CH_2Cl_2, rt, 63–95% yield.[13,14] This method was reported to be successful for epoxide containing substrates when other methods failed. Kaolinitic clay is also an effective catalyst except for phenols, which fail to react.[15] Spanish Speolite clay has also been used.[16]

10. Dihydropyran, $(TMSO)_2SO_2$, CH_2Cl_2, 92–100% yields.[17] Sulfuric acid is produced *in situ*. Sulfamic acid is also an effective catalyst.[18]

11. Dihydropyran, H_2SO_4-Silica gel, 5–10 min, CH_2Cl_2, 74–95% yield.[19]

12. Dihydropyran, TMSI, CH_2Cl_2, rt, 80–96% yield.[20]

13. Dihydropyran, I_2, 0.5–3.5 h, CH_2Cl_2, rt, 83–92% yield.[21] *In situ* generated HI is most likely the actual catalyst. This method modified by microwave heating has been used to monoprotect diols with modest selectivity.[22]

14. Dihydropyran, $(CH_3)_2SBr_2$, rt, 5 min to 3.5 h, 81–97% yield. HBr is generated *in situ*.[23] Phenols are also protected. Bu_4NBr_3 which also generates HBr *in situ* is similarly effective (75–97% yield).[24] NBS has been used similarly.[25]

15. Dihydropyran, acetonyltriphenylphosphonium bromide, CH_2Cl_2, 5 min, 80–97% yield. The EE and THF ethers are also formed using this reagent.[26]

16. Dihydropyran, trichloroisocyanuric acid, 60–80°C, neat, 75–95% yield. In the presence of methanol THP groups are removed. TCCA is known to react with alcohols to generate HCl, the likely catalyst.[27]

17. Dihydropyran, $PdCl_2(CH_3CN)_2$, THF, rt, 49–90% yield. Phenols do not react.[67] Dihydropyran, $Ph_3P·HBr$, 24 h, CH_2Cl_2, 88% yield.[28]

18. Dihydropyran, $LaCl_3$, CH_2Cl_2, rt, 4 h, 90% .[29] GaI_3 is similarly an effective catalyst (85–95% yield).[30]

19. Dihydropyran, Sc(OTf)$_3$, EtOAc, rt, 92–98% yield. THF ethers are formed with dihydrofuran.[31] The method is applicable to phenols. In(OTf)$_3$ can also be used (30 min, 64–85% yield).[32]

20. Dihydropyran, polystyrene supported AlCl$_3$, CH$_2$Cl$_2$, rt, 89–97% yield. Considerable selectivity can be achieved by this method. The more electron rich alcohols react in preference to electron poor derivatives, primary alcohols react faster than 2° alcohols, alkanols react in preference to phenols and diols can be monoprotected efficiently.[33] The catalyst AlCl$_3$·6H$_2$O under solvent-free conditions gives THP ethers of alcohols and phenols in 74–96% yield.[34]

21. Dihydropyran, CuSO$_4$·5H$_2$O, CH$_3$CN, rt, 40 min to 12h, 70–91% yield. Phenols are similarly derivatized.[35] Diols can be selectively monotetrahydropyranylated.

22. Dihydropyran, anhydrous FeSO$_4$, MW, 80–97% yield.[36] In the presence of water, THP groups are removed.

23. Dihydropyran, anhydrous Fe(ClO$_4$)$_3$, Et$_2$O, 75–98% yield. Fe(ClO)$_3$, MeOH is used to cleave the THP group.[37] **Note that metal perchlorates are generally hazardous.**

24. Dihydropyran, LiClO$_4$, Et$_2$O, 56–92% yield.[38] LiOTf[39] or LiPF$_6$[40] can be used as the catalyst to protect alkanols and phenols. 1,4- and 1,2-cyclohexanediols can be monoprotected in 83–87% yield with LiOTf.

25. Dihydropyran, InCl$_3$, [bmim]PF$_6$, 81–91% yield. THF ethers are formed with dihydrofuran.[41]

26. Polymer-bound dihydropyran, PPTS, 80°C.[42]

27. Dihydropyran, Al(PO$_4$)$_3$, reflux, 15 min, 97% yield.[43]

28. Dihydropyran, DDQ, CH$_2$Cl$_2$, 82–100% yield.[44]

29. 2-Tetrahydropyranyl phenyl sulfone, MgBr$_2$·Et$_2$O, NaHCO$_3$, THF, rt, 47–99% yield. [45]

30. Dihydropyran, H-Y Zeolite, hexane, reflux, 60–95% yield.[46] H-Rho Zeolite,[47] H-Beta Zeolite,[48] Zeolite HSZ-330 (dihydropyran, rt, 1.5h, 44–100% yield),[49] and Zeolite E4[50] can also be used as a catalyst.

31. Dihydropyran, H-MCM-41, ms, 69°C, 44–99% yield.[51]

32. Dihydropyran, H$_3$[PW$_{12}$O$_{40}$], CH$_2$Cl$_2$, rt, 64–96% yield. The same acid can be used to cleave the THP group if methanol is used as solvent.[52,53,] The similar K$_5$CoW$_{12}$O$_{40}$·3H$_2$O has also been used.[54]

33. Tetrahydropyran, (Bu$_4$N$^+$)$_2$S$_2$O$_8$$^{-2}$, reflux, 85–95% yield. These oxidative conditions do not affect thioethers.[55]

34. 3,4-(MeO)$_2$C$_6$H$_3$CH$_2$OTHF, DDQ, CH$_3$CN, 54–94% yield. These conditions can also be used for glycoside synthesis.[56]

35. Al$_2$(SO$_4$)$_3$-SiO$_2$ is a reasonable catalyst for the monotetrahydropyranylation of simple, symmetrical 1,ω-diols.[57]

36. Dihydropyran, Al$_2$O$_3$, ZnCl$_2$.[58]

37. Dihydropyran, CAN, CH$_3$CN, rt, 81–91% yield.[59]

38. Dihydropyran, CuCl, CH$_2$Cl$_2$, 75–93% yield.[60]

Cleavage

1. AcOH, THF, H_2O, (4:2:1), 45°C, 3.5 h.[1] MEM ethers are stable to these conditions.[61]

2. PPTS, EtOH, (pH 3.0), 55°C, 3 h, 95–100% yield.[2]

3. Amberlyst H-15, MeOH, 45°C, 1 h, 95% yield.[3] Dowex-50W-X8, 25°C, 1 h, MeOH, 99% yield.[62]

4. Boric acid, $EtOCH_2CH_2OH$, 90°C, 2 h, 80–95% yield.[63]

5. TsOH, MeOH, 25°C, 1 h, 94% yield.[64] The use of 2-propanol as solvent was found to enhance the selectivity for THP removal in the presence of a 1,3-TBDPS group.[65] TBDPS ethers are not affected by these conditions.[66]

6. H_2SO_4·silica gel, 5–10 min, MeOH, 78–92% yield.[19]

7. $K_5CoW_{12}O_{40}$·H_2O, MeOH, rt, 94–100% yield.[54]

8. MeOH, $(TMSO)_2SO_2$, 10–90 min, 93–100% yield.[6] This reagent forms H_2SO_4 *in situ.*

9. I_2, MeOH, rt, 3–6 h, 73–85% yield.[21]

10. $(CH_3)_2SBr_2$, rt, CH_2Cl_2, MeOH, 73–97% yield.[23]

11. CBr_4, MeOH, reflux, 89–96% yield. HBr is formed *in situ*. 1,3-dioxolanes are also cleaved.

12. Acetonyltriphenylphosphonium bromide, MeOH, rt, 90–99% yield.[26]

13. $PdCl_2(CH_3CN)_2$, wet CH_3CN, reflux, trace to 93% yield.[67] Phenolic THP ethers are cleaved also. The residual $PdCl_2$ found in some sources of Pd/C has been shown to catalyze cleavage of THP ethers during hydrogenation.[68]

14. $MgBr_2$, Et_2O, rt, 66–95% yield.[69] *t*-Butyldimethylsilyl and MEM ethers are not affected by these conditions, but the MOM ether is slowly cleaved. The THP derivatives of benzylic and tertiary alcohols give bromides.

15. $CuCl_2$·$2H_2O$, MeOH, rt, 68–95% yield.[70]

16. $TiCl_3$, CH_3CN, rt, 46–97% yield.[71]

17. $In(OTf)_3$, MeOH, H_2O, rt, 60–92% yield. In the presence of Ac_2O the THP is converted directly to an acetate.[32] InI_3 (EtOAc, reflux 12–15 h)[72] or $TiCl_4$ (CH_2Cl_2, Ac_2O, 0–25°C, 6 h, 72–90% yield)[73] also converts THP ethers directly to an acetate. The THP ether can be converted directly to an acetate by refluxing in AcOH/AcCl (91% yield).[74] These conditions would probably convert other related acetals to acetates as well.

18. Me_2AlCl, CH_2Cl_2, −25°C → rt, 1 h, 89–100% yield.[75]

19. $(NCSBu_2Sn)_2O$ 1%, THF, H_2O.[76] Acetonides and TMS ethers are also cleaved under these conditions, but TBDMS, MTM, and MOM groups are stable. This catalyst has also been used to effect transesterifications.[77]

20. MeOH, reagent prepared by heating Bu_2SnO and Bu_3SnPO_4, heat 2 h, 90% yield.[78] This method is effective for primary, secondary, tertiary, benzylic and allylic THP derivatives. The MEM group and ketals are inert to this reagent, but TMS and TBDMS ethers are cleaved.

21. 2,4,4,6-Tetrabromo-2,5-cyclohexadiene, Ph_3P in CH_2Cl_2 or CH_3CN converts THP ethers into bromides (78–99% yield).[79] $Ph_3P \cdot Br_2$, CH_2Cl_2, $-50°C$ to $35°C$, 85–94% yield.[80] Ethyl acetals and MOM groups are also cleaved with this reagent, but a THP ether can be selectively cleaved in the presence of a MOM ether. The use of this reagent at 0–10°C (16 h) will convert a THP ether directly into a bromide,[81] and with a slight modification of the reaction conditions, chlorides, nitriles, methyl ethers, and trifluoroacetates may also be directly produced.[82] THP ethers, when treated with the Viehe salt (CH_2Cl_2, 78–96% yield), are converted to chlorides.[83]

22. Bu_3SnSMe, $BF_3 \cdot Et_2O$, toluene, $-20°C$ to $0°C$, 1.5 h; H_3O^+, 70–97% yield. The intermediate stannanes from this reaction, when treated with various electrophiles, form benzyl and MEM ethers, benzoates, and tosylates, and when treated with PCC, they form aldehydes.[84,85]

23. Tonsil, a Mexican Bentonite, acetone, 30 min, rt, 60–95% yield. MOM and MEM groups are stable and phenolic THP ethers were also cleaved.[86]

24. Expansive graphite, MeOH, 40–50°C, 92–98% yield.[87]

25. TBDMSOTf, CH_2Cl_2 ; Me_2S, 95% yield. The THP group is converted directly into a TBDMS ether.[88]

26. $BH_3 \cdot THF$, 20°C, 24 h, 84% yield.[89]

27. DDQ, aq. MeOH, 81–98% yield.[90] DDQ in aqueous CH_3CN has also been used (42–95% yield), but since the medium was reported to be acidic (pH 3), the reaction probably occurs by simple acid catalysis. Benzylic, allylic, and primary THP derivatives are not efficiently cleaved.[91]

28. $NaCNBH_3$, $BF_3 \cdot Et_2O$, rt, 68–95% yield.[92]

29. LiCl, H_2O, DMSO, 90°C, 6 h, 81–92% yield.[93]

30. CAN, MeOH, 0°C, 0.5–3 h, 81–95% yield. TBDMS ethers are more easily cleaved, and thus a TBDMS ether is cleaved selectively in the presence of a THP ether (15 min, 95%).[94] An improved version of this method has been developed.[95] THF ethers are cleaved similarly.

31. $BF_3 \cdot Et_2O$, $HSCH_2CH_2SH$, CH_2Cl_2, 100% yield. A primary TBDMS ether was not affected.[96]

32. $SnCl_2$, MeOH, 80–95% yield.[97]

33. $CuSO_4 \cdot 5H_2O$, MeOH, rt, 2–6 h.[35]

34. β-cyclodextrin, H_2O, 50°C, 70–90% yield.[98] The phenolic derivative is also cleaved.

35. THP ethers can be converted directly to TBDMS and TES ethers using the silyl hydride and $Sn(OTf)_2$ or the silyl triflate (70–95% yield). The use of TMSOTf gives the free alcohols upon isolation.[99]

36. THP ethers can be converted directly to an acetate or formate by reaction with ethyl acetate, acetic acid, or ethyl formate and with $K_5CoW_{12}O_{40}\cdot3H_2O$ as the catalyst (20–98% yield). The transformation is most successful with primary THP ethers.[100]

37. In the presence of PhCHO, Et_3SiH, TMSOTf, CH_3CN, 0°C, 1 h, THP ethers are converted to benzyl ethers.[101]

38. Explosions have been reported on distillation of compounds containing a tetrahydropyranyl ether after a reaction with B_2H_6, H_2O_2, and OH^- and with 40% CH_3CO_3H:

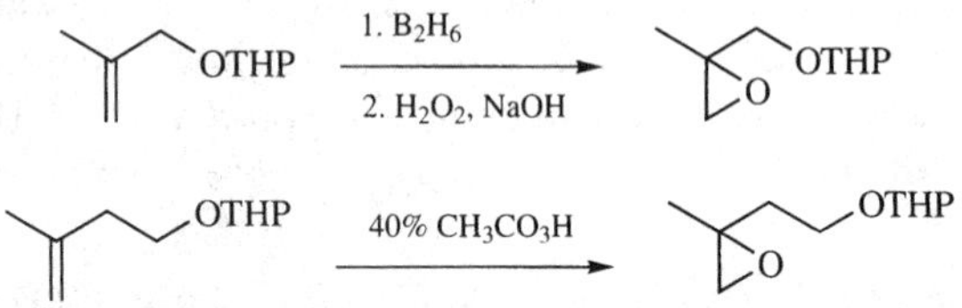

It was thought that the acetal might have reacted with peroxy reagents, forming explosive peroxides. It was suggested that this could also occur with compounds such as tetrahydrofuranyl acetals, 1,3-dioxolanes, and methoxymethyl ethers.[102]

Oxidative Deprotection

The THP or the TMS ether can be converted directly to an aldehyde or ketone using a variety of oxidative methods. In most of the examples the reagent cleaves the THP or TMS ether with acid or a liberated acid and then oxidizes the alcohol to the carbonyl derivative. The majority of examples are very simple and the generality of these methods in complex synthesis remains to be tested.

1. Montmorillonite K-10, $Fe(NO_3)_3$, MW, 80–90% yield.[103] Bis(trimethylsilyl) chromate[104] and ammonium chlorochromate/Montmorillonite K-10[105] as the oxidant gives similar results.

2. Clay supported $[Ce(NO_3)_3]_2CrO_4$ and $[Ce(NO_3)_3]_2HIO_6$, CH_2Cl_2, 65–90% yield.[106]

3. Ceric ammonium nitrate support on HNO_3/silica gel, MW, 6–10 min, 90–91% yield. The method only works for benzylic derivatives.[107]

4. Wet alumina-supported chromium(VI) oxide, CH_2Cl_2, 10–25 min, 83–93% yield.[108]

5. 3-Carboxypyridinium chlorochromate, CH_3CN or CH_2Cl_2, reflux, 0.1–2.5 h, 63–98% yield.[109]

6. $AgBrO_3(NaBrO_3)/AlCl_3$, CH_3CN, reflux, 0.6–3 h, 70–95% yield.[110]

7. 4-(Dimethylamino)pyridinium and 2,2′-bipyridinium chlorochromate, CH_3CN, 15–35 min, 25–95% yield[111] or tetramethylammonium chlorochromate (80–98% yield).[112]

8. K_2FeO_4/silica gel, CH_3CN, reflux, 2–14 h, 80–94% yield.[113]

9. $PhCH_2PPH_3HSO_5$, $BiCl_3$, CH_2Cl_2, MW, 80–99% yield.[114]

10. β-Cyclodextrin, NBS, H_2O, rt, 20–60 min, 74–98% yield.[115]

1. R. Schwalm, H. Binder, and D. Funhoff, *J. Appl. Polym. Sci.*, **78**, 208 (2000).

2. K. F. Bernady, M. B. Floyd, J. F. Poletto, and M. J. Weiss, *J. Org. Chem.*, **44**, 1438 (1979).

3. R. Azzouz, L. Bischoff, M.-H. Fouquet, and F. Marsais, *Synlett*, 2808 (2005).

4. M. Miyashita, A. Yoshikoshi, and P. A. Grieco, *J. Org. Chem.*, **42**, 3772 (1977).

5. R. D. Johnston, C.R. Marston, P. E. Krieger, and G. L. Goe, *Synthesis*, 393 (1988).

6. A. Bongini, G. Cardillo, M. Orena, and S. Sandri, *Synthesis*, 618 (1979).

7. T. Nishiguchi, M. Kuroda, M. Saitoh, A. Nishida, and S. Fujisaki, *J. Chem. Soc., Chem. Commun.*, 2491 (1995).

8. T. Nishiguchi, S. Hayakawa, Y. Hirasaka, and M. Saitoh, *Tetrahedron Lett.*, **41**, 9843 (2000).

9. R. J. Petroski, *Synth. Commum.*, **33**, 3251 (2003).

10. H. K. Patney, *Synth. Commun.*, **21**, 2329 (1991).

11. A. Sakar, O. S. Yemul, B. P. Bandgar, N. B. Gaikwad, and P. P. Wadgaonkar, *Org. Prep. Proced. Int.*, **28**, 613 (1996).

12. M. Curini, F. Epifano, M. C. Marcotullio, and O. Rosati, *Tetrahedron Lett.*, **39**, 8159 (1998).

13. S. Hoyer, P. Laszlo, M. Orlovic, and E. Polla, *Synthesis*, 655 (1986).

14. T. Taniguchi, K. Kadota, A. S. ElAzab, and K. Ogasawara, *Synlett*, 1247 (1999).

15. T. T. Upadhya, T. Daniel, A. Sudalai, T. Ravindranathan, and K. R. Sabu, *Synth. Commun.*, **26**, 4539 (1996).

16. J. M. Campelo, A. Garcia, F. Lafont, D. Luna, and J. M. Marinas, *Synth. Commun.*, **24**, 1345 (1994).

17. Y. Morizawa, I. Mori, T. Hiyama, and H. Nozaki, *Synthesis*, 899 (1981).

18. B. Wang, L.-M. Yang, and J.-S. Suo, *Synth. Commun.*, **33**, 3929 (2003).

19. M. M. Heravi, M. A. Bigdeli, N. Nahid, and D. Ajami, *Ind. J. Chem., Sect. B*, **38B**, 1285 (1999). D. M. Pore, U. V. Desai, R. B. Mane, and P. P. Wadgaonkar, *Synth. Commum.*, **34**, 2135 (2004).

20. G. A. Olah, A. Husain, and B. P. Singh, *Synthesis*, 703 (1985).

21. H. M. S. Kumar, B. V. S. Reddy, E. J. Reddy, and J. S. Yadav, *Chem. Lett.*, **28**, 857 (1999).

22. N. Deka and J. C. Sarma, *J. Org. Chem.*, **66**, 1947 (2001).

23. A. T. Khan, E. Mondal, B. M. Borah, and S. Ghosh, Eur. *J. Org. Chem.*, 4113 (2003).

24. S. Naik, R. Gopinath, and B. K. Patel, *Tetrahedron Lett.*, **42**, 7679 (2001). A. R. Hajipour, S. A. Pourmousavi, and A. E. Ruoho, *Synth. Commun.*, **35**, 2889 (2005).

25. B. Das, M. R. Reddy, N. Ravindranath, V. S. Reddy, and K. Venkateshwarlu, *Ind. J. Chem.*, **43B**, 1711 (2004).

26. Y.-S. Hon and C.-F. Lee, *Tetrahedron Lett.*, **40**, 2389 (1999). Y.-S. Hon, C.-F. Lee, R.-J. Chen, and P.-H. Szu, *Tetrahedron*, **57**, 5991 (2001).

27. H. Firouzabadi, N. Iranpoor, and H. Hazarkhani, *Synth. Commun.*, **34**, 3623 (2004).

28. V. Bolitt, C. Mioskowski, D.-S. Shin, and J. R. Falck, *Tetrahedron Lett.*, **29**, 4583 (1988).

29. V. Bhuma and M. L. Kantam, *Synth. Commun.*, **22**, 2941 (1992).

30. P.-P. Sun and Z.-X. Hu, *Chin. J. Chem.*, **22**, 1341 (2004).

31. T. Watahiki, H. Kikumoto, M. Matsuzaki, T. Suzuki, and T. Oriyama, *Bull. Chem. Soc. Jpn.*, **75**, 367 (2002).

32. T. Mineno, *Tetrahedron Lett.*, **43**, 7975 (2002).

33. B. Tamami and K. P. Borujeny, *Tetrahedron Lett.*, **45**, 715 (2004).

34. V. V. Namboodiri and R. S. Varma, *Tetrahedron Lett.*, **43**, 1143 (2002).

35. A. T. Khan, L. H. Choudhury, and S. Ghosh, *Tetrahedron Lett.*, **45**, 7891 (2004).

36. B. P. Bandgar and S. P. Kasture, *J. Chinese Chem. Soc. (Taipei, Taiwan)*, **48**, 877 (2001).

37. M. M. Heravi, F. K. Behbahani, H. A. Oskooie, and R. H. Shoar, *Tetrahedron Lett.*, **46**, 2543 (2005).

38. B. S. Babu and K. K. Balasubramanian, *Tetrahedron Lett.*, **39**, 9287 (1998).

39. B. Karimi and J. Maleki, *Tetrahedron Lett.*, **43**, 5353 (2002).

40. N. Hamada and T. Sato, *Synlett*, 1802 (2004).

41. J. Singh Yadav, B. V. Subba Reddy, and D. Gnaneshwar, *New J. Chem.*, **27**, 202 (2003).

42. L. A. Thompson and J. A. Ellman, *Tetrahedron Lett.*, **35**, 9333 (1994).

43. J. M. Campelo, A. Garcia, F. Lafont, D. Luna, and J. M. Marinas, *Synth. Commun.*, **22**, 2335 (1992).

44. K. Tanemura, T. Horaguchi, and T. Suzuki, *Bull., Chem. Soc. Jpn.*, **65**, 304 (1992).

45. D. S. Brown, S. V. Ley, S. Vile, and M. Thompson, *Tetrahedron*, **47**, 1329 (1991).

46. P. Kumar, C. U. Dinesh, R. S. Reddy, and B. Pandey, *Synthesis*, 1069 (1993).

47. D. P. Sabde, B. G. Naik, V. R. Hedge, and S. G. Hegde, *J. Chem. Res., Synop.*, 494 (1996).

48. J.-E. Choi and K.-Y. Ko, *Bull. Korean Chem. Soc.*, **22**, 1177 (2001).

49. R. Ballini, F. Bigi, S. Carloni, R. Maggi, and G. Sartori, *Tetrahedron Lett.*, **38**, 4169 (1997).

50. A. Hegedüs, I. Vigh, and Z. Hell, *Synth. Commum.*, **34**, 4145 (2004).

51. K. R. Kloetstra and H. van Bekkum, *J. Chem. Res. Synop.*, 26 (1995).

52. A. Molnar and T. Beregszaszi, *Tetrahedron Lett.*, **37**, 8597 (1996).

53. G. P. Romanelli, G. Baronetti, H. J. Thomas, and J. C. Autino, *Tetrahedron Lett.*, **43**, 7589 (2002).

54. M. H. Habibi, S. Tangestaninejad, I. Mohammadpoor-Baltork, V. Mirkhani, and B. Yadollahi, *Tetrahedron Lett.*, **42**, 2851 (2001).

55. H. C. Choi, K. I. Cho, and Y. H. Kim, *Synlett*, 207 (1995).

56. J. Inanaga, Y. Yokoyama, and T. Hanamato, *Chem. Lett.*, **22**, 85 (1993).

57. T. Nishiguchi and K. Kawamine, *J. Chem. Soc., Chem. Commun.*, 1766 (1990).

58. B. C. Ranu and M. Saha, *J. Org. Chem.*, **59**, 8269 (1994).

59. G. Maity and S. C. Roy, *Synth. Commun.*, **23**, 1667 (1993); K. Pachamuthu and Y. D. Vankar, *J. Org. Chem.*, **66**, 7511 (2001).

60. U. T. Bhalerao, K. J. Davis and B. V. Rao, *Synth. Commun.*, **26**, 3081 (1996).

61. E. J. Corey, R. L. Danheiser, S. Chandrasekaran, P. Siret, G. E. Keck and J.-L. Gras, *J. Am. Chem. Soc.*, **100**, 8031 (1978).

62. R. Beier and B. P. Mundy, *Synth. Commun.*, **9**, 271 (1979).

63. J. Gigg and R. Gigg, *J. Chem. Soc. C*, 431 (1967).

64. E. J. Corey, H. Niwa, and J. Knolle, *J. Am. Chem. Soc.*, **100**, 1942 (1978).

65. F. Almqvist and T. Frejd, *Tetrahedron: Asymmetry*, **6**, 957 (1995).

66. A. B. Shenvi and H. Gerlach, *Helv. Chim. Acta*, **63**, 2426 (1980).

67. Y.-G. Wang, X.-X. Wu, and Z.-Y. Jiang, *Tetrahedron Lett.*, **45**, 2973 (2004).

68. L. H. Kaisalo and T. A. Hase, *Tetrahedron Lett.*, **42**, 7699 (2001).

69. S. Kim and J. H. Park, *Tetrahedron Lett.*, **28**, 439 (1987).

70. K. J. Davis, U. T. Bhalerao, and B. V. Rao, *Ind. J. Chem., Sect. B*, **39B**, 860 (2000). J. Wang, C. Zhang, Z. Qu, Y. Hou, B. Chen, and P. Wu, *J. Chem. Res., Syn.*, 294 (1999).

71. A. Semwal and S. K. Nayak, *Synthesis*, 71 (2005).

72. B. C. Ranu and A. Hajra, *J. Chem. Soc. Perkin Trans. 1*, 355 (2001).

73. S. Chandrasekhar, T. Ramachandar, M. V. Reddy, and M. Takhi, *J. Org. Chem.*, **65**, 4729 (2000).

74. M. Jacobson, R. E. Redfern, W. A. Jones, and M. H. Aldridge, *Science*, **170**, 543 (1970); T. Bakos and I. Vincze, *Synth. Commun.*, **19**, 523 (1989).

75. Y. Ogawa and M. Shibasaki, *Tetrahedron Lett.*, **25**, 663 (1984).

76. J. Otera and H. Nozaki, *Tetrahedron Lett.*, **27**, 5743 (1986).

77. J. Otera, T. Yano, A. Kawabata, and H. Nozaki, *Tetrahedron Lett.*, **27**, 2383 (1986).

78. J. Otera, Y. Niibo, S. Chikada, and H. Nozaki, *Synthesis*, 328 (1988).

79. A. Tanaka and T. Oritani, *Tetrahedron Lett.*, **38**, 1955 (1997).

80. A. Wagner, M.-P. Heitz, and C. Mioskowski, *J. Chem. Soc., Chem. Commun.*, 1619 (1989).

81. M. Schwarz, J. E. Oliver, and P. E. Sonnet, *J. Org. Chem.*, **40**, 2410 (1975).

82. P. E. Sonnet, *Synth. Commun.*, **6**, 21 (1976).

83. T. Schlama, V. Gouverneur, and C. Mioskowski, *Tetrahedron Lett.*, **38**, 3517 (1997).

84. T. Sato, J. Otera, and H. Nozaki, *J. Org. Chem.*, **55**, 4770 (1990).

85. T. Sato, T. Tada, J. Otera, and H. Nozaki, *Tetrahedron Lett.*, **30**, 1665 (1989).

86. R. Cruz-Almanza, F. J. Peres-Flores, and M. Avila, *Synth. Commun.*, **20**, 1125 (1990).

87. Z.-H. Zhang, T.-S. Li, T.-S. Jin, and J.-X. Wang, *J. Chem. Res. (S)*, 152 (1998).

88. S. Kim and I. S. Kee, *Tetrahedron Lett.*, **31**, 2899 (1990).

89. J. Cossy, V. Bellosta, and M. C. Müller, *Tetrahedron Lett.*, **33**, 5045 (1992).

90. K. Tanemura, T. Suzuki, and T. Horaguchi, *Bull. Chem. Soc. Jpn.*, **67**, 290 (1994).

91. S. Raina and V. K. Singh, *Synth. Commun.*, **25**, 2395 (1995).

92. A. Srikrishna, J. A. Sattigeri, R.Viswajanani, and C. V. Yelamaggad, *J. Org. Chem.*, **60**, 2260 (1995).

93. G. Maiti and S. C. Roy, *J. Org. Chem.*, **61**, 6038 (1996).

94. A. DattaGupta, R. Singh, and V. K. Singh, *Synlett*, 69 (1996).

95. I. E. Markó, A. Ates, B. Augustyns, A. Gautier, Y. Quesnel, L. Turet, and M. Wiaux, *Tetrahedron Lett.*, **40**, 5613 (1999).

96. K. P. Nambiar and A. Mitra, *Tetrahedron Lett.*, **35**, 3033 (1994).

97. K. J. Davis, U. T. Bhalerao, and B. V. Rao, *Indian J. Chem., Sect. B*, **36B**, 211 (1997).

98. M. A. Reddy, L. R. Reddy, N. Bhanumathi, and K. R. Rao, *New J. Chem.*, **25**, 359 (2001).

99. T. Oriyama, K. Yatabe, S. Sugawara, Y. Machiguchi, and G. Koga, *Synlett*, 523 (1996).

100. E. Rafiee, S. Tangestaninejad, M. H. Habibi, I. Mohammadpoor-Baltork, and V. Mirkhani, *Russian J. Org. Chem.*, **41**, 393 (2005).

101. T. Suzuki, K. Ohashi, and T. Oriyama, *Synthesis*, 1561 (1999).

102. A. I. Meyers, S. Schwartzman, G. L. Olson, and H.-C. Cheung, *Tetrahedron Lett.*, **17**, 2417 (1976).

103. M. M. Heravi, D. Ajami, M. M. Mojtahedi, and M. Ghassemzadeh, *Tetrahedron Lett.*, **40**, 561 (1999).

104. M. M. Heravi and D. Ajami, *J. Chem. Res. (S)*, 718 (1998); M. M. Heravi and D. Ajami, *Monatsh. Chem.*, **130**, 709 (1999).

105. M. M. Heravi, R. Hekmatshoar, Y. S. Beheshtiha, and M. Ghassemzadeh, *Monatsh. Chem.*, **132**, 651 (2001).

106. M. M. Heravi, H. A. Oskooie, M. Ghassemzadeh, and F. F. Zameni, *Monatsh. Chem.*, **130**, 1253 (1999).

107. M. M. Heravi, P. Kazemian, H. A. Oskooie, and M. Ghassemzadeh, *J. Chem. Res.*, 105 (2005).

108. M. M. Heravi, D. Ajami, and K. Tabar-Heydar, *Monatsh. Chem.*, **130**, 337 (1999).

109. I. Mohammadpoor-Baltork and S. Pouranshirvani, *Synthesis*, 756 (1997).

110. I. Mohammadpoor-Baltork and A. R. Nourozi, *Synthesis*, 487 (1999).

111. I. Mohammadpoor-Baltork and B. Kharamesh, *J. Chem. Res. (S)*, 146 (1998).

112. A. R. Hajipour and A. E. Ruoho, *Synth. Commun.*, **33**, 871 (2003).

113. M. Tajbakhsh, M. M. Heravi, and S. Habibzadeh, *Phosphorus, Sulfur and Silicon and the Related Elements*, **176**, 191 (2001).

114. A. R. Hajipour, S. E. Mallapour, I. M. Baltork, and H. Adibi, *Synth. Commum.*, **31**, 1625 (2001).

115. M. Narender, M. S. Reddy, and K. R. Rao, *Synthesis*, 1741 (2004).

Fluorous Tetrahydropyranyl

This group was developed for the simple purification of small molecules by liquid/liquid extraction with CH_3CN/FC72.

Formation/Cleavage[1]

The related fluorous alkoxy ethyl ether $(C_8F_{17}CH_2CH_2)_2CHOCH(OR)CH_3$ has been prepared for the same purpose.[2]

1. P. Wipf and J. T. Reeves, *Tetrahedron Lett.*, **40**, 4649 (1999).
2. P. Wipf and J. T. Reeves, *Tetrahedron Lett.*, **40**, 5139 (1999).

3-Bromotetrahydropyranyl Ether: 3-BrTHP−OR

The 3-bromotetrahydropyranyl ether was prepared from a 17-hydroxy steroid and 2,3-dibromopyran (pyridine, benzene, 20°C, 24 h); it was cleaved by zinc/ethanol.[1]

1. A. D. Cross and I. T. Harrison, *Steroids*, **6**, 397 (1965).

Tetrahydrothiopyranyl Ether (Chart 1)

The tetrahydrothiopyranyl ether was prepared from a 3-hydroxy steroid and dihydro-thiopyran (CF_3COOH, $CHCl_3$, 35% yield); it can be cleaved under neutral conditions ($AgNO_3$, aq. acetone, 85% yield).[1]

1. L. A. Cohen and J. A. Steele, *J. Org. Chem.*, **31**, 2333 (1966).

1-Methoxycyclohexyl Ether[1]: A

4-Methoxytetrahydropyranyl Ether (MTHP−OR)[1]: B (Chart 1)

4-Methoxytetrahydrothiopyranyl Ether[2]: C (Chart 1)

4-Methoxytetrahydrothiopyranyl Ether S,S-Dioxide[2]: D

The above ethers have been examined as possible protective groups for the 2′-hydroxyl of ribonucleotides. The following rates of hydrolysis were found: A:B:C:D = 1:0.025:0.005:0.002.[3] These acetals can be prepared by the same methods used for the preparation of the THP derivative. Compounds B and C have been prepared from the vinyl ether and TMSCl as a catalyst.[4] Sulfoxide D was prepared from sulfide C by oxidation with m-ClC$_6$H$_4$CO$_3$H. These ethers have the advantage that they do not introduce an additional stereogenic center into the molecules as does the THP group. The 4-methoxytetrahydropyranyl group has seen extensive use in nucleoside synthesis, but still suffers from excessive acid lability when the 9-phenylxanthen-9-yl group is used to protect 5′-hydroxy functions in ribonucleotides.[5] The recommended conditions for removal of this group are 0.01 M HCl at room temperature. Little, if any, use of these groups has been made by the general synthetic community, but the wide range of selectivities observed in their acidic hydrolysis should make them useful for the selective protection of polyfunctional molecules.

1-[(2-Chloro-4-methyl)phenyl]-4-methoxypiperidin-4-yl Ether (CTMP−OR)[6]

This group was designed to have nearly constant acid stability with decreasing pH ($t_{1/2} = 80$ min at pH = 3.0, $t_{1/2} = 33.5$ min at pH = 0.5), which is in contrast to the MTHP group that is hydrolyzed faster as the pH is decreased ($t_{1/2} = 125$ min at pH = 3, $t_{1/2} = 0.9$ min at pH = 1.0). This group was reported to have excellent compatibility with the conditions used to remove the 9-phenylxanthen-9-yl group (5.5 eq. CF$_3$COOH, 16.5 eq. pyrrole, CH$_2$Cl$_2$, rt, 30 s, 95.5% yield).[3,7,8]

1-(2-Fluorophenyl)-4-methoxypiperidin-4-yl Ether (Fpmp−OR)

Formation

1-(2-Fluorophenyl)-4-methoxy-1,2,5,6-tetrahydropyridine, mesitylenesulfonic acid or TFA, CH$_2$Cl$_2$, 76–91% yield.[9–11]

Cleavage

Water, pH 2–2.5, 20 h. The $t_{1/2}$ for deblocking the 2′-Fpmp derivative of uridine is 166 min at pH 3 at 25°C, whereas it is 75 min for the bis-Fpmp r[UpU] derivative. The increased rate in the latter is assumed to be a result of internal phosphate participation.[12] The Fpmp group is ~1.3 times more stable than the related Ctmp group in the pH range 0.5–1.5. This added stability improves the selectivity for cleavage of the DMTr and pixyl groups in the presence of the Fpmp group during RNA synthesis.[11]

1-(4-Chlorophenyl)-4-methoxypiperidin-4-yl Ether (Cpep−OR)

The Cpep group, formed from the enol ether, has a rate of hydrolysis that is only 3.73 times slower at pH 3.75 than at pH 0.5. It is more stable than the Fpmp group at pH 0.5 and yet over twice as labile at pH 3.75. It has a nearly constant half-life between pH 0.5 and 2.5.[13]

1. C. B. Reese, R. Saffhill, and J. E. Sulston, *J. Am. Chem. Soc.*, **89**, 3366 (1967); *idem*, *Tetrahedron*, **26**, 1023 (1970).

2. J. H. van Boom, P. van Deursen, J. Meeuwse, and C. B. Reese, *J. Chem. Soc., Chem. Commun.*, 766 (1972).

3. C. B. Reese, H. T. Serafinowska, and G. Zappia, *Tetrahedron Lett.*, **27**, 2291 (1986).

4. H. C. P. F. Roelen, G. J. Ligtvoet, G. A. Van der Morel, and J. H. Van Boom, *Recl. Trav. Chim. Pays-Bas*, **106**, 545 (1987).

5. C. B. Reese and P. A. Skone, *Nucleic Acids Res.*, **13**, 5215 (1985).

6. For a large-scale preparation, see M. Faja, C. B. Reese, Q. Song, and P.-Z. Zhang, *J. Chem. Soc., Perkin Trans. 1*, 191 (1997).

7. For an improved preparation of the reagent, see C. B. Reese and E. A. Thompson, *J. Chem. Soc., Perkin Trans. 1*, 2881 (1988); M. Faja, C. B. Reese, Q. Song, and P.-Z. Zhang, *J. Chem. Soc. Perkin Trans. 1*, 191 (1997).

8. O. Sakatsume, M. Ohtsuki, H. Takaku, and C. B. Reese, *Nucleic Acid Symp. Ser.*, **20**, 77 (1988).

9. B. Beijer, I. Sulston, B. S. Sproat, P. Rider, A. I. Lamond, and P. Neuner, *Nucleic Acids Res*, **18**, 5143 (1990).

10. A. J. Lawrence, J. B. J. Pavey, I. A. O'Neil, and R. Cosstick, *Tetrahedron Lett.*, **36**, 6341 (1995).

11. V. M. Rao, C. B. Reese, V. Schehlmann, and P. S. Yu , *J. Chem. Soc., Perkin Trans. 1*, 43 (1993).

12. D. C. Capaldi and C. B. Reese, *Nucleic Acids Res.*, **22**, 2209 (1994).

13. W. Lloyd, C. B. Reese, Q. Song, A. M. Vandersteen, C. Visintin, and P.-Z. Zhang, *Perkin 1*, 165 (2000).

1,4-Dioxan-2-yl Ether

Formation

1,4-Dihydrodioxin, CuBr$_2$, THF, rt, 50–88% yield.[1]

Cleavage

6 N HCl, EtOH, reflux, 90% yield for cholesterol.[1] Although a direct stability comparison was not made, this group should be more stable than the THP group for the same reasons that the anomeric ethers of carbohydrates are more stable than their 2-deoxy counterparts.

1. M. Fetizon and I. Hanna, *Synthesis*, 806 (1985).

Tetrahydrofuranyl Ether (Chart 1)

Formation

1. 2-Chlorotetrahydrofuran, Et$_3$N, 30 min, 82–98% yield.[1] 2-Chlorotetrahydro-furan is readily prepared from THF with SO$_2$Cl$_2$ (25°C, 0.5 h, 85%).
2. Dihydrofuran, metallosalen catalyst, C$_6$H$_5$Cl or CH$_2$Cl$_2$, rt, 24 h, 81–100% yield. Since this reaction employs a chiral catalyst the derivatization proceeds with 71–86% ee or 40–99% de.[2]
3. Ph$_2$CHCO$_2$-2-tetrahydrofuranyl, 1% TsOH, CCl$_4$, 20°C, 30 min, 90–99% yield.[1,3] The authors report that formation of the THF ether by reaction with 2-chlorotetrahydrofuran avoids a laborious procedure[4] that is required when dihydrofuran is used. In addition, the use of dihydrofuran to protect the 2'-OH of a nucleotide gives low yields (24–42%).[5] The tetrahydrofuranyl ester is reported to be a readily available, stable solid. A tetrahydrofuranyl ether can be cleaved in the presence of a THP ether.[1]
4. THF, [Ce(Et$_3$NH)$_2$](NO$_3$)$_6$, 50–100°C, 8 h, 30–98% yield.[6] Hindered alcohols give the lower yields. The method was also used to introduce the THP group with tetrahydropyran.
5. THF, PhI(OAc)$_2$, 10–68% yield.[7] These results show that hypervalent iodine species should probably not be used in THF as a solvent.
6. THF, (n-Bu$_4$N$^+$)$_2$S$_2$O$_8{}^-$, reflux, 85% yield.[8] These oxidative conditions proved to be compatible with an aromatic thioether.

7. BrCCl$_3$, 60°C, 2,4,6-collidine, THF, 56–92% yield.[9] This method is not recommended for allylic and tertiary alcohols.

8. THF, CrCl$_2$, CCl$_4$, rt, 47–95% yield. The reaction proceed through *in situ* formation of 2-chlorotetrahydrofuran.[10] Phenols and tertiary alcohols give the ethers in only modest yields.

9. 1-*t*-Butylperoxy-1,2-benziodoxol-3(1H)-one, CCl$_4$, 50°C, THF, 10 h, K$_2$CO$_3$, 43–98% yield. Phenols and tertiary alcohols fail to react. 2-Chlorotetrahydrofuran is formed *in situ* by a free radical mechanism.[11]

Cleavage

1. AcOH, H$_2$O, THF, (3:1:1), 25°C, 30 min, 90% yield.[1]

2. 0.01 *N* HCl, THF (1:1), 25°C, 10 min, 50% yield.[1]

3. pH 5, 25°C, 3 h, 90% yield.[1]

1. C. G. Kruse, F. L. Jonkers, V. Dert, and A. van der Gen, *Recl. Trav. Chim. Pays-Bas*, **98** 371 (1979).

2. H. Nagano and T. Katsuki, *Chem. Lett.*, 782 (2002).

3. C. G. Kruse, E. K. Poels, F. L. Jonkers, and A. van der Gen, *J. Org. Chem.*, **43**, 3548 (1978).

4. E. L. Eliel, B. E. Nowak, R. A. Daignault, and V. G. Badding, *J. Org. Chem.*, **30**, 2441 (1965).

5. E. Ohtsuka, A. Yamane, and M. Ikehara, *Chem. Pharm. Bull.*, **31**, 1534 (1983).

6. A. M. Maione and A. Romeo, *Synthesis*, 250 (1987).

7. A. N. French, J. Cole, and T. Wirth, *Synlett*, 2291 (2004).

8. J. C. Jung, H. C. Choi, and Y. H. Kim, *Tetrahedron Lett.*, **34**, 3581 (1993).

9. J. M. Barks, B. C. Gilbert, A. F. Parsons, and B. Upeandran, *Tetrahedron Lett.*, **41**, 6249 (2000).

10. R. Baati, A. Valleix, C. Mioskowski, D. K. Barma, and J. R. Falck, *Org. Lett.*, **2**, 485 (2000).

11. M. Ochiai and T. Sueda, *Tetrahedron Lett.*, **45**, 3557 (2004).

Tetrahydrothiofuranyl Ether (Chart 1)

Formation

1. Dihydrothiofuran, CHCl$_3$, CF$_3$COOH, reflux, 6 days, 75% yield.[1]

2. cat. TsOH, CHCl$_3$, 20°C, 5 h, 85–95% yield.[2]

Cleavage

1. AgNO$_3$, acetone, H$_2$O, reflux, 90% yield.[1]
2. HgCl$_2$, CH$_3$CN, H$_2$O, 25°C, 10 min, quant.[2] Some of the methods used to cleave methylthiomethyl (MTM) ethers should also be applicable to the cleavage of tetrahydrothiofuranyl ethers.

1. L. A. Cohen and J. A. Steele, *J. Org. Chem.*, **31**, 2333 (1966).
2. C. G. Kruse, E. K. Poels, F. L. Jonkers, and A. van der Gen, *J. Org. Chem.*, **43**, 3548 (1978).

2,3,3a,4,5,6,7,7a-Octahydro-7,8,8-trimethyl-4,7-methanobenzofuran-2-yl Ether (RO−MBF)

Formation[1,2]

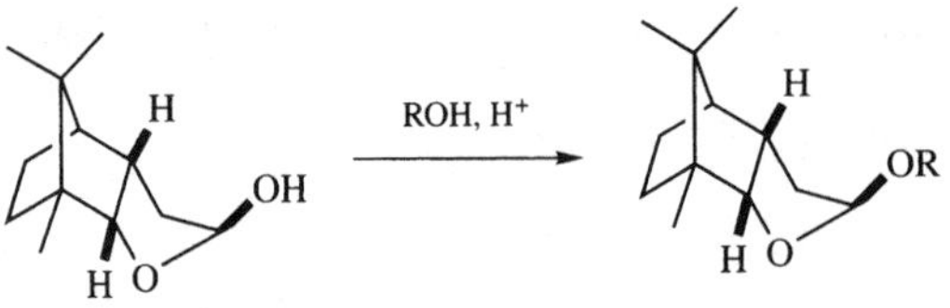

The advantage of this ketal is that unlike the THP group, only a single isomer is produced in the derivatization, but the disadvantage is that it is not commercially available. Conditions used to hydrolyze the THP group can be used to hydrolyze this acetal.[3] This group may also find applications in the resolution of racemic alcohols.

1. C. R. Noe, *Chem. Ber.*, **115**, 1576 1591 (1982); C. R. Noe, M. Knollmüller, G. Steinbauer, E. Jangg, and H. Völlenkle, *Chem. Ber.*, **121**, 1231 (1988).
2. U. Girreser and C. R. Noe, *Synthesis*, 1223 (1995).
3. K. Zimmermann, *Synth. Commun.*, **25**, 2959 (1995).

Substituted Ethyl Ethers

1-Ethoxyethyl Ether (EE−OR): ROCH(OC$_2$H$_5$)CH$_3$ (Chart 1)

Formation

1. Ethyl vinyl ether, HCl (anhydrous).[1]
2. Ethyl vinyl ether, TsOH, 25°C, 1 h.[2]
3. Ethyl vinyl ether, pyridinium tosylate (PPTS), CH$_2$Cl$_2$, rt, 0.5 h.[3]
4. The ethoxyethyl ether was selectively introduced on a primary alcohol in the presence of a secondary alcohol.[4]

5. $CH_3CH(Cl)OEt$, $PhNMe_2$, CH_2Cl_2, 0°C, 10–60 min.[5] These conditions are effective for extremely acid-sensitive substrates or where conditions **1** and **2** fail.

6. $CH_2=CHOEt$, $CoCl_2$, 65–91% yield.[6]

7. 2:1 Ethyl vinyl ether, CH_2Cl_2, PPTS, 25°C, 4 h, 84% yield. These conditions proved optimal for the protection of this acid-sensitive alcohol.[7]

Cleavage

1. 5% AcOH, 20°C, 2 h, 100% yield.[1]

2. 0.5 N HCl, THF, 0°C, 100% yield.[2] The ethoxyethyl ether is more readily cleaved by acidic hydrolysis than the THP ether, but it is more stable than the 1-methyl-1-methoxyethyl ether. TBDMS ethers are not affected by these conditions.[8]

3. Pyridinium tosylate, *n*-PrOH, 80–85% yield.[9] An acetonide was not affected by these conditions.

1. S. Chládek and J. Smrt, *Chem. Ind. (London)*, 1719 (1964).

2. A. I. Meyers, D. L. Comins, D. M. Roland, R. Henning, and K. Shimizu, *J. Am. Chem. Soc.*, **101**, 7104 (1979).

3. A. Fukuzawa, H. Sato, and T. Masamune, *Tetrahedron Lett.*, **28**, 4303 (1987).

4. M. F. Semmelhack and S. Tomoda, *J. Am. Chem. Soc.*, **103**, 2427 (1981).

5. W. C. Still, *J. Am. Chem. Soc.*, **100**, 1481 (1978).

6. J. Iqbal, R. R. Srivastava, K. B. Gupta, and M. A. Khan, *Synth. Commun.*, **19**, 901 (1989).

7. M. R. Hellberg, R. E. Conrow, N. A. Sharif, M. A. McLaughlin, J. E. Bishop, J. Y. Crider, W. D. Dean, K. A. DeWolf, D. R. Pierce, V. L. Sallee, R. D. Selliah, B. S. Severns, S. J. Sproull, G. W. Williams, P. W. Zinke, and P. G. Klimko, *Bioorg. & Med. Chem.*, **10**, 2031 (2002).

8. K. Zimmermann, *Synth. Commun.*, **25**, 2959 (1995).

9. M. A. Tius and A. H. Faug, *J. Am. Chem. Soc.*, **108**, 1035 (1986).

1-(2-Chloroethoxy)ethyl Ether (Cee−OR): $ROCH(CH_3)OCH_2CH_2Cl$

The Cee group was developed for the protection of the 2′-hydroxyl group of ribonucleosides.

Formation

CH$_2$=CHOCH$_2$CH$_2$Cl, PPTS, CH$_2$Cl$_2$, 80–83% yield.[1]

Cleavage

The relative rates of cleavage for a variety of uridine-protected acetals are given in the table below.

Relative Cleavage Rates for Various Uridine-Protected Acetals

	1.5% Cl$_2$CHCO$_2$H in CH$_2$Cl$_2$		0.01 N HCl (pH 2)	
Ether	$T_{1/2}$ (min)	T_∞ (min)	$T_{1/2}$ (min)	T_∞ (min)
ROCH(CH$_3$)OCH$_2$CH$_2$Cl	420	960	96	360
ROCH(CH$_3$)O−i-Pr	—	30 s	1	4
ROCH(CH$_3$)OBu	2	5	12	34
ROCH(CH$_3$)OEt	20 s	3	5	18
ROTHP	90	273	32	150
ROCTMP[a]	—	—	55	295

[a]CTMP = 1-[(2-chloro-4-methyl)phenyl]-4-methoxypiperidin-4-yl ether

The Cee group is stable under the acidic conditions used to cleave the DMTr group.[3]

1. S.-i. Yamakage, O. Sakatsume, E. Furuyama, and H. Takaku, *Tetrahedron Lett.*, **30**, 6361 (1989).
2. O. Sakatsume, T. Yamaguchi, M. Ishikawa, I. Ichiro, K. Miura, and H. Takaku, *Tetrahedron*, **47**, 8717 (1991).
3. O. Sakatsume, T. Ogawa, H. Hosaka, M. Kawashima, M. Takaki, and H. Takaku, *Nucleosides & Nucleotides*, **10**, 141 (1991).

2-Hydroxyethyl Ethers

Although not strictly used as a protective group, these ethers are often formed as a result of other transformations and thus block a hydroxyl. They are cleaved by the action of CAN to release the alcohol. What is unusual about this process is that even nonbenzylic ethers are cleaved as illustrated below.[1]

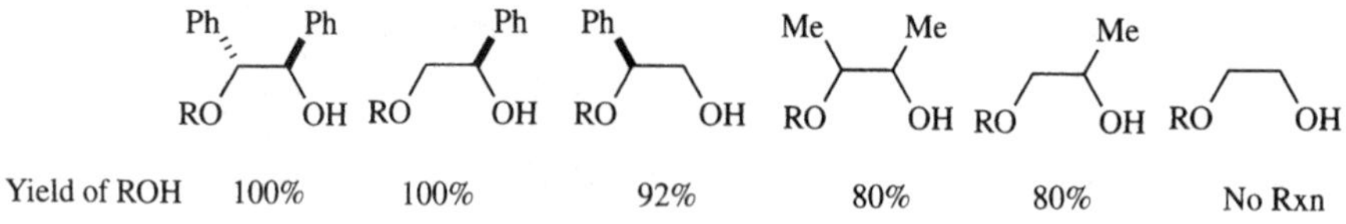

Yield of ROH	100%	100%	92%	80%	80%	No Rxn

1. H. Fujioka, Y. Ohba, H. Hirose, K. Murai, and Y. Kita, *Org. Lett.*, **7**, 3303 (2005).

2-Bromoethyl Ether: $BrCH_2CH_2OR$

The bromomethyl ether was used for the protection of the anomeric center in carbohydrate synthesis. It is readily introduced by normal glycosylation methodology. It is
cleaved by conversion to phenylsulfonylethyl ether, which, upon treatment with base,
releases the alcohol by an E-2 process.[1]

1. U. Ellervik, M. Jacobsson, and J. Ohlsson, *Tetrahedron*, **61**, 2421 (2005).

1-[2-(Trimethylsilyl)ethoxy]ethyl Ether (SEE−OR)

The chiral center produced upon derivatization of an alcohol may be a detriment to
this group.

Formation

2-TMSCH$_2$CH$_2$OCH=CH$_2$, CH$_2$Cl$_2$, PPTS, rt, 1–3 h, 76–96% yield. Phenols are
readily protected with this reagent.[1]

Cleavage

1. TBAF·H$_2$O, THF, 45°C, 20–24 h, 76–90% yield.
2. TsOH or PPTS, THF, H$_2$O, 4 h, rt.[1]
3. The section on the cleavage of the SEM ether should be consulted. The expectation is that this group is more easily cleaved by acid than the SEM group
   because the added stabilization the methyl group imparts to an intermediate
   carbenium ion.

1. J. Wu, B. K. Shull, and M. Koreeda, *Tetrahedron Lett.*, **37**, 3647 (1996).

1-Methyl-1-methoxyethyl Ether (MIP−OR): $ROC(OCH_3)(CH_3)_2$ (Chart 1)

This group can be used to protect the sensitive hydroperoxides.[1]

Formation

1. CH$_2$=C(CH$_3$)OMe, cat. POCl$_3$, 20°C, 30 min, 100% yield.[2]
2. CH$_2$=C(CH$_3$)OMe, neat, 20°C, TsOH.[3]

3. CH$_2$=C(CH$_3$)OMe has been used to protect a hydroperoxide.[4]

Cleavage

1. 20% AcOH, 20°C, 10 min.[1]
2. Pyridinium *p*-toluenesulfonate, 5°C, 1 h.[5] Similar selectivity can be achieved using a silica-alumina gel prepared by the sol–gel method.[6]

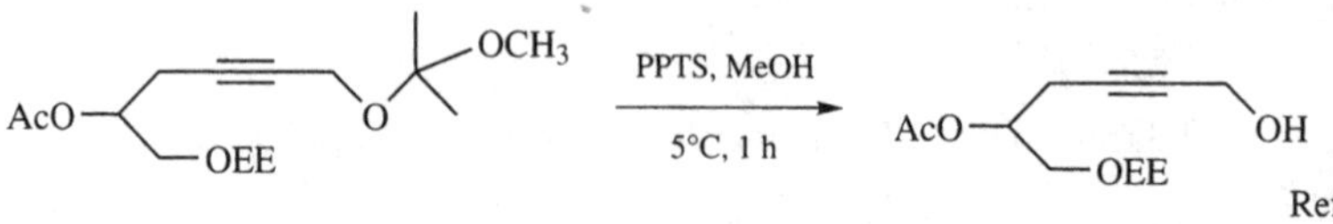

Ref. 5

In general, the MIP ether is very labile to acid and silica gel chromatography unless some TEA is used as part of the eluting solvent. The acid in the NMR solvent, CDCl$_3$, is sufficient to cleave the MIP ether.

1. P. H. Dussault and K. R. Woller, *J. Am. Chem. Soc.*, **119**, 3824 (1997).
2. A. F. Klug, K. G. Untch, and J. H. Fried, *J. Am. Chem. Soc.*, **94**, 7827 (1972).
3. P. L. Barili, G. Berti, G. Catelani, F. Colonna, and A. Marra, *Tetrahedron Lett.*, **27**, 2307 (1986).
4. P. H. Dussault and K. R. Woller, *J. Am. Chem. Soc.*, **119**, 3824 (1997).
5. G. Just, C. Luthe and M. T. P. Viet, *Can. J. Chem.*, **61**, 712 (1983).
6. Y. Matsumoto, K. Mita, K. Hashimoto, H. Iio, and T. Tokoroyama, *Tetrahedron*, **52**, 9387 (1996).

1-Methyl-1-benzyloxyethyl Ether (MBE−OR): ROC(OBn)(CH$_3$)$_2$

Formation

1. CH$_2$=C(OBn)(CH$_3$), PdCl$_2$(1,5-cyclooctadiene) [PdCl$_2$(COD)], 85–95% yield.[1]
2. CH$_2$=C(OBn)(CH$_3$), POCl$_3$ or TsOH, 61–98% yield.[1] It should be noted that these conditions do not afford a cyclic acetal with a 1,3-diol. This ketal is stable to LiAlH$_4$, diisobutylaluminum hydride, NaOH, alkyllithiums, and Grignard reagents.

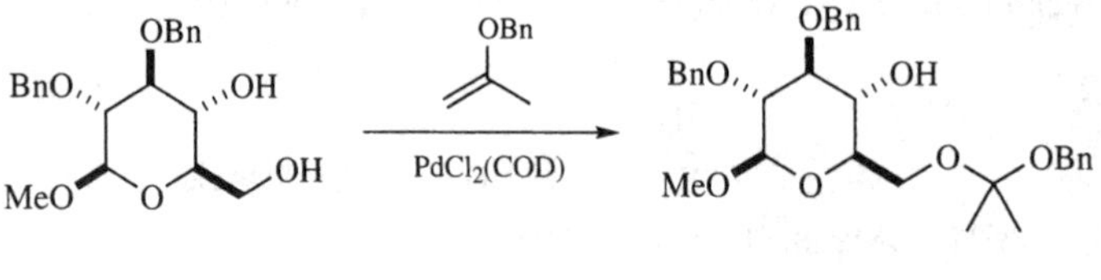

Cleavage

1. H$_2$, 5% Pd–C, EtOH, rt, 92–99% yield.[1]
2. 3 *M* AcOH, H$_2$O, THF.[2]

1-Methyl-1-benzyloxy-2-fluoroethyl Ether: $ROC(OBn)(CH_2F)(CH_3)$

The electron-withdrawing fluorine group should make this group more stable to acid than the MBE group.

Formation

CH_2=$C(OBn)CH_2F$, $PdCl_2(COD)$, CH_3CN, rt, 24 h, 89–100% yield.[2] Protic acids can also be used to introduce this group, but the yields are sometimes lower. A primary alcohol can be protected in the presence of a secondary alcohol. This reagent does not give cyclic acetals of 1,3-diols with palladium catalysis.

Cleavage

H_2, Pd–C, EtOH, 1 atm, 98–100% yield.[2] This group is stable to 3 *M* aqueous acetic acid at room temperature, conditions that cleave the TBDMS group and the 1-methyl-1-benzyloxyethyl ether.

1. T. Mukaiyama, M. Ohshima, and M. Murakami, *Chem. Lett.*, **13**, 265 (1984).
2. T. Mukaiyama, M. Ohishima, H. Nagaoka, and M. Murakami, *Chem. Lett.*, **13**, 615 (1984).

1-Methyl-1-phenoxyethyl Ether: $ROC(OPh)(CH_3)_2$

The electron-withdrawing phenyl group is expected to increase the stability of this group toward acid relative to its methyl counterpart.

Formation/Cleavage[1]

1. P. Zandbergen, H. M. G. Willems, G. A. Van der Marel, J. Brussee, and A. van der Gen, *Synth. Commun.*, **22**, 2781 (1992).

2,2,2-Trichloroethyl Ether: Cl_3CCH_2OR

The anomeric position of a carbohydrate is protected as its trichloroethyl ether. Cleavage is effected with Zn, AcOH, AcONa (3 h, 92%).[1]

1. R. U. Lemieux and H. Driguez, *J. Am. Chem. Soc.*, **97**, 4069 (1975).

1,1-Dianisyl-2,2,2-trichloroethyl Ether (DATE−OR)

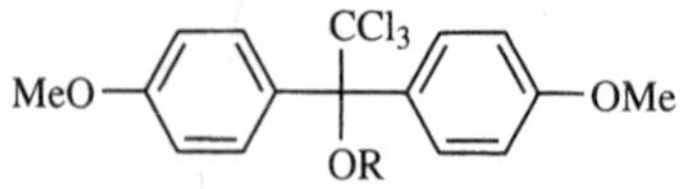

Formation

An$_2$(Cl$_3$C)CCl, AgOTf, CH$_3$CN, Pyr, rt, 12–18 h, 92% yield.[1]

Cleavage[1]

1. Li[Co(I)Pc], MeOH, 80–90% yield.
2. Zn, ZnBr$_2$, MeOH, Et$_2$O, or Zn, 80% AcOH–dioxane, 70–80% yield.
3. DATE ethers are stable to concd. HCl–MeOH–dioxane (1:2:2), Cl$_2$CHCO$_2$H–CH$_2$Cl$_2$ (3:97), and NH$_3$–dioxane (1:1).

1. R. M. Karl, R. Klösel, S. König, S. Lehnhoff, and I. Ugi, *Tetrahedron*, **51**, 3759 (1995).

1,1,1,3,3,3-Hexafluoro-2-phenylisopropyl Ether (HIP−OR):   Ph(CF$_3$)$_2$C−OR

This group is stable to strong acid and base, TMSI, Pd–C/H$_2$, DDQ, TBAF, and LAH at low temperatures, and thus has the potential to participate in a large number of orthogonal sets.[1]

Formation

1,1,1,3,3,3-Hexafluoro-2-phenylisopropyl alcohol, diethyl azodicarboxylate, PPh$_3$, benzene, 82–98% yield. Primary alcohols are effectively derivatized, but yields for secondary alcohols are low (46–65% yield).[1]

Cleavage

Lithium naphthalenide, <1 h, −78°C. The following protective groups can be cleaved in the presence of the HIP group: Tr, THP, MEM, Bn, MPM, TBDPS, Bz; all but the Bz group are stable to the conditions for the cleavage of the HIP group.[1]

1. H.-S. Cho, J. Yu, and J. R. Falck, *J. Am. Chem. Soc.*, **116**, 8354 (1994).

1-(2-Cyanoethoxy)ethyl Ether (CEE−OR):   ROCH(CH$_3$)OCH$_2$CH$_2$CN

This group was developed for the protection of ribonucleosides. The CEE group is stable to TEA·HF, 25% aq. NH$_3$, 25% aq. NH$_3$/EtOH and 2M NH$_3$/EtOH.

Formation

$CH_2=CHOCH_2CH_2CN$, dioxane, pTSA, 75–97% yield.[1]

Cleavage

1. 0.5M DBU, CH_3CN, $t_{1/2}$ = 240 min.
2. TBAF, THF, 1 min.

1. T. Umemoto and T. Wada, *Tetrahedron Lett.*, **45**, 9529 (2004).

2-Trimethylsilylethyl Ether: $Me_3SiCH_2CH_2OR$

Cleavage

1. $BF_3 \cdot Et_2O$, CH_2Cl_2, 0–25°C, 79% yield.[1]

2. CsF, DMF, 210°C, >65% yield.[2]

1. S. D. Burke, G. J. Pacofsky, and A. D. Piscopio, *Tetrahedron Lett.*, **27**, 3345 (1986).
2. L. A. Paquette, D. Backhaus, and R. Braun, *J. Am. Chem. Soc.*, **118**, 11990 (1996).

2-(Benzylthio)ethyl Ether: $BnSCH_2CH_2OR$

This ether, developed for protection of a pyranoside anomeric hydroxyl, is prepared via a Königs–Knorr reaction from the glycosyl bromide and 2-(benzylthio)ethanol in the presence of DIPEA. It is cleaved, after oxidation with dimethyldioxirane, by treatment with LDA or MeONa.[1]

1. T.-H. Chan and C. P. Fei, *J. Chem. Soc., Chem. Commun.*, 825 (1993).

2-(Phenylselenyl)ethyl Ether: ROCH₂CH₂SePh (Chart 1)

This ether was prepared from an alcohol and 2-(phenylselenyl)ethyl bromide (AgNO₃, CH₃CN, 20°C, 10–15 min, 80–90% yield); it is cleaved by oxidation (H₂O₂, 1 h; ozone; or NaIO₄), followed by acidic hydrolysis of the intermediate vinyl ether (dil. HCl, 65–70% yield).[1] The use of this group was crucial to the synthesis of lucilacaene which is not stable to acid, base or light.[2]

1. T.-L. Ho and T. W. Hall, *Synth. Commun.*, **5**, 367 (1975).
2. J. Yamaguchi, H. Kakeya, T. Uno, M. Shoji, H. Osada, and Y. Hayashi, *Angew. Chem. Int. Ed.*, **44**, 3110 (2005).

t-Butyl Ether: *t*-BuOR (Chart 1)

Formation

t-Butyl ethers can be prepared from a variety of alcohols, including allylic alcohols. They are stable to most reagents except strong acids. The *t*-butyl ether is probably one of the more under-used alcohol protective groups considering its stability, the ease and efficiency of introduction, and the ease of cleavage.

1. Isobutylene, BF₃·Et₂O, H₃PO₄, 100% yield.[1,2]

This method has been used for the preparation of the somewhat more hindered 2-ethyl-2-butyl ether (*t*-amyl ether); the introduction is selective for primary alcohols.[3]

2. Isobutylene, Amberlyst H-15, hexane.[4] Methylene chloride can also be used as solvent, and in this case a primary alcohol was selectively converted to the *t*-amyl ether in the presence of a secondary alcohol.[5]

3. Isobutylene, H_2SO_4.[6] Acyl migration has been observed using these conditions.[7]

4. Isobutylene, H_3PO_4, $BF_3 \cdot Et_2O$, $-72°C$, 3 h, $0°C$, 20 h, 79% yield.[8]

5. t-BuOC(=NH)CCl$_3$, $BF_3 \cdot Et_2O$, CH_2Cl_2, cyclohexane, 59–91% yield.[9]

6. BOC_2O, $Mg(ClO_4)_2$, CH_2Cl_2, $40°C$, 8–43 h, 65–95% yield.[10]

Cleavage

1. Anhydrous CF_3COOH, 0–$20°C$, 1–16 h, 80–90% yield.[2,4]

2. HBr, AcOH, $20°C$, 30 min.[11]

3. 4 N HCl, dioxane, reflux, 3 h.[12] In this case the t-butyl ether was stable to 10 N HCl, MeOH, 0–$5°C$, 30 h.

4. HCO_2H, rt, 24 h, >83% yield.[13]

5. Me_3SiI, CCl_4, or $CHCl_3$, $25°C$, <0.1 h, 100% yield.[14] Under suitable conditions, this reagent also cleaves many other ethers, esters, ketals, and carbamates.[15]

6. Ac_2O, $FeCl_3$, Et_2O, 76–93% yield.[4,16] These conditions give the acetate of the alcohol, which can then be cleaved by simple basic hydrolysis. The method is also effective for the conversion of t-butyl glycosides to acetates with retention of configuration (80–100% yield).[17]

7. $TiCl_4$, CH_2Cl_2, $0°C$, 1 min, 85% yield.[18]

8. TBDMSOTf, CH_2Cl_2, rt, 24 h, 82% yield. The use of a catalytic amount of the triflate will give the alcohol. If the triflate is used stoichiometrically and the reaction worked up with 2,6-lutidine the TBDMS ether is isolated (98% yield).[19]

9. $CeCl_3 \cdot 7H_2O$, CH_3CN, 93–98% yield.[10]

1 R. A. Micheli, Z. G. Hajos, N. Cohen, D. R. Parrish, L. A. Portland, W. Sciamanna, M. A. Scott, and P. A. Wehrli, *J. Org. Chem.*, **40**, 675 (1975).

2. H. C. Beyerman and G. L. Heiszwolf, *J. Chem. Soc.*, 755 (1963).

3. B. Figadère, X. Franck, and A. Cavè, *Tetrahedron Lett.*, **34**, 5893 (1993).

4. A. Alexakis and J. M. Duffault, *Tetrahedron Lett.*, **29**, 6243 (1988); A. Alexakis, M. Gardette, and S. Colin, *Tetrahedron Lett.*, **29**, 2951 (1988).

5. X. Franck, B. Figadère, and A. Cavé, *Tetrahedron Lett.*, **38**, 1413 (1997).

6. H. C. Beyerman and J. S. Bontekoe, *Proc. Chem. Soc.*, 249 (1961).

7. N. I. Simirskaya and M. V. Mavrov, *Zh. Org. Khim.*, **31**, 140 (1995); *Chem. Abstr.* **124**, 8220w (1996).

8. N. Cohen, W. F. Eichel, R. J. Lopresti, C. Neukom, and G. Saucy, *J. Org. Chem.*, **41**, 3505 (1976).

9. A. Armstrong, I. Brackenridge, R. F. W. Jackson, and J. M. Kirk, *Tetrahedron Lett.*, **29**, 2483 (1988).

10. G. Bartoli, M. Bosco, M. Locatelli, E. Marcantoni, P. Melchiorre, and L. Sambri, *Org. Lett.*, **7**, 427 (2005).

11. F. M. Callahan, G. W. Anderson, R. Paul, and J. E. Zimmerman, *J. Am. Chem. Soc.*, **85**, 201 (1963).

12. U. Eder, G. Haffer, G. Neef, G. Sauer, A. Seeger, and R. Wiechert, *Chem. Ber.*, **110**, 3161 (1977).

13. H. Paulsen and K. Adermann, *Liebigs Ann. Chem.*, 751 (1989).

14. M. E. Jung and M. A. Lyster, *J. Org. Chem.*, **42**, 3761 (1977).

15. A. H. Schmidt, *Aldrichimica Acta*, **14**, 31 (1981).

16. B. Ganem and V. R. Small, Jr., *J. Org. Chem.*, **39**, 3728 (1974).

17. N. Rakotomanomana, J. M. Lacombe, and A. A. Pavia, *J. Carbohydr. Chem.*, **9**, 93 (1990).

18. R. H. Schlessinger and R. A. Nugent, *J. Am. Chem. Soc.*, **104**, 1116 (1982).

19. X. Franck, B. Figadère, and A. Cavé, *Tetrahedron Lett.*, **36**, 711 (1995).

Cyclohexyl (Chx−OR) Ether: $C_6H_{11}OR$

The cyclohexyl group was developed as an alternative to the benzyl group for the protection of serine and threonine in BOC-based peptide synthesis because the benzyl group is partially lost upon deprotection of the BOC groups with TFA. It is about $20\times$ more stable to TFA than the benzyl group. Since a direct Williamson ether synthesis failed with cyclohexyl bromide, a two-step approach was used that relies on the greater reactivity of the cyclohexenyl bromide, which does undergo the S_N2 displacement in modest yield. The resulting allylic ether is then hydrogenated with PtO_2 to give the cyclohexyl ether. It is efficiently cleaved using 1 M TFMSA-thioanisole in TFA at rt for 30 min.[1]

1. Y. Nishiyama and K. Kurita, *Tetrahedron Lett.*, **40**, 927 (1999); Y. Nishiyama, S. Shikama, K.-i. Morita, and K. Kurita, *J. Chem. Soc., Perkin 1*, 1949 (2000).

1-Methyl-1′-cyclopropylmethyl (MCPM−OR) Ether: $C_3H_5CH(CH_3)OR$

This ether was developed as a protective group for carbohydrate synthesis. It has the disadvantage of having a chiral center which will complicate analysis. It is formed using the trichloroacetamidate method with Lewis acid catalysis ($BF_3\cdot Et_2O$, or AgOTf, 56% yield). It is somewhat more stable to TFA than the MPM ether. It is cleaved using 10% TFA, but was also cleaved with $Ac_2O/Sc(OTf)_3$.[1]

1. E. Eichler, F. Yan, J. Sealy, and D. M. Whitfield, *Tetrahedron*, **57**, 6679 (2001).

Allyl Ether (Allyl−OR): $CH_2{=}CHCH_2OR$ (Chart 1)

The use of allyl ethers for the protection of alcohols is common in the carbohydrate literature because allyl ethers are generally compatible with the various methods for glycoside formation.[1] Obviously the allyl ether is not compatible with powerful electrophiles such as bromine and catalytic hydrogenation, but it is stable to moderately acidic conditions (1 N HCl, reflux, 10 h).[2] The ease of formation, the many mild methods for its cleavage in the presence of numerous other protective groups, and its general stability have made it a mainstay of many orthogonal sets. The synthesis of perdeuteroallyl bromide and its use as a protective group in carbohydrates have been reported. The perdeutero derivative has the advantage that the allyl resonances in the NMR no longer obscure other, more diagnostic resonances, such as those of the anomeric carbon in glycosides.[3] The use of the allyl protective group primarily covering carbohydrate chemistry has been reviewed.[4]

Formation

1. CH_2=$CHCH_2Br$, NaOH, benzene, reflux, 1.5 h,[5] or NaH, benzene, 90–100% yield.[6]

2. CH_2=$CHCH_2OH$, [CpRu(CH_3CN)$_3$]PF_6 0.0005 eq., 2-quinolinecarboxylic acid, 70°C, 6 h, 87–98% yield.[7]

3. CH_2=$CHCH_2OC$(=NH)CCl_3, H^+.[8]

4. Bu_2SnO, toluene, THF; CH_2=$CHCH_2Br$, Bu_4NBr, 96% yield.[9] The crotyl ether has been introduced using similar methodology.[10]

5. CH_2=$CHCH_2OCO_2Et$, Pd_2(dba)$_3$, THF, 65°C, 4 h, 70–97% yield.[11]

Note the preferential reaction at the anomeric hydroxyl. The method is also effective for the protection of primary and secondary alcohols. A modification of this approach which uses t-$BuOCO_2CH_2CH$=CH_2 as the allyl source selectivity monoalkylates a tertiary hydroxyl in the erythronolide derivative.

The method is effective because *t*-BuOH does not compete effectively in the allylation process.[12]

6. MeO$_2$COCH$_2$CH=CH$_2$, Pd$_2$(dba)$_3$, CHCl$_3$, Ph$_2$P(CH$_2$)$_4$PPh$_2$, THF, 65°C, 95% yield.[13]

7. Allyl carbonates have been converted to allyl ethers with Pd(Ph$_3$P)$_4$.[14] The reaction also proceeds with Pd(OAc)$_2$ and Ph$_3$P (82% yield).[15] In the case below, acid- and base-catalyzed procedures failed because of the sensitivity of the [(*i*-Pr)$_2$Si]$_2$O group.

8. Allyl bromide, (RO)$_2$Mg.[16]

9. KF·alumina, allyl bromide, 80% yield. These conditions were developed because the typical strongly basic metal alkoxide-induced alkylation led to Beckmann fragmentation of the isoxazoline.[17]

10. Allyl bromide, DMF, BaO, rt.[18] This method is used in carbohydrates to prevent alkylation of an amide, which is a problem when NaH is used as the base.[19]

11. Allyl bromide, Al$_2$O$_3$, 1–10 days. These conditions were developed to alkylate selectively an alcohol in the presence of an amide.[20]

12. Allyl alcohol, DEAD, Ph$_3$P, THF, 69% yield. Other ethers were prepared but only ascorbic acid was used as a substrate.[21] The pK_a seems to determine the selectivity. pK_a of 2-OH ~ 8, 3-OH ~ 3–4, 5-OH ~ 12, 6-OH ~ 14, based on calculations using ACD software.

13. Allyl acetate, toluene, 100°C, [Ir(COD)$_2$]$^+$BF$_4^-$, 5 h, 62–98% yield. Phenols, acids, amines, and thiols are similarly allylated by this method in excellent yield.[22]

14. From an aldehyde: BiBr$_3$, Et$_3$SiH, CH$_3$CN, CH$_2$=CHCH$_2$OTBS, 92% yield. Other ethers can be prepared simply by changing the silyl ether. A propyl ether was prepared using this method on a 50-kg scale.[23] The BiBr$_3$ serves to generate HBr and TESBr *in situ.*

15. BrCH$_2$CH$_2$CH$_2$Br, NaH, THF, DMF, 2 h, 32–90% yield. This method is specific for the monoprotection of diols.

Cleavage

1. One of the primary methods for the cleavage of allyl ethers is through isomerization of the olefin to the vinyl ether. The vinyl ether can then be cleaved by a number of methods.

i. 0.1 *N* HCl, acetone–water, reflux, 30 min.[8]

ii. 0.1 eq. TsOH, MeOH, 25°C, 2.5 h, >86% yield.[25]

iii. KMnO$_4$, NaOH-H$_2$O, 10°C, 100% yield. These basic conditions avoid acid-catalyzed acetonide cleavage.[2]

iv. HgCl$_2$/HgO, acetone–H$_2$O, 5 min, 100% yield.[26]

v. Ozonolysis.[24,27]

vi. SeO$_2$, H$_2$O$_2$, 92% yield.[28]

vii. Me$_3$NO, OsO$_4$, CH$_2$Cl$_2$, >76% yield.[29]

viii. MCPBA, MeOH, H$_2$O.[30]

When the OAc group was a hydroxyl, the epoxidation selectivity was not very good, presumably because of the known directing effect of hydroxyl groups in peracid epoxidations.

    ix. NIS, CH_2Cl_2, H_2O.[31] Iodine can also be used.[32]

    x. $BF_3 \cdot Et_2O$, Bu_4NF, 0°C, 52–88% yield.[33]

    xi. $PdCl_2(MeCN)_2$, IPA, THF, or MeCN, 66–99% yield.[34]

2. Allyl group isomerization can also be performed using a variety of catalysts that have the advantage of being compatible with base-sensitive groups.

Allyl ethers are isomerized by $(Ph_3P)_3RhCl$, and t-BuOK/DMSO in the following order[35]:

$(Ph_3P)_3RhCl$:   allyl > 2-methylallyl > but-2-enyl

t-BuOK: but-2-enyl > allyl > 2-methylallyl

A variety of catalysts have been used to isomerize olefins and allyl ethers. It is possible to remove the allyl group in the presence of an allyloxycarbonyl (AOC, Alloc, or Aloc) group using an $[Ir(COD)(Ph_2MeP)_2]PF_6$-catalyzed isomerization, but the selectivity is not complete. The allyloxycarbonyl group can be removed selectively in the presence of an allyl group using a palladium or rhodium catalyst.[36] Hydrogen-activated $[Ir(COD)(Ph_2MeP)_2]PF_6$ is a better catalyst for allyl isomerization (91–100% yield) because there is no reduction of the alkene as is sometimes the case with $(Ph_3P)_3RhCl$.[37,38] Cationic iridium catalysts bearing σ-basic phosphines such as PCy_3 very efficiently isomerizes allylic ethers.[39,40] The preparation of a polymer-supported iridium catalyst that makes product isolation more facile has been reported.[41] When Wilkinson's catalyst is prereduced with BuLi, alkene reduction is not observed and high yields of enol ethers are obtained.[42] This method can also be used for isomerization of but-2-enyl ethers.[43] The iridium catalyst is also compatible with acetylenes.[44] Because the iridium catalyst can effect isomerization at room temperature, adjacent azides do not cycloadd to the allyl group during the isomerization reaction, as is the case when the isomerization must be performed at reflux.[29]

Ref. 27

Useful selectivity between allyl and 3-methylbut-2-enyl (prenyl) ethers has been achieved.[35]

i. $H_2Ru(PPh_3)_4$, EtOH, 95°C; 1.5 h, TsOH, MeOH, 2.5 h, 86% yield.[25]

ii. $RhH(Ph_3P)_4$, TFA, EtOH, 50°C, 30 min, 98% yield.[45]

iii. $RuH(CO)(Ph_3P)_3$, 60–80°C, 3 h.[46]

iv. $[CpRu(CH_3CN)]PF_6$, quinaldic acid, MeOH, 0.5–3 h, 41% to >99% yield.

v. $RhCl_3$, DABCO, EtOH, H_2O; H_3O^+, EtOH.[47]

vi. Polystyrene–CH_2NMe_4–$RhCl_4$ (EtOH, H_2O).[48]

vii. $RuCl_2(PPh)_3$ ($NaBH_4$, EtOH).[49]

viii. $Rh(diphos)(acetone)_2[ClO_4]_2$ (acetone, 25°C).[50]

ix. $Fe(CO)_5$ (xylene, 135°C, 8–15 h, 97% yield).[51] $Fe(CO)_5$, EtOH, H_2O, NaOH, reflux, 0.5 h, 63–96% yield. The isomerization is effective for a large variety of allyl ethers including the 2-methylpropenyl ether. An epoxide survives these conditions.[52]

x. trans-$Pd(NH_3)_2Cl_2$/t-BuOH isomerizes allyl ethers to vinyl ethers that can then be hydrolyzed in 90% yield, but in the presence of an α-hydroxy group the intermediate vinyl ether cyclizes to an acetal.[53] This reagent does not affect benzylidene acetals.

xi. Pd/C, H_2O, MeOH, cat. TsOH or $HClO_4$ 60–80°C, 24 h, 80–95% yield.[54] When TsOH is omitted the reaction gives the vinyl ether.[55]

xii. Pd/C, TFA, H_2O, dioxane, reflux 18 h, 70% yield.[56]

xiii. $Pd(Ph_3P)_4$, AcOH, 80°C, 10–60 min, 72–98% yield.[57]

xiv. $PdCl_2$, AcOH, H_2O, NaOAc, 89% yield.[27] This method has found application in complex carbohydrate synthesis.[58]

xv. Both the first and second generation Grubbs' olefin metathesis catalysts have been shown to isomerize allylic ethers to vinyl ethers that are readily hydrolyzed.[59] It is a decomposition product of the catalyst that was shown to be the isomerization catalyst.[60]

xvi. NiCl$_2$(diop), LiBHEt$_3$, THF, reflux, 2h, 80–87% yield. This catalyst selectively isomerizes allylic alcohols to the Z-vinyl ethers. RuCl$_2$(PPh$_3$)$_3$ reduced with LiBHEt$_3$ is also an effective isomerization catalyst, but in this case there is no E/Z selectivity.[61]

3. Allyl ethers can be cleaved using Pd(0) or Ni(0). In this case the π-allyl complex is intercepted with a good nucleophile.

   i. Pd(Ph$_3$P)$_4$, K$_2$CO$_3$, MeOH, reflux, 90% yield. If the reaction is performed at rt phenolic allyl ethers are cleaved selectively.[62]

   ii. Pd(Ph$_3$P)$_4$, PMHS–ZnCl$_2$, THF, rt, 85–94% yield. Additionally, allyl esters and allyl amines are cleaved, but a prenyl ether is stable.[63]

   iii. Pd(Ph$_3$P)$_4$, K$_2$CO$_3$, MeOH, reflux, 90% yield. If the reaction is performed at rt phenolic allyl ethers are cleaved selectively.[64]

   iv. Pd(Ph$_3$P)$_4$, RSO$_2$Na, CH$_2$Cl$_2$ or THF/MeOH, 70–99% yield. These conditions were shown to be superior to the use of sodium 2-ethylhexanoate. Methallyl, crotyl, and cinnamyl ethers, the alloc group, and allylamines are all efficiently cleaved by this method.[65] Using DME as solvent was found optimal for the deprotection of polymer bound allyl groups. Precipitated Pd can be removed by treatment with pyrrolidinedithiocarbamate in MeOH/THF.[66]

   v. Pd(Ph$_3$P)$_4$, N,N'-dimethylbarbituric acid, 90°C, 24 h, sealed tube, 78–100% yield. The prenyl groups along with other common ethers and esters are all stable.[67] Dimedone can be used as an allyl scavenger. In this example, deprotection of the SEM or PMB ethers was completely unsuccessful because of the sensitivity of the tetronate to base and oxidative reagents. The facile nature of this reaction is attributed to the increased acidity of the tetronate hydroxyl.[68]

vi. DIBAL, Et$_3$Al or NaBH$_4$, NiCl$_2$(dppp), toluene, CH$_2$Cl$_2$, THF, or ether, 80–97% yield.[69] These conditions are chemoselective for simple alkyl and phenolic allyl ethers. More highly substituted allyl ethers are unreactive. The following ethers and esters are stable: TBS, MPM, Bn, prenyl, MOM, THP, Ac, Bz, Pv.

vii. 1,2-Bis(4-methoxyphenyl)3,4-bis(2,4,6-tri-*tert*-butylphenylphosphinidiene)-cyclobutene, Pd(0), aniline, 84–99% yield. This is an excellent catalyst for the cleavage of allyl ethers, esters and carbamates.[70]

4. NBS, *hv*, CCl$_4$; base, 78–99% yield.[71]

5. Tetrabutylammonium peroxydisulfate, I$_2$, 25–50°C, CH$_3$CN, H$_2$O, 0.5–4 h, 81–95% yield.[72] When tetrabutylammonium peroxydisulfate is used alone the allyl group is oxidized to an ester which is then cleaved with MeONa/MeOH.[73]

6. NMO, OsO$_4$, then NaIO$_4$, dioxane, H$_2$O, 60°C, 18 h, 64–77% yield. Additionally, allyl amides are cleaved.[74]

7. *t*-BuOOH, cat. CuBr, *t*-BuOH, H$_2$O, 70°C, 60% yield at 90% conversion.[75]

8. DDQ, wet CH$_2$Cl$_2$, 70–92% yield. Anomeric and secondary allylic ethers could not be cleaved under these conditions.[76]

9. Pyridinium chlorochromate oxidation of an allyl ether or benzyl ether gives the enone (CH$_2$Cl$_2$, reflux, 84% yield).[77]

10. Protection for the double bond in the allyl protecting group may be achieved by epoxidation.

Regeneration of the allyl group occurs upon treatment with and TFA.[78]

11. Allyl groups are subject to oxidative deprotection with Chromiapillared Montmorillonite clay, *t*-BuOOH, CH$_2$Cl$_2$, isooctane, 85% yield.[79] Allylamines are cleaved in 84–90% yield and allyl phenyl ethers are cleaved in 80% yield.

12.

RO⟶ $\xrightarrow[\text{Reflux 1 h, 50\%}]{\text{SeO}_2,\ \text{AcOH}}$ ⟶ RO⟶OH ⟶ ROH

Ref. 80

13. PdCl$_2$, CuCl, DMF, O$_2$, 4 h, rt, 88–93% yield.[81]

14. Cp$_2$Zr prepared from CpZrCl$_2$, n-BuLi; H$_2$O, 50–98% yield. Allyl ethers are cleaved faster than allylamines that are also cleaved (66%).[82]

$\xrightarrow[\text{>87\%}]{\text{Cp}_2\text{ZrCl}_2,\ n\text{-BuLi}}$

Ref. 83

15. Li, naphthalene, THF, −78°C to 20°C, 1–12 h, 25–90% yield. Benzyl and PhMe$_2$Si ethers, sulfonamides, allyl sulfonamides sulfonyl amides, benzyl amides and some esters are also cleaved.[84]

16. SmI$_2$ (5 eq./allyl group), THF, i-PrNH$_2$ (20 eq.), H$_2$O (15 eq.), 80–99% yield. Phenolic allyl ethers are cleaved at a faster rate. An anomeric allyl ether is completely stable and other substituted allyl ethers along with allyl amines and allyl sulfides are also not cleaved.[85]

17. Ti(O-i-Pr)$_4$, n-BuMgCl, THF, rt, 69–97% yield. Methallyl and other substituted allyl ethers are not cleaved, but ester groups are partially removed as expected.[86]

18. TiCl$_3$, Mg, THF, 28–96% yield.[87]

19. Electrolysis, DMF, SmCl$_3$, (n-Bu)$_4$NBr, Mg anode, Ni cathode, 60–90% yield.[88]

20. Electrolysis, [Ni(bipyr)$_3$](BF$_4$)$_2$,Mg anode, DMF, rt, 25–99% yield.[89] Aryl halides are reduced.

21. Ac$_2$O, BF$_3$ ·Et$_2$O then MeONa/MeOH to hydrolyze the acetate.[90]

22. TMSCl, NaI, CH$_3$CN, 90–98% yield. Both alkyl and phenolic ethers were cleaved. This method generates TMSI in situ, which is known to cleave a large variety of ethers, ester, and carbamates.[91]

23. CoCl$_2$, AcCl, CH$_3$CN, rt, 8–12 h, 71–84% yield. Benzyl ethers and epoxides are among those that are also cleaved.[92]

24. RCO$_2$Br or RCO$_2$Cl, graphite, ClCH$_2$Cl$_2$Cl, reflux, 77% yield. Most other ethers are also cleaved.[93]

25. AlCl$_3$·PhNMe$_2$, CH$_2$Cl$_2$, 73–100% yield.[94] Benzyl ethers are also cleaved.

26. NaBH$_4$, I$_2$, THF, 0°C, 53–96% yield.[95] Methyl esters, an actonide, THP, TBDMS, and benzyl ethers were stable.

27. LiCl, NaBH$_4$, THF, 0–35°C, 70–92% yield.[96] Both alkyl and phenolic allyl ethers are cleaved.

28. I(CF$_2$)$_6$X (X = F or Cl), Na$_2$S$_2$O$_4$, NaHCO$_3$, CH$_3$CN (or DMF)/H$_2$O, rt, 30 min; Zn powder, NH$_4$Cl, EtOH, reflux, 15 min, ~87–93% yield. The reaction proceeds to give an iodohydrin ether, which is reductively cleaved with Zn.[97]

29. *t*-BuLi, pentane, $-78°C$ to rt, 1 h, 90–99% . The functional group compatibility of this method is somewhat limited, but TBS, THP, and Bn ethers were shown to be compatible.[98]

30. NaTeH, EtOH, AcOH, reflux, 2 h, 85–99% yield.[99]

31. $CeCl_3 \cdot 7H_2O$, NaI, CH_3CN, reflux, 69–95% yield. Phenolic and alkyl ethers are cleaved.[100] Another version of this method uses 1,3-propanethiol to scavenge formed allyl iodide. The relative rates for various allyl ethers are presented in the table below.[101] The following groups were unaffected by these conditions: TBS, Tr, and Alloc.

Cleavage of Substituted Allyl Octyl Ethers Promoted by $CeCl_3 \cdot 7H_2O$

Entry	Derivative	T	t (h)	% Yield	Solvent	Scavenger
1	Allyl	Reflux	109	24	CH_3CN	None
2	Allyl	Reflux	30	83	CH_3NO_2	$HS(CH_2)_3SH$
3	Prenyl	Reflux	10	17	CH_3NO_2	$HS(CH_2)_3SH$
4	Crotyl	Reflux	1.5	85	CH_3NO_2	$HS(CH_2)_3SH$
5	Cinnamyl	Reflux	9	63	CH_3NO_2	$HS(CH_2)_3SH$
6	β-Methallyl	Reflux	2.5	54	CH_3NO_2	$HS(CH_2)_3SH$

1. R. Gigg, *Am. Chem. Soc. Symp. Ser.*, **39**, 253 (1977); *ibid.*, **77**, 44 (1978); R. Gigg and R. Conant, *Carbohydr. Res.* **100**, C5 (1982).

2. J. Cunningham, R. Gigg, and C. D. Warren, *Tetrahedron Lett.*, **5**, 1191 (1964).

3. J. Thiem, H. Mohn, and A. Heesing, *Synthesis*, 775 (1985).

4. F. Guibé, *Tetrahedron*, **40**, 13509 (1997).

5. R. Gigg and C. D. Warren, *J. Chem. Soc. C*, 2367 (1969).

6. E. J. Corey and W. J. Suggs, *J. Org. Chem.*, **38**, 3224 (1973).

7. H. Saburi, S. Tanaka, and M. Kitawmura, *Angew. Chem. Int. Ed.*, **44**, 1730 (2005).

8. T. Iversen and D. R. Bundle, *J. Chem. Soc., Chem. Commun.*, 1240 (1981); H.-P. Wessel, T. Iversen, and D. R. Bundle, *J. Chem. Soc., Perkin Trans. I*, 2247 (1985).

9. S. Sato, S. Nunomura, T. Nakano, Y. Ito, and T. Ogawa, *Tetrahedron Lett.*, **29**, 4097 (1988).

10. A. K. M. Anisuzzaman, L. Anderson, and J. L. Navia, *Carbohydr. Res.*, **174**, 265 (1988).

11. R. Lakhmiri, P. Lhoste, and D. Sinou, *Tetrahedron Lett.*, **30**, 4669 (1989).

12. E. J. Stoner, M. J. Peterson, M. S. Allen, J. A. DeMattei, A. R. Haight, M. R. Leanna, S. R. Patel, D. J. Plata, R. H. Premchandran, and M. Rasmussen, *J. Org. Chem.*, **68**, 8847 (2003).

13. H. Oguri, S. Hishiyama, T. Oishi, and M. Hirama, *Synlett*, 1252 (1995).

14. J. J. Oltvoort, M. Kloosterman, and J. H. Van Boom, *Recl: J. R. Neth. Chem. Soc.*, **102**, 501 (1983); F. Guibe and Y. Saint M'Leux, *Tetrahedron Lett.*, **22**, 3591 (1981).

15. K.-i. Sato, S. Akai, A. Yoshitomo, and Y. Takai, *Tetrahedron Lett.*, **45**, 8199 (2004).

16. J.-M. Lin, H.-H. Li, and A.-M. Zhou, *Tetrahedron Lett.*, **37**, 5159 (1996).

17. H. Yin, R. W. Franck, S.-L. Chen, G. J. Quigley, and L. Todaro, *J. Org. Chem.*, **57**, 644 (1992).

18. J.-C. Jacquinet and P. Sinaÿ, *J. Org. Chem.*, **42**, 720 (1977).

19. O. Hindsgual, T. Norberg, J. Le pendu, and R. U. Lemieux, *Carbohydr. Res.*, **109**, 109 (1982).

20. I. A. Motorina, F. Parly, and D. S. Grierson, *Synlett*, 389 (1996).

21. H. Tahir and O. Hindsgaul, *J. Org. Chem.*, **65**, 911 (2000).

22. H. Nakagawa, T. Hirabayashi, S. Sakaguchi, and Y. Ishii, *J. Org. Chem.*, **69**, 3474 (2004).

23. J. S. Bajwa, X. Jiang, J. Slade, K. Prasad, O. Repic, and T. J. Blacklock, *Tetrahedron Lett.*, **43**, 6709 (2002).

24. J. Gigg and R. Gigg, *J. Chem. Soc. C*, 82 (1966).

25. K. C. Nicolaou, T. J. Caulfield, H. Kataoka, and N. A. Stylianides, *J. Am. Chem. Soc.*, **112**, 3693 (1990).

26. R. Gigg and C. D. Warren, *J. Chem. Soc. C*, 1903 (1968).

27. A. B. Smith, III, R. A. Rivero, K. J. Hale, and H. A. Vaccaro, *J. Am. Chem. Soc.*, **113**, 2092 (1991).

28. H. Yamada, T. Harada, and T. Takahashi, *J. Am. Chem. Soc.*, **116**, 7919 (1994).

29. C. Lamberth and M. D. Bednarski, *Tetrahedron Lett.*, **32**, 7369 (1991).

30. P. L. Barili, G. Berti, D. Bertozzi, G. Catelani, F. Colonna, T. Corsetti, and F. D'Andrea, *Tetrahedron*, **46**, 5365 (1990).

31. K. M. Halkes, T. M. Slaghek, H. J. Vermeer, J. P. Kamerling, and J. F. G. Vliegenthart, *Tetrahedron Lett.*, **36**, 6137 (1995).

32. S. Inamura, K. Fukase, and S. Kusumoto, *Tetrahedron Lett.*, **42**, 7613 (2001).

33. V. Gevorgyan and Y. Yamamoto, *Tetrahedron Lett.*, **36**, 7765 (1995).

34. H. Aoyama, M. Tokunaga, S.-i. Hiraiwa, Y. Shirogane, Y. Obora, and Y. Tsuji, *Org. Lett.*, **6**, 509 (2004); H. B. Mereyala and S. R. Lingannagaru, *Tetrahedron*, **53**, 17501 (1997).

35. P. A. Gent and R. Gigg, *J. Chem. Soc., Chem. Commun.*, 277 (1974); R. Gigg, *J. Chem. Soc., Perkin Trans. I*, 738 (1980).

36. P. Boullanger, P. Chatelard, G. Descotes, M. Kloosterman, and J. H. Van Boom, *J. Carbohydr. Chem.*, **5**, 541 (1986).

37. J. J. Oltvoort, C. A. A. van Boeckel, J. H. de Koning, and J. H. van Boom, *Synthesis*, 305 (1981).

38. For hydrogenation during isomerization, see C. D. Warren and R. W. Jeanloz, *Carbohydr. Res.*, **53**, 67 (1977); T. Nishiguchi, K. Tachi, and K. Fukuzumi, *J. Org. Chem.*, **40**, 237 (1975); C. A. A. van Boeckel and J. H. van Boom, *Tetrahedron Lett.*, 3561 (1979).

39. S. G. Nelson, C. J. Bungard, and K. Wang, *J. Am. Chem. Soc.*, **125**, 13000 (2003).

40. T. Higashino, S. Sakaguchi, and Y. Ishii, *Org. Lett.*, **2**, 4193 (2000).

41. I. R. Baxendale, A.-L. Lee, and S. V. Ley, *Synlett*, 516 (2002).

42. G.-J. Boons, A. Burton, and S. Isles, *J. Chem. Soc., Chem. Commun.*, 141 (1996).

43. G.-J. Boons, B. Heskamp, and F. Hout, *Angew. Chem., Int. Ed. Engl.*, **35**, 2845 (1996); G.-J. Boons and S. Isles, *J. Org. Chem.*, **61**, 4262 (1996).

44. J. Alzeer, C. Cai and A. Vasella, *Helv. Chim. Acta*, **78**, 242 (1995).

45. F. E. Ziegler, E. G. Brown, and S. B. Sobolov, *J. Org. Chem.*, **55**, 3691 (1990).

46. M. Urbala, N. Kuznik, S. Krompiec, and J. Rzepa, *Synlett*, 1203 (2004).

47. M. Dufour, J.-C. Gramain, H.-P. Husson, M.-E. Sinibaldi, and Y. Troin, *Tetrahedron Lett.*, **30**, 3429 (1989).

48. M. Setty-Fichman, J. Blum, Y. Sasson, and M. Eisen, *Tetrahedron Lett.*, **35**, 781 (1994).

49. H. Frauenrath, T. Arenz, G. Raube, and M. Zorn, *Angew. Chem., Int. Ed. Engl.*, **32**, 83 (1993).

50. S. H. Bergens and B. Bosnich, *J. Am. Chem. Soc.*, **113**, 958 (1991).

51. T. Arnold, B. Orschel, and H.-U. Reissig, *Angew. Chem., Int. Ed. Engl.*, **31**, 1033 (1992).

52. J. V. Crivello and S. Kong, *J. Org. Chem.*, **63**, 6745 (1998).

53. T. Bieg and W. Szeja, *J. Carbohydr. Chem.*, **4**, 441 (1985).

54. R. Boss and R. Scheffold, *Angew. Chem., Int. Ed. Engl.*, **15**, 558 (1976).

55. A. B. Smith, III, R. A. Rivero, K. J. Hale, and H. A. Vaccaro, *J. Am. Chem. Soc.*, **113**, 2092 (1991).

56. D. Liu, R. Chen, L. Hong, and M. J. Sofia, *Tetrahedron Lett.*, **39**, 4951 (1998).

57. K. Nakayama, K. Uoto, K. Higashi, T. Soga, and T. Kusama, *Chem. Pharm. Bull.*, **40**, 1718 (1992).

58. G. J. S. Lohman and P. H. Seeberger, *J. Org. Chem.*, **69**, 4081 (2004).

59. C. Cadot, P. I. Dalko, and J. Cossy, *Tetrahedron Lett.*, **43**, 1839 (2002); Y.-J. Hu, R. Dominique, S. K. Das and R. Roy, *Can. J. Chem.*, **78**, 838 (2000); T. R. Hoye and H. Zhao, *Org. Lett.*, **1**, 169 (1999).

60. S. H. Hong, M. W. Day, and R. H. Grubbs, *J. Am. Chem. Soc.*, **126**, 7414 (2004).

61. A. Wille, S. Tomm, and H. Frauenrath, *Synthesis*, 305 (1998).

62. D. R. Vutukuri, P. Bharathi, Z. Yu, K. Rajasekaran, M.-H. Tran, and S. Thayumanavan, *J. Org. Chem.*, **68**, 1146 (2003).

63. S. Chandrasekhar, C. Raji Reddy, and R. Jagadeeshwar Rao, *Tetrahedron*, **57**, 3435 (2001).

64. D. R. Vutukuri, P. Bharathi, Z. Yu, K. Rajasekaran, M.-H. Tran, and S. Thayumanavan, *J. Org. Chem.*, **68**, 1146 (2003).

65. M. Honda, H. Morita, and I. Nagakura, *J. Org. Chem.*, **62**, 8932 (1997).

66. T. Opatz and H. Kunz, *Tetrahedron Lett.*, **41**, 10185 (2000).

67. H. Tsukamoto and Y. Kondo, *Synlett*, 1061 (2003).

68. W. R. Roush and R. J. Sciotti, *J. Am. Chem. Soc.*, **120**, 7411 (1998).

69. T. Taniguchi and K. Ogasawara, *Angew. Chem., Int. Ed. Engl*, **37**, 1136 (1998).

70. H. Murakami, T. Minami, and F. Ozawa, *J. Org. Chem.*, **69**, 4482 (2004).

71. R. R. Diaz, C. R. Melgarejo, M. T. P. Lopez-Espinosa, and I. I. Cubero, *J. Org. Chem.*, **59**, 7928 (1994).

72. S. G. Yang, M. Y. Park, and Y. H. Kim, *Synlett*, 492 (2002).

73. F.-E. Chen, X.-H. Ling, Y.-P. He, and X.-H. Peng, *Synthesis*, 1772 (2001).

74. P. I. Kitov and D. R. Bundle, *Org. Lett.*, **3**, 2835 (2001).

75. R. Krähmer, L. Hennig, M. Findeisen, D. Muller, and P. Welzel, *Tetrahedron*, **54**, 10753 (1998).

76. J. S. Yadav, S. Chandrasekhar, G. Sumithra, and R. Kache, *Tetrahedron Lett.*, **37**, 6603 (1996).

77. J. Cossy, S. Bouzbouz, M. Lachgar, A. Hakiki, and B. Tabyaoui, *Tetrahedron Lett.*, **39**, 2561 (1998).

78. G. O. Aspinall, I. H. Ibrahim, and N. K. Khare, *Carbohydr. Res.*, **200**, 247 (1990); H. Paulsen, F. R. Heiker, J. Feldmann, and K. Heyns, *Synthesis*, 636 (1980).

79. B. M. Choudary, A. D. Prasad, V. Swapna, V. L. K. Valli, and V. Bhuma, *Tetrahedron*, **48**, 953 (1992).

80. K. Kariyone and H. Yazawa, *Tetrahedron Lett.*, **11**, 2885 (1970).

81. H. B. Mereyala and S. Guntha, *Tetrahedron Lett.*, **34**, 6929 (1993).

82. H. Ito, T. Taguchi, and Y. Hanzawa, *J. Org. Chem.*, **58**, 774 (1993).

83. E. Vedejs, B. N. Naidu, A. Klapars, D. L. Warner, V.-s. Li, Y. Na, and H. Kohn, *J. Am. Chem. Soc.*, **125**, 15796 (2003).

84. E. Alonso, D. J. Ramon, and M. Yus, *Tetrahedron*, **53**, 14355 (1997).

85. A. Dahlèn, A. Sundgren, M. Lahmann, S. Oscarson, and G. Hilmersson, *Org. Lett.*, **5**, 4085 (2003).

86. J. Lee and J. K. Cha, *Tetrahedron Lett.*, **37**, 3663 (1996).

87. S. M. Kadam, S. K. Nayak, and A. Banerji, *Tetrahedron Lett.*, **33**, 5129 (1992).

88. B. Espanet, E. Duñach, and J. Périchon, *Tetrahedron Lett.*, **33**, 2485 (1992).

89. S. Olivero and E. Duñach, *J. Chem. Soc., Chem. Commun.*, 2497 (1995).

90. C. F. Garbers, J. A. Steenkamp, and H. E. Visagie, *Tetrahedron Lett.*, **16**, 3753 (1975).

91. A. Kamal, E. Laxman, and N. V. Rao, *Tetrahedron Lett.*, **40**, 371 (1999).

92. J. Iqbal and R. R. Srivastava, *Tetrahedron*, **47**, 3155 (1991).

93. Y. Suzuki, M. Matsushima, and M. Kodomari, *Chem. Lett.*, **27**, 319 (1998).

94. T. Akiyama, H. Hirofuji, and S. Ozaki, *Tetrahedron Lett.*, **32**, 1321 (1991).

95. R. M. Thomas, G. H. Mohan, and D. S. Iyengar, *Tetrahedron Lett.*, **38**, 4721 (1997).

96. S. RajaRam, K. P. Chary, S. Salahuddin, and D. S. Iyengar, *Syn. Comm.*, **32**, 133 (2002).

97. B. Yu, B. Li, J. Zhang, and Y. Hui, *Tetrahedron Lett.*, **39**, 4871 (1998).

98. W. F. Bailey, M. D. England, M. J. Mealy, C. Thongsornkleeb, and L. Teng, *Org. Lett.*, **2**, 489 (2000).

99. N. Shobana and P. Shanmugam, *Indian J. Chem., Sect. B*, **25B**, 658 (1986).

100. R. M. Thomas, G. S. Reddy, and D. S. Iyengar, *Tetrahedron Lett.*, **40**, 7293 (1999).

101. G. Bartoli, G. Cupone, R. Dalpozzo, A. De Nino, L. Maiuolo, E. Marcantoni, and A. Procopio, *Synlett*, 1897 (2001).

Prenyl Ether (Pre): $(CH_3)_2C=CHCH_2OR$

Formation

Prenyl ethers can be formed using the typical Williamson ether synthesis—that is, by reacting the alcohol with a suitable base and a prenyl halide. Many of the methods used for the formation of allyl and benzyl ethers should be applicable.[1]

Cleavage

1. DDQ, CH_2Cl_2, H_2O, rt, 0.75 min to 9 h, 36–89% yield. The reaction can be run using catalytic DDQ with $Mn(OAc)_3$ as the reoxidant. Allyl, TBS, TBDPS, and a phenolic prenyl ether were stable to these conditions.[2]

2. *t*-BuOK, DMSO. In this case deprotection occurs by γ-elimination rather than isomerization as with the simple allyl group. Elimination is also faster than isomerization of the allyl group, but the rate difference is insufficient for good selectivity.[3] The crotyl group is removed similarly.

3. I_2, CH_2Cl_2, 3Å MS, 1–8 h, rt, 22–94% yield. The Bn, allyl, and TBDMS ethers are stable to these conditions, but TBS ether is partially cleaved.[4] Phenolic prenyl ethers react to give chromanes.

4. *p*-TSA, CH_2Cl_2, rt, 1–4 h, 76% yield. Phenolic prenyl ethers are also cleaved.[5]

5. $ZrCl_4$ (0.2 eq.), NaI (0.2 eq.), CH_3CN, reflux, 1–2 h, 79–94% yield. Allyl, crotyl, benzyl, and THP ethers and the acetate, Cbz, and BOC are not affected, but prenyl esters are cleaved efficiently (85–91% yield).[6]

6. $TiCl_4$, *n*-Bu_4NI, CH_2Cl_2, 0°C, 2 h, 64–100% yield.[7]

7. $Yb(OTf)_3$, CH_3NO_2, rt, 0.5–24 h, 55–85% yield. Prenyl esters and phenolic ethers are cleaved.[8]

8. $(PhSO_2)_2$, 10 mol %, 80°C, 88–93% yield.[9]

Approximate Half-Life of Various Allylic Ethers in Wet CD_2Cl_2 at 80°C with $(PhSO_2)_2$

				Bn	TBS	Ac
N.R.	120 h	21 h	6 h	N.R.	N. R.	N. R.

1. M. L. Fascio, A. Alvarez-Larena, and N. B. D'Accorso, *Carbohydr. Res.*, **337**, 2419 (2002).

2. J.-M. Vatèle, *Synlett*, 507 (2002). J.-M. Vatele, *Tetrahedron*, **58**, 5689 (2002).

3. R. Gigg, *J. Chem. Soc. Perkin Trans. 1*, 738 (1980).

4. J.-M. Vatèle, *Synlett*, 1989 (2001).

5. K. S. Babu, B. C. Raju, P. V. Srinivas, A. S. Rao, S. P. Kumar, and J. M. Rao, *Chem. Lett.*, **32**, 704 (2003).

6. G. V. M. Sharma, C. G. Reddy, and P. R. Krishna, *Synlett*, 1728 (2003).

7. T. Tsuritani, H. Shinokubo, and K. Oshima, *Tetrahedron Lett.*, **40**, 8121 (1999).

8. G. V. M. Sharma, A. Ilangovan, and A. K. Mahalingam, *J. Org. Chem.*, **63**, 9103 (1998).

9. D. Markovic and P. Vogel, *Org. Lett.*, **6**, 2693 (2004).

Cinnamyl Ether (Cin): $C_6H_5CH=CHCH_2OR$

Formation

1. The ether can be formed by the typical Williamson ether synthesis using a strong base and the cinnamyl bromide.[1] Many of the methods used for allyl ether synthesis should be applicable.

2. $PhCH=CHCH_2OAc$, 0.5 eq. Et_2Zn, 5% $Pd(Ph_3P)_4$, THF, rt, 56–99% yield.[2]

3. *n*-BuLi, Ph_2PCl; $PhCH=CHCH_2OH$, fluoranil, CH_2Cl_2, rt, 3 h 90% yield. This methods works for a variety of ethers.[3]

4. From a TMS ether: $PhCH=CHCHO$, TMSOTf, CH_2Cl_2, −86°C, Et_3SiH, 87% yield.[4]

5. 1-Phenylpropyne, $Pd(Ph_3P)_4$, benzoic acid, dioxane, 100°C, 66–89% yield. Acids react to give the esters, but phenols give a mixture of *O*- and C-alkylation products with C-alkylation predominating with prolonged reaction times.[5]

Cleavage

1. Electrolysis: −2.7 to −2.9 V, Hg electrode, 62–83% yield. The allyl group is unaffected.[6] Cinnamyl carbamates are cleaved.[7]

2. $CeCl_3 \cdot 7H_2O$, NaI, CH_3NO_2, reflux, 1,3-propanedithiol, 52–88% yield. Trityl, Alloc and TBDPS groups were stable, but benzyl and THP ethers were not.[8]

1. M. L. Fascio, A. Alvarez-Larena, and N. B. D'Accorso, *Carbohydr. Res.*, **337**, 2419 (2002).

2. H. Kim and C. Lee, *Org. Lett.*, **4**, 4369 (2002).

3. T. Shintou and T. Mukaiyama, *Chem. Lett.*, **32**, 984 (2003).

4. C.-C. Wang, J.-C. Lee, S.-Y. Luo, H.-F. Fan, C.-L. Pai, W.-C. Yang, L.-D. Lu, and S.-C. Hung, *Angew. Chem. Int. Ed.*, **41**, 2360 (2002).

5. W. Zhang, A. R. Haight, and M. C. Hsu, *Tetrahedron Lett.*, **43**, 6575 (2002).

6. A. Solis-Oba, T. Hudlicky, L. Koroniak, and D. Frey, *Tetrahedron Lett.*, **42**, 1241 (2001).

7. J. Hansen, S. Freeman, and T. Hudlicky, *Tetrahedron Lett.*, **44**, 1575 (2003).

8. G. Bartoli, G. Cupone, R. Dalpozzo, A. De Nino, L. Maiuolo, E. Marcantoni, and A. Procopio, *Synlett*, 1897 (2001).

2-Phenallyl Ether

$$Ph \diagdown \diagup OR$$

This ether is prepared by the Williamson ether synthesis from alcohols and phenols using α-bromomethylstyrene. It is cleaved by treating the ether in THF with *t*-BuLi at −78°C for 30 min (75–97% yield). The phenallyl ether can be cleaved in the presence of an allyl ether. Phenallyl amines and amides are cleaved similarly.[1] Cleavage occurs by an addition of the alkyllithium to the olefin followed by elimination.

1. J. Barluenga, F. J. Fananas, R. Sanz, C. Marcos, and J. M. Ignacio, *Chem. Commun.*, 933 (2005).

Propargyl Ethers: $HC \equiv CCH_2OR$

This group is smaller than an allyl group and has found value in directing the formation of β-mannosyl derivatives.

Formation

Propargyl ethers are readily formed from the alcohol by treatment with NaH, DMF, and propargyl bromide.[1] **Note that propargyl halides are explosive and shock-sensitive!**

Cleavage

1. Propargyl ethers are cleaved with $TiCl_3$–Mg in THF, 54–92% yield. Allyl and benzyl ethers were not cleaved; phenolic propargyl ethers are also cleaved.[2]
2. $(BnNEt_3)_2MoS_4$ (benzyltriethylammonium tetrathiomolybdate).[3,4]
3. *t*-BuOK for allene formation then OsO_4, *N*-methylmorpholine-*N*-oxide, 80–91% yield.[1]

1. D. Crich and P. Jayalath, *Org. Lett.*, **7**, 2277 (2005).
2. S. K. Nayak, S. M. Kadam, and A. Banerji, *Synlett*, 581 (1993).
3. V. M. Swamy, P. Ilankumaran, and S. Chandrasekaran, *Synlett*, 513 (1997).
4. K. R. Prabhu, N. Devan, and S. Chandrasekaran, *Synlett*, 1762 (2002).

p-Chlorophenyl Ether: *p*-ClC₆H₄-OR

Formation/Cleavage[1]

The *p*-chlorophenyl ether was used in this synthesis to minimize ring sulfonation during cyclization of a diketo ester with concentrated H_2SO_4/AcOH.[1] Cleavage occurs by reduction of the aromatic ring to form an enol ether which is hydrolyzed with acid.

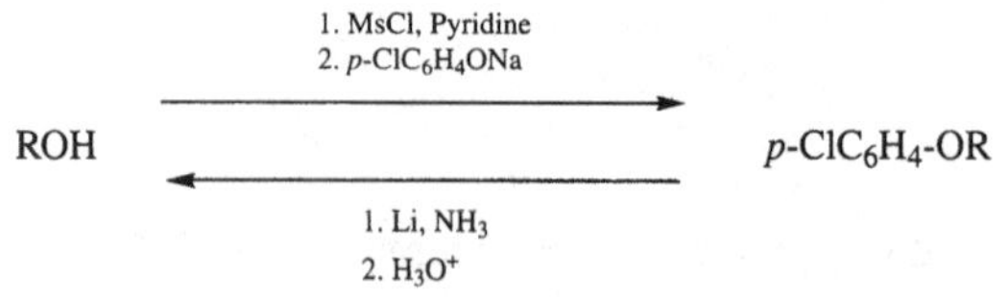

1. MsCl, Pyridine
2. p-ClC$_6$H$_4$ONa

ROH ⟶ p-ClC$_6$H$_4$-OR

1. Li, NH$_3$
2. H$_3$O$^+$

1. J. A. Marshall and J. J. Partridge, *J. Am. Chem. Soc.*, **90**, 1090 (1968).

p-Methoxyphenyl Ether (PMP-OR): p-MeOC$_6$H$_4$OR

This group is stable to 3 N HCl, 100°C; 3 N NaOH, 100°C; H$_2$, 1200 psi; O$_3$, MeOH, −78°C; RaNi, 100°C; LiAlH$_4$; Jones reagent and pyridinium chlorochromate (PCC). It has also been used for protection of the anomeric hydroxyl during oligosaccharide synthesis.[1]

Formation

1. From an alcohol: MeOC$_6$H$_4$BF$_3$$^-K^+$, Cu(OAc)$_2$, DMAP, CH$_2Cl_2$, MS4Å, rt, O$_2$, 24 h, quant.[2]
2. From an alcohol: MeOC$_6$H$_4$I, CuI, Cs$_2$CO$_3$, 1,10-phenanthroline, 18–24 h, 110°C, 64–93% yield.[3]
3. p-MeOC$_6$H$_4$OH, DEAD, Ph$_3$P, THF, 82–99% yield.[4,5] Using this method on a secondary alcohol would give inversion.

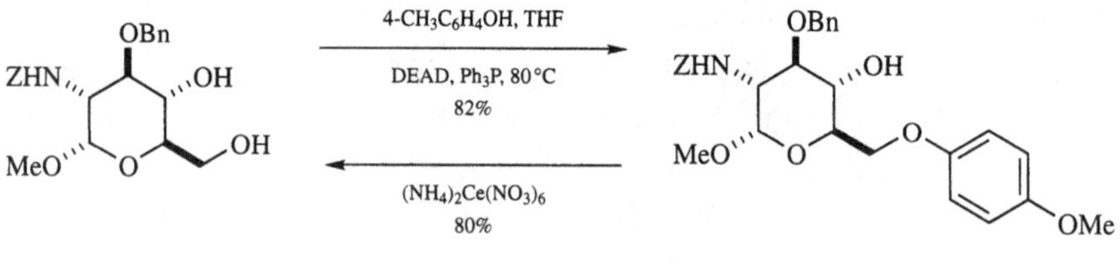

Z = benzyloxycarbonyl, DEAD = diethyl azodicarboxylate

4. From a mesylate: K$_2$CO$_3$, 18-crown-6, CH$_3$CN, reflux, 48 h, 81% yield.[6]
5. From a tosylate: p-MeOC$_6$H$_4$OH, DMF, NaH, 60°C, 14 h.[7]

Cleavage

1. Ceric ammonium nitrate, CH$_3$CN, H$_2$O (4:1), 0°C, 10 min, 80–85% yield[1,2] or CAN, Pyr, CH$_3$CN, H$_2$O, 0°C, 0.5 h, 96% yield.[6]
2. Anodic oxidation, CH$_3$CN, H$_2$O, Bu$_4$NPF$_6$, 20°C, 74–100% yield.[8]
3. Treatment of a PMP ether with Na/NH$_3$ results in the formation of an enol either, which in principle can be hydrolyzed to release the alcohol.[9]

1. Y. Matsuzaki, Y. Ito, Y. Nakahara, and T. Ogawa, *Tetrahedron Lett.*, **34**, 1061 (1993).
2. T. D. Quach and R. A. Batey, *Org. Lett.*, **5**, 1381 (2003).

3. M. Wolter, G. Nordmann, G. E. Job, and S. L. Buchwald, *Org. Lett.*, **4**, 973 (2002); G. F. Manbeck, A. J. Lipman, R. A. Stockland, Jr., A. L. Freidl, A. F. Hasler, J. J. Stone, and I. A. Guzei, *J. Org. Chem.*, **70**, 244 (2005).

4. T. Fukuyama, A. A. Laud, and L. M. Hotchkiss, *Tetrahedron Lett.*, **26**, 6291 (1985).

5. M. Petitou, P. Duchaussoy, and J. Choay, *Tetrahedron Lett.*, **29**, 1389 (1988).

6. Y. Masaki, K. Yoshizawa, and A. Itoh, *Tetrahedron Lett.*, **37**, 9321 (1996).

7. S. Takano, M. Moriya, M. Suzuki, Y. Iwabuchi, T. Sugihara, and K. Ogasawara, *Heterocycles*, **31**, 1555 (1990).

8. S. Iacobucci, N. Filippova, and M. d'Alarcao, *Carbohydr. Res.*, **277**, 321 (1995).

9. D. Qin, H.-S. Byun, and R. Bittman, *J. Am. Chem. Soc.*, **121**, 662–668 (1999).

p-Nitrophenyl Ether: $NO_2C_6H_4OR$

The *p*-nitrophenyl ether was used for the protection of the anomeric position of a pyranoside. It is installed using the Königs–Knorr process and can be cleaved by hydrogenolysis (Pd–C, H_2, Ac_2O), followed by oxidation with ceric ammonium nitrate (81–99% yield).[1]

1. K. Fukase, T. Yasukochi, Y. Nakai, and S. Kusumoto, *Tetrahedron Lett.*, **37**, 3343 (1996).

2,4-Dinitrophenyl Ether (RO−DNP): $2,4\text{-}(NO_2)_2\text{-}C_6H_3OR$

Formation

2,4-Dinitrofluorobenzene, DABCO, DMF, 85% yield.[1] When this group was used to protect an anomeric center of a carbohydrate, only the ß-isomer was formed, but this could be equilibrated to the α-isomer in 90% yield with K_2CO_3 in DMF.

1. H. J. Koeners, A. J. De Kok, C. Romers, and J. H. Van Boom, *Recl. Trav. Chim. Pays-Bas*, **99**, 355 (1980).

2,3,5,6-Tetrafluoro-4-(trifluoromethyl)phenyl Ether: $CF_3C_6F_4OR$

Treatment of a steroidal alcohol with perfluorotoluene [NaOH, (*n*-Bu)$_4$NHSO$_4$, CH_2Cl_2, 79%] gives the ether, which can be cleaved in 82% yield with NaOMe/ DMF.[1]

1. J. J. Deadman, R. McCague, and M. Jarman, *J. Chem. Soc., Perkin Trans. 1*, 2413 (1991).

Benzyl Ether (Bn−OR): PhCH₂OR (Chart 1)

The benzyl ether is one of the most robust of protecting groups and is orthogonal to a host of others, making it and its variants one of the most used of protecting groups, but it can participate in unwanted side reactions as the following illustrates.[1]

Formation

1. BnCl, powdered KOH, 130–140°C, 86% yield.[2]
2. BnBr, CsOH, TBAI, 4-Å MS, DMF, 23°C, 3 h, 73–97% yield.[3]
3. BnCl, Bu₄NHSO₄, 50% KOH, benzene.[4] This method was used to selectively monoprotect a diol.[5]

4. BnX (X=Cl, Br), Ag₂O, DMF, 25°C, good yields.[6] This method is very effective for the monobenzylation of diols.[7]
5. Ag₂O, BnBr, DMF, rt, 48 h, 76% yield.[8] In the following case all other methods failed.[9]

6. BnCl, Ni(acac)₂, reflux, 3 h, 80–90% .[10]
7. BnCl, Cu(acac)₂, reflux, 3–5 h, 65–92% when reaction is performed neat. Primary alcohols react preferentially and phenols fail to react. In THF the yields are much lower.[11]

8. $BnO-C(=NH)CCl_3$, CF_3SO_3H.[12–15]

9. OBn   MgO, CH_2Cl_2, reflux, 19–24 h, 45–84% yield.[16]

10. BnOH, $BiBr_3$, CCl_4, rt, 76–95% yield.[17]

11. NaH, THF, BnBr, Bu_4NI, 20°C, 3 h, 100%.[18] This method was used to protect a hindered hydroxyl group. Increased reactivity is achieved by the *in situ* generation of benzyl iodide.

12. The primary alcohol below was selectively benzylated using NaH and BnBr at −70°C.[19]

13. Note that in this case the primary alcohol was left unprotected.[20] This selectivity is probably due to the increased acidity of the secondary alcohol verses the primary alcohol.

14. BnI, NaH, rt, 90% yield.[21] Note that in this case the reaction proceeds without complication of the Payne rearrangement. This appears to be general.[22]

15. BnCl, NaH, $CuCl_2$, Bu_4NI, THF, reflux 25 h, 70% yield.[23]

16. (Bu₃Sn)₂O, toluene, reflux; BnBr, *N*-methylimidazole, 95% yield.[24] Equatorial alcohols are benzylated in preference to axial alcohols in diol-containing substrates. The application of the stannylene method for the selective protection of carbohydrates has been reviewed.[25]

17. Bu₂SnO, benzene; BnBr, DMF, heat, 80% yield.[26] This method has also been used to protect selectively the anomeric hydroxyl in a carbohydrate derivative.[27] The reaction can be accelerated using microwave heating.[28] The replacement of Bu₂SnO with Bu₂Sn(OMe)₂ improves this process procedurally.[29] The use of stannylene acetals for the regioselective manipulation of hydroxyl groups has been reviewed.[30]

18. PhCHN₂, HBF₄, −40°C, CH₂Cl₂, 66–92% yield.[31] Selective alcohol protection in the presence of amines is achieved under these conditions.[32]

19. Ph₂POBn, 2,6-dimethylquinone, CH₂Cl₂, rt, 0.5 h, 90–95% yield. This method is quite general and can be used to prepare a large variety of ethers (PMB, cinnamyl, *t*-Bu, etc.) and esters.[33]

20. From a TMS ether: PhCHO, TESH, TMSOTf, 96% yield.[34] This method is effective for the preparation of allyl ethers (85% yield). This method has been expanded to include the MPM, 2-Nap, cinnamyl, crotyl, and DMB ethers. Primary alcohols are derivatized in preference to secondary alcohols. The reaction is also regioselective.[35]

21. LiHMDS, TBAI, BnBr, THF, −78°C to 25°C, 72% . The use of other bases led to significant participation of the NHBOC group. LDA also proved unsatisfactory in this case.[36]

1. A. F. Petri, A. Bayer, and M. E. Maier, *Angew. Chem. Int. Ed.*, **43**, 5821 (2004).
2. H. G. Fletcher, *Methods Carbohydr. Chem.*, **II**, 166 (1963).

3. E. E. Dueno, F. Chu, S.-I. Kim, and K. W. Jung, *Tetrahedron Lett.*, **40**, 1843 (1999).

4. H. H. Freedman and R. A. Dubois, *Tetrahedron Lett.*, **16**, 3251 (1975).

5. D. Crich, W. Li, and H. Li, *J. Am. Chem. Soc.*, **126**, 15081 (2004).

6. R. Kuhn, I. Löw, and H. Trishmann, *Chem. Ber.*, **90**, 203 (1957).

7. A. Bouzide and G. Sauvé, *Tetrahedron Lett.*, **38**, 5945 (1997).

8. L. Van Hijfte and R. D. Little, *J. Org. Chem.*, **50**, 3940 (1985).

9. K. M. Brummond and H. S.-p, *J. Org. Chem.*, **70**, 907 (2005).

10. M. Yamashita and Y. Takegami, *Synthesis*, 803 (1977).

11. O. Sirkecioglu, B. Karliga, and N. Talinli, *Tetrahedron Lett.*, **44**, 8483 (2003).

12. T. Iversen and K. R. Bundle, *J. Chem Soc., Chem. Commun.*, 1240 (1981).

13. J. D. White, G. N. Reddy and G. O. Spessard, *J. Am. Chem. Soc.*, **110**, 1624 (1988).

14. U. Widmer, *Synthesis*, 568 (1987).

15. P. Eckenberg, U. Groth, T. Huhn, N. Richter, and C. Schmeck, *Tetrahedron*, **49**, 1619 (1993).

16. K. W. C. Poon, S. E. House, and G. B. Dudley, *Synlett*, 3142 (2005).

17. B. Boyer, E.-M. Keramane, J.-P. Roque, and A. A. Pavia, *Tetrahedron Lett.*, **41**, 2891 (2000).

18. S. Czernecki, C. Georgoulis, and C. Provelenghiou, *Tetrahedron Lett.*, 3535 (1976); K. Kanai, I. Sakamoto, S. Ogawa, and T. Suami, *Bull. Chem. Soc. Jpn.*, **60**, 1529 (1987).

19. A. Fukuzawa, H. Sato, and T. Masamune, *Tetrahedron Lett.*, **28**, 4303 (1987).

20. P. Grice, S. V. Ley, J. Pietruszka, H. W. M. Priepke, and S. L. Warriner, *J. Chem. Soc., Perkin Trans. 1*, 351 (1997).

21. E. E. van Tamelen, S. R. Zawacky, R. K. Russell, and J. G. Carlson, *J. Am. Chem. Soc.*, **105**, 142 (1983).

22. A. Furstner and O. R. Thiel, *J. Org. Chem.*, **65**, 1738 (2000).

23. B. Classon, P. J. Garegg, S. Oscarson, and A. K. Tidén, *Carbohydr. Res.*, **216**, 187 (1991).

24. C. Cruzado, M. Bernabe, and M. Martin-Lomas, *J. Org. Chem.*, **54**, 465 (1989). For an improved set of reaction conditions, see A. B. C. Simas, K. C. Pais and A. A. T. da Silva, *J. Org. Chem.*, **68**, 5426 (2003).

25. T. B. Grindley, *Adv. Carbohydr. Chem. Biochem*, **53**, 17 (1998).

26. W. R. Roush, M. R. Michaelides, D. F. Tai, B. M. Lesur, W. K. M. Chong, and D. J. Harris, *J. Am. Chem. Soc.*, **111**, 2984 (1989).

27. C. Bliard, P. Herczegh, A. Olesker, and G. Lukacs, *J. Carbohydr. Res.*, **8**, 103 (1989).

28. L. Ballell, J. A. F. Joosten, F. A. el Maate, R. M. J. Liskamp, and R. J. Pieters, *Tetrahedron Lett.*, **45**, 6685 (2004).

29. G. J. Boons, G. H. Castle, J. A. Clase, P. Grice, S. V. Ley, and C. Pinel, *Synlett*, 913 (1993).

30. S. David and S. Hanessian, *Tetrahedron*, **41**, 643 (1985); M. Pereyre, J.-P. Quintard, and A. Rahm, *Tin in Organic Synthesis*, Butterworths, London, 1987, pp. 261–285.

31. L. J. Liotta and B. Ganem, *Tetrahedron Lett.*, **30**, 4759 (1989).

32. L. J. Liotta and B. Ganem, *Isr. J. Chem.*, **31**, 215 (1991).

33. T. Shintou and T. Mukaiyama, *J. Am. Chem. Soc.*, **126**, 7359 (2004).

34. S. Hatakeyama, H. Mori, K. Kitano, H. Yamada, and M. Nishizawa, *Tetrahedron Lett.*, **35**, 4367 (1994).

35. C.-C. Wang, J.-C. Lee, S.-Y. Luo, H.-F. Fan, C.-L. Pai, W.-C. Yang, L.-D. Lu, and S.-C. Hung, *Angew. Chem. Int. Ed.*, **41**, 2360 (2002).

36. K. C. Nicolaou, Y.-K. Chen, X. Huang, T. Ling, M. Bella, and S. A. Snyder, *J. Am. Chem. Soc.*, **126**, 12888 (2004).

Cleavage: Reductively (Hydrogenolysis)

The following table shows how substituents can affect the relative rate of benzyl ether hydrogenolysis:

Relative Rates for Substituted Benzyl Ether Cleavage

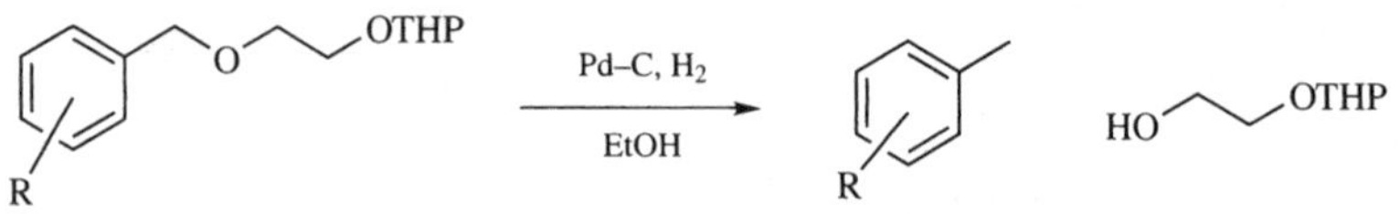

Substrate	k, $(M\ s^{21}) \times 10^{-6}$	Relative Rate
R = CF$_3$	0.080 ± 0.002	0.205
R = H	0.390 ± 0.008	1.00
R = 4-Me	3.07 ± 0.12	7.94
R = 3,5-Me$_2$	4.30 ± 0.22	11.01
R = 4-t-Bu	9.58 ± 0.78	24.78

1. H_2/Pd–C, EtOH, 95% yield.[1,2]

2. Pd is the preferred catalyst since the use of Pt results in ring hydrogenation.[1] Hydrogenolysis of the benzyl group of threonine in peptides containing tryptophan often results in reduction of tryptophan to the 2,3-dihydro derivative.[3] The presence of nonaromatic amines can retard *O*-debenzylation,[4,5] and the presence of Na$_2$CO$_3$ prevents benzyl group removal but allows double-bond reduction to occur.[6] Similarly, the ethylenediamine complex with Pd–C retards debenzylation except for benzyl esters, which are cleaved. Cbz group hydrogenolysis with this catalyst is strongly solvent-dependent with cleavage occurring in MeOH for aliphatic amine derivatives but not in THF where aromatic amines are released.[7] Epoxides[8] are stable to this catalyst and alkynes are cleanly reduced to Z-alkenes.[9] Although it is possible to effect benzyl ether cleavage in the presence of an isolated olefin (H$_2$/5% Pd–C, 97% yield),[10] in general, the degree of selectivity is dependent upon the substitution pattern and the level of steric hindrance. Good selectivity was achieved for hydrogenolysis of a benzyl group in the presence of a trisubstituted olefin conjugated to an ester.[11] Excellent selectivity has been observed in the hydrogenolysis (Pd–C, EtOAc, rt, 18 h) of a benzyl group in the presence of a *p*-methoxybenzyl group.[12] Hydrogenolysis of the benzyl group is solvent-dependent, as illustrated in the following table[13]:

Solvent Effect on the Hydrogenation of Benzyl Ether (1.1 bar H$_2$, 50°C)

Solvent	Relative Rate
Methanol	2.5
Ethanol	3.5
Propanol	7
Hexanol	12.5
Octanol	16
Acetic acid	17
THF	20
Hexane	3
Toluene	1

3. Ti–HMS modified Pd–C was found to accelerate the hydrogenolysis of simple benzyl ethers in the presence of acid-sensitive functional groups.[14] The use of benzyl protection for polymer-supported syntheses has been a problem because of trapping of the catalyst by the polymer. This problem is partially solved by the use of Pd nanoparticles which result in efficient benzyl group hydrogenolysis from polymer supports.[15]

4. In the following case, no hydrogenolysis of the benzyl groups occurs because the amino alcohol poisons the Pd–C or Pd(OH)$_2$.[16]

5. Hydrogenation of aromatic halides is often a problem,[17] but in the presence of an unprotected maleimide the catalyst is sufficiently poisoned that the chloride is retained.[18] Similarly, the presence of phthalimide has a poisoning effect on the Lindlar reduction of acetylenes.

6. Pd–C using transfer hydrogenation. A number of methods have been developed where hydrogen is generated *in situ*. These include the use of HCO$_2$H,[19] ammonium formate (MeOH, reflux, 91% yield),[20] isopropyl alcohol,[21] cyclohexene

(1–8 h, 80–90% yield),[22] and cyclohexadiene (25°C, 2 h, good yields).[23] PMB ethers are retained with these conditions when EtOAc is the solvent,[24] but further moderation of the catalyst with 2,6-lutidine was employed in the following case[25]:

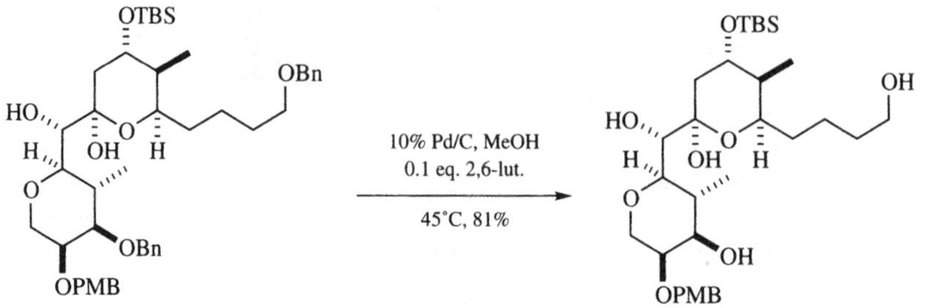

A benzylidene acetal is not cleaved when ammonium formate is used as the hydrogen source,[20] and a trisubstituted olefin is not affected when formic acid is used as a hydrogen source,[26] but the following groups are also cleaved under these conditions: N-Cbz, CO_2Bn, BOM(His), N-2-ClCbz, and PhOBn.[27] The use of hydrazinium monoformate was found advantageous for deprotection of N-Cbz, BnOR, N-2-ClCbz, N-2-BrCbz, and RCO_2Bn because the reaction could be run at rt in MeOH or AcOH rather than the usual refluxing conditions used with other hydrogen transfer agents.[28] A disubstituted olefin is retained when using the following conditions for cleavage of a primary benzyl ether (secondary BOM is also cleaved): 1-methyl-1,4-cyclohexadiene, Pd(OH)$_2$–C, CaCO$_3$, EtOH (90%).[29] In α-methyl 2,3-di-O-benzyl-4,6-O-benzylideneglucose the cleavage can be controlled to cleave the 2-benzyl group selectively (83%) when cyclohexene is used as the hydrogen source.[30] Hydrogenation was also shown to cleave only an anomeric benzyl group in perbenzylated galactose.[31] Benzyl ethers are stable to transfer hydrogenolysis with Pd–C, t-BuNH$_2$·BH$_3$/MeOH, whereas alkenes, alkynes, aryl halides, and benzyl esters are reduced.[32]

7. Pd–C, H$_2$, Cl$_3$CCO$_2$H (anhydrous), MeOH, 74–93% yield. These conditions were developed to retain the Troc group, which is normally incompatible with

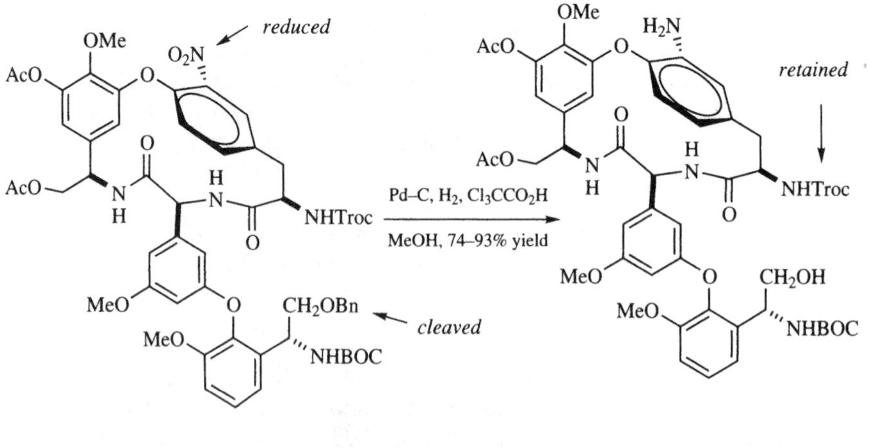

hydrogenolysis of benzyl ethers, thus solving a long-standing problem.[33] Trichloroacetic acid serves as a sacrificial Troc surrogate, thus preventing reduction of the Troc group.

8. Raney Nickel W2 or W4, EtOH, 85–100% yield.[34,35] Mono- and dimethoxy-substituted benzyl ethers, benzaldehyde and 4-methoxybenzaldehyde acetals are not cleaved under these conditions, and trisubstituted alkenes are not reduced.

9. PdCl$_2$, EtOH, H$_2$O, H$_2$, 79–99% yield. These conditions were used for the deprotection of peptides; the PdCl$_2$ was used stoichiometrically.[36]

10. Rh/Al$_2$O$_3$, H$_2$, 100% .[37]

Cleavage: Reductively (Single Electron)

11. Na/ammonia[38,39] or EtOH.[40]

Note that in this example the ester was not reduced. When the TBDMS group was replaced with an acetate, the benzyl cleavage reaction failed.[41] The reducing end hemiacetal of a polysaccharide is maintainable during a Birch debenzylation.[42]

12. Li, NH$_3$, THF, EtO–allyl. These conditions were used to prevent cleavage of an allylic ether. Presumably, the allyl ether serves as a sacrificial allyl ether, thus reducing the likelihood of reduction of the substrate allyl ether.[43]

A similar problem was encountered in the synthesis of okadaic acid, which contains a number of allylic ethers. In this case, successful debenzylation was achieved using LiDBB in THF (70% yield),[44] but in the case of a ciguatoxin synthesis, LiDBB did cleave an allylic ether. In this case, Na/NH$_3$,EtOH, THF, $-90°$C, 10 min resulted in successful deprotection albeit in only 30–40% yield.[45]

13. Lithium di-*tert*-butylbiphenyl (LiDBB), THF, $-78°$C, 3 h, 95% yield.[46] LiDBB has been found to cleave THF upon sonication.[47] A *p*-methoxybenzyl group is retained during benzyl cleavage with this reagent.[48]

14. Li, catalytic naphthalene, $-78°$C, THF, 68–99% yield. In addition, tosyl, benzyl and mesyl amides are cleaved with excellent efficiency.[49]

15. Lithium naphthalenide, THF, $-25°$C, 55–80 min, 73–98% yield.[50] These conditions will also cleave *N*-Ts, *N*-Ms, RCONRTs, RCONRMs, and the RCONRBn groups.[51]

16. Ca/NH$_3$, ether, or THF, 2 h; NH$_4$Cl, H$_2$O, 90% yield.[52] Acetylenes are **not** reduced under these conditions. One problem with the use of calcium is that the oxide coating makes it difficult to initiate the reaction. This is partially overcome by adding sand to the reaction mixture to abrade the surface of the calcium mechanically.

17. K (10 eq.), *t*-BuNH$_2$ (2 eq.), *t*-BuOH (2 eq.), 18-crown-6 (0.1 eq.), 90–99% . Benzylidine acetals are cleaved.[53]

18. Mg, HCO$_2$NH$_4$, methanol, rt, 88–90% yield. The following groups are cleaved similarly: *N*-Cbz, *N*-2-BrCbz, *N*-2-ClCbz, RCO$_2$Bn, His(BOM), *N*-Fmoc, 2,6-Cl$_2$BnOPh, and PhOBn.[54]

19. Zn, HCO$_2$NH$_4$, MeOH, rt, 79–82% yield.[55] Benzylthio ethers and benzylamines are also cleaved in excellent yield under these conditions.

20. Electrolytic reduction: -3.1 V, R$_4$NF, DMF.[56]

21. Lithium aluminum hydride will also cleave benzyl ethers, but this is seldom practical because of its high reactivity to other functional groups.[57]

22. DIBAL (150 eq.), PhCH$_3$, 50°C, 2 h, 82% yield, perbenzylated cyclodextrin as substrate.[58] The method is also applicable to the monodebenzylation of perbenzylated mono and disaccharides.[59] DIBAL in combination with triisobutylaluminum has also been used successfully to cleave benzyl groups from carbohydrates.[60]

Cleavage: Lewis Acid-Based

23. Me$_3$SiI, CH$_2$Cl$_2$, 25°C, 15 min, 100% yield.[61] This reagent also cleaves most other ethers and esters, but selectivity can be achieved with the proper choice of conditions.

24. Me$_2$BBr, ClCH$_2$CH$_2$Cl, 0°C to rt, 70–93% yield.[62] The reagent also cleaves phenolic methyl ethers; tertiary ethers and allylic ethers give the bromide rather than the alcohol.

25. FeCl$_3$, Ac$_2$O, 55–75% yield.[63] The relative rates of cleavage for the 6-, 3-, and 2-O-benzyl groups of a glucose derivative are 125:24:1. Sulfuric acid has also been used as a catalyst.[64] FeCl$_3$ (CH$_2$Cl$_2$, 0°C, rt, 64–88% yield) in the absence of acetic anhydride is also effective and was found to cleave secondary benzyl groups in the presence of a primary benzyl group.[65] This method has been used on complex polysaccharides.[66]

26. Ac$_2$O, H$_2$SO$_4$; MeOH, MeONa. A primary benzyl is removed from a perbenzylated galactose derivative.[67]

27. CrCl$_2$, LiI, EtOAc, H$_2$O, 80–89% yield. The relative reactivity of various benzyl ethers is as follows: DOB > DMB > PMB ~ Bn.[68,69]

28. Zn(OTf)$_2$, ClCH$_2$CH$_2$Cl, BzBr, rt, 10 min, 95–98% yield. TBDMS ether and acetonides are also cleaved by this method.[70]

29. BzBr, graphite, ClCH$_2$CH$_2$Cl, 50°C, 1–4 h, 67–91% . Allyl, alkyl, propargyl, and t-Bu ethers are also cleaved.[71]

30. Sc(CTf$_3$)$_3$ or Sc(NTf$_2$)$_3$, anisole, 100°C, 77–97% . These conditions also cleave the MPM ether, MPM amide, and the benzyl ester.[72]

31. PhSSiMe$_3$, Bu$_4$NI, ZnI$_2$, ClCH$_2$CH$_2$Cl, 60°C, 2 h, 75% yield.[73]

32. Ph$_3$CBF$_4$, CH$_2$Cl$_2$.[74]

33. t-BuMgBr, benzene, 80°C, 69% .[75] MeMgI fails in this reaction. In general, benzyl ethers are quite stable to Grignard reagents because these reactions are not usually run at such high temperatures.

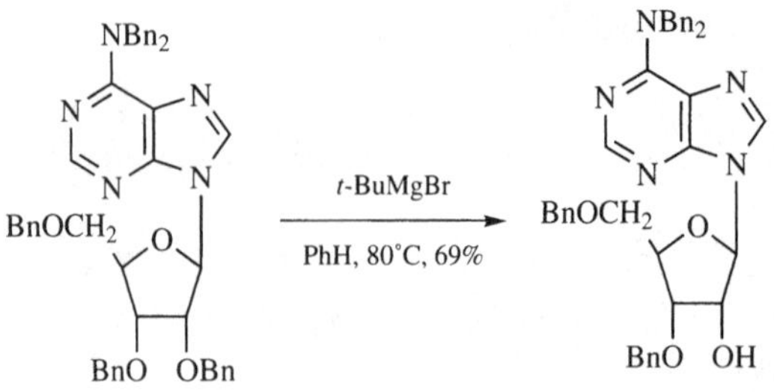

34. EtSH, $BF_3 \cdot Et_2O$, 63% yield.[76] Benzylamines are stable to these conditions, but $BF_3 \cdot Et_2O/Me_2S$ has been used to cleave an allylic benzyl ether.[77]

35. Et_2AlSPh, CH_2Cl_2, hexane, $-5°C$. This reagent causes partial cleavage of a benzyl ether.[78]

36. The fungus *Mortierella isabellina* NRRL 1757, 0–100% yield.[79]

37. $BF_3 \cdot Et_2O$, NaI, CH_3CN, 0°C, 1 h; rt, 7 h, 80% yield.[80]

38. BCl_3, CH_2Cl_2, $-78°C$ to 0°C; MeOH at $-78°C$, 77% yield.[81]

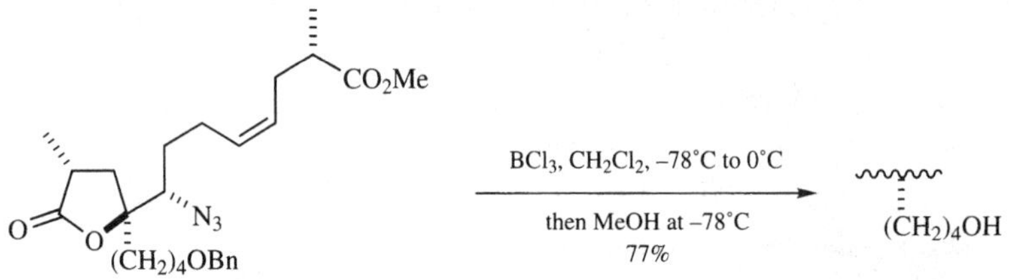

39. The following is an example of unexpected selectivity in the cleavage of a tribenzyl ether.[82] This selectivity is not general.

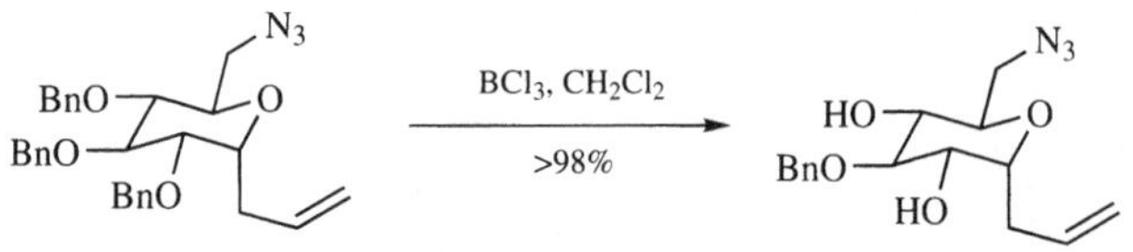

40. These conditions selectively remove an alkyl benzyl group in the presence of a phenolic benzyl group.[83]

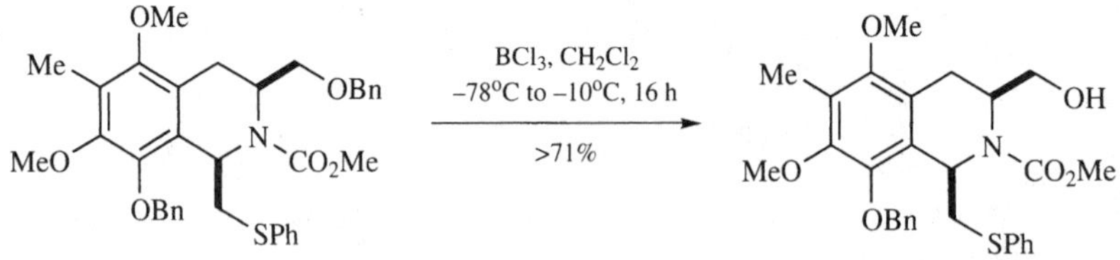

41. $BCl_3 \cdot DMS$, CH_2Cl_2, 5 min to 24 h, rt, 16–100%.[84] A trityl group is cleaved in preference to a benzyl group under these conditions. A phenolic benzyl ether is stable.[85]

42. BBr_3, 60% yield.[86] A SBn was not cleaved under these conditions.[87]

43. Me$_3$SiBr, thioanisole.[88] This reagent combination also cleaves a carbobenzoxy (Z) group, a 4-MeOC$_6$H$_4$CH$_2$SR group, and reduces sulfoxides to sulfides.

44. AlCl$_3$-aniline, CH$_2$Cl$_2$, rt, 80–96% yield.[89,90]

45. TMSOTf, Ac$_2$O, 10–15°C, 85% yield.[91] The acetate is produced that must then be hydrolyzed.

46. AcBr, SnBr$_2$ or Sn(OTf)$_2$, CH$_2$Cl$_2$, rt, 1–4 h, 76–97% yield.[92] These conditions convert a benzyl ether into an acetate.

47. ZnCl$_2$, Ac$_2$O, AcOH, rt, 80–94% yield. These conditions are selective for the cleavage of 6-O-benzylpyranosides.[93]

48. SnCl$_4$, CH$_2$Cl$_2$, rt, 30 min.[94]

Secondary benzyl ether is cleaved in preference to a primary benzyl ether.[95] TiCl$_4$ (CH$_2$Cl$_2$, rt, 30 min) has been used to cleave a secondary benzyl ether[96] and an α-methylbenzyl ether.[97] In carbohydrates where benzyl groups are used extensively for protection, their stability toward electrophilic reagents is increased by the presence of electron-withdrawing groups in the ring.[98]

Cleavage: Oxidative Methods

49. CrO$_3$/AcOH, 25°C, 50% yield [→ ROCOPh (→ ROH + PhCO$_2$H)].[99] This method was used to remove benzyl ethers from carbohydrates that contain functional groups sensitive to catalytic hydrogenation or dissolving metals. Esters are stable, but glycosides or acetals are cleaved.

50. RuO$_2$, NaIO$_4$, CCl$_4$, CH$_3$CN, H$_2$O, 54–96% yield.[100] The benzyl group is oxidized to a benzoate that can be hydrolyzed under basic conditions. In the following case, reductive conditions (Na/NH$_3$) failed.[101]

51. Ozone, 50 min, then NaOMe, 60–88% yield.[102]

52. $NaBrO_3$, $Na_2S_2O_4$, EtOAc, H_2O. Benzyl ethers are cleaved in preference to benzylcarbonates and 4,6-carbohydrate benzylidine acetals are unselectively cleaved to give a mixture of the primary and secondary monobenzoates.[103] This method was also found effective in complex carbohydrate synthesis.[104]

53. $(n\text{-}Bu_4N)_2S_2O_8$, CH_3CN, 5°C; MeONa, MeOH, 15°C, 85–90% yield. The benzyl ether is first oxidized to the benzoate and then cleaved by methanolysis.[105]

54. NBS, $h\nu$, $CaCO_3$, CCl_4, H_2O, 86% yield.[106]

55. NIS, 2.5 eq., CH_3CN, $h\nu$. This method cleaves a benzyl group in carbohydrates, provided that there is an adjacent hydroxyl. In some cases a benzylidine is formed.[107,108]

56. Electrolytic oxidation: 1.4–1.7 V, Ar_3N, CH_3CN, CH_2Cl_2, $LiClO_4$, lutidine.[109]

57. 4-Methoxy-TEMPO, CC_4, KBr, H_2O, $NaHSO_4$ to adjust pH to < 8.0, 0–5°C, NaOCl, 62–76% yield. These conditions oxidize the benzyl to a benzoate which can then be hydrolyzed by conventional means.[110]

58. Dimethyldioxirane, acetone, 48 h, rt, 85–93% yield.[111,112] p-Bromo-, p-cyano-, and 2-naphthylmethyl ethers and benzylidene acetals can also be deprotected.

59. PhI(OH)OTs, CH_3CN.[113]

60. DDQ, CH_2Cl_2, 58°C, 2 days, 52% yield.[114] In this example, conventional reductive methods failed. Anhydrous DDQ was used to prevent acid-promoted decomposition.

The removal of benzyl ethers in the presence of allylic ethers can be a problem, as illustrated in the synthesis of Ciguatoxin.[45] This method was found to prevent TIPS migration that occurred while attempting to remove a benzyl group with a variety of Lewis acids.[115]

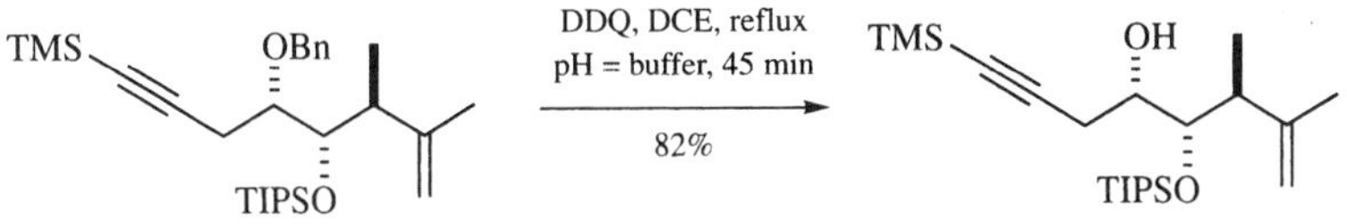

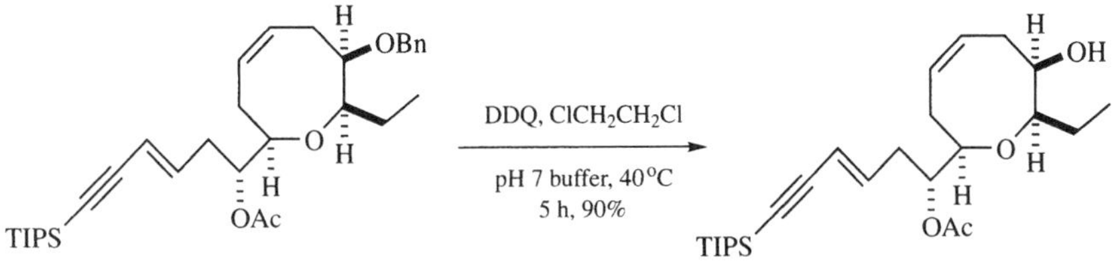

Ref. 116

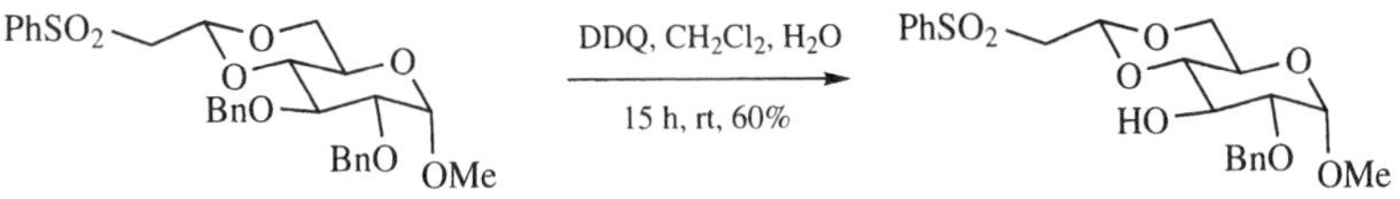

Ref. 117

Photolysis at 365 nm in CH$_3$CN improves the rate of DDQ promoted cleavage of benzyl ethers in that under these conditions cleavage occurs at rt. The MPM groups is still cleaved more rapidly and good selectivity can be achieved over benzyl ether cleavage. Unfortunately, olefins and acetylenes are incompatible with this protocol.[118]

61.

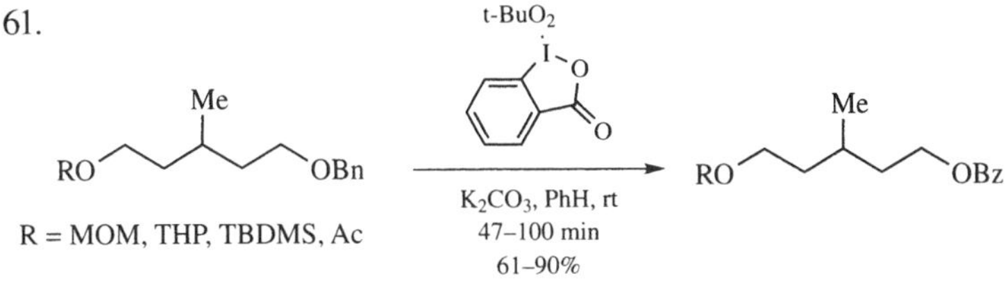

R = MOM, THP, TBDMS, Ac

Ref. 119

Allyl ethers are oxidized to acrylates with this reagent.

62. 25% MsOH/CHCl$_3$, 25°C, 84% yield.[120]

63. 6 N HCl, reflux, 92% yield. A N-Cbz group is also removed.[121]

64. P$_4$S$_{10}$, CH$_2$Cl$_2$, 88% yield.[122]

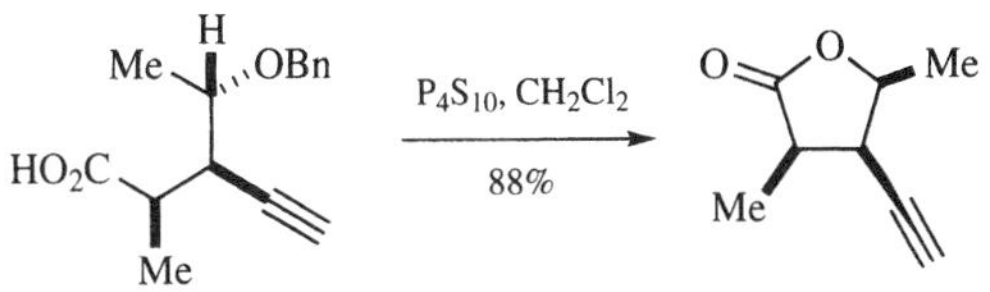

65. $ClO_2S-N=C=O$, K_2CO_3, CH_2Cl_2, reflux or $-78°C$; NaOH, MeOH, rt, 69–88% yield. PMB ethers can be cleaved in the presence of benzyl ethers; however, under more forcing conditions, benzyl ethers are cleaved.[123]

66. Although benzyl groups are considered robust and compatible with a myriad of transformations, they have been known to misbehave as in the following case where migration occurred unexpectedly.[124] The reaction presumably occurs through a bridged oxonium ion for which there is precedent.[125]

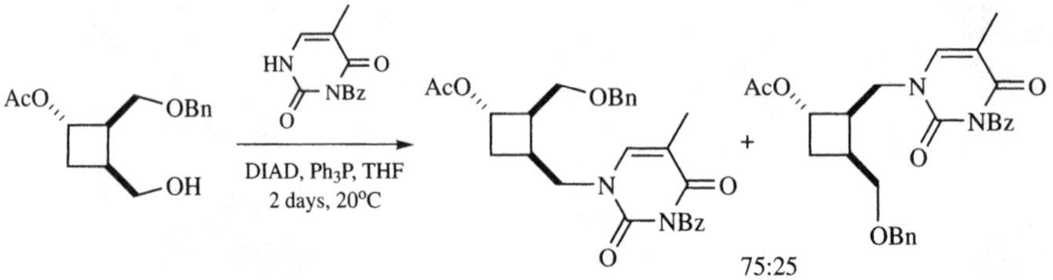

67. A special case that proceeds through ether formation followed by reductive cleavage is illustrated below.[126]

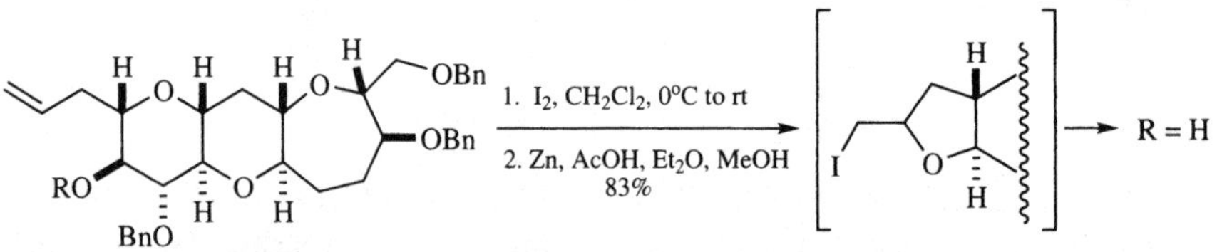

1. C. H. Heathcock and R. Ratcliffe, *J. Am. Chem. Soc.*, **93**, 1746 (1971).

2. W. H. Hartung and C. Simonoff, *Org. React.*, **7**, 263 (1953).

3. L. Kisfaludy, F. Korenczki, T. Mohacsi, M. Sajgo, and S. Fermandjian, *Int. J. Pept. Protein Res.*, **27**, 440 (1986).

4. B. P. Czech and R. A. Bartsch, *J. Org. Chem.*, **49**, 4076 (1984).

5. H. Sajiki and K. Hirota, *Tetrahedron*, **54**, 13981 (1998).

6. G. R. Cook, L. G. Beholz, and J. R. Stille, *J. Org. Chem.*, **59**, 3575 (1994).

7. H. Sajiki, K. Hattori, and K. Hirota, *J. Org. Chem.*, **63**, 7990 (1998); H. Sajiki and K. Hirota, *Tetrahedron*, **54**, 13981 (1998); H. Sajiki, Kuno, and K. Hirota, *Tetrahedron Lett.*, **39**, 13981 (1998).

8. H. Sajiki, K. Hattori, and K. Hirota, *Chem. Eur. J.*, **6**, 2200 (2000).

9. K. R. Campos, D. Cai, M. Journet, J. J. Kowal, R. D. Larsen, and P. J. Reider, *J. Org. Chem.*, **66**, 3634 (2001).

10. J. S. Bindra and A. Grodski, *J. Org. Chem.*, **43**, 3240 (1978).

11. D. Cain and T. L. Smith, Jr., *J. Am. Chem. Soc.*, **102**, 7568 (1980).

12. J. M. Chong and K. K. Sokoll, *Org. Prep. Proced. Int.*, **25**, 639 (1993).

13. S. Hawker, M. A. Bhatti, and K. G. Griffin, *Chim. Oggi*, **10**, 49 (1992).

14. A. Itoh, T. Kodama, S. Maeda, and Y. Masaki, *Tetrahedron Lett.*, **39**, 9461 (1998).

15. O. Kanie, G. Grotenbreg, and C.-H. Wong, *Angew. Chem. Int. Ed.*, **39**, 4545 (2000).

16. M. B.-U. Surfraz, M. Akhtar, and R. K. Allemann, *Tetrahedron Lett.*, **45**, 1223 (2004).

17. K. C. Nicolaou, H. J. Mitchell, N. F. Jain, N. Winssinger, R. Hughes, and T. Bando, *Angew. Chem. Int. Ed.*, **38**, 240 (1999).

18. M. M. Faul, L. L. Winneroski, and C. A. Krumrich, *J. Org. Chem.*, **64**, 2465 (1999).

19. B. El Amin, G. M. Anantharamaiah, G. P. Royer, and G. E. Means, *J. Org. Chem.*, **44**, 3442 (1979).

20. T. Bieg and W. Szeja, *Synthesis*, 76 (1985).

21. M. Del Carmen Cruzado and M. Martin-Lomias, *Tetrahedron Lett.*, **27**, 2497 (1986).

22. G. M. Anantharamaiah and K. M. Sivanandaiah, *J. Chem. Soc., Perkin Trans. 1*, 490 (1977); S. Hanessian, T. J. Liak, and B. Vanasse, *Synthesis*, 396 (1981).

23. A. M. Felix, E. P. Heimer, T. J. Lambros, C. Tzougraki, and J. Meienhofer, *J. Org. Chem.*, **43**, 4194 (1978).

24. D. A. Evans, D. H. B. Rippin, D. P. Halstead, and K. R. Campos, *J. Am. Chem. Soc.*, **121**, 6816, (1999).

25. A. B. Smith, III, W. Zhu, S. Shirakami, C. Sfouggatakis, V. A. Doughty, C. S. Bennett, and Y. Sakamoto, *Org. Lett.*, **5**, 761 (2003).

26. M. E. Jung, Y. Usui, and C. T. Vu, *Tetrahedron Lett.*, **28**, 5977 (1987).

27. D. C. Gowda, *Ind. J. Chem., Sect. B*, **41B**, 1064 (2002).

28. D. C. Gowda and B. Mahesh, *Prot. Pept. Lett.*, **9**, 225 (2002).

29. A. B. Smith III, V. A. Doughty, Q. Lin, L. Zhuang, M. D. McBriar, A. M. Boldi, W. H. Moser, N. Murase, K. Nakayama, and M. Sobukawa, *Angew. Chem. Int. Ed.*, **40**, 191 (2001).

30. D. Beaupere, I. Boutbaiba, G. Demailly, and R. Uzan, *Carbohydr. Res.*, **180**, 152 (1988).

31. T. Bieg and W. Szeja, *Carbohydr. Res.*, **205**, C10 (1990).

32. M. Couturier, B. M. Andresen, J. L. Tucker, P. Dube, S. J. Brenek, and J. T. Negri, *Tetrahedron Lett.*, **42**, 2763 (2001).

33. D. L. Boger, S. H. Kim, Y. Mori, J.-H. Weng, O. Rogel, S. L. Castle, and J. J. McAtee, *J. Am. Chem. Soc.*, **123**, 1862 (2001).

34. Y. Oikawa, T. Tanaka, K. Horita, and O. Yonemitsu, *Tetrahedron Lett.*, **25**, 5397 (1984); K. Horita, T. Yoshioka, T. Tanaka, Y. Oikawa, and O. Yonemitsu, *Tetrahedron*, **42**, 3021 (1986).

35. I. Paterson, H.-G. Lombart, and C. Allerton, *Org. Lett.*, **1**, 19, (1999); D. A. Evans, W. C. Trenkle, J. Zhang, and J. D. Burch, *Org. Lett.*, **7**, 3335 (2005).

36. A. J. Pallenberg, *Tetrahedron Lett.*, **33**, 7693 (1992).

37. Y. Oikawa, T. Tanaka, and O. Yonemitsu, *Tetrahedron Lett.*, **27**, 3647 (1986).

38. C. M. McCloskey, *Adv. Carbohydr. Chem.*, **12** 137 (1957); I. Schön, *Chem. Rev.*, **84**, 287 (1984).

39. K. D. Philips, J. Zemlicka, and J. P. Horowitz, *Carbohydr. Res.*, **30**, 281 (1973).

40. E. J. Reist, V. J. Bartuska, and L. Goodman, *J. Org. Chem.*, **29**, 3725 (1964).

41. M. M. Sulikowski, G. E. R. E. Davis, and A. B. Smith, III, *J. Chem. Soc., Perkin Trans. II*, 979 (1992).

42. V. Y. Dudkin, J. S. Miller, and S. J. Danishefsky, *J. Am. Chem. Soc.*, **126**, 736 (2004).

43. A. Zakarian, A. Batch, and R. A. Holton, *J. Am. Chem. Soc.*, **125**, 7822 (2003).

44. C. J. Forsyth, S. F. Sabes, and R. A. Urbanek, *J. Am. Chem. Soc.*, **119**, 8381 (1997).

45. M. Hirama, T. Oishi, H. Uehara, M. Inoue, M. Maruyama, H. Oguri, and M. Satake, *Science*, **294**, 1904 (2001).

46. S. J. Shimshock, R. E. Waltermire, and P. DeShong, *J. Am. Chem. Soc.*, **113**, 8791 (1991).

47. S. Streiff, N. Ribeiro, and L. Desaubry, *Chem. Commun.*, 346 (2004).

48. R. M. Owen and W. R. Roush, *Org. Lett.*, **7**, 3941 (2005).

49. E. Alonso, D. J. Ramón, and M. Yus, *Tetrahedron*, **53**, 14355 (1997).

50. H.-J. Liu, J. Yip, and K.-S. Shia, *Tetrahedron Lett.*, **38**, 2252 (1997).

51. E. Alonso, D. J. Ramon, and M. Yus, *Tetrahedron*, **53**, 14355–14368 (1997).

52. J. R. Hwu, V. Chua, J. E. Schroeder, R. E. Barrans, Jr., K. P. Khoudary, N. Wang, and J. M. Wetzel, *J. Org. Chem.*, **51** 4731 (1986); J. R. Hwu, Y. S. Wein, and Y.-J. Leu, *J. Org. Chem.*, **61**, 1493 (1996).

53. L. Shi, W. J. Xia, F. M. Zang, and Y. Q. Tu, *Synlett*, 1505 (2002).

54. D. C. Gowda, K. Abiraj, and P. Augustine, *Lett. Pept. Science*, **9**, 43 (2002). S. N. N. Babu, G. R. Srinivasa, D. C. Santhosh, and D. C. Gowda, *J. Chemical Res.*, 66 (2004).

55. G. R. Srinivasa, S. N. N. Babu, C. Lakshmi, and D. C. Gowda, *Synth. Commum.*, **34**, 1831 (2004).

56. V. G. Mairanovsky, *Angew. Chem., Int. Ed. Engl.*, **15**, 281 (1976).

57. J. P. Kutney, N. Abdurahman, C. Gletsos, P. LeQuesne, E. Piers, and I. Vlattas, *J. Am. Chem. Soc.*, **92**, 1727 (1970).

58. A. J. Pearce and P. Sinay, *Angew. Chem. Int. Ed.*, **39**, 3610 (2000).

59. T. Lecourt, A. Herault, A. J. Pearce, M. Sollogoub, and P. Sinay, *Chem. Eur. J.*, **10**, 2960 (2004).

60. B. C.-d. Roizel, E. Cabianca, P. Rollin, and P. Sinay, *Tetrahedron*, **58**, 9579 (2002).

61. M. E. Jung and M. A. Lyster, *J. Org. Chem.*, **42**, 3761 (1977).

62. Y. Guindon, C. Yoakim, and H. E. Morton, *Tetrahedron Lett.*, **24**, 2969 (1983).

63. K. P. R. Kartha, F. Dasgupta, P.P. Singh, and H. C. Srivastava, *J. Carbohydr. Chem.*, **5**, 437 (1986); J. I. Padron, and J. T. Vazquez, *Tetrahedron: Asymmetry*, **6**, 857 (1995).

64. J. Sakai, T. Takeda, and Y. Ogihara, *Carbohydr. Res.*, **95**, 125 (1981).

65. R. Rodebaugh, J. S. Debenham, and B. Fraser-Reid, *Tetrahedron Lett.*, **37**, 5477 (1996).

66. J. S. Debenham, R. Rodebaugh, and B. Fraser-Reid, *J. Org. Chem.*, **62**, 4591 (1997).

67. A. Dondoni, A. Marra, M. Mizuno, and P. P. Giovannini, *J. Org. Chem.*, **67**, 4186 (2002).

68. J. R. Falck, D. K. Barma, R. Baati, and C. Mioskowski, *Angew. Chem. Int. Ed.*, **40**, 1281 (2001).

69. J. R. Falck, D. K. Barma, S. K. Venkataraman, R. Baati, and C. Mioskowski, *Tetrahedron Lett.*, **43**, 963 (2002).

70. T. Polat and R. J. Linhardt, *Carbohydr. Res.*, **338**, 447 (2003).

71. Y. Suzuki, M. Matsushima, and M. Kodomari, *Chem. Lett.*, **27**, 319 (1998).

72. K. Ishihara, Y. Hiraiwa, and H. Yamamoto, *Synlett*, 80 (2000).

73. K. C. Nicolaou, M. R. Pavia, and S. P. Seitz, *J. Am. Chem. Soc.*, **104**, 2027 (1982).

74. T. R. Hoye, A. J. Caruso, J. F. Dellaria, Jr., and M. J. Kurth, *J. Am. Chem. Soc.*, **104**, 6704 (1982).

75. M. Kawana, *Chem. Lett.*, **10**, 1541 (1981).

76. S. M. Daly and R. W. Armstrong, *Tetrahedron Lett.*, **30**, 5713 (1989).

77. M. Ishizaki, O. Hoshino, and Y. Iitaka, *Tetrahedron Lett.*, **32**, 7079 (1991).

78. H. Imai, H. Uehara, M. Inoue, H. Oguri, T. Oishi, and M. Hirama, *Tetrahedron Lett.*, **42**, 6219 (2001).

79. H. L. Holland, M. Conn, P. C. Chenchaiah, and F. M. Brown, *Tetrahedron Lett.*, **29**, 6393 (1988).

80. Y. D. Vankar and C. T. Rao, *J. Chem. Res., Synop.*, 232 (1985).

81. D. R. Williams, D. L. Brown, and J. W. Benbow, *J. Am. Chem. Soc.*, **111**, 1923 (1989).

82. J. Xie, M. Menand and J.-M. Valery, *Carbohydr. Res.*, **340**, 481 (2005).

83. P. Magnus, K. S. Matthews, and V. Lynch, *Org. Lett.*, **5**, 2181 (2003).

84. M. S. Congreve, E. C. Davison, M. A. M. Fuhry, A. B. Holmes, A. N. Payne, R. A. Robinson, and S. E. Ward, *Synlett*, 663 (1993).

85. I. E. Wrona, A. E. Gabarda, G. Evano, and J. S. Panek, *J. Am. Chem. Soc.*, **127**, 15026 (2005).

86. D. E. Ward, Y. Gai, and B. F. Kaller, *J. Org. Chem.*, **60**, 7830 (1995).

87. K. Haraguchi, A. Nishikawa, E. Sasakura, H. Tanaka, K. T. Nakamura, and T. Miyasaka, *Tetrahedron Lett.*, **39**, 3713 (1998).

88. N. Fujii, A. Otaka, N. Sugiyama, M. Hatano, and H. Yajima, *Chem. Pharm. Bull.*, **35**, 3880 (1987).

89. T. Akiyama, H. Hirofuji, and S. Ozaki, *Tetrahedron Lett.*, **32**, 1321 (1991).

90. J. P. Vitale, S. A. Wolckenhauer, N. M. Do, and S. D. Rychnovsky, *Org. Lett.*, **7**, 3255 (2005).

91. J. Alzeer and A. Vasella, *Helv. Chim. Acta*, **78**, 177 (1995); P. Angibeaud and J.-P. Utille, *Synthesis*, 737 (1991).

92. T. Oriyama, M. Kimura, M. Oda, and G. Koga, *Synlett*, 437 (1993).

93. G. Yang, X. Ding, and F. Kong, *Tetrahedron Lett.*, **38**, 6725 (1997).

94. H. Hori, Y. Nishida, H. Ohrui, and H. Meguro, *J. Org. Chem.*, **54**, 1346 (1989).

95. K. S. Kim and S. D. Hong, *Tetrahedron Lett.*, **41**, 5909 (2000).

96. J.-P. Surivet and J.-M. Vatele, *Tetrahedron Lett.*, **39**, 9681 (1998).

97. M. Turks, M. C. Murcia, R. Scopelliti, and P. Vogel, *Org. Lett.*, **6**, 3031 (2004).

98. K. Jansson, G. Noori, and G. Magnusson, *J. Org. Chem.*, **55**, 3181 (1990).

99. S. J. Angyal and K. James, *Carbohydr. Res.*, **12**, 147 (1970).

100. P. F. Schuda, M. B. Cichowicz, and M. R. Heimann, *Tetrahedron Lett.*, **24**, 3829 (1983); P. F. Schuda and M. R. Heimann, *Tetrahedron Lett.*, **24**, 4267 (1983).

101. T. Ritter, P. Zarotti, and E. M. Carreira, *Org. Lett.*, **6**, 4371 (2004).

102. P. Angibeaud, J. Defaye, A. Gadelle, and J.-P. Utille, *Synthesis*, 1123 (1985).

103. M. Adinolfi, G. Barone, L. Guariniello, and A. Iadonisi, *Tetrahedron Lett.*, **40**, 8439 (1999). M. Adinolfi, G. Barone, A. Iadonisi, and M. Schiattarella, *Tetrahedron Lett.*, **42**, 5971 (2001). M. Adinolfi, L. Guariniello, A. Iadonisi, and L. Mangoni, *Synlett*, 1277 (2000).

104. Y. Du, M. Zhang and F. Kong, *Org. Lett.*, **2**, 3797 (2000).

105. F.-E. Chen, Z.-Z. Peng, H. Fu, G. Meng, Y. Cheng, and Y.-X. Lu, *Synlett*, 627 (2000).

106. R. W. Binkley and D. G. Hehemann, *J. Org. Chem.*, **55**, 378 (1990).

107. J. Madsen, C. Viuf, and M. Bols, *Chem. Eur. J.*, **6**, 1140 (2000).

108. J. Madsen and M. Bols, *Angew. Chem. Int. Ed.*, **37**, 3177 (1998).

109. W. Schmidt and E. Steckhan, *Angew. Chem., Int. Ed. Engl.*, **18**, 801 (1979); E. A. Mayeda, L.L. Miller, and J. F. Wolf, *J. Am. Chem. Soc.*, **94**, 6812 (1972).

110. N. S. Cho and C. H. Park, *Bull. Korean Chem. Soc.*, **15**, 924 (1994).

111. B. A. Marples, J. P. Muxworthy, and K. H. Baggaley, *Synlett*, 646 (1992).

112. R. Csuk and P. Dörr, *Tetrahedron*, **50**, 9983 (1994).

113. A. Kirschning, S. Domann, G. Dräger, and L. Rose, *Synlett*, 767 (1995).

114. N. Ikemoto and S. L Schreiber, *J. Am. Chem. Soc.*, **114**, 2524 (1992).

115. B. M. Trost and J. P. N. Papillon, *J. Am. Chem. Soc.*, **126**, 13618 (2004).

116. S. Baek, H. Jo, H. Kim, H. Kim, S. Kim, and D. Kim, *Org. Lett.*, **7**, 75 (2005).

117. E. Cabianca, A. Tatibouet, and P. Rollin, *Polish J. Chem.*, **79**, 317 (2005).

118. M. A. Rahim, S. Matsumura, and K. Toshima, *Tetrahedron Lett.*, **46**, 7307 (2005).

119. M. Ochiai, T. Ito, H. Takahashi, A. Nakanishi, M. Toyonari, T. Sueda, S. Goto, and M. Shiro, *J. Am. Chem. Soc.*, **118**, 7716 (1996).

120. D. S. Matteson, H.-W. Man, and O. C. Ho, *J. Am. Chem. Soc.*, **118**, 4560 (1996).

121. M. Katoh, R. Matsune, H. Nagase, and T. Honda, *Tetrahedron Lett.*, **45**, 6221 (2004).

122. P. A. Jacobi, J. Guo, and W. Zheng, *Tetrahedron Lett.*, **36**, 1197 (1995).

123. J. D. Kim, G. Han, O. P. Zee, and Y. H. Jung, *Tetrahedron Lett.*, **44**, 733 (2003).

124. Y. Marsac, A. Nourry, S. Legoupy, M. Pipelier, D. Dubreuil, and F. Huet, *Tetrahedron Lett.*, **45**, 6461 (2004).

125. V. K. Iyer and J. P. Horwitz, *J. Org. Chem.*, **47**, 644 (1982).

126. M. Sasaki, H. Fuwa, M. Ishikawa, and K. Tachibana, *Org. Lett.*, **1**, 1075 (1999).

Methoxy-Substituted Benzyl Ethers

Several methoxy-substituted benzyl ethers have been prepared and used as protective groups. Their utility lies in the fact that they are more readily cleaved oxidatively than the unsubstituted benzyl ethers. These ethers are not stable to methyl(trifluorom ethyl)dioxirane, which oxidizes the aromatic ring.[1] The related p-(dodecyloxy)benzyl ether has been prepared to facilitate chromatographic purification of carbohydrates on C_{18} silica gel.[2] The table below gives the relative rates of cleavage with dichloro-dicyanoquinone (DDQ).[3]

Cleavage of MPM, DMPM and TMPM Ethers with DDQ in CH_2Cl_2/H_2O at 20°C

Protective Group	Time (h)	Yield **ii** (%)	**iii**	Protective Group	Time (h)	Yield **ii** (%)	**iii**
3,4-DMPM	<0.33	86	84	2-MPM	3.5	93	70
4-MPM	0.33	89	86	3,5-DMPM	8	73	92
2,3,4-TMPM	0.5	60	75	2,3-DMPM	12.5	75	73
3,4,5-TMPM	1	89	89	3-MPM	24	80	94
2,5-DMPM	2.5	95	16	2,6-DMPM	27.5	80	95

From the table it is clear that there are considerable differences in the cleavage rates of the various ethers. These have been exploited in numerous syntheses.

p-Methoxybenzyl Ether (MPM-OR, PMB-OR): *p*-MeOC$_6$H$_4$CH$_2$OR

Formation

1. The section on the formation of benzyl ethers should also be consulted.
2. NaH, *p*-MeOC$_6$H$_4$CH$_2$Cl, THF, 81% yield.[4] For simple alcohols this is probably the most commonly used method.
3. NaH, *p*-MeOC$_6$H$_4$CH$_2$Br, DMF, −5°C, 1 h, 65%.[5,6] Additionally, other bases such as BuLi,[7] dimsyl potassium[8], CsOH or Cs$_2$CO$_3$,[9] and NaOH under phase transfer conditions[10] have been used to introduce the MPM group. The use of (*n*-Bu)$_4$NI for the *in situ* preparation of the very reactive *p*-methoxybenzyl iodide is often used for improving the protection of hindered alcohols.[11] In the following example, selectivity is probably achieved because of the increased acidity of the 2′-hydroxyl group.

Ref. 7

4. *p*-MeOC$_6$H$_4$CH$_2$Br (freshly distilled), THF, TEA, KHMDS, −78°C, 1 h then rt 2 h. The method was used to protect a secondary neopentyl alcohol.[12]
5. *p*-MeOC$_6$H$_4$CH$_2$OC(=NH)CCl$_3$, H$^+$, 52–84% yield[13–15] or with BF$_3$·Et$_2$O.[16] In addition, camphorsulfonic acid[14] and *p*-toluenesulfonic acid[15] have been used as a catalysts. La(OTf)$_3$ in toluene or acetonitrile is a superior catalyst giving the MPM ether in 87–93% yield of primary, secondary, and tertiary alcohols. It was necessary to use thioanisole as a carbocation scavenger for the protection of the epoxide of cinnamyl alcohol, 61% yield (34% without thioanisole).[17] The reagent is reported to be unstable which accounts for the low yields in some cases.
6.

Ref. 18

7. *p*-MeOC$_6$H$_4$CH$_2$OC(=NH)CF$_3$, PPTS or TfOH, CH$_2$Cl$_2$ or Et$_2$O, 70–88% yield. The trifluoroacetimidate is more stable than the trichloroacetimidate and can be chromatographed. A series of homologs were also prepared.[19] In the following example, basic conditions could not be used because of migration of the TBS group.[20]

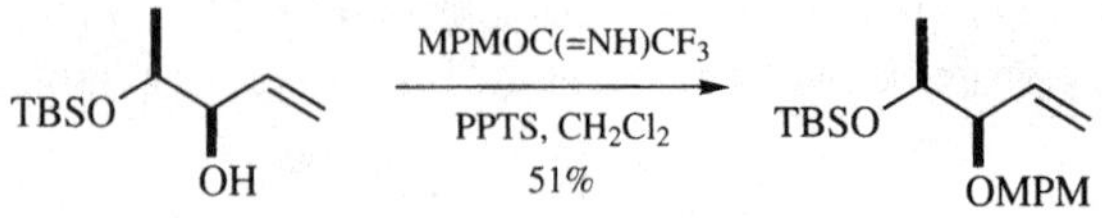

8. $p\text{-MeOC}_6H_4CH_2OC(O)S\text{-2-pyr}$, AgOTf, CH_2Cl_2, 72–88% yield.[21] In contrast to most other methods, the conditions are neutral.

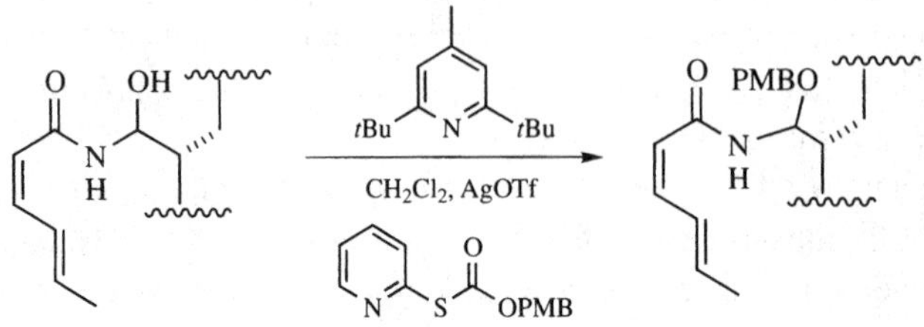

Ref. 22

9. PMBONPy, toluene, Et_2O, BTF or CH_2Cl_2, TMSOTf, or $TrB(C_6F_5)_4$, rt, 74–100% yield. This method has the advantage that the reagent is easily handled and quite stable.[23]

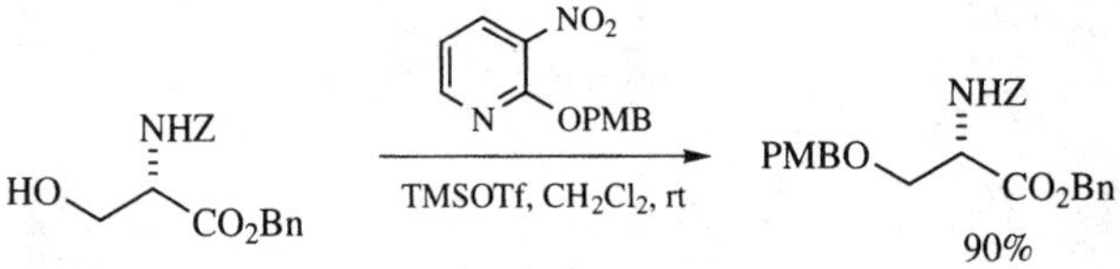

10. n-BuLi, Ph_2PCl; $p\text{-MeOC}_6H_4CH_2OH$, fluoranil, CH_2Cl_2, rt, 3 h 30–94% yield. This methods works for a variety of ethers.[24]

11. $p\text{-MeOC}_6H_4CH_2OH$, $Yb(OTf)_3$, CH_2Cl_2, rt, 60–88% yield.[25]

12. $p\text{-MeOC}_6H_4CH_2OH$, Al-MCM-41 zeolite, CH_3NO_3, 12–20 h, 32–75% yield. Primary alcohols are protected in preference to secondary alcohols.[26]

13.

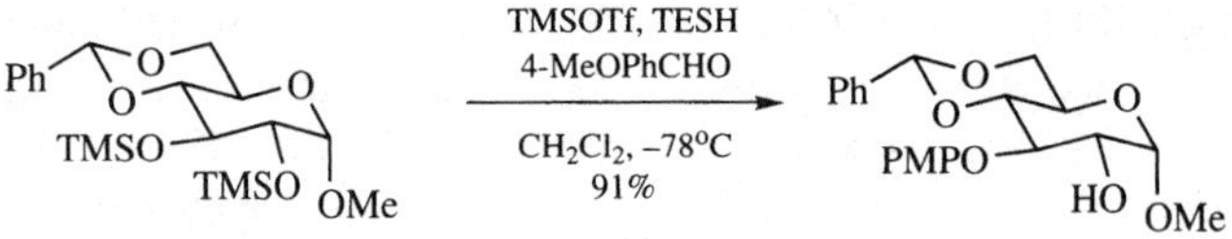

Ref. 27

Other ethers can be prepared similarly using this method.

14. $p\text{-MeOC}_6H_4CHN_2$, $SnCl_2$, ≈50% yield.[28] This method was used to introduce the MPM group at the 2′- and 3′-positions of ribonucleotides without selectivity for either the 2′- or 3′-isomer. The primary 5′-hydroxyl was not affected.

15.

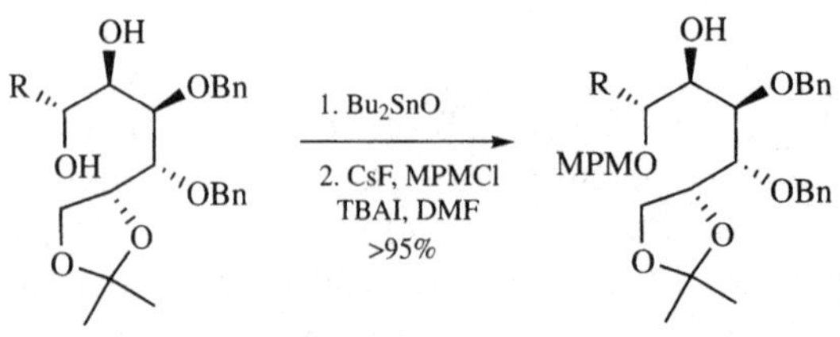

Ref. 29

16.

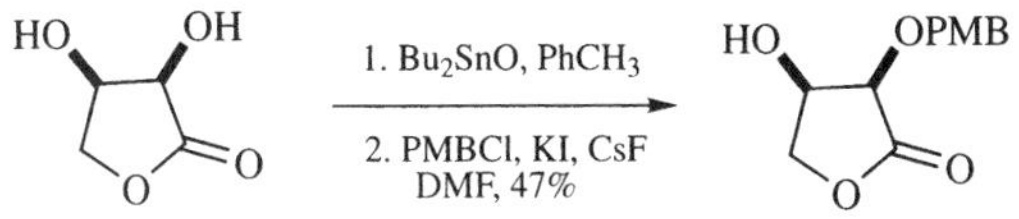

Ref. 30

17.

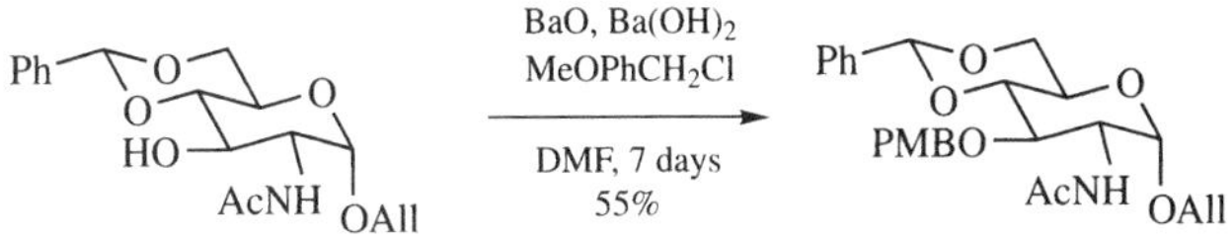

BaO, Ba(OH)$_2$
MeOPhCH$_2$Cl

DMF, 7 days
55%

Ref. 31

The authors do not indicate why these conditions were chosen over the more conventional, but it may be a result of competitive alkylation at the amide NH.

18. N-(4-Methoxybenzyl)-o-benzenedisulfonamide, NaH, THF, 57–78% yield.[32]

Cleavage

1. The section on the cleavage of benzyl ethers should also be consulted.
2. Electrolytic oxidation: Ar$_3$N, CH$_3$CN, LiClO$_4$, 20°C, 1.4–1.7 V, 80–90% yield.[33] Benzyl ethers are not affected by these conditions.
3. Dichlorodicyanoquinone (DDQ), CH$_2$Cl$_2$, H$_2$O, 40 min, rt, 84–93% yield.[34–36] This method normally does not cleave simple benzyl ethers, but forcing conditions will result in benzyl ether cleavage.[37] Surprisingly, a glycosidic TMS group was found to survive these conditions.[38] An O-MPM group can be cleaved in preference to an N-MPM protected amide[39] and a 2-naphthyl group (NAP).[40] The following groups are generally stable to these conditions: ketones, acetals, epoxides, alkenes, acetonides, tosylates, MOM, and MEM ethers, THP ethers, acetates, benzyloxymethyl (BOM) ethers, boronate, and TBDMS ethers, but exceptions do occur and will depend on the nature of the reaction conditions. MPM protected amide was shown to be stable to these conditions.[41] In this case the tertiary and electron-deficient MPM group is retained.[42]

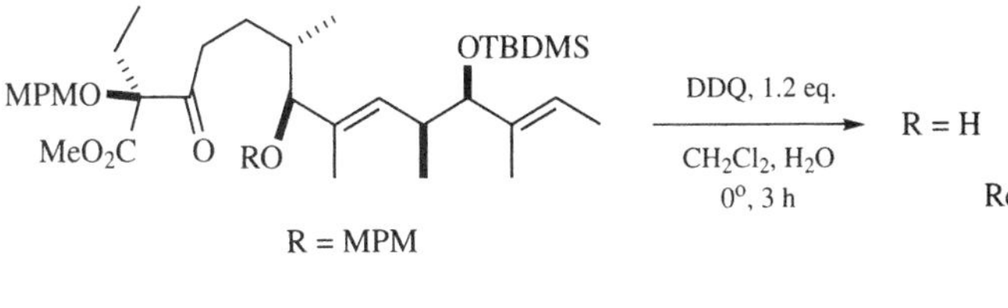

DDQ, 1.2 eq.
CH$_2$Cl$_2$, H$_2$O
0°, 3 h

R = H

Ref. 24

R = MPM

A very slow cleavage of an MPM protected adenosine was attributed to its reduced electron density as a result of π stacking with the adenine. Typically, these reactions are complete in <1 h, but in this case complete cleavage required 41 h.[43]

One problem that is encountered in the use of DDQ is that either 1,4-dienes[44,67] or 1,3-dienes[45] often interfere with deprotection, especially those that are have allylic heteroatoms. Trienes are even more problematic. The problem is less pronounced when there is an electron-withdrawing group conjugated to the diene.

A serendipitous deprotection of only one equatorial PMB group was observed with 1 eq. of DDQ (CH₂Cl₂, 0°C, 70%).[46] No explanation was offered for this result, but it may be that the electron withdrawing axial acetate deactivates the adjacent OPMB toward oxidation.

The hydroquinone produced from DDQ oxidations is fairly acidic and can interfere with acid sensitive glycals, but if the reaction is conducted in the presence of 2,6-di-*t*-butylpyridine glycals will survive.[47]

4. The following illustrates a rather surprising result where an allylic NHBOC was converted to a ketone during attempted PMB cleavage. As with dienic alcohols and ethers, this is probably a function of the diene.[48]

5. This example shows that overoxidation of allylic alcohols[48] may occur with DDQ.

6. In a rather unusual reaction oxidation of the PMB ether below with 2 eq. of DDQ affords the ortho ester.[50]

7. When MPM ethers bearing a proximal hydroxyl are treated DDQ acetals are formed.[51,52]

Placing 2 oxidatively removable groups adjacent to each other may not be the best synthetic strategy if they are both to be removed as in the following example where the desired diol could not be produced cleanly.[53]

R = PMB and R = H

Even a bis-PMB ether in a 1,3-relationship has been shown to form the 4-methoxybenzylidine acetal.[54]

An MPM group is readily cleaved in the presence of a 3-MPM.[55]

8. Catalytic DDQ, $FeCl_3$, CH_2Cl_2, H_2O, 62–94% yield.[56]

9. Catalytic DDQ (10%), $Mn(OAc)_3$ (3 eq.), CH_2Cl_2, H_2O, rt, 6–24 h, 61–90% yield.[57]

10. Ozone, acetone, −78°C, 42–82% yield.[58,59] PMB ethers are not stable during the ozonolysis of a monosubstituted alkene.[60]

11. Ceric ammonium nitrate (CAN), Br_2 or NBS, CH_2Cl_2, H_2O, 90% yield.[61] A PMB group is cleaved in preference to a 2-napthylmethyl group under these conditions, and it is also more efficient than when DDQ is used.[40]

Phth = phthalimido

12. Ph_3CBF_4 CH_2Cl_2 or CH_3CN, H_2O.[1,4] In one case the reaction with DDQ failed to go to completion. This was attributed to the reduced electron density on the aromatic ring because of its attachment at the more electron-poor anomeric center.

13. $hv > 280$ nm, H_2O, 1,4-dicyanonaphthalene, 70–81% yield.[63]

14. $Mg(ClO_4)_2$, hv, anthraquinone or dicyanoanthracene.[64] These conditions also cleave the DMPM group.

15. The following examples illustrate unusual and unexpected cleavage processes because of participation by nearby functionality.

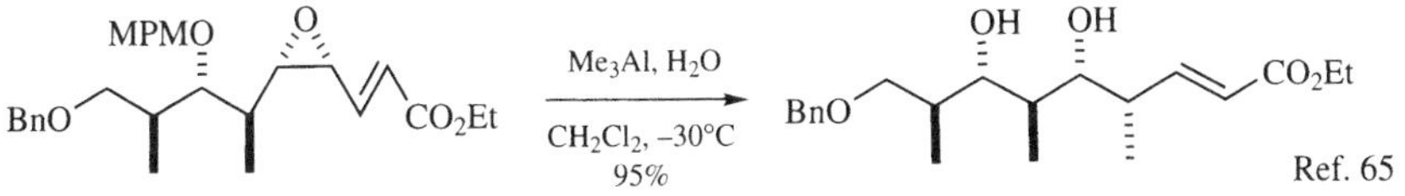

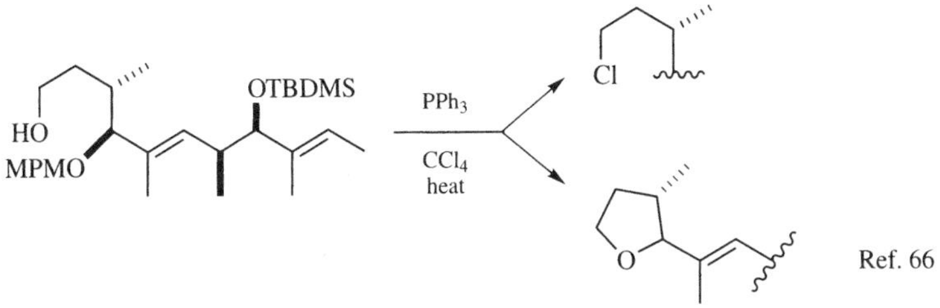

Ref. 65

Ref. 66

16. $MgBr_2 \cdot Et_2O$, Me_2S, CH_2Cl_2, 4–94 h, 75–96% yield.[67] The failure of this substrate to undergo cleavage with DDQ was attributed to the presence of the 1,3-diene. Acetonides and TBDMS ethers were found to be stable.

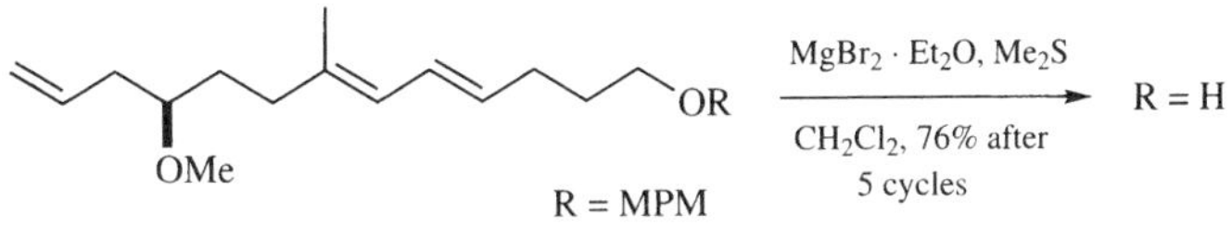

17. $AlCl_3$ or $SnCl_2$, EtSH, CH_2Cl_2, 73–97% yield.[68] Phenolic PMB ethers are also readily cleaved. In some cases the secondary ethers are cleaved faster than the primary PMB ether.[89]

18. $SnCl_4$, PhSH, CH_2Cl_2, −78°C to 50°C, 5 min to 1 h, 88–93% yield. Benzyl, allyl, and TBDMS ethers are stable along with various esters.[69] $BF_3 \cdot Et_2O$ can also be used as a Lewis acid (83% yield).[70]

19. $SnCl_4$ alone is capable of cleaving PMB ethers of carbohydrates with reasonable selectivity. The notable feature of this reaction is that the rate of cleavage of a primary benzyl ether is considerably faster than a secondary benzyl ether. In another example an axial derivative was cleaved faster than an equatorial PMB ether.[71]

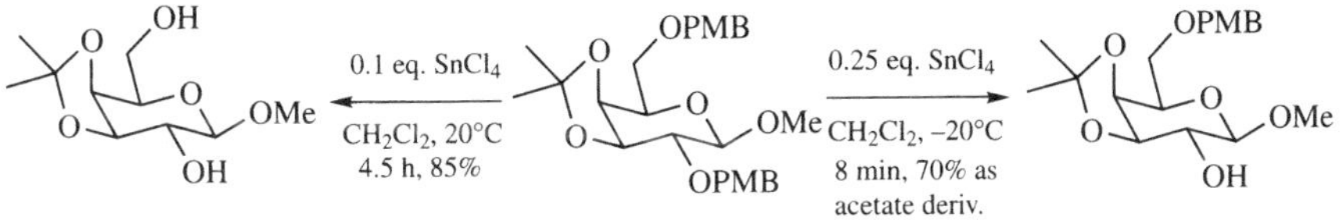

20. $ZrCl_4$ (20 mol %), CH_3CN, rt, 67–92% yield. The groups BOC, Ac, Bz, acetonide, THP, MEM, allyl, prenyl, and Bn were shown to be stable to these conditions, whereas the trityl group is cleaved. PMB esters are also cleaved.[72]

21. During the course of a dithiane-forming reaction, a PMB group was lost, which is consistent with a Lewis acid/thiol deprotection of the PMB group as in item 17.[73]

22. An allylic MPM ether has been converted directly to a bromide upon treatment with Me_2BBr (5 min, $-78°C$).[74] The reagent CBr_4/TPP (CH_2Cl_2, 0–30°C) is more general and converts alkyl, allyl, and benzyl PMB derivatives to bromides in 45–94% yield.[75]

23. BCl_3, dimethyl sulfide.[76,77] These conditions can remove a primary vs. a secondary PMB group.[78]

24. Me_2BBr, CH_2Cl_2, $-78°C$, 5 min, 100% yield.[16]

25. $SnBr_2$, AcBr, CH_2Cl_2, rt, 81–92% yield. These conditions, which also cleave alkyl and aryl benzyl ethers, produce an acetate that must then be hydrolyzed with base to release the alcohol.[79] When $SnCl_2/PhOCH_2COCl$ is used, only MPM ethers are cleaved, leaving benzyl ethers unaffected.

26. $CeCl_3 \cdot 7H_2O$, NaI, CH_3CN, reflux, 75–97% yield. PMB ether is selectively cleaved in the presence of a benzyl ether. TBDMS ethers are also cleaved.[80] Replacing $CeCl_3$ with $Ce(OTf)_3$ is a more efficient reagent for the deprotection of the MPM group (CH_3NO_2, reflux, 61–99% yield). It operates catalytically, but for aryl ethers a scavenging agent must be added to prevent Friedel–Crafts alkylation of the ring.[81] The trityl, THP, TBDPS, and benzyl ethers remain largely unaffected by this reagent.

27. TBDMSOTf, TEA, CH_2Cl_2, rt.[82] These conditions result in conversion of the MPM ether into a TBDMS ether.

28. TMSI, $CHCl_3$, 0.25 h, 25°C.[83]

29. $ClO_2S-N=C=O$, K_2CO_3, CH_2Cl_2, reflux or $-78°C$; NaOH, MeOH, rt, 72–88% yield. PMB ethers can be cleaved in the presence of benzyl ethers, but under more forcing conditions benzyl ethers are cleaved.[84]

30. TMSCl, anisole, $SnCl_2$, CH_2Cl_2, rt, 10–50 min, 78–96% yield.[85]

31. $BF_3 \cdot Et_2O$, $NaCNBH_3$, THF, reflux 4–24 h, 65–98% yield.[86] Functional groups such aryl ketones and nitro compounds are reduced and electron-rich

phenols tend to be alkylated with the released benzyl carbenium ion. The use of BF$_3$·Et$_2$O and triethylsilane as a cation scavenger is also effective.[87]

32. TFA, CH$_2$Cl$_2$, rt, 5–30 min, 84–99% yield.[88] An adamantyl glycoside was stable to these conditions. Secondary carbohydrate PMB ethers are cleaved faster than the primary PMB.[89] The reaction has also been performed in the presence of anisole to scavenge the liberated benzyl carbenium ion.[90] This method is probably preferred for the cleavage of two adjacent PMB ethers since competing benzylidine acetal formation is not a problem.[91]

33. AcOH, 90°C.[92] This method has been used for PMB cleavage in carbohydrates.[93]

34. 1 M HCl EtOH, reflux, 87% yield.[94]

35. 48% aq. HF, CH$_3$CN/CH$_2$Cl$_2$ (1:9), rt, 88% yield.[95] In this case it is possible that the released thiol assists in the cleavage of the PMB similar to the situation in entries 17 and 18.

36. TfOH (0.1 eq.), polymer–PhSO$_2$NH$_2$ or PhSO$_2$NH$_2$, dioxane, 64–98% yield. The benzylidine group interferes with deprotection because of sulfonimine formation. This can be prevented by using an N-methylsulfonamide as the PMB scavenger. Phenolic PMB groups are also readily cleaved, but benzyl groups are completely stable.[96]

37. Clay supported NH$_4$NO$_3$ (clayan), μW, 70–88% yield. The reaction is carried out neat since the use of a solvent resulted in incomplete deprotection.[97]

38. I$_2$, MeOH, reflux, 12–16 h, 75–91% yield. Benzyl ethers are stable to these conditions, but isopropylidenes are cleaved.[98]

39. AgO, HNO$_3$, 74% yield.[99]

40. Pd–C, H$_2$.[100]

41. Na, NH$_3$, 95% yield.[101] This is the method found most successful when DDQ oxidation fails.

42. The following surprising transformation indicates that the PMB ether may not always be such an innocent bystander.[102]

3,4-Dimethoxybenzyl Ether (DMPM−OR or DMP−OR):
3,4-(MeO)$_2$C$_6$H$_3$CH$_2$OR

Formation

1. 3,4-(MeO)$_2$C$_6$H$_3$CH$_2$OC(=NH)CCl$_3$, TsOH.[15] The dimethoxybenzyl ether has also been used for protection of the anomeric hydroxyl in carbohydrates.[103]
2. NaH, 3,4-(MeO)$_2$C$_6$H$_3$CH$_2$Br, DMF.[104]
3. Benzylidine acetals can be reduced selectively to give DMBN ethers.[105]

4. 2-(3,4-Dimethoxybenzyloxy)-3-nitropyridine, CSA, 85% yield.[106]

Cleavage

1. H$_2$, Pd/C, MeOH, >60–98% yield.[107]
2. This ether has properties similar to the *p*-methoxybenzyl (MPM) ether except that it can be removed from an alcohol with DDQ in the presence of an MPM group with 98% selectivity.[34–36] The selectivity is attributed to the lower oxidation potential of the DMPM group; 1.45 V for the DMPM versus 1.78 V for the MPM.

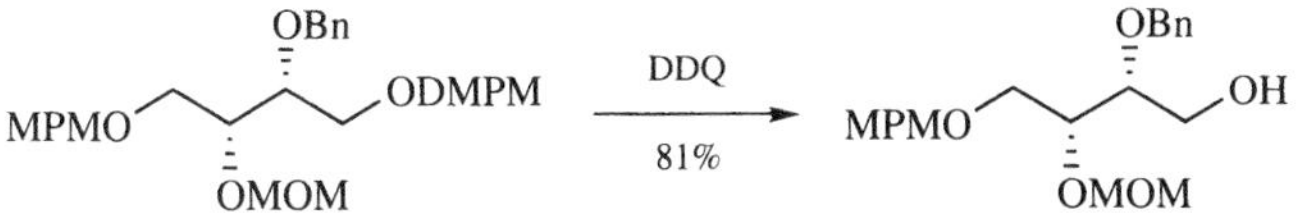

Ref. 34

As has been observed with MPM group, DDQ deprotection of a DMB group failed in the presence of a dienic allylic ether.[108] In the following case the DMB group was used successfully in the presence of allylic diene.[109]

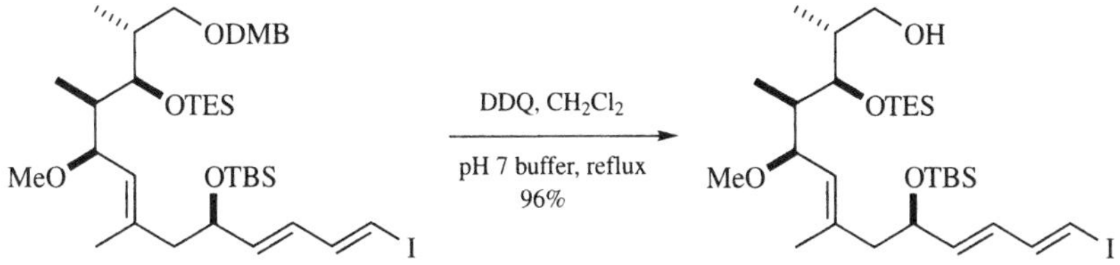

3. In the presence of a neighboring hydroxyl, DDQ cleavage results in the formation of a benzylidine acetal, which, upon extended treatment with DDQ, gives a hydroxy benzoate that can be hydrolyzed with LiOH (DDQ (4.0 eq.), CH_2Cl_2: buffer pH 7.0 (1:1), 0–25°C, 4 h: LiOH 2.0 eq.), MeOH, 25°C, 12 h, 85% over two steps).[110]

2,6-Dimethoxybenzyl Ether (DOB−OR): 2,6-$(MeO)_2C_6H_3CH_2OR$

Cleavage

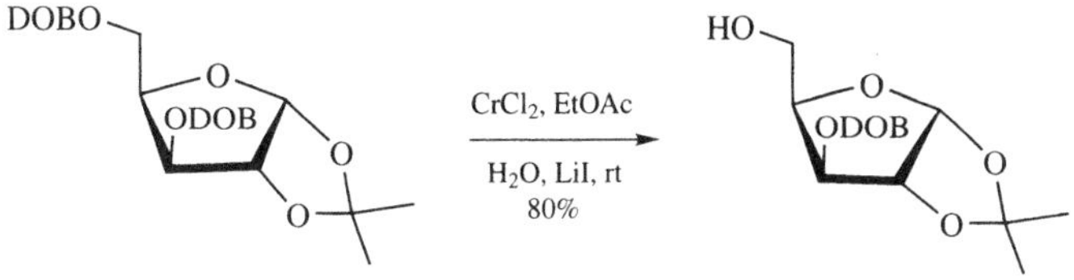

Ref. 111

The relative rates of benzyl either cleavage using these conditions is as follows: PMB > Bn (85%); DMBN > Bn (95%); DOB > Bn (98%); DOB > DMBN (85%). The reagent also does not cleave *N*-benzylamines. Benzyl groups are readily cleaved by hydrogenolysis in the presence of a DOB ether (Pd/C, EtOAc, hexane, 12 h, rt, H_2 (1 atm), >95% yield.[112]

1. L. A. Paquette, M. M. Kreilein, M. W. Bedore, and D. Friedrich, *Org. Lett.*, **7**, 4665 (2005).

2. V. Pozsgay, *Org. Lett.*, **1**, 477 (1999).

3. N. Nakajima, R. Abe, and O. Yonemitsu, *Chem. Pharm. Bull.*, **36**, 4244 (1988).

4. J. L. Marco and J. A. Hueso-Rodriquez, *Tetrahedron Lett.*, **29**, 2459 (1988).

5. H. Takaku and K. Kamaike, *Chem. Lett.*, **11**, 189 (1982).

6. H. Takaku, K. Kamaike, and H. Tsuchiya, *J. Org. Chem.*, **49**, 51 (1984).

7. H. Hoshi, T. Ohnuma, S. Aburaki, M. Konishi, and T. Oki, *Tetrahedron Lett.*, **34**, 1047 (1993).

8. N. Nakajima, T. Hamada, T. Tanaka, Y. Oikawa, and O. Yonemitsu, *J. Am. Chem. Soc.*, **108**, 4645 (1986).

9. E. E. Dueno, F. Chu, S.-I. Kim, and K. W. Jung, *Tetrahedron Lett.*, **40**, 1843 (1999).

10. P. J. Garegg, S. Oscarson, and H. Ritzen, *Carbohydr. Res.*, **181**, 89 (1988).

11. D. R. Mootoo and B. Fraser-Reid, *Tetrahedron*, **46**, 185 (1990).

12. P. Wipf and T. H. Graham, *J. Am. Chem. Soc.*, **126**, 15346 (2004).

13. H. Takaku, S. Ueda, and T. Ito, *Tetrahedron Lett.*, **24**, 5363 (1983); N. Nakajima, K. Horita, R. Abe, and O. Yonemitsu, *Tetrahedron Lett.*, **29**, 4139 (1988).

14. R. D. Walkup, R. R. Kane, P. D. Boatman, Jr., and R. T. Cunningham, *Tetrahedron Lett.*, **31**, 7587 (1990).

15. E. Adams, M. Hiegemann, H. Duddeck, and P. Welzel, *Tetrahedron*, **46**, 5975 (1990).

16. N. Hébert, A. Beck, R. B. Lennox, and G. Just, *J. Org. Chem.*, **57**, 1777 (1992).

17. A. N. Rai and A. Basu, *Tetrahedron Lett.*, **44**, 2267 (2003).

18. K. K. Reddy, M. Saady, and J. R. Falck, *J. Org. Chem.*, **60**, 3385 (1995).

19. N. Nakajima, M. Saito, and M. Ubukata, *Tetrahedron Lett.*, **39**, 5565 (1998).

20. Y. Matsuya, T. Kawaguchi, and H. Nemoto, *Org. Lett.*, **5**, 2939 (2003).

21. S. Hanessian and H. K. Huynh, *Tetrahedron Lett.*, **40**, 671 (1999).

22. A. B. Smith III, I. G. Safonov, and R. M. Corbett, *J. Am. Chem. Soc.*, **123**, 12426 (2001). A. B. Smith III, I. G. Safonov, and R. M. Corbett, *J. Am. Chem. Soc.*, **124**, 11102 (2002).

23. M. Nakano, W. Kikuchi, J.-i. Matsuo, and T. Mukaiyama, *Chem. Lett.*, **30**, 424 (2001).

24. T. Shintou and T. Mukaiyama, *Chem. Lett.*, **32**, 984 (2003).

25. G. V. M. Sharma and A. K. Mahalingam, *J. Org. Chem.*, **64**, 8943 (1999).

26. G. V. M. Sharma, S. Punna, A. Ratnamala, V. D. Kumari, and M. Subrahmanyam, *Org. Prep. & Proc. Int.*, **36**, 581 (2004).

27. C.-C. Wang, J.-C. Lee, S.-Y. Luo, H.-F. Fan, C.-L. Pai, W.-C. Yang, L.-D. Lu, and S.-C. Hung, *Angew. Chem. Int. Ed.*, **41**, 2360 (2002).

28. K. Kamaike, H. Tsuchiya, K. Imai, and H. Takaku, *Tetrahedron*, **42**, 4701 (1986).

29. A. Wei and Y. Kishi, *J. Org. Chem.*, **59**, 88 (1994).

30. S. Naito, M. Escobar, P. R. Kym, S. Liras, and S. F. Martin, *J. Org. Chem.*, **67**, 4200 (2002).

31. U. Schmid and H. Waldmann, *Chem. Eur. J.*, **4**, 494 (1998).

32. P. H. J. Carlsen, *Tetrahedron Lett.*, **39**, 1799 (1998).

33. W. Schmidt and E. Steckhan, *Angew. Chem., Int. Ed. Engl.*, **18**, 801 (1979). See also E. A. Mayeda, L. L. Miller, and J. F. Wolf, *J. Am. Chem. Soc.*, **94**, 6812 (1972); S. M. Weinreb, G. A. Epling, R. Comi, and M. Reitano, *J. Org. Chem.*, **40**, 1356 (1975).

34. K. Horita, T. Yoshioka, T. Tanaka, Y. Oikawa, and O. Yonemitsu, *Tetrahedron*, **42**, 3021 (1986); T. Tanaka, Y. Oikawa, T. Hamada, and O. Yonemitsu, *Tetrahedron Lett.*, **27**, 3651 (1986).

35. Y. Oikawa, T. Tanaka, K. Horita, and O. Yonemitsu, *Tetrahedron Lett.*, **25**, 5397 (1984).

36. Y. Oikawa, T. Yoshioka, and O. Yonemitsu, *Tetrahedron Lett.*, **23**, 885 (1982).

37. K. Horita, S. Nagato, Y. Oikawa, and O. Yonemitsu, *Tetrahedron Lett.*, **28**, 3253 (1987); N. Ikemoto and S. L. Schreiber, *J. Am. Chem. Soc.*, **114**, 2524 (1992).

38. K. Hiruma, T. Kajimoto, G. Weitz-Schmidt, I. Ollmann, and C.-H. Wong, *J. Am. Chem. Soc.*, **118**, 9265 (1996).

39. Y. Hamada, Y. Tanada, F. Yokokawa, and T. Shioiri, *Tetrahedron Lett.*, **32**, 5983 (1991).

40. J. A. Wright, J. Yu, and J. B. Spencer, *Tetrahedron Lett.*, **42**, 4033 (2001).

41. N. Chida, M. Ohtsuka, and S. Ogawa, *J. Org. Chem.*, **58**, 4441, (1993).

42. K. J. Hale and J. Cai, *Tetrahedron Lett.*, **37**, 4233 (1996).

43. H. Hotoda, M. Takahashi, K. Tanzawa, S. Takahashi, and M. Kaneko, *Tetrahedron Lett.*, **36**, 5037 (1995).

44. M. Hutchings, D. Moffat, and N. S. Simpkins, *Synlett*, 661 (2001).

45. I. R. Correa, Jr., and R. A. Pilli, *Angew. Chem. Int. Ed.*, **42**, 3017 (2003).

46. R. M. Conrad, M. J. Grogan, and C. R. Bertozzi, *Org. Lett.*, **4**, 1359 (2002).

47. H. M. Kim, I. J. Kim, and S. J. Danishefsky, *J. Am. Chem. Soc.*, **123**, 35 (2001).

48. S. M. Bauer and R. W. Armstrong, *J. Am. Chem. Soc.*, **121**, 6355 (1999).

49. B. M. Trost and J. Y. L. Chung, *J. Am. Chem. Soc.*, **107**, 4586 (1985).

50. D. A. Evans, D. H. B. Ripin, D. P. Halstead, and K. R. Campos, *J. Am. Chem. Soc.*, **121**, 6816 (1999).

51. R. Stürmer, K. Ritter, and R. W. Hoffmann, *Angew. Chem., Int. Ed. Engl.*, **32**, 101 (1993).

52. S. Hanessian, N. G. Cooke, B. DeHoff, and Y. Sakito, *J. Am. Chem. Soc.*, **112**, 5276 (1990).

53. A. B. Smith III, I. G. Safonov, and R. M. Corbett, *J. Am. Chem. Soc.*, **124**, 11102 (2002).

54. L. A. Paquette, T.-L. Shih, Q. Zeng, and J. E. Hofferberth, *Tetrahedron Lett.*, **40**, 3519 (1999).

55. E. Roulland and M. S. Ermolenko, *Org. Lett.*, **7**, 2225 (2005).

56. S. Chandrasekhar, G. Sumithra, and J. S. Yadav, *Tetrahedron Lett.*, **37**, 1645 (1996).

57. G. V. M. Sharma, B. Lavanya, A. K. Mahalingam, and P. R. Krishna, *Tetrahedron Lett.*, **41**, 10323 (2000).

58. M. Hirama and M. Shimizu, *Synth. Commun.*, **13**, 781 (1983).

59. P. Somfai, *Tetrahedron*, **50**, 11315 (1994).

60. W. Yu, Y. Mei, Y. Kang, Z. Hua, and Z. Jin, *Org. Lett.*, **6**, 3217 (2004).

61. B. Classon, P. J. Garegg, and B. Samuelsson, *Acta Chem. Scand. Ser. B*, **B38**, 419 (1984); R. Johansson and B. Samuelsson, *J. Chem. Soc., Perkin Trans. I*, 2371 (1984).

62. G. I. Georg, P. M. Mashava, E. Akgün, and M. W. Milstead, *Tetrahedron Lett.*, **32**, 3151 (1991); Y. Wang, S. A. Babirad, and Y. Kishi, *J. Org. Chem.*, **57**, 468 (1992).

63. G. Pandey and A. Krishna, *Synth. Commun.*, **18**, 2309 (1988).

64. A. Nishida, S. Oishi, and O. Yonemitsu, *Chem. Pharm. Bull.*, **37**, 2266 (1989).

65. M. Miyashita, M. Hoshino, and A. Yoshikoshi, *J. Org. Chem.*, **56**, 6483 (1991).

66. K. J. Hale, J. Cai, S. Manaviazar, and S. A. Peak, *Tetrahedron Lett.*, **36**, 6965 (1995).

67. T. Onoda, R. Shirai, and S. Iwasaki, *Tetrahedron Lett.*, **38**, 1443 (1997).

68. A. Bouzide and G. Sauvé, *Synlett*, 1153 (1997).

69. W. Yu, M. Su, X. Gao, Z. Yang, and Z. Jin, *Tetrahedron Lett.*, **41**, 4015 (2000).

70. K. C. Nicolaou, H. J. Mitchell, H. Suzuki, R. M. Rodriguez, O. Baudoin, and K. C. Fylaktakidou, *Angew. Chem. Int. Ed.*, **38**, 3334 (1999).

71. K. P. R. Kartha, M. Kiso, A. Hasegawa, and H. J. Jennings, *J. Carbohydr. Chem.*, **17**, 811 (1998).

72. G. V. M. Sharma, C. G. Reddy, and P. R. Krishna, *J. Org. Chem.*, **68**, 4574 (2003).

73. A. Vakalopoulos and H. M. R. Hoffmann, *Org. Lett.*, **3**, 177 (2001).

74. D. G. Hall and P. Deslongchamps, *J. Org. Chem.*, **60**, 7796 (1995).

75. J. S. Yadav and R. K. Mishra, *Tetrahedron Lett.*, **43**, 5419 (2002).

76. M. S. Congreve, E. C. Davison, M. A. M. Fuhry, A. B. Holmes, A. N. Payne, R. A. Robinson, and S. E. Ward, *Synlett*, 663 (1993).

77. J. W. Burton, J. S. Clark, S. Derrer, T. C. Stork, J. G. Bendall, and A. B. Holmes, *J. Am. Chem. Soc.*, **119**, 7483 (1997).

78. I. Paterson and I. Lyothier, *Org. Lett.*, **6**, 4933–4936 (2004).

79. T. Oriyama, M. Kimura, M. Oda, and G. Koga, *Synlett*, 437, (1993).

80. A. Cappa, E. Marcantoni, and E. Torregiani, *J. Org. Chem.*, **64**, 5696 (1999).

81. G. Bartoli, R. Dalpozzo, A. De Nino, L. Maiuolo, M. Nardi, A. Procopio, and A. Tagarelli, *Eur. J. Org. Chem.*, 2176 (2004).

82. T. Oriyama , K. Yatabe, Y. Kawada, and G. Koga, *Synlett*, 45, (1995).

83. D. M. Gordon and S. J. Danishefsky, *J. Am. Chem. Soc.*, **114**, 659 (1992).

84. J. D. Kim, G. Han, O. P. Zee, and Y. H. Jung, *Tetrahedron Lett.*, **44**, 733 (2003).

85. T. Akiyama, H. Shima, and S. Ozaki, *Synlett*, 415 (1992).

86. A. Srikrishna, R. Viswajanani, J. A. Sattigeri, and D. Vijaykumar, *J. Org. Chem.*, **60**, 5961 (1995).

87. Y. Morimoto, M. Iwahashi, K. Nishida, Y. Hayashi, and H. Shirahama, *Angew. Chem., Int. Ed. Engl.*, **35**, 904 (1996).

88. L. Yan and D. Kahne, *Synlett*, 523 (1995).

89. A. Bouzide and G. Sauve, *Tetrahedron Lett.*, **40**, 2883 (1999).

90. E. F. De Medeiros, J. M. Herbert, and R. J. K. Taylor. *J. Chem. Soc., Perkin Trans. 1*, 2725 (1991).

91. D. R. Li, C. Y. Sun, C. Su, G.-Q. Lin, and W.-S. Zhou, *Org. Lett.*, **6**, 4261 (2004).

92. K. J. Hodgetts and T. W. Wallace, *Synth. Commun.*, **24** 1151 (1994).

93. A. K. Misra, I. Mukherjee, B. Mukhopadhyay, and N. Roy, *Ind. J. Chem., Sect. B*, **38B**, 90 (1999).

94. D. J. Jenkins, A. M. Riley, and B. V. L. Potter, *J. Org. Chem.*, **61**, 7719 (1996).

95. A. B. Smith III, G. K. Friestad, J. Barbosa, E. Bertounesque, K. G. Hull, M. Iwashima, Y. Qiu, B. A. Salvatore, P. G. Spoors, and J. J.-W. Duan, *J. Am. Chem. Soc.*, **121**, 10468 (1999).

96. R. J. Hinklin and L. L. Kiessling, *Org. Lett.*, **4**, 113 (2002).

97. J. S. Yadav, H. M. Meshram, G. S. Reddy, and G. Sumithra, *Tetrahedron Lett.*, **39**, 3043 (1998).

98. A. R. Vaino and W. A. Szarek, *Synlett*, 1157 (1995).

99. S. Sisko, J. R. Henry, and S. M. Weinreb, *J. Org. Chem.*, **58**, 4945 (1993).

100. M. Hikota, H. Tone, K. Horita, and O. Yonemitsu, *J. Org. Chem.*, **55**, 7 (1990).

101. K. C. Nicolaou, J.-Y. Xu, S. Kim, T. Ohshima, S. Hosokawa, and J. Pfefferkorn, *J. Am. Chem. Soc.*, **119**, 11353 (1997).

102. L. A. Paquette and Q. Zeng, *Tetrahedron Lett.*, **40**, 3823 (1999).

103. S. J. Danishefsky, H. G. Selnick, R. E. Zelle, and M. P. DeNinno, *J. Am. Chem. Soc.*, **110**, 4368 (1988); A. De Mesmaeker, P. Hoffmann, and B. Ernst, *Tetrahedron Lett.*, **30**, 3773 (1989).

104. H. Takaku, T. Ito, and K. Imai, *Chem. Lett.*, 1005 (1986).

105. K. C. Nicolaou, Y. Li, K. C. Fylaktakidou, H. J. Mitchell, N.-X. Wei, and B. Weyershausen, *Angew. Chem. Int. Ed.*, **40**, 3849 (2001).

106. P. J. Mohr and R. L. Halcomb, *J. Am. Chem. Soc.*, **125**, 1712 (2003).

107. J. W. Lane and R. L. Halcomb, *Org. Lett.*, **5**, 4017 (2003).

108. K. A. Scheidt, T. D. Bannister, A. Tasaka, M. D. Wendt, B. M. Savall, G. J. Fegley, and W. R. Roush, *J. Am. Chem. Soc.*, **124**, 6981 (2002).

109. I. Paterson, R. D. M. Davies, A. C. Heimann, R. Marquez, and A. Meyer, *Org. Lett.*, **5**, 4477 (2003).

110. K. C. Nicolaou, Y. Li, K. C. Fylaktakidou, H. J. Mitchell, and K. Sugita, *Angew. Chem. Int. Ed.*, **40**, 3854 (2001).

111. J. R. Falck, D. K. Barma, S. K. Venkataraman, R. Baati, and C. Mioskowski, *Tetrahedron Lett.*, **43**, 963 (2002).

112. J. R. Falck, D. K. Barma, R. Baati, and C. Mioskowski, *Angew. Chem. Int. Ed.*, **40**, 1281 (2001).

o- and *p*-Nitrobenzyl Ether: *o*- and *p*-$NO_2C_6H_4CH_2OR$ (Chart 1)

The *o*-nitrobenzyl and *p*-nitrobenzyl ethers can be prepared and cleaved by many of the methods described for benzyl ethers.[1] In addition, the *o*-nitrobenzyl ether can be cleaved by irradiation (320 nm, 10 min, quant. yield of carbohydrate[2,3]; 280 nm, 95% yield of nucleotide[4]). This is one of the most important methods for cleavage of this ether. These ethers can also be cleaved oxidatively (DDQ or electrolysis) after reduction to the aniline derivative.[11] Clean reduction to the aniline is accomplished with Zn(Cu) (acetylacetone, rt, >93% yield).[12] Hydrogenolysis is also an effective means for cleavage.[5] A polymeric version of the *o*-nitrobenzyl ether has been prepared for oligosaccharide synthesis that is also conveniently cleaved by photolysis.[6] An unusual selective deprotection of a bis-*o*-nitrobenzyl ether has been observed.[7] The photochemical reaction of *o*-nitrobenzyl derivatives has been reviewed.[8]

A photodeprotection in a highly functionalized environment is illustrated with the deprotection of an intermediate in the synthesis of calicheamicin γ_1.[19]

p-Nitrobenzyl Ether: $p\text{-}NO_2\text{-}C_6H_4CH_2OR$

Formation

1. 4-NO_2BnBr, Ag_2O, CH_2Cl_2, reflux, 5 days, 58–84% yield.[10]
2. The p-nitrobenzyl ether is also prepared from an alcohol and p-nitrobenzyl alcohol (trifluoroacetic anhydride, 2,6-lutidine, CH_2Cl_2, 67% yield) or with the bromide and Ag_2O.[11,12]

Cleavage

Cleavage is generally accomplished by first reducing the nitro group and then removing the p-aminobenzyl ether with acid or oxidatively with DDQ.[11] Thus, conditions that reduce a nitro group should be applicable for the deprotection of this ether. Some of the methods that have been used specifically for the p-nitrobenzyl ether are as follows.

1. In, EtOH, H_2O, NH_4Cl, rt, 81–100% yield. These conditions generally reduce nitro groups.[10,13]
2. Electrolytic reduction (−1.1 V, DMF, R_4NX, 60% yield).[14,15]
3. Reduction with $Na_2S_2O_4$ (pH 8–9, 80–95% yield).[16]
4. Reduction by Zn/AcOH followed by acidolysis.[17]
5. Reduction with In, NH_4Cl, MeOH, IPA, 85°C, 73% yield.[18]

1. D. G. Bartholomew and A. D. Broom, *J. Chem. Soc., Chem. Commun.*, 38 (1975).
2. U. Zehavi, B. Ami, and A. Patchornik, *J. Org. Chem.*, **37**, 2281 (1972); U. Zehavi and A. Patchornik, *J. Org. Chem.*, **37**, 2285 (1972).
3. For reviews of photoremovable protective groups, see V. N. R. Pillai, *Synthesis*, 1 (1980); V. N. R. Pillai, *Org. Photochem.*, **9**, 225 (1987). C. G. Bochet, *J. Chem. Soc., Perkin Trans. 1*, 125 (2002). P. Pelliccioli Anna and J. Wirz, *Photochemical & Photobiological Sciences: Official Journal of the European Photochemistry Association and the European Society for Photobiology*, **1**, 441 (2002); A. Hasan, K.-P. Stengele, H. Giegrich, P. Cornwell, K. R. Isham, R. A. Sachleben, W. Pfleiderer, and R. S. Foote, *Tetrahedron*, **53**, 4247 (1997).
4. E. Ohtsuka, S. Tanaka, and M. Ikehara, *J. Am. Chem. Soc.*, **100**, 8210 (1978).

5. K. Khanbabaee and M. Grober, *Eur. J. Org. Chem.*, 2128 (2003).

6. K. C. Nicolaou, N. Winssinger, J. Pastor, and F. DeRoose, *J. Am. Chem. Soc.*, **119**, 449 (1997).

7. N. Katagiri, M. Makino, and C. Kaneko, *Chem. Pharm. Bull.*, **43**, 884 (1995).

8. Y. L. Chow, in *The Chemistry of Amino, Nitroso and Nitro Compounds and Their Derivatives.* Supplement F, Part 1, S. Patai, Ed., Wiley, New York, 1982. p. 181.

9. R. Groneberg, T. Miyazaki, N. A. Stylianides, T. J. Schulze, W. Stahl, E. P. Schreiner, T. Suzuki, Y. Iwabuchi, A. L. Smith, and K. C. Nicolaou, *J. Am. Chem. Soc.*, **115**, 7593 (1993); A. L. Smith, E. N. Pitsinos, C.-K. Hwang, Y. Mizuno, H. Saimoto, G. R. Scarlato, T. Suzuki, and K. C. Nicolaou, *J. Am. Chem. Soc.*, **115**, 7612 (1993); K. C. Nicolaou, C. W. Hummel, M. Nakada, K. Shibayama, E. N. Pitsinos, H. Saimoto, Y. Mizuno, K.-U. Baldenius, and A. L. Smith, *J. Am. Chem. Soc.*, **115**, 7625 (1993).

10. M. R. Pitts, J. R. Harrison, and C. J. Moody, *J. Chem. Soc., Perkin Trans. 1*, 955 (2001).

11. K. Fukase, H. Tanaka, S. Torii, and S. Kusumoto, *Tetrahedron Lett.*, **31**, 389 (1990).

12. K. Fukase, S. Hase, T. Ikenaka, and S. Kusumoto, *Bull. Chem. Soc. Jpn.*, **65**, 436 (1992).

13. C. J. Moody and M. R. Pitts, *Synlett*, 1575 (1999). M. R. Pitts, J. R. Harrison, and C. J. Moody, *J. Chem. Soc., Perkin Trans. 1*, 955 (2001).

14. V. G. Mairanovsky, *Angew. Chem., Int. Ed. Engl.*, **15**, 281 (1976).

15. K. Fukase, H. Tanaka, S. Torii, and S. Kusumoto, *Tetrahedron Lett.*, **31**, 381 (1990).

16. E. Guibe-Jampel and M. Wakselman, *Synth. Commun.*, **12**, 219 (1982).

17. T. Abiko and H. Sekino, *Chem. Pharm. Bull.*, **38**, 2304 (1990).

18. T. Nakatsuka, Y. Tomimori, Y. Fukuda, and H. Nukaya, *Bioorg. Med. Chem. Lett.*, **14**, 3201 (2002).

Pentadienylnitrobenzyl (PeNB−OR) Ether, Pentadienylnitropiperonyl (PeNP−OR) Ether:

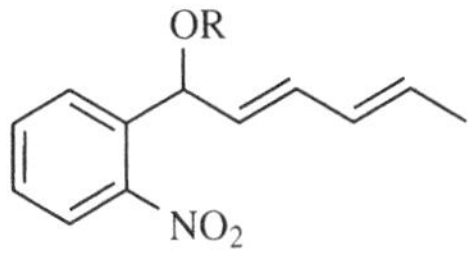

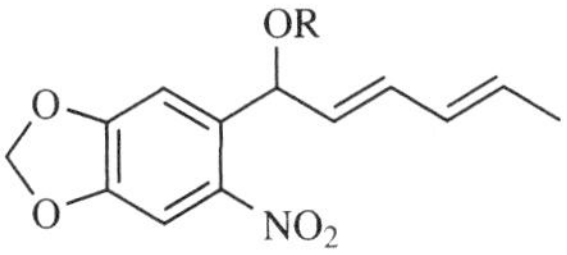

These groups were developed as photochemically cleavable protecting groups for alcohols and acids. They are cleaved by irradiation at 350 nm for 3 h in MeOH. The phenyl ethers required 254-nm irradiation. The photochemical deprotection does not produce a reactive by-product.[1]

1. M. C. Pirrung, Y. R. Lee, K. Park, and J. B. Springer, *J. Org. Chem.*, **64**, 5042 (1999).

Halobenzyl Ethers (X_n-PhCH_2-OR): $X_n-C_6H_{5-n}CH_2OR$

Halobenzyl ethers have been prepared to protect side-chain hydroxyl groups in amino acids. They are more stable to the conditions of acidic hydrolysis (50% CF_3COOH) than the unsubstituted benzyl ether; they are cleaved by HF (0°C, 10 min).[1] Deprotection can also be accomplished with Pearlman's catalyst,[2] Raney nickel W2, Li/NH$_3$,[3] or Na/NH$_3$.[4] These ethers also impart greater crystallinity, which often aids purification.[5] The electron-withdrawing effect can be used to advantage to stabilize the glycosidic bond toward acid[6] and the benzyl ether bond toward electrophilic reagents, as in the following case where the BrBn group (PPB) was used to prevent competition of the ether linkage with the carbonate group for the iodonium intermediate.[7]

The transformation of the PPB group to a more readily cleaved benzyl group has been exploited in carbohydrate synthesis. This transformation is accomplished with 4,4′-di-t-butylbiphenylide (LDBB).[8] Since the 4-ClBn group (PCB) is less reactive to Pd-catalyzed substitution with an amine, the PPB group can be selectively converted to a p-amine derivative which may then be cleaved with SnCl$_4$, dichloroacetic acid, TFA, ZnCl$_2$, TiCl$_4$ or CAN.[9] After derivatization of the alcohol as a propyl ether, the PCB group was removed similarly.

A similar strategy has been used where the PPB group is converted to a biphenyl group that can be removed oxidatively with DDQ. The DMPBn group is oxidatively cleaved at a rate similar to the MPM group, but it has a much greater stability toward acid, which allows cleavage of the PMB in the presence of the DMPBn with ZrCl$_4$ (catalytic, CH$_3$CN, 82% yield).[10]

The PPB group has been converted to a p-hydroxybenzyl group (PHB) that is readily cleaved with DDQ (CH$_2$Cl$_2$, H$_2$O, 97% yield). It has also been converted to a PMB group.[11]

The 2-bromobenzyl ether has been used as a self oxidizing protective group in the synthesis of the CP-225,917 core skeleton.[12] Other methods to oxidize this position all met with failure.

2,6-Dichlorobenzyl Ether: $2,6\text{-}Cl_2C_6H_3CH_2OR$ and

2,4-Dichlorobenzyl Ether: $2,4\text{-}Cl_2C_6H_3CH_2OR$

Formation[13]

The reaction proceeds without the complication of a Payne rearrangement.

Cleavage

This group is cleaved during an iodine-promoted tetrahydrofuran synthesis.[14] The 2,6-dichlorobenzyl ether (DCB) is sufficiently stable to DDQ that an MPM group can readily be cleaved in its presence. The DCB group is cleaved with TMSI generated in situ,[15] but dissolving metal reductions or hydrogenolysis should also cleave this group. The 2,4-dichlorobenzyl group has been cleaved with BCl$_3$ (CH$_2$Cl$_2$, $-78°$C to rt; aq. NaHCO$_3$, 59% yield).[16] The 2,4-dichlorobenzyl group has been used for the protection of a ribofuranosyl derivative. Selective cleavage at the 2-position was achieved with SnCl$_4$ as illustrated.[17]

$R = 2,4\text{-}Cl_2C_6H_3CH_2$

2,6-Difluorobenzyl Ether: $C_6H_3F_2CH_2OR$

This group was developed to prevent participation of the BnO bond during cationic reactions. It is formed from the bromide [C$_6$H$_3$F$_2$CH$_2$Br, Ba(OH)$_2$·8H$_2$O, DMF, 25 h, 94% yield][18] and cleaved by dissolving metal reduction (Ca, NH$_3$, 79% yield).[19] Hydrogenolysis, the process commonly used to cleave benzyl groups, is expected to be sluggish in comparison to the unsubstituted benzyl group.

1. D. Yamashiro, J. Org. Chem., **42**, 523 (1977).

2. J. D. White and J. D. Hansen, J. Org. Chem., **70**, 1963 (2005).

3. B. K. Goering, K. Lee, B. An, and J. K. Cha, J. Org. Chem., **58**, 1100 (1993).

4. J. D. White, P. R. Blakemore, N. J. Green, E. B. Hauser, M. A. Holoboski, L. E. Keown, C. S. N. Kolz, and B. W. Phillips, J. Org. Chem., **67**, 7750 (2002).

5. S. Koto, S. Inada, N. Morishima, and S. Zen, Carbohydr. Res., **87**, 294 (1980).

6. N. L. Pohl and L. L. Kiessling, Tetrahedron Lett., **38**, 6985 (1997).

7. A. B. Smith, III, L. Zhuang, C. S. Brook, Q. Lin, W. H. Moser, R. E. L. Trout, and A. M. Boldi, Tetrahedron Lett., **38**, 8671 (1997).

8. K. Fujiwara, A. Goto, D. Sato, Y. Ohtaniuchi, H. Tanaka, A. Murai, H. Kawai, and T. Suzuki, Tetrahedron Lett., **45**, 7011 (2004).

9. O. J. Plante, S. L. Buchwald, and P. H. Seeberger, J. Am. Chem. Soc., **122**, 7148 (2000)

10. X. Liu and P. H. Seeberger, Chem. Commun., 1708 (2004).

11. K. Fujiwara, Y. Koyama, K. Kawai, H. Tanaka, and A. Murai, Synlett, 1835 (2002).

12. K. C. Nicolaou, Y. He, K. C. Fong, W. H. Yoon, H.-S. Choi, Y.-L. Zhong, and P. S. Baran, Org. Lett., **1**, 63 (1999).

13. S. Hatakeyama, K. Sakurai, and S. Takano, Heterocycles, **24**, 633 (1986).

14. S. D. Rychnovsky and P. A. Bartlett, J. Am. Chem. Soc., **103**, 3963 (1981).

15. J. D. White, G. Wang, and L. Quaranta, Org. Lett., **5**, 4109 (2003).

16. A. M. Kawasaki, M. D. Casper, T. P. Prakash, S. Manalili, H. Sasmor, M. Manoharan, and P. D. Cook, *Tetrahedron Lett.*, **40**, 661 (1999).

17. P. Martia, *Helv. Chim. Acta*, **78**, 486 (1995).

18. R. Bürli and A. Vasella, *Helv. Chim. Acta*, **79**, 1159 (1996).

19. H. J. Borschberg, *Chimia*, **45**, 329 (1991).

p-Cyanobenzyl Ether: *p*-CN-C$_6$H$_4$CH$_2$OR

The *p*-cyanobenzyl ether, prepared from an alcohol and the benzyl bromide in the presence of sodium hydride (74% yield), can be cleaved by electrolytic reduction (-2.1 V, 71% yield)[1] or with Et$_3$GeNa, dioxane, HMPA, 50°C.[2] It is stable to electrolytic removal (-1.4 V) of a tritylone ether [i.e., 9-(9-phenyl-10-oxo)anthryl ether].

1. C. van der Stouwe and H. J. Schäfer, *Tetrahedron Lett.*, **20**, 2643 (1979); *idem, Chem. Ber.*, **114**, 946 (1981); J. P. Coleman, Naser-ud-din, H. G. Gilde, J. H. P. Utley, B. C. L. Weedon, and L. Eberson, *J.Chem. Soc., Perkin Trans. II*, 1903 (1973).

2. Y. Yokoyama, S. Takizawa, M. Nanjo, and K. Mochida, *Chem. Lett.*, **33**, 1032 (2004).

Fluorous Benzyl Ether (BnfOR): (C$_6$F$_{13}$CH$_2$CH$_2$)$_3$SiC$_6$H$_4$CH$_2$OR

The fluorous benzyl ether was prepared to take advantage of the fluorous synthesis technique. The Bnf ether is prepared using the conventional method: NaH, DMF, benzotrifluoride, TBAI. It is cleaved by hydrogenolysis: Pd(OH)$_2$, H$_2$, FC72.[1]

4-Fluorousalkoxybenzyl Ether: CF$_3$(CF$_2$)$_n$CH$_2$CH$_2$CH$_2$OC$_6$H$_4$CH$_2$OR, $n = 1,3,5,7$

This group was used to prepare a family of murisolin isomers that could be separated by fluorous chromatography.[2] As with the MPM, this group is cleaved using DDQ.

1. D. P. Curran, R. Ferritto, and Y. Hua, *Tetrahedron Lett.*, **39**, 4937 (1998).

2. C. S. Wilcox, V. Gudipati, H. Lu, S. Turkyilmaz, and D. P. Curran, *Angew. Chem. Int. Ed.*, **44**, 6938 (2005).

Trimethylsilylxylyl (TIX) Ether: TMSCH$_2$C$_6$H$_4$CH$_2$OR

The TIX group is not stable to TBAF or CsF because these reagents remove the silyl group, leaving a 4-methylbenzylether, but it is stable to HF·pyridine, BF$_3$·Et$_2$O, ZnCl$_2$, MgBr$_2$·DMS, LiBF$_4$ (CH$_3$CN, reflux), CeCl$_3$·7H$_2$O and NaI, CH$_3$CN, reflux.[1]

Formation

1. TMSCH$_2$C$_6$H$_4$CH$_2$OC(=NH)CCl$_3$, CH$_2$Cl$_2$, Sc(OTf)$_3$, 0°C to rt, 15 min, 78–95% yield.

2. TMSCH$_2$C$_6$H$_4$CH$_2$Br, NaH, THF, rt, 2–5 h, 78–87% yield.

Cleavage

1. TFA, CH_2Cl_2, 0°C, 0.25 h, 52% yield. It is stable to 5% TFA/CH_2Cl_2.
2. CAN, THF, H_2O, rt, 0.5 h, 62% .
3. DDQ, CH_2Cl_2, H_2O, rt, 15–60 min, 71–93% yield. At −10°C the TIX group is selectively cleaved over the PMB group in 74% yield and the PMB group is selectively cleaved over the TIX group with $ZrCl_4$ in 95% yield.

1. C. R. Reddy, A. G. Chittiboyina, R. Kache, J.-C. Jung, E. B. Watkins, and M. A. Avery, *Tetrahedron*, **61**, 1289 (2005).

p-Phenylbenzyl Ether: p-C_6H_5-$C_6H_4CH_2OR$

The section on the formation of benzyl ethers should be consulted. Such biphenyl-methyl ethers have also been prepared using a Suzuki coupling with a 4-bromobenzyl ether.[1] *p*-Phenylbenzyl ethers are more stable to acid than the PMB ethers (60°C in aq. AcOH or TFA, CH_2Cl_2, rt, several hours) [2]

Formation

1. PhBnBr, NaH, THF, 0°C, 24 h, 63% yield.[2]
2. PhBnOC(=NH)CCl₃, TfOH, CH_2Cl_2, rt, 4 h.[2]

Cleavage

1. $FeCl_3$, CH_2Cl_2, 2–3 min, 68% yield.[3] Benzyl ethers are cleaved in 15–20 min under these conditions. Methyl glycosides, acetates, and benzoates were not affected by this reagent.
2. $CrCl_2$, LiI, EtOAc, H_2O, 92% yield.[4]
3. DDQ, $Mn(OAc)_3$, CH_2Cl_2, 63–86% yield.[2]
4. Pd/C, H_2, EtOAc, >52% yield.[5] The *p*-phenylbenzyl ether is more easily cleaved by hydrogenolysis than normal benzyl ethers. This was used to great advantage in the deprotection of the vineomycinone intermediate shown below. The use of the *p*-methoxybenzyl ether proved unsuccessful in this application because it could not be removed by either hydrogenolysis or oxidatively with DDQ.

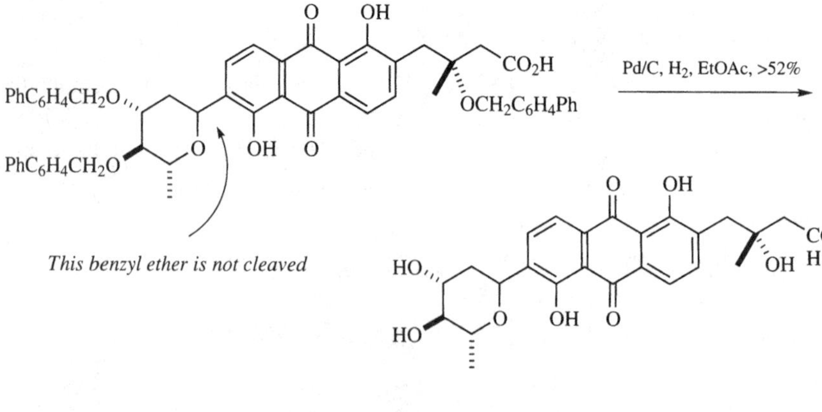

This benzyl ether is not cleaved

1. See the section on halobenzyl ethers.

2. G. V. M. Sharma and Rakesh, *Tetrahedron Lett.*, **42**, 5571 (2001).

3. M. H. Park, R. Takeda, and K. Nakanishi, *Tetrahedron Lett.*, **28**, 3823 (1987).

4. J. R. Falck, D. K. Barma, R. Baati, and C. Mioskowski, *Angew. Chem. Int. Ed.*, **40**, 1281 (2001).

5. V. Bollitt, C. Mioskowski, R. O. Kollah, S. Manna, D. Rajapaksa, and J. R. Falck, *J. Am. Chem. Soc.*, **113**, 6320 (1991).

2-Phenyl-2-propyl Ether (Pp−OR, Cumyl−OR): $C_6H_5C(CH_3)_2OR$

Formation

1. $PhCMe_2OH$, $BiBr_3$, CCl_4, 90–95% yield.[1]

2. $PhCMe_2OH$, dodecylbenzenesulfonic acid, H_2O, 83% yield.[2]

Cleavage

1. H_2, Pd/C, cat. $CHCl_3$, AcOEt, 94–97% yield.[3]

2. Ammonium formate, Pd–C, EtOH, 50°C, 2 h.

3. Na, NH_3, THF, 83% yield.[4]

4. 10% HCl, dioxane (1:1), rt 12 h, 87% yield.[4]

5. 50% TFA, CH_2Cl_2, rt. Benzyl ethers are stable to these conditions.

1. B. Boyer, E.-M. Keramane, J.-P. Roque, and A. A. Pavia, *Tetrahedron Lett.*, **41**, 2891 (2000).

2. S. Kobayashi, S. Iimura, and K. Manabe, *Chem. Lett.*, **31**, 10 (2002).

3. R. Muto and K. Ogasawara, *Tetrahedron Lett.*, **42**, 4143 (2001).

4. H. Nakashima, M Sato, T. Taniguchi, and K. Ogasawara, *Synlett*, 1754 (1999).

p-Acylaminobenzyl Ethers (PAB−OR): *p*-R′CONH-$C_6H_4CH_2OR$

The pivaloylamidobenzyl group was stable to acetic acid–water–90°C, MeOH–NaOMe, and iridium-induced allyl isomerization, as well as to many of the Lewis acids used in glycosylation.[1]

Formation

1. *p*-PvNH-$C_6H_4CH_2Cl$, $Ba(OH)_2$, BaO, DMF, 32 h, 58–99% yield.[1]

2. *p*-PvNH-$C_6H_4CH_2OC(=NH)CCl_3$, TfOH, CH_2Cl_2, 1.5 h, 82% yield.[1]

3. *p*-Acetamidobenzyl ether from a *p*-nitrobenzyl ether: Zn(Cu), acetylacetone; Ac_2O, 93% yield.[2]

4. *p*-Acetamidobenzyl ether from a *p*-nitrobenzyl ether: Pd black, H_2, HCO_2NH_4, or cyclohexadiene: Ac_2O, pyridine.[3]

Cleavage

1. DDQ oxidation.[1,2] Cleavage occurs selectively in the presence of a benzyl and *p*-nitrobenzyl group.
2. Hydrogenolysis.[1]

1. K. Fukase, T. Yoshimura, M. Hashida, and S. Kusumoto, *Tetrahedron Lett.*, **32**, 4019 (1991).
2. K. Fukase, S. Hase, T. Ikenaka, and S. Kusumoto, *Bull. Chem. Soc. Jpn.*, **65**, 436 (1992).
3. K. Fukase, H. Tanaka, S. Torii, and S. Kusumoto, *Tetrahedron Lett.*, **31**, 389 (1990).

p-Azidobenzyl Ether (Azb−OR): 4-N$_3$C$_6$H$_4$CH$_2$OR

This benzyl ether is partially stable to BF$_3$·Et$_2$O as used in glycosylation reactions and NaOMe, but it is not stable to TFA at rt for 30 min.

Formation

    p-N$_3$−C$_6$H$_4$CH$_2$Br, NaH, DMF, 92–98% yield.[1] The benzyl chloride may also be used.[2]

Cleavage[1]

1. H$_2$, Pd–C, 2. PPh$_3$, 3. DDQ, −5°C.
2. DDQ, rt, 90% yield. The reaction is slow.
3. PPh$_3$ then DDQ, 92% yield.

4-Azido-3-chlorobenzyl Ether (ClAzb−OR): 4-N$_3$-3-Cl-C$_6$H$_3$CH$_2$OR

The 3-chloro derivative was developed to impart greater acid stability to the azido-benzyl ether. It is formed using the benzyl bromide (NaH, DMF) and is much more stable to BF$_3$·Et$_2$O, but it is cleaved in neat TFA. Conditions used to cleave the azidobenzyl ether also cleave the 4-azido-3-chlorobenzyl ether (Ph$_3$P, THF; DDQ, H$_2$O, AcOH, rt, 1 h, 75% yield).[3,4] The ClAzb ether is inert to DDQ oxidation.[4]

1. K. Fukase, M. Hashida, and S. Kusumoto, *Tetrahedron Lett.*, **32**, 3557 (1991).
2. J. Sun, X. Han, and B. Yu, *Synlett* 437 (2005).
3. K. Egusa, K. Fukase, and S. Kusumoto, *Synlett*, 675 (1997).
4. K. Egusa, K. Fukase, Y. Nakai, and S. Kusumoto, *Synlett*, 27 (2000).

2- and 4-Trifluoromethylbenzyl Ethers: (2-, 4-CF$_3$-PhCH$_2$−OR), 2-, 4-CF$_3$-C$_6$H$_4$CH$_2$OR

The TfBn ethers are prepared by the standard method (NaH, DMF, CF$_3$BnX, 94–100%). They are oxidatively quite stable to NBS-promoted conversion of a

4,6-benzylidinepyranoside to the 6-bromo-4-benzoate. It can be quantitatively cleaved by simple hydrogenolysis with Pd–C and H$_2$.[1,2] It is completely stable to conditions used to deprotect a benzyl ether with DDQ.[3,4]

1. L. J. Liotta, K. L. Dombi, S. A. Kelley, S. Targontsidis, and A. M. Morin, *Tetrahedron Lett.*, **38**, 7833 (1997).
2. V. S. Kumar, D. L. Aubele, and P. E. Floreancig, *Org. Lett.*, **4**, 2489 (2002).
3. Y. Sakai, M. Oikawa, H. Yoshizaki, T. Ogawa, Y. Suda, K. Fukase, and S. Kusumoto, *Tetrahedron Lett.*, **41**, 6843 (2000).
4. H. Yoshizaki, N. Fukuda, K. Sato, M. Oikawa, K. Fukase, Y. Suda, and S. Kusumoto, *Angew. Chem. Int. Ed.*, **40**, 1475 (2001).

p-(Methylsulfinyl)benzyl Ether (Msib−OR):   *p*-(MeS(O))C$_6$H$_4$CH$_2$OR

Formation

CH$_3$S(O)C$_6$H$_4$CH$_2$Br, NaH.[1]

Cleavage

The cleavage of this group proceeds by initial reduction of the sulfoxide, which then makes the resulting methylthiobenzyl ether labile to trifluoroacetic acid. Thus, any method used to reduce a sulfoxide could be used to activate this group for deprotection.

1. SiCl$_4$, thioanisole, anisole, TFA, CH$_2$Cl$_2$, 25°C, 24 h, 82% yield.[2]
2. DMF·SO$_3$, ethanedithiol, rt, 36 h; 90% aq. TFA, 2-methylindole.[3]

1. S. Futaki, T. Yagami, T. Taike, T. Akita, and K. Kitagawa, *J. Chem. Soc., Perkin Trans. 1*, 653 (1990); Y. Kiso, S. Tanaka, T. Kimura, H. Itoh, and K. Akaji, *Chem. Pharm. Bull.*, **39**, 3097 (1991); S. Futaki, T. Taike, T. Akita, and K. Kitagawa, *J. Chem. Soc., Chem. Commun.*, 523 (1990).
2. Y. Kiso, T. Fukui, S. Tanaka, T. Kimura, and K. Akaji, *Tetrahedron Lett.*, **35**, 3571 (1994).
3. S. Futaki, T. Taike, T. Akita, and K. Kitagawa, *Tetrahedron*, **48**, 8899 (1992).

p-Siletanylbenzyl (PSB) Ether

The PSB ether can be prepared from the alcohol using the Mitsunobu reaction, from the bromide with base (K$_2$CO$_3$, TBAI, Cs$_2$CO$_3$ or NaH, DMF) or from the bromide with Ag_2O (CH_2Cl_2) in 38–96% yield. It is cleaved by oxidative removal of the silane (K_2CO_3, $KF \cdot H_2O$, 30% aq. H_2O_2, THF, MeOH, TBAF, t-BuOOH, DMF, 70°C) to form a 4-hydroxybenzyl ether, which is cleaved with base (85–99% yield). Alternatively, hydrogenolysis with Pd–C is also effective (88% yield). The PSB group is orthogonal to the MPM group in that it is stable to DDQ.[1]

1. H. Lam, S. E. House, and G. B. Dudley, *Tetrahedron Lett.*, **46**, 3283 (2005).

4-Acetoxybenzyl Ethers (PAB−OR): 4-AcOC₆H₄CH₂OR

4-(2-Trimethylsilyl)ethoxymethoxybenzyl Ether: 4-SEMOC₆H₄CH₂OR

These benzyl ethers were prepared to facilitate oligosaccharide synthesis. The PAB ether is introduced using either the trichloroacetamidate (TfOH, CH_2Cl_2, 67%) technology or from the bromide (AgOTf, CH_2Cl_2/hexane, 78%). Cleavage is effected by first hydrolyzing the acetate and then oxidatively cleaving the PHB group with either DDQ (CH_2Cl_2, 30 min, >95%), $FeCl_3$ (Et_2O, 5 min, 0°C, >95%), iodobenzene diacetate (CH_2Cl_2, 2 h, 20°C, 90%), or Ag_2CO_3/celite (CH_2Cl_2, 18 h, 20°C, 80%). The PHB group is also cleaved through a quinone methide with NaOMe/MeOH at 60°C (>95%). A PMB group can be cleaved in the presence of a PAB group with DDQ because the acetate is more electron-withdrawing than the methyl ether.[1]

The SEMOBn group is introduced with the bromide (NaH, DMF, 75%), and it is cleaved with fluoride (TBAF, DMF, 80°C, 48 h, 90%).[1] Other methods used to cleave SEM ethers should show similar effectiveness. Oxidative methods used to cleave the PMB group should also be applicable to this group.

1. L. Jobron and O. Hindsgaul, *J. Am. Chem. Soc.*, **121**, 5835 (1999).

2-Napthylmethyl Ether (Nap−OR): C₁₀H₇-2-CH₂OR

The 2-napthylmethyl group like the PMB group can be cleaved oxidatively or by hydrogenolysis, but it has the advantage that it is more acid stable than the PMB ether[1] and thus can resist conditions used to remove the isopropylidene group.[2]

Formation

The section on the formation of the benzyl group should be consulted since many of those methods should be applicable to the Nap group.

1. NapBr, NaH, DMF, 0°C to rt, 78% yield.[2] KH in THF has also been used.[3]

2.

3.

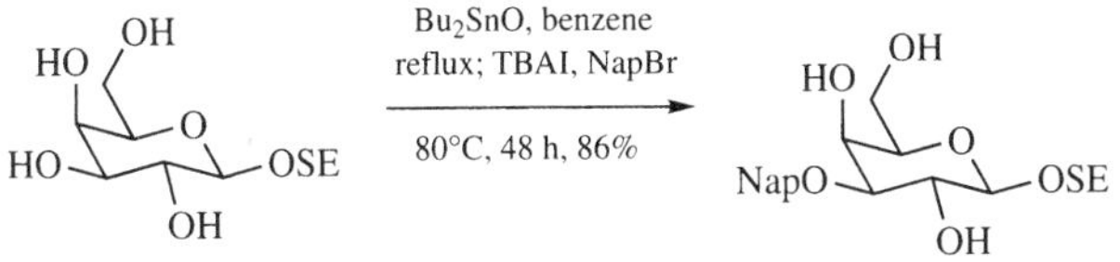

Bu$_2$SnO, benzene
reflux; TBAI, NapBr

80°C, 48 h, 86%

Ref. 5

4.

Bu$_2$SnO, benzene
reflux; TBABr, NapBr

80°C, 48 h, 69%

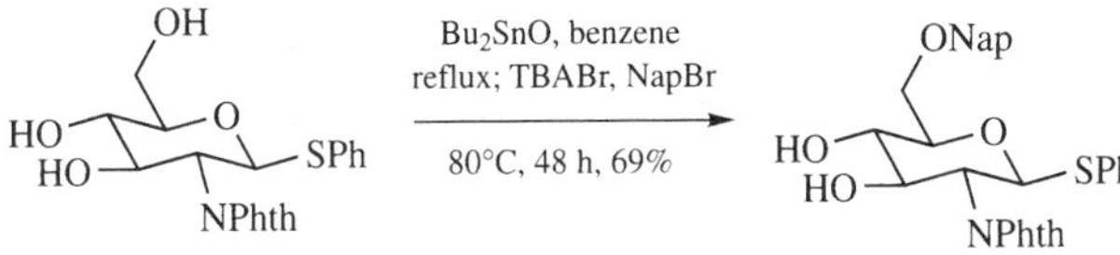

Ref. 6

Cleavage

1. DDQ, CH$_2$Cl$_2$, H$_2$O, rt 24 h, 58–80% yield.[3,6] Allylic ethers such as those in Ciguatoxin CTX3C, which are sometimes oxidized with DDQ, survived. In the presence of an adjacent hydroxyl the acetal can form as a by-product. This is the only product when using pure acetonitrile as the solvent.[7]

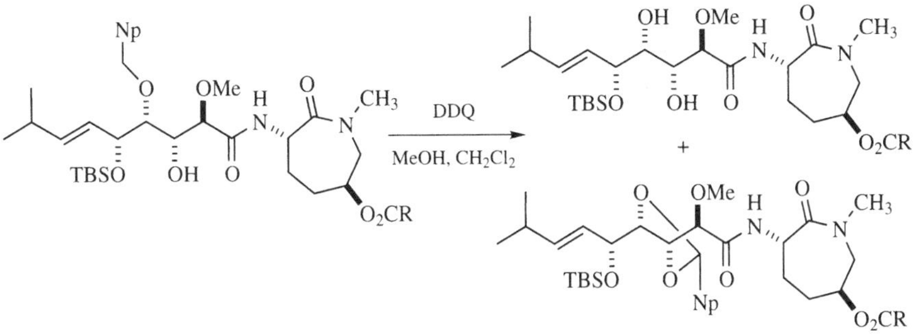

2. CAN, CH$_3$CN, H$_2$O, rt, 48 h, 65% yield.[3]
3. TFA, CH$_2$Cl$_2$, >1 h.[3]
4. Pd–C, EtOH, 96% yield. Hydrogenolysis of some common benyl groups occurs in the following order: NapOR > BnOR > PMPOR. The 2-methylnapthalene released during hydrogenolysis of the Nap group inhibits hydrogenolysis of the Bn group.[8] This may prove useful as a catalyst moderator.
5. Transfer hydrogenation: Pd–C, 1-methyl-1,4-cyclohexadiene, CaCO$_3$, EtOH, 98% yield. A disubstituted olefin survives these conditions.[9]

1. M. Inoue, H. Uehara, M. Maruyama, and M. Hirama, *Org. Lett.*, **4**, 4551 (2002).
2. M. Csavas, A. Borbás, L. Szilagyi, and A. Lipták, *Synlett*, 887 (2002); Z. B. Szabó, A. Borbas, I. Bajza, and A. Lipták, *Tetrahedron: Asymmetry*, **16**, 83 (2005).
3. H. Fuwa, S. Fujikawa, K. Tachibana, H. Takakura, and M. Sasaki, *Tetrahedron Lett.*, **45**, 4795, (2004).
4. C.-C. Wang, J.-C. Lee, S.-Y. Luo, H.-F. Fan, C.-L. Pai, W.-C. Yang, L.-D. Lu, and S.-C. Hung, *Angew. Chem. Int. Ed.*, **41**, 2360 (2002).

5. J. Xia, J. L. Alderfer, R. D. Locke, C. F. Piskorz, and K. L. Matta, *J. Org. Chem.*, **68**, 2752 (2003).

6. J. Xia, C. F. Piskorz, J. L. Alderfer, R. D. Locke, and K. L. Matta, *Tetrahedron Lett.*, **41**, 2773 (2000). J. Xia, J. L. Alderfer, C. F. Piskorz, and K. L. Matta, *Chem. Eur. J.*, **6**, 3442 (2000).

7. R. K. Boeckman, Jr., T. J. Clark, and B. C. Shook, *Helv. Chim. Acta*, **85**, 4532 (2002).

8. M. J. Gaunt, J. Yu, and J. B. Spencer, *J. Org. Chem.*, **63**, 4172 (1998).

9. A. B. Smith III, V. A. Doughty, C. Sfouggatakis, C. S. Bennett, J. Koyanagi, and M. Takeuchi, *Org. Lett.*, **4**, 783 (2002).

2- and 4-Picolyl Ether: $C_5H_4NCH_2OR$

Picolyl ethers are prepared from their chlorides by a Williamson ether synthesis (68–83% yield). Some selectivity for primary vs. secondary alcohols can be achieved (ratios = 4.3–4.6:1). They are cleaved electrolytically (-1.4 V, 0.5 M HBF$_4$, MeOH, 70% yield). Since picolyl chlorides are unstable as the free base, they must be generated from the hydrochloride prior to use.[1] These derivatives are relatively stable to acid (CF$_3$CO$_2$H, HF/anisole). Additionally, cleavage can be affected by hydrogenolysis in acetic acid.[2] The 2-picolyl ether was also found to be a participating group for the selective formation of 1,2-*trans* glycosides by participation of the nitrogen at the anomeric carbon.[3]

1. S. Wieditz and H. J. Schaefer, *Acta Chem. Scand. Ser. B.*, **B37**, 475 (1983); A. Gosden, R. Macrae, and G. T. Young, *J. Chem. Res., Synop.*, 22 (1977).

2. J. Rizo, F. Albericio, G. Romero, C. G. Esheverria, J. Claret, C. Muller, E. Giralt, and E. Pedroso, *J. Org. Chem.*, **53**, 5386 (1988).

3. J. T. Smoot, P. Pornsuriyasak, and A. V. Demchenko, *Angew. Chem. Int. Ed.*, **44**, 7123 (2005).

3-Methyl-2-picolyl *N*-Oxido Ether

The authors prepared a number of substituted 2-diazomethylene derivatives of picolyl oxide to use for monoprotection of the *cis*-glycol system in nucleosides. The 3-methyl derivative proved most satisfactory.[1]

Formation/Cleavage[1]

Ac$_2$O and BzCl/NaOH have been used to cleave this ether.[2]

1. Y. Mizuno, T. Endo, and K. Ikeda, *J. Org. Chem.*, **40**, 1385 (1975); Y. Mizuno, T. Endo and T. Nakamura, *J. Org. Chem.*, **40**, 1391 (1975).
2. Y. Mizuno, K. Ikeda, T. Endo, and K. Tsuchida, *Heterocycles*, **7**, 1189 (1977).

2-Quinolinylmethyl Ether (Qn−OR)

Formation[1,2]

Cleavage

1. CuCl$_2$·2H$_2$O, DMF, H$_2$O, air, 65°C, 56–80% yield.[1]
2. *h*ν, 61–85% yield.[2] In this case, cleavage results in simultaneous oxidation of the initially protected alcohol to give a ketone. The related 6-phenanthridinyl-methyl ethers similarly give ketones upon photochemical deprotection.[3]

1. L. Usypchuk and Y. Leblanc, *J. Org. Chem.*, **55**, 5344 (1990).
2. V. Rukachaisirikul, U. Koert, and R. W. Hoffmann, *Tetrahedron*, **48**, 4533 (1992).
3. V. Rukachaisirikul and R. W. Hoffmann, *Tetrahedron*, **48**, 10563 (1992).

6-Methoxy-2-(4-methylphenyl)-4-quinolinemethyl Ether

The ethers are formed by a Williamson ether synthesis (ROH, NaOH, DMF, 3 h, 70–93% yield) and are cleaved by photolysis at 350 nm in the presence of the radical scavengers sorbitol or dodecane thiol (IPA, 30–1440 min, 25–93% yield.[1]

1. G. A. Epling and A. A. Provatas, *Chem. Commun.*, 1036 (2002).

1-Pyrenylmethyl Ether

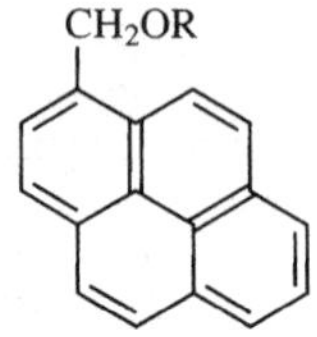

This is a fluorescent benzyl ether used for 2′-protection in nucleotide synthesis. It is introduced using 1-pyrenylmethyl chloride (KOH, benzene, dioxane, reflux, 2 h, >65% yield).[1] Most methods used for benzyl ether cleavage should be applicable to this ether.

1. K. Yamana, Y. Ohashi, K. Nunota, M. Kitamura, H. Nakano, O. Sangen, and T. Shimdzu, *Tetrahedron Lett.*, **32**, 6347 (1991).

Diphenylmethyl Ether (DPM−OR): Ph$_2$CHOR

Formation

1. (Ph$_2$CHO)$_3$PO, cat. CF$_3$COOH, CH$_2$Cl$_2$, reflux, 4–9 h, 65–92% yield.[1] This methodology has been applied to the protection of amino acid alcohols.[2]
2. Ph$_2$CHOH, concd. H$_2$SO$_4$, 12 h, 70% yield.[3] Acid-washed 4Å molecular sieves (52–86% yield),[4] Nafion H (35–92% yield),[5] Yb(OTf)$_3$, FeCl$_3$ (60–92% yield),[6,7] have been used as catalysts.
3. Ph$_2$CN$_2$, CH$_3$CN or benzene, 79–85% yield.[8]
4. Ph$_2$CHOC(=NH)CCl$_3$. TMSOTf, CH$_2$Cl$_2$, rt. 65–92% yield.[9] The 9-fluorenyl group is prepared similarly in 56–91% yield.
5. THP and silyl ethers can be converted directly to DPM ethers: Ph$_2$CHO$_2$CH, TMSOTf, silica gel, CH$_3$CN, 1 h, 74–94% yield.[10]

Cleavage

1. Pd–C, AlCl$_3$, cyclohexene, reflux, 24 h, 91% yield.[11] Simple hydrogenation also cleaves this ether (71–100% yield).[11]
2. Electrolytic reduction: −3.0 V, DMF, R$_4$NX.[3]

3. 10% CF_3COOH, anisole, CH_2Cl_2.[2] Anisole is present to scavenge the diphenylmethyl cation liberated during the cleavage reaction.

4. $TiCl_4$, low temperature, 77% yield.[12]

5. Aqueous HCl, THF, rt, 91% yield.[13]

4-Methoxydiphenylmethyl Ether (MDPM−OR): $(4\text{-}CH_3OC_6H_4)C_6H_5CH\text{-}OR$

4-Phenyldiphenylmethyl Ether (PPDPM−OR): $(4\text{-}C_6H_4C_6H_4)C_6H_5CH\text{-}OR$

Formation

$(4\text{-}CH_3OC_6H_4)C_6H_5CHOH$ or $(4\text{-}C_6H_4C_6H_4)C_6H_5CHOR$, $Yb(OTf)_3$, CH_2Cl_2, 59–84% yield.[14]

Cleavage

1. DDQ, CH_2Cl_2, rt, 72–84% yield. Both the MDPM ether and the PPDPM ether are cleaved by this method.

2. TFA, CH_2Cl_2, rt. This method only works for the MDPM ether with the PPDPM ether being stable to mild acid.

1. L. Lapatsanis, *Tetrahedron Lett.*, **19**, 3943 (1978).

2. C. Froussios and M. Kolovos, *Synthesis*, 1106 (1987); M. Kolovos and C. Froussios, *Tetrahedron Lett.*, **25**, 3909 (1984).

3. V. G. Mairanovsky, *Angew. Chem., Int. Ed. Engl.*, **15**, 281 (1976); R. Paredes and R. L. Perez, *Tetrahedron Lett.*, **39**, 2037 (1998).

4. M. Adinolfi, G. Barone, A. Iadonisi, and M. Schiattarella, *Tetrahedron Lett.*, **44**, 3733 (2003).

5. M. A. Stanescu and R. S. Varma, *Tetrahedron Lett.*, **43**, 7307 (2002).

6. G. V. M. Sharma, T. R. Prasad, and A. K. Mahalingam, *Tetrahedron Lett.*, **42**, 759 (2001).

7. V. V. Namboodiri and R. S. Varma, *Tetrahedron Lett.*, **43**, 4593 (2002).

8. G. Jackson, H. F. Jones, S. Petursson, and J. M. Webber, *Carbohydr. Res.*, **102**, 147 (1982).

9. I. A. I. Ali, E. S. H. El Ashry, and R. R. Schmidt, *Eur. J. Org. Chem.*, 4121 (2003).

10. T. Suzuki, K. Kobayashi, K. Noda, and T. Oriyama, *Syn. Comm.*, **31**, 2761 (2001).

11. G. A. Olah, G. K. S. Prakash, and S. C. Narang, *Synthesis*, 825 (1978).

12. M. B. Andrus, J. Liu, Z. Ye, and J. F. Cannon, *Org. Lett.*, **7**, 3861 (2005).

13. R. Martin, C. Murruzzu, M. A. Pericas, and A. Riera, *J. Org. Chem.*, **70**, 2325 (2005).

14. G. V. M. Sharma, T. R. Prasad, Rakesh, and B. Srinivas, *Synth. Commum.*, **34**, 941 (2004).

p,p'-Dinitrobenzhydryl Ether (RO−DNB): $ROCH(C_6H_4\text{-}p\text{-}NO_2)_2$

Formation/Cleavage[1]

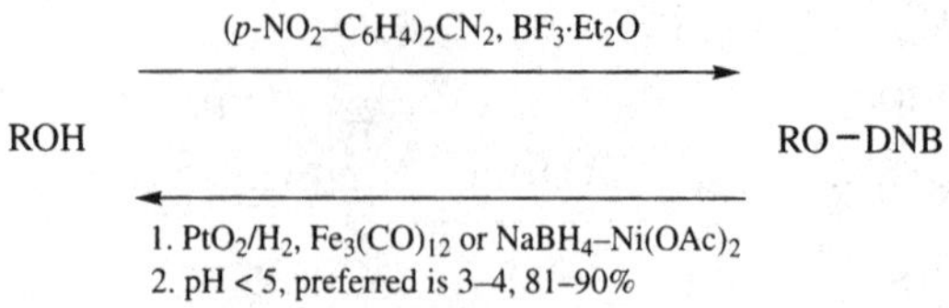

$$(p\text{-NO}_2\text{–C}_6\text{H}_4)_2\text{CN}_2, \text{BF}_3\cdot\text{Et}_2\text{O}$$

ROH $\xrightarrow{\hspace{4cm}}$ RO−DNB

$\xleftarrow{\hspace{4cm}}$

1. PtO_2/H_2, $Fe_3(CO)_{12}$ or $NaBH_4$–Ni(OAc)$_2$
2. pH < 5, preferred is 3–4, 81–90%

The cleavage proceeds by initial reduction of the nitro groups, followed by acid-catalyzed cleavage. The DNB group can be cleaved in the presence of allyl, benzyl, tetrahydropyranyl, methoxyethoxymethyl, methoxymethyl, silyl, trityl, and ketal protective groups.

1. G. Just, Z. Y. Wang, and L. Chan, *J. Org. Chem.*, **53**, 1030 (1988).

5-Dibenzosuberyl Ether

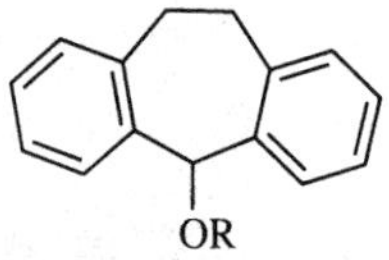

OR

The dibenzosuberyl ether is prepared from an alcohol and the suberyl chloride in the presence of triethylamine (CH_2Cl_2, 20°, 3 h, 75% yield). It is cleaved by acidic hydrolysis (1 *N* HCl/dioxane, 20°C, 6 h, 80% yield). This group has also been used to protect amines, thiols, and carboxylic acids. The alcohol derivative can be cleaved in the presence of a dibenzosuberylamine.[1]

1. J. Pless, *Helv. Chim. Acta*, **59**, 499 (1976).

Triphenylmethyl Ether (Tr−OR): $Ph_3C\text{-}OR$ (Chart 1)

Formation

1.

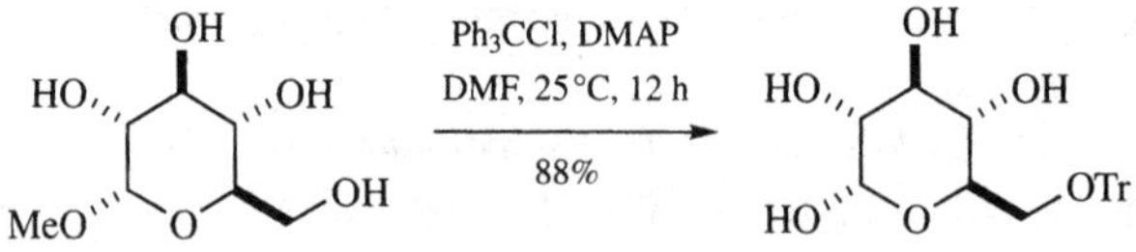

Ph$_3$CCl, DMAP
DMF, 25 °C, 12 h
88%

A secondary alcohol reacts more slowly (40–45°C, 18–24 h, 68–70% yield). In general, excellent selectivity can be achieved for primary alcohols in the presence of secondary alcohols.[1]

2. $C_5H_5N^+CPh_3BF_4^-$, CH_3CN, Pyr, 60–70°C, 75–90% yield.[2] Triphenylmethyl ethers can be prepared more readily with triphenylmethylpyridinium fluoroborate than with triphenylmethyl chloride/pyridine.

3. P-p-$C_6H_4Ph_2CCl$, Pyr, 25°C, 5 days, 90% [979] where P = styrene-divinylbenzene polymer. Triarylmethyl ethers of primary hydroxyl groups in glucopyranosides have been prepared using a polymeric form of triphenylmethyl chloride. Although the yields are not improved, the workup is simplified.

4. Ph_3CCl, 2,4,6-collidine, CH_2Cl_2, Bu_4NClO_4, 15 min, 97% yield.[4] This is an improved procedure for installing the trityl group on polymer-supported nucleosides. DBU is also a very effective base, and in this case secondary hydroxyls can be protected in good yield.[5]

5. $Me_2NC_5H_5NCPh_3^+Cl^-$, CH_2Cl_2, 25°C, 16 h, 95% yield.[6] In this case a primary alcohol is cleanly protected over a secondary alcohol. The reagent is a stable, isolable salt.[7] If the solvent is changed from CH_2Cl_2 to DMF, the amine of serine can be selectively protected.

6. $Ph_3COSiMe_3$, Me_3SiOTf, CH_2Cl_2, 0°C, 0.5 h, 73–97% yield.[8] These conditions also introduce the trityl group on a carboxyl group. The primary hydroxyl of persilylated ribose was selectively derivatized.

7. TrOTf, 2,6-lutidine, CH_2Cl_2, 0°C, >74% yield.[9]

8. $PhCH_2OCPh_3$, DDQ, MS4A, CH_2Cl_2, 46–99% yield. This method is effective for primary alcohols, but the yields for protection of secondary alcohols are only modest.[10]

9. The trityl group can migrate from one secondary center to another under acid catalysis.[11]

Cleavage

1. Formic acid, ether, 45 min, 88% yield.[12]

R = Ac   92%
R = TBDMS   88%
R = THP   60%/40% cleavage

2. CuSO$_4$ (anhydrous), benzene, heat, 89–100% yield.[13] In highly acylated carbohydrates, trityl removal proceeds without acyl migration.

3. Amberlyst 15-H, MeOH, rt, 5–10 min, 69–90%.[14]

4. AcOH, 56°C, 7.5 h, 96%.[15]

5. 90% CF$_3$COOH, t-BuOH, 20°C, 2–30 min, then Bio-Rad 1x2(OH$^-$) resin.[16] These conditions were used to cleave the trityl group from the 5'-hydroxyl of a nucleoside. Bio-Rad resin neutralizes the hydrolysis and minimizes cleavage of glycosyl bonds. TFA supported on silica gel will cleave trityl ethers (83–100% yield).[17]

6. CF$_3$COOH, TFAA, CH$_2$Cl$_2$. These conditions afford the trifluoroacetate, thus preventing retritylation that is sometimes a problem when a trityl group is cleaved with acid. A further advantage of these conditions was that a SEM group was completely stable. When TFAA was not used, traces of moisture resulted in partial SEM cleavage. The TFA group is easily cleaved with methanol and TEA.[18]

7. H$_2$/Pd, EtOH, 20°C, 14 h, 80% yield.[19]

8. HCl(g), CHCl$_3$, 0°C, 1 h, 91% yield.[20] Tritylthio ethers are stable during the deprotection of a primary trityl ether.[21]

9. TsOH, MeOH, 25°C, 5 h.[22]

10. NaHSO$_4$·SiO$_2$, CH$_2$Cl$_2$, MeOH, 2–2.5 hr, rt, 91–100% yield.[23] Trityl groups on amines are also cleaved.

11. Electrolytic reduction: −2.9 V, R$_4$NX, DMF.[24]

12.

CH$_3$CH(OCPh$_3$)(CH$_2$)$_4$CH$_2$OCPh$_3$ $\xrightarrow[\text{20°C, 15 min, 91\%}]{\text{Ph}_3\text{CBF}_4, \text{CH}_2\text{Cl}_2}$ CH$_3$CO(CH$_2$)CH$_2$OH

Since a secondary alcohol is oxidized in preference to a primary alcohol by Ph$_3$CBF$_4$, this reaction could result in selective protection of a primary alcohol.[25]

13. SnCl$_2$, Ac$_2$O, CH$_3$CN.[26] In this case a sulfoxide is also reduced.

14. Et$_2$AlCl, CH$_2$Cl$_2$, 3 min, 70–85% yield.[27] This method was used to remove the trityl group from various protected deoxyribonucleotides. The TBDPS group is stable to these conditions.

15. BiCl$_3$, CH$_3$CN, rt, 3–10 min, 89–95% yield.[28] BOC groups along with esters and THP and TBDMS ethers are unaffected.

16. CeCl$_3$·7H$_2$O, NaI, CH$_3$CN, 78–90% yield. DMTr ethers are also cleaved.[29]

17. Ce(OTf)$_4$, wet CH$_3$CN, 78–93% yield. DMTr ethers are cleaved similarly.[30] Yb(OTf)$_3$ can be used similarly.[31]

18. FeCl$_3$·6H$_2$O, CH$_2$Cl$_2$, rt, 1 h.[32]

19. BF$_3$·Et$_2$O, HSCH$_2$CH$_2$SH, 80% yield.[33]

20. BF$_3$·Et$_2$O, CH$_2$Cl$_2$, MeOH, 2 h, rt, 80% yield.[34]

21. ZnBr$_2$, MeOH, 100% yield.[35,36] TIP and TBDPS ethers are stable to these conditions.

22. BCl$_3$, CH$_2$Cl$_2$, −10°, 20 min, then cold NaHCO$_3$, 75–98% yield.[37,38] TBDMS ethers were stable to these conditions.

23. TESOTf, TESH, CH$_2$Cl$_2$, 88–99% yield.[39] In this case the trityl cation is reduced. Esters and Bn, MPM, TBDMS, and MOM ethers are stable.

24. Na, NH$_3$.[40] Additionally, benzyl groups are removed under these conditions.

25. Li, naphthalene, THF, 0°C, 80–92% yield. These conditions cleave a trityl ether in the presence of a tritylamine.

26. SiO$_2$, benzene, 25°C, 16 h, 81% yield.[41] This cleavage reaction is carried out on a column.

27. K-10 clay, MeOH, H$_2$O, 75°C, 95% yield.[42]

28. Ceric ammonium nitrate supported on silica gel, CH$_3$CN, 25°C, 90–98% yield. This reagent effectively removes the Tr, MMTr, and DMTr ethers from a variety of nucleosides and nucleotides and is more effective than CAN alone. It also cleaves the TBDMS group.[43] The reagent does not cause acyl migration during the removal of a trityl group.[44]

Ratio = 1.5:1: 2.5

29. I$_2$, MeOH. This reagent produces small amounts of HI by oxidizing the alcohol and it is the HI that cleaves the trityl group.[45]

30. CBr$_4$, MeOH, reflux, 88–93% yield.[46] Photolysis can also be used to activate reagent.[47]

31. Direct conversion to ester is possible by treating the trityl ether with an acid chloride in CH_2Cl_2 (12–100% yield).[48]

Tris(4-t-butylphenyl)methyl (^{s}Tr$-$OR) Ether: (4-t-BuC$_6$H$_4$)$_3$COR

The supertrityl group was originally prepared for use in the synthesis of rotaxanes by Stoddart.[49] Its bulkiness made it useful for the partial protection of cyclodextrins. It is introduced from the chloride, as is the typical trityl group, and can be cleaved with acid. It is somewhat less stable to acid than the trityl groups because of the additional stabilization of the carbenium ion imparted by the three t-Bu groups.[50]

α-Naphthyldiphenylmethyl Ether: RO$-$C(Ph)$_2$-α-C$_{10}$H$_7$ (Chart 1)

The α-naphthyldiphenylmethyl ether was prepared to protect, selectively, the 5′-OH group in nucleosides. It is prepared from α-naphthyldiphenylmethyl chloride in pyridine (65% yield) and cleaved selectively in the presence of a p-methoxyphenyl-diphenylmethyl ether with sodium anthracenide, **a** (THF, 97% , yield). The p-methoxyphenyldiphenylmethyl ether can be cleaved with acid in the presence of this group.[51]

a

p-Methoxyphenyldiphenylmethyl Ether (MMTr$-$OR): p-MeOC$_6$H$_4$(Ph)$_2$C$-$OR (Chart 1)

Di(p-methoxyphenyl)phenylmethyl Ether (DMTr$-$OR): (p-MeOC$_6$H$_4$)$_2$PhC$-$OR

Tri(p-methoxyphenyl)methyl Ether (TMTr$-$OR): (p-MeOC$_6$H$_4$)$_3$C$-$OR

These were originally prepared by Khorana from the appropriate chlorotriarylmethane in pyridine[52] or DMF[53] but can also be prepared from the corresponding triaryl tetrafluoroborate salts (80–98% yield for primary alcohols)[54] or by other less general methods.[55] They were developed to provide a selective protective group for the 5′-OH of nucleosides and nucleotides that is more acid-labile than the trityl group, because depurination is often a problem in the acid-catalyzed removal of the trityl group.[56] Introduction of p-methoxy groups increases the rate of hydrolysis by about one order of magnitude for each p-methoxy substituent. The monomethoxy derivative has been used for the selective protection of a primary allylic alcohol over a secondary allylic alcohol (MMTr, Pyr, $-10°C$).[57] The trimethoxy derivative is too labile for most applications, but the mono- and di-derivatives have been used extensively in the preparation of oligonucleotides and oligonucleosides. A series of triarylcarbinols has been prepared with similar acid stability, which upon acid treatment result in different colors. The use of these in oligonucleotide synthesis was demonstrated.[58]

Cleavage

1. For 5′-protected uridine derivatives in 80% AcOH, 20°C, the time for hydrolysis was as follows[52]:

   $(p\text{-MeOC}_6\text{H}_4)_n(\text{Ph})_m\text{COR}$

   $n = 0, m = 3, 48\,\text{h}$

   $n = 1, m = 2, 2\,\text{h}$

   $n = 2, m = 1, 15\,\text{min}$

   $n = 3, m = 0, 1\,\text{min}$

2. MMTr-OR: 1,1,1,2,2,2-Hexafluoro-2-propanol ($pK_a = 9.3$), 75–90% yield.[59]

3. The following is an example of the use of the MMTr group in a nonnucleoside setting where the usual trityl group was too stable.[60]

4. MMTr: $\text{Cl}_2\text{CCO}_2\text{H}$, Et_3SiH.[61]

5. MMTr: Sodium naphthalenide in HMPA (90% yield).[62] The MMTr group is not cleaved by sodium anthracenide, used to cleave α-naphthyldiphenylmethyl ethers.[51]

6. 3% $\text{CCl}_3\text{CO}_2\text{H}$ in 95:5 $\text{CH}_3\text{NO}_2/\text{MeOH}$ is recommended for removal of the DMTr group from the 5′-OH of deoxyribonucleotides because of reduced levels of depurination compared to $\text{Cl}_3\text{CO}_2\text{H}/\text{CH}_2\text{Cl}_2$, $\text{PhSO}_3\text{H}/\text{MeOH}/\text{CH}_2\text{Cl}_2$, and $\text{ZnBr}_2/\text{CH}_3\text{NO}_2$.[63]

7. MMTr: MeOH, CCl_4, ultrasound, 25–40°C, 1.5–12 h, 69–100% yield.[64]

8. MMTr: O-(Benzotriazol-1-yl)-N,N,N,N-tetramethyluronium tetrafluoroborate, CH_3CN, H_2O, 85–95% yield. The mechanism for cleavage is most likely the result of released acid from hydrolysis of the reagent. These conditions also cleave THP and TBDMS groups.[65]

4-(4′-Bromophenacyloxy)phenyldiphenylmethyl Ether:
$p\text{-}(p\text{-BrC}_6\text{H}_4\text{C(O)CH}_2\text{O})\text{C}_6\text{H}_4(\text{Ph})_2\text{C}-\text{OR}$

This group was developed for protection of the 5′-OH group in nucleosides. The derivative is prepared from the corresponding triarylmethyl chloride and is cleaved by reductive cleavage (Zn/AcOH) of the phenacyl ether to the p-hydroxyphenyldiphenylmethyl ether, followed by acidic hydrolysis with formic acid.[66]

4,4′,4″-Tris(4,5-dichlorophthalimidophenyl)methyl Ether (CPTr−OR)

The CPTr group was developed for the protection of the 5′-OH of ribonucleosides. It is introduced with CPTrBr/AgNO₃/DMF (15 min) in 80–96% yield and can be removed by ammonia, followed by 0.01 M HCl or 80% AcOH.[67] It can also be removed with hydrazine and acetic acid.[68,69]

4,4′,4″-Tris(levulinoyloxyphenyl)methyl Ether (TLTr-OR)

The TLTr group was developed for the protection of the 5′-OH of thymidine. It is introduced in 81% yield with TLTrBr/Pyr and is cleaved with hydrazine (3 min); Pyr–AcOH, 50°C, 3 min, 81% . The $t_{1/2}$ in 80% AcOH is 24 h.[70]

4,4′,4″-Tris(benzoyloxyphenyl)methyl Ether (TBTr−OR)

The TBTr group was prepared for 5′-OH protection in oligonucleotide synthesis. The group is introduced in >80% yield with TBTrBr/pyridine at 65°. It is five times more stable to 80% AcOH than the trityl group [$t_{1/2}$ (Tr) = 5 h; $t_{1/2}$ (TBTr) = 25 h]. The TBTr group is removed with 2 M NaOH. The di(4-methoxyphenyl)phenylmethyl (DMTr) group can be cleaved without affecting the TBTr derivative (80% AcOH, 95% yield).[71]

4,4′-Dimethoxy-3″-[N-(imidazolylmethyl)]trityl Ether (IDTr−OR)

4,4′-Dimethoxy-3″-[N-(imidazolylethyl)carbamoyl]trityl Ether (IETr−OR)

The IDTr group was developed to protect the 5′-OH of deoxyribonucleotides and to increase the rate of internucleotide bond formation through participation of the pendant imidazole group. Rate enhancements of ≈350 were observed except when (i-Pr)₂EtN was added to the reaction mixture, in which case reactions were complete

IDTr–OR   IETr–OR

in 30 s, as opposed to the usual 5–6 h without the pendant imidazole group. The group is efficiently introduced with the bistetrafluoroborate salt, IDTr–BBF, in DMF (70% yield). It is removed with 0.2 M Cl$_2$CHCO$_2$H or 1% CF$_3$COOH in CH$_2$Cl$_2$.[72]

The IETr group was developed for the same purpose, but found to be superior in its catalytic activity.[73]

Bis(4-methoxyphenyl)-1'-pyrenylmethyl Ether (Bmpm−OR)

This bulky group was developed as a fluorescent, acid-labile protective group for oligonucleotide synthesis. It has properties very similar to the DMTr group except that it can be detected down to 10^{-10} M on TLC plates with 360-nm ultraviolet light.[74]

4-(17-Tetrabenzo[*a,c,g,i*]fluorenylmethyl)-4',4''-dimethoxytrityl Ether (Tbf-DMTr−OR)

This group was developed for terminal protection of an oligonucleotide sequence for purposes of monitoring the purification by HPLC after a synthesis. It shows characteristic UV maxima at 365 and 380 nm. It is prepared from the chloride in pyridine and can be bound directly to the support-bound oligonucleotide.[75]

9-Anthryl Ether: 9-Anthryl−OR

This group is prepared by the reaction of the anion of 9-hydroxyanthracene and the tosylate of an alcohol. Since the formation of this group requires an S_N2 displacement on the alcohol to be protected, it is best suited for primary alcohols. It is cleaved by a novel singlet oxygen reaction followed by reduction of the endoperoxide with hydrogen and Raney nickel.[76]

9-(9-Phenyl)xanthenyl Ether (pixyl−OR)

The pixyl ether is prepared from the xanthenyl chloride in 68–87% yield. This group has been used extensively in the protection of the 5′-OH of nucleosides; it is readily cleaved by acidic hydrolysis (80% AcOH, 20°C, 8–15 min, 100% yield, or 3% trichloroacetic acid).[77] It can be cleaved under neutral conditions with $ZnBr_2$, thus reducing the extent of the often troublesome depurination of N-6-benzyloxyadenine residues during deprotection.[78] Photolysis in CH_3CN/H_2O also cleaves the pixyl group.[79] Acidic conditions that remove the pixyl group also partially cleave the THP group ($t_{1/2}$ for THP at 2′-OH of ribonucleoside = 560 s in 3% Cl_2CHCO_2H/CH_2Cl_2).[80,81] The pixyl group has advantages over the trityl group in that it produces derivatives with a greater tendency to be crystalline and that the UV extinction coefficients are ~100 times greater than for the trityl group. A series of pixyl derivatives has been prepared and the half-lives of TFA-induced cleavage determined.[82] Reaction conditions were TFA, CH_2Cl_2, EtOH, 22°C. Under these conditions the trityl group has an estimated $t_{1/2}$ of ~320 min.

R^1	R^2	R^3	Abbr.	$t^{1/2}$ (min)
OMe	H	H	—	0.3
Me	H	H	Tx	0.55
H	H	H	Px	1.37
H	CF_3	H	—	8.7
H	H	Br	—	244
H	CF_3	Br	—	1560

The addition of pyrrole as a cation scavenging agent has been recommended for use in deprotection during solid-phase DNA and RNA synthesis. The Px or Tx groups have been recommended as a better alternative to DMTr group in DNA and RNA synthesis because of their faster cleavage rates.[83]

Deprotection using photolysis at 254 or 300 nm in aqueous CH_3CN can also be used to cleave the pixyl group (83–97% yield).[84]

9-Phenylthioxanthyl (S-Px—OR, S-Pixyl—OR) Ether

The 9-Phenylthioxanthyl ether was developed as a photocleavable protective group for nucleosides and other alcohols. It is introduced from the chloride in dry pyridine (79–92% yield) and is cleaved by irradiation at 300 nm in aqueous CH_3CN or aqueous trifluoroethanol (75–97% yield).[85] The sulfoxide form is not ionized in 50% H_2SO_4 and thus serves as a protected form which upon reduction can readily be cleaved.[86]

9-(9-Phenyl-10-oxo)anthryl Ether (Tritylone Ether) (Chart 1)

The tritylone ether is used to protect primary hydroxyl groups in the presence of secondary hydroxyl groups. It is prepared by the reaction of an alcohol with 9-phenyl-9-hydroxyanthrone under acid catalysis (cat. TsOH, benzene, reflux, 55–95% yield).[87,88] It can be cleaved under the harsh conditions of the Wolff–Kishner reduction (H_2NNH_2, NaOH, 200°C, 88% yield)[51] and by electrolytic reduction (−1.4 V, LiBr, MeOH, 80–85% yield).[63] It is stable to 10% HCl, 55 h.[51]

1. S. K. Chaudhary and O. Hernandez, *Tetrahedron Lett.*, **20**, 95 (1979).

2. S. Hanessian and A. P. A. Staub, *Tetrahedron Lett.*, **14**, 3555 (1973).

3. J. M. J. Fréchet and K. E. Haque, *Tetrahedron Lett.*, **16**, 3055 (1975); K. Barlos, D. Gatos, J. Kallitsis, G. Papaphotiu, P. Sotiriuc, Y. Wenquig, and W. Schäfer, *Tetrahedron Lett.*, **30**, 3943 (1989).

4. M. P. Reddy, J. B. Rampal, and S. L. Beaucage, *Tetrahedron Lett.*, **28**, 23 (1987).

5. S. Colin-Messager, J.-P. Girard, and J.-C. Rossi, *Tetrahedron Lett.*, **33**, 2689 (1992).

6. O. Hernandez, S. K. Chaudhary, R. H. Cox, and J. Porter, *Tetrahedron Lett.*, **22**, 1491 (1981); R. P. Srivastava and J. Hajdu, *Tetrahedron Lett.*, **32**, 6525 (1991).

7. A. V. Bhatia, S. K. Chaudhary, and O. Hernandez, *Org. Synth.*, **75**, 184 (1997).

8. S. Murata and R.Noyori, *Tetrahedron Lett.*, **22**, 2107 (1981).

9. M. Hirama, T. Node, S. Yasuda, and S. Ito, *J. Am. Chem. Soc.*, **113**, 1830 (1991).

10. M. Oikawa, H. Yoshizaki, and S. Kusumoto, *Synlett*, 757 (1998); G. V. M. Sharma, A. K. Mahalingam, and T. R. Prasad, *Synlett*, **10**, 1479 (2000).

11. P. A. Bartlett and F. R. Green III, *J. Am. Chem. Soc.*, **100**, 4858 (1978).

12. M. Bessodes, D. Komiotis, and K. Antonakis, *Tetrahedron Lett.*, **27**, 579 (1986).

13. G. Randazzo, R. Capasso, M. R. Cicala, and A. Evidente, *Carbohydr. Res.*, **85**, 298 (1980).

14. C. Malanga, *Chem. Ind. (London)*, 856 (1987).

15. R. T. Blickenstaff, *J. Am. Chem. Soc.*, **82**, 3673 (1960).

16. M. MacCoss and D. J. Cameron, *Carbohydr. Res.*, **60**, 206 (1978).

17. A. K. Pathak, V. Paathak, L. E. Seitz, K. N. Tiwari, M. S. Akhtar, and R. C. Reynolds, *Tetrahedron Lett.*, **42**, 7755 (2001).

18. E. Krainer, F. Naider, and J. Becker, *Tetrahedron Lett.*, **34**, 1713 (1993).

19. R. N. Mirrington and K. J. Schmalzl, *J. Org. Chem.*, **37**, 2877 (1972); S. Hanessian and G. Rancourt, *Pure Appl. Chem.*, **49**, 1201 (1977).

20. Y. M. Choy and A. M. Unrau, *Carbohydr. Res.*, **17**, 439 (1971).

21. M. Maltese, *J. Org. Chem.*, **66**, 7615 (2001).

22. A. Ichihara, M. Ubukata, and S. Sakamura, *Tetrahedron Lett.*, **18**, 3473 (1977).

23. B. Das, G. Mahender, V. S. Kumar, and N. Chowdhury, *Tetrahedron Lett.*, **45**, 6709 (2004).

24. V. G. Mairanovsky, *Angew. Chem., Int. Ed. Engl.*, **15**, 281 (1976).

25. M. E. Jung and L. M. Speltz, *J. Am. Chem. Soc.*, **98**, 7882 (1976).

26. B. M. Trost and L. H. Latimer, *J. Org. Chem.*, **43**, 1031 (1978).

27. H. Köster and N. D. Sinha, *Tetrahedron Lett.*, **23**, 2641 (1982).

28. G. Sabitha, E. V. Reddy, R. Swapna, R. N. Mallikarjun, and J. S. Yadav, *Synlett*, 1276 (2004).

29. J. S. Yadav and B. V. S. Reddy, *Synlett*, 1275 (2000).

30. A. Khalafi-Nezhad and R. Fareghi Alamdari, *Tetrahedron*, **57**, 6805 (2001).

31. R. J. Lu, D. Liu, and R. W. Giese, *Tetrahedron Lett.*, **41**, 2817 (2000).

32. X. Ding, W. Wang, and F. Kong, *Carbohydr. Res.*, **303**, 445 (1997).

33. P.-E. Sum and L. Weiler, *Can. J. Chem.*, **56**, 2700 (1978).

34. D. Cabaret and M. Wakselman, *Can. J. Chem.*, **68**, 2253 (1990).

35. V. Kohli, H. Bloecker, and H. Koester, *Tetrahedron Lett.*, **21**, 2683 (1983).

36. T. F. S. Lampe and H. M. R. Hoffmann, *Tetrahedron Lett.*, **37**, 7695 (1996).

37. G. B. Jones, B. J. Chapman, R. S. Huber, and R. Beaty, *Tetrahedron: Asymmetry*, **5**, 1199 (1994).

38. G. B. Jones, G. Hynd, J. M. Wright, and A. Sharma, *J. Org. Chem.*, **65**, 263 (2000).

39. H. Imagawa, T. Tsuchihashi, R. K. Singh, H. Yamamoto, T. Sugihara, and M. Nishizawa, *Org. Lett.*, **5**, 153 (2003).

40. P. Kovác and S. Bauer, *Tetrahedron Lett.*, 2349 (1972); S. Hanessian, N. G. Cooke, B. Dehoff, and Y. Sakito, *J. Am. Chem. Soc.*, **112**, 5276 (1990).

41. J. Lehrfeld, *J. Org. Chem.*, **32**, 2544 (1967).

42. J.-i. Asakura, M. J. Robins, Y. Asaka, and T. H. Kim, *J. Org. Chem.*, **61**, 9026 (1996).

43. J. R. Hwu, M. L. Jain, F.-Y. Tsai, S.-C. Tsay, A. Balakumar, and G. H. Hakimelahi, *J. Org. Chem.*, **65**, 5077 (2000).

44. J. R. Hwu and K.-Y. King, *Curr, Sci.*, **81**, 1043 (2001).

45. J. L. Wahlstrom and R. C. Ronald, *J. Org. Chem.*, **63**, 6021 (1998).

46. J. S. Yadav and B. V. Subba Reddy, *Carbohydr. Res.*, **329**, 885 (2000).

47. M.-Y. Chen, L. N. Patkar, M.-D. Jan, A. S.-Y. Lee, and C.-C. Lin, *Tetrahedron Lett.*, **45**, 635 (2004). M.-Y. Chen, L. N. Patkar, K.-C. Lu, A. S.-Y. Lee, and C.-C. Lin, *Tetrahedron*, **60**, 11465 (2004).

48. S. C. Bergmeier and K. M. Arason, *Tetrahedron Lett.*, **41**, 5799 (2000).

49. P. R. Ashton, D. Philip, N. Spencer, and J. F. Stoddart, *J. Chem. Soc., Chem. Commun.*, 1124 (1992).

50. L. Poorters, D. Armspach, and D. Matt, *Eur. J. Org. Chem.*, **68**, 1377 (2003); D. Armspach and D. Matt, *Carbohydr. Res.*, **310**, 129 (1998).

51. R. L. Letsinger and J. L. Finnan, *J. Am. Chem. Soc.*, **97**, 7197 (1975).

52. H. G. Khorana, *Pure Appl. Chem.*, **17**, 349 (1968); M. Smith, D. H. Rammler, I. H. Goldberg, and H. G. Khorana, *J. Am. Chem. Soc.*, **84**, 430 (1962).

53. O. Hernandez, S. K. Chaudhary, R. H. Cox, and J. Porter, *Tetrahedron Lett.*, **22**, 1491 (1981).

54. C. Bleasdale, S. B. Ellwood, and B. T. Golding, *J. Chem. Soc., Perkin Trans. 1*, 803 (1990).

55. A. Khalafi-Nezhad and B. Mokhtari, *Tetrahedron Lett.*, **45**, 6737 (2004).

56. For a review in which the use of various trityl groups in nucleotide synthesis is discussed in the context of the phosphoramididite approach, see S. L. Beaucage and R. P. Iyer, *Tetrahedron*, **48**, 2223 (1992).

57. J. Adams and J. Rokach, *Tetrahedron Lett.*, **25**, 35 (1984).

58. E. F. Fisher and M. H. Caruthers, *Nucleic Acids Res.*, **11**, 1589 (1983).

59. N. J. Leonard and Neelima, *Tetrahedron Lett.*, **36**, 7833 (1995).

60. A. G. Myers and P. S. Dragovich, *J. Am. Chem. Soc.*, **114**, 5859 (1992).

61. V. T. Ravikumar, A. H. Krotz, and D. L. Cole, *Tetrahedron Lett.*, **36**, 6587 (1995).

62. G. L. Greene and R. L. Letsinger, *Tetrahedron Lett.*, **16**, 2081 (1975).

63. H. Takaku, K. Morita, and T. Sumiuchi, *Chem. Lett.*, **12**, 1661 (1983).

64. Y. Wang and C. McGuigan, *Synth. Commun.*, **27**, 3829 (1997).

65. K. S. Ramasamy and D. Averett, *Synlett*, 709 (1999).

66. A. T.-Rigby, Y.-H. Kim, C. J. Crosscup, and N. A. Starkovsky, *J. Org. Chem.*, **37**, 956 (1972).

67. M. Sekine and T. Hata, *J. Am. Chem. Soc.*, **108**, 4581 (1986).

68. M. D. Hagen, C. S.-Happ, E. Happ, and S. Chládek, *J. Org. Chem.*, **53**, 5040 (1988).

69. M. Sekine, J. Heikkilä, and T. Hata, *Bull. Chem. Soc. Jpn.*, **64**, 588 (1991).

70. M. Sekine and T. Hata, *Bull. Chem. Soc. Jpn.*, **58**, 336 (1985).

71. M. Sekine and T. Hata, *J. Org. Chem.*, **48**, 3011 (1983).

72. M. Sekine and T. Hata, *J. Org. Chem.*, **52**, 946 (1987).

73. M. Sekine, T. Mori, and T. Wada, *Tetrahedron Lett.*, **34**, 8289 (1993).

74. J. L. Fourrey, J. Varenne, C. Blonski, P. Dousset, and D. Shire, *Tetrahedron Lett.*, **28**, 5157 (1987).

75. R. Ramage and F. O. Wahl, *Tetrahedron Lett.*, **34**, 7133 (1993).

76. W. E. Barnett and L. L. Needham, *J. Chem. Soc., Chem. Commun.*, 1383 (1970); *idem, J. Org. Chem.*, **36**, 4134 (1971).

77. J. B. Chattopadhyaya and C. B. Reese, *J. Chem. Soc., Chem. Commun.*, 639 (1978).

78. M. D. Matteucci and M. H. Caruthers, *Tetrahedron Lett.*, **21**, 3243 (1980).

79. A. Misetic and M. K. Boyd, *Tetrahedron Lett.*, **39**, 1653 (1998).

80. C. Christodoulou, S. Agrawal, and M. J. Gait, *Tetrahedron Lett.*, **27**, 1521 (1986).

81. H. Tanimura and T. Imada, *Chem. Lett.*, **19**, 2081 (1990).

82. P. R. J. Gaffney, L. Changsheng, M. V. Rao, C. B. Reese, and J. C. Ward, *J. Chem. Soc., Perkin Trans. 1*, 1355 (1991); see Errata: *ibid., idem*, 1275 (1992).

83. C. B. Reese and H. Yan, *Tetrahedron Lett.*, **45**, 2567 (2004).

84. A. Misetic and M. K. Boyd, *Tetrahedron Lett.*, **39**, 1653 (1998).

85. M. P. Coleman and M. K. Boyd, *Tetrahedron Lett.*, **40**, 7911 (1999); M. P. Coleman and M. K. Boyd, *J. Org. Chem.*, **67**, 7641 (2002).

86. P. L. Bernad, Jr., S. Khan, V. A. Korshun, E. M. Southern, and M. S. Shchepinov, *Chem. Commun.*, 3466 (2005).

87. W. E. Barnett, L. L. Needham, and R. W. Powell, *Tetrahedron*, **28**, 419 (1972).

88. C. van der Stouwe and H. J. Schäfer, *Tetrahedron Lett.*, **20**, 2643 (1979).

1,3-Benzodithiolan-2-yl Ether (Bdt−OR)

Formation

1. BDTO-*i*-Am, H^+, dioxane, rt, 81% .[1]

2. Pyr, CH_2Cl_2, 95% .[1] The introduction of the Bdt group

proceeds under these rather neutral conditions; this proved advantageous for acid-sensitive substrates such as polyenes.[2] The Bdt group can also be reduced with Raney nickel to a methyl group or with Bu_3SnH followed by CH_3I to a [2-(methylthio)phenylthio]methyl ether (MTPM ether)[3,4] that can be cleaved with $AgNO_3$ ($DMF:H_2O$).[5]

Cleavage

1. 80% AcOH, 100°C, 30 min.[1]

2. 2% CF_3COOH, $CHCl_3$, 0°C, 20 min, 97% yield.[1]

Half-Lives for Cleavage of 5′-Protected Thymidine in 80% AcOH at 15°C

	DMTrT	mTHPT	Bdt-5′T	MMTrT	THPT	Bdt-3′T
$t_{1/2}$	3 min	23 min	38 min	48 min	3.5 h	2.5 h
$t_{complete}$	15 min	2.5 h	3 h	3 h	15 h	8 h

DMTrT = 5′-*O*-di-*p*-methoxytritylthymidine
mTHPT = 5′-*O*-(4-methoxytetrahydropyran-4-yl)thymidine
Bdt-5′T = 5′-*O*-(1,3-benzodithiolan-2-yl)thymidine
MMTrT = 5′-*O*-mono-*p*-methoxytritylthymidine
THPT = 5′-tetrahydropyranylthymidine
Bdt-3′T = 3′-*O*-(1,3-benzodithiolan-2-yl)thymidine

3. Dowex W50-1X, MeOH, 1.5 h, rt.[2]

1. M. Sekine and T. Hata, *J. Am. Chem. Soc.*, **105**, 2044 (1983); *idem, J. Org. Chem.*, **48**, 3112 (1983).
2. S. D. Rychnovsky and R. C. Hoye, *J. Am. Chem. Soc.*, **116**, 1753 (1994).
3. M. Sekine and T. Nakanishi, *J. Org. Chem.*, **54**, 5998 (1989).
4. M. Sekine and T. Nakanishi, *Nucleosides Nucleotides*, **11**, 679 (1992).
5. M. Sekine and T. Nakanishi, *Chem. Lett.*, **20**, 121 (1991).

4,5-Bis(ethoxycarbonyl)-[1,3]-dioxolan-2-yl Ether

This ether is introduced by an acid catalyzed orthoester exchange process with an alcohol. It was developed for protection of the 2′-hydroxyl in ribonucleotide synthesis. It is sufficiently stable to dichloroacetic acid, which is used for the cleavage of the dimethoxytrityl group.[1]

1. B. Karwowski, K. Seio, and M. Sekine, *Nucleosides & Nucleotides, and Nucleic Acids*, **24**, 1111 (2005).

Benzisothiazolyl *S,S*-Dioxido Ether

Formation/Cleavage[1]

1. H. Sommer and F. Cramer, *Chem. Ber.*, **107**, 24 (1974).

Silyl Ethers

Silyl ethers are among the most frequently used protective groups for the alcohol function.[1] This stems largely from the fact that their reactivity (both formation and cleavage) can be modulated by a suitable choice of substituents on the silicon atom. Both steric and electronic effects are the basic controlling elements that regulate the ease of cleavage in multiply functionalized substrates. In planning the selective deprotection, the steric environment around the silicon atom, as well as the environment of the protected molecular framework, must be considered. For example, it is normally quite easy to cleave a DEIPS group in the presence of a TBDMS group, but examples are known where the reverse is true. In these cases, the backbone structure provides additional steric encumbrance to reverse the selectivity. Differences in electronic factors are also used to achieve selectivity. For two alcohols of similar steric environments that have differing electron densities, the acid-catalyzed deprotection rates will vary substantially and can be used to advantage. This is especially true for phenolic vs. alkyl silyl ethers: The alkyl silyl ethers are more easily cleaved by acid, and the phenolic silyl ethers are more easily cleaved by base. The reduced basicity of the silyl oxygen can be used to change the course of Lewis acid-promoted reactions and help to provide selective deprotection.[2] Electron-withdrawing substituents on the silicon atom increase susceptibility toward basic hydrolysis, but decrease sensitivity toward acid. For some of the more common silyl ethers the stability toward acid increases in the following order: TMS (1) < TES (64) < TBDMS (20,000) < TIPS (700,000) < TBDPS (5,000,000), and the stability toward base increases in the following order: TMS (1) < TES (10–100) < TBDMS ~ TBDPS (20,000) < TIPS (100,000). Quantitative relationships have been developed[3] to examine the steric factors associated with nucleophilic attack on silicon and the solvolysis of silyl chlorides. Silyl ethers are also considered to be poor donor ligands for chelation-controlled reactions, and thus their use in reactions where stereoinduction is anticipated must be carefully considered.[4] One of the properties that has made silyl groups so popular is the fact that they are easily cleaved by fluoride ion, which is attributed to the high affinity that fluoride ion has for silicon. The Si–F bond strength is 30 kcal/mol greater than the Si–O bond strength.

Two excellent reviews that discuss the selective cleavage of numerous silyl derivatives are available.[5]

1. For a review on silylating agents, see *Silylating Agents*, G. van Look, G. Simchen, and J. Heberle, Fluka Chemie AG, 1995.
2. M. Oikawa, T. Ueno, H. Oikawa, and A. Ichihara, *J. Org. Chem.*, **60**, 5048 (1995).
3. N. Shimizu, N. Takesue, S. Yasuhara, and T. Inazu, *Chem. Lett.*, **22**, 1807 (1993); N. Shimizu, N. Takesue, A. Yamamoto, T. Tsutsumi, S. Yasuhara, and Y. Tsuno, *ibid.*, **21**, 1263 (1992).
4. L. Banfi, G. Guanti, and M. T. Zannetti, *Tetrahedron Lett.*, **37**, 521 (1996).
5. T. D. Nelson and R. D. Crouch, *Synthesis*, 1031 (1996); R. D. Crouch, *Tetrahedron*, **60**, 5833 (2004).

Migration of Silyl Groups

Silyl groups have found broad appeal as protective groups because their reactivity and stability can be tailored by varying the nature of the substituents on the silicon.

Their ability to migrate from one hydroxyl to another is a property that can be used to advantage,[1] but more often than not, it is a nuisance.[2] The migratory aptitude in nucleosides was found to be solvent-dependent, with migration proceeding fastest in protic solvents.[3] Migration usually occurs under basic conditions and proceeds intramolecularly through a pentacoordinate silicon,[4] but migrations do occur under acidic conditions.[5] The TBDMS group has been observed to migrate frequently,[2b, 6–11] while migration of the more stable TBDPS[12,13] and TIPS[14] groups occurs less frequently. The facile migration of the TBDMS residue is a severe problem in the synthesis of oligoribonucleotides.[3,15] Conditions favoring silyl migration are the presence of a strong base in protic solvents, but migrations in aprotic solvents are also observed.[3,16] Both 1,2-,[4] 1,3-,[17] and 1,5-migrations[22] have been observed, but if the topological features of a molecule are properly oriented, migrations that span many atoms have been observed. Such was the case during the attempted PMB ether formation in a cytovaricin synthesis where the C-32 DEIPS group migrated to the C-17 hydroxyl. In consonance with the fact that the larger, more stable silyl groups are not as prone to migration, the corresponding TIPS analog gave only the desired C-17 PMB ether.[18]

On the other hand, the TIPS group can readily migrate as was the case during the conversion of the iodide to the thioglycoside.[19] Migration may be driven by the preference of large silyl groups to assume axial orientations in sterically demanding environments. When the C-4 hydroxyl was protected as an acetate, the transformation proceeded as expected without TIPS migration.

Silyl migration can be used advantageously as in a disorazole C_1 synthesis by Meyers. Treatment of the hydroxyl with NaH results in TBS migration with concomitant liberation of an aldehyde which then reacts with the Horner–Emmons reagent to form the unsaturated ester.[20]

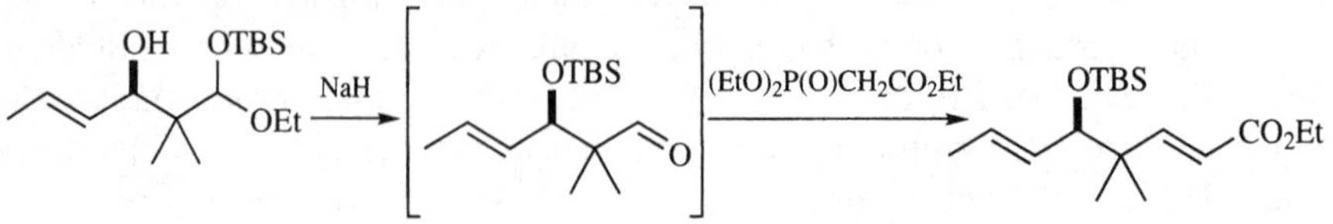

In Overman's synthesis of Alcyonin, silyl migration from the tertiary to the second-ary alcohol facilitated the deprotection of a hindered 3° TBS ether.[21]

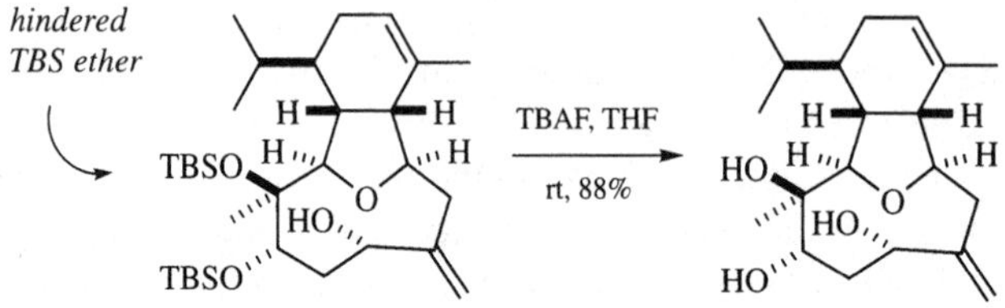

In essence, history has shown that placing negatively charged oxygen in proximity to a TBDMS ether will almost always result in some level of silyl migration, thus the planning of any synthesis should take this into account, especially since the degree of migration is largely unpredictable and is a function of spatial,[22] electronic, and steric effects. Moreover, as may be expected, the more acidic the hydroxyl, the less likely it is to bear the silyl group, as is illustrated below.[23]

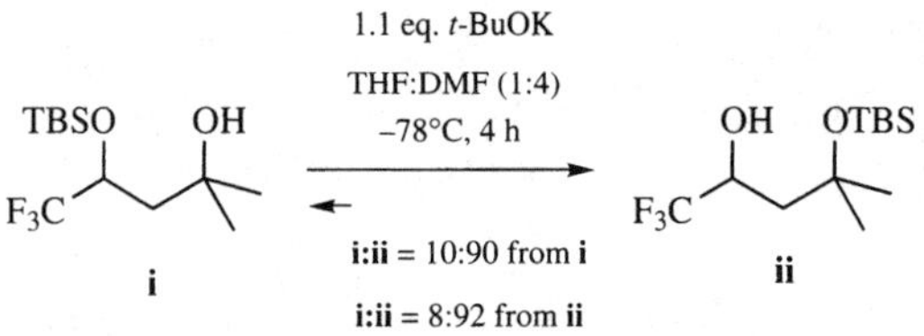

In consonance with this heuristic, a phenolic TBS derivative has been shown to mi-grate to a primary alcohol.[24] A pyranoside anomeric hydroxyl is more acidic than the 2-OH, and thus treatment of the disaccharide with NaH and BnBr results in migra-tion of the silyl group and protection of the anomeric center with a benzyl group.[25]

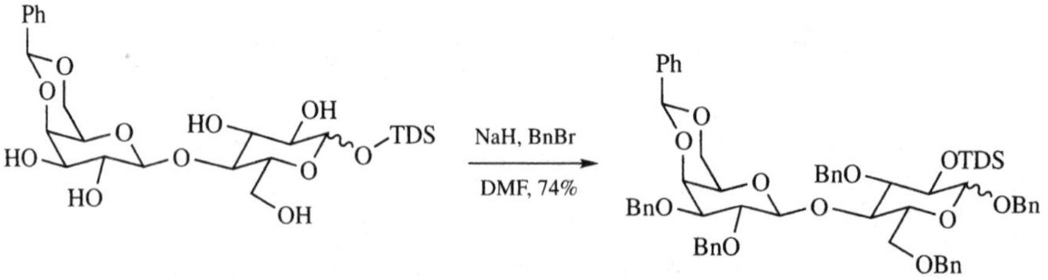

It appears that the counterion on the alkoxide has some remediating effects. For ex-ample, the NaBH$_4$ reduction of the lactol affords only the product of silyl migration whereas if CeCl$_3$ is included, no silyl migration was observed.[26] This case is also unusual because complete migration has occurred.

On the other hand, with a TBDPS group, CeCl$_3$ did not prevent migration; in this case, alcohol acidity seems to be an overriding factor, even at the expense of what is usually considered a sterically demanding situation.[12]

Initial product   *Product upon prolonged reaction*

Note that replacing the olefin with an epoxide, which is expected to reduce the acidity, drives the silyl group to the least hindered position.

In the following case, migration is complete because one alcohol is trapped by a Michael reaction preventing equilibrium.[27]

In the well-known Brook rearrangement,[28] silyl groups migrate from oxygen to carbon, but the following example is less obvious and not necessarily predictable.[29] This problem can be prevented by premixing ZnCl$_2$ with the iodide before t-BuLi addition.[30] Other cases of O-to-C migration have been observed.[31,32] This type of migration has been used to advantage for the preparation of 2-silylated benzyl alcohols.[33]

Although silyl migrations are usually acid- or base-catalyzed, they have been observed to occur thermally.[34]

1. G. A. Molander and S. Swallow, *J. Org. Chem.*, **59**, 7148 (1994); J. M. Lassaletta and R. R. Schmidt, Synlett, 925 (1995).

2. (a) C. A. A. Van Boeckel, S. F. Van Aelst, and T. Beetz, *Recl: J. R. Neth. Chem. Soc.*, **102**, 415 (1983); (b) P. G. M. Wuts and S. S. Bigelow, *J. Org. Chem.*, **53**, 5023 (1988); (c) F. Franke and R. D. Guthrie, Aust. J. Chem., **31** 1285 (1978); (d) Y. Torisawa, M. Shibasaki and S. Ikegami, *Tetrahedron Lett.*, **20**, 1865 (1979); (e) K. K. Ogilvie, S. L. Beaucage, A. L. Schifman, N. Y. Theriault and K. L. Sadana, *Can. J. Chem.*, **56**, 2768 (1978); (f) S. S Jones and C. B. Reese, *J. Chem. Soc., Perkin Trans. I*, 2762 (1979).

3. K. K. Ogilvie and D. W. Entwistle, *Carbohydr. Res.*, **89**, 203 (1981).

4. J. Mulzer and B. Schöllhorn, *Angew. Chem., Int. Ed Engl.*, **29**, 431 (1990).

5. J. A. Marshall and M. P. Bourbeau, *J. Org. Chem.*, **67**, 2751 (2002).

6. D. Crich and T. J. Ritchie, *Carbohydr. Res.*, **197**, 324 (1990) and ref. cit. therein.

7. R. W. Friesen and A. K. Daljeet, *Tetrahedron Lett.*, **31**, 6133 (1990).

8. M. T. Barros, C. D. Maycock, and M. R. Ventura, *Chem. Eur. J.*, **6**, 3991 (2000).

9. K. M. Gardinier and J. W. Leahy, *J. Org. Chem.*, **62**, 7098 (1997).

10. E. J. Jeong, E. J. Kang, L. T. Sung, S. K. Hong, and E. Lee, *J. Am. Chem. Soc.*, **124**, 14655 (2002).

11. A. B. Smith III, S. M. Pitram, and M. J. Fuertes, *Org. Lett.*, **5**, 2751 (2003).

12. J. A. Marshall and Y. Tang, *J. Org. Chem.*, **59**, 1457 (1994).

13. T. Ohgiya and S. Nishiyama, *Tetrahedron Lett.*, **45**, 8273 (2004).

14. F. Seela and T. Fröhlich, *Helv. Chim. Acta*, **77**, 399 (1994). Y. Li, D. Horton, V. Barberousse, S. Samreth, and F. Bellamy, *Carbohydr. Res.*, **316**, 104 (1999).

15. The issue of silyl migration in ribooligonucleotide synthesis has been reviewed in S. L. Beaucage and R. P. Iyer, *Tetrahedron*, **48**, 2223 (1992).

16. S. S. Jones and C. B. Reese, *J. Chem. Soc., Perkin Trans. 1*, 2762 (1979); W. Köhler and W. Pfleiderer, *Liebigs Ann. Chem.*, 1855 (1979).

17. U. Peters, W. Bankova and P. Welzel, *Tetrahedron*, **43**, 3803 (1987).

18. D. A. Evans, S. W. Kaldor, T. K. Jones, J. Clardy, and T. J. Stout. *J. Am. Chem. Soc.*, **112**, 7001 (1990).

19. O. Kwon and S. J. Danishefsky, *J. Am. Chem. Soc.*, **120**, 1588 (1998).

20. M. C. Hillier and A. I. Meyers, *Tetrahedron Lett.*, **42**, 5145 (2001); M. C. Hillier and A. I. Meyers, *Tetrahedron Lett.*, **42**, 5145 (2001).

21. O. Corminboeuf, L. E. Overman, and L. D. Pennington, *Org. Lett.*, **5**, 1543 (2003).

22. M. S. Arias-Perez, M. S. Lopez, and M. J. Santos, *J. Chem. Soc. Perkin Trans. 2*, 1549 (2002); S. Furegati, A. J. P. White, and A. D. Miller, *Synlett*, 2385 (2005).

23. T. Yamazaki, T. Ichige, and T. Kitazume, *Org. Lett*, **6**, 4073 (2004).

24. T.-Y. Ku, T. Grieme, P. Raje, P. Sharma, S. A. King, and H. E. Morton, *J. Am. Chem. Soc.*, **124**, 4282 (2002).

25. J. M. Lassaletta and R. R. Schmidt, *Synlett*, 925 (1995). See also D. L. Boger, S. Ichikawa and W. Zhong, *J. Am. Chem. Soc.*, **123**, 4161 (2001).

26. C. F. Masaguer, Y. Bleriot, J. Charlwood, B. G. Winchester, and G. W. J. Fleet, *Tetrahedron*, **53**, 15147 (1997).

27. T. J. Hunter and G. A. O'Doherty, *Org. Lett.*, **3**, 1049 (2001).

28. E. Colvin, *Silicon in Organic Synthesis*, Butterworths, Boston, Chapter 5, 1981.

29. A. B. Smith, III, Y. Qiu, D. R. Jones, and K. Kobayashi, *J. Am. Chem. Soc.*, **117**, 12011 (1995).

30. A. B. Smith, III, T. J. Beauchamp, M. J. LaMarche, M. D. Kaufman, Y. Qiu, H. Arimoto, D. R. Jones, and K. Kobayashi, *J. Am. Chem. Soc.*, **122**, 8654 (2000).

31. G. Simchen and J. Pfletschinger, *Angew. Chem., Int. Ed. Engl.*, **15**, 428 (1976); M. H. Hu, P. E. Fanwick, K. Wood, and M. Cushman, *J. Org. Chem.*, **60**, 5905 (1995); J. O. Karlsson, N. V. Nguyen, L. D. Foland, and H. W. Moore, *J. Am. Chem. Soc.*, **107**, 3392 (1985).

32. K. Sakaguchi, M. Fujita, H. Suzuki, M. Higashino, and Y. Ohfune, *Tetrahedron Lett.*, **41**, 6589 (2000).

33. Y. M. Hijji, P. F. Hudrlik, C. O. Okoro, and A. M. Hudrlik, *Synth. Commun.*, **27**, 4297 (1997).

34. M. K. Manthey, C. González-Bello, and C. Abell, *J. Chem. Soc., Perkin Trans. 1*, 625 (1997).

Trimethylsilyl Ether (TMS−OR): $ROSi(CH_3)_3$ (Chart 1)

A large number of silylating agents exist for the introduction of the trimethylsilyl group onto a variety of alcohols. In general, the sterically least hindered alcohols are the most readily silylated, but these are also the most labile to hydrolysis with either acid or base. Trimethylsilylation is used extensively for derivatization of most functional groups to increase volatility for gas chromatography and mass spectrometry.

Formation

1. Me_3SiCl, Et_3N, THF, 25°C, 8 h, 90% yield.[1]

2. Me_3SiCl, Li_2S, CH_3CN, 25°C, 12 h, 75–95% yield.[2] Silylation occurs under neutral conditions with this combination of reagents.

3. Me_3SiCl, Mg, DMF, rt, 70–99% yield. Tertiary alcohols are readily silylated. The TES and $PhMe_2Si$ ether have also been prepared by this method.[3]

4. $(Me_3Si)_2NH$, Me_3SiCl, Pyr, 20°C, 5 min, 100% yield.[4] ROH is a carbohydrate.

   Hexamethyldisilazane (HMDS) is one of the most common silylating agents and readily silylates alcohols, acids, amines, thiols, phenols, hydroxamic acids, amides, thioamides, sulfonamides, phosphoric amides, phosphites, hydrazines, and enolizable ketones. It works best in the presence of a catalyst such as X−NH−Y, where at least one of the groups X or Y is electron-withdrawing.[5] Saccharin is an excellent catalyst. Yttrium-based Lewis acids,[6] iodine,[7] zirconium sulfophenyl phosphonate,[8] $LiClO_4$,[9] $CuSO_4 \cdot 5H_2O$,[10] and tungstophosphoric acid[11] also serve as catalysts. $Cu(OTf)_2$ and I_2 have been used as catalysts for the silylation of α-hydroxyphosphonates.[12]

HO⌒⌒⌒OH →[0.5 eq. HMDS, THF / 1 drop TMSCl / Reflux, 92%] HO⌒⌒⌒OTMS

5. PhNHTMS, catalytic TBAF, DMF, 81–99% yield. This method efficiently silylates tertiary alcohols. The corresponding TES and TBS derivatives may be prepared with equal efficiency by the same method. These authors also report the following relative reactivity for various silylating agents.[14]

Reactivity for silylation of 1-octanol without TBAF catalysis

Reactivity for silylation of terpinen-4-ol with TBAF catalysis

6. $(Me_3Si)_2O$, PyH^+TsO^-, PhH, mol. sieves, reflux, 4 days, 80–90% yield.[15] These mildly acidic conditions are suitable for acid-sensitive alcohols.

7. Me_3SiNEt_2.[16] Trimethylsilyldiethylamine selectively silylates equatorial hydroxyl groups in quantitative yield (4–10 h, 25°C). The report indicated no reaction at axial hydroxyl groups. In the prostaglandin series the order of reactivity of trimethylsilyldiethylamine is $C_{11} > C_{15} \gg C_9$ (no reaction). These trimethylsilyl ethers are readily hydrolyzed in aqueous methanol containing a trace of acetic acid.[17] The reagent is also useful for the silylation of amino acids.[18]

8. $CH_3C(OSiMe_3)=NSiMe_3$, DMF, 78°C.[19] ROH is a C_{14}-hydroxy steroid. The sterically hindered silyl ether is stable to a Grignard reaction, but is hydrolyzed with 0.1 N HCl/10% aq. THF, 25°C.[19] The reagent also silylates amides, amino acids, phenols, carboxylic acids, enols, ureas, and imides.[20] Most active hydrogen compounds can be silylated with this reagent.

9. $Me_3SiCH_2CO_2Et$, cat. Bu_4NF, 25°C, 1–3 h, 90% yield. This reagent combination allows isolation of pure products under nonaqueous conditions. The reagent also converts aldehydes and ketones to trimethylsilyl enol ethers.[21] The analogous methyl trimethylsilylacetate has also been used.[22]

10. $Me_3SiNHSO_2OSiMe_3$, CH_2Cl_2, 30°C, 0.5 h, 92–98% yield. Higher yields of trimethylsilyl derivatives are realized by reaction of aliphatic, aromatic, and carboxylic hydroxyl groups with N,O-bis(trimethylsilyl)sulfamate than by reaction with N,O-bis(trimethylsilyl)acetamide.[23]

11. Me₃SiNHCO₂SiMe₃, THF, rapid, 80–95% yield. This reagent also silylates phenols and carboxyl groups.[24]

12. MeCH=C(OMe)OSiMe₃, CH₃CN, or CH₂Cl₂, 50°C, 30–50 min, 83–95% yield.[25] In addition, this reagent silylates phenols, thiols, amides, and carboxyl groups.

13. Me₃SiCH₂CH=CH₂, TsOH, CH₃CN, 70–80°C, 1–2 h, 90–95% yield.[26] This silylating reagent is stable to moisture. Allylsilanes can be used to protect alcohols, phenols, and carboxylic acids; there is no reaction with thiophenol except when CF₃SO₃H[27] is used as a catalyst. The method is also applicable to the formation of *t*-butyldimethylsilyl derivatives; the silyl ether of cyclohexanol was prepared in 95% yield from allyl-*t*-butyldimethylsilane. Iodine, bromine, trimethylsilyl bromide, and trimethylsilyl iodide have also been used as catalysts.[28] Nafion-H has been shown to be an effective catalyst.[29] The reaction of allyl trimethylsilane with TFA produces TMSOTf *in situ*; this can be trapped with pyridine to form a crystalline pyridinium salt, which serves as a powerful silylating reagent.[30]

14. (Me₃SiO)₂SO₂.[31] This is a powerful silylating reagent, but has seen little application in organic chemistry.

15. *N,O*-Bis(trimethylsilyl)trifluoroacetamide.[32] This reagent is suitable for the silylation of carboxylic acids, alcohols, phenols, amides, and ureas. It has the advantage over bis(trimethylsilyl)acetamide in that the by-products are more volatile. It has been used for the selective protection of 10-desacetylbaccatin III using LHMDS as a catalyst. The TES and TBDMS ethers were prepared similarly.[33] Conventional conditions using the silyl chloride results in silylation of the C-7 hydroxyl:

Entry	R	Reaction Conditions	% Yield
1	TMS	BTMSTFA, 08C, 5 h	91
2	TES	BTESTFA, rt, 24 h	85
3	TES	BTESTFA, THF, LHMDS (cat), 08C, 10 min	95
4	TBS	BTBSTFA, THF, LHMDS (cat), 08C, 5 h	70

16. *N,N'*-Bistrimethylsilylurea, CH₂Cl₂.[34] This reagent readily silylates carboxylic acids and alcohols. The by-product urea is easily removed by filtration. The use of this reagent has been reviewed.[35]

17. Me$_3$SiSEt.[36] Alcohols, thiols, amines, and carboxylic acids are silylated.

18. Nafion–TMS, Et$_3$N, CH$_2$Cl$_2$, 100% yield.[37]

19. Isopropenyloxytrimethylsilane.[38] In the presence of an acid catalyst, this reagent silylates alcohols and phenols. It also silylates carboxylic acids without added catalyst.

20. Methyl 3-trimethylsiloxy-2-butenoate.[39] This reagent silylates primary, secondary, and tertiary alcohols at room temperature without added catalyst.

21. N-Methyl-N-trimethylsilylacetamide.[40] This reagent has been used preparatively to silylate amino acids.[41]

22. Trimethylsilyl cyanide.[42] This reagent readily silylates alcohols, phenols, and carboxylic acids, but more slowly silylates thiols and amines. Amides and related compounds do not react with this reagent. The reagent has the advantage that a volatile gas (HCN is highly toxic) is the only by-product. In the following case the use of added base resulted in retro aldol condensation.[43]

23. TMSN$_3$, TBAB, 30°C. Primary, secondary and tertiary alcohols are all silylated in excellent yield.[44]

24. Me$_3$SiOC(O)NMe$_2$.[45] This reagent produces only volatile byproducts and autocatalytically silylates alcohols, phenols, and carboxylic acids.

25. Trimethylsilylimidazole, CCl$_4$, or THF, rt.[46] This is a powerful silylating agent for hydroxyl groups. Basic amines are not silylated with this reagent, but as the acidity increases, silylation can occur. TBAF has been used to catalyze trimethylsilylation with this reagent and other silylating agents of the general form R$_3$SiNR'$_2$.[47] A secondary aldol was readily silylated in the presence of a 3° hydroxyl.

26. Trimethylsilyl trichloroacetate, K$_2$CO$_3$, 18-crown-6, 100–150°C, 1–2 h, 80–90% yield.[48] This reagent silylates phenols, thiols, carboxylic acids, acetylenes, urethanes, and β-keto esters, producing CO$_2$ and chloroform as byproducts.

27. 3-Trimethylsilyloxazolidinone.[49] This reagent can be used to silylate most active hydrogen compounds.

28. Trimethylsilyl trifluoromethanesulfonate. This is an extremely powerful silylating agent, but probably is more useful for its many other applications in synthetic chemistry.[50] The following illustrates a recent case where conventional conditions failed.[51]

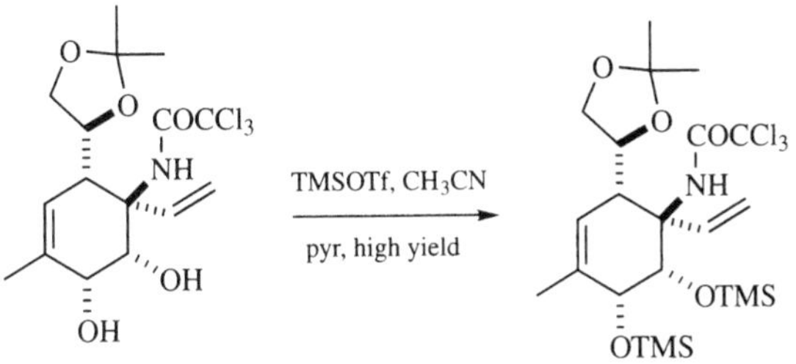

TMSOTf, CH₃CN

pyr, high yield

Cleavage

Trimethylsilyl ethers are quite susceptible to acid hydrolysis, but acid stability is quite dependent on the local steric environment. For example, the 17α-TMS ether of a steroid is quite difficult to hydrolyze. TMS ethers are readily cleaved with the numerous HF-based reagents. A polymer-bound ammonium fluoride is advantageous for isolation of small polar molecules.[52]

1. Bu₄NF, THF, aprotic conditions.[1]
2. H₂SiF₆.[53]
3. K₂CO₃, anhydrous MeOH, 0°C, 45 min, 100% yield.[54]

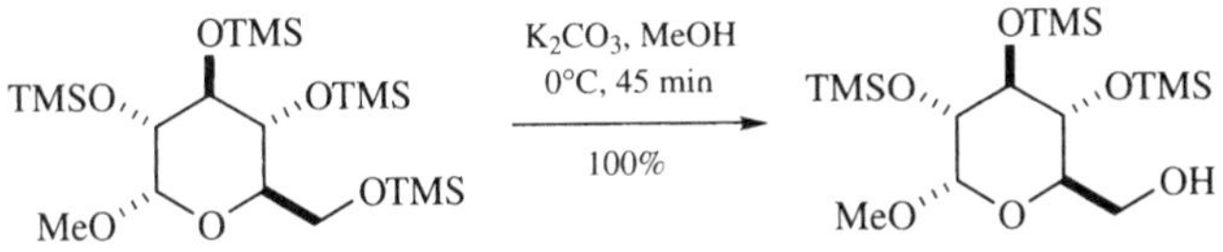

K₂CO₃, MeOH
0°C, 45 min

100%

4. Citric acid, MeOH, 20°C, 10 min, 100% yield.[55] For simple TMS ethers, almost any protic acid in an alcoholic solvent will remove the TMS group. It is only in highly functionalized and otherwise sensitive substrates that more specialized and unique methods are required.

5. Rexyn 101 (polystyrenesulfonic acid), 80–91% yield.[56] This method does not cleave the *t*-butyldimethylsilyl ether.

6. FeCl₃, CH₃CN, rt, 1 min.[57]

7. BF₃·Et₂O.[58]

8. DDQ, wet EtOAc.[59]

9. RedAl.[60]

10. Direct oxidative cleavage of the TMS ether to an aldehyde or ketone is possible and has been amply demonstrated only on relatively simple substrates. A large number of reagents are available to effect this conversion. The following are a sampling: (Ph₃SiO)₂CrO₂, *t*-BuOOH, CH₂Cl₂, rt, 42–98% yield,[61] Fe(NO₃)₃/montmorillonite clay, 70–95% yield,[62] NaBrO₃/NH₄Cl/aq. CH₃CN, 55–90% yield,[63] (*n*-BuPPh₃)₂S₂O₈/CH₃CN, 93–99% yield,[64] KMnO₄/AlCl₃/acetone/CH₃CN, 60–90% yield,[65] PdCl₂(PhCN)₂–CrO₃/clay–bis(trimeth

ylsilyl)chromate, 83–99% yield,[66] silica gel supported Dess–Martin peri-odane/CH_2Cl_2, 82–98% yield,[67] benzyltriphenylphosphonim chlorate/$AlCl_3$/CH_3CN, 20–100% yield,[68] tetrabutylammonium periodate/$AlCl_3$, 0–95% yield,[69] montmorillonoite-supported bis(trimethylsilyl)chromate/CH_2Cl_2, 82–93% yield,[70,71] benzyltriphenylphosphonium chlorochromate/$AlCl_3$/CH_3CN, 78–99% yield,[72] Zeofen/microwaves, 78–98% yield,[73] and wet alumina-supported CrO_3, 72–90% yield.[74]

1. E. J. Corey and B. B. Snider, *J. Am. Chem. Soc.*, **94**, 2549 (1972).

2. G. A. Olah, B. G. B. Gupta, S. C. Narang, and R. Malhotra, *J. Org. Chem.*, **44**, 4272 (1979).

3. I. Nishiguchi, Y. Kita, M. Watanabe, Y. Ishino, T. Ohno, and H. Mackawa, *Synlett*, **7**, 1025 (2000).

4. C. C. Sweeley, R. Bentley, M. Makita, and W. W. Wells, *J. Am. Chem. Soc.*, **85**, 2497 (1963).

5. C. A. Bruynes and T. K. Jurriens, *J. Org. Chem.*, **47**, 3966 (1982).

6. P. Kumar, G. C. G. Pais, and A. Keshavaraja, *J. Chem. Res., Synop.*, 376 (1996).

7. B. Karimi and B. Golshani, *J. Org. Chem.*, **65**, 7228 (2000).

8. M. Curini, F. Epifano, M. C. Marcotullio, O. Rosati, and U. Costantino, *Synth. Commum.*, **29**, 541 (1999).

9. B. P. Bandgar and S. P. Kasture, *Monatshefte fuer Chemie*, **132**, 1101 (2001); N. Azizi and M. R. Saidi, *Organometallics*, **23**, 1457 (2004).

10. B. Akhlaghinia and S. Tavakoli, *Synthesis*, 1775 (2005).

11. H. Firouzabadi, N. Iranpoor, K. Amani, and F. Nowrouzi, *J. Chem. Soc. Perkin Trans. 1*, 2601 (2002).

12. H. Firouzabadi, N. Iranpoor, S. Sobhani, S. Ghassamipour, and Z. Amoozgar, *Tetrahedron Lett.*, **44**, 891 (2003); H. Firouzabadi, N. Iranpoor, and S. Sobhani, *Tetrahedron Lett.*, **43**, 3653 (2002).

13. R. K. Kanjolia and V. D. Gupta, *Z. Naturforsch. B*, **35B**, 767 (1980).

14. A. Iiada, A. Horii, T. Misaki, and Y. Tanabe, *Synthesis*, 2677 (2005).

15. H. W. Pinnick, B. S. Bal, and N. H. Lajis, *Tetrahedron Lett.*, **19**, 4261 (1978); H. Matsumoto, Y. Hoshio, J. Nakabayashi, T. Nakano, and Y. Nagai, *Chem. Lett.*, **9**, 1475 (1980).

16. I. Weisz, K. Felföldi, and K. Kovács, *Acta Chim. Acad. Sci. Hung.*, **58**, 189 (1968).

17. E. W. Yankee, U. Axen, and G. L. Bundy, *J. Am. Chem. Soc.*, **96**, 5865 (1974); E. L. Cooper and E. W. Yankee, *J. Am. Chem. Soc.*, **96**, 5876 (1974).

18. K. Rühlmann, *J. Prakt. Chem.*, **9**, 315 (1959); K. Rühlmann, *Chem. Ber.*, **94**, 1876 (1961).

19. M. N. Galbraith, D. H. S. Horn, E. J. Middleton, and R. J. Hackney, *J. Chem. Soc., Chem. Commun.*, 466 (1968).

20. J. F. Klebe, H. Finkbeiner, and D. M. White, *J. Am. Chem. Soc.*, **88**, 3390 (1966).

21. E. Nakamura, T. Murofushi, M. Shimizu, and I. Kuwajima, *J. Am. Chem. Soc.*, **98**, 2346 (1976).

22. L. A. Paquette and T. Sugimura, *J. Am. Chem. Soc.*, **108**, 3841 (1986); T. Sugimura and L. A. Paquette, *J. Am. Chem. Soc.*, **109**, 3017 (1987).

23. B. E. Cooper and S. Westall, *J. Organomet. Chem.*, **118**, 135 (1976).

24. L. Birkofer and P. Sommer, *J. Organomet. Chem.*, **99**, C1 (1975).

25. Y. Kita, J. Haruta, J. Segawa, and Y. Tamura, *Tetrahedron Lett.*, **20**, 4311 (1979).

26. T. Morita, Y. Okamoto, and H. Sakurai, *Tetrahedron Lett.*, **21**, 835 (1980).

27. G. A. Olah, A. Husain, B. G. B. Gupta, G. F. Salem, and S. C. Narang, *J. Org. Chem.*, **46**, 5212 (1981).

28. H. Hosomi and H. Sakurai, *Chem Lett.*, **10**, 85 (1981).

29. G. A. Olah, A. Husain, and B. P. Singh, *Synthesis*, 892 (1983).

30. G. A Olah and D. A. Klumpp, *Synthesis*, 744, (1997).

31. L. H. Sommer, G. T. Kerr, and F. C. Whitmore, *J. Am. Chem. Soc.*, **70**, 445 (1948).

32. D. L. Stalling, C. W. Gehrke, and R. W. Zumalt, *Biochem. Biophys. Res. Commun.*, **31**, 616 (1968); M. G. Horning, E. A. Boucher, and A. M. Moss, *J. Gas Chromatogr.*, 297 (1967).

33. R. A. Holton, Z. Zhang, P. A. Clarke, H. Nadizadeh, and D. J. Procter, *Tetrahedron Lett*, **39**, 2883 (1998).

34. W. Verboom, G. W. Visser, and D. N. Reinhoudt, *Synthesis*, 807 (1981).

35. M. T. El Gihani and H. Heaney, *Synthesis*, 357 (1998).

36. E. W. Abel, *J. Chem. Soc.*, 4406 (1960); *idem, ibid.*, 4933 (1961).

37. S. Murata and R. Noyori, *Tetrahedron Lett.*, **21**, 767 (1980).

38. M. Donike and L. Jaenicke, *Angew. Chem., Int. Ed. Engl.*, **8**, 974 (1969).

39. T. Veysoglu and L. A. Mitscher, *Tetrahedron Lett.*, **22**, 1303 (1981).

40. L. Birkofer and M. Donike, *J. Chromatogr.*, **26**, 270 (1967).

41. H. R. Kricheldorf, *Justus Liebigs Ann. Chem.*, **763**, 17 (1972).

42. K. Mai and G. Patil, *J. Org. Chem.*, **51**, 3545 (1986).

43. E. J. Corey and Y.-J. Wu, *J. Am. Chem. Soc.*, **115**, 8871 (1993).

44. D. Amantini, F. Fringuelli, F. Pizzo, and L. Vaccaro, *J. Org. Chem.*, **66**, 6734 (2001).

45. D. Knausz, A. Meszticzky, L. Szakacs, B. Csakvari, and K. D. Ujszaszy, *J. Organomet. Chem.*, **256**, 11 (1983); D. Knausz, A. Meszticzky, L. Szakacs, and B. Csakvari, *J. Organomet. Chem.*, **268**, 207 (1984).

46. S. Torkelson and C. Ainsworth, *Synthesis* 722 (1976).

47. Y. Tanabe, M. Murakami, K. Kitaichi, and Y. Yoshida, *Tetrahedron Lett.*, **35**, 8409, (1994); Y. Tanabe, H. Okumura, A. Maeda, and M. Murakami, *ibid.*, **35**, 8413 (1994).

48. J. M. Renga and P.-C. Wang, *Tetrahedron Lett.*, **26**, 1175 (1985).

49. C. Palomo, *Synthesis*, 809 (1981); J. M. Aizpurua, C. Palomo, and A. L. Palomo, *Can. J. Chem.*, **62**, 336 (1984).

50. Review: H. Emde, D. Domsch, H. Feger, U. Frick, A. Götz, H. H. Hergott, K. Hofmann, W. Kober, K. Krägeloh, T. Oesterle, W. Steppan, W. West, and G. Simchen, *Synthesis*, 1 (1982).

51. T. Nishikawa, M. Asai, and M. Isobe, *J. Am. Chem. Soc.*, **124**, 7847 (2002).

52. C. Li, Y. Lu, W. Huang, and B. He, *Synth. Commun.*, **21**, 1315 (1991).

53. J. A. Marshall and M. P. Bourbeau, *J. Org. Chem.*, **67**, 2751 (2002).

54. D. T. Hurst and A. G. McInnes, *Can. J. Chem.*, **43**, 2004 (1965).

55. G. L. Bundy and D. C. Peterson, *Tetrahedron Lett.*, **19**, 41 (1978).

56. R. A. Bunce and D. V. Hertzler, *J. Org. Chem.*, **51**, 3451 (1986).

57. A. D. Cort, *Synth. Commun.*, **20**, 757 (1990); P. Saravanan and V. K. Singh, *J. Ind. Chem. Soc.*, **75**, 565 (1998).

58. L. Pettersson and T. Frejd, *J. Chem. Soc., Chem. Commun.*, 1823 (1993).

59. A. Oku, M. Kinugasa, and T. Kamada, *Chem. Lett.*, **22**, 165 (1993).

60. S.-H. Chen, V. Farina, D. M. Vyas, T. W. Doyle, B. H. Long, and C. Fairchild, *J. Org. Chem.*, **61**, 2065 (1996).

61. J. Muzart and A. N. Ajjou, *Synlett*, 497 (1991).

62. M. M. Mojtahedi, M. R. Saidi, M. Bolourtchian, and M. M. Heravi, *Synth. Commum.*, **29**, 3283 (1999).

63. A. Shaabani and A.-R. Karimi, *Synth. Commum.*, **31**, 759 (2001).

64. I. Mohammadpoor-Baltork, A. R. Hajipour, and M. Aghajari, *Syn. Comm.*, **32**, 1311 (2002).

65. H. Firouzabadi, S. Etemadi, B. Karimi, and A. S. Jarrahpour, *Phosphorus, Sulfur and Silicon*, **152**, 141 (1999).

66. M. M. Heravi, D. Ajami, D. Aghapoor, and M. Ghassemzadeh, *Phosphorus, Sulfur and Silicon*, **158**, 151 (2000).

67. H. A. Oskooie, M. Khalilpoor, A. Saednia, N. Sarmad, and M. M. Heravi, *Phosphorus, Sulfur and Silicon*, **166**, 197 (2000).

68. A. R. Hajipour, S. E. Mallakpour, I. Mohammadpoor-Baltork, and M. Malakoutikhah, *Tetrahedron*, **58**, 143 (2002).

69. H. Firouzabadi, H. Badparva, and A. R. Sardarian, *Iran. J. Chem. & Chem. Eng.*, **17**, 33 (1998).

70. M. M. Heravi, D. Ajami, K. Tabar-Heydar, and M. M. Mojtahedi, *J. Chem. Res. (S)*, 620 (1998).

71. M. M. Heravi, D. Ajami, and K. Tabar-Heydar, *Synth. Commum.*, **29**, 1009 (1999).

72. A. R. Hajipour, S. E. Mallakpour, I. M. Baltork, and H. Backnezhad, *Org. Prep. Proc. Int.*, **34**, 169 (2002).

73. M. M. Heravi, D. Ajami, M. Ghassemzadeh, and K. Tabar-Hydar, *Synth. Commum.*, **31**, 2097 (2001).

74. M. M. Heravi, D. Ajami, and M. Ghassemzadeh, *Synthesis*, 393 (1999).

Triethylsilyl Ether (TES−OR): Et₃SiOR

Formation

1. Et₃SiCl, Pyr. Triethylsilyl chloride is by far the most common reagent for the introduction of the TES group.[1] Silylation also occurs with imidazole and DMF,[2] and with dimethylaminopyridine as a catalyst.[3] Phenols,[4] carboxylic acids,[5] and amines[6] have also been silylated with TESCl.

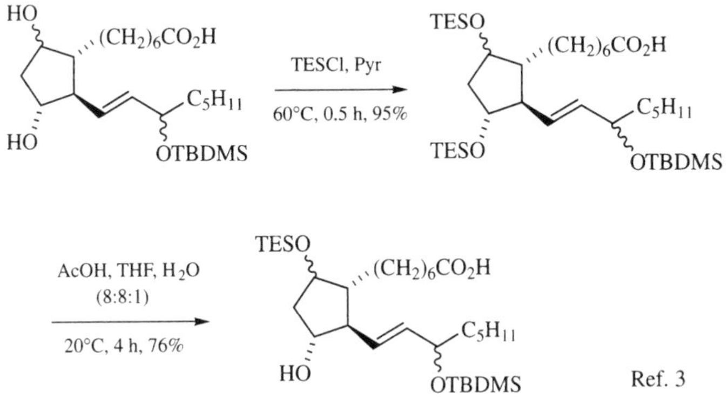

$$\text{AcOH, THF, H}_2\text{O} \atop (8:8:1)$$

$$20°C, 4 \text{ h}, 76\%$$

Ref. 3

More acidic conditions [AcOH, THF, H$_2$O (6:1:3), 45°C, 3 h] cleave all the protective groups, 76% yield.

2. TESCl, 2,6-lutidine, CH$_2$Cl$_2$, −78°C, 97% yield. Lutidine was crucial to getting selectivity for the primary hydroxyl at C-38 over C-24 and the carboxyl group. The use of imidazole as base resulted in over silylation.[7]

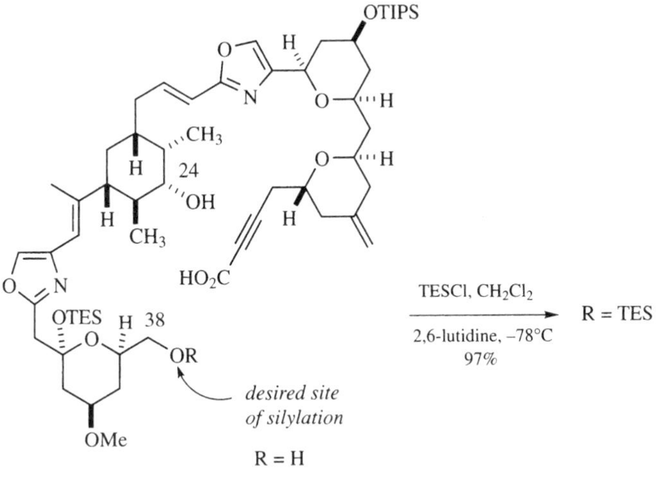

TESCl, CH$_2$Cl$_2$

2,6-lutidine, −78°C
97%

R = TES

desired site of silylation

R = H

3. Triethylsilyl triflate.[8] This has become a popular reagent for the preparation of TES ethers. Commonly used bases are pyridine and 2,6-lutidine.[9] The most frequently used solvent is CH$_2$Cl$_2$, but others such as CH$_3$CN have also been used.

4. Triethylsilane, catalytic B(C$_6$H$_5$)$_3$, hexane, or CH$_2$Cl$_2$, 86–95% yield. Primary alcohols can be reduced with this reagent. Alcohols and phenols are readily silylated, but under suitable conditions some alcohols and ethers are reduced.[10,11]

5. Triethylsilane, t-BuOCu, DTBM-Xantphos, toluene, 84–95% yield. This method will also introduce other silyl groups such as PhMe$_2$Si, Ph$_3$Si, t-BuPh$_2$Si, and t-BuMeSi groups. Primary alcohols can be protected selectively in the presence of secondary alcohols.[12]

6. Triethylsilane, [RuCl$_2$(p-cym)]$_2$, CH$_2$Cl$_2$, 50°C, 6 h, >95% yield.[13]

7. Triethylsilane, Cl$_2$(PCy$_3$)$_2$Ru=CHPh, 45°C, 6 h, 95% yield.[14] Aldehydes are reduced with this reagent. The method can be used to prepare a variety of other silyl ethers. Rh$_2$(pfb)$_4$ can also be used as a catalyst.[15]

8. Triethylsilane, CsF, imidazole.[16]

9. Triethylsilane, CH$_2$Cl$_2$, 1% Rh$_2$(pfb)$_4$ (rhodium perfluorobutyrate), 2 h, 88% yield.[17]

10. N-Methyl-N-triethylsilyltrifluoroacetamide.[18]

11. Allyltriethylsilane.[19]

12. N-Triethylsilylacetamide.[20]

13. Triethylsilyldiethylamine.[21]

14. 1-Methoxy-1-triethylsiloxypropene.[22]

15. 1-Methoxy-2-methyl-1-triethylsiloxypropene.[23]

16. Triethylsilyl perchlorate.[24] This reagent represents an **explosion** hazard.

17. Triethylsilyl cyanide.[25]

Cleavage

The triethylsilyl ether is approximately 10–100 times more stable[5] than the TMS ether and thus shows a greater stability to many reagents. Although TMS ethers can be cleaved in the presence of TES ethers, steric factors will play an important role in determining selectivity. The TES ether can be cleaved in the presence of a t-butyldimethylsilyl ether using 2% HF in acetonitrile.[26] In general, methods used to cleave the TBDMS ether are effective for cleavage of the TES ether.[27]

Pv = Pivaloyl

1. H$_2$SiF$_6$, IPA, −40°C, 88% yield. A primary TES group was removed in the presence of TBS and TIPS ether.[28]

2. DDQ, CH$_3$CN or THF, H$_2$O, 86–100% yield.[29] TBDMS ethers are not usually cleaved.

3. AcOH, TFA, H$_2$O, 80% yield. This procedure was developed to remove the 7-TES group from 7-TES Paclitaxel while retaining the C10 acetate.[30]

4. MeOH, 1-chloroethylchloroformate, 86–99% yield. This method produces HCl in situ. These conditions will cleave the TES group in the presence of TBDMS, THP, Tr, MOM, MEM, and Ts groups. They may also be be used to cleave a TBDMS group in the presence larger silyl ethers and the MOM and MEM ethers.[31]

5. Ph$_3$P·HBr, MeOH, CH$_2$Cl$_2$, 0°C, 80% yield.[32]

6. Iodoxybenzoic acid, DMSO, 20°C, 30 min, 62–93% yield. Primary TES groups are cleaved in the presence of TBDMS ethers. The drawback to this reagent is that some oxidation of the alcohol to an aldehyde occurs.[33]

7. Mesoporous silica (MCM-41), MeOH, rt, 2 h, 80–97% yield. TES groups are cleaved in preference to TBDMS groups.[34]

8. $ZnBr_2$, CH_2Cl_2, H_2O, >80% yield. This reagent is not selective; TES, TBDMS, and TIPS ethers are also cleaved, but phenolic TBDMS ethers are stable.[35]

9. $BiOClO_4 \cdot xH_2O$, CH_2Cl_2, 32–92% , TES, TBDMS, TIPS and TBDPS ethers are all cleaved.[36]

10. DMSO, $(COCl)_2$, CH_2Cl_2, TEA, −70°C, 64–86% yield. These conditions selectively convert a primary TES group to an aldehyde without effecting secondary TES ethers.[37] TMS ethers react similarly.

11. NaOH, DMPU, H_2O, 60% yield.[38]

12. Pd/C, MeOH, H_2.[39,40] There have been many instances where a silyl ether has been lost during a hydrogenation, which has led to speculation that silyl ethers can be cleaved by hydrogenolysis. It has been determined that the real mechanism for silyl ether loss is really a simple acid-catalyzed process that results from residual acid in the catalyst or acid that is formed from $PdCl_2$ used to prepare some forms of Pd–C. The only case where a true hydrogenolysis seems to cleave a silyl ether is the TES group. The reaction has a strong steric dependence.[41] Phenolic TES ethers are cleaved at a much slower rate than the alkyl counterpart.

1. T. W. Hart, D. A. Metcalfe, and F. Scheinmann, *J. Chem. Soc., Chem. Commun.*, 156 (1979).

2. W. Oppolzer, R. L. Snowden, and D. P. Simmons, *Helv. Chim. Acta*, **64**, 2002 (1981).

3. W. R. Roush and S. Russo-Rodriguez, *J. Org. Chem.*, **52**, 598 (1987).

4. T. L. McDonald, *J. Org. Chem.*, **43**, 3621 (1978).

5. C. E. Peishoff and W. L. Jorgensen, *J. Org. Chem.*, **48**, 1970 (1983).

6. R. West, P. Nowakowski, and P. Boudjouk, *J. Am. Chem. Soc.*, **98**, 5620 (1976).

7. D. A. Evans and D. M. Fitch, *Angew. Chem. Int. Ed.*, **39**, 2536 (2000).

8. C. H. Heathcock, S. D. Young, J. P. Hagen, R. Pilli, and U. Badertscher, *J. Org. Chem.*, **50**, 2095 (1985).

9. D. Seebach, H.-F. Chow, R. F. W. Jackson, M. A. Sutter, S. Thaisrivongs, and J. Zimmermann, *Liebigs Ann. Chem.*, 1281, (1986).

10. V. Gevorgyan, J.-X. Liu, M. Rubin, S. Benson, and Y. Yamamoto, *Tetrahedron Lett.*, **40**, 8919 (1999).

11. V. Gevorgyan, M. Rubin, S. Benson, J.-X. Liu, and Y. Yamamoto, *J. Org. Chem.*, **65**, 6179 (2000).

12. H. Ito, A. Watanabe, and M. Sawamura, *Org. Lett.*, **7**, 1869 (2005).

13. R. L. Miller, S. V. Maifeld, and D. Lee, *Org. Lett.*, **6**, 2773 (2004).

14. S. V. Maifeld, R. L. Miller, and D. Lee, *Tetrahedron Lett.*, **43**, 6363 (2002).

15. A. Biffis, M. Braga, and M. Basato, *Adv. Synth. Catal.*, **346**, 451 (2004).

16. L. Horner and J. Mathias, *J. Organomet. Chem.*, **282**, 175 (1985).

17. M. P. Doyle, K. G. High, V. Bagheri, R. J. Pieters, P. J. Lewis, and M. M. Pearson, *J. Org. Chem.*, **55**, 6082 (1990).

18. M. Donike and J. Zimmermann, *J. Chromatogr.*, **202**, 483 (1980).

19. A. Hosomi and H. Sakurai, *Chem. Lett.*, **10**, 85 (1981).

20. J. Dieckman and C. Djerassi, *J. Org. Chem.*, **32**, 1005 (1967); J. Dieckman, J. B. Thompson, and C. Djerassi, *ibid.*, 3904 (1967).

21. A. R. Bassindale and D. R. M. Walton, *J. Organomet. Chem.*, **25**, 389 (1970).

22. Y. Kita, J. Haruta, J. Segawa, and Y. Tamura, *Tetrahedron Lett.*, **20**, 4311 (1979).

23. E. Yoshii and K. Takeda, *Chem. Pharm. Bull.*, **31**, 4586 (1983).

24. T. J. Barton and C. R. Tully, *J. Org. Chem.*, **43**, 3649 (1978); D. B. Collum, J. H. McDonald, III, and W. C. Still, *J. Am. Chem. Soc.*, **102**, 2117 (1980). For *O*-silylation of esters, see C. S. Wilcox and R. E. Babston, *Tetrahedron Lett.*, **25**, 699 (1984).

25. K. Mai and P. Patil, *J. Org. Chem.*, **51**, 3545 (1986).

26. D. Boschelli, T. Takemasa, Y. Nishitani, and S. Masamune, *Tetrahedron Lett.*, **26**, 5239 (1985).

27. For an extensive review on selective silyl ether cleavage, see T. D. Nelson and R. D. Crouch, *Synthesis*, 1031 (1996).

28. J. A. Lafontaine, D. P. Provencal, C. Gardelli, and J. W. Leahy, *J. Org. Chem.*, **68**, 4215 (2003).

29. K. Tanemura, T. Suzuki, and T. Horaguchi, *J. Chem. Soc., Perkin Trans. 1*, 2997 (1992).

30. A. K. Singh, R. E. Weaver, G. L. Powers, V. W. Rosso, C. Wei, D. A. Lust, A. S. Kotnis, F. T. Comezoglu, M. Liu, K. S. Bembenek, B. D. Phan, D. J. Vanyo, M. L. Davies, R. Mathew, V. A. Palaniswamy, W.-S. Li, K. Gadamsetti, C. J. Spagnuolo, and W. J. Winter, *Organic Process Research & Development*, **7**, 25 (2003).

31. C.-E. Yeom, Y. J. Kim, S. Y. Lee, Y. J. Shin, and B. M. Kim, *Tetrahedron*, **61**, 12227 (2005).

32. C. H. Heathcock, M. McLaughlin, J. Medina, J. L. Hubbs, G. A. Wallace, R. Scott, M. M. Claffey, C. J. Hayes, and G. R. Ott, *J. Am. Chem. Soc.*, **125**, 12844 (2003).

33. Y. Wu, J.-H. Huang, X. Shen, Q. Hu, C.-J. Tang, and L. Li, *Org. Lett.*, **4**, 2141 (2002).

34. A. Itoh, T. Kodama, and Y. Masaki, *Synlett*, 357 (1999).

35. R. D. Crouch, J. M. Polizzi, R. A. Cleiman, J. Yi, and C. A. Romany, *Tetrahedron Lett.*, **43**, 7151 (2002).

36. R. D. Crouch, C. A. Romany, A. C. Kreshock, K. A. Menconi, and J. L. Zile, *Tetrahedron Lett.*, **45**, 1279 (2004).

37. A. Rodriguez, M. Nomen, B. W. Spur, and J. J. Godfroid, *Tetrahedron Lett.*, **40**, 5161 (1999).

38. A. B. Smith, III, Q. Lin, V. A. Doughty,L. Zhuang, M. D. McBriar, J. K. Kerns, C. S. Brook, N. Murase, and K. Nakayama, *Angew. Chem. Int. Ed.*, **40**, 196 (2001).

39. T. Ikawa, H. Sajiki, and K. Hirota, *Tetrahedron*, **60**, 6189 (2004).

40. T. Ikawa, K. Hattori, H. Sajiki, and K. Hirota, *Tetrahedron*, **60**, 6901 (2004).

41. D. Rotulo-Sims and J. Prunet, *Org. Lett.*, **4**, 4701 (2002).

Triisopropylsilyl Ether (TIPS$-$OR)[1]: $(i\text{-Pr})_3\text{SiOR}$ (Chart 1)

The greater bulkiness of the TIPS group makes it more stable than the *t*-butyldimethylsilyl (TBDMS) group, but not as stable as the *t*-butyldiphenylsilyl (TBDPS) group to acidic hydrolysis. The TIPS group is more stable to basic hydrolysis than the TBDMS group and the TBDPS group.[2] TIPS group introduction onto primary hydroxyls proceeds selectively over secondary hydroxyls.[3] The TIPS group has been used to prevent chelation with Grignard reagents during additions to carbonyls.[4] As a note of caution, some lots of the reagent are contaminated with varying quantities of diisopropyl(*n*-propyl)silyl chloride and as such it would be prudent to check the quality of the reagent prior to use.[5]

Formation

1. TIPSCl, imidazole, DMF, 82% yield.[2]

2. TIPSCl, imidazole, DMAP[6] or TEA[7], CH_2Cl_2.

3. TIPSCl, pyridine, $AgNO_3$ or $Pb(NO_3)_2$, >90% yield.[8] These conditions cleanly introduce the hindered TIPS group onto the 3′-position of thymidine.

4. TIPSCl, $AgNO_3$, 78% yield. This method was used when the typical conditions failed.[9]

5. TIPSH, CsF, imidazole.[10]

6. TIPSOTf, NaH, THF, rt, 2 h, 24–85% yield. This method was used to persilylate a variety of glucose derivatives.[11] When the reaction was attempted with TIPSCl, no product was isolated. The TBS group can be introduced similarly.

7. $TIPSOSO_2CF_3$, 2,6-lutidine,[12] TEA or DIPEA,[13] CH_2Cl_2.

8. *N*-Triisopropylsilylpyridinium triflate, CH_2Cl_2, 84% yield.[14]

9. The sluggishness of the reaction of TIPSOTf with tertiary alcohols can be exploited to advantage as was the case in Magnus' strychnine synthesis.[27] The equilibrium favors the tertiary hemiketal, but silylation favors the primary alcohol.

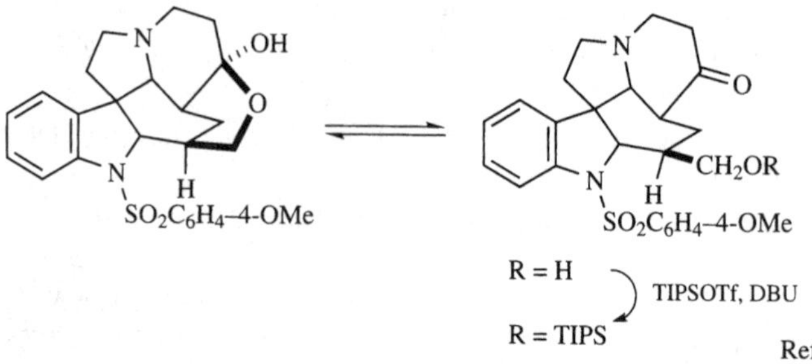

R = H
R = TIPS ⟩ TIPSOTf, DBU

Ref. 27

10. Unusual and unexpected things do happen with TIPSOTf as in the case below, but the problem was simply solved by using a more sterically demanding pivalate rather than an acetate to prevent orthoester formation.[5]

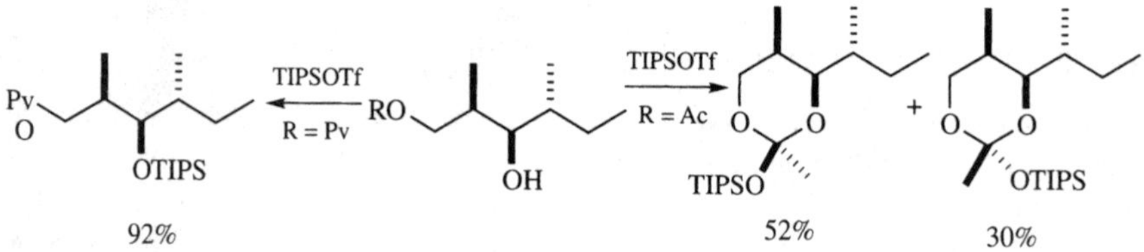

92% 52% 30%

Cleavage

1. HF, CH$_3$CN.[15] In certain sensitive substrates it may be advisable to run this reaction in a polypropylene vessel as was the case in Schreiber's synthesis of FK-506 where the yield increased from 35% to 73% when switching from the standard glass vessel.[16] This is presumably because of the fluorosilicic acid formed when HF reacts with glass.

2. 40% aqueous HF in THF.[17]

3. Pyr·HF, THF.[18]

4. Et$_3$N·HF, 25°C, 9 d, 79% yield. A 2° TIPS group was removed in the presence of a more hindered 2° TBS group.[19] The TBS group was later removed with Pyr·HF indicating that this is a more reactive reagent.

5. NH$_4$F, MeOH, rt, 9 h, 35–61% yield.[20,21,]

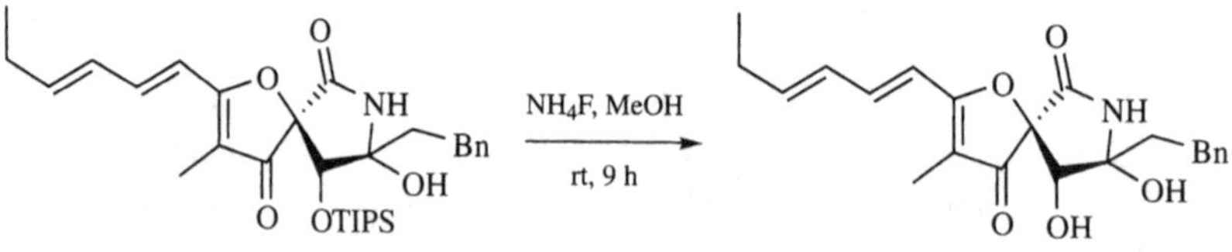

6. Bu$_4$NF, THF.[22] TBAF buffered with acetic acid is used to remove a TIPS and prevent acyl migration which is often prevalent with more basic reagents.[23,24]

7. SiF$_4$, CH$_2$Cl$_2$, CH$_3$CN, 0°C, >72% yield.[25]

8. TAS-F, DMA, 100°C, 85% yield.[26] The following example cleaves a very hindered neopentyl derivative.

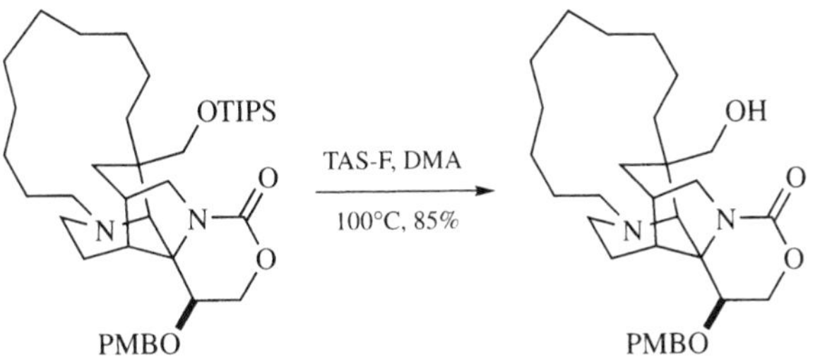

TAS-F, DMA

100°C, 85%

9. 0.01 N HCl, EtOH, 90°C, 15 min, 100% yield.[2] HCl in a variety of other concentrations has also been used to cleave the TIPS ether.[27]

10. HCl, EtOAc, −30°C to 0°C.[28]

11. 80% AcOH, H₂O.[29]

12. TFA, THF, H₂O.[30]

13. The following examples illustrate how the local steric electronic environment can reverse the expected selectivity for the deprotection of TIPS ether verses a TBS ether. The allylic TBS ether is also less nucleophilic relative to the TIPS ether because of electron withdrawing character of the olefin.[31–33]

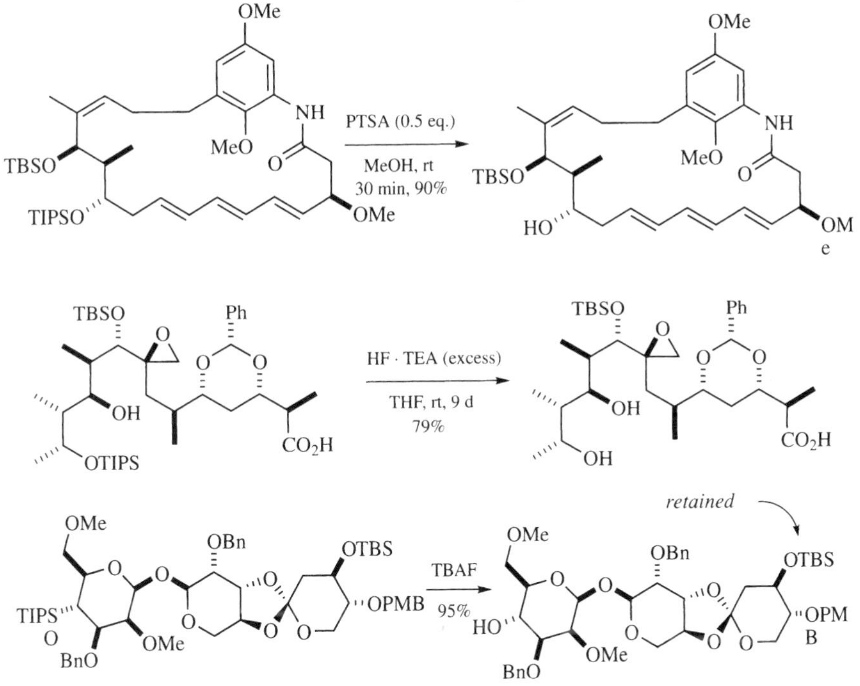

PTSA (0.5 eq.)

MeOH, rt
30 min, 90%

HF · TEA (excess)

THF, rt, 9 d
79%

retained

TBAF

95%

14. 40% KOH, MeOH, reflux, 18 h.[14]

15. NO₂BF₄.[35]

16. FeCl₃, CH₃CN, 70% yield. In this case deprotection occurs during an oxidative coupling in which HCl maybe released.[36]

1. For an extensive review of the chemistry of the triisopropylsilyl group, see C. Rücker, *Chem. Rev.*, **95**, 1009 (1995).

2. R. F. Cunico and L. Bedell, *J. Org. Chem.*, **45**, 4797 (1980).

3. K. K. Ogilvie, E. A. Thompson, M. A. Quilliam, and J. B. Westmore, *Tetrahedron Lett.*, **15**, 2865 (1974).

4. S. V. Frye and E. L. Eliel, *Tetrahedron Lett.*, **27**, 3223 (1986).

5. D. J. Barden and I. Fleming, *Chem. Commun.*, 2366 (2001); I. Fleming and A. K. Mandal, *J. Indian Chem. Soc.*, **77**, 593 (2000).

6. M. Ohwa and E. L. Eliel, *Chem. Lett.*, **16**, 41 (1987).

7. P. R. Maloney and F. G. Fang, *Tetrahedron Lett.*, **35**, 2823 (1994).

8. S. Nishino, Y. Nagato, H. Yamamoto, and Y. Ishido, *J. Carbohydr. Chem.*, **5**, 199 (1986).

9. D. R. Li, C. Y. Sun, C. Su, G.-Q. Lin, and W.-S. Zhou, *Org. Lett.*, **6**, 4261 (2004).

10. L. Horner and J. Mathias, *J. Organomet. Chem.*, **282**, 175 (1985).

11. H. Abe, S. Shuto, S. Tamura, and A. Matsuda, *Tetrahedron Lett.*, **42**, 6159 (2001).

12. E. J. Corey, H. Cho, C. Rücker, and D. H. Hua, *Tetrahedron Lett.*, **22**, 3455 (1981); K. Tanaka, H. Yoda, Y. Isobe, and A. Kaji, *J. Org. Chem.*, **51**, 1856 (1986).

13. D. W. Knight, D. Shaw, and G. Fenton, *Synlett*, 295 (1994).

14. G. A. Olah and D. A. Klumpp, *Synthesis*, 744 (1997).

15. J. L. Mascareñas, A. Mouriño, and L. Castedo, *J. Org. Chem.*, **51**, 1269 (1986).

16. M. Nakatsuka, J. A. Ragan, T. Sammakia, D. B. Smith, D. B. Uehling, and S. L. Schreiber, *J. Am. Chem. Soc.*, **112**, 5583 (1990).

17. J. Cooper, D. W. Knight, and P. T. Gallagher, *J. Chem. Soc., Chem. Commun.*, 1220 (1987).

18. P. Wipf and H. Kim, *J. Org. Chem.*, **58**, 5592 (1993).

19. D. A. Evans, A. S. Kim, R. Metternich, and V. J. Novack, *J. Am. Chem. Soc.*, **120**, 5921 (1998).

20. Y. Hayashi, M. Shoji, J. Yamaguchi, K. Sato, S. Yamaguchi, T. Mukaiyama, K. Sakai, Y. Asami, H. Kakeya, and H. Osada, *J. Am. Chem. Soc.*, **124**, 12079 (2002).

21. Y. Hayashi, M. Shoji, S. Yamaguchi, T. Mukaiyama, J. Yamaguchi, H. Kakeya, and H. Osada, *Org. Lett.*, **5**, 2287 (2003).

22. J. C.-Y. Cheng, U. Hacksell, and G. P. Daves, Jr., *J. Org. Chem.*, **51** 4941 (1986).

23. L. Han and R. K. Razdan, *Tetrahedron Lett.*, **40**, 1631 (1999).

24. J. B. Schwarz, S. D. Kuduk, X.-T. Chen, D. Sames, P. W. Glunz, and S. J. Danishefsky, *J. Am. Chem. Soc.*, **121**, 2662 (1999).

25. I. Kadota, H. Takamura, K. Sato, A. Ohno, K. Matsuda, M. Satake, and Y. Yamamoto, *J. Am. Chem. Soc.*, **125**, 11893 (2003).

26. C. J. Douglas, S. Hiebert, and L. E. Overman, *Org. Lett.*, **7**, 933 (2005).

27. P. Magnus, M. Giles, R. Bonnert, G. Johnson, L. McQuire, M. Deluca, A. Merritt, C. S. Kim, and N. Vicker, *J. Am. Chem. Soc.*, **115**, 8116 (1993). P. Magnus, M. Giles, R. Bonnert, C. S. Kim, L. McQuire, A. Merritt, and N. Vicker, *ibid.*, **114**, 4403 (1992). H. Yoda, K. Shirakawa, and K. Takabe, *Tetrahedron Lett.*, **32**, 3401 (1991).

28. W.-R. Li, W. R. Ewing, B. D. Harris, and M. M. Joullie, *J. Am. Chem. Soc.*, **112**, 7659 (1990).

29. K. K. Ogilvie, S. L. Beaucage, D. W. Entwistle, E. A. Thompson, M. A. Quilliam, and J. B. Westmore, *J. Carbohyd., Nucleosides, Nucleotides*, **3**, 197 (1976).

30. C. Eisenberg and P. Knochel, *J. Org. Chem.*, **59**, 3760 (1994).

31. C. E. Masse, M. Yang, J. Solomon, and J. S. Panek, *J. Am. Chem. Soc.*, **120**, 4123 (1998).

32. D. A. Evans, A. S. Kim, R. Metternich, and V. J. Novak, *J. Am. Chem. Soc.*, **120**, 5921 (1998).

33. K. C. Nicolaou, H. J. Mitchell, K. C. Fylaktakidou, R. M. Rodriguez, and H. Suzuki, *Chem. Eur. J.*, **6**, 3116 (2000).

34. L. E. Overman and S. R. Angle, *J. Org. Chem.*, **50**, 4021 (1985).

35. N. Hussain, D. O. Morgan, C. R. White, and J. A. Murphy, *Tetrahedron Lett.*, **35**, 5069 (1994).

36. C. A. Merlic, C. C. Aldrich, J. Albaneze-Walker, and A. Saghatelian, *J. Am. Chem. Soc.*, **122**, 3224 (2000).

Dimethylisopropylsilyl Ether (IPDMS−OR): ROSiMe$_2$−*i*-Pr (Chart 1)

Formation

1. (*i*-PrMe$_2$Si)$_2$NH, *i*-PrMe$_2$SiCl, 25°C, 48 h, 98% yield.[1]
2. *i*-PrMe$_2$SiCl, imidazole, DMF, 26°C, 2 h, 65% yield.[2]

Cleavage

1. AcOH/H$_2$O, (3:1), 35°C, 10 min, 100% yield.[1] An IPDMS ether is more easily cleaved than a THP ether. It is not stable to Grignard or Wittig reactions or to Jones oxidation.

2. Many of the fluoride based reagents such as TBAF will cleave this ether.

1. E. J. Corey and R. K. Varma, *J. Am. Chem. Soc.*, **93**, 7319 (1971).
2. K. Toshima, K. Tatsuta, and M. Kinoshita, *Tetrahedron Lett.*, **27**, 4741 (1986).

Diethylisopropylsilyl Ether (DEIPS−OR): ROSiEt$_2$−*i*-Pr

This group is more labile to hydrolysis than the TBDMS group and has been used to protect an alcohol where the TBDMS group was too resistant to cleavage. The DEIPS group is ≈90 times more stable than the TMS group to acid hydrolysis and 600 times more stable than the TMS group to base-catalyzed solvolysis.

Formation

1. Diethylisopropylsilyl chloride, imidazole, CH_2Cl_2, 25°C, 1 h.[1]
2. $Et_2(i\text{-}Pr)SiOTf$, CH_2Cl_2, 2,6-lutidine, rt.[2]

Cleavage

1. 3:1:3 AcOH, H_2O, THF.[1] Any of the methods used to cleave the TBDMS ether also cleave the DEIPS ether.

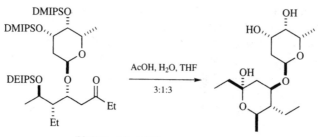

DMIPS = $Me_2i\text{-}PrSi$

2. AcOH, KF·HF, THF, H_2O, 30°C, 46 h, 94% yield.[3] These conditions did not affect a secondary OTBDMS group.
3. H_2, $Pd(OH)_2$.[4] When the cleavage is performed in dioxane, the DEIPS group is stable and benzyl ethers are selectively removed, whereas if MeOH is used as solvent, both the DEIPS and the benzyl ether are cleaved.
4. RMgX.[5]
5. HF·Pyr, Pyr, THF, 74% yield.[6]

1. K. Toshima, K. Tatsuta, and M. Kinoshita, *Tetrahedron Lett.*, **27**, 4741 (1986).
2. K. Toshima, S. Mukaiyama, M. Kinoshita, and K. Tatsuta, *Tetrahedron Lett.*, **30**, 6413 (1989).
3. K. Toshima, M. Misawa, K. Ohta, K. Tatsuta, and M. Kinoshita, *Tetrahedron Lett.*, **30**, 6417 (1989).
4. K. Toshima, K. Yanagawa, S. Mukaiyama, and K. Tatsuta, *Tetrahedron Lett.*, **31**, 6697 (1990).
5. Y. Watanabe, T. Fujimoto, and S. Ozaki, *J. Chem. Soc., Chem. Commun.*, 681 (1992).
6. D. A. Evans, S. W. Kaldor, T. K. Jones, J. Clardy, and T. J. Stout, *J. Am. Chem. Soc.*, **112**, 7001 (1990).

Dimethylthexylsilyl Ether (TDS−OR): $(CH_3)_2CHC(CH_3)_2Si(CH_3)_2OR$

Both TDSCl and $TDSOSO_2CF_3$ are used to introduce the TDS group. In general, conditions similar to those used to introduce the TBDMS group are effective. This group is slightly more hindered than the TBDMS group, and the chloride has the advantage of being a liquid, which is useful when handling large quantities of material. Cleavage of this group can be accomplished by the same methods used to cleave the TBDMS group, but it is two to three times slower because of its increased steric bulk.[1] A disadvantage is that the NMR spectrum is not as simple as in the case when the similar TBDMS group is used.

1. H. Wetter and K. Oertle, *Tetrahedron Lett.*, **26**, 5515 (1985).

2-Norbornyldimethylsilyl (NDMS−OR):

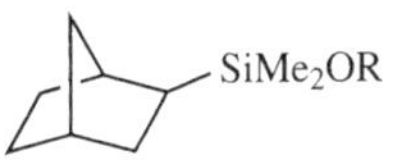

This silyl ether was developed as an economical alternative to the TBDMS ether. It can be introduced using either the silyl chloride or the triflate under conventional conditions. Its stability is intermediate to that of isopropyldimethylsilyl (IPDMS) group and the TBDMS group. It is stable to KF in MeOH at 25°C, but is cleaved in 7 h at 65°C conditions where the TBDMS ether is stable. It is cleaved with TBAF in <1 min.[1] The corresponding silyl ester has also been prepared. Unfortunately, this group carries the liability of a chiral center.

1. D. K. Heldmann, J. Stohrer, and R. Zauner, *Synlett*, 1919 (2002).

t-Butyldimethylsilyl Ether (TBS−OR, TBDMS−OR): *t*-BuMe$_2$SiOR (Chart 1)

The TBDMS ether has become one of the most popular silyl protective groups used in chemical synthesis. It is easily introduced with a variety of reagents, has the virtue of being quite stable to a variety of organic reactions, and is readily removed under conditions that do not attack other functional groups. It was also shown to withstand 230°C.[1] It is approximately 10^4 times more stable to basic hydrolysis than the trimethylsilyl (TMS) group. It has excellent stability toward base but is relatively sensitive to acid. The ease of introduction and removal of the TBDMS ether are influenced by steric factors that often allow for its selective introduction in polyfunctional, sterically differentiated molecules. It is relatively easy to introduce a primary TBDMS group in the presence of a secondary alcohol. One problem that has been encountered with the TBDMS group is that it can be metalated on the silyl methyl with *t*-BuLi.[2] Surprisingly, it was shown to be stable to a Tamao oxidation, which uses fluoride ion.[3]

Formation

1. TBDMSCl, imidazole, DMF, 25°C, 10 h, high yields.[4] This is the most common method for the introduction of the TBDMS group on alcohols with low steric demand. The method works best when the reactions are run in very concentrated solutions. This combination of reagents also silylates phenols,[5] hydroperoxides,[6] and hydroxylamines,[7] but under suitable conditions it is

possible to silylate a primary alcohol in preference to a phenol.[8] Thiols, amines, and carboxylic acids are not effectively silylated under these conditions.[9] Tertiary alcohols can be silylated with the phosphoramidate catalyst **i**.[10]

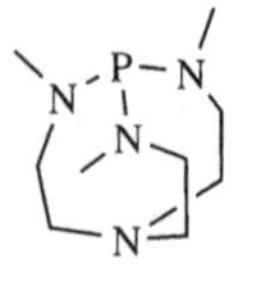

i

Although, silylation using these conditions normally proceeds uneventfully, the following scheme shows that reactions are not always straightforward.[11]

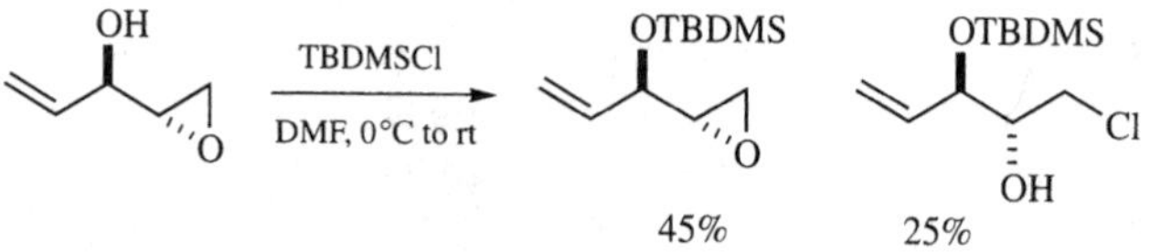

45%   25%

Ionic liquids have been used to replace DMF as a solvent.[12]

2. TBDMSCl, Li$_2$S, CH$_3$CN, 25°C, 5–8 h, 75–95% yield.[13] This reaction occurs under nearly neutral conditions.

3. TBDMSCl, DMAP, Et$_3$N, DMF, 25°C, 12 h.[14] These conditions were used to silylate selectively a primary over a secondary alcohol.[15] In the silylation of carbohydrates, it was shown that these conditions inhibit silyl migration whereas the use of imidazole as base causes migration.[16] Besides DMAP, other catalysts such as 1,1,3,3-tetramethylguanidine,[17] 1,8-diazabicyclo[5.4.0]undec-7-ene (83–99%),[9] 1,5-diazabicyclo[4.3.0]non-5-ene,[18] and ethyldiisopropyl-amine have also been used.[19] A chiral guanidine has been used to give modest kinetic resolution of chiral secondary alcohols with TBDMSCl and TIPSCl.[20]

4. TBDMSCl, KH, 18-crown-6, THF, 0°C to rt, 78% yield.[21] This combination of reagents is very effective in silylating extremely hindered alcohols.

5. Since the Si–N bond is much weaker than the Si–O bond, even if silylation occurs on nitrogen it will generally transfer to the oxygen.

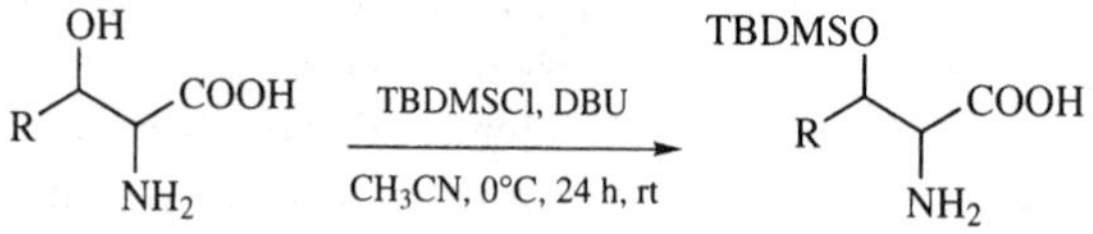

These conditions were chosen specifically to facilitate the silylation of hydroxylated amino acids.[22]

6. (a) Bu$_2$SnO, MeOH. (b) TBDMSCl, CH$_2$Cl$_2$. These conditions selectively protect the equatorial alcohol of a *cis*-diol on a pyranoside ring.[23] In the case of β-lactosides, the primary TBDMS ether is formed in 96% yield.[24] Butane-1,2,4-triol shows unusual selectivity in that the stannylene methods give the 4-TBDMS derivative, whereas benzylation, acetylation, and tosylation give the 1-substituted derivatives.[25]

7. Heating an alcohol and TBDMSCl in DMF to 120°C without added base will form the silyl ether, but HCl is also formed, which must be considered in the context of the rest of the molecule.[26]

8. TBDMSOClO$_3$, CH$_3$CN, Pyr, 20 min, 100% yield.[27] This reagent works well, but it has the disadvantage of being **explosive** and has been supplanted by TBDMSOSO$_2$CF$_3$.

9. TBDMSOSO$_2$CF$_3$, CH$_2$Cl$_2$, 2,6-lutidine, 0–25°C.[28] This is one of the most powerful methods for introducing the TBDMS group. Other bases such as triethylamine,[29] ethyldiisopropylamine,[30] and pyridine[31] have also been used successfully. In the presence of an ester or ketone, it is possible simultaneously to form a silyl enol ether while silylating a hydroxyl group.[27] Not all protections proceed as expected, as illustrated with the following glutarimide.[32]

10. A secondary alcohol was selectively protected in the presence of a secondary allylic alcohol with TBDMSOTf, 2,6-lutidine at −78°C.[33]

t-Butyl or *t*-amyl ethers are converted to TBDMS ethers with this reagent. If the lutidine is not present, cleavage to the alcohol occurs.[34] Silyl migration has been observed during protection of an alcohol with a proximal silyl ether using TBDMSOTf-2,6-lutidine.[35] See section on silyl migration.

1:1 ratio

The following case shows a very interesting solvent effect that was not explained by the authors,[36] but it has been shown by others that the 3-hydroxyl is typically the kinetic product and the 2-hydroxyl is the thermodynamic product, thus implicating possible silyl migration.

11. From a THP ether: TBDMSOTf, Me$_2$S, CH$_2$Cl$_2$, $-50°C$, 24–97% yield. Allylic THP ethers are converted inefficiently.[37]

12. TBDMSCH$_2$CH=CH$_2$, TsOH, CH$_3$CN, 70–80°C, 2.5 h, 95% yield.[38]

13. Methallyl-TBDMS and Sc(OTf)$_3$, CH$_3$CH$_2$CN, rt, 85–98% yield. Tertiary alcohols and phenols can be silylated using this method. The TES and TBDPS ether are also prepared by this method.[39]

14. 4-t-Butyldimethylsiloxy-3-penten-2-one, DMF, TsOH, rt, 83–92% yield.[40]

15. 1-(t-Butyldimethylsilyl)imidazole.[41,42]

16. N-t-Butyldimethylsilyl-N-methyltrifluoroacetamide, CH$_3$CN, 5 min, 97–100% yield.[43] This reagent also silylates thiols, amines, amides, carboxylic acids, and enolizable carbonyl groups.

17. 1-(t-Butyldimethylsiloxy)-1-methoxyethene, CH$_3$CN, 91–100% yield.[44] This reagent also silylates thiols and carboxylic acids.

18. TBDMSCN, 80°C, 5 min, 95% yield.[45]

19. From a THP ether: TBDMSH, CH$_2$Cl$_2$, Sn(OTf)$_2$, rt, 1 h, 78% yield. TIPS ethers are prepared analogously.[46]

20. TBDMSONO$_2$.[47]

21. N,N-Bis-TBDMSdimethylhydantoin, cat. TBAF.[48] Primary alcohols are selectively protected.

22. CH$_3$C(OTBDMS)=NTBDMS, TBAF, NMP (N-methylpyrrolidinone), 76–99% yield.[49]

23. PhC(OTBDMS)=NPh, (Si-BEZA) catalytic pyridinium triflate, THF or benzotrifluoride, 25–50°C, 5–2400 min, 23–99% yield. This is a general method and can be used to prepare TMS, TES, TBDPS, and TIPS ethers even from 3° alcohols and phenols.[50]

24. TBDMSH, 10% Pd–C.[51] This method has been used to study the disilylation of a variety of monosacharrides. The major isomer is the 3,6-bis-TBDMS derivative, with the remainder being primarily the 2,6-derivative.[52]

25. TBDMSH, [RuCl$_2$(p-cym)]$_2$, CH$_2$Cl$_2$, 50°C, 6 h, >95% yield.[53]

26. TBDMSH, Cl$_2$(PCy$_3$)$_2$Ru=CHPh, 45°C, 6 h, 95% yield.[54]

27. TBDMSOH, Ph$_3$P, DEAD, THF, $-78°C$, 68–85% yield.[55]

28. Ph$_2$P-TBDMS, DEAD, CH$_2$Cl$_2$, rt, 5 min, 68–95% yield. The method works for 1°, 2°, and 3° alcohols and phenols. It can also be used to introduce the TES and the TIPS groups.[56]

29. TBDMSH, THF, TBAF, rt, 1 h, 97% yield. Other silanes react similarly.[57]

30. The following schemes represent some interesting examples where the TBDMS
group is introduced selectively on compounds with more than one alcohol.

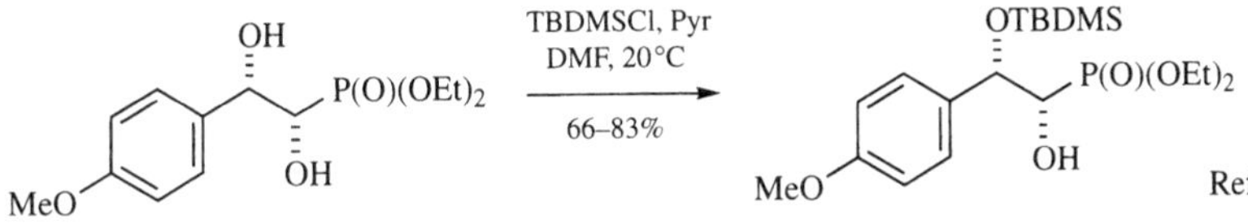

Ref. 58

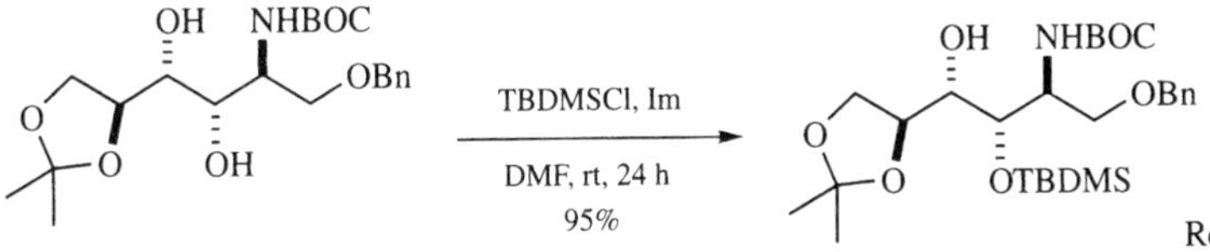

Ref. 59

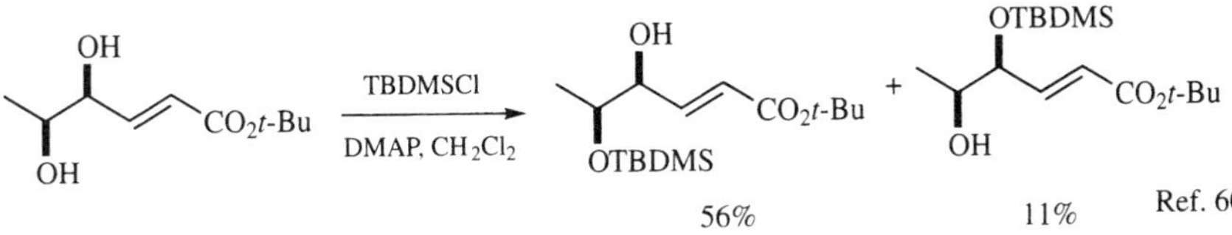

56%   11%   Ref. 60

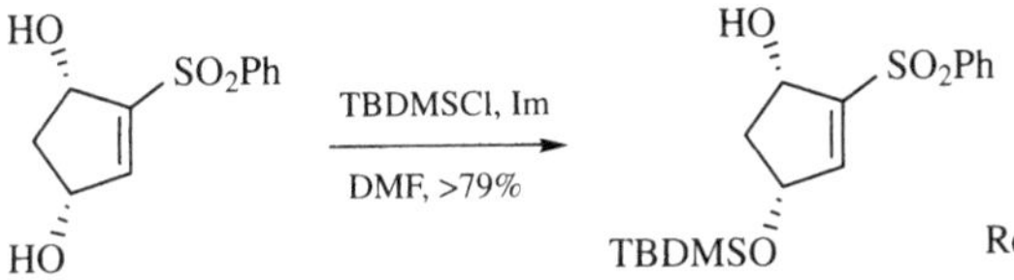

Ref. 61

From these examples, it appears that with the reagent TBDMSCl–Im–DMF,
the acidity of the alcohol plays an important role in determining the regiochemical preference of hydroxyl protection. In the case of 1,2-diols with similar steric requirements, it appears that when using imidazole as a base, the
least acidic hydroxyl is silylated. This may not be the kinetic result, since
imidazole has been shown to cause silyl migration.[16] The use of less basic
amines tends to give the kinetic result because these are not as prone to pro-
mote silyl migration. The section on the migration of silyl groups should be
consulted. Given this, the following result is counterintuitive, but it may be
conformationally driven.[62]

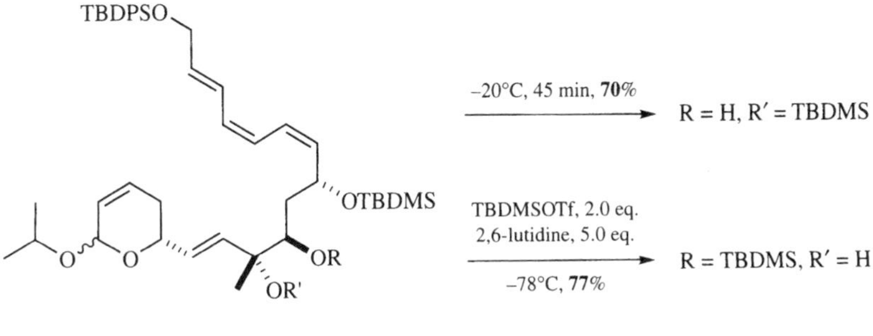

$$HO(CH_2)_nOH \xrightarrow[\text{54 – 90\%}]{\begin{array}{l}\text{1. NaOH, THF}\\\text{2. TBDMSCl}\end{array}} HO(CH_2)_nOTBDMS$$

Ref. 63, 64

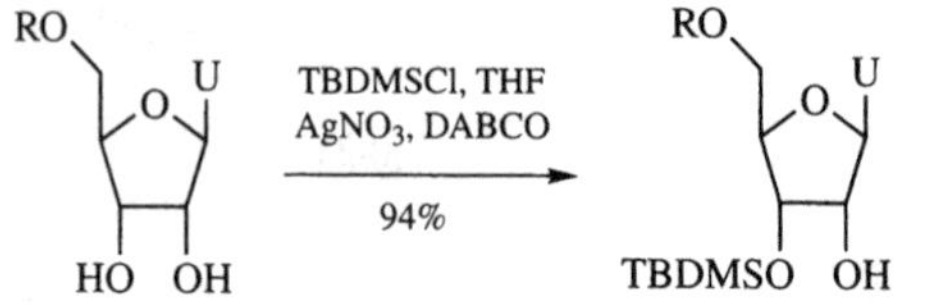

R = DMTr or TBDMS

Ref. 65

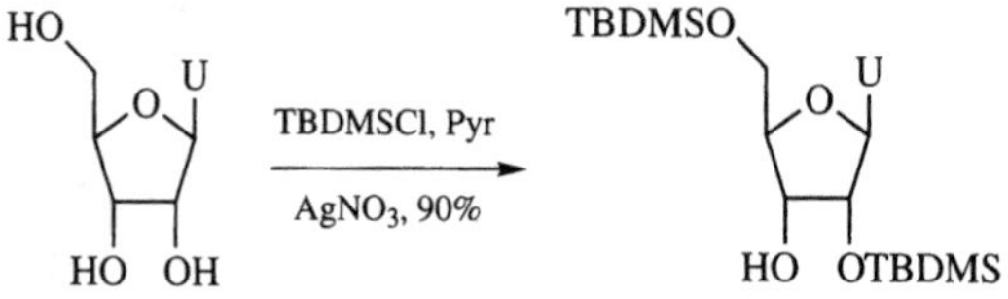

Ref. 65

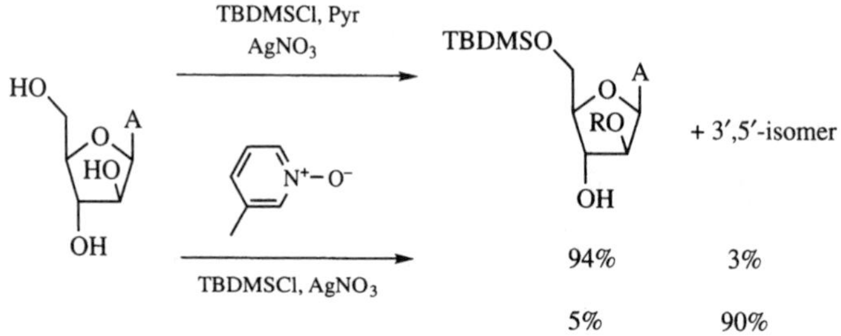

+ 3′,5′-isomer

94%   3%

5%   90%   Ref. 66

The following alcohol could not be silylated using conventional conditions. The use of AgNO$_3$ made silylation possible.[67]

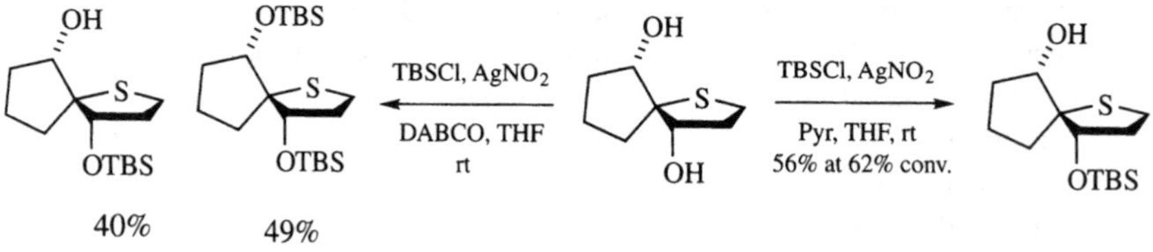

40%   49%

56% at 62% conv.

1. L. G. Monovich, Y. L. Huérou, M. Rönn, and G. A. Molander, *J. Am. Chem. Soc.*, **122**, 52 (2000).

2. R. W. Friesen and L. A. Trimble, *J. Org. Chem.*, **61**, 1165 (1996).

3. M. R. Elliott, A.-L. Dhimane, and M. Malacria, *J. Am. Chem. Soc.*, **119**, 3427 (1997).

4. E. J. Corey and A. Venkateswarlu, *J. Am. Chem. Soc.*, **94**, 6190 (1972).

5. D. W. Hansen, Jr., and D. Pilipauskas, *J. Org. Chem.*, **50**, 945 (1985).

6. G. R. Clark, M. M. Nikaido, C. K. Fair, and J. Lin, *J. Org. Chem.*, **50**, 1994 (1985).

7. J. F. W. Keana, G. S. Heo, and G. T. Gaughan, *J. Org. Chem.*, **50**, 2346 (1985).

8. M. Sefkow and H. Kaatz, *Tetrahedron Lett.*, **40**, 6561 (1999).

9. J. M. Aizpurua and C. Palomo, *Tetrahedron Lett.*, **26**, 475 (1985).

10. B. A. D'Sa and J. G. Verkade, *J. Am. Chem. Soc.*, **118**, 12832 (1996); B. A. D'Sa, D. McLeod, and J. G. Verkade, *J. Org. Chem.*, **62**, 5057 (1997).

11. J. Jin and S. M. Weinreb, *J. Am. Chem. Soc.*, **119**, 5773 (1997).

12. Z.-Y. Xu, D.-Q. Xu, B.-Y. Liu, and S.-P. Luo, *Synth. Commum.*, **33**, 4143 (2003).

13. G. A. Olah, B. G. B. Gupta, S. C. Narang, and R. Malhotra, *J. Org. Chem.*, **44**, 4272 (1979).

14. S. K. Chaudhary and O. Hernandez, *Tetrahedron Lett.*, **20**, 99 (1979).

15. K. K. Ogilvie, A. L. Shifman, and C. L. Penney, *Can. J. Chem.*, **57**, 2230 (1979); W. Kinzy and R. R. Schmidt, *Liebigs Ann. Chem.*, 407 (1987).

16. T. Halmos, R. Montserrat, J. Filippi, and K. Anonakis, *Carbohydr. Res.*, **170**, 57 (1987).

17. S. Kim and H. Chang, *Synth. Commun.*, **14**, 899 (1984).

18. S. Kim and H. Chang, *Bull. Chem. Soc. Jpn.*, **58**, 3669 (1985).

19. L. Lombardo, *Tetrahedron Lett.*, **25**, 227 (1984).

20. T. Isobe, T. Fukuda, Y. Araki, and T. Ishikawa, *Chem. Commun.*, 243 (2001).

21. T. F. Braish and P. L. Fuchs, *Synth. Commun.*, **16**, 111 (1986).

22. F. Orsini, F. Pelizzoni, M. Sisti, and L. Verotta, *Org. Prep. Proced. Int.*, **21**, 505 (1989).

23. P. J. Garegg, L. Olsson, and S. Oscarson, *J. Carbohydr. Chem.*, **12**, 955 (1993).

24. A. Glen, D. A. Leigh, R. P. Martin, J. P. Smart, and A. M. Truscello, *Carbohydr. Res.*, **248**, 365 (1993).

25. D. A. Leigh, R. P. Martin, J. P. Smart, and A. M. Truscello, *J. Chem. Soc., Chem. Commun.*, 1373 (1994).

26. B. Hatano, S. Toyota, and F. Toda, *Green Chem.*, **3**, 140 (2001).

27. T. J. Barton and C. R. Tully, *J. Org. Chem.*, **43**, 3649 (1978).

28. E. J. Corey, H. Cho, C. Rücker, and D. H. Hua, *Tetrahedron Lett.*, **22**, 3455 (1981).

29. L. N. Mander and S. P. Sethi, *Tetrahedron Lett.*, **25**, 5953 (1984).

30. D. Boschelli, T. Takemasa, Y. Nishitani, and S. Masamune, *Tetrahedron Lett.*, **26**, 5239 (1985).

31. P. G. Gassman and L. M. Haberman, *J. Org. Chem.*, **51**, 5010 (1986).

32. W. J. Vloon, J. C. van den Bos, N. P. Willard, G.-J. Koomen, and U. K. Pandit, *Recl. Trav. Chim. Pays-Bas*, **108**, 393 (1989).

33. D. Askin, D. Angst, and S. Danishefsky, *J. Org. Chem.*, **52**, 622 (1987).

34. X. Franck, B. Figadere, and A. Cavé, *Tetrahedron Lett.*, **36**, 711 (1995).

35. D. Seebach, H. F. Chow, R. F. W. Jackson, M. A. Sutter, S. Thaisrivongs, and J. Zimmermann, *Liebigs Ann. Chem.*, 1281 (1986).

36. K. C. Nicolaou, H. J. Mitchell, K. C. Fylaktakidou, R. M. Rodriguez, and H. Suzuki, *Chem. Eur. J.*, **6**, 3116 (2000).

37. S. Kim and I. S. Kee, *Tetrahedron Lett.*, **31**, 2899 (1990).

38. T. Morita, Y. Okamoto, and H. Sakurai, *Tetrahedron Lett.*, **21**, 835 (1980).

39. T. Suzuki, T. Watahiki, and T. Oriyama, *Tetrahedron Lett.*, **41**, 8903 (2000). T. Suzuki, T. Watahiki, and T. Oriyama, *Tennen Yuki Kagobutsu Toronkai Koen Yoshishu*, **42nd**, 625 (2000).

40. T. Veysoglu and L. A. Mitscher, *Tetrahedron Lett.*, **22**, 1299 (1981).

41. M. T. Reetz and G. Neumeier, *Liebigs Ann. Chem.*, 1234 (1981).

42. G. R. Martinez, P. A. Grieco, E. Williams, K.-i. Kanai, and C. V. Srinivasan, *J. Am. Chem. Soc.*, **104**, 1436 (1982).

43. T. P. Mawhinney and M. A. Madson, *J. Org. Chem.*, **47**, 3336 (1982).

44. Y. Kita, J.-i. Haruta, T. Fujii, J. Segawa, and Y. Tamura, *Synthesis* 451 (1981).

45. K. Kai and G. Patil, *J. Org. Chem.*, **51**, 3545 (1986).

46. T. Oriyama, K. Yatabe, S. Sugawara, Y. Machiguchi, and G. Koga, *Synlett*, 523 (1996).

47. B. K. Goering, K. Lee, B. An, and J. K. Cha, *J. Org. Chem.*, **58**, 1100 (1993).

48. Y. Tanabe, M. Murakami, K. Kitaichi, and Y. Yoshida, *Tetrahedron Lett.*, **35**, 8409 (1994).

49. D. A. Johnson and L. M. Taubner, *Tetrahedron Lett.*, **37**, 605 (1996).

50. T. Misaki, M. Kurihara, and Y. Tanabe, *Chem. Commun.*, 2478 (2001).

51. K. Yamamoto and M. Takemae, *Bull. Chem. Soc. Jpn.*, **62**, 2111 (1989).

52. M.-K. Chung, G. Orlova, J. D. Goddard, M. Schlaf, R. Harris, T. J. Beveridge, G. White, and F. R. Hallett, *J. Am. Chem. Soc.*, **124**, 10508 (2002).

53. R. L. Miller, S. V. Maifeld, and D. Lee, *Org. Lett.*, **6**, 2773 (2004).

54. S. V. Maifeld, R. L. Miller, and D. Lee, *Tetrahedron Lett.*, **43**, 6363 (2002).

55. D. L. J. Clive and D. Kellner, *Tetrahedron Lett.*, **32**, 7159 (1991).

56. M. Hayashi, Y. Matsuura, and Y. Watanabe, *Tetrahedron Lett.*, **45**, 1409 (2004).

57. Y. Tanabe, H. Okumura, A. Maeda, and M. Murakami, *Tetrahedron Lett.*, **35**, 8413 (1994).

58. T. Yokomatsu, K. Suemune, T. Yamagishi, and S. Shibuya, *Synlett*, 847 (1995).

59. L. Ermolenko, N. A. Sasaki, and P. Potier, *Tetrahedron Lett.*, **40**, 5187 (1999).

60. T. Sunazuka, T. Hirose, Y. Harigaya, S. Takamatsu, M. Hayashi, K. Komiyama, and S. Omura, *J. Am. Chem. Soc.*, **119**, 10247 (1997).

61. R. E. Donaldson and P. L. Fuchs, *J. Am. Chem. Soc.*, **103**, 2108 (1981).

62. D. L. Boger, S. Ichikawa, and W. Zhong, *J. Am. Chem. Soc.*, **123**, 4161 (2001).

63. P. G. McDougal, J. G. Rico, Y.-I. Oh, and B.D. Condon, *J. Org. Chem.*, **51**, 3388 (1986).

64. M. Achmatowicz and L. S. Hegedus, *J. Org. Chem.*, **69**, 2229 (2004).

65. (a) G. H. Hakimelahi, Z. A. Proba, and K. K. Ogilvie, *Tetrahedron Lett.*, **22**, 5243 (1981); (b) *idem, ibid.*, **22**, 4775 (1981).

66. K. K. Ogilvie, G. H. Hakimelahi, Z. A. Proba, and D. P. C. McGee, *Tetrahedron Lett.*, **23**, 1997 (1982); K. K. Ogilvie, D. P. C. McGee, S. M. Boisvert, G. H. Hakimelahi, and Z. A. Proba, *Can. J. Chem.*, **61**, 1204 (1983).

67. S. Dong and L. A. Paquette, *J. Org. Chem.*, **70**, 1580 (2005).

Cleavage

The following tables give a comparison of the stability of various silyl ethers to acid, base, and TBAF. The reported half-lives vary as a function of environment and acid or base concentration, but they serve to help define the relative stabilities of these silyl groups.

Half-Lives of Hydrolysis of Primary Silyl Ethers[1]

Silyl Ether	Half-Lives 5% NaOH–95% MeOH	Half-Lives 1% HCl–MeOH, 25°C
n-C_6H_{13}OTMS	≤1 min	≤1 min
n-C_6H_{13}OSi-i-BuMe₂	2.5 min	≤1 min
n-C_6H_{13}OTBDMS	Stable for 24 h	≤1 min
n-C_6H_{13}OMDPS	≤1 min	14 min
n-C_6H_{13}OTIPS	Stable for 24 h	55 min
n-C_6H_{13}OTBDPS	Stable for 24 h	225 min

Half-Lives of Hydrolysis of Primary Silyl Ethers[2]: Comparison of Trialkylsilyl vs. Alkoxysilyl Ethers

Ether	Half-Lives with Bu₄NF	Half-Lives with 0.1 M HClO₄
n-$C_{12}H_{25}$OTBDMS	140 h	1.4 h
n-$C_{12}H_{25}$OTBDPS	375 h	≤200 h
n-$C_{12}H_{25}$OSiPh₂(O-i-Pr)	<0.03 h	0.7 h
n-$C_{12}H_{25}$OSiPh₂(O-t-Bu)	5.8 h	17.5 h
n-$C_{12}H_{25}$OPh(t-Bu)(OMe)	22 h	200 h

1. Bu₄NF, THF, 25°C, 1 h, >90% yield.[3] Fluoride ion is very basic, especially under anhydrous conditions, and thus may cause side reactions with base-sensitive substrates.[4] The strong basicity can be moderated by the addition of acetic acid to the reaction, as was the case in the following reaction, where all others methods failed to remove the TBDMS group.[5]

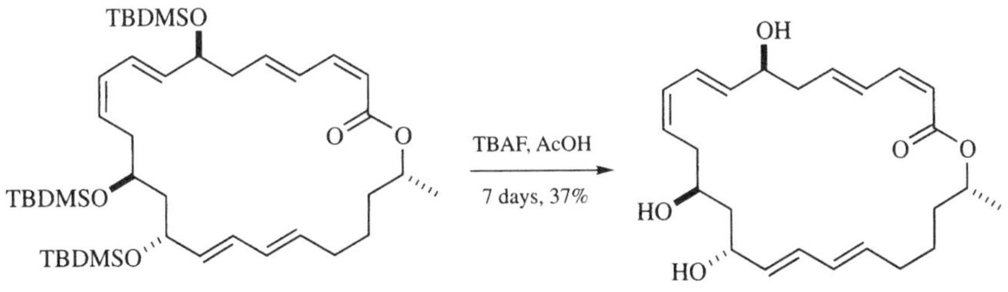

Commercial TBAF is known to contain water, but the water content seems to vary from lot to lot. This variation in water concentration was determined to be the cause for the often ineffective cleavage of TBDMS groups of ribosyl pyrimidine nucleosides. Interestingly, the cleavage of ribosyl purine nucleoside is not affected by the water content. In order to ensure consistency in deprotection in this case, the reaction should be run with molecular sieve-treated TBAF, which results in a water content of 2.3%.[6] It is also known that the addition of 4-Å ms increases the rate of TBAF-induced deprotection[7] and occasionally prevents decomposition.[8] No attempt should be made to dehy-drate TBAF, because it results in decomposition to tributylamine and HF_2^-, but anhydrous TBAF can be prepared by the addition of Bu_4NCN to hexafluo-robenzene in THF, CH_3CN, or DMSO at or below rt.[9] ArOTBDMS ethers can be cleaved in the presence of alkylOTBDMS ethers a process that is covered in two excellent reviews.[10] Similarly, allyl TBDMS ethers have been cleaved in the presence of alkyl TBDMS ethers.[11] The insolubility of Bu_4NClO_4 in water has been used to advantage in the workup of reactions that use large quanti-ties of TBAF.[12] Long-range stereoelectronic effects are seen in the rate of silyl ether cleavage, as shown by the TBAF-induced cleavage rates for the following three ethers[13]:

15.4×10^{-3} min^{-1} 4.3×10^{-3} min^{-1} 1.3×10^{-3} min^{-1}

2. 4-Methoxysalicylaldehyde·BF_3, CH_2Cl_2, 25°C. This method generates HF *in situ.*[14] The following table gives the relative rates of silyl cleavage for three different reagents (TIBS = triisobutylsilyl).

Relative Rates of Silyl Ether Cleavage

Protective Group	BF_3·Et_2O CH_2Cl_2, rt	TBAF THF, rt	BF_3·Et_2O Aldehyde, CH_2Cl_2
TBDMS	45 min	20 min	10 min
TIPS	45 min	15 min	10 min
TIBS	1 h	15 min	15 min
ThxDMS	1.5 h	25 min	15 min
TPS	15 h	2.5 h	20 min
TBDPS	NR	50 min	20 min

3. TBAF, NH_4F, THF, rt, 30 min, 63% yield. Ammonium fluoride was used to buffer the basicity of TBAF.[15]

4. $(Me_2N)_3S^+$ $F_2SiMe_3^-$ (TAS-F)[16], DMF, 73–98% yield. This is a very promis-ing method that was demonstrated on a variety of complex and base-sensitive substrates.[17] This reagent also does not have the liability associated with re-moving the n-Bu_4N^+ from reaction mixtures. Teoc groups are also cleaved. The addition of water is used to moderate the basicity of the reagent.[18]

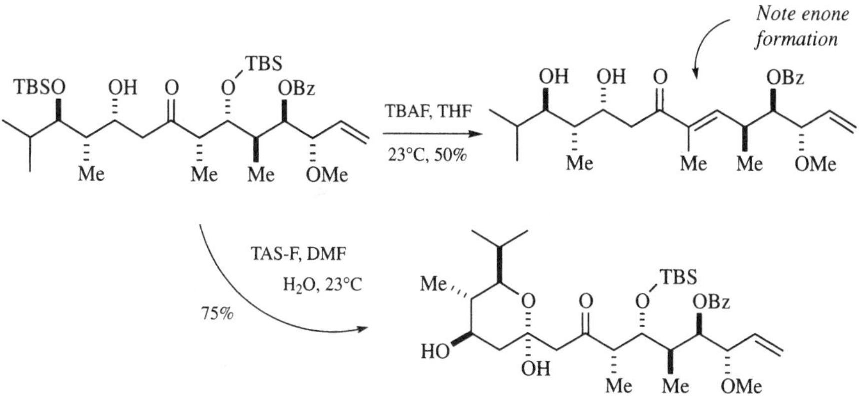

Note enone formation

TBAF, THF

23°C, 50%

TAS-F, DMF
H₂O, 23°C

75%

5. KF, 18-crown-6.[19]
6. KF·Al₂O₃, CH₃CN, 0°C, 2 h, 60% yield.[20]

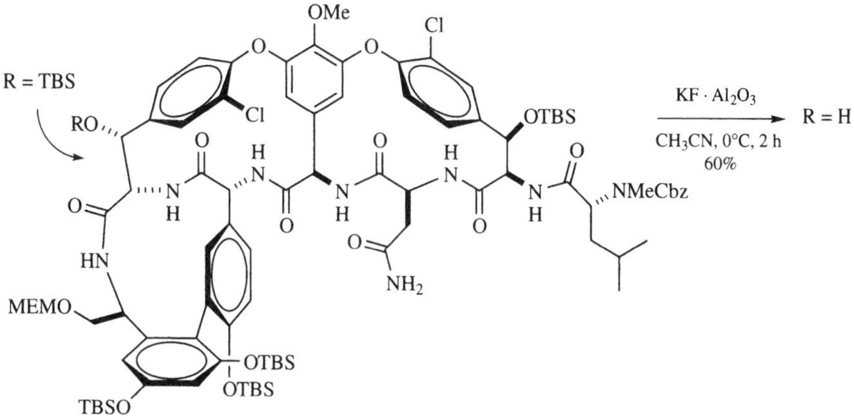

R = TBS

KF · Al₂O₃
───────────→ R = H
CH₃CN, 0°C, 2 h
60%

7. Bu₄NCl, KF·H₂O, CH₃CN, 25°C, 4 h, 95% yield.[21] This method generates TBAF *in situ* and is reported to be suitable for reactions that normally require anhydrous conditions.

8. Aq. HF, CH₃CN (5:95), 20°C, 1–3 h, 90–100% yield.[22] This reagent will cleave ROTBDMS ethers in the presence of ArOTBDMS ethers.[10] This reagent can be used to remove TBDMS groups from prostaglandins.

9. Pyridine·HF, THF, 0–25°C, 70% yield.[23] Cyclic acetals and THP derivatives were found to be stable to these conditions.[24] A primary TBDMS can be cleaved in the presence of a secondary TBDMS.[25] In the following reaction, if excess pyridine was not included as a buffer, some acyl transfer was observed.[26]

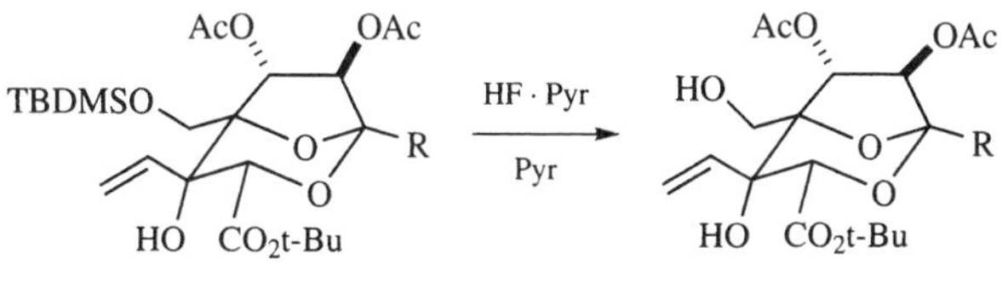

HF · Pyr
──────→
Pyr

10. 57% HF in urea.[27]

11. Et$_3$N·HF, cyclohexane, rt, 30 min.[28] The use of Et$_3$N·3HF was recommended for the desilylation of nucleosides and nucleotides.[29]

12. NH$_4$F·HF, DMF, NMP, 20°C, 90–98% yield. These conditions were developed to remove the TBDMS group from the sensitive carbapenems.[30]

13. NH$_4$F, MeOH, H$_2$O, 60–65°C, 65% yield.[31,62] Selectivity for primary TBDMS ethers has been observed with this reagent.[33]

14. Selectivity in the cleavage of a primary allylic TBDMS group was achieved with HF/CH$_3$CN in the presence of a more hindered secondary TBDMS group.[34]

15. Selective cleavage of a secondary TBDMS ether in the presence of a somewhat more hindered one was achieved with Bu$_4$NF in THF.[35]

16. SiF$_4$, CH$_3$CN, 23°C, 20 min, 94% yield. This reaction is faster in CH$_3$CN; tertiary and phenolic TBDMS groups react much more slowly,[36,37] but can be cleaved with this reagent.[38] In another example a 3° TBDMS ether was cleaved.[39]

17. H$_2$SiF$_6$, TEA, CH$_3$CN, >70% yield. TIPS groups are fairly stable to these conditions.[40]

18. (BF$_3$·Et$_2$O)−Bu$_4$NF. This reagent is selective for TBDMS ethers in the presence of TIPS and TBDPS ethers.[41]

19. CsF, CH$_3$CN, H$_2$O, reflux.[42]

20. Zn(BF$_4$)$_2$, H$_2$O, rt, 2–24 h, 80–96% yield. Phenolic ethers required heating for cleavage to occur and the TBDPS ether was completely stable.[43]

21. AcOH, H_2O, THF (3:1:1), 25–80°C, 15 min to 5 h.[3] Selective cleavage of a primary TBDMS group was achieved with acetic acid in the presence of a secondary TBDMS group.[44]

22. Dowex 50W-X8, MeOH, 20°C.[45] Dowex 50W-X8 is a carboxylic acid resin, H^+ form.

23. Low-loading alkylated polystyrene-supported sulfonic acid, water, 40°C, 12–24 h, 76–94% yield. A tertiary TBDMS ether was not cleaved. A TBDMS can be cleaved in the presence of a TBDPS ether. TIPS, TBDPS, OTr, OMOM ethers, and an acetate can all be cleaved, but the authors do not indicate relative rates.[46]

24. TsOH (0.1 eq.), THF, H_2O (20:1), 65% yield.[47]

25. Pyridinium *p*-toluensulfonate, EtOH, 22–55°C, 1.2–2 h, 80–92% yield.[48] These conditions were used to remove cleanly a TBDMS group in the presence of a TBDPS group or a primary TBDMS group in the presence of secondary.[49]

26. HCO_2H, THF, H_2O, 82% yield. In this case all fluoride-based methods failed.[50] This may be do to the potential for this system to undergo a retro Claisen condensation with the often basic fluoride reagents.

In the case of oligonucleotides, the phosphate has been shown to increase the rate of formic acid induced TBDMS hydrolysis by internal phosphate participation.[51]

27. 1% concd. HCl in EtOH.[27,52]

28. 1 *N* aq. periodic acid in THF was found effective when numerous other methods failed.[53]

29. H_2SO_4.[54] A silica based sulfonic acid has also been developed.[55]

30. Oxone, 50% aqueous MeOH, 75–92% yield. This method is selective for primary TBDMS ethers.[56]

31. $NaIO_4$, THF, H_2O, rt, 1–2 h, 90–94% yield. This method also removes the TMS, TES, TIPS, and TPS groups effectively, but does not cleave a TBDPS group cleanly.[57]

32. Trifluoroacetic acid, H_2O (9:1), CH_2Cl_2, rt, 96 h.[58] In the following riboside the selectivity is more likely the result of the reduced basicity of the OTBDMS group adjacent to the carbonyl oxygen rather than steric differences associated with the two ethers.[59] Similarly, a glycosidic TBDMS group was retained, whereas a primary TBDMS group was cleaved with TFA. In that case also, the glycosidic oxygen is less basic and would be less susceptible to acid-catalyzed cleavage.[60] The use of TFA:H_2O:THF in a ratio of 1:1:4 was recommended for primary TBDMS removal in multisilylated nucleosides (85–99% yield).[61]

Trichloroacetic acid similarly deprotects the primary 5′-TBDMS in the presence of the secondary TBDMS ethers.[62]

33. 0.5% Phosphomolybdic acid supported on silica gel, THF, rt, 92–99% yield. Phenolic TBS ethers are cleaved much more slowly.[63]

34. Nafion-H, NaI, MeOH, 73–99% yield.[64]

35. AcCl, MeOH, 0–5°C, rt, 3–15 min, 80–98% yield.[65] TBDPS ethers are also cleaved but much more slowly (2–4 h) This combination of reagents is well known to produce HCl.

36. $NiCl_2$, $HSCH_2CH_2SH$, MeOH, CH_2Cl_2, rt, 65–99% yield.[66]

37. TMSCl, wet CH_3CN, 2–21 h, rt, 78–94% yield. Phenolic TBDMS ethers are unaffected.[67]

38. $SbCl_5$, wet CH_3CN, rt, 85–95% yield. Phenolic TBDMS ethers are cleaved along with TBDMS esters and amines.[68]

39. Decaborane, THF, MeOH, 1–12 h, rt, 90–98% yield. A triphenylsilyl (TPS) ether is cleaved, but TBDPS, TIPS, and Tr ethers were stable.[69]

40. $BiBr_3$, wet CH_3CN, rt, 72–94% yield. Phenolic TBDMS ethers were stable to these conditions.[70]

41. (n-Bu)$_4$NBr$_3$ (0.1 eq.), MeOH, rt to reflux, 92–99% yield. Phenolic ethers required heating to reflux to get cleavage. The relative order of stability for various ethers is as follows: phenolic TBDMS > 1° TBDMS > 2° TBDPS > 2° OTHP > 1° OTHP > 1° TBDMS > 1° ODMT.[71]

42. NBS, DMSO, H_2O, rt, 17 h.[72] A trisubstituted steroidal alkene was not affected by these conditions. These conditions have been used to cleave a primary TBDMS ether in the presence of a secondary TBDMS ether.[73]

43. Bromine, MeOH, 20–360 min, reflux, 64–99% yield. TBDPS ethers are also cleaved but can be retained if the reaction is conducted at rt.[74]

44. IBr, MeOH, 1–12 min, 80–95% yield. The TBDPS was stable.[75]

45. CBr_4, MeOH, reflux, 83–95% yield. TIPS and TBDMS ethers are also cleaved.[76] Using photolysis[77] at rt or using sonication,[78] primary TBDMS ethers were efficiently cleaved in the presence of secondary TBDMS ethers. This method also removes O-trityl groups.

46. Methanol, CCl_4, ultrasonication, 40–50°, 90–96% yield.[79] Phenolic TBDMS and TBDPS ethers are stable.

47. Acetonyltriphenylphosphonium bromide, MeOH, 7 min to 6 h, 70–95% yield. Phenolic TBDMS ethers are preserved during cleavage of alkyl TBDMS ethers.[80]

48. I_2, MeOH, 65°C, 12 h, 90% yield.[81] PMB ethers are also cleaved, but benzyl ethers are stable. Phenolic TBDMS ethers are stable.[82]

49. Catalytic NIS, MeOH, rt, 69–100% yield. Phenolic TBDMS ethers are inert.[83]

50. $Sc(OTf)_3$, CH_3CN, H_2O, rt, 1 h, 91–98% yield. Phenolic TBDMS ethers were stable to these conditions.[84] TBDPS and TIPS ethers could be cleaved if the reaction time was extended to 24 h.

51. $Ce(OTf)_4$, THF, H_2O, 38–95% yield. Phenolic derivatives are slowly cleaved, but phenolic TBDPS ether is stable.[85]

52. $CeCl_3 \cdot 7H_2O$, NaI, CH_3CN, rt or reflux, 87–99% yield. Secondary derivatives are cleaved at reflux, whereas primary derivatives are cleaved at rt. The TBDPS and TIPS ethers are cleaved more slowly.[86]

53. $BiCl_3$, NaI, CH_3CN, rt, 30–120 min, rt, 70–86% yield.[87] The phenolic TBDMS ether is stable.

54. $Bi(OTf)_3$, MeOH, 90–95% yield. The use of $BiCl_3$ or $Bi(TFA)_3$ does not cleave the TBDMS group, but they do cleave the TMS group.[88]

55. $InCl_3$, wet CH_3CN, reflux, 75–93% yield. Phenolic TBDMS, TBDPS, and alkyl TBDPS ethers are stable.[89]

56. $CuCl_2 \cdot H_2O$, acetone, H_2O, reflux, 80–99% yield. A TBDPS ether was also cleaved.[90]

57. $BF_3 \cdot Et_2O$, $CHCl_3$, 0–25°C, 15 min to 3 h, 70–90% yield.[91] CH_3CN is also an effective solvent.[92] This method has been used when TBAF and HF/CH_3CN failed do to ester hydrolysis.[93]

58. $Bu_4Sn_2O(NCS)_2$, MeOH, reflux, 16 h, 70% yield.[94] This reagent also cleaves ketals and acetals, 77–97% yield.

59. i-Bu_2AlH, CH_2Cl_2, 25°C, 1–2 h, 84–95% yield.[95]

60. $ZrCl_4$, dry CH_3CN, rt, 20–45 min, 76–95% yield.[96] In the presence of Ac_2O acetates are formed and THP ethers are also converted.[97] The TBDMS group is cleaved selectively in the presence of the TBDPS group.[98]

61. $BH_3 \cdot DMS$, TMSOTf, CH_2Cl_2, $-78°C$, 70% yield.[99] Esters and acetals also react with this combination of reagents.

62. $SnCl_2$, $FeCl_3$, $Cu(NO_3)_2$ or $Ce(NO_3)_3$, CH_3CN, rt, 5 min, 95% yield.[100] TBDPS ethers can also be cleaved with prolonged reaction times (3 h, 85–93% yield), but can be retained during the cleavage of a primary TBS ether. With $SnCl_2 \cdot 2H_2O$, primary TBS ethers are cleaved in the presence of secondary derivatives and phenolic TBS ethers are retained during the cleavage of a primary TBS ether.[101]

63. Me_2BBr.[102]

64. BCl_3, THF, 65–83% yield. The primary TBDMS ether was selectively cleaved from a series of persilylated carbohydrate derivatives.[103]

65. $LiBF_4$, CH_3CN, CH_2Cl_2, 40–86% yield.[104] In this case, Bu_4NF or acid failed to remove a primary TBDMS group from a steroid.

66. LiBr, 18-crown-6.[105] Selectivity for primary derivatives was achieved.

67. TMSOTf, CH_2Cl_2, 0°C, 5 min, then neutral alumina, 92% yield.[106,107] TBDPS groups are stable to these conditions.

68. LiCl, H_2O, DMF, 90°C, 81–98% yield.[108]

69. DMSO, $P(MeNCH_2CH_2)_3N$, 80°C, 19–36 h, 68–94% yield.[109] Phenolic derivatives are also cleaved.[109]

70. KO_2, DMSO, DME, 18-crown-6, 50–85% yield.[110]

71. LiOH, dioxane, EtOH, H_2O, 90°C, 83% yield.[111]

72. The loss of the TBDMS group during $LiAlH_4$ reductions has been observed in cases where there is an adjacent amine or hydroxyl.[112]

73. In this case, cleavage of the primary TBDMS group is attributed to the presence of the 2'-hydroxyl, since in its absence the cleavage reaction does not proceed.[113]

R = TBDMS or TBDPS

74. The oxidative deprotection of silyl ethers such as the TBDMS ether has been reviewed for years prior to 1997.[114]

75. *N*-Hydroxyphthalimide, O_2, $Co(O_2C(CH_2)_8CH_3$, CH_3CN, 86–95% yield. This method converts either a TBDMS or a TMS ether directly to an aldehyde or ketone.[115]

76. DDQ, CH_3CN, H_2O.[116] These conditions normally cleave the PMB group selectively in the presence of a TBDMS group,[117] but in the case of an allylic derivative below the alcohol was oxidized directly to an aldehyde.[118] This reaction has some generality in that other electron-rich substrates as well as a TES ether are similarly oxidized. It is also selective in that PMB ethers survive.[119] It should be noted that in the presence of protic solvents, DDQ forms acidic adducts which are probably responsible for the hydrolysis.[120]

77. Quinolinium fluorochromate, DMF, rt, 15 h, 64–92% yield.[121]

78. 3 eq. *t*-BuOOH, 1.2 eq. $MoO_2(acac)_2$, CH_2Cl_2, 50–87% yield.[122]

79. 0.01 eq. $PdCl_2(CH_3CN)_2$, acetone, rt, 99% yield.[123,124] Additionally, acetals are cleaved with this reagent, but the TBDPS, MEM, and THP groups are completely stable.

80. Ceric ammonium nitrate, MeOH, 0°C, 15 min, 82–95% yield.[125] Dioxolanes and some THP ethers are not affected, but in general, with extended reaction times, THP ethers are cleaved. Silica gel-supported CAN was found to be advantageous for the deprotection of nucleosides and nucleotides with primary TBS groups cleaved in preference to secondary derivatives. The TIPS group can also be cleaved by this method.[126] This method was found effective where more traditional methods failed.[127]

81. Ph_3CBF_4, CH_3CN, CH_2Cl_2, rt, 60 h.[128]

82. During an attempt to metalate a glycal with t-BuLi, it was discovered by deuterium labeling that a TBDMS ether can be deprotonated.[129,130]

83. Lewatit 500, MeOH, 96% yield.[131]

84. DMSO, H_2O, 90°C, 79–87% yield. These conditions are only effective for primary allylic and homoallylic, primary benzylic, and aryl TBDMS ethers.[132]

85. Al_2O_3, H_2O, hexanes, 81–98% yield. These conditions are selective for the primary derivative. TBDPS and TMS ethers are also cleaved.[133] The use of alumina in a microwave oven is also effective (68–93% yield).[134]

86. PdO, cyclohexene, methanol, 30 min for a primary ROH, 90–95% yield. Secondary alcohols require longer times. The primary TBDPS and TIPS groups are cleaved much more slowly (18–21 h). Benzylic TBDMS ethers are cleaved without hydrogenolysis.[135]

87. Pd–C, MeOH, H_2, 71–99% yield. In solvents other than MeOH, TBDMS ethers are quite stable, but the addition of H_2O does increase the rate of cleavage. TES and TPS ethers are also cleaved, but TIPS and TBDPS ethers are stable. A phenolic TBDMS ether is also stable even with MeOH as the solvent.[136] With Pd–C(ethylenediamine) as a catalyst, TBDMS ether cleavage is completely suppressed.[137] This has led to a study of a variety of Pd–C catalysts which has shown that the likely mechanism for cleavage of silyl ethers is a result of residual acid in the catalyst. Stirring a variety of Pd–C catalysts in H_2O results in a pH range of 2.88–6.28. This would also account for the variability observed in the literature for the hydrogenolysis of various silyl ethers. *Only with the TES ether is there any indication that the cleavage occurs by hydrogenolysis, with others being the result of acid catalyzed hydrolysis.*[138]

Conversion of the TBDMS Group to Other Derivatives

1. AcBr, CH_2Cl_2, rt, 20 min, 90% yield. These conditions convert the TBDMS ether into the acetate. Benzyl and TBDPS ethers are stable, except when $SnBr_2$ is included in the reaction mixture, in which case these groups are also converted to acetates in excellent yield.[139]

2. Ac_2O, $Cu(OTf)_2$, CH_2Cl_2, 2–24 h, rt, 60–93% yield. THP and TBDMS groups are converted to acetates. MEM groups react but do not give clean products.[140] The ionic liquid, [bmim]Cl and $FeCl_3$ in the presence of Ac_2O, has been used to convert a TBDMS ether into an acetate.[141]

3. BzBr, $Zn(OTf)_2$, $ClCH_2CH_2Cl$, 10–30 min, 9–98% yield. The benzoate is formed from TBDMS, Bn, and anomeric 4-methoxyphenyl ethers.[142]

4. Treatment of a primary TBDMS group with Ph_3P and Br_2 converts it to a primary bromide.[143]

5. Silica chloride, NaI, CH_3CN, rt, 76–92% yield. This method converts TMS, THP, and TBDMS ethers directly to the iodide.[144]

6. $POCl_3$, DMF, 3–14 h, 0°C, 60–98% yield.[145] TES ethers are also converted.

1. J. S. Davies, L. C. L. Higginbotham, E. J. Tremeer, C. Brown, and R. S. Treadgold, *J. Chem. Soc., Perkin Trans. 1*, 3043 (1992).

2. J. W. Gillard, R. Fortin, H. E. Morton, C. Yoakim, C. A. Quesnelle, S. Daignault, and Y. Guindon, *J. Org. Chem.*, **53**, 2602 (1988).

3. E. J. Corey and A. Venkateswarlu, *J. Am. Chem. Soc.*, **94**, 6190 (1972).

4. J. H. Clark, *Chem. Rev.*, **80**, 429 (1980).

5. A. B. Smith, III, and G. R. Ott, *J. Am. Chem. Soc.*, **118**, 13095 (1996).

6. R. I. Hogrefe, A. P. McCaffrey, L. U. Borozdina, E. S. McCampbell, and M. M. Vaghefi, *Nucleic Acids Res.*, **21**, 4739 (1993).

7. H. C. Kolb, S. V. Ley, A. M. Z. Slawin, and D. J. Williams, *J. Chem. Soc., Perkin Trans. 1*, 2735 (1992).

8. M. D. B. Fenster and G. R. Dake, *Org. Lett.*, **5**, 4313 (2003).

9. H. Sun and S. G. DiMagno, *J. Am. Chem. Soc.*, **127**, 2050 (2005).

10. R. D. Crouch, *Tetrahedron*, **60**, 5833–5871 (2004), T. D. Nelson and R. D. Crouch, *Synthesis*, 1031, (1996).

11. K. C. Nicolaou, S. Ninkovic, F. Sarabia, D. Vourloumis, Y. He, H. Vallberg, M. R. V. Finlay, and Z. Yang, *J. Am. Chem. Soc.*, **119**, 7974 (1997).

12. J. C. Craig and E. T. Everhart, *Synth. Commun.*, **20**, 2147 (1990).

13. P. M. F. M. Bastiaansen, R. V. A. Orrû, J. B. P. A. Wijnberg, and A. de Groot, *J. Org. Chem.*, **60**, 6154 (1995).

14. S. Mabic and J.-P. Lepoittevin, *Synlett*, 851 (1994).

15. A. Fürstner and H. Weintritt, *J. Am. Chem. Soc.*, **120**, 2817 (1998)

16. W. J. Middleton, *Org. Synth.*, **64**, 221 (1985).

17. K. A. Scheidt, H. Chen, B. C. Follows, S. R. Chemler, D. S. Coffey, and W. R. Roush, *J. Org. Chem.*, **63**, 6436 (1998).

18. K. A. Scheidt, T. D. Bannister, A. Tasaka, M. D. Wendt, B. M. Savall, G. J. Fegley, and W. R. Roush, *J. Am. Chem. Soc.*, **124**, 6981 (2002).

19. G. Stork and P. F. Hudrlik *J. Am. Chem. Soc.*, **60**, 4462 4464 (1968); C. L. Liotta and H. P. Harris, *J. Am. Chem. Soc.*, **96**, 2250 (1974).

20. K. C. Nicolaou, H. J. Mitchell, N. F. Jain, T. Bando, R. Hughes, N. Winssinger, S. Natarajan, and A. E. Koumbis, *Chem. Eur. J.*, **5**, 2648 (1999).

21. L. A. Carpino and A. C. Sau, *J. Chem. Soc., Chem. Commun.*, 514 (1979).

22. R. F. Newton, D. P. Reynolds, M. A. W. Finch, D. R. Kelly, and S. M. Roberts, *Tetrahedron Lett.*, **20**, 3981 (1979).

23. K. C. Nicolaou and S. E. Webber, *Synthesis*, 453 (1986).

24. S. Masamune, L. D.-L. Lu, W. P. Jackson, T. Kaiho, and T. Toyoda, *J. Am. Chem. Soc.*, **104**, 5523 (1982).

25. P. Ruiz, J. Murga, M. Carda, and J. A. Marco, *J. Org. Chem.*, **70**, 713 (2005).

26. E. M. Carreira and J. Du Bois, *J. Am. Chem. Soc.*, **117**, 8106 (1995).

27. H. Wetter and K. Oertle, *Tetrahedron Lett.*, **26**, 5515 (1985).

28. J.-E. Nyström, T. D. McCanna, P. Helquist, and R. S. Iyer, *Tetrahedron Lett.*, **26**, 5393 (1985).

29. M. C. Pirrung, S. W. Shuey, D. C. Lever, and L. Fallon, *Biorg. Med. Chem. Lett.*, **4**, 1345 (1994).

30. M. Seki, K. Kondo, T. Kuroda, T. Yamanaka, and T. Iwasaki, *Synlett*, 609 (1995).

31. J. D. White, J. C. Amedio, Jr., S. Gut, and L. Jayasinghe, *J. Org. Chem.*, **54**, 4268 (1989).

32. W. Zhang and M. J. Robins, *Tetrahedron Lett.*, **33**, 1177 (1992).

33. D. Crich and F. Hermann, *Tetrahedron Lett.*, **34**, 3385 (1993).

34. S. J. Danishefsky, D. M. Armistead, F. E. Wincott, H. G. Selnick, and R. Hungate, *J. Am. Chem. Soc.*, **109**, 8117 (1987).

35. T. Nakaba, M. Fukui, and T. Oishi, *Tetrahedron Lett.*, **29**, 2219 2223 (1988).

36. E. J. Corey and K. Y. Ki, *Tetrahedron Lett.*, **33**, 2289 (1992).

37. K. C. Nicolaou, K. R. Reddy, G. Skokotas, F. Sato, and X.-Y. Xiao, *J. Am. Chem. Soc.*, **114**, 7935 (1992).

38. H. W. B. Johnson, U. Majumder, and J. D. Rainier, *J. Am. Chem. Soc.*, **127**, 848 (2005).

39. I. Kadota, H. Takamura, K. Sato, A. Ohno, K. Matsuda, M. Satake, and Y. Yamamoto, *J. Am. Chem. Soc.*, **125**, 11893 (2003).

40. A. S. Pilcher, D. K. Hill, S. J. Shimshock, R. E. Waltermire, and P. DeShong, *J. Org. Chem.*, **57**, 2492 (1992); S. J. Shimshock, R. E. Waltermire, and P. Deshong, *J. Am. Chem. Soc.*, **113**, 8791 (1991).

41. S.-i. Kawahara, T. Wada, and M. Sekine, *Tetrahedron Lett.*, **37**, 509 (1996).

42. P. F. Cirillo and J. S. Panek, *J. Org. Chem.*, **55**, 6071 (1990).

43. B. C. Ranu, U. Jana, and A. Majee, *Tetrahedron Lett.*, **40**, 1985 (1999).

44. A. Kawai, O. Hara, Y. Hamada, and T. Shiari, *Tetrahedron Lett.*, **29**, 6331 (1988).

45. E. J. Corey, J. W. Ponder, and P. Ulrich, *Tetrahedron Lett.*, **21**, 137 (1980).

46. S. Iimura, K. Manabe, and S. Kobayashi, *J. Org. Chem.*, **68**, 8723 (2003).

47. E. J. Thomas and A. C. Williams, *J. Chem. Soc., Chem. Commun.*, 992 (1987).

48. C. Prakash, S. Saleh, and I. A. Blair, *Tetrahedron Lett.*, **30**, 19 (1989).

49. J. A. Marshall and K. C. Ellis, *Org. Lett.*, **5**, 1729 (2003).

50. A. S. Kende, I. Liu, I. Kaldor, G. Dorey, and K. Koch, *J. Am. Chem. Soc.*, **117**, 8258 (1995).

51. S.-i. Kawahara, T. Wada, and M. Sekine, *J. Am. Chem. Soc.*, **118**, 9461 (1996).

52. R. F. Cunico and L. Bedell, *J. Org. Chem.*, **45**, 4797 (1980).

53. G. Kim, M. Y. Chu-Moyer, S. J. Danishefsky, and G. K. Schulte, *J. Am. Chem. Soc.*, **115**, 30 (1993).

54. F. Franke and R. D. Guthrie, *Aust. J. Chem.*, **31**, 1285 (1978).

55. B. Karimi and D. Zareyee, *Tetrahedron Lett.*, **46**, 4661 (2005).

56. G. Sabitha, M. Syamala, and J. S. Yadav, *Org. Lett.*, **1**, 1701 (1999).

57. M. Wang, C. Li, D. Yin, and X.-T. Liang, *Tetrahedron Lett.*, **43**, 8727 (2002).

58. R. Baker, W. J. Cummings, J. F. Hayes, and A. Kumar, *J. Chem. Soc., Chem. Commun.*, 1237 (1986).

59. M. J. Robins, V. Samano, and M. D. Johnson, *J. Org. Chem.*, **55**, 410 (1990).

60. S. F. Martin, J. A. Dodge, L. E. Burgess, and M. Hartmann, *J. Org. Chem.*, **57**, 1070 (1992).

61. X.-F. Zhu, H. J. Williams, and A. I. Scott, *J. Chem. Soc. Perkin Trans. 1*, 2305 (2000).

62. X.-F. Zhu, H. J. Williams, and A. I. Scott, *Synth. Commum.*, **33**, 2011 (2003).

63. G. D. K. Kumar and S. Baskaran, *J. Org. Chem.*, **70**, 4520 (2005).

64. S. Rani, J. L. Babu, and Y. D. Vankar, *Synth. Commum.*, **33**, 4043 (2003).

65. A. T. Khan and E. Mondal, *Synlett*, 694 (2003).

66. A. T. Khan, S. Islam, L. H. Choudhury, and S. Ghosh, *Tetrahedron Lett.*, **45**, 9617 (2004).

67. P. A. Grieco and C. J. Markworth, *Tetrahedron Lett.*, **40**, 665 (1999).

68. P. M. C. Glória, S. Prabhakar, A. M. Lobo, and M. J. S. Gomes, *Tetrahedron Lett.*, **44**, 8819 (2003).

69. Y. J. Jeong, J. H. Lee, E. S. Park, and C. M. Yoon, *J. Chem. Soc. Perkin Trans. 1*, 1223 (2002).

70. J. S. Bajwa, J. Vivelo, J. Slade, O. Repic, and T. Blacklock, *Tetrahedron Lett.*, **41**, 6021 (2000).

71. R. Gopinath, S. J. Haque, and B. K. Patel, *J. Org. Chem.*, **67**, 5842 (2002).

72. R. J. Batten, A. J. Dixon, R. J. K. Taylor, and R. F. Newton, *Synthesis*, 234, (1980).

73. N. Tsukada, T. Shimada, Y. S. Gyoung, N. Asao, and Y. Yamamoto, *J. Org. Chem.*, **60**, 143 (1995).

74. M. T. Barros, C. D. Maycock, and C. Thomassigny, *Synlett*, 1146 (2001).

75. K. P. R. Kartha and R. A. Field, *Synlett*, 311 (1999).

76. A. S.-Y. Lee, H.-C. Yeh, and J.-J. Shie, *Tetrahedron Lett.*, **39**, 5249 (1998).

77. M.-Y. Chen, K.-C. Lu, A. S.-Y. Lee, and C.-C. Lin, *Tetrahedron Lett.*, **43**, 2777 (2002). M.-Y. Chen, L. N. Patkar, K.-C. Lu, A. S.-Y. Lee, and C.-C. Lin, *Tetrahedron*, **60**, 11465 (2004).

78. A. S. Balnaves, G. McGowan, P. D. P. Shapland, and E. J. Thomas, *Tetrahedron Lett.*, **44**, 2713 (2003).

79. A. S.-Y. Lee, H.-C. Yeh, and M.-H. Tsai, *Tetrahedron Lett.*, **36**, 6891 (1995).

80. A. T. Khan, S. Ghosh, and L. H. Choudhury, *Eur. J. Org. Chem.*, 2198 (2004).

81. A. R. Vaino and W. A. Szarek, *J. Chem. Soc., Chem. Commun.*, 2351 (1996).

82. B. H. Lipshutz and J. Keith, *Tetrahedron Lett.*, **39**, 2495 (1998).

83. B. Karimi, A. Zamani, and D. Zareyee, *Tetrahedron Lett.*, **45**, 9139 (2004).

84. T. Oriyama, Y. Kobayashi, and K. Noda, *Synlett*, 1047 (1998).

85. G. Bartoli, G. Cupone, R. Dalpozzo, A. De Nino, L. Maiuolo, A. Procopio, L. Sambri, and A. Tagarelli, *Tetrahedron Lett.*, **43**, 5945 (2002).

86. G. Bartoli, M. Bosco, E. Marcantoni, L. Sambri, and E. Torregiani, *Synlett*, 209 (1998).

87. G. Sabitha, R. S. Babu, E. V. Reddy, R. Srividya, and J. S. Yadav, *Adv. Synth. Catal.*, **343**, 169 (2001).

88. H. Firouzabadi, I. Mohammadpoor-Baltork, and S. Kolagar, *Synth. Commun.*, **31**, 905 (2001).

89. J. S. Yadav, B. V. S. Reddy, and C. Madan, *New J. Chem.*, **24**, 853 (2000).

90. Z. P. Tan, L. Wang, and J. B. Wang, *Chin. Chem. Lett.*, **11**, 753 (2000).

91. D. R. Kelly, S. M. Roberts, and R. F. Newton, *Synth. Commun.*, **9**, 295 (1979); M. Eggen, S. K. Nair, and G. I. Georg, *Org. Lett.*, **3**, 1813 (2001).

92. S. A. King, B. Pipik, A. S. Thompson, A. DeCamp, and T. R. Verhoeven, *Tetrahedron Lett.*, **36**, 4563 (1995).

93. A. Fürstner and I. Konetzki, *J. Org. Chem.*, **63**, 3072 (1998).

94. J. Otera and H. Nozaki, *Tetrahedron Lett.*, **27**, 5743 (1986).

95. E. J. Corey and G. B. Jones, *J. Org. Chem.*, **57**, 1028 (1992).

96. G. V. M. Sharma, B. Srinivas, and P. R. Krishna, *Tetrahedron Lett.*, **44**, 4689 (2003).

97. C. S. Reddy, G. Smitha, and S. Chandrasekhar, *Tetrahedron Lett.*, **44**, 4693 (2003).

98. G. V. M. Sharma, B. Srinivas, and P. R. Krishna, *Letters in Organic Chemistry*, **2**, 57 (2005).

99. R. Hunter, B. Bartels, and J. F. Michael, *Tetrahedron Lett.*, **32**, 1095 (1991).

100. A. D. Cort, *Synth. Commun.*, **20**, 757 (1990).

101. J. Hua, Z. Y. Jiang, and Y. G. Wang, *Chin. Chem. Lett.*, **15**, 1430 (2004).

102. Y. Guindon, C. Yoakim, and H. E. Morton, *J. Org. Chem.*, **49**, 3912 (1984).

103. Y.-Y. Yang, W.-B. Yang, C.-F. Teo, and C.-H. Lin, *Synlett*, **11**, 1634 (2000).

104. B. W. Metcalf, J. P. Burkhart, and K. Jund, *Tetrahedron Lett.*, **21**, 35 (1980).

105. M. Tandon and T. P. Begley, *Synth. Commun.*, **27**, 2953 (1997).

106. V. Bou and J. Vilarrasa, *Tetrahedron Lett.*, **31**, 567 (1990).

107. R. Hunter, W. Hinz, and P. Richards, *Tetrahedron Lett.*, **40**, 3643 (1999).

108. J. Farras, C. Serra, and J. Vilarrasa, *Tetrahedron Lett.*, **39**, 327 (1998).

109. Z. Yu and J. G. Verkade, *J. Org. Chem.*, **65**, 2065 (2000).

110. Y. Torisawa, M. Shibasaki, and S. Ikegami, *Chem. Pharm. Bull.*, **31**, 2607 (1983).

111. P. Wipf and S. Lim, *J. Am. Chem. Soc.*, **117**, 558 (1995).

112. J. N. Glushka and A. S. Perlin, *Carbohydr. Res.*, **205**, 305 (1990); P. A. Wender, F. C. Bi, M. A. Brodney, and F. Gosselin, *Org. Lett.*, **3**, 2105–2108 (2001); E. F. J. De Vries, J. Brussee, and A. van der Gen, *J. Org. Chem.*, **59**, 7133 (1994).

113. L. L. H. de Fallois, J.-L. Décout, and M. Fontecave, *Tetrahedron Lett.*, **36**, 9479 (1995).

114. J. Muzart, *Synthesis*, 11 (1993); S. Chandrasekhar, P. K. Mohanty, and M Takhi, *J. Org. Chem.*, **62**, 2628 (1997).

115. B. Karimi and J. Rajabi, *Org. Lett.*, **6**, 2841 (2004).

116 K. Tanemura, T. Suzuki, and T. Horaguchi, *J. Chem. Soc., Perkin Trans. 1*, 2997 (1992); K. Tanemura, T. Suzuki, and T. Horaguchi, *Bull. Chem. Soc. Jpn.*, **67**, 290 (1994).

117. A. B. Smith, III, Y. Qiu, D. R. Jones, and K. Kobayashi, *J. Am. Chem. Soc.*, **117**, 12011 (1995); J. A. Marshall and M. P. Bourbeau, *J. Org. Chem.*, **67**, 2751 (2002).

118. I. Paterson, M. D. Woodrow, and C. J. Cowden, *Tetrahedron Lett.*, **39**, 6041 (1998).

119. I. Paterson, C. J. Cowden, V. S. Rahn, and M. D. Woodrow, *Synlett*, 915 (1998).

120. K. Tanemura, Y. Nishida, T. Suzuki, K. Satsumabayashi, and T. Horaguchi, *J. Chem. Res. (S)*, 40, (1999).

121. S. Chandrasekhar, P. K. Mohanty, and M. Takhi, *J. Org. Chem.*, **62**, 2628 (1997).

122. T. Hanamoto, T. Hayama, T. Katsuki, and M. Yamaguchi, *Tetrahedron Lett.*, **28**, 6329 (1987).

123. B. H. Lipshutz, D. Pollart, J. Monforte, and H. Kotsuki, *Tetrahedron Lett.*, **26**, 705 (1985).

124. N. S. Wilson and B. A. Keay, *J. Org. Chem.*, **61**, 2918 (1996).

125. A. DattaGupta, R. Singh, and V. K. Singh, *Synlett*, 69 (1996).

126. J. R. Hwu, M. L. Jain, F.-Y. Tsai, S.-C. Tsay, A. Balakumar, and G. H. Hakimelahi, *J. Org. Chem.*, **65**, 5077 (2000).

127. P. A. Wender, S. G. Hegde, R. D. Hubbard, and L. Zhang, *J. Am. Chem. Soc.*, **124**, 4956 (2002).

128. T. J. Barton and C. R. Tully, *J. Org. Chem.*, **43**, 3649 (1978).

129. R. W. Frieser and L. A. Trimble, *J. Org. Chem.*, **61**, 1165 (1996).

130. J. D. White and M. Kawasaki, *J. Am. Chem. Soc.*, **112**, 4991 (1990).

131. L. F. Tietze, C. Schneider, and A. Grote, *Chem.-Eur. J.*, **2**, 139 (1996).

132. G. Maiti and S. C. Roy, *Tetrahedron Lett.*, **38**, 495 (1997).

133. J. Feixas, A. Capdevila, and A. Guerrero, *Tetrahedron*, **50**, 8539 (1994).

134. R. S. Varma, J. B. Lamture, and M. Varma, *Tetrahedron Lett.*, **34**, 3029 (1993).

135. J. F. Cormier, M. B. Isaac, and L.-F. Chen, *Tetrahedron Lett.*, **34**, 243 (1993).

136. T. Ikawa, K. Hattori, H. Sajiki, and K. Hirota, *Tetrahedron*, **60**, 6901 (2004).

137. K. Hattori, H. Sajiki, and K. Hirota, *Tetrahedron*, **57**, 2109 (2001). K. Hattori, H. Sajiki and K. Hirota, *Tetrahedron Lett.*, **41**, 5711 (2000).

138. T. Ikawa, H. Sajiki, and K. Hirota, *Tetrahedron*, **60**, 6189 (2004).

139. T. Oriyama, M. Oda, J. Gono, and G. Koga, *Tetrahedron Lett.*, **35** 2027 (1994).

140. K. L. Chandra, P. Saravanan, and V. K. Singh, *Tetrahedron Lett.*, **42**, 5309 (2001).

141. J. R. Harjani, S. J. Nara, M. M. Salunkhe, and Y. S. Sanghvi, *Nucleosides & Nucleotides, and Nucleic Acids*, **24**, 819 (2005).

142. T. Polat and R. J. Linhardt, *Carbohydr. Res.*, **338**, 447 (2003).

143. P. R. Ashton, R. Königer, J. F. Stoddart, D. Alker, and V. D. Harding, *J. Org. Chem.*, **61**, 903 (1996).

144. H. Firouzabadi, N. Iranpoor, and H. Hazarkhani, *Tetrahedron Lett.*, **43**, 7139 (2002).

145. S. Koeller and J.-P. Lellouche, *Tetrahedron Lett.*, **40**, 7043 (1999).

t-Butyldiphenylsilyl Ether (TBDPS−OR): *t*-BuPh$_2$SiOR (Chart 1)

The TBDPS group is considerably more stable ($\approx$100 times) than the TBDMS group toward acidic hydrolysis. The TBDPS group is less stable to base than the TBDMS group. The TBDPS group shows greater stability than the TBDMS group to many reagents with which the TBDMS group is incompatible. The TBDMS group is less prone to undergo migration under basic conditions.[1] TBDPS ethers are stable to K_2CO_3/CH_3OH, to 9 M NH_4OH, 60°C, 2 h, and to $NaOCH_3$ (cat.)/CH_3OH, 25°C, 24 h. The ether is stable to 80% AcOH, used to cleave TBDMS, triphenylmethyl, and tetrahydropyranyl ethers. It is also stable to HBr/AcOH, 12°C, 2 min, to 25–75% HCO_2H, 25°C, 2–6 h, and to 50% aq. CF_3CO_2H, 25°C, 15 min (conditions used to cleave acetals).[2] It was the only protective group stable to *B*-I-9-BBN in an iodoboration of an acetylene.[3]

Formation

1. TBDPSCl, imidazole, DMF, rt.[2] This is the original procedure used to introduce this group and is also the most widely employed method.

2. TBDPSCl, DMAP, Pyr.[4] Selective silylation of a primary hydroxyl was achieved under these conditions.

3. TBDPSCl, N(CH$_2$CH$_2$NMe)$_3$P, DMF or CH$_3$CN, TEA, 37–99% yield. This system was effective at silylating hindered alcohols.[5]

4. TBDPSCl, DMAP, triethylamine, CH$_2$Cl$_2$.[6] This combination of reagents was shown to be very selective for the silylation of a primary hydroxyl in the presence of a secondary hydroxyl.

5. TBDPSCl, poly(vinylpyridine), HMPT, CH$_2$Cl$_2$.[7]

6. TBDPSCl, CH$_2$Cl$_2$, DIPEA, rt, 2 h, 95% yield.[8] The selective monosilylation can also be achieved in DMF as the solvent; in this the DIPEA is only partially soluble and slowly delivers the base to the reaction mixture.[9]

7. TBDPSCl, NH$_4$NO$_3$, DMF, 72–96% yield.[10] This reagent can be used to avoid benzoyl group migration that can occur under more basic conditions.[11]

8. TBDPSOTf, 2,6-lutidine, CH$_2$Cl$_2$.[12]

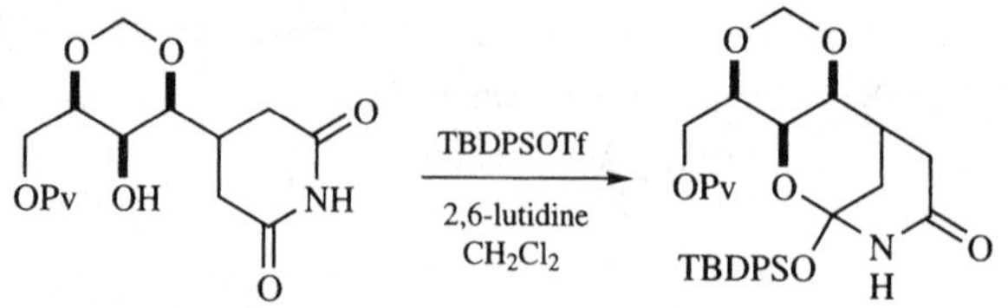

9. TBDPSCl, AgNO$_3$, Pyr, THF, rt, 3 h, 70% yield.[13,14] The addition of AgNO$_3$ increases the rate of silylation. It appears that the more acidic alcohol is the most reactive by this method.

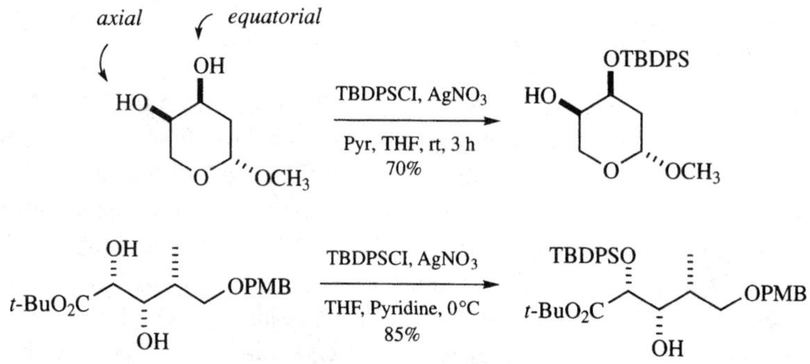

10. It is possible for the TBDPS group to participate in cationic reactions by a phenyl transfer as illustrated.[15]

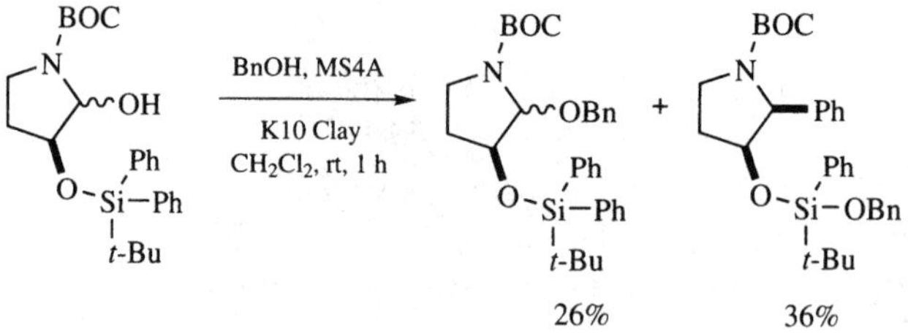

Cleavage

1. Bu$_4$NF, THF, 25°C, 1–5 h, >90% yield.[2]

2. Bu$_4$NF, AcOH, H$_2$O, DMF, 89% yield. These conditions cleave a TBDMS ether in the presence of a TBS ether.[16]

3. NH$_4$F.[17]

4. Pyr·HF, THF.[18] When the reaction is conducted under high pressure (1.0 GPa), it proved to be very effective for cleaving hindered TBDPS ethers.[19]

5. HF, CH$_3$CN.[20]

6. [(Me$_2$N)$_3$S][Me$_3$SiF$_2$], CH$_3$CN, reflux, quant. or (Bu$_4$N)(Ph$_3$SiF$_2$), CH$_3$CN, reflux, 84% yield. Use of HF·pyridine resulted in formyl acetal formation by participation of an adjacent MOM ether.[21]

7. Amberlite 26 F$^-$ form.[7]

8. 3% methanolic HCl, 25°C, 3 h, 71% yield.[1] In benzoyl-protected carbohydrates this method gives clean deprotection without acyl migration.[22]

9. Br$_2$, MeOH, reflux, 64–99% yield. TBS ether are cleaved at rt in preference to TBDMS ethers.[23]

10. BF$_3$·Et$_2$O, 4-methoxysalicylaldehyde.[24] The relative rate of cleavage of the TBDPS ethers of the following alcohols is PhCH$_2$CH$_2$O-, propargylO-, BnO-, menthol, PhO- (20 min, 45 min, 1.5 h, 5 h, 8 h).

11. 5 N NaOH, EtOH, 25°C, 7 h, 93% yield.[1] TBDMS ethers are stable[25,26] and in some cases a sterically congested TES group will also survive NaOH (DMPU, H$_2$O, 60% yield).[27]

In the following case, there was no indication of any Payne rearrangement of the epoxy alcohol.[28]

12. 10% KOH, CH_3OH.[29]

13. KO_2, DMSO, 18-crown-6.[1]

14. $LiAlH_4$ has resulted in the cleavage of a TBDPS group, but generally,[30,31] TBDPS ethers are not affected by $LiAlH_4$.

15. NaH, HMPA, 0°C, 5 min; H_2O, 83–84% yield.[32] These conditions selectively cleave a TBDPS ether in the presence of a t-butyldimethylsilyl ether.

16. Alumina.[33]

1. Y. Torisawa, M. Shibasaki, and S. Ikegami, *Chem. Pharm. Bull.*, **31**, 2607 (1983); W. W. Wood and A. Rashid, *Tetrahedron Lett.*, **28**, 1933 (1987).

2. S. Hanessian and P. Lavallee, *Can. J. Chem.*, **53** 2975 (1975); *idem, ibid.*, **55**, 562 (1977).

3. M. D. Chappell, C. R. Harris, S. D. Kuduk, A. Balog, Z. Wu, F. Zhang, C. B. Lee, S. J. Stachel, S. J. Danishefsky, T.-C. Chou, and Y. Guan, *J. Org. Chem.*, **67**, 7730 (2002).

4. R. E. Ireland and D. M. Obrecht, *Helv. Chim. Acta.*, **69**, 1273 (1986); D. M. Clode, W. A. Laurie, D. McHale and J. B. Sheridan, *Carbohydr. Res.*, **139**, 161 (1985).

5. B. A. D'Sa, D. McLeod, and J. G. Verkade, *J. Org. Chem.*, **62**, 5057 (1997).

6. S. K. Chaudhary and O. Hernandez, *Tetrahedron Lett.*, **20**, 99 (1979); Y. Guindon, C. Yoakim, M. A. Bernstein, and H. E. Morton, *ibid.*, **26**, 1185 (1985).

7. G. Cardillo, M. Orena, S. Sandri, and C. Tomasihi, *Chem. Ind. (London)*, 643 (1983).

8. F. Freeman and D. S. H. L. Kim, *J. Org. Chem.*, **57**, 1722 (1992).

9. C. Yu, B. Liu, and L. Hu, *Tetrahedron Lett.*, **41**, 4281 (2000).

10. S. A. Hardinger and N. Wijaya, *Tetrahedron Lett.*, **34**, 3821 (1993).

11. L. A. Paquette, J. Chang, and Z. Liu, *J. Org. Chem.*, **69**, 6441 (2004).

12. P. A. Grieco, K. J. Henry, J. J. Nunes, and J. E. Matt, Jr., *J. Chem. Soc., Chem. Commun.*, 368 (1992).

13. R. K. Bhatt, K. Chauhan, P. Wheelan, R. C. Murphy, and J. R. Falck, *J. Am. Chem. Soc.*, **116**, 5050 (1994).

14. T.-P. Loh and L.-C. Feng, *Tetrahedron Lett.*, **42**, 6001 (2001).

15. K. Tomooka, A. Nakazaki, and T. Nakai, *J. Am. Chem. Soc.*, **122**, 408 (2000).

16. S. Higashibayashi, K. Shinko, T. Ishizu, K. Hashimoto, H. Shirahama, and M. Nakata, *Synlett*, 1306 (2000).

17. W. Zhang and M. J. Robins, *Tetrahedron Lett.*, **33**, 1177 (1992).

18. K. C. Nicolaou, S. P. Seitz, M. R. Pavia, and N. A. Petasis, *J. Org. Chem.*, **44**, 4011 (1979); K. C. Nicolaou, S. P. Seitz, and M. R. Pavia, *J. Am. Chem. Soc.*, **103**, 1222 (1981).

19. I. Matsuo, M. Wada, and Y. Ito, *Tetrahedron Lett.*, **43**, 3273 (2002).

20. Y. Ogawa, M. Nunomoto, and M. Shibasaki, *J. Org. Chem.*, **51**, 1625 (1986).

21. C. Aïssa, R. Riveiros, J. Ragot, and A. Furstner, *J. Am. Chem. Soc.*, **125**, 15512 (2003).

22. E. M. Nashed and C. P. J. Glaudemans, *J. Org. Chem.*, **52**, 5255 (1987).

23. M. T. Barros, C. D. Maycock, and C. Thomassigny, *Synlett*, 1146 (2001).

24. S. Mabic and J. P. Lepoittevin, *Synlett*, 851 (1994).

25. S. Hatakeyama, H. Irie, T. Shintani, Y. Noguchi, H. Yamada, and M. Nishizawa, *Tetrahedron*, **50**, 13369 (1994).

26. T.-P. Loh and L.-C. Feng, *Tetrahedron Lett.*, **42**, 3223 (2001).

27. A. B. Smith III, Q. Lin, V. A. Doughty, L. Zhuang, M. D. McBriar, J. K. Kerns, C. S. Brook, N. Murase, K. Nakayama, and M. Sobukawa, *Angew. Chem. Int. Ed.*, **40**, 196 (2001).

28. D. R. Williams and K. G. Meyer, *J. Am. Chem. Soc.*, **123**, 765 (2001).

29. A. A. Malik, R. J. Cormier, and C. M. Sharts, *Org. Prep. Proced. Int.*, **18**, 345 (1986).

30. B. Rajashekhar and E. T. Kaiser, *J. Org. Chem.*, **50**, 5480 (1985).

31. J. C. McWilliams and J. Clardy, *J. Am. Chem. Soc.*, **116**, 8378 (1994).

32. M. S. Shekhani, K. M. Khan, K. Mahmood, P. M. Shah, and S. Malik, *Tetrahedron Lett.*, **31**, 1669 (1990).

33. J. Feixas, A. Capdevila, and A. Guerrero, *Tetrahedron* **50**, 8539 (1994).

Tribenzylsilyl Ether: $ROSi(CH_2C_6H_5)_3$ (Chart 1)

Tri-*p*-xylylsilyl Ether: $ROSi(CH_2C_6H_4-p-CH_3)_3$

To control the stereochemistry of epoxidation at the 10,11-double bond in intermediates in prostaglandin synthesis, a bulky protective group was used for the $C_{15}-OH$ group. Epoxidation of the tribenzylsilyl ether yielded 88% α-oxide; epoxidation of the tri-*p*-xylylsilyl ether was less selective.[1]

Formation

$ClSi(CH_2C_6H_4-p-Y)_3$ (Y = H or CH_3), DMF, 2,6-lutidine, −20°C, 24–36 h, 90–100% yield.[1]

Cleavage

1. AcOH, THF, H_2O, (3:1:1), 26°C, 6 h → 45°C, 3 h, 85% yield.[1]
2. Many of the fluoride-based reagents found in the TBDMS section will cleave this ether.

1. E. J. Corey and H. E. Ensley, *J. Org. Chem.*, **38**, 3187 (1973).

Triphenylsilyl Ether (TPS−OR): $ROSiPh_3$

The stability of the TPS group to basic hydrolysis is similar to that of the TMS group, but its stability to acid hydrolysis is about 400 times greater than the TMS group.[1]

Formation

1. Ph$_3$SiCl, Pyr.[2]
2. Ph$_3$SiBr, Pyr, −40°C, 15 min.[3]
3. Ph$_3$SiH, cat.[4] KOH, 18-crown-6 has been used as a catalyst (57–100% yield).[5] B(C$_6$H$_5$)$_3$ is a very effective catalyst for this transformation.[6] It has also been applied to the formation of other silyl ethers.

Cleavage

1. AcOH:H$_2$O:THF (3:1:1), 70°C, 3 h, 70% yield.[3]
2. Bu$_4$NF.[7]
3. NaOH, EtOH.[2]
4. HCl.[8]
5. HF·Pyr, THF, rt, 99% yield.[9]
6. NaBF$_4$ or NaPF$_6$, 0.5–16 h, 92–96% yield.[10]
7. Li, naphthalene, THF, 0°C. This system also works for other phenyl substituted silyl ethers.[11]

1. L. H. Sommer, *Stereochemistry, Mechanism and Silicon: An Introduction to the Dynamic Stereochemistry and Reaction Mechanisms of Silicon Centers*, McGraw-Hill, New York, 1965, p. 126.
2. S. A. Barker, J. S. Brimacombe, M. R. Harnden, and J. A. Jarvis, *J. Chem Soc.*, 3403 (1963).
3. H. Nakai, N. Hamanaka, H. Miyake and M. Hayashi, *Chem Lett.*, **8**, 1499 (1979).
4. E. Lukevics and M. Dzintara, *J. Organomet. Chem.*, **271**, 307 (1984); L. Horner and J. Mathias, *J. Organomet. Chem.*, **282**, 175 (1985).
5. F. L. Bideau, T. Coradin, J. Henique, and E. Samuel, *Chem. Commun.*, 1408 (2001).
6. J. M. Blackwell, K. L. Foster, V. H. Beck, and W. E. Piers, *J. Org. Chem.*, **64**, 4887 (1999).
7. K. Maruoka, M. Hasegawa, H. Yamamoto, K. Suzuki, M. Shimazaki, and G.-i. Tsuchihashi, *J. Am. Chem. Soc.*, **108**, 3827 (1986).
8. R. G. Neville, *J. Org. Chem.*, **26**, 3031 (1961).
9. A. Balog, D. Meng, T. Kamenecka, P. Bertinato, D.-S. Su, E. J. Sorensen, and S. J. Danishefsky, *Angew. Chem., Int. Ed . Engl.*, **35**, 2801 (1996).
10. O. Farooq, *J. Chem. Soc. Perkin Trans. 1*, 661 (1998).
11. C. Behloul, D. Guijarro, and M. Yus, *Tetrahedron*, **61**, 6908 (2005).

Diphenylmethylsilyl Ether (DPMS−OR): Ph$_2$MeSiOR

The DPMS group has stability intermediate between the TMS and TES (triethylsilyl) groups. It is incompatible with base, acid, BuLi, LiAlH$_4$, pyridinium chlorochromate, pyridinium dichromate, and CrO_3/pyridine. It is stable to Grignard reagents, Wittig reagents, m-chloroperoxybenzoic acid, and silica gel chromatography.[1]

Formation

1. $Ph_2MeSiCl$, DMF, imidazole, 83–92% yield.[1]
2. Ph_2MeSiH, $Cl_2(PCy_3)_2Ru=CHPh$, 25–35°C, 3, 95% yield.[2]
3. Ph_2MeSiH, $[RuCl_2(p\text{-cym})]_2$, CH_2Cl_2, 25°C, 6 h, 95% yield.[3]

Cleavage

1. It can be cleaved with mild acid, fluoride ion or base.[1]
2. NaN_3, DMF, 40°C, 80–93% yield.[4]
3. Photolysis at 254 nm, CH_3OH, CH_2Cl_2, phenanthrene, 51–84% yield. These conditions are selective for allylic and benzylic alcohols. In the absence of the phenanthrene, TBDMS ethers are also cleaved.[5]

1. S. E. Denmark, R. P. Hammer, E. J. Weber, and K. L. Habermas, *J. Org. Chem.*, **52**, 165 (1987).
2. S. V. Maifeld, R. L. Miller, and D. Lee, *Tetrahedron Lett.*, **43**, 6363 (2002).
3. R. L. Miller, S. V. Maifeld, and D. Lee, *Org. Lett.*, **6**, 2773 (2004).
4. S. J. Monger, D. M. Parry, and S. M. Roberts, *J. Chem. Soc., Chem. Commun.*, 381 (1989).
5. O. Piva, A. Amougay, and J.-P. Pete, *Synth. Commun.*, **25**, 219 (1995).

Di-t-butylmethylsilyl Ether (DTBMS−OR): $(t\text{-Bu})_2MeSiOR$

Formation

1. $DTBMSClO_4$, MeCN, Pyr, 100% yield.[1]
2. DTBMSOTf, 2,6-lutidine, DMAP, 70°C, 87% yield.[2,3]

Cleavage

1. $BF_3 \cdot Et_2O$, CH_2Cl_2; $NaHCO_3$, H_2O, 0°C, 30 min, 94% yield. CsF in DMSO fails to cleave this group.[1]
2. 49% Aqueous HF, $MeNO_2$, 0°C, 24 h, 30% yield.[2]

1. T. J. Barton and C. R. Tully, *J. Org. Chem.*, **43**, 3649 (1978).
2. K. C. Nicolaou, E. W. Yue, S. La Greca, A. Nadin, Z. Yang, J. E. Leresche, T. Tsari, Y. Naniwa, and F. De Riccardis, *Chem. Eur. J.*, **1**, 467 (1995).
3. R. S. Bhide, B. S. Levison, R. B. Sharma, S. Ghosh, and R. G. Salomon, *Tetrahedron Lett.*, **27**, 671 (1986).

Bis(*t*-butyl)-1-pyrenylmethoxysilyl Ether

This group was developed as a fluorescent silyl protective group for oligonucleotide synthesis. It has excitation and emission wavelengths of 346 nm and 390 nm, respectively, which are outside the range of the DNA-damaging wavelength of 254–260 nm. It is prepared from the *in situ* prepared silyl chloride. It is stable to 0.01 M HCl and 30% ammonia. It is cleaved with 0.1 M TBAF in 3 min at rt.[1]

1. S. Tripathi, K. Misra, and Y. S. Sanghvi, *Nucleosides & Nucleotides, and Nucleic Acids*, **24**, 1345 (2005).

Sisyl Ether [Tris(trimethylsilyl)silyl Ether]: $[(CH_3)_3Si]_3SiOR$

The sisyl ether is stable to Grignard and Wittig reagents, oxidation with Jones' reagent, KF/18-crown-6. CsF, and strongly acidic conditions (TsOH, HCl) that cleave most other silyl groups. It is not stable to alkyllithiums or $LiAlH_4$.

Formation

$[(CH_3)_3Si]_3SiCl$, CH_2Cl_2, DMAP, 70–97% yield.[1]

Cleavage

1. TBAF, THF.[2]
2. Photolysis, MeOH, CH_2Cl_2, 62–95% yield.[1]
3. Relative rates for acidic hydrolysis of silyl ethers (aqueous THF and AcOH)[3]

SiR_3	$PhCH_2CH_2OSiR_3$	$PhCH_2OSiR_3$	$C_5H_9OSiR_3$
$Si(SiMe_3)_3$	6.2	5.5	3.7
$SiMe_2t$-Bu	1	1	1

1. M. A. Brook, C. Gottardo, S. Balduzzi, and M. Mohamed, *Tetrahedron Lett.*, **38**, 6997 (1997).
2. K. J. Kulicke and B. Giese, *Synlett*, 91 (1990).
3. M. A. Brook, S. Baladuzzi, M. Mohamed, and C. Gottardo, *Tetrahedron*, **55**, 10027 (1999).

(2-Hydroxystyryl)dimethylsilyl Ether (HSDMS−OR) and
(2-Hydroxystyryl)diisopropylsilyl Ether (HSDIS−OR)

Formation

The reagent is readily prepared by the addition of Me$_2$NLi to the silyl chloride.[1]

Formation scheme: R'$_2$Si(NMe$_2$)-styryl with ortho-OH; ROH, THF, rt, or reflux; 72–95% yield; R' = Me or *i*-Pr → R'$_2$Si(OR)-styryl with ortho-OH

Cleavage[1]

Photolysis at 254 nm, rt, 30 min, CH$_3$CN, 75–92% yield. Cleavage occurs by *trans* to *cis* isomerization followed by hydroxyl exchange to release the alcohol. Cleavage of the naphthyl analog occurs at 350 nm.[2]

naphthyl-Si(OR) with ortho-OH; *hv*, 350 nm, MeOH; 86–94% → cyclic product with O–Si−*l*-Pr / *l*-Pr + ROH

1. M. C. Pirrung and Y. R. Lee, *J. Org. Chem.*, **58**, 6961 (1993).
2. M. C. Pirrung, L. Fallon, J. Zhu, and Y. R. Lee, *J. Am. Chem. Soc.*, **123**, 3638 (2001).

t-Butylmethoxyphenylsilyl Ether (TBMPS−OR): *t*-Bu(CH$_3$O)PhSiOR

The TBMPS group has a greater sensitivity to fluoride ion than the TBDMS and TBDPS groups, which allows for the selective cleavage of the TBMPS group in the presence of the latter two. The TBMPS group is also 140 times more stable to 0.01 N HClO$_4$ than the TBDMS group, thus allowing selective hydrolysis of the TBDMS group. The group can be introduced onto primary, secondary, and tertiary hydroxyls in excellent yield when DMF is used as the solvent, and it can be selectively introduced onto primary hydroxyls when CH$_2$Cl$_2$ is used as solvent. The main problem with this group is that when it is introduced onto chiral molecules, diastereomers result that may complicate NMR interpretation.[1]

Formation/Cleavage[1]

ROH   *t*-BuPhMeOSiBr, Et$_3$N, CH$_2$Cl$_2$ or DMF, 71–100% → ROTBDMS
       ← Bu$_4$NF, THF or acid

In the following case, the TBMPS group was used to advantage to get reasonable acid stability during the cleavage of 2° TBS group earlier in the synthesis and yet allow removal under mild treatment with TBAF.[2]

1. Y. Guindon, R. Fortin, C. Yoakim, and J. W. Gillard, *Tetrahedron Lett.*, **25**, 4717 (1984); J. W. Gillard, R. Fortin, H. E. Morton, C. Yoakim, C. A. Quesnelle, S. Daignault, and Y. Guindon, *J. Org. Chem.*, **53**, 2602 (1988).
2. D. R. Williams, M. P. Clark, U. Emde, and M. A. Berliner, *Org. Lett.*, **2**, 3023 (2000).

t-Butoxydiphenylsilyl Ether (DPTBOS−OR): Ph₂(*t*-BuO)SiOR

The DPTBOS group is considered a low-cost alternative to the TBDMS group with comparable acid stability and retained sensitivity to fluoride ion.

Formation

DPTBOSCl, TEA, CH_2Cl_2, rt, 98% yield.[1]

Cleavage

1. 0.01 *M* $HClO_4$.[2]
2. TBAF.[2]
3. $Na_2S \cdot 9H_2O$, EtOH, rt, 12 h, 70% yield.[3]
4. TAS−F, H_2O, DMF, 85% yield. In this case the TBS ether could not be cleaved at a reasonable rate.[4]

1. L. F. Tietze, C. Schnieder, and A Grote, *Chem. Eur. J.*, **2**, 139 (1996).
2. J. W. Gillard, R. Fortin, H. E. Morton, C. Yoakim, C. A. Quesnelle, S. Daignault, and Y. Guindon, *J. Org. Chem.*, **53**, 2602 (1988).
3. T. Schmittberger and D. Uguen, *Tetrahedron Lett.*, **36**, 7445 (1995).
4. D. A. Evans, H. A. Rajapakse, A. Chiu, and D. Stenkamp, *Angew. Chem. Int. Ed.*, **41**, 4573 (2002).

1,1,3,3-Tetraisopropyl-3-[2-(triphenylmethoxy)ethoxy]disiloxane-1-yl Ether

This group was developed for the protection of the 5′-hydroxyl for solid-phase RNA synthesis. It is introduced with the silyl chloride, and pyridine and can be cleaved with TBAF in THF. The trityl group introduces a chromophore for analytical purposes.[1]

1. I. Hirao, M. Koizumi, Y. Ishido, and A. Andrus, *Tetrahedron Lett.*, **39**, 2989 (1998).

Fluorous Silyl Ethers: $(C_6F_{13}CH_2CH_2)_3Si-OR$, $C_6F_{13}CH_2CH_2(i\text{-}Pr)_2Si-OR$, $C_8F_{17}CH_2CH_2(Ph)(t\text{-}Bu)Si-OR$, $(C_8F_{17}CH_2CH_2)_2CHO)(Ph)(Me)Si-OR$, $(C_8F_{17}CH_2CH_2)_2CHO)(Ph)_2Si-OR$, $(C_8F_{17}CH_2CH_2)_2CHO)(Ph)(t\text{-}Bu)Si-OR$

These ethers have been prepared to use the "fluorous synthesis" technique. They are introduced using the standard methods and can be cleaved with TBAF in THF.[1–5]

1. H. Nakamura, B. Linclau, and D. P. Curran, *J. Am. Chem. Soc.*, **123**, 10119 (2001).
2. L. Manzoni and R. Castelli, *Org. Lett.*, **6**, 4195 (2004).
3. S. Röver and P. Wipf, *Tetrahedron Lett.*, **40**, 5667 (1999).
4. Z. Luo, Q. Zhang, Y. Oderaotoshi, and D. P. Curran, *Science*, **291**, 1766 (2001).
5. S. Tripathi, K. Misra, and Y. S. Sanghvi, *Org. Prep. & Proc. Int.*, **37**, 257 (2005).

Conversion of Silyl Ethers to Other Functional Groups

The ability to convert a protective group to another functional group directly without first performing a deprotection is a potentially valuable transformation. Silyl-protected alcohols have been converted directly to aldehydes,[1,2] ketones,[3] bromides,[4] acetates[5] and ethers[6] without first liberating the alcohol in a prior deprotection step. The smaller sterically less demanding silyl ethers can often be oxidized to aldehydes and ketones with reagents such as pyridinium chlorochromate.

1. G. A. Tolstikov, M. S. Miftakhov, N. S. Vostrikov, N. G. Komissarova, M. E. Adler, and O. Kuznetsov, *Zh. Org. Khim.*, **24**, 224 (1988); *Chem. Abstr.*, **110**, 7162c (1989).
2. I. Mohammadpoor-Baltork and S. Pouranshirvani, *Synthesis*, 756 (1997).
3. F. P. Cossio, J. M. Aizpurua, and C. Palomo, *Can. J. Chem.*, **64**, 225 (1986).
4. H. Mattes and C. Benezra, *Tetrahedron Lett.*, **28**, 1697 (1987); S. Kim and J. H. Park, *J. Org. Chem.*, **53**, 3111 (1988); J. M. Aizpurua, F. P. Cossio, and C. Palomo, *J. Org. Chem.*, **51**, 4941 (1986).
5. S. J. Danishefsky and N. Mantlo, *J. Am. Chem. Soc.*, **110**, 8129 (1988); B. Ganem and V. R. Small, Jr., *J. Org. Chem.*, **39**, 3728 (1974); S. Kim and W. J. Lee, *Synth. Commun.*, **16**, 659 (1986); E.- F. Fuchs and J. Lehmann, *Chem. Ber.*, **107**, 721 (1974).
6. D. G. Saunders, *Synthesis*, 377 (1988).

ESTERS

See also Chapter 5, on the preparation of esters as protective groups for carboxylic acids.

Formate Ester: ROCHO (Chart 2)

Formation

1. 85% HCOOH, 60°C, 1 h, 93% yield.[1] This method can be used to selectively protect only the primary alcohol of a pyranoside.[2]
2. 70% HCOOH, cat. $HClO_4$, 50–55°C, good yields.[3]
3. $CH_3COOCHO$, Pyr, −20°C, 80–100% yield.[4–6] The related $(CH_3)_3CCO_2$-C(O)H has been used similarly and has the advantage that no pivalate was formed as is sometimes the case with the acetyl derivative.[7]
4. Me_2N^+=CHOBz Cl$^-$, Et_2O, overnight; dil. H_2SO_4, 60–96% yield.[8]
5. DMF, Cs_2CO_3, TBAI, 100°C, 20 h, cyclohexyl bromide, 86% yield.[9]
6. 2,4,6-trichloro-1,3,5-triazine, DMF, LiF, CH_2Cl_2, rt, 15 min to 4 h, 76–100% yield. Primary alcohols are formylated in the presence of secondary alcohols.[10]
7. HCO_2H, $BF_3 \cdot 2MeOH$, 90% yield.[11]
8. Ethyl formate, $Ce(SO_4)_2$–silica gel, reflux 0.5–24 h, 90–100% yield.[12]
9. Methyl formate, HBr, 88% yield.[13]
10. β-Oxopropyl formate, DBN, 50–70°C, 3 h, THF, 70–82% yield.[14]
11. From a silyl ether (TES, TBDMS, TBDPS, TIPS): Vilsmeier–Haack reagents, 10–85% yield.[15] TIPS ethers give low yields.

Cleavage

1. $KHCO_3$, H_2O, MeOH, 20°C, 3 days.[3]
2. Dil. NH_3, pH 11.2, 22°C, 62% yield.[16] A formate ester can be cleaved selectively in the presence of an acetate (MeOH, reflux)[5] or dil. NH_3 (formate is 100 times faster than an acetate)[16] or benzoate ester (dil. NH_3).[16]

1. H. J. Ringold, B. Löken, G. Rosenkranz, and F. Sondheimer, *J. Am. Chem. Soc.*, **78**, 816 (1956).
2. L. X. Gan and R. L. Whistler, *Carbohydr. Res.*, **206**, 65 (1990).
3. I. W. Hughes, F. Smith, and M. Webb, *J. Chem. Soc.*, 3437 (1949).
4. F. Reber, A. Lardon, and T. Reichstein, *Helv. Chim. Acta*, **37**, 45 (1954).
5. J. Zemlicka, J. Beránek, and J. Smrt, Collect. *Czech. Chem. Commun.*, **27**, 2784 (1962).
6. For a review on acetic formic anhydride, see P. Strazzolini, A. G. Giumanini, and S. Cauci, *Tetrahedron*, **46**, 1081 (1990).
7. E. Vedejs and S. M. Duncan, *J. Org. Chem.*, **65**, 6073 (2000).
8. J. Barluenga, P. J. Campos, E. Gonzalez-Nunez, and G. Asensio, *Synthesis*, 426 (1985).

9. F. Chu, E. E. Dueno, and K. W. Jung, *Tetrahedron Lett.*, **40**, 1847 (1999).

10. L. De Luca, G. Giacomelli, and A. Porcheddu, *J. Org. Chem.*, **67**, 5152 (2002).

11. M. Dymicky, *Org. Prep. Proced. Int.*, **14**, 177 (1982).

12. T. Nishiguchi and H. Taya, *J. Chem. Soc., Perkin Trans. I*, 172 (1990).

13. H. Hagiwara, K. Morohashi, H. Sakai, T. Suzuki, and M. Ando, *Tetrahedron*, **54**, 5845 (1998).

14. A. Kabouche and Z. Kabouche, *Tetrahedron Lett.*, **40**, 2127 (1999).

15. J.-P. Lellouche and V. Kotlyar, *Synlett*, 564 (2004); S. Koeller and J.-P. Lellouche, *Tetrahedron Lett.*, **40**, 7043 (1999).

16. C. B. Reese and J. C. M. Stewart, *Tetrahedron Lett.*, **9**, 4273 (1968).

Benzoylformate Ester: ROCOCOPh

The benzoylformate ester can be prepared from the 3′-hydroxy group in a deoxyribonucleotide by reaction with benzoyl chloroformate (anhydrous pyridine, 20°C, 12 h, 86% yield); it is cleaved by aqueous pyridine (20°C, 12 h, 31% yield), conditions that do not cleave an acetate ester.[1]

1. R. L. Letsinger and P. S. Miller, *J. Am. Chem. Soc.*, **91**, 3356 (1969).

Acetate Ester (ROAc): CH_3CO_2R (Chart 2)

Formation

Methods Based on Base Catalysis

1. Ac_2O, Pyr, 20°C, 12 h, 100% yield.[1] This is one of the most common methods for acetate introduction. By running the reaction at lower temperatures, good selectivity can be achieved for primary alcohols over secondary alcohols.[2] Tertiary alcohols are generally not acylated under these conditions.

2. Ac_2O or AcCl, Pyr, DMAP, 24–80°C, 1–40 h, 72–95% yield.[3] The use of DMAP increases the rate of acylation by a factor of 10^4. These conditions will acylate most alcohols, including tertiary alcohols. Although DMAP is a great catalyst, the modifications embodied in catalysts **2** and **3** make them superior.[4] The relative rates for the catalysts **1**, **2**, and **3** are 1:2.4:6.

1 (DMAP)   2 (4-PPY)   3   Cat.

The use of DMAP (4-*N*,*N*-dimethylaminopyridine) as a catalyst to improve the rate of esterification is quite general and works for other esters as well, but it is not effective with hindered anhydrides such as pivalic anhydride.

3. The phosphine **i**[5] (48–99% yield) and Bu$_3$P[6] have been developed as active acylation catalysts for acetates and benzoates.

i

4. Ac$_2$O, pyridine–alumina, microwave heating, no solvent, 54–100% yield.[7] Phenols, thiols, and amines are also acylated.

5. CH$_3$COCl, CH$_2$Cl$_2$, collidine, 91% yield. A primary acetate was formed selectively in the presence of a secondary. These conditions are suitable for a variety of other esters.[8]

6. CH$_2$=C=O, *t*-BuOK, THF.[9] The 17α-hydroxy group of a steroid was acetylated by this method.

7. AcCl, Ag$_2$O, cat. KI, CH$_2$Cl$_2$, 40°C, 60–99% yield. In some cases, this method gives results that are complementary to the stannylene method. Selectivity, in the esterification is dependent upon the configuration at the anomeric position of a pyranoside.[10] Benzoates give similar results, but with tosylates the regioselectivity is reversed in some cases.

8. NaH, 93% yield.[11] Primary alcohols are selectively acylated.

Methods Based on Acid Catalysis

1. CH_3COCl neat or in CH_2Cl_2, $ZrOCl_2 \cdot 8H_2O$,[12] or $BiOCl$[13] 86–98% yield. Phenols, thiols, and amines are all readily acylated.

2. CH_3COCl, 25°C, 16 h, 67–79% yield.[14]

3. The direct conversion of a THP-protected alcohol to an acetate is possible, thus avoiding a deprotection step.[15]

4. Ac-imidazole, $PtCl_2(C_2H_4)$, 23°C, 0.5–144 h, 51–87% yield.[16] Platinum(II) acts as a template to catalyze the acetylation of the pyridinyl alcohol, $C_5H_4N(CH_2)_nCH_2OH$. Normally acylimidazoles are not very reactive acylating agents with alcohols.

5. Ac_2O, CH_2Cl_2, 15 kbar (1.5 GPa), 79–98% yield.[17] This high-pressure technique also works to introduce benzoates and TBDMS ethers onto highly hindered tertiary alcohols.

6. The monoacetylation of alpha–omega diols can be accomplished in excellent yield.[18]

$$HOCH_2(CH_2)_nCH_2OH \xrightarrow[\substack{30 \text{ h to } 1\text{wk} \\ 60–90\%}]{AcOH, H_2SO_4, H_2O} AcOCH_2(CH_2)_nCH_2OH$$

A monoacetate can be isolated by continuous extraction with organic solvents such as cyclohexane/CCl_4. Monoacylation can also be achieved by ion exchange resin,[19] HY-Zeolite,[20] or acid-catalyzed[21] transesterification.

7. AcOH, TMSCl, 81% yield.[22]

8. AcOH, $FeCl_3$, CH_2Cl_2, 81–99% yield. Acetonides, THP, TBDMS and TPS ethers are converted directly to acetates.[23]

9. $Sc(OTf)_3$, AcOH, p-nitrobenzoic anhydride[24] or $Sc(OTf)_3$, Ac_2O, 66% to >95% yield. The lower yields are obtained with allylic alcohols, but propargylic alcohols give high yields. Phenols are effectively acylated with this catalyst, but at a much slower rate than simple aliphatic alcohols.[25] The method was shown to be superior to most other methods for macrolactonization with minimum diolide formation.

10. Ac_2O, cat. TMSOTf, CH_2Cl_2, 0°C, 0.5–60 min, 71–100% yield. This is a more reactive combination of reagents than DMAP/Ac_2O. Phenols are also efficiently acylated by this method.[26]

11. Ac_2O, $BF_3 \cdot Et_2O$, THF, 0°C.[27] These conditions give good chemoselectivity for the most nucleophilic hydroxyl group. Alcohols are acetylated in the presence of phenols.

12. Ac_2O, HBF_4 absorbed on silica gel, neat, rt, 75–100% yield. Phenols, thiols, and amines are also readily acylated.[28]

13. Ac_2O, polystyrene-bound $C_6F_4CH(Tf)_2$, <1 h, >99% yield. Benzoyl esters are formed when using Bz_2O.[29]

14. A large number of metal salts have been used to activate Ac_2O for the acylation of alcohols and phenols. At least with the triflates, a dual mechanism has been demonstrated. In the first process, TfOH generated *in situ* serves as a very effective catalyst for very rapid acylation of the alcohol; the second, but slower, process is catalyzed by the metal triflate.[30] Although it is not clear how far this can be extrapolated to the numerous other metal salts that have been used to catalyze ester formation, it is likely that these too will participate in an acid-induced catalytic cycle. The following is a compilation of many of the metal salts that have been used for ester formation with Ac_2O and Bz_2O and other anhydrides: $Sc(NTf_2)_3$ (CH_3CN, 0°C, 1 h, 90–99% yield),[31,32] $Bi(OTf)_3$ (CH_3CN, 15 min to 3 h, 80–92% yield),[33-37] $Cu(OTf)_2$ (0°C to rt, 66–99% yield, a racemization free method[112]),[38,39] LiOTf (neat, rt, 44–97% yield),[40] $In(OTf)_3$ (CH_3CN, rt, 95–98% yield),[41] $LiClO_4$ (neat, rt, 4–48 h, 84–100% yield),[42] $Mg(ClO_4)_2$ (neat, 1 min 7.5 h, 92–99% yield),[43] $BiOClO_4$ (CH_3CN, 10 min to 2 h, 79–100% yield),[44] $AlPW_{12}O_{40}$ (neat, rt, 88–98% yield),[45] $TaCl_5$ (CH_2Cl_2, rt, 40–80% yield),[46] $Sc(OTf)_3$ (neat, rt, 88–99% yield),[47] $Ce(OTf)_3$ (CH_3CN, rt, 73–98% yield),[48] $RuCl_3$ (CH_3CN, rt, 81–95% yield),[49] $CoCl_2$ (69–100% yield). This method does not work for 3° alcohols).[50,51] TMS ethers can be converted directly to acetates using $Sc(OTf)_3$ and Ac_2O.[52]

Ref. 53

15. Ac_2O, Amberlyst 15, 77% yield. These conditions introduce an acetyl group on oxygen in preference to the normally more reactive primary amine.[54] The amine is protonated, thereby reducing its reactivity. A number of other solid acids have been used to catalyze acylations: yttria–zirconia (CH_3CN, reflux, 71–99% yield),[55] Montmorillonite clay (CH_2Cl_2, 28–98% yield),[56] Zeolite H-FER (neat, 75°C, 45–99% yield).[57] Amines and thiols are also acylated. Zeolite HSZ-360 (neat, 60°C, 1–8 h, 84–100% yield),[58] Nafion-H (CH_2Cl_2, 2–24 h, 75–99% yield),[59] 4-Å molecular sieves (neat, 1–24 h, 56–98% yield).[60]

16. Ac_2O, $YbCl_3$, THF, 64–100% yield of the monoacetate from 1,2-diols.[61]

17. VO(OTf)$_2$, Ac$_2$O, CH$_2$Cl$_2$, 75–100% yield. Other esters can be formed by using other anhydrides. Thiols and amines and phenols are also acylated, but tertiary alcohols are not reactive.[62]

18. Ac$_2$O, I$_2$, 85–100% yield.[63] Phenols and 3° alcohols are also efficiently acylated.

19. Ac$_2$O, NBS, CH$_2$Cl$_2$, 84–98% yield.[64]

Methods Based on Transesterification

1. AcOC$_6$F$_5$, Et$_3$N, DMF, 80°C, 12–60 h, 72–95% yield.[65] This reagent reacts with amines (25°C, no Et$_3$N) selectively in the presence of alcohols to form N-acetyl derivatives in 80–90% yield.

2. Vinyl acetate or 2-propenyl acetate, toluene, Cp*$_2$Sm(THF)$_2$, rt, 3 h, 88–99% yield. Other esters can also be prepared by this method.[66] Iminophosphorane bases also serve as excellent transesterification catalysts with vinyl acetate (74–99% yield).[67]

3. Vinyl acetate, PdCl$_2$, CuCl$_2$, toluene, rt, 58–96% yield. Phenols, amines, and tertiary alcohols are not acylated with this method.[68]

4. Isopropenyl acetate, Y$_5$(O$-i$-Pr)$_{13}$O, 72–99% yield. Esters are formed in the presence of phenols and amines.[69]

5. Ethyl acetate, Ce(SO$_4$)$_2$·silica gel, reflux, 91–99% yield.[70]

6. 1,3-Disubstituted tetraalkyldistannoxanes, Ac$_2$O, EtOAc, or vinyl acetate, 17–99% yield. Primary alcohols are acylated selectively over secondary alcohols.[71]

7. AcOEt, Al$_2$O$_3$, 75–80°, 24 h, 45–69% yield.[72] This method is selective for primary alcohols. Phenols do not react under these conditions. The use of SiO$_2$-NaHSO$_4$ as a solid support was also found to be effective.[73]

8. Ph$_3$P, CBr$_4$, EtOAc, 51–100% yield.[74]

9. AcOMe, N-hetereocyclic carbene catalyst, molecular sieves, 25°C, 56–92% yield.[75]

Biotransformations

1. The use of biocatalysts for the selective introduction and cleavage of esters is vast and has been extensively reviewed.[76] Therefore, only a few examples of the types of transformations that are encountered in the area of protective group chemistry will be illustrated to show some of the basic transformations that have appeared in the literature. The selective protection or deprotection of symmetrical intermediates to give enantio-enriched products has also been used extensively.

2. AcOCH$_2$CF$_3$, porcine pancreatic lipase, THF, 60 h, 77% yield.[77] This enzymatic method was used to acetylate selectively the primary hydroxyl group of a variety of carbohydrates. The selective enzymatic acylation of carbohydrates has been partially reviewed.[78]

3. $AcOCH_2CCl_3$, pyridine, porcine pancreatic lipase, 85% yield.[79] These studies examined the selective acylation of carbohydrates. Mannose is acylated at the 6-position in 85% yield in one example.

4. Lipase Fp from Amano, vinyl acetate, 4 h, 90% yield.[80,81] This method can also be used for the selective introduction of other esters such as the methoxyacetyl, phenoxyacetyl, and phenylacetyl groups in excellent yield.

(a)

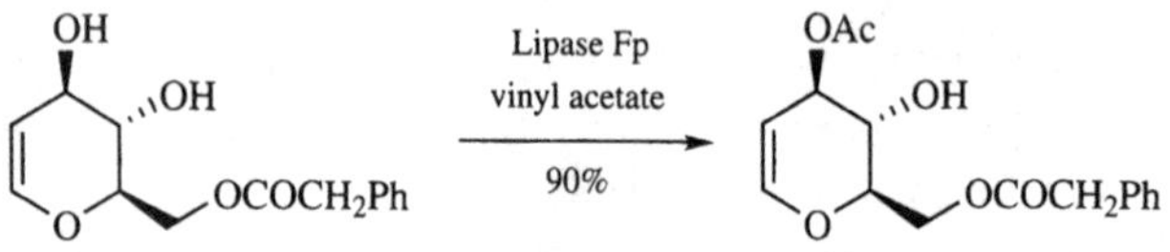

(b)

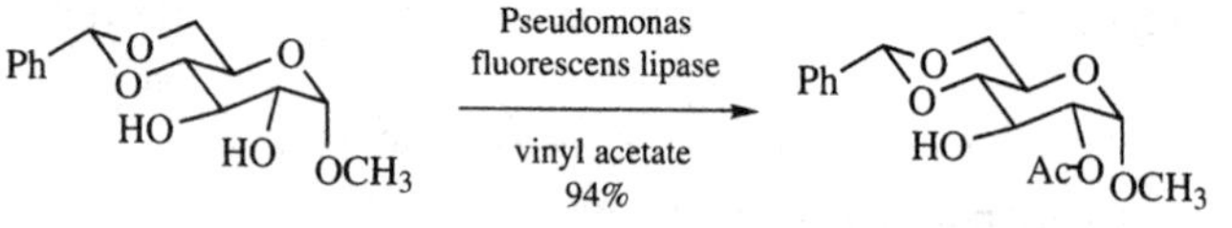

Ref. 82

(c)

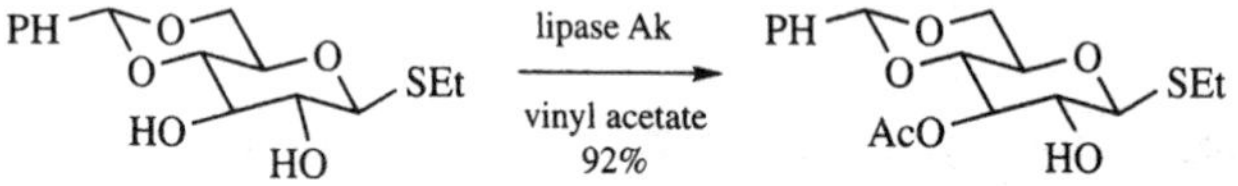

Refs. 83, 84

(d)

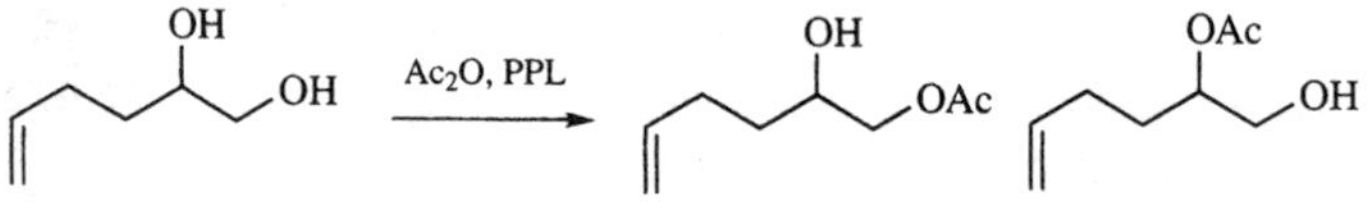

98:2   Ref. 85

(e) Carbohydrates with their multiple hydroxyl groups can often be selectively protected more easily using lipases than by conventional esterifications.[86]

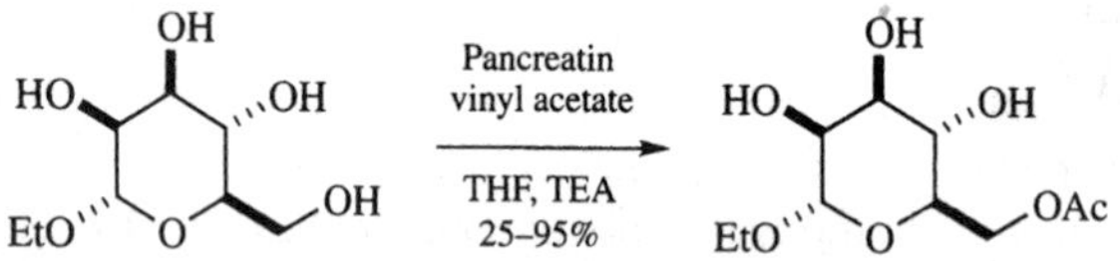

Ref. 87

(f) Desymmetrization of alcohols is useful not only in that a diol is selectively protected but resolution of the alcohol is also observed. 1-Ethoxyvinyl 2-furoate was found to be superior to vinyl acetate in these reactions giving monoprotected alcohols in 82–99% ee.[88]

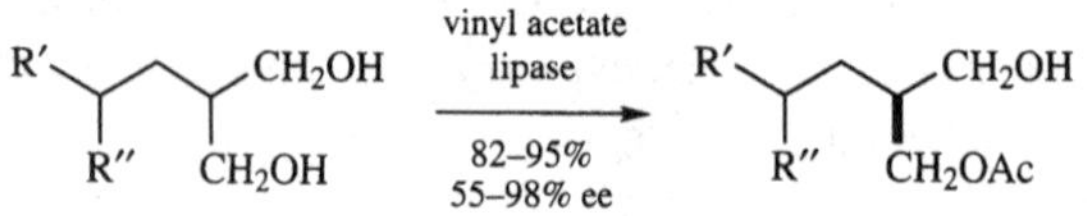

Ref. 89

(g)

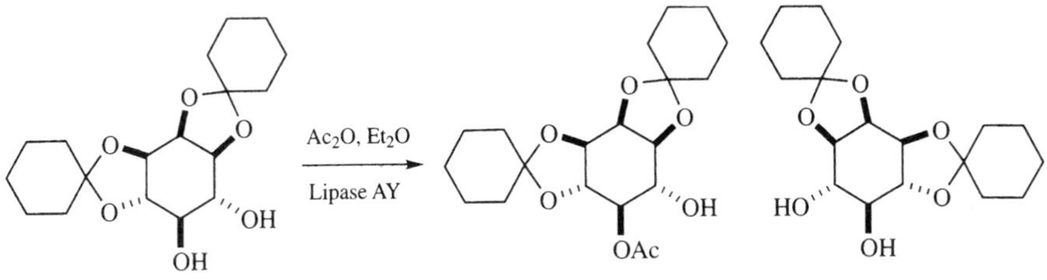

(+/-)   44% (100% ee)   54% (88% ee)

Ref. 90

(h)

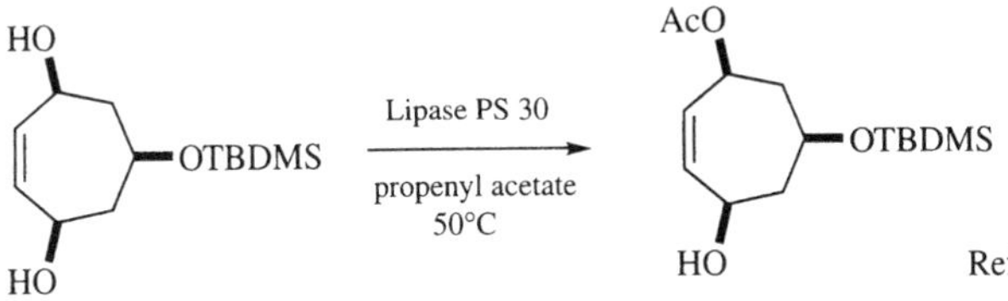

Ref. 91

This lipase has been used to selectively acetylate the 3'-hydroxyl of 2'-deoxynucleosides and ribonucleosides in the presence of the free 5'-hydroxyl.[92]

Miscellaneous Methods

1. Bu_2SnO, $PhCH_3$, 110°C, 2 h; AcCl, CH_2Cl_2, 0°C, 30 min, 84% yield.[93]

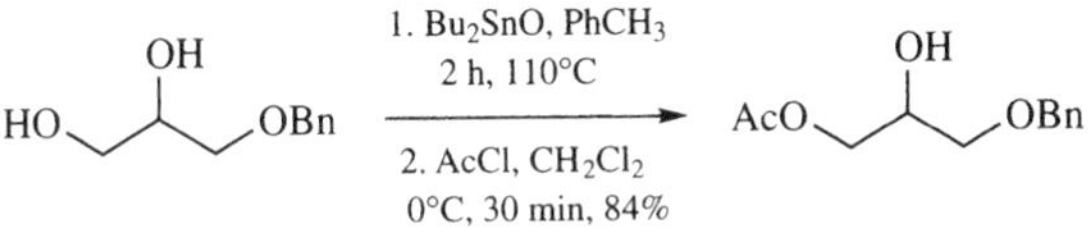

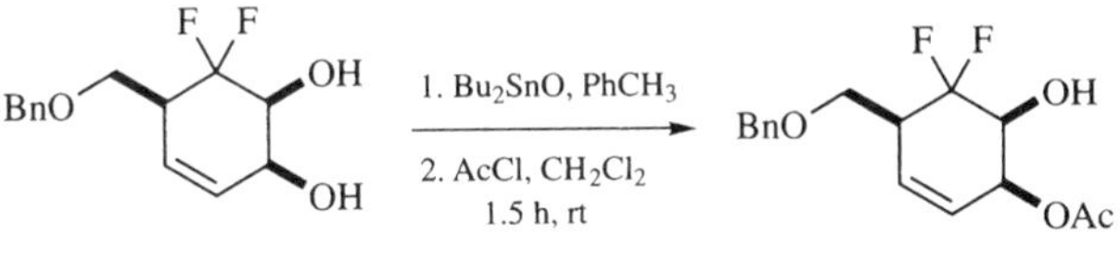

Ref. 94

2.

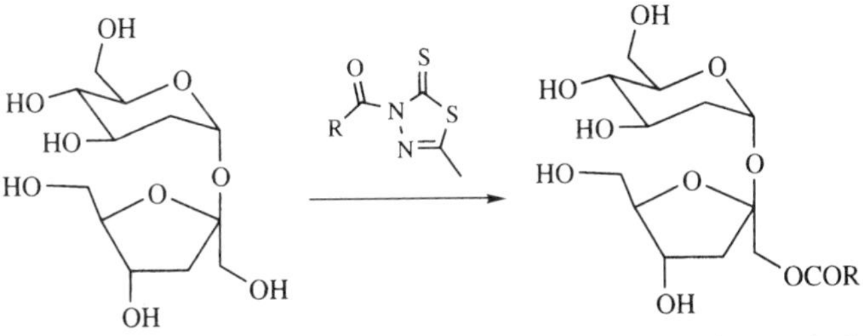

Ref. 95

3. An acyl thiazolidone is also effective for the selective acylation (Ac, Pv, Bz) of primary alcohols.[96]

4. $Me(OMe)_3$, TsOH, 1.5 h, then H_2O for 30 min.[97] When TMSCl is used as a catalyst simple alcohols are acylated in preference to phenols (70–88% yield).[98]

When the reaction was run in CH_3CN migration of the EtS group to the 2-position was observed. This is attributed to episulfonium salt formation with resultant addition of acetate at the anomeric position.[99]

5.

Enantioselective Acetylation not Using Enzymes

One form of protecting group selectivity is selectivity for a single enantiomer of a racemic alcohol. A number of catalytic systems have been developed that give good to excellent results for the selective acylation of a single enantiomer.[101]

Cleavage

1. K_2CO_3, MeOH, H_2O, 20°C, 1 h, 100% yield.[102] When catalytic NaOMe is used as the base in methanol, the method is referred to as the Zemplén de-O-acetylation. Acetyl groups are known to migrate under these conditions, but a recent study indicated that acyl migration is reduced with decreasing solvent polarity (6:1 chloroform/MeOH vs. MeOH).[103]

2. Phase transfer catalysis: TBAH, NaOH, THF, or CH_2Cl_2, rt, 51–96% yield.[104]

3. KCN, 95% EtOH, 20°C to reflux, 12 h, 93% yield.[105,106] Potassium cyanide is a mild transesterification catalyst, suitable for acid- or base-sensitive compounds. When used with 1,2-diol acetates hydrolysis proceeds slowly until the first acetate is removed.[107]

4. Guanidine, EtOH, CH_2Cl_2, rt, 85–100% yield.[108] Acetamides, benzoates and pivaloates are stable under these conditions. Phenolic acetates can be removed in the presence of primary and secondary acetates with excellent selectivity.

5. 50% NH_3, MeOH, 20°C, 2.5 h, 85% yield.[109] The 3′-acetate is removed from cytosine in the presence of a 5′-benzoate. If the reaction time is extended to 2 days the benzoate is removed as well as the benzoyl protection on nitrogen.

6. Bu₃SnOMe, ClCH₂CH₂Cl, 1 h, 77% yield.[110] These conditions selectively cleave the anomeric acetate of a glucose derivative in the presence of other acetates.

7. BF₃·Et₂O, wet CH₃CN, 96% yield.[111]

8. Sc(OTf)₃, MeOH, H₂O, 88% yield. This method is good for systems that are prone to racemization as in the following case.[112]

9. Yb(OTf)₃, IPA, reflux, 8–78 h, 51–97% yield. Phenolic acetates are cleaved somewhat faster, and some selectivity for primary over secondary acetates was achieved.[113]

10. 1,8-Diazabicyclo[5.4.0]undec-7-ene (DBU), benzene, 60°C, 45 h, 47–97% yield.[114] Benzoates are not cleaved under these conditions.

11. Tris(2,4,6-trimethoxyphenyl)phosphine, MeOH, 20°C, 7.5–48 h, 73–99% yield.[115] Note that axial acetates are cleaved much more slowly.

12. CH₃ONa, La(OTf)₃, MeOH, 97–100% yield. This method was developed specifically for the isomerization free cleavage of 6-*exo*-acetoxybicyclo-[2.2.2]octan-2-ones.[116] Isomerization can occur through a retro aldol process in the presence of base.

13. Sm, I₂, MeOH, rt, 3–60 min, 95–100% yield. Tertiary alcohols were not affected. As the reaction time and temperature are increased benzoates and carbonates can also be cleaved.[117]

14. I_2, MeOH, 68–80°C, 5–40 h, 38–69% yield. The method was used to selectively cleave the primary acetate from peracetylated nucleosides. Lower
yields were obtained for substrates having a thioether.[118]

15. HBF_4, MeOH, 23°C, 48 h, 83% yield. This system cleaves acetate groups
in the presence of benzoate groups.[119,120] HCl in methanol can also be used,
and this method will cleave a primary acetate in the presence of secondary
benzoates.[121–123]

16. $LiEt_3BH$, THF, −78°C, 2 h, 98% yield.[124] An anomeric acetate can be selectively cleaved in the presence of a secondary acetate.

17. Distannoxanes, MeOH or EtOH in $CHCl_3$, CH_2Cl_2, PhH, or THF. 1-ω di-
acetates are selectively cleaved, but the selectivity goes down as the chain
length increases.[125]

18. $[t\text{-}Bu_2SOH(Cl)]_2$, MeOH, 47–96% yield. The primary acetate is selectively
removed in a multitude of carbohydrate polyacetates.[126]

19. Bu_2SnO, toluene, 80–110°C, 1.5–27 h, 15–92% yield.[127]

20. Mg, MeOH or $Mg(OMe)_2$ in MeOH. The acetate is cleaved in the presence
of the benzoate and pivalate (76–96% yield).[128] The relative rates of cleavage
are: p-nitrobenzoate > acetate > benzoate > pivalate ≫ acetamide. Tertiary
acetates are not cleaved.[129]

21. $Ti(O\text{-}i\text{-}Pr)_4$, THF, rt, 10–18 h, 75–92% yield.[130]

22. H_2O_2, $NaHCO_3$, THF. The 10-acetate, which is an α-keto acetate, is cleaved
in the presence of the taxol side chain that is prone to hydrolysis with other
reagents.[131]

23. H_2NNH_2, MeOH, 92% yield. An anomeric acetate was cleaved selectively in
the presence of an axial secondary acetate.[132] Hydrazine will also selectively
remove the C2 acetate or benzoate in the presence of other acetates or benzoates in a variety of pyranosides.[133]

24. MeOH, 4-Å molecular sieves, quantitative.[134] This method was developed to
deacylate acetylated carbohydrates.

Enzymatic hydrolysis

25. Deprotection using enzymes can be quite useful. An added benefit is that a
racemic or meso substrate can often be resolved with excellent enantioselectivity.[135] Numerous examples of this process are described in the literature.
Although acetates are the most common substrates in enzymatic reactions,
other aliphatic esters have been examined with good success.[76] Enzymatic
transformations in nucleoside chemistry have been reviewed.[136]

26. *Candida Cylindracea*, phosphate buffer pH 7, Bu_2O.[137] The 6-*O*-acetyl of α-methyl peracetylglucose was selectively removed. Porcine pancreatic lipase will also hydrolyze acetyl groups from other carbohydrates. These lipases are not specific for acetate, since they hydrolyze other esters as well. In general, selectivity is dependent upon the ester and the substrate.[77,138]

27. *Rhodosporidium toruloides*, 54–88% yield. A number of peracetylated glycosides were hydrolyzed selectively at the 6-hydroxyl. These derivatives when treated with acetic acid undergo acetyl migration to give the C4-deprotected monosaccharide.[139]

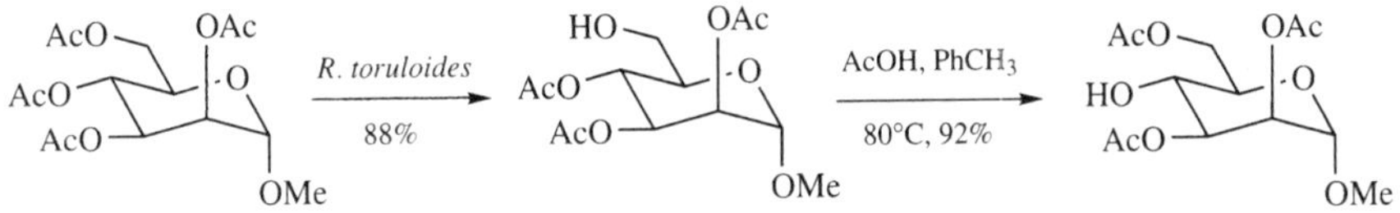

28.

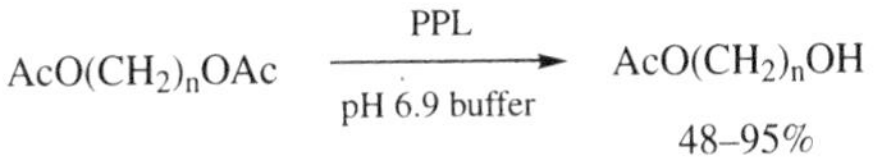

$$AcO(CH_2)_nOAc \xrightarrow[\text{pH 6.9 buffer}]{\text{PPL}} AcO(CH_2)_nOH$$

48–95%

Ref. 140

Larger n gives lower yield

29.

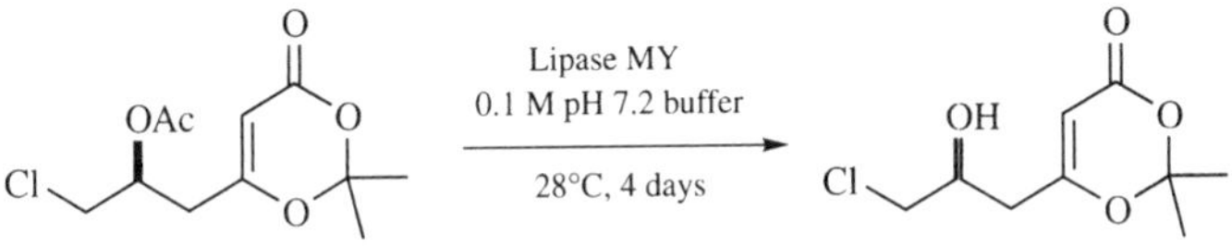

In this case, chemical methods were unsuccessful.[141]

30.

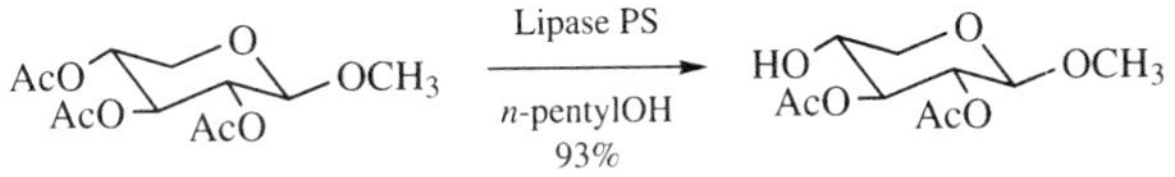

Ref. 142

31.

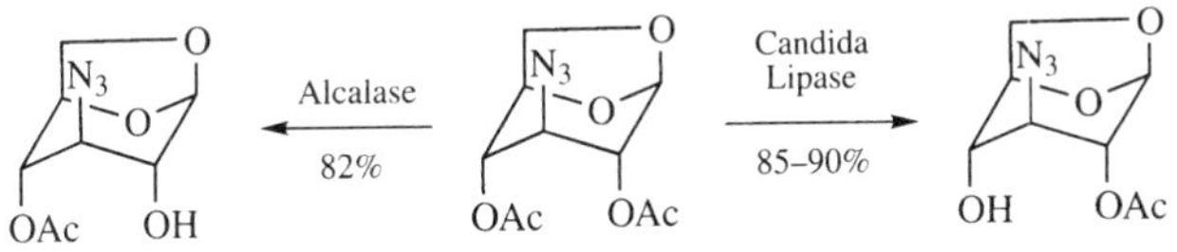

Ref. 143

32.

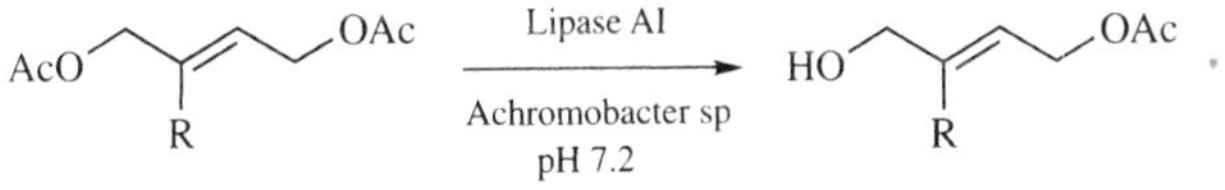

Selectivity depends upon R   Ref. 144

33. Guanidine, guanidinium nitrate, MeOH, CH_2Cl_2, 91–99% yield. These conditions were designed to be compatible with the *N*-Troc group. The

tetrachlorophthalimido, *N*-Fmoc, and *O*-Troc groups were unstable in the presence of this reagent. Benzoates are cleaved, but 20 × more slowly.[145]

1. H. Weber and H. G. Khorana, *J. Mol. Biol.*, **72**, 219 (1972); R. I. Zhdanov and S. M. Zhenodarova, *Synthesis*, 222 (1975).

2. G. Stork, T. Takahashi, I. Kawamoto, and T. Suzuki, *J. Am. Chem. Soc.* **100**, 8272 (1978).

3. G. Höfle, W. Steglich, and H. Vorbrüggen, *Angew. Chem., Inter. Ed. Engl.*, **17**, 569 (1978).

4. For a brief review of this family of catalysts, see A. C. Spivey and S. Arseniyadis, *Angew. Chem. Int. Ed.*, **43**, 5436 (2004).

5. B. A. D'Sa and J. G. Verkade, *J. Org. Chem.*, **61**, 2963 (1996).

6. E. Vedejs and S. T. Diver, *J. Am. Chem. Soc.*, **115**, 3358 (1993).

7. S. Paul, P. Nanda, R. Gupta, and A. Loupy, *Tetrahedron Lett.*, **43**, 4261 (2002).

8. K. Ishihara, H. Kurihara, and H. Yamamoto, *J. Org. Chem.*, **58**, 3791 (1993).

9. J. N. Cardner, T. L. Popper, F. E. Carlon, O. Gnoj, and H. L Herzog, *J. Org. Chem.*, **33**, 3695 (1968).

10. H. Wang, J. She, L.-H. Zhang, and X.-S. Ye, *J. Org. Chem.*, **69**, 5774 (2004).

11. S. Yamada, *J. Org. Chem.*, **57**, 1591 (1992).

12. R. Ghosh, S. Maiti, and A. Chakraborty, *Tetrahedron Lett.*, **46**, 147 (2005).

13. R. Ghosh, S. Maiti, and A. Chakraborty, *Tetrahedron Lett.*, **45**, 6775 (2004).

14. D. Horton, *Org. Synth., Collect. Vol. V*, 1 (1973).

15. M. Jacobson, R. E. Redfern, W. A. Jones, and M. H. Aldridge, *Science*, **170**, 542 (1970).

16. J. C. Chottard, E. Mulliez, and D. Mansuy, *J. Am. Chem. Soc.*, **99**, 3531 (1977).

17. W. G. Dauben, R. A. Bunce, J. M. Gerdes, K. E. Henegar, A. F. Cunningham, Jr., and T. B. Ottoboni, *Tetrahedron Lett.*, **24**, 5709 (1983).

18. J. H. Babler and M. J. Coghlan, *Tetrahedron Lett.*, **20**, 1971 (1979).

19. T. Nishiguchi, S. Fujisaki, Y. Ishii, Y. Yano, and A. Nishida, *J. Org. Chem.*, **59**, 1191 (1994).

20. K. V. N. S. Srinivas, I. Mahender, and B. Das, *Synlett*, 2419 (2003).

21. T. Nishiguchi and H. Taya, *J. Am. Chem. Soc.*, **111**, 9102 (1989).

22. R. Nakao, K. Oka, and T. Fukomoto, *Bull. Chem. Soc. Jpn.*, **54**, 1267 (1981).

23. G. V. M. Sharma, A. K. Mahalingam, M. Nagarajan, A. Llangovan, and P. Radhakrishna, *Synlett*, 1200 (1999).

24. I. Shiina and T. Mukaiyama, *Chem. Lett.*, **23**, 677 (1994); J. Izumi, I. Shiina, and T. Mukaiyama, *Chem. Lett.*, **24**, 141 (1995).

25. K. Ishihara, M. Kubota, H. Kurihara, and H. Yamamoto, *J. Org. Chem.*, **61**, 4560 (1996).

26. P. A. Procopiou, S. P. D. Baugh, S. S. Flack, and G. G. A. Inglis, *J. Chem. Soc., Chem. Commun.*, 2625 (1996); idem, *J. Org. Chem.*, **63**, 2342 (1998).

27. Y. Nagao, E. Fujita, T. Kohno, and M. Yagi, *Chem. Pharm. Bull.*, **29**, 3202 (1981).

28. A. K. Chakraborti and R. Gulhane, *Tetrahedron Lett.*, **44**, 3521 (2003).

29. K. Ishihara, A. Hasegawa, and H. Yamamoto, *Angew. Chem. Int. Ed.*, **40**, 4077 (2001).

30. R. Dumeunier and I. E. Marko, *Tetrahedron Lett.*, **45**, 825 (2004).

31. K. Ishihara, M. Kubota, and H. Yamamoto, *Synlett*, 265 (1996).

32. W. R. Roush and D. A. Barda, *Tetrahedron Lett.*, **38**, 8785 (1997).

33. M. D. Carrigan, D. A. Freiberg, R. C. Smith, H. M. Zerth, and R. S. Mohan, *Synthesis*, 2091 (2001).

34. I. Mohammadpoor-Baltork, H. Aliyan, and A. R. Khosropour, *Tetrahedron*, **57**, 5851 (2001).

35. A. Orita, C. Tanahashi, A. Kakuda, and J. Otera, *J. Org. Chem.*, **66**, 8926 (2001).

36. For a review, see C. LeRoux, and J. Dubac, *Synlett*, 181 (2002).

37. A. Orita, C. Tanahashi, A. Kakuda, and J. Otera, *Angew. Chem. Int. Ed.*, **39**, 2877 (2000).

38. C.-A. Tai, S. S. Kulkarni, and S.-C. Hung, *J. Org. Chem.*, **68**, 8719 (2003).

39. K. L. Chandra, P. Saravanan, R. K. Singh, and V. K. Singh, *Tetrahedron*, **58**, 1369 (2002).

40. B. Karimi and J. Maleki, *J. Org. Chem.*, **68**, 4951 (2003).

41. K. K. Chauhan, C. G. Frost, I. Love, and D. Waite, *Synlett*, 1743 (1999).

42. Y. Nakae, I. Kusaki, and T. Sato, *Synlett*, 1584 (2001); K.-C. Lu, S.-Y. Hsieh, L. N. Patkar, C.-T. Chen, and C.-C. Lin, *Tetrahedron*, **60**, 8967 (2004).

43. G. Bartoli, M. Bosco, R. Dalpozzo, E. Marcantoni, M. Massaccesi, S. Rinaldi, and L. Sambri, *Synlett*, 39 (2003).

44. A. K. Chakraborti, R. Gulhane, and Shivani, *Synlett*, 1805 (2003).

45. H. Firouzabadi, N. Iranpoor, F. Nowrouzi, and K. Amani, *Chem. Commun.*, 764 (2003).

46. S. Chandrasekhar, T. Ramachander, and M. Takhi, *Tetrahedron Lett.*, **39**, 3263 (1998).

47. J.-C. Lee, C.-A. Tai, and S.-C. Hung, *Tetrahedron Lett.*, **43**, 851 (2002).

48. R. Dalpozzo, A. De Nino, L. Maiuolo, A. Procopio, M. Nardi, G. Bartoli, and R. Romeo, *Tetrahedron Lett.*, **44**, 5621 (2003).

49. S. K. De, *Tetrahedron Lett.*, **45**, 2919 (2004).

50. J. Iqbal and R. R. Srivastava, *J. Org. Chem.*, **57**, 2001 (1992).

51. S. Velusamy, S. Borpuzari, and T. Punniyamurthy, *Tetrahedron*, **61**, 2011 (2005).

52. W. Ke and D. M. Whitfield, *Carbohydr. Res.*, **339**, 2841 (2004).

53. B. B. Metaferia, J. Hoch, T. E. Glass, S. L. Bane, S. K. Chatterjee, J. P. Snyder, A. Lakdawala, B. Cornett, and D. G. I. Kingston, *Org. Lett.*, **3**, 1461 (2001); R. A. Holton, Z. Zhang, P. A. Clarke, H. Nadizadeh, and D. J. Procter, *Tetrahedron Lett.*, **39**, 2883 (1998); E. W. P. Damen, L. Braamer, and H. W. Scheeren, *Tetrahedron Lett.*, **39**, 6081 (1998).

54. V. Srivastava, A.Tandon, and S. Ray, *Synth. Commun.*, **22**, 2703 (1992).

55. P. Kumar, R. K. Pandey, M. S. Bodas, and M. K. Dongare, *Synlett*, 206 (2001).

56. T.-S. Li and A.-X. Li, *J. Chem. Soc. Perkin Trans. 1*, 1913 (1998); B. M. Choudary, V. Bhaskar, M. L. Kantam, K. K. Rao, and K. V. Raghavan, *Green Chem.*, **2**, 67 (2000).

57. S. P. Chavan, R. Anand, K. Pasupathy, and B. S. Rao, *Green Chem.*, **3**, 320 (2001).

58. R. Ballini, G. Bosica, S. Carloni, L. Ciaralli, R. Maggi, and G. Sartori, *Tetrahedron Lett.*, **39**, 6049 (1998).

59. R. Kumareswaran, K. Pachamuthu, and Y. D. Vankar, *Synlett*, 1652 (2000).

60. M. Adinolfi, G. Barone, A. Iadonisi, and M. Schiattarella, *Tetrahedron Lett.*, **44**, 4661 (2003).

61. P. A. Clarke, R. A. Holton, and N. E. Kayaleh, *Tetrahedron Lett.*, **41**, 2687 (2000). P. Clarke, *Tetrahedron Lett.*, **43**, 4761 (2002). P. A. Clarke, N. E. Kayaleh, M. A. Smith, J. R. Baker, S. J. Bird, and C. Chan, *J. Org. Chem.*, **67**, 5226 (2002).

62. C.-T. Chen, J.-H. Kuo, C.-H. Li, N. B. Barhate, S.-W. Hon, T.-W. Li, S.-D. Chao, C.-C. Liu, Y.-C. Li, I.-H. Chang, J.-S. Lin, C.-J. Liu, and Y.-C. Chou, *Org. Lett.*, **3**, 3729 (2001).

63. P. Phukan, *Tetrahedron Lett.*, **45**, 4785 (2004); K. P. R. Kartha, and R. A. Field, *Tetrahedron*, **53**, 11753 (1997).

64. B. Karimi and H. Seradj, *Synlett*, 519 (2001).

65. L. Kisfaludy, T. Mohacsi, M. Low, and F. Drexler, *J. Org. Chem.*, **44**, 654 (1979).

66. Y. Ishii, M. Takeno, Y. Kawasaki, A. Muromachi, Y. Nishiyama, and S. Sakaguchi, *J. Org. Chem.*, **61**, 3088 (1996).

67. P. Ilankumaran and J. G. Verkade, *J. Org. Chem.*, **64**, 9063 (1999).

68. J. W. J. Bosco and A. K. Saikia, *Chem. Commun.*, 1116 (2004).

69. M.-H. Lin and T. V. RajanBabu, *Org. Lett.*, **2**, 997 (2000).

70. T. Nishiguchi and H. Taya, *J. Chem. Soc., Perkin Trans. 1*, 172 (1990).

71. A. Orita, K. Sakamoto, Y. Hamada, A. Mitsutome, and J. Otera, *Tetrahedron*, **55**, 2899 (1999); A. Orita, A. Mitsutome, and J. Otera, *J. Org. Chem.*, **63**, 2420 (1998).

72. G. H. Posner and M. Oda, *Tetrahedron Lett.*, **22**, 5003 (1981); S. S. Rana, J. J. Barlow, and K. L. Matta, *Tetrahedron Lett.*, **22**, 5007 (1981).

73. T. Nishiguchi and H. Taya, *J. Am. Chem. Soc.*, **111**, 9102 (1989).

74. H. Hagiwara, K. Morohashi, H. Sakai, T. Suzuki, and M. Ando, *Tetrahedron*, **54**, 5845 (1998).

75. R. Singh, R. M. Kissling, M.-A. Letellier, and S. P. Nolan, *J. Org. Chem.*, **69**, 209 (2004).

76. (a) C.-S. Chen and C. J. Sih, "General Aspects and Optimization of Enantioselective Biocatalysis in Organic Solvents—The Use of Lipases," *Angew. Chem., Int. Ed. Engl.*, **28**, 695 (1989). (b)D. H. G Crout and M. Christen, "Biotransformations in Organic Synthesis," *Mod. Synth. Methods* **5**, 1 (1989). U. Hanefeld, *Org. Biomol. Chem.*, **1**, 2405 (2003). (c) H. G. Davies, R. H. Green, D. R. Kelly, and S. M. Roberts, *Biotransformations in Preparative Organic Chemistry: The Use of Isolated Enzymes and Whole Cell Systems*, Academic Press, New York, 1989. (d) "Chiral Synthons by Ester Hydrolysis Catalysed by Pig Liver Esterase," M. Ohno and M. Otsuka, *Org. React.*, **37**, 1 (1989). (e) C.-H. Wong, "Enzymatic Catalysts in Organic Synthesis," *Science*, **244**, 1145 (1989). (f) C. J. Sih and S. H. Wu, "Resolution of Enantiomers via Biocatalysis," *Top. Stereochem.*, **19**, 63 (1989). (g) N. Turner, "Recent Advances in the Use of Enzyme-Catalysed Reactions in Organic Synthesis," *Nat. Prod. Rep.* **6**, 625 (1989). (h) L. Zhu and M. C. Tedford, "Applications of Pig Liver Esterases (PLE) in Asymmetric Synthesis," *Tetrahedron*, **46**, 6587 (1990). (i) A. M. Klibanov, "Asymmetric Transformations Catalysed by Enzymes in Organic Solvents," *Acc. Chem. Res.*, **23**, 114 (1990). (j) D. G. Drueckhammer, W. J. Hennen, R. L. Pederson, C. F. Barbas, III, C. M. Gautheron, T. Krach, and C.-H. Wong, "Enzyme Catalysis in Synthetic Carbohydrate Chemistry," *Synthesis*, 499 (1991). (k) "Esterolytic and Lipolytic Enzymes in Organic Synthesis," W. Boland, C. Frössl and M. Lorenz, *Synthesis*, 1049 (1991). (l) "Enzymic Methods in Preparative Carbohydrate Chemistry," S. David, C. Augé and C. Gautheron, *Adv. Carbohydr. Chem. Biochem.*, **49**, 175 (1992). (m) "Enzymic Protecting Group Techniques," H. Waldmann, *Kontakte (Darmstadt)*, **2**,

33 (1991). (n) "The Biocatalytic Approach to the Preparation of Enantiomerically Pure Chiral Building Blocks," E. Santaniello, P. Ferraboschi, P. Grisenti, A. Manzocchi, *Chem. Rev.*, **92**, 1071 (1992). (o) L. Poppe and L. Novak, *Selective Biocatalysis. A Synthetic Approach.* VCH: Weinheim 1992. (p) K. Farber, *Biotransformations in Organic Chemistry*, Springer-Verlag: Berlin 1992. (q) "Enzymic Protecting Group Techniques in Bioorganic Synthesis," A. Reidel and H. Waldmann, *J. Prakt. Chem./Chem.-Ztg.*, **335**, 109 (1993). (r) H. Waldmann and D. Sebastian, "Enzymatic Protecting Group Techniques," *Chem. Rev.*, **94**, 911 (1994). (s) K. Drauz and H. Waldmann, Eds., *Enzyme Catalysis in Organic Chemistry: A Comprehensive Handbook*, VCH, Weinheim, 1995.

77. W. J. Hennen, H. M. Sweers, Y.-F. Wang, and C.-H. Wong, *J. Org. Chem.*, **53**, 4939 (1988). See also E. W. Holla, *Angew Chem., Int. Ed. Engl.*, **28**, 220 (1989).

78. N. B. Bashir, S. J. Phythian, A. J. Reason, and S. M. Roberts, *J. Chem. Soc., Perkin Trans. I*, 2203 (1995).

79. M. Therisod and A. M. Klibanov, *J. Am. Chem. Soc.* **108**, 5638 (1986); H. M. Sweers and C.-H. Wong, *J. Am. Chem. Soc.*, **108**, 6421 (1986).

80. E. W. Holla, *J. Carbohydr. Chem.*, **9**, 113 (1990).

81. V. Framis, F. Camps, and P. Clapes, *Tetrahedron Lett.*, **45**, 5031 (2004).

82. G. Iacazio and S. M. Roberts, *J. Chem. Soc., Perkin Trans. I*, 1099 (1993); M. J. Chinn, G. Iacazio, D. G. Spackman, N. J. Turner, and S. H. Roberts, *J. Chem. Soc., Perkin Trans. I*, 661 (1992).

83. I. Matsuo, M. Isomura, R. Walton, and K. Ajisaka, *Tetrahedron Lett.*, **37**, 8795 (1996).

84. J. J. Gridley, A. J. Hacking, H. M. I. Osborn, and D. Spackman, *Synlett*, 1397 (1997).

85. S. Ramaswamy, B. Morgan, and A. C. Oehlschager, *Tetrahedron Lett.*, **31**, 3405 (1990).

86. J. J. Gridley, A. J. Hacking, H. M. I. Osborn, and D. G. Spackman, *Synlett*, 1397 (1997); N. Boissiere-Junot, C. Tellier, and C. Rabiller, *J. Carbohydr. Chem.*, **17**, 99 (1998); B. Danieli, M. Luisetti, G. Sampognaro, G. Carrea, and S. Riva, *J. Mol. Catal. B: Enzymatic*, **3**, 193 (1997).

87. F. Theil and H. Schick, *Synthesis*, 533 (1991).

88. S. Akai, T. Naka, T. Fujita, Y. Takebe, T. Tsujino, and Y. Kita, *J. Org. Chem.*, **67**, 411 (2002).

89. Y. Terao, M. Akamatsu, and K. Achiwa, *Chem. Pharm. Bull.*, **39**, 823 (1991).

90. L. Ling, Y. Watanabe, T. Akiyama, and S. Ozaki, *Tetrahedron Lett.*, **33**, 1911 (1992).

91. C. R. Johnson, A. Golebiowski, T. K. McGill, and D. H. Steensma, *Tetrahedron Lett.*, **32**, 2597 (1991).

92. R. V. Nair and M. M. Salunkhe, *Synth. Commum.*, **30**, 3115 (2000).

93. F. Aragozzini, E. Maconi, D. Potenza, and C. Scolastico, *Synthesis*, 225 (1989). For a review on the use of Sn–O derivatives to direct regioselective acylation and alkylation, see S. David and S. Hanessian, *Tetrahedron*, **41**, 643 (1985).

94. C. Audouard, J. Fawcett, G. A. Griffith, E. Kerouredan, A. Miah, J. M. Percy, and H. Yang, *Org. Lett.*, **6**, 4269 (2004).

95. C. Chauvin and D. Plusquellec, *Tetrahedron Lett.*, **32**, 3495 (1991).

96. S. Yamada, *J. Org. Chem.*, **57**, 1591 (1992).

97. M. Oikawa, A. Wada, F. Okazaki, and S. Kusumoto, *J. Org. Chem.*, **61**, 4469 (1996).

98. G. Sabitha, B. V. S. Reddy, G. S. K. K. Reddy, and J. S. Yadav, *New J. Chem.*, **24**, 63 (2000).

99. F. I. Auzanneau and D. R. Bundle, *Carbohydr. Res.*, **212**, 13 (1991).

100. P. C. Zhu, J. Lin, and C. U. Pittman, Jr., *J. Org. Chem.*, **60**, 5729 (1995).

101. S. J. Miller, G. T. Copeland, N. Papaioannou, T. E. Horstmann, and E. M. Ruel, *J. Am. Chem. Soc.*, **120**, 1629 (1998); T. Sano, K. Imai, K. Ohashi, and T. Oriyama, *Chem. Lett.*, **28**, 265 (1999); M. M. Vasbinder, E. R. Jarvo and S. J. Miller, *Angew. Chem. Int. Ed.*, **40**, 2824 (2001); B. Tao, J. C. Ruble, D. A. Hoic, and G. C. Fu, *J. Am. Chem. Soc.*, **121**, 5091 (1999); M.-H. Lin and T. V. Rajanbabu, *Org. Lett.*, **4**, 1607 (2002). E. Vedejs, O. Daugulis, L. A. Harper, J. A. MacKay, and D. R. Powell, *J. Org. Chem.*, **68**, 5020 (2003).

102. J. J. Plattner, R. D. Gless, and H. Rapoport, *J. Am. Chem. Soc.*, **94**, 8613 (1972).

103. B. Reinhard and H. Faillard, *Liebigs Ann. Chem.*, 193 (1994).

104. R. D. Crouch, J. S. Burger, K. A. Zietek, A. B. Cadwallader, J. E. Bedison, and M. M. Smielewska, *Synlett*, 991 (2003).

105. K. Mori, M. Tominaga, T. Takigawa, and M. Matsui, *Synthesis*, 790 (1973).

106. K. Mori and M. Sasaki, *Tetrahedron Lett.*, **20**, 1329 (1979).

107. J. Herzig, A. Nudelman, H. E. Gottlieb, and B. Fischer, *J. Org. Chem.*, **51**, 727 (1986).

108. N. Kunesch, C. Meit, and J. Poisson, *Tetrahedron Lett.*, **28**, 3569 (1987).

109. T. Neilson and E. S. Werstiuk, *Can. J. Chem.*, **49**, 493 (1971).

110. A. Nudelman, J. Herzig, H. E. Gottlieb, E. Keinan, and J. Sterling, *Carbohydr. Res.*, **162**, 145 (1987).

111. D. Askin, C. Angst, and S. Danishefsky, *J. Org. Chem.*, **52**, 622 (1987).

112. A. S. Demir and O. Sesenoglu, *Org. Lett.*, **4**, 2021 (2002); H. Kajiro, S. Mitamura, A. Mori, and T. Hiyama, *Bull. Chem. Soc. Jpn.*, **72**, 1553 (1999); H. Kajiro, S. Mitamura, A. Mori, and T. Hiyama, *Tetrahedron Lett.*, **40**, 1689 (1999).

113. G. V. M. Sharma and A. Ilangovan, *Synlett*, 1963 (1999).

114. L. H. B. Baptistella, J. F. dos Santos, K. C. Ballabio, and A. J. Marsaioli, *Synthesis*, 436 (1989).

115. K. Yoshimoto, H. Kawabata, N. Nakamichi, and M. Hayashi, *Chem. Lett.*, **30**, 934 (2001).

116. S. Di Stefano, F. Leonelli, B. Garofalo, L. Mandolini, R. M. Bettolo, and L. M. Migneco, *Org. Lett.*, **4**, 2783 (2002).

117. R. Yanada, N. Negoro, K. Bessho, and K. Yanada, *Synlett*, 1261 (1995).

118. B. Ren, L. Cai, L.-R. Zhang, Z.-J. Yang, and L.-H. Zhang, *Tetrahedron Lett.*, **46**, 8083 (2005).

119. V. Pozsgay, *J. Am. Chem. Soc.*, **117**, 6673 (1995).

120. A. G. González, I. Brouard, F. Leon, J. I. Padron, and J. Bermejo, *Tetrahedron Lett.*, **42**, 3187 (2001).

121. N. Yamamoto, T. Nishikawa, and M. Isobe, *Synlett*, 505 (1995).

122. D. Solomon, M. Fridman, J. Zhang, and T. Baasov, *Org. Lett.*, **3**, 4311 (2001).

123. C.-E. Yeom, S. Y. Lee, Y. J. Kim, and B. M. Kim, *Synlett*, 1527 (2005).

124. S. V. Ley, A. Armstrong, D. Diez-Martin, M. J. Ford, P. Grice, J. G. Knight, H. C. Kolb, A. Madin, C. A. Marby, S. Mukherjee, A. N. Shaw, A. M. Z. Slawin, S. Vile, A. D. White, D. J. Williams, and M. Woods, *J. Chem. Soc., Perkin Trans. I*, 667 (1991).

125. J. Otera, N. Dan-oh, and H. Nozaki, *Tetrahedron*, **49**, 3065 (1993).

126. A. Orita, Y. Hamada, T. Nakano, S. Toyoshima, and J. Otera, *Chem. Eur. J.*, **7**, 3321 (2001); A. Orita, A. Sakamoto, Y. Hamada, and J. Otera, *Synlett*, 140 (2000).

127. M. G. Perez and M. S. Maier, *Tetrahedron Lett.*, **36**, 3311 (1995); S.-M. Wang, W.-Z. Ge, H.-M. Liu, D.-P. Zou, and X.-B. Yan, *Steroids*, **69**, 599 (2004).

128. Y.-C. Xu, A. Bizuneh, and C. Walker, *Tetrahedron Lett.*, **37**, 455 (1996).

129. Y.-C. Xu, A. Bizuneh, and C. Walker *J. Org. Chem.*, **61**, 9086 (1996).

130. B. C. Ranu, S. K. Guchhait, and M. Saha, *J. Indian Chem. Soc.*, **76**, 547(1999).

131. Q. Y. Zheng, L. G. Darbie, X. Cheng, and C. K. Murray, *Tetrahedron Lett.*, **36**, 2001 (1995).

132. W. R. Roush and X.-F. Lin, *J. Am. Chem. Soc.*, **117**, 2236 (1995).

133. J. Li and Y. Wang, *Synth. Commum.*, **34**, 211 (2004).

134. K. P. R. Kartha, B. Mukhopadhyaya, and R. A. Field, *Carbohydr. Res.*, **339**, 729 (2004).

135. Y.-F. Wang, C.-S. Chen, G. Girdaukas, and C. J. Sih, in *Enzymes in Organic Synthesis (Ciba Foundation Symposium*, Vol. 111), 128 (1985); K. Tsuji, Y. Terao, and K. Achiwa, *Tetrahedron Lett.*, **30**, 6189 (1989); R. Csuk and B. I. Glaenzer, *Z. Naturforsch. B, Chem. Sci.*, **43**, 1355 (1988). For examples in a cyclic series, see K. Laumen and M. Schneider, *Tetrahedron Lett.*, **26**, 2073 (1985); K. Naemura, N. Takahashi, and H. Chikamatsu, *Chem. Lett.*, **17**, 1717 (1988); C. R. Johnson and C. H. Senanayake, *J. Org. Chem.*, **54**, 735 (1989); D. R. Deardorff, A. J. Matthews, D. S. McMeekin, and C. L. Craney, *Tetrahedron Lett.*, **27**, 1255 (1986); N. W. Boaz, *Tetrahedron Lett.*, **30**, 2061 (1989).

136. M. Ferrero and V. Gotor, *Monatsh. Chem.*, **131**, 585 (2000).

137. M. Kloosterman, E. W. J. Mosuller, H. E. Schoemaker, and E. M. Meijer, *Tetrahedron Lett.*, **28**, 2989 (1987).

138. Y. Kodera, K. Sakurai, Y. Satoh, T. Uemura, Y. Kaneda, H. Nishimura, M. Hiroto, A. Matsushima, and Y. Inada, *Biotechnol. Lett.*, **20**, 177 (1998).

139. T. Horrobin, C. H. Tran, and D. Crout, *J. Chem. Soc. Perkin Trans. 1*, 1069 (1998). G. Fernandez-Lorente, J. M. Palomo, J. Cocca, C. Mateo, P. Moro, M. Terreni, R. Fernandez-Lafuente, and J. M. Guisan, *Tetrahedron*, **59**, 5705 (2003).

140. O. Houille, T. Schmittberger, and D. Uguen, *Tetrahedron Lett.*, **37**, 625 (1996).

141. J. Sakaki, H. Sakoda, Y. Sugita, M. Sato, and C. Kaneto, *Tetrahedron: Asymmetry*, **2**, 343 (1991).

142. R. Lopez, E. Montero, F. Sanchez, J. Cañada, and A. Fernandez-Mayoralas, *J. Org. Chem.*, **59**, 7027 (1994).

143. E. W. Holla, V. Sinnwell, and W. Klaffke, *Synlett*, 413 (1992).

144. T. Itoh, A. Uzu, N. Kanda, and Y. Takagi, *Tetrahedron Lett.*, **37**, 91 (1996).

145. U. Ellervik and G. Magnusson, *Tetrahedron Lett.*, **38**, 1627 (1997).

Chloroacetate Ester: $ClCH_2CO_2R$

Formation

1. $(ClCH_2CO)_2O$, Pyr, 0°C, 70–90% yield.[1]
2. $ClCH_2COCl$, Pyr, ether, 87% yield.[2]
3. PPh_3, DEAD, $ClCH_2CO_2H$, 73% yield.[3] In this case the esterification proceeds with inversion of configuration at the alcoholic center.

4. Vinyl chloroacetate, $Cp^*_2Sm(THF)_2$, toluene, rt, 99% yield. With SmI_2 as catalyst the yield is 79% .[4]

5. Bu_2SnO, MeOH, 65°C, 2 h, then $ClCH_2COCl$, 89% yield.[5]

Cleavage

The chloroacetate group has been observed to migrate during silica gel chromatography.[6] In general, cleavage of chloroacetates can be accomplished in the presence of other esters such as acetates and benzoates because of the large difference in the hydrolysis rates for esters bearing electron-withdrawing groups. A study comparing the half-lives for hydrolysis of a variety of esters of 5'-O-acyluridines gave the following results.[7]

Half-Lives for Hydrolysis of Various Esters

	$t_{1/2}$ min	
Acyl Group	Reagent I	Reagent II
CH_3CO-	191	59
$MeOCH_2CO-$	10.4	2.5
$PhOCH_2CO-$	3.9	$<1^a$
Formyl-	0.4	$(0.22)^b$
$ClCH_2CO-$	0.28	$(0.17)^b$

Reagent I = 155 mM NH_3/H_2O; reagent II = $NH_3/MeOH$.
[a]Reaction is too fast to measure.
[b]Time for complete solvolysis of the substrate.

The relative rates of alkaline hydrolysis of acetate, chloro-, dichloro-, and trichloroacetates have been compared and give the following relative rates: $1 : 760 : 1.6 \times 10^4 : 10^5$.[8]

Cleavage

1. $HSCH_2CH_2NH_2$ or $H_2NCH_2CH_2NH_2$ or o-phenylenediamine, Pyr, Et_3N, 1 h, rt.[1]
2. Thiourea, $NaHCO_3$, EtOH, 70°C, 5 h, 70% yield.[2]
3. H_2O, Pyr, pH 6.7, 20 h, 100% yield.[9]
4. MeOH, TEA, 96% yield.[10]
5. $NH_2NHC(S)SH$, lutidine, AcOH, 2–20 min, rt, 88–99% yield.[11,12] This method is superior to the use of thiourea in that it proceeds at lower temperatures and affords much higher yields. This reagent also serves to remove the related bromoacetyl esters that under these conditions are 5–10 times more labile. Cleavage occurs cleanly in the presence of an acetate.[13]

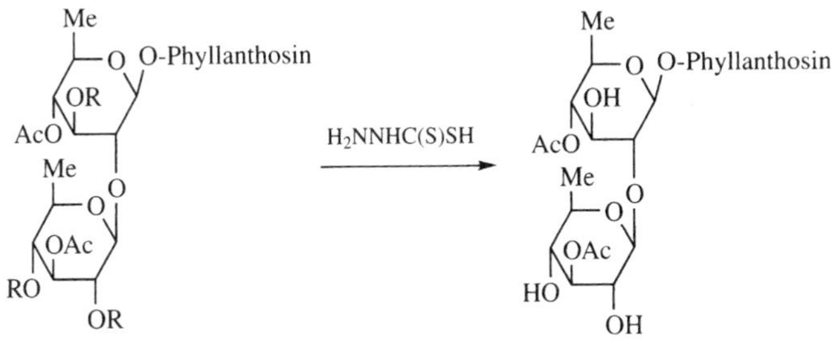

R = ClCH₂CO–

6. Hydrazinedithiocarbonate, DMF.[16] "Hydrazine acetate."[14]

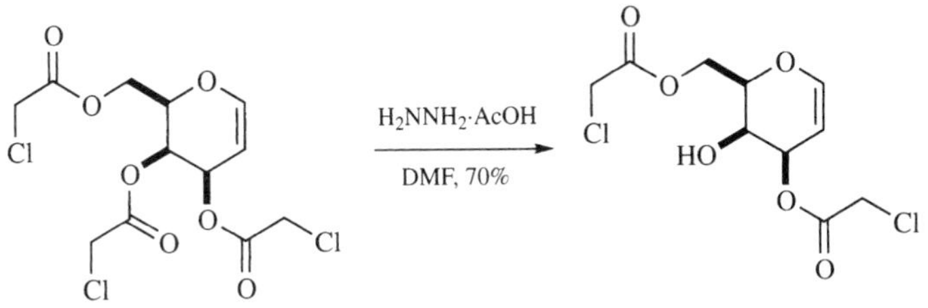

Ref. 15

7. DABCO, ethanol, pyridine, 20–70°C, >94% yield. This method is faster than the thiourea method by a factor of about 9. It does not cause benzoyl migration in the carbohydrates examined.[17]

8. The lipase from *Pseudomonas sp* K10 has also been used to cleave the chloroacetate, resulting in resolution of a racemic mixture since only one enantiomer was cleaved.[18]

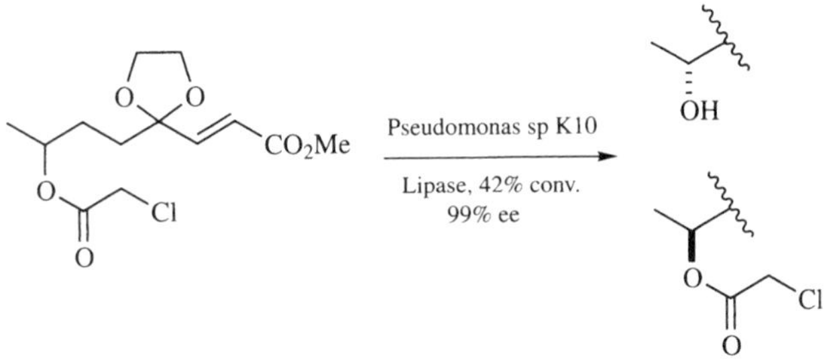

9. *N,N*-Pentamethylenethiourea, TEA, dioxane, 70°C, 3 h.[19]

10. NH₃, THF, −50°C to −40°C, 2.5 h. The use of hydrazine failed in this case.[20]

1. A. F. Cook and D. T. Maichuk, *J. Org. Chem.*, **35**, 1940 (1970).

2. M. Naruto, K. Ohno, N. Naruse, and H. Takeuchi, *Tetrahedron Lett.*, **20**, 251 (1979).

3. M. Saiah, M. Bessodes, and K. Antonakis, *Tetrahedron Lett.*, **33**, 4317 (1992); B. Lipshutz and T. A. Miller, *Tetrahedron Lett.*, **31**, 5253 (1990).

4. Y. Ishii, M. Takeno, Y. Kawasaki, A. Muromachi, Y. Nishiyama, and S. Sakaguchi, *J. Org. Chem.*, **61**, 3088 (1996).

5. A. Liakatos, M. J. Kiefel, and M. von Itzstein, *Org. Lett.*, **5**, 4365 (2003).

6. V. Pozsgay, *J. Am. Chem. Soc.*, **117**, 6673 (1995).

7. C. B. Reese, J. C. M. Stewart, J. H van Boom, H. P. M. de Leeuw, J. Nagel, and J. F. M. de Rooy, *J. Chem Soc., Perkin Trans. I*, 934 (1975).

8. N. S. Isaacs, *Physical Organic Chemistry*, 2nd. ed.; John Wiley & Sons, New York, 1995; p. 515.

9. F. Johnson, N. A. Starkovsky, A. C. Paton, and A. A. Carlson, *J. Am. Chem. Soc.*, **86**, 118 (1964).

10. K. C. Nicolaou, H. J. Mitchell, K. C. Fylaktakidou, R. M. Rodriguez, and H. Suzuki, *Chem. Eur. J.*, **6**, 3116 (2000).

11. C. A. A. van Boeckel and T. Beetz, *Tetrahedron Lett.*, **24**, 3775 (1983).

12. A. S. Cambell and B. Fraser-Reid, *J. Am. Chem. Soc.*, **117**, 10387 (1995).

13. A. B. Smith III, K. J. Hale, and H. A. Vaccaro, *J. Chem. Soc., Chem. Commun.*, 1026 (1987).

14. U. E. Udodong, C. S. Rao, and B. Fraser-Reid, *Tetrahedron*, **48**, 4713 (1992).

15. S. Bouhroum and P. J. A. Vottero, *Tetrahedron Lett.*, **31**, 7441 (1990).

16. S. Manabe and Y. Ito, *J. Am. Chem. Soc.*, **124**, 12638 (2002); C. A. A. Boechel, T. Beetz, *Tetrahedron Lett.*, **24**, 3775 (1983); J. G. Allen and B. Fraser-Reid, *J. Am. Chem. Soc.*, **121**, 468 (1999).

17. D. J. Lefeber, J. P. Kamerling, and J. F. G. Vliegenthart, *Org. Lett.*, **2**, 701 (2000).

18. T. K. Ngooi, A. Scilimati, Z.-W. Guo, and C. J. Sih, *J. Org. Chem.*, **54**, 911 (1989).

19. U. Schmidt, M. Kroner, and H. Griesser, *Synthesis*, 294 (1991).

20. J. C. McWilliams and J. Clardy, *J. Am. Chem. Soc.*, **116**, 8378 (1994).

Dichloroacetate Ester: Cl_2CHCO_2R

Formation

1. $Cl_2CHCOCl$.[1]

2. $(Cl_2CHCO)_2O$, Pyr, CH_2Cl_2.[2] This reagent is more reactive than Ac_2O and was used for the protection of a very hindered alcohol where silyl groups and a simple acetate could not be introduced.[3]

3. Cl$_2$CHCOCCl$_3$, DMF, 56% yield.[4] This reagent was used to acylate selectively the 6-position of an α-methyl glucoside.

Cleavage

1. pH 9–9.5, 20°C, 30 min.[1]
2. NH$_3$, MeOH.[4,5]
3. KOH, t-BuOH, H$_2$O, THF.[2]

1. J. R. E. Hoover, G. L. Dunn, D. R. Jakas, L. L. Lam, J. J. Taggart, J. R. Guarini, and L. Phillips, *J. Med. Chem.*, **17**, 34 (1974).
2. S. Masamune, W. Choy, F. A. J. Kerdesky, and B. Imperiali, *J. Am. Chem. Soc.* **103**, 1566 (1981).
3. K. C. Nicolaou, Y. Li, K. Sugita, H. Monenschein, P. Guntupalli, H. J. Mitchell, K. C. Fylaktakidou, D. Vourloumis, P. Giannakakou, and A. O'Brate, *J. Am. Chem. Soc.*, **125**, 15443 (2003).
4. A. H. Haines and E. J. Sutcliffe, *Carbohydr. Res.*, **138**, 143 (1985).
5. C. B. Reese, J. C. M. Stewart, J. H van Boom, H. P. M. de Leeuw, J. Nagel, and J. F. M. de Rooy, *J. Chem Soc., Perkin Trans. I*, 934 (1975).

Trichloroacetate Ester: RO$_2$CCCl$_3$ (Chart 2)

Formation

1. Cl$_3$CCOCl, Pyr, DMF, 20°C, 2 days, 60–90% yield.[1]

2. From a TBDMS or TIPS ether: trichloroacetic anhydride, 3HF·TEA, 80°C, 2 h, 90–93% yield.[3]

Cleavage

1. NH$_3$, EtOH, CHCl$_3$, 20°C, 6 h, 81% yield.[1] Cleavage of the trichloroacetate occurs selectively in the presence of an acetate.
2. KOH, MeOH, 72% yield.[1] A formate ester was not hydrolyzed under these conditions.

1. V. Schwarz, *Collect. Czech. Chem. Commun.*, **27**, 2567 (1962).
2. S. Bailey, A. Teerawutgulrag, and E. J. Thomas, *J. Chem. Soc., Chem. Commun.*, 2519 (1995).
3. S. D. Stamatov, M. Kullberg, and J. Stawinski, *Tetrahedron Lett.*, **46**, 6855 (2005).

Trichloroacetamidate: $Cl_3CC(=NH)OR$

Typically, the trichloacetamidate group is used as an activating group for the introduction of ethers such as the benzyl and MPM ether, among others, and for activation of the anomeric position in glycoside synthesis. Thus the use of this group as a protective group must be carefully considered, since it is expected to be unstable to strong acids and Lewis acids. It is formed from the alcohol, trichloroacetonitrile, and DBU as a strong base. It is cleaved by acid hydrolysis (TsOH, H_2O, MeOH, CH_2Cl_2), DBU (MeOH by exchange), and Zn (NH_4Cl, EtOH, reflux, 5 min). Yields range from 73–100% .[1]

1. B. Yu, H. Yu, Y. Hui, and X. Han, *Synlett*, 753 (1999).

Trifluoroacetate Ester (RO-TFA): CF_3CO_2R

Formation

1. $(CF_3CO)_2O$, Pyr.[1]
2. Even with this highly reactive reagent, excellent selectivity was achieved for one of two very similar alcohols.[2]

3. 2-Pyridyl trifluoroacetate, ether, 20°C, 30 min, 99% yield.[3]
4. CF_3CO_3H, 20°C, 4 h, 83% yield. [4] In this case, a hindered alcohol was converted to the TFA derivative.
5. *N*-(Trifluoroacetyl)succinimide, THF or toluene, reflux, 86–99% yield. Phenols and amines react to give the phenolic ester and TFA amides respectively.[5]

Cleavage

A series of nucleoside trifluoroacetates were rapidly hydrolyzed in 100% yield at 20°C, pH 7.[6] In general, these are easily hydrolyzed under mildly basic conditions.

1. A. Lardon and T. Reichstein, *Helv. Chim. Acta*, **37**, 443 (1954).

2. P. T. Lansbury, T. E. Nickson, J. P. Vacca, R. D. Sindelar, and J. M. Messinger, II, *Tetrahedron*, **43**, 5583 (1987).

3. T. Keumi, M. Shimada, T. Morita, and H. Kitajima, *Bull. Chem. Soc. Jpn.*, **63**, 2252 (1990).

4. G. W. Holbert and B. Ganem, *J. Chem. Soc., Chem. Commun.*, 248 (1978).

5. A. R. Katritzky, B. Yang, G. Qiu, and Z. Zhang, *Synthesis*, 55 (1999).

6. F. Cramer, H. P. Bär, H. J. Rhaese, W. Sänger, K. H. Scheit, G. Schneider, and J. Tennigkeit, *Tetrahedron Lett.*, **4**, 1039 (1963).

Methoxyacetate Ester: $MeOCH_2CO_2R$

Formation

1. $MeOCH_2COCl$, Pyr.[1]

2. $(MeOCH_2CO)_2O$, DIPEA, CH_2Cl_2.[2] In this case the methoxyacetate was used because attempts to deprotect the primary acetate in the presence of a β-acetoxy ketone lead to its elimination.

Cleavage

1. NH_3/MeOH or NH_3/H_2O, 78% yield.[1] In nucleoside derivatives the methoxyacetate is cleaved 20 times faster than an acetate. It can be cleaved in the presence of a benzoate.

2. $Yb(OTf)_3$, MeOH, 0–25°C, 92–99% yield. Acetates, benzoates, THP, TBDMS, TBDPS, and MEM ethers are not affected by this reagent.[3]

3. Ethanolamine, IPA, reflux, 21 h, >50% yield. These conditions did not affect the C-10 acetate or the C-13 side chain of a taxol derivative.[4]

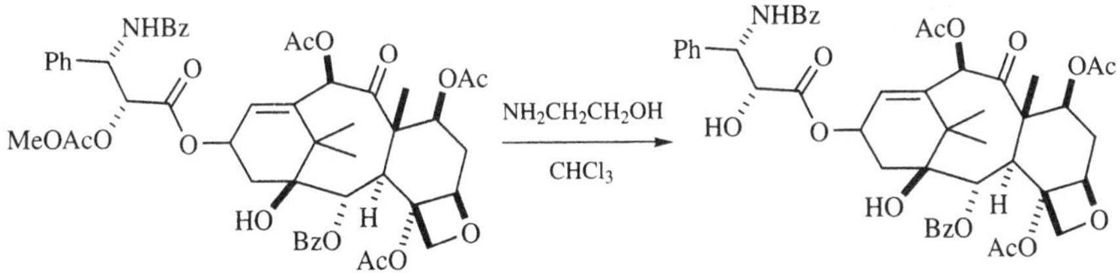

1. C. B. Reese and J. C. M. Stewart, *Tetrahedron Lett.*, **9**, 4273 (1968).

2. D. A. Evans, B. W. Trotter, B. Cote, P. J. Coleman, L. C. Dias, and A. N. Tyler, *Angew. Chem. Int. Ed.*, **36**, 2744 (1997).

3. T. Hanamoto, Y. Sugimato, Y. Yokoyama, and J. Inanaga, *J. Org. Chem.*, **61**, 4491 (1996).

4. R. B. Greenwald, A. Pendri, and D. Bolikal, *J. Org. Chem.*, **60**, 331 (1995).

Triphenylmethoxyacetate Ester: ROCOCH$_2$OCPh$_3$

The triphenylmethoxyacetate was prepared in 53% yield from a nucleoside and the sodium acetate (Ph$_3$COCH$_2$CO$_2$Na, i-Pr$_3$C$_6$H$_2$SO$_2$Cl, Pyr) as a derivative that could be easily detected on TLC (i.e., it has a distinct orange-yellow color after it is sprayed with ceric sulfate). It is readily cleaved by NH$_3$/MeOH (100% yield).[1]

1. E. S. Werstiuk and T. Neilson, *Can. J. Chem.*, **50**, 1283 (1972).

Phenoxyacetate Ester: PhOCH$_2$CO$_2$R (Chart 2)

Formation

1. (PhOCH$_2$CO)$_2$O, Pyr.[1,2]
2. (PhOCH$_2$CO)$_2$O, Pyr, DMAP, CH$_2$Cl$_2$, 0°C.[3]

3. PhOCH$_2$CO$_2$Cl, pyridine, 81% yield.[4]

Cleavage

1. t-BuNH$_2$, MeOH.[2] Methylamine is similarly effective.[4]
2. NH$_3$ in H$_2$O or MeOH.[1] The phenoxyacetate is 50 times more labile to aqueous ammonia than is an acetate.
3. Er(OTf)$_3$, MeOH, rt, 68% yield.[5]
4. 0.001 M K$_2$CO$_3$, MeOH, CH$_2$Cl$_2$, 86% yield. A cinnamyl ester was retained.[6]

1. C. B. Reese and J. C. M. Stewart, *Tetrahedron Lett.*, **9**, 4273 (1968).
2. T. Kamimura, T. Masegi, and T. Hata, *Chem. Lett.*, **11**, 965 (1982).
3. R. B. Woodward and 48 co-workers, *J. Am. Chem. Soc.* **103**, 3210 (1981).
4. K. Pekari and R. R. Schmidt, *J. Org. Chem.*, **68**, 1295 (2003).
5. K. Shimada, Y. Kaburagi, and T. Fukuyama, *J. Am. Chem. Soc.*, **125**, 4048 (2003).
6. H. I. Duynstee, M. C. de Koning, H. Ovaa, G. A. van der Marel, and J. H. van Boom, *Eur. J. Org. Chem.*, 2623 (1999).

p-Chlorophenoxyacetate Ester: ROCOCH$_2$OC$_6$H$_4$-p-Cl

The p-chlorophenoxyacetate, prepared to protect a nucleoside by reaction with the acetyl chloride, is cleaved by 0.2 M NaOH, dioxane-H$_2$O, 0°C, 30 s.[1] The presence of the electron-withdrawing group facilitates ester cleavage over the parent phenoxy-acetate.

1. S. S. Jones and C. B. Reese, *J. Am. Chem. Soc.*, **101**, 7399 (1979).

Phenylacetate Ester: $PhCH_2CO_2R$

Formation

1. *Lipase Fp*, $PhCH_2CO_2CH=CH_2$, 84–88% yield.[1]
2. $PhCH_2CO_2H$, DCC, DMAP.[3]

Cleavage

Penicillin G Acylase.[1,2] This method was used to cleave a phenylacetate in the presence of an acetate.[3]

p-P-Phenylacetate Ester: $ROCOCH_2C_6H_4-p-P$ (P = polymer)

Monoprotection of a symmetrical diol can be affected by reaction with a polymer-supported phenylacetyl chloride. The free hydroxyl group is then converted to an ether and the phenylacetate cleaved by aqueous ammonia-dioxane, 48 h.[4]

$$HO(CH_2)_nOH \; + \; p\text{-}P-C_6H_4CH_2COCl \; \xrightarrow{\text{Pyr}} \; HO(CH_2)_nOCOCH_2C_6H_4-p\text{-}P$$

1. E. W. Holla, *J. Carbohydr. Chem.*, **9**, 113 (1990).
2. H. Waldmann, A. Heuser, P. Braun, M. Schulz, and H. Kunz, in *Microbial Reagents in Organic Synthesis*; S. Servi, Ed., Kluwer Academic, Dordrecht, 1992, pp. 113–122.
3. R. S. Coleman, T. E. Richardson, and A. J. Carpenter, *J. Org. Chem.*, **63**, 5738 (1998).
4. J. Y. Wong and C. C. Leznoff, *Can. J. Chem.*, **51**, 2452 (1973).

Diphenylacetate Ester (DPA-OR): Ph_2CHCO_2R

The DPA ester is formed from the acid chloride in pyridine (40–96% yield). It is cleaved oxidatively by treatment with NBS followed by thiourea (40–88% yield).[1]

1. F. Santoyo-Gonzalez, F. Garcia-Calvo-Flores, J. Isac-Garcia, R. Robles-Diaz, and A. Vargas-Berenguel, *Synthesis*, 97 (1994).

3-Phenylpropionate Ester: $ROCOCH_2CH_2Ph$

The 3-phenylpropionate ester has been used in nucleoside synthesis.[1] It is cleaved by α-chymotrypsin (37°C, 8–16 h, 70–90% yield),[2] and it can be cleaved in the presence of an acetate.[3]

1. H. S. Sachdev and N. A. Starkovsky, *Tetrahedron Lett.*, **10**, 733 (1969).
2. A. T.-Rigby, *J. Org. Chem.*, **38**, 977 (1973).
3. Y. Y. Lin and J. B. Jones, *J. Org. Chem.*, **38**, 3575 (1973).

Bisfluorous Chain Type Propanoyl (Bfp−OR) Ester

This group was used to protect carbohydrates for fluorous based synthesis.[1] The ester is prepared using DCC (CH_2Cl_2, DMAP, 87% yield) as a coupling agent. It is cleaved by methanolysis with NaOMe (2 h, rt, 93% yield).[2] A similarly functionalized benzoyl ester has been prepared and tested as a protective group in fluorous based synthesis.[3]

1. "Handbook of Fluorous Chemistry," J. A. Gladysz, D. P. Curran, and I. T. Horváth, Eds., Wiley-VCH, Weinheim, 2004.
2. T. Miura, K. Goto, H. Waragai, H. Matsumoto, Y. Hirose, M. Ohmae, H.-k. Ishida, A. Satoh, and T. Inazu, *J. Org. Chem.*, **69**, 5348 (2004); T. Miura, Y. Hirose, M. Ohmae, and T. Inazu, Org. Lett., **3**, 3947 (2001); T. Miura and T. Inazu, *Tetrahedron Lett.*, **44**, 1819 (2003).
3. T. Miura, A. Satoh, K. Goto, Y. Murakami, N. Imai, and T. Inazu, Tetrahedron: Asymmetry, **16**, 3 (2005).

4-Pentenoate Ester: $CH=CHCH_2CH_2CO_2R$

Formation

$CH=CHCH_2CH_2COCl$.[1] This group was used for the protection of anomeric hydroxyl groups.

Cleavage

NBS, 1% H_2O, CH_3CN. 36–85% yield.[1]

1. J. C. Lopez and B. Fraser-Reid, *J. Chem. Soc., Chem. Commun.*, 159 (1991).

4-Oxopentanoate (Levulinate) Ester (Lev−OR): ROCOCH₂CH₂COCH₃

The levulinate is less prone to migrate than the benzoate and acetate.[1]

Formation

1. $(CH_3COCH_2CH_2CO)_2O$, Pyr, 25°C, 24 h, 70–85% yield.[2]
2. $CH_3COCH_2CH_2CO_2H$, DCC, DMAP, 96% yield.[3]

3. 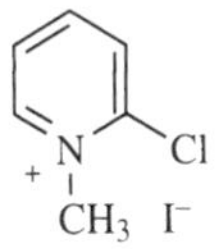 Cl CMPI, $CH_3COCH_2CH_2CO_2H$, DABCO, 86% yield.[4]

    (CMPI = 2-chloro-1-methylpyridinium iodide)

4. *Candida antarctica* Lipase, trifluoroethyl levulinate, THF, 40°C, 4 days, 65–83% yield. The method was used for the selective protection of the primary alcohol of the galactose saccharide.[5]

Cleavage

1. $NaBH_4$, H_2O, pH 5–8, 20°C, 20 min, 80–95% yield.[1] The by-product, 5-methyl-γ-butyrolactone, is water-soluble and thus easily removed.
2. 0.5 *M* H_2NNH_2, H_2O, Pyr, AcOH, 2 min, 100% yield.[2] Normal esters are not cleaved under these conditions.[6]
3. MeMgI, 0°C, 2 h, 93% yield.[7] A levulinate is cleaved in preference to a benzoate.
4. $NaHSO_3$, THF, CH_3CN or EtOH, 86–90% yield.[8]

4,4-(Ethylenedithio)pentanoate Ester (Levulinoyl Dithioacetal Ester): RO-LevS

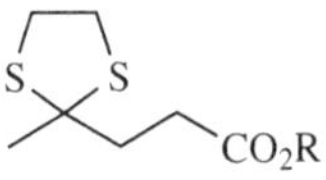

Formation

1. 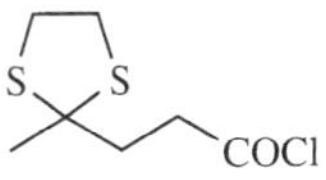 2,6-lutidine, 0°C, 70% yield.[9]

2. 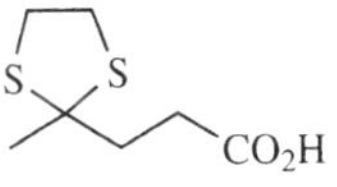 CMPI, DABCO, dioxane, 2 h, 20°C, 96% yield.[3] (CMPI = 2-chloro-1-methylpyridinium iodide)

Cleavage

The LevS group is converted to the Lev group with $HgCl_2/HgO$ (acetone/H_2O, 4 h, 20°C, 74% yield). It can then be hydrolyzed using the conditions that remove the

Lev group.[9] The LevS group is stable to the conditions used for glycoside formation [HgBr$_2$, Hg(CN)$_2$].

1. J. N. Glushka and A. S. Perlin, *Carbohydr. Res.*, **205**, 305 (1990); R. N. Rej, J. N. Glushka, W. Chew, and A. S. Perlin, *Carbohydr. Res.*, **189**, 135 (1989).
2. A. Hassner, G. Strand, M. Rubinstein, and A. Patchornik, *J. Am. Chem. Soc.*, **97**, 1614 (1975).
3. J. H. van Boom and P. M. J. Burgers, *Tetrahedron Lett.*, **17**, 4875 (1976).
4. H. J. Koeners, J. Verhoeven, and J. H. van Boom, *Tetrahedron Lett.*, **21**, 381 (1980).
5. A. Rencurosi, L. Poletti, L. Panza, and L. Lay, *J. Carbohydr. Chem.*, **20**, 761 (2001).
6. N. Jeker and C. Tamm, *Helv. Chim. Acta*, **71**, 1895, 1904 (1988).
7. Y. Watanabe, T. Fujmoto, and S. Ozaki, *J. Chem. Soc., Chem. Commun.*, 681 (1992).
8. M. Ono and I. Itoh, *Chem. Lett.*, **17**, 585 (1988).
9. H. .J. Koeners, C. H. M. Verdegaal, and J. H. Van Boom, *Recl. Trav. Chim. Pays-Bas*, **100**, 118 (1981).

5-[3-Bis(4-methoxyphenyl)hydroxymethylphenoxy]levulinic Acid Ester

This ester is formed from the anhydride in pyridine and is quantitatively cleaved with H$_2$NNH$_2$·H$_2$O, Pyr-AcOH. The sensitivity of detection of this ester is high with its absorbance maximum of 513 nm and extinction coefficient of 78,600 in 5% Cl$_2$CHCO$_2$H/CH$_2$Cl$_2$ where it forms the trityl cation.[1]

1. E. Leikauf and H. Köster, *Tetrahedron*, **51**, 5557 (1995).

Pivaloate Ester (Pv—OR): (CH$_3$)$_3$CCO$_2$R, (Chart 2)

Formation

1. PvCl, Pyr, 0–75°C, 2.5 days, 99% yield.[1] In general, such extended reaction times are not required to obtain complete reaction. This is an excellent reagent for selective acylation of a primary alcohol over a secondary alcohol.[2–4] Microwave heating has been used to accelerate the esterification for more sterically demanding alcohols.[5]

Ref. 6

2. Selective acylation can be obtained for one of two primary alcohols having slightly different steric environments.[7,8]

Some reactions are not so easily explained as in the following case where the seemingly more hindered alcohol was acylated in preference to the less hindered alcohol.[9]

Good selectivity among secondary carbohydrate alcohols has been achieved, but the regiochemistry is structure-dependent.[10] α-Methylglucoside can be selectively acylated at the 2,6-positions in 89% yield and α-methyl 4,6-*O*-benzylidineglucoside can be selectively acylated at the 2-position in 77% yield.[11]

3. Vinyl pivaloate, Cp*Sm(THF)₂, toluene, 3 h, 99% yield.[12]

4. Pivaloic anhydride, Sc(OTf)₃, CH₃CN, −20°C, 4 h.[13,14]

5. Pivaloic anhydride, VO(OTf)₂, CH₂Cl₂, 85–100% yield. Amines thiols and phenols also react.[15]

6. Pivaloic anhydride, MgBr₂, TEA, CH₂Cl₂, rt, 99% yield.[13]

7.

Thiazolidine-2-thione shows excellent selectivity for primary alcohols over secondary alcohols (>20:1).[16] A chiral version of this reagent gives moderate enantioselectivity in the acylation of racemic alcohols.[17]

8. Pivaloic anhydride, MoO_2Cl_2, CH_2Cl_2, 91–99% yield.[18] This method is quite general and can be used to prepare esters from a large variety of anhydrides. It is also suitable for the preparation of amides and thioesters.

Cleavage

1. Bu_4NOH, 20°C, 4 h.[19]

2. aq. $MeNH_2$, 20°C, $t_{1/2}$ = 3 h.[20] In this case the 5′-position of uridine was deprotected. Acetates can be cleaved selectively in the presence of a pivaloate with NH_3/MeOH. Pivaloates are not cleaved by hydrazine in refluxing ethanol, conditions that cleave phthalimides.[21]

3. 0.5 N NaOH, EtOH, H_2O, 20°C, 12 h, 58% yield.[22]

4. K_2CO_3, MeOH, reflux, 48 h, 63% yield.[23] The survival of the lactone is probably the result of a conformational effect that increases the steric hindrance around the carbonyl.

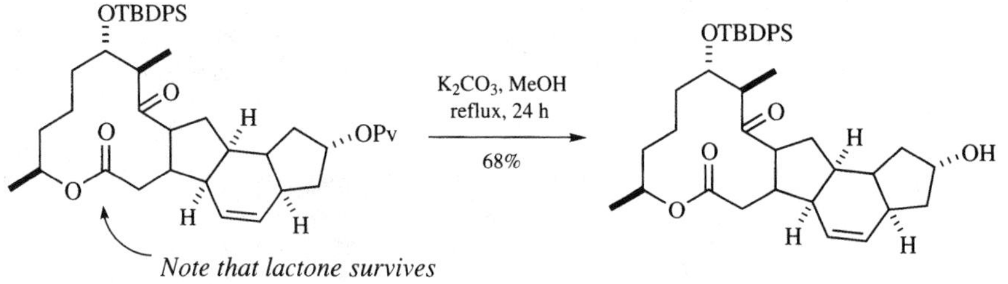

Note that lactone survives

5. NaOMe, MeOH, 90% yield.[24]

6. MeLi, Et_2O, 20°C.[25]

7. t-BuOK, H_2O (8:2), 20°C, 3 h, 94% yield.[26]

8. i-Bu_2AlH, CH_2Cl_2, −78°C, 95% yield.[2] i-Bu_2AlH, CH_2Cl_2, toluene, 84% yield. Three pivaloates were cleaved from a zaragozic acid intermediate. The use of THF or ether as solvent failed to remove all three.[27]

9. Fungus, *Currulania lunata*, 6 h, 64% yield.[28] In this case, a 21-pivaloate was removed from a steroid.

10. KEt_3BH, THF, −78°C, 78% yield.[29]

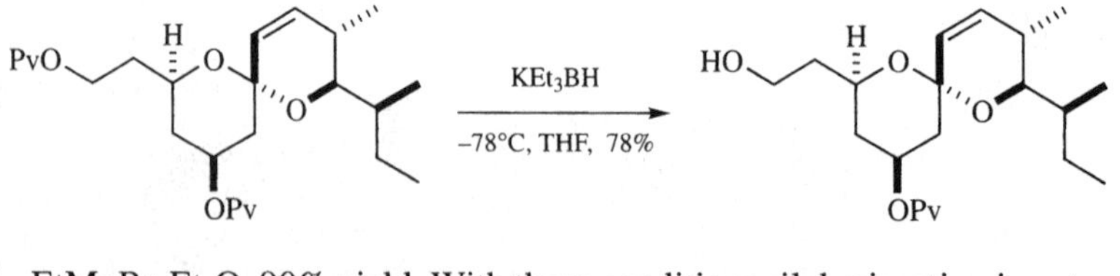

11. EtMgBr, Et_2O, 90% yield. With these conditions silyl migration is not a problem as it was when the typical hydrolysis conditions were used.[30]

12.

PvO$\diagdown$$\diagup$O$\diagdown$$\underset{\underset{O}{|}}{\overset{\overset{O}{\parallel}}{P}}$$-$OPh $\xrightarrow{\text{esterase}}$ $\left[\text{HO}\diagdown\diagup\text{O}\diagdown\underset{\underset{O}{|}}{\overset{\overset{O}{\parallel}}{P}}-\text{OPh} \right]$ $\longrightarrow$ $(HO)_2\overset{\overset{O}{\parallel}}{P}-OPh$

                                                                            Ref. 31

13. Al$_2$O$_3$, microwaves, 12 min, 93% yield.[32] Cleavage of acetates occurs similarly.

14. Esterase from rabbit serum, 53–95% yield.[33]

15. Li, NH$_3$, Et$_2$O; NH$_4$Cl, 70–85% yield.[34]

16. 3 *M* HCl, dioxane, reflux, 18 h, 80% yield.[35]

17. Sm, I$_2$, MeOH, 24 h reflux, 95% yield.[36] Troc, Ac, and Bz groups are also cleaved.

1. M. J. Robins, S. D. Hawrelak, T. Kanai, J.-M. Siefert, and R. Mengel, *J. Org. Chem.*, **44**, 1317 (1979).

2. K. C. Nicolaou and S. E. Webber, *Synthesis*, 453 (1986).

3. D. Boschelli, T. Takemasa, Y. Nishitani, and S. Masamune, *Tetrahedron Lett.*, **26**, 5239 (1985).

4. H. Nagaoka, W. Rutsch, G. Schmid, H. Ilio, M. R. Johnson, and Y. Kishi, *J. Am. Chem. Soc.* **102**, 7962 (1980).

5. E. Söderberg, J. Westman, and S. Oscarson, *J. Carbohydr. Chem.*, **20**, 397 (2001).

6. P. Jütten and H. D. Scharf, *J. Carbohydr. Chem.*, **9**, 675 (1990). Z. Yang, E. L.-M. Wong, T. Y.-T. Shum, C.-M. Che, and Y. Hui, *Org. Lett.*, **7**, 669 (2005).

7. N. Kato, H. Kataoka, S. Ohbuchi, S. Tanaka, and H. Takeshita, *J. Chem. Soc., Chem. Commun.*, 354 (1988).

8. P. F. Schuda and M. R. Heimann, *Tetrahedron Lett.*, **24**, 4267 (1983).

9. M. Yu, D. L. J. Clive, V. S. C. Yeh, S. Kang, and J. Wang, *Tetrahedron Lett.*, **45**, 2879 (2004).

10. L. Jiang and T.-H. Chan, *J. Org. Chem.*, **63**, 6035 (1998).

11. S. Tomic-Kulenovic and D. Keglevic, *Carbohydr. Res.*, **85**, 302 (1980).

12. Y. Ishii, M. Takeno, Y. Kawasaki, A. Muromachi, Y. Nishiyama, and S. Sakaguchi, *J. Org. Chem.*, **61**, 3088 (1996).

13. E. Vedejs and O. Daugulis, *J. Org. Chem.*, **61**, 5702 (1996).

14. Z.-H. Peng and K. A. Woerpel, *J. Am. Chem. Soc.*, **125**, 6018 (2003).

15. C.-T. Chen, J.-H. Kuo, C.-H. Li, N. B. Barhate, S.-W. Hon, T.-W. Li, S.-D. Chao, C.-C. Liu, Y.-C. Li, I.-H. Chang, J.-S. Lin, C.-J. Liu, and Y.-C. Chou, *Org. Lett.*, **3**, 3729 (2001).

16. S. Yamada, *Tetrahedron Lett.*, **33**, 2171 (1992); S. Yamada, T. Sugaki, and K. Matsuzaki, *J. Org. Chem.*, **61**, 5932 (1996).

17. S. Yamada and H. Katsumata, *J. Org. Chem.*, **64**, 9365 (1999).

18. C-T. Chen, J.-H. Kuo, V. D. Pawar, Y. S. Munot, S.-S. Weng, C.-H. Ku, and C.-Y. Liu, *J. Org. Chem.*, **70**, 1188 (2005).

19. C. A. A. van Boeckel and J. H. van Boom, *Tetrahedron Lett.*, **20**, 3561 (1979).

20. B. E. Griffin, M. Jarman, and C. B. Reese, *Tetrahedron*, **24**, 639 (1968).

21. T. Nakano, Y. Ito, and T. Ogawa, *Carbohydr. Res.*, **243**, 43 (1993).

22. K. K. Ogilvie and D. J. Iwacha, *Tetrahedron Lett.*, **14**, 317 (1973).

23. L. A. Paquette, I. Collado, and M. Purdie, *J. Am. Chem. Soc.*, **120**, 2553 (1998).

24. K. C. Nicolaou, T. J. Caulfield, H. Kataoka, and N. A. Stylianides, *J. Am. Chem. Soc.*, **112**, 3693 (1990).

25. B. M. Trost, S. A. Godleski, and J. L. Belletire, *J. Org. Chem.*, **44**, 2052 (1979).

26. P. G. Gassman and W. N. Schenk, *J. Org. Chem.*, **42**, 918 (1977).

27. E. M. Carreira and J. Du Bois, *J. Am. Chem. Soc.*, **117**, 8106 (1995).

28. H. Kosmol, F. Hill, U. Kerb, and K. Kieslich, *Tetrahedron Lett.*, **11**, 641 (1970).

29. S. J. Danishefsky, D. M. Armistead, F. E. Wincott, H. G. Selnick, and R. Hungate, *J. Am. Chem. Soc.*, **111**, 2967 (1989).

30. Y. Watanabe, T. Fujimoto, and S. Ozaki, *J. Chem. Soc., Chem. Commun.*, 681 (1992).

31. D. Farquhar, S. Khan, M. C. Wilkerson, and B. S. Andersson, *Tetrahedron Lett.*, **36**, 655 (1995).

32. S. V. Ley and D. M. Mynett, *Synlett*, 793 (1993).

33. S. Tomic, A. Tresec, D. Ljevakovic, and J. Tomasic, *Carbohydr. Res.*, **210**, 191 (1991); D. Ljevakovic, S. Tomic, and J. Tomasic, *Carbohydr. Res.*, **230**, 107 (1992).

34. H. W. Pinnick and E. Fernandez, *J. Org. Chem.*, **44**, 2810 (1979).

35. A.-M. Fernandez, J.-C. Plaquevent, and L. Duhamel, *J. Org. Chem.*, **62**, 4007 (1997).

36. R. Yanada, N. Negoro, K. Bessho, and K. Yanada, *Synlett*, 1261 (1995).

1-Adamantoate Ester: ROCO-1-adamantyl (Chart 2)

The adamantoate ester is formed selectively from a primary hydroxyl group (e.g., from the 5'-OH in a ribonucleoside) by reaction with adamantoyl chloride, Pyr (20°C, 16 h). It is cleaved by alkaline hydrolysis (0.25 N NaOH, 20 min), but is stable to milder alkaline hydrolysis (e.g., NH_3, MeOH), conditions that cleave an acetate ester.[1] Its steric properties are similar to that of the pivalate.

1. K. Gerzon and D. Kau, *J. Med. Chem.*, **10**, 189 (1967).

Crotonate Ester: $ROCOCH=CHCH_3$

4-Methoxycrotonate Ester: $ROCOCH=CHCH_2OCH_3$

The crotonate esters, prepared to protect a primary hydroxyl group in nucleosides, are cleaved by hydrazine (MeOH, Pyr, 2 h). The methoxycrotonate is 100-fold more reactive to hydrazinolysis and 2-fold less reactive to alkaline hydrolysis than the corresponding acetate.[1]

1. R. Arentzen and C. B. Reese, *J. Chem. Soc., Chem. Commun.*, 270 (1977).

Benzoate Ester (Bz−OR): PhCO$_2$R (Chart 2)

The benzoate ester is one of the more common esters used to protect alcohols. Benzoates are less readily hydrolyzed than acetates, and the tendency for benzoate migration to adjacent hydroxyls, in contrast to acetates, is not nearly as strong,[1] but they can be forced to migrate to a thermodynamically more stable position.[2,3] For the most part, this migration is a major annoyance,[4] but it has been used to advantage.[3,5] The p-methoxybenzoate is even less prone to migrate than the benzoate.[6] Migration from a secondary to primary alcohol has also been induced with AgNO$_3$, KF, Pyr, H$_2$O at 100°C.[7]

The use of TBAF, a fairly basic reagent, for silyl ether cleavage can result in ester migration as the following example illustrates.[8]

Formation

1. BzCl or Bz$_2$O, Pyr, 0°C. Benzoyl chloride is the most common reagent for the introduction of the benzoate group. Reaction conditions vary, depending on the nature of the alcohol to be protected. Cosolvents such as CH$_2$Cl$_2$ are often used with pyridine. Benzoylation in a polyhydroxylated system is much more selective than acetylation.[1] A primary alcohol is selectively protected over a secondary allylic alcohol,[9] and an equatorial alcohol can be selectively protected in preference to an axial alcohol,[10] but this has been shown to be solvent dependant in some cases.[11] A cyclic secondary alcohol was selectively protected in the presence of a secondary acyclic alcohol.[12]

With pyridine the ratio is: **21:1:1.5**

With DMF/TEA the ratio is: **1:45:0.4**

The use of chiral amines will selectively monobenzoylate a diol and simultaneously generate a chiral product with reasonable ee's.[13]

2. BzCl, TMEDA, CH_2Cl_2, MS 4 Å, −78°C, 95–96% yield. The use of TMEDA as a base greatly accelerates the esterification in comparison to the use of more conventional bases.[14] TMEDA also improves the formation of carbonates from chloroformates.

3. BzCl, $LiClO_4$, THF, 5–10 h, 70–87% yield. An acetate and a pivaloate have been prepared correspondingly.[15]

4. Regioselective benzoylation of methyl 4,6-*O*-benzylidene-α-galactopyranoside can be effected by phase transfer catalysis (BzCl, Bu_4NCl, 40% NaOH, PhH, 69% yield of 2-benzoate; BzCl, Bu_4NCl, 40% NaOH, HMPA, 62% yield of 3-benzoate).[16]

5. NBz Et₃N, DMF, 20°C, 15 min, 90% yield.[17] The 2-hydroxyl of methyl 4,6-*O*-benzylidine-α-glucopyranoside was selectively protected.[18]

6. Benzoyloxybenzotriazole (BzOBt), CH_2Cl_2, TEA, rt, 89% yield. An anomeric hydroxyl was selectively acylated in the presence of a secondary hydroxyl.[19] This reagent selectively acylates primary alcohols in the presence of secondary alcohols and will selectively acylate the 2-hydroxyl in a 4,6-protected glucose derivative.[20]

7. BzCN, Et₃N, CH_3CN, 5 min to 2 h, >80% yield.[21,22] This reagent selectively acylates a primary hydroxyl group in the presence of a secondary hydroxyl group.[23]

8. $BzOCF(CF_3)_2$, TMEDA, 20°C, 30 min, 90% yield.[24] This reagent also reacts with amines to form benzamides in high yields.

9. $BzOSO_2CF_3$, −78°C, CH_2Cl_2, few min.[25,26] With acid-sensitive substrates pyridine is used as a cosolvent. This reagent also reacts with ketals, epoxides,[25] and aldehydes.[27] This reagent works where BzCl fails to give complete reaction.[28]

10. PhCO$_2$H, DIAD, Ph$_3$P, THF, 84% yield.[29]

The Mitsunobu reaction is usually used to introduce an ester with inversion of configuration. The use of this methodology on an anomeric hydroxyl was found to give only the β-benzoate, whereas other methods gave mixtures of anomers.[30] Improved yields are obtained in the Mitsunobu esterification when p-nitrobenzoic acid is used as the nucleophile,[31] and bis(dimethylamino) azodicarboxylate as an activating agent was found to be advantageous for hindered esters.[32] Bu$_3$P=CHCN was introduced as an alternative activating agent for the Mitsunobu reaction.[33]

11. BzOH, Al$_2$O$_3$/MeSO$_3$H, neat, 80–92% yield. This method was found to be excellent for the monoesterification of diols, but remotely oriented diols tend to give diesters as well. Amino alcohols are also selectively esterified.[34] In this case the nitrogen is protected by protonation, but under basic conditions O to N migration will occur.

12. NBz CHCl$_3$, reflux, 10 h.[35]

13. An alcohol can be selectively benzoylated in the presence of a primary amine if steric diminish its reactivity.[36]

14. BuLi, BzCl; 10% Na$_2$CO$_3$, H$_2$O, 82% yield.[37] These conditions were used to monoprotect 1,4-butanediol.

15. BzOOBz, Ph$_3$P, CH$_2$Cl$_2$, 1 h, rt ≈80% yield.[38] When these conditions are applied to unsymmetrical 1,2-diols the benzoate of the kinetically and thermodynamically less stable isomer is formed.

16. (Bu$_3$Sn)$_2$O; BzCl.[39,40] The use of microwaves accelerates this reaction.[41] Bu$_2$Sn(OMe)$_2$ is reported to work better than Bu$_2$SnO in the monoprotection of diols.[42] The monoprotection of diols at the more hindered position can be accomplished through the stannylene if the reaction is quenched with PhMe$_2$SiCl (45–77% yield).[43] A cautionary note concerning this method is

that in some cases a temperature-induced post-acylation migration may occur to give unexpected mixtures.[44]

The reaction can also be run using catalytic amounts of a tin reagent which results in acylation of the least hindered alcohol or monoacylation of symmetrical diols is also possible.[45] The use of a chiral tin reagent gives modest levels of kinetic resolution of racemic diols.[46]

17.

Ref. 47

18.

The selectivity here relies on the fact that the β-benzoate is the thermodynamically more stable ester. A mixture of esters is formed upon hydrolysis of the ortho ester and then equilibrated with DBU.[48] Carbohydrates are selectively protected with this methodology.[49]

19. Bz$_2$O, MgBr$_2$, TEA, CH$_2$Cl$_2$, rt, 95% yield. Tertiary alcohols are readily acylated.[50]

20. Bz$_2$O, K$_2$CO$_3$, acetone, 90% yield. Note that the secondary hydroxyl was not esterified.[51]

21. Bz$_2$O, Sc(NTf$_2$)$_3$, CH$_3$CN, 25°C, 1.5–3 h, 90–98% yield. Phenols are also acylated efficiently.[52]

22. Vinyl benzoate, Cp$^*_2$Sm(THF)$_2$, toluene, rt, 3 h, 99% yield.[53]

23. N-Benzoyl-4-(dimethylamino)pyridinium chloride, CH$_2$Cl$_2$, TEA.[54]

24. As with acetates, enzymatic methods can be used to regioselectively introduce a primary benzoate in the presence of a secondary alcohol (Cal-B, vinyl benzoate, THF, 60°C, 89–96% yield).[55]

25. (R,R)-P-box-CuCl$_2$, BzCl, 0.5 eq., DIPEA, CH$_2$Cl$_2$, 0°C, 77–99% ee.[56]

Cleavage

The section on the cleavage of acetates should be consulted, since many of the methods presented there are applicable to benzoates.

1. 1% NaOH, MeOH, 20°C, 50 min, 90% yield.[57]

2. Et$_3$N, MeOH, H$_2$O (1:5:1), reflux, 20 h, 86% yield.[58]

3. MeOH, KCN.[59]

4. A benzoate ester can be cleaved in 60–90% yield by electrolytic reduction at −2.3 V.[60]

5. The following example illustrates the selective cleavage of a 2′-benzoate in a nucleotide derivative.[62] This selectivity is achieved because the hydroxyl at the 2′-position is the most acidic of the three.

The use of hydrazine was also found very effective in the deprotection of a complex glycopeptide where conventional methods failed to give complete deprotection.[63]

6. Ammonia, MeOH, 65–70% yield. This method was developed to selectively cleave secondary benzoates in the presence of the primary benzoate.[64] This method was also successful for the cleavage of secondary benzoates in the presence of a primary benzoate of pyranosides.

7. Ammonia, 87% yield. In this case an anomeric benzoate was deprotected in the presence of a primary benzoate which shows that benzoates of more acidic hydroxyls are cleaved more rapidly.[65]

8. $BF_3 \cdot Et_2O$, Me_2S.[66]

9. Mg, MeOH, rt, 13 h, 91% yield. Esters are cleaved selectively in the order *p*-nitrobenzoate > acetate > benzoate > pivalate ≫ trifluoroacetamide.[67]

10. EtMgBr, Et_2O, rt, 1 h, 90–100% yield.[68,69] These conditions were used to prevent a neighboring silyl ether from migrating. Ethylmagnesium chloride is much more reactive; thus the reaction can be run at $-42°C$ giving a 90% yield of the alcohol. Acetates and pivaloates are also cleaved.

1 A. H. Haines, *Adv. Carbohydr. Chem. Biochem.*, **33**, 11 (1976).

2. S. J. Danishefsky, M. P. DeNinno, and S.-h. Chen, *J. Am. Chem. Soc.* **110**, 3929 (1988).

3. A. Graziani, P. Passacantilli, G. Piancatelli, and S. Tani, *Tetrahedron Lett.*, **42**, 3857 (2001).

4. M. Chandrasekhar, K. L. Chandra, and V. K. Singh, *J. Org. Chem.*, **68**, 4039 (2003); T. Nukada, A. Berces, and D. M. Whitfield, *J. Org. Chem.*, **64**, 9030 (1999).

5. S. Jaracz, K. Nakanishi, A. A. Jensen, and K. Stromgaard, *Chem. Eur. J.*, **10**, 1507 (2004).

6. E. J. Corey, A. Guzman-Perez, and M. C. Noe, *J. Am. Chem. Soc.*, **117**, 10805 (1995).

7. Z. Zhang and G. Magnusson, *J. Org. Chem.*, **61**, 2383 (1996).

8. K. C. Nicolaou, H. J. Mitchell, K. C. Fylaktakidou, R. M. Rodriguez, and H. Suzuki, *Chem. Eur. J.*, **6**, 3116 (2000).

9. R. H. Schlessinger and A. Lopes, *J. Org. Chem.*, **46**, 5252 (1981).

10. A. P. Kozikowski, X. Yan, and J. M. Rusnak, *J. Chem. Soc., Chem. Commun.*, 1301 (1988).

11. M. Flores-Mosquera, M. Martin-Lomas, and J. L. Chiara, *Tetrahedron Lett.*, **39**, 5085 (1998).

12. K. Furuhata, K. Takeda, and H. Ogura, *Chem. Pharm. Bull.*, **39**, 817 (1991).

13. T. Oriyama, K. Imai, T. Hosoya, and T. Sano, *Tetrahedron Lett.*, **39**, 397 (1998); T. Oriyama, K. Imai, T. Sano, and T. Hosoya, *Tetrahedron Lett.*, **39**, 3529 (1998); S. Mizuta, M. Sadamori, T. Fujimoto, and I. Yamamoto, *Angew. Chem. Int. Ed.*, **42**, 3383 (2003).

14. T. Sano, K. Ohashi, and T. Oriyama, *Synthesis*, 1141 (1999).

15. B. P. Bandgar, V. T. Kamble, V. S. Sadavarte, and L. S. Uppalla, *Synlett*, 735 (2002).

16. W. Szeja, *Synthesis*, 821 (1979).

17. J. Stawinski, T. Hozumi, and S. A. Narang, *J. Chem. Soc., Chem. Commun.*, 243 (1976).

18. S. Kim, H. Chang, and W.J. Kim, *J. Org. Chem.*, **50**, 1751 (1985).

19. S.-C. Hung, S. R. Thopate, F.-C. Chi, S.-W. Chang, J.-C. Lee, C.-C. Wang, and Y.-S. Wen, *J. Am. Chem. Soc.*, **123**, 3153 (2001).

20. S. Kim, H. Chang, and W. J. Kim, *J. Org. Chem.*, **50**, 1751 (1985).

21. M. Havel, J. Velek, J. Pospíšek, and M. Soucek, *Collect. Czech. Chem. Commun.*, **44**, 2443 (1979).

22. A. Holý and M. Soucek, *Tetrahedron Lett.*, **12**, 185 (1971).

23. R. M. Soll and S. P. Seitz, *Tetrahedron Lett.*, **28**, 5457 (1987); C. Gege, J. Vogel, G. Bendas, U. Rothe, and R. R. Schmidt, *Chem. Eur. J.*, **6**, 111 (2000).

24. N. Ishikawa and S. Shin-ya, *Chem. Lett.*, **5**, 673 (1976).

25. L. Brown and M. Koreeda, *J. Org. Chem.*, **49**, 3875 (1984).

26. A. Liakatos, M. J. Kiefel, and M. von Itzstein, *Org. Lett.*, **5**, 4365 (2003).

27. K. Takeuchi, K. Ikai, M. Yoshida, and A. Tsugeno, *Tetrahedron*, **44**, 5681 (1988).

28. A. Liakatos, M. J. Kiefel, and M. von Itzstein, *Org. Lett.*, **5**, 4365 (2003).

29. A. B. Smith III and K. J. Hale, *Tetrahedron Lett.*, **30**, 1037 (1989).

30. A. B. Smith, III, R. A. Rivero, K. J. Hale, and H. A. Vaccaro, *J. Am. Chem. Soc.*, **113**, 2092 (1991).

31. S. F. Martin and J. A. Dodge, *Tetrahedron Lett.*, **32**, 3017 (1991).

32. T. Tsunoda, Y. Yamamiya, Y. Kawamura, and S. Ito, *Tetrahedron Lett.*, **36**, 2529 (1995).

33. T. Tsunoda, F. Ozaki, and S. Ito, *Tetrahedron Lett.*, **35**, 5081 (1994).

34. H. Sharghi and M. H. Sarvari, *Tetrahedron*, **59**, 3627 (2003).

35. C. L. Brewer, S. David, and A. Veyriérs, *Carbohydr. Res.*, **36**, 188 (1974).

36. Y. Ito, M. Sawamura, E. Shirakawa, K. Hayashizaki, and T. Hayashi, *Tetrahedron*, **44**, 5253 (1988). See also T.-Y. Luh and Y. H. Chong, *Synth. Commun.*, **8**, 327 (1978).

37. A. J. Castellino and H. Rapoport, *J. Org. Chem.*, **51**, 1006 (1986).

38. A. M. Pautard and S. A. Evans, Jr., *J. Org. Chem.*, **53**, 2300 (1988).

39. S. Hanessian and R. Roy, *Can. J. Chem.*, **63**, 163 (1985).

40. For a mechanistic study of the tin-directed acylation, see S. Roelens, *J. Chem. Soc., Perkin Trans. II*, 2105 (1988).

41. B. Herradón, A. Morcuende, and S. Valverde, *Synlett*, 455 (1995). A. Morcuende, S. Valverde, and B. Herradón, *Synlett*, 89, (1994).

42. G. J. Boons, G. H. Castle, J. A. Clase, P. Grice, S. V. Ley, and C. Pinel, *Synlett*, 913 (1993).

43. G. Reginato, A. Ricci, S. Roelens, and S. Scapecchi, *J. Org. Chem.*, **55**, 5132 (1990).

44. M. W. Bredenkamp, and H. S. C. Spies, *Tetrahedron Lett.*, **41**, 543 (2000).

45. T. Maki, F. Iwasaki, and Y. Matsumura, *Tetrahedron Lett.*, **39**, 5601 (1998); F. Iwasaki, T. Maki, O. Onomura, W. Nakashima, and Y. Matsumura, *J. Org. Chem.*, **65**, 996 (2000).

46. F. Iwasaki, T. Maki, W. Nakashima, O. Onomura, and Y. Matsumura, *Org. Lett.*, **1**, 969 (1999).

47. H. Yamda, T. Harada, and T. Takahashi, *J. Am. Chem. Soc.*, **116**, 7919 (1994).

48. J. W. Lampe, P. F. Hughes, C. K. Biggers, S. H. Smith, and H. Hu, *J. Org. Chem.*, **61**, 4572 (1996).

49. F. I. Auzanneau and D. R. Bundle, *Carbohydr. Res.*, **212**, 13 (1991).

50. E. Vedejs and O. Daugulis, *J. Org. Chem.*, **61**, 5702 (1996).

51. T. Ritter, P. Zarotti, and E. M. Carreira, *Org. Lett.*, **6**, 4371 (2004).

52. K. Ishihara, M. Kubota, and H. Yamamoto, *Synlett*, 265 (1996).

53. Y. Ishii, M. Takeno, Y. Kawasaki, A. Muromachi, Y. Nishiyama, and S. Sakaguchi, *J. Org. Chem.*, **61**, 3088 (1996).

54. M. S. Wolfe, *Synth. Commun.*, **27**, 2975 (1997).

55. J. García, S. Fernandez, M. Ferrero, Y. S. Sanghvi, and V. Gotor, *Tetrahedron Lett.*, **45**, 1709 (2004).

56. Y. Matsumura, T. Maki, S. Murakami, and O. Onomura, *J. Am. Chem. Soc.*, **125**, 2052 (2003).

57. K. Mashimo and Y. Sato, *Tetrahedron*, **26**, 803 (1970).

58. K. Tsuzuki, Y. Nakajima, T. Watanabe, M. Yanagiya, and T. Matsumoto, *Tetrahedron Lett.*, **19**, 989 (1978).

59. J. Herzig, A. Nudelman, H. E. Gottlieb, and B. Fischer, *J. Org. Chem.*, **51**, 727 (1986).

60. V. G. Mairanovsky, *Angew. Chem., Inter. Ed. Engl.*, **15**, 281 (1976).

61. J.-P. Pulicani, D. Bézard, J.-D. Bourzat, H. Bouchard, M. Zucco, D. Deprez, and A. Commercon, *Tetrahedron Lett.*, **35**, 9717 (1994).

62. Y. Ishido, N. Nakazaki, and N. Sakairi, *J. Chem. Soc., Perkin Trans. 1*, 2088 (1979).

63. P. W. Glunz, S. Hintermann, J. B. Schwarz, S. D. Kuduk, X.-T. Chen, L. J. Williams, D. Sames, S. J. Danishefsky, V. Kudryashov, and K. O. Lloyd, *J. Am. Chem. Soc.*, **121**, 10636 (1999).

64. R. Zerrouki, V. Roy, A. Hadj-Bouazza, and P. Krausz, *J. Carbohydr. Chem.*, **23**, 299 (2004).

65. J.-C. Lee, S.-W. Chang, C.-C. Liao, F.-C. Chi, C.-S. Chen, Y.-S. Wen, C.-C. Wang, S. S. Kulkarni, R. Puranik, Y.-H. Liu, and S.-C. Hung, *Chem. Eur. J.*, **10**, 399 (2004).

66. K. Fuji, T. Kawabata, and E. Fujita, *Chem. Pharm. Bull.*, **28**, 3662 (1980).

67. Y.-C. Xu, E. Lebeau, and C. Walker, *Tetrahedron Lett.*, **35**, 6207 (1994); Y.-C. Xu, A. Bizuneh, and C. Walker, *Tetrahedron Lett.*, **37**, 455 (1996).

68. Y. Watanabe, T. Fujimoto, and S. Ozaki, *J. Chem. Soc., Chem. Commun.*, 681 (1992).

69. Y. Watanabe, T. Fujimoto, T. Shinohara, and S. Ozaki, *J. Chem. Soc., Chem. Commun.*, 428 (1991).

p-Phenylbenzoate Ester: $ROCOC_6H_4$-*p*-C_6H_5

The *p*-phenylbenzoate ester was prepared to protect the hydroxyl group of a prostaglandin intermediate by reaction with the benzoyl chloride (Pyr, 25°C, 1 h, 97% yield). It was a more crystalline, more readily separated derivative than 15 other esters that were investigated.[1] It can be cleaved with K_2CO_3 in MeOH in the presence of a lactone.[2]

1. E. J. Corey, S. M. Albonico, U. Koelliker, T. K. Schaaf, and R. K. Varma, *J. Am. Chem. Soc.*, **93**, 1491 (1971).
2. T. V. RaganBabu, *J. Org. Chem.*, **53**, 4522 (1988).

2,4,6-Trimethylbenzoate (Mesitoate) Ester: $2,4,6\text{-}Me_3C_6H_2CO_2R$ (Chart 2)

Formation

1. $Me_3C_6H_2COCl$, Pyr, $CHCl_3$, 0°C, 14 h → 23°C, 1 h, 95% yield.[1]
2. $Me_3C_6H_2CO_2H$, $(CF_3CO)_2O$, PhH, 20°C, 15 min.[2]

Cleavage

1. $LiAlH_4$, Et_2O, 20°C, 2 h.[2]
2. *t*-BuOK, H_2O (8:1) "anhydrous hydroxide," 20°C, 24–72 h, 50–72% yield.[3] A mesitoate ester is exceptionally stable to base: 2 *N* NaOH, 20°C, 20 h; 12 *N* NaOH, EtOH, 50°C, 15 min.

1. E. J. Corey, K. Achiwa, and J. A. Katzenellenbogen, *J. Am. Chem. Soc.*, **91**, 4318 (1969).
2. I. J. Bolton, R. G. Harrison, B. Lythgoe, and R. S. Manwaring, *J. Chem. Soc. C*, 2944 (1971).
3. P. G. Gassman and W. N. Schenk, *J. Org. Chem.*, **42**, 918 (1977).

4-Bromobenzoate: $4\text{-}BrC_6H_4CO_2R$

The 4-bromobenzoate[1] is often used in place of a benzoate because it tends to impart crystallinity to a molecule which makes x-ray structure determinations possible.[2] It is prepared and cleaved by the same methods as the benzoate.[3]

1. K. Ohmori, S. Nishiyama, and S. Yamamura, *Tetrahedron Lett.*, **36**, 6519 (1995).
2. G. Zhou, Q.-Y. Hu, and E. J. Corey, *Org. Lett.*, **5**, 3979 (2003).
3. T. Yoshimura, T. Bando, M. Shindo, and K. Shishido, *Tetrahedron Lett.*, **45**, 9241 (2004); P. G. Reddy and S. Baskaran, *J. Org. Chem.*, **69**, 3093 (2004).

2,5-Difluorobenzoate: $2,5\text{-}F_2C_6H_3CO_2R$

The 2,5-difluorobenzoyl group was developed for the protection of *O*-linked glycopeptides. In contrast to the use of acetates and benzoates, this group does not result in the formation of orthoesters or transfer the ester to the alcohol being glycosylated as is the case with an acetate. It can be cleaved using conditions that do not result in elimination of the serine or threonine to dehydropeptides with loss of the glycoside, as is the case with the benzoate. The ester is formed from the acid chloride using pyridine with DMAP catalysis (91% yield). It can be cleaved with LiOH/MeOH

(0.5 h) or with NH_3/MeOH (2 h). Of the four fluorinated esters tested, the rate of cleavage is as follows: 2,5-difluor > 3-fluoro > 2-fluoro > 4-fluorobenzoyl derivative.[1] Only the 2,5-derivative was found satisfactory for glycopeptide synthesis.

1. P. Sjölin and J. Kihlberg, *J. Org. Chem.*, **66**, 2957 (2001).

p-Nitrobenzoate (*p*NBz−OR or PNB−OR) Ester: 4-$NO_2C_6H_4CO_2R$

Formation

1. *p*-Nitrobenzoyl chloride, imidazole, >52% yield.[1, 2]
2. *p*-Nitrobenzoic acid, Ph_3P, DEAD, THF^3.[4] This method results in inversion of configuration when using secondary alcohols.

Cleavage

1. NaOH, dioxane, H_2O, >97% yield.[1]
2. NaN_3, MeOH, 40°C, 52–100% yield. This method is sufficiently mild that Aldol esters are not eliminated during cleavage.

1. R. Carter, K. Hodgetts, J. McKenna, P. Magnus, and S. Wren, *Tetrahedron*, **56**, 4367 (2000); J. W. C. Cheing, W. P. D. Goldring, and G. Pattenden, *Chem. Commun.*, 2788 (2003).
2. C. Kolar, K. Dehmel, H. Moldenhauer, and M. Gerken, *J. Carbohydr.Chem.*, **9**, 873 (1990).
3. J. A. Dodge, J. I. Trujillo, and M. Presnell, *J. Org. Chem.*, **59**, 234 (1994).
4. D. L. Hughes and R. A. Reamer, *J. Org. Chem.*, **61**, 2967 (1996); T. Haack, K. Haack, W. E. Diederich, B. Blackman, S. Roy, S. Pusuluri, and G. I. Georg, *J. Org. Chem.*, **70**, 7592 (2005).

Picolinate (Pic) Ester

Formation

1. Via the Mitsunobu reaction: Pyridyl-2-CO_2H, Ph_3P, DIAD, 20°C, 3 h, rt, 16 h, 67–94% yield.[1]
2. The picolinate is readily prepared from the commercially available acid chloride and an alcohol or phenol. [2]

Cleavage

1. $Cu(OAc)_2$, MeOH, or $CHCl_3$/MeOH, 79–95% yield. This hydrolysis was successful where the hydrolysis of the 4-nitrobenzoate or benzoate resulted in elimination.

2. $Zn(OAc)_2 \cdot 2H_2O$ in CH_2Cl_2, MeOH at rt in 1.5–4 h in 89–97% yield.[2]

1. T. Sammakia and J. S. Jacobs, *Tetrahedron Lett.*, **40**, 2685 (1999).
2. J. Y. Baek, Y.-J. Shin, H. B. Jeon, and K. S. Kim, *Tetrahedron Lett.*, **46**, 5143 (2005).

Nicotinate Ester

Formation

3-Pyridylcarboxylic acid anhydride, 93–99% yield.[1]

Cleavage

MeI followed by hydroxide, 55–98% yield. Quaternization of the pyridine increases the rate of hydrolysis of the ester.

1. S. Ushida, *Chem. Lett.*, **18**, 59 (1989).

Proximity-Assisted Deprotection for Ester Cleavage

The following derivatives represent protective groups that contain an auxilliary functionality, which when chemically modified, results in intramolecular, assisted cleavage, thus increasing the rate of cleavage over simple basic hydrolysis. In general, this allows for their removal in the presence of other esters that would normally be cleaved using conventional hydrolytic methods.

2-(Azidomethyl)benzoate Ester (AZMB−OR): 2-$(N_3CH_2)C_6H_4CO_2R$

This ester was developed as a participating group in glycosylations that could be removed in the presence of other esters. It is introduced using the acid chloride

(CH_2Cl_2, DMAP, rt, 87% yield or pyridine). It is cleaved by reduction of the azide with Bu_3P or $MePPh_2$ (THF, H_2O, 76–96% yield) which causes facile intramolecular amide formation with release of the protected alcohol.[1] Other conditions that reduce azides to amines such as hydrogenation (NH_4HCO_2, Pd–C, MeOH, rt) or $NaBH_4$ reduction will cleave this ester (86–98% yield).[2,3]

4-Azidobutyrate Ester: $N_3(CH_2)_3CO_2R$

The 4-azidobutyrate ester is introduced via the acid chloride. Cleavage occurs by pyrrolidone formation after the azide is reduced by hydrogenation, H_2S or Ph_3P.[4,5]

(2-Azidomethyl)phenylacetate Ester (AMPA−OR): $2\text{-}(N_3CH_2)C_6H_4CH_2CO_2R$

This group is similar to the AZMB group. It is introduced from the acid using DCC as a coupling agent (73–92% yield). It is cleaved by reduction with Lindlar catalyst but should be cleavable by the same methods used to cleave the AZMB group. As expected, NaOMe/MeOH also hydrolyzes this ester.[6]

2-{[(Tritylthio)oxy]methyl}benzoate Ester (TOB−OR)

2-{[(4-Methoxytritylthio)oxy]methyl}benzoate Ester (MOB−OR)

2-{[Methyl(tritylthio)amino]methyl}benzoate Ester (TAB−OR)

2-{{[(4-methoxytrityl)thio]methylamino}-methyl}benzoate (MAB−OR) Ester

X = H, 4-MeO

Y = O or NH

These groups were developed for the protection of the 5′-hydroxyl in nucleoside synthesis. Its advantage is that it can be cleaved using the same conditions that oxidize the phosphite to the phosphate (I_2, pyridine) thus taking one step out of the synthesis. It is cleaved with 3% trichloroacetic acid and was stable to the following reagents: Ac_2O/pyridine/DMAP, t-butyl hydroperoxide, 1,2-benzodithiol-3-one 1,1-dioxide, N,N,N,N-tetramethylthioruram disulfide.[7] Introduction of these selectively at the 5′-hydroxyl of a nucleoside did prove problematic because it requires protection of the 3′-hydroxyl. The TAB group is induced using BOPCl (DMAP, pyridine, 64% yield) as a coupling agent. It is also cleaved oxidatively with I_2.

2-(Allyloxy)phenylacetate (APAC−OR) Ester: $2\text{-}(CH_2{=}CHCH_2O)C_6H_4CO_2R$

This ester is a participating group in glycosylations. It is introduced using DCC/DMAP as a coupling agent (almost quantitative yield). It is cleaved by lactone formation upon allyl group removal with $(Ph_3P)_4Pd$ (proton sponge, EtOH, H_2O, reflux, 2–7 h, almost quantitative yield). For other potential methods of deprotection the sections on allyl group cleavage should be consulted. This group was shown to be orthogonal to the acetate and levulinate esters.[8]

2-(Prenyloxymethyl)benzoate Ester (POMB): $((CH_3)_2C=CHCH_2O)C_6H_4CO_2R$

The ester is prepared using DCC/DMAP (90–97% yield). It is cleaved in a two-step process wherein the prenyl ether is removed with DDQ in CH_2Cl_2/H_2O to reveal an alcohol that is induced to lactonize with $Yb(OTf)_3 \cdot H_2O$ releasing the protected alcohol 90–92% yield.[9]

6-(Levulinyloxymethyl)-3-methoxy-2 and 4-nitrobenzoate Ester (LMMo(p)NBz−OR)

This group was developed for 5′ protection in oligonucleotide synthesis. It is introduced using triisopropylbenzenesulfonyl chloride/pyridine (55–76% yield).[10] It is cleaved with hydrazine. Other methods used to cleave the levulinate groups should also be applicable. The PAC$_{LEV}$ group is another levulinate-based protected protective group.[24]

4-Benzyloxybutyrate Ester (BOB): $C_6H_5CH_2OCH_2CH_2CH_2CH_2CO_2R$

This ester is prepared by condensing the acid and alcohol with EDC (DMAP, CH_2Cl_2, 58–99% yield). It is cleaved by hydrogenolysis followed by t-BuOK treatment.[11]

4-Trialkylsilyloxybutyrate Ester (SOB): $4\text{-}(t\text{-Bu}(CH_3)_2SiO)CH_2CH_2CH_2CH_2CO_2R$

This ester was developed as a BOB replacement because the BOB could not be efficiently removed by hydrogenolysis. It is prepared from the acid (TsCl, DMAP, THF, 0°C to rt, 98% yield). It is cleaved with TBAF (THF, rt, 75% yield).[12]

4-Acetoxy-2,2-dimethylbutyrate Ester (ADMB): $CH_3CO_2CH_2CH_2C(CH_3)_2CO_2R$

This group was developed for C-2 protection of carbohydrates. It selectively directs glycosylation to give primarily the β-glycoside. This group has the advantage over the pivalate, which has a similar directing effect in that it is easily cleaved with catalytic DBU in MeOH.[13]

2,2-Dimethyl-4-pentenoate Ester: $CH_2=CH(CH_3)_2CCO_2R$

This group is a pivalate ester equivalent that still has the steric advantage associated with pivalic acid but can be removed after the olefin is converted to an alcohol by hydroboration.[14]

2-Iodobenzoate Ester: $2\text{-I-}C_6H_4CO_2R$

The 2-iodobenzoate is introduced by acylation of the alcohol with the acid (DCC, DMAP, CH_2Cl_2, 25°C, 96% yield); it is removed by oxidation with Cl_2 (MeOH, H_2O, Na_2CO_3, pH >7.5).[15]

4-Nitro-4-methylpentanoate Ester

Formation/Cleavage[16]

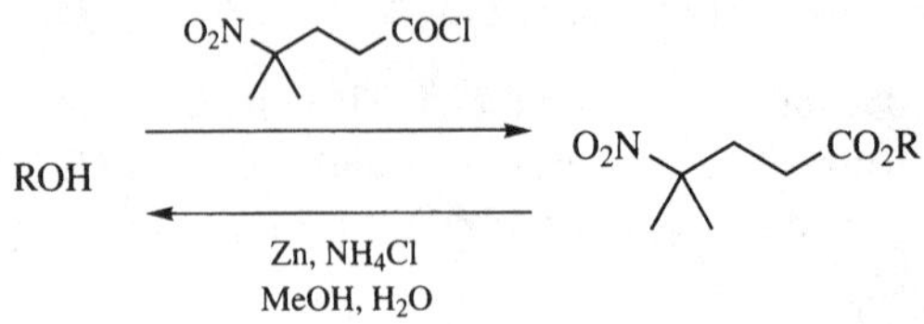

o-(Dibromomethyl)benzoate Ester: o-$(Br_2CH)C_6H_4CO_2R$

The o-(dibromomethyl)benzoate, prepared to protect nucleosides by reaction with the benzoyl chloride (CH_3CN, 65–90% yield), can be cleaved under nearly neutral conditions. The cleavage involves conversion of the $-CHBr_2$ group to $-CHO$ by silver ion-assisted hydrolysis. The benzoate group, *ortho* to the $-CHO$ group, now is rapidly hydrolyzed by neighboring group participation (the morpholine and hydroxide ion-catalyzed hydrolyses of methyl 2-formylbenzoate are particularly rapid).[17,18]

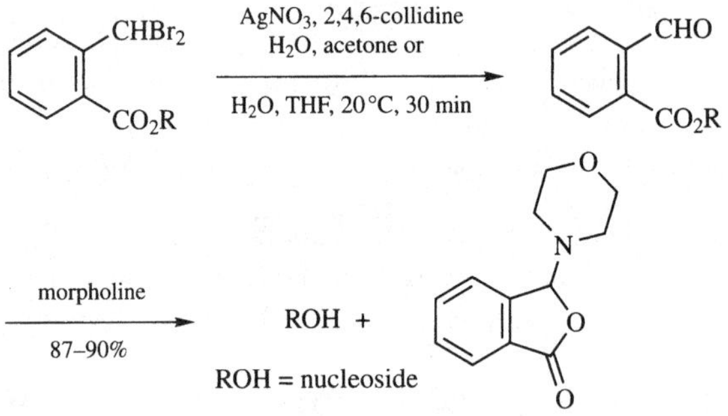

2-Formylbenzenesulfonate Ester

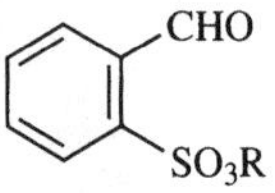

This sulfonate is prepared by reaction with the sulfonyl chloride. Cleavage occurs with 0.05 M NaOH (acetone, H_2O, 25°C, 5 min, 83–93% yield). Here also, cleavage is facilitated by intramolecular participation through the hydrate of the aldehyde.[19]

4-(Methylthiomethoxy)butyrate Ester (MTMB−OR): $CH_3SCH_2O(CH_2)_3CO_2R$

Formation

4-$(CH_3SCH_2O)(CH_2)_3CO_2H$, 2,6-dichlorobenzoyl chloride, Pyr, CH_3CN, 70% yield.[20] The MTMB group was selectively introduced onto the 5′-OH of thymidine.

Cleavage

Hg(ClO$_4$)$_2$, THF, H$_2$O, collidine, rt, 5 min; 1 M K$_2$CO$_3$ (10 min) or TEA (30 min).[7] Hg(II) cleaves the MTM group, liberating a hydroxyl group that assists in the cleavage of the ester.

2-(Methylthiomethoxymethyl)benzoate Ester (MTMT−OR): 2-(CH$_3$SCH$_2$OCH$_2$)C$_6$H$_4$CO$_2$R

This group was introduced and removed using the same conditions as the MTMB group. The half-lives for ammonolysis of acetate, MTMB, and MTMT are 5 min, 15 min, and 6 h, respectively.[7]

2-(Chloroacetoxymethyl)benzoate Ester (CAMB−OR)

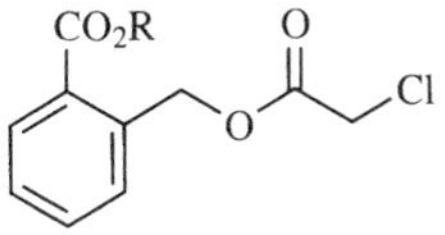

This ester was designed as a protective group for the 2-position in glycosyl donors. It has the stability of the benzoate during glycosylation, but has the ease of removal of the chloroacetate. It is readily introduced through the acid chloride (CH$_2$Cl$_2$, Pyr, 71–88% yield) and is cleaved with thiourea to release the alcohol that closes to the phthalide, releasing the carbohydrate.[21] Its use for nitrogen protection was unsuccessful.

2-[(2-Chloroacetoxy)ethyl]benzoate Ester (CAEB−OR)

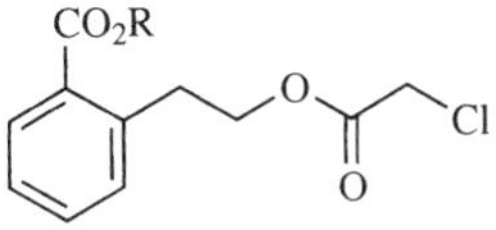

The CAEB group is similar to the CAMB group except that the final deprotection requires acid treatment to initiate ring closure and cleavage.[22] It is introduced through the acid chloride (Pyr, CH$_2$Cl$_2$, 72 h, 61–91% yield) and is cleaved with thiourea (DMF, 55°C, 8–17 h; TsOH, 120 h, 83% yield). This group is reported to be stable to hydrogenolysis.

2-[2-(Benzyloxy)ethyl]benzoate Ester (PAC$_H$−OR) and

2-[2-(4-Methoxybenzyloxy)ethyl]benzoate Ester (PAC$_M$−OR)

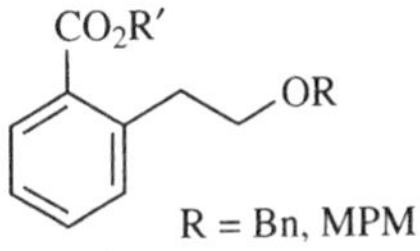

R = Bn, MPM

These groups were designed for use in the synthesis of phosphatidylinositol phosphates where it was desirable to be able to cleave a benzoate without cleaving a glyceryl ester.[23]

Formation

PAC—OH, DCC, CH_2Cl_2, DMAP, rt, ~4 h, 87–100% yield.[23]

Cleavage

1. R = H, H_2, Pd–C, AcOEt then *t*-BuOK or *t*-BuMgCl, 85–96% yield.[23] When $Pd(OH)_2$ is used as the catalyst, base treatment is not required because lactonization occurs spontaneously.[24]

2. R = OMe, DDQ, CH_2Cl_2, H_2O, 0°C or rt and then *t*-BuOK or *t*-BuMgCl, 82–98% yield.[23]

3. R = OMe, $AlCl_3$, $PhNMe_2$, CH_2Cl_2, rt and then *t*-BuOK or *t*-BuMgCl, 88–91% yield.[23]

1. K. R. Love, R. B. Andrade, and P. H. Seeberger, *J. Org. Chem.*, **66**, 8165 (2001).

2. T. Wada, A. Ohkubo, A. Mochizuki, and M. Sekine, *Tetrahedron Lett.*, **42**, 1069 (2001).

3. W. Peng, J. Sun, F. Lin, X. Han, and B. Yu, *Synlett*, 259 (2004).

4. S. Kusumoto, K. Sakai, and T. Shiba, *Bull. Chem. Soc. Jpn.*, **59**, 1296 (1986).

5. S. Velarde, J. Urbina, and M. R. Pena, *J. Org. Chem.*, **61**, 9541 (1996).

6. J. Xu and Z. Guo, *Carbohydr. Res.*, **337**, 87 (2002).

7. K. Seio, E. Utagawa, and M. Sekine, *Helv. Chim. Acta*, **87**, 2318 (2004).

8. E. Arranz and G.-J. Boons, *Tetrahedron Lett.*, **42**, 6469 (2001).

9. J.-M. Vatéle, *Tetrahedron Lett.*, **46**, 2299 (2005).

10. K. Kamaike, H. Takahashi, K. Morohoshi, N. Kataoka, T. Kakinuma, and Y. Ishido, *Acta Biochimica Polonica*, **45**, 949 (1998); K. Kamaike, T. Namiki, Y. Kayama, and E. Kawashima, *Nucleic Acids Research Supplement*, 151 (2002); K. Kamaike, T. Namiki, and E. Kawashima, *Nucleosides, Nucleotides & Nucleic Acids*, **22**, 1011 (2003); K. Kamaike, K. Takahashi, T. Kakinuma, K. Morohoshi, and Y. Ishido, *Tetrahedron Lett.*, **38**, 6857 (1997).

11. M. A. Clark and B. Ganem, *Tetrahedron Lett.*, **41**, 9523 (2000).

12. P. Renton, D. Gala, and G. M. Lee, *Tetrahedron Lett.*, **42**, 7141 (2001).

13. H. Yu, D. L. Williams, and H. E. Ensley, *Tetrahedron Lett.*, **46**, 3417 (2005).

14. M. T. Crimmins, C. A. Carroll, and A. J. Wells, *Tetrahedron Lett.*, **39**, 7005 (1998).

15. R. A. Moss, P. Scrimin, S. Bhattacharya, and S. Chatterjee, *Tetrahedron Lett.*, **28**, 5005 (1987).

16. T.-L. Ho, *Synth. Commun.*, **10**, 469 (1980).

17. J. B. Chattopadhyaya, C. B. Reese, and A. H. Todd, *J. Chem. Soc., Chem. Commun.*, 987 (1979); J. B. Chattopadhyaya and C. B. Reese, *Nucleic Acids Res.*, **8**, 2039 (1980).

18. K. Zegelaar-Jaarsveld, H. I. Duynstee, G. A. van der Marel, and J. H. van Boom, *Tetrahedron*, **52**, 3575 (1996).

19. M. S. Shashidhar and M. V. Bhatt, *J. Chem. Soc., Chem. Commun.*, 654 (1987).

20. J. M. Brown, C. Christodoulou, C. B. Reese, and G. Sindona, *J. Chem. Soc., Perkin Trans. I*, 1785 (1984).

21. T. Ziegler and G. Pantkowski, *Liebigs Ann. Chem.*, 659 (1994).

22. T. Ziegler and G. Pantkowski, *Tetrahedron Lett.*, **36**, 5727 (1995).

23. Y. Watanabe, M. Ishimaru, and S. Ozaki, *Chem. Lett.*, **23**, 2163 (1994).

24. Y. Watanabe and T. Nakamura, *Nat. Prod. Lett.*, **10**, 275 (1997).

Miscellaneous Esters

The following miscellaneous esters have been prepared as protective groups, but they have not been widely used. Therefore, they are simply listed for completeness, rather than described in detail.

1. 2,6-Dichloro-4-methylphenoxyacetate ester[1]
2. 2,6-Dichloro-4-(1,1,3,3-tetramethylbutyl)phenoxyacetate ester[1]
3. 2,4-Bis(1,1-dimethylpropyl)phenoxyacetate ester[1]
4. Chlorodiphenylacetate ester[2]
5. Isobutyrate ester[3] (Chart 2)
6. Monosuccinoate ester[4]
7. (*E*)-2-Methyl-2-butenoate (Tigloate) ester[5]
8. *o*-(Methoxycarbonyl)benzoate ester[6]
9. *p*-***P***-Benzoate ester[7] ***P*** = polymer
10. α-Naphthoate ester[8]
11. Nitrate ester[9] (Chart 2)
12. Alkyl *N,N,N',N'*-tetramethylphosphorodiamidate: $[(CH_3)_2N]_2P(O)OR$[10]
13. 2-Chlorobenzoate ester.[11]

1. C. B. Reese, *Tetrahedron*, **34**, 3143 (1978).

2. A. F. Cook and D. T. Maichuk, *J. Org. Chem.*, **35**, 1940 (1970).

3. H. Büchi and H. G. Khorana, *J. Mol. Biol.*, **72**, 251 (1972).

4. P. L. Julian, C. C. Cochrane, A. Magnani, and W. J. Karpel, *J. Am. Chem. Soc.*, **78**, 3153 (1956).

5. S. M. Kupchan, A. D. J. Balon, and E. Fujita, *J. Org. Chem.*, **27**, 3103 (1962).

6. G. Losse and H. Raue, *Chem. Ber.*, **98**, 1522 (1965).

7. R. D. Guthrie, A. D. Jenkins, and J. Stehlicek, *J. Chem. Soc. C*, 2690 (1971).

8. I. Watanabe, T. Tsuchiya, T. Takase, S. Umezawa, and H. Umezawa, *Bull. Chem. Soc. Jpn.*, **50**, 2369 (1977).

9. J. Honeyman and J. W. W. Morgan, *Adv. Carbohydr. Chem.*, **12**, 117 (1957); J. F. W. Keana, in *Steroid Reactions*, C. Djerassi, Ed., Holden-Day, San Franscisco, 1963, pp. 75–76; R. Boschan, R. T. Merrow, and R. W. Van Dolah, *Chem. Rev.*, **55**, 485 (1955); R. W. Binkley and D. J. Koholic, *J. Org. Chem.*, **44**, 2047 (1979); R. W. Binkley and D. J. Koholic, *J. Carbohydr. Chem.*, **3**, 85 (1984).

10. R. E. Ireland, D. C. Muchmore, and U. Hengartner, *J. Am. Chem. Soc.* **94**, 5098 (1972).

11. E. Rozners, R. Renhofa, M. Petrova, J. Popelis, V. Kumpins, and E. Bizdena, *Nucleosides & Nucleotides*, **11**, 1579 (1992).

Sulfonates, Sulfenates, and Sulfinates as Protective Groups for Alcohols

Sulfonate protective groups have largely been restricted to carbohydrates where they serve to protect the 2-OH with a nonparticipating group so that coupling gives predominately 1,2-*cis* glycosides.

Sulfate: $ROSO_3^-$

Formation[1]/Cleavage[2]

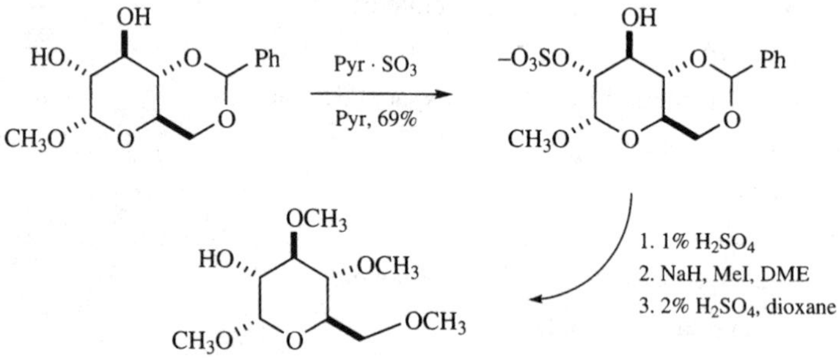

The α-anomer gives better selectivity for the 2-OH than does the β-anomer (3:2). Note that the conditions used to remove the 4,6-*O*-benzylidene group are sufficiently mild to retain the sulfate.[2]

Allylsulfonate (Als−OR): $CH_2=CHCH_2SO_3R$

The allylsulfonate was developed for the protection of carbohydrates.

Formation

Allylsulfonyl chloride, Pyr, CH_2Cl_2, 55–71% yield.[3]

Cleavage

THF, morpholine, 35% aq. formaldehyde, $(Ph_3P)_4Pd$, >85% yield.[3]

Methanesulfonate (Mesylate) (RO−Ms): $MeSO_3R$

Formation

1. MsCl, Et_3N, CH_2Cl_2, 0°C, generally >90% yield.[4]

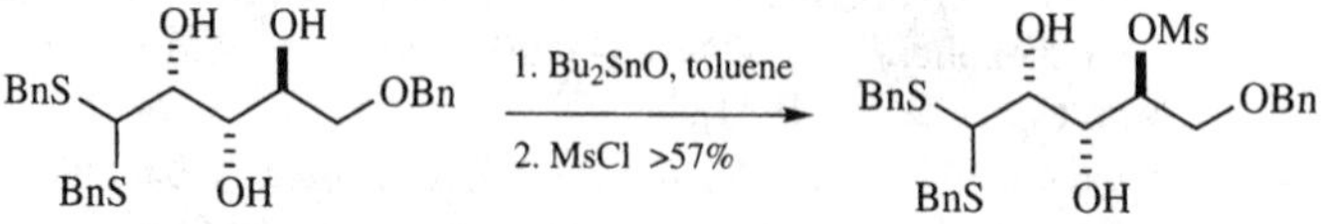

2. MsCl, TEA, Me_3NHCl, toluene, 0–5°C, 1 h, 87–94% yield.[26]

Cleavage

1. Na(Hg), 2-propanol, 84–98% yield.[6] The use of methanol or ethanol gives very slow reactions. Benzyl groups are not affected by these conditions.
2. Photolysis, KI, MeOH.[7] The triflates are also cleaved, but the products are partitioned between cleavage and reduction.[8]
3. MeMgBr, THF, 90% yield.[9,10]
4. MeLi, THF.[11]
5. LiAlH$_4$, THF, 50°C, 15 h.[12]

Benzylsulfonate: ROSO$_2$Bn

Formation

BnSO$_2$Cl, 2,6-lutidine, CH$_2$Cl$_2$, >72% yield.[13]

Cleavage

NaNH$_2$, DMF, 67–95% yield.[3, 14]

Tosylate (Ts−OR): CH$_3$C$_6$H$_4$SO$_3$R

Formation

1. TsCl, Pyr.[15] Some interesting selectivity has been obtained.[16]

2. Ts−N⁺N−Me TfO⁻ This reagent selectively protects a primary alcohol in the presence of a secondary alcohol.[17]
3. Bu$_2$SnO, toluene reflux; TsCl, CHCl$_3$, 36–99% yield. The primary alcohol of a 1,2-diol is selectively tosylated, but when hexamethylene stannylene acetals are used, selectivity is reversed and the secondary diol is preferentially tosylated.[18, 19] This method has been made catalytic in Bu$_2$SnO to rapidly sulfonate the primary alcohol of 1,2-diols and to selectively monotosylate internal 1,2-diols.[20] A fluorous version of this process has been developed which allows for the simple recycling of the tin species.[21]

4. TsCl, DABCO, CH$_2$Cl$_2$, MTBE or AcOEt, 45–97% yield. In many cases these conditions were found to be superior to the use of pyridine as a base. DABCO is also less toxic than pyridine, which may prove useful in a commercial setting.[22]

5. TsCl, Me$_2$N(CH$_2$)$_n$NMe$_2$, n = 3 and 6, TEA, toluene or CH$_3$CN, 0–5°C, 87–95% yield. Attempts at using TMEDA result in the formation of TsNMe$_2$.[23] Almost no chloride formation is observed under these conditions.

6. TsCl, Ag$_2$O, cat. KI, CH$_2$Cl$_2$, 40°C, 60–99% yield. Nosylates and Mesylates can also be formed by this method.[24] In some cases this method gives results that are complementary to the stannylene method. Selectivity is also dependent upon the substituent at the anomeric position of a pyranoside, but not the configuration.[25] Acetates and benzoates give similar results.

7. TsCl, TEA, Me$_3$NHCl, toluene, 80–97% yield. With this method allylic and propargylic alcohols can be tosylated without chloride formation.[26]

8. Ts$_2$O, Yb(OTf)$_3$, CH$_2$Cl$_2$, rt, 10 min to 24 h, 76–89% yield.[27] With this method the conversion of a tosylate to the chloride is avoided.

9. TsOH, ZrCl$_4$, CH$_2$Cl$_2$, reflux, 6–14 h, 51–95% yield. Tertiary alcohols fail to form tosylates.[28] CoCl$_2$·2H$_2$O (26–95% yield)[29] and silica chloride (0–95% yield)[30] have also been used successfully as catalysts.

Cleavage

1. $h\nu$, 90% CH$_3$CN/H$_2$O, 1,5-dimethoxynaphthalene, NH$_2$NH$_2$ or NaBH$_4$ or Pyr·BH$_3$, 59–97% yield.[31]

2. $h\nu$, Et$_3$N, MeOH, 12 h, 91% yield.[32]

3. The tosyl group has also been removed by reductive cleavage with Na/NH$_3$ (65–73% yield),[33] Na/naphthalene (50–87% yield),[34] and Na(Hg)/MeOH (96.7% yield).[35]

4. TiCl$_3$, Li, THF, rt, 18 h, 43–76% yield.[36]

5. NaBH$_4$, DMSO, 140°C, 71% yield.[37]

6. LiAlH$_4$, ether.[38]

7. Mg, MeOH, 4–6 h, 80–95% yield.[39] Phenolic tosylates are also cleaved efficiently.

8. KF·Al$_2$O$_3$, Microwave, 85–90% yield. This method uses no solvent and is likely to be difficult to scale.[40] Phenolic tosylates and sulfonamides are also cleaved.

9. NaOMe, MeOH, reflux, 12 h, 99% yield. This reaction is successful because the sulfonates can not eliminate to form olefins and displacement is hindered by the axial substituents.[41]

2-[(4-Nitrophenyl)ethyl]sulfonate (Npes−OR): $4\text{-}NO_2C_6H_4CH_2CH_2SO_3R$

Formation

$NO_2C_6H_4CH_2CH_2SO_2Cl$, Pyr, 70–90% yield.[42]

Cleavage

0.1 M DBU, CH_3CN, 2 h.[43] The Npes group is more labile to base than the Npeoc and Npe groups. It is not very rapidly removed by fluoride ions. K_2CO_3, MeOH can be used for acetate cleavage in the presence of a Npes ester.[44]

2-Trifluoromethylbenzenesulfonate: $2\text{-}CF_3C_6H_4SO_2\text{-}OR$

This group was developed to improve the β-selectivity in the glycosylation of rhamnose and mannose thioglycosides. It is prepared from the sulfonyl chloride and cleaved using Na(Hg) in isopropanol (61–80% yield).[45]

4-Monomethoxytritylsulfenate (MMTrS−OR): $4\text{-}CH_3OC_6H_4(C_6H_5)_2CS\text{-}OR$

This group was developed for 5′ protection in acid free oligonucleotide synthesis. It is introduced by the reaction of the sulfenyl chloride with the lithium anion generated from LiHMDS in THF at rt. It is cleaved with I_2/CH_3CN–pyridine–H_2O conditions that simultaneously oxidize phosphite to phosphate. Unlike the 2,4-dinitrobenzenesulfenyl group, it is completely compatible with tervalent phosphorous.[46]

1. A. Liav and M. B. Goren, *Carbohydr. Res.*, **131**, C8 (1984).

2. M. B. Goren, and M. E. Kochansky, *J. Org. Chem.*, **38**, 3510 (1973); A. Liav and M. B. Goren, *Carbohydr. Res.*, **127**, 211 (1984).

3. W. K. D. Brill and H. Kunz, *Synlett*, 163 (1991).

4. A. Fürst and F. Koller, *Helv. Chim. Acta*, **30**, 1454 (1947).

5. M. W. Bredenkamp, C. W. Holzapfel, and A. D. Swanepoel, *Tetrahedron Lett.*, **31**, 2759 (1990).

6. K. T. Webster, R. Eby, and C. Schuerch, *Carbohydr. Res.*, **123**, 335 (1983).

7. R. W. Binkley and X. Liu, *J. Carbohydr. Chem.*, **11**, 183 (1992).

8. X. G. Liu, R. W. Binkley, and P. Yeh, *J. Carbohydr. Chem.*, **11**, 1053 (1992).

9. M. E. Jung and D. Sun, *Tetrahedron Lett.*, **40**, 8343 (1999).

10. J. Cossy, J.-L. Ranaivosata, V. Bellosta, and R. Wietzke, *Synth. Commun.*, **25**, 3109 (1995).

11. M. E. Jung and C. P. Lee, *Org. Lett.*, **3**, 333 (2001). L. Qiao, Y. Hu, F. Nan, G. Powis, and A. P. Kozikowski, *Org. Lett.*, **2**, 115 (2000).

12. E. Bozó, S. Boros, J. Kuszmann, and E. Gács-Baitz, *Tetrahedron*, **55**, 8095 (1999).

13. L. F. Awad, El S. H. Ashry, and C. Schuerch, *Bull. Chem. Soc. Jpn.*, **59**, 1587 (1986).

14. A. A.-H. Abdel-Rahman, S. Jonke, E. S. H. El Ashry, and R. R. Schmidt, *Angew. Chem. Int. Ed.*, **41**, 2972 (2002).

15. L. F. Fieser and M. Fieser, *Reagents for Organic Synthesis*, Vol. **1**, Wiley, New York, 1967, p. 1179.

16. K. M. Sureshan, M. S. Shashidhar, T. Praveen, R. G. Gonnade, and M. M. Bhadbhade, *Carbohydr. Res.*, **337**, 2399 (2002).

17. M. Gerspacher and H. Rapoport, *J. Org. Chem.*, **56**, 3700 (1991).

18. X. Kong and T. B. Grindley, *Can. J. Chem.*, **72**, 2396 (1994).

19. Y. Tsuda, M. Nishimura, T. Kobayashi, Y. Sato, and K. Kanemitsu, *Chem. Pharm. Bull. Tokyo*, **39**, 2883 (1991).

20. M. J. Martinelli, N. K. Hayyar, E. D. Moher, U. P. Dhokte, J. M. Pawlak, and R. Vaidyanathan, *Org. Lett.*, **1**, 447 (1999); M. J. Martinelli, R. Vaidyanathan, and V. V. Khau, *Tetrahedron Lett.*, **41**, 3773 (2000); M. J. Martinelli, R. Vaidyanathan, J. M. Pawlak, N. K. Nayyar, U. P. Dhokte, C. W. Doecke, L. M. H. Zollars, E. D. Moher, V. V. Khau, and B. Kosmrlj, *J. Am. Chem. Soc.*, **124**, 3578 (2002).

21. B. Bucher and D. P. Curran, *Tetrahedron Lett.*, **41**, 9617 (2000).

22. J. Hartung, S. Hünig, R. Kneuer, M. Schwaz, and H. Wenner, *Synthesis*, 1433 (1997).

23. Y. Yoshida, K. Shimonishi, Y. Sakakura, S. Okada, N. Aso, and Y. Tanabe, *Synthesis*, 1633 (1999).

24. A. Bouzide, N. LeBerre, and G. Sauve, *Tetrahedron Lett.*, **42**, 8781 (2001).

25. H. Wang, J. She, L.-H. Zhang, and X.-S. Ye, *J. Org. Chem.*, **69**, 5774 (2004). A. Bouzide and G. Sauve, *Org. Lett.*, **4**, 2329 (2002).

26. Y. Yoshida, Y. Sakakura, N. Aso, S. Okada, and Y. Tanabe, *Tetrahedron*, **55**, 2183 (1999).

27. S. Comagic and R. Schirrmacher, *Synthesis*, 885 (2004).

28. B. Das and V. S. Reddy, *Chem. Lett.*, **33**, 1428 (2004).

29. S. Velusamy, J. S. K. Kumar, and T. Punniyamurthy, *Tetrahedron Lett.*, **45**, 203 (2004).

30. B. Das, V. S. Reddy, and M. R. Reddy, *Tetrahedron Lett.*, **45**, 6717 (2004).

31. A. Nishida, T. Hamada, and O. Yonemitsu, *J. Org. Chem.*, **53**, 3386 (1988); *idem.*, *Chem. Pharm. Bull.*, **38**, 2977 (1990).

32. R. W. Binkley and D. J. Koholic, *J. Org. Chem.*, **54**, 3577 (1989).

33. M. A. Miljkovic, M. Pesic, A. Jokic, and E. A. Davidson, *Carbohydr. Res.*, **15**, 162 (1970); J. Kovar, *Can. J. Chem.*, **48**, 2383 (1970).

34. H. C. Jarrell, R. G. S. Ritchie, W. A. Szarek, and J. K. N. Jones, *Can. J. Chem.*, **51**, 1767 (1973). E. Lewandowska, V. Neschadimenko, S. F. Wnuk, and M. J. Robins, *Tetrahedron*, **53**, 6295 (1997).

35. R. S. Tipson, *Methods Carbohydr. Chem.*, **II**, 250 (1963).

36. S. K. Nayak, *Synthesis*, 1575 (2000).
37. V. Pozsgay, E. P. Dubois, and L. Pannell, *J. Org. Chem.*, **62**, 2832 (1997).
38. H. B. Borén, G. Ekborg, and J. Lönngren, *Acta. Chem. Scand., Ser B*, **B29**, 1085 (1975).
39. M. Sridhar, B. A. Kumar, and R. Narender, *Tetrahedron Lett.*, **39**, 2847 (1998).
40. G. Sabitha, S. Abraham, B. V. S. Reddy, and J. S. Yadav, *Synlett*, 1745 (1999).
41. M. P. Sarmah, M. S. Shashidhar, K. M. Sureshan, R. G.Gonnade, and M. M. Bhadbhade, *Tetrahedron*, **61**, 4437 (2005).
42. M. Pfister, H. Schirmeister, M. Mohr, S. Farkas, K.-P. Stengele, T. Reiner, M. Dunkel, S. Gokhale, R. Charubala, and W. Pfleiderer, *Helv. Chim. Acta*, **78**, 1705 (1995).
43. H. Schirmeister and W. Pfleiderer, *Helv. Chim. Acta*, **77**, 10 (1994); R. Charubala, W. Pfleiderer, R. W. Sobol, S. W. Li, and R. J. Suhadolnik, *Helv. Chim. Acta*, **72**, 1354 (1989).
44. C. Hörndler and W. Pfleiderer, *Helv. Chim. Acta*, **79**, 798 (1996).
45. D. Crich and J. Picione, *Org. Lett.*, **5**, 781 (2003).
46. K. Seio and M. Sekine, *Tetrahedron Lett.*, **42**, 8657 (2001).

Alkyl 2,4-Dinitrophenylsulfenate: $ROSC_6H_3$-2,4-$(NO_2)_2$ (Chart 2)

A nitrophenylsulfenate, cleaved by nucleophiles under very mild conditions, was developed as protection for an hydroxyl group during solid-phase nucleotide synthesis.[1] The sulfenate ester is stable to the acidic hydrolysis of acetonides.[2]

Formation

1. 2,4-$(NO_2)_2C_6H_3SCl$, Pyr, DMF or CH_2Cl_2, 20°C, 1 h, 70–85% yield.[1]

Cleavage

1. Nu^-, MeOH, H_2O, 25°C, 4 h, 63–80% yield.[1]
2. $Nu^- = Na_2S_2O_3$, pH 8.9; NaCN, pH 8.9; Na_2S, pH 6.6; PhSH, pH 11.8.[1]
3. H_2, Raney Ni, 54% yield.[1]
4. Al, Hg(OAc)$_2$, MeOH, 5 h, 67% yield.[2]
5. An *o*-nitrophenylsulfenate is cleaved by electrolytic reduction (−1.0 V, DMF, R_4NX).[4]
6. PhSH, Pyr, THF, 83% yield.[3]
7. Photolysis, >280 nm, Et_3N, CH_2Cl_2. Cleavage is believed to occur by an electron transfer from TEA to the sulfenate.[5]

1. R. L. Letsinger, J. Fontaine, V. Mahadevan, D. A. Schexnayder, and R. E. Leone, *J. Org. Chem.*, **29**, 2615 (1964).

2. K. Takiura, S. Honda, and T. Endo, *Carbohydr. Res.* **21**, 301 (1972).

3. P. Magnus, G. F. Miknis, N. J. Press, D. Grandjean, G. M. Taylor, and J. Harling, *J. Am. Chem. Soc.*, **119**, 6739 (1997).

4. V. G. Mairanovsky, *Angew. Chem., Int. Ed. Engl.*, **15**, 281 (1976).

5. K. Wakamatsu, M. Kouda, K. Shimaoka, and H. Yamada, *Tetrahedron Lett.*, **45**, 6395 (2004).

2,2,5,5-Tetramethylpyrrolidin-3-one-1-sulfinate

This group was developed for 5′-hydroxyl protection in oligonucleotide synthesis. It is stable to the conditions for nucleotide coupling using the phosphoramidite approach. It is not stable to acid or to I_2/pyridine/THF, conditions used for phosphite oxidation. It has been used to prepare a 20-mer.[1]

1. V. Marchan, J. Cieslak, V. Livengood, and S. L. Beaucage, *J. Am. Chem. Soc.*, **126**, 9601 (2004).

Borate Ester: $(RO)_3B$

Formation

1. $BH_3 \cdot Me_2S$, 25°C, 1 h, 80–90% yield.[1]
2. $B(OH)_3$, benzene, $-H_2O$, 100% yield.[2,3]

Cleavage

Simple borate esters are readily hydrolyzed with aqueous acid or base. More sterically hindered borates such as pinanediol derivatives are quite stable to hydrolysis.[4] Some hindered borates are stable to anhydrous acid and base, to HBr/BzOOBz, to NaH, and Wittig reactions.[3]

1. C. A. Brown and S. Krishnamurthy, *J. Org. Chem.*, **43**, 2731 (1978).
2. W. I. Fanta and W. F. Erman, *J. Org. Chem.*, **37**, 1624 (1972).

3. W. I. Fanta and W. F. Erman, *Tetrahedron Lett.*, **10**, 4155 (1969).

4. D. S. Matteson and R. Ray, *J. Am. Chem. Soc.*, **102**, 7590 (1980).

Dimethylphosphinothioyl Ester: $(CH_3)_2P(S)OR$

The dimethylphosphinothioyl group has been used to protect hydroxyl groups in carbohydrates. It is prepared from the alcohol and $Me_2P(S)Cl$ (cat. DMAP, DBU). It is not prone to undergo "acyl" migration as are carboxylate esters. It is stable to the acidic conditions used to cleave acetonides and trityl groups, to DBU/MeOH, Bu_4NF, Bu_3SnH, Grignard reagents and cat. NaOMe/MeOH. The dimethylphosphinothioyl group is cleaved with $BnMe_3NOH$. It can also be cleaved by Bu_4NF after conversion to the dimethylphosphonyl group with *m*-chloroperoxybenzoic acid.[1]

1. T. Inazu and T. Yamanoi, *Noguchi Kenkyusho Jiho*, 43 (1988); *Chem. Abstr.*, **111**, 7685w (1989).

Carbonates

Carbonates, like esters, can be cleaved by basic hydrolysis, but generally are much less susceptible to hydrolysis because of the resonance effect of the second oxygen. In general, carbonates are cleaved by taking advantage of the properties of the second alkyl substituent (e.g., zinc reduction of the 2,2,2-trichloroethyl carbonate). The reagents used to introduce the carbonate onto alcohols react readily with amines as well. As expected, basic hydrolysis of the resulting carbamate is considerably more difficult than basic hydrolysis of a carbonate.

Alkyl Methyl Carbonate: $ROCO_2CH_3$ (Chart 2)

Carbonates are not always the innocent bystander and can function as leaving groups under some conditions.[1]

Formation

1.

2. $(CH_3)_2C=NOCO_2CH_3$, CAL, dioxane, 60°C, 3 d, 45% yield. Only a primary alcohol is protected.[3]

3. $BtOCO_2CH_3$, Pyr, DMAP, rt, 70–99% yield. This reagent proved effective for hindered alcohols where methyl chloroformate failed. Severely hindered alcohols such as the 13-hydroxyl of Baccatin III fail to react.[4]

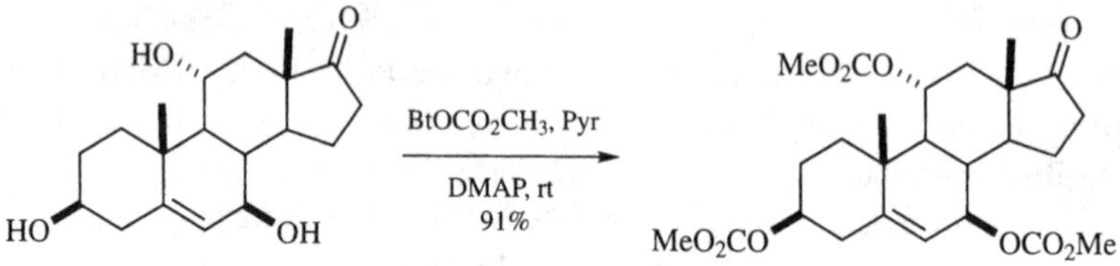

Cleavage

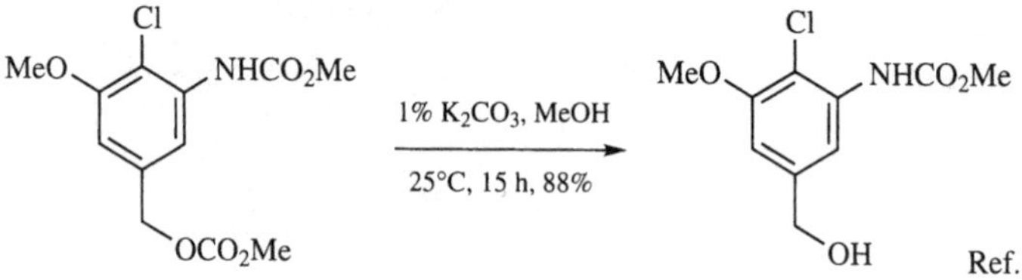

Ref. 2

1. M. D. Ganton and M. A. Kerr, *Org. Lett.*, **7**, 4777 (2005).

2. A. I. Meyers, K. Tomioka, D. M. Roland, and D. Comins, *Tetrahedron Lett.*, **19**, 1375 (1978).

3. R. Pulido and V Gator, *J. Chem. Soc., Perkin Trans. I*, 589 (1993).

4. P. G. M. Wuts, S. W. Ashford, A. M. Anderson, and J. R. Atkins, *Org. Lett.*, **5**, 1483 (2003).

Methoxymethyl Carbonate: $CH_3OCH_2OCO_2R$

Formation

1. K_2CO_3, $ClCH_2OMe$, DMF, −20°C, 28–95% yield.[1]
2. $AgCO_3$, $ClCH_2OMe$, DMF, −15°C, 15–67% yield.[2]

Cleavage

1. K_2CO_3, MeOH, H_2O, 30 min, 20°C, 19–93% yield.[2]
2. TFA, MeOH, 30 h, 20°C, 79–93% yield.[1, 2]

1. K. Teranishi, A. Komoda, M. Hisamatsu, and T. Yamada, *Bull. Chem. Soc. Jpn.*, **68**, 309 (1995).

2. K. Teranishi, H. Nakao, A. Komoda, M. Hisamatsu, and T. Yamada, *Synthesis*, 176 (1995).

Alkyl 9-Fluorenylmethyl Carbonate (Fmoc−OR)

Formation

1. FmocCl, Pyr, 20°C, 40 min, 81–96% yield.[1] TMEDA is a very effective base for this transformation.[2]

2.

Cleavage

Et$_3$N, Pyr, 2 h, 83–96% yield (half life = 20 min).[1]

1. C. Gioeli and J. B. Chattopadhyaya, *J. Chem. Soc., Chem. Commun.*, 672 (1982).
2. M. Adinolfi, G. Barone, L. Guariniello, and A. Iadonisi, *Tetrahedron Lett.*, **41**, 9305 (2000).
3. K. Takeda, K. Tsuboyama, M. Hoshino, M. Kishino, and H. Ogura, *Synthesis*, 557 (1987).

Alkyl Ethyl Carbonate: ROCO$_2$Et

An ethyl carbonate, prepared and cleaved by conditions similar to those described for a methyl carbonate, was used to protect a hydroxyl group in glucose.[1] Ethyl chloroformate in pyridine or CH$_2$Cl$_2$/TEA is the most common method of preparation for this carbonate. The carbonate may be prepared by exchange with diethyl carbonate in the presence of a MgLa mixed oxide catalyst.[2] The carbonates of 2-hydroxycarboxylic acids may also be prepared by the reaction of 2-ethoxy-1-(ethoxycarbonyl)-1,2-dihydroquinoline (EEDQ).[3] These carbonates can also be cleaved enzymatically with Lipase B from *Candida antarctica* (phosphate buffer, pH 7, 30–60°C).[4]

1. F. Reber and T. Reichstein, *Helv. Chim. Acta*, **28**, 1164 (1945).
2. B. Veldurthy and F. Figueras, *Chem. Commun.*, 734 (2004).
3. M. H. Hyun, M. H. Kang, and S. C. Han, *Tetrahedron Lett.*, **40**, 3435 (1999).
4. M. Capello, M. Gonzalez, S. D. Rodriguez, L. E. Iglesias, and A. M. Iribarren, *J. Mol. Catal. B: Enzymatic*, **36**, 36 (2005).

Alkyl Bromoethyl Carbonate (BEC−OR): $BrCH_2CH_2OCO_2R$

A bromoethyl carbonate of a primary alcohol was prepared from the chloroformate and DMAP. This group was used in place of the desired Alloc group, so that an oxidative cleavage of an olefin with OsO_4 could be performed. The BEC group was later converted to the desired Alloc group by treatment with allyl alcohol and MeMgBr/THF.[1] It should be possible to cleave this group with Zn/AcOH or other reducing systems.

1. L. D. Julian, J. S. Newcom, and W. R. Roush, *J. Am. Chem. Soc.*, **127**, 6186 (2005).

Alkyl 2-(Methylthiomethoxy)ethyl Carbonate (MTMEC−OR): $CH_3SCH_2OCH_2CH_2OCO_2R$

Formation

$CH_3SCH_2OCH_2CH_2OCOCl$, 1-methylimidazole, CH_3CN, 1 h, >72% yield.[1]

Cleavage

$Hg(ClO_4)_2$, 2,4,6-collidine, acetone, H_2O (9:1), 5 h; NH_3, dioxane, H_2O (1:1).[6] In this case, Hg(II) is used to cleave the MTM group liberating a hydroxyl group, which assists in the cleavage of the carbonate upon treatment with ammonia. Cleavage by ammonia is 500 times faster for this hydroxy derivative than for the initial MTM derivative.

1. S. S. Jones, C. B. Reese, and S. Sibanda, *Tetrahedron Lett.*, **22**, 1933 (1981).

Alkyl 2,2,2-Trichloroethyl Carbonate (Troc−OR): $ROCO_2CH_2CCl_3$ (Chart 2)

Formation

Cl_3CCH_2OCOCl, Pyr, 20°C, 12 h.[1] The trichloroethyl carbonate can be introduced selectively onto a primary alcohol in the presence of a secondary alcohol.[2] DMAP has been used to catalyze this acylation.[3] TMEDA is probably the best amine to use for the formation of carbonates.

Cleavage

1. Zn, AcOH, 20°C, 1–3 h, 80% yield.[1]
2. Zn, MeOH, reflux, short time.[1]
3. Zn–Cu, AcOH, 20°C, 3.5 h, 100% yield.[4] A 2,2,2-tribromoethyl carbonate is cleaved by Zn–Cu/AcOH 10 times faster than trichloroethyl carbonate.
4. Electrolysis, −1.65 V, MeOH, LiClO$_4$, 80% yield.[5]
5. Sm, I$_2$, MeOH, rt, 5 min, 100% yield.[6, 7]
6. In, NH$_4$Cl, H$_2$O, MeOH, 0.5–1.5 h, 82–98% yield.[8]

1. T. B. Windholz and D. B. R. Johnston, *Tetrahedron Lett.*, **8**, 2555 (1967).
2. M. Imoto, N. Kusunose, S. Kusumoto, and T. Shiba, *Tetrahedron Lett.*, **29**, 2227 (1988).
3. S. Hanessian and R. Roy, *Can. J. Chem.*, **63**, 163 (1985).
4. A. F. Cook, *J. Org. Chem.*, **33**, 3589 (1968).
5. M. F. Semmelhack and G. E. Heinsohn, *J. Am. Chem. Soc.*, **94**, 5139 (1972).
6. R. Yanada, N. Negoro, K. Bessho, and K. Yanada, *Synlett*, 1261 (1995).
7. C. B. Lee, T.-C. Chou, X.-G. Zhang, Z.-G. Wang, S. D. Kuduk, M. D. Chappell, S. J. Stachel, and S. J. Danishefsky, *J. Org. Chem.*, **65**, 6525 (2000).
8. M. Valluri, T. Mineno, R. M. Hindupur, and M. A. Avery, *Tetrahedron Lett.*, **42**, 7153 (2001).

1,1-Dimethyl-2,2,2-trichloroethyl Carbonate (TCBOC−OR):
Cl$_3$CC(CH$_3$)$_2$OCO$_2$R

Formation

Cl$_3$CC(CH$_3$)$_2$OCOCl, base, solvent.[1]

Cleavage

(Et$_3$NH)Sn(SPh)$_3$, tetrabutylammonium cobalt(II)phthalocyanine-5,12,19,26-tetrasulfonate, CH$_3$CN, MeOH, 20°C, 1 h, 90% yield.[1]

1. S. Lehnhoff, R. M. Karl, and I. Ugi, *Synthesis*, 309 (1991).

Alkyl 2-(Trimethylsilyl)ethyl Carbonate (TMSEC−OR, Teoc−OR):
Me$_3$SiCH$_2$CH$_2$OCO$_2$R

Formation

1. TMSCH$_2$CH$_2$OCOCl, Pyr, 65–97% yield.[1]
2. TMSCH$_2$CH$_2$OCO-imidazole, DBU, benzene, 54% yield.[2]

Cleavage

1. 0.2 M Bu_4NF, THF, 20°C, 10 min, 87–94% yield.[1]
2. $ZnCl_2$, CH_2Cl_2 or CH_3NO_2, 20°C, 81–90% yield.[1]
3. $ZnBr_2$, CH_2Cl_2 or CH_3NO_2, 20°C, 65–92% yield.[1]

1. C. Gioeli, N. Balgobin, S. Josephson, and J.B. Chattopadhyaya, *Tetrahedron Lett.*, **22**, 969 (1981).
2. W. R. Roush and T. A. Bizzard, *J Org. Chem.*, **49**, 4332 (1984).

2-[Dimethyl(2-naphthylmethyl)silyl]ethyl Carbonate (NSEC-OR)

This group was developed as a UV-active group for carbohydrate synthesis. It is introduced with the chloroformate (DMAP, CH_2Cl_2, rt, 15 h, 59–66% yield). As with the Teoc group, it is cleaved with TBAF, which can be done in the presence of a variety of esters. It can not be cleaved in the presence of the Fmoc group even with AcOH buffered TBAF.[1]

1. S. Bufali, A. Holemann, and P. H. Seeberger, *J. Carbohydr. Chem.*, **24**, 441 (2005).

Alkyl 2-(Phenylsulfonyl)ethyl Carbonate (Psec−OR): $PhSO_2CH_2CH_2OCO_2R$

Formation

$PhSO_2CH_2CH_2OCOCl$, Pyr, 20°C, 74–99% yield.[1]

Cleavage

1. Et_3N, Pyr, 20 h, rt, 85–99% yield.[1]
2. NH_3, dioxane, H_2O (9:1), 7 min.[1]
3. K_2CO_3 (0.04 M) 1 min.[1]
4. 4-Substituted phenylsulfonyl analogs (4-$RC_6H_4SO_2CH_2CH_2OCOR'$) of this protective group have also been prepared and their relative rates of cleavage studied in TEA/Pyr at 20°C).[2]

Cleavage rates for 4-Substituted Psec Derivatives

R	Relative Rate, $T_{1/2}$ (min)
H	180
Me	1140
Cl	60
NO_2	10

1. N. Balgobin, S. Josephson, and J. B. Chattopadhyaya, *Tetrahedron Lett.*, **22**, 3667 (1981).
2. S. Josephson, N. Balgobin, and J. Chattopadhyaya, *Tetrahedron Lett.*, **22**, 4537 (1981).

Alkyl 2-(Triphenylphosphonio)ethyl Carbonate (Peoc−OR):
$Ph_3P^+CH_2CH_2OCO_2R\ Cl^-$

Formation

$Ph_3P^+CH_2CH_2OCOCl\ Cl^-$, Pyr, CH_2Cl_2, 4 h, 0°C, 65–94% yield.[1]

Cleavage

Me_2NH, MeOH, 0°C, 75% yield.[1] *t*-Butyl esters could be cleaved with HCl without affecting the Peoc group.

1. H. Kunz and H.-H. Bechtolsheimer, *Synthesis*, 303 (1982).

Cis-[4-[[(-Methoxytrityl)sulfenyl]oxy]tetraydrofuran-3-yl]oxy Carbonate (MTFOC−OR)

This group was developed as an oxidatively cleavable group for 5′-protection in oligonucleotide synthesis. It is prepared either from the carbonylimidozolide or the 4-nitrophenyl carbonate. Alternatively the alcohol to be protected can be treated with carbonyl diimidazole followed by sulfenyl protected diol. Yields range from 70% to 93%. The MTFOC group is cleaved upon oxidation with I_2, which releases the alcohol that in the presence of pyridine cyclizes to form a carbonate with release of the nucleotide. The oxidation step is fast (<1 min), and the cyclization to form the carbonate has a half-life of 51 min.[1]

1. E. Utagawa, K. Seio, and M. Sekine, *Nucleosides & Nucleotides, and Nucleic Acids*, **24** 927 (2005).

Alkyl Isobutyl Carbonate: $ROCO_2CH_2CH(CH_3)_2$

An isobutyl carbonate was prepared by reaction with isobutyl chloroformate (Pyr, 20°C, 3 days, 73% yield), to protect the 5′-OH group in thymidine. It was cleaved by acidic hydrolysis (80% AcOH, reflux, 15 min, 88% yield).[1]

1. K. K. Ogilvie and R. L. Letsinger, *J. Org. Chem.*, **32**, 2365 (1967).

Alkyl *t*-Butyl Carbonate (BOC): $(CH_3)_3COCO_2R$

Formation

1. BOC_2O, methylimidazole or DMAP, solvent, 0°C. The formation of a BOC carbonate under these conditions is very dependent upon the alcohol. Only acidic alcohols give clean conversion. The usual product from the reaction is a dialkyl carbonate mixed with the desired BOC carbonate.[1] Although there are cases that give the expected products,[2] in this case the cyclic carbonate does not form because of the trans relationship of the two alcohols.

2. BOC-Im, toluene, 60°C. The reagent reacts selectively with primary alcohols, 96–98% yield. 1,2-diols give the cyclic carbonate and 2° alcohols fail to react.[3]
3. BOC_2O, $CeCl_3$, THF, 24 h, 25°C, 94% yield.[4] $V(O)(OTf)_2$ can also be used as a catalyst.[5]

Cleavage

The section on the cleavage of BOC amines should be consulted, since many of those methods should be applicable to the cleavage of the carbonate. TFA, CH_2Cl_2, rt, >73% yield.[2]

1. Y. Basel and A. Hassner, *J. Org. Chem.*, **65**, 6368 (2000).
2. K. Tomooka, M. Kikuchi, K. Igawa, M. Susuki, P.-H. Keong, and T. Nakai, *Angew. Chem. Int. Ed.*, **39**, 4502 (2000).
3. S. P. Rannard and N. J. Davis, *Org. Lett.*, **1**, 933 (1999).
4. R. A. Holton, Z. Zhang, P. A. Clarke, H. Nadizadeh, and D. J. Procter, *Tetrahedron Lett.*, **39**, 2883 (1998).
5. C.-T. Chen, J.-H. Kuo, C.-H. Li, N. B. Barhate, S.-W. Hon, T.-W. Li, S.-D. Chao, C.-C. Liu, Y.-C. Li, I.-H. Chang, J.-S. Lin, C.-J. Liu, and Y.-C. Chou, *Org. Lett.*, **3**, 3729 (2001).

Alkyl Vinyl Carbonate: $ROCO_2CH=CH_2$

Formation

$CH_2=CHOCOCl$, Pyr, CH_2Cl_2, 93% yield.[1]

Cleavage

Na_2CO_3, H_2O, dioxane, warm, 97% yield.[1] Phenols can be protected under similar conditions. Amines are converted by these conditions to carbamates that are stable to alkaline hydrolysis with sodium carbonate. Carbamates are cleaved by acidic hydrolysis (HBr, MeOH, CH_2Cl_2, 8h), conditions that do not cleave alkyl or aryl vinyl carbonates.

1. R. A. Olofson and R. C. Schnur, *Tetrahedron Lett.*, **18**, 1571 (1977).

Alkyl Allyl Carbonate (Alloc−OR): $ROCO_2CH_2CH=CH_2$ (Chart 2)

Formation

1. $CH_2=CHCH_2OCOCl$, Pyr, THF, 0–20°C, 2h, 90% yield.[1]
2. $CH_2=CHCH_2OCOCl$, TMEDA, CH_2Cl_2, 0°C, 20 min, 95% yield. The use of TMEDA greatly improves formation of carbonates from the respective chloroformates. The method was also applied to the preparation of Bn, Fm, and CCl_3CH_2 carbonates, all in excellent yield.[2]
3.

This reaction[3] showed a remarkable selectivity with respect to the solvent and base used. In THF and EtOAc using TEA as the base, a 1:1 mixture of the allylic carbonate and bisacylated products is obtained, but when CH_2Cl_2 is used as solvent the reaction favors the allylic alcohol by a factor of 97:3 (mono/bis). In THF or MTBE, use of TMEDA as the base also results in a 97:3 mono/bis ratio.[3]

4. Diallyl carbonate, $Pd(OAc)_2$, Ph_3P. Conventional methods failed to protect this hindered 12-α-hydroxycholestane derivative.[4] This reaction is unusual in that the carbonate was formed rather than the expected allyl ether.

5. $CH_2=CHCH_2OCO_2N=C(CH_3)_2$, CAL, dioxane, 60°C, 3 days.[5]

6. DMAP, THF, 65% yield. This reaction is selective for pri-

mary alcohols.[6] Benzyl, isobutyl, and ethyl carbonates are also prepared using this method (63–85% yield).

7. Allylbromide, Cs_2CO_3, TBAI, DMF, CO_2, 23°C, 91% yield. This is a general method for the preparation of carbonates.[7]

Cleavage

1. $Ni(CO)_4$, TMEDA, DMF, 55°C, 4 h, 87–95% yield.[1] Because of the toxicity associated with nickelcarbonyl, this method is rarely used and has largely been supplanted by palladium-based reagents.
2. $Pd(Ph_3P)_4$, HCO_2NH_4.[8]
3. $Pd(Ph_3P)_4$, Bu_3SnH, 90–100% yield.[9]
4. $PdCl_2(Ph_3P)_2$, dimedone, 91% yield.[10]
5. $Pd(OAc)_2$, TPPTS, Et_2NH, CH_3CN, H_2O, 51–100% yield. If the reaction is run in a biphasic system using butyronitrile as the solvent, a dimethylallyl carbamate can be retained; however, in a homogeneous system using CH_3CN, both groups are cleaved quantitatively.[11,12]
6. $Pd(dba)_2$, dppe, Et_2NH, THF, 15 min–5 h, 96–100% yield.[13]
7. $Pd(Ph_3P)_4$, $NaBH_4$, ethanol, >88% yield.[3]
8. $Pd(OAc)_2$, TPPTS, Et_2NH, $CH_3CN–H_2O$ or $Et_2O–H_2O$, 94–98% yield.[14]
9. Lithium naphthalenide, THF, 0°C, 1–2 h, 71–99% yield. Cbz carbonates, thio-carbonates, and carbamates are also cleaved under these conditions.[15]

Cinnamyl Carbonate: $PhCH=CHCH_2OCO_2R$

A cinnamyl carbonate is cleaved electrochemically (−2.3 V, Hg, CH_3CN) in preference to the cinnamyl carbamate.[16]

1. E. J. Corey and J. W. Suggs, *J. Org. Chem.*, **38**, 3223 (1973).
2. M. Adinolfi, G. Barone, L. Guariniello, and A. Iadonisi, *Tetrahedron Lett.*, **41**, 9305 (2000).

3. R. J. Cvetovich, D. H. Kelly, L. M. DiMichele, R. F. Shuman, and E. J. J. Grabowski, *J. Org. Chem.*, **59**, 7704 (1994).

4. A. P. Davis, B. J. Dorgan, and E. R. Mageean, *J. Chem. Soc., Chem. Commun.*, 492 (1993).

5. R. Pulido and V. Gotor, *J. Chem. Soc., Perkin Trans. I*, 589 (1993).

6. M. Allainmat, P. L'Haridon, L. Toupet, and D. Plusquellec, *Synthesis*, 27 (1990).

7. S.-I. Kim, F. Chu, E. E. Dueno, and K. W. Jung, *J. Org. Chem.*, **64**, 4578 (1999).

8. Y. Hayakawa, H. Kato, M. Uchiyama, H. Kajino, and R. Noyori, *J. Org. Chem.*, **51**, 2400 (1986).

9. F. Guibe and Y. Saint M'Leux, *Tetrahedron Lett.*, **22**, 3591 (1981).

10. H. X. Zhang, F. Guibé, and G. Balavoine, *Tetrahedron Lett.*, **29**, 623 (1988).

11. S. Lemaire-Audoire, M. Savignac, E. Blart, G. Pourcelot, J. P. Genét, and J. M. Bernard, *Tetrahedron Lett.*, **35**, 8783 (1994).

12. J. P. Genét, E. Blart, M. Savignac, S. Lemeune, S. Lemaire-Audoire, J. M. Paris, and J. M. Bernard, *Tetrahedron*, **50**, 497 (1994).

13. J. P. Genét, E. Blart, M. Savignac, S. Lemeune, S. Lemaire-Audoire, and J.-M. Bernard, Synlett, 680 (1993).

14. J. P. Genét, E. Blart, M. Savignac, S. Lemeune, and J.-L. Paris, *Tetrahedron Lett.*, **34**, 4189 (1993).

15. C. Behloul, D. Guijarro, and M. Yus, *Tetrahedron*, **61**, 9319 (2005).

16. P. Cankar, D. Dubas, S. C. Banfield, M. Chahma, and T. Hudlicky, *Tetrahedron Lett.*, **46**, 6851 (2005).

Propargyl (Poc) Carbonate: $HC\equiv CCH_2OCO_2R$

This group was developed for the protection of carbohydrates. Orthogonality was demonstrated to the following groups: Cbz, Alloc, Lev, acetate, Bn, benzylidene.

Formation/Cleavage[1]

These cleavage conditions can be used to cleave the carbonate in the presence of the Poc carbamate in 78–90% yield.[2]

1. P. R. Sridhar and S. Chandrasekaran, *Org. Lett.*, **4**, 4731 (2002).

2. R. Ramesh, R. G. Bhat, and S. Chandresekaran, *J. Org. Chem.*, **70**, 837 (2005).

Alkyl *p*-Chlorophenyl Carbonate (CPC−OR): $4\text{-}ClC_6H_4OCO_2R$

This group was developed for the protection of carbohydrates and is a participating group during glycosylation. It is prepared from the chloroformate (CH_2Cl_2, pyridine,

DMAP, 85–95% yield). It was shown to be orthogonal to the Bz, Pv, All, and PMB groups. It is cleaved with LiOOH in THF/H_2O at 0°C.[1]

1. K. R. Love and P. H. Seeberger, *Synthesis*, 317 (2001).

Alkyl *p*-Nitrophenyl Carbonate: $ROCOOC_6H_4$–*p*-NO_2 (Chart 2)

Formation/Cleavage[1]

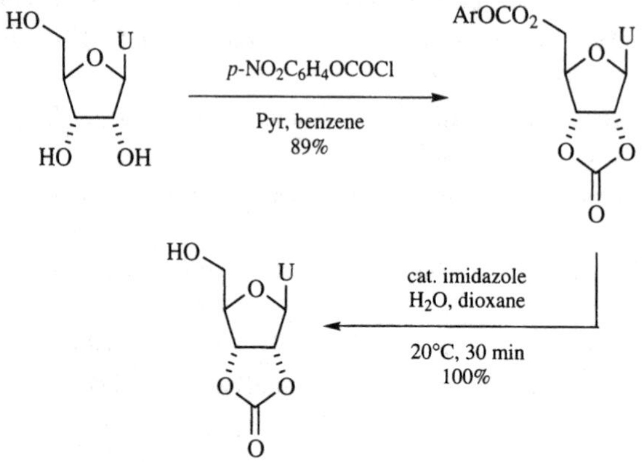

Acetates, benzoates, and cyclic carbonates are stable to these hydrolysis conditions. [Cyclic carbonates are cleaved by more alkaline conditions (e.g., dil. NaOH, 20°C, 5 min, or aq. Pyr, warm, 15 min, 100% yield).][1] The cleavage process can be monitored by the release of the yellow *p*-nitrophenol anion.

1. R. L. Letsinger and K. K. Ogilvie, *J. Org. Chem.*, **32**, 296 (1967).

Alkyl 4-Ethoxy-1-naphthyl Carbonate

Formation/Cleavage[1]

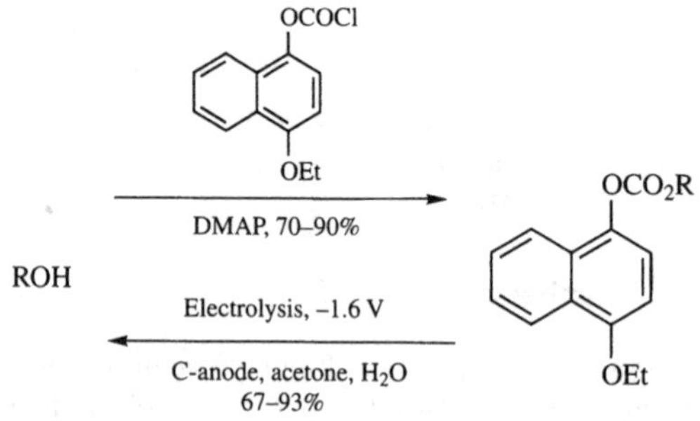

Amines can also be protected by this reagent. Cleavage must be carried out in acidic media to avoid amine oxidation. The by-product naphthoquinone can be removed by extraction with basic hydrosulfite. Ceric ammonium nitrate also serves as an oxidant for deprotection, but the yields are much lower.

1. R. W. Johnson, E. R. Grover, and L. J. MacPherson, *Tetrahedron Lett.*, **22**, 3719 (1981).

Alkyl 6-Bromo-7-hydroxycoumarin-4-ylmethyl Carbonate (Bhcmoc)

The Bhcmoc group was developed as a photochemically removable protective group for caged compounds. Among the series tested this one showed the highest photochemical efficiency in its release of an alcohol.[1]

1. A. Z. Suzuki, T. Watanabe, M. Kawamoto, K. Nishiyama, H. Yamashita, M. Ishii, M. Iwamura, and T. Furuta, *Org. Lett.*, **5**, 4867 (2003).

Alkyl Benzyl Carbonate: $ROCO_2Bn$ (Chart 2)

Formation

1. $BnOCOCl$, CH_2Cl_2, TMEDA, 0°C, 82–91% yield.[1] TMEDA is a superior base to TEA or pyridine. The use of DMAP/DABCO results in selective carbonate formation at the C-2 hydroxyl of a glucose and galactose derivative, whereas the mannose derivative selectively reacts at the C-3 position.[2]
2. $BnOCO_2Bt$, DMF, Pyr, DMAP. The reagent is a stable easily handled solid. This method is good for relatively unhindered carbonates.[3] Its use with hindered alcohols results in disproportionation to give the benzyl ether of HOBt.
3. A benzyl carbonate was prepared in 83% yield from the sodium alkoxide of glycerol and benzyl chloroformate (20°C, 24 h).[4]
4. Lipase catalyzed ester exchange with allyl benzyl carbonate.[5]
5. BnCl, TBAI, CO_2, Cs_2CO_3, DMF, 94–97% yield.[6] The MPM carbonate is prepared by the same method.

Cleavage

1. Hydrogenolysis: H_2/Pd-C, EtOH, 20°C, 2 h, 2 atm, 76% yield.[1] Good selectivity can be obtained in the presence of a phenyl aminal and a nitrile.[7]

2. Transfer hydrogenation: cyclohexadiene, 10% Pd/C, DMF, 90 min, 99% yield. This method was developed for deprotection of nucleoside derivatives because conventional hydrogenolysis often results in over reduction of the nucleobase.[8]

3. Electrolytic reduction: -2.7 V, R_4NX, DMF, 70% yield.[9]

4. As with most other carbonates, cleavage with aqueous base is also an option, but confers little advantage because esters are also hydrolyzed. The only advantage may be that they are more resistant to hydrolysis than are typical esters.

5. Ceric ammonium nitrate, TBAF, TFA, HBr, and HCl have been reported to cleave Cbz-protected carbohydrates, but no details were provided.[10]

6. $NaBrO_3$, $Na_2S_2O_4$, EtOAc, H_2O, 95% yield. A sterically hindered benzyl carbonate was not cleaved and benzyl ethers are cleaved much more readily.[11]

1. M. Adinolfi, G. Barone, L. Guariniello, and A. Iadonisi, *Tetrahedron Lett.*, **41**, 9305 (2000).

2. A. Morère, F. Mouffouk, A. Jeanjean, A. Leydet, and J.-L. Montero, *Carbohydr. Res.*, **338**, 2409 (2003).

3. P. G. M. Wuts, unpublished results.

4. B. F. Daubert and C. G. King, *J. Am. Chem. Soc.*, **61**, 3328 (1939).

5. M. Pozo, R. Pulido, and V. Gotor, *Tetrahedron*, **48**, 6477 (1992).

6. R. N. Salvatore, F. Chu, A. S. Nagle, E. A. Kapxhiu, R. M. Cross, and K. W. Jung, *Tetrahedron*, **58**, 3329 (2002).

7. N. Langlois and B. K. L. Nguyen, *J. Org. Chem.*, **69**, 7558 (2004).

8. D. C. Johnson, II and T. S. Widlanski, *Org. Lett.*, **6**, 4643 (2004).

9. V. G. Mairanovsky, *Angew. Chem., Inter. Ed. Engl.*, **15**, 281 (1976).

10. F. Mouffouk, A. Morere, S. Vidal, A. Leydet, and J.-L. Montero, *Synth. Commum.*, **34**, 303 (2004).

11. M. Adinolfi, L. Guariniello, A. Iadonisi, and L. Mangoni, *Synlett*, 1277 (2000).

Alkyl *o*-Nitrobenzyl Carbonate: $ROCO_2CH_2C_6H_4-o\text{-}NO_2$

Alkyl *p*-Nitrobenzyl Carbonate: $ROCO_2CH_2C_6H_4-p\text{-}NO_2$ (Chart 2)

The nitrobenzyl carbonates were prepared to protect a secondary hydroxyl group in a thienamycin precursor. The *o*-nitrobenzyl carbonate was prepared from the chloroformate (DMAP, CH_2Cl_2, 0–20°C, 3 h) and cleaved by photolysis, pH 7.[1] Cleavage occurs by an internal redox process to liberate 2-nitrosobenzaldehyde. The *p*-nitrobenzyl carbonate was prepared from the chloroformate (-78°C, *n*-BuLi, THF, 85% yield) and cleaved by hydrogenolysis (H_2/Pd–C, dioxane, H_2O, EtOH, K_2HPO_4)[2] or by electrolytic reduction.[3]

1. L. D. Cama and B. G. Christensen, *J. Am. Chem. Soc.*, **100**, 8006 (1978).
2. D. B. R. Johnston, S. M. Schmitt, F. A. Bouffard, and B. G. Christensen, *J. Am. Chem. Soc.*, **100**, 313 (1978).
3. V. G. Mairanovsky, *Angew. Chem., Inter. Ed., Engl.*, **15**, 281 (1976).

Alkyl *p*-Methoxybenzyl Carbonate: p-$MeOC_6H_4CH_2OCO_2R$

Alkyl 3,4-Dimethoxybenzyl Carbonate: 3,4-$(MeO)_2C_6H_3CH_2OCO_2R$

These carbonates are formed from the chloroformates but can also be formed from the alcohol from CO_2 (Cs_2CO_3, benzyl halide, TBAI, DMF, 3h, 92–94% yield).[1] These groups are readily cleaved with Ph_3CBF_4, 0°C, 6min, 90% yield; 0°C, 15min, 90% yield. It should also be possible to cleave these carbonates with DDQ like the corresponding methoxy- and dimethoxyphenylmethyl ethers, although the reactions are expected to be slower because of the reduced electron density imparted by the carbonyl group.[2] These carbonates are expected to be susceptible to strong acids.

1. S.-I. Kim, F. Chu, E. E. Dueno, and K. W. Jung, *J. Org. Chem.*, **64**, 4578 (1999).
2. D. H. R. Barton, P. D. Magnus, G. Smith, G. Streckert, and D. Zurr, *J. Chem. Soc., Perkin Trans. I*, 542 (1972).

Alkyl Anthraquinon-2-ylmethyl Carbonate (Aqmoc−OR)

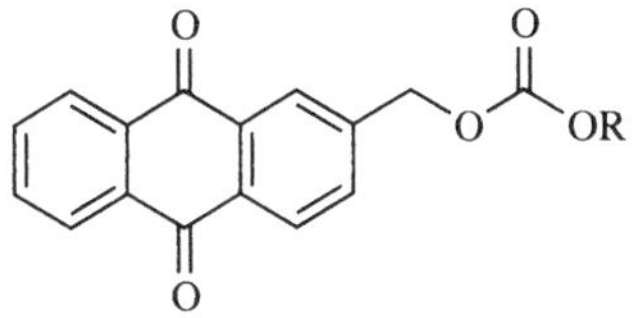

The anthraquinon-2-ylmethyl carbonate is prepared by reaction of anthraquinon-2-ylmethanol with the 4-nitrophenylcarbonate of the alcohol to be derivatized. It is cleaved by photolysis at 350nm in THF/H_2O with a quantum yield of 0.10 and a rate constant of $10^6\ s^{-1}$ in 91% yield for adenosine.[1]

1. T. Furuta, Y. Hirayama, and M. Iwamura, *Org. Lett.*, **3**, 1809 (2001).

Alkyl 2-Dansylethyl Carbonate (Dnseoc−OR)

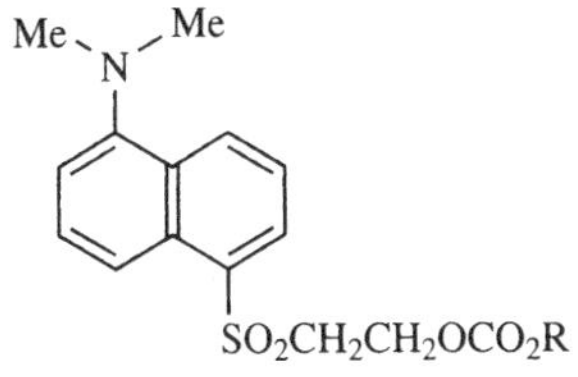

Formation

When the Dnseoc group is used in nucleoside synthesis, the coupling yields are determined by measuring the absorbance at 350 nm of each eluate from the Dnseoc-deprotection steps containing the 5-(dimethylamino)naphthalene-1yl-vinyl sulfone or by measuring the fluorescence at 530 nm. [1]

Cleavage

DBU, CH_3CN, 140 s.[2] The 2-(4-nitrophenyl)ethyl (Npe) phosphate protective group and the 2-(4-nitrophenyl)ethoxycarbonyl (Npeoc) group are stable to these conditions, but the cyanoethyl group is not.

Alkyl 2-(4-Nitrophenyl)ethyl Carbonate (Npeoc−OR):
$4-NO_2C_6H_4CH_2CH_2OCO_2R$

The incorporation of the additional methylene unit serves to substantially increase the rate of photochemical deprotection vs *o*-nitrobenzyl carbonate. Introduction of an additional methyl group in the α-position further increase the rate of deprotection.[3]

Formation

1. $4-NO_2C_6H_4CH_2CH_2OCOCl$, Pyr, CH_2Cl_2, −10°C, 3 h, >70% yield.[4]
2. 3-Methyl-1-[2-(4-nitrophenyl)ethoxycarbonyl]-1*H*-imidazol-3-ium chloride, CH_2Cl_2, DMAP, rt, 100% yield.[4]

Cleavage

1. 0.5 *M* DBU in dry pyridine.[4]
2. K_2CO_3, MeOH, 69–75% yield.[5]

Alkyl 2-(2,4-Dinitrophenyl)ethyl Carbonate (Dnpeoc−OR):
$2,4-NO_2C_6H_3CH_2CH_2OCO_2R$

Formation

$2,4-NO_2C_6H_3CH_2CH_2OCOCl$, Pyr, CH_2Cl_2, −10°C, 3 h, >75% yield.[4]

Cleavage

TEA, MeOH, dioxane.[4]

Alkyl 2-(2-Nitrophenyl)propyl carbonate (NPPOC−OR)

Alkyl 2-(3,4-Methylenedioxy-6-nitrophenylpropyl carbonate (MNPPOC−OR)

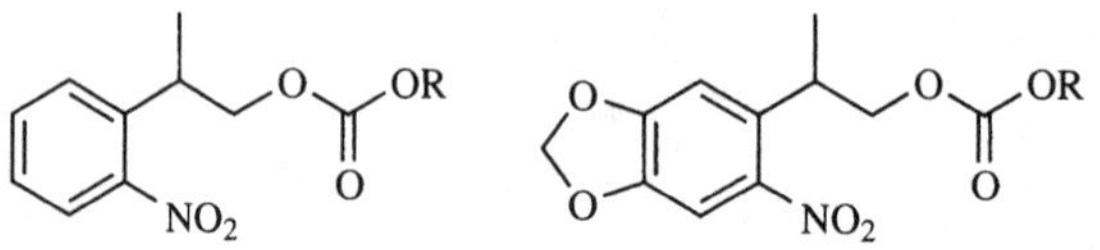

These groups were developed for automated DNA synthesis.[6-8] They are introduced with the acid chloride (0°C to rt, pyridine, 88–92% yield). Cleavage is affected by photolysis at 365 nm, in MeOH/H_2O in 95–99% yield and proceeds by a β-elimination mechanism in contrast to the 2-nitrobenzyl carbonate which is cleaved by an internal redox process.[9,10] Pfleiderer has done an exhaustive substituent effect study on the 2-(2-nitrophenyl)propyl template and has shown that addition of a phenyl group at the 4-position gives improved cleavage rates and purities during deprotection of the 5′-thymidine derivative.[11] Deprotection can be accelerated a factor of 3 by using a sensitizer such as 9H-thioxanthen-9-one.[12] Alternatively, the following derivative was developed having a built-in triplet sensitizer to improve the absorption coefficient at 366 nm in the presence of oxygen[13]:

Olefinic and saturated versions were also prepared.

Alkyl 2-Cyano-1-phenylethyl Carbonate (Cpeoc−OR): NCCH₂CH(C₆H₅)OCO₂R

This group was developed as a 5′-protective group in nucleoside synthesis that is compatible with the 2-(4-nitrophenyl)ethyl (npe) and 2-(4-nitrophenyl)ethoxycarbonyl (npeoc) groups. It is introduced using the chloroformate (3–83% yield) and is rapidly cleaved with 0.1 M DBU in CH₃CN with half-lives of 7–14 s.[14]

1. F. Bergmann and W. Pfleiderer, *Helv. Chim. Acta*, **77**, 203 (1994).

2. F. Bergmann and W. Pfleiderer, *Helv. Chim. Acta*, **77**, 988 (1994).

3. A. Hasan, K.-P. Stengele, H. Giegrich, P. Cornwell, K. R. Isham, R. A. Sachleben, W. Pfleiderer, and R. S. Foote, *Tetrahedron*, **53**, 4247 (1997).

4. H. Schirmeister, F. Himmelsbach, and W. Pfleiderer, *Helv. Chim. Acta*, **76**, 385 (1993).

5. M. Wasner, R. J. Suhadolnik, S. E. Horvath, M. E. Adelson, N. Kon, M.-X. Guan, E. E. Henderson, and W. Pfleiderer, *Helv. Chim. Acta*, **79**, 619 (1996).

6. M. C. Pirrung, L. Wang and M. P. Montague-Smith, *Org. Lett.*, **3**, 1105 (2001).

7. S. Bühler, H. Giegrich, and W. Pfleiderer, *Nucleosides & Nucleotides*, **18**, 1281 (1999).

8. P. Berroy, M. L. Viriot, and M. C. Carre, *Sensors and Actuators, B: Chemical*, **B74**, 186 (2001).

9. H. Giegrich, S. Eisele-Bühler, C. Hermann, E. Kvasyuk, R. Charubala, and W. Pfleiderer, *Nucleosides & Nucleotides*, **17**, 1987 (1998).

10. P. Berroy, M. L. Viriot, and M. C. Carre, *Sensors and Actuators, B: Chemical*, **B74**, 186 (2001).

11. S. Bühler, I. Lagoja, H. Giegrich, K.-P. Stengele, and W. Pfleiderer, *Helv. Chim. Acta*, **87**, 620 (2004); S. Walbert, W. Pfleiderer, and U. E. Steiner, *Helv. Chim. Acta*, **84**, 1601 (2001).

12. D. Wöll, S. Walbert, K.-P. Stengele, T. J. Albert, T. Richmond, J. Norton, M. Singer, R. D. Green, W. Pfleiderer, and U. E. Steiner, *Helv. Chim. Acta*, **87**, 28 (2004).

13. J. Smirnova, D. Woll, W. Pfleiderer, and U. E. Steiner, *Helv. Chim. Acta*, **88**, 891 (2005).

14. U. Münch and W. Pfleiderer, *Nucleosides & Nucleotides*, **16**, 801 (1997).

Alkyl 2-(2-Pyridyl)amino-1-phenylethyl Carbonate and

Alkyl 2-[*N*-Methyl-*N*-(2-pyridyl)]amino-1-phenylethyl Carbonate

These groups were evaluated as thermolytically labile protective groups for 5′-hydroxyl protection in nucleoside synthesis; however, because of the 60 min required to get complete deprotection at 90°C, they were deemed impractical for this application.[1]

1. K. Chmielewski Marcin, V. Marchan, J. Cieslak, A. Grajkowski, V. Livengood, U. Munch, A. Wilk, and L. Beaucage Serge, *J. Org. Chem.*, **68**, 10003 (2003).

Alkyl Phenacyl Carbonate

Phenacyl carbonates can be cleaved by photolysis at 320–390 nm in the presence of an aromatic triplet sensitizer such as 9,10-dimethylanthracene or *N*-methylcarbazole (61–91% yield). Phenacyl phosphates and esters are cleaved similarly.[1]

1. A. Banerjee, K. Lee, and D. E. Falvey, *Tetrahedron*, **55**, 12699 (1999).

Alkyl 3′,5′-Dimethoxybenzoin Carbonate (DMB−O₂COR)

Formation/Cleavage

The dimethoxybenzoin group has an advantage over the *o*-nitrobenzyl group because it produces a nonreactive benzofuran upon photolysis, whereas the o-nitrobenzyl group gives a reactive nitroso aldehyde upon photolytic cleavage. The DMB group is also cleaved much more rapidly and with greater quantum efficiency than the o-nitrobenzyl group.[1] A convenient procedure for preparation of DMB has been reported.[2]

1. M. C. Pirrung and J.-C. Bradley, *J. Org. Chem.*, **60**, 1116 (1995).
2. M. H. B. Stowell, R. S. Rock, D. C. Rees, and S. I Chan, *Tetrahedron Lett.*, **37**, 307 (1996).

Alkyl Methyl Dithiocarbonate: CH_3SCSOR

Formation[1]

Most attempts to differentiate these hydroxyl groups with conventional derivatives resulted in the formation of a tetrahydrofuran. The dithiocarbonate can also be prepared by phase transfer catalysis (Bu_4NHSO_4, 50% $NaOH/H_2O$, CS_2, MeI, rt, 1.5 h).[2]

Cleavage

These esters can be deoxygenated with Bu_3SnH[3] or, as in the above example, with $LiAlH_4$.

1. R. H. Schlessinger and J. A. Schultz, *J. Org. Chem.*, **48**, 407 (1983).
2. A. W. M. Lee, W. H. Chan, H. C. Wong, and M. S. Wong, *Synth. Commun.*, **19**, 547 (1989).
3. D. H. R. Barton and S. W. McCombie, *J. Chem. Soc., Perkin Trans. I*, 1574 (1975).

Alkyl S-Benzyl Thiocarbonate: $ROCOSCH_2Ph$ (Chart 2)

Formation

$PhCH_2SCOCl$, Pyr, 65–70% yield.[1]

Cleavage

H_2O_2, AcOH, AcOK, $CHCl_3$, 20°C, 4 days, 50–55% yield.[1]

1. J. J. Willard, *Can. J. Chem.*, **40**, 2035 (1962).

Carbamates

Alkyl Dimethylthiocarbamate (DMTC): $(CH_3)_2NC(=S)-OR$

This group has excellent stability to a wide variety of reagents. Orthogonality has been demonstrated to the following groups: TBDMS, TBDPS, PMB, MOM, THP, MEM, Ac, Bn.[1]

Formation

1. From the Na salt of an alcohol: $Me_2NC(=S)Cl$, NaI, THF, 0°C, 89–99% yield.
2. From the alcohol: $Im_2C=S$, CH_2Cl_2, DMAP then dimethylamine, 96% yield.

Cleavage

1. $NaIO_4$, H_2O, MeOH, 92–95% yield.
2. NaOH, H_2O_2, THF or CH_3CN, 18 h, 90% yield.

1. D. K. Barma, A. Bandyopadhyay, J. H. Capdevila, and J. R. Falck, *Org. Lett.*, **5**, 4755 (2003).

Alkyl N-Phenylcarbamate: ROCONHPh (Chart 2)

Phenyl isocyanates are generally more reactive than alkyl isocyanates in their reactions with alcohols, but with CuCl catalysis even alkyl isocyanates will react readily with primary, secondary, or tertiary alcohols (45–95% yield).[1]

Formation

$PhN=C=O$, Pyr, 20°C, 2–3 h, 100% yield.[2] This method was used to protect selectively the primary hydroxyl group in several pyranosides.[3]

Cleavage

1. MeONa, MeOH, reflux, 1.5 h, good yield.[4]
2. LiAlH$_4$, THF, or dioxane, reflux, 3–4 h, 90% yield.[3]
3. Cl$_3$SiH, Et$_3$N, CH$_2$Cl$_2$, 4–48 h, 25–80°C, 80–95% yield.[5] Primary, secondary, tertiary, allylic, propargylic, or benzylic derivatives are cleaved by this method.
4. Bu$_4$NNO$_2$, Ac$_2$O, pyridine, 40°C, 79–100% yield. Deprotection proceeds by nitrosation of the amine which facilitates nucleophilic addition to the carbonyl.[6] A similar process is used to hydrolyze some amides.

1. M. E. Duggan and J. S. Imagire, *Synthesis*, 131 (1989).
2. K. L. Agarwal and H. G. Khorana, *J. Am. Chem. Soc.*, **94**, 3578 (1972).
3. D. Plusquellec and M. Lefeuvre, *Tetrahedron Lett.*, **28**, 4165 (1987).
4. H. O. Bouveng, *Acta Chem. Scand.*, **15**, 87, 96 (1961).
5. W. H. Pirkle and J. R. Hauske, *J. Org. Chem.*, **42**, 2781 (1977).
6. S. Akai, N. Nishino, Y. Iwata, J.-i. Hiyama, E. Kawashima, K.-i. Sato, and Y. Ishido, *Tetrahedron Lett.*, **39**, 5583 (1998).

Alkyl *N*-Methyl-*N*-(*o*-nitrophenyl) Carbamate

This carbamate is prepared from the carbamoyl chloride (CH$_2$Cl$_2$, DMAP, TEA or RONa, 88–94% yield). It is cleaved by photolysis at 248–365 nm in EtOH, H$_2$O, (91–100% yield) to afford the alcohol and 2-nitrosoaniline.[1]

1. S. Loudwig and M. Goeldner, *Tetrahedron Lett.*, **42**, 7957 (2001).

PROTECTION FOR 1,2- AND 1,3-DIOLS

The prevalence of diols in synthetic planning and in natural sources (e.g., in carbohydrates, macrolides, and nucleosides) has led to the development of a number of protective groups of varying stability to a substantial array of reagents. Dioxolanes and dioxanes are the most common protective groups for diols.

In some cases the formation of a dioxolane or dioxane can result in the generation of a new stereogenic center, either with complete selectivity or as a mixture of the two possible isomers. Although the new stereogenic center is removed on deprotection, this center often causes problems because it complicates NMR interpretation.

Cyclic carbonates and cyclic boronates have also found considerable use as protective groups. In contrast to most acetals and ketals, the carbonates are cleaved with a strong base and sterically unencumbered boronates are readily cleaved by water.

Some of the protective groups for diols are listed in Reactivity Chart 3.

Cyclic Acetals and Ketals

Methylene Acetal (Chart 3)

Methylene acetals are the most stable acetals to acid hydrolysis. Difficulty in their removal is probably the reason that these compounds have not seen much use. Cleavage usually occurs under strongly acidic or Lewis acidic conditions.

Formation

1. 40% CH_2O, concd. HCl, 50°C, 4 days, 68% yield.[1] The trismethylenedioxy derivative of a carbohydrate was formed.
2. Paraformaldehyde, H_2SO_4, AcOH, 90°C, 1 h, good yield.[2]
3. DMSO, NBS, 50°C, 12 h, 62% yield.[3]
4. CH_2Br_2, NaH, DMF, 0–30°C, 40 h, 46% yield.[4]
5. $(MeO)_2CH_2$, 2,6-lutidine, TMSOTf, 0°C, 15 min.[5] Similar conditions have been used to introduce MOM ethers on alcohols.

6. $(MeO)_2CH_2$, LiBr, TsOH, CH_2Cl_2, 23°C, 83% yield.[6] In this case a 1,3-methylene acetal is formed in preference to a 1,2-methylene acetal from a 1,2,3-triol. These conditions also protect simple alcohols as their MOM derivatives.
7. CH_2Br_2, NaOH, CH_2Cl_2, cetyl-NMe$_3$Br, heat, 81% yield.[7] This method is effective for both *cis*- and *trans*-1,2-diols.
8. DMSO, TMSCl, 36–72 h.[8]
9. DMSO, POCl$_3$ or SOCl$_2$, 30–120 min, 10–95% yield.[9] With *trans*-1,2-diols, 1,3,5-trioxapanes are formed.

In some examples, the trioxaheptane system could be hydrolyzed with acid to give the diol. The trioxaheptane may also release formaldehyde upon heating.

10. CH_2Br_2, powdered KOH, DMSO, rt, 49% yield.[10]

11. HCHO, cat. SO_2.[11]

12. From a *bis*-MEM ether: $ZnBr_2$, EtOAc, rt.[12]

13. 1,1'-Thiocarbonyldiimidazole, solvent, rt, then reduce with Ph_3SnH, AIBN, toluene, reflux, 36–90% yield.[13]

Cleavage

1. BCl_3, CH_2Cl_2, −80°C, 30 min, warm to 20°C, 61% yield; isolated as the acetate derivative.[1]

2. 2 N HCl, 100°C, 3 h.[2]

3. AcOH, Ac_2O, H_2SO_4, 2 h, 0°C, 91.5% yield.[14]

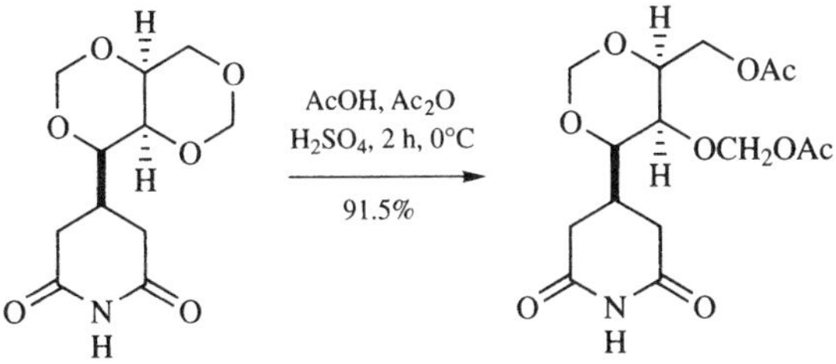

4. NaI, $SiCl_4$, rt, 20–60 min, 78% yield. Cleavage results in subsequent formation of a diiodide, but this is not a general process. For the most part ketals are cleaved to give the ketone, while catechol methylene acetals return the catechol.[15]

5. Ph_3CBF_4, CH_2Cl_2, reflux, 48 h; HCl, rt, 17.5 h, 86% yield.[16] Cleavage occurs by hydride abstraction.

6. $(CF_3CO)_2O$, AcOH, CH_2Cl_2, 21°C; MeOH, K_2CO_3, 92% yield.[17]

7. HF, EtOH, THF, 0–5°C, 14 h.[18]

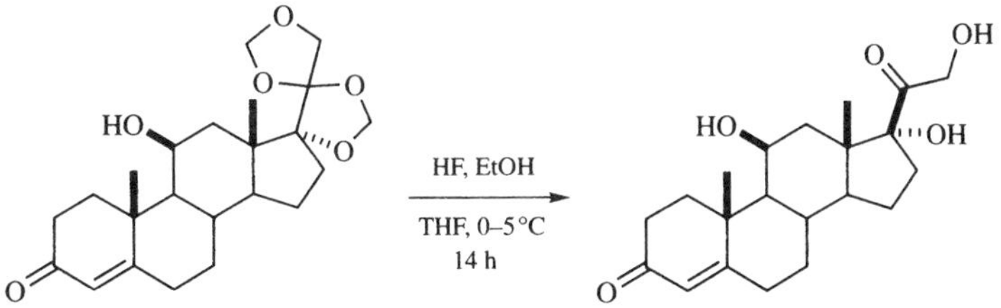

8. AcCl, $ZnCl_2$, Et_2O; ROH, 75–97% yield.[19, 20] When methanol is replaced with benzyl alcohol or methoxyethanol the BOM or MEM groups are formed, respectively.

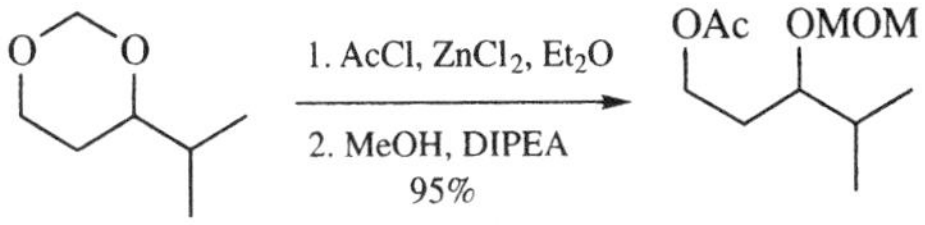

1. T. G. Bonner, *Methods Carbohydr. Chem.*, **II**, 314 (1963).

2. L. Hough, J. K. N. Jones, and M. S. Magson, *J. Chem. Soc.*, 1525 (1952).

3. S. Hanessian, G. Y.-Chung, P. Lavallee, and A. G. Pernet, *J. Am. Chem. Soc.*, **94**, 8929 (1972).

4. J. S. Brimacombe, A. B. Foster, B. D. Jones, and J. J. Willard, *J. Chem. Soc. C*, 2404 (1967).

5. F. Matsuda, M. Kawasaki, and S. Terashima, *Tetrahedron Lett.*, **26**, 4639 (1985).

6. J. L. Gras, R. Nouguier, and M. Mchich, *Tetrahedron Lett.*, **28**, 6601 (1987).

7. D. G. Norman, C. B. Reese, and H. T. Serafinowska, *Synthesis*, 751 (1985). For a similar method, see K. S. Kim and W. A. Szarek, *Synthesis*, 48 (1978).

8. B. S. Bal and H. W. Pinnick, *J. Org. Chem.*, **44**, 3727 (1979); Z. Gu, L. Zeng, X.-p. Fang, T. Colman-Saizarbitoria, M. Huo, and J. L. Mclaughlin, *J. Org. Chem.*, **59**, 5162 (1994); E. F. Queiroz, E. L. M. Silva, F. Roblot, R. Hocquemiller, and B. Figadere, *Tetrahedron Lett.*, **40**, 697 (1999).

9. M. Guiso, C. Procaccio, M. R. Fizzano, and F. Piccioni, *Tetrahedron Lett.*, **38**, 4291 (1997).

10. A. Liptak, V. A. Oláh and J. Kerékgyártó, *Synthesis*, 421 (1982).

11. B. Burczyk, *J. Prakt. Chem.*, **322**, 173 (1980).

12. J. A. Boynton and J. R. Hanson, *J. Chem. Res., Synop.*, 378 (1992).

13. F. De Angelis, M. Marzi, P. Minetti, D. Misiti, and S. Muck, *J. Org. Chem.*, **62**, 4159 (1997).

14. M. J. Wanner, N. P. Willard, G. J. Kooman, and U. K. Pandet, *Tetrahdron*, **43**, 2549 (1987).

15. S. S. Elmorsy, M. V. Bhatt, and A. Pelter, *Tetrahedron Lett.*, **33**, 1657 (1992).

16. H. Niwa, O. Okamoto, and K. Yamada, *Tetrahedron Lett.*, **29**, 5139 (1988).

17. J.-L. Gras, H. Pellissier, and R. Nouguier, *J. Org. Chem.*, **54**, 5675 (1989).

18. H. Shibasaki, T. Furuta, and Y. Kasuya, *Steroids*, **57**, 13 (1992).

19. W. F. Bailey, L. M. J. Zarcone, and A. D. Rivera, *J. Org. Chem.*, **60**, 2532 (1995).

20. W. F. Bailey, M. W. Carson, and L. M. J. Zarcone, *Org. Syn.*, **75**, 177 (1997).

Ethylidene Acetal: (Chart 3)

Formation

1. CH_3CHO, $CH_3CH(OMe)_2$, or paraldehyde, concd. H_2SO_4, 2–3 h, 60% yield.[1]

2. In the following example the ethylidene acetal was used because attempts to make the acetonide led to formation of a 1:1 mixture of the 1,3- and 1,4-acetonide.[2]

3. Diborane reduction of an ortho ester that is prepared from a triol with
   $CH_3C(OEt)_3$, PPTS.[3]

Cleavage

1. 0.67 N H_2SO_4, aq. acetone, reflux, 7 h.[1]
2. Ac_2O, cat. H_2SO_4, 20°C, 5 min, 60% yield.[1] The ethylidene acetal is cleaved to
   form an acetate that can be hydrolyzed with base.
3. 80% AcOH, reflux, 1.5 h.[4]
4. O_3, CH_2Cl_2, 75% yield.[3]

1. T. G. Bonner, *Methods Carbohydr. Chem.*, **II**, 309 (1963); D. M. Hall, T. E. Lawler, and
   B. C. Childress, *Carbohydr. Res.*, **38**, 359 (1974).
2. A. G. Brewster and A. Leach, *Tetrahedron Lett.*, **27**, 2539 (1986).
3. G. Stork and S. D. Rychnovsky, *J. Am. Chem. Soc.* **109**, 1565 (1987).
4. J. W. Van Cleve and C. E. Rist, *Carbohydr. Res.*, **4**, 82 (1967).

t-Butylmethylidene Acetal: t-BuCH(OR)$_2$[1]

1-t-Butylethylidene Ketal: t-BuC(CH$_3$)(OR)$_2$[2]

1-Phenylethylidene Ketal: Ph(CH$_3$)C(OR)$_2$[2]

1-t-Butylethylidene and 1-phenylethylidene ketals were prepared selectively from
the C$_4$–C$_6$, 1,3-diol in glucose by an acid-catalyzed *trans*-ketalization reaction [e.g.,
Me$_3$CC(OMe)$_2$CH$_3$, TsOH/DMF, 24 h, 79% yield; PhC(OMe)$_2$Me, TsOH, DMF,
24 h, 90% yield, respectively]. They are cleaved by acidic hydrolysis: AcOH, 20°C,
90 min, 100% yield, and AcOH, 20°C, 3 days, 100% yield, respectively.[2] Ozonolysis
of the t-butylmethylidene ketal affords a hydroxy ester, albeit with poor regiocontrol,
but a more sterically differentiated derivative may give better selectivity as was observed with the ethylidene ketal.[1]

Ref. 1

1. S. D. Rychnovsky and N. A. Powell, *J. Org. Chem.*, **62**, 6460 (1997).

2. M. E. Evans, F. W. Parrish, and L. Long, Jr., *Carbohydr. Res.*, **3**, 453 (1967).

2-(Methoxycarbonyl)ethylidene (Mocdene) or 2-(*t*-Butylcarbonyl)ethylidene (Bocdene) Acetals

These acetals are prepared by reaction of a 1,2-diol with the corresponding propynoic ester in CH_3CN and DMAP in 90–95% yields. The reaction fails with 1,3-diols because vinyl ethers are formed instead. These acetals are exceptionally stable to strong acids and thus cannot be deprotected by acid hydrolysis. The preferred method for deprotection is by heating in neat pyrrolidine which returns the diol in 93–94% yield by an elimination addition mechanism.[1]

1. X. Ariza, A. M. Costa, M. Faja, O. Pineda, and J. Vilarrasa, *Org. Lett.*, **2**, 2809 (2000).

Phenylsulfonylethylidene Acetal (PSE): $PhSO_2CH_2CH_2CH(OR)_2$

The phenylsulfonylethylidene derivative is an exceptionally stable diol-protective group in that it is stable to strong bases such as DBU and strong acids such as 6 *N* HCl. It is readily prepared from the diethyl acetal with Amberlyst 15 in refluxing toluene (69–87% yield). It also introduced by a double Micheal addition with 1,2-bis(phenylsulfonyl)ethylene in DMF using *t*-BuOK as the base in generally good yields (70–99%). It can be cleaved with $LiNH_2$ in liquid ammonia, BuLi/rt or Na-naphthalenide/$-78°C$/4h (72–86% yield)[1] or reductively with alane.[2,3]

1. S. Chandrasekhar, C. Srinivas, and P. Srihari, *Synth. Commum.*, **33**, 895 (2003).

2. F. Chery, P. Rollin, O. De Lucchi, and S. Cossu, *Tetrahedron Lett.*, **41**, 2357 (2000).

3. F. Chery, P. Rollin, O. De Lucchi, and S. Cossu, *Synthesis*, 286 (2001).

2,2,2-Trichloroethylidene Acetal

Formation

1. Trichloroacetaldehyde (chloral) reacts with glucose in the presence of sulfuric acid to form two mono- and four diacetals. [1]

2.

note inversion of
configuration

Ref. 2, 3

Cleavage

Cleavage occurs by prior conversion to the ethylidene acetal with RaNi or Bu₃SnH and then the normal acid hydrolysis.[2,3] The trichloro acetal is cleaved by reduction (H₂, Raney Ni, 50% NaOH, EtOH, 15 min).[3] The trichloro acetal can probably be cleaved with Zn/AcOH [cf. ROCH(R′)OCH₂CCl₃ cleaved by Zn/AcOH, AcONa, 20°C, 3 h, 90% yield[4]].

1. S. Forsén, B. Lindberg, and B.-G. Silvander, *Acta Chem. Scand.*, **19**, 359 (1965).
2. R. Miethchen and D. Rentsch, *J. Prakt. Chem.*, **337**, 422 (1995).
3. R. Miethchen and D. Rentsch, *Synthesis*, 827 (1994).
4. R. U. Lemieux and H. Driguez, *J. Am. Chem. Soc.*, **97**, 4069 (1975).

3-(Benzyloxy)propyl Acetal

The 3-(benzyloxy)propyl acetal was developed to be deprotected in two stages: hydrogenolysis of the benzyl group followed by mild acid treatment to cleave the acetal by intramolecular transketalization. Prolonged hydrogenolysis over Pd–C also resulted in acetal cleavage,[1] but this is most likely the result of residual acid in the catalyst—a well-known problem.[2]

1. N. A. Powell and S. D. Rychnovsky, *J. Org. Chem.*, **64**, 2026 (1999).
2. See the sections on TES and TBDMS ether deprotection.

Acrolein Acetal: $CH_2=CHCH(OR)_2$

Formation

Bu_2SnO, toluene, reflux, 4h; $Pd(Ph_3P)_4$, THF, $CH_2=CHCH(OAc)_2$, rt, 1 h 80–89% yield. In pyranoside protection, selectivity for 1,3-dioxane formation is generally observed, but dioxolanes are often formed.

Cleavage

1. $(Ph_3P)_3RhCl$, EtOH, with or without TFA, 90% yield.
2. 1% H_2SO_4, refluxing dioxane, >80% yield.[1]
3. Reductive cleavage of the acrolein acetal proceeds similarly to that of the benzylidene acetals.[2]

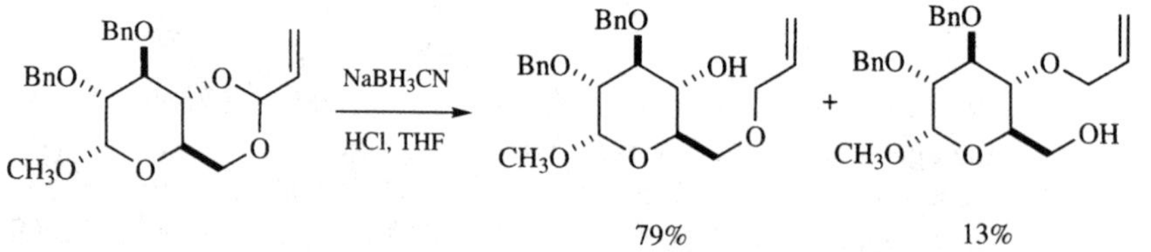

79%   13%

1. C. W. Holzapfel, J. J. Huyser, T. L. Van der Merwe, and F. R. Van Heerden, *Heterocycles*, **32**, 1445 (1991).
2. P. J. Garegg, *Acc. Chem. Res.*, **25**, 575 (1992).

Acetonide (Isopropylidene Ketal) (Chart 3)

Acetonide formation is the most commonly used protection for 1,2- and 1,3-diols. The acetonide has been used extensively in carbohydrate chemistry to mask selectively the hydroxyls of the many different sugars.[1] In preparing acetonides of triols, the 1,2-derivative is generally favored over the 1,3-derivative and a 1,3-derivative is favored over the 1,4-derivative,[2] but the extent to which the 1,2-acetonide is favored is dependent upon structure.[3–6] Note that the 1,2-selectivity for the ketal from 3-pentanone is better than that from acetone.[7] Its greater lipophilicity also improves the isolation of the ketals of small alcohols such as glycerol.[8]

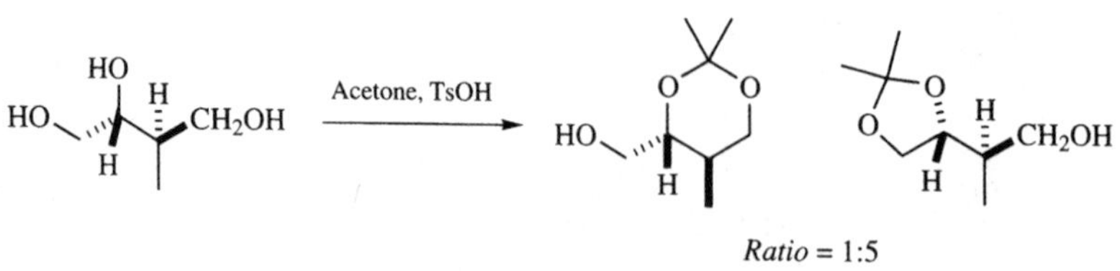

Ratio = 1:5

In cases where two 1,2-acetonides are possible, the thermodynamically more favored one prevails. Secondary alcohols have a greater tendency to form cyclic acetals than do primary alcohols,[7,9] but an acetonide from a primary alcohol is preferred over an acetonide from two *trans*, secondary alcohols.

Below, **i** is isomerized to **ii** producing a *trans* derivative, but acetonide **iii** fails to isomerize to the internal derivative because the less favorable *cis* product would be formed.[11]

The following unusual and unexpected isomerization has been observed indicating that steric effects play an important role in determining thermodynamic stability. In this case the placement of two very large substituents in a *cis* relationship prevents the expected formation of the five-membered ring.[12]

Trityltetrafluoroborate has been observed to equilibrate ketals.[13] This may have broader implications in synthesis because it occurs in the absence of a protic acid.

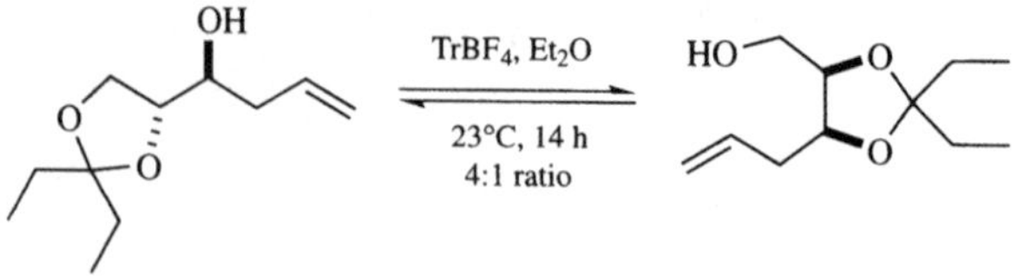
OH
$TrBF_4$, Et_2O
23°C, 14 h
4:1 ratio
HO
O
O

Acetonides may also participate in unexpected reactions, such as in the chlorination and iodination shown below.[14–16]

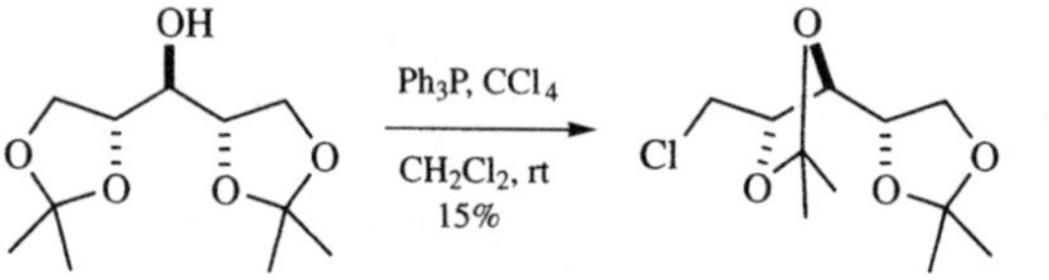
OH
Ph_3P, CCl_4
CH_2Cl_2, rt
15%
Cl
O

Ref. 10

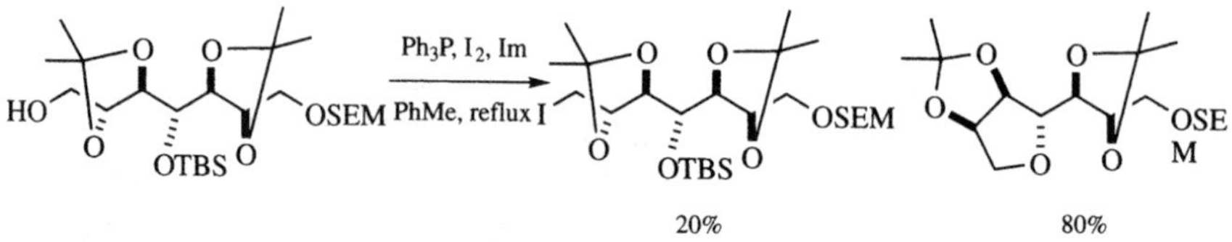
Ph_3P, I_2, Im
PhMe, reflux
HO
OSEM
OTBS
I
OSE
M
20%
80%

The attempted allylation of the aldehyde shown in the following equation resulted in unanticipated tetrahydrofuran formation.[17]

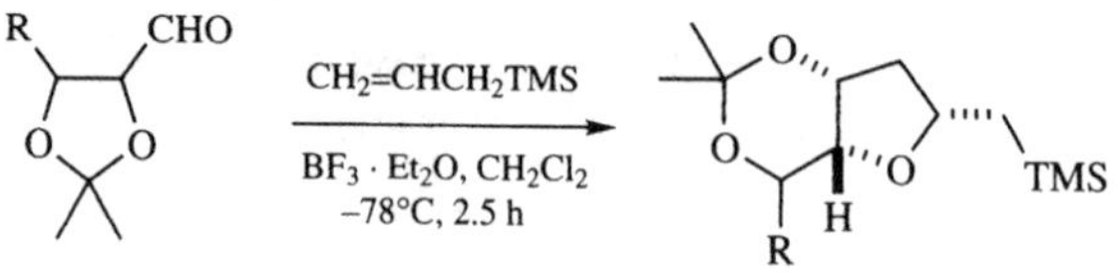
R
CHO
CH_2=$CHCH_2TMS$
$BF_3 \cdot Et_2O$, CH_2Cl_2
−78°C, 2.5 h
TMS
H
R

In the following case, it was anticipated that the nitrogen would participate in the iodocyclization but instead the acetonide proved more reactive.[18]

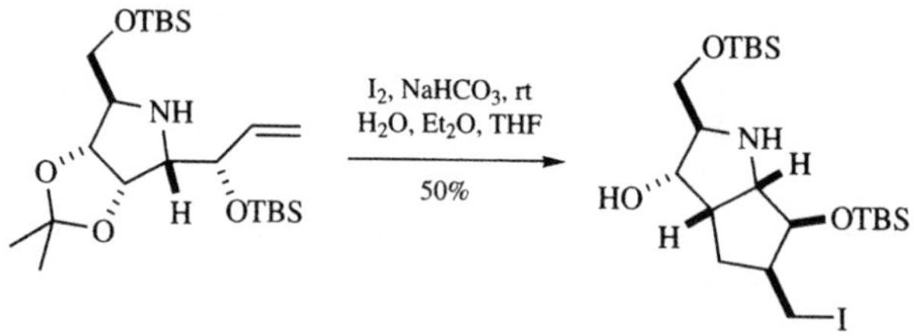
OTBS
NH
H
I_2, $NaHCO_3$, rt
H_2O, Et_2O, THF
50%
HO
I

These examples serve to illustrate the fact that in reactions where carbenium ions are formed in proximity to the acetal lone pairs, unexpected rearrangements may occur.

Formation

1. The classical method for acetonide formation is by reaction of a diol with acetone and an acid catalyst.[19,20]

2. $CH_3C(OCH_3)=CH_2$, dry HBr, CH_2Cl_2, 0°C, 16 h, 75% yield.[21]

Under these conditions, 2-methoxypropene reacts to form the kinetically-controlled 1,3-O-isopropylidene, instead of the thermodynamically more stable 1,2-O-isopropylidene.[22]

3. TsOH, DMF, $Me_2C(OMe)_2$, 24 h.[23,24] This method has become one of the most popular methods for the preparation of acetonides. It generally gives high yields and is compatible with acid-sensitive protective groups such as the TBDMS group.

4. $Me_2C(OMe)_2$, DMF, pyridinium p-toluenesulfonate (PPTS).[25] The use of PPTS for acid-catalyzed reactions has been quite successful and is particularly useful when TsOH acid is too strong an acid for the functionality in a given substrate. TBDMS groups are stable under these conditions.[26]

5. Anhydrous acetone, $FeCl_3$, 36°C, 5 h, 60–70% yield.[27]

6. $Me_2C(OMe)_2$, di-p-nitrophenyl hydrogen phosphate, 3–5 h, 90–100% yield.[28]

7. $Me_2C(OMe)_2$, $SnCl_2$, DME, 30 min, 54% yield. This reaction has been used to prepare the bisacetonide of mannitol on a 100-kg scale.[29]

8. $MeC(OEt)=CH_2$, cat. HCl, DMF, 25°C, 12 h, 90–100% yield.[30] This method is subject to solvent effects. In the formation of a *trans*-acetonide, the use of CH_2Cl_2 did not give the acetonide, but when the solvent was changed to THF, acetonide formation proceeded in 90% yield.[31] These conditions are used to obtain the kinetic acetonide.[32]

9. $MeC(OTMS)=CH_2$, concd. HCl or TMSCl, 10–30 min, 80–85% yield.[33] This method is effective for the formation of *cis*- or *trans*-acetonides of 1,2-cyclohexanediol.

10. Acetone, I_2, 70–85% yield, rt or reflux.[34]

11. Acetone, $CuSO_4$, H_2SO_4, 90% yield.[35] If PPTS replaces H_2SO_4 as the acid, the acetonide can be formed in the presence of a trityl group.[36] $CuSO_4$ serves as a dehydrating agent.

These conditions were used when dimethoxypropane was ineffective because of lactone opening as a result of the released methanol.[37]

12. Trimethylsilylated diols are converted to acetonides with acetone and TMSOTf, −78°C, 3.5 h, >76% yield.[38]

13. Acetone, AlCl₃, Et₂O, rt, 3.5 h, 80% yield.[39] Other methods failed in this sterically demanding case.

14. CH₃CCl(OMe)CH₃, DMF, 92% yield.[40]

15. Conversion of silyl ethers to acetonides without prior cleavage of the silyl ether is possible (acetone, AcOH, CuSO₄, 81% yield),[41] but is dependent upon the conditions of the reaction.[11]

Compare the following examples:

16. Lactone methanolysis followed by acetonide formation has also been observed.[42]

17. Conversion of an epoxide directly to an acetonide is accomplished with acetone and SnCl₄ (81–86% yield)[43] or with N-(4-methoxybenyl)-2-cyanopyridinium hexafluoroantimonate [N-(4-MeOC₆H₄CH₂)-2-CN-PyrSbF₆] (59–100% yield).[44]

18. (CH₃)₂C(OCH₂CH₂CH₂CH=CH₂)₂, NBS, TESOTf, 94% yield.[45]

19. Acetone, K-10 clay.[46]

20. Acetone, FeCl₃, reflux, 20 min, 77% yield.[47]

21. From an epoxide: Er(OTf)$_3$, acetone, rt, 29–99% yield. The lower yields are obtained from epoxides such as glycidol ethers bearing Bn, propargyl, and phenyl ethers. Benzylidene groups are also cleaved in the process.[48]

Cleavage

Cleavage rates for 1,3-dioxanes are greater than for 1,3-dioxolanes,[49] but hydrolysis of a *trans*-fused dioxolane is faster than the dioxane. In substrates having more than one acetonide, the least hindered and more electron-rich acetonide can be hydrolyzed selectively.[50] In a classic example, 1,2-5,6-diacetoneglucofuranose is hydrolyzed selectively at the 5,6-acetonide. *Trans*-acetonides are generally cleaved faster than *cis*-acetonides.[51]

1. Dowex 50-W (H$^+$), water, 70°C, excellent yield.[52] Amberlyst 15 has been used to cleave an acetonide from an acid-sensitive substrate.[53]
2. 1 *N* HCl, THF (1:1), 20°.[7]
3. 2 *N* HCl, 80°C, 6 h.[54] 2 *N* HCl has been used to selectively hydrolyze the acetonide of an *anti* acetonide in the presence of a *syn*-acetonide.[55]

4. 60–80% AcOH, 25°C, 2 h, 92% yield of *cis*-1,2-diol.[56] MOM groups are stable to these conditions.[57]
5. 80% AcOH, reflux, 30 min, 78% yield of *trans*-1,2-diol.[56]

6. $NaHSO_3 \cdot SiO_2$, CH_2Cl_2, rt, 82–100% yield. Ether, ester, and sulfonate protective groups were compatible with this method, but silyl and trityl ethers were not because of low selectivity.[58] $HClO_4 \cdot SiO_2$ also cleaves acetonides and trityl ethers in excellent yield.[59]

7. TsOH, MeOH, 25°C, 5 h.[60] These conditions failed to cleave the acetonide of a 2′,3′-ribonucleoside.[61]

8. TFA, CH_2Cl_2, rt, 2–11 h, 77–92% yield. These conditions cleave ribosyl acetonides in the presence of a MOM group in the absence of a proximal oxygen that can direct the cleavage.[62]

9. CF_3CO_2H, THF, H_2O, 83% yield.[63]

10. 40% aqueous HF, CH_3CN, >56% yield.[64]

11. Phosphomolybdic acid (PMA) supported on silica gel, CH_3CN, rt, 4–7 min, 89–95% yield.[65] Esters, benzyl, allyl, silyl, propargyl, and MOM ethers are all compatible with this method.

12. MeOH, PPTS, heat, high yield.[66] The conditions cleave a terminal acetonide in the presence of an internal acetonide.[53]

13. EtSH, TsOH, $CHCl_3$, >76% yield.[3]

14. BCl_3, 25°C, 2 min, 100% yield.[67]

15. $MgBr_2$, benzene, reflux, 70–80% yield. Ether must be removed from the $MgBr_2$ to get reasonable rates for the deprotection.[68]

16. PdCl$_2$(CH$_3$CN)$_2$, CH$_3$CN, H$_2$O, rt.[69] When the solvent is changed to wet acetone the reagent cleaves an ethylene glycol ketal from ketones in 82–100% yield. TBDPS and MEM groups are stable, but TBDMS and THP groups are cleaved under these conditions.

17. (Bu$_2$SnNCS)$_2$O, diglyme, H$_2$O, 100°C, 82% yield.[70] The THP group is also cleaved by this reagent.

18. FeCl$_3$·SiO$_2$, CHCl$_3$, 74% yield.[71] When used in acetone, this reagent cleaves the trityl and TBDMS groups. These conditions also cleave THP and TMS groups, but TBDMS, MTM, and MOM groups are not affected when CHCl$_3$ is used as solvent. The use of polyvinylpyridine supported FeCl$_3$ is similarly effective giving high yields (CH$_3$CN, CH$_2$Cl$_2$, 87–94% yield). A secondary TMS ether, a vinyl ether, and a THP group were all stable to these conditions.[72]

19. La(NO$_3$)$_3$·6H$_2$O, CH$_3$CN, reflux, 4–6 h, 81–96% yield. Terminal acetonides are cleaved in preference to internal acetonides. The following ethers and esters were stable to these conditions: Tr, TMS, TBDMS, THP, Ac, Bz, Bn, Me.[73]

20. CeCl$_3$·7H$_2$O, oxalic acid, CH$_3$CN, rt, 64–98% yield. Neither reagent alone would cleave the acetonide. The method is compatible with the following protective groups: Tr, Ts, TBDMS, TBDPS, PMB, and esters.[74]

21. Zn(NO$_2$)$_2$·6H$_2$O, CH$_3$CN, 6–8 h, 82–88% yield.[75] This method will selectively remove a terminal acetonide in the presence of an internal acetonide.

22. BiCl$_3$, CH$_3$CN, or CH$_2$Cl$_2$, 10–30 min, 79–93% yield. BOC groups, esters, THP, and TBS ethers are unaffected by this reagent.[76] VCl$_3$/MeOH has been used for this and related transformations.[77]

23. SnCl$_2$, CH$_3$NO$_2$, H$_2$O, >80% yield.[78]

24. HSCH$_2$CH$_2$CH$_2$SH, BF$_3$·Et$_2$O, CH$_2$Cl$_2$, 0°C, 89% yield. A primary TBDMS group was not affected.[79] TiCl$_4$ can also be used as a catalyst, but in this case a PMB ether is also cleaved.[80]

25. Me$_2$BBr, CH$_2$Cl$_2$, −78°C, ~50%.[81]

26. SO$_2$, H$_2$O, 40°C, >67% yield.[82]

27. Cat. I_2, MeOH, rt, 24 h, >80% yield.[83] Benzylidene ketals and thioketals are also cleaved under these conditions. The use of I_2 in CH_3CN/H_2O cleaves terminal acetonides (90–95% yield) but does not cleave MOM, PMB, Bn, allyl, and propargyl ethers. Silyl ethers are cleaved to some extent.[84]

28. Br_2, Et_2O.[20]

29. 5% CBr_4, MeOH, photolysis, 5–48 h, 72–93% yield.[85] A terminal acetonide is cleaved in the presence of an internal derivative. TBS and TBDPS ethers are unaffected by these conditions, but trityl groups are cleaved.

30. Ceric ammonium nitrate, pyridine, acetone, water. Benzylidene acetals are also cleaved. The pH of the system is 4.4, making this method compatible with acid-sensitive substrates.[86]

31. Polymer supported dicyanoketene acetal, CH_3CN, H_2O, rt, 2 h, 73–96% yield. This reagent also cleaves dioxolanes and THP and TBS ethers.[87]

32. In the following examples the acetonide protective group is selectively converted to one of two *t*-butyl groups. The reaction appears to be general, but the alcohol bearing the *t*-butyl group varies with structure.[88] Benzylidene ketals are also cleaved. The reaction of acetonides with MeMgI proceeds similarly and the selectivity is driven by chelation.[89]

An analogy to the above process is when $TMSCH_2MgCl$ is substituted for MeMgI deprotection of the acetonide and takes[90] place probably through a Peterson olefination process. *Trans*-acetonides react in preference to the *cis* derivatives.

33. Although acetonides are generally considered stable to reagents like BH_3, they can on occasion undergo unexpected side reactions, such as the cleavage observed during a hydroboration.[91,92] Changing the solvent to THF completely prevents the aberrant cleavage process.

34. The rather unusual conversion of an acetonide to an isopropenyl ether was developed to differentiate a terminal acetonide from several internal ones. It was, in turn, converted to the 1-methylcyclopropyl ether that was later cleaved with NBS or DDQ.[93,94] The intermediate isopropenyl group can be removed with I_2 ($NaHCO_3$, THF, H_2O, rt, 78 yield).

1. For a review, see D. M. Clode, *Chem. Rev.*, **79**, 491 (1979).

2. M. R. Kotecha, S. V. Ley, and S. Mantegani, *Synlett*, 395 (1992).

3. D. R. Williams and S.-Y. Sit, *J. Am. Chem. Soc.*, **106**, 2949 (1984).

4. P. Lavallee, R. Ruel, L. Grenier, and M. Bissonnette, *Tetrahedron Lett.*, **27**, 679 (1986).

5. A. I. Meyers and J. P. Lawson, *Tetrahedron Lett.*, **23**, 4883 (1982).

6. S. Hanessian, *Aldrichimica Acta*, **22**, 3 (1989).

7. S. J. Angyal and R. J. Beveridge, *Carbohydr. Res.*, **65**, 229 (1978).

8. C. R. Schmid and D. A. Bradley, *Synthesis*, 587 (1992).

9. P. A. Grieco, Y. Yokoyama, G. P. Withers, F. J. Okuniewicz, and C.-L. J. Wang, *J. Org. Chem.*, **43**, 4178 (1978).

10. S. Nishiyama, Y. Ikeda, S. Yoshida, and S. Yamamura, *Tetrahedron Lett.*, **30**, 105 (1989).

11. J. W. Coe and W. R. Roush, *J. Org. Chem.*, **54**, 915 (1989); C. Mukai, M. Miyakawa, and M. Hanaoka, *J. Chem. Soc., Perkin Trans. 1*, 913 (1997); J. D. White, P. Hrnciar, and A. F. T. Yokochi, *J. Am. Chem. Soc.*, **120**, 7359 (1998).

12. T. Ritter, P. Zarotti, and E. M. Carreira, *Org. Lett.*, **6**, 4371 (2004).

13. S. A. Frank and W. R. Roush, *J. Org. Chem.*, **67**, 4316 (2002).

14. J. S. Edmonds, Y. Shibata, F. Yang, and M. Morita, *Tetrahedron Lett.*, **38**, 5819 (1997).

15. J.-C. Lee, S.-W. Chang, C.-C. Liao, F.-C. Chi, C.-S. Chen, Y.-S. Wen, C.-C. Wang, S. S. Kulkarni, R. Puranik, Y.-H. Liu, and S.-C. Hung, *Chem. Eur. J.*, **10**, 399 (2004).

16. S.-K. Chang and L. A. Paquette, *Synlett*, 2915 (2005).

17. K. Osumi and H. Sugimura, *Tetrahedron Lett.*, **36**, 5789 (1995).

18. T. J. Donohoe, H. O. Sintim, and J. Hollinshead, *J. Org. Chem.*, **70**, 7297 (2005).

19. O. Th. Schmidt, *Methods Carbohydr. Chem.*, **II**, 318 (1963).

20. A. N. de Belder, *Adv. Carbohydr. Chem.*, **20**, 219 (1965).

21. E. J. Corey, S. Kim, S. Yoo, K. C. Nicolaou, L. S. Melvin, Jr., D. J. Brunelle, J. R. Falck, E. J. Trybulski, R. Lett, and P. W. Sheldrake, *J. Am. Chem. Soc.*, **100**, 4620 (1978).

22. E. Fanton, J. Gelas, and D. Horton, *J. Chem. Soc., Chem. Commun.*, 21 (1980).

23. M. E. Evans, F. W. Parrish, and L. Long, Jr., *Carbohydr. Res.*, **3**, 453 (1967).

24. B. H. Lipshutz and J. C. Barton, *J. Org. Chem.* **53**, 4495 (1988).

25. M. Kitamura, M. Isobe, Y. Ichikawa, and T. Goto, *J. Am. Chem. Soc.* **106**, 3252 (1984).

26. K. Mori and S. Maemoto, *Liebigs Ann. Chem.*, 863 (1987).

27. P. P. Singh, M. M. Gharia, F. Dasgupta, and H. C. Srivastava, *Tetrahedron Lett.*, **18**, 439 (1977).

28. A. Hampton, *J. Am. Chem. Soc.*, **83**, 3640 (1961).

29. C. R. Schmid, J. D. Bryant, M. Dowlatzedah, J. L. Phillips, D. E. Prather, R. D. Schantz, N. L. Sear, and C. S. Vianco, *J. Org. Chem.*, **56**, 4056 (1991).

30. S. Chládek, and J. Smrt, *Collect. Czech. Chem. Commun.*, **28**, 1301 (1963).

31. J. Cai, B. E. Davison, C. R. Ganellin, and S. Thaisrivongs, *Tetrahedron Lett.*, **36**, 6535 (1995).

32. J. Gelas and D. Horton, *Heterocycles*, **16**, 1587 (1981).

33. G. L. Larson and A. Hernandez, *J. Org. Chem.*, **38**, 3935 (1973).

34. K. P. R. Kartha, *Tetrahedron Lett.*, **27**, 3415 (1986).

35. P. Rollin and J.-R. Pougny, *Tetrahedron*, **42**, 3479 (1986).

36. T. Nakata, M. Fukui, and T. Oishi, *Tetrahedron Lett.*, **29**, 2219 (1988).

37. K. W. Hering, K. Karaveg, K. W. Moremen, and W. H. Pearson, *J. Org. Chem.*, **70**, 9892 (2005).

38. S. D. Rychnovsky, *J. Org. Chem.*, **54**, 4982 (1989).

39. B. Lal, R. M. Gidwani, and R. H. Rupp, *Synthesis*, 711 (1989).

40. A. Kilpala, M. Lindberg, T. Norberg, and S. Oscarson, *J. Carbohydr. Chem.*, **10**, 499 (1991).

41. D. Schinzer, A. Limberg, and O. M. Böhm, *Chem. Eur. J.*, **2**, 1477 (1996).

42. J. P. Férézou, M. Julia, Y. Li, L. W. Liu, and A. Pancrazi, *Synlett*, 766 (1990).

43. R. Stürmer, *Liebigs Ann. Chem.*, 311 (1991).

44. S. B. Lee, T. Takata, and T. Endo, *Chem. Lett.*, **19**, 2019 (1990).

45. R. Madsen and B. Fraser-Reid, *J. Org. Chem.*, **60**, 772 (1995).

46. J.-I. Asakura, Y. Matsubara, and M. Yoshihara, *J. Carbohydr. Chem.*, **15**, 231 (1996).

47. Y. Cai, C.-C. Ling, and D. R. Bundle, *Org. Lett.*, **7**, 4021 (2005).

48. A. Procopio, R. Dalpozzo, A. De Nino, L. Maiuolo, M. Nardi, and B. Russo, *Adv. Synth. Catal.*, **347**, 1447 (2005).

49. S.-K. Chun and S.-H. Moon, *J. Chem. Soc., Chem. Commun.*, 77 (1992).

50. K.-H. Park, Y. J. Yoon, and S. G. Lee, *Tetrahedron Lett.*, **35**, 9737 (1994); S. D. Burke, K. W. Jung, J. R. Phillips, and R. E. Perri, *Tetrahedron Lett.*, **35**, 703 (1994); M. Gerspacher and H. Rapoport, *J. Org. Chem.*, **56**, 3700 (1991).

51. K. S. Ravikumar and D. Farquhar, *Tetrahedron Lett.*, **43**, 1367 (2002).

52. P.-T. Ho, *Tetrahedron Lett.*, **19**, 1623 (1978).

53. J. Zhu and D. Ma, *Angew. Chem. Int. Ed.*, **42**, 5348 (2003).

54. T. Ohgi, T. Kondo, and T. Goto, *Tetrahedron Lett.*, **18**, 4051 (1977).

55. S. E. Bode, M. Muller, and M. Wolberg, *Org. Lett.*, **4**, 619 (2002).

56. M. L. Lewbart and J. J. Schneider, *J. Org. Chem.*, **34**, 3505 (1969).

57. S. Hanessian, D. Delorme, P. C. Tyler, G. Demailly, and Y. Chapleur, *Can. J. Chem.* **61**, 634 (1983).

58. G. Mahender, R. Ramu, C. Ramesh, and B. Das, *Chem. Lett.*, **32**, 734 (2003).

59. A. Agarwal and Y. D. Vankar, *Carbohydr. Res.*, **340**, 1661 (2005).

60. A. Ichihara, M. Ubukata, and S. Sakamura, *Tetrahedron Lett.*, **18**, 3473 (1977).

61. J. Kimura and O. Mitsunobu, *Bull. Chem. Soc. Jpn.*, **51**, 1903 (1978).

62. R. D. Wakharkar, M. B. Sahasrabuddhe, H. B. Borate, and M. K. Gurjar, *Synthesis*, 1830 (2004).

63. Y. Leblanc, B. J. Fitzsimmons, J. Adams, F. Perez, and J. Rokach, *J. Org. Chem.*, **51**, 789 (1986).

64. K.-G. Liu, S. Yan, Y.-L. Wu, and Z.-J. Yao, *Org. Lett.*, **6**, 2269 (2004).

65. J. S. Yadav, S. Raghavendra, M. Satyanarayana, and E. Balanarsaiah, *Synlett*, 2461 (2005).

66. R. Van Rijsbergen, M. J. O. Anteunis, and A. De Bruyn, *J. Carbohydr. Chem.*, **2**, 395 (1983).

67. T. J. Tewson and M. J. Welch, *J. Org. Chem.*, **43**, 1090 (1978).

68. G. G. Haraldsson, T. Stefansson, and H. Snorrason, *Acta Chem. Scand.*, **52**, 824 (1998).

69. B. H. Lipshutz, D. Pollart, J. Monforte, and H. Kotsuki, *Tetrahedron Lett.*, **26**, 705 (1985); C. Schmeck, and L. S. Hegadus, *J. Am. Chem. Soc.*, **116**, 9927 (1994).

70. J. Otera and H. Nozaki, *Tetrahedron Lett.*, **27**, 5743 (1986).

71. K. S. Kim, Y. H. Song, B. H. Lee, and C. S. Hahn, *J. Org. Chem.*, **51**, 404 (1986).

72. M. A. Chari and K. Syamasundar, *Synthesis*, 708 (2005).

73. S. M. Reddy, Y. V. Reddy, and Y. Venkateswarlu, *Tetrahedron Lett.*, **46**, 7439 (2005).

74. X. Xiao and D. Bai, *Synlett*, 535 (2001).

75. S. Vijayasaradhi, J. Singh, and I. S. Aidhen, *Synlett*, 110 (2000); L. Chabaud, Y. Landais, and P. Renaud, *Org. Lett.*, **7**, 2587 (2005).

76. N. R. Swamy and Y. Venkateswarlu, *Tetrahedron Lett.*, **43**, 7549 (2002).

77. G. Sabitha, G. S. K. K. Reddy, K. B. Reddy, N. M. Reddy, and J. S. Yadav, *J. Mol. Catal. A: Chemical*, **238**, 229 (2005).

78. K. A. Ahrendt and R. M. Williams, *Org. Lett.*, **6**, 4539 (2004).

79. T. Konosu and S. Oida, *Chem. Pharm. Bull.*, **39**, 2212 (1991).

80. K. C. Nicolaou, Y. He, K. C. Fong, W. H. Yoon, H.-S. Choi, Y.-L. Zhong, and P. S. Baran, *Org. Lett.*, **1**, 63 (1999).

81. S. E. de Laszlo, M. J. Ford, S. V. Ley, and G. N. Maw, *Tetrahedron Lett.*, **31**, 5525 (1990).

82. A. Dondoni and D. Perrone, *J. Org. Chem.*, **60**, 4749 (1995).

83. W. A. Szarek, A. Zamojski, K. N. Tiwari, and E. R. Isoni, *Tetrahedron Lett.*, **27**, 3827 (1986).

84. J. S. Yadav, M. Satyanarayana, S. Raghavendra, and E. Balanarsaiah, *Tetrahedron Lett.*, **46**, 8745 (2005).

85. M.-Y. Chen, L. N. Patkar, M.-D. Jan, A. S.-Y. Lee, and C.-C. Lin, *Tetrahedron Lett.*, **45**, 635 (2004); T. Chandra and K. L. Brown, *Tetrahedron Lett.*, **46**, 8617 (2005).

86. G. Barone, E. Bedini, A. Iadonisi, E. Manzo, and M. Parrilli, *Synlett*, 1645 (2002); E. Manzo, G. Barone, E. Bedini, A. Iadonisi, L. Mangoni, and M. Parrilli, *Tetrahedron*, **58**, 129 (2002).

87. Y. Masaki, T. Yamada, and N. Tanaka, *Synlett*, 1311 (2001).

88. S. Takano, T. Ohkawa, and K. Ogasawara, *Tetrahedron Lett.*, **29**, 1823 (1988).

89. T.-Y. Luh, *Synlett*, 201 (1996). W.-L. Cheng, S.-M. Yeh, and T. -Y. Luh, *J. Org. Chem.*, **58**, 5576 (1993); W.-L. Cheng, Y.-J. Shaw, S.-M. Yeh, P. P. Kanakamma, Y.-H. Chen, C. Chen, J.-C. Shieu, S.-J. Yiin, G.-H. Lee, Y. Wang, and T.-Y. Luh, *J. Org. Chem.*, **64**, 532 (1999).

90. C.-C. Chiang, Y.-H. Chen, Y.-T. Hsieh, and T.-Y. Luh, *J. Org. Chem.*, **65**, 4694 (2000).

91. L. D. Coutts, C. L. Cywin, and J. Kallmerten, *Synlett*, 696 (1993).

92. G. Casiraghi, F. Ulgheri, P. Spanu, G. Rassu, L. Pinna, G. Gasparri Fava, M. Belicchi Ferrari, and G. Pelosi, *J. Chem. Soc., Perkin Trans. 1*, 2991 (1993).

93. S. D. Rychnovsky and J. Kim, *Tetrahedron Lett.*, **32**, 7219 (1991).

94. S. D. Rychnovsky and R. C. Hoye, *J. Am. Chem. Soc.*, **116**, 1753 (1994).

Cyclopentylidene Ketal, i

Cyclohexylidene Ketal, ii

Cycloheptylidene Ketal, iii

Compounds **i**, **ii**, and **iii** can be prepared by an acid-catalyzed reaction of a diol and the cycloalkanone in the presence of triethyl orthoformate and mesitylenesulfonic acid.[1] The relative ease of acid-catalyzed hydrolysis [0.53 M H_2SO_4, H_2O, PrOH (65:35), 20°C] for compounds **i**, **iii**, acetonide, and **ii** is $C_5 \approx C_7$ > acetonide $\gg C_6$ (e.g., $t_{1/2}$ for 1,2-O-alkylidene-α-D-glucopyranoses of C_5, C_7, acetonide, and C_6 derivatives are 8, 10, 20, and 124 h, respectively[1])[2]. The efficiency of cleavage seems to be dependent upon the electronic environment about the ketal.[3]

The cyclohexylidene ketal has been prepared from dimethoxycyclohexane and TsOH[4]; HC(OEt)$_3$, cyclohexanone, TsOH, EtOAc, heat, 5 h, 78%; 1-(trimethylsiloxy)-cyclohexene, concd. HCl, 20°C, 10–30 min, 70–75% yield,[5] cyclohexanone, TsOH, CuSO$_4$,[6] and 1-ethoxycyclohexene, TsOH, DMF.[7] The cyclohexylidene derivative of a *trans*-1,2-diol has been prepared.[8] Cyclohexylidene ketals may also be prepared directly from an epoxide with MTO catalysis.[9]

Cyclohexylidene derivatives are cleaved by acidic hydrolysis: 10% HCl, Et$_2$O, 25°C, 5 min[3]; TFA, H$_2$O, 20°C, 6 min to 2 h, 65–85% yield[10]; 0.1 N HCl, dioxane[8]; BCl$_3$, CH$_2$Cl$_2$, −80°C, 15 h, 90% yield.[11] The cyclohexylidene derivative is also subject to cleavage with Grignard reagents, but under harsh reaction conditions (MeMgI, PhH, 85°C, 58% yield).[12] *trans*-Cyclohexylidene ketals are preferentially cleaved in the presence of *cis*-cyclohexylidene ketals.[13] Selective cleavage of the less substituted of two cyclohexylidenes is possible.[14,15] The rather water-insoluble cyclohexanone that is formed during deprotection can reketalize a diol unless provision is made for its removal. Hexane extraction from a methanolic reaction has been found effective in removing the cyclohexanone.[16]

A cyclohexylidene acetal can be cleaved with Py(HF)$_n$, (CHCl$_3$, 0°C, to rt, 89% yield) in the presence of the fluoride labile TIPS protective group.[17]

In addition, the cyclopentylidene ketal has been prepared from dimethoxycyclopentane, TsOH, CH$_3$CN,[18] or cyclopentanone (PTSA, CuSO$_4$ >70% yield)[19] and can be cleaved with 2:1 AcOH/H$_2$O, rt, 2 h.[20] Certain epoxides can be converted directly to cyclopentylidene derivatives as illustrated. [21]

PNB = *p*-nitrobenzyl

The 1,2-position of a 6-deoxyglucose derivative has been protected using this reagent, giving primarily the pyranose form. These can be cleaved by alcoholysis with allyl alcohol (benzene, CSA, Δ, 29 h, 82–96%).[22] Methoxycyclopentene (PPTS, CH$_2$Cl$_2$, rt, 100%) has been used to introduce this group.[23] The following example shows that a cyclopentylidene can be hydrolyzed in the presence of a p-methoxybenzaldehyde ketal. The ketal is first deactivated toward acid hydrolysis by formation of a charge transfer complex with trinitrotoluene.[24]

A five-membered cyclopentylidene can be cleaved in the presence of a six-membered derivative.[25]

1. W. A. R. van Heeswijk, J. B. Goedhart, and J. F. G. Vliegenthart, *Carbohydr. Res.*, **58**, 337 (1977).

2. J. M. J. Tronchet, G. Zosimo-Landolfo, F. Villedon-Denaide, M. Balkadjian, D. Cabrini, and F. Barbalat-Rey, *J. Carbohydr. Chem.*, **9**, 823 (1990).

3. J. D. White, J. H. Cammack, K. Sakuma, G. W. Rewcastle, and R. K. Widener, *J. Org. Chem.*, **60**, 3600 (1995).

4. C. Kuroda, P. Theramongkol, J. R. Engebrecht, and J. D. White, *J. Org. Chem.*, **51**, 956 (1986). J. Haddad, L. P. Kotra, B. Llano-Sotelo, C. Kim, E. F. Azucena, Jr., M. Liu, S. B. Vakulenko, C. S. Chow, and S. Mobashery, *J. Am. Chem. Soc.*, **124**, 3229 (2002).

5. G. L. Larson and A. Hernandez, *J. Org. Chem.*, **38**, 3935 (1973).

6. W. R. Rousch, M. R. Michaelides, D. F. Tai, B. M. Lesur, W. K. M. Chong, and D. J. Harris, *J. Am. Chem. Soc.*, **111**, 2984 (1989).

7. H. B. Mereyala and M. Pannala, *Tetrahedron Lett.*, **36**, 2121 (1995).

8. D. Askin, C. Angst, and S. Danishefsky, *J. Org. Chem.*, **50**, 5005 (1985).

9. Z. Zhu and J. H. Espenson, *Organometallics*, **16**, 3658 (1997).

10. S. L. Cook and J. A. Secrist, *J. Am. Chem. Soc.*, **101**, 1554 (1979).

11. S. D. Géro, *Tetrahedron Lett.*, **7**, 591 (1966).

12. M. Kawana and S. Emoto, *Bull. Chem. Soc. Jpn.*, **53**, 230 (1980).

13. Y.-C. Liu and C.-S. Chen, *Tetrahedron Lett.*, **30**, 1617 (1989). J. E. Innes, P. J. Edwards, and S. V. Ley, *J. Chem. Soc., Perkin Trans. 1.*, 795 (1997). T. Suzuki, S. Tanaka, I. Yamada, Y. Koashi, K. Yamada, and N. Chida, *Org. Lett.*, **2**, 1137 (2000).

14. D. P. Stamos and Y. Kishi, *Tetrahedron Lett.*, **37**, 8643 (1996).

15. M. S. Wolfe, B. L. Anderson, D. R. Borcherding, and R. T. Borchardt, *J. Org. Chem.*, **55**, 4712 (1990).

16. D. A. Evans and J. D. Burch, *Org. Lett.*, **3**, 503 (2001).

17. Y. Watanabe, Y. Kiyosawa, A. Tatsukawa, and M. Hayashi, *Tetrahedron Lett.*, **42**, 4641 (2001).

18. K. Ditrich, *Liebigs Ann. Chem.*, 789 (1990).

19. D. B. Collum, J. H. McDonald III, and W. C. Still, *J. Am. Chem. Soc.*, **102**, 2118 (1980).

20. C. B. Reese and J. G. Ward, *Tetrahedron Lett.*, **28**, 2309 (1987).

21. A. B. Smith, III, J. Kingery-Wood, T. L. Leenay, E. G. Nolen, and T. Sunazuka, *J. Am. Chem. Soc.*, **114**, 1438 (1992).

22. A. B. Smith, III, R. A. Rivero, K. J. Hale, and H. A. Vaccaro, *J. Am. Chem. Soc.*, **113**, 2092 (1991).

23. R. M. Soll and S. P. Seitz, *Tetrahedron Lett.*, **28**, 5457 (1987).

24. R. Stürmer, K. Ritter, and R. W. Hoffmann, *Angew. Chem., Int. Ed. Engl.*, **32**, 101 (1993).

25. T. K. M. Shing and V. W.-F. Tai, *J. Org. Chem.*, **64**, 2140 (1999).

Benzylidene Acetal (Chart 3)

A benzylidene acetal is a commonly used protective group for 1,2- and 1,3-diols. In the case of a 1,2,3-triol, the 1,3-acetal is the preferred product—in contrast to the acetonide, which gives the 1,2-derivative. It has the advantage that it can be removed under neutral conditions by hydrogenolysis or by acid hydrolysis. Benzyl groups[1] and isolated olefins[2] have been hydrogenated in the presence of 1,3-benzylidene acetals. Benzylidene acetals of 1,2-diols are more susceptible to hydrogenolysis than are those of 1,3-diols. In fact, the former can be removed in the presence of the latter.[3] A polymer-bound benzylidene acetal has also been prepared.[4]

Formation

1. PhCHO, $ZnCl_2$, 28°C, 4 h.[5]
2. PhCHO, DMSO, concd. H_2SO_4, 25°C, 4 h.[6]
3. X = FSO_3^- or BF_4^-, K_2CO_3 or Pyr, CH_2Cl_2, 25°C, 16 h, 45–82% yield.[7]
   This method is suitable for the protection of 1,2-, 1,3-, and 1,4-diols.
4. PhCHO, TsOH, reflux, $-H_2O$, 72% yield.[8]
5. $PhCHBr_2$, Pyr.[9]
6. $PhCH(OMe)_2$, HBF_4, Et_2O, DMF, 97% yield.[10,11] 1,3-Diols are protected in preference to 1,2-diols.[12]

7. PhC(OMe)$_2$, SnCl$_2$, DME, heat, 45 min.[13] A modification of this procedure that uses Sn(OTf)$_2$ has been reported to be superior.[14]

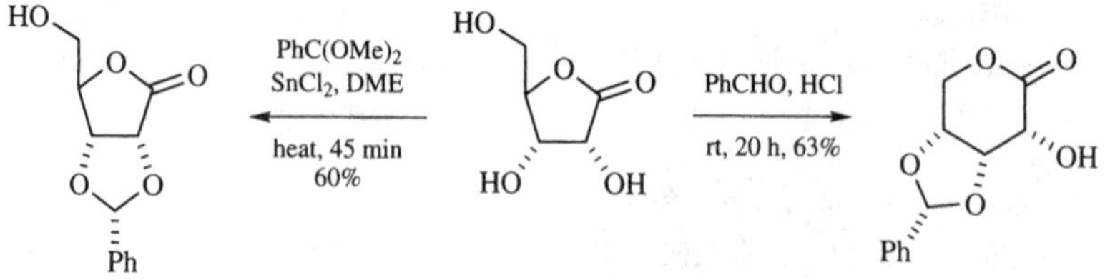

8.

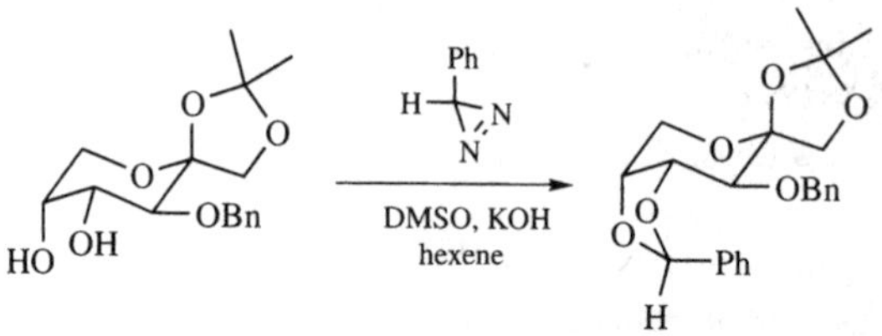

Ref. 15

9. PhCH(OCH$_2$CH$_2$CH=CH$_2$)$_2$, CSA, NBS. Standard methods failed because of cleavage of the dispiroketal (dispoke) protective group.[16,17]

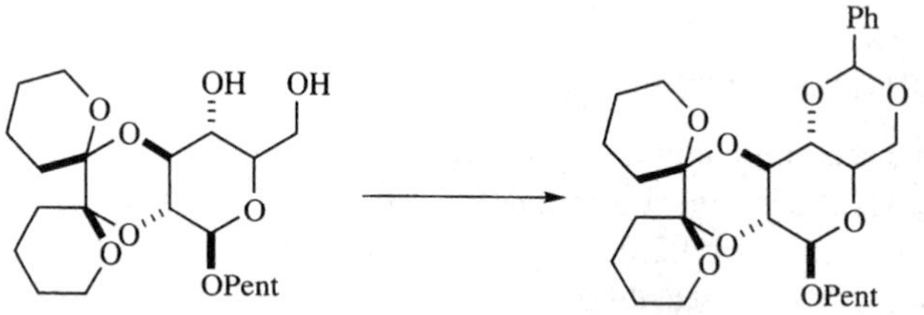

10. By an intramolecular Michael addition.[18]

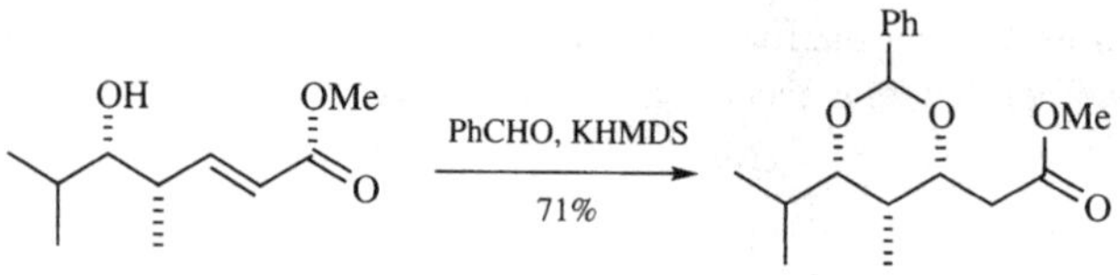

98:2 diastereoselectivity

Cleavage

1. H$_2$/Pd–C, AcOH, 25°C, 30–45 min, 90% yield.[19]
2. Na, NH$_3$, 85% yield.[20]
3. The benzylidene acetal is cleaved by acidic hydrolysis (e.g., 0.01 N H$_2$SO$_4$, 100°C, 3 h, 92% yield[21]; 80% AcOH, 25°C, $t_{1/2}$ for uridine = 60 h[22]), conditions that do not cleave a methylenedioxy group.[22] The rate of acid-catalyzed hydrolysis of benzylidene acetals increases as the size of the substituent R increases. The second-order rate constant k_H, on going from R = Me to R = t-amyl, increases about 100-fold, indicating that steric effects play a large role in determining hydrolysis rates.[23]

4. Electrolysis: -2.9 V, R_4NX, DMF.[24]

5. BCl_3, 100% yield. This reagent also cleaves a number of other ketal-type protective groups.[25]

6. I_2, MeOH, 85% yield.[26]

7. $FeCl_3$, CH_2Cl_2, 3–30 min, 68–85% yield.[27] Benzyl groups are also cleaved by this reagent.

8. $Pd(OH)_2$, cyclohexene, 98% yield.[1]

9. Pd-C, hydrazine, MeOH.[28] In this case a 1,2-benzylidene acetal was cleaved in the presence of a 1,3-benzylidene acetal.

10. Pd–C, HCO_2NH_4, 97% yield.[3]

11. EtSH, $NaHCO_3$, $Zn(OTf)_2$, CH_2Cl_2, rt, 5 h, 90% yield.[29] In the following case, these conditions were the only ones that retained the acetonide and the TBS ether.[30]

12. $SnCl_2$, CH_2Cl_2, rt, 3–12 h, 86–95% yield.[31]

Partial Cleavage of Benzylidene Acetals to Give Benzyl Ethers

Reductive Methods

Benzylidene acetals have the useful property that one of the two C–O bonds can be selectively cleaved. The direction of cleavage is dependent upon steric and electronic factors as well as on the nature of the cleavage reagent. This transformation has been reviewed in the context of carbohydrates.[32]

1. $(i\text{-}Bu)_2AlH$, CH_2Cl_2 or $PhCH_3$, 0°C to rt, yields generally $>80\%$.[33,34] With this reagent, cleavage occurs to give the least hindered alcohol. The cleavage of 1,2-benzylidene acetals with this reagent has been studied.

Coordinating groups such as a sulfone[35] or a MOM[36] group can be used to direct the regiochemical cleavage with DIBAH.

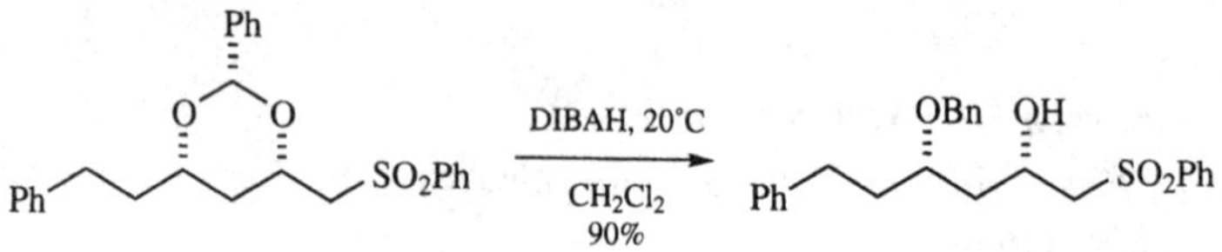

In general, the direction of this cleavage process is a function of the electron density on the two oxygens in the ring.[37]

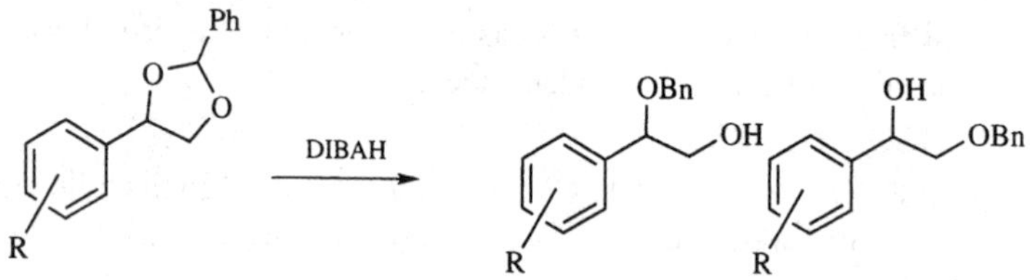

R = 3-CF$_3$ Ratio = 1:3.9

R = 4-MeO Ratio = 3:1

2.

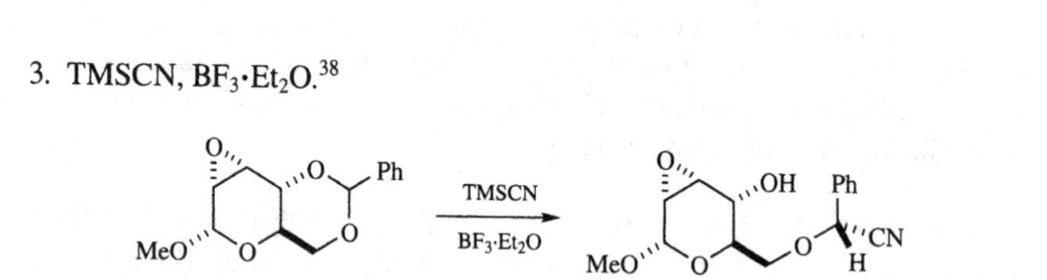

R = Me

R = H

Ref. 38

3. TMSCN, BF$_3$·Et$_2$O.[38]

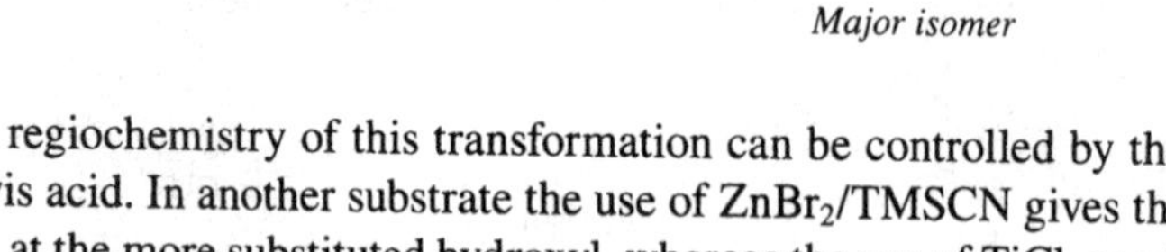

Major isomer

The regiochemistry of this transformation can be controlled by the choice of Lewis acid. In another substrate the use of ZnBr$_2$/TMSCN gives the cyanohydrin at the more substituted hydroxyl, whereas the use of TiCl$_4$ as a Lewis acid places the cyanohydrin at the least substituted hydroxyl.[39]

TiCl₄	250 : 1
ZnBr₂	<1 : 250

4. The reaction of a benzylidene acetal with allyltrimethylsilane and AlCl₃ or TMSOTf gives an allyl-substituted benzyl ether that can then be cleaved.[40]

5. Zn(BH₄)₂, TMSCl, Et₂O, 25°C, 45 min, 77–97% yield.[41] Reduction occurs to form a monobenzyl derivative of a diol.

6. NaBH₃CN, TiCl₄, CH₃CN, rt, 3 h, 83% yield.[42] NaBH₃CN, THF, ether/HCl converts a 4,6-benzylidene to a 6-O-benzylpyranoside,[43] as does NaBH₃CN/ TMSCl/CH₃CN.[44] The use of triflic acid improves this process because the stoichiometry is more conveniently controlled.[45]

These methods have been applied to 1,2-O-benzylidene sugars and, in general good selectivity, can be achieved for cleavage at the anomeric side of the acetal to give the benzyl ether at the 2-position.[46]

7. Me₂BBr, TEA, BH₃·THF, −78°C warm to −20°C over 1 h, 70–97% yield. These conditions cleave the benzylidene acetal to leave the least hindered alcohol as a free hydroxyl. If diborane is omitted from the reaction mixture and the reaction is quenched with PhSH and TEA, the benzylidene group is cleaved to give an O,S-acetal [ROCH(SPh)Ph]. Acetonides are cleaved similarly.[47]

8. AlCl₃, BH₃·TEA, THF, 60°C, 96% yield.[48] In a 2-aminoglucose derivative the 6-O-benzyl derivative was formed selectively. The use of Me₃N·BH₃ in THF

gives the 6-*O*-benzyl derivative, but when the solvent is changed to toluene or CH_2Cl_2 the 4-*O*-benzyl ether is produced.[49] A mechanistic study on the reductive cleavage of acetals has been published.[50] The addition of some water was reported to improve the ring opening process.[51]

9. $V(O)(OTf)_2$, $BH_3 \cdot THF$, CH_2Cl_2, rt, 74–94% yield.[53]

R = H and X = OMe   86% yield
R = OMe and X = SEt   86% yield

10. Bu_2BOTf, $BH_3 \cdot THF$, CH_2Cl_2, 0°C, 70–91% yield. In a variety of pyranosides, cleavage occurs primarily to give the primary alcohol with the secondary alcohol protected as the benzyl ether.[54] The method has been successfully employed on a pentasaccharide.[55] A stereochemical dependence in selectivity has been observed under these conditions for five-membered rings.[56]

11. TFA, Et_3SiH, CH_2Cl_2.[57] 6-*O*-Benzylpyranosides are formed from a 4,6-benzylidenepyranoside in 80–98% yield. $BF_3 \cdot Et_2O$ has also been used as a catalyst for this type of cleavage.[58]

Comparing the use of a protic acid vs. a Lewis acid on the same substrate results in reversal of the cleavage process with TESH as the reductant.[59]

BnO, HO, AcO — AcO OMe ←(TfOH, TESH, CH₂Cl₂, rt, 90%)— Ph O O AcO AcO OMe —(PhBCl₂, TESH, CH₂Cl₂, rt, 91%)→ HO, Bn–O, AcO — AcO OMe

12. Cu(OTf)₂, Me₂EtSiH, or BH₃·THF, CH₂Cl₂.[60]

BnO, HO, BnO — BnO OMe ←(Cu(OTf)₂, Me₂EtSiH, CH₂Cl₂, rt, 84%)— Ph O O BnO BnO OMe —(Cu(OTf)₂, BH₃·THF, CH₂Cl₂, rt, 94%)→ HO, Bn–O, BnO — BnO OMe

13. BH₃·SMe₂, CH₂Cl₂, 0°C, 1 h, then BF₃, 5 min.[61] Simple benzylidene acetals are cleaved efficiently without hydroboration of alkenes that may be present, and acetonides are converted to the hydroxy isopropyl ethers.

Ph—(O,O acetal) OTBDMS, OH —(1. BH₃·SMe₂ 2. BF₃·Et₂O, 87%)→ BnO OTBDMS, HO OH

A related hydroxyl-directed cleavage has been observed using LiBF₄/BH₃ or LiBH₄/BF₃.[62]

Oxidative Methods

1. t-BuOOH, CuCl₂, benzene, 50°C, 15 h, 87% yield. [63] Additionally, Pd(OAc)₂, FeCl₂, PdCl₂, and NiCl₂ were found to be active catalysts in this transformation, but in each case a mixture of benzoates was formed from a 4,6-benzylidene glucose derivative.[64]

2. Ozonolysis, Ac₂O, NaOAc, −78°C, 1 h, 95% yield.[65] In this case the benzylidene acetal is converted to a diester.

3. NaBO₃·4H₂O, Ac₂O, 67–95% yield.[66] Cleavage occurs to give a monobenzoate. A similar process using NaBO₃/Na₂S₂O₄ gives a mixture of the primary and the secondary benzoate.[67]

4. t-BuOOH, Pd(TFA)(t-BuOOH), 26–78% yield.[68] Palladium acetate can also be used.[69] Cleavage occurs to give a monobenzoate.

5. t-BuOOH, t-butylperoxy-λ^3-iodane, K₂CO₃, benzene, rt, 57% yield.[70]

6. 2,2′-bypyridinium chlorochromate, m-chloroperoxybenzoic acid, CH₂Cl₂, rt, 48–72% yield.[71]

Ph—(O,O acetal on cyclohexane) —(BPCC, mCPBA, CH₂Cl₂, 67%)→ BzO, HO (cyclohexane)

7. Ph₃CBF₄, CH₃CN, 25°C, 8 h, 80% yield.[72] A 1:1 mixture of diol monobenzoates is formed.

8. NBS, CCl_4, H_2O, 75% yield.[73] Mechanistically, the reaction proceeds by initial benzylic bromide formation, which then fragments by a polar pathway.[74]

In this type of cleavage reaction, it appears that the axial benzoate is the preferred product. If water is excluded from the reaction, a bromo benzoate is obtained.[75] The highly oxidizing medium of 2,2′-bipyridinium chlorochromate and MCPBA in CH_2Cl_2 at rt for 36h effects a similar conversion of benzylidene acetals to hydroxy benzoates in 25–72% yield.[76]

9. $BrCCl_3$, CCl_4, hv, 30 min, 100% yield.[77]

10. Molecular oxygen, *N*-hydroxyphthalimide, $Co(OAc)_2$, 66–91% yield. A monobenzoate is formed without regioselectivity.[78]

11. DDQ, CH_2Cl_2, H_2O, >71% yield.[79]

12. The following redox rearrangement of a benzylidene acetal has been reported.[80]

1. S. Hanessian, T. J. Liak, and B. Vanasse, *Synthesis*, 396 (1981).

2. A. B. Smith III and K. J. Hale, *Tetrahedron Lett.*, **30**, 1037 (1989); I. Kadota, H. Takamura, and Y. Yamamoto, *Tetrahedron Lett.*, **42**, 3649 (2001).

3. T. Bieg and W. Szeja, *Carbohydr. Res.*, **140**, C7 (1985).

4. J. M. J. M. Fréchet and G. Pellé, *J. Chem. Soc., Chem. Commun.*, 225 (1975).

5. H. G. Fletcher, Jr., *Methods Carbohydr. Chem.*, **II**, 307 (1963).

6. R. M. Carman and J. J. Kibby, *Aust. J. Chem.*, **29**, 1761 (1976).

7. R. M. Munavu and H. H. Szmant, *Tetrahedron Lett.*, **16**, 4543 (1975).

8. D. A. McGowan and G. A. Berchtold, *J. Am. Chem. Soc.*, **104**, 7036 (1982).

9. R. N. Russell, T. M. Weigel, O. Han, and H.-w. Liu, *Carbohydr. Res.*, **201**, 95 (1990).

10. R. Albert, K. Dax, R. Pleschko, and A. Stütz, *Carbohydr. Res.*, **137**, 282 (1985); T. Yamanoi, T. Akiyama, E. Ishida, H. Abe, M. Amemiya, and T. Inazu, *Chem. Lett.*, **18**, 335 (1989); M. El Sous and M. A. Rizzacasa, *Tetrahedron Lett.*, **46**, 293 (2005).

11. M. T. Crimmins, W. G. Hollis, Jr., and G. J. Lever, *Tetrahedron Lett.*, **28**, 3647 (1987).

12. Y. Morimoto, A. Mikami, S.-i. Kuwabe, and H. Shirahama, *Tetrahedron Lett.*, **32**, 2909 (1991).

13. S. Y. Han, M. M. Joullie, N. A. Petasis, J. Bigorra, J. Cobera, J. Font, and R. M. Ortuno, *Tetrahedron*, **49**, 349 (1992).

14. C. C. Joseph, B. Zwanenburg, and G. J. F. Chittenden, *Synth. Commum.*, **33**, 493 (2003).

15. C. Li and A. Vasella, *Helv. Chim. Acta*, **76**, 211 (1993).

16. C. W. Andrews, R. Rodebaugh, and B. Fraser-Reid, *J. Org. Chem.*, **61**, 5280 (1996).

17. R. Madsen and B. Fraser-Reid, *J. Org. Chem.*, **60**, 772 (1995).

18. D. A. Evans and J. A. Gauchet-Prunet, *J. Org. Chem.*, **58**, 2446 (1993).

19. W. H. Hartung and R. Simonoff, *Org. React.*, **7**, 263 (1953); see pp. 271, 284, 302.

20. M. Zaoral, J. Jezek, R. Straka, and K. Masek, *Collect. Czech Chem. Commun.*, **43**, 1797 (1978).

21. R. M. Hann, N. K. Richtmyer, H. W. Diehl, and C. S. Hudson, *J. Am. Chem. Soc.*, **72**, 561 (1950).

22. M. Smith, D. H. Rammler, I. H. Goldberg, and H. G. Khorana, *J. Am. Chem. Soc.*, **84**, 430 (1962).

23. A. T. N. Belarmino, S. Froehner, and D. Zanette, *J. Org. Chem.*, **68**, 706 (2003).

24. V. G. Mairanovsky, *Angew. Chem., Int. Ed. Engl.*, **15**, 281 (1976).

25. T. G. Bonner, E. J. Bourne, and S. McNally, *J. Chem. Soc.*, 2929 (1960).

26. W. A. Szarek, A. Zamojski, K. N. Tiwari, and E. R. Ison *Tetrahedron Lett.*, **27**, 3827 (1986).

27. M. H. Park, R. Takeda, and K. Nakanishi, *Tetrahedron Lett.*, **28**, 3823 (1987).

28. T. Bieg and W. Szeja, *Synthesis*, 317 (1986).

29. K. C. Nicolaou, C. A. Veale, C.-K. Hwang, J. Hutchinson, C. V. C. Prasad, and W. W. Ogilvie, *Angew. Chem., Int. Ed. Engl.*, **30**, 299 (1991).

30. T. Hu, N. Takenaka, and J. S. Panek, *J. Am. Chem. Soc.*, **124**, 12806 (2002).

31. J. Xia and Y. Hui, *Synth. Commun.*, **26**, 881 (1996).

32. P. J. Garegg, "Regioselective Cleavage of *O*-Benzylidene Acetals to Benzyl Ethers," in *Preparative Carbohydrate Chemistry*, S. Hanessian, Ed., Marcel Dekker, New York, 1997, pp. 53–65.

33. S. Takano, M. Akiyama, S. Sato, and K. Ogasawara, *Chem. Lett.*, **12** 1593 (1983); S. Hatakeyama, K. Sakurai, K. Saijo, and S. Takano, *Tetrahedron Lett.*, **26**, 1333 (1985).

34. S. L. Schreiber, Z. Wang, and G. Schulte, *Tetrahedron Lett.*, **29**, 4085 (1988).

35. L. Grimaud, D. Rotulo, R. Ros-Perez, L. Guitry-Azam, and J. Prunet, *Tetrahedron Lett.*, **43**, 7477 (2002).

36. G. A. Molander and F. Dehmel, *J. Am. Chem. Soc.*, **126**, 10313 (2004).

37. D. R. Gauthier, Jr., R. H. Szumigala, Jr., J. D. Armstrong III, and R. P. Volante, *Tetrahedron Lett.*, **42**, 7011 (2001).

38. F. G. De las Heras, A. San Felix, A. Calvo-Mateo, and P. Fernandez-Resa, *Tetrahedron*, **41**, 3867 (1985).

39. R. C. Corcoran, *Tetrahedron Lett.*, **31**, 2101 (1990).

40. F. Peri, L. Cipolla, and F. Nicotra, *Carbohydr. Lett.*, **4**, 21 (2000).

41. H. Kotsuki, Y. Ushio, N. Yoshimura, and M. Ochi, *J. Org. Chem.*, **52**, 2594 (1987).

42. G. Adam and D. Seebach, *Synthesis*, 373 (1988).

43. P. J. Garegg and H. Hultberg, *Carbohydr. Res.*, **93**, C10 (1981). L. Qiao and J. C. Vederas, *J. Org. Chem.*, **58**, 3480 (1993).

44. M. Ghosh, R. G. Dulina, R. Kakarla, and M. J. Sofia, *J. Org. Chem.*, **65**, 8387 (2000).

45. N. L. Pohl and L. L. Kiessling, *Tetrahedron Lett.*, **38**, 6985 (1997).

46. K. Suzuki, H. Nonaka, and M. Yamaura, *Tetrahedron Lett.*, **44**, 1975 (2003).

47. Y. Guindon, Y. Girard, S. Berthiaume, V. Gorys, R. Lemieux, and C. Yoakim, *Can. J. Chem.*, **68**, 897 (1990).

48. B. Classon, P. J. Garegg, and A.-C. Helland, *J. Carbohydr. Chem.*, **8**, 543 (1989).

49. P. J. Garegg, *Acc. Chem. Res.*, **25**, 575 (1992).

50. K. Ishihara, A. Mori, and H. Yamamoto, *Tetrahedron*, **46**, 4595 (1990).

51. A. A. Sherman, Y. V. Mironov, O. N. Yudina, and N. E. Nifantiev, *Carbohydr. Res.*, **338**, 697 (2003).

52. H. Ando, Y. Koike, S. Koizumi, H. Ishida, and M. Kiso, *Angew. Chem. Int. Ed.*, **44**, 6759 (2005).

53. C.-C. Wang, S.-Y. Luo, C.-R. Shie, and S.-C. Hung, *Org. Lett.*, **4**, 847 (2002).

54. L. Jiang and T.-H. Chan, *Tetrahedron Lett.*, **39**, 355 (1998).

55. M. Mandal, V. Y. Dudkin, X. Geng, and S. J. Danishefsky, *Angew. Chem. Int. Ed.*, **43**, 2557 (2004).

56. S. N. Lam and J. Gervay-Hague, *J. Org. Chem.*, **70**, 8772 (2005).

57. M. P. DeNinno, J. B. Etienne, and K. C. Duplantier, *Tetrahedron Lett.*, **36**, 669 (1995); A. Arasappan and B. Fraser-Reid, *J. Org. Chem.*, **61**, 2401 (1996).

58. A. Aravind and S. Baskaran, *Tetrahedron Lett.*, **46**, 743 (2005); S. D. Debenham and E. J. Toone, *Tetrahedron: Asymmetry*, **11**, 385 (2000).

59. M. Sakagami and H. Hamana, *Tetrahedron Lett.*, **41**, 5547 (2000).

60. C.-R. Shie, Z.-H. Tzeng, S. S. Kulkarni, B.-J. Uang, C.-Y. Hsu, and S.-C. Hung, *Angew. Chem. Int. Ed.*, **44**, 1665 (2005).

61. S. Saito, A. Kuroda, K. Tanaka, and R. Kimura, *Synlett*, 231 (1996).

62. B. Delpech, D. Calvo, and R. Lett, *Tetrahedron Lett.*, **37**, 1015 (1996).

63. K. Sato, T. Igarashi, Y. Yanagisawa, N. Kawauchi, H. Hashimoto, and J. Yoshimura, *Chem. Lett.*, **17**, 1699 (1988).

64. F. E. Ziegler and J. S. Tung, *J. Org. Chem.*, **56**, 6530 (1991).

65. P. Deslongchamps, C. Moreau, D. Fréhel, and R. Chênevert, *Can. J. Chem.*, **53**, 1204 (1975).

66. S. Bhat, A. R. Ramesha, and S. Chandrasekaran, *Synlett*, 329 (1995).

67. M. Adinolfi, G. Barone, L. Guariniello, and A. Iadonisi, *Tetrahedron Lett.*, **40**, 8439 (1999).

68. T. Hosokawa, Y. Imada, and S. I. Murahashi, *J. Chem. Soc., Chem. Commun.*, 1245 (1983).

69. P. H. G. Wiegerinck, L. Fluks, J. B. Hammink, S. J. E. Mulders, F. M. H. de Groot, H. L. M. van Rozendaal, and H. W. Scheeren, *J. Org. Chem.*, **61**, 7092 (1996).

70. T. Sueda, S. Fukuda, and M. Ochial, *Org. Lett.*, **3**, 2387 (2001).

71. F. A. Luzzio and R. A. Bobb, *Tetrahedron Lett.*, **38**, 1733 (1997).

72. S. Hanessian and A. P. A. Staub, *Tetrahedron Lett.*, **14**, 3551 (1973).

73. R. W. Binkley, G. S. Goewey, and J. C. Johnston, *J. Org. Chem.*, **49**, 992 (1984).

74. J. McNulty, J. Wilson, and A. C. Rochon, *J. Org. Chem.*, **69**, 563 (2004).

75. O. Han and H.-w. Liu, *Tetrahedron Lett.*, **28**, 1073 (1987); F. Chretien, M. Khaldi, and Y. Chapleur, *Synth. Commun.*, **20**, 1589 (1990).

76. F. A. Luzzio and R. A. Bobb, *Tetrahedron Lett.*, **38**, 1733 (1997).

77. P. M. Collins, A. Manro, E. C. Opara-Mottah, and M. H. Ali, *J. Chem. Soc., Chem. Commun.*, 272 (1988).

78. Y. Chen and P. G. Wang, *Tetrahedron Lett.*, **42**, 4955 (2001).

79. Y.-G. Suh, J.-K. Jung, S.-Y. Seo, K.-H. Min, D.-Y. Shin, Y.-S. Lee, S.-H. Kim, and H.-J. Park, *J. Org. Chem.*, **67**, 4127 (2002).

80. B. P. Roberts and T. M. Smits, *Tetrahedron Lett.*, **42**, 137 (2001).

p-Methoxybenzylidene Acetal (Chart 3)

The *p*-methoxybenzylidene acetal is a versatile protective group for diols that undergoes acid hydrolysis 10 times faster than the benzylidene group.[1] As with the benzylidene derivative, the 1,3-derivative is thermodynamically favored over the 1,2-derivative.[2] Because of its acid sensitivity, it has been observed to migrate in during chromatography on silica gel.[3]

The following example shows that the methoxybenzylidene acetal is not always an innocent bystander. During an attempted Barton deoxygenation the benzylidene acetal participated in a 1,5-hydrogen shift when the reaction was run under dilute conditions, but this could be obviated by running the reaction in neat Bu_3SnH.[4]

Concentration	Ratio a:b:c	Yield a
0.03 M toluene, 1.1 eq. Bu_3SnH	1:2:2	20%
0.003 M toluene, 1.1 eq. Bu_3SnH	0:1:1	NA
0.03 M toluene, neat Bu_3SnH	1:0:0	84%

Formation

1. p-MeOC$_6$H$_4$CHO, acid, 70–95% yield.[1, 5] The thermodynamic isomer is favored.

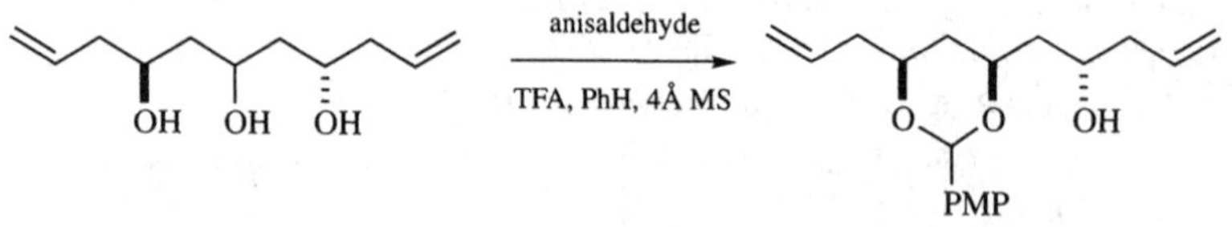

2. From a trimethylsilylated triol: p-MeOC$_6$H$_4$CHO, TMSOTf, CH$_2$Cl$_2$, −78°C, 5 h, 96% yield.[6]

3. p-MeOC$_6$H$_4$CH$_2$OMe, DDQ, CH$_2$Cl$_2$, rt, 30 min, 49–82% yield.[7,8]

4. p-MeOC$_6$H$_4$CHO, ZnCl$_2$.[9]

5. p-MeOC$_6$H$_4$CH(OMe)$_2$, acid.[10] The related o-methoxybenzylidene acetal has been prepared by this method.[11] Useful diol selectivity has been achieved as in the following illustration.[12]

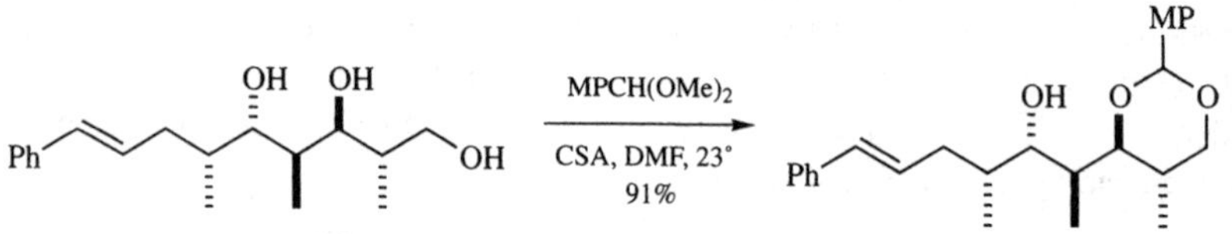

MP = p-methoxyphenyl

6. The p-methoxybenzylidene ketal can be prepared by DDQ oxidation of a p-methoxybenzyl group that has a neighboring hydroxyl.[13] This methodology has been used to advantage in a number of syntheses.[14,15] In one case, to prevent an unwanted acid-catalyzed acetal isomerization, it was necessary to recrystallize the DDQ and use molecular sieves.[16] The following examples serve to illustrate the reaction.[17,18]

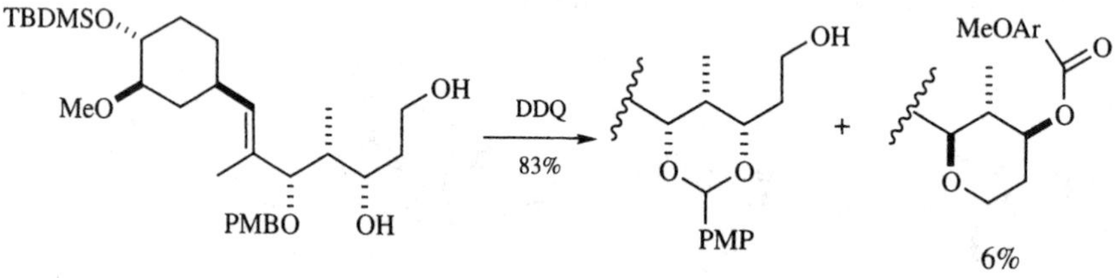

Ref. 18

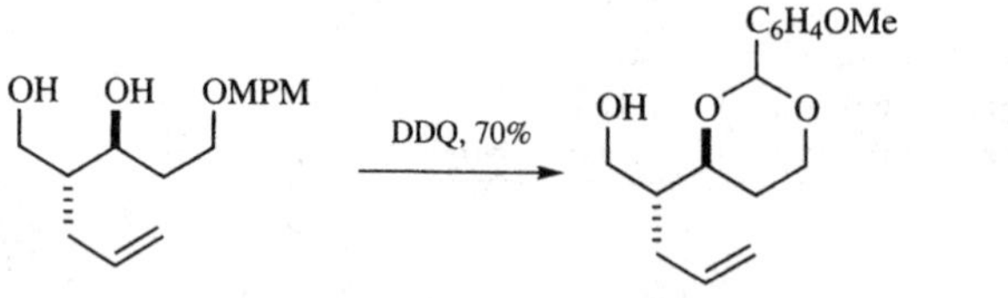

MPM = PMB = p-methoxybenzyl   Ref. 17

Cleavage

1. 80% AcOH, 25°C, 10 h, 100% yield.[1] Mesitylene acetals have been found to be stable during the acid (pH = 1)-catalyzed cleavage of *p*-methoxybenzylidene acetals.[19]

2. The PMP acetal is quite susceptible to acid-catalyzed cleavage. In the following case a normally readily cleaved cyclopentylidene group could not be cleaved in preference to the PMP acetal. In a very creative move the authors prepared a charge transfer complex with the extremely electron-deficient trinitrotoluene and the electron-rich PMP groups to suppress protonation of the oxygens of these acetals and allow hydrolysis of the cyclopentylidene group.[20]

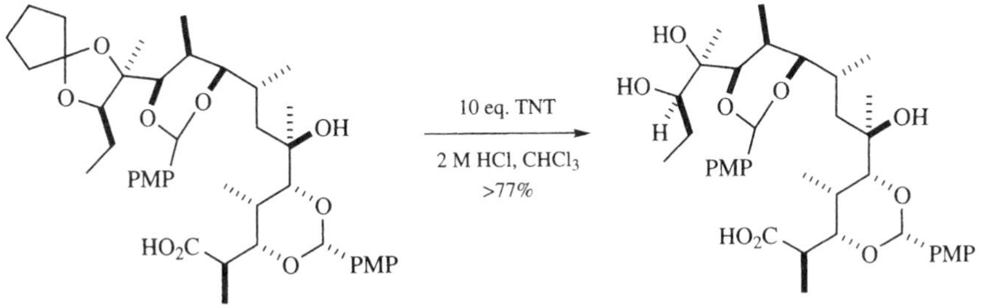

3. Pd(OH)₂, 25°C, 2 h, H₂, >95% yield.[21]
4. EtSH, Zn(OTf)₂, NaHCO₃, 100% yield.[22]

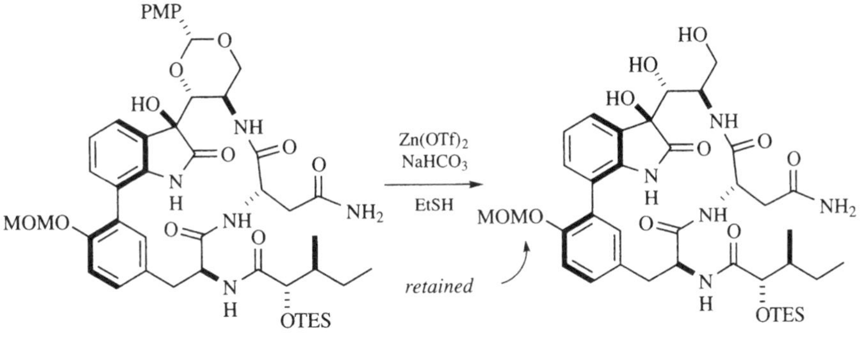

5. Ce(NH₄)₂(NO₃)₆, CH₃CN, H₂O.[23]

   As with the benzylidene group, a variety of methods shown below have been developed to effect cleavage of one of the two C–O bonds in this acetal.

6. (*i*-Bu)₂AlH, PhCH₃, 75% yield.[8,10,24] This reagent generally gives the product that results from reduction at the least hindered position,[25] but neighboring groups such as a carbonyl that can coordinate to DIBAH can change the course of the reaction to give the secondary alcohol.[26]

7. DDQ, water, 87% yield.[7] This method results in the formation of a mixture of the two possible monobenzoates.[27]

$LiAlH_4/AlCl_3$,[28,29] $BH_3 \cdot NMe_3/AlCl_3$,[4] $BH_3 \cdot THF/heat$,[29] $BH_3 \cdot THF/TMSOTf/ CH_2Cl_2$,[30] or $NaBH_3CN/TMSCl$, CH_3CN[11] result in cleavage at the least hindered side of the ketal, giving the more hindered ether, whereas $NaBH_3CN/ HCl$[4] or $NaBH_3CN/TFA/DMF$[11] results in formation of an MPM ether at the least hindered alcohol.

8. BH_3, Bu_2OTf, THF. In this case the direction of cleavage is temperature-dependent.[31] The allyl group is compatible with the low-temperature conditions.

9. Bu$_3$SnH, MgBr$_2$·Et$_2$O, CH$_2$Cl$_2$. This method results in the formation of a primary PMB ether when chelation control is possible; otherwise it gives the secondary ether.[32]

10. PhBCl$_2$, Et$_3$SiH, 4-Å MS, Et$_2$O, $-78°C$ to $-40°C$, 90% yield.[33]

11. DDQ, CH$_2$Cl$_2$, Bu$_4$NCl, ClCH$_2$CH$_2$Cl, 96% yield. When CuBr$_2$/ Bu$_4$NBr is used the 6-Br derivative is produced in 93% yield. [27]

MP = *p*-methoxyphenyl

12. Ozone.[34] Most acetals are subject to cleavage with ozone giving a mono ester of the original diol.

13. PDC, *t*-BuOOH, 0°C, 4–8 h.[35] Other acetals are similarly cleaved.

14. Selectfluor, CH$_3$CN, 5% H$_2$O, 5 h, rt, 87–92% yield. This reagent also cleaves dithianes and THP ethers.[36]

1. M. Smith, D. H. Rammler, I. H. Goldberg, and H. G. Khorana, *J. Am. Chem. Soc.*, **84**, 430 (1962).

2. K. Takebuchi, Y. Hamada, and T. Shioiri, *Tetrahedron Lett.*, **35**, 5239 (1994); T. Gustafsson, M. Schou, F. Almqvist, and J. Kihlberg, *J. Org. Chem.*, **69**, 8694 (2004).

3. A. Arefolov and J. S. Panek, *Org. Lett.*, **4**, 2397 (2002).

4. D. A. Evans, A. S. Kim, R. Metternich, and V. J. Novack, *J. Am. Chem. Soc.*, **120**, 5921 (1998).

5. J. N. Shepherd and D. C. Myles, *Org. Lett.*, **5**, 1027 (2003).

6. P. Breuilles, G. Oddon, and D. Uguen, *Tetrahedron Lett.*, **38**, 6607 (1997).

7. Y. Oikawa, T. Nishi, and O. Yonemitsu, *Tetrahedron Lett.*, **24**, 4037 (1983).

8. Y. Ito, Y. Ohnishi, T. Ogawa, and Y. Nakahara, *Synlett*, 1102 (1998).

9. S. Hanessian, J. Kloss, and T. Sugawara, *ACS Symp. Ser.* **386**, "Trends in Synth. Carbohydr. Chem., 64 (1989).

10. M. Kloosterman, T. Slaghek, J.P.G. Hermans, and J. H. Van Boom, *Recl: J. R. Neth. Chem. Soc.*, **103**, 335 (1984).

11. V. Box, R. Hollingsworth, and E. Roberts, *Heterocycles*, **14**, 1713 (1980).

12. D. A. Evans and H. P. Ng, *Tetrahedron Lett.*, **34**, 2229 (1993).

13. Y. Oikawa, T. Yoshioka, and O. Yonemitsu, *Tetrahedron Lett.*, **23**, 889 (1982).

14. A. F. Sviridov, M. S. Ermolenko, D. V. Yaskunsky, V. S. Borodkin, and N. K. Kochetkov, *Tetrahedron Lett.*, **28**, 3835 (1987).

15. J. S. Yadav, M. C. Chander, and B. V. Joshi, *Tetrahedron Lett.*, **29**, 2737 (1988).

16. R. Stürmer, K. Ritter, and R. W. Hoffman, *Angew. Chem., Int. Ed. Engl.*, **32**, 101 (1993).

17. A. B. Jones, M. Yamaguchi, A. Patten, S. J. Danishefsky, J. A Ragan, D. B. Smith, and S. L. Schreiber, *J. Org. Chem.*, **54**, 17 (1989); A. B. Smith III, K. J. Hale, L. M. Laakso, K. Chen, and A. Riera, *Tetrahedron Lett.*, **30**, 6963 (1989).

18. J. A. Marshall and S. Xie, *J. Org. Chem.*, **60**, 7230 (1995).

19. S. F. Martin, T. Hida, P. R. Kym, M. Loft, and A. Hodgson, *J. Am. Chem. Soc.*, **119**, 3193 (1997).

20. R. Stürmer, K. Ritter, and R. W. Hoffmann, *Angew. Chem. Int. Ed.* **32**, 101 (1993).

21. K. Toshima, S. Murkaiyama, T. Yoshida, T. Tamai, and K. Tatsuta, *Tetrahedron Lett.*, **32**, 6155 (1991).

22. M. Inoue, H. Sakazaki, H. Furuyama, and M. Hirama, *Angew. Chem. Int. Ed.*, **42**, 2654 (2003).

23. R. Johansson and B. Samuelsson, *J. Chem. Soc., Chem. Commun.*, 201 (1984).

24. E. Marotta, I. Pagani, P. Righi, G. Rosini, V. Bortolasi, and A. Medici, *Tetrahedron: Asymmetry*, **6**, 2319 (1995).

25. R. Munakata, H. Katakai, T. Ueki, J. Kurosaka, K.-i. Takao, and K.-i. Tadano, *J. Am. Chem. Soc.*, **126**, 11254 (2004).

26. J. Mulzer, A. Mantoulidis, and E. Ohler, *J. Org. Chem.*, **65**, 7456 (2000).

27. Z. Zhang and G. Magnusson, *J. Org. Chem.*, **61**, 2394 (1996).

28. I. Sato, Y. Akahori, K.-i. Iida, and M. Hirama, *Tetrahedron Lett.*, **37**, 5135 (1996).

29. T. Tsuri and S. Kamata, *Tetrahedron Lett.*, **26**, 5195 (1985).

30. J. M. Hernández-Torres, S.-T. Liew, J. Achkar, and A. Wei, *Synthesis*, 487 (2002).

31. J. M. Hernández-Torres, J. Achkar, and A. Wei, *J. Org. Chem.*, **69**, 7206 (2004).

32. B.-Z. Zheng, M. Yamauchi, H. Dei, S.-i. Kusaka, K. Matsui, and O. Yonemitsu, *Tetrahedron Lett.*, **41**, 6441 (2000).

33. A. Dilhas and D. Bonnaffe, *Tetrahedron Lett.*, **45**, 3643 (2004).

34. P. Deslongchamps, P. Atlani, D. Fréhel, A. Malaval, and C. Moreau, *Can. J. Chem.*, **52**, 3651 (1974).

35. N. Chidambaram, S. Bhat, and S. Chandrasekaran, *J. Org. Chem.*, **57**, 5013 (1992).

36. J. Liu and C.-H. Wong, *Tetrahedron Lett.*, **43**, 4037 (2002).

1-(4-Methoxyphenyl)ethylidene Ketal

Formation/Cleavage[1]

PPTS = pyridinium p-toluenesulfonate

1. B. H. Lipshutz and M. C. Morey, *J. Org. Chem.*, **46**, 2419 (1981).

2,4-Dimethoxybenzylidene Acetal: $2,4\text{-}(CH_3O)_2C_6H_3CH(OR)_2$

This acetal is stable to hydrogenation with W4-Raney Ni, which was used to cleave a benzyl group in 99% yield.[1]

Formation

$2,4\text{-}(MeO)_2C_6H_3CHO$, benzene, TsOH, heat, >81% yield.[2]

Cleavage

As with the benzylidine acetal the DMP derivative can be selectively reduced with DIBAL to give an alcohol and a protected alcohol.[3]

1. K. Horita, T. Yoshioka, T. Tanaka, Y. Oikawa, and O. Yonemitsu, *Tetrahedron*, **42**, 3021 (1986).

2. M. Smith, D. H. Rammler, I. H. Goldberg, and H. G. Khorana, *J. Am. Chem. Soc.*, **84**, 430 (1962).

3. A. B. Smith III, V. A. Doughty, Q. Lin, L. Zhuang, M. D. McBriar, A. M. Boldi, W. H. Moser, N. Murase, K. Nakayama, and M. Sobukawa, *Angew. Chem. Int. Ed.*, **40**, 191 (2001).

3,4-Dimethoxybenzylidene Acetal

Formation

Treatment of a 3,4-dimethoxybenzyl ether containing a free hydroxyl with DDQ (benzene, 3-Å molecular sieves, rt) affords the 3,4-dimethoxybenzylidene acetal.[1]

Cleavage[2]

The acetal can also be cleaved with DDQ (CH_2Cl_2, H_2O, 66% yield) to afford the monobenzoate. Treatment with DIBAL (CH_2Cl_2, 0°C, 91% yield) affords the hydroxy ether.[3]

1. K. Nozaki and H. Shirahama, *Chem. Lett.*, **17**, 1847 (1988).

2. M. J. Wanner, N. P. Willard, G. J. Koomen, and U. K. Pandet, *Tetrahedron*, **43**, 2549 (1987).

3. A. B. Smith, III, Q. Lin, K. Nakayama, A. M. Boldi, C. S. Brook, M. D. McBriar, W. H. Moser, M. Sobukawa, and L. Zhuang, *Tetrahedron Lett.*, **38**, 8675 (1997).

p-Acetoxybenzylidene Acetal

Formation

p-AcOC$_6$H$_4$CHO, ZnCl$_2$, CH$_2$Cl$_2$, rt, 18 h, 85% yield.[1]

Cleavage

1. As with the 4-TBSObenzylidene acetal, treatment with base should cleave this group.
2. HCl, Et$_2$O, NaCNBH$_3$, THF.[1]

PAB = *p*-Acetoxybenzyl

1. L. Jobron and O. Hindsgaul, *J. Am. Chem. Soc.*, **121**, 5835 (1999).

4-(*t*-Butyldimethylsilyloxy)benzylidene Acetal

The 4-(*t*-butyldimethylsilyloxy)benzylidene acetal was developed for protection of 1,2-diols in situations where strong acid conditions could not be used for deprotection.

Formation

1. From the bis TMS ether: TBSOC$_6$H$_4$CHO, TMSOTf, CH$_2$Cl$_2$, −78°C, 5 min, 91–94% yield.[1]
2. From the diol: TBSOC$_6$H$_4$CH(OMe)$_2$, CSA, DMF, rt – 50°C, 84–96% yield.[1]

Cleavage

1. K$_2$CO$_3$, NH$_2$OH·HCl, CsF, MeOH, H$_2$O, 70°C, 91–93% yield. The inclusion of CsF improves the rate of deprotection, but its absence does not prevent deprotection. These conditions could not be used with substrates containing esters because of their hydrolysis.[1]
2. A 2-step process: TBAF, THF or (HF)$_3$·TEA, THF to remove the TBS group followed by AcOH, THF, H$_2$O at rt.[1,2]

A comparison of hydrolysis rates of various benzylidene acetals with AcOH/ H_2O showed that the *p*-hydroxybenzylidene group was removed in about 1 h vs. 2.5 h for the benzylidene acetal and 2 h for the *p*-methoxybenzylidene acetal.

1. Y. Kaburagi, H. Osajima, K. Shimada, H. Tokuyama, and T. Fukuyama, *Tetrahedron Lett.*, **45**, 3817 (2004).
2. K. Shimada, Y. Kaburagi, and T. Fukuyama, *J. Am. Chem. Soc.*, **125**, 4048 (2003).

2-Nitrobenzylidene Acetal

The 2-nitrobenzylidene acetal has been used to protect carbohydrates. It can be cleaved by photolysis (45 min, MeOH; CF_3CO_3H, CH_2Cl_2, 0°C, 95% yield) to form primarily axial 2-nitrobenzoates from diols containing at least one axial alcohol.[1] As with other benzylidene acetals the ring can be opened to give a benzyl ether and an alcohol.[2] The resulting benzyl ethers can be removed photochemically.

4-Nitrobenzylidene Acetal

Formation

1. $4\text{-NO}_2\text{PhCH(OMe)}_2$, TsOH, DMF, benzene, heat. Used to protect a 4,6-glucopyranoside.[3]
2. $4\text{-NO}_2\text{PhCHO}$, TMS_2O, TMSOTf, Et_3SiH, THF, 96% yield.[4]

1. P. M. Collins and V. R. N. Munasinghe, *J. Chem. Soc., Perkin Trans. I*, 921 (1983).
2. S. Watanabe, T. Sueyoshi, M. Ichihara, C. Uehara, and M. Iwamura, *Org. Lett.*, **3**, 255 (2001).
3. W. Guenther and H. Kunz, *Carbohydr. Res.*, **228**, 217 (1992).
4. Y. Fukase, S.-Q. Zhang, K. Iseki, M. Oikawa, K. Fukase, and S. Kusumoto, *Synlett*, 1693 (2001).

Mesitylene Acetal: MesCH(OR)$_2$

Formation

MesCH(OR)$_2$, CSA, CH$_2$Cl$_2$, 61–91% yield.[1,2]

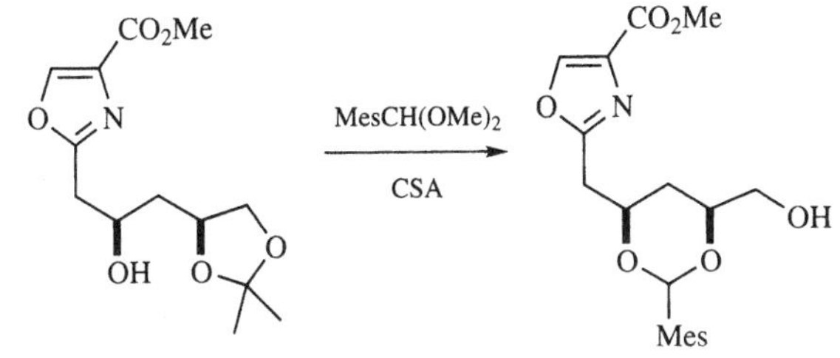

Cleavage

Cleavage of the mesitylene acetal is facilitated by the steric compression induced by the two ortho-methyl groups which raise the ground state energy of the acetal.

1. Pd(OH)$_2$, H$_2$, EtOH, rt, 12 h.[2] A BOM group can be removed by hydrogenolysis (10% Pd–C, MeOH, THF, 83% yield) in the presence of the mesitylene and 4-methoxyphenyl acetals.[1]

2. 50% Aq. AcOH, 35°C, >70% yield.[1] In the following illustration, methoxy-substituted benzylidene acetals could not be hydrolyzed,[3] which implies that the mesitylene acetal is more stable, but this was

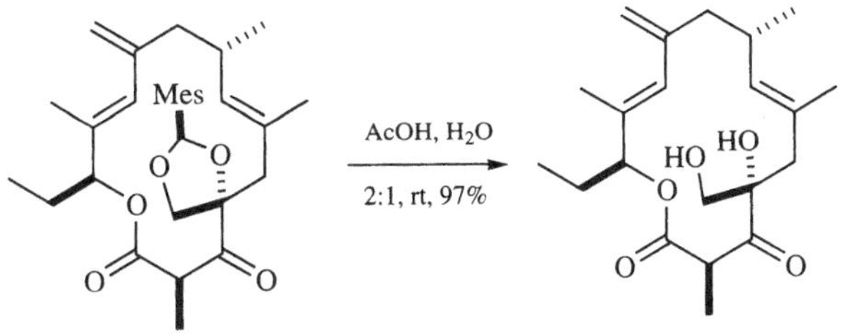

contradicted by the following example where the PMP acetal is cleaved in preference to the mesitylene derivative.[4]

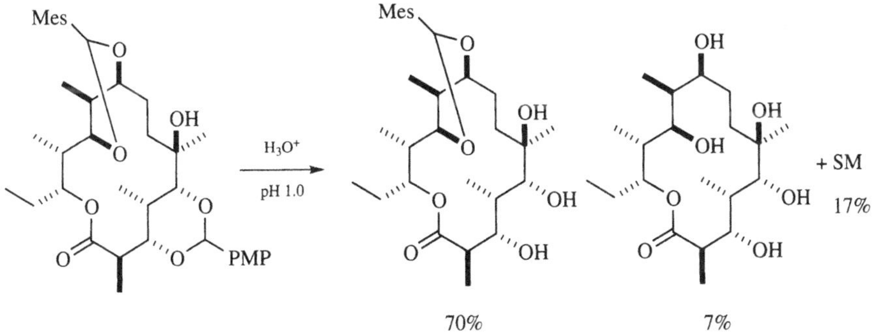

1. S. F. Martin, T. Hida, P. R. Kym, M. Loft, and A. Hodgson, *J. Am. Chem. Soc.*, **119**, 3193 (1997). P. J. Hergenrother, A. Hodgson, A. S. Judd, W.-C. Lee, and S. F. Martin, *Angew. Chem. Int. Ed.*, **42**, 3278 (2003).

2. M. Hikota, H. Tone, K. Horita, and O. Yonemitsu, *J. Org. Chem.*, **55**, 7 (1990).

3. B. Tse, *J. Am. Chem. Soc.*, **118**, 7094 (1996).

4. S. F. Martin, T. Hida, P. R. Kym, M. Loft, and A. Hodgson, *J. Am. Chem. Soc.*, **119**, 3193 (1997).

6-Bromo-7-hydroxycoumarin-2-yl-methylidene Acetal

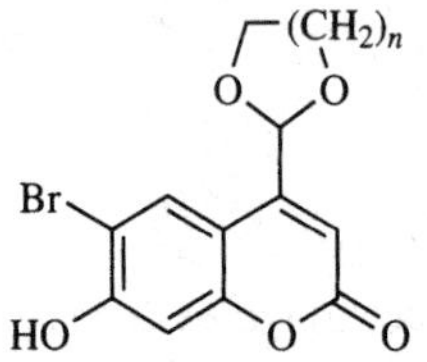

This photolabile protective group was developed for the protection of diols, which could release caged biologically active molecules in biological systems. The acetal is prepared from the aldehyde and a diol (PPTS, toluene, $MgSO_4$, reflux) and is cleaved by photolysis at 348 nm in a pH 7.4 buffer.[1]

1. W. Lin and D. S. Lawrence, *J. Org. Chem.*, **67**, 2723 (2002).

1-Naphthaldehyde Acetal: $C_{10}H_7CH(OR)_2$

This acetal was prepared to confer crystallinity on the intermediates in the synthesis of the lysocellin antibiotics.[1]

Formation

$C_{10}H_7CHO$, trichloroacetic acid, PhH, >74% yield.

Cleavage

1. Pd/C, H_2O, $(COOH)_2$, EtOAc, 0°C, 61% yield.
2. 2:1 THF 1 M H_2SO_4, 45°C, 81% yield.

2-Naphthaldehyde acetal: $C_{10}H_7CH(OR)_2$

Formation

1. 2-(dimethoxymethyl)naphthalene, PTSA, DMF, rt, overnight, 90–97% yield.[2]
2. 2-naphthaldehyde, CH_3CN, DMF, PTSA, 2 days, 90–97% yield.[2]

Cleavage

1. DDQ, CH_3CN, H_2O, 2–3 h, 95–97% yield.[2]
2. The naphthylidene acetal can be selectively cleaved in a manner similar to the benzylidene acetal.[2] $VO(OTf)_2/BH_3 \cdot THF$ can be used as a substituted for $AlCl_3/LiAlH_4$.[3,4]

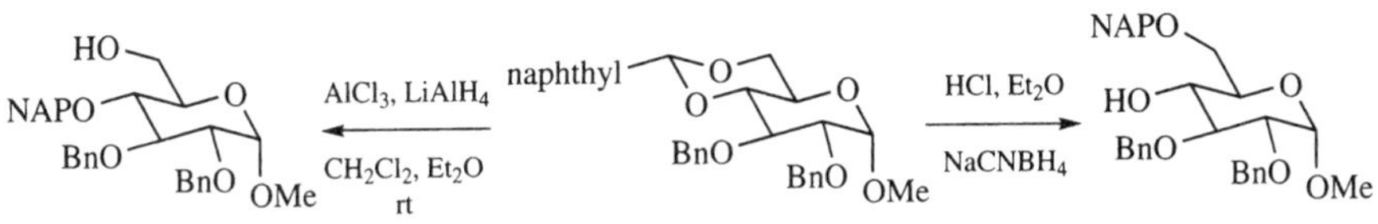

1. D. A. Evans, R. P. Polniaszek, K. M. DeVries, D. E. Guinn, and D. J. Mathre, *J. Am. Chem. Soc.*, **113**, 7613 (1991).
2. A. Lipták, A. Borbas, L. Janossy, and L. Szilagyi, *Tetrahedron Lett.*, **41**, 4949 (2000). K. Fujiwara, A. Goto, D. Sato, Y. Ohtaniuchi, H. Tanaka, A. Murai, H. Kawai, and T. Suzuki, *Tetrahedron Lett.*, **45**, 7011 (2004).
3. J.-C. Lee, X.-A. Lu, S. S. Kulkarni, Y.-S. Wen, and S.-C. Hung, *J. Am. Chem. Soc.*, **126**, 476 (2004).
4. A. Borbás, Z. B. Szabo, L. Szilágyi, A. Bényei, and A. Lipták, *Tetrahedron*, **58**, 5723 (2002).

9-Anthracene Acetal

The 9-anthracene acetal was developed as a fluorescent protective group to facilitate purification and reaction monitoring on solid supports. These acetals are also very crystalline.[1]

Formation

Anthracene-9-CH(OMe)$_2$, CH_3CN, PTSA, 3 h, 94–96% yield. Deprotection is more facile than the related benzylidene acetal.

Cleavage

1. 80% AcOH, H_2O, 90°C, 2 h, 94–97% yield.
2. $NaBH_3CN$, THF, Et_2O, HCl, 91% yield.

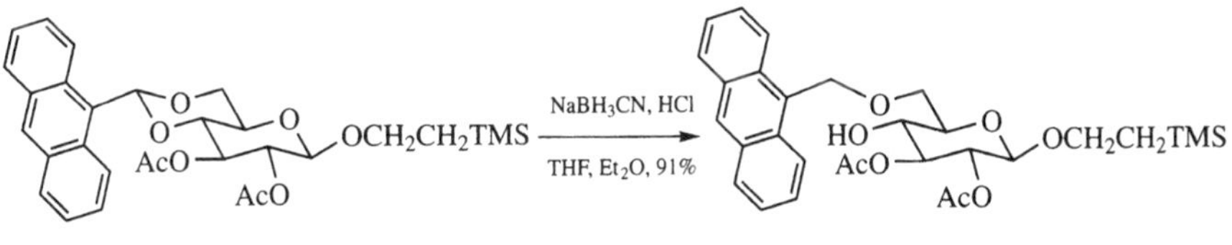

1. U. Ellervik, *Tetrahedron Lett.*, **44**, 2279 (2003).

Benzophenone Ketal: $Ph_2C(OR)_2$

Formation

1. $Ph_2C(OMe)_2$, H_2SO_4.[1]
2. $Ph_2C(OMe)_2$, DMF, TsOH, 50°C, vacuum to remove MeOH, 40–72% yield.[2]
3. Ph_2CCl_2, Pyr.[3]

Cleavage

1. AcOH, H_2O.[4]
2. Hydrochloric acid, 80% dioxane/water.[5] Cleavage rates for various ring sizes were examined.

1. T. Yoon, M. D. Shair, S. J. Danishefsky, and G. K. Shulte *J. Org. Chem.*, **59**, 3752 (1994).
2. L. Di Donna, A. Napoli, C. Siciliano, and G. Sindona, *Tetrahedron Lett.*, **40**, 1013 (1999).
3. A. Borbas, J. Hajko, M. Kajtar-Peredy, and A. Liptak, *J. Carbohydr. Chem.*, **12**, 191 (1993).
4. K. S. Feldman and A. Sambandam, *J. Org. Chem.*, **60**, 8171 (1995).
5. T. Oshima, S.-y. Ueno, and T. Nagai, *Heterocycles*, **40**, 607 (1995).

Di-(p-anisyl)methylidene, Xanthen-9-ylidene, 2,7-Dimethylxanthen-9-ylidene Ketals

These groups were prepared to examine the relative acid lability to the classic iso-propylidene group. They are formed from the corresponding dimethyl ketals in acetonitrile with CSA as a catalyst in 95%, 88%, 70% yield, respectively. The relative rates for the hydrolysis of the uridine derivatives in $TFA/H_2O/MeOH$ at 30° were examined and the results are reported in the following table.[1]

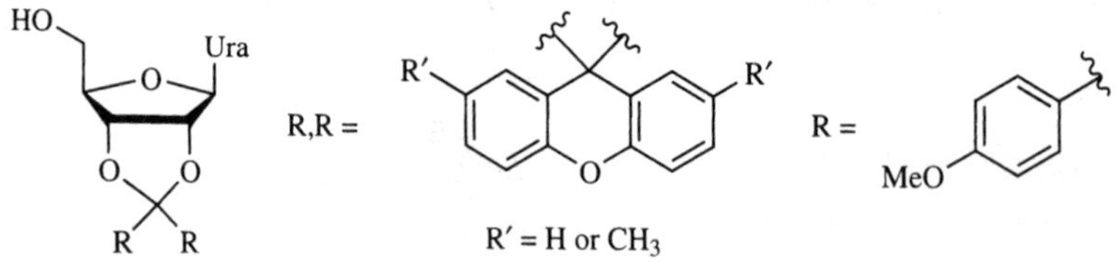

$R,R =$ $R' =$ H or CH_3 $R =$ MeO

Acidic Hydrolysis of 2′, 3′-Protected Uridine Derivatives

Entry	Substrate	Half-Life ($t_{1/2}$) min
1	2′, 3′-*O*-Isoprotylideneuridine (R=Me)	178
2	2′, 3′-*O*-[Di-(p-anisyl)methylene]uridine (R = MP)	56.7
3	2′, 3′-*O*-(Xanthen-9-ylidene)uridine	31.7
4	2′, 3′-*O*-(2, 7-Dimethylxanthen-9-ylidene)uridine	8.6

The xanthen-9-ylidene groups were also examined for the protection of glycerol derivatives.[2] In this case the xanthen-9-ylidene group was removed by reaction with pyrrole in dichloroacetic acid, which forms a bis-pyrrole that is removed with $FeCl_3/Et_2O$.

1. C. B. Reese, Q. Song, and H. Yan, *Tetrahedron Lett.*, **42**, 1789 (2001).
2. C. B. Reese and H. Yan, *J. Chem. Soc. Perkin Trans. 1*, 1807 (2001).

Chiral Ketones

The use of chiral ketones for protection of diols serves two purposes: First, diol protection is accomplished, and second, symmetrical intermediates are converted to chiral derivatives that can be elaborated further so that when the diol is deprotected the molecule retains chirality.[1]

Camphor Ketal

Formation

1. Camphor dimethyl ketal, TMSOTf, DMSO, 90°C, 3 h, 25% yield.[2]
2. Camphor dimethyl ketal, H_2SO_4, DMSO, 70°C, 3 h.[3]

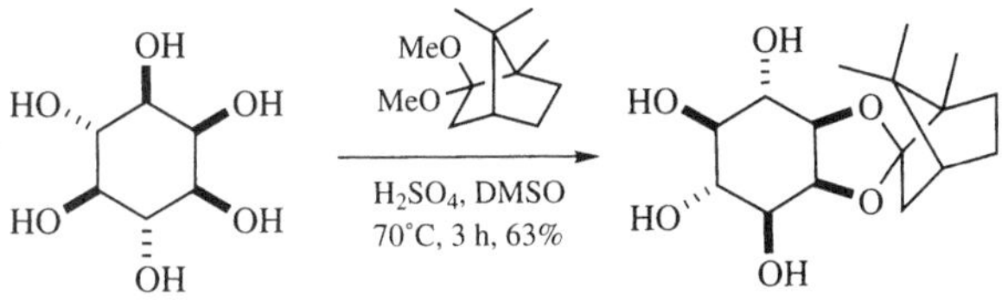

3. Camphor, TsOH, 65–70% yield.[4]

Cleavage

AcOH, H_2O, >88% yield.[4]

Menthone Ketal

Formation

1. Menthone TMS enol ether, TfOH, THF, −40°C, 2 h, 51–91% yield.[5]

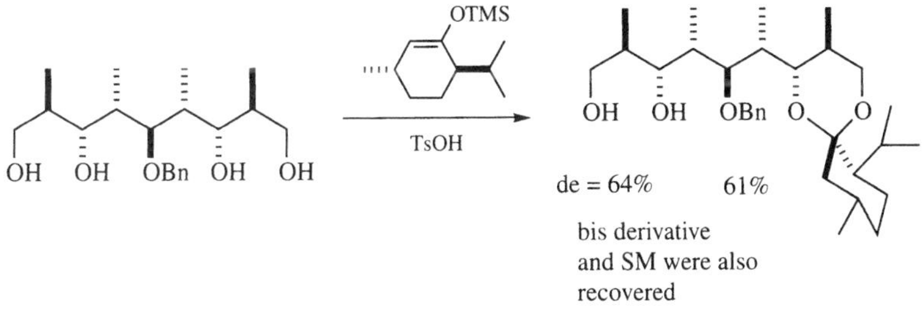

de = 64%   61%

bis derivative
and SM were also
recovered

2. From a TMS protected triol using (−)-menthone.[6]

Cleavage

1. CSA, MeOH, 2 days, rt, 89–90% yield.[6]
2. CHCl$_3$ saturated with 9 N HCl, 85% yield.[6]

1. T. Harada and A. Oku, *Synlett*, 95 (1994).
2. K. S. Bruzik and M.-D. Tsai, *J. Am. Chem. Soc.*, **114**, 6361 (1992).
3. Y. Takahashi, H. Nakayama, K. Katagiri, K. Ichikawa, N. Ito, T. Takita, T. Takeuchi, and T. Miyake, *Tetrahedron Lett.*, **42**, 1053 (2001); J. Lindberg, L. Ohberg, P. J. Garegg, and P. Konradsson, *Tetrahedron*, **58**, 1387 (2002).
4. G. M. Salamonczyk and K. M. Pietrusiewicz, *Tetrahedron Lett.*, **32**, 4031 (1991).
5. T. Harada, Y. Kagamihara, S. Tanaka, K. Sakamoto, and A. Oku, *J. Org. Chem.*, **57**, 1637 (1992); T. Harada, S. Tanaka, and A. Oku, *Tetrahedron*, **48**, 8621 (1992); T. Harada, T. Shintani, and A. Oku, *J. Am. Chem. Soc.*, **117**, 12346 (1995); R. Chenevert and Y. S. Rose, *J. Org. Chem.*, **65**, 1707 (2000).
6. N. Adjé, P. Breuilles and D. Uguen, *Tetrahedron Lett.*, **33**, 2151 (1992).

Cyclic Ortho Esters

A variety of cyclic ortho esters,[1,2] including cyclic orthoformates, have been developed to protect *cis*-1,2-diols. Cyclic ortho esters are more readily cleaved by acidic hydrolysis (e.g., by a phosphate buffer, pH 4.5–7.5, or by 0.005–0.05 M HCl)[3] than are acetonides. Careful hydrolysis or reduction can be used to prepare selectively monoprotected diol derivatives.

Methoxymethylene and Ethoxymethylene Acetal (Chart 3)

Formation

1. HC(OMe)$_3$ or HC(OEt)$_3$, acid catalyst, 77% or 45–80% yields, respectively.[4–6] The reaction is selective for *cis*-diols when there is a choice.[7]

2. Ceric ammonium nitrate, HC(OMe)$_3$, CH$_2$Cl$_2$.[8]

Cleavage

1. 98% formic acid or HCl at pH 2, 20°C.[4]
2. 80% AcOH, rt, 2 h, >80% yield.[9] This method is selective for the inside alcohol of 1,2-diols.[10]

3. Reduction with $(i\text{-Bu})_2\text{AlH}$ affords a diol with one hydroxyl group protected as a MOM group. In general, the more substituted hydroxyl bears the MOM group.[11]

1. C. B. Reese, *Tetrahedron*, **34**, 3143 (1978).

2. V. Amarnath and A. D. Broom, *Chem. Rev.*, **77**, 183 (1977).

3. M. Ahmad, R. G. Bergstrom, M. J. Cashen, A. J. Kresge, R. A. McClelland, and M. F. Powell, *J. Am. Chem. Soc.*, **99**, 4827 (1977).

4. B. E. Griffin, M. Jarman, C. B. Reese, and J. E. Sulston, *Tetrahedron*, **23**, 2301 (1967).

5. J. Zemlicka, *Chem. Ind. (London)*, 581 (1964); F. Eckstein and F. Cramer, *Chem. Ber.*, **98**, 995 (1965).

6. R. M. Ortuño, R. Mercé, and J. Font, *Tetrahedron Lett.*, **27**, 2519 (1986).

7. A.-L. Chauvin, S. A. Nepogodiev, and R. A. Field, *J. Org. Chem.*, **70**, 960 (2005).

8. M. J. Comin, E. Elhalem, and J. B. Rodriguez, *Tetrahedron*, **60**, 11851 (2004).

9. D.-S. Hsu, T. Matsumoto, and K. Suzuki, *Synlett*, 801 (2005).

10. M. Ikejiri, K. Miyashita, T. Tsunemi, and T. Imanishi, *Tetrahedron Lett.*, **45**, 1243 (2004).

11. M. Takasu, Y. Naruse, and H. Yamamoto, *Tetrahedron Lett.*, **29**, 1947 (1988).

2-Oxacyclopentylidene Ortho Ester

This ortho ester does not form a monoester upon deprotection as do acyclic ortho esters, thus avoiding a hydrolysis step.[1]

1. R. M. Kennedy, A. Abiko, T. Takemasa, H. Okumoto, and S. Masamune, *Tetrahedron Lett.*, **29**, 451 (1988).

The following ortho esters have been prepared to protect the diols of nucleosides. They are readily hydrolyzed with mild acid to afford monoester derivatives, generally as a mixture of positional isomers.

Dimethoxymethylene Ortho Ester[1] (Chart 3)

1-Methoxyethylidene Ortho Ester[2]

1-Ethoxyethylidene Ortho Ester[3]

Formation

1. $CH_2=C(OMe)_2$, DMF, TsOH, $<5°C.[4]$ These conditions will completely protect certain triols.[5]

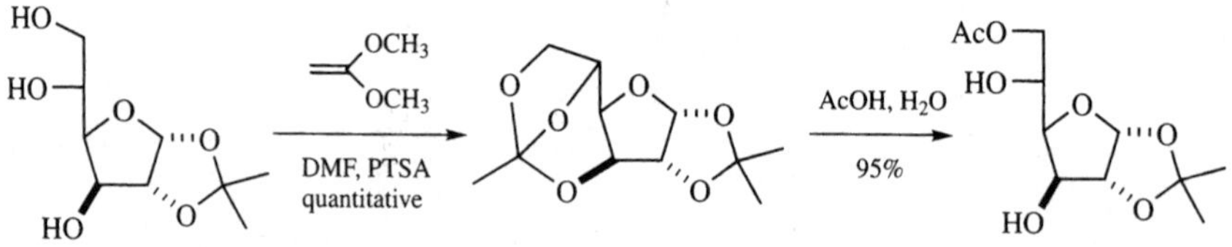

2. $CH_3C(OEt)_3$.[6b] With this ortho ester good selectivity for the axial alcohol is achieved in the acidic hydrolysis of a pyranoside derivative.[4,7]

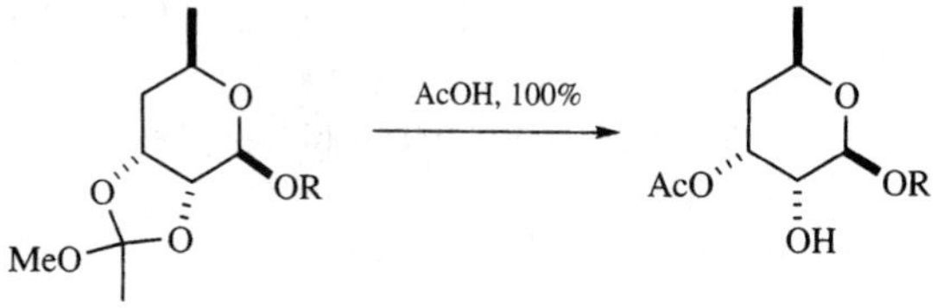

Ref. 7

Methylidene Ortho Ester

Formation[8]

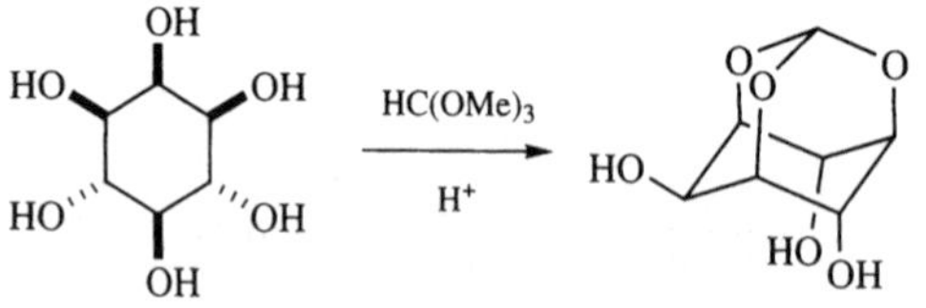

Cleavage

1. TFA, H_2O, rt, 40 h, 85% yield.[9]

2.

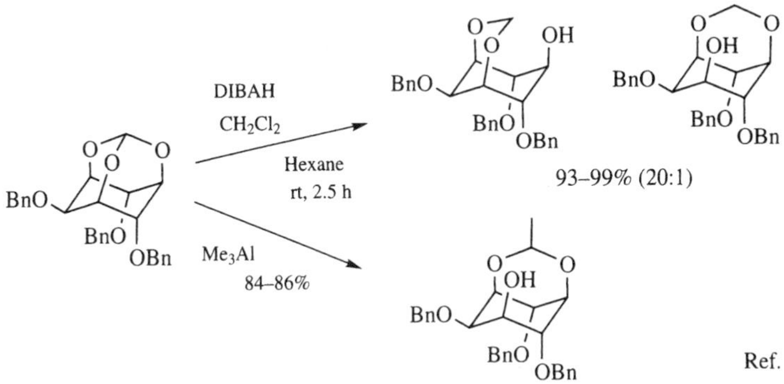
DIBAH
CH_2Cl_2
Hexane
rt, 2.5 h
Me_3Al
84–86%
93–99% (20:1)
BnO
OBn
OH

Ref. 10

Phthalide Ortho Ester

Formation/Cleavage[11]

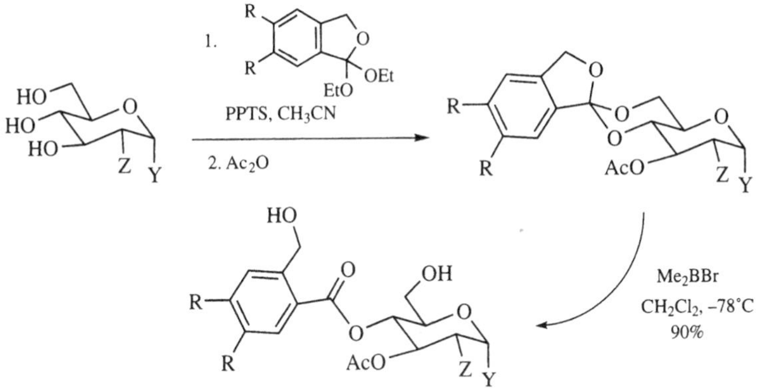
1.
PPTS, CH_3CN
2. Ac_2O
Me_2BBr
CH_2Cl_2, −78°C
90%
OEt
EtO
AcO
HO
OH

1,2-Dimethoxyethylidene Ortho Ester[12]

α-Methoxybenzylidene Ortho Ester[2]

1-(*N,N*-Dimethylamino)ethylidene Derivative[13]

α-(*N,N*-Dimethylamino)benzylidene Derivative[13]

1. G. R. Niaz and C. B. Reese, *J. Chem. Soc., Chem. Commun.*, 552 (1969).
2. C. B. Reese and J. E. Sulston, *Proc. Chem. Soc.*, 214 (1964).
3. V. P. Miller, D.-y. Yang, T. M. Weigel, O. Han, and H.-w. Liu, *J. Org. Chem.*, **54**, 4175 (1989).
4. M. Bouchra, P. Calinaud, and J. Gelas, *Carbohydr. Res.*, **267**, 227 (1995).
5. M. Bouchra, P. Calinaud, and J. Gelas, *Synthesis*, 561 (1995).

6. (a) H. P. M. Fromageot, B. E. Griffin, C. B. Reese, and J. E. Sulston, *Tetrahedron*, **23**, 2315 (1967); (b) U. E. Udodong, C. S. Rao, and B. Fraser-Reid, *Tetrahedron*, **48**, 4713 (1992).

7. S. Hanessian and R. Roy, *Can. J. Chem.*, **63**, 163 (1985).

8. J. P. Vacca, S. J. De Solms, S. D. Young, J. R. Huff, D. C. Billington, R. Baker, J. J. Kulagowski, and I. M. Mawer, *ACS Symp. Ser.*, **463** (Inositol Phosphates Deriv.) 66 (1991).

9. A. M. Riley, M. F. Mahon, and B. V. L. Potter, *Angew. Chem., Int. Ed. Engl.*, **36**, 1472 (1997).

10. S.-M. Yeh, G. H. Lee, Y. Wang, and T.-Y. Luh, *J. Org. Chem.*, **62**, 8315 (1997).

11. A. Arasappan and P. L. Fuchs, *J. Am. Chem. Soc.*, **117**, 177 (1995).

12. J. H. van Boom, G. R. Owen, J. Preston, T. Ravindranathan, and C.B. Reese, *J. Chem. Soc. C*, 3230 (1971).

13. S. Hanessian and E. Moralioglu, *Can. J. Chem.*, **50**, 233 (1972).

Butane-2,3-bisacetal (BBA)

This family of bisacetals has been reviewed in the context of their application in organic synthesis.[1] Note that these selectively protect *trans*-diols in preference to *cis*-diols.

Formation

1. EtOAc, BF$_3$·Et$_2$O, 65% yield.[2]

2. 2,3-Butanedione, TMOF (trimethyl orthoformate), CSA, MeOH, 60–82% yield.[3, 4, ,5]

3. 2,3-Butanedione, TMOF, BF$_3$·Et$_2$O.[6]

4. 2,3-Butanedione, TMSOMe, TMSOTf, CH_2Cl_2, 0°C, 97% yield.[7]

5. 2,2,3,3-Tetramethoxybutane, TMOF, MeOH, CSA, 54–91% yield. *trans*-Diols are protected in preference to *cis*-diols in contrast to acetonide formation which prefers protection of *cis*-diols.[8]

6. 2,3-Dimethoxybutadiene, $Ph_3P \cdot HBr$, CH_2Cl_2, 24 h the $BF_3 \cdot Et_2O$, 63–93% yield.[9]

Cleavage

1. PTSA, MeOH, reflux, 2 h, 94% yield.[10] HCl may also be used as the acid.[11]

2. TFA, H_2O, quantitative.[5]

1. S. V. Ley, D. K. Baeschlin, D. J. Dixon, A. C. Foster, S. J. Ince, H. W. M. Priepke, and D. J. Reynolds, *Chem. Rev.*, **101**, 53 (2001).

2. U. Berens, D. Leckel, and S. C. Oepen, *J. Org. Chem.*, **60**, 8204 (1995).

3. N. L. Douglas, S. V. Ley, H. M. I. Osborn, D. R. Owen, H. W. M. Priepke, and S. L. Warriner, *Synlett*, 793 (1996).

4. A. Hense, S. V. Ley, H. M. I. Osborn, D. R. Owen, J.-R. Poisson, S. L. Warriner, and D. E. Wesson, *J. Chem. Soc., Perkin Trans. I*, 2023 (1997).

5. N. Armesto, M. Ferrero, S. Fernandez, and V. Gotor, *Tetrahedron Lett.*, **41**, 8759 (2000).

6. C.-H. Chou, C.-S. Wu, C.-H. Chen, L.-D. Lu, S. S. Kulkarni, C.-H. Wong, and S.-C. Hung, *Org. Lett.*, **6**, 585 (2004).

7. E. Lence, L. Castedo, and C. Gonzalez, *Tetrahedron Lett.*, **43**, 7917 (2002).

8. J.-L. Montchamp, F. Tian, M. E. Hart, and J. W. Frost, *J. Org. Chem.*, **61**, 3897 (1996).

9. S. V. Ley and P. Michel, *Synlett*, 1793 (2001).

10. S. V. Ley, P. Michel, and C. Trapella, *Org. Lett.*, **5**, 4553 (2003).

11. D. J. Dixon, S. V. Ley, A. Polara, and T. Sheppard, *Org. Lett.*, **3**, 3749 (2001).

Cyclohexane-1,2-diacetal (CDA)

Formation

1. 1,1,2,2-Tetramethoxycyclohexane,[1] CSA, MeOH, trimethyl orthoformate.[2] This reagent selectively protects *trans*-1,2-diols.

2. 1,2-Cyclohexanedione, trimethyl orthoformate, CSA, MeOH, 61 yield.[3] 9,10-Phenanthrenequinone and 2,3-butanedione were similarly converted to diacetals by this method.[4]

Cleavage

TFA, H$_2$O, 5 min, 81% per CDA unit.[2]

1. S. V. Ley, H. M. I. Osborn, H. W. M. Priepke, and S. L. Warriner, *Org. Synth.*, **75**, 170 (1997).

2. P. Grice, S. V. Ley, J. Pietruszka, and H. M. W. Priepke, *Angew. Chem., Int. Ed. Engl.*, **35**, 197 (1996); P. Grice, S. V. Ley, J. Pietruszka, H. M. W. Priepke, and S. L.Warriner, *J. Chem. Soc., Perkin Trans. 1*, 351 (1997).

3. N. L. Douglas, S. V. Ley, H. M. I. Osborn, D. R. Owen, H. W. M. Priepke, and S. L. Warriner, *Synlett*, 793 (1996).

4. A. Hense, S. V. Ley, H. M. I. Osborn, D. R. Owen, J.-F. Poisson, S. L. Warriner, and K. E. Wesson, *J. Chem. Soc., Perkin Trans. 1*, 2023 (1997).

Dispiroketals

Formation

1. Bisdihydropyran[1], CSA, toluene, reflux, 36–98% yield.[2]

2. 2,2′-Bis(phenylthiomethyl)dihydropyran, CSA, CHCl$_3$, 54–93% yield. This dihydropyran can be used for resolution of racemic diols or regioselective protection. The regioselective protection is directed by the chirality of the dihydropyran.[3,4]

Other 2,2′-substituted bisdihydropyrans that can be cleaved by a variety of methods are available and their use in synthesis has been reviewed.[5]

TBDMSO ... SEt, OH, OH, HO → CSA, CH₂Cl₂, reflux overnight, 64% → TBDMSO ... SEt, HO

Cleavage

The simplest of the dispiroketals is cleaved with TFA and H_2O.[6] The 2,2'-bis(phenylthiomethyl) dispiroketal (dispoke) derivative is cleaved by oxidation to the sulfone followed by treatment with $LiN(TMS)_2$.[3] The related bromo and iodo derivatives are cleaved reductively with LDBB (lithium 4,4'-di-t-butylbiphenyl-ide) or by elimination with the P4-t-butylphosphazene base and acid hydrolysis of the enol ether.[5] The 2,2-diphenyl dispiroketal is cleaved with $FeCl_3$ (CH_2Cl_2, rt, overnight)[7]. The dimethyl dispiroketal is cleaved with TFA,[8] and the allyl deriva-tive is cleaved by ozonolysis followed by elimination.[2]

1. S. V. Ley and H. M. I. Osborn, *Org. Synth.*, **77**, 212 (2000).
2. For a review, see S. V. Ley, R. Downham, P. J. Edwards, J. E. Innes, and M. Woods, *Con-temp. Org. Synth.*, **2**, 365 (1995).
3. S. V. Ley, S. Mio, and B. Meseguer, *Synlett*, 791 (1996); W. A. Greenberg, E. S. Priestley, P. S. Sears, P. B. Alper, C. Rosenbohm, M. Hendrix, S.-C. Hung, and C.-H. Wong, *J. Am. Chem. Soc.*, **121**, 6527 (1999).
4. D. K. Baeschlin, A. R. Chaperon, L. G. Green, M. G. Hahn, S. J. Ince, and S. V. Ley, *Chem. Eur. J.*, **6**, 172 (2000).
5. S. V. Ley and S. Mio, *Synlett*, 789 (1996).
6. M. G. Banwell, N. L. Hungerford, and K. A. Jolliffe, *Org. Lett.*, **6**, 2737 (2004).
7. D. A. Entwistle, A. B. Hughes, S. V. Ley, and G. Visentin, *Tetrahedron Lett.*, **35**, 777 (1994).
8. R. Downham, P. J. Edwards, D. A. Entwistle, A. B. Huges, K. S. Kim, and S. V. Ley, *Tetrahedron: Asymmetry*, **6**, 2403 (1995).

Silyl Derivatives

Di-t-butylsilylene Group (DTBS(OR)₂)

The DTBS group is probably the most useful of the bifunctional silyl ethers. Dimeth-ylsilyl and diisopropylsilyl derivatives of diols are very susceptible to hydrolysis, even in water, and therefore are of limited use, unless other structurally imposed steric effects provide additional stabilization.

Formation

1. (*t*-Bu)$_2$SiCl$_2$, CH$_3$CN, TEA, HOBt, 65°C.[1,2] Tertiary alcohols do not react under these conditions. The reagent is effective for both 1,2- and 1,3-diols, but 1,3-derivatives are preferred over the 1,2-derivatives at least in the carbohydrate manifold.[3]

2. (*t*-Bu)$_2$Si(OTf)$_2$, 2,6-lutidine, 0–25°C, CHCl$_3$.[4] This reagent readily silylates 1,2-, 1,3- and 1,4-diols even when one of the alcohols is tertiary. THP and PMB protected diols are converted to the silylene derivative with this reagent.[5] 1,3-Diols are preferably protected over *cis*- or *trans*- 1,2-diols.[6]

3. The di-*t*-butylsilylene group has been used to connect a diene and a dienophile to control the intramolecular Diels–Alder reaction.[7]

4. (*t*-Bu)$_2$SiCl$_2$, AgNO$_3$, Pyr, DMF, >84% yield.[8]

5. DMF is the only solvent that works in this transformation.[9]

6. (*t*-Bu)$_2$SiHCl, *n*-BuLi, THF, −78°C to rt, 84–94% yield.[10,11]

Cleavage

Derivatives of 1,3- and 1,4-diols are stable to pH 4–10 at 22°C for several hours, but derivatives of 1,2-diols undergo rapid hydrolysis under basic conditions (5:1 THF, pH 10 buffer, 22°C, 5 min) to form monosilyl ethers of the parent diol.

1. 48% aq. HF, CH$_3$CN, 25°C, 15 min, 95% yield.[3]
2. Bu$_3$NHF, THF.[12]
3. Pyr·HF, THF, 25°C, 85–92% yield.[1]

Note the retention of the TBDMS group

4. TBAF, ZnCl$_2$, ms, rt to 65°C, 3 h.[13]

5. TBAF, THF, rt, 96% yield.[6]

6. TBAF, AcOH, 60°C, 12 h, 45% yield.[14]

7. Tris(dimethylamino)sulfonium difluorotrimethylsilicate (TSAF), THF, 0°C, 5 h, 64% yield. A TES and a two phenolic TIPS groups were also cleaved.[15]

8. BF$_3$·Et$_2$O, allyltrimethylsilane, toluene, 85°C, 95% yield.[16] This is a general method for the selective ring opening of the DTBS derivative to give silyl ethers of the more hindered alcohol. The silylene derivatives of tertiary or benzylic alcohols result in elimination.

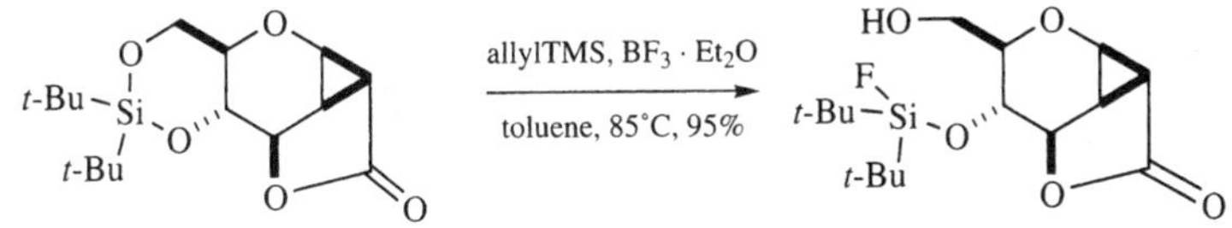

9. Reaction with n-BuLi/TMEDA results in the formation of a penta-co-ordinate intermediate that cleaves to give regioselectively the secondary silyl ether.[17]

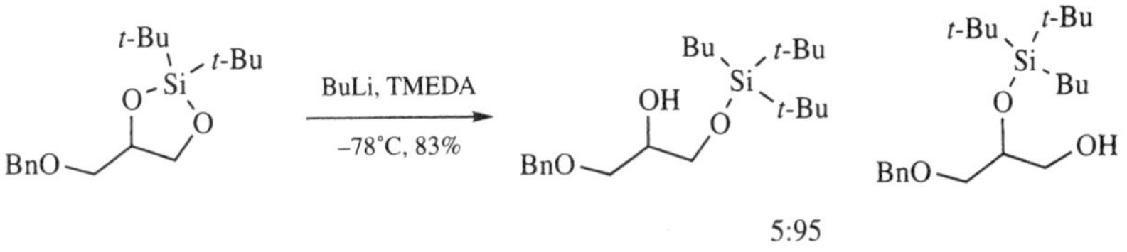

5:95

Dialkylsilylene Groups

Three different silylene derivatives were used to achieve selective protection of a more hindered diol during a taxol synthesis. Treatment of the silylene with MeLi opens the ring to afford the more hindered silyl ether.[18,19]

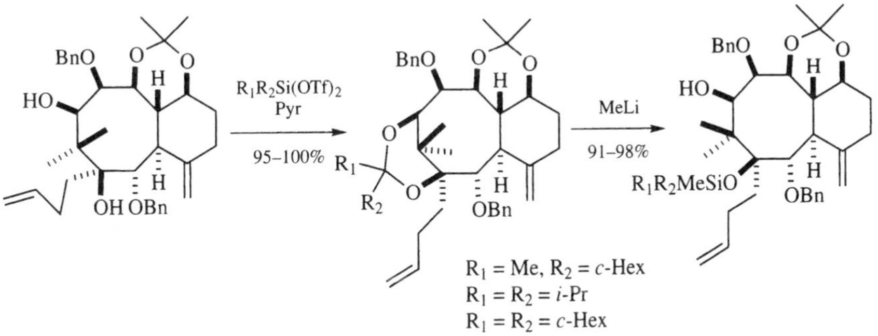

R$_1$ = Me, R$_2$ = c-Hex
R$_1$ = R$_2$ = i-Pr
R$_1$ = R$_2$ = c-Hex

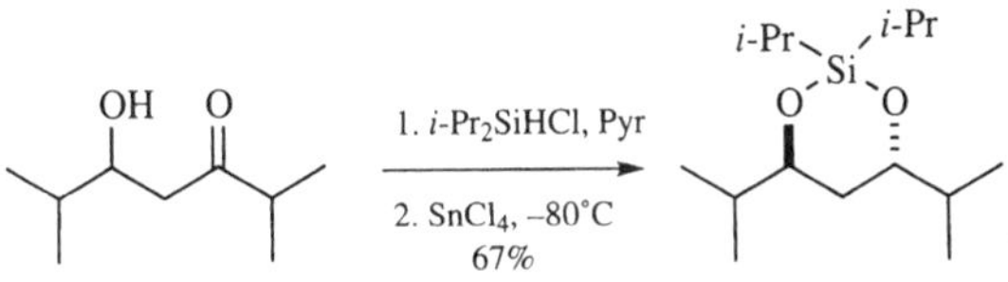

1. i-Pr$_2$SiHCl, Pyr

2. SnCl$_4$, –80°C
67%

Ref. 20

1. B. M. Trost and C. G. Caldwell, *Tetrahedron Lett.*, **22**, 4999 (1981).

2. B. M. Trost, C. G. Caldwell, E. Murayama, and D. Heissler, *J. Org. Chem.*, **48**, 3252 (1983).

3. D. Kumagai, M. Miyazaki, and S.-I. Nishimura, *Tetrahedron Lett.*, **42**, 1953 (2001).

4. E. J. Corey and P. B. Hopkins, *Tetrahedron Lett.*, **23**, 4871 (1982).

5. T. Oriyama, K. Yatabe, S. Sugawara, Y. Machiguchi, and G. Koga, *Synlett*, 523 (1996).

6. B. Delpech, D. Calvo, and R. Lett, *Tetrahedron Lett.*, **37**, 1019 (1996); K. A. Parker and A. T. Georges, *Org. Lett.*, **2**, 497 (2000).

7. J. W. Gillard, R. Fortin, E. L. Grimm, M. Maillard, M. Tjepkema, M. A. Bernstein, and R. Glaser, *Tetrahedron Lett.*, **32**, 1145 (1991).

8. C. W. Gundlach, IV, T. R. Ryder, and G. D. Glick, *Tetrahedron Lett.*, **38**, 4039 (1997).

9. K. Furusawa, K. Ueno, and T. Katsura, *Chem. Lett.*, **19**, 97 (1990).

10. K. Tanino, T. Shimizu, M. Kuwahara, and I. Kuwajima, *J. Org. Chem.*, **63**, 2422 (1998).

11. K. Morihira, R. Hara, S. Kawahara, T. Nishimori, N. Nakamura, H. Kusama, and I. Kuwajima, *J. Am. Chem. Soc.*, **120**, 12980 (1998).

12. K. Furusawa, *Chem. Lett.*, **18**, 509 (1989).

13. R. Van Speybroeck, H. Guo, J. van der Eycken, and M. Vandewalle, *Tetrahedron*, **47**, 4675 (1991).

14. K. Toshima, H. Yamaguchi, T. Jyojima, Y. Noguchi, M. Nakata, and S. Matsumura, *Tetrahedron Lett.*, **37**, 1073 (1996).

15. D. A. Johnson and L. M. Taubner, *Tetrahedron Lett.*, **37**, 605 (1996).

16. M. Yu and B. L. Pagenkopf, *J. Org. Chem.*, **67**, 4553 (2002).

17. K. Tanikno, T. Shimizu, M. Kuwahara, and I. Kuwajima, *J. Org. Chem.*, **63**, 2422 (1998).

18. I. Shiina, T. Nishimura, N. Ohkawa, H. Sakoh, K. Nishimura, K. Saitoh, and T. Mukaiyama, *Chem. Lett.*, **26**, 419 (1997).

19. T. Mukaiyama, Y. Ogawa, K. Kuroda, and J.-i. Matsuo, *Chem. Lett.*, **33**, 1412 (2004).

20. S. Anwar and A. P. Davis, *J. Chem. Soc., Chem. Commun.*, 831 (1986). See also reference 18.

1,3-(1,1,3,3-Tetraisopropyldisiloxanylidene) Derivative (TIPDS(OR)₂)

Formation

1. TIPDSCl₂, DMF, imidazole.[1-3] This reagent is primarily used in carbohydrate protection, but occasionally it proves valuable in other circumstances.[4] Its use in natural product synthesis has been reviewed.[5]

2. TIPDSCl$_2$, Pyr.[6–8] In polyhydroxylated systems the regiochemical outcome is determined by initial reaction at the sterically less hindered alcohol.[9]

3. TIPDSCl$_2$, AgOTf, *sym*-collidine DMF, 45% yield.[10]

4. (*i*-Pr)$_2$SiH)$_2$O, PdCl$_2$, CCl$_4$, 60°C, 2 h, then substrate in pyridine. This method produces the silyl chloride *in situ.*[11]

Cleavage

1. Bu$_4$NF, THF.[1,5,12] When Bu$_4$NF is used to remove the TIPDS group, ester groups can migrate because of the basic nature of fluoride ion. Migration can be prevented by the addition of Pyr·HCl.[13]

2. TBAF, AcOH, THF.[14]

3. TEA·HF.[15]

4. 0.2 *M* HCl, dioxane, H$_2$O, or MeOH.[1]

5. 0.2 *M* NaOH, dioxane, H$_2$O.[1]

6. TMSI, CH$_2$Cl$_2$, 0°C, 0.5 h, 83% yield.[16]

7. Ac$_2$O, AcOH, H$_2$SO$_4$.[2]

8. The TIPDS derivative can be induced to isomerize from the thermodynamically less stable eight-membered ring to the more stable seven-membered ring derivative.[6,17] The isomerization occurs only in DMF.

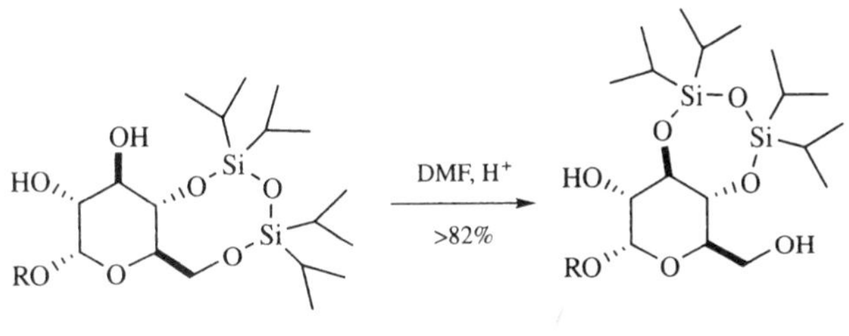

9. NH$_4$F, MeOH, 60°C, 3 h, 99% yield.[18]

10. CsF, NH$_3$, MeOH.[19] The TIPDS group is partially cleaved with MeOH/NH$_3$ in an attempt to remove an acetyl group.[20]

11. KF·2H$_2$O, 18-Crown-6, DMF or THF, rt, 55–81% yield.[21]

12. Treatment of a TIPDS group with methyl pyruvate (TMSOTf, 0°C to rt, 69–99% yield) converts it to the pyruvate acetal.[10]

13. (HF)$_n$ pyridine, rt.[22]

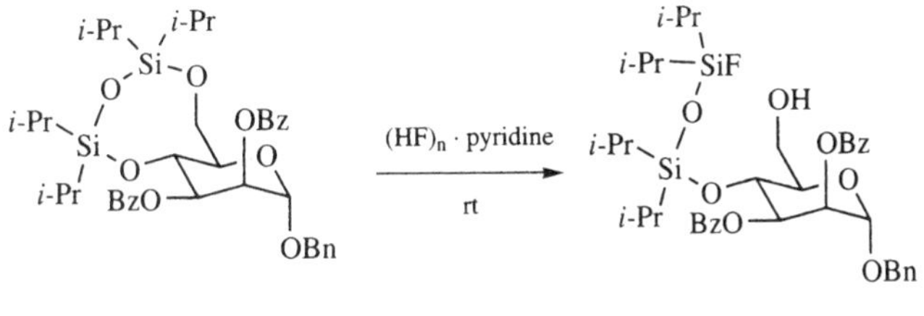

14. 1 *N* HCl, dioxane, 88% yield.[23] Aqueous TFA in THF also efficiently carries out this transformation.[24]

1,1,3,3-Tetra-*t*-butoxydisiloxanylidene Derivative (TBDS(OR)₂)

Formation

1,3-Dichloro-1,1,3,3-tetra-*t*-butoxydisiloxane, Pyr, rt, 50–87% yield.[25]

B = pyrimidine or purine residue

Cleavage

Bu₄NF, THF, 2 min.[25] This group is less reactive toward triethylammonium fluoride than the TIPDS group. It is stable to 2 *M* HCl, aq. dioxane, overnight. Treatment with 0.2 *M* NaOH, aq. dioxane leads to cleavage of only the Si–O bond at the 5′-position of the uridine derivative. The TBDS derivative is 25 times more stable than the TIPDS derivative to basic hydrolysis.

Methylene-bis-(diisopropylsilanoxanylidene) (MDPS(OR)₂)

This group was developed to retain the properties of the TIPDS group but to have improved base stability by replacing the connecting oxygen with the robust methylene group. It is introduced with the dichloride (DMF, imidazole, 79% yield) and is cleaved with TBAF (97% yield) although more slowly than the TIPDS group.[26]

1,1,4,4-Tetraphenyl-1,4-disilanylidene (SIBA(OR)₂

This group was developed as a passive *O*-2 protective group that could be removed in the presence of an acid sensitive target molecule after affecting an α-selective glycosylation. It is introduced with the dichloride (DMF, imidazole, 1 h, 92% yield) and can be removed with Bu₄NF (THF, 20°C, 99% yield).[27]

1. W. T. Markiewicz, *J. Chem. Res. Synop.*, 24 (1979).

2. J. P. Schaumberg, G. C. Hokanson, J. C. French, E. Smal, and D. C. Baker, *J. Org. Chem.*, **50**, 1651 (1985).

3. E. Ohtsuka, M. Ohkubo, A. Yamane, and M. Ikebara, *Chem. Pharm. Bull.*, **31**, 1910 (1983).

4. A. G. Myers, P. M. Harrington, and E. Y. Kuo, *J. Am. Chem. Soc.*, **113**, 694 (1991).

5. T. Ziegler, R. Dettmann, F. Bien, and C. Jurisch, *Trends in Organic Chemistry*, **6**, 91 (1997).

6. C. A. A. van Boeckel and J. H. van Boom, *Tetrahedron*, **41**, 4545, 4557 (1985).

7. J. Thiem, V. Duckstein, A. Prahst, and M. Matzke, *Liebigs Ann. Chem.*, 289 (1987).

8. J. S. Davies, E. J. Tremeer, and R. C. Treadgold, *J. Chem. Soc., Perkin Trans. I*, 1107 (1987).

9. W. T. Markiewicz, N. Sh. Padyukova, S. Samek, and J. Smrt, *Collect. Czech. Chem. Commun.*, **45**, 1860 (1980).

10. T. Ziegler, E. Eckhardt, K. Neumann, and V. Birault, *Synthesis*, 1013 (1992).

11. C. Ferreri, C. Costantino, R. Romeo, and C. Chatgilialoglu, *Tetrahedron Lett.*, **40**, 1197 (1999).

12. M. D. Hagen, C. S.-Happ, E. Happ, and S. Chládek, *J. Org. Chem.*, **53**, 5040 (1988).

13. J. J. Oltvoort, M. Kloosterman, and J. H. Van Boom, *Recl: J. R. Neth. Chem. Soc.*, **102**, 501 (1983).

14. W. Pfleiderer, M. Pfister, S. Farkas, H. Schirmeister, R. Charubala, K. P. Stengele, M. Mohr, F. Bergmann, and S. Gokhale, *Nucleosides & Nucleotides*, **10**, 377 (1991).

15. W. T. Markiewicz, E. Biala, and R. Kierzek, *Bull. Pol. Acad. Sci.,Chem.*, **32**, 433 (1984).

16. T. Tatsuoka, K. Imao, and K. Suzuki, *Heterocycles*, **24**, 617 (1986).

17. C. H. M. Verdegaal, P. L. Jansse, J. F. M. de Rooij, and J. H. Van Boom, *Tetrahedron Lett.*, **21**, 1571 (1980).

18. W. Zhang and M. J. Robins, *Tetrahedron Lett.*, **33**, 1177 (1992).

19. J. R. McCarthy, D. P. Matthews, D. M. Stemerick, E. W. Huber, P. Bey, B. J. Lippert, R. D. Snyder, and P. S. Sunkara, *J. Am. Chem. Soc.*, **113**, 7439 (1991).

20. K. Haraguchi, N. Shiina, Y. Yoshimura, H. Shimada, K. Hashimoto, and H. Tanaka, *Org. Lett.*, **6**, 2645 (2004).

21. J. N. Kremsky and N. D. Sinha, *Bioorg. Med. Chem. Lett.*, **4**, 2171 (1994).

22. T. Ziegler, R. Dettmann, M. Duszenko, and V. Kolb, *J.Carbohydr. Chem.*, **13**, 81 (1994); J. J. Oltvoort, M. Klosterman, and J. H. van Boom, *Recl. Trav. Chim Pays-Bas*, **102**, 501 (1983).

23. S. Hanessian, S. Marcotte, R. Machaalani, and G. Huang, *Org. Lett.*, **5**, 4277 (2003).

24. X.-F. Zhu, H. J. Williams, and A. I. Scott, *Tetrahedron Lett.*, **41**, 9541 (2000).

25. W. T. Markiewicz, B. Nowakowska, and K. Adrych, *Tetrahedron Lett.*, **29**, 1561 (1988); W. T. Markiewicz and A.-R. Katarzyna, *Nucleosides & Nucleotides*, **10**, 415 (1991).

26. K. Wen, S. Chow, Y. S. Sanghvi, and E. A. Theodorakis, *J. Org. Chem.*, **67**, 7887 (2002).

27. H. Wehlan, M. Dauber, M.-T. M. Fernaud, J. Schuppan, R. Mahrwald, B. Ziemer, M.-E. J. Garcia, and U. Koert, *Angew. Chem. Int. Ed.*, **43**, 4597 (2004).

o-Xylyl Ether

This derivative is formed from the diol and 1,2-di(bromomethyl)benzene (NaH, THF, HMPA, 0°C, 66% yield). It is cleaved by hydrogenolysis [Pd(OH)$_2$, EtOH, H$_2$, 89–99% yield].[1]

1. A. J. Poss and M. S. Smyth, *Synth. Commun.*, **19**, 3363 (1989).

3,3'-Oxybis(dimethoxytrityl) Ether (O-DMT)

The 3,3'-oxybis(dimethoxytrityl) group was developed for protection of ribonucleosides, but unexpectedly both the 2',5'- and 3',5'-derivatives are formed.[1] The group is introduced using the bis trityl chloride (2,4,6-collidine, AgClO$_4$, pyridine, 65°C, 1 h). Acid catalysis is used to remove it.

1,2-Ethylene-3,3-bis(4'4''-dimethoxytrityl) Ether (E-DMT)

The E-DMT group is similar to the O-DMT group except that there is a two-carbon spacer joining the aryl rings. It is introduced using the bischloride in pyridine and will protect thymidine in 65% yield.[2]

1. N. Oka, Y. S. Sanghvi, and E. A. Theodorakis, *Bioorg. Med. Chem. Lett.*, **14**, 3241 (2004).

2. N. Oka, Y. S. Sanghvi, and E. A. Theodorakis, *Synlett*, 823 (2004).

Cyclic Carbonates(Chart 3)

Cyclic carbonates[1,2] are very stable to acidic hydrolysis (AcOH, HBr, and H_2SO_4/MeOH) and are more stable to basic hydrolysis than esters.

Formation

1. Phosgene, pyridine, 20°C, 1 h.[3]
2. The related thionocarbonate is prepared from thiophosgene (pyridine, DMAP, 78% yield).[4]
3. p-$NO_2C_6H_4OCOCl$, Pyr, 20°C, 5 days, 72% yield.[5]
4. *N,N'*-Carbonyldiimidazole, PhH, heat, 12 h to 4 days, 90% yield.[6,7]
5. Cl_3CCOCl, pyridine, 1 h, rt, >80% yield.[8]

6. $Cl_3COCO_2CCl_3$ (triphosgene), pyridine, CH_2Cl_2, 84–99% yield.[9] Triphosgene is a much safer source of phosgene and is an easily handled solid. A 1,2,3-triol was selectively protected at the 1,2-position with this reagent.[10] Reactions using triphosgene often need to be run at higher temperatures because it is not as reactive as phosgene.
7. CO, S, Et_3N, 80°C, 4 h; $CuCl_2$, rt, 18 h, 66–100% yield.[11]
8. Ethylene carbonate, $NaHCO_3$, 120°C, 80% yield.[12]
9. Cyclic carbonates are prepared directly from epoxides with LiBr, CO_2, NMP (1-methyl-2-pyrrolidinone), 100°C.[13]

Cleavage

1. $Ba(OH)_2$, H_2O, 70°C.[14]
2. Pyridine, H_2O, reflux, 15 min, 100% yield.[4] These conditions were used to remove the carbonate from uridine.
3. 0.5 *M* NaOH, 50% aq. dioxane, 25°C, 5 min, 100% yield.[4] K_2CO_3 is a similarly effective base.[15]
4. 0.1 *M* MeONa, MeOH, quantitative yield.[16]
5. As with the benzylidene ketals, the carbonate can be opened to give a mono-protected diol.[17]

6. In the following case a carbonate could not be removed in the presence of the diolide using hydrolytic conditions. It was found that treatment with the bifunctional Grignard reagent cleaved the carbonate in 65% yield by taking advantage of the intramolecularity of the second addition.[18]

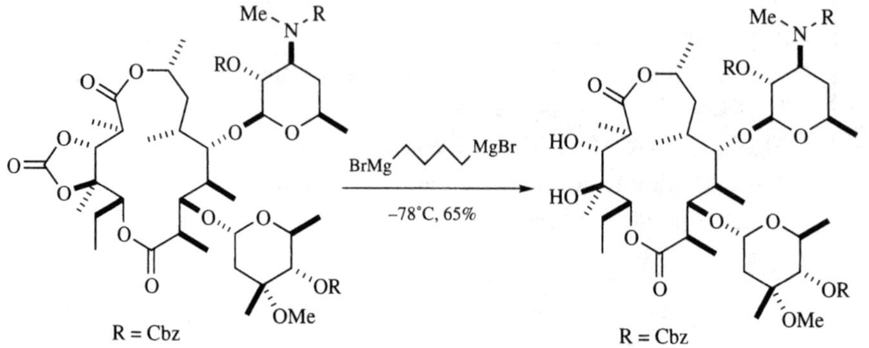

R = Cbz   R = Cbz

7. Enzymatic cleavage: PPL was found to cleave carbonates bearing an unsaturated substituent. This also results in the resolution of the diol and the remaining carbonate, since only one enantiomer is hydrolyzed preferentially. The yields and enantiomeric excesses depend on the level of conversion. This method may be useful for the hydrolysis of carbonates that cannot be treated with base.[19]

8. During the course of the preparation of a vinyl iodide using Schwartz's reagent, a carbonate was unexpectedly cleaved.[20]

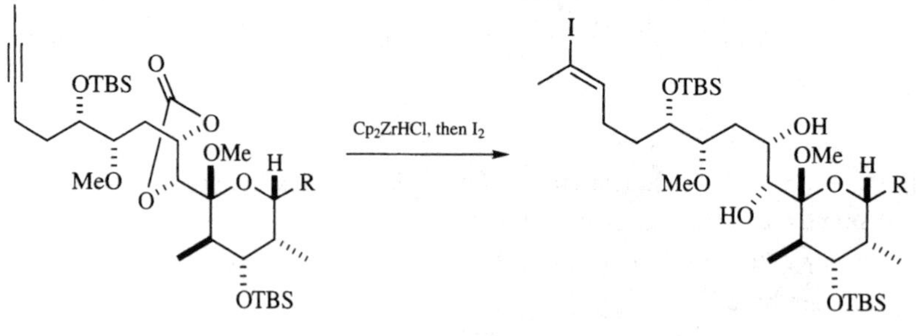

9. Reaction of a cyclic carbonate with ammonia results in the selective ring-opening to give a carbamate.[21]

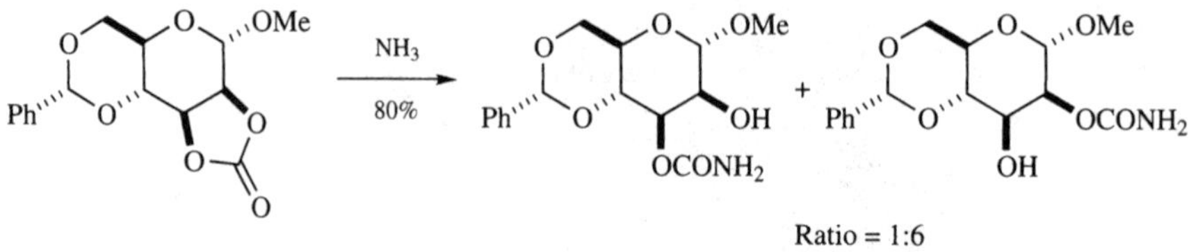

Ratio = 1:6

1. L. Hough, J. E. Priddle, and R. S. Theobald, *Adv. Carbohydr. Chem.*, **15**, 91 (1960).
2. V. Amarnath and A. D. Broom, *Chem. Rev.*, **77**, 183 (1977).
3. W. N. Haworth and C. R. Porter, *J. Chem. Soc.*, 151 (1930).
4. S. Y. Ko, *J. Org. Chem.*, **60**, 6250 (1995).
5. R. L. Letsinger and K. K. Ogilvie, *J. Org. Chem.*, **32**, 296 (1967).
6. J. P. Kutney and A. H. Ratcliffe, *Synth. Commun.*, **5**, 47 (1975).
7. K. Narasaka, *ACS Symp. Ser.* **386**, "Trends in Synth. Carbohydr. Chem.," p. 290 (1989).
8. K. Tatsuta, K. Akimoto, M. Annaka, Y. Ohno, and M. Kinoshita, *Bull. Chem. Soc. Jpn.*, **58**, 1699 (1985).
9. R. M. Burk and M. B. Roof, *Tetrahedron Lett.*, **34**, 395 (1993).
10. S. K. Kang, J. H. Jeon, K. S. Nam, C. H. Park, and H. W. Lee, *Synth. Commun.*, **24**, 305 (1994).
11. T. Mizuno, F. Nakamura, Y. Egashira, I. Nishiguchi, T. Hirashima, A. Ogawa, N. Kambe, and N. Sonoda, *Synthesis*, 636 (1989)
12. T. Desai, J. Gigg, and R. Gigg, *Carbohydr. Res.*, **277**, C5 (1995).
13. N. Kihara, Y. Nakawaki, and T. Endo, *J. Org. Chem.*, **60**, 473 (1995).
14. W. G. Overend, M. Stacey, and L. F. Wiggins, *J. Chem. Soc.*, 1358 (1949).
15. M. Yamashita, N. Ohta, T. Shimizu, K. Matsumoto, Y. Matsuura, I. Kawasaki, T. Tanaka, N. Maezaki, and S. Ohta, *J. Org. Chem.*, **68**, 1216 (2003).
16. P. Kosma, G. Schulz, and F. M. Unger, *Carbohydr. Res.*, **180**, 19 (1988).
17. K. C. Nicolaou, C. F. Claiborne, K. Paulvannan, M. H. D. Postema, and R. K. Guy, *Chem. Eur. J.*, **3**, 399 (1997).
18. T. Ohara, M. Kume, Y. Narukawa, K. Motokawa, K. Uotani, and H. Nakai, *J. Org. Chem.*, **67**, 9146 (2002).
19. K. Matsumoto, Y. Nakamura, M. Shimojo, and M. Hatanaka, *Tetrahedron Lett.*, **43**, 6933 (2002).
20. K. C. Nicolaou, Y. Li, K. Sugita, H. Monenschein, P. Guntupalli, H. J. Mitchell, K. C. Fylaktakidou, D. Vourloumis, P. Giannakakou, and A. O'Brate, *J. Am. Chem. Soc.*, **125**, 15443 (2003).
21. D. L. Boger and T. Honda, *J. Am. Chem. Soc.*, **116**, 5647 (1994).

Cyclic Boronates

Although boronates are quite susceptible to hydrolysis, they have been useful for the protection of carbohydrates.[1,2] It should be noted that as the steric demands of the diol increase, the rate of hydrolysis decreases. For example, pinacol boronates are rather difficult to hydrolyze; in fact, they can be isolated from aqueous systems with no hydrolysis. The section on the protection of boronic acids should be consulted. The use of boron acids as protective agents has been reviewed.[3] Boric acid has been used to transiently protect diols.[4]

Methyl and Ethyl Boronate[5] (Chart 3)

Formation

1.

Ref. 6

2. $[t\text{-}C_4H_9CO_2B(C_2H_5)]_2O$, Pyr; then concentrate under reduced pressure.[7]
3. $EtB(OMe)_2$, ion exchange resin, 85% yield.[8]
4. $LiEt_3BH$, THF, 0°C to rt, 98% yield.[9]
5. $(MeBO)_3$, pyridine, rt, 0.5 h, 77% yield.[10]

Cleavage

1. Pinacol, DMAP, benzene, rt. This method proceeds by ester exchange to form the more stable pinacolate ester.[10]
2. MeOH or 2,4-dihydroxy-4-methylpentane, >82% yield.[11]

Phenyl Boronate

Formation

1. $PhB(OH)_2$, PhH,[12] or pyridine.[13] A polymeric version of the phenyl boronate has been developed.[14] The phenyl boronates are stable to the conditions of

stannylation and have been used for selective sulfation to produce monosulfated monosaccharides.[15] Phenyl boronates were found to be stable to oxidation with PCC.[16] Syn-1,2-diols can be selectively protected in the presence of anti-1,2-diols.[17]

2. PhB(OH)$_2$, benzene, MeOH, reflux, and distill out the MeOH.[18]

3. From a benzylidene acetal: PhB(OH)$_2$, (EtO)$_3$B, heat.[19]

Cleavage

1. 1,3-Propanediol, acetone.[1] This method removes the boronate by exchange. 2-Methylpentane-2,5-diol in acetic acid cleaves a phenyl boronate (85% yield).[20] Pinacol is also very effective for removing the boronate.[21]

2. Acetone, H$_2$O (4:1), 30 min, 83% yield.[10]

3. H$_2$O$_2$, EtOAc, >80% yield.[22,23]

4. Ac$_2$O, Pyr, 99% yield. In this case the boronate is converted to an acetate.[24]

5. Treatment of the boronate with BuI, AgO affords the monoalkylated diol in a manner similar to stannylene-directed monoalkylation and acylation.[25]

o-Acetamidophenyl Boronate: [2,6-(AcNH)$_2$C$_6$H$_3$B(OR)$_2$]

This boronate was developed to confer added stability toward hydrolysis. It was shown to be substantially more stable to hydrolysis than the simple phenyl boronate because of coordination of the ortho acetamide to the boronate.[26]

1. R. J. Ferrier, Adv. Carbohydr. Chem. Biochem., 35, 31–80 (1978).

2. W. V. Dahlhoff and R. Köster, Heterocycles, 18, 421 (1982).

3. P. J. Duggan and E. M. Tyndall, J. Chem. Soc. Perkin Trans. 1, 1325 (2002).

4. H.-R. Bjørsvik, H. Priebe, J. Cervenka, A. W. Aabye, T. Gulbrandsen, and A. C. Bryde, Org. Proc. Res. Dev., 5, 472 (2001).

5. W. V. Dahloff and R. Köster, J. Org. Chem., 41, 2316 (1976), and references cited therein.

6. W. V. Dahlhoff, A. Geisheimer, and R. Köster, *Synthesis*, 935 (1980); W. V. Dahlhoff and R. Köster, *Synthesis*, 936 (1980).

7. R. Köster, K. Taba, and W. V. Dahlhoff, *Liebigs Ann. Chem.*, 1422 (1983).

8. W. V. Dahlhoff, W. Fenzl, and R. Köster, *Liebigs Ann. Chem.*, 807 (1990).

9. L. Garlaschelli, G. Mellerio, and G. Vidari, *Tetrahedron Lett.*, **30**, 597 (1989).

10. H. Kusama, R. Hara, S. Kawahara, T. Nishimori, H. Kashima, N. Nakamura, K. Morihira, and I. Kuwajima, *J. Am. Chem. Soc.*, **122**, 3811 (2000).

11. J. Gu, M. E. Ruppen, and P. Cai, *Org. Lett.*, **7**, 3945 (2005).

12. R. J. Ferrier, *Methods Carbohydr. Chem.*, **VI**, 419 (1972).

13. J. M. J. Fréchet, L. J. Nuyens, and E. Seymour, *J. Am. Chem. Soc.*, **101**, 432 (1979).

14. N. P. Bullen, P. Hodge, and F. G. Thorpe, *J. Chem. Soc., Perkin Trans. I*, 1863 (1981).

15. S. Langston, B. Bernet, and A. Vasella, *Helv. Chim. Acta*, **77**, 2341 (1994).

16. C. Lifjebris, B. M. Nilsson, B. Resul, and U. Hacksell, *J. Org. Chem.*, **61**, 4028 (1996).

17. M. Journet, D. Cai, L. M. DiMichele, D. L. Hughes, R. D. Larsen, T. R. Verhoeven, and P. J. Reider, *J. Org. Chem.*, **64**, 2411 (1999).

18. G. G. Cross and D. M. Whitfield, *Synlett*, 487 (1998).

19. H. H. Seltzman, D. N. Fleming, G. D. Hawkins, and F. I. Carroll, *Tetrahedron Lett.*, **41**, 3589 (2000).

20. E. Bertounesque, J.-C. Florent, and C. Monneret, *Synthesis*, 270 (1991).

21. Q. Wang and A. Padwa, *Org. Lett.*, **6**, 2189 (2004).

22. D. A. Evans and R. P. Polniaszek, *Tetrahedron Lett.*, **27**, 5683 (1986).

23. D. A. Evans, R. P. Polniaszek, D. M. DeVries, D. E. Guinn, and D. J. Mathre, *J. Am. Chem. Soc.*, **113**, 7613 (1991).

24. A. Flores-Parra, C. Paredes-Tepox, P. Joseph-Nathan, and R. Contreras, *Tetrahedron*, **46**, 4137 (1990).

25. K. Oshima, E.-i. Kitazono, and Y. Aoyama, *Tetrahedron Lett.*, **38**, 5001 (1997).

26. S. X. Cai and J. F. W. Keana, *Bioconjugate Chem.*, **2**, 317 (1991).

3

PROTECTION FOR PHENOLS AND CATECHOLS

Greene's Protective Groups in Organic Synthesis, Fourth Edition, by Peter G. M. Wuts and Theodora W. Greene
Copyright © 2007 John Wiley & Sons, Inc.

The phenolic hydroxyl group occurs widely in plant and animal life, both terrestrial and pelagic, as demonstrated by the vast number of natural products that contain this group. In developing a synthesis of any phenol-containing product, protection is often mandatory to prevent reaction with oxidizing agents and electrophiles or reaction of the nucleophilic phenoxide ion with even mild alkylating and acylating agents. Many of the protective groups developed for alcohol protection are also applicable to phenol protection, and thus the chapter on alcohol protection should also be consulted. Ethers are the most widely used protective groups for phenols and in general they are more easily cleaved than the analogous ethers of simple alcohols.[1] Esters are also important protective groups for phenols, but are not as stable to hydrolysis as the related alcohol derivatives. Simple esters are easily hydrolyzed with mild base (e.g., NaHCO$_3$/aq. MeOH, 25°C), but more sterically demanding esters (e.g., pivalate) require harsher conditions to effect hydrolysis. Catechols can be protected in the presence of phenols as cyclic acetals or ketals or cyclic esters.

Some of the more important phenol and catechol protective groups are included in Reactivity Chart 4.[2]

---

1. For a review on ether cleavage, see M. V. Bhatt and S. U. Kulkarni, *Synthesis*, 249 (1983).

2. See also E. Haslam, "Protection of Phenols and Catechols," in *Protective Groups in Organic Chemistry*, J. F. W. McOmie, Ed., Plenum, New York and London, 1973, pp. 145–182.

PROTECTION FOR PHENOLS

Ethers

Historically, simple *n*-alkyl ethers formed from a phenol and a halide or sulfonate were cleaved under rather drastic conditions (e.g., refluxing HBr). Newer methods of alkyl ether cleavage have been developed that do not rely on harshly acidic conditions. New ether protective groups have been developed that are removed under much milder conditions (e.g., via nucleophilic displacement, hydrogenolysis of benzyl ethers, or mild acid hydrolysis of acetal-type ethers) that often do not affect other functional groups in a molecule. When exploring methods for phenol protection, the section on protection of alcohols should also be consulted, since in many cases those methods are applicable to phenols. The difference between the two groups is their pK_a's, which will effect both the deprotection and cleavage process.

Methyl Ether: $ArOCH_3$ (Chart 4)

Deuteromethyl ethers have been used to protect phenols to prevent the methyl hydrogens from participating in free radical reactions.[1]

Formation

1. MeI, K_2CO_3, acetone, reflux, 6 h.[2,3] This is a very common and often very efficient method for the preparation of phenolic methyl ethers. The method is also applicable to the formation of phenolic benzyl ethers. Stronger bases are not required because of the increased acidity of a phenol versus a typical alcohol. In the following case the ortho OH is more acidic by about 1 pK_a unit therefore more reactive.[4]

2. Me_2SO_4, NaOH, EtOH, reflux, 3 h, 71–74% yield.[2]

3. Li_2CO_3, MeI, DMF, 55°C, 18 h, 54–90%.[5]

This method selectively protects phenols with p$K_a \leq 8$ as a result of electron withdrawing *ortho*- or *para*-substituents.

4. $LiOH \cdot H_2O$, Me_2SO_4 0.5 eq., THF, 70–100% yield. This method results in the transfer of both methyl groups, does not isomerize amino acid derivatives, and is selective for a PhOH in the presence of an amide.[6]

5. RX, or R'_2SO_4, NaOH, CH_2Cl_2, H_2O, $PhCH_2N^+Bu_3Br^-$, 25°C, 2–13 h, 75–95% yield.

    Ar=simple; 2- or 2,6-disubstituted[7,8] The phase transfer approach is probably the simplest method to scale up.

    R=Me, allyl, $\overset{\triangleright}{\underset{O}{}}$—$CH_2^-$, n-Bu, c-C_5H_{11}, $PhCH_2$,—CH_2CO_2Et, R'=Me, Et

6. Phenols protected as t-$BuMe_2Si$ ethers can be converted directly to methyl or benzyl ethers (MeI or BnBr, KF, DMF, rt, >90% yield).[9]

7. Methyl, ethyl, and benzyl ethers have been prepared in the presence of tetraethylammonium fluoride as a Lewis base (alkyl halide, DME, 20°C, 3 h, 60–85% yields).[8]

8. Diazomethane[10]

$$p\text{-}NO_2\text{-}C_6H_4ONa + MeN(NO)CONH_2 \xrightarrow{DME,\ 0-25°C,\ 6\ h} [p\text{-}NO_2^-C_6H_4O^- + CH_2N_2]$$

$$\xrightarrow{\hspace{3cm}} [p\text{-}NO_2^-C_6H_4OCH_3, >90\%$$

9. Diazomethane, ether, 80% yield.[11]

10. $TMSCHN_2$, MeOH, MeCN, rt, DIPEA, 31–100% yield.[12] The following illustrates the power of the method.[13] $TMSCHN_2$ is much less hazardous than diazomethane especially on scale.

11. Dimethyl carbonate, $(Bu_2N)_2C=NMe$, 180°C, 4.5 h, 54–99% yield.[14] In the presence of this guanidine, aromatic methyl carbonates are converted to methyl ethers with loss of CO_2. The reaction can also be carried out with K_2CO_3 at 140°C in triglyme or DMF, 60–81% yield[15] or with Cs_2CO_3 at 120°C in neat dimethyl carbonate.[16] In the latter case, simple alcohols are converted to methyl carbonates. DBU can be used as a base in this process, either at 90°C or with

microwave heating.[17] Phase transfer conditions have been shown to be effective on a limited number of cases (Bu$_4$NBr, DMC, K$_2$CO$_3$, 93°C, 95–99% yield).[18]

12. MeOH, 1,2-bis(diphenylphosphino)ethane, diisopropylazidodicarboxylate, 20°C. This method is selective for the phenolic OH in the presence of acidic NH groups where conventional base promoted conditions result in *O*- and *N*-alkylation.[19]

Cleavage

Nucleophilic Methods

1. EtSNa, DMF, reflux, 3 h, 94–98% yield.[20,21] Potassium thiophenoxide has been used to cleave an aryl methyl ether without causing migration of a double bond.[22] Sodium benzylselenide (PhCH$_2$SeNa) and sodium thiocresolate (*p*-CH$_3$C$_6$H$_4$SNa) cleave dimethoxyaryl compounds regioselectively, reportedly because of steric factors in the former case[23] and electronic factors in the latter case.[24]

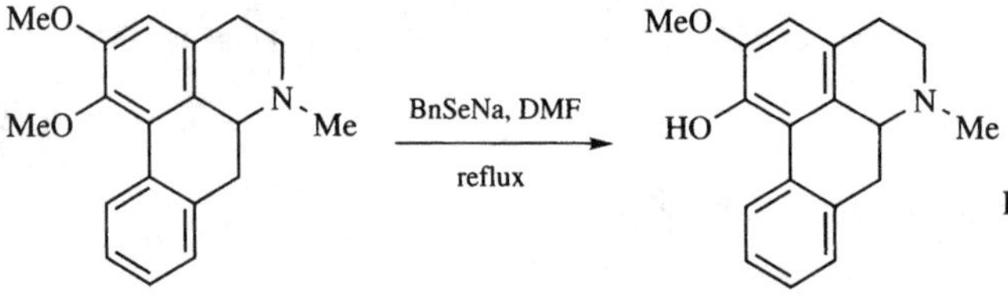

Ref. 23

2. PhSH, catalytic K$_2$CO$_3$, NMP, 60–97% yield.[25]
3. Sodium ethanethiolate has been examined for the selective cleavage of aryl methyl ethers. Methyl ethers *para* to an electron withdrawing group are cleaved preferentially.[26]

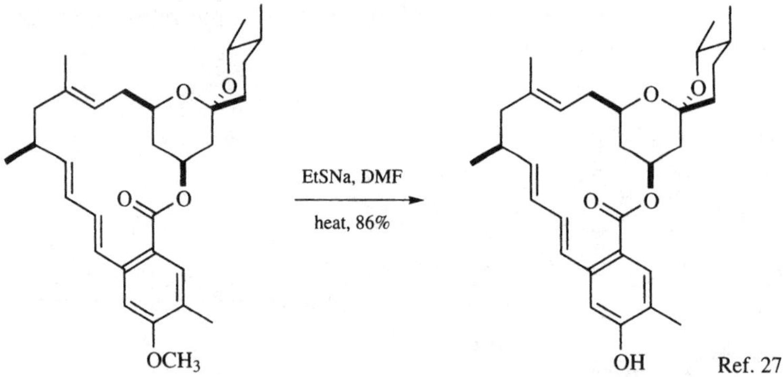

Ref. 27

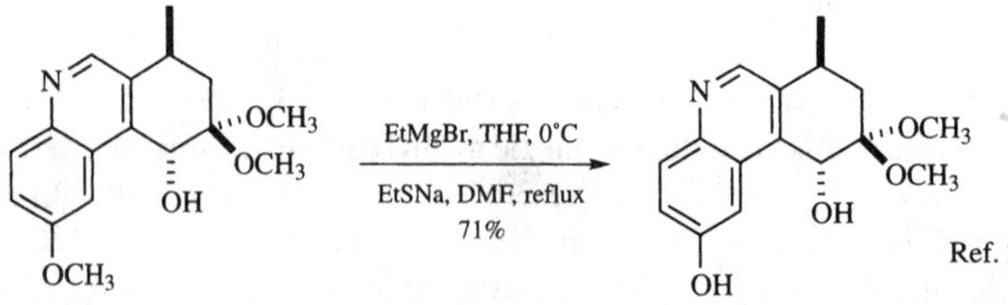

Ref. 28

In this case the magnesium alkoxide protects the ketal from cleavage.[28]

4. PhSPh, Na, NMP, 65–100% yield. This method generates the phenylthiolate ion *in situ*.[29]

5. 4-MePhSLi, HMPA, toluene reflux, 57%. The sodium salt failed to give complete deprotection and acidic reagents could not be used because of the sensitive cyclpropane and olefin.[30]

6. Sodium sulfide in *N*-methylpyrrolidone, NMP, (140°C, 2–4 h) cleaves aryl methyl ethers in 78–85% yield.[31]

7. Me₃SiSNa, DMPU, 185°C, 78–95% yield.[32]

8. (TMS)₂NNa or LDA, THF, DMPU, 185°C, 80–91% yield.[33]

9. DMSO, NaCN, 125–180°C, 5–48 h, 65–90% yield.[34] This cleavage reaction is successful for aromatic systems containing ketones, amides, and carboxylic acids; mixtures are obtained from nitro-substituted aromatic compounds; there is no reaction with 5-methoxyindole (180°C, 48 h).

10. LiI, collidine, reflux, 10 h, quant.[35] Aryl ethyl ethers are cleaved more slowly; dialkyl ethers are stable to these conditions.

11. LiI, quinoline, 140–180°C, 10–30 min, 65–88% yield.[36]

12. Sodium *N*-methylanilide, xylene, HMPA, 60–120°C, 70–95% yield. Methyl ethers of polyhydric phenols are cleaved to give the monophenol.[37] Benzyl ethers are also cleaved. Halogenated phenols are not effectively cleaved because of competing aromatic substitution.

13. Lithium diphenyphosphide (THF, 25°C, 2 h; HCl, H₂O, 87% yield) selectively cleaves an aryl methyl ether in the presence of an aryl ethyl ether.[38] It also cleaves a phenyl benzyl ether and a phenyl allyl ether to the phenol in 88% and 78% yield, respectively.[39,40]

14. L-Selectride or Super Hydride, 67°C, 88–92% yield.[41] Other methods to convert thebaine to oripavine have not been successful.[42]

15. xs MeMgI, 155–165°C, 15 min, 80% yield.[43] In the following case the use of AlBr$_3$/EtSH which was successful in a vancomycin synthesis was not successful.[44]

16.

Ref. 45, 46

The loss of the ethyl group probably occurs by an E-2 elimination whereas methyl cleavage occurs by an S$_N$2 process.

17. LiCl, DMF, heat, 4–72 h.[47]

18. Piperizine, DMA, 150°C, 52–96% yield. This method only works for *o*-anisic acids.[48]

Lewis Acid-Based Methods

1. Me$_3$SiI, CHCl$_3$, 25–50°C, 12–140 h.[49] Iodotrimethylsilane in quinoline (180°C, 70 min) selectively cleaves an aryl methyl group, in 72% yield, in the presence of a methylenedioxy group.[50] Me$_3$SiI cleaves esters more slowly than ethers and cleaves alkyl aryl ethers (48 h, 25°C) more slowly than alkyl alkyl ethers (1.3–48 h, 25°), but benzyl, trityl, and *t*-butyl ethers are cleaved quite rapidly (0.1 h, 25°C).[49] In the following case the reaction fails with the methyl esters do to elimination.[51]

2. *t*-Bu$_2$Si(OTf)$_2$, TEA, MeI, DMF, 100% yield.[52] This method probably produces a silyl iodide *in situ*, which is the real cleaving agent. It was used to prevent loss of the di-*t*-butylsilylene group.

3. AlBr$_3$, EtSH, 25°C, <1 h, 94% yield.[53] Both methyl aryl and methyl alkyl ethers are cleaved under these conditions. A methylenedioxy group, used to protect a catechol, is cleaved under similar conditions in satisfactory yields; methyl and ethyl esters and amides are stable (0–20°C, 2 h).[53]

4. AlCl$_3$, HSCH$_2$CH$_2$SH.[54] *t*-BuSH has been used similarly when the dithiol failed because of reaction at the C12 ketone in the following case.[55]

5. AlCl$_3$, 3 h, 0°C, 75% yield.[56,57] A selectivity study on the demethylation of polymethoxy substituted acetophenones has been performed using AlCl$_3$ in

CH$_3$CN.[58]

AlCl$_3$, 0°C, 3 h
75%

6. AlBr$_3$, CH$_3$CN.[59]

a: R″ = Br   b: R″ = H

7. AlCl$_3$, 1-ethyl-3-methylimidazolium iodide (ionic liquid), BzCl, 25% yield of the benzoate. This method can also be used to cleave other ethers.[60]

8. BBr$_3$, CH$_2$Cl$_2$, −80°C → 20°C, 12 h, 77–86% yield.[61] Methylenedioxy groups and diphenyl ethers are stable to these cleavage conditions. Benzyloxycarbonyl and t-butoxycarbonyl groups, benzyl esters[62] and 1,3-dioxolanes are cleaved with this reagent. Boron tribromide is reported to be more effective than iodotrimethylsilane for cleaving aryl methyl ethers.[63]

9. Boron triiodide rapidly cleaves methyl ethers of o-, m-, or p-substituted aromatic aldehydes (0°C, 25°C; 0.5–5 min; 40–86% yield).[64] BI$_3$ complexed with N,N-diethylaniline is similarly effective, but benzyl ethers are converted to the iodide.[65]

10. BBr$_3$·S(CH$_3$)$_2$, ClCH$_2$CH$_2$Cl, 83°C, 50–99% yield.[66] The advantage of this method is that the reagent is a stable, easily-handled solid. Methylenedioxy groups are also cleaved by this reagent.

11. BF$_3$·Me$_2$S, CH$_2$Cl$_2$, 0°C to rt, 5 min to 3 h, 80–95% yield. These conditions also cleave phenolic allyl ethers.[67]

12. 9-Bromo-9-borabicyclo[3.3.0]nonane (9-Br-BBN), CH$_2$Cl$_2$, reflux, 87–100% yield.[68] 9-Br-BBN also cleaves dialkyl ethers, allyl aryl ethers and methylenedioxy groups. 9-Iodo-9-borabicyclo[3.3.0]nonane has also been used effectively and does not cause haloboration of an alkene.[69]

13. BH$_2$Cl·DMS, toluene, reflux, 95% yield. Acetonides and THP ethers are cleaved and epoxides are converted to the chlorohydrin.[70]

14. Me$_2$BBr, CH$_2$Cl$_2$, 70°C, 30–36 h, 72–96% yield.[71] Alkyl methyl ethers are also cleaved, but tertiary methyl ethers are converted to the bromide.

15. 2-Bromo-1,3,2-benzodioxaborole, CH$_2$Cl$_2$ (cat. BF$_3$·Et$_2$O), 25°C, 0.5–36 h, 95–98% yield.[72] Aryl benzyl ethers, methyl esters, and aromatic benzoates are also cleaved.[72]

16. BCl$_3$, CH$_2$Cl$_2$, −20°C, 94% yield.[73]

Either an aryl methyl ether or a methylenedioxy group can be cleaved with boron trichloride under various conditions.[74] BCl$_3$ in the presence of Bu$_4$NI is more effective than BCl$_3$ alone and the reaction can be run at much lower temperatures.[75] The following case shows that some selectivity is achievable. In this case, coordination probably facilitates the cleavage of the methyl ethers ortho to the carbonyl goups.[76]

17. (C$_6$F$_5$)$_3$B, Et$_3$SiH, CH$_2$Cl$_2$, >99% yield. This method also cleaves a large variety of other ethers.[77] TES ethers are produced in this reaction.

18. MgI$_2$, THF, 92% yield.[78] This method is selective for methyl ethers ortho to a carbonyl group.

19. SiCl$_4$, LiI, BF$_3$, CH$_3$CN, toluene, 45 min to 15 h, 82–98% yield. BF$_3$ was required to get good yields. Benzyl and allyl ethers are cleaved similarly, but methyl thioethers are stable.[79]

20. NbCl$_5$, CH$_2$Cl$_2$, reflux, 3.5 h.[80]

21. $CeCl_3 \cdot 7H_2O$, NaI, CH_3CN, 80–90% yield.[81]

Methods Based on a Brønsted acid

1. CF_3SO_3H, PhSMe, 0–25°C.[82,83] In this case, O-methyltyrosine was deprotected without evidence of O→C migration, which is often a problem when removing protective groups from tyrosine.

2. TFA, thioanisole, TfOH, 2 h, 0°C, 87% yield.[92] Triflic acid alone with microwave heating will cleave phenolic methyl ethers.[84]

3. H_2SO_4, 70°C, 14 h, 52% yield.[85]

4. Methanesulfonic acid, methionine, 20°C, 40 h, 90% yield.[86] Methionine serves to scavenge the methyl group.

5. Regioselective cleavage of dimethoxyaryl derivatives with methanesulfonic acid/methionine has been reported.[93]

6. Pyr·HCl, 220°C, 6 min, 34% yield of morphine from codeine.[87]

7. 48% HBr, AcOH, reflux, 30 min, 85%.[88] The efficiency of this method is significantly improved if a phase transfer catalyst (n-$C_{16}H_{33}PBu_3Br$) is added to the mixture.[89] Methods that use HBr for ether cleavage can give bromides in the presence of benzylic alcohols.[90]

8. 48% HBr, Bu_4NBr, 100°C, 6 h, 80–98% yield.[91]

9. Use of the ionic liquid, [bmim]BF_4 in the presence of a strong protic acid such as HBr or TsOH results in clean phenolic ether cleavage at 115°C, 80–95% yield. Alkyl ethers are also cleaved but in poor yield.

10. HBr, NaI, 90–94°C, sealed tube, 90% yield.[92]

Miscellaneous Methods

1. Ceric ammonium nitrate converts, a 1,4-dimethoxy aromatic compound to the quinone, which is reduced with sodium dithionite to give a deprotected hydroquinone.[94]

2.

Ref. 95

3. Toluene, potassium, 18-crown-6, 100% yield.[96] Tetrahydrofuran can also be used as the solvent in this process.[97]

4. Sodium, liquid ammonia.[98] The utility of this method depends on the nature of the substituents on the aromatic ring. Rings containing electron-withdrawing groups will be reduced, as in the classic Birch reduction.

5. Li, ethylenediamine, THF, −10°C, 34–90% yield. Allyl and benzyl ethers are cleaved similarly, and the method is not compatible with reducible groups such as halides and esters.[99]

6. Microbial O-demethylation has been reported in a few examples. This is a rather specialized method and not necessarily predictable as are most of the chemical methods.[100]

1. D. L. J. Clive, M. Cantin, A. Khodabocus, X. Kong, and Y. Tao, *Tetrahedron*, **49**, 7917 (1993); D. L. J. Clive, A. Khodabocus, M. Cantin, and Y. Tao, *J. Chem. Soc., Chem. Commun.*, 1755 (1991).

2. G. N. Vyas and N. M. Shah, *Org. Synth., Coll. Vol. IV*, 836 (1963).

3. A. R. MacKenzie, C. J. Moody, and C. W. Rees, *Tetrahedron*, **42**, 3259 (1986).

4. D. L. Boger, J. Hong, M. Hikota, and M. Ishida, *J. Am. Chem. Soc.*, **121**, 2471 (1999).

5. W. E. Wymann, R. Davis, J. W. Patterson, Jr., and J. R. Pfister, *Synth. Commun.*, **18**, 1379 (1988).

6. A. Basek, M. K. Nayak, and A. K. Chakraborti, *Tetrahedron Lett.*, **39**, 4883 (1998).

7. A. McKillop, J.-C. Fiaud, and R. P. Hug, *Tetrahedron*, **30**, 1379 (1974).

8. J. M. Miller, K. H. So, and J. H. Clark, *Can. J. Chem.*, **57**, 1887 (1979).

9. A. K. Sinhababu, M. Kawase, and R. T. Borchardt, *Tetrahedron Lett.*, **28**, 4139 (1987).

10. S. M. Hecht, and J. W. Kozarich, *Tetrahedron Lett.*, **14**, 1307 (1973).

11. F. Bracher and B. Schulte, *J. Chem. Soc., Perkin Trans. I*, 2619 (1996).

12. T. Aoyama, S. Terasawa, K. Sudo, and T. Shioiri, *Chem. Pharm. Bull.*, **32**, 3759 (1984).

13. B. M. Crowley, Y. Mori, C. C. McComas, D. Tang, and D. L. Boger, *J. Am. Chem. Soc.*, **126**, 4310 (2004).

14. G. Barcelo, D. Grenouillat, J. P. Senet, and G. Sennyey, *Tetrahedron*, **46**, 1839 (1990).

15. A. Perosa, M. Selva, P. Tundo, and F. Zordan, *Synlett*, 272 (2000).

16. Y. Lee and I. Shimizu, *Synlett*, 1063 (1998).

17. W.-C. Shieh, S. Dell, and O. Repic, *Org. Lett.*, **3**, 4279 (2001).

18. S. Ouk, S. Thiebaud, E. Borredon, P. Legars, and L. Lecomte, *Tetrahedron Lett.*, **43**, 2661 (2002).

19. M. Attolini, T. Boxus, S. Biltresse, and J. Marchand-Brynaert, *Tetrahedron Lett.*, **43**, 1187 (2002).

20. G. I. Feutrill and R. N. Mirrington, *Tetrahedron Lett.*, 1327 (1970); *idem, Aust. J. Chem.*, **25**, 1719, 1731 (1972).

21. A. S. Kende and J. P. Rizzi, *Tetrahedron Lett.*, **22**, 1779 (1981).

22. J. W. Wildes, N. H. Martin, C. G. Pitt, and M. E. Wall, *J. Org. Chem.*, **36**, 721 (1971).

23. R. Ahmad, J. M. Saá, and M. P. Cava, *J. Org. Chem.*, **42**, 1228 (1977).

24. C. Hansson and B. Wickberg, *Synthesis*, 191 (1976).

25. M. K. Nayak and A. K. Chakraborti, *Tetrahedron Lett.*, **38**, 8749 (1997); A. K. Chakraborti, L. Sharma, and M. K. Nayak, *J. Org. Chem.*, **67**, 6406 (2002).

26. J. A. Dodge, M. G. Stocksdale, K. J. Fahey, and C. D. Jones, *J. Org. Chem.*, **60**, 739 (1995).

27. A. B. Smith, III, S. R. Schow, J. D. Bloom, A. S. Thompson, and K. N. Winzenberg, *J. Am. Chem. Soc.*, **104**, 4015 (1982).

28. A. G. Myers, N. J. Tom, M. E. Fraley, S. B. Cohen, and D. J. Mader, *J. Am. Chem. Soc.*, **119**, 6072 (1997).

29. A. K. Chakraborti, M. K. Nayak, and L. Sharma, *J. Org. Chem.*, **67**, 1776 (2002).

30. T. Tanaka, H. Mikamiyama, K. Maeda, and C. Iwata, *J. Org. Chem.*, **63**, 9782 (1998).

31. M. S. Newman, V. Sankaran, and D. R. Olson, *J. Am. Chem. Soc.*, **98**, 3237 (1976).

32. J. R. Hwu and S.-C. Tsay, *J. Org. Chem.*, **55**, 5987 (1990).

33. J. R. Hwu, F. F. Wong, J.-J. Huang, and S.-C. Tsay, *J. Org. Chem.*, **62**, 4097 (1997).

34. J. R. McCarthy, J. L. Moore,, and R. J. Crege, *Tetrahedron Lett.*, 5183 (1978).

35. I. T. Harrison, *J. Chem. Soc., Chem. Commun.*, 616 (1969).

36. K. Kirschke and E. Wolff, *J. Prakt. Chem./Chem. Ztg.*, **337**, 405 (1995).

37. B. Loubinoux, G. Coudert, and G. Guillaumet, *Synthesis*, 638 (1980).

38. R. E. Ireland and D. M. Walba, *Org. Synth., Coll.Vol. VI*, 567 (1988).

39. F. G. Mann and M. J. Pragnell, *Chem. Ind. (London)*, 1386 (1964).

40. H. Meier and U. Dullweber, *Tetrahedron Lett.*, **37**, 1191 (1996).

41. G. Majetich, Y. Zhang, and K. Wheless, *Tetrahedron Lett.*, **35**, 8727 (1994).

42. A. Coop, J. W. Lewis, and K. C. Rice, *J. Org. Chem.*, **61**, 6774 (1996).

43. R. Mechoulam and Y. Gaoni, *J. Am. Chem. Soc.*, **87**, 3273 (1965).

44. T. R. Hoye, P. E. Humpal, and B. Moon, *J. Am. Chem. Soc.*, **122**, 4982 (2000).

45. A. S. Radhakrishna, K. R. K. P. Rao, S. K. Suri, K. Sivaprakash, and B. B. Singh, *Synth. Commun.*, **21**, 379 (1991).

46. A.Oussaïd, L. N. Thach, and A. Loupy, *Tetrahedron Lett.*, **38**, 2451 (1997).

47. A. M. Bernard, M. R. Ghiani, P. P. Piras, and A. Rivoldini, *Synthesis*, 287 (1989).

48. H. Nishioka, M. Nagasawa, and K. Yoshida, *Synthesis*, 243 (2000).

49. M. E. Jung and M. A. Lyster, *J. Org. Chem.*, **42**, 3761 (1977).

50. J. Minamikawa and A. Brossi, *Tetrahedron Lett.*, **19**, 3085 (1978).

51. S. J. O'Malley, K. L. Tan, A. Watzke, R. G. Bergman, and J. A. Ellman, *J. Am. Chem. Soc.*, **127**, 13496 (2005).

52. Y. Kita, K. Iio, K. Kawaguchi, N. Fukuda, Y. Takeda, H. Ueno, R. Okunaka, K. Higuchi, T. Tsujino, H. Fujioka, and S. Akai, *Chem. Eur. J.*, **6**, 3897 (2000).

53. M. Node, K. Nishide, K. Fuji, and E. Fujita, *J. Org. Chem.*, **45**, 4275 (1980).

54. T. Inaba, I. Umezawa, M. Yuasa, T. Inoue, S. Mihashi, H. Itokawa, and K. Ogura, *J. Org. Chem.*, **52**, 2957 (1987).

55. Z. Fei and F. E. McDonald, *Org. Lett.*, **7**, 3617 (2005).

56. K. A. Parker and J. J. Petraitis, *Tetrahedron Lett.*, **22**, 397 (1981).

57. T.-t. Li and Y. L. Wu, *J. Am. Chem. Soc.*, **103**, 7007 (1981).

58. Y. Kawamura, H. Takatsuki, F. Torii, and T. Horie, *Bull. Chem. Soc. Jpn.*, **67**, 511 (1994).

59. T. Horie, T. Kobayashi, Y. Kawamura, I. Yoshida, H. Tominaga, and K. Yamashita, *Bull. Chem. Soc. Jpn.*, **68**, 2033 (1995).

60. L. Green, I. Hemeon, and R. D. Singer, *Tetrahedron Lett.*, **41**, 1343 (2000).

61. J. F. W. McOmie and D. E. West, *Org. Synth., Coll. Vol. V*, 412 (1973).

62. A. M. Felix, *J. Org. Chem.*, **39**, 1427 (1974).

63. E. H. Vickery, L. F. Pahler, and E. J. Eisenbraun, *J. Org. Chem.*, **44**, 4444 (1979).

64. J. M. Lansinger and R. C. Ronald, *Synth. Commun.*, **9**, 341 (1979).

65. C. Narayana, S. Padmanabhan, and G. W. Kabalka, *Tetrahedron Lett.*, **31**, 6977 (1990).

66. P. G. Williard and C. B. Fryhle, *Tetrahedron Lett.*, **21**, 3731 (1980).

67. M. T. Konieczny, G. Maciejewski, and W. Konieczny, *Synthesis*, 1575 (2005).

68. M. V. Bhatt, *J. Organomet. Chem.*, **156**, 221 (1978).

69. A. Fürstner, and G. Seidel, *J. Org. Chem.*, **62**, 2332 (1997).

70. P. Bovicelli, E. Mincione, and G. Ortaggi, *Tetrahedron Lett.*, **32**, 3719 (1991).

71. Y. Guindon, C. Yoackim, and H. E. Morton, *Tetrahedron Lett.*, **24**, 2969 (1983).

72. P. F. King and S. G. Stroud, *Tetrahedron Lett.*, **26**, 1415 (1985).

73. H. Nagaoka, G. Schmid, H. Iio, and Y. Kishi, *Tetrahedron Lett.*, **22**, 899 (1981).

74. M. Gerecke, R. Borer, and A. Brossi, *Helv. Chim. Acta*, **59**, 2551 (1976).

75. P. R. Brooks, M. C. Wirtz, M. G. Vetelino, D. M. Rescek, G. F. Woodworth, B. P. Morgan, and J. W. Coe, *J. Org. Chem.*, **64**, 9719 (1999).

76. M. Kitamura, K. Ohmori, T. Kawase, and K. Suzuki, *Angew. Chem. Int. Ed.*, **38**, 1229 (1999).

77. V. Gevorgyan, M. Rubin, S. Benson, J.-X. Liu, and Y. Yamamoto, *J. Org. Chem.*, **65**, 6179 (2000).

78. S. Yamaguchi, M. Nedachi, H. Yokoyama, and Y. Hirai, *Tetrahedron Lett.*, **40**, 7363 (1999).

79. D. Zewge, A. King, S. Weissman, and D. Tschaen, *Tetrahedron Lett.*, **45**, 3729 (2004).

80. S. Arai, Y. Sudo, and A. Nishida, *Synlett*, 1104 (2004).

81. J. S. Yadav, B. V. S. Reddy, C. Madan, and S. R. Hashim, *Chem. Lett.*, **29**, 738 (2000).

82. Y. Kiso, S. Nakamura, K. Ito, K. Ukawa, K. Kitagawa, T. Akita, and H. Moritoki, *J. Chem. Soc., Chem. Commun.*, 971 (1979).

83. Y. Kiso, K. Ukawa, S. Nakamura, K. Ito, and T. Akita, *Chem. Pharm. Bull.*, **28**, 673 (1980).

84. A. Fredriksson and S. Stone-Elander, *J. Labelled Compd. Radiopharm.*, **45**, 529 (2002).

85. C. Li, E. Lobkovsky and J. J. A. Porco, *J. Am. Chem. Soc.*, **122**, 10484 (2000).

86. D. G. Melillo, R. S. Larsen, D. J. Mathre, W. F. Shukis, A. W. Wood, and J. R. Colleluori, *J. Org. Chem.*, **52**, 5143 (1987); N. Fujii, H. Irie, and H. Yajima, *J. Chem. Soc. Perkin Trans. 1*, 2288 (1977).

87. M. Gates and G. Tschudi, *J. Am. Chem. Soc.*, **78**, 1380 (1956).

88. I. Kawasaki, K. Matsuda, and T. Kaneko, *Bull. Chem. Soc. Jpn.*, **44**, 1986 (1971).

89. D. Landini, F. Montanari, and F. Rolla, *Synthesis*, 771 (1978).

90. A. Kamai and N. L. Gayatri, *Tetrahedron Lett.*, **37**, 3359 (1996).

91. K. Hwang and S. Park, *Synth. Commun.*, **23**, 2845 (1993).

92. G. Li, D. Patel and V. J. Hruby, *Tetrahedron Lett.*, **34**, 5393 (1993).

93. N. Fujii, H. Irie, and H. Yajima, *J. Chem . Soc., Perkin Trans. I*, 2288 (1977).

94. M. Kawaski, F. Matsuda, and S. Terashima, *Tetrahedron*, **44**, 5713 (1988).

95. P. Deslongchamps, A. Bélanger, D. J. F. Berney, H. J. Borschberg, R. Brousseau, A. Doutheau, R. Durand, H. Katayama, R. Lapalme, D. M. Leturc, C.-C. Liao, F. N. MacLachan, J.-P. Maffrand, F. Marazza, R. Martino, C. M. L. Ruest, L. Saint-Laurent, and R. Saintonge, and P. Soucy, *Can. J. Chem.*, **68**, 115 (1990).

96. T. Ohsawa, K. Hatano, K. Kayoh, J. Kotabe, and T. Oishi, *Tetrahedron Lett.*, **33**, 5555 (1992).

97. U. Azzena, T. Denurra, G. Melloni, E. Fenude, and G. Rassa, *J. Org. Chem.*, **57**, 1444 (1992).

98. A. J. Birch, *Q. Rev.*, **4**, 69 (1950).

99. T. Shindo, Y. Fukuyama, and T. Sugai, *Synthesis*, 692 (2004).

100. G. S. Wu, A. Gard, and J. P. Rosazza, *J. J. Antibiotics*, **33**, 705 (1980). H. Kanatani, C. Sakakibara, M. Tanaka, K. Niitsu, Y. Ikeya, T. Wakamatsu, and M. Maruno, *Biosci. Biotech. Biochem.*, **58**, 1054 (1994).

Methoxymethyl Ether (MOM Ether): $ArOCH_2OCH_3$ (Chart 4)

Formation

1. $ClCH_2OCH_3$, CH_2Cl_2, NaOH-H_2O, Adogen (phase transfer cat.), 20°C, 20 min, 80–95% yield.[1,2] This method has been used to protect selectively a phenol in the presence of an alcohol.[3]

2. $ClCH_2OCH_3$, CH_3CN, 18-crown-6, 80% yield.[4]

3. $ClCH_2OCH_3$, acetone or DMF, K_2CO_3, 86% yield.[5,6] In the following example the selectivity is attributed to the hydrogen bonding of the peri OH with the carbonyl thus reducing its activity.

4. $ClCH_2OCH_3$, DMF, NaH, 93% yield.[5]

5. $CH_3OCH_2OCH_3$ TsOH, CH_2Cl_2, molecular sieves, N_2, reflux, 12 h, 60–80% yield.[7] This method of formation avoids the use of the **carcinogen chloromethyl methyl ether**.

6. MOM-2-pyridylsulfide, AgOTf, NaOAc, THF, 14–98% yield. Alkanols are similarly derivatized, but electron-deficient alcohols such as 4-nitrophenol give low yields.[8]

7. The ethoxymethyl ether (EOM ether) can be used as a replacement for the MOM group.[9]

Cleavage

1. HCl, *i*-PrOH, THF, 25°C, 12 h, quant.[7]

2. 2 *N* HOAc, 90°C, 40 h, high yield.[10] The group has been used in a synthesis of 13-desoxydelphonine from *o*-cresol, a synthesis that required the group to be stable to many reagents.[11]

3. CF_3CO_2H, CH_2Cl_2, 0°C, 3 h, 99% yield.[12] The method was selective for a phenolic MOM group.

4. Montmorillonite clay, CH_2Cl_2 or benzene, 25–50°C, 0.5–5 h, 74–96% yield. This method only works for systems that contain ortho heteroatoms.[13] Other systems give very low yield or do not react.

5. 1-Fluoro-3,5-dichloropyridinium triflate, CH_2Cl_2, 0°C, 2 h, 69% yield. The authors indicate that the MOM group is cleaved by fluorination of the methylene followed by hydrolysis.[14] An alternative explanation is that triflic acid is generated during the oxidation of the A-ring, which cleaves the MOM group by conventional acid hydrolysis.

6. NaHSO$_4$, SiO$_2$, CH$_2$Cl$_2$, rt, 1–1.5 h, 90–100% yield.[15] This method also cleaves MOM esters.

7. NaI, acetone, cat. HCl, 50°C, 85% yield.[16]

8. P$_2$I$_4$, CH$_2$Cl$_2$, 0°C to rt, 30 min, 70–90% yield.[17] This method is also effective for removal of the SEM and MEM groups.

9. (EtO)$_3$SiCl, NaI, CH$_3$CN, CH$_2$Cl$_2$, −5°C, 0.5 h, 74% yield. This method was reported to work better than TMSI.[18] TBDPS groups were not affected by this reagent.

10. TMSBr, CH$_2$Cl$_2$, 30°C to 0°C, 87% yield.[19]

11. CBr$_4$, Ph$_3$P, ClCH$_2$CH$_2$Cl, 40°C, 90–99% yield.[20]

1. F. R. van Heerden, J. J. van Zyl, G. J. H. Rall, E. V. Brandt, and D. G. Roux, *Tetrahedron Lett.*, **19**, 661 (1978).

2. W. R. Roush, D. S. Coffey, and D. J. Madar, *J. Am. Chem. Soc.*, **119**, 11331 (1997).

3. T. R. Kelly, C. T. Jagoe, and Q. Li, *J. Am. Chem. Soc.*, **111**, 4522 (1989).

4. G. J. H. Rall, M. E. Oberholzer, D. Ferreira, and D. G. Roux, *Tetrahedron Lett.*, **17**, 1033 (1976).

5. M. Süsse, S. Johne, and M. Hesse, *Helv. Chim. Acta*, **75**, 457 (1992).

6. A. Scopton and T. R. Kelly, *Org. Lett.*, **6**, 3869 (2004).

7. J. P. Yardley and H. Fletcher III, *Synthesis*, 244 (1976).

8. B. F. Marcune, S. Karady, U.-H. Dolling, and T. J. Novak, *J. Org. Chem.*, **64**, 2446 (1999).

9. E. Moulin, S. Barluenga, and N. Winssinger, *Org. Lett.*, **7**, 5637 (2005).

10. M. A. A.-Rahman, H. W. Elliott, R. Binks, W. Küng, and H. Rapoport, *J. Med. Chem.*, **9**, 1 (1966).

11. K. Wiesner, *Pure Appl. Chem.*, **51**, 689 (1979).

12. M. Kitamura, K. Ohmori, T. Kawase, and K. Suzuki, *Angew. Chem. Int. Ed.*, **38**, 1229 (1999).

13. J. P. Deville and V. Behar, *J. Org. Chem.*, **66**, 4097 (2001).

14. E. J. Martinez and E. J. Corey, *Org. Lett.*, **1**, 75 (1999).

15. C. Ramesh, N. Ravindranath, and B. Das, *J. Org. Chem.*, **68**, 7101 (2003).

16. D. R. Williams, B. A. Barner, K. Nishitani, and J. G. Phillips, *J. Am. Chem. Soc.*, **104**, 4708 (1982).
17. H. Saimoto, Y. Kusano, and T. Hiyama, *Tetrahedron Lett.*, **27**, 1607 (1986).
18. J. R. Falck, K. K. Reddy, and S. Chandrasekhar, *Tetrahedron Lett.*, **38**, 5245 (1997).
19. J. W. Huffman, X. Zhang, M.-J. Wu, H. H. Joyner, and W. T. Pennington, *J. Org. Chem.*, **56**, 1481 (1991).
20. Y. Peng, C. Ji, Y. Chen, C. Huang, and Y. Jiang, *Synth. Commum.*, **34**, 4325 (2004).

Benzyloxymethyl Ether (BOM Ether): $C_6H_5CH_2OCH_2OAr$

Formation

BOMCl, NaH, DMF, >81% yield.[1]

Cleavage

1. MeOH, Dowex 50W-X8 (H^+), 90% yield.[1]
2. RaNi, THF, EtOH, 57% yield.[2]
3. Pd catalyzed hydrogenolysis should also be effective for the cleavage of this ether.

1. W. R. Roush, M. R. Michaelides, D. F. Tai, B. M. Lesur, W. K. M. Chong, and D. J. Harris, *J. Am. Chem. Soc.*, **111**, 2984 (1989).
2. W. R. Roush, R. A. Hartz, and D. J. Gustin, *J. Am. Chem. Soc.*, **121**, 1990 (1999).

Methoxyethoxymethyl Ether (MEM Ether): $ArOCH_2OCH_2CH_2OCH_3$ (Chart 4)

In an attempt to metalate a MEM-protected phenol with BuLi, the methoxy group was eliminated forming the vinyloxymethyl ether. This was attributed to intramolecular proton abstraction.[1] A 2-methoxyethoxymethyl ether was used to protect one phenol group during a total synthesis of gibberellic acid.[2]

Formation

1. NaH, THF, 0°C; $MeOCH_2CH_2OCH_2Cl$, 0–25°C, 2 h, 75% yield.[2]
2. $MeOCH_2CH_2OCH_2Cl$, DIPEA.[3]

Cleavage

1. CF_3CO_2H, CH_2Cl_2, 23°C, 1 h, 74% yield.[2]
2. $(Ipc)_2BCl$, THF, 0°C, 80 h. Cleavage occurred during the reduction of an acetophenone.[3]
3. For other methods of cleavage, the chapter on alcohol protection should be consulted.

1. J. Mayrargue, M. Essamkaoui, and H. Moskowitz, *Tetrahedron Lett.*, **30**, 6867 (1989).

2. E. J. Corey, R. L. Danheiser, S. Chandrasekaran, P. Siret, G. E. Keck, and J.-L. Gras, *J. Am. Chem. Soc.*, **100**, 8031 (1978).

3. E. T. Everhart and J. C. Craig, *J. Chem. Soc., Perkin Trans. I*, 1701 (1991).

2-(Trimethylsilyl)ethoxymethyl Ether (SEM Ether):
$(CH_3)_3SiCH_2CH_2OCH_2OAr$

Formation

1. SEMCl, DMAP, Et_3N, benzene, reflux, 3 h, 98% yield.[1]

2. SEMCl, $(i\text{-}Pr)_2NEt$, CH_2Cl_2, 97% yield.[3]

Cleavage

1. Bu_4NF, HMPA, 40°C, 2 h, >23–51% yield.[2]

2. H_2SO_4, MeOH, THF, 90% yield.[1]

3. P_2I_4, CH_2Cl_2, 0°C to rt, 30 min, 62–86% yield.[3,4] These conditions also cleave methoxymethyl and methoxyethoxymethyl ethers.

4. In the following case the SEM group served as a good leaving group because of its ability to stabilize positive charge.[5]

5. $MgBr_2$, Et_2O, CH_2Cl_2, 70% yield.[6] In this case previous attempts to cleave the phenolic EOM groups (ethoxymethyl ether) with acid all failed because of epoxide opening.

1. T. L. Shih, M. J. Wyvratt, and H. Mrozik, *J. Org. Chem.*, **52**, 2029 (1987).

2. A. Leboff, A.-C. Carbonnelle, J.-P. Alazard, C. Thal, and A. S. Kende, *Tetrahedron Lett.*, **28**, 4163 (1987).

3. H. Saimoto, Y. Kusano, and T. Hiyama, *Tetrahedron Lett.*, **27**, 1607 (1986).

4. H. Saimoto, S.-i. Ohrai, H. Sashiwa, Y. Shigemasa, and T. Hiyama, *Bull. Chem. Soc. Jpn*, **68**, 2727 (1995).

5. L. K. Casillas and C. A. Townsend, *J. Org. Chem.*, **64**, 4050 (1999).

6. E. Moulin, S. Barluenga, and N. Winssinger, *Org. Lett.*, **7**, 5637 (2005).

Methylthiomethyl Ether (MTM Ether): $ArOCH_2SCH_3$ (Chart 4)

Formation

NaOH, $ClCH_2SMe$, HMPA, 25°C, 16 h, 91–94% yield.[1]

Cleavage

1. $HgCl_2$, $CH_3CN–H_2O$, reflux, 10 h, 90–95% yield.[1] Aryl methylthiomethyl ethers are stable to the conditions used to hydrolyze primary alkyl MTM ethers (e.g., $HgCl_2/CH_3CN–H_2O$, 25°C, 6 h). They are moderately stable to acidic conditions (95% recovered from $HOAc/THF–H_2O$, 25°C, 4 h).

2. Ac_2O, Me_3SiCl, 25 min, rt, 95% yield.[2]

1. R. A. Holton and R. G. Davis, *Tetrahedron Lett.*, **18**, 533 (1977).

2. N. C. Barua, R. P. Sharma, and J. N. Baruah, *Tetrahedron Lett.*, **24**, 1189 (1983).

Phenylthiomethyl Ether (PTM Ether): $C_6H_5SCH_2OAr$

Formation

NaI, $PhSCH_2Cl$, NaH, HMPA, 87–94% yield.[1]

Cleavage

$CH_3CN:H_2O$ (4:1), $HgCl_2$, 24 h, 90–94% yield. The methylthiomethyl ether group can be removed in the presence of the phenylthiomethyl ether.[1]

1. R. A. Holton and R. V. Nelson, *Synth. Commun.*, **10**, 911 (1980).

Azidomethyl Ether (Azm−OAr): N_3CH_2OAr

The azidomethyl ether, used to protect phenols and prepared by displacement of azide on the chloromethylene group, is cleaved reductively with $LiAH_4$, by hydrogenolysis (Pd–C, H_2) or reduction with $SnCl_2/PhSH/TEA$.[1] It is stable to strong acids, permanganate, and free-radical bromination.[2]

1. T. Young and L. L. Kiessling, *Angew. Chem. Int. Ed.*, **41**, 3449 (2002).

2. B. Loubinoux, S. Tabbache, P. Gerardin, and J. Miazimbakana, *Tetrahedron*, **44**, 6055 (1988).

Cyanomethyl Ether: $ArOCH_2CN$

The cyanomethyl ether, formed from bromoacetonitrile (acetone, K_2CO_3, 97–100% yield), is cleaved by hydrogenation of the nitrile with PtO_2 in EtOH, 98% yield.[1] The method has also been used for the protection of amines and carbamates.

1. A. Benarab, S. Boye, L. Savelon, and G. Guillaumet, *Tetrahedron Lett.*, **34**, 7567 (1993).

2,2-Dichloro-1,1-difluoroethyl Ether: $CHCl_2CF_2OAr$

Formation/Cleavage

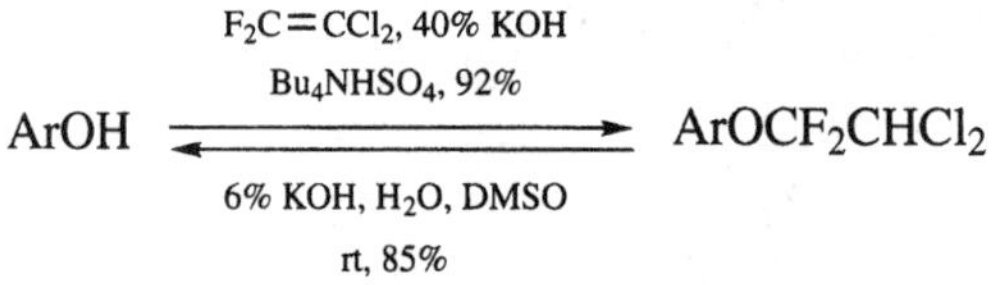

$$ArOH \underset{\underset{rt,\ 85\%}{6\%\ KOH,\ H_2O,\ DMSO}}{\overset{\overset{F_2C=CCl_2,\ 40\%\ KOH}{Bu_4NHSO_4,\ 92\%}}{\rightleftharpoons}} ArOCF_2CHCl_2$$

This group decreases the electron density on the aromatic ring and thus inhibits solvolysis of the tertiary alcohol **i** and the derived acetate **ii**.[1]

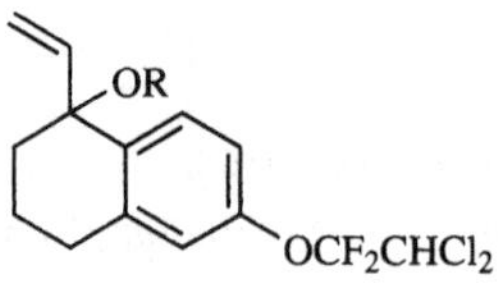

i R = H
ii R = Ac

1. S. G. Will, P. Magriotis, E. R. Marinelli, J. Dolan, and F. Johnson, *J. Org. Chem.*, **50**, 5432 (1985).

2-Chloro- and 2-Bromoethyl Ether: XCH_2CH_2OAr, X=Cl, Br

These ethers can be removed from naphthohydroquinones, either by elimination to the vinyl ether followed by hydrolysis or by Finklestein reaction with iodide followed by reduction with zinc.[1]

1. H. Laatsch, *Z. Naturforsch., B: Anorg. Chem., Org. Chem.*, **40b**, 534 (1985).

t-Butyldiphenylsilylethyl Ether (TBDPSE–OAr)

This group was developed as an alternative to the TMSE group, which can only be introduced via the Mitsunobu reaction in low yield because of competing *O*-silylation. The TBDPSE group is introduced using the Mitsunobu reaction (TB-DMSCH$_2$CH$_2$OH, DIAD, PPh$_3$, 57–98% yield). It is stable to mild acid (5% TFA), base, hydrogenolysis, and lithium halogen exchange. It is cleaved with strong acid (50% TFA, CH$_2$Cl$_2$) or TBAF/THF (75–92% yield).[1]

1. B. S. Gerstenberger and J. P. Konopelski, *J. Org. Chem.*, **70**, 1467 (2005).

Tetrahydropyranyl Ether (THP Ether): ArO-2-tetrahydropyranyl

The tetrahydropyranyl ether, prepared from a phenol and dihydropyran (HCl/EtOAc, 25°C, 24 h), is cleaved by aqueous oxalic acid (MeOH, 50–90°C, 1–2 h)[1] or other acidic reagents such as oxone[2] or TMSI.[3] Tonsil, Mexican Bentonite earth,[4] HSZ Zeolite,[5] and H$_3$[PW$_{12}$O$_{40}$][6] have also been used for the tetrahydropyranylation of phenols. The use of [Ru(ACN)$_3$(triphos)](OTf)$_2$ in acetone selectively removes the THP group from a phenol in the presence of an alkyl THP group. Ketals of aceto-phenones are also cleaved.[7]

1-Ethoxyethyl Ether (EE): ArOCH(OC$_2$H$_5$)CH$_3$

The ethoxyethyl ether is prepared by acid catalysis from a phenol and ethyl vinyl ether and is cleaved by acid-catalyzed methanolysis.[8]

1. H. N. Grant, V. Prelog, and R. P. A. Sneeden, *Helv. Chim. Acta*, **46**, 415 (1963).
2. I. Mohammadpoor-Baltork, M. K. Amini, and S. Farshidipoor, *Bull. Chem. Soc. Jpn.*, **73**, 2775 (2000).
3. N. Foy, E. Stephan, and G. Jaouen, *J. Chem. Res.(S)*, 518 (2001).
4. R. Cruz-Almanza, F. J. Pérez-Floress, and M. Avila, *Synth. Commun.*, **20**, 1125 (1990).
5. R. Ballini, F. Bigi, S. Carloni, R. Maggi, and G. Sartori, *Tetrahedron Lett.*, **38**, 4169 (1997).
6. A. Moinar and T. Beregszaszi, *Tetrahedron Lett.*, **37**, 8597 (1996).
7. S. Ma and L. M. Venanzi, *Tetrahedron Lett.*, **34**, 8071 (1993).
8. J. H. Rigby and M. E. Mateo, *J. Am. Chem. Soc.*, **119**, 12655 (1997).

Phenacyl Ether: ArOCH$_2$COC$_6$H$_5$ (Chart 4)

4-Bromophenacyl Ether: ArOCH$_2$COC$_6$H$_4$-4-Br

Formation

BrCH$_2$COPh, K$_2$CO$_3$, acetone, reflux, 1–2 h, 85–95% yield.[1]

Cleavage

Zn, HOAc, 25°C, 1 h, 88–96% yield.[1] Phenacyl and *p*-bromophenacyl ethers of phenols are stable to 1% ethanolic alkali (reflux, 2 h) and to 5 *N* sulfuric acid in ethanol–water. The phenacyl ether, prepared from β-naphthol, is cleaved in 82% yield by 5% ethanolic alkali (reflux, 2 h).

1. J. B. Hendrickson and C. Kandall, *Tetrahedron Lett.*, **11**, 343 (1970).

Cyclopropylmethyl Ether: $ArOCH_2\text{-}c\text{-}C_3H_5$

For a particular phenol, the authors required a protective group that would be stable to reduction (by complex metals, catalytic hydrogenation, and Birch conditions) and that could be easily and selectively removed.

Formation

t-BuOK, DMF, 0°C, 30 min; *c*-$C_3H_5CH_2Br$, 20°C, 20 min to 40°C, 6 h, 80% yield.[1]

Cleavage

aq. HCl, MeOH, reflux, 2 h, 94% yield.[1]

1. W. Nagata, K. Okada, H. Itazaki, and S. Uyeo, *Chem. Pharm. Bull.*, **23**, 2878 (1975).

Allyl Ether: $ArOCH_2CH=CH_2$ (Chart 4)

Formation

1. Allyl ethers can be prepared by reaction of a phenol and the allyl bromide in the presence of base.[1] The use of KOH in EtOH with allyl bromide is an excellent method.
2. AllylOH, $Pd(OAc)_2$, PPh_3, $Ti(O-i\text{-}Pr)_4$, 73–87% yield.[2]
3. The section on allyl ethers of alcohols should be consulted.

Cleavage

1. The section on the cleavage of allyl ethers of alcohols should also be consulted.
2. *t*-BuOK, DMSO, 92% yield; MeOH, HCl, >75% yield.[3] This reaction proceeds by isomerization to the enol ether followed by hydrolysis.
3. EtOH, $RhCl_3$, reflux, 86% yield.[1] Cleavage proceeds by isomerization and enol ether hydrolysis. See the section on alkyl allyl ether cleavage for other methods to perform the isomerization.
4. Pd–C, TsOH, H_2O or MeOH; 60–80°C, 6 h, > 95% yield.[4]

5. 10% Pd–C, 10% KOH, MeOH, rt, 8 h, 71–100% yield. Other allyl ethers such as prenyl, cinnamyl, cyclohexenyl and 2-methylpropenyl ethers are cleaved similarly.[5]

6. $Ph_3P/Pd(OAc)_2$, HCOOH, 90°C, 1 h.[6]

7. Pd° cat., Bu_3SnH, AcOH, p-NO_2−phenol.[7] The crotyl ether has been cleaved by a similar method.[8] In the following case, isomerization methods failed presumably because of the MTM group, which may poison the catalysts.[9]

8. $Pd(Ph_3P)_4$, $LiBH_4$, THF, 88% yield.[10] $NaBH_4$ can also be used as an allyl scavenging agent.[11]

9. $Pd(Ph_3P)_4$, Et_3SiH, AcOH, toluene, 92% yield.

10. $Pd(Ph_3P)_4$, $PhSiH_3$, 20–40 min, 74–100% yield.[12]

11. $Pd(Ph_3P)_4$, K_2CO_3, MeOH, reflux, 6–12 h, 85–97% yield.[13]

12. Bis(benzonitrile)palladium(II) chloride, benzene, reflux, 16–20 h, 86% yield.[14]

13. 1,2-Bis(4-methoxyphenyl)3,4-bis(2,4,6-tri-*tert*-butylphenylphosphinidiene) cyclobutene, Pd(0), aniline, 84–99% yield. This is an excellent catalyst for the cleavage of allyl ethers, esters, and carbamates.[15]

14. $LiPPh_2$, THF, 4 h, reflux, 78% yield.[16] Cleavage proceeds by an S_N2' process.

15. $NaAlH_2(OCH_2CH_2OCH_3)_2$, $PhCH_3$, reflux, 10 h, 62% yield.[17] An aryl allyl ether is selectively cleaved by this reagent (which also cleaves aryl benzyl ethers) in the presence of an N-allylamide.

16. $SiCl_4$, NaI, CH_2Cl_2, CH_3CN, 8 h, 84% yield.[18]

17. $NaBH_4$, I_2, THF, 0°C, 84–95% yield.[19]

18. I_2, DMSO, 130°C, 30 min, 85–97% yield.[20] Iodine probably also causes the required oxidation that is observed.

19. Electrolysis: $PdCl_2$, bipyridine, DMF, Bu_4NBF_4, Mg/stainless steel electrodes, 20°C, 73–99% yield.[21]

20. Electrolysis, DMF, Bu_4NBr, $SmCl_3$, Mg anode, 67–90% yield.[22]

21. Electrogenerated elemental nickel, NaOAc, DMF, 18 h, rt, 72–100% yield. The presence of aryl iodides results in low yields.[23]

22. Electrolysis, $[Ni(bipy)_3](BF_3)$, Mg anode, DMF, rt, 40–99% yield.[24] Aryl bromides and iodides are reduced under these conditions.

23. Chromium-pillared clay, t-BuOOH, CH_2Cl_2, 10 h, 80% yield. Simple allyl ethers are cleaved to give ketones, and allylamines are also deprotected (84–90% yield).[25]

24. SeO_2/HOAc, dioxane, reflux, 1 h, 40–75% yield.[26]

25. Li, naphthalene, THF, 51–91% yield.[27]

26. $TiCl_3$, Mg, THF, reflux, 3 h, 70% yield.[28]

1. See for example: S. F. Martin, and P. J. Garrison, *J. Org. Chem.*, **47**, 1513 (1982).

2. T. Satoh, M. Ikeda, M. Miura, and M. Nomura, *J. Org. Chem.*, **62**, 4877 (1997).

3. F. Effenberger and J. Jäger, *J. Org. Chem.*, **62**, 3867 (1997).

4. R. Boss and R. Scheffold, Angew. Chem., *Int. Ed. Engl.*, **15**, 558 (1976).

5. M. Ishizaki, M. Yamada, S.-i. Watanabe, O. Hoshino, K. Nishitani, M. Hayashida, A. Tanaka, and H. Hara, *Tetrahedron*, **60**, 7973 (2004).

6. H. Hey and H.-J. Arpe, *Angew. Chem., Int. Ed. Engl.*, **12**, 928 (1973).

7. P. Four and F. Guibe, *Tetrahedron Lett.*, **23**, 1825 (1982).

8. W. R. Roush, R. A. Hartz, and D. J. Gustin, *J. Am. Chem. Soc.*, **121**, 1990 (1999).

9. H. Yin, R. W. Franck, S.-L. Chen, G. J. Quigley, and L. Todaro, *J. Org. Chem.*, **57**, 644 (1992).

10. M. Bois-Choussy, L. Neuville, R. Beugelmans, and J. Zhu, *J. Org. Chem.*, **61**, 9309 (1996).

11. R. Beugelmans, S. Bourdet, A. Bigot, and J. Zhu, *Tetrahedron Lett.*, **35**, 4349 (1994).

12. M. Dessolin, M.-G. Guillerez, N. Thieriet, F. Guibé, and A. Loffet, *Tetrahedron Lett.*, **36**, 5741 (1995).

13. D. R. Vutukuri, P. Bharathi, Z. Yu, K. Rajasekaran, M.-H. Tran, and S. Thayumanavan, *J. Org. Chem.*, **68**, 1146 (2003).

14. J. M. Bruce and Y. Roshan-Ali, *J. Chem. Res., Synop.*, 193 (1981).

15. H. Murakami, T. Minami, and F. Ozawa, *J. Org. Chem.*, **69**, 4482 (2004).

16. F. G. Mann and M. J. Pragnell, *J. Chem. Soc.*, 4120 (1965).

17. T. Kametani, S.-P. Huang, M. Ihara, and K. Fukumoto, *J. Org. Chem.*, **41**, 2545 (1976).

18. M. V. Bhatt and S. S. El-Morey, *Synthesis*, 1048 (1982).

19. R. M. Thomas, G. H. Mohan, and D. S. Iyengar, *Tetrahedron Lett.*, **38**, 4721 (1997).

20. P. D. Lokhande, S. S. Sakate, K. N. Taksande, and B. Navghare, *Tetrahedron Lett.*, **46**, 1573 (2005).

21. D. Franco, D. Panyella, M. Rocamora, M. Gomez, J. C. Clinet, G. Muller, and E. Duñach, *Tetrahedron Lett.*, **40**, 5685 (1999).

22. B. Espanet, E. Duñach, and J. Perichon, *Tetrahedron Lett.*, **33**, 2485 (1992).

23. A. Yasuhara, A. Kasano, and T. Sakamoto, *J. Org. Chem.*, **64**, 4211 (1999).

24. S. Olivero and E. Duñach, *J. Chem. Soc., Chem. Commun.*, 2497 (1995).

25. B M. Choudary, A. D. Prasad, V. Swapna, V. L. K. Valli, and V Bhuma, *Tetrahedron*, **48**, 953 (1992).

26. K. Kariyone and H. Yazawa, *Tetrahedron Lett.*, **11**, 2885 (1970).

27. E. Alonso, D. J. Ramon, and M. Yus, *Tetrahedron*, **42**, 14355 (1997).

28. S. M. Kadam, S. K. Nayak, and A. Banerji, *Tetrahedron Lett.*, **33**, 5129 (1992).

Prenyl Ether: $(CH_3)_2C=CHCH_2OR$

Formation

The section on the formation of allyl ethers should be consulted, since many of those methods are applicable to the prenyl ether. One difference is that the phenolic OH is more acidic, thus weaker bases may be used in methods that rely on an S_N2 process.

Cleavage

1. $TiCl_4$, n-Bu$_4$NI, CH_2Cl_2, $-78°C$, 30 min, 81–100% yield. Alkyl prenyl ethers are not cleaved under these conditions. Their cleavage occurs at higher temperatures and longer reaction times. Selectivity can be obtained in the presence of a coordinating group. Phenolic crotyl ethers are stable.[1]

2. p-TSA, CH_2Cl_2, rt, 70–98% yield. Allyl ethers are not cleaved.[2]
3. $ZrCl_4$, NaI, CH_3CN, reflux, 1–2 h, 94% yield.[3]
4. $ZrCl_4$, $NaBH_4$, CH_2Cl_2, 1.5–4 h, 70–96% yield. Prenyl esters are retained.[4]
5. $CeCl_3\cdot7H_2O$, NaI, CH_3CN, reflux, 80–90% yield. Phenolic allyl and benzyl ethers are stable, but methyl ethers are cleaved.[5]

6. $Yb(OTf)_3$, CH_3NO_2, rt, 0.5–12 h 72–90 yield. The rate is dependent upon the nature of the substituents on the ring. Electron poor aromatics are cleaved more slowly.[6]

1. T. Tsuritani, H. Shinokubo, and K. Oshima, *Tetrahedron Lett.*, **40**, 8121 (1999).
2. K. S. Babu, B. C. Raju, P. V. Srinivas, A. S. Rao, S. P. Kumar, and J. M. Rao, *Chem. Lett.*, **32**, 704 (2003).
3. G. V. M. Sharma, C. G. Reddy, and P. R. Krishna, *Synlett*, 1728 (2003).
4. K. S. Babu, B. C. Raju, P. V. Srinivas, and J. M. Rao, *Tetrahedron Lett.*, **44**, 2525 (2003).
5. J. S. Yadav, B. V. S. Reddy, C. Madan, and S. R. Hashim, *Chem. Lett.*, **29**, 738 (2000).
6. G. V. M. Sharma, A. Ilangovan, and A. K. Mahalingam, *J. Org. Chem.*, **63**, 9103 (1998).

Cyclohex-2-en-1-yl Ether

The cyclohexenyl ether is prepared from the bromide and K_2CO_3 in acetone. It is cleaved with HCl in ether (92–98% yield)[1] and with 10% Pd/C, 10% KOH, MeOH.[5]

1. P. Carato, G. Laconde, C. Ladjel, P. Depreux, and J.-P. Henichart, *Tetrahedron Lett.*, **43**, 6533 (2002).

Propargyl Ether: $HC\equiv CCH_2OAr$

Formation

Propargyl ethers are generally formed using some variant of the Williamson ether synthesis. See section on alcohol protection.

Cleavage

1. Electrolysis, $Ni(bipyr)_3(BF_4)_2$, Mg anode, DMF, rt, 77–99% yield. This method is not compatible with halogenated phenols because of competing halogen cleavage.[1] Propargyl esters are also cleaved.
2. $TiCl_3$, Mg, THF, 54–92% yield.[2]
3. BBr_3, CH_2Cl_2, rt, 72–99% yield. Benzyl ethers are cleaved more rapidly and methyl ethers are also cleaved, but the propargyl ether is cleaved in preference to the methyl ether if steric factors are similar.[3]
4. $PdCl_2(Ph_3P)_2$, TEA, DMF, H_2O, 2–3 h, 45–78% yield. Propargyl anilines are cleaved similarly but in generally low yields.[4]

1. S. Olivero and E. Duñach, *Tetrahedron Lett.*, **38**, 6193 (1997).
2. S. K. Nayak, S. M. Kadam, and A. Banerji, *Synlett*, 581 (1993).
3. S. Punna, S. Meunier, and M. G. Finn, *Org. Lett.*, **6**, 2777 (2004).
4. M. Pal, K. Parasuraman, and K. R. Yeleswarapu, *Org. Lett.*, **5**, 349 (2003).

Isopropyl Ether: $ArOCH(CH_3)_2$

An isopropyl ether was developed as a phenol protective group that would be more stable to Lewis acids than an aryl benzyl ether.[1] The isopropyl group has been tested for use in protection of the phenolic oxygen of tyrosine during peptide synthesis.[2]

Formation

Me$_2$CHBr, K$_2$CO$_3$, DMF, acetone, 20°C, 19 h.[1]

Cleavage

1. BCl$_3$, CH$_2$Cl$_2$, 0°C, rapid; or TiCl$_4$, CH$_2$Cl$_2$, 0°C, slower.[1] There was no reaction with SnCl$_4$.[1]
2. SiCl$_4$, NaI, 14 h, CH$_2$Cl$_2$, CH$_3$CN, 80% yield.[3]
3. AlCl$_3$, CH$_2$Cl$_2$, rt, 80–96% yield. The isopropyl group is selectively cleaved in the presence of a phenolic methyl ether.[4]
4. TMSOTf, Ac$_2$O, CH$_3$CN, 68–98% yield.[5] These conditions convert the ether to an acetate.

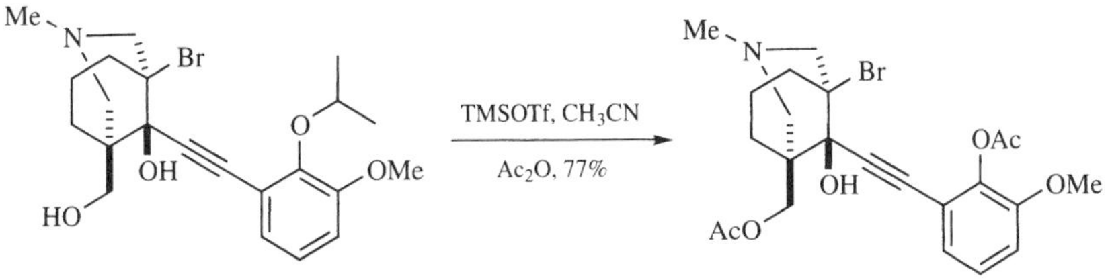

1. T. Sala and M. V. Sargent, *J. Chem. Soc., Perkin Trans. I*, 2593 (1979).
2. See cyclohexyl ether in this section: M. Engelhard and R. B. Merrifield, *J. Am. Chem. Soc.*, **100**, 3559 (1978).
3. M. V. Bhatt and S. S. El-Morey, *Synthesis*, 1048 (1982).
4. M. G. Banwell, B. L. Flynn, and S. G. Stewart, *J. Org. Chem.*, **64**, 9139 (1998). Erratum: M. G. Banwell, B. L. Flynn, and S. G. Stewart, *J. Org. Chem.*, **64**, 6118 (1999).
5. C. M. Williams and L. N. Mander, *Tetrahedron Lett.*, **45**, 667 (2004).

Cyclohexyl Ether: ArO-c-C$_6$H$_{11}$ (Chart 4)

Formation[1]

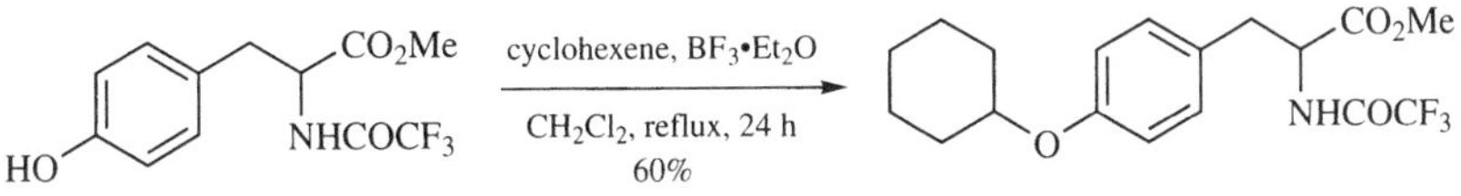

Cleavage

1. HF, 0°C, 30 min, 100% yield.[1]
2. 5.3 N HBr/AcOH, 25°C, 2 h, 99% yield. An ether that would not undergo rearrangement to a 3-alkyl derivative during acid-catalyzed removal of −NH protective groups was required to protect the phenol group in tyrosine. Four compounds were investigated: *O*-cyclohexyl-, *O*-isobornyl-, *O*-[1-(5-pentamethylcyclopentadienyl)ethyl]-, and *O*-isopropyltyrosine.[1]

The O-isobornyl- and O-[1-(5-pentamethylcyclopentadienyl)ethyl]- derivatives do not undergo rearrangement to form alkyl tyrosine derivatives, but are very labile in trifluoroacetic acid (100% cleaved in 5 min). The cyclohexyl, isopropyl, and 3-pentyl[2] derivatives are more stable to acid, but undergo some rearrangement. The cyclohexyl and 3-pentyl groups combine minimal rearrangement with ready removal.[1] A comparison has been made with several other common protective groups for tyrosine and the degree of alkylation ortho to the phenolic OH decreases in the order: Bn > 2-ClC$_6$H$_4$CH$_2$ > 2,6-Cl$_2$C$_6$H$_3$CH$_2$ > cyclohexyl > t-Bu ~ benzyloxycarbonyl ~ 2-Br-benzyloxycarbonyl.[3]

1. M. Engelhard and R. B. Merrifield, *J. Am. Chem. Soc.*, **100**, 3559 (1978).
2. J. Bodi, Y. Nishiuchi, H. Nishio, T. Inui, and T. Kimura, *Tetrahedron Lett.*, **39**, 7117 (1998).
3. J. P. Tam, W. F. Heath, and R. B. Merrifield, *Int. J. Pept. Protein Res.*, **21**, 57 (1983).

t-Butyl Ether: ArOC(CH$_3$)$_3$ (Chart 4)

The section on t-butyl ethers of alcohols should also be consulted.

Formation

1. Isobutylene, cat. concd. H$_2$SO$_4$, CH$_2$Cl$_2$, 25°C, 6–10 h, 93% yield.[1] These conditions also convert carboxylic acids to t-Bu esters.
2. Isobutylene, CF$_3$SO$_3$H, CH$_2$Cl$_2$, −78°C, 70–90% yield.[2] These conditions will protect a phenol in the presence of a primary alcohol.
3. t-Butyl halide, Pyr, 20–30°C, few h, 65–95% yield.[3]

Cleavage

1. Anhydrous CF$_3$CO$_2$H, 25°C, 16 h, 81% yield.[1]
2. CF$_3$CH$_2$OH, CF$_3$SO$_3$H, −5°C, 60 s, 100% yield.[2]

1. H. C. Beyerman and J. S. Bontekoe, *Recl. Trav. Chim. Pays-Bas*, **81**, 691 (1962).
2. J. L. Holcombe and T. Livinghouse, *J. Org. Chem.*, **51**, 111 (1986).
3. H. Masada and Y. Oishi, *Chem. Lett.*, **7**, 57 (1978).

Benzyl Ether: ArOCH$_2$C$_6$H$_5$ (Chart 4)

Formation

1. In general, benzyl ethers are prepared from a phenol by treating an alkaline solution of the phenol with a benzyl halide.[1] In the following cases, hydrogen bonding of the ortho OH with the carbonyl reduces its reactivity which leads

to benzylation of the remaining hydroxyl.[2]

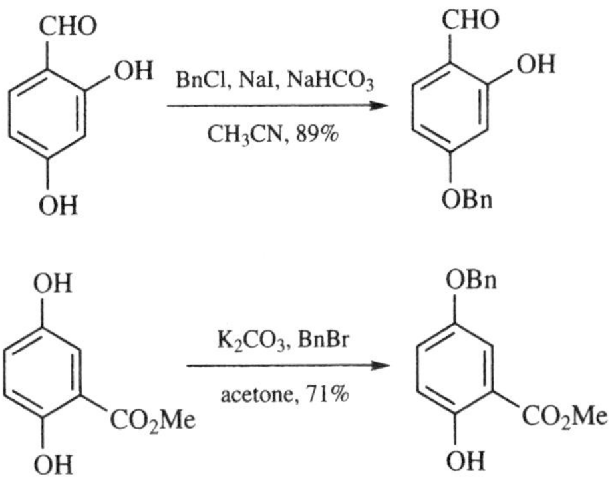

Ref. 3

2. The greater acidity of the phenolic hydroxyls makes them more reactive than simple alkanols.[4]

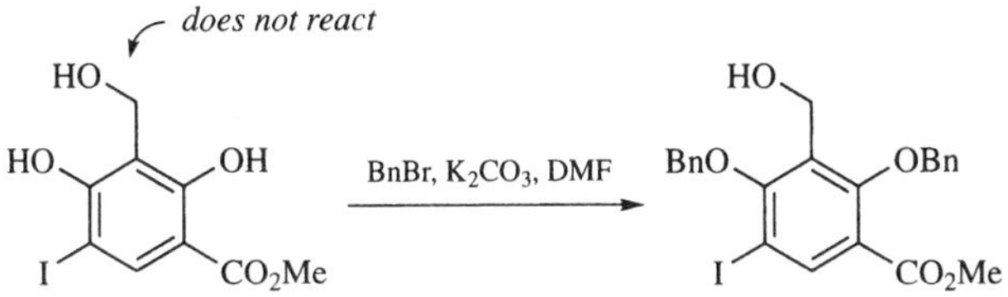

3. CHCl$_3$, MeOH, K$_2$CO$_3$, BnBr, 4 h, heat.[5] In this case, some (5:1) selectivity was achieved for a less hindered phenol in the presence of a more hindered one.

4. KF·alumina, DME, 80% yield. Both a phenol and an amide nitrogen are benzylated.[6]

5. Benzyl ethers of phenols can also be prepared by reaction with phenyldiazomethane.

6. (BnO)$_2$CO, DMF, 155°C, 2 h, 80% yield. Active methylenes are also benzylated.[7]

7. Ph$_2$POBn, 2,6-dimethylquinone, CH$_2$Cl$_2$, rt, 0.5 h, 70–92% yield. This method is quite general and can be used to prepare a large variety of ethers using either alkynols or phenols.[8]

Cleavage

The section on the cleavage of alkyl benzyl ethers should be consulted, since many of those methods are applicable to phenolic benzyl ethers. It should be noted that phenolic benzyl ethers can be retained during the hydrogenation of olefins and the hydrogenolysis of the Cbz group by the addition of 2,2'-dipyridyl as an additive.[9] There is also a solvent dependence with aromatic solvents allowing olefin reduction in the presence of a phenolic benzyl ether. Methanol as solvent gives both reduction and cleavage.[10]

1.

Catalytic hydrogenation in acetic anhydride–benzene removes the aromatic benzyl ether and forms a monoacetate; hydrogenation in ethyl acetate removes the aliphatic benzyl ether to give, after acetylation, the diacetate.[11] Trisubstituted alkenes can be retained during the hydrogenolysis of a phenolic benzyl ether.[12]

2. 5% Pd–C, H_2-balloon, Pyr (0.5 eq.), 24 h. The use of pyridine poisoned catalyst allows for the hydrogenation of benzyl ether in the presence of a phenolic PMB ether. Good selectivity is also obtained for the dimethyl and trimethyl-benzyl ethers.[13]

3. Pd–C, 1,4-cyclohexadiene, 25°C, 1.5 h, 95–100% yield.[14] This method has been used for the deprotection of a variety of benzyl-based protective groups in peptides.[15]

Ref. 16

4. Palladium black, a more reactive catalyst than Pd–C, must be used to cleave the more stable aliphatic benzyl ethers.[14] The retention of aryl halides can be a problem during the hydrogenolysis of benzyl groups. In a synthesis of the putative structure of Diazonamide A, an aryl chloride is retained.[17] This selectivity may be the result of catalyst poisoning by the heterocyclic amines. It is known that amines moderate the activity of Pd catalysts. Note that a Z group was also cleaved. Dehalogenation of aryl chlorides can be suppressed by the inclusion of chloride into the reaction mixture. Hydrochloric acid is effective because dehalogenation is faster under basic conditions. The dielectric constant of the

solvent also has a profound effect, with solvents of low dielectric constant giving less dechlorination.[18]

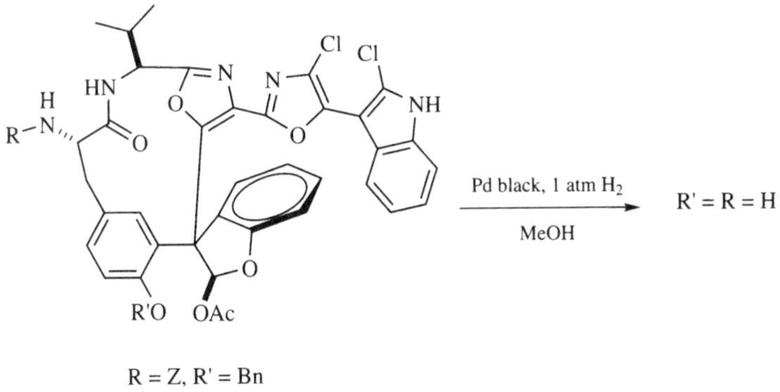

R = Z, R' = Bn

5. The following case illustrates a very unusual Pd-catalyzed oxidation.[6] Mechanistically, this was postulated to involve coordination of the Pd with the released OH and NH followed by a β-hydride elimination. The second oxidation proceeded similarly but through a hemiaminal.

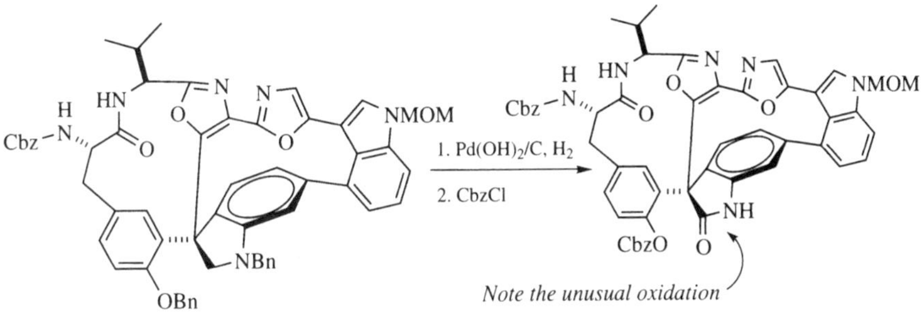

Note the unusual oxidation

6. PdCl$_2$, Et$_3$SiH, CH$_2$Cl$_2$, TEA, 66–71% yield for halogen containing phenols. The level of dehalogenation is dependent upon the steric environment and the halogen with chlorides being stable to reduction.[19]
7. Pd/BaSO$_4$, H$_2$, >75% yield.[20]

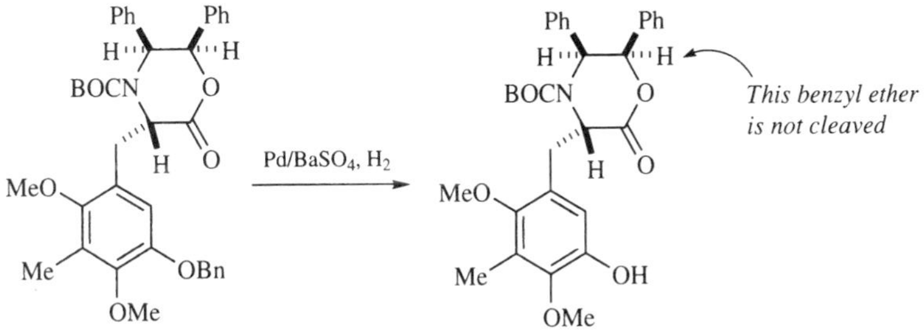

This benzyl ether is not cleaved

8. Pd–C encapsulated in POEPOP$_{1500}$, MeOH, H$_2$O, 25°C, 40 bar.[21]
9. Raney nickel, K$_2$CO$_3$, ethanol, EtOAc, 60°C, 70% yield.[22]

10. Na, *t*-BuOH, 70–80°C, 2 h, 78%.[23]

Note the reduction of the conjugated olefin

In this example, sodium in *t*-butyl alcohol cleaves two aryl benzyl ethers and reduces a double bond that is conjugated with an aromatic ring; nonconjugated double bonds are stable.

11. Calcium, ammonia, 95% yield.[24] For this method to work the oxide coating on the Ca must be removed. This is sometimes accomplished by stirring with sand.

12. BF$_3$·Et$_2$O, EtSH, 25°C, 40 min, 80–90% yield.[25] Addition of sodium sulfate prevents hydrolysis of a dithioacetal group present in the compound; replacement of ethanethiol with ethanedithiol prevents cleavage of a dithiolane group.

13. CF$_3$OSO$_2$F or CH$_3$OSO$_2$F, PhSCH$_3$, CF$_3$CO$_2$H, 0°C, 30 min, 100% yield.[26] Thioanisole suppresses acid-catalyzed rearrangement of the benzyl group to form 3-benzyltyrosine. The more acid-stable 2,6-dichlorobenzyl ether is cleaved in a similar manner.

14. Me$_3$SiI, CH$_3$CN, 25–50°C, 100% yield.[27] Selective removal of protective groups is possible with this reagent since a carbamate, =NCOOCMe$_3$, is cleaved in 6 min at 25°C; an aryl benzyl ether is cleaved in 100% yield, with no formation of 3-benzyltyrosine, in 1 h at 50°C, at which time a methyl ester begins to be cleaved.

15. 2-Bromo-1,3,2-benzodioxaborole, CH$_2$Cl$_2$, 95% yield.[28]

16. BBr$_3$, CH$_2$Cl$_2$, rt, 15 min, 75% yield.[29]

17. NaI, BF$_3$·Et$_2$O, 0°C, 45 min, rt, 15 min, 75–90% yield.[30]

18. CF$_3$CO$_2$H, PhSCH$_3$, 25°C, 3 h.[31] The use of dimethyl sulfide or anisole as a cation scavenger was not as effective because of side reactions. Benzyl ethers of serine and threonine were slowly cleaved (30% in 3 h; complete cleavage in 30 h). The use of pentamethylbenzene has been shown to increase the rate of deprotection of *O*-Bn-Tyrosine.[32] The use of pentamethylbenzene was developed to minimize the formation of 3-benzyltyrosine during the acidolysis of benzyl-protected tyrosine.[33]

19. PhNMe$_2$, AlCl$_3$, CH$_2$Cl$_2$, 78–91% yield.[34]

20. MgBr$_2$, benzene, Et$_2$O, reflux, 24 h, 63–95% yield.[35] Coordination facilitates selective cleavage.

21. Dimethyldioxirane, acetone, 20°C, 45 h, 69% yield.[36]
22. SnBr$_2$, AcBr, CH$_2$Cl$_2$, rt, 5–24 h, 76–86% yield. These conditions convert a benzyl ether to the acetate and are effective for alkyl benzyl ethers as well.[37]
23. TiCl$_3$, Mg, THF, reflux, 28–96% yield.[38]

1. For example, M. C. Venuti, B. E. Loe, G. H. Jones, and J. M. Young, *J. Med. Chem.*, **31**, 2132 (1988).

2. W. L. Mendelson, M. Holmes, and J. Dougherty, *Synth. Commun.*, **26**, 593 (1996).

3. N. R. Kotecha, S. V. Ley, and S. Montégani, *Synlett*, 395 (1992).

4. J. B. Shotwell, E. S. Krygowski, J. Hines, B. Koh, E. W. D. Huntsman, H. W. Choi, J. J. S. Schneekloth, J. L. Wood, and C. M. Crews, *Org. Lett.*, **4**, 3087 (2002).

5. H. Schmidhammer and A. Brossi, *J. Org. Chem.*, **48**, 1469 (1983).

6. K. C. Nicolaou, J. Hao, M. V. Reddy, P. B. Rao, G. Rassias, S. A. Snyder, H. Huang, D. Y.-K. Chen, W. E. Brenzovich, N. Giuseppone, P. Giannakakou, and A. O'brate, *J. Am. Chem. Soc.*, **126**, 12897 (2004).

7. M. Selva, C. A. Margues, and P. Tundo, *J. Chem. Soc., Perkin Trans. I*, 1889 (1995).

8. T. Shintou and T. Mukaiyama, *J. Am. Chem. Soc.*, **126**, 7359 (2004).

9. H. Sajiki, H. Kuno, and K. Hirota, *Tetrahedron Lett.*, **39**, 7127 (1998).

10. S. Maki, M. Okawa, R. Matusi, T. Hirano, and H. Niwa, *Synlett*, 1590 (2001).

11. G. Büchi and S. M. Weinreb, *J. Am. Chem. Soc.*, **93**, 746 (1971).

12. A. F. Barrero, E. J. Alvarez-Manzaneda, and R. Chahboun, *Tetrahedron Lett.*, **38**, 8101 (1997).

13. H. Sajiki, H. Kuno, and K. Hirota, *Tetrahedron Lett.*, **38**, 399 (1997).

14. A. M. Felix, E. P. Heimer, T. J. Lambros, C. Tzougraki, and J. Meienhofer, *J. Org. Chem.*, **43**, 4194 (1978).

15. D. C. Gowda and K. Abiraj, *Letters in Peptide Science*, **9**, 153 (2003).

16. D. E. Thurston, V. S. Murty, D. R. Langley, and G. B. Jones, *Synthesis*, 81 (1990).

17. J. Li, X. Chen, A. W. G. Burgett, and P. G. Harran, *Angew. Chem. Int. Ed.*, **40**, 2682 (2001).

18. J. Li, S. Wang, G. A. Crispino, K. Tenhuisen, A. Singh, and J. A. Grosso, *Tetrahedron Lett.*, **44**, 4041 (2003).

19. R. S. Coleman and J. A. Shah, *Synthesis*, 1399 (1999).

20. J. W. Lane, Y. Chen, and R. M. Williams, *J. Am. Chem. Soc.*, **127**, 12684 (2005).

21. A. M. Jansson, M. Grotli, K. M. Halkes, and M. Meldal, *Org. Lett.*, **4**, 27 (2002).

22. M. K. Schwaebe, T. J. Moran, and J. P. Whitten, *Tetrahedron Lett.*, **46**, 827 (2005).

23. B. Loev and C. R. Dawson, *J. Am. Chem. Soc.*, **78**, 6095 (1956). S. I. Odejinmi and D. F. Weimer, *Tetrahedron Lett.*, **46**, 3871 (2005).

24. J. R. Hwu, Y. S. Wein, and Y.-J. Leu, *J. Org. Chem.*, **61**, 1493 (1996).

25. K. Fuji, K. Ichikawa, M. Node, and E. Fujita, *J. Org. Chem.*, **44**, 1661 (1979).

26. Y. Kiso, H. Isawa, K. Kitagawa, and T. Akita, *Chem. Pharm. Bull.*, **26**, 2562 (1978).

27. R. S. Lott, V. S. Chauhan, and C. H. Stammer, *J. Chem. Soc., Chem. Commun.*, 495 (1979).

28. P. F. King and S. G. Stroud, *Tetrahedron Lett.*, **26**, 1415 (1985).

29. E. Paliakov and L. Strekowski, *Tetrahedron Lett.*, **45**, 4093 (2004).

30. Y. D. Vankar and C. T. Rao, *J. Chem. Res., Synop.*, 232 (1985).

31. Y. Kiso, K. Ukawa, S. Nakamura, K. Ito, and T. Akita, *Chem. Pharm. Bull.*, **28**, 673 (1980).

32. H. Yoshino, Y. Tsuchiya, I. Saito, and M. Tsujii, *Chem. Pharm. Bull.*, **35**, 3438 (1987).

33. H. Yoshino, M. Tsujii, M. Kodama, K. Komeda, N. Niikawa, T. Tanase, N. Asakawa, K. Nose, and K. Yamatsu, *Chem. Pharm. Bull.*, **38**, 1735 (1990).

34. T. Akiyama, H. Hirofuji, and S. Ozaki, *Tetrahedron Lett.*, **32**, 1321 (1991).

35. J. E. Baldwin and G. G. Haraldsson, *Acta. Chem. Scand., Ser. B*, **B40**, 400 (1986).

36. B. A. Marples, J. P. Muxworthy, and K. H. Baggaley, *Synlett*, 646 (1992).

37. T. Oriyama, M. Kimura, M. Oda, and G. Koga, *Synlett*, 437 (1993).

38. S. M. Kadam, S. K. Nayak, and A. Banerji, *Tetrahedron Lett.*, **33**, 5129 (1992).

2,4-Dimethylbenzyl Ether: $2,4\text{-}(CH_3)_2C_6H_3CH_2OAr$

The 2,4-dimethylbenzyl ether is considerably more stable to hydrogenolysis than the benzyl ether. It has a half-life of 15 h at 1 atm of hydrogen in the presence of Pd–C, whereas the benzyl ether has a half-life of ~45 min. This added stability allows hydrogenation of azides, nitro groups, and olefins in the presence of a dimethylbenzyl group.[1]

1. R. Davis and J. M. Muchowski, *Synthesis*, 987 (1982).

4-Methoxybenzyl Ether (MPM–OAr or PMB–OAr): $4\text{-}CH_3OC_6H_4CH_2OAr$

Formation

1. $MeOC_6H_4CH_2Cl$, Bu_4NI, K_2CO_3, acetone, 55°C, 96% yield.[1] Sodium iodide can be used in place of Bu_4NI.[2]

2. $MeOC_6H_4CH_2Br$, $(i\text{-}Pr)_2NEt$, CH_2Cl_2, rt, 80% yield.[3]

Cleavage

1. CF_3CO_2H, CH_2Cl_2, 85% yield.[1]

2. Camphorsulfonic acid, $(CH_3)_2C(OCH_3)_2$, rt.[3]

3. Dowex 50WX8-100, H_2O.[4]

4. $BF_3 \cdot Et_2O$, $NaCNBH_3$, THF, reflux, 6–10 h, 65–77% yield.[5]

5. 18-Crown-6, toluene, K, 2–3 h, 81–96% yield.[6]

6. Acetic acid, 90°C, 89–96% yield.[7] Benzyl groups are not affected by these conditions.

7. DDQ, 35% yield.[8] The DDQ-promoted cleavage of phenolic MPM ethers can be complicated by overoxidation, especially with electron-rich phenolic compounds.

8. 5% Pd–C, H$_2$. In the presence of pyridine, hydrogenolysis of the MPM group is suppressed.[9]

9. Formation of a mesylate resulted in cleavage of a PMB group by a solvolytic process.[10]

1. J. D. White and J. C. Amedio, Jr., *J. Org. Chem.*, **54**, 736 (1989).

2. I. A. McDonald, P. L. Nyce, M. J. Jung, and J. S. Sabol, *Tetrahedron Lett.*, **32**, 887 (1991).

3. H. Nagaoka, G. Schmid, H. Iio, and Y. Kishi, *Tetrahedron Lett.*, **22**, 899 (1981).

4. J. M. Pletcher and F. E. McDonald, *Org. Lett.*, **7**, 4749 (2005).

5. A. Srikrishna, R. Viswajanani, J. A. Sattigeri, and D. Vijaykumar, *J. Org. Chem.*, **60**, 5961 (1995).

6. T. Ohsawa, K. Hatano, K. Kayoh, J. Kotabe, and T. Oishi, *Tetrahedron Lett.*, **33**, 5555 (1992).

7. K. J. Hodgetts and T. W. Wallace, *Synth. Commun.*, **24**, 1151 (1994).

8. O. P. Vig, S. S. Bari, A.Sharma, and M. A. Sattar, *Indian J. Chem., Sect. B*, **29B**, 284 (1990).

9. H. Sajiki, H. Kuno, and K. Hirota, *Tetrahedron Lett.*, **38**, 399 (1997).

10. M. A. Zajac and E. Vedejs, *Org. Lett.*, **6**, 237 (2004); E. Vedejs and M. A. Zajac, *Org. Lett.*, **3**, 2451 (2001).

o-**Nitrobenzyl Ether:** o-NO$_2$–C$_6$H$_4$CH$_2$OAr (Chart 4)

An *o*-nitrobenzyl ether can be cleaved by photolysis. In tyrosine this avoids the use of acid-catalyzed cleavage and the attendant conversion to 3-benzyltyrosine.[1] (Note that this unwanted conversion can also be suppressed by the addition of thioanisole; see benzyl ether cleavage.)

p-**Nitrobenzyl Ether:** $o\text{-}NO_2\text{-}C_6H_4CH_2OAr$

Formation

4-NO_2BnBr, Ag_2O, CH_2Cl_2, reflux, 5 days, 58–84% yield.[2]

Cleavage

1. Indium, EtOH, H_2O, NH_4Cl, rt, 81–100% yield. These conditions generally reduce nitro groups.[2] Thus other conditions that reduce nitro groups should cleave this ether.

2. Mg, MeOH, 90% yield.[3]

1. B. Amit, E. Hazum, M. Fridkin, and A. Patchornik, *Int. J. Pept. Protein Res.*, **9**, 91 (1977).

2. M. R. Pitts, J. R. Harrison, and C. J. Moody, *J. Chem. Soc., Perkin Trans. 1*, 955 (2001).

3. W. Huang, X. Zhang, H. Liu, J. Shen, and H. Jiang, *Tetrahedron Lett.*, **46**, 5965 (2005).

2,6-Dichlorobenzyl Ether: $ArOCH_2C_6H_3\text{-}2,6\text{-}Cl_2$

This group is readily cleaved by a mixture of CF_3SO_3H, $PhSCH_3$, and CF_3CO_2H.[1,2] Of the common benzyl protecting groups used to protect the hydroxyl of tyrosine, the 2,6-dichlorobenzyl shows a low incidence of alkylation at the 3-position of tyrosine during cleavage with HF/anisole. A comparative study on deprotection of X-Tyr in HF/anisole gives the following percentages of side reactions for various X groups: Bn, 24.5; 2-ClBn, 9.8; 2,6-Cl_2Bn, 6.5; cyclohexyl, 1.5; *t*-Bu, <0.2; Cbz, 0.5; 2-Br-Cbz, 0.2.[3] As with most other benzyl groups, hydrogenolysis (ammonium formate, Pd–C, MeOH, rt, 90% yield) can be used to cleave this ether.[4]

3,4-Dichlorobenzyl Ether: $3,4\text{-}Cl_2C_6H_3CH_2OAr$

As with the 2,6-dichlorobenzyl ether the electron-withdrawing chlorine atoms confer greater acid stability to this group than the usual benzyl group. It is cleaved by hydrogenolysis (Pd–C, H_2).[5]

1. Y. Kiso, M. Satomi, K. Ukawa, and T. Akita, *J. Chem. Soc., Chem. Commun.*, 1063 (1980).

2. J. Deng, Y. Hamada, and T. Shioiri, *Tetrahedron Lett.*, **37**, 2261 (1996).

3. J. P. Tam, W. F. Heath, and R. B. Merrifield, *Int. J. Pept. Protein Res.*, **21**, 57 (1983).

4. D. C. Gowda, B. Rajesh, and S. Gowda, *Ind. J. Chem., Sect. B*, **39B**, 504 (2000).

5. D. A. Evans, C. J. Dinsmore, D. A. Evrard, and K. M. DeVries, *J. Am. Chem. Soc.*, **115**, 6426 (1993).

4-(Dimethylamino)carbonylbenzyl Ether: $(CH_3)_2NCOC_6H_4CH_2OAr$

The 4-(dimethylamino)carbonylbenzyl ether has been used to protect the phenolic hydroxyl of tyrosine. It is stable to CF_3CO_2H (120 h), but not to HBr/AcOH (complete cleavage in 16 h). It can also be cleaved by hydrogenolysis $(H_2/Pd-C)$.[1]

1. V. S. Chauhan, S. J. Ratcliffe, and G. T. Young, *Int. J. Pept. Protein Res.*, **15**, 96 (1980).

4-Methylsulfinylbenzyl Ether (Msib−OR): $CH_3S(O)C_6H_4CH_2OAr$

The Msib group has been used for the protection of tyrosine. It is cleaved by reduction of the sulfoxide to the sulfide, which is then deprotected with acid. Reduction is achieved with $DMF-SO_3/HSCH_2CH_2SH$ or Bu_4NI[1] or with $SiCl_3/TFA$.[2]

1. S. Futaki, T. Yagami, T. Taike, T. Ogawa, T. Akita, and K. Kitagawa, *Chem. Pharm. Bull.*, **38**, 1165 (1990).
2. Y. Kiso, S. Tanaka, T. Kimura, H. Itoh, and K. Akaji, *Chem. Pharm. Bull.*, **39**, 3097 (1991).

9-Anthrylmethyl Ether: $ArOCH_2$-9-anthryl (Chart 4)

9-Anthrylmethyl ethers, formed from the sodium salt of a phenol and 9-anthrylmethyl chloride in DMF can be cleaved with CH_3SNa (DMF, 25°C, 20 min, 85–99% yield). They are also cleaved by CF_3CO_2H/CH_2Cl_2 (0°C, 10 min, 100% yield); they are stable to CF_3CO_2H/dioxane (25°C, 1 h).[1]

1. N. Kornblum and A. Scott, *J. Am. Chem. Soc.*, **96**, 590 (1974).

4-Picolyl Ether: $ArOCH_2$−4-pyridyl (Chart 4)

Formation[1]/Cleavage[1,2]

An aryl 4-picolyl ether is stable to trifluoroacetic acid, used to cleave an *N-t*-butoxycarbonyl group.[2]

1. A. Gosden, D. Stevenson, and G. T. Young, *J. Chem. Soc., Chem. Commun.*, 1123 (1972).
2. P. M. Scopes, K. B. Walshaw, M. Welford, and G. T. Young, *J. Chem. Soc.*, 782 (1965).

Heptafluoro-*p*-tolyl and Tetrafluoro-4-pyridyl Ethers: $ArOC_6F_4$-CF_3, $ArOC_5F_4N$

Formation/Cleavage[1-3]

Ref. 4

If 2 eq. of reagent are used, both hydroxyls can be protected and the phenolic hydroxyl can be selectively cleaved with NaOMe. The tetrafluoropyridyl derivative is introduced under similar conditions. The use of this methodology has been reviewed.[5]

1. M. Jarman and R. McCague, *J. Chem. Soc., Chem. Commun.*, 125 (1984).
2. M. Jarman and R. McCague, *J. Chem. Res., Synop.*, 114 (1985).
3. J. J. Deadman, R. McCague, and M. Jarman, *J. Chem. Soc., Perkin Trans. I*, 2413 (1991).
4. S. Singh and R. A. Magarian, *Chem. Lett.*, **23**, 1821 (1994).
5. M. Jarman, *J. Fluorine Chem.*, **42**, 3 (1989).

Silyl Ethers

Aryl and alkyl trimethylsilyl ethers can often be cleaved by refluxing in aqueous methanol, an advantage for acid- or base-sensitive substrates. The ethers are stable to Grignard and Wittig reactions, and to reduction with lithium aluminum hydride at $-15°C$. Aryl *t*-butyldimethylsilyl ethers and other sterically more demanding silyl ethers require acid- or fluoride ion-catalyzed hydrolysis for removal. Increased steric bulk also improves their stability to a much harsher set of conditions. Two excellent reviews on the selective deprotection of alkyl silyl ethers and aryl silyl ethers have been published.[1]

1. T. D. Nelson and R. D. Crouch, *Synthesis*, 1031 (1996); R. D. Crouch, *Tetrahedron*, **60**, 5833 (2004).

Trimethylsilyl Ether (TMS Ether): $ArOSi(CH_3)_3$

Formation

1. Me_3SiCl, Pyr, 30–35°C, 12 h, satisfactory yield.[1]
2. $(Me_3Si)_2NH$, cat. concd. H_2SO_4, reflux, 2 h, 97% yield.[2]
3. A large number of other silylating agents have been described for the derivatization of phenols, but the two listed above are among the most common.[3]

Cleavage

Trimethylsilyl ethers are readily cleaved by fluoride ion, mild acids, and mild bases. If the TMS derivative is somewhat hindered, it also becomes less susceptible to cleavage. A phenolic TMS ether can be cleaved in the presence of an alkyl TMS ether [Dowex 1-x8 (HO⁻), EtOH, rt, 6 h, 78% yield].[4]

1. Cl. Moreau, F. Roessac, and J. M. Conia, *Tetrahedron Lett.*, **11**, 3527 (1970).
2. S. A. Barker and R. L. Settine, *Org. Prep. Proced. Int.*, **11**, 87 (1979).
3. G. van Look, G. Simchen, and J. Heberle, *Silylating Agents*, Fluka Chemie, AG, (1995).
4. Y. Kawazoe, M. Nomura, Y. Kondo, and K. Kohda, *Tetrahedron Lett.*, **28**, 4307 (1987).

t-Butyldimethylsilyl Ether (TBDMS, TBS Ether): $ArOSi(CH_3)_2C(CH_3)_3$ (Chart 4)

The section on alcohol protection should be examined since many of the methods for formation and cleavage of TBDMS ethers are similar. The primary difference is that phenolic TBDMS ethers are much less susceptible to acid hydrolysis because of the reduced basicity of the oxygen, but are more susceptible to basic reagents because phenol is a much better leaving group than a simple alcohol.[1] The monodeprotection of mixed aryl and alkyl silyl ethers has been reviewed.[2]

Formation

1. *t*-BuMe₂SiCl, DMF, imidazole, 25°C, 3 h, 96% yield.[3,4]
2.

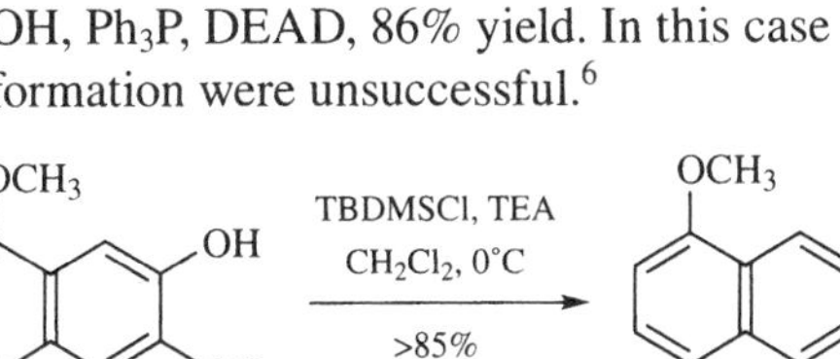

Ref. 5

3. *t*-BuMe₂SiOH, Ph₃P, DEAD, 86% yield. In this case the standard methods for silyl ether formation were unsuccessful.[6]
4.

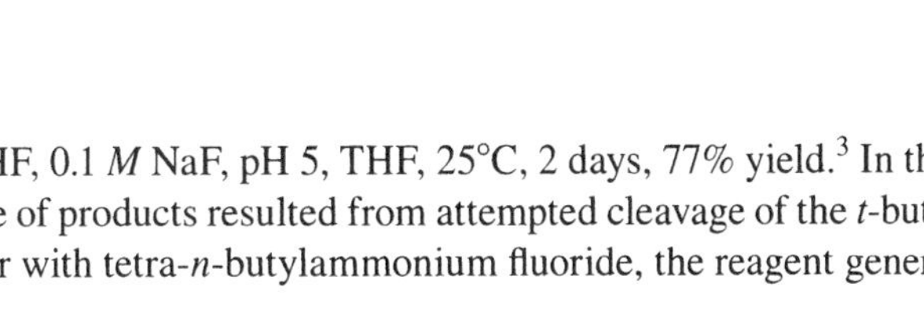

Ref. 7

Cleavage

1. 0.1 *M* HF, 0.1 *M* NaF, pH 5, THF, 25°C, 2 days, 77% yield.[3] In this substrate a mixture of products resulted from attempted cleavage of the *t*-butyldimethylsilyl ether with tetra-*n*-butylammonium fluoride, the reagent generally used.[8]

2. KF, 48% aq. HBr, DMF, rt, 91% yield.[9]

The use of Bu$_4$NF results in decomposition of this substrate.

3. KF/Al$_2$O$_3$, DME, or dioxane, 16 h, 25°C, 94% yield. These conditions do not cleave a TBDPS group.[10]

4.

Ref. 11

5. LiOH, DMF, rt, 1–16 h, 76–97% yield. Alkyl TBDMS ethers are stable to these conditions. The rate is dependent upon the substituents with electron-withdrawing groups increasing the rate.[12]

6. DIBALH, THF, hexane, −78°C, 45 min.[13]

7. THF, MeOH, Borax buffer (1:1:1), 40–50°C, 8 h, >90% yield.[13]

8. PdCl$_2$(CH$_3$CN)$_2$, aq. acetone, 75°C, 10–96% yield.[14]

9. BF$_3$·Et$_2$O, CH$_2$Cl$_2$, rt, 8 h.[15]

10. K$_2$CO$_3$, Kriptofix 222, CH$_3$CN, 55°C, 2 h, 70–95% yield.[16,17] Phenolic silyl ethers are cleaved selectively, but when TsOH or BF$_3$·Et$_2$O is used, alkyl TBDMS groups are cleaved in preference to phenolic derivatives.

11. Amberlite IRA-400 fluoride form, CH$_2$Cl$_2$ or DMF; then elute with aq. HCl, 80–90% yield.[18]

12. Ultrasound, MeOH, CCl$_4$, 45-98% yield. This method is specific for cleavage of TBDMS ethers ortho to a carbonyl group.[19]

13. DMSO, H$_2$O, 90°C, 82% yield. Selective cleavage of a phenolic TBDMS ether occurs in the presence of the alkyl ether.[20]

The table below gives the relative half-life to acid or base hydrolysis of a number of silylated p-cresols.[21]

Susceptibility of Silylated Cresols to Hydrolysis

	Half-Life ($t_{1/2}$ min) at 25°C	
Substrate	Acid Hydrolysis 1% HCl in 95% MeOH	Base Hydrolysis 5% NaOH, in 95% MeOH
p-MeC$_6$H$_4$OSiEt$_3$	≤1[a]	≤1[a]
p-MeC$_6$H$_4$OSi$-i$-BuMe$_2$	≤1[a]	≤1[a]
p-MeC$_6$H$_4$OSi$-t$-BuMe$_2$	273	3.5
p-MeC$_6$H$_4$OSi$-t$-BuPh$_2$	100 (h)	6.5
p-MeC$_6$H$_4$OSi$-i$-Pr$_3$	100 (h)	188

[a]A $t_{1/2}$ of 1 min is a minimum value because of sampling methods.

1. E. W. Collington, H. Finch, and I. J. Smith, *Tetrahedron Lett.*, **26**, 681, (1985).

2. T. D. Nelson and R. D. Crouch, *Synthesis*, 1031 (1996); R. D. Crouch, *Tetrahedron*, **60**, 5833 (2004).

3. P. M. Kendall, J. V. Johnson, and C. E. Cook, *J. Org. Chem.*, **44**, 1421 (1979).

4. R. C. Ronald, J. M. Lansinger, T. S. Lillie, and C. J. Wheeler, *J. Org. Chem.*, **47**, 2541 (1982).

5. A. Liu, K. Dillon, R. M. Campbell, D. C. Cox, and D. M. Huryn, *Tetrahedron Lett.*, **37**, 3785 (1996).

6. D. L. J. Clive and D. Kellner, *Tetrahedron Lett.*, **32**, 7159 (1991).

7. A. Kojima, T. Takemoto, M. Sodeoka, and M. Shibasaki, *J. Org. Chem.*, **61**, 4876 (1996).

8. E. J. Corey and A. Venkateswarlu, *J. Am. Chem. Soc.*, **94**, 6190 (1972).

9. A. K. Sinhababu, M. Kawase, and R. T. Borchardt, *Synthesis*, 710 (1988).

10. B. E. Blass, C. L. Harris, and D. E. Portlock, *Tetrahedron Lett.*, **42**, 1611 (2001).

11. E. A. Schmittling and J. S. Sawyer, *Tetrahedron Lett.*, **32**, 7207 (1991).

12. S. V. Ankala and G. Fenteany, *Tetrahedron Lett.*, **43**, 4729 (2002).

13. I. Tichkowsky and R. Lett, *Tetrahedron Lett.*, **43**, 3997 (2002).

14. N. S. Wilson and B. A. Keay, *Tetrahedron Lett.*, **37**, 153 (1996).

15. S. Mabic and J.-P. Lepoittevin, *Synlett*, 851 (1994).

16. C. Prakash, S. Saleh, and I. A. Blair, *Tetrahedron Lett.*, **35**, 7565 (1994).

17. N. S. Wilson and B. A. Keay, *Tetrahedron Lett.*, **38**, 187 (1997).

18. B. P. Bandgar, S. D. Unde, D. S. Unde, V. H. Kulkarni, and S. V. Patil, *Indian J. Chem., Sect. B*, **33B**, 782 (1994).

19. A. H. De Groot, R. A. Dommisse, and G. L. Lemiére, *Tetrahedron*, **56**, 1541 (2000).

20. G. Maiti and S. C. Roy, *Tetrahedron Lett.*, **38**, 495 (1997).

21. J. S. Davies, C. L. Higginbotham, E. J. Tremeer, C. Brown, and R. C. Treadgold, *J. Chem. Soc., Perkin Trans. I*, 3043 (1992).

t-Butyldiphenylsilyl Ether (TBDPS-OAr)

The TBDPS ether has been used for the monoprotection of a catechol (TBDPSCl, Im, DMF, 5 h, 83% yield)[1] or simple phenol protection. It is cleaved with Bu$_4$NF (THF, 94% yield).[2]

1. J. C. Kim and W.-W. Park, *Org. Prep. Proced. Int.*, **26**, 479 (1994).
2. A. B. Smith, III, J. Barbosa, W. Wong, and J. L. Wood, *J. Am. Chem. Soc.*, **118**, 8316 (1996).

Triisopropylsilyl Ether (TIPS—OAr)

The bulk of the TIPS group, introduced with TIPSCl (DMF, Im, 92% yield), directs metalation away from the silyl group as illustrated.[1] Cleavage is accomplished with 3HF·TEA, THF.[2]

1. J. J. Landi, Jr., and K. Ramig, *Synth. Commun.*, **21**, 167 (1991).
2. C. Visintin, A. E. Aliev, D. Riddall, D. Baker, M. Okuyama, P. M. Hoi, R. Hiley, and D. L. Selwood, *Org. Lett.*, **7**, 1699 (2005).

Esters

Aryl esters, prepared from the phenol and an acid chloride or anhydride in the presence of base, are readily cleaved by saponification. In general, they are more readily cleaved than the related esters of alcohols, thus allowing selective removal of phenolic esters. Steric factors play a significant role in that hindered esters are much slower to hydrolyze. 9-Fluorenecarboxylates and 9-xanthenecarboxylates are also cleaved by photolysis. To permit selective removal, a number of carbonate esters have been investigated: Aryl benzyl carbonates can be cleaved by hydrogenolysis; aryl 2,2,2-trichloroethyl carbonates by Zn/THF-H$_2$O. Esters of electron deficient phenols are good acylating agents for alcohols and amines.

Aryl Formate: HCO$_2$Ar

The formate ester of phenol is rarely formed, but can be prepared from the phenol, formic acid, and DCC, 94–99% yield, or from the mixed anhydride, HCO$_2$OAc (pyridine, CH$_2$Cl$_2$).[1] The formate ester is not very stable to basic conditions or to other good nucleophiles.[2]

1. A. G. Schultz and A. Wang, *J. Am. Chem. Soc.*, **120**, 8259 (1998).
2. J. Huang and H. K. Hall, Jr., *J. Chem. Res., Synop.*, 292 (1991).

Aryl Acetate: ArOCOCH$_3$ (Chart 4)

Formation

1. AcCl, NaOH, dioxane, Bu$_4$NHSO$_4$, 25°C, 30 min, 90% yield.[1] Phase transfer catalysis with tetra-*n*-butylammonium hydrogen sulfate effects acylation of sterically hindered phenols and selective acylation of a phenol in the presence of an aliphatic secondary alcohol.

2. NaH, Ac$_2$O, DMF, 66% yield.[2]

3. 1-Acetyl-*v*-triazolo[4,5-*b*]pyridine, THF, 1 *N* NaOH, 30 min.[3]

This method is also effective in the selective introduction of a benzoate ester.

4. IPA, NaOH, Ac$_2$O, pH 7.8. Phenols are selectively esterified in the presence of other alcohols.[4] These authors also showed that an alcohol could be acetylated in the presence of an amine using Ac$_2$O and Amberlyst 15 resin.

5. *Chromobacterium viscosum* lipase, cyclohexane, vinyl acetate, THF, 40°C.[5]

6. Ac$_2$O in the presence of Lewis acids such as Mg(ClO$_4$)$_2^6$ or InCl$_3^7$ serve as catalysts for the acylation of phenols.

7. I$_2$, Ac$_2$O, microwaves, 2–4 min, 94–98% yield. The method is good for very hindered phenols such as 2,6-di-*tert*-butylhydroquinone.[8]

Cleavage

Aryl acetates are very easily cleaved by even the mildest of bases in alcoholic solvents.

1. NaHCO$_3$/aq. MeOH, 25°C, 0.75 h, 94% yield.[9] Ammonium acetate[10] and NaBO$_3^{11}$ have also been used as a base.

2. aq. NH$_3$, 0°C, 48 h.[16]

3. NaBH$_4$, HO(CH$_2$)$_2$OH, 40°C, 18 h, 87% yield.[12] Lithium aluminum hydride can be used to affect efficient ester cleavage if no other functional group is present that can be attacked by this strong reducing agent.[13]

4. NaBH$_4$, LiCl, diglyme. A diacylated guanidine was not deacetylated under these conditions, whereas the usual basic conditions for acetate hydrolysis also resulted in guanidine deacylation.[14]

5. Sm, I_2, EtOH, 82–100% yield. Esters of other alcohols are similarly deacylated.[15]

6. 3 N HCl, acetone, reflux, 2 h.[16]

The following conditions selectively remove a phenolic acetate in the presence of a normal alkyl acetate.

1. TsOH, SiO_2, toluene, 80°C, 6–40 h, 79–100% yield.[17] Ammonium formate supported on silica can also be used.[18]

2. Amberlyst-15 or iodine, MeOH, 48–100% yield.[19]

3. Kaolinitic clay, MeOH, 25°C, 88–96% yield.[20]

4. $(NH_2)_2C=NH$, MeOH, 50°C, 95% yield.[21]

5. $Me_2NCH_2C(O)N(OH)Me$, MeOH or THF/H_2O, 84% yield.[22]

6. Zn, MeOH, 91–100% yield.[23]

7. Neutral alumina, microwaves, 82–96% yield.[24]

8. Bi(III)-mandelate, DMSO, 80–125°C, 44–96% yield. Phenolic acetates with strong electron withdrawing groups are hydrolyzed the fastest.[25]

9. $LiClO_4 \cdot 2H_2O$, MeOH, rt, 3 h, 52–71% yield.[26]

10. Porcine pancreatic lipase, 28–30°C, 95% yield.[27]

11. *Candida cylindracea* lipase, BuOH, hexane, 3 h, 25°C, 40–100% yield.[28]

12. *Pseudomonas cepacia* PS lipase, acetone, pH 7 phosphate buffer, 25°C.[29]

1. V. O. Illi, *Tetrahedron Lett.*, **20**, 2431 (1979).

2. T. Ritter, P. Zarotti, and E. M. Carreira, *Org. Lett.*, **6**, 4371 (2004).

3. M. P. Paradisi, G. P. Zecchini, and I. Torrini, *Tetrahedron Lett.*, **27**, 5029 (1986).

4. V. Srivastava, A. Tandon, and S. Ray, *Synth. Commun.*, **22**, 2703 (1992).

5. G. Nicolosi, M. Piattelli, and C. Sanfilippo, *Tetrahedron*, **48**, 2477 (1992).

6. A. K. Chakraborti, L. Sharma, R. Gulhane, and Shivani, *Tetrahedron*, **59**, 7661 (2003).

7. A. K. Chakraborti and R. Gulhane, *Tetrahedron Lett.*, **44**, 6749 (2003).

8. N. Deka, A.-M. Mariotte, and A. Boumendjel, *Green Chem.*, **3**, 263 (2001).

9. For example, see G. Büchi and S. M. Weinreb, *J. Am. Chem. Soc.*, **93**, 746 (1971).

10. C. Ramesh, G. Mahender, N. Ravindranath, and B. Das, *Tetrahedron*, **59**, 1049 (2003).

11. B. P. Bandgar, L. S. Uppalla, V. S. Sadavarte, and S. V. Patil, *New J. Chem.*, **26**, 1273 (2002).

12. J. Quick and J. K. Crelling, *J. Org. Chem.*, **43**, 155 (1978); B. P. Bandgar and V. T. Kamble, *J. Chem. Research (S)*, 54 (2001).

13. H. Mayer, P. Schudel, R. Rüegg and O. Isler, *Helv. Chim. Acta*, **46**, 650 (1963).

14. D. Huber, G. Leclerc, and G. Andermann, *Tetrahedron Lett.*, **27**, 5731 (1986).

15. R. Yanada, N. Negoro, K. Bessho, and K. Yanada, *Synlett*, 1261 (1995).

16. E. Haslam, G. K. Makinson, M. O. Naumann, and J. Cunningham, *J. Chem. Soc.*, 2137 (1964).

17. G. Blay, M. L. Cardona, M. B. Garcia, and J. P. Pedro, *Synthesis*, 438 (1989).

18. C. Ramesh, G. Mahender, N. Ravindranath and B. Das, *Green Chem.*, **5**, 68 (2003).

19. B. Das, J. Banerjee, R. Ramu, R. Pal, N. Ravindranath, and C. Ramesh, *Tetrahedron Lett.*, **44**, 5465 (2003).

20. B. P. Bandgar and S. P. Kasture, *Green Chem.*, **2**, 154 (2000).

21. N. Kunesch, C. Miet, and J. Poisson, *Tetrahedron Lett.*, **28**, 3569 (1987).

22. M. Ono and I. Itoh, *Tetrahedron Lett.*, **30**, 207 (1989).

23. A. G. González, Z. D. Jorge, H. L. Dorta, and F. R. Luis, *Tetrahedron Lett.*, **22**, 335 (1981).

24. R. S. Varma, M. Varma, and A. K. Chatterjee, *J. Chem. Soc., Perkin Trans. I*, 999 (1993).

25. V. Le Boisselier, M. Postel, and E. Duñch, *Tetrahedron Lett.*, **38**, 2981 (1997).

26. F. Rajabi and M. R. Saidi, *Synth. Commum.*, **35**, 483 (2005).

27. V. S. Parmar, A. Kumar, K. S. Bisht, S. Mukherjee, A. K. Prasad, S. K. Sharma, J. Wengel, and C. E. Olsen, *Tetrahedron*, **53**, 2163 (1997).

28. G. Pedrocchi-Fantoni and S. Servi, *J. Chem. Soc., Perkin Trans. I*, 1029 (1992); P. Ciuffreda, S. Casati, and E. Santaniello, *Tetrahedron*, **56**, 317 (2000).

29. P. Allevi, P. Ciuffreda, A. Longo, and M. Anastasia, *Tetrahedron: Asymmetry*, **9**, 2915 (1998).

Aryl Levulinate: $CH_3COCH_2CH_2CO_2Ar$

Cleavage[1]

1. M. Ono and I. Itoh, *Chem. Lett.*, **17**, 585 (1988).

Aryl Pivaloate (ArOPv): $(CH_3)_3CCO_2Ar$ (Chart 4)

Formation

1. Pivaloyl chloride reacts selectively with the less hindered phenol group.[1]

2. NaH, THF, 99% yield.[2] This method works well for the esterification of a phenol in the presence of an aniline. When the thiazolidone is reacted with a hydroxyaniline in the absence of base, only the nitrogen is derivatized to form a pivalamide.[3]

Cleavage

1. 50% aqueous KOH, EtOH, reflux, 64 h, 87% yield.[1]
2. PhSH, K_2CO_3, NMP, reflux, 15–30 min, 70–90% yield.[4]
3. Polymer-SK, MeOH, THF, 40°C, 99% yield.[5] This method also cleaves pivalates from aryl amines and alcohols.

1. L. K. T. Lam and K. Farhat, *Org. Prep. Proced. Int.*, **10**, 79 (1978).
2. K. C. Nicolaou and W.-M. Dai, *J. Am. Chem. Soc.*, **114**, 8908 (1992).
3. W.-M. Dai, Y. K. Cheung, K. W. Tang. P. Y. Choi, and S. L. Chung, *Tetrahedron*, **51**, 12263 (1995).
4. A. K. Chakraborti, M. K. Nayak, and L. Sharma, *J. Org. Chem.*, **64**, 8027 (1999).
5. R. N. MacCoss, D. J. Henry, C. T. Brain, and S. V. Ley, *Synlett*, 675 (2004).

Aryl Benzoate: $ArOCOC_6H_5$ (Chart 4)

Aryl benzoates, stable to alkylation conditions using K_2CO_3/Me_2SO_4, are cleaved by more basic hydrolysis (KOH).[1] They are stable to anhydrous hydrogen chloride,[2] but are cleaved by hydrochloric acid.[3]

Formation

1. $(ClCO)_2$, Me_2NCHO, PhCOOH; Pyr, 20°C, 2 h, 90% yield.[4]

2. SBz aq. $NaHCO_3$ or aq. NaOH, 80% yield.[5] This reagent forms aryl benzoates under aqueous conditions. (It also acylates amines and carboxylic acids.)

3. Monoesterification of a symmetrical dihydroxy aromatic compound can be effected by reaction with polymer-bound benzoyl chloride (Pyr, benzene, reflux, 15 h) to give a polymer-bound benzoate, which can be alkylated with diazomethane to form, after basic hydrolysis (0.5 M NaOH, dioxane, H_2O, 25°C, 20 h, or 60°C, 3 h), a monomethyl ether.[6]

4. $Fe_2(SO_4)_3 \cdot SiO_2$, methyl benzoate, 97% yield.[7]

Cleavage

1. Under anhydrous conditions, cesium carbonate or bicarbonate quantitatively cleaves an aryl dibenzoate or diacetate to the monoester; yields are considerably lower with potassium carbonate.[8] K_2CO_3 in NMP at 100°C results in selective cleavage of aryl benzoates and acetates, but does not hydrolyze other nonphenolic esters.[9]

2. $BuNH_2$, benzene, rt, 1–24 h, >85% yield.[10] This method is generally selective for phenolic esters.

3. 2-Bromo-1,3,2-benzodioxaborole, CH_2Cl_2 (cat. $BF_3 \cdot Et_2O$), 25°C, 0.25 h, 71% yield.[11]

4. Aryl benzoates are subject to acyl migration under basic conditions.[12]

1. M. Gates, *J. Am. Chem. Soc.*, **72**, 228 (1950).

2. D. D. Pratt and R. Robinson, *J. Chem. Soc.*, 1577 (1922).

3. A. Robertson and R. Robinson, *J. Chem. Soc.*, 1710 (1927).

4. P. A. Stadler, *Helv. Chim. Acta*, **61**, 1675 (1978).

5. M. Yamada, Y. Watabe, T. Sakakibara, and R. Sudoh, *J. Chem. Soc., Chem. Commun.*, 179 (1979).

6. C. C. Leznoff and D. M. Dixit, *Can. J. Chem.*, **55**, 3351 (1977).

7. T. Nishiguchi and H. Taya, *J. Chem. Soc., Perkin Trans. I*, 172 (1990).

8. H. E. Zaugg, *J. Org. Chem.*, **41**, 3419 (1976).
9. A. K. Chakraborti, L. Sharma, and U. Sharma, *Tetrahedron*, **57**, 9343 (2001).
10. K. H. Bell, *Tetrahedron Lett.*, **27**, 2263 (1986).
11. P. F. King and S. G. Stroud, *Tetrahedron Lett.*, **26**, 1415 (1985).
12. C. Pugh, *Org. Lett.*, **2**, 1329 (2000).

Aryl 9-Fluorenecarboxylate: (Chart 4)

Aryl 9-fluorenecarboxylates (designed to be cleaved photolytically) were prepared from the phenol and the acid chloride (9-fluorenecarbonyl chloride, Pyr, C_6H_6, 25°C, 1 h, 65% yield) and cleaved by photolysis ($h\nu$, Et_2O, reflux, 4 h, 60% yield). The related aryl **xanthenecarboxylates, i,** were prepared and cleaved in the same way.[1]

i

1. D. H. R. Barton, Y. L. Chow, A. Cox, and G. W. Kirby, *J. Chem. Soc.*, 3571 (1965).

Carbonates

Aryl Methyl Carbonate: $ArOCO_2CH_3$ (Chart 4)

In an early synthesis a methyl carbonate, prepared by reaction of a phenol with methyl chloroformate, was cleaved selectively in the presence of a phenyl ester.[1] In this case the ester is partially protected by formation of an ammonium salt, which reduces the leaving group ability of the phenol.

An ethyl carbonate was cleaved by refluxing in acetic acid for 6 h.[2]

1. E. Fischer and H. O. L. Fischer, *Ber.* **46**, 1138 (1913).
2. E. Haslam, R. D. Haworth, and G. K. Makinson, *J. Chem. Soc.*, 5153 (1961).

t-Butyl Carbonate (BOC-OAr): $(CH_3)_3COCO_2Ar$

The BOC derivative of phenols can be prepared using a phase transfer protocol $(BOC_2O, Bu_4NHSO_4$ or 18-crown-6, NaOH, CH_2Cl_2, 80% yield)[1] or by direct acylation with BOC_2O and DMAP as a catalyst (79–100% yield).[2] The unusual process of protecting a phenol in the presence of the more nucleophilic amine has been accomplished with 1-*tert*-butoxy-*tert*-butoxycarbonyl-1,2-dihydroquinoline.[3] Chemoselectivity is controlled by the solvent.

Cleavage is achieved by refluxing a mixture of the carbonate with 3 *M* HCl in dioxane. The use of TFA for cleavage often results in *t*-butylation of the phenol.[2] This can be prevented by adding a cation scavenger to the reaction mixture. Basic hydrolysis (NaOH/MeOH or piperidine/CH_2Cl_2) is also very effective at removing the BOC group from a phenol.[4]

1. F. Houlihan, F. Bouchard, J. M. J. Frechet, and C. G. Willson, *Can. J. Chem.*, **63**, 153 (1985).
2. M. M. Hansen and J. R. Riggs, *Tetrahedron Lett.*, **39**, 2705 (1998).
3. H. Ouchi, Y. Saito, Y. Yamamoto, and H. Takahata, *Org. Lett.*, **4**, 585 (2002).
4. K. Nakamura, T. Nakajima, H. Kayahara, E. Nomura, and H. Taniguchi, *Tetrahedron Lett.*, **45**, 495 (2004).

1-Adamantyl Carbonate (Adoc−OAr)

The adamantyl carbonate is prepared from $Adoc_2CO_3$ (DMAP, CH_3CN, >79% yield)[1] or in the case of electron-deficient phenols, the fluoroformate (THF, Pyr, 54–95% yield).[2] It is somewhat more stable to TFA than the adamantyl carbamate.

1. B. Nyasse and U. Ragnarsson, *Acta Chem. Scand.*, **47**, 374 (1993).
2. I. Niculescu-Duvaz and C. J. Springer, *J. Chem. Res., Synop.*, 242 (1994).

2,4-Dimethylpent-3-yl Carbonate (Doc−OAr): $(i$-Pr$)_2CHOCO_2Ar$

The Doc group, used for the protection of the phenolic hydroxyl group in tyrosine, is introduced with the chloroformate (DIPEA, CH_3CN). The Doc group has a half-life in 20% piperidine/DMF of 8 h, which compares to 30 s for the 2-BrZ (2-BrCbz) group, making it about 1000 times more stable. The 2-BrZ group is only slightly more stable to acid than the Doc group. The Doc group is completely cleaved by HF.[1] When used in peptide synthesis, the Doc group results in much lower levels of alkylation by-products during the deprotection process.[2]

1. K. Rosenthal, A. Karlström, and A. Undén, *Tetrahedron Lett.*, **38**, 1075 (1997).
2. A. Karlström, K. Rosenthal, and A. Undén, *J. Pept. Res.*, **55**, 36 (2000).

Allyl Carbonate (Alloc−OAr): $CH_2=CHCH_2OCO_2Ar$

Allyl chloroformate was used to protect both the phenolic hydroxyl and the amine of a series of amino acids (85–98% yield) with the aim of using a single protective group that was readily cleaved from the phenol (20% piperidine/DMF) but retained on the amine.[1] Many of the Pd based methods discussed in the alcohol section should be applicable.

1. A. D. Morley, *Tetrahedron Lett.*, **41**, 7401 (2000).

4-Methylsulfinylbenzyl Ether (Msib−OR): $CH_3S(O)C_6H_4CH_2OAr$

The Msib group has been used for the protection of tyrosine. It is cleaved by reduction of the sulfoxide to the sulfide, which is then deprotected with acid. Reduction is achieved with $DMF\text{-}SO_3/HSCH_2CH_2SH$ or Bu_4NI[1] or with $SiCl_3/TFA$.[2]

1. S. Futaki, T. Yagami, T. Taike, T. Ogawa, T. Akita, and K. Kitagawa, *Chem. Pharm. Bull.*, **38**, 1165 (1990).
2. Y. Kiso, S. Tanaka, T. Kimura, H. Itoh, and K. Akaji, *Chem. Pharm. Bull.*, **39**, 3097 (1991).

Aryl 2,2,2-Trichloroethyl Carbonate: $ArOCOOCH_2CCl_3$ (Chart 4)

Formation

Cl_3CCH_2OCOCl, Pyr or aq. NaOH, 25°C, 12 h.[1]

Cleavage

1. Zn, HOAc, 25°C, 1–3 h, or Zn, CH_3OH, heat, few min.[1]
2. Zn, THF–H_2O, pH 4.2, 25°C, 4 h.[2] The authors suggest that selective cleavage should be possible by this method, since, at pH 4.2 and 25°C, 2,2,2-trichloroethyl esters are cleaved in 10 min, 2,2,2-trichloroethyl carbamates are cleaved in 30 min, and the 2,2,2-trichloroethyl carbonate of estrone, formed in 87% yield from estrone and the acid chloride, is cleaved in 4 h (97% yield).

1. T. B. Windholz and D. B. R. Johnston, *Tetrahedron Lett.*, **8**, 2555 (1967).
2. G. Just and K. Grozinger, *Synthesis*, 457 (1976).

Aryl Vinyl Carbonate: $ArOCO_2CH=CH_2$ (Chart 4)

Formation

$CH_2=CHOCOCl$, Pyr, 95% yield.[1]

Cleavage

Na_2CO_3, warm aq. dioxane, 96% yield. Selective protection of an aryl $-OH$ or an amine $-NH$ group is possible by reaction of the compound with vinyl chloroformate. Vinyl carbamates ($RR'NCO_2CH=CH_2$) are stable to the basic conditions (Na_2CO_3) used to cleave vinyl carbonates. Conversely, vinyl carbonates are stable to the acidic conditions ($HBr/CH_3OH/CH_2Cl_2$) used to cleave vinyl carbamates. Vinyl carbonates are cleaved by more acidic conditions: $2 N$ anhydrous HCl/ dioxane, 25°C, 3 h, 10% yield; HBF_4, 25°C, 12 h, 30% yield; $2 N$ HCl/CH_3OH-H_2O(4:1), 60°C, 8 h, 100% yield.[1]

1. R. A. Olofson and R. C. Schnur, *Tetrahedron Lett.*, **18**, 1571 (1977).

Aryl Benzyl Carbonate: $ArOCOOCH_2C_6H_5$ (Chart 4)

The related *o*-bromobenzyl carbonates have been developed for use in solid-phase peptide synthesis. An aryl *o*-bromobenzyl carbonate is stable to acidic cleavage (CF_3CO_2H) of a *t*-butyl carbamate; a benzyl carbonate is cleaved. The *o*-bromo derivative is quantitatively cleaved with hydrogen fluoride (0°C, 10 min).[1]

Formation

$PhCH_2OCOCl$, Pyr, CH_2Cl_2, THF.[2]

Cleavage

H_2/Pd–C, EtOH, 20°C.[2]

1. D. Yamashiro and C. H. Li, *J. Org. Chem.*, **38**, 591 (1973).
2. M. Kuhn and A. von Wartburg, *Helv. Chim. Acta*, **52**, 948 (1969).

Aryl Carbamates: ArOCONHR

Formation

RNCO (R=Ph, *i*-Bu), 60°C, 2 h, 65–85% yield.[1]

Cleavage

1. $2 N$ NaOH, 20°C, 2 h, 78% yield.[1]
2. $H_2NNH_2 \cdot H_2O$, DMF, 20°C, 3 h, 59–87% yield.[1]

1. G. Jäger, R. Geiger, and W. Siedel, *Chem. Ber.*, **101**, 2762 (1968).

Phosphinates

Dimethylphosphinyl Ester (Dmp−OAr Ester): $(CH_3)_2P(O)OAr$

Formation

$Me_2P(O)Cl$, Et_3N, $CHCl_3$, 76% yield.[1] The Dmp group was used to protect tyrosine for use in peptide synthesis. It is stable to 1 M HCl/MeOH, 1 M HCl/AcOH, CF_3CO_2H, HBr/AcOH, and H_2/Pd–C.

Cleavage

The Dmp group can be cleaved by the following reagents: liq. HF (0°C, 1 h); 1 M Et_3N/MeOH (rt, 7 h); 0.1 M NaOH (rt, <5 min); 5% aq. $NaHCO_3$ (rt, 5 h); 20% hydrazine/MeOH (rt, <5 min); 50% pyridine/DMF (rt, 6 h); Bu_4NF (rt, <5 min).[1]

Dimethylphosphinothioyl Ester (Mpt−OAr): $(CH_3)_2P(S)OAr$

Formation

MptCl, CH_2Cl_2, Et_3N, 66% yield.[2]

Cleavage

The *O*-Mpt group is quite stable to acidic conditions (HBr/AcOH, CF_3CO_2H, 1 M HCl/AcOH), but is slowly cleaved under basic conditions (1 M NaOH/MeOH, 5 min; 1 M Et_3N/MeOH, reflux, 12 h). In contrast, the *N*-Mpt group is readily cleaved with acid (CF_3CO_2H, 60 min; 1 M HCl/AcOH, 15 min; HBr/AcOH, 5 min), but not with base. The Mpt group was used to protect tyrosine during peptide synthesis.[2] The Mpt group can be removed with aq. $AgNO_3$ or $Hg(OAc)_2^3$ or fluoride ion.[4]

Diphenylphosphinothioyl Ester (Dpt−OAr): $(C_6H_5)_2P(S)OAr$

The diphenylphosphinothioyl ester, used to protect a tryptophan, is cleaved with $Bu_4NF·3H_2O$/DMF.[5]

1. M. Ueki, Y. Sano, I. Sori, K. Shinozaki, H. Oyamada, and S. Ikeda, *Tetrahedron Lett.*, **27**, 4181 (1986).
2. M. Ueki and T. Inazu, *Bull. Chem. Soc. Jpn.*, **55**, 204 (1982).
3. M. Ueki and K. Shinozaki, *Bull. Chem. Soc. Jpn.*, **56**, 1187 (1983).
4. M. Ueki and K. Shinozaki, *Bull. Chem. Soc. Jpn.*, **57**, 2156 (1984).
5. Y. Kiso, T. Kimura, Y. Fujiwara, M. Shimokura, and A. Nishitani, *Chem. Pharm. Bull.*, **36**, 5024 (1988).

Sulfonates

An aryl methane- or toluenesulfonate ester is stable to reduction with lithium aluminum hydride, to the acidic conditions used for nitration of an aromatic ring $(HNO_3/HOAc)$[1], and to the high temperatures (200–250°C) of an Ullmann reaction. Aryl sulfonate esters, formed by reaction of a phenol with a sulfonyl chloride in pyridine or aqueous sodium hydroxide, are cleaved by warming in aqueous sodium hydroxide.[2]

1. E. M. Kampouris, *J. Chem. Soc.*, 2651 (1965).
2. F. G. Bordwell and P. J. Boutan, *J. Am. Chem. Soc.*, **79**, 717 (1957).

Aryl Methanesulfonate: $ArOSO_2CH_3$ (Chart 4)

In a synthesis of decinine, a phenol was protected as a methanesulfonate that was stable during an Ullmann coupling reaction and during a condensation, catalyzed by calcium hydroxide, of an amine with an aldehyde. Aryl methanesulfonates are cleaved by warm sodium hydroxide solution,[1,2] with LDA (THF, −78°C to rt, 57–95% yield)[3] or with TMSOK/CH$_3$CN.[4] An aryl methanesulfonate was cleaved to a phenol by phenyllithium or phenylmagnesium bromide;[5,6] it was reduced to an aromatic hydrocarbon by sodium in liquid ammonia.[7]

1. I. Lantos and B. Loev, *Tetrahedron Lett.*, **16**, 2011 (1975).
2. J. E. Rice, N. Hussain, and E. J. LaVoie, *J. Labelled Compd. Radiopharm.*, **24**, 1043 (1987).
3. T. Ritter, K. Stanek, I. Larrosa, and E. M. Carreira, *Org. Lett.*, **6**, 1513 (2004).
4. K. Mori, K. Rikimaru, T. Kan, and T. Fukuyama, *Org. Lett.*, **6**, 3095 (2004).
5. J. E. Baldwin, D. H. R. Barton, I. Dainis, and J. L. C. Pereira, *J. Chem. Soc. C*, 2283 (1968).
6. E. J. Corey and S. E. Lazerwith, *J. Am. Chem. Soc.*, **120**, 12777 (1998).
7. G. W. Kenner and N. R. Williams, *J. Chem. Soc.*, 522 (1955).

Aryl Trifluoromethanesulfonate (ArO-Tf): CF_3SO_2-OAr

Phenolic triflates are formed with 4-nitrophenyl triflate in the presence of K_2CO_3 in DMF or with Triflic anhydride in the presence of an amine base.[1] It can be cleaved with Et$_4$NOH in dioxane or with TBAF in THF (70–99% yield). Et$_4$NOH will also cleave phenolic mesylates and tosylates.[2] These triflates are also substrates for a variety of Pd-catalyzed coupling reactions.

1. J. Zhu, A. Bigot, M. Elise, and T. H. Dau, *Tetrahedron Lett.*, **38**, 1181 (1997).
2. T. Ohgiya and S. Nishiyama, *Tetrahedron Lett.*, **45**, 6317 (2004).

Aryl Toluenesulfonate: $ArOSO_2C_6H_4$-p-CH_3

An aryl toluenesulfonate is stable to lithium aluminum hydride (Et_2O, reflux, 4 h) and to p-toluenesulfonic acid ($C_6H_5CH_3$, reflux, 15 min).[1]

Formation[1]

1.

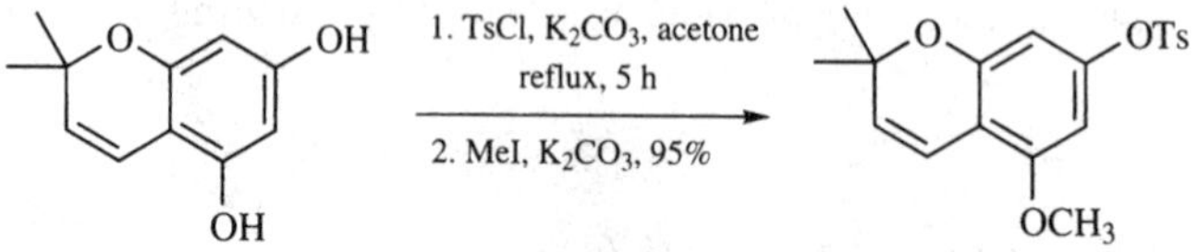

1. TsCl, K_2CO_3, acetone
   reflux, 5 h

2. MeI, K_2CO_3, 95%

2. *o*-Aminophenol can be selectively protected as a sulfonate or a sulfonamide.[2]

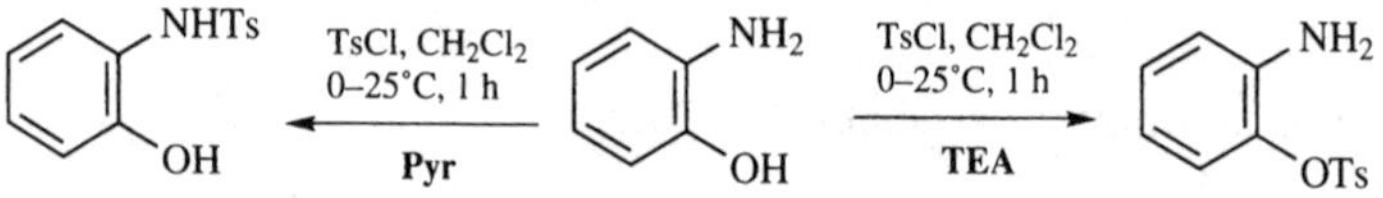

TsCl, CH_2Cl_2
0–25°C, 1 h

Pyr

TsCl, CH_2Cl_2
0–25°C, 1 h

TEA

Cleavage

1.

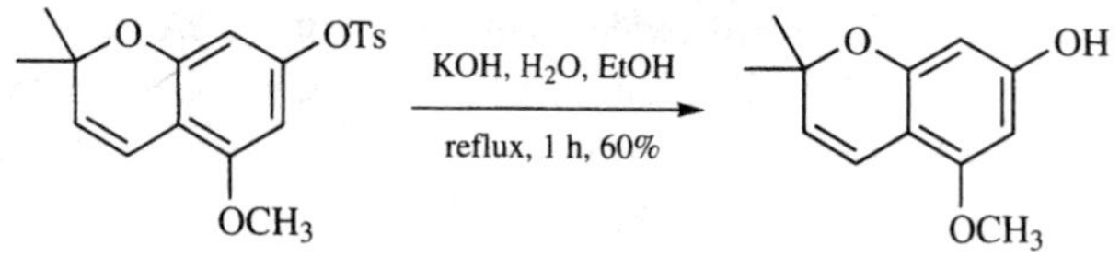

KOH, H_2O, EtOH

reflux, 1 h, 60%

2. KOTMS, CH_3CN, >87% yield.[3]

TMSOK, CH_3CN

>87%

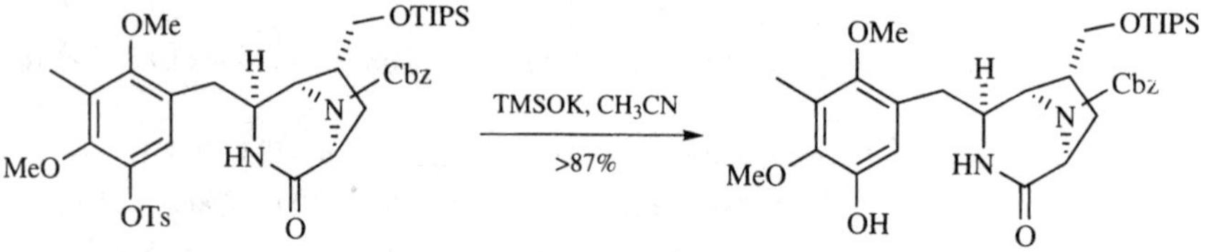

3. PhSH, K_2CO_3, NMP, reflux, 60 min, 60–95% yield. Aryl esters are cleaved similarly, but faster.[4]

4. TBAF, THF, −5°C to 2°C or KF, DME, 80% yield.[5]

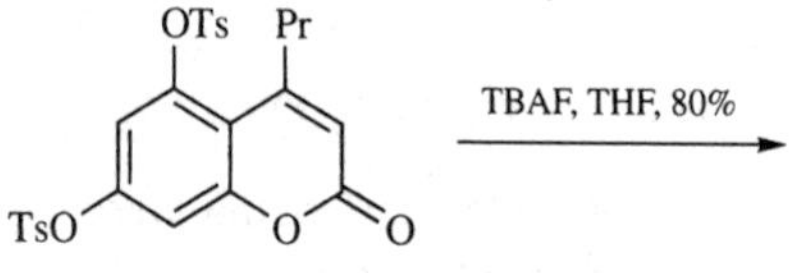

TBAF, THF, 80%

5. Electrolysis: Hg anode, Pt cathode, DMF, O_2, cyclohexene, Bu_4NBr, 62% yield.[6]

6.

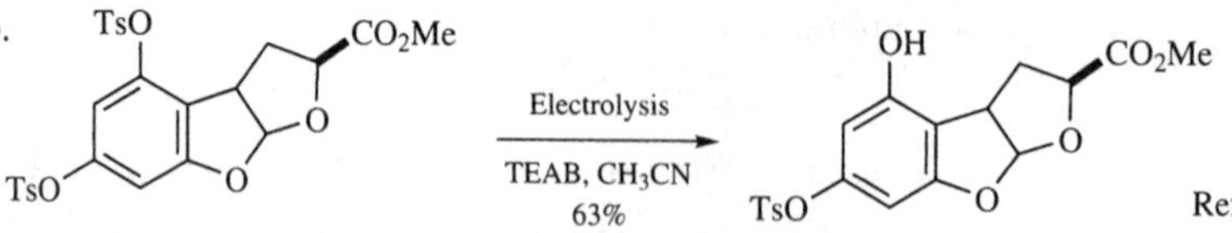

Electrolysis

TEAB, CH_3CN
63%

Ref. 7

7. TiCl$_3$, Li, THF, rt, 68–91% yield. The toluenesulfonamide of an aniline can also be cleaved.[8]

8. Na(Hg), MeOH, 96.7% yield.[9]

9. Mg, MeOH, 4–6 h, 90–95% yield.[10,11]

10. SmI$_2$, THF, H$_2$O, 0°C, 94% yield.[12]

1. M. L. Wolfrom, E. W. Koos, and H. B. Bhat, *J. Org. Chem.*, **32**, 1058 (1967).

2. K. Kurita, *Chem. Ind. (London)*, 345 (1974).

3. E. R. Ashley, E. G. Cruz, and B. M. Stoltz, *J. Am. Chem. Soc.*, **125**, 15000 (2003).

4. A. K. Chakraborti, M. K. Nayak, and L. Sharma, *J. Org. Chem.*, **64**, 8027 (1999).

5. M. E. Fox, I. C. Lennon, and G. Meek, *Tetrahedron Lett.*, **43**, 2899 (2002).

6. S. Dwivedi and R. A. Misra, *Indian J. Chem., Sect. B*, **B31**, 282 (1992).

7. E. R. Civitello and H. Rapoport, *J. Org. Chem.*, **59**, 3775 (1994).

8. S. K. Nayak, *Synthesis*, 1575 (2000).

9. R. S. Tipson, *Methods Carbohydr. Chem.*, **2**, 250 (1963).

10. M. Sridhar, B. A. Kumar, and R. Narender, *Tetrahedron Lett.*, **39**, 2847 (1998).

11. P. S. Baran, C. A. Guerrero, N. B. Ambhaikar, and B. D. Hafensteiner, *Angew. Chem. Int. Ed.*, **44**, 606 (2005).

12. G. E. Keck, T. T. Wager, and J. F. D. Rodriquez, *J. Am. Chem. Soc.*, **121**, 5176 (1999).

Aryl 2-Formylbenzenesulfonate

The formylbenzenesulfonate prepared from a phenol (2-CHO-C$_6$H$_4$SO$_2$Cl, Et$_3$N) can be cleaved with NaOH (aq. acetone, rt, 5 min) in the presence of a hindered acetate.[1]

1. M. S. Shashidhar and M. V. Bhatt, *J. Chem. Soc., Chem. Commun.*, 654 (1987).

Aryl Benzylsulfonate (Bns−OAr): $PhOSO_2CH_2C_6H_5$

The aryl benzylsulfonate, introduced with the sulfonyl chloride (THF, TEA, 100% yield), is stable to hydrogenolysis with Pd, Rh, or Ru, but is readily cleaved with Raney nickel (H_2, EtOH, 99% yield). Single electron reduction with LiDTBB (THF, 0°C, 50–88% yield) is a reasonably effective method for cleaving this group. These reducing conditions were compatible with aryl halides, esters, nitro groups and aldehydes.[1] It is removed with strong base such as KOH, NaOH, or K_2CO_3, but Grignard reactions can be performed in its presence.[2]

1. F. Alonso, Y. Moglie, C. Vitale, G. Radivoy, and M. Yus, *Synthesis*, 1971 (2005).
2. A. Briot, C. Baehr, R. Brouillard, A. Wagner, and C. Mioskowski, *Tetrahedron Lett.*, **44**, 965 (2003).

PROTECTION FOR CATECHOLS (1,2-Dihydroxybenzenes)

Catechols can be protected as diethers or diesters by methods that have been described to protect phenols. However, formation of cyclic acetals and ketals (e.g., methylenedioxy, acetonide, cyclohexylidenedioxy, and diphenylmethylenedioxy derivatives) or cyclic esters (e.g., borates or carbonates) selectively protects the two adjacent hydroxyl groups in the presence of isolated phenol groups.

Cyclic Acetals and Ketals

Methylene Acetal (Chart 4)

The methylenedioxy group, often present in natural products, is stable to many reagents including Grignard and alkyllithium reagents.[1] Efficient methods for both formation and removal of the group are available.

Formation

1. CH_2Br_2, NaOH, H_2O, Adogen, reflux, 3h, 76–86% yield[2] [Adogen= $R_3N^+CH_3Cl^-$, phase transfer catalyst (R=C_8–C_{10} straight-chain alkyl groups)]. Earlier methods required anhydrous conditions and aprotic solvents.
2. CH_2X_2 (X=Br, Cl), DMF, KF or CsF, 110°C, 1.5 h, 70–98% yield.[3]
3. $BrCH_2Cl$, DMF, Cs_2CO_3, 70–110°C, 86–97% yield.[4]
4. CH_2Cl_2, CsF, DMF, reflux, 91% yield.[5]
5. CH_2I_2, KF, DMF, 110°C, overnight, 84% yield.[6]

Cleavage

1. AlBr$_3$, EtSH, 0°C, 0.5–1 h, 73–78% yield.[7] Aluminum bromide cleaves aryl and alkyl methyl ethers in high yield; methyl esters are stable.

2. PCl$_5$, CH$_2$Cl$_2$, reflux; H$_2$O; reflux, 3 h, 61% yield.[8]

3. BCl$_3$, CH$_3$SCH$_3$, ClCH$_2$CH$_2$Cl, 83°C, 98% yield.[9] Selective cleavage of an aryl methylenedioxy group, or an aryl methyl ether, by boron trichloride has been investigated.[10–12]

4. 9-Br-BBN, 24 h, 40°C, CH$_2$Cl$_2$.[13]

5. A 4-nitro-1,2-methylenedioxybenzene has been cleaved to a catechol with 2 *N* NaOH, 90°C, 30 min[14]; a similar compound substituted with a 4-nitro or 4-formyl group has been cleaved by NaOCH$_3$/DMSO, 150°C, 2.5 min (13–74% catechol, 6–60% recovered starting material).[15]

6. Pb(OAc)$_4$, benzene, 50°C, 8 h.[16]

7. (TMS)$_2$NNa or LDA, THF, DMPU, 93–99% yield.[17]

8. AlBr$_3$, EtSH, 0°C, 93% yield.[18]

9. Et$_3$SiH, B(C$_6$H$_5$)$_3$, CH$_2$Cl$_2$, 79% yield. These conditions will cleave a variety of ethers to give the TES derivative.[19]

1. P. Zhang and R. E. Gawley, *J. Org. Chem.*, **58**, 3223 (1993); M. L. Pedersen and D. B. Berkowitz, *J. Org. Chem.*, **58**, 6966 (1993).

2. A. P. Bashall and J. F. Collins, *Tetrahedron Lett.*, **16**, 3489 (1975).

3. J. H. Clark, H. L. Holland, and J. M. Miller, *Tetrahedron Lett.*, **17**, 3361 (1976).

4. R. E. Zelle and W. J. McClellan, *Tetrahedron Lett.*, **32**, 2461 (1991); B. Zhou, J. Guo, and S. J. Danishefsky, *Org. Lett.*, **4**, 43 (2002).

5. T. Geller, J. Jakupovic, and H.-G. Schmalz, *Tetrahedron Lett.*, **39**, 1541 (1998).

6. A. Alam, Y. Takaguchi, H. Ito, T. Yoshida, and S. Tsuboi, *Tetrahedron*, **61**, 1909 (2005).

7. M. Node, K. Nishide, M. Sai, K. Ichikawa, K. Fuji, and E. Fujita, *Chem. Lett.*, **8**, 97 (1979).

8. G. L. Trammell, *Tetrahedron Lett.*, **19**, 1525 (1978).

9. P. G. Williard and C. B. Fryhle, *Tetrahedron Lett.*, **21**, 3731 (1980).

10. M. Gerecke, R. Borer, and A. Brossi, *Helv. Chim. Acta*, **59**, 2551 (1976).

11. S. Teitel, J. O'Brien, and A. Brossi, *J. Org. Chem.*, **37**, 3368 (1972).

12. F. M. Dean, J. Goodchild, L. E. Houghton, J. A. Martin, R. B. Morton, B. Parton, A. W. Price, and N. Somvichien, *Tetrahedron Lett.*, **7**, 4153 (1966).

13. M. V. Bhatt, *J. Organomet. Chem.*, **156**, 221 (1978).

14. E. Haslam and R. D. Haworth, *J. Chem. Soc.*, 827 (1955).

15. S. Kobayashi, M. Kihara, and Y. Yamahara, *Chem. Pharm. Bull.*, **26**, 3113 (1978).

16. Y. Ikeya, H. Taguchi, and I. Yoshioka, *Chem. Pharm. Bull.*, **29**, 2893 (1981).

17. J. R. Hwu, F. F. Wong, J.-J. Huang, and S.-C. Tsay, *J. Org. Chem.*, **62**, 4097 (1997).

18. Y.-Z. Hu and D. L. J. Clive, *J. Chem. Soc., Perkin Trans. I*, 1421 (1997).

19. V. Gevorgyan, M. Rubin, S. Benson, J.-X. Liu, and Y. Yamamoto, *J. Org. Chem.*, **65**, 6179 (2000).

Pivaldehyde Acetal

The acetal is prepared from a catechol and pivaldehyde with TMSCl catalysis.[1]

1. Y. Nishida, M. Abe, H. Ohrui, and H. Meguro, *Tetrahedron: Asymmetry*, **4**, 1431 (1993).

2-BOC-ethylidene (Bocdene) and 2-Moc-ethylidene (Mocdene) Acetals

Formation/Cleavage

If the *t*-Bu group is cleaved with TFA, pyrrolidine will no longer remove the Bocdene group.[1]

1. X. Ariza, O. Pineda, J. Vilarrasa, G. W. Shipps, Jr., Y. Ma, and X. Dai, *Org. Lett.*, **3**, 1399 (2001).

Acetonide Derivative (Chart 4)

A catechol can be protected as an acetonide (acetone, 70% yield). It is cleaved with 6 *N* HCl (reflux, 2 h, high yield)[1] or by refluxing in acetic acid/H_2O (100°C, 18 h, 90% yield).[2]

1. K. Ogura and G.-i. Tsuchihashi, *Tetrahedron Lett.*, **12**, 3151 (1971).

2. E. J. Corey and S. D. Hurt, *Tetrahedron Lett.*, **18**, 3923 (1977).

Cyclohexylidene Ketal

The cyclohexylidene ketal, prepared from a catechol and cyclohexanone (Al_2O_3/ TsOH, CH_2Cl_2, reflux, 36 h),[1] is stable to metalation conditions (RX/BuLi) that cleave aryl methyl ethers.[2] The ketal is cleaved by acidic hydrolysis (concd. HCl/ EtOH, reflux, 1.5 h, → 20°C, 12 h); it is stable to milder acidic hydrolysis that cleaves tetrahydropyranyl ethers (1 N HCl/EtOH, reflux, 5 h, 91% yield).[3]

1. G. Schill and E. Logemann, *Chem. Ber.*, **106**, 2910 (1973).
2. G. Schill and K. Murjahn, *Chem. Ber.*, **104**, 3587 (1971).
3. J. Boeckmann and G. Schill, *Chem. Ber.*, **110**, 703 (1977).

Diphenylmethylene Ketal (Chart 4)

The diphenylmethylene ketal prepared from a catechol (Ph_2CCl_2, Pyr, acetone, 12 h),[1] (Ph_2CCl_2, neat, 170°C, 5 min, 59%),[2] or [$Ph_2C(OMe)_2$, H_2SO_4, CH_2Cl_2, 40°C, >83% yield][3] can be cleaved by hydrogenolysis (H_2/Pd–C, THF).[4,5] It has also been prepared from a 1,2,3-trihydroxybenzene (Ph_2CCl_2, 160°C, 5 min, 80% yield) and cleaved by acidic hydrolysis (HOAc, reflux, 7 h[6,7] or with TFA, rt, 30 min).[8] This group is stable to bromination conditions where the cyclic ethylorthoformate and the 4-methoxyphenyl acetal were not.[9]

1. W. Bradley, R. Robinson, and G. Schwarzenbach, *J. Chem. Soc.*, 793 (1930).
2. S. Bengtsson and T. Högberg, *J. Org. Chem.*, **54**, 4549 (1989).
3. M. D. Shair, T. Y. Yoon, K. K. Mosny, T. C. Chou, and S. J. Danishefsky, *J. Am. Chem. Soc.*, **118**, 9509 (1996).
4. E. Haslam, R. D. Haworth, S. D. Mills, H. J. Rogers, R. Armitage, and T. Searle, *J. Chem. Soc.*, 1836 (1961).
5. K. S. Feldman, S. M. Ensel, and R. D. Minard, *J. Am. Chem. Soc.*, **116**, 1742 (1994).
6. L. Jurd, *J. Am. Chem. Soc.*, **81**, 4606 (1959).
7. T. R. Kelly, A. Szabados, and Y.-J. Lee, *J. Org. Chem.*, **62**, 428 (1997).
8. Y. Kita, M. Arisawa, M. Gyoten, M. Nakajima, R. Hamada, H. Tohma, and T. Takada, *J. Org. Chem.*, **63**, 6625 (1998).
9. A. Alam, Y. Takaguchi, H. Ito, T. Yoshida, and S. Tsuboi, *Tetrahedron*, **61**, 1909 (2005).

Cyclic Ethyl Orthoformate (Ceof)

The Ceof group was developed for protection of L-DOPA in peptide synthesis using the Fmoc strategy.[1]

Formation

1. $HC(OEt)_3$, TsOH, 4-Å molecular sieves, benzene, reflux, 3 days, 80% yield.
2. $HC(OEt)_3$, Amberlyst 15E, benzene, reflux, 15 h, 99% yield.

Cleavage

1. 1 M TMSBr, TFA, thioanisole, m-cresol and EDT, 0°C, 60 min. These conditions are overkill for this hydrolysis, but were used because deprotection was part of a global peptide deprotection.
2. TsOH or HCl, MeOH, H_2O, rt, 16 h, 80–88% yield.[2,3]

1. B.-H. Hu and P. B. Messersmith, *Tetrahedron Lett.*, **41**, 5795 (2000).
2. A. Merz and M. Rauschel, *Synthesis*, 797 (1993).
3. A. Alam, Y. Takaguchi, H. Ito, T. Yoshida, and S. Tsuboi, *Tetrahedron*, **61**, 1909 (2005).

Diisopropylsilylene Derivative: $[(CH_3)_2CH]_2Si(OR)_2$

The diisopropylsilylene, formed from a catechol with $(i$-$Pr)_2Si(OTf)_2$ and 2,6-lutidine in 96% yield, is cleaved with KF (MeOH, 2 eq. HCl).[1]

1. E. J. Corey and J. O. Link, *Tetrahedron Lett.*, **31**, 601 (1990).

Cyclic Esters

Cyclic Borate (Chart 4)

A cyclic borate can be used to protect a catechol group during base-catalyzed alkylation or acylation of an isolated phenol group; the borate ester is then readily hydrolyzed by dilute acid.[1]

Formation[1]

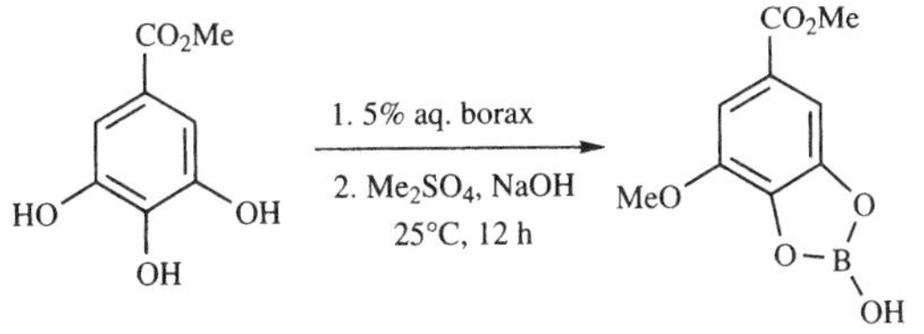

Cleavage[1]

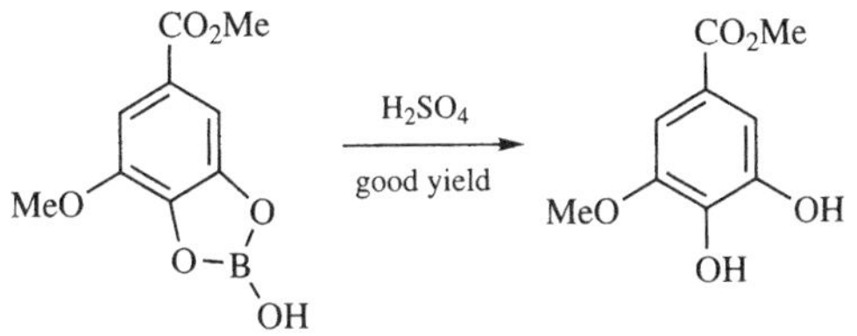

1. R. R. Scheline, *Acta Chem. Scand.*, **20**, 1182 (1966).

Cyclic Carbonate (Chart 4)

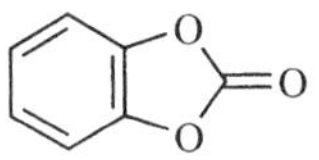

Cyclic carbonates have been used to a limited extent only (since they are readily hydrolyzed) to protect the catechol group in a polyhydroxy benzene.

Formation[1,2]

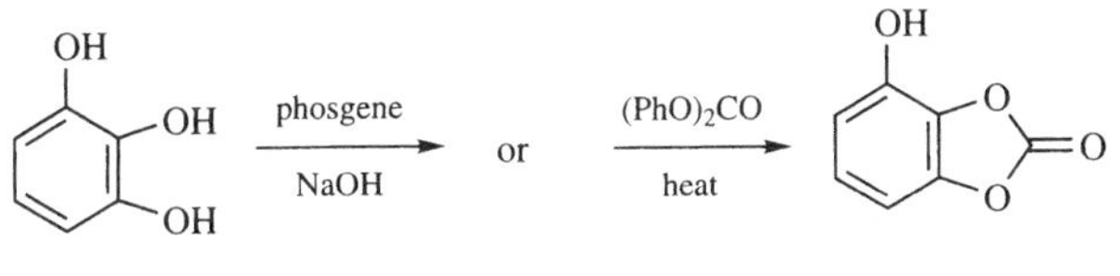

Cleavage

The cyclic carbonate is easily cleaved by refluxing in water for 30 min.[3] It can be converted to the 1,2-dimethoxybenzene derivative (aq. NaOH, Me$_2$SO$_4$, reflux, 3 h).[4]

1. A. Einhorn, J. Cobliner, and H. Pfeiffer, *Ber.*, **37**, 100 (1904).
2. S. M. O. Van Dyck, G. L. F. Lemiere, T. H. M. Jonckers, and R. Dommise, *Molecules [Electronic Publication]*, **5**, 153 (2000).
3. H. Hillemann, *Ber.*, **71**, 34 (1938).
4. W. Baker, J. A. Godsell, J. F. W. McOmie, and T. L. V. Ulbricht, *J. Chem. Soc.*, 4058 (1953).

PROTECTION FOR 2-HYDROXYBENZENETHIOLS

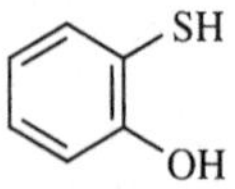
SH
OH

Two derivatives have been prepared that may prove useful as protective groups for 2-hydroxybenzenethiols. The methylene acetal is expected to be quite stable, whereas the orthoester derivative should be much more labile and cleavable by acid hydrolysis.

Formation

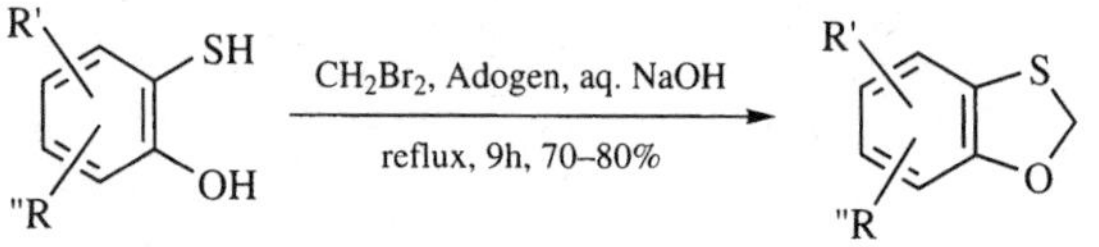
R'
SH
OH
"R
CH_2Br_2, Adogen, aq. NaOH
reflux, 9h, 70–80%
R'
S
O
"R

R', R" = H, Me, Cl
Adogen = MeR_3NCl, phase transfer catalyst
R = C_8-C_{10} straight chain alkyl groups

Ref. 1

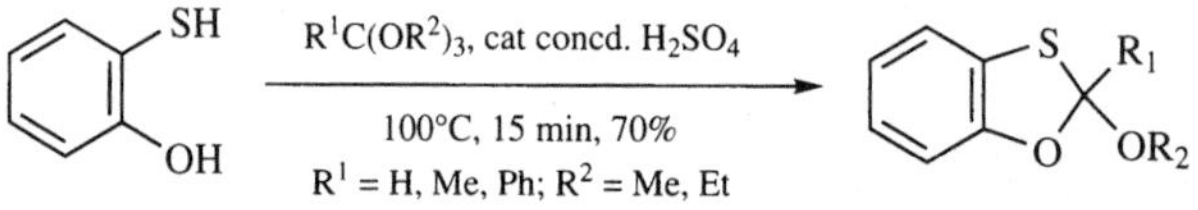
SH
OH
$R^1C(OR^2)_3$, cat concd. H_2SO_4
100°C, 15 min, 70%
R^1 = H, Me, Ph; R^2 = Me, Et
S
O
R_1
OR_2

Ref. 2

1. S. Cabiddu, S. Melis, L. Bonsignore, and M. T. Cocco, *Synthesis*, 660 (1975).
2. S. Cabiddu, A. Maccioni, and M. Secci, *Synthesis*, 797 (1976).

4

PROTECTION FOR THE CARBONYL GROUP

Greene's Protective Groups in Organic Synthesis, Fourth Edition, by Peter G. M. Wuts and
Theodora W. Greene
Copyright © 2007 John Wiley & Sons, Inc.

During a synthetic sequence a carbonyl group may have to be protected against attack by various reagents such as strong or moderately strong nucleophiles, including organometallic reagents; acidic, basic, catalytic, or hydride reducing agents; and some oxidants. Because of the order of reactivity of the carbonyl group [e.g., aldehydes (aliphatic > aromatic) > acyclic ketones and cyclohexanones > cyclopentanones > α,β-unsaturated ketones or α,α-disubstituted ketones $\gg$ aromatic ketones], it may be possible to protect a reactive carbonyl group selectively in the presence of a less reactive one. In keto steroids the order of reactivity to ketalization is C_3 or Δ^4-$C_3 > C_{17} > C_{12} > C_{20} > C_{17,21-(OH)_2} C_{20} > C_{11}$.[1] A review discusses the relative rates of hydrolysis of acetals, ketals, and ortho esters which are most commonly used to protect ketones and aldehydes.[2]

The most useful protective groups are the acyclic and cyclic acetals or ketals, and the acyclic or cyclic thioacetals or ketals. The protective group is introduced by treating the carbonyl compound in the presence of acid with an alcohol, diol, thiol, or dithiol. Cyclic and acyclic acetals and ketals are stable to aqueous and nonaqueous bases, to nucleophiles including organometallic reagents, and to hydride reduction. A 1,3-dithiane or 1,3-dithiolane, prepared to protect an aldehyde, is converted by strong base (such as BuLi) to an anion. The oxygen derivatives are stable to neutral and basic catalytic reduction, as well as to reduction by sodium in ammonia. Although the sulfur analogs poison hydrogenation catalysts, they can be cleaved by Raney Ni and by sodium/ammonia. The oxygen derivatives are stable to most oxidants; the sulfur derivatives are cleaved by a wide range of oxidants. The oxygen, but not the sulfur, analogs are readily cleaved by acidic hydrolysis. Sulfur derivatives are cleaved under neutral conditions by mercury(II), silver(I), or copper(II) salts as well as a variety of oxidants; oxygen analogs are stable to those conditions. The properties of oxygen and sulfur derivatives are combined in the cyclic 1,3-oxathianes and 1,3-oxathiolanes.

The carbonyl group forms a number of other very stable derivatives. They are less used as protective groups because of the greater difficulty involved in their removal or because of stability issues. Such derivatives include cyanohydrins, hydrazones, imines, oximes, and semicarbazones. Enol ethers are used to protect one carbonyl group in a 1,2- or 1,3-dicarbonyl compound.

Although IUPAC no longer uses the term "ketal," we have retained it to indicate compounds formed from ketones.

Derivatives of carbonyl compounds that have been used as protective groups in synthetic schemes are described in this chapter; some of the more important protective groups are listed in Reactivity Chart 5.[3-5]

1. H. J. E. Loewenthal, *Tetrahedron*, **6**, 269 (1959).
2. E. H. Cordes and H. G. Bull, *Chem. Rev.*, **74**, 581 (1974).
3. See also H. J. E. Loewenthal, "Protection of Aldehydes and Ketones," in *Protective Groups in Organic Chemistry*, J. F. W. McOmie, Ed., Plenum, New York and London, 1973, pp. 323–402.
4. J. F. W. Keana, in *Steroid Reactions*, C. Djerassi, Ed., Holden-Day, San Francisco, 1963, pp. 1–66, 83–87.
5. P. J. Kociensk, *Protecting Groups*, 3rd ed., G. Thieme, New York, 2004, Chapter 2.

ACETALS AND KETALS

Acyclic Acetals and Ketals

Methods similar to those used to form and cleave dimethyl acetal and ketal derivatives can be used for other dialkyl acetals and ketals.[1]

Dimethyl Acetals and Ketals: $R_2C(OCH_3)_2$ (Chart 5)

Formation

The formation of dimethyl acetals is relatively easy. In most cases, the reaction of an aldehyde with an acid in the presence of a water scavenger such as trimethylorthoacetate or trimethylorthoformate will give the acetal in excellent yield.

1. MeOH, dry HCl, 2 min.[2]

Photochemically generated HCl from chloranil has been shown to be an effective catalyst system for the formation of dimethyl acetals but less so for ketals.[3]

2. MeOH, pyridinium tosylate, 3 h, 55°C, 89% yield.[4] In this case the steric crowding imposed by the ethyl group drives the selectivity.

3. DCC-SnCl$_4$; ROH, (CO$_2$H)$_2$, 90% yield.[5]
4. CH(OMe)$_3$, MeNO$_2$, CF$_3$COOH, reflux, 4 h, 81–93% yield.[6] This procedure was reported to be particularly effective for the preparation of ketals of diaryl ketones.
5. MeOH, LaCl$_3$, (MeO)$_3$CH, 25°C, 10 min, 80–100% yield.[7] Dimethyl acetals can be prepared efficiently under neutral conditions by catalysis with lanthanide halides, but the results of the reaction with ketones are unpredictable.
6. LiBF$_4$, ROH, (MeO)$_3$CH, reflux, 72–100% yield. Aromatic ketones and aldehydes react more slowly, but are efficiently derivatized.[8]
7. Cu(BF$_4$)$_2$·xH$_2$O, MeOH, trimethylorthoformate, rt, 78–95% yield. Aldehydes are more reactive than ketones but with insufficient chemoselectivity to be useful.[9]
8. Me$_3$SiOCH$_3$, Me$_3$SiOTf, CH$_2$Cl$_2$, −78°C, 86% yield.[10] The use of TMSOFs to catalyze this transformation has also been demonstrated.[11] A norbornyl ketone was not ketalized under these conditions.
9. (MeO)$_3$CH, anhydrous MeOH, TsOH, reflux, 2 h.[12] Diethyl ketals have been prepared under similar conditions (EtOH, TsOH, 0–23°C, 15 min to 6 h, 80–95% yield) in the presence of molecular sieves to shift the equilibrium by adsorbing water.[13] Amberlyst-15,[14] sulfamic acid,[15] or graphite bisulfate[16] and (EtO)$_3$CH have been used to prepare diethyl ketals.

In the following example a mixture of the *cis*- and *trans*-decalones is converted completely to the *cis*- isomer, in general the thermodynamically less favored isomer.[17]

Trimethylorthoformate in MeOH under 0.8 GPa has been used to prepare dimethylacetals with out the aid of an acid catalyst.[18]

10. MeOH, (MeO)$_4$Si, dry HCl, 25°C, 3 days.[19]

11. MeOH, acidic ion-exchange resin, 7–86% yield.[20]

12. (MeO)$_3$CH, Montmorillonite clay K-10, 5 min to 15 h, >90% yield.[21] Diethyl ketals have been prepared in satisfactory yield by reaction of the carbonyl compound and ethanol in the presence of Kaolinitic clay.[22] SO$_3$H–silica has been used as a solid acid catalyst.[23]

13. MeOH, Ce$^+$-exchanged Montmorillonite clay, 25°C, 0.5–12 h, 18–99% yield. Aldehydes can be selectively protected in the presence of ketones.[24]

14. MeOH, NH$_4$Cl, reflux, 1.5 h, 66% yield.[25]

15. Hydrogenation of enones in MeOH with Pd–C resulted in acetal formation. This is most likely due to the fact that some forms of Pd–C contain PdCl$_2$, which, upon reduction with hydrogen, releases HCl, which actually catalyzes ketal formation (see section on TBDMS and TES ethers). When ethylene glycol/THF is used as solvent, the related dioxolane is formed in 86% yield.[26] Ketal formation is probably caused by the now well-documented residual acid or PdCl$_2$ in some lots of Pd–C that is converted to HCl by hydrogenation.

16. I$_2$, MeOH, rt, 80–99% yield. As in the above case, the cyclohexanone which is sterically less encumbered reacts preferentially.[27]

17. MeOH, PhSO$_2$NHOH, 25°C, 15 min, 75–85% yield.[28]

18. Allyl bromide, Sb(OEt)$_3$, 80°C, 2–6 h, 85–98% yield.[29] This method is chemoselective for aldehydes in the presence of ketones.

19. Sc(NTf)$_3$, HC(OCH$_3$)$_3$ (TMOF), toluene, 0°C, 0.5 h, 92% yield.[30]

20. CeCl$_3$·7H$_2$O, MeOH, TMOF.[31]

21. WCl$_6$, MeOH rt, neat, 35–96% yield.[32]

22. CoCl$_2$, MeOH, reflux, 52–96% yield. 67–97% yield. Aldehydes are protected in the presence of ketones.[33] The use of RuCl$_3$,[34] or TiO$_2$/SO$_4$[2,35] give similar results.

23. Me_2SO_4, $2N$ NaOH, MeOH, H_2O, reflux, 30 min, 85% yield.[36] In this case the hemiacetal of phthaldehyde is alkylated with methyl sulfate; this use is probably restricted to cases that are stable to the strongly basic conditions.

24.

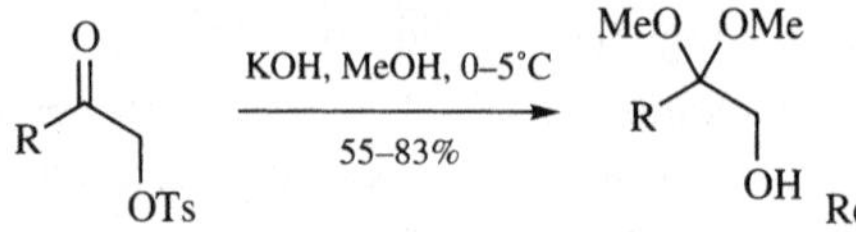

KOH, MeOH, 0–5°C

55–83%

Ref. 37

Cleavage

The acid-catalyzed cleavage of acetals and ketals is greatly influenced by the substitution on the acetal or ketal carbon atom. The following values for k_H^+ illustrate the magnitude of the effect[38]:

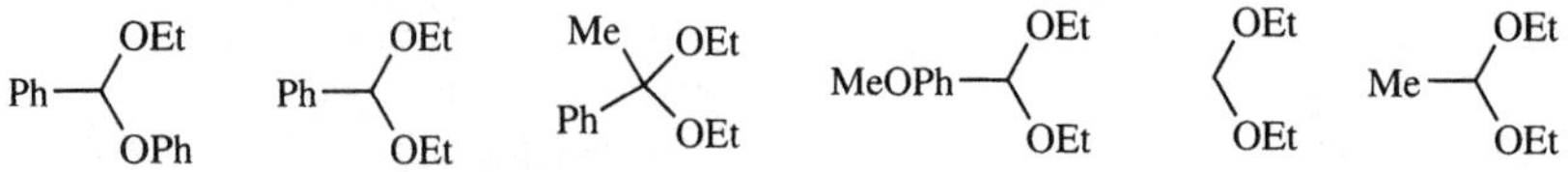

| 41 | 160 | 6×10^3 | 5×10^3 | 1.5×10^{-4} | 1.6 |

1. 50% CF_3COOH, $CHCl_3$, H_2O, 0°C, 90 min, 96% yield.[39]

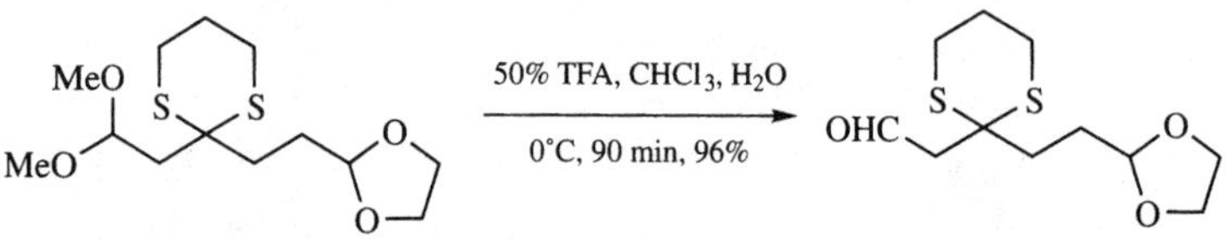

50% TFA, $CHCl_3$, H_2O

0°C, 90 min, 96%

2. TsOH, acetone.[40]

3. $LiBF_4$, wet CH_3CN, 96% yield. Unsubstituted 1,3-dioxolanes are hydrolyzed only slowly, but substituted dioxolanes are completely stable.[41] This reagent proved excellent for hydrolysis of the dimethyl ketal in the presence of the acid-sensitive oxazolidine[42] and polyene.[43]

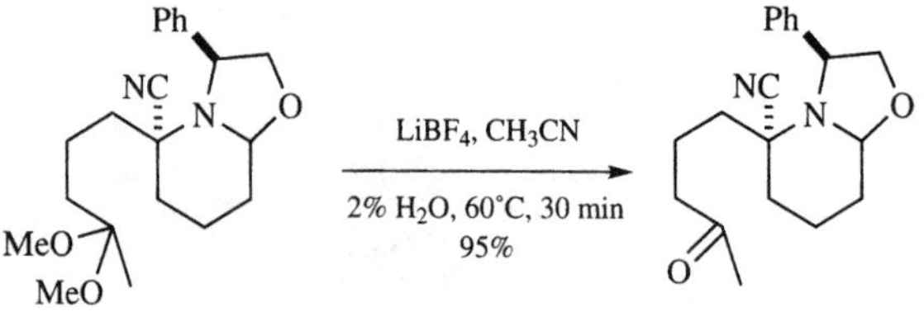

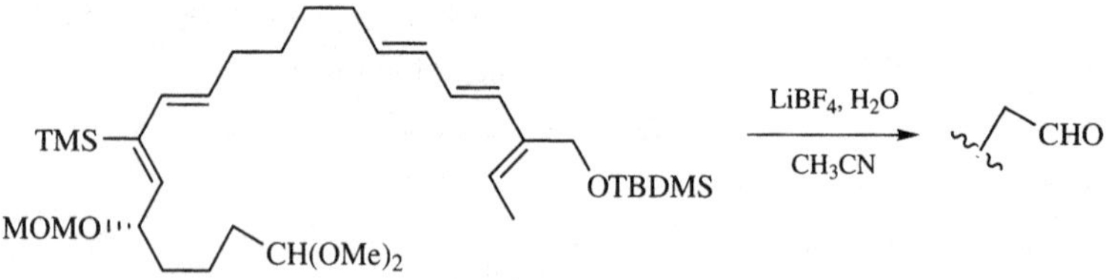

$LiBF_4$, CH_3CN

2% H_2O, 60°C, 30 min
95%

$LiBF_4$, H_2O

CH_3CN

4. HCO_2H, pentane, 1 h, 20°C.[44] Under these conditions a β-γ-double bond does not migrate into conjugation.

5. Amberlyst-15, acetone, H_2O, 20h.[45] Aldehyde acetals conjugated with electron withdrawing groups tend to be slow to hydrolyze. The use of HCl/THF or PPTS/acetone in the case below was slow and caused considerable isomerization. A TBDMS group is stable under these conditions.[46]

6. 70% H_2O_2, Cl_3CCO_2H, CH_2Cl_2, t-BuOH; dimethyl sulfide, 80% yield.[47] Other methods cleaved the epoxide. This method also cleaves the THP and trityl groups.

7. CF_3COOH, rt; $NaHCO_3$, 98% yield.[48]

8. AcOH, H_2O, 89% yield.[49] A factor of 400 in the relative rate of hydrolysis is attributed to a conformational effect where the lone pair on oxygen in the silyl ketals does not overlap with the incipient cation during hydrolysis. Hydrolysis of the second ketal is retarded by the enone, which destabilizes the intermediate carbenium ion.

9. Oxalic acid, THF, H_2O, rt, 12 min, 72% yield.[50]

10. 10% H_2O, silica gel, CH_2Cl_2, 18h, rt.[51] In this example attempts to use HCl resulted in THP cleavage followed by cyclization to form a furan.

E-isomer is also formed

11. DMSO, H_2O, dioxane, reflux, 12 h, 65–99% yield.[52] These conditions cleave a dimethyl ketal in the presence of a t-butyldimethylsilyl ether.

12. The direct conversion of dimethyl ketals to other carbonyl protected derivatives is also possible. Treatment of a dimethyl ketal with $HSCH_2CH_2SH$, $TeCl_4$, $ClCH_2CH_2Cl$ gives the dithiolane in 99% yield.[53]

13. [Ru(ACN)$_3$(triphos)](OTf)$_2$, acetone, rt, 5 h 99% yield.[54] Dioxolanes are also cleaved when not conjugated as in the case below. Nonphenolic THP groups and dioxolane ketals are stable.

14. DDQ, MeCN, H_2O, rt, 75–92% yield.[55] It was shown that this reaction does not proceed through acid catalysis by the hydroquinone.

15. Me_3SiI, CH_2Cl_2, 25°C, 15 min, 85–95% yield.[56] Under these cleavage conditions 1,3-dithiolanes, alkyl and trimethylsilyl enol ethers, and enol acetates are stable. 1,3-Dioxolanes give complex mixtures. Alcohols, epoxides, trityl, t-butyl, and benzyl ethers and esters are reactive. Most other ethers and esters, amines, amides, ketones, olefins, acetylenes, and halides are expected to be stable.

16. $ISiCl_3$, rt, 20–30 min, 74–95% yield.[57] Esters and phenolic methyl ethers are reported to survive, whereas with the related TMSI they are cleaved.

17. SiH_2I_2, CH_3CN, −42°C, 3–40 min, 90–100% yield. Other ketals are also cleaved under these conditions.[58]

18. $ZnCl_2$, Me_2S, AcCl, THF, 89% yield.[59] A dimethyl acetal is chemoselectively cleaved in the presence of a dioxolane acetal.

19. $FeCl_3 \cdot SiO_2$, acetone, rt, 50 min, 80% yield.[60]

20. $Na_2S_2O_4$, THF, H_2O, 90% yield.[61]

21. Me_2BBr, CH_2Cl_2, $-78°C$, 45 min, 100% yield. These conditions were chosen when conventional acid-catalyzed hydrolysis resulted in aldehyde epimerization during a kainic acid synthesis.[62]

22. I_2, acetone, rt, 5–45 min, 93–98% yield. A t-Bu ether is stable to these conditions.[63]

23. $Bi(NO_3)_3 \cdot 5H_2O$, CH_2Cl_2, 76–98% yield. This method works for ketals and acetals that can delocalize a positive charge such as aromatic acetals.[64]

24. Decaborane in aqueous THF, >92% yield. The method only works for acetals that are electron-rich. Aromatic acetals with electron withdrawing groups fail to react thus providing some chemoselectivity.[65] Decaborane can also be used for the formation of dimethyl acetals.

25. $TMSN(SO_2F)_2$, CH_2Cl_2, $-78°C$, 79–96% yield. The reaction proceeds by a unique mechanism with methyl ether as the by product. Dioxolanes are also cleaved but the reaction requires 0°C to go to completion thus a selective deprotection is in principle possible.[66]

26. TESOTf, 2,6-lutidine, CH_2Cl_2, 0°C, 5 min, 50–93% yield. This is an unusual method in that deprotection occurs under basic conditions. The reaction is selective for the cleavage of acetals over ketals with excellent chemoselectivity. Similar selectivity is achieved with dioxolanes.[67]

27. The following miscellaneous reagents have been used to cleave dimethyl acetals, but these have not been extensively tested in large molecule synthesis and as such are listed here for completeness. In most cases for the simple systems

studied, the yields tend to be high. Vanadyl(IV) acetate,[68] Er(OTf)$_3$,[69] polymer-supported π-acid,[70] Montmorillonite K10,[71] HM-zeolite,[72] hexagonal mesoporous molecular sieves[73] and titanium cation-exchanged Montmorillonite clay,[74] Mo$_2$(acac)$_2$,[75] acetyl chloride, SmCl$_3$,[76] β-cyclodextrin/H$_2$O,[77] SiO$_2$ and oxalic or sulfuric acid,[78] SnCl$_2$·2H$_2$O, C$_{60}$,[79] TiCl$_4$, LiI,[80] BF$_3$·Et$_2$O, Et$_4$NI.[81]

1. F. A. J. Meskens, *Synthesis*, 501 (1981).

2. A. F. B. Cameron, J. S. Hunt, J. F. Oughton, P. A. Wilkinson, and B. M. Wilson, *J. Chem. Soc.*, 3864 (1953).

3. H. J. P. de Lijser and N. A. Rangel, J. Org. Chem., **69**, 8315 (2004).

4. J. D. White and Y. Choi, *Org. Lett.*, **2**, 2373 (2000).

5. N. H. Andersen and H.-S. Uh, *Synth. Commun.*, **3**, 125 (1973).

6. A. Thurkauf, A. E. Jacobson, and K. C. Rice, *Synthesis*, 233 (1988).

7. A. L. Gemal and J.-L. Luche, *J. Org. Chem.*, **44**, 4187 (1979).

8. N. Hamada, K. Kazahaya, H. Shimizu, and T. Sato, *Synlett*, 1074 (2004).

9. R. Kumar and A. K. Chakraborti, *Tetrahedron Lett.*, **46**, 8319 (2005).

10. M. Vandewalle, J. Van der Eycken, W. Oppolzer, and C. Vullioud, *Tetrahedron*, **42**, 4035 (1986).

11. B. H. Lipshutz, J. Burgess-Henry, and G. P. Roth, *Tetrahedron Lett.*, **34**, 995 (1993).

12. E. Wenkert and T. E. Goodwin, *Synth. Commun.*, **7**, 409 (1977).

13. D. P. Roelofsen, E. R. J. Wils, and H. Van Bekkum, *Recl. Trav. Chim. Pays-Bas*, **90**, 1141 (1971).

14. S. A. Patwardhan and S. Dev, *Synthesis*, 348 (1974).

15. W. Gong, B. Wang, Y. Gu, L. Yan, L. Yang, and J. Suo, *Synth. Commun.*, **34**, 4243 (2004).

16. J. P. Alazard, H. B. Kagan, and R. Setton, *Bull. Soc. Chim. Fr.*, 499 (1977).

17. J. B. P. A. Wijnberg, R. P. W. Kesselmans, and A. de Groot, *Tetrahedron Lett.*, **27**, 2415 (1986).

18. K. Kumamoto, Y. Ichikawa, and H. Kotsuki, *Synlett*, 2254 (2005).

19. W. W. Zajac and K. J. Byrne, *J. Org. Chem.*, **35**, 3375 (1970).

20. N. B. Lorette, W. L. Howard, and J. H. Brown, Jr., *J. Org. Chem.*, **24**, 1731 (1959).

21. E. C. Taylor and C.-S. Chiang, *Synthesis*, 467 (1977). Montmorillonite clay is activated Al$_2$O$_3$/SiO$_2$/H$_2$O. V. M. Thuy and P. Maitte, *Bull. Soc. Chim. Fr.*, 2558 (1975).

22. D. Ponde, H. B. Borate, A. Sudalai, T. Ravindranathan, and V. H. Deshpande, *Tetrahedron Lett.*, **37**, 4605 (1996).

23. K.-i. Shimizu, E. Hayashi, T. Hatamachi, T. Kodama, and Y. Kitayama, *Tetrahedron Lett.*, **45**, 5135 (2004).

24. J.-i. Tateiwa, H. Horiuchi, and S. Uemura, *J. Org. Chem.*, **60**, 4039 (1995).

25. J. I. DeGraw, L. Goodman, and B. R. Baker, *J. Org. Chem.*, **26**, 1156 (1961).

26. P. Hudson and P. J. Parsons, *Synlett*, 867 (1992).

27. M. K. Basu, S. Samajdar, F. F. Becker, and B. K. Banik, *Synlett*, 319 (2002).

28. A. Hassner, R. Wiederkehr, and A. J. Kascheres, *J. Org. Chem.*, **35**, 1962 (1970).

29. Y. Liao, Y.-Z. Huang, and F.-H. Zhu, *J. Chem. Soc., Chem. Commun.*, 493 (1990).

30. K. Ishihara, Y. Karumi, M. Kubota, and H. Yamamoto, *Synlett* 839 (1996).

31. A. B. Smith, III, M. Fukui, H. A. Vaccaro, and J. R. Empfield, *J. Am. Chem. Soc.*, **113**, 2071 (1991).

32. H. Firouzabadi, N. Iranpoor, and B. Karimi, *Synth. Commum.*, **29**, 2255 (1999).

33. S. Velusamy and T. Punniyamurthy, *Tetrahedron Lett.*, **45**, 4917 (2004).

34. S. K. De and R. A. Gibbs, *Tetrahedron Lett.*, **45**, 8141 (2004).

35. Y.-R. Ma, T.-S. Jin, S.-X. Shi, and T.-S. Li, *Synth. Commum.*, **33**, 2103 (2003).

36. E. Schmitz, *Chem. Ber.*, **91**, 410 (1958).

37. O. Prakash, N. Saini, and P. K. Sharma, *J. Chem. Res., Synop.*, 430 (1993).

38. D. P. N. Satchell and R. S. Satchell, *Chem. Soc. Rev.*, **19**, 55 (1990).

39. R. A. Ellison, E. R. Lukenbach, and C.-W. Chiu, *Tetrahedron Lett.*, **16**, 499 (1975).

40. E. W. Colvin, R. A. Raphael, and J. S. Roberts, *J. Chem. Soc., Chem. Commun.*, 858 (1971).

41. B. H. Lipshutz and D. F. Harvey, *Synth. Commun.*, **12**, 267 (1982).

42. M. Bonin, J. Royer, D. S. Grierson, and H.-P. Husson, *Tetrahedron Lett.*, **27**, 1569 (1986).

43. W. R. Roush and R. J. Sciotti, *J. Am. Chem. Soc.*, **116**, 6457 (1994).

44. F. Barbot and P. Miginiac, *Synthesis*, 651 (1983).

45. G. M. Cappola, *Synthesis*, 1021 (1984).

46. A. E. Greene, M. A. Teixeira, E. Barreiro, A. Cruz, and P. Crabbé, *J. Org. Chem.*, **47**, 2553 (1982).

47. A. G. Meyers, M. A. M. Fundy, and P. A. Linstrom, Jr., *Tetrahedron Lett.*, **29**, 5609 (1988).

48. J. J. Tufariello and K. Winzenberg, *Tetrahedron Lett.*, **27**, 1645 (1986).

49. A. J. Stern and J. S. Swenton, *J. Org. Chem.*, **54**, 2953 (1989).

50. D. A. Evans, S. P. Tanis, and D. J. Hart, *J. Am. Chem. Soc.*, **103**, 5813 (1981).

51. L. Crombie and D. Fisher, *Tetrahedron Lett.*, **26**, 2477 (1985).

52. T. Kametani, H. Kondoh, T. Honda, H. Ishizone, Y. Suzuki, and W. Mori, *Chem. Lett.*, **18**, 901 (1989); K. R. Muralidharan, M. K. Mokhallalati, and L. N. Pridgen, *Tetrahedron Lett.*, **35**, 7489 (1994).

53. H. Tani, K. Masumoto, and T. Inamasu, *Tetrahedron Lett.*, **32**, 2039 (1991).

54. S. Ma and L. M. Venanzi, *Tetrahedron Lett.*, **34**, 8071 (1993).

55. K. Tanemura, T. Suzuki, and T. Horaguchi, *J. Chem. Soc., Chem. Commun.*, 979 (1992); A. Oku, M. Kinugasa, and T. Kamada, *Chem. Lett.*, **22**, 165 (1993); B. Karimi and A. M. Ashtiani, *Chem. Lett.*, **28**, 1199 (1999).

56. M. E. Jung, W. A. Andrus, and P. L. Ornstein, *Tetrahedron Lett.*, **18**, 4175 (1977).

57. S. S. Elmorsy, M. V. Bhatt, and A. Pelter, *Tetrahedron Lett.*, **33**, 1657 (1992).

58. E. Keinan, D. Perez, M. Sahai, and R. Shvily, *J. Org. Chem.*, **55**, 2927 (1990).

59. C. Chang, K. C. Chu, and S. Yue, *Synth. Commun.*, **22**, 1217 (1992).

60. T. Nishimata, Y. Sato, and M. Mori, *J. Org. Chem.*, **69**, 1837 (2004).

61. K. A. Parker and D.-S. Su, *J. Org. Chem.*, **61**, 2191 (1996).

62. S. Hanessian and S. Ninkovic, *J. Org. Chem.*, **61**, 5418 (1996).

63. J. Sun, Y. Dong, L. Cao, X. Wang, S. Wang, and Y. Hu, *J. Org. Chem.*, **69**, 8932 (2004).

64. K. J. Eash, M. S. Pulia, L. C. Wieland, and R. S. Mohan, *J. Org. Chem.*, **65**, 8399 (2000).

65. S. H. Lee, J. H. Lee, and C. M. Yoon, *Tetrahedron Lett.*, **43**, 2699 (2002).

66. G. Kaur, A. Trehan, and S. Trehan, *J. Org. Chem.*, **63**, 2365 (1998).

67. H. Fujioka, Y. Sawama, N. Murata, T. Okitsu, O. Kubo, S. Matsuda, and Y. Kita, *J. Am. Chem. Soc.*, **126**, 11800 (2004).

68. M. L. Kantam, V. Neeraja, and P. Sreekanth, *Catal. Commun.*, **2**, 301 (2001).

69. R. Dalpozzo, A. De Nino, L. Maiuolo, M. Nardi, A. Procopio, and A. Tagarelli, *Synthesis*, 496 (2004).

70. N. Tanaka and Y. Masaki, *Synlett*, 1960 (1999).

71. E. C. L. Gautier, A. E. Graham, A. McKillop, S. P. Standen, and R. J. K. Taylor, *Tetrahedron Lett.*, **38**, 1881 (1997).

72. M. N. Rao, P. Kumar, A. P. Singh, and R. S. Reddy, *Synth. Commun.*, **22**, 1299 (1992).

73. K.-Y. Ko, S.-T. Park, and M.-J. Choi, *Bull. Korean Chem.l Soc.*, **21**, 951 (2000).

74. T. Kawabata, M. Kato, T. Mizugaki, K. Ebitani, and K. Kaneda, *Chem. Lett.*, **32**, 648 (2003).

75. M. L. Kantam, V. Swapna and P. L. Santhi, *Synth. Commun.*, **25**, 2529 (1995).

76. S.-H. Wu and Z.-B. Ding, *Synth. Commun.*, **24**, 2173 (1994).

77. N. S. Krishnaveni, K. Surendra, M. A. Reddy, Y. V. D. Nageswar, and K. R. Rao, *J. Org. Chem.*, **68**, 2018 (2003).

78. F. Huet, A. Lechevallier, M. Pellet, and J. M. Conia, *Synthesis*, 63 (1978).

79. K. L. Ford and E. J. Roskamp, *J. Org. Chem.*, **58**, 4142 (1993); K. L. Ford and E. J. Roskamp, *Tetrahedron Lett.*, **33**, 1135 (1992).

80. G. Balme and J. Goré, *J. Org. Chem.*, **48**, 3336 (1983).

81. A. K. Mandal, P. Y. Shrotri, and A. D. Ghogare, *Synthesis*, 221 (1986).

Diisopropyl Acetal: $(i\text{-PrO})_2\text{CHR}$

Formation

$\text{CH}(\text{O}i\text{-Pr})_3$, CSA, IPA, removal of i-PrOH by distillation, 3 h, 68–92% yield.[1,2]

Cleavage

Formic acid, THF, H_2O, 20°C, 100% yield. This acetal was chosen to prevent conjugation of a double bond during hydrolysis, which occurred when the corresponding dimethyl acetal was hydrolyzed.[1]

1. J. Sandri and J. Viala, *Synthesis*, 271 (1995).

2. A. Pommier, J.-M. Pons, and P. J. Kocienski, *J. Org. Chem.*, **60**, 7334 (1995).

Bis(2,2,2-trichloroethyl) Acetals and Ketals: $R_2C(\text{OCH}_2\text{CCl}_3)_2$ (Chart 5)

Formation[1]

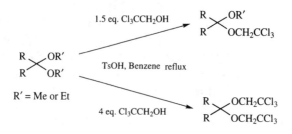

It is more efficient to prepare this ketal by an exchange reaction with the dimethyl or diethyl ketal than directly from the carbonyl compound. Hydrolysis can also be affected by acid catalysis.

Cleavage

Zn/EtOAc or THF, reflux, 3–12 h, 40–100% yield.[1]

1. J. L. Isidor and R. M. Carlson, *J. Org. Chem.*, **38**, 554 (1973).

Dibenzyl Acetals and Ketals: $R_2C(OCH_2Ph)_2$

Formation

1. From a thioacetal:[1]

2. $BnOSiMe_3$, $FeCl_3$, 2 h, 0°C, CH_2Cl_2, 20–97% yield.[2]

Cleavage

1. Cleavage is accomplished by hydrogenolysis (Pd–C, MeOH, 3 h).[1]
2. Acid-catalyzed hydrolysis may also be used to regenerate the aldehyde or ketone.

1. J. H. Jordaan and W. J. Serfontein, *J. Org. Chem.*, **28**, 1395 (1963).
2. T. Watahiki, Y. Akabane, S. Mori, and T. Oriyama, *Org. Lett.*, **5**, 3045 (2003).

Bis(2-nitrobenzyl) Acetals and Ketals: $R_2C(OCH_2C_6H_4-2-NO_2)_2$

Formation

$2-NO_2C_6H_4CH_2OSiMe_3$, Me_3SiOTf, −78°C, 78–95% yield.[1]

Cleavage

Photolysis at 350 nm, 85–95% yield.[1]

1. D. Gravel, S. Murray, and G. Ladouceur, *J. Chem. Soc., Chem. Commun.*, 1828 (1985).

Diacetyl Acetals and Ketals: $R_2C(OAc)_2$

Although there are numerous methods for the protection and deprotection of diacetyl acetals, these are rarely used in synthesis as protective groups, but have been used as starting materials for palladium-catalyzed alkylations.[1]

Formation

Acylals are, in general, easily formed by the reaction of an aldehyde with Ac_2O and a Brønsted or Lewis acid. The protection process usually proceeds at rt in yields ranging from about 50–99%. The following catalysts have been used for the preparation of acylals: 1 drop concd. H_2SO_4,[2] $ZnCl_2$,[3] $Zn(BF_4)_2$,[4] $FeCl_3$,[5,6] PCl_3,[7] Nafion H,[8] expansive Graphite,[9] β-Zeolite,[10] Environcat EPZG,[11] HY-zeolite,[12] Amberlyst-15,[13] I_2,[14] $Cu(BF_4)_2$,[15] $Cu(OTf)_2$,[16] $Sc(OTf)_3$,[17] $Bi(OTf)_3$,[18] $Bi(NO_3)_3$,[19] $AlPW_{12}O_{40}$,[20] $InBr_3$,[21] $InCl_3$,[22] $LiBF_4$,[23] $Zr(CH_3PO_3)_{1.2}(O_3PC_6H_4SO_3H)_{0.8}$,[24] $Zr(SO_4)_2 \cdot 4H_2O/SiO_2$,[25] Wells–Dawson acid,[26] Mo/TiO_2–ZrO_2,[27] ceric ammonium nitrate,[28] N-bromosuccinimide,[29] In general, these methods are aldehyde selective with ketones being unreactive.

Cleavage

As with the acetate group, acylals are readily hydrolyzed with base and the reagents used to cleave an acetate for the most part should cleave an acylal. Cleavage reactions are quite efficient with yields generally exceeding 80%. The use of enzymes for the hydrolysis of acylals is effective and in the case of racemic derivatives some enantioenrichment of the aldehyde is possible.[30] The following reagents have been used for the cleavage of acylals: NaOH or K_2CO_3,[5] alumina,[31] $AlCl_3$,[32] $BiCl_3$,[33] potassium 3-dimethylaminophenoxide,[34] expansive Graphite,[35] Zeolite Y,[36] Envirocat EPZG,[37] CAN, silica gel,[38] Montmorillonite clay K 10 or KSF,[39] $InBr_3$/polyethylene glycol,[40] $Fe_2(SO_4)_3 \cdot xH_2O$,[41] Well–Dawson heteropolyacid,[42] CBr_4,[43] $SnCl_2 \cdot 2H_2O$,[44] $NaHSO_4$, polyethylene glycol.[45]

1. B. M. Trost and C. B. Lee, *J. Am. Chem. Soc.*, **123**, 3671 (2001).

2. M. Tomita, T. Kikuchi, K. Bessho, T. Hori, and Y. Inubushi, *Chem. Pharm. Bull.*, **11**, 1484 (1963).

3. I. Scriabine, *Bull. Soc. Chim. Fr.*, 1194 (1961).

4. B. C. Ranu, J. Dutta and A. Das, *Chem. Lett.*, **32**, 366 (2003).

5. K. S. Kochhar, B. S. Bal, R. P. Deshpande, S. N. Rajadhyaksha, and H. W. Pinnick, *J. Org. Chem.*, **48**, 1765 (1983).

6. J. Kula, *Synth. Commun.*, **16**, 833 (1986).

7. J. K. Michie and J. A. Miller, *Synthesis*, 824 (1981).

8. G. A. Olah and A. K. Mehrotra, *Synthesis*, 962 (1982).

9. T.-S. Jin, G.-Y. Du, Z.-H. Zhang, and T.-S. Li, *Synth. Commun.*, **27**, 2261 (1997).

10. P. Kumar, V. R. Hedge, and J. T. P. Kumar, *Tetrahedron Lett.*, **36**, 601 (1995).

11. B. P. Bandgar, N. P. Mahajan, D. P. Mulay, J. L. Thote, and P. P. Wadgaonkar, *J. Chem. Res., Synop.*, 470 (1995).

12. C. Pereira, B. Gigante, M. J. Marcelo-Curto, H. Carreyre, G. Pérot, and M. Guisnet, *Synthesis*, 1077 (1995).

13. A. V. Reddy, K. Ravinder, V. L. N. Reddy, V. Ravikanth, and Y. Venkateswarlu, *Synth. Commun.*, **33**, 1531 (2003).

14. N. Deka, D. J. Kalita, R. Borah, and J. C. Sarma, *J. Org. Chem.*, **62**, 1563 (1997).

15. A. K. Chakraborti, R. Thilagavathi, and R. Kumar, *Synthesis*, 831 (2004).

16. K. L. Chandra, P. Saravanan, and V. K. Singh, *Synlett*, 359 (2000).

17. V. K. Aggarwal, S. Fonquerna, and G. P. Vennall, *Synlett*, 849 (1998).

18. M. D. Carrigan, K. J. Eash, M. C. Oswald, and R. S. Mohan, *Tetrahedron Lett.*, **42**, 8133 (2001).

19. D. H. Aggen, J. N. Arnold, P. D. Hayes, N. J. Smoter, and R. S. Mohan, *Tetrahedron*, **60**, 3675 (2004).

20. H. Firouzabadi, N. Iranpoor, F. Nowrouzi, and K. Amani, *Tetrahedron Lett.*, **44**, 3951 (2003).

21. L. Yin, Z.-H. Zhang, Y.-M. Wang, and M.-L. Pang, *Synlett*, 1727 (2004).

22. M. Salavati-Niasari and S. Hydarzadeh, *J. Mol. Catal. A: Chemical*, **237**, 254 (2005).

23. N. Sumida, K. Nishioka, and T. Sato, *Synlett*, 1921 (2001); J. S. Yadav, B. V. S. Reddy, C. Venugopal, and T. Ramalingam, *Synlett*, 604 (2002).

24. M. Curini, F. Epifano, M. C. Marcotullio, O. Rosati, and M. Nocchetti, *Tetrahedron Lett.*, **43**, 2709 (2002).

25. T. Jin, G. Feng, M. Yang, and T. Li, *Synth. Commum.*, **34**, 1645 (2004).

26. G. P. Romanelli, H. J. Thomas, G. T. Baronetti, and J. C. Autino, *Tetrahedron Lett.*, **44**, 1301 (2003).

27. B. M. Reddy, P. M. Sreekanth, and A. Khan, *Synth. Commum.*, **34**, 1839 (2004).

28. S. C. Roy and B. Banerjee, *Synlett*, 1677 (2002).

29. B. Karimi, H. Seradj, and G. R. Ebrahimian, *Synlett*, 623 (2000).

30. Y. S. Angelis and I. Smonou, *Tetrahedron Lett.*, **38**, 8109 (1997).

31. R. S. Varma, A. K. Chatterjee, and M. Varma, *Tetrahedron Lett.*, **34**, 3207 (1993).

32. G. Sabitha, S. Abraham, T. Ramalingam, and J. S. Yadav, *J. Chem. Res. (S)*, 144 (2002). I. Mohammadpoor-Baltork, and H. Aliyan, *J. Chem. Res. (S)*, 272 (1999).

33. I. Mohammadpoor-Baltork and H. Aliyan, *Synth. Commum.*, **29**, 2741 (1999).

34. Y.-Y. Ku, R. Patel, and D. Sawick, *Tetrahedron Lett.*, **34**, 8037 (1993).

35. T.-S. Jin, Y.-R. Ma, Z.-H. Zhang, and T.-S. Li, *Synth. Commun.*, **27**, 3379 (1997).

36. R. Ballini, M. Bordoni, G. Bosica, R. Maggi, and G. Sartori, *Tetrahedron Lett.*, **39**, 7587 (1998).

37. B. P. Bandgar, S. P. Kasture, K. Tidke, and S. S. Makone, *Green Chem.*, **2**, 152 (2000).

38. P. Cotelle and J.-P. Catteau, *Tetrahedron Lett.*, **33**, 3855 (1992).

39. T.-S. Li, Z.-H. Zhang, and C.-G. Fu, *Tetrahedron Lett.*, **38**, 3285 (1997).

40. Z.-H. Zhang, L. Yin, Y.-M. Wang, J.-Y. Liu, and Y. Li, *Green Chem.*, **6**, 563 (2004); Z.-H. Zhang, L. Yin, Y. Li, and Y.-M. Wang, *Tetrahedron Lett.*, **46**, 889 (2005).

41. L. Li, X. Zhang, G. Zhang, and G. Qu, *J. Chemical Res.*, 39 (2004); T. S. Jin, Y. R. Ma, Z. H. Zhang, and T. S. Li, *Org. Prep. Proc. Int.*, **30**, 463 (1998).

42. G. P. Romanelli, J. C. Autino, G. Baronetti, and H. J. Thomas, *Synth. Commum.*, **34**, 3909 (2004).

43. T. Ramalingam, R. Srinivas, B. V. S. Reddy, and J. S. Yadav, *Synth. Commum.*, **31**, 1091 (2001).

44. I. Mohammadpoor-Baltork and H. Alivan, *Ind. J. Chem., Sect. B*, **38B**, 1223 (1999).

45. Z.-H. Zhang, *Monatsh. Chem.*, **136**, 1191 (2005).

Cyclic Acetals and Ketals

Ring size plays a significant role in the hydrolysis rates (hydrolysis in 0.003 M HCl in 7:3 dioxane–H_2O, 30°C).[1]

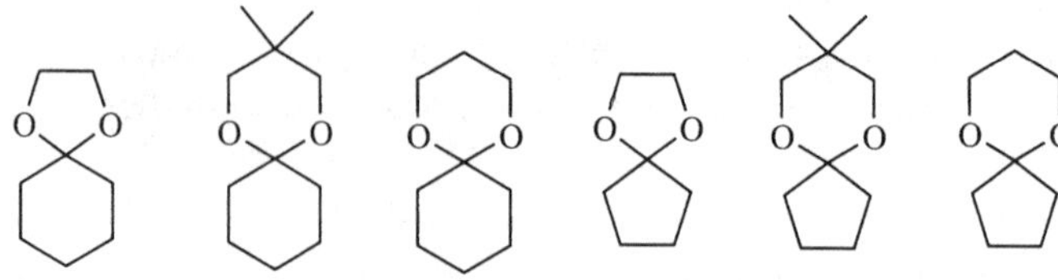

| Relative rate = 1.0 | 2.0 | 30.6 | 13.0 | 15.5 | 172 |

Formation

$$HOCH_2C(CH_3)_2CH_2OH > HO(CH_2)_2OH > HO(CH_2)_3OH$$

Cleavage

For acid-catalyzed hydrolysis the following generalizations apply.

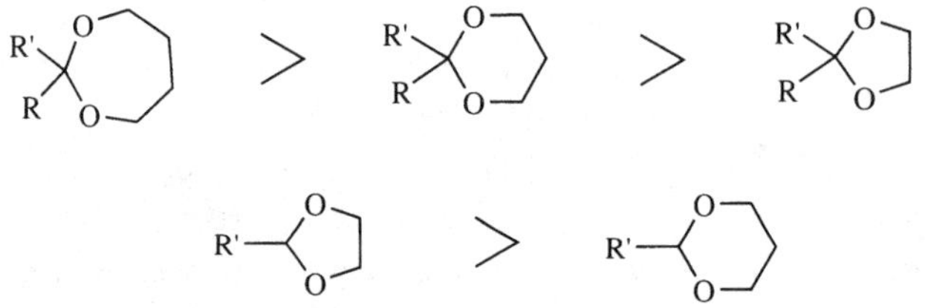

The relative rates of acid-catalyzed hydrolysis of some dioxolanes [dioxolane: aq. HCl (1:1)] are: 2,2-dimethyldioxolane: 2-methyldioxolane: dioxolane, 50,000:5000:1.[2] The following table gives the relative hydrolysis rates for 5α-androstane cyclic ketals in 0.02 N HCl at 37°C.[3]

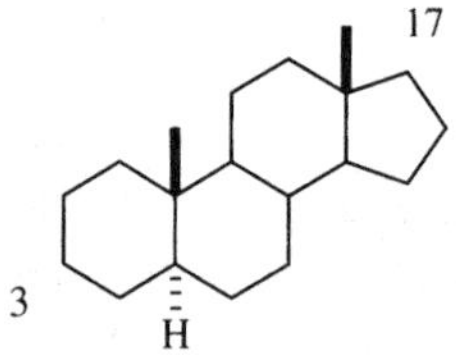

Relative Hydrolysis Rates of α-Androstane Cyclic Ketals in 0.02 N HCl at 37°C

Glycol	3-Ketal	17-Ketal	3-Ketal-17-one	17-ketal-3-one
Ethylene glycol	1.00	1.64	1.06	1.51
1,3-Propanediol	14.5	40.5	13.8	48.3
2,2-Dimethyl-1,3-propanediol	1.52	6.90	1.26	5.24
2,2-Diethyl-1,3-propanediol	0.75	2.63	0.47	2.09

These results show that unsubstituted dioxanes hydrolyze faster than dioxolanes, but that substitution reduces the rate of hydrolysis and that cyclopentanone ketals hydrolyze faster than cyclohexanone derivatives.

A review[4] discusses the condensation of aldehydes and ketones with glycerol to give 1,3-dioxanes and 1,3-dioxolanes. The chemistry of O/O and O/S acetals has been reviewed,[5] and a recent monograph discusses this area of protective groups in a didactic sense.[6]

1. M. S. Newman and R. J. Harper, *J. Am. Chem. Soc.*, **80**, 6350 (1958); S. W. Smith and M. S. Newman, *J. Am. Chem. Soc.*, **90**, 1249, 1253 (1968).

2. P. Salomaa and A. Kankaanperä, *Acta Chem. Scand.*, **15**, 871 (1961).

3. S. W. Smith and M. S. Newman, *J. Am. Chem. Soc.*, **90**, 1249 (1968).

4. A. J. Showler and P. A. Darley, Chem. Rev., **67**, 427 (1967).

5. H. Hagemann and D. Klamann, Eds., *O/O-und O/S-Acetale* [*Methoden Der Organishen Chemie*, Houben-Weyl)] 4th ed., G. Thieme, Stuttgart, 1991, Band E 14a/1.

6. P. J. Kocienski, "Carbonyl Protecting Groups," in *Protecting Groups*, 3rd ed., Thieme Medical Publishers, New York, 2004, Chapter 2.

1,3-Dioxanes (Chart 5)

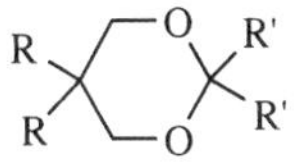

R = H, CH$_3$

The section on the formation of 1,3-dioxolanes should be consulted since many of the methods are also applicable to the formation of 1,3-dioxanes.

Formation

1. HO(CH$_2$)$_3$OH, TsOH, benzene, reflux.[1-3]

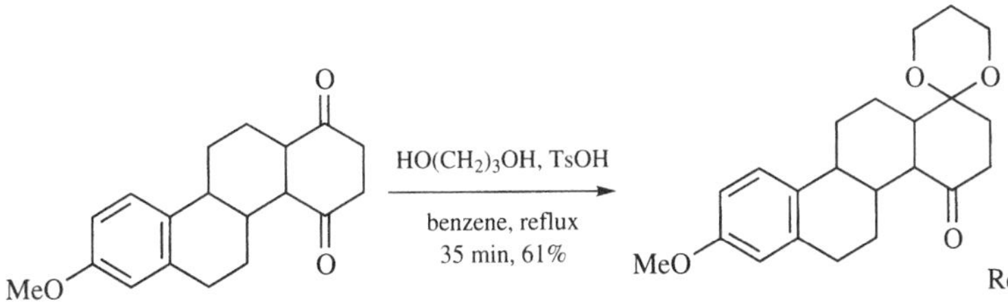

Ref. 1

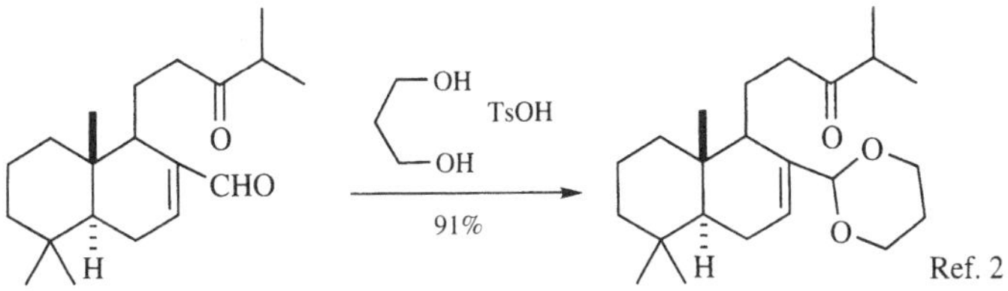

Ref. 2

In the first example selective protection was more successful with 1,3-propanediol than with ethylene glycol.[1]

2. 1,3-Propanediol, THF, Amberlyst-15, 5 min, 50–70% yield.[4] This method is also effective for the preparation of 1,3-dioxolanes.

3. HOCH$_2$C(CH$_3$)$_2$CH$_2$OH, Sc(NTf$_2$)$_3$, toluene, 0°C, 3 h, 87–92% yield.[5]

4.

Other methods for ketalization met with failure.[6]

5. HOCH$_2$CH$_2$CH$_2$OH, (EtO)$_3$CH, NBS, MeOH, CH$_2$Cl$_2$, rt, 6 h, 25–97% yield. As is usually the case, aldehydes are protected faster than ketones.[7]

6. 2-Methoxy-5,5-dimethyl-1,3-dioxane, HOCH$_2$C(CH$_3$)$_2$CH$_2$OH, TsOH, 97% yield.[8]

This method is also effective for the unsubstituted derivative.[9] Protection and TES group hydrolysis occurs without competing dehydration.

7. HOCH$_2$C(CH$_3$)$_2$CH$_2$OH, N-4-methoxybenzyl-2-cyanopyridinium hexafluoroantimonate, toluene, reflux, 1.5–3.7 h, 85–99% yield.[10]

8. TMSOCH$_2$C(CH$_3$)$_2$CH$_2$OTMS, TMSOTf, Pyr, 75% yield.[11] These are kinetically controlled conditions. Iodine[12] and NBS[13] can also be used as a catalyst with this protected diol.

9. HOCH$_2$CH$_2$CH$_2$OH, Ru(CH$_3$CN)$_3$(triphos)(OTf)$_2$, 94–99% yield.[14]

10. HOCH$_2$C(CH$_3$)$_2$CH$_2$OH, sulfated zirconia, benzene, reflux, 88–97% yield.[15]

11. HOCH$_2$C(CH$_3$)$_2$CH$_2$OH, yttria–zirconia, rt, CHCl$_3$, 75–96% yield.[16]

12. From a dithiane: NBS, 1,3-propanediol, DABCO, CH$_2$Cl$_2$, rt, 5 min, 30–97% yield.[17] The method is also applicable to other thioacetals.

13. From a dimethylacetal.[18] This acetal was used because it improved a subsequent epoxidation of the enone. It was later cleaved with 48% HF/CH$_3$CN in 92% yield.

14. HOCH$_2$CH$_2$CH$_2$OH, ZrCl$_4$, (EtO)$_3$CH, rt, CH$_2$Cl$_2$, 52–98% yield. Aldehydes react faster than ketones.[19]

Cleavage

1. For the most part, some form of aqueous acid will cleave these acetals and ketals. The section on the cleavage of 1,3-dioxolanes should be consulted, since a majority of the methods available are applicable to 1,3-dioxanes as well.

2. TMSCl, SmCl$_3$, THF, 71–99% yield. Ketals are cleaved faster than acetals.[20]

1. J. E. Cole, W. S. Johnson, P. A. Robins, and J. Walker, *J. Chem. Soc.*, 244 (1962).

2. H. Okawara, H. Nakai, and M. Ohno, *Tetrahedron Lett.*, **23**, 1087 (1982).

3. For examples on the use of the related 4,4-dimethyl-1,3-dioxane, see E. Piers, J. Banville, C. K. Lau, and I. Nagakura, *Can. J. Chem.*, **60**, 2965 (1982); M. A. Avery, C. Jennings-White, and W. K. M. Chong, *Tetrahedron Lett.*, **28**, 4629 (1987).

4. A. E. Dann, J. B. Davis, and M. J. Nagler, *J. Chem. Soc., Perkin Trans. I*, 158 (1979).

5. K. Ishihara, Y. Karumi, M. Kubota, and H. Yamamoto, *Synlett*, 839 (1996).

6. L. A. Paquette and S. Borrelly, *J. Org. Chem.*, **60**, 6912 (1995).

7. B. Karimi, G. R. Ebrahimian, and H. Seradj, *Org. Lett.*, **1**, 1737 (1999).

8. J. D. White, F. W. J. Demnitz, H. Oda, C. Hassler, and J. P. Snyder, *Org. Lett.*, **2**, 3313 (2000).

9. D. S. Coffey, A. I. McDonald, L. E. Overman, and F. Stappenbeck, *J. Am. Chem. Soc.*, **121**, 6944 (1999).

10. S.-B. Lee, S.-D. Lee, T. Takata, and T. Endo, *Synthesis*, 368 (1991).

11. C. K. F. Chiu, L. N. Mander, A. D. Stuart, and A. C. Willis, *Aust. J. Chem.*, **45**, 227 (1992).

12. B. Karimi and B. Golshani, *Synthesis*, 784 (2002).

13. B. Karimi, H. Hazarkhani, and J. Maleki, *Synthesis*, 279 (2005).

14. S. Ma and L. M. Venanzi, *Synlett*, 751 (1993).

15. A. Sakar, O. S. Yemul, B. P. Bandgar, N. B. Gaikwad, and P. P. Wadgaonkar, *Org. Prep. Proced. Int.*, **28**, 613 (1996).

16. G. C. G. Pals, A. Keshavaraja, K. Saravanan, and P. Kumar, *J. Chem. Res., Synop.*, 426 (1996).

17. B. Karimi, H. Seradj, and J. Maleki, *Tetrahedron*, **58**, 4513 (2002).

18. C. Li, E. A. Pace, M.-C. Liang, E. Lobkovsky, T. D. Gilmore, and J. J. A. Porco, *J. Am. Chem. Soc.*, **123**, 11308 (2001).

19. H. Firouzabadi, N. Iranpoor, and B. Karimi, *Synlett*, 321 (1999).

20. Y. Ukaji, N. Koumoto, and T. Fujisawa, *Chem. Lett.*, **18**, 1623 (1989).

5-Methylene-1,3-dioxane (Chart 5)

Formation[1]

$CH_2=C(CH_2OH)_2$, TsOH, benzene, reflux, 90% yield.

Cleavage[1]

The rhodium-catalyzed isomerization can also be carried out with the chiral catalyst, $Ru_2Cl_4(diop)_3$ (H_2, 20–80°C, 1–6 h, 47–90% yield) or with $NiBr_2Diop/LiB$-HEt_3.[2] In this case, optically enriched enol ethers are obtained.[3] The section on allyl ethers should be consulted for other methods of isomerization.

1. E. J. Corey and J. W. Suggs, *Tetrahedron Lett.*, **16**, 3775 (1975).

2. S. Flock and H. Frauenrath, *Synlett*, 839 (2001).

3. H. Frauenrath and M. Kaulard, *Synlett*, 517 (1994).

5-Trimethylsilyl-1,3-dioxane (cyclo-SEM)

Formation

TMSCH(CH$_2$OH)$_2$, CSA, 3-Å ms, rt, 45–97% yield. Attempts to force recalcitrant reactions to completion by heating fails as a result of diol decomposition through the Peterson olefination process.[1]

Cleavage

1. BF$_3$·Et$_2$O, THF.
2. LiBF$_4$, THF, 66°C, reflux, 71–93% yield. The use of LiBH$_4$, CH$_3$CN was found not to be selective because these conditions will cleave 1,3-dioxanes and dioxolanes. Other fluoride sources that fail to cleave the cyclo-SEM group include TBAF, CsF, and Bu$_4$NBF$_4$.

1. B. H. Lipshutz, P. Mollard, C. Lindsley, and V. Chang, *Tetrahedron Lett.*, **38**, 1873 (1997).

5,5-Dibromo-1,3-dioxane (Chart 5)

Formation

Br$_2$C(CH$_2$OH)$_2$, TsOH, benzene, heat for several hours, 84–94% yield.[1]

Cleavage

Zn–Ag, THF, AcOH, 25°C, 1 h, ~90% yield.[1]

1. E. J. Corey, E. J. Trybulski, and J. W. Suggs, *Tetrahedron Lett.*, **17**, 4577 (1976).

5-(2-Pyridyl)-1,3-dioxane

Formation/Cleavage[1]

This group is stable to 0.1 *M* HCl.

1. A. R. Katritzky, W.-Q. Fan, and Q.-L. Li, *Tetrahedron Lett.*, **28**, 1195 (1987).

Salicylate Acetals

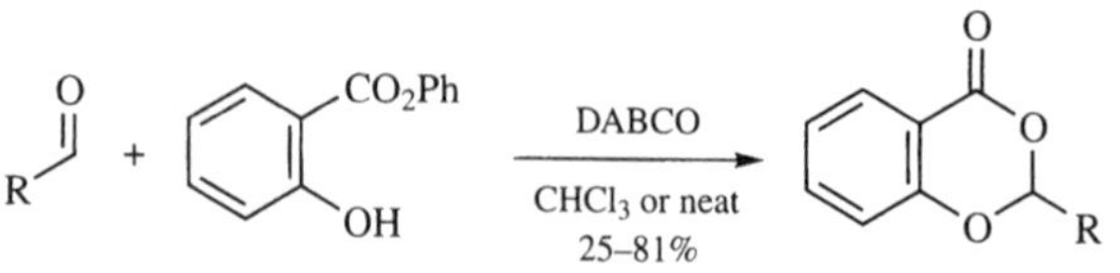

Although aromatic aldehydes failed to react, this is one of the few methods available for the preparation of acetals under basic conditions.[1,2]

1. P. Perlmutter and E. Puniani, *Tetrahedron Lett.*, **37**, 3755 (1996).
2. A. A. Khan, N. D. Emslie, S. E. Drewes, J. S. Field, and N. Ramesar, *Chem. Ber.*, **126**, 1477 (1993).

1,3-Dioxolanes (Chart 5)

The 1,3-dioxolane group is probably the most widely used carbonyl protective group. For the protection of carbonyls containing other acid-sensitive functionality, one should use acids of low acidity or pyridinium salts. In general, a molecule containing two similar ketones can be selectively protected at the less hindered carbonyl, assuming that neither or both of the carbonyls are conjugated to an alkene.[1]

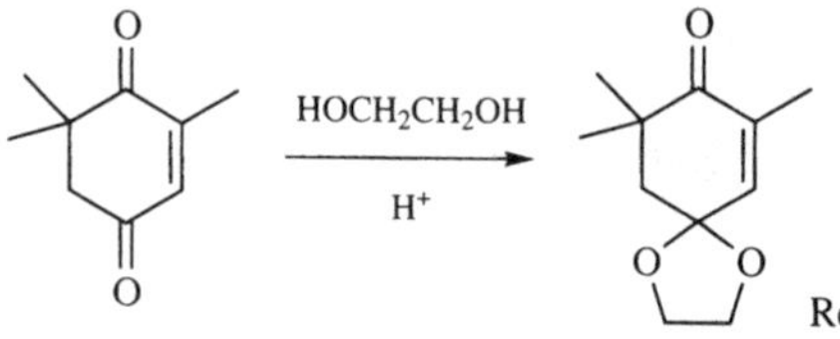

Ref. 1b

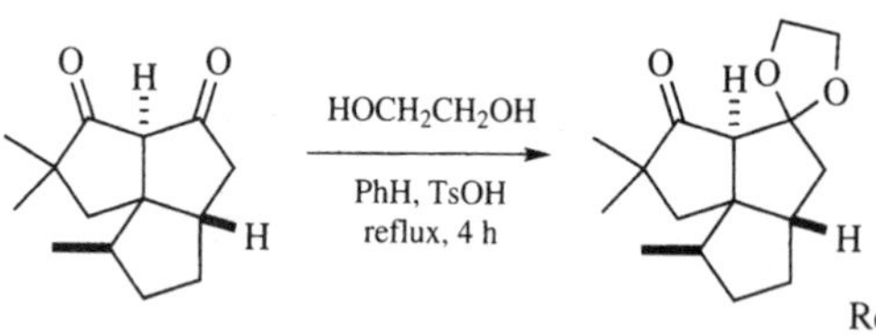

Ref. 1a

If one carbonyl is conjugated with a double bond, the unconjugated carbonyl is selectively protected. This generalization appears to be independent of ring size.[2] Simple aldehydes are generally selectively protected over simple ketones.[3] In the formation of 1,3-dioxolanes of enones, control of the olefin regiochemistry is determined by the acidity of the acid catalyst. Acids of high acidity ($pK_a \sim 1$) may cause the double bond to migrate to the β,γ-position, whereas acids of low acidity ($pK_a \sim 3$) do not cause double-bond migration (see table below).[4] In addition, the use of the bistrimethylsilyl derivative of ethylene glycol and Me_3SiOTf (CH_2Cl_2, $-78°C$, 20h, pyridine quench, 92%) for the protection of enones proceeds without double bond migration.[5,6] A similar result was obtained with the Wieland–Miescher ketone using stoichiometric amounts of TsOH.[7]

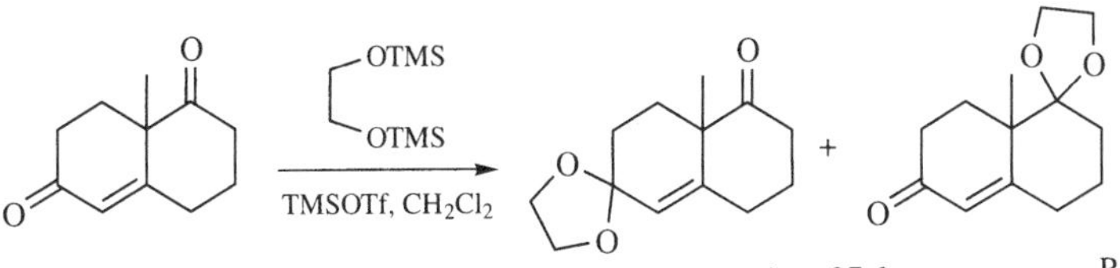

ratio = 27:1   Ref. 5

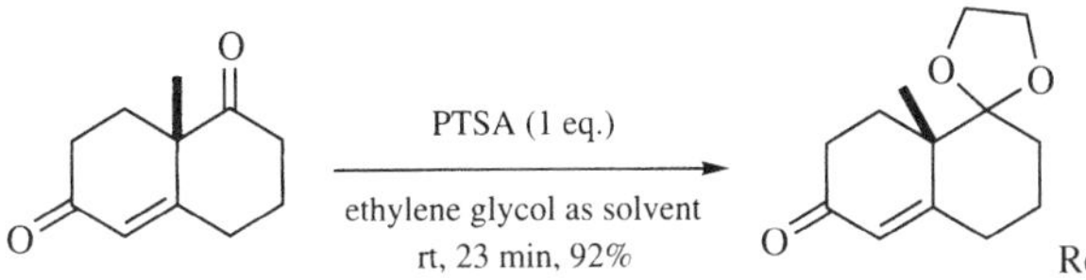

Ref. 7

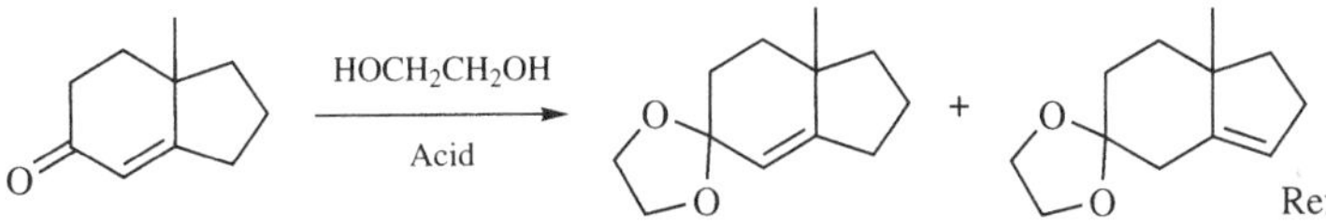

Ref. 4

Olefin Isomerization as a Function of Acid pK_a

Acid	pK_a	% α,β	% β,γ	% Conversion
Fumaric acid	3.03	100	0	90
Phthalic acid	2.89	70	30	90
Oxalic acid[8]	1.23	80	20	93
TsOH acid	<1.0	0	100	100

The following is an interesting example of selective protection.[9] The selectivity is probably the result of greater steric compression associated with the ketal of the cyclopentanone.

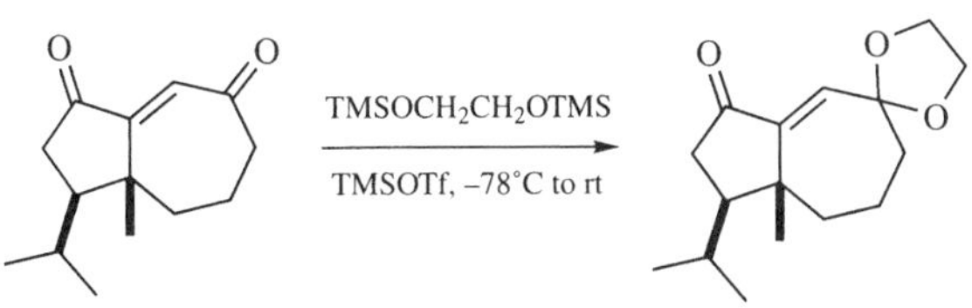

A polymer-supported 1,2-diol has also been developed for use in carbonyl protection.[10]

Formation

The most common method to prepare a ketal is to treat the carbonyl compound with ethylene glycol and an acid at reflux with a solvent that will azeotrope water using a Dean–Stark trap. For substrates that can not tolerate high temperatures, a dehydrating agent such as trimethylorthoformate is often used to scavenge the water.

1. HO(CH$_2$)$_2$OH, C$_5$H$_5$N·TsOH, C$_6$H$_6$, reflux, 1–3 h, 90–95% yield.[11] This is a commonly used, mild and general method for dioxolane formation.
2. HO(CH$_2$)$_2$OH, TsOH, C$_6$H$_6$, reflux, 75–85% yield.[12]
3. HO(CH$_2$)$_2$OH, TsOH, (EtO)$_3$CH, 25°C, 65% yield.[13]
4. HO(CH$_2$)$_2$OH, BF$_3$·Et$_2$O, HOAc, 35–40°C, 15 min, 90% yield.[14]
5. HO(CH$_2$)$_2$OH, HCl, 25°C, 12 h, 55–90% yield.[15]
6. HO(CH$_2$)$_2$OH, Tetrabutylammonium tribromide, triethylorthoformate, 21–97% yield. This method produces HBr *in situ* and can be use to prepare both cyclic and acyclic acetals.[16]
7. HO(CH$_2$)$_2$OH, Me$_3$SiCl, MeOH or CH$_2$Cl$_2$.[17] HCl is produced *in situ*.
8. HO(CH$_2$)$_2$OH, Al$_2$O$_3$, PhCH$_3$ or CCl$_4$, heat, 24 h, 80–100% yield.[3] These conditions are selective for the formation of acetals from aldehydes in the presence of ketones.
9. HO(CH$_2$)$_2$OH, 0.1 eq. CuCl$_2$·H$_2$O, 80°C, 30 min, 82–100% yield.[18] The use of 5 eq. of CuCl$_2$ results in the formation of the α-chloro ketal.

10. HO(CH$_2$)$_2$OH, oxalic acid, CH$_3$CN, 25°C, 95% yield.[19] Note that ketals prepared with oxalic acid from enones tend to retain the olefin regiochemistry.[8]
11. HO(CH$_2$)$_2$OH, adipic acid, C$_6$H$_6$, reflux, 17–24 h, 10–85% yield.[20]
12.

On a large scale, isomerization occurs

With the dimethyl derivative, isomerization is prevented

Ref. 21

13. HO(CH$_2$)$_2$OH, SeO$_2$, CHCl$_3$, 28°C, 4 h, 60% yield.[22]

14. HO(CH$_2$)$_2$OH, C$_5$H$_5$N·HCl, C$_6$H$_6$, reflux, 6 h, 85% yield.[23]

15. HO(CH$_2$)$_n$OH (n = 2,3)/MeOCH$^+$NMe$_2$ MeOSO$_3^-$, 0–25°C, 2 h, 40–95% yield.[24]

16. HO(CH$_2$)$_n$OH (n = 2,3)/column packed with an acid ion-exchange resin, 5 min, 50–90% yield.[25]

17. HOCH$_2$CH$_2$OH, (EtO)$_3$CH, p-TsOH, 83% yield.[26]

18. 2-Methoxy-1,3-dioxolane/TsOH, C$_6$H$_6$, 40–50°C, 4 h, 85% yield.[27]

19. 2-Ethoxy-1,3-dioxolane, pyridinium tosylate (PPTS), benzene, heat, 8 h, 89% yield.[28] In this case, protection of an enone proceeds without double-bond migration.

20. 2-Ethyl-2-methyl-1,3-dioxolane/TsOH, reflux, 75% yield.[29,30] These conditions selectively protect a ketone in the presence of an enone.

21. 2,2-Dimethyl-1,3-dioxolane, microwave irradiation, montmorillonite KSF, 38–95% yield.[31] Titanium cation-exchanged montmorillonite has also been used.[32]

22. 2-Dimethylamino-1,3-dioxolane/cat. HOAc, CH$_2$Cl$_2$, 83% yield.[33] 2-Dimethylamino-1,3-dioxolane protects a reactive ketone under mild conditions: It reacts selectively with a C$_3$-keto steroid in the presence of a Δ^4-3-keto steroid. C$_{12}$- and C$_{20}$-keto steroids do not react.

23. Diethylene orthocarbonate, C(−OCH$_2$CH$_2$O−)$_2$/TsOH or wet BF$_3$·Et$_2$O, CHCl$_3$, 20°C, 70–95% yield.[34]

24. 1,3-Dioxolanes have been prepared from a carbonyl compound and an epoxide (e.g., ketone/SnCl$_4$, CCl$_4$, 20°C, 4 h, 53% yield[35] or aldehyde, Et$_4$NBr, 125–220°C, 2–4 h, 20–85% yield[36]). Perhaloketones can be protected by reaction with ethylene chlorohydrin under basic conditions (K$_2$CO$_3$, pentane, 25°C, 2 h, 85% yield[37] or NaOH, EtOH–H$_2$O, 95% yield[38]).

25. Ethylene oxide, BF$_3$·Et$_2$O, >120 min, CH$_2$Cl$_2$, 25°C, 47–95% yield.[39]

26. HO(CH$_2$)$_2$OH, I$_2$, 30–90% yield. HI is formed *in situ*.[40]

27. HO(CH$_2$)$_2$OH, PhH, catalyst, quant.[41]

4.7% of the 17-ketal and 8.3% of the diketal are also obtained.

28. HOCH$_2$CH$_2$OH, BuSnCl$_3$, 0°C, 10 min, 75–92% yield.[42]

29. HO(CH$_2$)$_2$OH, ZrOCl$_2$·8 H$_2$O, aq. NaOH, 65–98% 74% yield.[43]

30. HO(CH$_2$)$_2$OH, PhH, N-benzylpyridinium hexafluoroantimonate, 1.5–9 h, reflux, 72–91% yield.[44] It is also possible to form the 4,4-dimethyldioxane (85–99% yield) under these conditions.

31. HO(CH$_2$)$_2$OH, [Ru(MeCN)$_3$(Ph$_3$P)](OTf)$_2$, PhH, azeotropic distillation, 87–99% yield.[45]

32. HOCH$_2$CH$_2$OH, (*i*-PrO)$_3$CH, RhCl$_3$(*triphos*), [*triphos* = H$_3$CC(CH$_2$PPh$_2$)$_3$], rt, reflux, 80–100% yield.[46] Benzophenone, which normally does not react well, can be ketalized using this method.

33. HOCH$_2$CH$_2$OH or other alcohols, RuCl$_3$·3H$_2$O, rt, 45–95% yield. Ketones do not react.[47]

34. HOCH$_2$CH$_2$OH, [(dppb)Pt(μ-OH)](BF$_4$)$_2$, 82°C, ClCH$_2$CH$_2$Cl, 10–83% yield. The method works for acrolein where pTSA does not because of competing Michael addition.[48] Unsaturated ketones give low yields.

35. From a tosylhydrazone: ethylene glycol, 200°C, 89% yield.[49]

36. HO(CH$_2$)$_n$OH, n = 2,3, Fe or Al, rt, 52–99% yield.[50]

37. Selective ketone protection: The −CHO group is converted in Step 1 to a siloxysulfonium salt [R′CH((OTMS)S$^+$Me$_2$ $^-$OTf] that is reconverted to an aldehyde group in Step 3.[51]

38. Me$_3$SiOCH$_2$CH$_2$OSiMe$_3$, Me$_3$SiOTf, 15 Kbar (1.5 GPa), 40°C, 48 h.[52] These conditions were used to prepare the ketal of fenchone, which cannot be done under normal acid-catalyzed conditions. This method was found useful for the protection of α-haloketones for which there are otherwise few methods.[53]

39. TMSOCH$_2$CH$_2$OTMS, TfOH or FsOH (fluorosulfonic acid), BTMSA [bis(trimethylsilyl)acetamide] or BTMSU [bis(trimethylsilyl)urea], 76–97% yield.[54]

40. HO(CH$_2$)$_n$OH, n = 2,3, *i*-PrOTMS, TMSOTf, CH$_2$Cl$_2$, −20°C, 3 h, 84–99% yield.[55]

41. HOCH$_2$CH$_2$OH, MgSO$_4$, PhH, L-tartaric acid, reflux, 20 h, 97% yield. These conditions were optimized for protection of unsaturated aldehydes to prevent double bond migration.[56]

42.

Ref. 57

43. HOCH$_2$CH$_2$OH, Bi(OTf)$_3$·4H$_2$O, toluene or fluorobenzene, trimethylortho-formate, reflux, 56–79% yield. Dimethyl acetals are prepared similarly in good yields.[58]

44. The following is a rare example of ketal formation using basic conditions.[59] When the carbonyl group is very electron-deficient, thus stabilizing the hemi-acetal, a dioxolane can be prepared under basic conditions.[37,60]

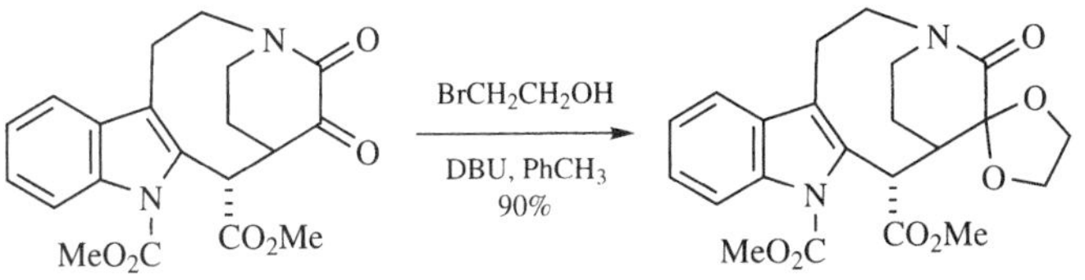

45. Microwaves[61] and ionic liquids[62] have been used to induce acetal formation, but the methods have not been broadly tested on significant substrates.

Cleavage

1,3-Dioxolanes can be cleaved by acid-catalyzed exchange dioxolanation, acid-cata-lyzed hydrolysis, or oxidation. Many different forms of acid have been used to cleave 1,3-dioxolanes. Some representative examples are shown below. Many of the reports give only simple examples, so it is not clear how they will stand up to the rigors of multifunctional substrates.

1. Pyridinium tosylate (PPTS), acetone, H$_2$O, heat, 100% yield.[11,63] Microwaves have been used to accelerate this cleavage reaction.[64]

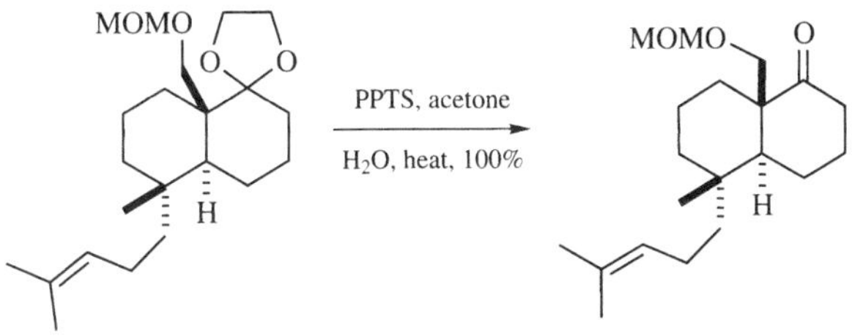

2. Acetone, TsOH, 20°C, 12 h.[65] The reactant is a 3,6,17-tris(ethylenedioxy) steroid; the product has carbonyl groups at C-6 and C-17.

3. 5% HCl, THF, 25°C, 20 h.[66]

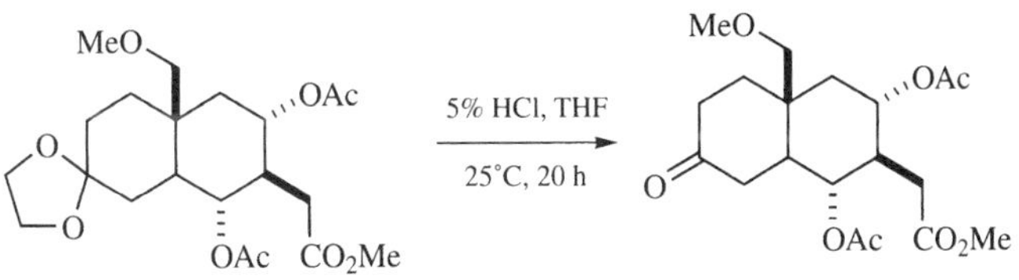

4. 1 *M* HCl, THF, 0–25°C, 13 h, 71% yield. Note that the acetonide survives these conditions.[67] Some variations have been reported in this system (including the use of 30% AcOH, 90°C, high yield).[68]

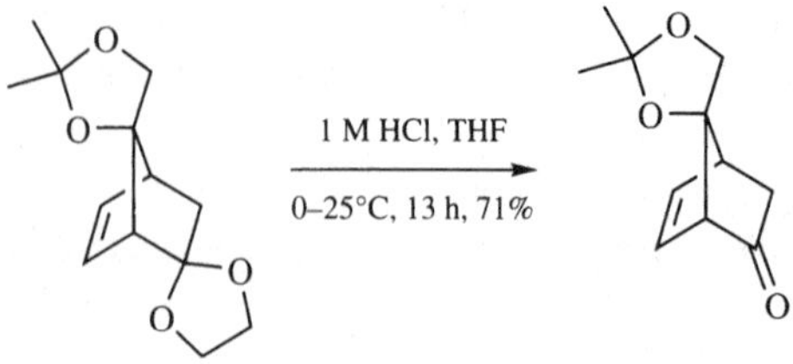

5. 80% AcOH, 65°C, 5 min, 85% yield.[69]

6. Wet magnesium sulfate (C_6H_6, 20°C, 1 h) effects selective, quantitative cleavage of an α,β-unsaturated 1,3-dioxolane in the presence of a 1,3-dioxolane.[20]

7. Perchloric acid (79% $HClO_4$/CH_2Cl_2, 0°, 1 h → 25°C, 3 h, 87% yield)[70] and periodic acid (aq. dioxane, 3 h, quant. yield)[71] cleave 1,3-dioxolanes; the latter drives the reaction to completion by oxidation of the ethylene glycol that forms. Yields are substantially higher from cleavage with perchloric acid (3 N $HClO_4$/THF, 25°C, 3 h, 80% yield) than with hydrochloric acid (HCl/HOAc, 65% yield).[72]

8. SiO_2, H_2O, CH_2Cl_2, oxalic acid, 90–95% yield.[73] These conditions selectively cleave α,β-unsaturated ketals.

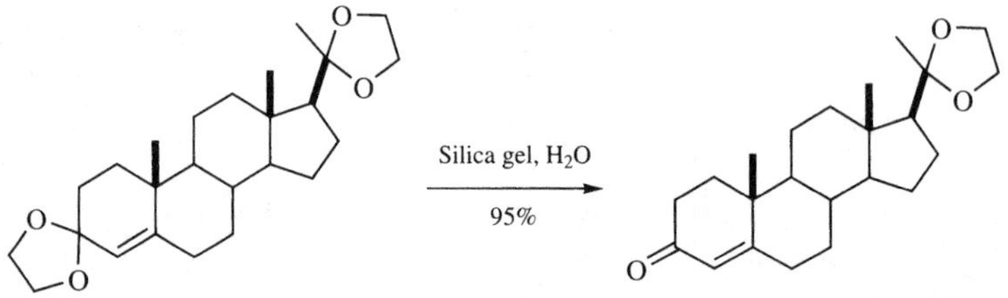

9. Ph_3CBF_4, CH_2Cl_2, 25°C, 60–100% yield.[74,75] 1,3-Dithiolanes are not affected by these conditions, but a 1,3-oxathiolane is cleaved (100% yield).[76]

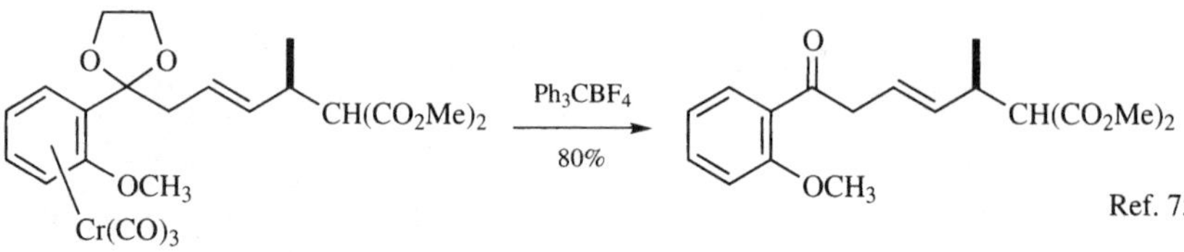

Ref. 75

10. Me_2BBr, CH_2Cl_2, −78°C, 90–97% yield.[77] This reagent also cleaves MTM, MEM and MOM ethers (87–95% yield).

11. $PdCl_2(CH_3CN)_2$, acetone, H_2O, 82–100% yield.[78]

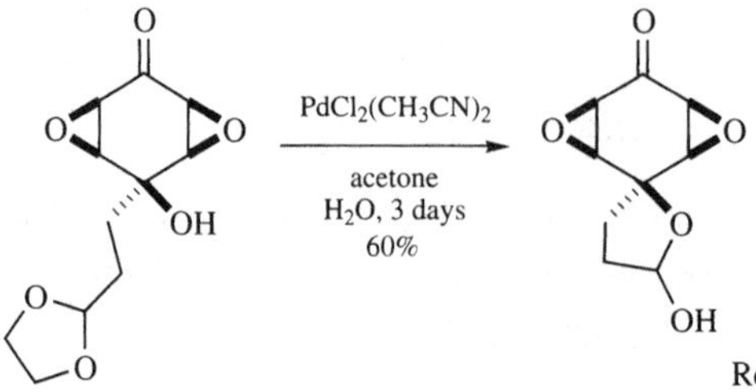

Ref. 79

12. LiBF$_4$, wet CH$_3$CN.[80] Unsubstituted 1,3-dioxolanes are cleaved slowly under these conditions (40% in 5 h). The 4,5-dimethyl- and 4,4,5,5-tetramethyl-dioxolane and 1,3-dioxane are inert under these conditions. Dimethyl ketals are readily cleaved.

13. Dimethyl sulfoxide, 180°C, H$_2$O, 10 h, 89% yield. A diethyl acetal can be cleaved in the presence of a 1,3-dioxolane under these conditions. TBDMS, THP, and MOM groups are stable. The use of refluxing DMSO/dioxane is also effective.[81]

14. Hydrothermal conditions cleave a 1,3-dioxolane, but the reaction must be conducted under pressure and uses a catalytic amount of CaCl$_2$ (453°K, 1.02 MPa, 20 min).[82] It is likely that acid is generated *in situ*.

15. NaTeH, EtOH, 25°C, 30 min; air, 80–85% yield.[83]

16. H$_2$SiI$_2$, CDCl$_3$, −42°C, 1–10 min, 100% yield.[84] Aromatic ketals are cleaved faster than the corresponding aliphatic derivatives, and cyclic ketals are cleaved more slowly than the acyclic analogues such as dimethyl ketals. Substituted ketals such as those derived from butane-2,3-diol, which react only slowly with Me$_3$SiI, can also be cleaved with H$_2$SiI$_2$. If the reaction is run at 22°C, ketals and acetals are reduced to iodides in excellent yield. The related Me$_3$SiI also cleaves 1,3-doxolanes.[85]

17. Me$_3$SiNEt$_2$, MeI is a synthetic equivalent to TMSI that will open dioxolanes to enol ethers.[86]

18. CuSO$_4$·SiO$_2$, CH$_2$Cl$_2$, 20–80 h, 70–90% yield.[87]

19. DDQ, CH$_3$CN, H$_2$O, 68–95% yield.[88]

20. *t*-BuOOH, Pd(OOCCF$_3$)(OO-*t*-Bu), benzene, 50°C, 12 h, 60–80% yield.[89] In this case, an acetal is oxidized to the ester of ethylene glycol (RCO$_2$CH$_2$CH$_2$OH). A similar process that uses H$_2$O$_2$ as the oxidant has been developed for 1,3-dioxolanes and dimethyl acetals.[90] α,β-Unsaturated acetals gave poor yields.

21. V$_2$O$_5$, H$_2$O$_2$, CH$_3$CN, 92–96% yield. If MeOH is used as the solvent, esters are obtained rather than aldehydes (82–95% yields).[91]

22. O$_3$, AcOEt, −78°C, 94% yield. These conditions are used to convert an acetal to an ester.[92] Oxone[93] and dimethyldioxirane[94] can also be used to generate esters from 1,3-dioxolanes, but oxone does not always result in oxidation.[95]

23. Dimethyldioxirane, acetone, CH$_2$Cl$_2$, 0°C, 24 h, >95% yield.[96] Although ketone dioxolanes are cleaved to ketones, aldehyde dimethyl acetals will gives the ester, but the generality of the later process has not been established beyond the acetal of benzaldehyde. Ethers are also oxidized under these conditions.

24. 3 mol% Ceric ammonium nitrate, CH_3CN, borate buffer, pH = $8, 60°C, 100\%$ yield. This method also cleaves dimethyl acetals and the THP group.[97] This method can be used to cleave a dioxolane in the presence of an enol triflate.[98]

25. NO_2, silica gel, CCl_4, 30°C, 40 min, 88–100% yield.[99]

26. PPh_3, CBr_4, THF, 0°C, 96% yield.[100] CBr_4 alone has also been used.[101]

27. $SmCl_3$, TMSCl, THF, 92% yield. A ketal is cleaved in preference to an acetal.[102]

28. 2,4,6-Triphenylpyrilium tetrafluoroborate, H_2O, CH_2Cl_2, 3 h, $h\nu$, 67–88% yield.[103]

29. $RuCl_3 \cdot nH_2O$, t-BuOH, PhH, 1 h, rt, 46–86% yield. In this case the acetal is cleaved with simultaneous oxidation to an ethylene glycol ester.[104]

30. NaI, $CeCl_3 \cdot 7H_2O$, CH_3CN, rt, 0.5–21 h, 84–96% yield.[105] Chemoselective cleavage of ketone derivatives is observed in the presence of aldehyde derivatives, and enone ketals are cleaved in the presence of simple ketone ketals.

31. Thiourea, EtOH, H_2O, reflux, 82–89% yield. This method also cleaves acetonides (64–93% yield).[106]

32. Some of the other miscellaneous reagents that have been examined for their ability to cleave dioxolanes—and in some cases other acetals and ketals— are as follows. Their scope and utility have not been examined in complex scenarios. $Ce(OTf)_3$,[107] $InCl_3$,[108] WCl_6,[109] $CuCl_2 \cdot 2H_2O$,[110] $AgBrO_3/AlCl_3$,[111] $FeCl_3 \cdot 6H_2O$,[112] $BiCl_3$[113] or $Bi(OTf)_3$,[114] AlI_3,[115] $TiCl_4/LiI$,[116] Pt–Mo/ZrO_2,[117] polyaniline-supported sulfuric acid,[118] $LiCl/H_2O/DMSO$,[119] wet-SiO_2,[120] $BnPh_3P^+HSO_5^-/BiCl_3$,[121] $K_5CoW_{12}O_{40} \cdot 3H_2O$,[122] $(PhCH_2PPh_3)_2S_2O_8$,[123] and Magtrieve™.[124]

1. For two examples, see (a) M. T. Crimmins and J. A. DeLoach, *J. Am. Chem. Soc.*, **108**, 800 (1986); (b) M. G. Constantino, P. M. Donate, and N. Petragnani, *J. Org. Chem.*, **51**, 253 (1986).

2. For a variety of examples with varying ring sizes, see Y. Ohtsuka and T. Oishi, *Tetrahedron Lett.*, **27**, 203 (1986); C. Iwata, Y. Takemoto, M. Doi, and T. Imanishi, *J. Org. Chem.*, **53**, 1623 (1988); S. D. Burke, C. W. Murtiashaw, J. O. Saunders, and M. S. Dike, *J. Am. Chem. Soc.*, **104**, 872 (1982); P. A. Wender, M. A. Eisenstat, and M. P. Filosa, *J. Am. Chem. Soc.*, **101**, 2196 (1979); A. A. Devreese, P. J. de Clercq, and M. Vandewalle, *Tetrahedron Lett.*, **21**, 4767 (1980); P. G. Baraldi, A. Barco, S. Benetti, G. P. Pollini, E. Polo, and D. Simoni, *J. Org. Chem.*, **50**, 23 (1985); M. P. Bosch, F. Camps, J. Coll, A. Guerrero, T. Tatsuoka, and J. Meinwald, *J. Org. Chem.*, **51**, 773 (1986).

3. Y. Kamitori, M. Hojo, R. Masuda, and T. Yoshida, *Tetrahedron Lett.*, **26**, 4767 (1985).

4. J. W. De Leeuw, E. R. De Waard, T. Beetz, and H. O. Huisman, *Recl. Trav. Chim. Pays-Bas*, **92**, 1047 (1973).

5. J. R. Hwu and J. M. Wetzel, *J. Org. Chem.*, **50**, 3946 (1985); J. R. Hwu, L.-C. Leu, J. A. Robl, D. A. Anderson, and J. M. Wetzel, *J. Org. Chem.*, **52**, 188 (1987).

6. T. Tsunoda, M. Suzuki, and R. Noyori, *Tetrahedron Lett.*, **21**, 1357 (1980).

7. P. Ciceri and F. W. J. Demnitz, *Tetrahedron Lett.*, **38**, 389 (1997).

8. G. H. Posner and G. L. Loomis, *Tetrahedron Lett.*, 4213 (1978).

9. B. Shi, N. A. Hawryluk, and B. B. Snider, *J. Org. Chem.*, **68**, 1030 (2003).

10. P. Hodge and J. Waterhouse, *J. Chem. Soc., Perkin Trans. I*, 2319 (1983); Z. H. Xu, C. R. McArthur, and C. C. Leznoff, *Can. J. Chem.*, **61**, 1405 (1983).

11. R. Sterzycki, *Synthesis*, 724 (1979).

12. R. A. Daignault and E. L. Eliel, *Org. Synth., Collect. Vol. V*, 303 (1973).

13. F. F. Caserio, Jr., and J. D. Roberts, *J. Am. Chem. Soc.*, **80**, 5837 (1958).

14. L. F. Fieser and R. Stevenson, *J. Am. Chem. Soc.*, **76**, 1728 (1954).

15. E. G. Howard and R. V. Lindsey, *J. Am. Chem. Soc.*, **82**, 158 (1960).

16. R. Gopinath, S. J. Haque, and B. K. Patel, *J. Org. Chem.*, **67**, 5842 (2002).

17. T. H. Chan, M. A. Brook, and T. Chaly, *Synthesis*, 203 (1983).

18. J. Y. Satoh, C. T. Yokoyama, A. M. Haruta, K. Nishizawa, M. Hirose, and A. Hagitani, *Chem. Lett.*, **3**, 1521 (1974); P. Saravanan, M. Chandrasekhar, R. V. Anand, and V. K. Singh, *Tetrahedron Lett.*, **39**, 3091 (1998).

19. N. H. Andersen and H.-S. Uh, *Synth. Commun.*, **3**, 125 (1973).

20. J. J. Brown, R. H. Lenhard, and S. Bernstein, *J. Am. Chem. Soc.*, **86**, 2183 (1964).

21. T. Ohshima, K. Kagechika, M. Adachi, M. Sodeoka, and M. Shibasaki, *J. Am. Chem. Soc.*, **118**, 7108 (1996).

22. E. P. Oliveto, H. Q. Smith, C. Gerold, L. Weber, R. Rausser, and E. B. Hershberg, *J. Am. Chem. Soc.*, **77**, 2224 (1955).

23. F. T. Bond, J. E. Stemke, and D. W. Powell, *Synth. Commun.*, **5**, 427 (1975).

24. W. Kantlehner and H.-D. Gutbrod, *Liebigs Ann. Chem.*, 1362 (1979).

25. A. E. Dann, J. B. Davis, and M. J. Nagler, *J. Chem. Soc., Perkin Trans. I*, 158 (1979); K. Ishihara, A. Hasegawa, and H. Yamamoto, *Synlett*, 1296 (2002).

26. M. Koreeda and L. Brown, *J. Org. Chem.*, **48**, 2122 (1983).

27. B. Glatz, G. Helmchen, H. Muxfeldt, H. Porcher, R. Prewo, J. Senn, J. J. Stezowski, R. J. Stojda, and D. R. White, *J. Am. Chem. Soc.*, **101**, 2171 (1979).

28. R. A. Holton, R. M. Kennedy, H.-B. Kim, and M. E. Krafft, *J. Am. Chem. Soc.*, **109**, 1597 (1987).

29. H. J. Dauben, B. Löken, and H. J. Ringold, *J. Am. Chem. Soc.*, **76**, 1359 (1954).

30. H. Hagiwara and H. Uda, *J. Org. Chem.*, **53**, 2308 (1988); Y. Tamai, H. Hagiwara, and H. Uda, *J. Chem. Soc., Perkin Trans. I*, 1311 (1986).

31. B. Pério, M.-J. Dozias, P. Jacquault, and J. Hamelin, *Tetrahedron Lett.*, **38**, 7867 (1997).

32. T. Kawabata, T. Mizugaki, K. Ebitani, and K. Kaneda, *Tetrahedron Lett.*, **42**, 8329 (2001).

33. H. Vorbrueggen, *Steroids*, **1**, 45 (1963).

34. D. H. R. Barton, C. C. Dawes, and P. D. Magnus, *J. Chem. Soc., Chem. Commun.*, 432 (1975).

35. J. L. E. Erickson and F. E. Collins, *J. Org. Chem.*, **30**, 1050 (1965).

36. F. Nerdel, J. Buddrus, G. Scherowsky, D. Klamann, and M. Fligge, *Liebigs Ann. Chem.*, **710**, 85 (1967).

37. H. E. Simmons and D. W. Wiley, *J. Am. Chem. Soc.*, **82**, 2288 (1960).

38. R. J. Stedman, L. D. Davis, and L. S. Miller, *Tetrahedron Lett.*, **8**, 4915 (1967).

39. D. S. Torok, J. J. Figueroa, and W. J. Scott, *J. Org. Chem.*, **58**, 7274 (1993).

40. B. K. Banik, M. Chapa, J. Marquez, and M. Cardona, *Tetrahedron Lett.*, **46**, 2341 (2005).

41. J. Otera, N. Danoh, and H. Nozaki, *Tetrahedron*, **48**, 1449 (1992).

42. D. Marton, P. Slaviero, and G. Taglianini, *Gazz. Chim. Ital.*, **119**, 359 (1989).

43. M. Shibagaki, K. Takahashi, H. Kuno, and H. Matsushita, *Bull. Chem. Soc. Jpn.*, **63**, 1258 (1990).

44. S.-B. Lee, S.-D. Lee, T. Takata, and T. Endo, *Synthesis*, 368 (1991).

45. S. Ma and L. M. Venanzi, *Synlett*, 751 (1993).

46. J. Ott, G. M. Ramos Tombo, B. Schmid, L. M. Venanzi, G. Wang, and T. R. Ward, *Tetrahedron Lett.*, **30**, 6151 (1989); M. Sülü and L. M. Venanzi, *Helv. Chim. Acta*, **84**, 898 (2001).

47. J.-Y. Qi, J.-X. Ji, C.-H. Yueng, H.-L. Kwong, and A. S. C. Chan, *Tetrahedron Lett.*, **45**, 7719 (2004).

48. E. Nieddu, M. Cataldo, F. Pinna, and G. Strukul, *Tetrahedron Lett.*, **40**, 6987 (1999).

49. Z. Paryzek and J. Martynow, *J. Chem. Soc., Perkin Trans. I*, 243 (1991).

50. W. Wang, L. Shi, and Y. Huang, *Tetrahedron*, **46**, 3315 (1990).

51. S. Kim, Y. G. Kim, and D.-i. Kim, *Tetrahedron Lett.*, **33**, 2565 (1992).

52. W. G. Dauben, J. M. Gerdes, and G. C. Look, *J. Org. Chem.*, **51**, 4964 (1986); H. Eibisch, *Z. Chem.*, **26**, 375 (1986).

53. R. Carlson, H. Gautun, and A. Westerlund, *Adv. Synth. Catal.*, **344**, 57 (2002).

54. M. El Gihani and H. Heaney, *Synlett*. 433 (1993). *idem, ibid.*, 583 (1993).

55. M. Kurihara and N. Miyata, *Chem. Lett.*, 263 (1995); M. Kurihara and W. Hakamata, *J. Org. Chem.*, **68**, 3413 (2003).

56. T.-J. Lu, J.-F. Yang, and L.-J. Sheu, *J. Org. Chem.*, **60**, 2931 (1995).

57. A. A. Haaksma, B. J. M. Jansen, and A. de Groot, *Tetrahedron*, **48**, 3121 (1992).

58. N. M. Leonard, M. C. Oswald, D. A. Freilberg, B. A. Nattier, R. C. Smith, and R. S. Mohan, *J. Org. Chem.*, **67**, 5202 (2002).

59. P. Magnus, M. Giles, R. Bonnert, C. S. Kim, L. McQuire, A. Merritt, and N. Vicker, *J. Am. Chem. Soc.*, **114**, 4403 (1992).

60. G. R. Newkome, J. D. Sauer, and C. L. McClure, *Tetrahedron Lett.*, **14**, 1599 (1973).

61. D. J. Kalita, R. Borah, and J. C. Sarma, *Tetrahedron Lett.*, **39**, 4573 (1998). D. D. Laskar, D. Prajapati, and J. S. Sandhu, *Chem. Lett.*, 1283 (1999); J. S. Yadov, B. V. S. Reddy, R. Srinivas, and T. Ramalingam, *Synlett*, 701 (2000); F. M. Moghaddam, A. A. Oskoui, and H. Z. Boinee, *Letters in Organic Chemistry*, **2**, 151 (2005).

62. H.-H. Wu, F. Yang, P. Cui, J. Tang, and M.-Y. He, *Tetrahedron Lett.*, **45**, 4963 (2004); D. Li, F. Shi, J. Peng, S. Guo, and Y. Deng, *J. Org. Chem.*, **69**, 3582 (2004).

63. H. Hagiwara and H. Uda, *J. Chem. Soc., Chem. Commun.*, 1351 (1987).

64. Y. He, M. Johansson, and O. Sterner, *Synth. Commum.*, **34**, 4153 (2004).

65. G. Bauduin, D. Bondon, Y. Pietrasanta, and B. Pucci, *Tetrahedron*, **34**, 3269 (1978).

66. P. A. Grieco, M. Nishizawa, T. Oguri, S. D. Burke, and N. Marinovic, *J. Am. Chem. Soc.*, **99**, 5773 (1977).

67. P. A. Grieco, Y. Yokoyama, G. P. Withers, F. J. Okuniewicz, and C.-L. J. Wang, *J. Org. Chem.*, **43**, 4178 (1978).

68. P. A. Grieco, Y. Ohfune, and G. Majetich, *J. Am. Chem. Soc.*, **99**, 7393 (1977).

69. J. H. Babler, N. C. Malek, and M. J. Coghlan, *J. Org. Chem.*, **43**, 1821 (1978).

70. P. A. Grieco, T. Oguri, S. Gilman, and G. R. DeTitta, *J. Am. Chem. Soc.*, **100**, 1616 (1978).

71. H. M. Walborsky, R. H. Davis, and D. R. Howton, *J. Am. Chem. Soc.*, **73**, 2590 (1951).

72. J. A. Zderic and D. C. Limon, *J. Am. Chem. Soc.*, **81**, 4570 (1959).

73. F. Huet, A. Lechevallier, M. Pellet, and J. M. Conia, *Synthesis*, 63 (1978).

74. D. H. R. Barton, P. D. Magnus, G. Smith, and D. Zurr, *J. Chem. Soc., Chem. Commun.*, 861 (1971).

75. M. Uemura, T. Minami, and Y. Hayashi, *Tetrahedron Lett.*, **29**, 6271 (1988).

76. D. H. R. Barton, P. D. Magnus, G. Smith, G. Streckert, and D. Zurr, *J. Chem. Soc., Perkin Trans. I*, 542 (1972).

77. Y. Guindon, H. E. Morton, and C. Yoakim, *Tetrahedron Lett.*, **24**, 3969 (1983).

78. B. H. Lipshutz, D. Pollart, J. Monforte, and H. Kotsuki, *Tetrahedron Lett.*, **26**, 705 (1985).

79. A. McKillop, R. J. K. Taylor, R. J. Watson, and N. Lewis, *Synlett*, 1005 (1992).

80. B. H. Lipshutz and D. F. Harvey, *Synth. Commun.*, **12**, 267 (1982).

81. T. Kametani, H. Kondoh, T. Honda, H. Ishizone, Y. Suzuki, and W. Mori, *Chem. Lett.*, **18**, 901 (1989).

82. K. Sato, T. Kishimoto, M. Morimoto, H. Saimoto, and Y. Shigemasa, *Tetrahedron Lett.*, **44**, 8623 (2003).

83. P. Lue, W.-Q. Fan, and X.-J. Zhou, *Synthesis*, 692 (1989).

84. E. Keinan, D. Perez, M. Sahai, and R. Shvily, *J. Org. Chem.*, **55**, 2927 (1990).

85. M. E. Jung, W. A. Andrus, and P. L. Ornstein, *Tetrahedron Lett.*, **18**, 4175 (1977).

86. A. Iwata, H. Tang, and A. Kunai, *J. Org. Chem.*, **67**, 5170 (2002).

87. G. M. Caballero and E. G. Gros, *Synth. Commun.*, **25**, 395 (1995).

88. K. Tanemura, T. Suzuki, and T. Horaguchi, *J. Chem. Soc., Chem. Commun.*, 979 (1992).

89. T. Hosokawa, Y. Imada, and S.-i. Murahashi, *J. Chem. Soc., Chem. Commun.*, 1245 (1983).

90. T. Takeda, H. Watanabe, and T. Kitahara, *Synlett*, 1149 (1997).

91. R. Gopinath, A. R. Paital, and B. K. Patel, *Tetrahedron Lett.*, **43**, 5123 (2002).

92. M. Fernandez and R. Alonso, *Org. Lett.*, **5**, 2461 (2003).

93. M. Curini, F. Epifano, M. C. Marcotullio, and O. Rosati, *Synlett*, 777 (1999).

94. M. Frigerio, M. Santagostino, and S. Sputore, *Synlett*, 833 (1997).

95. D. S. Bose, B. Jayalakshmi, and A. V. Narsaiah, *Synthesis*, 67 (2000).

96. R. Curci, L. D'Accolti, M. Fiorentino, C. Fusco, W. Adam, M. E. Gonzalez-Nunez, and R. Mello, *Tetrahedron Lett.*, **33**, 4225 (1992).

97. I. E. Markó, A. Ates, B. Augustyns, A. Gautier, Y. Quesnel, L. Turet, and M. Wiaux, *Tetrahedron Lett.*, **40**, 5613 (1999); V. Nair, L. G. Nair, L. Balagopal, and R. Rajan, *Ind. J. Chem., Sect. B*, **38B**, 1234 (1999); I. E. Markó, A. Ates, A. Gautier, B. Leroy, J.-M. Plancher, Y. Quesnel, and J.-C. Vanherck, *Angew. Chem. Int. Ed.*, **38**, 3207 (1999); A. Ates, A. Gautier, B. Leroy, J.-M. Plancher, Y. Quesnel, J.-C. Vanherck, and I. E. Markó, *Tetrahedron*, **59**, 8989 (2003).

98. N. Maulide and I. E. Marko, *Synlett*, 2195 (2005).

99. T. Nishiguchi, T. Ohosima, A. Nishida, and S. Fujisaki, *J. Chem. Soc., Chem. Commun.*, 1121 (1995).

100. C. Johnstone, W. J. Kerr, and J. S. Scott, *J. Chem. Soc., Chem. Commun.*, 341 (1996).

101. A. S.-Y. Lee and C.-L. Cheng, *Tetrahedron*, **53**, 14255 (1997).

102. Y. Ukaji, N. Koumoto, and T. Fujisawa, *Chem. Lett.*, **18**, 1623 (1989).

103. H. Garcia, S. Iborra, M. A. Miranda, and J. Primo, *New J. Chem.*, **13**, 805 (1989).

104. S. Murahashi, Y. Oda, and T. Naota, *Chem. Lett.*, **21**, 2237 (1992).

105. E. Marcantoni, F. Nobili, G. Bartoli, M. Bosco, and L. Sambri, *J. Org. Chem.*, **62**, 4183 (1997); O. Arjona, R. Menchaca, and J. Plumet, *Org. Lett.*, **3**, 107 (2001).

106. S. Majumdar and A. Bhattacharjya, *J. Org. Chem.*, **64**, 5682 (1999).

107. R. Dalpozzo, A. De Nino, L. Maiuolo, A. Procopio, A. Tagarelli, G. Sindona, and G. Bartoli, *J. Org. Chem.*, **67**, 9093 (2002).

108. B. C. Ranu, R. Jana and S. Samanta, *Adv. Synth. Catal.*, **346**, 446 (2004).

109. H. Firouzabadi, N. Iranpoor, and B. Karimi, *J. Chem. Res. (S)*, 664 (1998).

110. P. Saravanan, M. Chandrasekhar, R. V. Anand, and V. K. Singh, *Tetrahedron Lett.*, **39**, 3091 (1998).

111. I. Mohammadpoor-Baltork and A. R. Nourozi, *Synthesis*, 487 (1999).

112. S. E. Sen, S. L. Roach, J. K. Boggs, G. J. Ewing, and J. Magrath, *J. Org. Chem.*, **62**, 6684(1997).

113. G. Sabitha, R. S. Babu, E. V. Reddy, and J. S. Yadav, *Chem. Lett.*, **29**, 1074 (2000).

114. M. D. Carrigan, D. Sarapa, R. C. Smith, L. C. Wieland, and R. S. Mohan, *J. Org. Chem.*, **67**, 1027 (2002).

115. P. Sarmah and N. C. Barua, *Tetrahedron Lett.*, **30**, 4703 (1989).

116. G. Balme and J. Goré, *J. Org. Chem.*, **48**, 3336 (1983).

117. B. M. Reddy, V. M. Reddy, and D. Giridhar, *Synth. Commum.*, **31**, 1819 (2001).

118. S. Palaniappan, P. Narender, C. Saravanan, and V. J. Rao, *Synlett*, 1793 (2003).

119. P. K. Mandal, P. Dutta, and S. C. Roy, *Tetrahedron Lett.*, **38**, 7271 (1997).

120. B. F. Mirjalili, M. A. Zolfigol, A. Bamoniri, and A. Hazar, *Bull. Korean Chem. Soc.*, **25**, 1075 (2004).

121. A. R. Hajipour, S. E. Mallakpour, I. Mohammadpoor-Baltork, and H. Adibi, *Phosphorus, Sulfur and Silicon and the Related Elements*, **165**, 155 (2000).

122. M. H. Habibi, S. Tangestaninejad, I. Mohammadpoor-Baltork, V. Mirkhani, and B. Yadollahi, *Tetrahedron Lett.*, **42**, 6771 (2001).

123. M. Tajbakhsh, I. Mohammadpoor-Baltork, and F. Ramzanian-Lehmali, *J. Chem. Res., Syn.*, 185 (2001).

124. K.-Y. Ko and S.-T. Park, *Tetrahedron Lett.*, **40**, 6025 (1999).

4,4,5,5-Tetramethyl-1,3-dioxolane

The acetal is readily formed from an aldehyde upon treatment with pinacol and PTSA in toluene (95% yield). This group was used to protect an aldehyde during metalation and boronic acid formation when the dithiane group proved unsuccessful. It was removed by transacetalization with 1,3-propanedithiol and $BF_3 \cdot Et_2O$ to give the dithiane (95% yield).[1]

1. G. J. McKiernan and R. C. Hartley, *Org. Lett.*, **5**, 4389 (2003).

4-Bromomethyl-1,3-dioxolane (Chart 5)

This ketal is stable to several reagents that react with carbonyl groups (e.g., m-ClC$_6$H$_4$CO$_3$H, NH$_3$, NaBH$_4$, and MeLi). It is cleaved under neutral conditions.[1]

Formation

HOCH$_2$CH(OH)CH$_2$Br, TsOH, benzene, reflux, 5 h, 93–98% yield.

Cleavage

Activated Zn, MeOH, reflux, 12 h, 89–96% yield.

1. E. J. Corey and R. A. Ruden, *J. Org. Chem.*, **38**, 834 (1973).

4-Phenylsulfonylmethyl-1,3-dioxolane

This derivative is prepared from the readily available diol under standard conditions (PPTS, benzene, reflux, 90%). It is cleaved with DBU (CH$_2$Cl$_2$, rt, 12–36 h, 70–90%) yield.[1]

1. S. Chandrasekhar and S. Sarkar, *Tetrahedron Lett.*, **39**, 2401 (1998).

4-(3-Butenyl)-1,3-dioxolane

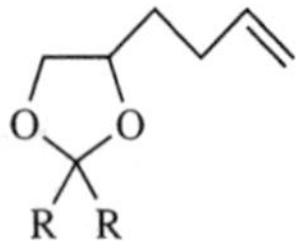

Formation/Cleavage[1]

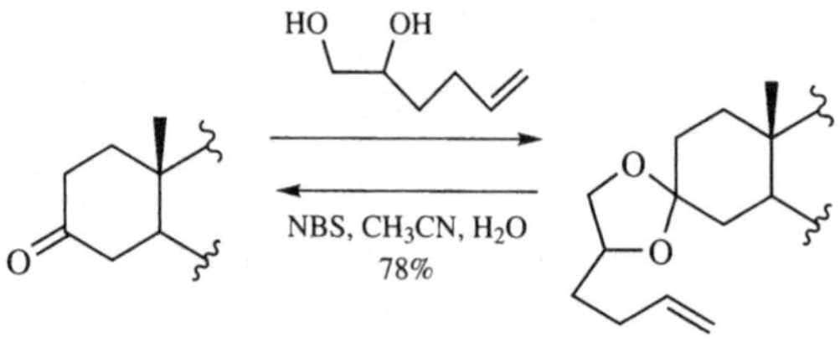

NBS, CH$_3$CN, H$_2$O
78%

1. Z. Wu, D. R. Mootoo, and B. Fraser-Reid, *Tetrahedron Lett.*, **29**, 6549 (1988).

4-Phenyl-1,3-dioxolane

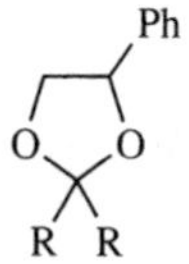

Cleavage[1]

1. Electrolysis: LiClO$_4$, H$_2$O, Pyr, CH$_3$CN, *N*-hydroxyphthalimide, 0.85 V SCE, 22–90% yield.
2. Pd/C, H$_2$.[2]

1. M. Masui, T. Kawaguchi, and S. Ozaki, *J. Chem. Soc., Chem. Commun.*, 1484 (1985).
2. S. Chandrasekhar, B. Muralidhar, and S. Sarkar, *Synth. Commun.*, **27**, 2691 (1997).

4-(4-Methoxyphenyl)-1,3-dioxolane

This protective group can be removed oxidatively in excellent yields.[1] The section on the cleavage of the *p*-methoxybenzyl ether should be consulted, since a number of the methods presented there are should be applicable to this derivative.

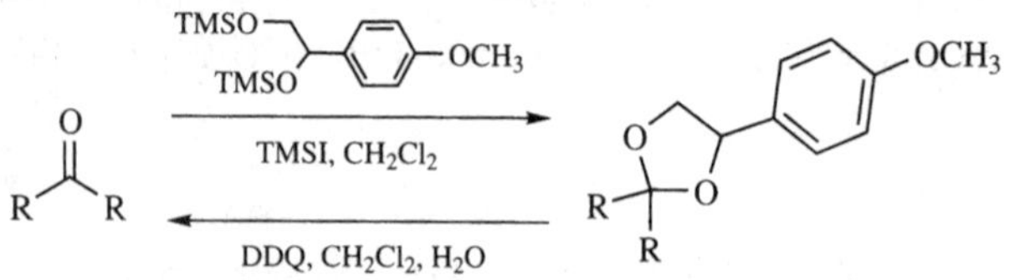

TMSI, CH$_2$Cl$_2$

DDQ, CH$_2$Cl$_2$, H$_2$O

1. C. E. McDonald, L. E. Nice, and K. E. Kennedy, *Tetrahedron Lett.*, **35**, 57 (1994).

4-(2-Nitrophenyl)-1,3-dioxolane (Chart 5)

This dioxolane is readily formed from the glycol (TsOH, benzene, reflux, 70–95% yield); it is cleaved by irradiation (350 nm, benzene, 25°C, 6 h, 75–90% yield). The rate of cleavage is decreased with increasing steric bulk.[1] This group is stable to 5% HCl/THF; 10% AcOH/THF; 2% oxalic acid/THF; 10% aq. H_2SO_4/THF; 3% aq. TsOH/THF.[2]

4-(4-Nitrophenyl)-1,3-dioxolane

This derivative is prepared from the diol by standard acid catalyzed ketal formation. It is cleaved by electrochemical reduction at a Hg electrode.[3]

1. L. Ceita, A. K. Maiti, R. Mestres, and A. Tortajada, *J. Chem. Res. (S)*, 403 (2001).

2. J. Hébert and D. Gravel, *Can. J. Chem.*, **52**, 187 (1974); D. Gravel, J. Hébert, and D. Thoraval, *Can. J. Chem.*, **61**, 400 (1983).

3. J. M. Chapuzet, C. Gru, R. Labrecque, and J. Lessard, *J. Electroanalyt. Chem.*, **507**, 22 (2001); R. Labrecque, J. Mailhot, B. Daoust, J. M. Chapuzet, and J. Lessard, *Electrochimica Acta*, **42**, 2089 (1997).

4-Fluorous Acetal derivatives

$R_f = (CF_2)_nCF_3$ $n = 5$ or 7

This and other fluorous acetal derivatives are prepared from diols bearing 13 or more F atoms to make them soluble in fluorinated hydrocarbons. They are prepared by the standard methods of heating the ketone with the diol in the presence of an acid such as TsOH or pyridinium tosylate. As with most 1,3-dioxanes and 1,3-dioxolanes, they can be cleaved with aqueous acid.[1,2]

1. Y. Huang and F.-L. Qing, *Tetrahedron*, **60**, 8341 (2004).

2. R. W. Read and C. Zhang, *Tetrahedron Lett.*, **44**, 7045 (2003).

4-[6-Bromo-7-hydroxycoumar-4-yl]-1,3-dioxolane (Bhc-diol) Ketal

The ketal is prepared in low yield from the diol (PPTS, MgSO$_4$, toluene, BuOH, 110°C, 22–57%) and is cleaved by irradiation at 365 nm at pH 7.2. It was developed for releasing aldehydes and ketones by a one- or two-photon excitation under physiological conditions.[1]

1. M. Lu, O. D. Fedoryak, B. R. Moister, and T. M. Dore, *Org. Lett.*, **5**, 2119 (2003).

4-Trimethylsilylmethyl-1,3-dioxolane

Formation/Cleavage[1]

Hindered ketones and enones fail to form the ketal because of competing decomposition of the silyl reagent. This occurs via a Peterson olefination process.

1. B. M. Lillie and M. A. Avery, *Tetrahedron Lett.*, **35**, 969 (1994).

O,O′-Phenylenedioxy Ketal

The phenylenedioxy ketal is prepared from catechol (TsOH, 90°C, 30 h, 85% yield) or KSF or K-10 clay (benzene, reflux)[1] and is cleaved with $5N$ HCl (dioxane, reflux, 6 h). It is more stable to acid than the ethylene ketal.[2,3]

1. T.-S. Li, L.-J. Li, B. Lu, and F. Yang, *J. Chem. Soc. Perkin Trans. 1*, 3561 (1998); B. List, D. Shabat, C. F. Barbos, III, and R. A. Lerner, *Chem. Eur. J.*, **4**, 881 (1998).

2. M. Rosenberger, D. Andrews, F. DiMaria, A. J. Duggan, and G. Saucy, *Helv. Chim. Acta*, **55**, 249 (1972).

3. M. Rosenberger, A. J. Duggan, and G. Saucy, *Helv. Chim. Acta*, **55**, 1333 (1972).

1,3-Dioxapane

Medium ring cyclic acetals are much more labile than either the 1,3-dioxolane or 1,3-dioxane. They can be formed by some of the same methods used for the preparation of other acetals. The following are the relative cleavage rates for various benzophenone ketals.[1]

1	4.9	14	34.9

Formation

1. HO(CH$_2$)$_4$OH, HC(OEt)$_3$, EtOH, 2,4,4,6-tetrabromo-2,5-cyclohexadienone (TABCO), 73% yield.[2]
2. HO(CH$_2$)$_4$OH, PPTS, benzene, reflux, 92% yield.[3]

Cleavage

0.1 M HCl, acetone, H$_2$O, rt, 3 h, 75% yield.[4] The dioxolane could not be cleaved from this substrate.

1. T. Oshima, S.-y. Ueno, and T. Nagai, *Heterocycles*, **40**, 607 (1995). See also J.-Y. Conan, A. Natat, and D. Priolet, *Bull. Soc. Chim.*, 1935 (1976).

2. H. Firouzabadi, N. Iranpoor, and H. R. Shaterian, *Bull. Chem. Soc. Jpn.*, **75**, 2195 (2002).

3. K. M. Brummond and J. Lu, *Org. Lett.*, **3**, 1347 (2001).

4. B. B. Snider and H. Lin, *Org. Lett.*, **2**, 643 (2000).

1,5-Dihydro-3*H*-2,4-benzodioxepin

Formation[1,2]

1.

Ref. 1

Camphor cannot be protected with this reagent, indicating that steric factors will prevent its use in very hindered systems.

2. 1,2-Dihydroxymethylbenzene, $CH(OCH_3)_3$, TsOH, 80% yield.[3,4]

3. From a methyl enol ether: 1,2-dihydroxymethylbenzene, Amberlyst H^+, 85% yield.[5]

4. 1,2-Dihydroxymethylbenzene, sulfonated charcoal or TsOH, PhH, reflux, 88–98% yield.[6]

5. 1,2-Ditrimethylsiloxymethylbenzene, TMSOTf, CH_2Cl_2, −78°C, 96% yield.[7]

6. 1,2-Dihydroxymethylbenzene, H-Y Zeolite, CH_2Cl_2, reflux, 3–12 h, 46–95% yield.[8]

7. 1,2-Dihydroxymethylbenzene, Environcat EPZG, toluene, reflux, 93–99% yield. Ketones were not reactive under these conditions.[9]

Cleavage

1. H_2, PdO, THF, rt, 0.5 h, 100% yield.[1]

2. 5% Pd–C, H_2, 95% yield.[10]

1. N. Machinaga and C. Kibayashi, *Tetrahedron Lett.*, **30**, 4165 (1989).

2. K. Mori, T. Yoshimura, and T. Sugai, *Liebigs Ann. Chem.*, 899 (1988).

3. R. Oi and K. B. Sharpless, *Tetrahedron Lett.*, **33**, 2095 (1992).

4. S. D. Burke and D. N. Deaton, *Tetrahedron Lett.*, **32**, 4651 (1991).

5. L. Schmitt, B. Spiess, and G. Schlewer, *Tetrahedron Lett.*, **33**, 2013 (1992).

6. H. K. Patney, *Tetrahedron Lett.*, **32**, 413 (1992).

7. S. V. D'Andrea, J. P. Freeman, and J. Szmuszkovicz, *Org. Prep. Proced. Int.*, **23**, 432 (1991).

8. T. P. Kumar, K. R. Reddy, and R. S. Reddy, *J. Chem. Res., Synop.*, 394 (1994).

9. B. P. Bandgar, M. M. Kulkarni, and P. P. Wadgaonkar, *Synth. Commun.*, **27**, 627 (1997).

10. R. K. Boeckman, Jr., J. Zhang, and M. R. Reeder, *Org. Lett.*, **4**, 3891 (2002).

7,7-Dimethyl-1,2,4-trioxepane

These acetals are remarkably stable to acid. They are also stable to the following conditions: toluene, 110°C, Ph_3P, $Pd(Ph_3P)_4$, $NaBH_4$, H_2CrO_4, DDQ, TEA, Me_2NH,

TEA, CuCl, NaH, DMSO, 10% aq. HCl/THF, 10% NaOH/MeOH, TsOH/MeOH, t-BuOK/THF, Pt/H$_2$, LiAlH$_4$. This group is not stable to BuLi. A 1,3-dioxolane can be cleaved in the presence of the trioxepane group.[1]

Formation

The require peroxide is easily prepare and can be used in crude form.

Cleavage

Zn, AcOH or Mg, MeOH, 40–100% yield.[1]

3,3-Dialkyl-6-(1-phenylvinyl)-1,2,4-trioxane

These derivatives are prepared from the readily prepared hydroperoxide by the standard acid catalyzed ketal formation. Cleavage is achieved under basic conditions by treatment with Triton B in THF at rt, 62–87% yield. This group is stable to Grignard reagents, the Wadsworth–Emmons reaction, and reductive amination with NaBH(OAc)$_3$.[2]

1. A. Ahmed and P. H. Dussault, *Org. Lett.*, **6**, 3609 (2004).
2. C. Singh and H. Malik, *Org. Lett.*, **7**, 5673 (2005).

Chiral Acetals and Ketals

Chiral protecting groups, although less frequently used in synthesis, provide sought-after protection, diastereochemical control, and enantioselectivity, and can improve the chemical characteristics of a molecule to facilitate a synthesis.[1]

(4R,5R)-Diphenyl-1,3-dioxolane

Formation

1. (1*R*,2*R*)-Diphenyl-1,2-ditrimethylsiloxyethane, TMSOTf, 66% yield.[2]
2. (1*R*,2*R*)-Diphenyl-1,2-ethanediol, PPTS, 80°C.[3]

Cleavage

1. 2.7 *N* HCl, MeOH, 25°C, 90% yield.[3]
2. Pd(OH)$_2$, H$_2$, EtOAc, quant.[2]

4,5-Dimethyl-1,3-dioxolane

Formation

1. 2,3-Bistrimethylsiloxybutane, TMSOTf, CH$_2$Cl$_2$, 66% yield. An enone does not migrate out of conjugation.[4]
2. 2,3-Butanediol, benzene, PPTS, reflux, 66% yield.[5]
3.

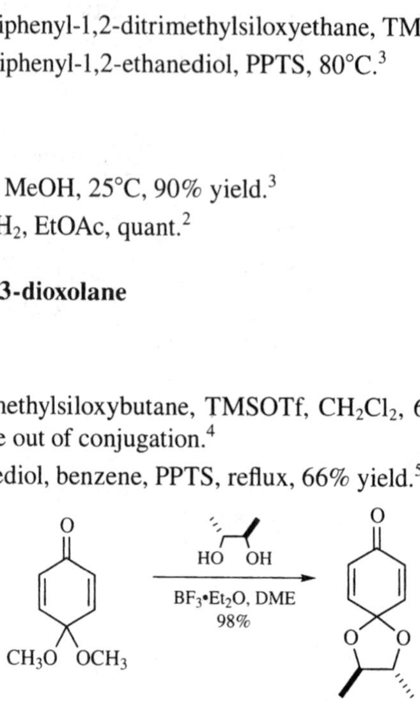

Refs. 6, 7

This reaction also works to form the related dioxane, but the yields are lower.[6]

trans-1,2-Cyclohexanediol Ketal

Formation

trans-1,2-Cyclohexanediol, *i*-PrOTMS, TMSOTf, CH$_2$Cl$_2$, −20°C, 3h, 85% yield.[8]

trans-4,6-Dimethyl-1,3-dioxane

Formation

1. 2,3-Pentanediol, PPTS, >95% yield.[8,9]
2.

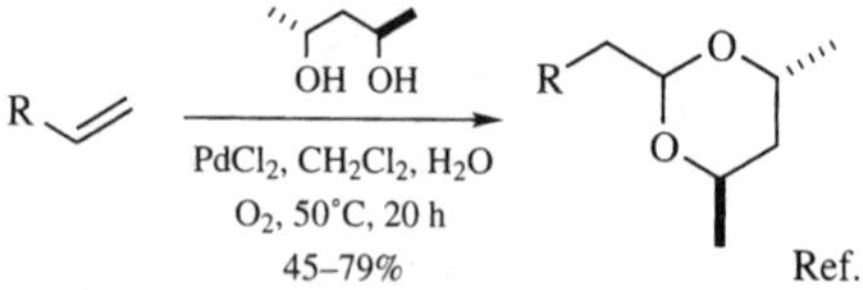

Ref. 10

3. 2,3-Pentanediol, Sc(OTf)$_3$, rt, 13 h to 2 days, benzene, THF or CH$_2$Cl$_2$, 59–100%. This method is also effective for formation of a 4,5-dimethyldioxolane.[11]

Cleavage

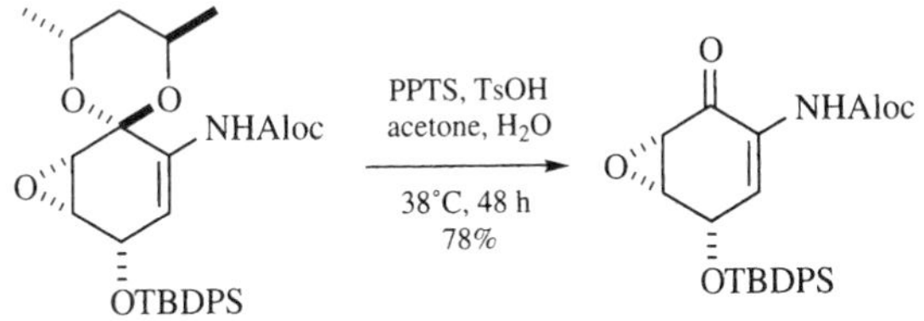
NHAloc
OTBDPS
PPTS, TsOH
acetone, H_2O
38°C, 48 h
78%
NHAloc
OTBDPS

Hydrolysis is facilitated by the increased level of strain imparted by the axial methyl group, thus allowing cleavage under conditions to which the product is stable.[12]

4,5-Bis(dimethylaminocarbonyl)-1,3-dioxolane

This chiral protective group was developed for use in the synthesis of optically active alcohols.[13]

Formation[13]

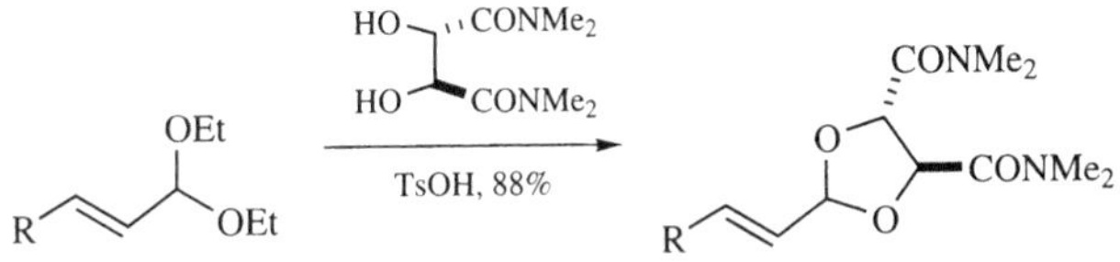
OEt
OEt
$CONMe_2$
$CONMe_2$
TsOH, 88%
$CONMe_2$
$CONMe_2$

Cleavage[13]

6 *M* HCl, dioxane, >92% yield.

4,5-Dicarbomethoxy-1,3-dioxolane

Formation

1. Dimethyl tartrate, $Sc(OTf)_3$, MeCN, rt, 3 h, 95% yield.[14]
2.

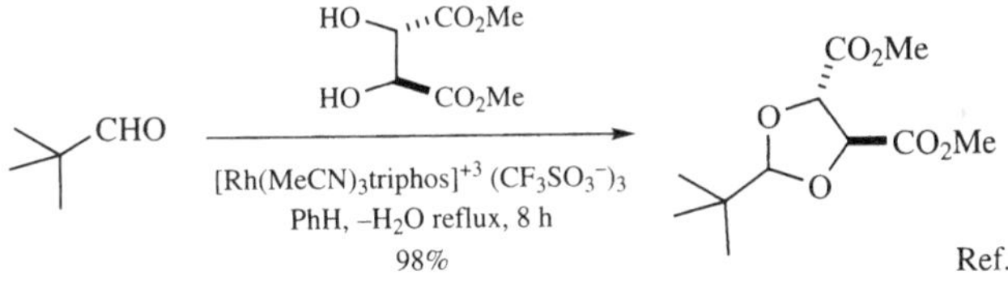
CHO
CO_2Me
CO_2Me
$[Rh(MeCN)_3triphos]^{+3}$ $(CF_3SO_3^-)_3$
PhH, $-H_2O$ reflux, 8 h
98%
CO_2Me
CO_2Me

Ref.15

4,5-Dimethoxymethyl-1,3-dioxolane

Formation/Cleavage[16]

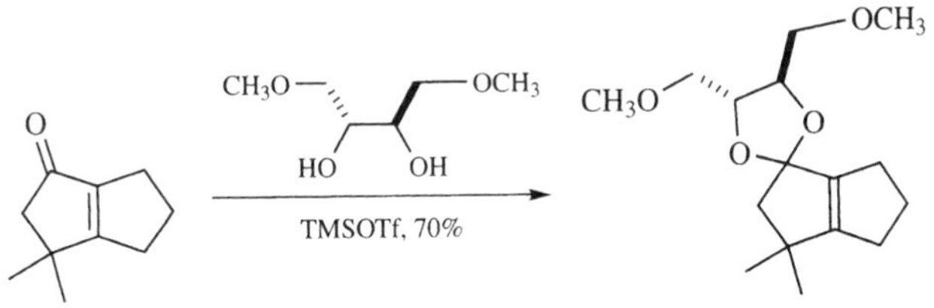
CH_3O
OCH_3
HO
OH
TMSOTf, 70%
OCH_3
CH_3O

This protective group was used to direct the selective cyclopropanation of a variety of enones. Hydrolysis (HCl, MeOH, H_2O, rt, 94% yield) affords optically active cyclopropyl ketones.

2,2-Dialkyl-4,5-bis(2-nitrophenyl)-1,3-dioxolane

Formation

Bis(o-nitrophenyl)ethanediol, benzene, reflux, PPTS, 67–92% yield.[17]

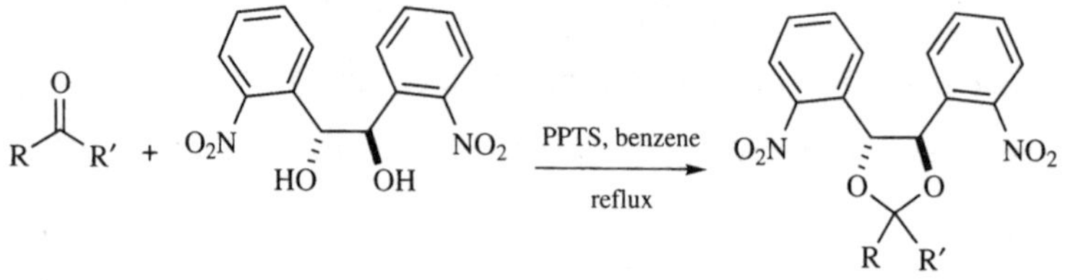

Cleavage

$h\nu$ 350 nm, CH_3CN or CH_2Cl_2, 1–2 h, 69–97% yield by GC or NMR.

4,5-Bis(2-nitro-4,5-dimethoxyphenyl)-1,3-dioxolane:

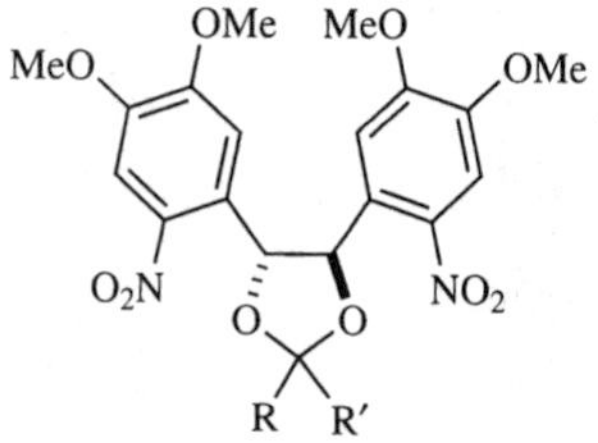

This dioxolane was developed as a photochemically removable dioxolane for ketones. It is formed from a ketone and the diol in benzene with PTSA catalysis in 55–95% yield. The ketal is stable to dilute acid, $2N$ NaOH, NaH, $LiAlH_4$, t-BuOK, $NaBH_4$, DDQ, TBAF, and CAN. Cleavage is accomplished by irradiation at 350 nm in 68–92% yield.[18]

1. A review: A. Alexakis and P. Mangeney, *Tetrahedron: Asymmetry*, **1**, 477 (1990).
2. C. N. Eid, Jr., and J. P. Konopelski, *Tetrahedron Lett.*, **32**, 461 (1991).
3. J. Cossy and S. BouzBouz, *Tetrahedron Lett.*, **37**, 5091 (1996).
4. E. A. Mash and S. B. Hemperly, *J. Org. Chem.*, **55**, 2055 (1990).
5. M. Toyota, Y. Nishikawa, and K. Fukumoto, *Tetrahedron*, **52**, 10347 (1996).
6. M. C. Pirrung and D. S. Nunn, *Tetrahedron Lett.*, **33**, 6591 (1992).
7. P. de March, M. Escoda, M. Figueredo, J. Font, A. Alvarez-Larena, and J. F. Piniella, *J. Org. Chem.*, **60**, 3895 (1995).
8. M. Kurihara and N. Miyata, *Chem. Lett.*, **24**, 263 (1995).
9. A. Mori and H. Yamamoto, *J. Org. Chem.*, **50**, 5444 (1985).

10. T. Hosokawa, T. Ohta, S. Kanayama, and S. I. Murahashi, *J. Org. Chem.*, **52**, 1758 (1987).

11. S.-i. Fukuzawa, T. Tsuchimoto, T. Hotaka, and T. Hiyama, *Synlett*, 1077 (1995).

12. P. Wipf, Y. Kim and H. Jahn, *Synthesis*, 1549 (1995).

13. J. Fujiwara, Y. Fukutani, M. Hasegawa, K. Maruoka, and H. Yamamoto, *J. Am. Chem. Soc.*, **106**, 5004 (1984).

14. K. Ishihara, Y. Karumi, M. Kubota, and H. Yamamoto, *Synlett*, 839 (1996).

15. J. Ott, G. M. Ramos Tombo, B. Schmid, L. M. Venanzi, G. Wang, and T. R. Ward, *Tetrahedron Lett.*, **30**, 6151 (1989).

16. E. A. Mash, S. K. Math, and C. J. Flann, *Tetrahedron Lett.*, **29**, 2147 (1988).

17. A. Blanc and C. G. Bochet, *J. Org. Chem.*, **68**, 1138 (2003).

18. S. Kantevari, C. V. Narasimhaji, and H. B. Mereyala, *Tetrahedron*, **61**, 5849 (2005).

Dithio Acetals and Ketals

A carbonyl group can be protected as a dithio acetal or ketal, 1,3-dithiane, or 1,3-dithiolane by reaction of the carbonyl compound in the presence of an acid catalyst with a thiol or dithiol. The derivatives are, in general, cleaved by reaction with Lewis acids or oxidation; acidic hydrolysis is unsatisfactory. The acyclic derivatives are formed and hydrolyzed much more readily than their cyclic counterparts. Representative examples of formation and cleavage are shown below.

Acyclic Dithio Acetals and Ketals

S,S'-Dimethyl Acetals and Ketals: $RR'C(SCH_3)_2$ (Chart 5)

S,S'-Diethyl Acetals and Ketals: $RR'C(SC_2H_5)_2$

S,S'-Dipropyl Acetals and Ketals: $RR'C(SC_3H_7)_2$

S,S'-Dibutyl Acetals and Ketals: $RR'C(SC_4H_9)_2$

S,S'-Dipentyl Acetals and Ketals: $RR'C(SC_5H_{11})_2$

S,S'-Diphenyl Acetals and Ketals: $RR'C(SC_6H_5)_2$

S,S'-Dibenzyl Acetals and Ketals: $RR'C(SCH_2C_6H_5)_2$

General Methods of Formation

1. RSH, concd. HCl, 20°C, 30 min.[1] These conditions were used to protect an aldose as the methyl or ethyl thioketal.

2. $RSSiMe_3$, ZnI_2, Et_2O, 0–25°C, 70–95% yield.[2] This method is satisfactory for a variety of aldehydes and ketones and is also suitable for the preparation of 1,3-dithianes. Methacrolein gives the product of Michael addition rather than the thioacetal. The less hindered of two ketones is readily protected using this methodology.[3]

3. RSH, Me$_3$SiCl, CHCl$_3$, 20°C, 1 h, >80% yield.[4]

4. B(SR)$_3$, reflux, 2 h or 25°C, 18 h, 75–85% yield.[5]

5. Al(SPh)$_3$, 25°C, 1 h, 65% yield.[6] This method also converts esters to thioesters.

6. PhSH, BF$_3$·Et$_2$O, CHCl$_3$, 0°C, 10 min, 86% yield.[7] ZnCl$_2$[8] and MgBr$_2$[9] have also been used as catalysts. With MgBr$_2$ acetals can be converted to thioacetals in the presence of ketones.

7. RSH, LiBr, 75–80°C, 80–99% yield. This method is also effective for the preparation of dithianes.[10]

8. Sc(OTf)$_3$, EtSH, ionic liquid, 7–15 min, 90–95% yield.[11]

9. RSH, SO$_2$, benzene, 54–81% yield.[12]

10. EtSH, TiCl$_4$, CHCl$_3$, 6–12 h, rt, 90–98% yield.[13]

11. P-PPh$_2$·I$_2$, RSH, Et$_3$N, CH$_3$CN; K$_2$CO$_3$, H$_2$O, 80–98% yield.[14] This method is also effective for the formation of dioxolanes and dithiolanes.

12. RSSR (R = Me, Ph, Bu), Bu$_3$P, rt, 15–83% yield. This reagent also reacts with epoxides to form 1,2-dithioethers.[15,16]

13. H-Y or H-M zeolite, hexane or CH$_2$Cl$_2$, EtSH, reflux, 0.75–144 h, 50–96% yield.[17]

14. NaHSO$_4$·SiO$_2$, CH$_2$Cl$_2$, rt, 5–10 min, 5–10 h. 75–98% yield. Aldehydes are selectively protected over ketones. In the presence of water, this reagent will cleave dithioacetals and in the presence of a diol it will convert a dithioacetal to an acetal.[18]

General Methods of Cleavage

1. AgNO$_3$/Ag$_2$O, CH$_3$CN-H$_2$O, 0°C, 2 h, 85% yield.[19]

This method has also been used to cleave dithianes and dithiolanes.[20] The
S,S'-dibutyl group is stable to acids (e.g., HOAc/H_2O-THF, 45°C, 3 h; TsOH/
CH_2Cl_2, 0°C, 0.5 h).[19]

2. $AgClO_4$, H_2O, C_6H_6, 25°C, 4 h, 80–100% yield.[22]

3. $FeCl_3$·$6H_2O$, CH_2Cl_2, rt, 15 min, 80–98% yield.[23]

4. $Bi(OTf)_3$,·xH_2O, CH_2Cl_2, H_2O, rt, 10 min, 80–95% yield.[23]

5. $GaCl_3$, CH_2Cl_2, H_2O, rt, 20 min.[24] Thioketals are cleaved in preference to
   thioacetals and dithianes, which do not react.

6. $HgCl_2$, $CdCO_3$, aq. acetone[25] or $HgCl_2$, $CaCO_3$, CH_3CN, H_2O.[26] In a case
   where this combination of reagents was not effective, HgO/BF_3·Et_2O was
   found to work.[27]

7. $HgCl_2$, HgO, 80% CH_3CN, H_2O, 30 min, rt, 96% yield.[28]

8. $Tl(NO_3)_3$, CH_3OH, H_2O, 25°C, 5 min, 73–98% yield.[7] These conditions are
   also effective for the cleavage of dithiolanes and dithianes.

9. SO_2Cl_2, SiO_2·H_2O, CH_2Cl_2, 25°C, 2–3 h, 90–100% yield.[29,30]

10. The dithioacetal can be converted to an O,S-acetal.[31] The mixed acetals were
    then used to prepare furanosides.

11. In the presence of dibromantin and an alcohol dithioacetals are converted to
    the acetal (85–90% yield) and in the presence of a 1,2-diol they are converted
    to dioxolanes (75–80% yield).[32]

12. DMSO, 140–160°C, 4–5 h, 79–94% yield.[33]

13. I_2, $NaHCO_3$, dioxane, H_2O, 25°C, 4.5 h, 80–95% yield.[34]

14. I_2, MeOH, reflux, 2 h, 79%; $HClO_4$, H_2O, 25°C, 16 h, 87% yield.[35] These con-
    ditions also cleave acetonides and benzylidene acetals.[36]

15. Cetyltrimethylammonium tribromide, CH_2Cl_2, 0–5°C, 5–30 min, 65–95%
    yield.[37]

16. H_2O_2, aq. acetone or $NaIO_4$/H_2O, 25°C; g HCl/$CHCl_3$, 0°C, 50–70% yield.[38]

17. O_2, $h\nu$, hexane, Ph_2CO, 2–5 h, 60–80% yield.[39] 1,3-Oxathiolanes and dithi-
    olanes are also cleaved by these conditions.

18. CuCl, CuO, H_2O, acetone, 2 h, 20°C, 61–73% yield.[40]

19. MCPBA, CF_3COOH, CH_2Cl_2, 0°C.[41]

20. Ph_3CClO_4, Ph_3COMe, CH_2Cl_2, −45°C, 2.5 h; aq. $NaHCO_3$, 84–96% yield.[42]
    A diethyl thioketal could be cleaved in the presence of a diphenyl thioketal.

21. DDQ, CH_3CN, H_2O, 80°C, 43–95% yield.[43] These conditions also resulted in
    cleavage of acetyl groups; a dithiolane was stable to these conditions.

22. $Me_2CH(CH_2)_2ONO$, CH_2Cl_2; 25°C, 15 min, H_2O, 63–93% yield.[44] Isoamyl
    nitrite cleaves aromatic dithioacetals in preference to aliphatic dithioacetals,
    and dithioacetals in preference to dithioketals. It also cleaves 1,3-oxathiolanes
    (1 h, 65–90% yield).

23. Clay supported NH_4NO_3, CH_2Cl_2, rt, 76–90% yield.[45]

24. N-Chlorosuccinimide, $AgNO_3$, CH_3CN, H_2O, 0°C, >68% yield.[46]

1. H. Zinner, *Chem. Ber.*, **83**, 275 (1950).

2. D. A. Evans, L. K. Truesdale, K. G. Grimm, and S. L. Nesbitt, *J. Am. Chem. Soc.*, **99**, 5009 (1977).

3. D. A. Evans, K. G. Grimm, and L. K. Truesdale, *J. Am. Chem. Soc.*, **97**, 3229 (1975).

4. B. S. Ong and T. H. Chan, *Synth. Commun.*, **7**, 283 (1977).

5. F. Bessette, J. Brault, and J. M. Lalancette, *Can. J. Chem.*, **43**, 307 (1965).

6. T. Cohen and R. E. Gapinski, *Tetrahedron Lett.*, **19**, 4319 (1978).

7. E. Fujita, Y. Nagao, and K. Kaneko, *Chem. Pharm. Bull.*, **26**, 3743 (1978).

8. W. E. Truce and F. E. Roberts, *J. Org. Chem.*, **28**, 961 (1963).

9. J. H. Park and S. Kim, *Chem. Lett.*, **18**, 629 (1989).

10. H. Firouzabadi, N. Iranpoor, and B. Karimi, *Synthesis*, 58 (1999).

11. A. Kamal and G. Chouhan, *Tetrahedron Lett.*, **44**, 3337 (2003).

12. B. Burczyk and Z. Kortylewicz, *Synthesis*, 831 (1982).

13. V. Kumar and S. Dev, *Tetrahedron Lett.*, **24**, 1289 (1983).

14. R. Caputo, C. Ferreri, and G. Palumbo, *Synthesis*, 386 (1987).

15. M. Tazaki and M. Takagi, *Chem. Lett.*, **8**, 767 (1979).

16. M. Inoue, S. Yamashita, A. Tatami, K. Miyazaki, and M. Hirama, *J. Org. Chem.*, **69**, 2797 (2004).

17. P. Kumar, R. S. Reddy, A. P. Singh, and B. Pandey, *Synthesis*, 67 (1993).

18. B. Das, R. Ramu, M. R. Reddy, and G. Mahender, *Synthesis*, 250 (2005).

19. E. J. Corey, M. Shibasaki, J. Knolle, and T. Sugahara, *Tetrahedron Lett.*, **18**, 785 (1977).

20. C. H. Heathcock, M. J. Taschner, T. Rosen, J. A. Thomas, C. R. Hadley, and G. Popják, *Tetrahedron Lett.*, **23**, 4747 (1982); R. Zamboni and J. Rokach, *Tetrahedron Lett.*, **23**, 4751 (1982).

21. T. Mukaiyama, S. Kobayashi, K. Kamio, and H. Takei, *Chem. Lett.*, **1**, 237 (1972).

22. A. Kamal, E. Laxman, and P. S. M. M. Reddy, *Synlett*, 1476 (2000).

23. A. Kamal, P. S. M. M. Reddy, and D. R. Reddy, *Tetrahedron Lett.*, **44**, 2857 (2003).

24. K. Saigo, Y. Hashimoto, N. Kihara, H. Umehara, and M. Hasegawa, *Chem. Lett.*, **19**, 831 (1990).

25. J. English, Jr., and P. H. Griswold, Jr., *J. Am. Chem. Soc.*, **67**, 2039 (1945).

26. A. I. Meyers, D. L. Comins, D. M. Roland, R. Henning, and K. Shimizu, *J. Am. Chem. Soc.*, **101**, 7104 (1979).

27. P. Norris, D. Horton, and B. R. Levine, *Tetrahedron Lett.*, **36**, 7811 (1995).

28. V. E. Amoo, S. De Bernardo, and M. Weigele, *Tetrahedron Lett.*, **29**, 2401 (1988).

29. M. Hojo and R. Masuda, *Synthesis*, 678 (1976).

30. Y. Kamitori, M. Hojo, R. Masuda, T. Kimura, and T. Yoshida, *J. Org. Chem.*, **51**, 1427 (1986).

31. J. C. McAuliffe and O. Hindsgaul, *J. Org. Chem.*, **62**, 1234 (1997).

32. S. K. Madhusudan and A. K. Misra, *Carbohydr. Res.*, **340**, 497 (2005).

33. Ch. S. Rao, M. Chandrasekharam, H. Ila, and H. Junjappa, *Tetrahedron Lett.*, **33**, 8163 (1992).

34. G. A. Russell and L. A. Ochrymowycz, *J. Org. Chem.*, **34**, 3618 (1969).

35. B. M. Trost, T. N. Salzmann, and K. Hiroi, *J. Am. Chem. Soc.*, **98**, 4887 (1976).

36. W. A. Szarek, A. Zamojski, K. N. Tiwari, and E. R. Ison, *Tetrahedron Lett.*, **27**, 3827 (1986).

37. E. Mondal, G. Bose, and A. T. Khan, *Synlett*, 785 (2001).

38. H. Nieuwenhuyse and R. Louw, *Tetrahedron Lett.*, **12**, 4141 (1971).

39. T. T. Takahashi, C. Y. Nakamura, and J. Y. Satoh, *J. Chem. Soc., Chem. Commun.*, 680 (1977).

40. B. Cazes and S. Julia, *Tetrahedron Lett.*, **19**, 4065 (1978).

41. J. Cossy, *Synthesis*, 1113 (1987).

42. M. Ohshima, M. Murakami, and T. Mukaiyama, *Chem. Lett.*, **15**, 1593 (1986).

43. J. M. Garcia Fernandez, C. O. Mellet, A. M. Marin, and J. Fuentes, *Carbohyd. Res.* **274**, 263 (1993).

44. K. Fuji, K. Ichikawa and E. Fujita, *Tetrahedron Lett.*, **19**, 3561 (1978).

45. H. M. Meshram, G. S. Reddy, and J. S. Yadav, *Tetrahedron Lett.*, **38**, 8891 (1997).

46. M. Naruto, K. Ohno, and N. Naruse, *Chem. Lett.*, **7**, 1419 (1978).

S,S'-Diacetyl Acetals and Ketals: $R_2C(SCOCH_3)_2$

Formation[1]

Cleavage[1]

The formyl group was lost during attempted protection with ethylene glycol, TsOH.

1. T. Kametani, Y. Kigawa, K. Takahashi, H. Nemoto, and K. Fukumoto, *Chem. Pharm. Bull.*, **26**, 1918 (1978).

Cyclic Dithio Acetals and Ketals

1,3-Dithiane Derivative ($n = 3$): (Chart 5)

1,3-Dithiolane Derivative ($n = 2$): (Chart 5)

The popularity of the dithiane group stems largely from its ability to be deprotonated by n-BuLi to form an anion that reacts with a variety of reagents to form a carbon–carbon bond. It is exceptionally acid stable when compared to the 1,3-dioxolane or 1,3-dioxane groups. As with most sulfur-containing molecules, its downfall is the stench associated with the reagents used to introduce it and the by-products that result from its deprotection. Because of its unique position as a conjunctive unit in synthesis, it is nonetheless a frequently used protective group.[1] Although numerous methods are available for deprotection of this group, most have not been tested during the rigors of complex synthesis. The majority of examples published tend to be simple unfunctionalized substrates. A review that covers the synthesis and cleavage of 1,3-dithiolanes has been published.[2] The role of dithianes in natural product synthesis has been extensive and has been reviewed.[3]

General Methods of Formation

Lewis Acid-Catalyzed Methods

1. $HS(CH_2)_nSH$, BF_3·Et_2O, CH_2Cl_2, 25°C, 12h, high yield, $n = 2$[4], $n = 3$[5]. In α,β-unsaturated ketones the olefin does not migrate to the β,γ-position as occurs when an ethylene ketal is prepared.[6] Aldehydes are selectively protected in the presence of ketones, except when large steric factors disfavor the aldehyde group, as in the example below.[7] A TBDMS group is not stable to these conditions.[8] Oxazolidines are converted to the dithiane in 70% yield under these conditions,[9] but the use of methanesulfonic acid as a catalyst is equally effective.[10]

2.

   R = Cl or Ph   CHCl_3, 25°C, 2 h, 90–100% yield.[11]

When R = Ph, the reaction is selective for unhindered ketones. Diaryl ketones, generally unreactive compounds, react rapidly when R = Cl.

3. Me$_3$SiSCH$_2$CH$_2$SSiMe$_3$, ZnI$_2$, Et$_2$O, 0–25°C, 12–24 h, high yields.[12] Less hindered ketones can be selectively protected in the presence of more hindered ketones. α,β-Unsaturated ketones are selectively protected (94:1, 94:4) in the presence of saturated ketones by this reagent.[13]

4. HS(CH$_2$)$_2$SH, TiCl$_4$, −10–25°C, 96% yield.[14]

5. HS(CH$_2$)$_n$SH, MeCN, ScCl$_3$, or CoCl$_2$, rt, 2 h, 70–93% yield. Aldehydes react chemoselectively in the presence of ketones.[15]

6. HSCH$_2$CH$_2$SH, SnCl$_2$·H$_2$O, THF, reflux, 10–240 min, 51–96% yield.[16] Under these conditions, aldehydes react faster than ketones. Dimethyl ketals, which react faster than dimethyl acetals, are also converted to dithianes and dithiolanes under these conditions (75–100% yield).[17]

7. HSCH$_2$CH$_2$SH, MgI$_2$, Et$_2$O, rt, 8 h, 95–96% yield.[18] Aryl ketones are not efficiently protected.

8. HS(CH$_2$)$_n$SH, MeCN, SmI$_3$, 62–92% yield.[19]

9. HSCH$_2$CH$_2$SH, Zn(OTf)$_2$ or Mg(OTf)$_2$, ClCH$_2$CH$_2$Cl, heat, 16 h, 85–99% yield.[20,21] Excellent selectivity can be achieved between a hindered and an unhindered ketone.[22] α,β-Unsaturated ketones such as carvone are not cleanly converted to ketals because of Michael addition of the thiol.[20]

In this case other methods failed because of β-elimination.

10. HS(CH$_2$)$_n$SH, 40% aq. Zn(BF$_4$)$_2$, CH$_2$Cl$_2$, 5 min to 15 h, 70–95% yield. Acyclic ketones are unreactive.[23]

11. Sc(OTf)$_3$, HS(CH$_2$)$_n$SH, CH$_2$Cl$_2$, rt, 55–94% yield. Aldehydes react in preference to ketones.[24]

12. HS(CH$_2$)$_3$SH, Al(OTf)$_3$, ClCH$_2$CH$_2$Cl, rt, 50–98% yield.[25]

13. HS(CH$_2$)$_n$SH, Lu(OTf)$_3$, rt, CH$_3$CN, 68–90% yield. Aldehydes react in preference to ketones.[26] Y(OTf)$_3$ as a catalyst gives similar results.[27]

14. HSCH$_2$CH$_2$SH, LiClO$_4$, ether, 70–95% yield.[28] Lithium triflate is a similarly effective catalyst.[29]

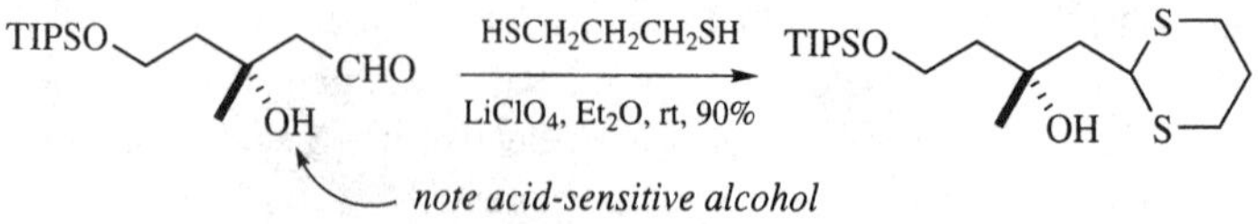

— *note acid-sensitive alcohol*

15. HS(CH$_2$)$_3$SH, LiBF$_4$, neat, 25°C, 74–100% yield.[30]
16. 1,3-Dioxolanes[31,32] and 1,3-dioxanes[33] are readily converted to 1,3-dithiolanes and 1,3-dithianes in good to excellent yields.

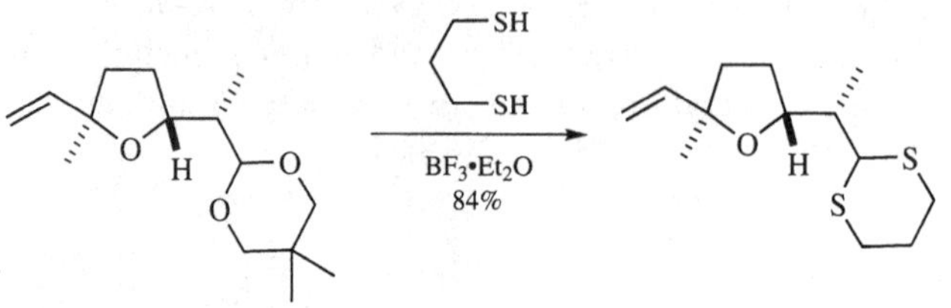

Ref. 33

17.

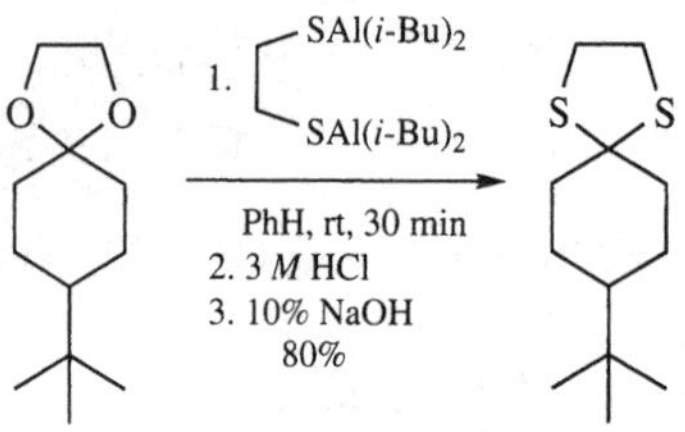

Ref. 32

18. 2,2-Dimethyl-2-sila-1,3-dithiane, BF$_3$·Et$_2$O, CH$_2$Cl$_2$, 0°C, 82–99% yield.[34] This method was reported to be superior to the conventional synthesis because cleaner products are formed. Aldehydes are selectively protected in the presence of ketones, which do not react competitively with this reagent.

19. 2,2-Dibutyl-2-stanna-1,3-dithiane, Bu$_2$Sn(OTf)$_2$, ClCH$_2$CH$_2$Cl, 35°C, 1 h, 77–94% yield.[35] TBDMS, TBDPS, THP, and OAc groups are not affected by these conditions.

20. HS(CH$_2$)$_n$SH, ClCH$_2$CH$_2$Cl, TeCl$_4$, rt, 80–99% yield.[36] This method is also effective for converting dimethyl acetals to the thioacetal and for selectively protecting an aldehyde in the presence of a ketone.

21. HSCH$_2$CH$_2$SH, CH$_2$Cl$_2$, LaCl$_3$, 1–96 h, 25–93% yield.[37]

22. InBr$_3$, InCl$_3$, or In(OTf)$_3$; HS(CH$_2$)$_n$SH; CH$_2$Cl$_2$ or H$_2$O, 33–98% yield.[38] InCl$_3$ will convert acetals and ketals to the dithianes and dithiolanes.[39]

23. HSCH$_2$CH$_2$SH, VO(OTf)$_2$, CH$_3$CN, rt, 72–95% yield. Aldehydes are protected selectively in the presence of ketones. Acyclic thioacetals are formed similarly.[40] This author has also used RuCl$_3$ to affect this transformation.[41]

24. HS(CH$_2$)$_n$SH or HSCH$_2$CH$_2$OH, MoO$_2$(acac)$_2$, CH$_3$CN, rt, 1.5–4 h, 78–98% yield.[42]

25. HS(CH$_2$)$_n$SH, MoCl$_5$, CH$_2$Cl$_2$, rt, 2 min to 36 h, 70–98% yield. This method selectively converts open chain acetals to dithiolanes in the presence of

the cyclic analog. In the presence of DMSO this reagent will also cleave thioacetals.[43]

26. From *N,N*-dialkylhydrazones: HSCH$_2$CH$_2$SH, CH$_2$Cl$_2$, BF$_3$·Et$_2$O, 84–98% yield. With electronically deficient derivatives the reaction can require days to complete.[44]

27. From an enol either: HSCH$_2$CH$_2$SH, CH$_2$Cl$_2$, TMSOTf, −78°C, 4 h, 76–94% yield.[45]

28. From an acetylenic ketone by Michael addition.[46]

29. The following method is one that does not use a malodorous reagent to introduce a dithiane. The reaction can also be done in water in the presence of a surfactant.[47]

30. HSCH$_2$CH$_2$SH, *p*-dodecylbenzenesulfonic acid, H$_2$O, 40°C, 4 h, 74–94% yield.[48]

31. Dithiol or thiol, tungstophosphoric acid, 89–94% yield. Hindered ketones were effectively derivatized. In an unusual reaction, anthrone was reduced to anthracene under these conditions.[49]

Solid-Supported Reagents

1. HS(CH$_2$)$_n$SH, Montmorillonite KSF clay, without solvent, 85–90% yield.[50]

2. From an acetal or ketal or oxime: $HS(CH_2)_nSH$, Kaolinitic clay, CCl_4, reflux, 50–94% yield.[51]

3. H-Y Zeolite, hexane, or CH_2Cl_2, $HSCH_2CH_2SH$, 0.75–144 h, 50–96% yield.[52]

4. $HSCH_2CH_2SH$, PhMe, activated Bentonite, 5 h, 99% yield.[53]

5. H-Rho-zeolite, hexane, reflux, 85–94% yield.[54]

6. $HSCH_2CH_2SH$, $FeCl_3$–SiO_2, CH_2Cl_2, <1 min to 7 h.[55] Montmorillonite clay can also be used as a support medium for the ferric ion (75–98%). In this case the reaction is chemoselective for aldehydes.[56]

7. $HSCH_2CH_2SH$, CH_2Cl_2, $(TMSO)_2SO_2$–silica, 75–99% yield.[57]

8. $HS(CH_2)_nSH$, $SOCl_2$-SiO_2, 88–100% yield.[58] Aldehydes are selectively protected in the presence of ketones. This reagent also converts acetals and ketals directly to thioacetals.[59]

9. $HSCH_2CH_2SH$, CH_2Cl_2, $CoBr_2$–silica, rt, 3 min to 24 h, 87–99% yield.[60]

10. $HSCH_2CH_2SH$, $ZrCl_4$–silica, CH_2Cl_2, rt, 3 h, 98% yield. Unreactive ketones such as benzophenone are efficiently protected. $ZrCl_4$ alone is also an effective catalyst.[61]

11. $HSCH_2CH_2SH$, $AlCl_3$–SiO_2, $ClCH_2CH_2Cl$, reflux, 8–95% yield. Aryl ketones are unreactive.[62]

12. $HSCH_2CH_2SH$, polyphosphoric acid on silica gel, CH_2Cl_2, 45–100% yield. Ketones react less efficiently than aldehydes.[63]

13. $HSCH_2CH_2SH$, Dowex-50W-X8 acidified with HCl, Et_2O, 35–200 min, 60–90% yield.[64]

14. $HSCH_2CH_2SH$, Amberlyst 15, 83–100% yield.[65]

Methods that Form an Acid *In Situ*

1. $HS(CH_2)_nSH$, neat, $Me_2S \cdot Br_2$, 65–98% yield. HBr, probably generated *in situ* by oxidation of the dithiol, is probably the true catalyst in this reaction. Aldehydes react selectively in the presence of ketones. This catalyst has also been used to prepare 1,3-dioxolanes.[66] Tetrabutylammonium tribromide similarly serves as a catalyst.[67]

2. $HSCH_2CH_2SH$, I_2, Al_2O_3, CH_2Cl_2, reflux, 85–95% yield. Aldehydes react in preference to ketones.[68]

3. From an aldehyde, acetal or ketal: $HS(CH_2)_nSH$, CH_2Cl_2, NBS, rt, 57–91% yield.[69] This method was also used to prepare oxathiolanes.

4. $HS(CH_2)_nSH$, $NiCl_2$, CH_2Cl_2, MeOH, rt, 75–97% yield.[70]

5. $HS(CH_2)_nSH$, CH_2Cl_2, zirconium sulfophenyl phosphonate, reflux, 69–95% yield.[71]

6. $HSCH_2CH_2SH$, THF, $CuSO_4$, 40–96% yield.[72]

7. $HSCH_2CH_2SH$, MeCN, rt, $Bi_2(SO_4)_3$, air, 2.5 h, 93–100% yield.[73] $Bi(NO_3)_3$ also serves as a catalyst and can be used to catalyze the formation of acetals and ketals.[74]

8. $HS(CH_2)_nSH$, $CHCl_3$, trichloroisocyanuric acid, 40–95% yield. Acetals and ketals are also converted and aldehydes react in preference to ketones.[75]

9. $HS(CH_2)_nSH$, AcCl, rt, 68–98% yield.[76]

General Methods of Cleavage[77]

Methods Based on Oxidation

1. $AgNO_3$, EtOH, H_2O, 50°C, 20 min, 55% yield.[78]

Attempted cleavage using Hg(II) salts gave material that could not be distilled. 1,3-Dithiolanes can also be cleaved with Ag_2O (MeOH, H_2O, reflux, 16 h to 4 days, 75–85% yield).[79]

2. For ($n = 3$): NCS, $AgNO_3$, CH_3CN, H_2O, 25°C, 5–10 min, 70–100% yield.[80,81]

3. For ($n = 3$): NBS, $AgClO_4$, acetone, H_2O, 0°C, 1 min, 90% yield.[82]

4. For ($n = 2$): NBS, aq. acetone, 0°C, 20 min, 80% yield.[83]

5. $AgNO_3$, I_2, THF, H_2O, 53–100% yield[84]

6. 1,3-Dithiolanes, 1,3-dithianes, and 1,3-oxathiolanes in the presence of a diol and NBS are converted to acetals and ketals, 30–96% yield.[85]

7. For ($n = 3$): NCS or 2,4,4,6-tetrabromo-2,5-cyclohexadien-1-one (TABCO) or trichlorocyanuric acid, DMSO, $CHCl_3$, 4–70 min, 87–98% yield. Other thioacetals are similarly cleaved.[86]

8. For ($n = 2,3$): $Tl(NO_3)_3$, CH_3OH, 25°C, 5 min, 73–99% yield. These conditions have been used to effect selective cleavage of α,β-unsaturated thioketals.[87] In this case $Hg(OAc)_2$ was found not to be reliable.

9. For ($n = 2,3$): Tl(OCOCF$_3$)$_3$, THF, 25°C, 1 min, 83–95% yield.[88] Tl(TFA)$_3$, Et$_2$O, H$_2$O, 94% yield.[89] α,β-Unsaturated 1,3-dithiolanes are selectively cleaved in the presence of saturated 1,3-dithiolanes [Tl(NO$_3$)$_3$, 5 min, 97% yield].[90]

10. For ($n = 2,3$): ZnCr$_2$O$_7$·3H$_2$O,[91] or 2,6-dicarboxypyridinium chlorochromate[92] CH$_3$CN, rt, 85–94% yield.

11. For ($n = 2,3$): SO$_2$Cl$_2$, SiO$_2$, CH$_2$Cl$_2$, H$_2$O, 0–25°C, 90–100% yield.[93]

12. For ($n = 2,3$): SiO$_2$Cl$_2$, CH$_2$Cl$_2$, DMSO, rt, 88–96% yield. For carbonyl derivatives that have enolizable hydrogens, the reaction proceeds to give ring-expanded products.[94]

13. For ($n = 2$): I$_2$, DMSO, 90°C, 1 h, 75–85% yield.[95]

14. I$_2$, NaHCO$_3$, CH$_3$CN, 0°C, >89% yield.[96,97] A variation of the method recycles the iodine by reoxidation with TaCl$_5$/H$_2$O$_2$ (81–100% yield). With this method ketone derivatives are cleaved more rapidly than aldehyde derivatives.[98]

15. Diiodohydantoin, −20°C, 5:5:1 acetone: THF:H$_2$O.[13]

16. ZnBr$_2$, CH$_2$Cl$_2$, MeOH, rt, 4 h, 93% yield.[99] This method is specific for systems that have hydroxyl groups that can direct the hydrolysis.

(i) $ZnBr_2$ (20 eq.), CH_2Cl_2, MeOH, rt, 20 h, 95%

(ii) $ZnBr_2$ (20 eq.), CH_2Cl_2, MeOH, rt, 4 h, 95%

17. For ($n = 3$): DMSO, dioxane, 1.8 M HCl, 90–96% yield.[100]

18. For ($n = 2$[101], 3[102]): p-MeC$_6$H$_4$SO$_2$N(Cl)Na, aq. MeOH, 75–100% yield. 1,3-Oxathiolanes are also cleaved by Chloramine-T.[102]

19. For ($n = 2,3$): N-Chlorobenzotriazole, CH_2Cl_2, $-80°C$; NaOH, 50% yield.[103] 1,3-Dithianes and 1,3-dithiolanes, used in this example to protect C$_3$-keto steroids, were not cleaved by HgCl$_2$–CdCO$_3$.

20. During the course of an aldehyde oxidation with $NaClO_2$, it was observed that a dithiane was cleaved during the reaction. Optimization of the conditions led to a cleavage process that gave 61–97% yields of ketones and aldehydes.[104]

21. For ($n = 2,3$): (PhSeO)$_2$O, THF or CH_2Cl_2, 25°C, 30 min to 50 h, 63–78% yield.[105]

22. For ($n = 3$): Me$_2$CH(CH$_2$)$_2$ONO, CH_2Cl_2, reflux, 2.5 h, 65% yield.[106] 1,3-Oxathiolanes are also cleaved by isoamyl nitrite.

23. NO$^+$HSO$_4^-$, CH_2Cl_2, 25°C, 45 min; H_2O, 56–82% yield.[107]

24. For ($n = 2,3$): Nitrogen oxides, CH_2Cl_2, 40–96%, yield.[108]

25. Cu(NO$_3$)$_2$·N$_2$O$_4$, CCl$_4$, rt, 83–95% yield. This reagent and its iron analog also cleave TBDMS, THP, and TMS ethers to give aldehydes and ketones.[109]

26. For ($n = 2,3$): Ce(NH$_4$)$_2$(NO$_3$)$_6$, aq. CH$_3$CN, 3 min, 70–87% yield.[110]

27. For ($n = 2$): Me$_2$S·Br$_2$, CH_2Cl_2, 25°C, 1 h → reflux, 8 h, followed by H_2O, 55–91% yield.[111]

28. (CF$_3$CO$_2$)$_2$IPh, H_2O, CH$_3$CN, 85–99% yield.[112] This reagent produces TFA and thus some silyl-protective groups, and some olefins have been found

incompatible with this method. In the presence of ethylene glycol the dithiane can be converted to a dioxolane (91% yield)[112] or in the presence of methanol to the dimethyl acetal.[113] The reaction conditions are not compatible with primary amides. Thioesters are not affected.[112] A phenylthio ester is stable to these conditions, but some amides are not. The hypervalent iodine derivative 1-(*t*-butylperoxy)-1,2-benziodoxol-3(1*H*)-one[114] or *o*-iodoxybenzoic acid (IBX)[115] similarly cleaves thioketals. IBX in DMSO/trace H_2O selectively cleaves benzylic and allylic dithianes.[116] $(CF_3CO_2)_2IPh$ is effective at the deprotection of dithiane containing alkaloids which often react with many of the other available methods.[117] In this procedure the amine is protected by protonation, thus preventing oxidation.

Dess–Martin Periodinane (CH_3CN, H_2O, CH_2Cl_2, 68–99% yield), which liberates AcOH rather than TFA during the reaction, was found to be an excellent replacement for $(CF_3CO_2)_2IPh$ in substrates containing silyl groups and olefins.[118] The following case could not be deprotected with $(CF_3CO_2)_2IPh$ directly without significant decomposition. When the reaction was run in MeOH, a dimethyl ketal was produced that could be hydrolyzed with $AcOH/H_2O$.[96]

29. $PhI(O_2CCl_3)_2$, CH_3CN, H_2O, rt, 5 min, >95% yield.[119]
30. MCPBA; Ac_2O, Et_3N, H_2O, THF, 28–37% yield. Subsequent use of this method has resulted in much higher yields.[120] The deprotection proceeds by sulfoxide formation followed by a Pummer-like rearrangement to release the ketone.

31. For (n = 3): MCPBA, TFA, CH_2Cl_2, 0°C, 75–96% yield.[121]

32. Pyr·HBr·Br$_2$, CH_2Cl_2, pyridine, Bu$_4$NBr, 0°C to rt, 2 h, 80–90% yield.[122] The deprotection proceeds without olefin or aromatic ring bromination.

33. PhOP(O)Cl$_2$, DMF, NaI, 1 h, rt, 71–94% yield.[123]

34. MeP(Ph)$_3^+$Br$^-$, CH_2Cl_2, H_2O, NaH$_2$PO$_4$, Na$_2$HPO$_4$, 0–100% yield.[124]

35. For (n = 2): Me$_3$SiI or Me$_3$SiBr, DMSO, 65–99% yield.[125]

36. For (n = 3): Me$_3$S$^+$SbCl$_6^-$, −77°C; Na$_2$CO$_3$, H_2O, 95–97% yield.[100]

37. DMSO, 140–160°C, 4–5 h.[126]

38. For (n = 3): NaNO$_2$, AcCl, H_2O, CH_2Cl_2, rt, 0°C, 82–97% yield. This method also cleaves oxathiolanes.[127]

39. For (n = 2,3): Bi(NO$_3$)$_3$·5H$_2$O, H_2O, CH_3CN or CH_2Cl_2, rt, air, 72–98% yield.[128] Oxathiolanes are also cleaved by this method.

40. For (n = 2): SeO$_2$, AcOH, rt, 0.5–2 h, 90–98% yield.[129]

41. For (n = 2, 3): H$_5$IO$_5$, ether, THF, 77–99% yield.[130] This method also cleaves oxathioacetals, but did not affect the acid sensitive acetonide or 1,3-dioxolane. It should be noted that ethereal periodic acid has been used to cleave terminal acetonides with subsequent glycol cleavage.[131]

42. 1-Benzyl-4-aza-1-azaoniabicyclo[2.2.2]octane periodate, AlCl$_3$ neat, 85–96% yield.[132] This method proceeds in the solid state and as such it is probably not very practical because there is no way to dissipate heat or to achieve adequate mixing on scale.

43. An anomolous cleavage of a dithiolane was observed during an attempted hydroboration.[133]

Ref. 133

44. DDQ, BF$_3$·Et$_2$O, CH_2Cl_2, air, H_2O, >90% yield.[134]

45. DDQ, CH_3CN, photolysis or reflux, 1.5–2 h, 90–95% yield.[135]

46. DDQ, CH_3CN, H_2O (9:1), 0.5–6 h, 30–88% yield.[136] Dithiane derivatives of aromatic aldehydes give thioesters in low yields; dithiolanes are not effectively cleaved.

47. Ceric ammonium nitrate, acetone, H_2O, rt, 12, 99% yield.[137] This method has resulted in over oxidation to give an enone.[138]

48. NaTeH; H_2O, air, 80–85% yield.[139]

49. $SbCl_5$, N_2, CH_2Cl_2, 0°C, 10 min; aq. $NaHCO_3$, 0°C, 10 min, 63–100% yield.[140]

50. $GaCl_3$, MeOH, O_2, CH_2Cl_2, rt, 24 h, 71–99% yield.[141]

51. N-Fluoro-2,4,6-trimethylpyridinium trifluoromethanesulfonate, −10°C, CH_2Cl_2, THF, H_2O, 68–91% yield.[142]

52. Selectfluor™, CH_3CN or CH_3NO_2, 5% H_2O, <5 min, 80–95% yield.[143] The THP and p-methoxybenzylidene groups are also cleaved in excellent yield with this reagent.

53. Oxone, wet alumina, $CHCl_3$, reflux, 15–180 min, 70–96% yield.[144]

54. Pe(phen)$_3$(PF$_6$), CH_3CN, H_2O, 43–75% yield. Hydroxyl and THP groups are not compatible with these conditions.[145]

55. Clayfen, microwave, 87–97%. The reaction is done in the solid state.[146]

56. Fe(NO$_3$)$_3$, silica gel, hexane, 40–50°C, 3–30 min, 86–100% yield.[147] Fe(NO$_3$)$_3$ and Montmorillonite K10 clay in hexane[148] and Fe(NO$_3$)$_3$/basic alumina are also effective.[149] Kaolinitic clay which contains Fe$_2$O$_3$ is also effective.[150]

57. FeCl$_3$, KI, methanol, reflux, 88–91% yield. CeCl$_3$ will replace FeCl$_3$ in this method to cleave dithiolanes and oxathiolanes.[151]

58. CuCl$_2$, CuO, acetone, reflux, 90 min, 85% yield.[152]

59. For (n = 2): CuCl$_2$·2H$_2$O, SiO$_2$, CH$_2$Cl$_2$, H$_2$O, 50–94% yield.[72]

60. Clay-supported ammonium nitrate, CH$_2$Cl$_2$, 16–27 h, 75–90% yield.[153]

61. t-BuOOH, MeOH, reflux, 70–93% yield.[154]

62. NaBO$_3$·H$_2$O, AcOH, Na$_2$CO$_3$, 25°C, 80–97% yield.[155]

63. V$_2$O$_3$, H$_2$O$_2$, NH$_4$Br, CH$_2$Cl$_2$, H$_2$O, 0–5°C, 65–95% yield. Dialkyl thioacetals are also cleaved.[156]

64. 48% HBr, 30% H$_2$O$_2$, CH$_3$CN, rt, 70–96% yield.[157]

Methods Based on Alkylation

1. For (n = 2,3): MeOSO$_2$F, C$_6$H$_6$, 25°C, 1 h, 62–88% yield[158] or liq. SO$_2$, 70–85% yield.[159]

2. For ($n = 2$): MeI, aq. MeOH, reflux, 2–20h, 60–80% yield.[159]
3. For ($n = 3$): MeI, aq. CH_3CN, 25°C.[160]
4. For ($n = 2$): EtI, $CaCO_3$, CH_3CN, H_2O, 81% yield.[161]

5. For ($n = 2$): Et_3OBF_4, followed by 3% aq. $CuSO_4$, 81% yield.[162]
6. 1-Benzenesulfinyl piperidine (BSP), Tf_2O, 2,4,6-tri-*t*-butylpyrimidine (TTBP), CH_2Cl_2, −60°C, 76–91% yield. The TTBP is only required for acid sensitive substrates.[163]

Methods Based on Acetal Exchange

1. Deprotection of a thioketal can occur with HF, which usually does not affect this group, when neighboring group participation occurs as in the case below.[164]

R = TBDMS   PMB = *p*-methoxybenzyl

Note the unusual cleavage of the PMB ether as well.[165]

2. Dowex 50W, acetone, paraformaldehyde, reflux, 50–90% yield.[166]
3. Amberlyst 15, acetone, CH_2O, H_2O, 80°C, 10–25h, 50–80% yield.[167]
4. OHCCOOH, HOAc, 25°C, 15min to 20h, 60–90% yield.[168]
5. TMSOTf, CH_2Cl_2, $NO_2C_6H_4CHO$, rt, 95% yield.[169] Diphenylthio acetals are also cleaved in high yield. This reagent system proved useful in scavenging PhSH that is produced in an electrophilic cyclization.[170]
6. Layered zirconium sulfophenyl phosphonate, glycolic acid monohydrate, 60°C, 79–95% yield.[171]

Mercury-Based Methods

The use of Hg(II) to cleave a dithiane is among the oldest methods to accomplish dithiane deprotection, but because of the environmental issues associated with this toxic element, it should be avoided where possible.

1. $Hg(ClO_4)_2$, MeOH, $CHCl_3$, 25°C, 5 min, 93% yield.[172,173]

2. A 1,3-dithiane is stable to the conditions ($HgCl_2$, $CaCO_3$, CH_3CN-H_2O, 25°C, 1–2 h) used to cleave a methylthiomethyl (MTM) ether (i.e., a monothio acetal).[174]
3. HgO, $BF_3\cdot Et_2O$.[175]
4. $HgCl_2$, HgO, MeOH; $LiBF_4$, H_2O, CH_3CN, 89–91% yield.[175]

Photochemical Methods

1. For ($n = 2,3$): Visible light, methylene green, CH_3CN, H_2O, 86–97% yield.[176]
2. hv, sen., O_2, CH_3CN or CH_2Cl_2, 62–96% yield.[177,178]
3. For ($n = 2,3$): 2,4,6-Triphenylpyrylium perchlorate, hv, O_2, CH_2Cl_2, 13–95% yield.[179,180]
4. hv, benzophenone, CH_3CN, 1.5-3 h, 35–97% yield.[181]
5. For ($n = 2$): O_2, hv, 4.5 h, 60–80% yield.[182] 1,3-Oxathiolanes are also cleaved by O_2/hv.

Methods Based on Electrolysis

1. Electrolysis: 1.5 V, CH_3CN, H_2O, $LiClO_4$ or Bu_4NClO_4, 50–75% yield.[183,184] 1,3-Dithiolanes were not cleaved efficiently by electrolytic oxidation. This method has been applied to dithiane deprotection to produce α-diketones.[185]
2. Electrolysis: 1 V, (p-$CH_3C_6H_4$)$_3$N, CH_3CN, H_2O, $NaHCO_3$, 70–95% yield.[186]

1. For a review on the use of 1,3-dithianes in natural product synthesis, see M. Yus, C. Najera, and F. Foubelo, *Tetrahedron*, **59**, 6147 (2003).
2. A. K. Banerjee and M. S. Laya, *Russ. Chem. Rev.*, **69**, 947 (2000).
3. M. Yus, C. Najera, and F. Foubelo, *Tetrahedron*, **59**, 6147 (2003).
4. R. P. Hatch, J. Shringarpure, and S. M. Weinreb, *J. Org. Chem.*, **43**, 4172 (1978).

5. J. A. Marshall and J. L. Belletire, *Tetrahedron Lett.*, **12**, 871 (1971).

6. F. Sondheimer and D. Rosenthal, *J. Am. Chem. Soc.*, **80**, 3995 (1958).

7. W.-S. Zhou, *Pure Appl. Chem.*, **58**, 817 (1986).

8. T. Nakata, S. Nagao, N. Mori, and T. Oishi, *Tetrahedron Lett.*, **26**, 6461 (1985).

9. A. Pasquarello, G. Poli, and C. Scolastico, *Synlett*, 93 (1992).

10. I. Hoppe, D. Hoppe, R. Herbst-Irmer, and E. Egert, *Tetrahedron Lett.*, **31**, 6859 (1990).

11. D. R. Morton and S. J. Hobbs, *J. Org. Chem.*, **44**, 656 (1979).

12. D. A. Evans, L. K. Truesdale, K. G. Grimm, and S. L. Nesbitt, *J. Am. Chem. Soc.*, **99**, 5009 (1977).

13. E. J. Corey, M. A. Tius and J. Das, *J. Am. Chem. Soc.*, **102**, 7612 (1980).

14. V. Kumar and S. Dev, *Tetrahedron Lett.*, **24**, 1289 (1983).

15. S. K. De, *Synthesis*, 828 (2004); S. K. De, *Tetrahedron Lett.*, **45**, 1035 (2004).

16. N. B. Das, A. Nayak and R. P. Sharma, *J. Chem. Res., Synop.*, 242 (1993).

17. T. Sato, J. Otera and H. Nozaki, *J. Org. Chem.*, **58**, 4971 (1993).

18. P. K. Chowdhury, *J. Chem. Res., Synop.*, 124 (1993).

19. Y. Zhang, Y. Yu, and R. Lin, *Org. Prep. Proced. Int.*, **25**, 365 (1993).

20. E. J. Corey and K. Shimoji, *Tetrahedron Lett.*, **24**, 169 (1983).

21. M. E. Kuehne, W. G. Bornmann, W. G. Earley, and I. Marko, *J. Org. Chem.*, **51**, 2913 (1986).

22. B. M. Trost and J. R. Parquette, *J. Org. Chem.*, **59**, 7568 (1994).

23. S. Islam, A. Majee, T. Mandal, and A. T. Khan, *Synth. Commum.*, **34**, 2911 (2004).

24. A. Kamal and G. Chouhan, *Tetrahedron Lett.*, **43**, 1347 (2002).

25. H. Firouzabadi, N. Iranpoor, and G. Kohmareh, *Synth. Commum.*, **33**, 167 (2003).

26. S. K. De, *Synth. Commum.*, **34**, 4401 (2004).

27. S. K. De, *Tetrahedron Lett.*, **45**, 2339 (2004).

28. V. G. Saraswathy and S. Sankaraman, *J. Org. Chem.*, **59**, 4665 (1994); L. F. Tietze, B. Weigand, and C. Wulff, *Synthesis*, 69 (2000).

29. H. Firouzabadi, B. Karimi, and S. Eslami, *Tetrahedron Lett.*, **40**, 4055 (1999); H. Firouzabadi, S. Eslami, and B. Karimi, *Bull. Chem. Soc. Jpn.*, **74**, 2401 (2001).

30. K. Kazahaya, S. Kiyoshi, and T. Sato, *Synlett*, 1640 (2004).

31. R. A. Moss and C. B. Mallon, *J. Org. Chem.*, **40**, 1368 (1975).

32. T. Satoh, S. Uwaya, and K. Yamakawa, *Chem. Lett.*, **12**, 667 (1983).

33. Y. Honda, A. Ori, and G. Tsuchihashi, *Chem. Lett.*, **16**, 1259 (1987).

34. J. A. Soderquist and E. I. Miranda, *Tetrahedron Lett.*, **27**, 6305 (1986).

35. T. Sato, E. Yoshida, T. Kobayashi, J. Otera, and H. Nozaki, *Tetrahedron Lett.*, **29**, 3971 (1988).

36. H. Tani, K. Masumoto, T. Inamasu, and H. Suzuki, *Tetrahedron Lett.*, **32**, 2039 (1991).

37. L. Garlaschelli and G. Vidari, *Tetrahedron Lett.*, **31**, 5815 (1990).

38. M. A. Ceschi, L. d. A. Felix, and C. Peppe, *Tetrahedron Lett.*, **41**, 9695 (2000); S. Muthusamy, S. A. Babu, and C. Gunanathan, *Tetrahedron*, **58**, 7897 (2002); S. Muthusamy, S. A. Babu, and C. Gunanathan, *Tetrahedron Lett.*, **42**, 359 (2001).

39. B. C. Ranu, A. Das, and S. Samanta, *Synlett*, 727 (2002).

40. S. K. De, *J. Mol. Catal. A: Chemical*, **226**, 77 (2005).

41. S. K. De, *Adv. Synth. Catal.*, **347**, 673 (2005).

42. K. K. Rana, C. Guin, S. Jana, and S. C. Roy, *Tetrahedron Lett.*, **44**, 8597 (2003).

43. H. Firouzabadi and B. Karimi, *Phosphorus, Sulfur and Silicon and the Related Elements*, **175**, 207 (2001).

44. E. Diez, A. M. Lopez, C. Pareja, E. Martin, R. Fernandez, and J. M. Lassaletta, *Tetrahedron Lett.*, **39**, 7955 (1998).

45. A. Martel, S. Chewchanwuttiwong, G. Dujardin, and E. Brown, *Tetrahedron Lett.*, **44**, 1491 (2003).

46. M. Ball, M. J. Gaunt, D. F. Hook, A. S. Jessiman, S. Kawahara, P. Orsini, A. Scolaro, A. C. Talbot, H. R. Tanner, S. Yamanoi, and S. V. Ley, *Angew. Chem. Int. Ed.*, **44**, 5433 (2005).

47. D. Dong, Y. Ouyang, H. Yu, Q. Liu, J. Liu, M. Wang, and J. Zhu, *J. Org. Chem.*, **70**, 4535 (2005); Q. Liu, G. Che, H. Yu, Y. Liu, J. Zhang, Q. Zhang, and D. Dong, *J. Org. Chem.*, **68**, 9148 (2003); H. Yu, Q. Liu, Y. Yin, Q. Fang, J. Zhang, and D. Dong, *Synlett*, 999 (2004).

48. S. Kobayashi, S. Iimura, and K. Manabe, *Chem. Lett.*, **31**, 10 (2002).

49. H. Firouzabadi, N. Iranpoor, and K. Amani, *Synthesis*, 59 (2002).

50. D. Villemin, B. Labiad, and M. Hammadi, *J. Chem. Soc., Chem. Commun.*, 1192 (1992).

51. G. K. Jnaneshwara, N. B. Barhate, A. Sudalai, V. H. Deshpande, R. D. Wakharkar, A. S. Gajare, M. S. Shingare, and R. Sukumar, *J. Chem. Soc. Perkin Trans. 1*, 965 (1998).

52. P. Kumar, R. S. Reddy, A. P. Singh, and B. Pandey, *Synthesis*, 67 (1993); *idem, Tetrahedron Lett.*, **33**, 825 (1992).

53. R. Miranda, H. Cervantes, and P. Joseph-Nathan, *Synth. Commun.*, **20**, 153 (1990).

54. D. P. Sabde, B. G. Naik, V. R. Hedge, and S. G. Hegde, *J. Chem. Res., Synop.*, 494 (1996).

55. H. K. Patney, *Tetrahedron Lett.*, **32**, 2259 (1991); M. Hirano, K. Ukawa, S. Yakabe and T. Morimoto, *Org. Prep. Proced. Int.*, **29**, 480 (1997).

56. B. M. Choudary and Y. Sudha, *Synth. Commun.*, **26**, 2993 (1996).

57. H. K. Patney, *Tetrahedron Lett.*, **34**, 7127 (1993).

58. Y. Kamitori, M. Hojo, R. Masuda, T. Kimura, and T. Yoshida, *J. Org. Chem.*, **51**, 1427 (1986).

59. H. Firouzabadi, N. Iranpoor, B. Karimi, and H. Hazarkhani, *Synlett*, 263 (2000).

60. H. K. Patney, *Tetrahedron Lett.*, **35**, 5717 (1994).

61. H. K. Patney and S. Margan, *Tetrahedron Lett.*, **37**, 4621 (1996); H. Firouzabadi, N. Iranpoor, and B. Karimi, *Synlett*, 319 (1999).

62. B. Tamami and K. P. Borujeny, *Synth. Commun.*, **33**, 4253 (2003).

63. T. Aoyama, T. Takido, and M. Kodomari, *Synlett*, 2307 (2004).

64. A. K. Maiti, K. Basu, and P. Bhattacharyya, *J. Chem. Res., Synop.*, 108 (1995).

65. R. B. Perni, *Synth. Commun.*, **19** 2383 (1989); B. Ku and D. Y. Oh, *ibid.*, **19**, 433 (1989).

66. A. T. Khan, E. Mondal, S. Ghosh, and S. Islam, *Eur. J. Org. Chem.*, 2002 (2004).

67. S. Naik, R. Gopinath, M. Goswami, and B. K. Patel, *Org. Biomol. Chem.*, **2**, 1670 (2004).

68. N. Deka and J. C. Sarma, *Chem. Lett.*, **30**, 794 (2001); S. Samajdar, M. K. Basu, F. F. Becker, and B. K. Banik, *Tetrahedron Lett.*, **42**, 4425 (2001).

69. A. Kamal, G. Chouhan, and K. Ahmed, *Tetrahedron Lett.*, **43**, 6947 (2002); A. Kamal and G. Chouhan, *Synlett*, 474 (2002).

70. A. T. Khan, E. Mondal, P. R. Sahu, and S. Islam, *Tetrahedron Lett.*, **44**, 919 (2003).

71. M. Curini, F. Epifano, M. C. Marcotullio, and O. Rosati, *Synlett*, 1182 (2001).

72. A. Nayak, B. Nanda, N. B. Das, and R. P. Sharma, *J. Chem. Res., Synop.*, 100 (1994).

73. N. Komatsu, M. Uda, and H. Suzuki, *Synlett*, 984 (1995).

74. N. Sriivastava, S. K. Dasgupta, and B. K. Banik, *Tetrahedron Lett.*, **44**, 1191 (2003).

75. H. Firouzabadi, N. Iranpoor, and H. Hazarkhani, *Synlett*, 1641 (2001).

76. A. T. Khan and E. Mondal, *Ind. J. Chem.*, **44B**, 844 (2005).

77. Mechanisms of hydrolysis of thioacetals: D. P. N. Satchell and R. S. Satchell, *Chem. Soc. Rev.*, **19**, 55 (1990).

78. C. A. Reece, J. O. Rodin, R. G. Brownlee, W. G. Duncan, and R. M. Silverstein, *Tetrahedron*, **24**, 4249 (1968).

79. D. Gravel, C. Vaziri, and S. Rahal, *J. Chem. Soc., Chem. Commun.*, 1323 (1972).

80. E. J. Corey and B. W. Erickson, *J. Org. Chem.*, **36**, 3553 (1971).

81. A. V. Rama Rao, G. Venkatswamy, S. M. Javeed, V. H. Deshpande, and B. R. Rao, *J. Org. Chem.*, **48**, 1552 (1983).

82. K. C. Nicolaou, K. Ajito, A. P. Patron, H. Khatuya, P. K. Richter, and P. Bertinato, *J. Am. Chem. Soc.*, **118**, 3059 (1996).

83. E. N. Cain and L. L. Welling, *Tetrahedron Lett.*, **16**, 1353 (1975).

84. K. Nishide, K. Yokota, D. Nakamura, T. Sumiya, M. Node, M. Ueda, and K. Fuji, *Tetrahedron Lett.*, **34**, 3425 (1993).

85. B. Karimi, H. Seradj, and J. Maleki, *Tetrahedron*, **58**, 4513 (2002).

86. N. Iranpoor, H. Firouzabadi, and H. R. Shaterian, *Tetrahedron Lett.*, **44**, 4769 (2003).

87. P. S. Jones, S. V. Ley, N. S. Simpkins, and A. J. Whittle, *Tetrahedron*, **42**, 6519 (1986).

88. T.-L. Ho and C. M. Wong, *Can. J. Chem.*, **50**, 3740 (1972).

89. W. O. Moss, R. H. Bradbury, N. J. Hales, and T. Gallagher, *J. Chem. Soc., Perkin Trans. I*, 1901 (1992).

90. R. A. J. Smith and D. J. Hannah, *Synth. Commun.*, **9**, 301 (1979).

91. H. Firouzabadi, N. Iranpoor, H. Hassani, and S. Sobhani, *Synth. Commun.*, **34**, 1967 (2004).

92. R. Hosseinzadeh, M. Tajbakhsh, A. Shakoori, and M. Y. Niaki, *Monatsh. Chem.*, **135**, 1243 (2004).

93. M. Hojo and R. Masuda, *Synthesis*, 678 (1976).

94. H. Firouzabadi, N. Iranpoor, H. Hazarkhani, and B. Karimi, *J. Org. Chem.*, **67**, 2572 (2002).

95. J. B. Chattopadhyaya and A. V. Rama Rao, *Tetrahedron Lett.*, **14**, 3735 (1973).

96. M. J. Gaunt, D. F. Hook, H. R. Tanner, and S. V. Ley, *Org. Lett.*, **5**, 4815 (2003); M. J. Gaunt, A. S. Jessiman, P. Orsini, H. R. Tanner, D. F. Hook, and S. V. Ley, *Org. Lett.*, **5**, 4819 (2003).

97. J. Ishihara and A. Murai, *Synlett*, 363 (1996).

98. M. Kirihara, A. Harano, H. Tsukiji, R. Takizawa, T. Uchiyama, and A. Hatano, *Tetrahedron Lett.*, **46**, 6377 (2005).

99. A. Vakalopoulos and H. M. R. Hoffmann, *Org. Lett.*, **3**, 2185 (2001).

100. M. Prato, U. Quintily, G. Scorrano, and A. Sturaro, *Synthesis*, 679 (1982).

101. W. F. J. Huurdeman, H. Wynberg, and D. W. Emerson, *Tetrahedron Lett.*, **12**, 3449 (1971).

102. D. W. Emerson and H. Wynberg, *Tetrahedron Lett.*, **12**, 3445 (1971).

103. P. R. Heaton, J. M. Midgley, and W. B. Whalley, *J. Chem. Soc., Chem. Commun.*, 750 (1971).

104. T. Ichige, A. Miyake, N. Kanoh, and M. Nakata, *Synlett*, 1686 (2004).

105. D. H. R. Barton, N. J. Cussans, and S. V. Ley, *J. Chem. Soc., Chem. Commun.*, 751 (1977).

106. K. Fuji, K. Ichikawa, and E. Fujita, *Tetrahedron Lett.*, **19**, 3561 (1978).

107. G. A. Olah, S. C. Narang, G. F. Salem, and B. G. B. Gupta, *Synthesis*, 273 (1979).

108. G. Mehta and R. Uma, *Tetrahedron Lett.*, **37**, 1897 (1996).

109. H. Firouzabadi, N. Iranpoor, and M. A. Zolfigol, *Bull. Chem. Soc. Jpn.*, **71**, 2169 (1998).

110. T.-L. Ho, H. C. Ho, and C. M. Wong, *J. Chem. Soc., Chem. Commun.*, 791 (1972).

111. G. A. Olah, Y. D. Vankar, M. Arvanaghi, and G. K. S. Prakash, *Synthesis*, 720 (1979).

112. G. Stork and K. Zhao, *Tetrahedron Lett.*, **30**, 287 (1989).

113. M. Nakatsuka, J. A. Ragan, T. Sammakia, D. B. Smith, D. E. Uehling, and S. L. Schreiber, *J. Am. Chem. Soc.*, **112**, 5583 (1990).

114. M. Ochiai, A. Nakanishi, and T. Ito, *J. Org. Chem.*, **62**, 4253 (1997).

115. K. C. Nicolaou, C. J. N. Mathison, and T. Montagnon, *Angew. Chem. Int. Ed.*, **42**, 4077 (2003); K. C. Nicolaou, C. J. N. Mathison, and T. Montagnon, *J. Am. Chem. Soc.*, **126**, 5192 (2004).

116. Y. Wu, X. Shen, J.-H. Huang, C.-J. Tang, H.-H. Liu, and Q. Hu, *Tetrahedron Lett.*, **43**, 6443 (2002).

117. F. F. Fleming, L. Funk, R. Altundas, and Y. Tu, *J. Org. Chem.*, **66**, 6502 (2001).

118. N. F. Langille, L. A. Dakin, and J. S. Panek, *Org. Lett.*, **5**, 575 (2003).

119. M. H. B. Stowell, R. S. Rock, D. C. Rees, and S. I. Chan, *Tetrahedron Lett.*, **37**, 307 (1996).

120. A. B. Smith, III, B. D. Dorsey, M. Visnick, T. Maeda, and M. S. Malamas, *J. Am. Chem. Soc.*, **108**, 3110 (1986); R. M. Garbaccio, and S. J. Danishefsky, *Org. Lett.*, **2**, 3127 (2000); R. M. Garbaccio, S. J. Stachel, D. K. Baeschlin, and S. J. Danishefsky, *J. Am. Chem. Soc.*, **123**, 10903 (2001).

121. J. Cossy, *Synthesis*, 1113 (1987).

122. G. S. Bates and J. O'Doherty, *J. Org. Chem.*, **46**, 1745 (1981).

123. H.-J. Liu and V. Wiszniewski, *Tetrahedron Lett.*, **29**, 5471 (1988).

124. H.-J. Cristau, A. Bazbouz, P. Morand, and E. Torreilles, *Tetrahedron Lett.*, **27**, 2965 (1986).

125. G. A. Olah, S. C. Narang, and A. K. Mehrotra, *Synthesis*, 965 (1982).

126. CH. S. Rao, M. Chandrasekharam, H. Ila, and H. Junjappa, *Tetrahedron Lett.*, **33**, 8163 (1992).

127. A. T. Khan, E. Mondal, and P. R. Sahu, *Synlett*, 377 (2003).

128. N. Komatsu, A. Taniguchi, S. Wada, and H. Suzuki, *Adv. Synth. Catal.*, **343**, 473 (2001).

129. S. A. Haroutounian, *Synthesis*, 39 (1995).

130. X.-X. Shi, S. P. Khanapure, and J. Rokach, *Tetrahedron Lett.*, **37**, 4331 (1996).

131. W.-L. Wu and Y.-L. Wu, *J. Org. Chem.*, **58**, 3586 (1993).

132. A. R. Hajipour and A. E. Ruoho, *Org. Prep. Proc. Int.*, **37** 298–303 (2005).

133. C. D'Alessandro, S. Giacopello, A. M. Seldes, and M. E. Deluca, *Synth. Commun.*, **25**, 2703 (1995).

134. J. P. Collman, D. A. Tyvoll, L. L. Chng, and H. T. Fish, *J. Org. Chem.*, **60**, 1926 (1995).

135. L. Mathew and S. Sankararaman, *J. Org. Chem.*, **58**, 7576 (1993).

136. K. Tanemura, H. Dohya, M. Imamura, T. Suzuki, and T. Horaguchi, *Chem. Lett.* **23**, 965 (1994). *idem, J. Chem. Soc., Perkin Trans. I*, 453 (1996).

137. A. Okada, T. Minami, Y. Umezu, S. Nishikawa, R. Mori, and Y. Nakayama, *Tetrahedron: Asymmetry*, **2**, 667 (1991).

138. J. J. La Caire, P. T. Lansbury, B. Zhi, and K. Hoogsteen, *J. Org. Chem.*, **60**, 4822 (1995).

139. P. Lue, W.-Q. Fan, and X.-J. Zhou, *Synthesis*, 692 (1989).

140. M. Kamata, H. Otogawa, and E. Hasegawa, *Tetrahedron Lett.*, **32**, 7421 (1991).

141. K. Saigo, Y. Hashimoto, N. Kihara, H. Umehara, and M. Hasegawa, *Chem. Lett.* **19**, 831 (1990).

142. A. S. Kiselyov, L. Strekowski, and V. V. Semenov, *Tetrahedron*, **49**, 2151 (1993).

143. J. Liu and C.-H. Wong, *Tetrahedron Lett.*, **43**, 4037 (2002).

144. P. Ceccherelli, M. Curini, M. C. Marcotullio, F. Epifano, and O. Rosati, *Synlett*, 767 (1996).

145. M. Schmittel and M. Levis, *Synlett*, 315 (1996).

146. R. S. Varma and R. K. Saini, *Tetrahedron Lett.*, **38**, 2623 (1997).

147. M. Hirano, K. Ukawa, S. Yakabe, and T. Morimoto, *Synth. Commun.*, **27**, 1527 (1997).

148. M. Hirano, K. Ukawa, S. Yakabe, J. H. Clark, and T. Morimoto, *Synthesis*, 858 (1997).

149. P. Wipf and M. J. Soth, *Org. Lett.*, **4**, 1787 (2002).

150. B. P. Bandgar and S. P. Kasture, *Green Chem.*, **2**, 154 (2000).

151. J. S. Yadav, B. V. S. Reddy, S. Raghavendra, and M. Satyanarayana, *Tetrahedron Lett.*, **43**, 4679 (2002).

152. P. Stütz and P. A. Stadler, *Org. Synth., Collect. Vol. VI*, 109 (1988).

153. H. M. Meshram, G. S. Reddy, and J. S. Yadav, *Tetrahedron Lett.*, **38**, 8891 (1997).

154. N. B. Barhate, P. D. Shinde, V. A. Mahajan, and R. D. Wakharkar, *Tetrahedron Lett.*, **43**, 6031 (2002).

155. B. P. Bandgar, S. A. Kulkarni, and J. N. Nigal, *OPPI Briefs*, **30**, 706 (1998).

156. E. Mondal, G. Bose, P. R. Sahu, and A. T. Khan, *Chem. Lett.*, **30**, 1158 (2001).

157. N. C. Ganguly and M. Datta, *J. Chem. Res*, 218 (2005).

158. T.-L. Ho and C. M. Wong, *Synthesis*, 561 (1972).

159. M. Fetizon and M. Jurion, *J. Chem. Soc., Chem. Commun.*, 382 (1972).

160. S. Takano, S. Hatakeyama, and K. Ogasawara, *J. Chem. Soc., Chem. Commun.*, 68 (1977).

161. P. A. Clarke and A. P. Cridland, *Org. Lett.*, **7**, 4221 (2005).

162. T. Oishi, K. Kamemoto, and Y. Ban, *Tetrahedron Lett.*, **13**, 1085 (1972).

163. D. Crich and J. Picione, *Synlett*, 1257 (2003).

164. P. G. Steet and E. J. Thomas, *J. Chem. Soc., Perkin Trans. I*, 371 (1997).

165. A. B. Smith III, J. J.-W. Duan, K. G. Hull, and B. A. Salvatore, *Tetrahedron Lett.*, **32**, 4855 (1991).

166. V. S. Giri and P. J. Sankar, *Synth. Commun.*, **23**, 1795 (1993).

167. R. Ballini and M. Petrini, *Synthesis*, 336 (1990).

168. H. Muxfeldt, W.-D. Unterweger, and G. Helmchen, *Synthesis*, 694 (1976).

169. T. Ravindranathan, S. P. Chavan, R. B. Tejwani, and J. P. Varghese, *J. Chem. Soc., Chem. Commun.*, 1750 (1991).

170. S. P. Chavan, R. B. Tejwani, and T. Ravindranathan, *J. Org. Chem.*, **66**, 6197 (2001).

171. M. Curini, M. C. Marcotullio, E. Pisani, and O. Rosati, *Synlett*, 769 (1997).

172. E. Fujita, Y. Nagao, and K. Kaneko, *Chem. Pharm. Bull.*, **26**, 3743 (1978).

173. B. H. Lipshutz, R. Moretti, and R. Crow, *Tetrahedron Lett.*, **30**, 15 (1989).

174. E. J. Corey and M. G. Bock, *Tetrahedron Lett.*, **16**, 2643 (1975).

175. J. A. Soderquist and E. L. Miranda, *J. Am. Chem. Soc.*, **114**, 10078 (1992).

176. G. A. Epling and Q. Wang, *Synlett*, 335 (1992).

177. M. Kamata, M. Sato, and E. Hasagawa, *Tetrahedron Lett.*, **33**, 5085 (1992); M. Kamata, Y. Murakami, Y. Tamagawa, Y. Kato, and E. Hasegawa, *Tetrahedron*, **50**, 12821 (1994).

178. E. Fasani, M. Freccero, M. Mella, and A. Albini, *Tetrahedron*, **53**, 2219 (1997).

179. M. Kamata, Y. Murakami, Y. Tamagawa, M. Kato, and E. Hasegawa, *Tetrahedron*, **50**, 12821 (1994).

180. E. Fasani, M. Freccero, M. Mella, and A. Albini, *Tetrahedron*, **53**, 2219 (1997).

181. W. A. McHale and A. G. Kutateladze, *J. Org. Chem.*, **63**, 9924 (1998).

182. T. T. Takahashi, C. Y. Nakamura, and J. Y. Satoh, *J. Chem. Soc., Chem. Commun.*, 680 (1977).

183. Q. N. Porter and J. H. P. Utley, *J. Chem. Soc., Chem. Commun.*, 255 (1978).

184. H. J. Cristau, B. Chabaud, and C. Niangoran, *J. Org. Chem.*, **48**, 1527 (1983).

185. A.-M. Martre, G. Mousset, R. B. Rhlid, and H. Veschambre, *Tetrahedron Lett.*, **31**, 2599 (1990).

186. M. Platen and E. Steckhan, *Tetrahedron Lett.*, **21** 511 (1980); *idem, Chem. Ber.*, **117**, 1679 (1984).

1,5-Dihydro-3*H*-2,4-benzodithiepin Derivative:

Dithiepin derivatives, prepared in high yield (FeCl$_3$·SiO$_3$, CH$_2$Cl$_2$, rt, 84–99%)[1] from 1,2-bis(mercaptomethyl)benzenes, are cleaved by HgCl$_2$ (80% yield). Neither reagents nor products have unpleasant odors.[2]

1. H. K. Patney, *Synth. Commun.*, **23**, 1829 (1993).

2. I. Shahak and E. D. Bergmann, *J. Chem. Soc. C*, 1005 (1966).

Monothio Acetals and Ketals

Acyclic Monothio Acetals and Ketals

Acyclic monothio acetals and ketals can be prepared directly from a carbonyl compound or by *trans*-ketalization, a reaction that does not involve a free carbonyl group, from a 1,3-dithiane or 1,3-dithiolane. They are cleaved by acidic hydrolysis or Hg(II) salts. One of their primary liabilities is that with ketones a new chiral center is introduced which may complicate product analysis.

O-Trimethylsilyl-*S*-alkyl Acetals and Ketals: $R_2C(SR')OSiMe_3$

Formation

1. $RSSiMe_3$, ZnI_2, 25°C, 30 min, 80–90% yield.[1]
2. Me_3SiCl, R'SH, Pyr, 25°C, 3 h, 75–90% yield.[2]
3. TMS-Imidazole, RSH, 90 min, 81–94% yield.[3]

Cleavage

1. Dilute HCl.[2]
2. In ether or tetrahydrofuran organolithium reagents cleave the silicon–oxygen bond; in hexamethylphosphoramide, they react at the carbon atom.[2]

1. D. A. Evans, L. K. Truesdale, K. G. Grimm, and S. L. Nesbitt, *J. Am. Chem. Soc.*, **99**, 5009 (1977).
2. T. H. Chan and B. S. Ong, *Tetrahedron Lett.*, **17**, 319 (1976).
3. M. B. Sassaman, G. K. S. Prakash, and G. A. Olah, *Synthesis*, 104 (1990).

O-Alkyl-*S*-alkyl or -*S*-phenyl Acetals and Ketals: $R_2C(OR')SR''$

Formation

Monothioacetals are generally formed by *trans*-ketalization of simple acetals.

1. From a dimethyl acetal: Et_2AlSPh, 0°C, 78% yield.[1]
2. From a dimethyl acetal: $BCl_3·Et_2O$, −45°C, CH_3SH, 73% yield.[2]
3. From a dialkyl acetal: Bu_3SnSPh, $BF_3·Et_2O$, toluene, −78° to 0°C, 64–100% yield.[3] These conditions also convert MOM and MEM groups to the corresponding phenylthiomethyl groups in 64–77% yield. Reaction of α,β-unsaturated acetals results in the formation of a vinyl ether.

4. From a dialkyl acetal: $MgBr_2$, Et_2O, rt, PhSH, 91% yield.[4] MOM groups are converted to phenylthiomethyl groups, 75% yield, but MEM groups do not react.

5. ROTMS (R = 4-MeBn, 4-MeOBn, 2-butenyl), PhSTMS, $CHCl_3$, TMSOTf, $-75°C$, 37–93%.[5]

6. From a dimethyl ketal: cat. , PhSTMS, DMF, 0–60°C, 62–90% yield.[6]

7. RSH, LiBr, toluene, 0–80°C, 70–99% yield. MOM and MEM groups as well as furanose and pyranone acetals all react to give the monothioacetal, but simple dimethylacetals and dimethylketals react faster than the furanose and pyranose acetals.[7]

Cleavage

1. The mechanisms for hydrolysis of *O,S*-acetals have been reviewed. The following acid-catalyzed cleavage rates show that the *O,S*-acetals have a stability that lies between thioacetals and acetals.[8]

3.5×10^{-4}	1.3	41	160

An extensive review of the chemistry of *O,S*-acetals has been published.[9]

2. Electrolysis: Pt electrode, KOAc, AcOH, 10 V, 18–20°C; K_2CO_3, MeOH, 81–91% yield.[10] These cleavage conditions could, in principle, be used to cleave the MTM group.

3. $HgCl_2$, H_2O, $HClO_4$.[11] The section on MTM ethers should be consulted.

4. V_2O_5, H_2O_2, NH_4Br, CH_2Cl_2, H_2O, 0–5°C, 68–96% yield.[12]

1. Y. Masaki, Y. Serizawa, and K. Kaji, *Chem. Lett.*, **14**, 1933 (1985).

2. F. Nakatsubo, A. J. Cocuzza, D. E. Keely, and Y. Kishi, *J. Am. Chem. Soc.*, **99**, 4835 (1977).

3. T. Sato, T. Kobayashi, T. Gojo, E. Yishida, J. Otera, and H. Nozaki, *Chem. Lett.*, **16**, 1661 (1987); T. Sato, J. Otera, and H. Nozaki, *Tetrahedron*, **45**, 1209 (1989).

4. S. Kim, J. H. Park and S. Lee, *Tetrahedron Lett.*, **30**, 6697 (1989).

5. A. Kusche, R. Hoffmann, I. Münster, P. Keiner, and R. Brückner, *Tetrahedron Lett.*, **32**, 467 (1991).

6. T. Miura and Y. Masaki, *Tetrahedron*, **51**, 10477 (1995); *idem, Tetrahedron Lett.*, **35**, 7961 (1994).

7. F. Ono, R. Negoro, and T. Sato, *Synlett*, **10**, 1581 (2001).

8. D. P. N. Satchell and R. S. Satchell, *Chem. Soc. Rev.*, **19**, 55 (1990).

9. P. Wimmer, "O/S Acetale," in *O/O- und O/S-Acetale [Methoden der Organischen Chemie]* (*Houben-Weyl*), Band E14a/1, H. Hagemann and D. Klamann, Eds., G. Theime Stuttgart, 1991, p. 785.

10. T. Mandai, H. Irei, M. Kuwada, and J. Otera, *Tetrahedron Lett.*, **25**, 2371 (1984).

11. J. L. Jensen, D. F. Maynard, G. R. Shaw, and T. W. Smith, Jr., *J. Org. Chem.*, **57**, 1982 (1992).

12. E. Mondal, P. R. Sahu, G. Bose, and A. T. Khan, *J. Chem. Soc. Perkin Trans. 1*, 1026 (2002).

O-Methyl-*S*-2-(methylthio)ethyl Acetals and Ketals: $R_2C(OMe)SCH_2CH_2SMe$

These derivatives are less susceptible to oxidation and hydrogenolysis than are the 1,3-dithiane and 1,3-dithiolane precursors.

Formation[1]

Cleavage

$HgCl_2$, $CaCO_3$, THF, H_2O, 0°C, rapid.[1]

1. E. J. Corey and T. Hase, *Tetrahedron Lett.*, **16**, 3267 (1975).

Cyclic Monothio Acetals and Ketals

1,3-Oxathiolanes: (Chart 5)

Formation

1. $HSCH_2CH_2OH$, $ZnCl_2$ AcONa, dioxane, 25°C, 20 h, 60–90% yield.[1,2]

2. $HSCH_2CH_2OH$, $LiBF_4$, CH_3CN, rt, 80–95% yield. Ketones fail to react. Dithiolanes can also be prepared by this method.[3]

3. $HSCH_2CH_2OH$, $ZrCl_4$, CH_2Cl_2, 55–97% yield. Aldehydes react much faster than ketones.[4] Indium triflate can be used as a catalyst (70–92% yield).[5]

4. $HSCH_2CH_2OH$, TMSOTf, 10 min, 50–78% yield.[6]

5. $HSCH_2CH_2OH$, ionic liquid: [bmim]BF_4, rt, 70–90% yield. Dithiolanes can also be prepared by this method, but the method is selective for reaction of aldehydes.[7]

6. $HSCH_2CH_2OH$, *n*-Bu_4NBr_3 0.01–0.1 eq., CH_2Cl_2, 60–98% yield. HBr is probably generated *in situ* by oxidation of the thiol to a disulfide. $Me_2S \cdot Br_2$ has also been used as a catalyst.[8] Using 0.5 eq. of *n*-Bu_4NBr_3 can be used to cleave a 1,3-oxathiolane.[9]

7. Polymer supported ammonium chloride (APSG·HCl), MeOH, rt, HSCH$_2$-CH$_2$OH, TMOF, 54–91% yield. This method was developed specifically for the protection of α,β-unsaturated aldehydes and ketones.[10]

Cleavage

The section on the cleavage of 1,3-dithianes and 1,3-dithiolanes should be consulted since many of the methods described there are also applicable to the cleavage of oxathiolanes. The cleavage of O, S-acetals has been reviewed.[11]

1. HgCl$_2$, AcOH, AcOK, 100°C, 1 h, 83% yield.[12]
2. HgCl$_2$, NaOH, EtOH, H$_2$O, 25°C, 30 min, 91% yield.[12]
3. Raney Ni, AcOH, AcOK, 100°C, 90 min, 92% yield.[12]
4. HCl, AcOH, reflux, 22 h, 60% yield.[13]
5. AgNO$_3$, NCS, 80% CH$_3$CN, H$_2$O.[14]
6. 0.1 eq. VOCl$_3$, O$_2$, CF$_3$CH$_2$OH, reflux, then H$_2$O, 73–100% yield. The reaction proceeds through a trifluoroethyl acetal that is hydrolyzed with water. Dithianes react much more slowly.[15]
7. V$_2$O$_5$, H$_2$O$_2$, NH$_4$Br, CH$_2$Cl$_2$, H$_2$O, 0–5°C, 68–96% yield. This system generates Br$_2$ *in situ*. The method was compatible with the presence of allylic ethers.[16] H$_2$MoO$_4$·H$_2$O is also a good catalyst that can be used in deprotection of oxathiolanes.[17]
8. 30% H$_2$O$_2$, CH$_3$CN, reflux, 71–100% yield.[18]
9. Phenyliodo(III) bistrifluoroacetate, NaI, CH$_2$Cl$_2$, 15 min. 84–92% yield. Iodine is generated *in situ* by this method.[19]
10. N-Bromosuccinimide, DABCO, 75% aq. Acetone, rt, 84–94% yield.[20]
11. Benzyne, ClCH$_2$CH$_2$Cl, 49–100% yield.[21]
12. 4-Nitrobenzaldehyde, TMSOTf, CH$_2$Cl$_2$, rt, 75–97% yield.[22] Dithiolanes are stable to these conditions.
13. Glycolic acid, Amberlyst 15, neat, 80–94% yield. This method proceeds by an exchange process.[23]
14. MeI, aq. acetone, reflux, 91% yield.[24]

1. J. Romo, G. Rosenkranz, and C. Djerassi, *J. Am. Chem. Soc.*, **73**, 4961 (1951).
2. V. K. Yadav and A. G. Fallis, *Tetrahedron Lett.*, **29**, 897 (1988).
3. J. S. Yadav, B. V. S. Reddy, and S. K. Pandey, *Synlett*, 238 (2001).
4. B. Karimi and H. Seradj, *Synlett*, 805 (2000).
5. K. Kazahaya, N. Hamada, S. Ito, and T. Sato, *Synlett*, 1535 (2002).
6. T. Ravindranathan, S. P. Chavan, and S. W. Dantale, *Tetrahedron Lett.*, **36**, 2285 (1995).
7. J. S. Yadav, B. V. S. Reddy, and G. Kondaji, *Chem. Lett.*, **32**, 672 (2003).
8. A. T. Khan, P. R. Sahu, and A. Majee, *J. Mol. Catal. A: Chemical*, **226**, 207 (2005).
9. E. Mondal, P. R. Sahu, G. Bose, and A. T. Khan, *Tetrahedron Lett.*, **43**, 2843 (2002).

10. S. Kerverdo, L. Lizzani-Cuvelier, and E. Duñach, *Tetrahedron*, **58**, 10455 (2002).

11. *O,S*-Acetals. P. Wimmer, in *O/O- und O/S-Acetale*; H. Hagemann and D. Klamann, Eds., *Houben-Weyl*, 4th ed., Vol. E14a/1, Thieme, Stuttgart, 1991, pp. 785–831.

12. C. Djerassi, M. Shamma, and T. Y. Kan, *J. Am. Chem. Soc.*, **80**, 4723 (1958).

13. R. H. Mazur and E. A. Brown, *J. Am. Chem. Soc.*, **77**, 6670 (1955).

14. S. V. Frye and E. L. Eliel, *Tetrahedron Lett.*, **26**, 3907 (1985).

15. M. Kirihara, Y. Ochiai, N. Arai, S. Takizawa, T. Momose, and H. Nemoto, *Tetrahedron Lett.*, **40**, 9055 (1999).

16. E. Mondal, P. R. Sahu, G. Bose, and A. T. Khan, *J. Chem. Soc. Perkin Trans. 1*, 1026 (2002).

17. E. Mondal, P. R. Sahu, and A. T. Khan, *Synlett*, 463 (2002).

18. S. P. Chavan, S. W. Dantale, K. Pasupathy, R. B. Tejwani, S. K. Kamat, and T. Ravindra-nathan, *Green Chem.*, **4**, 337 (2002).

19. L.-C. Chen and H.-M. Wang, *Org. Prep. Proc. Int.*, **31**, 562 (1999).

20. B. Karimi, H. Seradj, and M. H. Tabaei, *Synlett*, 1798 (2000).

21. J. Nakayama, H. Sugiura, A. Shiotsuki, and M. Hoshino, *Tetrahedron Lett.*, **26**, 2195 (1985).

22. T. Ravindranathan, S. P. Chaven, J. P. Varghese, S. W. Dantale, and R. B. Tejwani, *J. Chem. Soc., Chem. Commun.*, 1937 (1994); T. Ravindranathan, S. P. Chavan, and M. M. Awachat, *Tetrahedron Lett.*, **35**, 8835 (1994).

23. S. P. Chavan, P. Soni, and S. K. Kamat, *Synlett*, 1251 (2001).

24. E. J. Corey and M. G. Bock, *Tetrahedron Lett.*, **16**, 2643 (1975).

Diseleno Acetals and Ketals: $R_2C(SeR')_2$

Selenium compounds are generally highly toxic.

Formation

1. RSeH, $ZnCl_2$, N_2, CCl_4, 20°C, 3 h, 70–95% yield.[1]
2. From a ketal: $(PhSe)_3B$, CF_3COOH, $CHCl_3$, 20°C, 20 min to 24 h.[2]

Cleavage

Diseleno acetals and ketals are cleaved more rapidly than their dithio counterparts; a methyl derivative is cleaved more rapidly than a phenyl derivative. Methyl iodide or ozone converts diseleno acetals and ketals to vinyl selenides.[1]

1. $HgCl_2$, $CaCO_3$, CH_3CN, H_2O, 20°C, 2–4 h, 65–80% yield.[1]
2. $CuCl_2$, CuO, acetone, H_2O, 20°C, 5 min to 2 h, 73–99% yield.[1]
3. H_2O_2, THF, 0°C, 15 min to 20°C, 3 h, 60–65% yield.[1]
4. $(PhSeO)_2O$, THF, 20°C or 60°C, 5 min to 6 h, 60–90% yield.[1]
5. Clay-supported ferric nitrate (Clayfen) or clay-supported cupric nitrate (Claycop), pentane, rt, 60–97% yield.[3]

1. A. Burton, L. Hevesi, W. Dumont, A. Cravador, and A. Krief, *Synthesis*, 877 (1979).
2. D. L. J. Clive and S. M. Menchen, *J. Org. Chem.*, **44**, 4279 (1979).
3. P. Laszlo, P. Pennetreau, and A. Krief, *Tetrahedron Lett.*, **27**, 3153 (1986).

MISCELLANEOUS DERIVATIVES

O-Substituted Cyanohydrins

O-Acetyl Cyanohydrin: $R_2C(CN)OAc$

Formation

1. $Me_2C(CN)OH$, Et_3N, 25°C, 2 h, 82% yield; Ac_2O, Pyr, 25°C, 40 h, 82% yield.[1]
2. From a cyanohydrin: Ac_2O, $FeCl_3$, 25–92% yield.[2] Other anhydrides are also effective in this conversion.
3. AcCN, K_2CO_3, CH_3CN, 79–96% yield.[3]

Cleavage

Li(O-*t*-Bu)$_3$AlH, THF; KOH, CH_3OH, H_2O, 25°C, 5 min, 84% yield.[1]

O-Methoxycarbonyl Cyanohydrin: $R_2C(CN)OCO_2CH_3$

This derivative is prepared by reaction of a ketone with CH_3O_2CCN, diisopropyl-amine in THF at rt for 16–18 h (15–98% yield). From the two examples provided, it appears that ketones conjugated to either an aromatic ring or an olefin tend to give low yields.[4] This group is stable to acids, oxidants, and Lewis acids. It reacts with nucleophilic reagents.

O-Trimethylsilyl Cyanohydrin: $R_2C(CN)OSiMe_3$ (Chart 5)

Formation

1. The following results indicate that there are essentially two modes by which these cyanohydrins form. The first is a Lewis acid-catalyzed mode which presumably activates the carbonyl toward addition, and the second is a nu-cleophilic mode whereby the nucleophile reacts with TMSCN to release CN^- which adds to the carbonyl followed by silylation of the oxygen. There is also a large body of literature on the preparation of chiral cyanohydrins.[5]
2. Me_3SiCN, cat. KCN or Bu_4NF, 18-crown-6, 75–95% yield.[6]
3. Me_3SiCN, Ph_3P, CH_3CN, 0°C, 1 h, 100% yield.[7]

4. Me$_2$C(CN)OSiMe$_3$, KCN, 130°C.[8]

5. Me$_3$SiCl, KCN, Amberlite XAD-4, CH$_3$CN, 60°C, 8 h, 81–97% yield.[9]

6. Me$_3$SiCl, KCN, NaI, Pyr, CH$_3$CN, 50–77% distilled yields, 100% by NMR.[10]

7. R$_3$SiCl, KCN, ZnI$_2$, CH$_3$CN, 86–98% yield.[11] This method was used to prepare the t-BuPh$_2$Si, t-BuMe$_2$Si and i-Pr$_3$Si cyanohydrins.

8. TMSCN, TEA, 91–100% yield.[12] K$_2$CO$_3$ has also been used effectively as a base.[13] A polymer-supported amine is also an effective catalyst.[14]

9. TMSCN, P(RNCH$_2$CH$_2$)$_3$N, THF, rt, 59–95% yield. These conditions also give excellent results with TBSCN, giving the TBS protected cyanohydrins (99% yield except for camphor which gave a 43% yield).[15]

10. LiO(CH$_2$CH$_2$O)$_3$Me, TMSCN, THF, 91–98% yield. Bicyclic systems show good endo selectivity.[16]

11. N-Methylmorpholine N-oxide, TMSCN, CH$_2$Cl$_2$, 86–99% yield.[17] Triethanolamine N-oxide is also effective.[18]

12. TMSCN, THF, Yb(CN)$_3$, 0°C to rt, 84–99% yield.[19]

13. TMSCN, CH$_2$Cl$_2$, Yb(OTf)$_3$, 55–95% yield. Aromatic ketones fail to react.[20]

14. TMSCN, CH$_2$Cl$_2$, −40°C, Eu(fod)$_3$, 45–95% yield.[21]

15. TMSCN, CH$_3$CN, reflux, 2 h, 89–95% yield.[22] These conditions are selective for aldehydes.

16. TMSCN, MgAlCO$_3$, heptane, 90–99% yield.[23]

17. TMSCN, (−)-DIPT [diisopropyl L-tartrate], Ti(i-PrO)$_4$, CH$_2$Cl$_2$, 0°C, 6 h, rt, 12 h, 95% yield. These conditions afford chiral cyanohydrins.[24]

18. (R)-BINOL-Ti(O−i-Pr)$_2$, TMSCN, CH$_2$Cl$_2$. Enantioselectivity of up to 75% is obtained.[25]

19. Chiral (salene)Ti(IV) complexes, TMSCN. This system is selective for aldehydes; the asymmetric induction is dependent upon aldehyde structure.[26,27]

20. Pybox-AlCl$_3$, [(S,S)-2,6-bis(4′-isopropyloxazolin-2′-yl)pyridine], TMSCN. Mandelonitrile was formed in 92% yield (>90% ee).[28]

21. Ti(Oi-Pr)$_4$, sulfoximines, TMSCN.[29]

22. TMSCN, Zr(KPO$_4$)$_2$, CH$_2$Cl$_2$, reflux, 83–98% yield.[30]

23. Bu$_2$SnCl$_2$ or Ph$_2$SnCl$_2$, TMSCN, 71–97% yield.[31]

24. TMSCN, I$_2$, CH$_2$Cl$_2$, rt, 30 min, 85–93% yield.[32]

Cleavage

1. AgF, THF, H$_2$O, 25°C, 2.5 h, 77% yield.[7]

2. Dilute acid or base.[33]

3. (*S*)-Hydroxynitrile lyase can be used for the decomposition of cyanohydrins with some level of enantioselectivity.[34]

O-1-Ethoxyethyl Cyanohydrin: $R_2C(CN)OCH(OC_2H_5)CH_3$

The ethoxyethyl cyanohydrin was prepared (NaCN, HCl, THF, 0°C, 75% yield, followed by EtOCH=CH₂, HCl, 50% yield) to convert an aldehyde ultimately to a protected ketone. It was cleaved by hydrolysis (0.01 *N* HCl, MeOH, 25°C, followed by NaOH, 0°C, 85% yield).[35] Butyl vinyl ether can be used similarly.

O-Tetrahydropyranyl Cyanohydrin: $R_2C(CN)O$-THP

The tetrahydropyranyl cyanohydrin was prepared from a steroid cyanohydrin (dihydropyran, TsOH, reflux, 1.5 h) and cleaved by hydrolysis (cat. concd. HCl, acetone, reflux, 15 min, followed by aq. pyridine, reflux, 1 h).[36]

1. P. D. Klimstra and F. B. Colton, *Steroids*, **10**, 411 (1967).

2. T. Hiyama, H. Oishi, and H. Saimoto, *Tetrahedron Lett.*, **26**, 2459 (1985).

3. M. Okimoto and T. Chiba, *Synthesis*, 1188 (1996).

4. D. Poirier, D. Berthiaume, and R. P. Boivin, *Synlett*, 1423 (1999); D. Berthiaume and D. Poirier, *Tetrahedron*, **56**, 5995 (2000).

5. Tetrahedron Symposium in Print Number 109, M. North, Ed., *Tetrahedron*, **60**, 10379 (2004).

6. D. A. Evans, J. M. Hoffman, and L. K. Truesdale, *J. Am. Chem. Soc.*, **95**, 5822 (1973).

7. D. A. Evans and R. Y. Wong, *J. Org. Chem.*, **42**, 350 (1977).

8. D. A. Evans and L. K. Truesdale, *Tetrahedron Lett.*, **14**, 4929 (1973).

9. K. Sukata, *Bull. Chem. Soc. Jpn.*, **60**, 3820 (1987).

10. F. Duboudin, Ph. Cazeau, F. Moulines, and O. Laporte, *Synthesis*, 212 (1982).

11. V. H. Rawal, J. A. Rao, and M. P. Cava, *Tetrahedron Lett.*, **26**, 4275 (1985).

12. S. Kobayashi, Y. Tsuchiya, and T. Mukaiyama, *Chem. Lett.*, **20**, 537 (1991).

13. B. He, Y. Li, X. Feng, and G. Zhang, *Synlett*, 1776 (2004).

14. M. L. Kantam, P. Sreekanth, and P. L. Santhi, *Green Chem.*, **47** (2000).

15. B. M. Fetterly and J. G. Verkade, *Tetrahedron Lett.*, **46**, 8061 (2005).

16. H. S. Wilkinson, P. T. Grover, C. P. Vandenbossche, R. P. Bakale, N. N. Bhongle, S. A. Wald, and C. H. Senanayake, *Org. Lett.*, **3**, 553 (2001).

17. S. S. Kim, D. W. Kim, and G. Rajagopal, *Synthesis*, 213 (2004).

18. H. Zhou, F.-X. Chen, B. Qin, X. Feng, and G. Zhang, *Synlett*, 1077 (2004).

19. S. Matsubara, T. Takai, and K. Utimoto, *Chem. Lett.*, **20**, 1447 (1991).

20. Y. Yang and D. Wang, *Chem. Lett.*, **26**, 1379 (1997).

21. J. H. Gu, M. Okamoto, M. Terada, K. Mikami, and T. Nakai, *Chem. Lett.*, **21**, 1169 (1992).

22. K. Manju and S. Trehan, *J. Chem. Soc., Perkin Trans. I*, 2383 (1995).

23. B. M. Choudary, N. Narender, and V. Bhuma, *Synth. Commun.*, **25**, 2829 (1995).

24. M. C. Pirrung and S. W. Shuey, *J. Org. Chem.*, **59**, 3890 (1994).

25. M. Mori, H. Imma, and T. Nakai, *Tetrahedron Lett.*, **38**, 6229 (1997).

26. Y. Belokon, M. Flego, N. Ikonnikow, M. Moscalenko, M. North, C. Orizu, V. Tararov, and M. Tasinazzo, *J. Chem. Soc., Perkin Trans. I*, 1293 (1997).

27. Y. Jiang, X. Zhou, W. Hu, L. Wu, and A. Mi, *Tetrahedron: Asymmetry*, **6**, 405 (1995).

28. I. Iovel, Y. Popelis, M. Fleisher, and E. Lukevics, *Tetrahedron: Asymmetry*, **8**, 1279 (1997).

29. C. Bolm, P. Mueller, and K. Harms, *Acta Chem. Scand.*, **50**, 305 (1996); C. Bolm and P. Mueller, *Tetrahedron Lett.*, **36**, 1625 (1995).

30. M. Curini, F. Epifano, M. C. Marcotullio, O. Rosati, and M. Rossi, *Synlett*, 315 (1999).

31. J. K. Whitesell and R. Apodaca, *Tetrahedron Lett.*, **37**, 2525 (1996).

32. J. S. Yadav, B. V. S. Reddy, M. S. Reddy, and A. R. Prasad, *Tetrahedron Lett.*, **43**, 9703 (2002).

33. D. A. Evans, L. K. Truesdale, and G. L. Carroll, *J. Chem. Soc., Chem. Commun.*, 55 (1973).

34. M. Schmidt, S. Herve, N. Klempier, and H. Griengl, *Tetrahedron*, **52**, 7833 (1996).

35. G. Stork and L. Maldonado, *J. Am. Chem. Soc.*, **93**, 5286 (1971).

36. P. deRuggieri and C. Ferrari, *J. Am. Chem. Soc.* **81**, 5725 (1959).

Substituted Hydrazones

N,N-**Dimethylhydrazone:** $RR'C=NN(CH_3)_2$ (Chart 5)

Although *N,N*-dimethylhydrazones are used as protective groups their use is not nearly as ubiquitous as the acetal and ketal. This is likely a result of the fact that these can still be deprotonated with strong base and are susceptible to nucleophilic reagents.

Formation

1. H_2NNMe_2, EtOH–HOAc, reflux, 24 h, 90–94% yield.[1]
2. $Me_2AlNHNMe_2$, $PhCH_3$, reflux, 3–5 h, 77–99% yield.[2]
3. H_2NNMe_2, TMSCl, 25°C, 36 h, 92% yield.[3]

Cleavage

The cleavage of *N,N*-dialkylhydrazones in connection with the synthesis of natural products has been reviewed.[4] Most of the methods presented below have not been rigorously tested for their functional group compatibility.

1. Aqueous $NH_4H_2PO_4$, THF, 77–99% yield.[5] Cyclic acetals are compatible with this method.
2. $NaIO_4$, MeOH, pH 7, 2–3 h, 90% yield.[6]
3. $Cu(OAc)_2$, H_2O, THF, pH 5.4, 25°C, 15 min, 97% yield.[7]
4. $CuCl_2$, THF, HPO_4^-, → pH 7, 85–100% yield.[7,8]
5. CH_3I, 95% EtOH, reflux, 80–90% yield.[9]

6. O_3, CH_2Cl_2, $-78°C$, 60–100% yield.[10]

7. O_2, hv, Rose Bengal, MeOH, $-78°C$ to $-20°C$, followed by Ph_3P or Me_2S, 48–88% yield.[11]

8. N_2O_4, $-40°C$ to $0°C$, CH_3CN, THF, $CHCl_3$, CCl_4, ~10 min, 75–95% yield.[12] This method is also effective for the regeneration of ketones from oximes (45–95% yield).

9. $NaBO_3 \cdot 4H_2O$, t-BuOH, pH 7, 60°C, 24 h, 70–95% yield.[13]

10. AcOH, THF, H_2O, AcONa, 25°C, 24 h, 95% yield.[14]

N,N-Dimethylhydrazones are stable to CrO_3/H_2SO_4 (0°C, 3 min), to $NaBH_4$ (EtOH, 25°C), to $LiAlH_4$ (THF, 25°C), and to B_2H_6 followed by H_2O_2/OH^-. They are cleaved by CrO_3/Pyr and by p-$NO_2C_6H_4CO_3H/CHCl_3$, 25°C.[9]

11. Silica gel, THF, H_2O, rt, 3–10 h, 60–74% yield[15] or silica gel, CH_2Cl_2, 77–100% yield.[16]

12. $BF_3 \cdot Et_2O$, acetone, H_2O, 93–100% yield.[17]

13. MCPBA, DMF, $-63°C$, 100% yield.[18] Hydrazones of aldols are cleaved without elimination under these conditions.[19] An axial α-methyl group on a cyclohexanone does not epimerize under these conditions.[18]

14. $MMPP \cdot 6H_2O$ (magnesium monoperoxyphthalate), pH 7 buffer, MeOH, 0°, 5–120 min, 76–99% yield.[20] These conditions were used to cleave the related SAMP hydrazone in the presence of 2 trisubstituted alkenes in 46% yield.[21]

15. Peracetic acid.[22]

16. Dimethyldioxirane, acetone, 89% yield.[23]

17. $NOBF_4$, CH_2Cl_2, Pyr, 59–86% yield. Oximes are cleaved similarly in 55–82% yield.[24]

18. $Pd(OAc)_2$, $SnCl_2$, DMF, H_2O, 53–100% yield. This is a catalytic procedure for the cleavage of dimethylhydrazones.[25]

19. $[(n\text{-}Bu)_4N]_2S_2O_8$, $ClCH_2CH_2Cl$, reflux, 0.6 h, 89–97% yield.[26]

20. $MeReO_3$, H_2O_2, CH_3CN, AcOH, 85–93% yield.[27]

21. $(NMe_4)_2[Ni(Me_2opba)] \cdot 4H_2O$, pivaldehyde, N-methylimidazole, fluorobenzene, O_2, 46–95% yield.[28] Oximes and tosylhydrazones are also cleaved with this method.

22. $FeSO_4 \cdot 7H_2O$, $CHCl_3$, rt, 20–60 min, 86–94% yield. Phenylhydrazones are also cleaved.[29]

23. $FeCl_3 \cdot SiO_2$, CH_2Cl_2, 82–93% yield. Oximes and tosylhydrazones are also cleaved.[30]

24. $CeCl_3 \cdot 7H_2O/SiO_2$, microwaves, 88–91% yield.[31]

25. Porcine pancreatic lipase, acetone, H_2O, 11–96% yield.[32]

26. TMSCl, NaI, CH_3CN, 87–95% yield.[33]

27. CoF_3 ($CHCl_3$, reflux, 67–93% yield);[34] $MoOCl_3$ or MoF_6 (H_2O, THF, 25°C, 4 h, 80–90% yield);[35] WF_6 ($CHCl_3$, 0–25°C, 1 h, 84–95% yield)[36]; UF_6 (50–95% yield)[37] $[Ni(en)_3]S_2O_3$, $Hg([Co(SCN)_4$ or $Mn(acac)_3$, ($CHCl_3$, 88–98% yield).[38]

1. G. R. Newkome and D. L. Fishel, *Org. Synth., Collect. Vol. VI*, 12 (1988).

2. B. Bildstein and P. Denifl, *Synthesis*, 158 (1994).

3. D. A. Evans, R. P. Polniaszek, K. M. DeVries, D. E. Guinn, and D. J. Mathre, *J. Am. Chem. Soc.*, **113**, 7613 (1991).

4. D. Enders, L. Wortmann, and R. Peters, *Acc. Chem. Res.*, **33**, 157 (2000).

5. T. Ulven and P. H. J. Carlsen, *Eur. J. Org. Chem.*, 3971 (2000).

6. E. J. Corey and D. Enders, *Tetrahedron Lett.*, **17**, 11 (1976).

7. E. J. Corey and S. Knapp, *Tetrahedron Lett.*, **17**, 3667 (1976).

8. A. Sadeghi-Khomami, A. J. Blake, C. Wilson, and N. R. Thomas, *Org. Lett.*, **7**, 4891 (2005).

9. M. Avaro, J. Levisalles, and H. Rudler, *J. Chem. Soc., Chem. Commun.*, 445 (1969).

10. R. E. Erickson, P. J. Andrulis, J. C. Collins, M. L. Lungle, and G. D. Mercer, *J. Org. Chem.*, **34**, 2961 (1969).

11. E. Friedrich, W. Lutz, H. Eichenauer, and D. Enders, *Synthesis*, 893 (1977).

12. S. B. Shim, K. Kim, and Y. H. Kim, *Tetrahedron Lett.*, **28**, 645 (1987).

13. D. Enders and V. Bhushan, *Z. Naturforsch. B: Chem. Sci.*, **42**, 1595 (1987).

14. E. J. Corey and H. L. Pearce, *J. Am. Chem. Soc.*, **101**, 5841 (1979).

15. R. B. Mitra and G. B. Reddy, *Synthesis*, 694 (1989).

16. H. Kotsuki, A. Miyazaki, I. Kadota, and M. Ochi, *J. Chem. Soc., Perkin Trans. I*, 429 (1990).

17. D. Enders, H. Dyker, G. Raabe, and J. Runsink, *Synlett*, 901 (1992).

18. M. Duraisamy and H. M. Walborsky, *J. Org. Chem.*, **49**, 3410 (1984).

19. M. M. Claffey and C. H. Heathcock, *J. Org. Chem.*, **61**, 7646 (1996).

20. D. Enders and A. Plant, *Synlett*, 725 (1990).

21. K. C. Nicolaou, F. Sarabia, M. R. V. Finlay, S. Ninkovic, N. P. King, D. Vourloumis, and Y. He, *Chem. Eur. J.*, **3**, 1971 (1997).

22. L. Horner and H. Fernekess, *Chem. Ber.*, **94**, 712 (1961).

23. A. Altamura, R. Curci, and J. O. Edwards, *J. Org. Chem.*, **58**, 7289 (1993).

24. G. A. Olah and T.-L. Ho, *Synthesis*, 610 (1976).

25. T. Mino, T. Hirota, and M. Yamashita, *Synlett*, 999 (1996); T. Mino, T. Hirota, N. Fujita, and M. Yamashita, *Synthesis*, 2024 (1999).

26. H. C. Choi and Y. H. Kim, *Synth. Commun.*, **24**, 2307 (1994).

27. S. Stankovic and J. H. Espenson, *J. Org. Chem.*, **65**, 2218 (2000).

28. G. Blay, E. Benach, I. Fernandez, S. Galletero, J. R. Pedro, and R. Ruiz, *Synthesis*, 403 (2000).

29. A. Nasreen and S. R. Adapa, *Org. Prep. Proc. Int.*, **31**, 573 (1999).

30. D. S. Bose, A. V. Narsaiah, and P. R. Goud, *Ind. J. Chem., Sect. B*, **40B**, 719 (2001).

31. J. S. Yadov, B. V. S. Reddy, M. S. K. Reddy, and G. Sabitha, *Synlett*, **7**, 1134 (2001).

32. T. Mino, T. Matsuda, D. Hiramatsu, and M. Yamashita, *Tetrahedron Lett.*, **41**, 1461 (2000).

33. A. Kamal, K. V. Ramana, and M. Arifuddin, *Chem. Lett.*, **28**, 827 (1999).

34. G. A. Olah, J. Welch, and M. Henninger, *Synthesis*, 308 (1977).

35. G. A. Olah, J. Welch, G. K. S. Prakash, and T.-L. Ho, *Synthesis*, 808 (1976).

36. G. A. Olah and J. Welch, *Synthesis*, 809 (1976).

37. G. A. Olah, J. Welch, and T.-L. Ho, *J. Am. Chem. Soc.*, **98**, 6717 (1976).

38. A. Kamal, M. Arifuddin, and M. V. Rao, *Synlett*, 1482 (2000).

Phenylhydrazone: $C_6H_5NHN=CR_2$

Formation

$PhNHNH_2$, AcOH, EtOH.[1] This is a standard method that works well for a large variety of substrates. The cationic ion exchange resin Dowex 50-X8 is also a good catalyst for this reaction.[2]

Cleavage

1. $PhI(OTFA)_2$, CH_3CN, H_2O, 82–90% yield or $PhI(OH)OTs$, $CDCl_3$, rt, 2h, 74–98% yield.[3] Mild oxidative regeneration of ketones occurs in good yields.

2. $(NH_4)_2S_2O_8$, clay, microwaves or ultrasound, 62–90% yield.[4]

3. Wet silica supported $KMnO_4$, 70–98 yield.[5]

4. Wet silica gel, $SiBr_4$, 79–91% yield.[6] This method probably produces HBr *in situ*, which is probably the real catalyst. Oximes and semicarbazones are also hydrolyzed.

1. R. L. Shriner, R. C. Fuson, D. Y. Curtin, and T. C. Morrill, *The Systematic Identification of Organic Compounds: A Laboratory Manual*, 6th ed., Wiley, New York, 1980, p. 165.

2. K. Niknam, A. R. Kiasat, and S. Karimi, *Synth. Commun.*, **35**, 2231 (2005).

3. D. H. R. Barton, J. Cs. Jaszberenyi, and T. Shinade, *Tetrahedron Lett.*, **34**, 7191 (1993).

4. R. S. Varma and H. M. Meshram, *Tetrahedron Lett.*, **38**, 7973 (1997).

5. A. R. Hajipour, H. Adibi, and A. E. Ruoho, *J. Org. Chem.*, **68**, 4553 (2003).

6. S. K. De, *Tetrahedron Lett.*, **44**, 9055 (2003).

2,4-Dinitrophenylhydrazone (2,4-DNP Group): $R_2C=NNHC_6H_3$-2,4-$(NO_2)_2$
(Chart 5)

Formation

2,4-$(NO_2)_2C_6H_3NHNH_2 \cdot H_2SO_4$, EtOH, H_2O, 25°C, 10 min, 80% yield.[1]

In a synthesis of sativene a carbonyl group was protected as a 2,4-DNP while a double bond was hydrated with $BH_3/H_2O_2/OH^-$. Attempted protection of the carbonyl group as a ketal caused migration of the double bond; protection as an oxime or oxime acetate was unsatisfactory, since they would be reduced with BH_3.

Cleavage

2,4-Dinitrophenylhydrazones are cleaved by various oxidizing and reducing agents, and by exchange reactions. Some of the methods used for the cleavage of oximes should be applicable for DNP cleavage.

1. O_3, EtOAc, $-78°C$, 70% yield.[1]
2. $TiCl_3$, DME, H_2O, N_2, reflux, 80–95% yield.[2]
3. Acetone, sealed tube, 75°C, 20 h, 80–85% yield.[3]

1. J. E. McMurry, *J. Am. Chem. Soc.*, **90**, 6821 (1968).
2. J. E. McMurry and M. Silvestri, *J. Org. Chem.*, **40**, 1502 (1975).
3. S. R. Maynez, L. Pelavin, and G. Erker, *J. Org. Chem.*, **40**, 3302 (1975).

Tosylhydrazone: $CH_3C_6H_4SO_2NHN=CR_2$

Formation

$TsNHNH_2$, AcOH, EtOH.[1]

Cleavage

1. TS-1(titanium silicate molecular sieve), H_2O_2, MeOH, reflux, 4–18 h, 60–64% yield.[2]
2. Dimethyldioxirane, acetone, 95% yield.[3]
3. $Zr(O_3PCH_3)_{1.2}(O_3PC_6H_4SO_3H)_{0.8}$, acetone, H_2O, reflux, 70–95% yield.[4]
4. $KHSO_5$, aq. CH_3CN, 63–99% yield.[5]
5. Dimethyldioxirane, acetone, pH 6, 10–144 h, 67–99% yield.[6]
6. 70% t-Butyl hydroperoxide, CCl_4, reflux, 4–18 h, 50–100% yield.[7] Cleavage is only effective for aromatic tosylhydrazones.
7. Na_2O_2, pentane, H_2O, reflux, 6 h, 69–72% yield.[8]
8. DDQ, CH_2Cl_2, H_2O, 80–95% yield.[9]

1. R. H. Shapiro, *Org. React.*, **23**, 405 (1976).
2. P. Kumar, V. R. Hegde, B. Paudey, and T. Ravindranathan, *J. Chem. Soc., Chem. Commun.*, 1553 (1993).
3. A. Altamura, R. Curci, and J. O. Edwards, *J. Org. Chem.*, **58**, 7289 (1993).
4. M. Curini, O. Rosati, and E. Pisani, *Synlett*, 333 (1996).

5. Y. H. Kim, J. C. Jung, and K. S. Kim, *Chem. Ind. (London)* 31 (1992).

6. J. C. Jung, K. S. Kim, and Y. H. Kim, *Synth. Commun.*, **22**, 1583 (1991).

7. N. B. Barhate, A. S. Gajare, R. D. Wakharkar, and A. Sudalai, *Tetrahedron Lett.*, **38**, 653 (1997).

8. T.-L. Ho and G. A. Olah, *Synthesis*, 611 (1976).

9. S. Chandrasekhar, C. R. Reddy, and M. V. Reddy, *Chem. Lett.*, **29**, 430 (2000).

Semicarbazone ($NH_2CONHN=CR_2$)

Formation

$NH_2CONHNH_2$, NaOAc, MeOH.[1]

Cleavage

1. $PhI(OAc)_2$, CH_3CN, H_2O, 70–83 yield.[2]
2. $(Bu_4N^+)_2S_2O_8^{-2}$, $ClCH_2CH_2Cl$, reflux, 89–97% yield.[3]
3. Pyruvic acid, acetic acid, 43–61% $CHCl_3$.[4]
4. $CuCl_2·2H_2O$, CH_3CN, reflux, 10–390 min, 7–97% yield.[5]
5. TMSCl, $NaNO_2$ or $NaNO_3$, Aliquat 366, 3–5 h, CH_2Cl_2, 75–95% yield.[6]

Diphenylmethylsemicarbazone ($Ph_2CHNHCONHN=CR_2$)

This derivative was used to improve the solubility characteristic of an argininal semicarbazone for solution phase peptide synthesis.

Formation

$Ph_2CHNHCONHNH_2$, NaOAc, EtOH, H_2O, reflux, 1 h, 78% yield.[7]

Cleavage

Since hydrogenolysis resulted in a only 20% yield of the free aldehyde, a two-step procedure was developed in which the diphenylmethyl group was first cleaved with HF/anisole and then the unsubstituted semicarbazone was cleaved with formalin in 40–60% overall yield.

1. R. L. Shriner, R. C. Fuson, D. Y. Curtin, and T. C. Morrill, *The Systematic Identification of Organic Compounds*, 6th ed., Wiley, New York, 1980, p. 179.

2. D. W. Chen and Z. C. Chen, *Synthesis*, 773 (1994).

3. H. C. Choi and Y. H. Kim, *Synth. Commun.*, **24**, 2307 (1994).

4. H. Hosoda, K. Osanai, I. Fukasawa, and T. Nambara, *Chem. Pharm. Bull.*, **38**, 1949 (1990).

5. R. N. Ram and K. Varsha, *Tetrahedron Lett.*, **32**, 5829 (1991).

6. R. H. Khan, R. K. Mathur, and A. C. Ghosh, *J. Chem. Res., Synop.*, 506 (1995).

7. R. Dagnino, Jr., and T. R. Webb, *Tetrahedron Lett.*, **35**, 2125 (1994).

Oxime Derivatives: $R_2C=NOH$

The use of oximes for carbonyl protection has become quite rare. This may be do to the fact that oximes still contain an acidic hydrogen and a somewhat reactive C=N.

Formation

1. $H_2NOH \cdot HCl$, Pyr, 60°C. This is the standard method for the preparation of oximes. Ethanol or methanol can be used as cosolvents.
2. $H_2NOH \cdot HCl$, DABCO, MeOH, rt, 87% for a camphor derivative.[1] This method was reported to be better than when pyridine was used as the solvent and base.
3. TMSNHOTMS, KH, 100% yield.[2]
4. $H_2NOH \cdot HCl$, Amberlyst A21, EtOH, 1–10 h, 70–97% yield.[3]

Cleavage

Oximes are cleaved by oxidation, reduction, or hydrolysis in the presence of another carbonyl compound. Some synthetically useful methods are shown below. The cleavage of oximes has been reviewed.[4] Most of the methods have not been tested in significant synthetic endeavors and as such their functional group compatibility is uncertain.

1. $CH_3CO(CH_2)_2COOH$, 1 N HCl, 25°C, 3 h, 94% yield.[5] Pyruvic acid (HOAc, reflux, 1–3 h, 77% yield),[6] acetone (80–100 h, 72% yield),[7] and glycolic acid[8] effect cleavage in a similar manner.
2. $TiCl_4$, NaI, CH_3CN, rt, 63–97% yield.[9]
3. $Zr(O_3PCH_3)_{1.2}(O_3PC_6H_4SO_3H)_{0.8}$, acetone, water, reflux 30 min to 24 h, 70–95% yield. Semicarbazones, tosylhydrazones and hydrazones are also cleaved.[10] $Zr(HSO_4)_4$ also serves as a good catalyst.[11]
4. $BiCl_3$, microwave irradiation, 2 min, THF, 70–96% yield. α,β-Unsaturated systems were not effectively cleaved under these conditions.[12] $BiCl_3$, $Bi(OTf)_3$[13] or $Bi(NO_2)_3$[14] can also be used.
5. Ionic liquid/silica gel, acetone, water, 89–96% yield.[15]
6. $Na_2S_2O_4$, H_2O, 25°C, 12 h or 40°C, few hours ~95% yield.[16]
7. $NaHSO_3$, EtOH, H_2O, reflux, 2–16 h; dil. HCl, 30 min, 85% yield.[17,18]
8. $Mg(HSO_4)_2$, wet SiO_2, rt, 72–96% yield. These conditions also cleave simple semicarbazones and phenylhydrazones.[19]
9. Ac_2O, 20°C; $Cr(OAc)_2$, THF, H_2O, 25–65°C, 75–95% yield.[20] Chromous acetate also cleaves unsubstituted oximes, but the reaction is slow and requires high temperatures.
10. $TiCl_3$, H_2O, rt, 1 h, 85% yield.[21] This is an excellent reagent that works when cleavage of a methoxy oxime with chromous ion fails.
11. VCl_2, H_2O, THF, 8 h, rt, 75–92% yield.[22]

12. Fe, HCl, MeOH, H_2O, reflux, 30 min, 80–94% yield.[23]

13. Baker's yeast, pH 7.2, H_2O, EtOH, 62–95% yield with sonication.[24]

14. $Ru_3(CO)_{12}$, CO, 20 atm, 4 h, 100°C. These conditions reduce the oxime to an imine that is easily hydrolyzed with water.[25] Aldehyde oximes give low yields of nitriles.

15. $Mo(CO)_6$, CH_3CN, H_2O, 59–94% yield.[26] $Co_2(CO)_8$/TEA is similarly effective.[27]

16. $NaNO_2$, 1 N HCl, CH_3OH, H_2O, 0°C, 3 h, 76% yield.[28] In the last step of a synthesis of erythronolide A, acid-catalyzed hydrolysis of an acetonide failed because the carbonyl-containing precursor was unstable to acidic hydrolysis (3% MeOH, HCl, 0°C, 30 min, conditions developed for the synthesis of erythronolide B). Consequently, the carbonyl group was protected as an oxime, the acetonide was cleaved, and the carbonyl group was regenerated.

17. NOCl, Pyr, −20°C; H_2O, reflux, 70–90% yield.[29] Olefins were not affected under these conditions. The related nitrosyl tetrafluoroborate has also been used.[30]

18. $Et_3N \cdot HCl \cdot CrO_3$, $ClCH_2CH_2Cl$, 2 h, rt, 60–90% yield.[31] This reagent was reported to work better than PCC (pyridinium chlorochromate[32]). Trimethylsilyl chlorochromate,[33] 2,6-dicarboxypyridinium chlorochromate,[34] bistetrabutylammonium dichromate,[35] imidazolium dichromate,[36] and CrO_3/silica gel[37] are also effective.

19. t-BuONO, t-BuOK; H_2O, NaOH; acidify, 40°C.[38]

20. TMSCl, $NaNO_2$, CCl_4, 5% Aliquat 336, rt, 3–5 h, 64–98% yield.[39]

21. NaOCl, MeCN, rt, 23–99% yield.[40]

22. t-Butylhypoiodite, CCl_4, rt, ~20 min, 93–96% yield.[41]

23. Zinc bismuthate, $PhCH_3$ or CH_3CN, reflux, 0.5–2 h, 56–85% yield.[42]

24. MnO_2, hexane or CH_2Cl_2, rt, 70–92% yield.[43] The oximes of pyruvates and O-alkyl oximes are not cleaved under these conditions.

25. $PhICl_2$, Pyr, $CHCl_3$, 3 h, 10°C, 65–80% yield.[44]

26. Dess-Martin periodinane, CH_2Cl_2, rt, 20 min, 90–100% yield.[45]

27. I_2, water, SDS, 25–40°C, 67–90% yield.[46]

28. $(NH_4)_2S_2O_8$·silica gel, microwave irradiation, 59–83% yield.[47] $AgNO_3$ will catalyze the oxidative cleavage with this reagent.[48] Benzyltriphenylphosphonium peroxodisulfate has also been used.[49]

29. $(PhSeO)_2$/THF, 50°C, 1–3 h, 80–95% yield.[50] An O-methyl oxime is stable to phenylselenic anhydride.

30. TS-1 zeolite, H_2O_2, acetone, reflux, 65–86% yield.[51]

31. $MoO_2(acac)_2$,[52] sodium tungstate,[53] or $VO(acac)_2$,[54] H_2O_2, acetone, 73–94% yield.

32. Dimethyldioxirane, acetone, 0°C or rt, 80–100% yield.[55]

33. $Cu(NO_3)_2$, Bentonite, hexane, acetone, 60–97% yield.[56] When silica gel is used as the support, tosylhydrazones and thioketals are also cleaved in excellent yield.[57]

34. $Fe(NO_3)_3$ or $Bi(NO_3)$ activated with $H_3PW \cdot 6H_2O$, neat, 40–45°C, 45–95% yield.[58]

35. $KMnO_4$, CH_3CN, H_2O, rt, 25–96% yield.[59] Alumina supported permanganate[60] and $KMnO_4$–MnO_2[61] are similarly effective. $KMnO_4$ also cleaves semicarbazones and phenylhydrazones.

36. $Mn(OAc)_3$, benzene, reflux, 1–2 h, 86–96% yield.[62]

37. 70% *t*-Butyl hydroperoxide, CCl_4, reflux, 4–18 h, 30–100% yield.[63]

38. NBS, CCl_4, 25°C, 80–96% yield.[64] *N*-bromosaccharin,[65] *N,N'*-dibromo-*N,N'*-1,2-ethanediylbis(*p*-toluenesulfphonamide),[66] *N,N*-dibromobenzenesulfonamide,[67] and poly[4-vinyl-*N,N*-dichlorobenzenesulfonamide][68] can be used similarly.

39. Wet $NaIO_4 \cdot$ silica, microwave, 68–93% yield.[69]

40. HIO_3, CH_2Cl_2, rt, 72–97% yield.[70]

41. $KHSO_5$, AcOH, 70–88% yield.[71]

42. Bu_3P, PhSSPh, THF, 85% yield.[72]

43. Platinum(II) terpyridyl acetylide complex, *h*v, CH_3CN 10-94% yield.[73]

44. Chloranil, *h*v, CH_3CN, 5–66% yield. In some cases a nitrile is formed under these conditions.[74]

1. R. V. Stevens, F. C. A. Gaeta, and D. S. Lawrence, *J. Am. Chem. Soc.*, **105**, 7713 (1983).

2. R. V. Hoffman and G. A. Buntain, *Synthesis*, 831 (1987).

3. R. Ballini, L. Barboni, and P. Filippone, *Chem. Lett.*, **26**, 475 (1997).

4. A. R. Hajipour, S. Khoee, and A. E. Ruoho, *Org. Prep. & Proc. Int.*, **35**, 529 (2003).

5. C. H. Depuy and B. W. Ponder, *J. Am. Chem. Soc.*, **81**, 4629 (1959).

6. E. B. Hershberg, *J. Org. Chem.*, **13**, 542 (1948).

7. S. R. Maynez, L. Pelavin, and G. Erker, *J. Org. Chem.*, **40**, 3302 (1975).

8. S. P. Chavan and P. Soni, *Tetrahedron Lett.*, **45**, 3161 (2004).

9. R. Balicki and L. Kaczmarek, *Synth. Commun.*, **21**, 1777 (1991).

10. M. Curini, O. Rosati, and E. Pisani, *Synlett*, 333 (1996).

11. F. Shirini, M. A. Zolfigol, A. Safari, I. Mohammadpoor-Baltork, and B. F. Mirjalili, *Tetrahedron Lett.*, **44**, 7463 (2003).

12. A. Boruah, B. Baruah, D. Prajapati, and J. S. Sandhu, *Tetrahedron Lett.*, **38**, 4267 (1997).

13. J. N. Arnold, P. D. Hayes, R. L. Kohaus, and R. S. Mohan, *Tetrahedron Lett.*, **44**, 9173 (2003).

14. T. T. Niaki, H. A. Oskooiee, M. M. Heravi, and B. Miralaee, *J. Chem. Res.*, 488 (2004); M. M. Mojtahedi and M. M. Heravi, *Ind. J. Chem.*, **44B**, 831 (2005).

15. D. Li, F. Shi, S. Guo, and Y. Deng, *Tetrahedron Lett.*, **45**, 265 (2004).

16. P. M. Pojer, *Aust. J. Chem.*, **32**, 201 (1979).

17. S. H. Pines, J. M. Chemerda, and M. A. Kozlowski, *J. Org. Chem.*, **31**, 3446 (1966).

18. Y. Watanabe, S. Morimoto, T. Adachi, M. Kashimura, and T. Asaka, *J. Antibiot.*, **46**, 647 (1993).

19. F. Shirini, M. A. Zolfigol, B. Mallakpour, S. E. Mallakpour, A. R. Hajipour, and I. M. Baltork, *Tetrahedron Lett.*, **43**, 1555 (2002).

20. E. J. Corey and J. E. Richman, *J. Am. Chem. Soc.*, **92**, 5276 (1970).

21. G. H. Timms and E. Wildsmith, *Tetrahedron Lett.*, **12**, 195 (1971).

22. G. A. Olah, M. Arvanaghi, and G. K. S. Prakash, *Synthesis*, 220 (1980).

23. P. K. Pradhan, S. Dey, P. Jaisankar, and V. S. Giri, *Synth. Commum.*, **35**, 913 (2005).

24. A. Kamal, M. V. Rao, and H. M. Meshram, *J. Chem. Soc., Perkin Trans. 1*, 2056 (1991).

25. M. Akazome, Y. Tsuji, and Y. Watanabe, *Chem. Lett.*, **19**, 635 (1990).

26. F. Geneste, N. Racelma, and A. Moradpour, *Synth. Commun.*, **27**, 957 (1997).

27. C. Mukai, I. Nomura, O. Kataoka, and M. Hanaoka, *Synthesis*, 1872 (1999).

28. E. J. Corey, P. B. Hopkins, S. Kim, S. Yoo, K. P. Nambiar, and J. R. Falck, *J. Am. Chem. Soc.*, **101**, 7131 (1979).

29. C. R. Narayanan, P. S. Ramaswamy, and M. S. Wadia, *Chem. Ind. (London)*, 454 (1977).

30. G. A. Olah and T. L. Ho, *Synthesis*, 609 (1976).

31. C. Gundu Rao, A. S. Radhakrishna, B. Bali Singh, and S. P. Bhatnagar, *Synthesis*, 808 (1983); G.-S. Zhang, D.-H. Yang, and M.-F. Chen, *OPPI Briefs*, **30**, 713 (1998).

32. N. C. Ganguly, M. Datta, and P. De, *J. Indian Chem. Soc.*, **81**, 308 (2004).

33. J. M. Aizpurua, M. Juarista, B. L. Lecea, and C. Palomo, *Tetrahedron*, **41**, 2903 (1985).

34. R. Hosseinzadeh, M. Tajbakhsh, and M. Y. Niaki, *Tetrahedron Lett.*, **43**, 9413 (2002).

35. R. Murugan and B. S. R. Reddy, *Chem. Lett.*, **33**, 1038 (2004).

36. S. K. De, *Synth. Commum.*, **34**, 2751 (2004).

37. P. M. Bendale and B. M. Khadilkar, *Tetrahedron Lett.*, **39**, 5867 (1998).

38. E. J. Corey, M. Narisada, T. Hiraoka, and R. A. Ellison, *J. Am. Chem. Soc.*, **92**, 396 (1970).

39. J. G. Lee, K. H. Kwak, and J. P. Hwang, *Tetrahedron Lett.*, **31**, 6677 (1990).

40. J. M. Khurana, A. Ray, and P. K. Sahoo, *Bull. Chem. Soc. Jpn.*, **67**, 1091 (1994).

41. V. N. Telvekar, *Synth. Commun.*, **35**, 2827 (2005).

42. H. Firouzabadi and I. Mohammadpoor-Baltork, *Synth. Commun.*, **24**, 489 (1994).

43. T. Shinada and K. Yoshihara, *Tetrahedron Lett.*, **36**, 6701 (1995).

44. A. S. Radhakrishna, A. Beena, K. Sivaprakash, and B. B. Singh, *Synth. Commun.*, **21**, 1473 (1991).

45. S. S. Chaudhari and K. G. Akamanchi, *Tetrahedron Lett.*, **39**, 3209 (1998); S. S. Chaudhari and K. G. Akamanchi, *Synthesis*, 760 (1999).

46. P. Gogoi, P. Hazarika, and D. Konwar, *J. Org. Chem.*, **70**, 1934 (2005).

47. R. S. Varma and H. M. Meshram, *Tetrahedron Lett.*, **38**, 5427 (1997).

48. M. Hirano, K. Kojima, S. Yakabe, and T. Morimoto, *J. Chem. Res., Synop.*, 277 (2001).

49. I. Mohammadpoor-Baltork, A. R. Hajipour, and H. Mohammadi, *Bull. Chem. Soc. Jpn.*, **71**, 1649 (1998); I. Mohammadpoor-Baltork, A. R. Hajipour, and R. Haddadi, *J. Chem. Res., Synop.*, 102 (1999).

50. D. H. R. Barton, D. J. Lester, and S. V. Ley, *J. Chem. Soc., Chem. Commun.*, 445 (1977).

51. R. Joseph, A Sudalai, and T. Ravindranathan, *Tetrahedron Lett.*, **35**, 5493 (1994).

52. S. K. De, *J. Chem. Res.*, 78 (2004).

53. A. Manjula, G. N. Reddy, and B. V. Rao, *Synth. Commum.*, **33**, 3455 (2003).

54. S. K. De, *Synth. Commum.*, **34**, 4409 (2004).

55. G. A. Olah, Q. Liao, C. S. Lee, and G. K. S. Prakash, *Synlett*, 427 (1993).

56. R. Sanabria, P. Castañeda, R. Miranda, A. Tubón, F. Delgado, and L. Velasco, *Org. Prep. Proced. Int.*, **27**, 480 (1995).

57. J. G. Lee and J. P. Hwang, *Chem. Lett.*, **24**, 507 (1995).

58. H. Firouzabadi, N. Iranpoor, and K. Amani, *Synth. Commum.*, **34**, 3587 (2004). B. A. Nattier, K. J. Eash, and R. S. Mohan, *Synthesis*, 1010 (2001).

59. A. Wali, P. A. Ganeshpure, and S. Satish, *Bull. Chem. Soc. Jpn.*, **66**, 1847 (1993).

60. W. Chrisman, M. J. Blankinship, B. Taylor, and C. E. Harris, *Tetrahedron Lett.*, **42**, 4775 (2001); G. H. Imanzadeh, A. R. Hajipour, and S. E. Mallakpour, *Synth. Commum.*, **33**, 735 (2003).

61. A. Shaabani, S. Naderi, A. Rahmati, Z. Badri, M. Darvishi, and D. G. Lee, *Synthesis*, 3023 (2005).

62. A. S. Demir, C. Tanyeli, and E. Altinel, *Tetrahedron Lett.*, **38**, 7267 (1997).

63. N. B. Barhate, A. S. Gajare, R. D. Wakharkar, and A. Sudalai, *Tetrahedron Lett.*, **38**, 653 (1997).

64. B. P. Bandgar, . L. B. Kunde, and J. L. Thote, *Synth. Commun.*, **27**, 1149 (1997); B. P. Bandgar and S. S. Makone, *OPPI Briefs*, **32**, 391 (2000).

65. A. Khazaei and A. A. Manesh, *Synthesis*, 1739 (2004).

66. A. Khazaei, R. G. Vaghei, and M. Tajbakhsh, *Tetrahedron Lett.*, **42**, 5099 (2001).

67. M. Tajbakhsh, A. Khazaei, M. Shabani-Mahalli, and R. Ghorbani-Vaghai, *J. Chem. Res.*, 141 (2004).

68. A. Khazaei and R. G. Vaghei, *Tetrahedron Lett.*, **43**, 3073 (2002).

69. R. S. Varma, R. Dahiya, and R. K. Saini, *Tetrahedron Lett.*, **38**, 8819 (1997).

70. S. Chandrasekhar and K. Gopalaiah, *Tetrahedron Lett.*, **43**, 4023 (2002).

71. D. S. Bose and P. Srinivas, *Synth. Commun.*, **27**, 3835 (1997).

72. D. H. R. Barton, W. B. Motherwell, E. S. Simon, and S. Z. Zard, *J. Chem. Soc., Chem. Commun.*, 337 (1984).

73. Y. Yang, D. Zhang, L.-Z. Wu, B. Chen, L.-P. Zhang, and C.-H. Tung, *J. Org. Chem.*, **69**, 4788 (2004).

74. H. J. P. de Lijser, F. H. Fardoun, J. R. Sawyer, and M. Quant, *Org. Lett.*, **4**, 2325 (2002).

O-Methyl Oxime: $R_2C=NOCH_3$

Formation

MeONH$_2$·HCl, Pyr, MeOH, 23°C, 30 min, 81% yield.[1]

Cleavage

This method was developed because conventional procedures failed to cleave the oxime.[1] Cleavage occurs by reduction of the oxime to the imine which is then readily hydrolyzed.

1. E. J. Corey, K. Niimura, Y. Konishi, S. Hashimoto, and Y. Hamada, *Tetrahedron Lett.*, **27**, 2199 (1986).

O-Benzyl Oxime: $R_2C=NOCH_2Ph$

The reactions shown below were used in a synthesis of perhydohistrionicotoxin; the carbonyl groups were protected as an oxime and an *O*-benzyl oxime.[1]

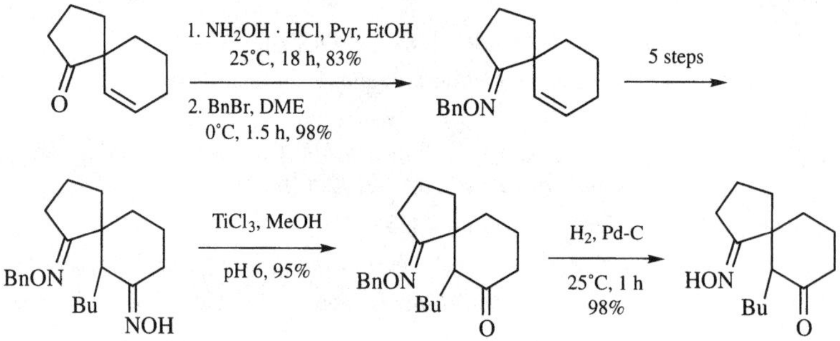

The 2-chlorobenzyl group has been used in the protection of an oxime during the modification of erythromycin A.[2]

1. E. J. Corey, M. Petrzilka, and Y. Ueda, *Helv. Chim. Acta*, **60**, 2294 (1977).
2. Y. Watanabe, S. Morimoto, T. Adachi, M. Kashimura, and T. Asaka, *J. Antibiot.*, **46**, 647 (1993).

O-Phenylthiomethyl Oxime: $R_2C=NOCH_2SC_6H_5$ (Chart 5)

In a prostaglandin synthesis a carbonyl group was protected as an oxime that had its hydroxyl group protected against Collins oxidation by the phenylthiomethyl group. The phenylthiomethyl group is readily removed to give an oxime that is then cleaved to the carbonyl compound.[1]

Formation

$PhSCH_2ONH_2$, Pyr, 25°C, 24 h, 100% yield.[1]

Cleavage

$HgCl_2$, HgO, AcOH, AcOK, 25–50°C, 0.5–48 h, 75% yield; K_2CO_3, MeOH, 25°C, 5 min, 100% yield. These conditions remove the $PhSCH_2-$ group from the oxime, which is then cleaved with AcOH/NaNO$_2$ (10°C, 1 h). This group was also stable to acid, base and LiAlH$_4$.[1]

1. I. Vlattas, L. Della Vecchia, and J. J. Fitt, *J. Org. Chem.*, **38**, 3749 (1973).

1,2-Adducts to Aldehydes and Ketones

Diethylamine Adduct: R$_2$C[OTi(NEt$_2$)$_3$]NEt$_2$

Titanium tetrakis(diethylamide) selectively adds to aldehydes in the presence of ketones and to the least hindered ketone in compounds containing more than one ketone. The protection is *in situ*, which thus avoids the usual protection/deprotection sequence. Selective aldol and Grignard additions are readily performed employing this protection methodology.[1]

N-Methoxy-N-methylamine Adduct: [R$_2$C(OLi)N(OMe)Me]

The use of various amine adducts of carbonyl compounds as a method of carbonyl protection has been reviewed.[2,3]

Pyrrole Carbinol

The pyrrole carbinol first prepared in 1934 is easily prepared from an aldehyde by reaction with the lithium anion of pyrrole in THF. The unprotected carbinol is relatively stable but as with the imidazolide it may be protected as the TBS ether to improve its stability. The pyrrole carbinol is sufficiently stable as the lithium salt that aryl halides may be metalated with BuLi. These derivatives may also be converted directly to α,β-unsaturated esters using the Wadsworth–Horner–Emmons olefination using the Masamune–Roush protocol. Deprotection is accomplished with catalytic DBU or NaOMe.[7]

1-Methyl-2-(1'-hydroxyalkyl)imidazoles

Formation/Cleavage[8]

This protective group is stable to 1 N KOH/MeOH, 70°C, 7 h; 20% H_2SO_4, 70°C, 7 h; H_2, Pd–C, EtOH, 1 atm, 18 h; $NaBH_4$, $LiAlH_4$, CF_3COOH, Al_2O_3/MeOH.

O-Silylimidazolyl Aminals

Formation

1. Imidazole, TBDMSCl, DMF, rt, 88–96% yield.[9] This group was stable to $NaBH_4$, MeMgCl, and thioketal formation with $HSCH_2CH_2SH/BF_3\cdot Et_2O$.

2. TMS-imidazole, 35°C, CH_2Cl_2, >82% yield.[10] This derivative was used to protect the aldehyde during a $LiAlH_4$ reduction.

Cleavage

48% HF, CH$_3$CN, 88–96% yield for the TBDMS derivative.[9]

Sodium Bisulfite Adducts: RCH(OH)SO$_3$Na

Sodium bisulfite adducts are readily formed from aldehydes by reaction with NaHSO$_3$. These derivatives are often crystalline and thus serve as a convenient method for purification of aldehydes. Reversion to the aldehyde usually is accomplished by treatment with aqueous acid or base. TMSCl can be used to regenerate the aldehyde under nonaqueous conditions.[11]

o-**Carborane**

Formation/Cleavage[12]

The carboranyl alcohol can also be prepared from the stannyl carborane and an aldehyde using Pd$_2$(dba)$_3$–CHCl$_3$/dppe. The carborane is stable to Brønsted and Lewis acids and to LiAlH$_4$.

Amino Nitrile Derivatives

These were prepared to protect an aldehyde of an α-amino aldehyde and thus prevent racemization. A variety of amines were examined, and it was found that the morpholine derivative was the most stable and the ammonia derivative the least stable. The iminium ion could be regenerated upon treatment with ZnCl$_2$, but regeneration of the aldehyde was not reported.[13] The method was used to advantage in a (−)-Saframycin A synthesis.[14]

1. M. T. Reetz, B. Wenderoth, and R. Peter, *J. Chem. Soc., Chem. Commun.*, 406 (1983).

2. D. L. Comins, *Synlett*, 615 (1992).

3. F. Roschangar, J. C. Brown, B. E. Cooley, Jr., M. J. Sharp, and R. T. Matsuoka, *Tetrahedron*, **58**, 1657 (2002).

4. R. W. Hoffmann and I. Münster, *Tetrahedron Lett.*, **36**, 1431 (1995).

5. D. A. Evans, R. P. Polniaszek K. M. DeVries, D. E. Guinn, and D. J. Mathre, *J. Am. Chem. Soc.*, **113**, 7613 (1991).

6. N. Brémand, P. Mangeney, and J. F. Normant, *Tetrahedron Lett.*, **42**, 1883 (2001).

7. D. J. Dixon, M. S. Scott, and C. A. Luckhurst, *Synlett*, 2317 (2003).

8. S. Ohta, S. Hayakawa, K. Nishimura, and M. Okamoto, *Tetrahedron Lett.*, **25**, 3251 (1984).

9. L. G. Quan and J. K. Cha, *Synlett*, 1925 (2001).

10. M. Kim and E. Vedejs, *J. Org. Chem.*, **69**, 7262 (2004).

11. D. P. Kjell, B. J. Slattery, and M. J. Semo, *J. Org. Chem.*, **64**, 5722 (1999).

12. H. Nakamura, K. Aoyagi, and Y. Yamamoto, *J. Org. Chem.*, **62**, 780 (1997); H. Nakamura, K. Aoyagi and Y. Yamamoto, *J. Organomet. Chem.*, **574**, 107 (1999).

13. A. G. Myers, D. W. Kung, B. Zhong, M. Movassaghi, and S. Kwon, *J. Am. Chem. Soc.*, **121**, 8401 (1999).

14. A. G. Myers and D. W. Kung, *J. Am. Chem. Soc.*, **121**, 10828 (1999).

Cyclic Derivatives

N,N'-Dimethylimidazolidine and *N,N'*-Diarylimidazolidine

R' = Me, Ar

The imidazolidine was prepared from an aldehyde with *N,N'*-dimethyl-1,2-ethylene-diamine (benzene, heat, 78% yield) and cleaved with MeI (Et$_2$O; H$_2$O, 92% yield) or aqueous HCl.[1] Derivatization is chemoselective for aldehydes. The imidazolidine is stable to BuLi and LDA[2-4] and Li/NH$_3$.[5] The diphenylimidazolidine has been prepared analogously and can be cleaved with aqueous HCl.[6,7] Alternatively it can be prepared using thionyl chloride (Pyr, CH$_2$Cl$_2$, 0–25°C, 7 h, 93% yield).[8] A chiral version using *N,N'*-dimethyl-1*S*,2*S*-diphenyl-1,2-ethylenediamine has been used for protection as well as asymmetric induction.[9,10]

Ref. 6

The related bis-*N,N'*-(3,5-dichlorophenyl)imidazolidine has been used to protect an aldehyde. It is prepared from bis-*N,N'*-(3,5-dichlorophenyl)-1,2-diaminoethane (CSA,

DMF, rt, 18 h, 72% yield) and is cleaved with aq. AcOH (rt, overnight, 98% yield).[11] Similarly (1*R*,2*R*)-bismethylamino cyclohexane has been used as a protecting group for an aldehyde and concomitantly served to induce chirality in a conjugate addition.[12]

1. A. Alexakis, N. Lensen, and P. Mangeney, *Tetrahedron Lett.*, **32**, 1171 (1991).

2. A. J. Carpenter and D. J. Chadwick, *Tetrahedron*, **41**, 3803 (1985).

3. M. Gray and P. J. Parsons, *Synlett*, 729 (1991).

4. A. Couture, E. Deniau, P. Grandelaudon, and C. Hoarau, *J. Org. Chem.*, **63**, 3128 (1998).

5. L. E. Overman, D. J. Ricca, and V. D. Tran, *J. Am. Chem. Soc.*, **119**, 12031 (1997).

6. H.-W. Wanzlick and W. Löchel, *Chem. Ber.*, **86**, 1463 (1953).

7. A. Giannis, P. Münster, K. Sandhoff, and W. Steglich, *Tetrahedron*, **44**, 7177 (1988).

8. J. J. Vanden Eynde, A. Mayence, and A. Maquestiau, *Bull. Soc. Chim. Belg.*, **101**, 233 (1992).

9. A. Alexakis, N. Lensen, and P. Mangeney, *Tetrahedron Lett.*, **32**, 1171 (1991).

10. I. Marek, A. Alexakis, and J.-F. Normant, *Tetrahedron Lett.*, **32**, 5329 (1991).

11. A. Ono, T. Okamoto, M. Inada, H. Nara, and A. Matsuda, *Chem. Pharm. Bull.*, **42**, 2231 (1994).

12. L. F. Frey, R. D. Tillyer, A.-S. Caille, D. M. Tschaen, U.-H. Dolling, E. J. J. Grabowski, and P. J. Reider, *J. Org. Chem.*, **63**, 3120 (1998).

2,3-Dihydro-1,3-benzothiazole

The benzothiazole group is introduced by heating 2-methylaminobenzenethiol with a carbonyl compound in ethanol (70–93% yield).[1] An enone is selectively protected over a ketone, and aldehydes react faster than ketones. Cleavage is effected with $AgNO_3$ (CH_3CN, H_2O, pH 7, 83–93% yield)[2] or by heating in Ac_2O followed by aqueous hydrolysis (HCl, $CHCl_3$, 50°C, 1 h, 40% yield) of the resulting enamide.[3] Nonaromatic thiazolidines have also been used as protective groups. They can be cleaved by basic hydrolysis (NaOH, 25°C, 95% yield).[4]

1. H. Chikashita, N. Ishimoto, S. Komazawa, and K. Itoh, *Heterocycles*, **23**, 2509 (1985).

2. H. Chikashita, S. Komazawa, N. Ishimoto, K. Inoue, and K. Itoh, *Bull. Chem. Soc. Jpn.*, **62**, 1215 (1989).

3. G. Trapani, A. Reho, A. Latrofa, and G. Liso, *Synthesis*, 84 (1988).

4. K. Ueno, F. Ishikawa, and T. Naito, *Tetrahedron Lett.*, **10**, 1283 (1969).

Protection of the Carbonyl Group as an Enolate Anions, Enol Ethers, Enamines, and Imines

Lithium Diisopropylamide (LDA)

A 17-steroidal ketone was deprotonated by LDA to protect it from reduction during a lithium naphthalenide cleavage of a benzyl ether.[1]

Trimethylsilyl Enol Ethers

Trimethylsilyl enol ethers can be used to protect ketones, but in general are not used for this purpose because they are reactive under both acidic and basic conditions. More highly hindered silyl enol ethers are much less susceptible to acid and base. A less hindered silyl enol ether can be hydrolyzed in the presence of a more hindered one.[2]

The preparation of silyl enol ethers has been reviewed.[3-5] A nontraditional approach to their preparation involves a dehydrogenative silylation using a silane, a metal catalyst, and an amine.[6]

Enamines

The use of enamines as protective groups seems largely to be confined to steroid chemistry where they serve (in their protonated form) to protect the A–B enone system from bromination[7] and reduction.[8] A large body of literature exists on the preparation and chemistry of enamines[9]; they are easily hydrolyzed with water or aqueous acid.

Imines

In general, imines are too reactive to be used to protect carbonyl groups. In a synthesis of juncusol,[10] however, a bromo- and an iodocyclohexylimine of two identical aromatic aldehydes were coupled by an Ullmann coupling reaction modified by Ziegler.[11] The imines were cleaved by acidic hydrolysis (aq. oxalic acid, THF, 20°C, 1 h, 95% yield). Imines of aromatic aldehydes have also been prepared to protect the aldehyde during ring metalation with s-BuLi.[12] Imines have been used successfully to protect amines and are stable to phase transfer alkylations.

1. H.-J. Liu, J. Yip, and K.-S. Shia, *Tetrahedron Lett.*, **38**, 2253 (1997).

2. H. Urabe, Y. Takano, and I. Kuwajima, *J. Am. Chem. Soc.*, **105**, 5703 (1983).

3. E. Colvin, *Silicon in Organic Synthesis*, Butterworths, Boston, 1981, pp. 198–287.

4. W. P. Weber, *Silicon Reagents for Organic Synthesis*, Springer-Verlag, New York, 1983, pp. 255–272.

5. J. Hydrio, P. Van de Weghe, and J. Collin, *Synthesis*, 68 (1997).

6. M. Igarashi, Y. Sugihara, and T. Fuchikami, *Tetrahedron Lett.*, **40**, 711 (1999).

7. N. I. Carruthers, S. Garshasb, and A. T. McPhail, *J. Org. Chem.*, **57**, 961 (1992).

8. J. A. Hogg, *Steroids*, **57**, 593 (1992).

9. Enamine review: *Enamines: Synthesis, Structure and Reactions*, A. G. Cook, Ed., 2nd ed., Marcel Dekker, New York, 1988.

10. A. S. Kende and D. P. Curran, *J. Am. Chem. Soc.*, **101**, 1857 (1979).

11. F. E. Ziegler, K. W. Fowler, and S. Kanfer, *J. Am. Chem. Soc.*, **98**, 8282 (1976).

12. B. A. Keay and R. Rodrigo, *J. Am. Chem. Soc.*, **104**, 4725 (1982).

Substituted Methylene Derivatives: $RR'C=C(CN)R''$ (Chart 5) $RR' =$ substituted pyrrole; $R'' = -CN,^1 -CO_2Et^2$

The substituted methylene derivative, prepared from a 2-formylpyrrole and a malonic acid derivative, was used in a synthesis of chlorophyll.[1] It is cleaved under drastic conditions (concd. alkali).[1,2]

1. R. B. Woodward and 17 co-workers, *J. Am. Chem. Soc.*, **82**, 3800 (1960).

2. J. B. Paine, R. B. Woodward, and D. Dolphin, *J. Org. Chem.*, **41**, 2826 (1976).

Methylaluminum Bis(2,6-di-t-butyl-4-methylphenoxide) (MAD) Complex

This approach to carbonyl protection uses the relative differences in basicity and the differences in steric effects to protect selectively either the more basic carbonyl group or the less hindered carbonyl group from reactions with nucleophiles such as DIBAH[1] and MeLi.[2]

1. K. Maruoka, Y. Araki, and H. Yamamoto, *J. Am. Chem. Soc.*, **110**, 2650 (1988).

2. K. Maruoka, H. Imoto, and H. Yamamoto, *Synlett*, 441 (1994).

MONOPROTECTION OF DICARBONYL COMPOUNDS

Selective Protection of α- and β- Diketones

α- and β-Diketones can be protected as enol ethers, thioenol ethers, enol acetates, and enamines.

Enamines: and

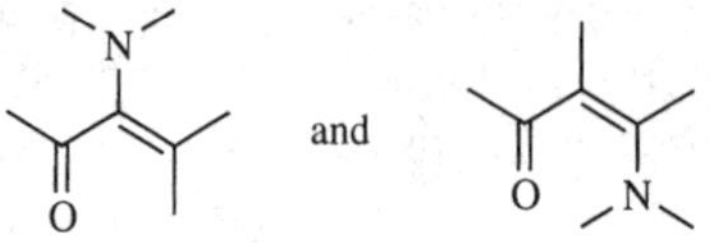

Enol Acetates: and

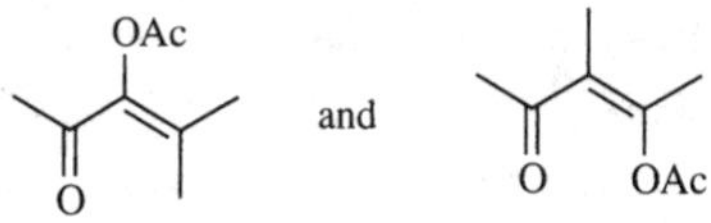

Enol Ethers: and

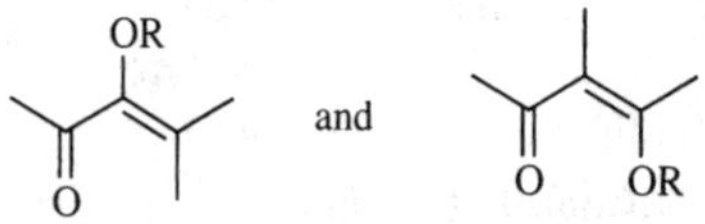

Methyl Enol Ether, Ethyl Enol Ether, *i*-Butyl Enol Ether

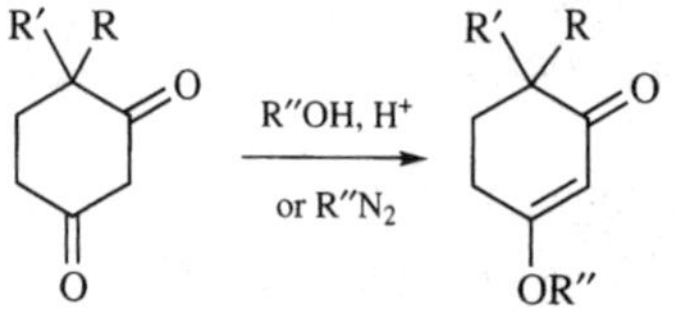

R"OH: R" = Me (HCl, 25°C, 8 h, 83% yield).[1]

R" = Et (TsOH, benzene, reflux, 6–8 h, 70–75% yield).[2]

R" = $(CH_3)_2CHCH_2$ (*i*-BuOH, benzene, reflux, TsOH, 16 h, 100% yield).[3]   In this case, 2-methyl-1,3-cyclopentanedione was monoprotected.

R" = Me ($TiCl_4$, MeOH, 1 h, rt, then TEA, MeOH, 80–97% yield).[4]

R" = various alcohols, I_2, rt, 3–7 min, 65–96% yield.[5]

R" = various alcohols, $B(C_6F_5)_3$, rt, 5–10 min, 89–96% yield.[6]

Methoxyethoxymethyl (MEM) Enol Ether

Formation

   Triethylamine, MEMCl, 92% yield.[7]

Methoxymethyl (MOM) Enol Ether

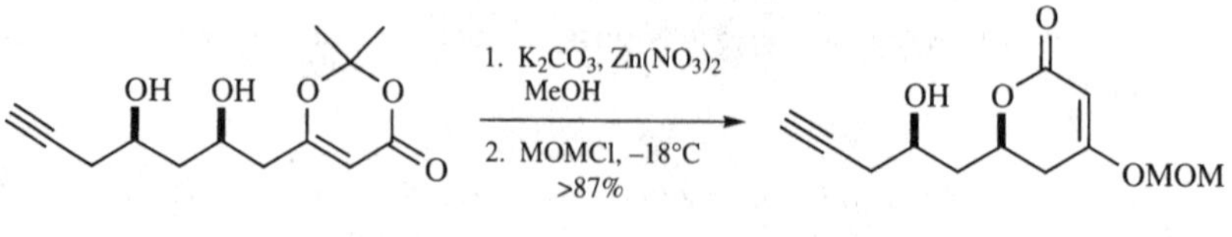

Ref. 8

The best method found for cleavage was MgBr$_2$·Et$_2$O, EtSH, Et$_2$O, rt. Without EtSH, the released formaldehyde reacts with the β-keto ester.
Ethyl vinyl ether has been used to prepare a related acetal.[9]

Enamino Derivatives (Vinylogous Amides)

1. R$'_2$NH = piperidine, TsOH, benzene, reflux, 92% yield.[10]
2. R$'_2$NH = morpholine, TsOH, PhCH$_3$, reflux, 4–5 h, 72–80% yield.[11]
3. R$'_2$NH = various, 300 MPa, with or without Yb(OTf)$_3$, 0–99% yield.[12]
4. R$'_2$NH = various, K10 clay or SiO$_2$, 1–10 min, microwave, 35–99% yield.[13]
5. R$'_2$NH = various, BF$_3$·Et$_2$O, benzene, reflux, 4–6 h, 82–96% yield.[14]
6. R$'_2$NH = various, Montmorillonite or alumina, 20–100°C, 1–5 h, 85–99% yield.[15,16]
7. R$'_2$NH = various, Bi(OTf)$_3$, H$_2$O, rt, 63–98% yield.[17]
8. R$'_2$NH = various, Zn(ClO$_4$)$_2$·6H$_2$O, CH$_2$Cl$_2$, 71–99% yield.[18]
9. R$'_2$NH = various, AcOH, ultrasound, rt, 60–98% yield.[19]

4-Methyl-1,3-dioxolanyl Enol Acetate

Pyrrolidinyl Enamine

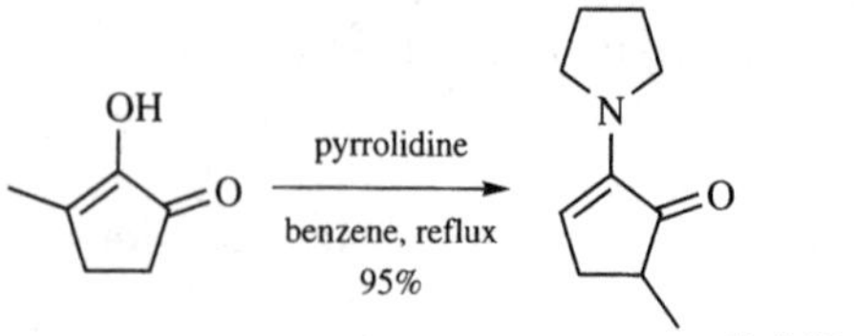

Ref. 21

Benzyl Enol Ether

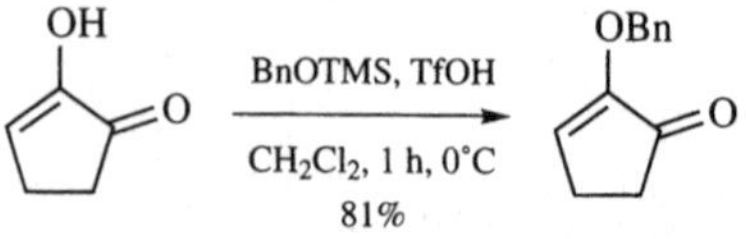

Ref. 22

Butyl Thioenol Ether

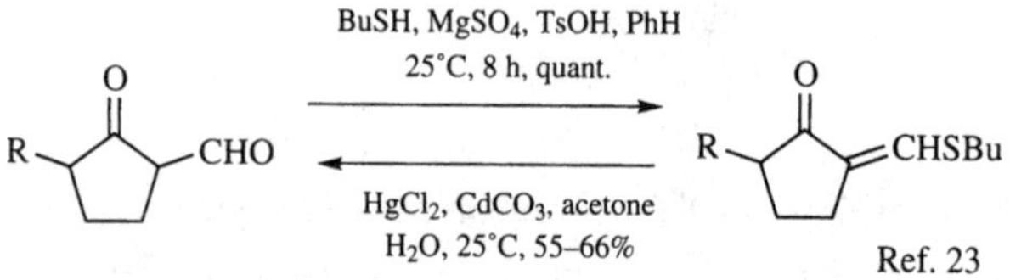

Ref. 23

Protection of Tetronic acids

1. R′ = Me (MeI, CsF, DMF, 45–81% yield).[24]
2. R′ = Bn, allyl, Me, TMSCH$_2$CH$_2$, t-Bu, etc. (R′OH, Ph$_3$P, DEAD, 31–100% yield).[25]

1. H. O. House and G. H. Rasmusson, *J. Org. Chem.*, **28**, 27 (1963).
2. W. F. Gannon and H. O. House, *Org. Synth., Collect. Vol V*, 539 (1973).
3. M. Rosenberger and P. J. McDougal, *J. Org. Chem.*, **47**, 2134 (1982).
4. A. Clerici, N. Pastori, and O. Porta, *Tetrahedron*, **57**, 217 (2001).

5. R. S. Bhosale, S. V. Bhosale, S. V. Bhosale, T. Wang, and P. K. Zubaidha, *Tetrahedron Lett.*, **45**, 7187 (2004).

6. S. Chandrasekhar, Y. S. Rao, and N. R. Reddy, *Synlett*, 1471 (2005).

7. A. J. H. Klunder, G. J. A. Ariaans, E. A. R. M. v. d. Loop, and B. Zwanenburg, *Tetrahedron*, **42**, 1903 (1986).

8. R. Munakata, H. Katakai, T. Ueki, J. Kurosaka, K.-i. Takao, and K.-i. Tadano, *J. Am. Chem. Soc.*, **126**, 11254 (2004).

9. K. Watanabe, K. Iwasaki, T. Abe, M. Inoue, K. Ohkubo, T. Suzuki, and T. Katoh, *Org. Lett.*, **7**, 3745 (2005).

10. P. Kloss, *Chem. Ber.*, **97**, 1723 (1964).

11. S. Hünig, E. Lücke, and W. Brenninger, *Org. Synth., Collect. Vol. V*, 808 (1973).

12. G. Jenner, *Tetrahedron Lett.*, **37**, 3691 (1996).

13. B. Rechsteiner, F. Texier-Boullet, and J. Hamelin, *Tetrahedron Lett.*, **34**, 5071 (1993).

14. M. Azzaro, S. Geribaldi, and B. Videau, *Synthesis*, 880 (1981).

15. F. Texier-Boullet, B. Klein, and J. Hamelin, Synthesis, 409 (1986).

16. M. E. F. Braibante, H. S. Braibante, L. Missio, and A. Andricopulo, *Synthesis*, 898 (1994).

17. A. R. Khosropour, M. M. Khodaei, and M. Kookhazadeh, *Tetrahedron Lett.*, **45**, 1725 (2004).

18. G. Bartoli, M. Bosco, M. Locatelli, E. Marcantoni, P. Melchiorre, and L. Sambri, *Synlett*, 239 (2004).

19. C. A. Brandt, A. C. M. P. da Silva, C. G. Pancote, C. L. Brito, and M. A. B. da Silveira, *Synthesis*, 1557 (2004).

20. J. L. E. Erickson and F. E. Collins, Jr., *J. Org. Chem.*, **30**, 1050 (1965).

21. E. Gordon, F. Martens, and H. Gault, *C. R. Hebd. Seances Acad. Sci., Ser. C*, **261**, 4129 (1965).

22. A. A. Ponaras and Md. Y. Meah, *Tetrahedron Lett.*, **27**, 4953 (1986).

23. P. R. Bernstein, *Tetrahedron Lett.*, **20**, 1015 (1979).

24. T. Sato, K. Yoshimatsu, and J. Otera, *Synlett*, 845 (1995).

25. J. S. Bajwa and R. C. Anderson, *Tetrahedron Lett.*, **31**, 6973 (1990).

Cyclic Ketals, Monothio and Dithio Ketals

Cyclohexane-1,2-dione reacts with ethylene glycol (TsOH, benzene, 6 h) to form the diprotected compound. Monoprotected 1,3-oxathiolanes and 1,3-dithiolanes are isolated on reaction under similar conditions with 2-mercaptoethanol and ethanedithiol, respectively.[1]

Bismethylenedioxy Derivatives: (Chart 5)

Formation/Cleavage[2,3]

This derivative is stable to TsOH/benzene at reflux, and to CrO_3/H^+.[4] It is stable to NBS/*hv*.[5] In the formation of a related derivative, formaldehyde from formalin (containing methanol) converted a C_{11}-hydroxyl group to the C_{11}-methoxymethyl ether. Paraformaldehyde can be used as a source of methanol-free formaldehyde to avoid formation of the ethers.[6]

Tetramethylbismethylenedioxy Derivatives

A bismethylenedioxy group in a 4-chloro or 11-keto steroid is stable to cleavage by formic acid or glacial acetic acid (100°C, 6h), whereas the tetramethyl derivative is readily hydrolyzed (50% AcOH, 90°C, 3–4h, 80–90% yield).[7]

1. R. H. Jaeger and H. Smith, *J. Chem. Soc.*, **160**, 646 (1955).
2. R. E. Beyler, F. Hoffman, R. M. Moriarty, and L. H. Sarett, *J. Org. Chem.*, **26**, 2421 (1961).
3. Y. Nishiguchi, N. Tagawa, F. Watanabe, T. Kiguchi, and I. Ninomiya, *Chem. Pharm. Bull.*, **38**, 2268 (1990).
4. J. F. W. Keana, in *Steroid Reactions*, C. Djerassi, Ed., Holden-Day, San Francisco, 1963, pp. 56–61.
5. D. Duval, R. Condom, and R. Emiliozzi, *C. R. Hebd. Seances Acad. Sci., Ser C*, **285**, 281 (1977).
6. J. A. Edwards, M. C. Calzada, and A. Bowers, *J. Med. Chem.*, **7**, 528 (1964).
7. A. Roy, W. D. Slaunwhite, and S. Roy, *J. Org. Chem.*, **34**, 1455 (1969).

5

PROTECTION FOR THE CARBOXYL GROUP

Greene's Protective Groups in Organic Synthesis, Fourth Edition, by Peter G. M. Wuts and Theodora W. Greene
Copyright © 2007 John Wiley & Sons, Inc.

Carboxylic acids are protected for a number of reasons: (1) to mask the acidic proton so that it does not interfere with base-catalyzed reactions; (2) to mask the carbonyl group to prevent nucleophilic addition reactions; and (3) to improve the handling of the molecule in question (e.g., to make the compound less water soluble, to improve its NMR characteristics, or to make it more volatile so that it can be analyzed by gas chromatography). Besides stability to a planned set of reaction conditions, the protective group must also be removed without affecting other functionality in the molecule. For this reason, a large number of protective groups for acids have been developed that are removed under a variety of conditions even though most can readily be cleaved by simple hydrolysis. Hydrolysis is an important means of deprotection, and the rate of hydrolysis is, of course, dependent upon steric and electronic factors that help to achieve differential deprotection in polyfunctional substrates. An approximate order of reactivity for some esters is as follows: $OEt < OBn < OMe < OPh < SPh < OCH_2CN < O\text{-}4\text{-}$nitrophenyl $< OSu < OC_6Cl_5 < OC_6F_5$.[1] These factors are also important in the selective protection of compounds containing two or more carboxylic acids. Hydrolysis using HOO^- is about 400 times faster than simple hydrolysis with hydroxide (phenyl acetate = substrate).[2]

Polymer-supported esters[3] are widely used in solid-phase peptide synthesis, and extensive information for this specialized protection is reported annually.[4] Some activated esters that have been used as macrolide precursors and some that have been used in peptide synthesis are also described in this chapter; the many activated esters that are used in peptide synthesis are discussed elsewhere.[4] A useful list, with references, of many protected amino acids (e.g., $-NH_2$, COOH, and side chain-protected compounds) has been compiled.[5] Some general methods for the preparation of esters are provided at the beginning of this chapter[6]; conditions that are unique to a protective group are described with that group.[7] Some esters that have been used as protective groups are included in Reactivity Chart 6.

1. G. Szókán, M. Almás, A. Kátai, and A. R. Khlafulla, *J. Chin. Chem. Soc. (Taipei)*, **44**, 519 (1997).

2. W. P. Jencks and M. Gilchrist, *J. Am. Chem. Soc.*, **90**, 2622 (1968).

3. See reference 22 (peptides) in Chapter 1. See also: P. Hodge, "Polymer-Supported Protecting Groups," *Chem. Ind. (London)*, 624 (1979); R. B. Merrifield, G. Barany, W. L. Cosand, M. Engelhard, and S. Mojsov, "Some Recent Developments in Solid Phase Peptide Synthesis," in *Peptides: Proceedings of the Fifth American Peptide Symposium*, M. Goodman and J. Meienhofer, Eds., Wiley, New York, 1977, pp. 488–502; J. M. J. Fréchet, "Synthesis and Applications of Organic Polymers as Supports and Protecting Groups," *Tetrahedron*, **37**, 663 (1981).

4. *Specialist Periodical Reports: Amino-Acids, Peptides, and Proteins*, Royal Society of Chemistry: London, Vols. 1–16, (1969–1983); Amino Acids and Peptides, Vols. 17–28 (1984–1998).

5. G. A. Fletcher and J. H. Jones, *Int. J. Pept. Protein Res.*, **4**, 347 (1972).

6. For classical methods, see C. A. Buehler and D. E. Pearson, *Survey of Organic Syntheses*, Wiley-Interscience, New York, 1970, Vol. 1, pp. 801–830; 1977, Vol. 2, pp. 711–726.

7. See also E. Haslam, "Recent Developments in Methods for the Esterification and Protection of the Carboxyl Group," *Tetrahedron*, **36** 2409–2433 (1980); E. Haslam, "Activation and Protection of the Carboxyl Group," *Chem. Ind. (London)*, 610–617 (1979); E. Haslam, "Protection of Carboxyl Groups," in *Protective Groups in Organic Chemistry*, J. F. W. McOmie, Ed., Plenum, New York and London, 1973, pp.183–215; P. J. Kocienski, *Protecting Groups*, G. Thieme, New York, 2004, p. 393. H. J. Kohlbau, R. Thurmer, W. Voelter, "Protection for the Carboxyl Group," in *Synthesis of Peptides and Peptidomimetics*, M. Goodman, Ed., *Houben-Weyl*, 4th ed., Vol. 22a, Thieme, Stuttgart, 2002, pp. 193–259; B. M. Trost and I. Fleming, Ed., *Comprehensive Organic Synthesis*, Vol. 6, Pergamon Press, Elmsford, NY, 1991, pp. 324–380.

ESTERS

General Preparations of Esters[1]

The preparation of esters can be classified into two main categories: (1) carboxylate activation with a good leaving group and (2) nucleophilic displacement of a carboxylate on an alkyl halide or sulfonate. For simple esters, acid-catalyzed esterification with azeotropic removal of water is also very effective, but limited to simple systems for the most part. The nucleophilic approach is generally not suitable for the preparation of esters if the halide or tosylate is sterically hindered, but there has been some success with simple secondary halides[2] and tosylates (ROTs, DMF, K_2CO_3, 69–93% yield).[3] The section on transesterification should also be consulted, since this technology can be quite useful for the preparation of esters from other esters.

1. The most commonly used method for the preparation of an ester is to react an acid chloride or anhydride with an alcohol in the presence of a base such as

pyridine or triethyl amine in a suitable solvent. With hindered alcohols the re-action is often slow, but can be accelerated by the addition of dimethylamino-pyridine (DMAP). The classic method for the preparation of the acid chloride is to react the acid with $SOCl_2$, $POCl_3$ at reflux. A milder process involves the reaction of the acid with oxalyl chloride in the presence of a catalytic amount of DMF in CH_2Cl_2 at rt or below.

2. RCO_2H, R′OH, DCC/DMAP, Et_2O, 25°C, 1–24 h, 70–95% yield. This method is suitable for a large variety of hindered and unhindered acids and alcohols.[4] The use of $Sc(OTf)_3$ as a cocatalyst improves the esterification of 3° alcohols.[5] Carboxylic acids that can form ketenes with DCC react preferentially with ali-phatic alcohols in the presence of phenols whereas those that do not show the opposite selectivity.[6] In some sterically congested situations the O-acyl urea will migrate to an unreactive N-acyl urea in competition with esterification. Carbodi-imide I was developed to make the urea by-product water soluble and thus easily washed out.[7] Isoureas are prepared from a carbodiimide and an alcohol which upon reaction with a carboxylic acid give esters in excellent yield. A polymer supported version of this process has been developed.[8] This process has been reviewed.[9] **Note that DCC is a potent skin irritant in some individuals.**

I

3. RCO_2H, R′OH, 2-chloro-1,3-dimethylimidazolinium chloride, 76–96% yield. The reagent is a powerful dehydrating agent which has a number of other uses such as the conversion of amides to nitriles, acids to anhydrides, etc.[10]

4. RCO_2H, R′OH, (Chlorophenylthiomethylene)dimethylammonium chloride, DIPEA, CH_2Cl_2, 75–100% yield. This coupling reagent can also be used to prepare amides from acids.[11]

5. RCO_2H, desired alcohol as solvent, 2-ethoxy-1-ethoxy-1-(ethoxycarbonyl)-1, 2-dihydroquinoline (EEDQ), 5 h to overnight, rt, reflux, 56–95% yield. Amino acids are not racemized.[12]

6. RCO_2H, R′OH, MeTHF, Me_3SiCl, (or Me_2SiCl_2, $MeSiCl_3$ or $SiCl_4$), rt, 15 min to 100 h, 90–97% yield.[13,14] In this case, both R and R′ can be hindered. Since the reaction conditions generate HCl, the substrates should be stable to strong acid. MeTHF is not as water-soluble as THF, thus facilitating an aqueous ex-traction. It also makes an azeotrope with water. HCl has also been generated photochemically using CCl_4.[15]

7. RCO_2H, R′OH, $NaHSO_4·SiO_2$, 5–15 h. 42–96%.[16] Aliphatic acids are esteri-fied in the presence of aromatic acids.

8. RCO_2H, R′OH, $HfCl_4·2THF$, toluene, reflux, azeotrope out H_2O, 91–99% yield. This method will only work for acids and alcohols that are higher boil-ing than toluene. A primary alcohol can be esterified in the presence of a secondary alcohol.[17,18]

9. RCO_2H, $B(OH)_3$, ROH, rt, 18 h, 65–99% yield. This method is specific for α-hydroxy acids.[19] The catalyst N-alkyl-4-boronopyridinium chloride is a better catalyst than boric acid.[20]

10. $(RCO_2)O$, R′OH, Bu_3P, excellent yields.[21] The nearly neutral esterification proceeds without the need for basic additives.

11. RCO_2H, R′OH, BOP-Cl, Et_3N, CH_2Cl_2, 23°, 2 h, 71–99% yield.[22] This is an excellent general method for the preparation of esters.

12. RCO_2H, R′OH, (a) $2,4,6-Cl_3C_6H_2COCl$, Et_3N, THF,[23] (b). R′OH, DMAP, >95% yield. This method is best suited to the preparation of relatively unhindered esters; otherwise some esterification of the benzoic acid may occur at the expense of the acid to be esterified. This method has also been used extensively for macrolide synthesis.

13. RCO_2H, TsCl, N-methylimidazole, CH_3CN or CH_2Cl_2, 0–5°C, 30 min, 82–96% yield. This method has the advantage over the mixed anhydride method in that the activating sulfonate does not form an ester in competition with the reacting acid. The method is also good for the preparation of thio esters and amides.[24]

14.

MeOTf is highly toxic.

15. RCO_2H (a) TsCl, K_2CO_3, TEBAC ($Et_3N^+CH_2Ph\ Cl^-$), 40°C reflux, 5–60 min (a) R′OH, reflux, 5–120 min, 80–90%.[26]

16. RCO_2H, $ClCO_2R'$, CH_2Cl_2, 0°C, Et_3N, DMAP, 89–98%.[27] This reaction is not suitable for hindered carboxylic acids, since considerable symmetrical anhydride formation (52% with pivalic acid) results. Symmetrical anhydride

formation can sometimes be suppressed by the use of stoichiometric quantities of DMAP.

17. $RCO_2H + R'X$, DBU, benzene, 25–80°C, 1–10 h, 70–95% yield.[28] RCO_2H = alkyl, aryl, hindered acids, R' = Et, n- and s-Bu, CH_3SCH_2, X = Cl, Br, I. The reaction also proceeds well in acetonitrile, allowing lower temperatures (25°C) and shorter times.[29]

18. $RCH(NHPG)CO_2H$, Cs_2CO_3, R'X, DMF pH 7, 6 h.[30] R' = Me, 80%; $PhCH_2$, 70–90%; o-$NO_2C_6H_4CH_2$, 90%; p-$MeOC_6H_4CH_2$, 70%; Ph_3C, 40–60%; t-Bu, 14%; PhCOCH(Me), 80%; N-phthalimidomethyl, 80% yield. A study of relative rates of this reaction indicates that $Cs^+ > K^+ > Na^+ > Li^+$; $I^- \gg Br^- \gg Cl^-$; HMPA > DMSO > DMF.[31]

19. $RCH(NHPG)CO_2H$, R'X, $NaHCO_3$, DMF, 25°C, 24 h, 90–95%.[32] R' = Et, n-Bu, s-Bu, X = Br, I

20. $RCH(NHPG)CO_2H$, R'X, $(C_8H_{17})_3N$ MeCl, aq. $NaHCO_3$, CH_2Cl_2 25°C, 3–24 h, 70–95%.[33]

21. $RCO_2H + R'_3OBF_4$, EtN-i-Pr_2, CH_2Cl_2, 20°, 1–24 h, 70–95%.[34] RCO_2H = hindered acids, R' = Me, Et.

22. RCO_2H, $Me_2NCH(OR')_2$, 25–80°C, 1–36 h, 80–95%.[35] RCO_2H = Ph, 2,4,6-$Me_3C_6H_2$-, N-protected amino acids, R' = Me, Et, $PhCH_2$, s-Bu

23. RCO_2H, $CH_3C(OEt)_3$, 30 min to 5 h, 80°C, [bmim]PF_6, 91–98%. The ionic liquid was compared with other solvents and found to be superior.[36]

24. $RCO_2H + R'OH$, t-BuNC, 0–20°C, 24 h, 36–98%.[37] RCO_2H = amino, dicarboxylic acids; ≠$PhCO_2H$, R' = Me, Et, t-Bu.

25. $RCO_2H + R'OH$, $Ph_3P(OSO_2CF_3)_2$, CH_2Cl_2, 25°C, 12 h, 75–85%.[38] R = aryl, R' = Et. A polymer supported version of this reagent has been developed.[39]

26. $RCO_2H + R'X$, Electrolysis: pyrrolidone, DMF, R''_4NX, rt, 80–99%.[40] This method is based on the generation of the tetraalkylammonium salt of pyrrolidone, which acts as a base. The method is compatible with a large variety of carboxylic acids and alkylating agents. The method is effective for the preparation of macrolides.

27. $RCH(NHPG)CO_2H$, isopropenyl chloroformate, DMAP, CH_2Cl_2, 0°C, R'OH, 60–96%.[41]

28. R'OH, TiCl(OTf)$_3$, $(Me_2SiO)_4$, 50°C, 12–48 h, 50–99%.[42]

29. RCO_2H, R'OH, $TiCl_4$, $AgClO_4$, $(ArCO)_2O$, TMSCl, CH_2Cl_2, rt, 0.5–17 h, 90–99%.[43]

30. RCO_2TMS, R'OTMS, $TiCl_4$, $AgClO_4$, $(ArCO)_2O$, CH_2Cl_2, rt, 80–99%. $Sn(OTf)_2$ has also been used as an effective catalyst.[44]

31. $Cl^\ominus$ RCO_2H, R'OH, 2,6-lutidine, 39–84%.[45]

32. RCO_2H, R'OH, EEDQ, 56–95%.[46]

33.

Esterification proceeds with inversion   Ref. 47

34. The Mitsunobu reaction is used to convert an alcohol and an acid into an ester by formation of an activated alcohol (Ph$_3$P, diethyl diazodicarboxylate), which then undergoes displacement with inversion by the carboxylate.[48] Although this reaction works very well, it suffers from the fact that large quantities of by-products are produced, which generally require removal by chromatography.

35. The following is a very general method that works for a variety of acids and sterically demanding alcohols.[49] This methodology has been reviewed.[50] In the case of chiral secondary alcohols, the ester is obtained with perfect inversion of configuration.

36. RCO$_2$H, 2-thienyl carbonate, DMAP, then R'OH and I$_2$, 81–93%.[51]

37. RCO$_2$H, O,O-di(2-pyridyl)thiocarbonate (DPTC), DMAP, toluene, 79–99% yield. This method has been used to prepare Taxol from the phenylisoserine side chain and protected Baccatin III in 95% yield, an esterification that is generally considered difficult.[52]

38. RCO$_2$H, (RCO$_2$)O, Mg(ClO$_4$)$_2$, 87–99% yield. The method was tested for methyl, benzyl and t-Butyl esters.[53]

39. RCO$_2$H, R'OH, 2-methyl-6-nitrobenzoic anhydride, TEA, DMAP, CH$_2$Cl$_2$, rt, 72–100% yield. Other aryl anhydrides are also effective.[54]

40. Tetrabutylammonium hydrogensulfate, KF·2H$_2$O, RX, THF rt, 3–24 h, 51–99% yield.[55] Trialkylsilyl esters can be converted similarly.[56]

41. Polymer-OC$_6$H$_4$N=N-NHR, rt, 90–96% yield. R=Me, Bn, n-Bu, 2-pyridylethyl.[57]

42. RCO$_2$H, R'OH, 1-t-butoxy-2-t-butoxycarbonyl-1,2-dihydroisoquinoline (BBDI), dioxane, rt, 51–96% yield.[58]

43. RCO$_2$H, R'OH, Di-2-thienyl carbonate, I$_2$, DMAP, 57–91% yield. This reagent is also suitable for macrolactonization.[59]

44. From a diazoderivative.[60]

General Cleavage of Esters[61]

1. The simplest and most frequently used method for the hydrolysis of esters is through the use of hydroxide in an organic aqueous medium such as MeOH/ H_2O. In the case of proximal diesters, hydroxide will selectively cleave only one of the esters.[62]

2. $RCO_2R' + Nu^- \xrightarrow{\text{aprotic solvent}^{63}} RCO_2H$

   In this method, cleavage occurs by nucleophilic displacement of the carboxylate.

   Nu^- = LiS-n-Pr: HMPA, 25°C, 1 h, ca. quant. yield[64]
   = NaSePh: HMPA-THF, reflux, 7 h, 90–100% yield[65]
   = LiCl: DMF or Pyr, reflux, 1–18 h, 60–90% yield[66]
   = KO-t-Bu: DMSO, 50–100°C, 1–24 h, 65–95% yield[67]
   = NaCN (for decarboxylation of malonic esters): DMSO, 160°C, 4 h, 70–80% yield[68]
   = NaTeH from Te, DMF, t-BuOH, NaBH$_4$, 80–90°C, 15 min, 85–98% yield[69]
   = KO$_2$: 18-crown-6, benzene, 25°C, 8–72 h, 80–95% yield[70]
   = LiI: EtOAc, reflux, 26–98% yield.[71] Bn, PMB, PNB, t-Bu and Me esters are all cleaved.
   = PhSH, KF, N-methylpyrrolidone, 190°C, 10 min, 50–100% yield.[72]

3. Hydrolysis of RCO_2R': TMSCl, NaI, CH_3CN reflux, 5–35 h, 70–90% yield.[73–75] RCO_2H = alkyl, aryl, hindered acids, R′ = Me, Et, i-Pr, t-Bu, PhCH$_2$. This method generates Me$_3$SiI *in situ*. The reagent also cleaves a number of other protective groups.

4. Hydrolysis of RCO_2R': MgI$_2$, toluene, 1–3 days, 41–96%.[76] RCO_2H = alkyl, aryl, hindered acids, R′ = Me, Et, cHex, 1-Ad, 2-Ad, t-Bu, PhCH$_2$

5. aq. NaOH, DMF; HCl, 15–60 min, 36–98% yield.[77]

6. Hydrolysis of RCO_2R': KO-t-Bu/H_2O (4:1), 25°C, 2–48 h, 80–100% yield.[78] RCO_2H = Ph, aryl, hindered acids, R′ = Me, t-Bu, alkyl, "anhydrous hydroxide," which is formed under these conditions also cleaves tertiary amides.

7. $RCH(NHPG)CO_2R'$: BBr$_3$, CH$_2$Cl$_2$, −10°C, 1 h → 25°C, 2 h, 60–85% yield.[79] R′ = Me, Et, t-Bu, PhCH$_2$, PG = − CO$_2$CH$_2$Ph, − CO$_2$-t-Bu; OMe, OEt, O-t-Bu, OCH$_2$Ph side-chain ethers.

8. Hydrolysis of RCO_2R': AlX$_3$ (X = Cl, Br), R″SH, 25°C, 5–50 h, 70–95% yield.[80,81] R = Ph, steroid side-chain,... R′ = Me, Et, PhCH$_2$, R″ = Et, HO(CH$_2$)$_2$−.

9. Hydrolysis of RCO_2R': xs $(Bu_3Sn)_2O$, $80°C$, benzene, 1–30 h, 40–95% yield.[82–84] $R' = CH_2O_2CC(CH_3)_3$, Me, Et, Ph.

10. $RCH(NHPG)CO_2Me$: (i) CH_2O, TsOH, (ii) $NaHCO_3$, MeOH, H_2O, reflux 5–10 min, 25–90% yield.[85] PG = Cbz, Boc, Fmoc.

11. $KF \cdot Al_2O_3$, microwave heating, 90–98% yield. The method was tested on a series of trivial esters.[86]

12. Isopropyl esters and carbamates are selectively cleaved in the presence of their methyl counterparts with $AlCl_3$ in CH_3NO_2 (0–50°C, 1–24 h, 78–92% yield).[87]

1. For a recent review, see J. Otera, *Esterification*, Wiley-VCH, Weinheim, 2003.

2. T. Shono, O. Ishige, H. Uyama, and S. Kashimura, *J. Org. Chem.*, **51**, 546 (1986).

3. W. L. Garbrecht, G. Marzoni, K. R. Whitten, and M. L. Cohen, *J. Med. Chem.*, **31**, 444 (1988).

4. A. Hassner and V. Alexanian, *Tetrahedron Lett.*, **19**, 4475 (1978).

5. H. Zhao, A. Pendri, and R. B. Greenwald, *J. Org. Chem.*, **63**, 7559 (1998).

6. R. Shelkov, M. Nahmany, and A. Melman, *Org. Biomol. Chem.*, **2**, 397 (2004).

7. F. S. Gibson, M. S. Park, and H. Rapoport, *J. Org. Chem.*, **59**, 7503 (1994).

8. S. Crosignani, P. D. White, and B. Linclau, *J. Org. Chem.*, **69**, 5897 (2004).

9. L. J. Mathias, *Synthesis*, 561 (1979).

10. T. Isobe and T. Ishikawa, *J. Org. Chem.*, **64**, 6984 (1999). For a related reagent, see T. Fujisawa, T. Mori, K. Fukumoto, and T. Sato, *Chem. Lett.*, 1891 (1982).

11. L. Gomez, S. Ngouela, F. Gellibert, A. Wagner, and C. Mioskowski, *Tetrahedron Lett.*, **43**, 7597 (2002).

12. B. Zacharie, T. P. Connolly, and C. L. Penney, *J. Org. Chem.*, **60**, 7072 (1995).

13. R. Nakao, K. Oka, and T. Fukomoto, *Bull. Chem. Soc. Jpn.*, **54**, 1267 (1981).

14. M. A. Brook and T. H. Chan, *Synthesis*, 201 (1983).

15. J. R. Hwu, C.-Y. Hsu, and M. L. Jain, *Tetrahedron Lett.*, **45**, 5151 (2004).

16. B. Das, B. Venkataiah, and P. Madhusudhan, *Synlett*, 59 (2000).

17. K. Ishihara, M. Nakayama, S. Ohara, and H. Yamamoto, *Synlett*, 1117 (2001).

18. J. Otera, *Angew. Chem. Int. Ed.*, **40**, 2044 (2001).

19. T. A. Houston, B. L. Wilkinson, and J. T. Blanchfield, *Org. Lett.*, **6**, 679 (2004).

20. T. Maki, K. Ishihara, and H. Yamamoto, *Org. Lett.*, **7**, 5047 (2005).

21. E. Vedejs, N. S. Bennett, L. M. Conn, S. T. Diver, M. Gingras, S. Lin, P. A. Oliver, and M. J. Peterson, *J. Org. Chem.*, **58**, 7286 (1993).

22. J. Diago-Meseguer, A. L. Palomo-Coll, J. R. Fernández-Lizarbe, and A. Zugaza-Bilbao, *Synthesis*, 547 (1980).

23. J. Inanaga, K. Hirata, T. Saeki, T. Katsuki, and M. Yamaguchi, *Bull. Chem. Soc. Jpn.*, **52**, 1989 (1979).

24. K. Wakasugi, A. Iida, T. Misaki, Y. Nishii, and Y. Tanabe, *Adv. Synth. Catal.*, **345**, 1209 (2003). See also: K. Wakasugi, A. Nakamura, and Y. Tanabe, *Tetrahedron Lett.*, **42**, 7427 (2001). K. Wakasugi, A. Nakamura, A. Iida, Y. Nishii, N. Nakatani, S. Fukushima, and Y. Tanabe, *Tetrahedron*, **59**, 5337 (2003).

25. G. Ulibarri, N. Choret, and D. C. H. Bigg, *Synthesis*, 1286 (1996).

26. Z. M. Jászay, I. Petneházy, and L. Töke, *Synthesis*, 745 (1989).

27. S. Kim, Y. C. Kim, and J. I. Lee, *Tetrahedron Lett.*, **24**, 3365 (1983).

28. N. Ono, T. Yamada, T. Saito, K. Tanaka, and A. Kaji, *Bull. Chem. Soc. Jpn.*, **51**, 2401 (1978).

29. C. G. Rao, *Org. Prep. Proced. Int.*, **12**, 225 (1980).

30. S.-S. Wang, B. F. Gisin, D. P. Winter, R. Makofske, I. D. Kulesha, C. Tzougraki, and J. Meienhofer, *J. Org. Chem.*, **42**, 1286 (1977).

31. P. E. Pfeffer and L. S. Silbert, *J. Org. Chem.*, **41**, 1373 (1976).

32. V. Bocchi, G. Casnati, A. Dossena, and R. Marchelli, *Synthesis*, 961 (1979).

33. V. Bocchi, G. Casnati, A. Dossena, and R. Marchelli, *Synthesis*, 957 (1979).

34. D. J. Raber, P. Gariano, A. O. Brod, A. Gariano, W. C. Guida, A. R. Guida, and M. D. Herbst, *J. Org. Chem.*, **44**, 1149 (1979).

35. H. Brechbühler, H. Büchi, E. Hatz, J. Schreiber, and A. Eschenmoser, *Helv. Chim. Acta*, **48**, 1746 (1965).

36. T. Yoshino and H. Togo, *Synlett*, 1604 (2004).

37. D. Rehn and I. Ugi, *J. Chem. Res., Synop.*, 119 (1977).

38. J. B. Hendrickson and S. M. Schwartzman, *Tetrahedron Lett.*, 277 (1975).

39. K. E. Elson, I. D. Jenkins, and W. A. Loughlin, *Tetrahedron Lett.*, **45**, 2491 (2004).

40. T. Shono, O. Ishige, H. Uyama, and S. Kashimura, *J. Org. Chem.*, **51**, 546 (1986).

41. P. Jouin, B. Castro, C. Zeggaf, A. Pantaloni, J. P. Senet, S. Lecolier, and G. Sennyey, *Tetrahedron Lett.*, **28**, 1661 (1987).

42. J. Izumi, I. Shiina, and T. Mukaiyama, *Chem. Lett.*, **24**, 141 (1995).

43. I. Shiina, S. Miyoshi, M. Miyashita, and T. Mukaiyama, *Chem. Lett.*, **23**, 515 (1994).

44. T. Mukaiyama, I. Shiina, and M. Miyashita, *Chem. Lett.*, **21**, 625 (1992).

45. J. J. Folmer and S. M. Weinreb, *Tetrahedron Lett.*, **34**, 2737 (1993).

46. B. Zacharie, T. P. Connolly, and C. L. Penney, *J. Org. Chem.*, **60**, 7072 (1995).

47. J. Boivin, E. Henriet, and S. Z. Zard, *J. Am. Chem. Soc.*, **116**, 9739 (1994).

48. D. L. Hughes, *Org. React.*, **42**, 335 (1992); O. Mitsunobu, *Synthesis*, 1 (1981).

49. T. Mukaiyama, W. Kikuchi, and T. Shintou, *Chem. Lett.*, **32**, 300 (2003); T. Mukaiyama, T. Shintou, and K. Fukumoto, *J. Am. Chem. Soc.*, **125**, 10538 (2003), T. Shintou, W. Kikuchi, and T. Mukaiyama, *Bull. Chem. Soc. Jpn.*, **76**, 1645 (2003).

50. T. Mukaiyama, *Angew. Chem. Int. Ed.*, **43**, 5590 (2004).

51. Y. Oohashi, K. Fukumoto, and T. Mukaiyama, *Chem. Lett.*, **33**, 968 (2004); T. Mukaiyama, Y. Oohashi, and K. Fukumoto, *Chem. Lett.*, **33**, 552 (2004).

52. K. Saitoh, I. Shiina, and T. Mukaiyama, *Chem. Lett.*, **27**, 679 (1998).

53. L. Gooßen and A. Dohring, *Adv. Synth. Catal.*, **345**, 943 (2003).

54. I. Shiina, M. Kubota, H. Oshiumi, and M. Hashizume, *J. Org. Chem.*, **69**, 1822 (2004); I. Shiina, R. Ibuka, and M. Kubota, *Chem. Lett.*, **31**, 286 (2002).

55. T. Ooi, H. Sugimoto, K. Doda, and K. Maruoka, *Tetrahedron Lett.*, **42**, 9245 (2001).

56. T. Ooi, H. Sugimoto, and K. Maruoka, *Heterocycles*, **54**, 593 (2001).

57. J. Rademann, J. Smerdka, G. Jung, P. Grosche, and D. Schmid, *Angew. Chem. Int. Ed.*, **40**, 381 (2001).

58. Y. Saito, T. Yamaki, F. Kohashi, T. Watanabe, H. Ouchi, and H. Takahata, *Tetrahedron Lett.*, **46**, 1277 (2005).

59. Y. Oohashi, K. Fukumoto, and T. Mukaiyama, *Bull. Chem. Soc. Jpn.*, **78**, 1508 (2005).

60. M. E. Furrow and A. G. Myers, *J. Am. Chem. Soc.*, **126**, 12222 (2004).

61. For a review, see C. J. Salomon, E. G. Mata, and O. A. Mascaretti, *Tetrahedron*, **49**, 3691 (1993).

62. S. Niwayama, *J. Org. Chem.*, **65**, 5834 (2000).

63. J. McMurry, "Ester Cleavages via S^N2-Type Dealkylation," *Org. React.*, **24**, 187 (1976); A. Krapcho, *Synthesis*, **805**, 893, (1982).

64. P. A. Bartlett and W. S. Johnson, *Tetrahedron Lett.*, **11**, 4459 (1970).

65. D. Liotta, W. Markiewicz, and H. Santiesteban, *Tetrahedron Lett.*, **18**, 4365 (1977).

66. F. Elsinger, J. Schreiber, and A. Eschenmoser, *Helv. Chim. Acta*, **43**, 113 (1960).

67. F. C. Chang and N. F. Wood, *Tetrahedron Lett.*, **5**, 2969 (1964).

68. A. P. Krapcho, G. A. Glynn, and B. J. Grenon, *Tetrahedron Lett.*, **8**, 215 (1967).

69. J. Chen, and X. J. Zhou, *Synthesis*, 586 (1987).

70. J. San Filippo, L. J. Romano, C.-I. Chern, and J. S. Valentine, *J. Org. Chem.*, **41**, 586 (1976).

71. J. W. Fisher and K. L. Trinkle, *Tetrahedron Lett.*, **35**, 2505 (1994).

72. M. K. Nayak and A. K. Chakraborti, *Chem. Lett.*, **27**, 297 (1998).

73. M. E. Jung and M. A. Lyster, *J. Am. Chem. Soc.*, **99**, 968 (1977).

74. T. Morita, Y. Okamoto, and H. Sakurai, *J. Chem. Soc., Chem. Commun.*, 874 (1978).

75. G. A. Olah, S. C. Narang, B. G. B. Gupta, and R. Malhotra, *J. Org. Chem.*, **44**, 1247 (1979).

76. A. G. Martinez, J. O. Barcina, G. H. del Veccio, M. Hanack, and L. R. Subramanian, *Tetrahedron Lett.*, **32**, 5931 (1991).

77. J. M. Khurana and A. Sehgal, *Org. Prep. Proced. Int.*, **26**, 580 (1994).

78. P. G. Gassman and W. N. Schenk, *J. Org. Chem.*, **42**, 918 (1977).

79. A. M. Felix, *J. Org. Chem.*, **39**, 1427 (1974).

80. M. Node, K. Nishide, M. Sai, and E. Fujita, *Tetrahedron Lett.*, **19**, 5211 (1978).

81. M. Node, K. Nishide, M. Sai, K. Fuji, and E. Fujita, *J. Org. Chem.*, **46**, 1991 (1981).

82. C. J. Salomon, E. G. Mata, and O. A. Mascaretti, *Tetrahedron Lett.*, **32**, 4239 (1991).

83. C. J. Salomon, E. G. Mata, and O. A. Mascaretti, *J. Org. Chem.*, **59**, 7259 (1994).

84. E. G. Mata and O. A. Mascaretti, *Tetrahedron Lett.*, **29**, 6893 (1988).

85. P. Allevi, and M. Anastasia, *Tetrahedron Lett.*, **44**, 7663 (2003).

86. G. W. Kabalka, L. Wang, and R. M. Pagni, *Green Chem.*, **3**, 261 (2001).

87. G.-L. Chee, *Synlett*, 1593 (2001).

Transesterification

The process of transesterification is an important way to prepare a large number of esters from more complex or simple esters without passing through the carboxylic acid. Transesterification can be used to convert one type of ester to another type removable under a different set of conditions. This section describes many of the methods that have been found effective for ester metathesis.[1] In many cases, in order to get good conversion, a large excess of one of the components is required. This is not a problem with low-molecular-weight alcohols and esters that are easily removed by distillation during the isolation process.

1. ROH, DBU, LiBr. When a large excess of the alcohol is undesirable, the reaction can be run in THF/CH_2Cl_2 in the presence of 5-Å ms. The combination of DBU–LiBr is required, since neither reagent is effective alone.[2]

2. $P(RNCH_2CH_2)_3N$ (R = Me, i-Pr), alcohol as solvent, 4–24 h, 81–100% yield. Acetates are formed from an alcohol, vinyl acetate, or isopropenyl acetate, and this catalyst in excellent yield. The isopropyl derivative results in less racemization of amino acid esters than does the methyl derivative.[3]

3. Alkali metal alkoxides, t-butyl acetate neat, 45°C, 30 min, 98% yield of t-butyl ester from methyl benzoate. The rate constant for the reaction increases with increasing ionic radius of the metal and with decreasing solvent polarity. Equilibrium for the reaction is achieved in <10 s. Other examples are presented.[4–6] This method has been improved by changing the catalyst from t-BuONa to a 1:3 mixture of t-BuONa and t-BuC_6H_4ONa. Equilibration times are fast, and t-Bu esters can be prepared efficiently from methyl and ethyl esters (55–99% yield).[7] In this case the mixed aggregate remains in solution whereas without the phenolic component the alkali methoxide precipitates from solution. The low-molecular-weight alcohol is removed by distillation. K_2CO_3 has been tested as a catalyst but was found rather ineffective.[8]

4. The reduction of β-keto esters with $NaBH_4$ concomitantly causes transesterification of the remaining ester in modest yield.[9]

5. M(O−i-Pr)₃; M = La,[10] Nd, Gd, Yb.[11]

6. The use of 1,3-disubstituted 1,1,3,3-tetraalkyldistannoxanes for ester metathesis has been reviewed.[12,13] A "fluorous" version of this catalyst has been developed that allows one to utilize the concept of "fluorous synthesis."[14] The "fluorous" version requires 150°C to induce the transesterification, which may limit this process to simple substrates.

7. BuSn(O)OH, toluene, reflux, 19–64 h, 46–90% yield. Tertiary alcohols do not react.[15]

8. Bu_2SnO, MeOH, reflux, 5–12 h, 77–96% yield. Phenols do not react and chiral substrates are not isomerized.[16,17]

9. Ti(O−i-Pr)₄, ROH, 50–90% yield.[18–20] This method has been expanded to include sterically hindered secondary alcohols, but not tertiary alcohols.[21]

10. $Ti(O)(acac)_2$, toluene, reflux, 70–100% yield. Methyl esters are converted to a variety of other esters. The method was partially successful in converting a methyl ester into an *N*-acyl oxazolidinone and a thioester.[22]

11. Mg, MeOH.[23]

12. $Ce(SO_4)_2 \cdot SiO_2$, ROH, reflux, 0.25–2h.[24] $Ce(OTf)_4$ can be used to prepare acetates and formates with yields ranging from 80–92%. $Ce(OTf)_4$ also catalyzes the direct esterification of acids and alcohols.[25]

13. Indium metal, I_2, alcohol solvent, 4.56–32h, 68–90% yield. *t*-Bu esters may be prepared by this method from methyl esters.[26]

14. I_2, alcohol solvent, 15–20h, reflux, 45–94% yield.[27] These conditions also convert acids and alcohols to esters, 0–95% yield.

15.
$$RCO_2R' + R''OH \xrightarrow[\text{Toluene >88\% yield}]{\text{Bu}_2\text{Sn(OH)OSn(NCS)Bu}_2 \text{ cat.}} RCO_2R'' + R'OH$$

This method is not effective for tertiary alcohols. It has a strong rate dependence on solvent polarity with less polar solvents giving faster rates.[28]

16. *N*-heterocyclic carbenes, vinyl acetate, 4Å, ~1h, rt, THF, 95–100% yield. In this case the reaction is driven to completion by the release of acetaldehyde. More acidic alcohols like benzyl alcohol react faster than 2-butanol.[29] Transesterifications of simple esters and alcohols are also catalyzed by these carbenes.

17. Diphenylammonium triflate, toluene, 80°C, 33–97% yield. This catalyst can also be used to prepare esters from carboxylic acids and alcohols, 78–96% yield.[30]

18. *N*-Acyloxazolidinones are transesterified with a Lewis acid in MeOH, 70–98% yield.[31]

19. From a methyl ester: Tetracyanoethylene, ROH, 60°C, 48h, 40–100% yield.[32]

Methods for the Transesterification of β-keto Esters

1. ROH, toluene, reflux, 95% yield. The reaction in this case is proposed to proceed through a ketene intermediate.[33] Similar conditions with catalytic sodium perborate give esters in 58–90% yields.[34]

2. ROH, sulfated SnO_2, 50–97% yield.[35]

3. Various clays (smectite, atapulgite, vermiculite, K-10)[36] or Kalolinitic clay,[37] toluene, reflux, 48h, 0–98% yield.

4. *N,N*-Diethylaminopropylated silica gel, refluxing xylene, 56–97% yield.[38]

5. Yttria–zirconia-based Lewis acid catalyst, toluene, reflux, 35–99% yield.[39]

6. ZnSO₄, toluene, 60–80°C, 66–97% yield. This method works for allylic alcohols, which will often undergo the Carrol rearrangement followed by decarboxylation. The method can also be used to prepare esters of 3° alcohols.[40]

7. Zn (2 eq.), I₂ (0.5 eq.), toluene reflux, 45–89% yield.[41] When the reaction is performed with phenols as the alcohol, coumarins are produced in modest yields (25–78% yield).

8. LiClO₄, toluene, 100°C, distillation to remove low boiling alcohol, 57–94% yield. Cinnamyl alcohols were prepared without Carrol rearrangement and a trityl ester was prepared, but this most likely proceeds by an alternative mechanism.[42]

9. Sodium perborate, toluene, reflux, 2–10 h, 81–91% yield. Even trityl alcohol will participate in this reaction in moderate yield.[43]

10. Catalytic NBS, toluene, 90–100°C, 52–94% yield.[44] It is likely that the reaction is actually HBr catalyzed. It is noteworthy that normal esters fail to react, which implicates a mechanism that may involve a ketene intermediate.

1. J. Otera, *Chem. Rev.*, **93**, 1449 (1993); G. A. Grasa, R. Singh, and S. P. Nolan, *Synthesis*, 971 (2004).

2. D. Seebach, A. Thaler, D. Blaser, and S. Y. Ko, *Helv. Chim. Acta*, **74**, 1102 (1991).

3. P. Ilankumaran and J. G. Verkade, *J. Org. Chem.*, **64**, 3086 (1999).

4. M. G. Stanton and M. R. Gagné, *J. Am. Chem. Soc.*, **119**, 5075 (1997); V. A. Vasin and V. V. Razin, *Synlett*, 658 (2001).

5. M. G. Stanton and M. R. Gagné, *J. Org. Chem.*, **62**, 8240 (1997).

6. R. M. Kissling and M. R. Gagné, *J. Org. Chem.*, **66**, 9005 (2001).

7. R. M. Kissling and M. R. Gagné, *Org. Lett.*, **2**, 4209 (2000); M. G. Stanton, C. B. Allen, R. M. Kissling, A. L. Lincoln, and M. R. Gagne, *J. Am. Chem. Soc.*, **120**, 5981 (1998).

8. D. Janczewski, L. Synoradzki, and M. Wlostowski, *Synlett*, 420 (2003).

9. S. K. Padhi and A. Chadha, *Synlett*, **5**, 639 (2003).

10. T. Okano, K. Miyamoto, and J. Kiji, *Chem. Lett.*, **24**, 246 (1995).

11. T. Okano, Y. Hayashizaki, and J. Kiji, *Bull. Chem. Soc. Jpn.*, **66**, 1863 (1993).

12. J. Otera, *Adv. Detailed React. Mech.*, **3**, 167 (1994).

13. O. A. Mascaretti and R. L. E. Furlan, *Aldrichimica Acta*, **30**, 55 (1997).

14. J. Xiang, A. Orita, and J. Otera, *Angew. Chem. Int. Ed.*, **41**, 4117 (2002).

15. R. L. E. Furlán, E. G. Mata, and O. A. Mascaretti, *Tetrahedron Lett.*, **39**, 2257 (1998).

16. P. Baumhof, R. Mazitschek, and A. Giannis, *Angew. Chem. Int. Ed.*, **40**, 3672 (2001).

17. K. C. Nicolaou, B. S. Safina, M. Zak, A. A. Estrada, and S. H. Lee, *Angew. Chem. Int. Ed.*, **43**, 5087 (2004).

18. D. Seebach, E. Hungerbühler, R. Naef, P. Schnurrenberger, B. Weidmann, and M. Züger, *Synthesis*, 138 (1982).

19. U. D. Lengweiler, M. G. Fritz, and D. Seebach, *Helv. Chim. Acta*, **79**, 670 (1996).

20. For a review of titanium compounds as catalysts for transesterification, see M. I. Siling and T. N. Laricheva, *Russ. Chem. Rev.*, **65**, 279 (1996).

21. P. Krasik, *Tetrahedron Lett.*, **39**, 4223 (1998).

22. C.-T. Chen, J.-H. Kuo, C.-H. Ku, S.-S. Weng, and C.-Y. Liu, *J. Org. Chem.*, **70**, 1328 (2005).

23. Y.-C. Xu, E. Lebeau, and C. Walker, *Tetrahedron Lett.*, **35**, 6207 (1994).

24. T. Nishiguchi and H. Taya, *J. Chem. Soc., Perkin Trans. I*, 172 (1990).

25. N. Iranpoor and M. Shekarriz, *Bull. Chem. Soc. Jpn.*, **72**, 455 (1999).

26. B. C. Ranu, P. Dutta, and A. Sarkar, *J. Org. Chem.*, **63**, 6027 (1998); B. C. Ranu, P. Dutta, and A. Sarkar, *J. Chem. Soc. Perkin Trans. 1*, 2223 (2000).

27. K. Ramalinga, P. Vijayalakshmi, and T. N. B. Kaimal, *Tetrahedron Lett.*, **43**, 879 (2002).

28. J. Otera, T. Yano, A. Kawabata, and H. Nozaki, *Tetrahedron Lett.*, **27**, 2383 (1986); J. Otera, S. Ioka, and H. Nozaki, *J. Org. Chem.*, **54**, 4013 (1989).

29. G. A. Grasa, R. M. Kissling, and S. P. Nolan, *Org. Lett.*, **4**, 3583 (2002). See also G. W. Nyce, J. A. Lamboy, E. F. Connor, R. M. Waymouth, and J. L. Hedrick, *Org. Lett.*, **4**, 3587 (2002). G. A. Grasa, T. Guveli, R. Singh, and S. P. Nolan, *J. Org. Chem.*, **68**, 2812 (2003).

30. K. Wakasugi, T. Misaki, K. Yamada, and Y. Tanaabe, *Tetrahedron Lett.*, **41**, 5249 (2000).

31. A. Orita, Y. Nagano, J. Hirano, and J. Otera, *Synlett*, 637 (2001).

32. Y. Masaki, N. Tanaka, and T. Miura, *Chem. Lett.*, **26**, 55 (1997).

33. A. G. Myers, N. J. Tom, M. E. Fraley, S. B. Cohen, and D. J. Madar, *J. Am. Chem. Soc.*, **119**, 6072 (1997).

34. B. P. Bandgar, V. S. Sadavarte, and L. S. Uppalla, *Chem. Lett.*, **30**, 894 (2001).

35. S. P. Chavan, P. K. Zubaidha, S. W. Dantale, A. Keshavaraja, A. V. Ramaswany, and T. Ravindranathan, *Tetrahedron Lett.*, **37**, 233 (1996).

36. F. C. da Silva, V. F. Ferreira, R. S. Rianelli, and W. C. Perreira, *Tetrahedron Lett.*, **43**, 1165 (2002).

37. D. E. Ponde, V. H. Deshpande, V. J. Bulbule, A. Sudalai, and A. S. Gajare, *J. Org. Chem.*, **63**, 1058 (1998).

38. H. Hagiwara, A. Koseki, K. Isobe, K.-i. Shimizu, T. Hoshi, and T. Suzuki, *Synlett*, 2188 (2004).

39. P. Kumar, and R. K. Pandey, *Synlett*, 251 (2000).

40. B. P. Bandgar, S. S. Pandit, and L. S. Uppalla, *OPPI Briefs*, **35**, 219 (2003).

41. S. P. Chavan, K. Shivasankar, R. Sivappa, and R. Kale, *Tetrahedron Lett.*, **43**, 8583 (2002).

42. B. P. Bandgar, V. S. Sadavarte, and L. S. Uppalla, *Synlett*, 1338 (2001).

43. B. P. Bandgar, V. S. Sadavarte, and L. S. Uppalla, *Chem. Lett.*, **30**, 894 (2001).

44. B. P. Bandgar, L. S. Uppalla, and V. S. Sadavarte, *Synlett*, 1715 (2001).

Enzymatically Cleavable Esters

The enzymatic cleavage of esters is a vast and extensively reviewed area of chemistry.[1] More recently, several new esters have been examined primarily for the preparation of peptides and glycopeptides.

Heptyl Esters: $C_7H_{15}O_2CR$

The heptyl ester was developed as an enzymatically removable protective group for C-terminal amino acid protection.

Formation

1. Heptyl alcohol, TsOH, benzene, reflux, 66–92% yield.[2]
2. Many of the standard methods for ester formation are certainly applicable to heptyl ester formation.

Cleavage

1. Lipase from *Rhizopus niveus*, pH 7, rt, 50–96% yield.[3]
2. Lipase from *Aspergillus niger*, 0.2 *M* phosphate buffer, acetone, pH 7, 37°C, 50–96% yield. This lipase was used in the cleavage of phosphopeptide heptyl esters. These conditions are sufficiently mild to prevent elimination of phosphorylated serine and threonine residues.[4]
3. Lipase M (*Mucor javanicus*), pH 7, 37°C, 70–88% yield. In this case, α- and β-glycosidic peptide derivatives were deprotected. Acetates on the pyranosides were not affected.[5]
4. Newlase F, pH 7, 30°C.[6]

2-*N*-(Morpholino)ethyl Ester (MoEtO$_2$CR)

The ester was developed to impart greater hydrophilicity in C-terminal peptides which contain large hydrophobic amino acids, since the velocity of deprotection with enzymes often was reduced to nearly useless levels. Efficient cleavage is achieved with the lipase from *R. niveus* (pH 7, 37°C, 16h, H_2O, acetone, 78–91% yield).[7]

Choline Ester: $Me_3N^+CH_2CH_2O_2CR\ Br^-$

The choline ester is prepared by treating the 2-bromoethyl ester with trimethylamine.[8] The ester is cleaved with butyrylcholine esterase (pH 6, 0.05 M phosphate buffer, rt, 50–95% yield). As with the morpholinoethyl ester, it imparts greater solubility to the C-terminal end of very hydrophobic peptides, thus improving the ability to enzymatically cleave the C-terminal ester.[9–11]

(Methoxyethoxy)ethyl Ester (Mee Ester): $CH_3OCH_2CH_2OCH_2CH_2O_2CR$

Because O-glycoproteins are susceptible to strong base and anomerization with acid, their preparation presents a number of difficulties, among which is the issue of mild and selective deprotection. Although in many cases the heptyl group was found quite useful because of the mild conditions associated with its enzymatic cleavage, in some cases the enzymatic cleavage would not proceed because the high level of hydrophobicity reduced solubility enough that the cleavage velocity approached zero. Increasing the hydrophilicity of the C-terminal protective group by incorporating some oxygen in the chain as in the Mee ester, allows for the reasonably facile cleavage with the lipase M from *M. javanicus* or papain. The pyranosidic acetates were not cleaved with these enzymes, but they could be cleaved with lipase WG.[12]

Methoxyethyl Ester (ME$-O_2$CR): $CH_3OCH_2CH_2O_2CR$

The advantages of the methoxyethyl ester over some of the other water solubilizing esters are that many of the amino acid esters are crystalline and thus easily purified; they are cleaved with a number of readily available lipases and are useful for the synthesis of *N*-linked glycopeptides.[13]

1. (a) K. Faber and S. Riva, *Synthesis*, 895 (1992); (b) H. Waldmann and D. Sebastian, *Chem. Rev.*, **94**, 911 (1994); (c) *Enzyme Catalysis in Organic Synthesis: A Comprehensive Handbook*, K. Drauz, and H. Waldmann, Eds., VCH, New York, 1995; (d) T. Pohl, E. Nägele, and H. Waldmann, *Catal. Today*, **22**, 407 (1994); (e) H. Waldmann, P. Braun, and H. Kunz, *Chem. Pept. Proteins*, **5/6** (Pt. A), 227 (1993); (f) A. Reidel and H. Waldmann, *J. Prakt. Chem./Chem.-Ztg.*, **335**, 109 (1993); (g) C.-H. Wong and G. M. Whitesides, *Enzymes in Synthetic Organic Chemistry*, Pergamon, Oxford, (1994).

2. P. Braun, H. Waldmann, W. Vogt, and H. Kunz, *Synlett*, 105 (1990); *idem, Liebigs Ann. Chem.*, 165 (1991).

3. P. Braun, H. Waldmann, W. Vogt, and H. Kunz, *Liebigs Ann. Chem.*, 165 (1991).

4. D. Sebastian and H. Waldmann, *Tetrahedron Lett.*, **38**, 2927 (1997).

5. H. Waldmann, A. Heuser, P. Braun, and H. Kunz, *Indian J. Chem., Sect. B* **31B**, 799 (1992); H. Waldmann, P. Braun, and H. Kunz, *Biomed. Biochim. Acta*, **50** S, 243 (1991); P. Braun, H. Waldmann, and H. Kunz, *Synlett*, 39 (1992).

6. Z.-Z. Chen, Y.-M. Li, X. Peng, F.-R. Huang, and Y.-F. Zhao, *J. Mol. Cat. B: Enzymatic*, **18**, 243 (2002); Z.-Z. Chen, Y.-M. Li, and Y.-F. Zhao, *J. Chem. Res., Synop.*, 101 (2003).

7. G. Braum, P. Braun, D. Kowalczyk, and H. Kunz, *Tetrahedron Lett.*, **34**, 3111 (1993).

8. J. Sander and H. Waldmann, *Chem. Eur. J.*, **6**, 1564 (2000).

9. M. Schelhaas, S. Glomsda, M. Hänsler, H.-D. Jakubke, and H. Waldmann, *Angew. Chem., Int. Ed. Engl.*, **35**, 106 (1996).

10. M. Schelhaas, E. Nagele, N. Kuder, B. Bader, J. Kuhlmann, A. Wittinghofer, and H. Waldmann, *Chem. Eur. J.*, **5**, 1239 (1999).

11. K. Kuhn and H. Waldmann, *Tetrahedron Lett.*, **40**, 6369 (1999).

12. J. Eberling, P. Braun, D. Kowalczyk, M. Schultz, and H. Kunz, *J. Org. Chem.*, **61**, 2638 (1996); S. Flohr, V. Jungmann, and H. Waldmann, *Chem. Eur. J.*, **5**, 669 (1999).

13. M. Gewehr and H. Kunz, *Synthesis*, 1499 (1997).

Methyl Ester: RCO_2CH_3 (Chart 6)

Formation

The section on general methods should also be consulted.

1. Dimethylsulfate, LiOH·H_2O, THF, reflux, 66–100% yield.[1] K_2CO_3 in acetone can effectively be used as base and solvent with dimethylsulfate to form esters. A polymer supported methyl sulfate also effectively esterifies carboxylic acids (K_2CO_3, CH_3CN, reflux, 72–99% yield). This reagent also alkylates thiols, phenols, phosphates, and amines.[2]

2. KHCO_3,[3] Na_2CO_3[4] or Cs_2CO_3,[5] MeI or dimethylsulfate, DMF, excellent yields. This is a general method that works with a variety of other carbonates and solvents such as acetone.

3. $MeSO_2Cl$, pyridine, 0°C, 65–83% yield.[6] Although the yields are moderate compared to more conventional methods, this reaction is important in that these conditions are often used to prepare mesylates of alcohols which indicates that some caution must be exercised with free acids during reactions with alcohols.

4. Dimethyl carbonate, DBU, reflux, 98–99% yield.[7]

5. H_2NCON(NO)Me, KOH, DME, H_2O, 0°C, 75% yield. This method generates diazomethane *in situ*.[8] **N-Methyl-N-nitrosourea is a proven carcinogen.**

6. Me_3SiCHN_2, MeOH, benzene, 20°C.[9,10] This reagent also reacts with phenols. This is a safe alternative to the use of diazomethane. A detailed, large scale preparation of this useful reagent has been described.[11] The reagent reacts with various maleic anhydrides in the presence of an alcohol to form diesters (70–96% yield).[12]

7. Me_2C(OMe)_2, cat. HCl, 25°C, 18 h, 80–95% yield.[13] These reaction conditions were used to prepare methyl esters of amino acids.

8. (MeO)_2NH, heat, 98% yield.[14] Amines are also alkylated.

9. MeOH, H_2SO_4, 0°C, 1 h; 5°C, 18 h, 98% yield.[15]

Ratio = 4:1

10. MeOH, HBF$_4$, Na$_2$SO$_4$, 25–60°C, 15 h, 45–94% yield.[16] The selectivity observed here is also observed for Et, i-Pr, Bn and cyclohexyl esters ($n = 1, 2$).

$$R = CH_3, Et, i\text{-}Pr, Bn, cyclohexyl$$

11. CBr$_4$, MeOH, $h\nu$ (30 min), stir at rt 2–24 h, 90–99% yield. This method is selective for carboxylates attached to sp^3 centers. Carboxylates attached to sp^2 centers react substantially slower allowing almost complete selectivity for the saturated systems.[17] It would seem that HBr is generated which actually catalyzes the reaction.

12. TMSCl, MeOH, 2,2-dimethoxypropane, rt, 95–99% yield. As with the above case, aromatic acids are not esterified by this method which generates HCl *in situ*.[18] In general, it is more difficult to prepare aromatic esters by acid catalyzed esterification than aliphatic esters because aromatic acids are not as easily protonated. BCl$_3$ in MeOH has been used to prepare methyl esters and this combination of reagents also produces HCl.[19]

13. From a t-Bu ester: CSA, MeOH, sealed tube, 105°C, 100% yield.[20]

14. NiCl$_2$·6H$_2$O, 10 mol%, MeOH, reflux, 9–93% yield.[21] Aromatic and conjugated acids are not effectively esterified under these conditions.

15. 1-Methyl-p-tolyltriazene, ether, 70–90% yield.[22]

16. Polymer supported methyltriazine, CH$_2$Cl$_2$, rt, 34–100% yield. The process is effective for both aromatic and alkyl acids. Ethyl and benzyl esters have also been prepared by this method. Acidic phenols such as 4-nitrophenol can be methylated by this method but more electron-rich phenols give excruciatingly slow reactions.[23] The rate of reaction is pK_a-dependent.

17. O-Alkylisoureas (Me, Bn, p-MeOBn), microwaves, THF, 75–98% yield.[24]

18. For Boc protected amino acids: Ceric ammonium nitrate, MeOH, rt, 38–83% yield. When the reaction is conducted at reflux Boc cleavage is accompanied by esterification.[25]

Cleavage

Under normal circumstances, methyl esters are readily cleaved by alkali metal hydroxides and carbonates in an aqueous/organic solvent mixture.

1. LiOH, CH$_3$OH, H$_2$O (3:1), 5°C, 15 h.[26]

2. LiOH, H$_2$O$_2$, THF, H$_2$O, 25°C, 6 h, 97% yield.[27] In the following case, LiOH resulted in an unusual amide cleavage that is probably the result of rotation about the amide bond which removes the usual amide resonance, thus making it more susceptible to cleavage by base.

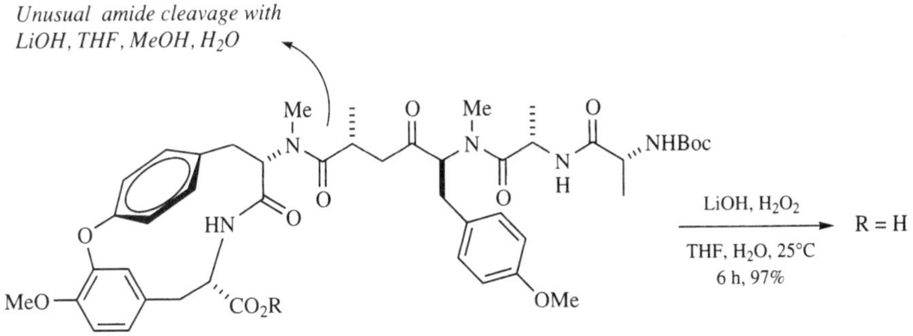

Unusual amide cleavage with LiOH, THF, MeOH, H$_2$O

LiOH, H$_2$O$_2$

THF, H$_2$O, 25°C
6 h, 97%

R = H

3. Ba(OH)$_2$·8H$_2$O, MeOH, rt, 7 h, 72% yield.[28] A nonaqueous workup procedure has been developed for this method.[29]

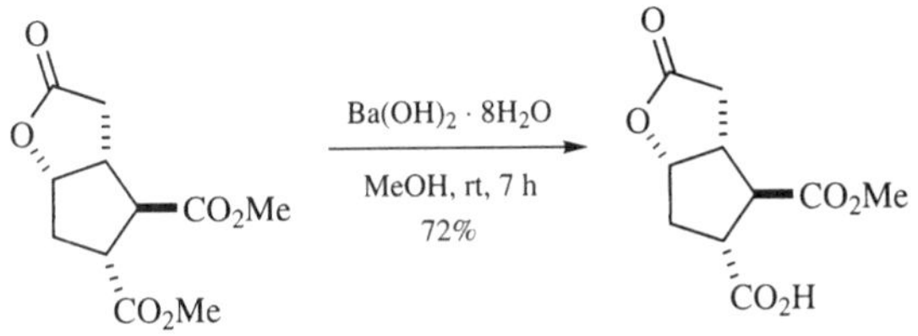

Ba(OH)$_2$ · 8H$_2$O

MeOH, rt, 7 h
72%

These conditions gave excellent selectivity for an external methyl dienoate in the presence of a more hindered internal dienoate during a synthesis of the complex macrolide swinholide.[30] These conditions are also mild enough to prevent retroaldol condensation during ester hydrolysis.[31] In general, the barium salts may also be removed by precipitation with CO$_2$ to form BaCO$_3$ which is readily filtered off, a method that is especially useful for water-soluble substrates.

4. In the following case the authors propose that the selectivity is due to participation of the hydroxyl group.[32]

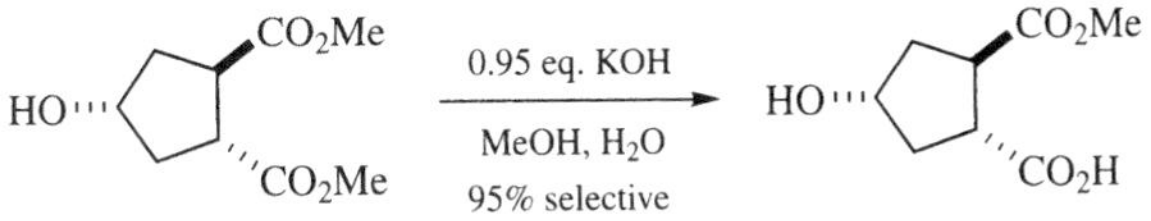

0.95 eq. KOH

MeOH, H$_2$O
95% selective

5. AlBr$_3$, tetrahydrothiophene, rt, 62 h, 99% yield.[33]

6. AlCl$_3$, DMA, CH$_2$Cl$_2$, reflux, 78–98% yield.[34] This method cleaves the methyl ester from Fmoc protected amino acids.

7. AlCl$_3$, Me$_2$S, >29% yield. Deprotection proceeds without isomerization at C2 and C9.[35]

8. BCl$_3$, 0°C, 5–6h, 90% yield.[36] In this example a phenolic methyl group, normally cleaved with boron trichloride, was not affected.

9. NaBH$_4$, I$_2$, 3 h, rt.[37]

10. NaCN, HMPA, 75°C, 24 h, 75–92% yield.[38] Ethyl esters are not cleaved under these conditions.

11. LiCl (5 eq.), H$_2$O (1.5 eq.), HMPA, 100°C, 2h, 88% yield.[39] In general, nucleophilic cleavage of β-ketoesters and sulfones results in decarboxylation.

12. Me$_4$NOAc, HMPA, 100°C, 17 h, 71% yield.[40]

13. Cs$_2$CO$_3$, PhSH, DMF, 85°C, 3 h, 91% yield. A methyl carbonate was cleaved simultaneously.[41]

14. H$_2$NC$_6$H$_4$SH, Cs$_2$CO$_3$, DMF, 85°C, 1–3 h.[42]

15. Catalytic KF,[43] or K$_2$CO$_3$[44] PhSH, NMP, 190°C, 50–100% yield. The method was only tested on aromatic esters, which include ethyl and benzyl esters as well as methyl esters. Aromatic nitro groups and aryl chlorides are compatible in that they do not give products of substitution.

16. n-PrSLi, HMPA, rt, 94% yield.[45]

17. Ph$_3$SiSH, Cs$_2$CO$_3$, 2,6-di-t-butylcresol, DMF, 80°C, 96% yield.[46]

18. LiI, Pyr, reflux, 91% yield.[47,48]

19. (CH$_3$)$_3$SiOK, ether[49] or THF, 4 h, 61–95% yields as the acid salt.[50] This has become a very popular and effective method for the cleavage of methyl esters,[51] often when conventional hydrolysis fails. It was even found effective for cleavage of an ethyl ester when other methods failed.[52,53] Hindered esters are cleaved with this reagent.[54]

20. [MeTeAlMe$_2$]$_2$, toluene, 23°C, 12 h, >89% yield. This method was developed when all other conventional methods failed to effect cleavage.[55] Note that in a very similar case which is less sterically encumbered, conventional NaOH hydrolysis was effective.[56]

21.

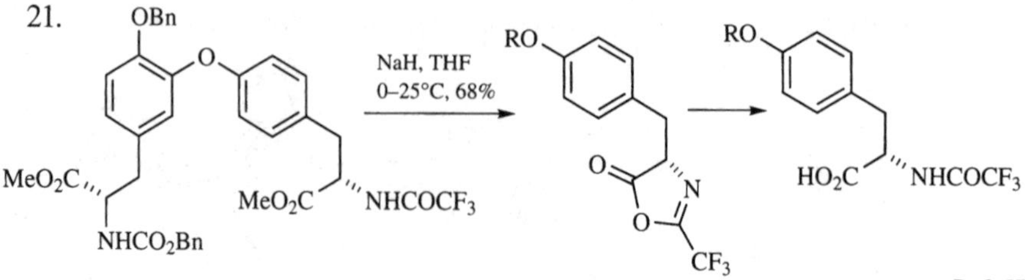

Ref. 57

22. (Bu$_3$Sn)$_2$O, benzene, 80°C, 2–24 h, 73–100% yield.[58] Only relatively unhindered esters are cleaved with this reagent. Acetates of primary and secondary alcohols and phenols are also cleaved efficiently.[59]

23. Me$_3$SnOH, 1,2-dichloroethane, 80°C, 1 h, 100°C.[60]

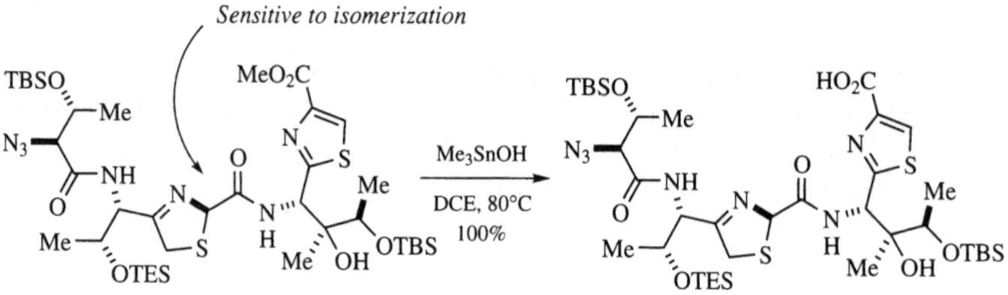

24. NaOCH$_2$CH$_2$CN, THF, 0–23°C, 10 min, 93% yield. This method was used to prevent formation of coumarin **i**.[61]

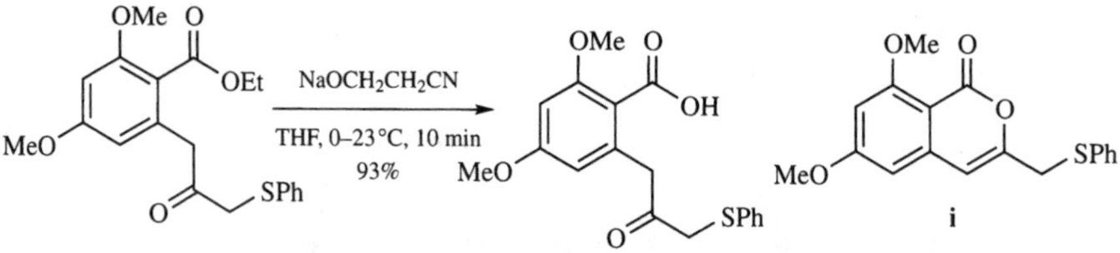

25. CuCO$_3$, Cu(OH)$_2$; H$_2$S workup, 50–60°C.[62]

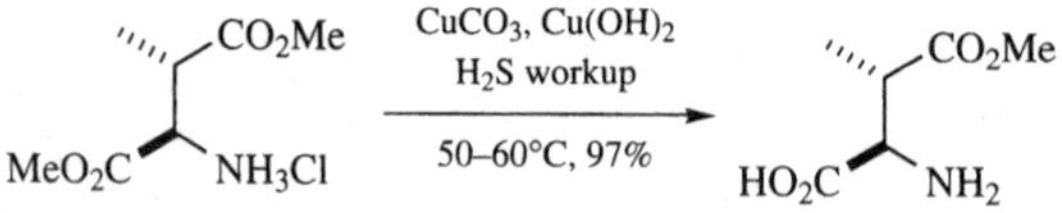

26. Pig liver esterase is particularly effective in cleaving one ester of a symmetrical pair.[63–65]

27.

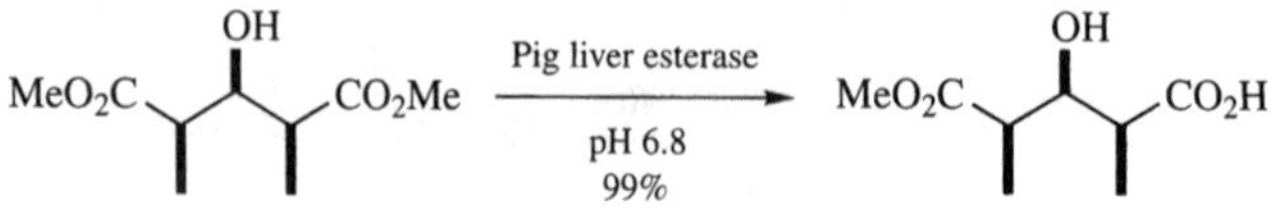

Ref. 66

28.

Pig liver esterase

98% chemical
96% ee

Ref. 67

29.

Pig liver esterase

$E = 21.5$ (enatiomeric ratio)

Ref. 68

30. Carbonic anhydrase, H_2O, 23–83% yield. This enzyme was used for the selective hydrolysis of the D-form of methyl N-acetyl α-amino acids.[69]

31. Porcine pancreatic lipase, pH 7.5, 23°C 4.5 h, 55% yield. These conditions were used to suppress facile racemization of 2-chlorocyclohexenone.[70]

32. Thermitase, pH 7.5, 55°C, 50% DMSO, 3–140 min. This method was used to avoid degradation of base-sensitive side chains during peptide synthesis. The method is compatible with the Fmoc group.[71]

1. A. K. Chakraborti, A. Basak-Nandi, and V. Grover, *J. Org. Chem.*, **64**, 8014 (1999).

2. T. Yoshino and H. Togo, *Synlett* 517 (2005).

3. D. S. Karanewsky, M. F. Malley, J. Z. Goutoutas, *J. Org. Chem.*, **56**, 3744 (1991).

4. G. Stork, S. Rychnovsky, *J. Am. Chem. Soc.*, **109**, 1565 (1987).

5. K. Luthman, M. Orbe, T. Waglund, and A. Claesson, *J. Org. Chem.*, **52**, 3777 (1987).

6. B. S. Siddiqui, F. Begum, and S. Begum, *Tetrahedron Lett.*, **42**, 9059 (2001).

7. W.-C. Shieh, S. Dell, and O. Repic, *J. Org. Chem.*, **67**, 2188 (2002); W.-C. Shieh, S. Dell, and O. Repic, *Tetrahedron Lett.*, **43**, 5607 (2002); F. Rajabi and M. R. Saidi, *Synth. Commum.*, **34**, 4179 (2004).

8. For example, see S. M. Hecht, and J. W. Kozarich, *Tetrahedron Lett.*, **4**, 1397 (1973).

9. N. Hashimoto, T. Aoyama, and T. Shioiri, *Chem. Pharm. Bull.*, **29**, 1475 (1981).

10. Y. Hirai, T. Aida, and S. Inoue, *J. Am. Chem. Soc.*, **111**, 3062 (1989).

11. T. Shioiri, T. Aoyama, and S. Mori, *Org. Synth.*, **68**, 1 (1990).

12. S. C. Fields, W. H. Dent, III, F. R. Green, III, and E. G. Tromiczak, *Tetrahedron Lett.*, **37**, 1967 (1996).

13. J. R. Rachele, *J. Org. Chem.*, **28**, 2898 (1963).

14. V. F. Rudchenko, S. M. Ignator, and R. G. Kostyanovsky, *J. Chem. Soc., Chem. Commun.*, 261 (1990).

15. S. Danishefsky, M. Hirama, K. Gombatz, T. Harayama, E. Berman, and P. Schuda, *J. Am. Chem. Soc.*, **100**, 6536 (1978); *idem, ibid.*, **101**, 7020 (1979).

16. R. Albert, J. Danklmaier, H. Hönig, and H. Kandolf, *Synthesis*, 635 (1987).

17. A. S.-Y. Lee, H.-C. Yang, and F.-Y. Su, *Tetrahedron Lett.*, **42**, 301 (2001).

18. A. Rodriguez, M. Nomen, B. W. Spur, and J. J. Godfroid, *Tetrahedron Lett.*, **39**, 8563 (1998).

19. C. A. Dyke and T. A. Bryson, *Tetrahedron Lett.*, **42**, 3959 (2001).
20. S. D. Burke and G. M. Sametz, *Org. Lett.*, **1**, 71 (1999).
21. R. N. Ram and I. Charles, *Tetrahedron*, **53**, 7335 (1997).
22. E. H. White, A. A. Baum, and D. E. Eitel, *Org. Syn.*, **48**, 102 (1968).
23. B. Erb, J.-P. Kucma, S. Mourey, and F. Struber, *Chem. Eur. J.*, **9**, 2582 (2003).
24. S. Crosignani, P. D. White, and B. Linclau, *Org. Lett.*, **4**, 2961 (2002); S. Crosignani, P. D. White, and B. Linclau, *Org. Lett.*, **4**, 1035 (2002).
25. A. Kuttan, S. Nowshudin, and M. N. A. Rao, *Tetrahedron Lett.*, **45**, 2663 (2004).
26. E. J. Corey, I. Székely, and C. S. Shiner, *Tetrahedron Lett.*, **18**, 3529 (1977).
27. D. L. Boger, D. Yohannes, J. Zhou, and M. A. Patane, *J. Am. Chem. Soc.*, **115**, 3420 (1993).
28. K. Inoue and K. Sakai, *Tetrahedron Lett.*, **18**, 4063 (1977).
29. M. C. Anderson, J. Moser, J. Sherrill, and R. K. Guy, *Synlett*, 2391 (2004).
30. I. Paterson, K.-S. Yeung, R. A. Ward, J. D. Smith, J. G. Cumming, and S. Lamboley, *Tetrahedron*, **51**, 9467 (1995).
31. M. Nambu and J. D. White, *J. Chem. Soc., Chem. Commun.*, 1619 (1996).
32. M. Honda, K. Hirata, H. Sueoka, T. Katsuki, and M. Yamaguchi, *Tetrahedron Lett.*, **22**, 2679 (1981).
33. A. E. Greene, M.-J. Luche, and J.-P. Deprés, *J. Am. Chem. Soc.*, **105**, 2435 (1983).
34. M. L. Di Gioia, A. Leggio, A. Le Pera, A. Liguori, F. Perri, and C. Siciliano, *Eur. J. Org. Chem.*, 4437 (2004). M. L. Di Gioia, A. Leggio, A. Le Pera, C. Siciliano, G. Sindona, and A. Liguori, *J. Peptide Res.*, **63**, 383 (2004).
35. M. Kawasaki, T. Shinada, M. Hamada, and Y. Ohfune, *Org. Lett.*, **7**, 4165 (2005).
36. P. S. Manchand, *J. Chem. Soc., Chem. Commun.*, 667 (1971).
37. D. H. R. Barton, L. Bould, D. L. J. Clive, P.D. Magnus, and T. Hase, *J. Chem. Soc. C*, 2204 (1971).
38. P. Müller and B. Siegfried, *Helv. Chim. Acta*, **57**, 987 (1974).
39. R. M. Williams, T. Glinka, E. Dwast, H. Coffman, and J. K. Sille, *J. Am. Chem. Soc.*, **112**, 808 (1990).
40. A. S. Kende, J. S. Mendoza, and Y. Fujii, *Tetrahedron*, **49**, 8015 (1993).
41. D. Eren and E. Keinan, *J. Am. Chem. Soc.*, **110**, 4356 (1988); S. Bouzbouz, and B. Kirschleger, *Synthesis*, 714 (1994).
42. E. Keinan and D. Eren, *J. Org. Chem.*, **51**, 3165 (1986).
43. M. K. Nayak and A. K. Chakraborti, *Chem. Lett.*, **27**, 297 (1998); A. K. Chakraborti, L. Sharma, and M. K. Nayak, *J. Org. Chem.*, **67**, 2541 (2002).
44. L. Sharma, M. K. Nayak, and A. K. Chakraborti, *Tetrahedron*, **55**, 9595 (1999).
45. B. M. Trost, H. Yang, and G. D. Probst, *J. Am. Chem. Soc.*, **126**, 48 (2004).
46. A. Ishiwata and Y. Ito, *Synlett*, 1339 (2003).
47. P. Magnus and T. Gallagher, *J. Chem. Soc., Chem. Commun.*, 389 (1984).
48. O. Lepage, E. Kattnig, and A. Fürstner, *J. Am. Chem. Soc.*, **126**, 15970 (2004).
49. E. D. Laganis and B. L. Chenard, *Tetrahedron Lett.*, **25**, 5831 (1984).
50. C. Rasset-Deloge, P. Martinez-Fresneda, and M. Vaultier, *Bull. Soc. Chim. Fr.*, **129**, 285 (1992).
51. I. Paterson, V. A. Doughty, M. D. McLeod, and T. Trieselmann, *Angew. Chem. Int. Ed.*, **39**, 1308 (2000); G. Böttcher and H.-U. Reissig, *Synlett*, 725 (2000); T. B. Durham, N.

Blanchard, B. M. Savall, N. A. Powell, and W. R. Roush, *J. Am. Chem. Soc.*, **126**, 9307 (2004); A. G. M. Barrett, M. Pena, and J. A. Willardsen, *J. Org. Chem.*, **61**, 1082 (1996).

52. J. A. Lafontaine, D. P. Provencal, C. Gardelli, and J. W. Leahy, *J. Org. Chem.*, **68**, 4215 (2003).

53. A. Fettes and E. M. Carreira, *J. Org. Chem.*, **68**, 9274 (2003).

54. J. Rachon, V. Goedken, and H. M. Walborsky, *J. Org. Chem.*, **54**, 1006 (1989).

55. L. R. Reddy, J.-F. Fournier, B. V. S. Reddy, and E. J. Corey, *J. Am. Chem. Soc.*, **127**, 8974 (2005); L. R. Reddy, J.-F. Fournier, B. V. S. Reddy, and E. J. Corey, *Org. Lett.*, **7**, 2699 (2005).

56. L. R. Reddy, P. Saravanan, J.-F. Fournier, B. V. S. Reddy, and E. J. Corey, *Org. Lett.*, **7**, 2703 (2005); B. V. S. Reddy, L. R. Reddy, and E. J. Corey, *Tetrahedron Lett.*, **46**, 4589 (2005).

57. D. L. Boger and D. Yohannes, *J. Org. Chem.*, **54**, 2498 (1989).

58. C. J. Saloman, E. G. Mata, and O. A. Masceretti, *J. Chem. Soc. Perkin Trans. 1*, 995 (1996); E. G. Mata and O. A. Mascaretti, *Tetrahedron Lett.*, **29**, 6893 (1988).

59. C. J. Salomon, G. E. Mata, and O. A. Mascaretti, *J. Org. Chem.*, **59**, 7259 (1994).

60. K. C. Nicolaou, M. Nevalainen, M. Zak, S. Bulat, M. Bella, and B. S. Safina, *Angew. Chem. Int. Ed.*, **42**, 3418 (2003); K. C. Nicolaou, B. S. Safina, M. Zak, A. A. Estrada and S. H. Lee, *Angew. Chem. Int. Ed.*, **43**, 5087 (2004); K. C. Nicolaou, M. Zak, B. S. Safina, S. H. Lee, and A. A. Estrada, *Angew. Chem. Int. Ed.*, **43**, 5092 (2004).

61. S. Barluenga, E. Moulin, P. Lopez, and N. Winssinger, *Chem. Eur. J.*, **11**, 4935 (2005).

62. J. M. Humphrey, J. B. Aggen, and A. R. Chamberlin, *J. Am. Chem. Soc.*, **118**, 11759 (1996).

63. M. Ohno, Y. Ito, M. Arita, T. Shibata, K. Adachi, and H. Sawai, *Tetrahedron*, **40**, 145 (1984).

64. E. Alvarez, T. Cuvigny, C. Hervé du Penhoat, and M. Julia, *Tetrahedron*, **44**, 119 (1988).

65. K. Adachi, S. Kobayashi, and M. Ohno, *Chimia*, **40**, 311 (1986).

66. D. S. Holmes, U. C. Dyer, S. Russell, J. A. Sherringham, and J. A. Robinson, *Tetrahedron Lett.*, **29**, 6357 (1988).

67. S. Kobayashi, K. Kamiyama, T. Iimori, and M. Ohno, *Tetrahedron Lett.*, **25**, 2557 (1984).

68. P. Mohr, L. Rösslein, and C. Tamm, *Tetrahedron Lett.*, **30**, 2513 (1989).

69. R. Chênevert, R. B. Rhlid, M. Létourneau, R. Gagnon, and L. D'Astous, *Tetrahedron: Asymmetry*, **4**, 1137 (1993).

70. H. Wild, *J. Org. Chem.*, **59**, 2748 (1994).

71. S. Reissmann and G. Greiner, *Int. J. Pept. Protein Res.*, **40**, 110 (1992).

Substituted Methyl Esters

9-Fluorenylmethyl (Fm) Ester

9-Fluorenylmethyl esters of *N*-protected amino acids were prepared using the DCC/DMAP method (50–89% yield),[1] by imidazole-catalyzed transesterification of protected amino acid active esters with FmOH[2] or by reaction with Fmoc-Cl (DIPEA, DMAP, 0°C, 30 min, 25–84% yield).[3] Cleavage is accomplished with either diethylamine or piperidine in CH_2Cl_2 at rt for 2 h. No racemization was observed during formation or cleavage of the Fm esters.[1] The Fm ester is cleaved slowly by hydrogenolysis,[4] but complete selectivity for hydrogenolysis of benzyloxycarbonyl group could not be obtained. Fm esters also improved the solubility of protected peptides in organic solvents.[2]

1. H. Kessler and R. Siegmeier, *Tetrahedron Lett.*, **24**, 281 (1983).
2. M. A. Bednarek and M. Bodanszky, *Int. J. Pept. Protein Res.*, **21**, 196 (1983).
3. S. A. M. Mérette, A. P. Burd, and J. J. Deadman, *Tetrahedron Lett.*, **40**, 753 (1999).
4. A. Lender, W. Yao, P. A. Sprengeler, R. A. Spanevello, G. T. Furst, R. Hirschmann, and A. B. Smith, III, *Int. J. Pept. Protein Res.*, **42**, 509 (1993).

Methoxymethyl Ester (MOM Ester): $RCOOCH_2OCH_3$ (Chart 6)

In general, MOM esters are not nearly as stable as are the ether counterparts. They are often not stable to silica gel chromatography.

Formation

The section on the formation of MOM ethers should be consulted, since many of the methods described there should also be applicable to the formation of MOM esters.

1. CH_3OCH_2Cl, Et_3N, DMF, 25°C, 1 h.[1]
2. $CH_3OCH_2OCH_3$, $Zn/BrCH_2CO_2Et$, 0°C; CH_3COCl, 0–20°C, 2 h, 75–85%.[2] A number of methoxymethyl esters were prepared by this method, which avoids the use of the *carcinogen chloromethyl methyl ether.*

Cleavage

1. R'_3SiBr, trace MeOH. Methoxymethyl ethers are stable to these cleavage conditions.[3] Methoxymethyl esters are unstable to silica gel chromatography, but are stable to mild acid (0.01 *N* HCl, EtOAc, MeOH, 25°C, 16 h).[4]
2. $MgBr_2$, Et_2O. MEM, MTM and SEM ethers are cleaved as well.[5]
3. Solvolysis in $MeOH/H_2O$ at 21°C. This method was developed for a series of penicillin derivatives where conventional cleavage methods resulted in partial β-lactam cleavage.[6]
4. $AlCl_3$, $PhNMe_2$, 80–99% yield. MEM, MTM, Me, Bn and SEM esters are cleaved similarly.[7]
5. Pyridine, H_2O.[8]

6. CBr$_4$, IPA, reflux, 82°C, 91–95% yield.[9] This method most likely generates HBr *in situ* and thus is incompatible with acid sensitive groups like the TBS group. MEM esters are cleaved similarly.

7. NaHSO$_4$, SiO$_2$, CH$_2$Cl$_2$, rt, 1–1.5 h, 90–100% yield.[10] These conditions have also been used for the cleavage of MOM, MEM and TBS ethers.

1. A. B. A. Jansen and T. J. Russell, *J. Chem. Soc.*, 2127 (1965).

2. F. Dardoize, M. Gaudemar, and N. Goasdoue, *Synthesis*, 567 (1977).

3. S. Masamune, *Aldrichimica Acta*, **11**, 23 (1978), see p. 30.

4. L. M. Weinstock, S. Karady, F. E. Roberts, A. M. Hoinowski, G. S. Brenner, T. B. K. Lee, W. C. Lumma, and M. Sletzinger, *Tetrahedron Lett.*, **16**, 3979 (1975).

5. S. Kim, Y. H. Park, and I. S. Kee, *Tetrahedron Lett.*, **32**, 3099 (1991).

6. S. Vanwetswinkel, V. Carlier, J. Marchand-Brynaert, and J. Fastrez, *Tetrahedron Lett.*, **37**, 2761 (1996).

7. T. Akiyama, H. Hirofuji, A. Hirose, and S. Ozaki, *Synth. Commun.*, **24**, 2179 (1994).

8. M. Shimano, H. Nagaoka, and Y. Yamada, *Chem. Pharm. Bull.*, **38**, 276 (1990).

9. A. S.-Y. Lee, Y.-J. Hu, and S.-F. Chu, *Tetrahedron*, **57**, 2121 (2001).

10. C. Ramesh, N. Ravindranath, and B. Das, *J. Org. Chem.*, **68**, 7101 (2003).

Methoxyethoxymethyl Ester (MEM Ester): RCO$_2$CH$_2$OCH$_2$CH$_2$OCH$_3$

In an attempt to synthesize the macrolide antibiotic chlorothricolide, an unhindered −COOH group was selectively protected, in the presence of a hindered −COOH group, as a MEM ester that was then reduced to an alcohol group.[1]

Formation

MeOCH$_2$CH$_2$OCH$_2$Cl, *i*-Pr$_2$NEt, CH$_2$Cl$_2$, 0°C 2 h, high yield.[2]

Cleavage

1. 3 *N* HCl, THF, 40°C, 12 h.[2]

2. MgBr$_2$, Et$_2$O, rt, 12 h.[3,4] These conditions also cleaved a THP group and MTM, MEM and MOM esters. The MEM ester is cleaved the slowest.[5]

3. AlCl$_3$-dimethylaniline[6]

1. R. E. Ireland and W. J. Thompson, *Tetrahedron Lett.*, **20**, 4705 (1979).

2. A. I. Meyers and P. J. Reider, *J. Am. Chem. Soc.*, **101**, 2501 (1979).

3. J. A. O'Neill, S. D. Lindell, T. J. Simpson, and C. L. Willis, *J. Chem. Soc., Perkin Trans. 1*, 637 (1996).

4. A. J. Pearson and H. Shin, *J. Org. Chem.*, **59**, 2314 (1994).

5. S. Kim, Y. H. Park, and I. S. Kee, *Tetrahedron Lett.*, **32**, 3099 (1991).

6. T. Akiyama, H. Hirofuji, A. Hirose, and S. Ozaki, *Synth. Commun.*, **24**, 2179 (1994).

Methylthiomethyl Ester (MTM Ester): $RCOOCH_2SCH_3$ (Chart 6)

Formation

1. From RCO_2K: CH_3SCH_2Cl, NaI, 18-crown-6, C_6H_6, reflux, 6 h, 85–97% yield.[1]
2. $Me_2S^+ClX^-$, Et_3N, 0.5 h, $-70°C$ to $25°C$, 80–85% yield.[2]
3. t-BuBr, DMSO, $NaHCO_3$, 62–98% yield.[3,4] This method was used to prepare the MTM esters of N-protected amino acids.

Cleavage

1. $HgCl_2$, CH_3CN, H_2O, reflux, 6 h; H_2S, $20°C$, 30 min, 82–98% yield.[1]
2. MeI, acetone, reflux, 24 h; 1 N NaOH, 87–97% yield.[5]
3. CF_3COOH, $25°C$, 15 min, 80–90% yield.[6]
4. HCl, Et_2O, 6 h, 83–88% yield.[4] Acidic deprotection of the BOC group could not be achieved with complete selectivity in the presence of an MTM ester. The trityl and NPS (2-nitrophenylsulfenyl) groups were the preferred nitrogen protective groups.
5. H_2O_2, $(NH_4)_6Mo_7O_{24}$; NaOH, pH 11, 97% yield.[5] The MTM ester is converted to the much more base labile methylsulfonylmethyl ester. It is possible to hydrolyze the methylsulfonylmethyl ester in the presence of the MTM ester.
6. MCPBA converts the MTM ester to a methylsulfonylmethyl ester (78–98% yield), which can be hydrolyzed enzymatically with rabbit serum (pH 4.5 phosphate buffer, EtOH, $25–28°C$, 1 h, 84% yield).[7]

1. L. G. Wade, J. M. Gerdes, and R. P. Wirth, *Tetrahedron Lett.*, **19**, 731 (1978).
2. T.-L. Ho, *Synth. Commun.*, **9**, 267 (1979).
3. A. Dossena, R. Marchelli, and G. Casnati, *J. Chem. Soc., Perkin Trans. I*, 2737 (1981).
4. A. Dossena, G. Palla, R. Marchelli, and T. Lodi, *Int. J. Pept. Protein Res.*, **23**, 198 (1984).
5. J. M. Gerdes and L. G. Wade, *Tetrahedron Lett.*, 689 (1979).
6. T.-L. Ho and C. M. Wong, *J. Chem. Soc., Chem. Commun.*, 224 (1973).
7. A. Kamal, *Synth. Commun.*, **21**, 1293 (1991).

Tetrahydropyranyl Ester (THP Ester): RCOO-2-tetrahydropyranyl (Chart 6)

The THP ester is readily formed from dihydropyran (TsOH, CH_2Cl_2, $20°C$, 1.5 h, quant.). It is cleaved under mildly acidic conditions (AcOH, THF, H_2O (4:2:1), $45°C$, 3.5 h).[1]

1. K. F. Bernady, M. B. Floyd, J. F. Poletto, and M. J. Weiss, *J. Org. Chem.*, **44**, 1438 (1979).

Tetrahydrofuranyl Ester: RCO$_2$-2-tetrahydrofuranyl

Formation/Cleavage[1]

1. C. G. Kruse, N. L. J. M. Broekhof, and A. van der Gen, *Tetrahedron Lett.*, **17**, 1725 (1976).

2-(Trimethylsilyl)ethoxymethyl Ester (SEM Ester):
RCO$_2$CH$_2$OCH$_2$CH$_2$Si(CH$_3$)$_3$

The SEM ester was used to protect a carboxyl group where DCC-mediated esterification caused destruction of the substrate.[2] It is formed from the acid and SEM chloride (THF, TEA, 0°C, 80% yield). The SEM group can be introduced on an acid in the presence of a diol.[1]

In the following case, the SEM group was removed by solvolysis. The ease of removal in this case was attributed to anchimeric assistance by the phosphate group.[2]

Normally SEM groups are cleaved by treatment with fluoride ion. Note that in this case the SEM group is removed considerably faster than the phenyl groups from the phosphate. Additionally, cleavage is affected with MgBr$_2$ in ether (61–100% yield),[3] HF in acetonitrile,[4] or neat HF.[5]

1. T. Motozaki, K. Sawamura, A. Suzuki, K. Yoshida, T. Ueki, A. Ohara, R. Munakata, K.-i. Takao, and K.-i. Tadano, *Org. Lett.*, **7**, 2261 (2005).
2. E. W. Logusch, *Tetrahedron Lett.*, **25**, 4195 (1984).

3. W.-C. Chen, M. D. Vera, and M. M. Joullié, *Tetrahedron Lett.*, **38**, 4025 (1997).

4. W.-R. Li, W. R. Ewing, B. D. Harris, and M. M. Joullié, *J. Am. Chem. Soc.*, **112**, 7659 (1990).

5. G. Jou, I. Gonzalez, F. Albericio, P. Lloyd-Williams, and E. Giralt, *J. Org. Chem.*, **62**, 354 (1997).

Benzyloxymethyl Ester (BOM Ester): $RCOOCH_2OCH_2C_6H_5$ (Chart 6)

Formation[1]

$$RCOONa + PhCH_2OCH_2Cl \xrightarrow{\text{HMPA, 25°C, 70%}} RCO_2BOM$$

Cleavage[1]

1. H_2/Pd–C, EtOH, 25°C, 70–100% yield.
2. Aqueous HCl, THF, 25°C, 2 h, 75–95% yield.

1. P. A. Zoretic, P. Soja, and W. E. Conrad, *J. Org. Chem.*, **40**, 2962 (1975).

Triisopropylsilyloxymethyl Ester (TIPSOCH$_2$O$_2$CR)

Formation

$TIPSOCH_2SEt$, $CuBr_2$, Bu_4NBr, 4-Å molecular sieves, CH_2Cl_2, 89–98% yield.[1] This method can also be used to prepare a variety of other formyl acetals and esters.

Cleavage

1. Conditions used to cleave TIPS ethers can be used to cleave this group.
2. Since this is an ester, simple hydrolysis with base can also be used to cleave this group.

1. D. Sawada and Y. Ito, *Tetrahedron Lett.*, **42**, 2501 (2001).

Pivaloyloxymethyl ester (POM−O$_2$CR): $(CH_3)_3CCO_2CH_2OR$

The ester is prepared from the acid with $PvOCH_2I$ and Ag_2CO_3 in DMF.[1] It is cleaved with $(Bu_3Sn)_2O$ (Et$_2$O, 3 h, 25°C, 56% yield).[2,3]

1. D. V. Patel, E. M. Gordon, R. J. Schmidt, H. N. Weller, M. G. Young, R. Zahler, M. Barbacid, J. M. Carboni, J. L. Gullo-Brown, L. Hunihan, C. Ricca, S. Robinson, B. R. Seizinger, A. V. Tuomari, and V. Manne, *J. Med. Chem.*, **38**, 435 (1995).

2. J. Salomon, E. G. Mata, and O. A. Mascaretti, *Tetrahedron Lett.*, **32**, 4239 (1991); C. J. Salomon, E. G. Mata, and O. A. Mascaretti, *J. Org. Chem.*, **59**, 7259 (1994).
3. E. G. Mata and O. A. Mascaretti, *Tetrahedron Lett.*, **29**, 6893 (1988).

Phenylacetoxymethyl Ester: $PhCH_2CO_2CH_2O_2CR$

The ester is conveniently formed from a penicillinic acid with $PhCH_2CO_2CH_2Cl$ and TEA. Cleavage is accomplished by enzymatic hydrolysis with penicillin G acylase in 70–90% yield.[1,2]

1. E. Baldaro, C. Fuganti, S. Servi, A. Tahliani, and M. Terreni, in *Microbial Reagents in Organic Synthesis*, S. Servi, Ed.; Kluwer Academic Publishers, Dordrecht, 1992, pp. 175ff.
2. E. Baldaro, D. Faiardi, C. Fuganti, P. Grasselli, and A. Lazzzarini, *Tetrahedron Lett.*, **29**, 4623 (1988).

Triisopropylsilylmethyl Ester: $(i\text{-}Pr_3SiCH_2O_2R)$

Formation

$i\text{-}Pr_3SiCHN_2$, 76–96% yield.[1] In contrast, when $TMSCHN_2$ is used to prepare an ester the methyl ester is formed.

Cleavage

3 N NaOH, EtOH, 6 h, reflux. These cleavage conditions indicate that this ester is quite hindered and resists addition of nucleophiles to the carbonyl group.

1. J. A. Soderquist and E. I. Miranda, *Tetrahedron Lett.*, **34**, 4905 (1993).

Cyanomethyl Ester: RCO_2CH_2CN

Formation

1. $ClCH_2CN$, TEA, 78–96% yield.[1]
2.

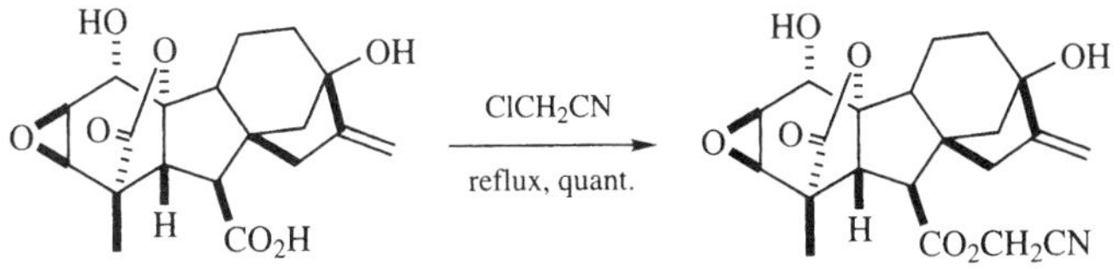

Ref. 2

Cleavage

Na$_2$S, acetone, water, 74–90% yield.[1]

1. H. M. Hugel, K. V. Bhaskar, and R. W. Longmore, *Synth. Commun.*, **22**, 693 (1992).
2. S. Findlow, P. Gaskin, P. A. Harrison, J. R. Lenton, M. Penny, and C. L. Willis, *J. Chem. Soc., Perkin Trans. I*, 751 (1997).

Acetol Ester: CH$_3$COCH$_2$O$_2$CR

Developed as an acid protecting group for peptide synthesis because of its stability to hydrogenolysis and acidic conditions, the acetol (hydroxy acetone) ester is prepared by DCC coupling (68–92% yield) of the acid with acetol. It is cleaved with TBAF in THF.[1]

1. B. Kundu, *Tetrahedron Lett.*, **33**, 3193 (1992).

Phenacyl Ester: RCOOCH$_2$COC$_6$H$_5$ (Chart 6)

Formation

1. PhCOCH$_2$Br, Et$_3$N, EtOAc, 20°C, 12 h, 83% yield.[1]
2. PhCOCH$_2$Br, KF/DMF, 25°C, 10 min, 90–99% yield.[2] Hindered acids are protected at 100°C.
3. From the K salt: PhCOCH$_2$Br, Bu$_4$NBr, CH$_3$CN, rt, dibenzo-18-crown-6, 86–98% yield.[3]

Cleavage

A phenacyl ester is much more readily cleaved by nucleophiles than are other esters such as the benzyl ester. Phenacyl esters are stable to acidic hydrolysis (e.g., concd. HCl[1]; HBr/HOAc[1]; 50% CF$_3$COOH/CH$_2$Cl$_2$[4]; HF, 0°C, 1 h[4]).

1. Zn/HOAc, 25°C, 1 h, 90% yield.[5]
2. Zn, acetylacetone, Pyr, DMF, 35°C, 0.6 h, 90–98% yield.[6]
3. Mg, MeOH, DMF, AcOH, 60–100 min. No racemization was observed for a variety of amino acids.[7]
4. H$_2$/Pd–C, aq. MeOH, 20°C, 1 h, 72% yield.[1]
5. PhSNa, DMF, 20°C, 30 min, 72% yield.[1]
6. CuCl$_2$, O$_2$, DMF, H$_2$O, 23–92% yield.[8]
7. Photolysis, sensitizer, CH$_3$CN, 2 h, 76–100% yield.[9,10] Irradiation of buffered solutions of *p*-hydroxyphenacyl esters releases the acid.[11]

8. PhSeH, DMF, rt, 48 h, 79% yield.[12] Under basic coupling conditions an aspartyl peptide that has a β-phenacyl ester is converted to a succinimide.[13] The use of PhSeH prevents the α,β-rearrangement of the aspartyl residue during deprotection.

9. TBAF, THF or DMSO or DMF, 72–98% yield. 4-Nitrobenzyl and trichloroethyl esters of amino acids are also cleaved.[14]

10. $(Bu_3Sn)_2O$ or Me_3SnOH, $ClCH_2CH_2Cl$, reflux 15–25 h, 45–100% yield. This method was used to cleave various BOC protected amino acids from polystyrene–phenacyl esters.[15]

1. G. C. Stelakatos, A. Paganou, and L. Zervas, *J. Chem. Soc. C*, 1191 (1966).

2. J. H. Clark and J. M. Miller, *Tetrahedron Lett.*, **18**, 599 (1977).

3. S. J. Jagdale, S. V. Patil, and M. M. Salunkhe, *Synth. Commun.*, **26**, 1747 (1996).

4. C. C. Yang and R. B. Merrifield, *J. Org. Chem.*, **41**, 1032 (1976).

5. J. B. Hendrickson and C. Kandall, *Tetrahedron Lett.*, **11**, 343 (1970).

6. D. Hagiwara, M. Neya, and M. Hashimoto, *Tetrahedron Lett.*, **31**, 6539 (1990).

7. S. Kokinaki, L. Leondiadis, and N. Ferderigos, *Org. Lett.*, **7**, 1723 (2005).

8. R. N. Ram and L. Singh, *Tetrahedron Lett.*, **36**, 5401 (1995).

9. A. Banerjee and D. E. Falvey, *J. Org. Chem.*, **62**, 6245 (1997). A. Banerjee and D. E. Falvey, *J. Am. Chem. Soc.*, **120**, 2965 (1998).

10. A. Banerjee, K. Lee, and D. E. Falvey, *Tetrahedron*, **55**, 12699 (1999); A. Banerjee, K. Lee, Q. Yu, A. G. Fang, and D. E. Falvey, *Tetrahedron Lett.*, **39**, 4635 (1998).

11. R. S. Givens, A. Jung, C.-H. Park, J. Weber, and W. Bartlett, *J. Am. Chem. Soc.*, **119**, 8369 (1997); R. S. Givens, J. F. W. Weber, P. G. Conrad, II, G. Orosz, S. L. Donahue, and S. A. Thayer, *J. Am. Chem. Soc.*, **122**, 2687 (2000).

12. J. L. Morell, P. Gaudreau, and E. Gross, *Int. J. Pept. Protein Res.*, **19**, 487 (1982).

13. M. Bodanszky and J. Martinez, *J. Org. Chem.*, **43**, 3071 (1978).

14. M. Namikoshi, B. Kundu, and K. L. Rinehart, *J. Org. Chem.*, **56**, 5464 (1991).

15. R. L. E. Furlan, E. G. Mata, and O. A. Mascaretti, *J. Chem. Soc. Perkin Trans. 1*, 355 (1998).

p-Bromophenacyl Ester: $RCOOCH_2COC_6H_4-p-Br$

In a penicillin synthesis, the carboxyl group was protected as a *p*-bromophenacyl ester that was cleaved by nucleophilic displacement (PhSK, DMF, 20°C, 30 min, 64% yield). Hydrogenolysis of a benzyl ester was difficult (perhaps because of catalyst poisoning by sulfur present in the penicillin); basic hydrolysis of methyl or ethyl esters led to attack at the β-lactam ring.[1] The phenacyl ester may also be cleaved by photolysis in the presence of 9,10-dimethylanthracene.[2]

1. P. Bamberg, B. Eckström, and B. Sjöberg, *Acta Chem. Scand.*, **21**, 2210 (1967).

2. A. Banerjee, K. Lee, and D. E. Falvey, *Tetrahedron*, **55**, 12699 (1999).

α-Methylphenacyl Ester: $RCO_2CH(CH_3)COC_6H_5$

***p*-Methoxyphenacyl Ester:** $RCO_2CH_2COC_6H_4$-*p*-OCH_3

3,4,5-Trimethoxyphenacyl Ester: $RCO_2CH_2COC_6H_2$-3,4,5-$(OCH_3)_3$

These phenacyl esters can be prepared from the phenacyl bromide, a carboxylic acid and potassium fluoride as base.[1] These phenacyl esters can be cleaved by irradiation (313 nm, dioxane or EtOH, 20°C, 6 h, 80–95% yield, R = amino acids;[2] >300 nm, 30°C, 8 h, R = a gibberellic acid, 36–62% yield[3]). The 3,4,5-trimethoxyphenacyl ester has been prepared and can be cleaved by irradiation at 350 nm.[4] Thioketal and ketal protected versions of this ester are photochemically stable until deprotected using conventional means. Another phenacyl derivative, $RCO_2CH(COC_6H_5)C_6H_3$-3,5-$(OCH_3)_2$, cleaved by irradiation, has also been reported.[5] It is stable during the photochemical cleavage of the 2-nitro-4,5-dimethoxybenzyl ester (cleaved at 420 nm).[6]

1. F. S. Tjoeng and G. A. Heavner, *Synthesis*, 897 (1981).
2. J. C. Sheehan and K. Umezawa, *J. Org. Chem.*, **38**, 3771 (1973).
3. E. P. Serebryakov, L. M. Suslova, and V. K. Kucherov, *Tetrahedron*, **34**, 345 (1978).
4. A. Shaginian, M. Patel, M.-H. Li, S. T. Flickinger, C. Kim, F. Cerrina, and P. J. Belshaw, *J. Am. Chem. Soc.*, **126**, 16704 (2004).
5. J. C. Sheehan, R. M. Wilson, and A. W. Oxford, *J. Am. Chem. Soc.*, **93**, 7222 (1971); Y. Shi, J. E. T. Corrie, and P. Wan, *J. Org. Chem.*, **62**, 8278 (1997).
6. C. G. Bochet, *Angew. Chem. Int. Ed.*, **40**, 2071 (2001).

2,5-Dimethylphenacyl (DMP) Ester

The DMP ester can be photochemically removed (>254 nm) without the presence of a sensitizer (51–95% yield).[1,2] The by-product from the reaction is an indanone. Quantum yields increase with increasing temperature.[3]

1. P. Klán, A. P. Pelliccioli, T. Pospisil, and J. Wirz, *Photochem. Photobiol. Sci.*, **1**, 920 (2002); R. Ruzicka, M. Zabada and P. P. Klán, *Synth. Commun.*, **32**, 2581 (2002).
2. M. Zabadal, A. P. Pelliccioli, P. Klan, and J. Wirz, *J. Phys. Chem. A*, **105**, 10329 (2001).
3. J. Literák, S. Relich, P. Kulhanek, and P. Klán, *Molecular Diversity*, **7**, 265 (2003)

Desyl Ester

Formation

Desyl bromide, DBU, benzene, reflux, 57–95% yield.[1] A polymer-supported version of this ester has been prepared.[2]

Cleavage

Photolysis at 350 nm, CH_3CN, H_2O. The by-product from the reaction is 2-phenylbenzo[*b*]furan. Cleavage with TBAF and $PhCH_2SH$ has been demonstrated (70–94% yield).[3] The related 3,5-dimethoxybenzoin analog is cleaved with a rate constant of $>10^{10}$ s^{-1}.[4] Photolytic cleavage occurs by heterolytic bond dissociation.[5,6]

1. K. R. Gee, L. W. Kueper III, J. Barnes, G. Dudley, and R. S. Givens, *J. Org. Chem.*, **61**, 1228 (1996).
2. A. Routledge, C. Abell, and S. Balasubramanian, *Tetrahedron Lett.*, **38**, 1227 (1997).
3. M. Ueki, H. Aoki, and T. Katoh, *Tetrahedron Lett.*, **34**, 2783 (1993).
4. M. H. B. Stowell, R. S. Rock, D. C. Rees, and S. I. Chan, *Tetrahedron Lett.*, **37**, 307 (1996).
5. Y. Shi, J. E. T. Corrie, and P. Wan, *J. Org. Chem.*, **62**, 8278 (1997).
6. New Photoprotecting Groups: R. S. Givens, J. F. W. Weber, A. H. Jung, C.-H. Park, "Desyl and *p*-Hydroxyphenyacyl Phosphate and Carboxylate Esters," in *Methods in Enzymology: Caged Compounds*, Vol. 291, G. Marriott, Ed., Academic Press, San Diego, 1998, pp. 1–29.

Carboxamidomethyl Ester (Cam Ester): $RCO_2CH_2CONH_2$

The carboxamidomethyl ester was prepared for use in peptide synthesis. It is formed from the cesium salt of an *N*-protected amino acid and α-chloroacetamide (60–85% yield). It is cleaved with 0.5 *M* NaOH or $NaHCO_3$ in DMF/H_2O. It is stable to the conditions required to remove BOC, Cbz, Fmoc, and *t*-butyl esters. It cannot be selectively cleaved in the presence of a benzyl ester of aspartic acid.[1]

1. J. Martinez, J. Laur, and B. Castro, *Tetrahedron*, **41**, 739 (1985); *idem.*, *Tetrahedron Lett.*, **24**, 5219 (1983); R. J. Bergeron, C. Ludin, R. Muller, R. E. Smith, and O. Phanstiel, IV, *J. Org. Chem.*, **62**, 3285 (1997).

p-Azobenzenecarboxamidomethyl Ester: $C_6H_5N=NC_6H_4NHC(O)CH_2O_2CR$

This ester was developed for C-terminal amino acids during solution phase peptide synthesis. Purification of intermediates can be monitored colorimetrically or visually. Protection is achieved by reacting the sodium salt of the *N*-protected amino acid with the bromoacetamide derivative to give the ester in 70–95% yield. Cleavage is affected by simple hydrolysis with K_2CO_3 or NH_4OH.[1] A related chromogenic ester, the p-(p-(dimethylamino)phenylazo)benzyl ester, has also been used for the same purpose, except that it can be cleaved by hydrogenolysis.[2]

1. V. G. Zhuravlev, A. A. Mazurov, and S. A. Andronati, *Collect. Czech. Chem. Commun.*, **57**, 1495 (1992).
2. G. D. Reynolds, D. R. K. Harding, and W. S. Hancock, *Int. J. Pept. Protein Res.*, **17**, 231 (1981).

6-Bromo-7-hydroxycoumarin-4-ylmethyl Ester

This group was developed for the photochemical release of bioactive messengers. They are introduced by displacement of the carboxylate on the chloromethyl derivative. Release is accomplished by a single or two-photon process, the latter allows for spatial resolution in tissue.[1]

1. T. Furuta, S. S.-H. Wang, J. L. Dantzker, T. M. Dore, W. J. Bybee, E. M. Callaway, W. Denk, and R. Y. Tsien, *Proc. Natl. Acad. Sci. USA*, **96**, 1193 (1999).

N-Phthalimidomethyl Ester (Chart 6)

Formation

RCO_2H + XCH_2-N-phthalimido

X = OH: Et_2NH, EtOAc, 37°C, 12 h, 70–80% yield.[1]
X = Cl: $(c\text{-}C_6H_{11})_2NH$, DMF or DMSO, 60°C, few minutes, 70–80% yield.[1]
X = Cl, Br: KF, DMF, 80°C, 2 h, 65–75% yield.[2]

Cleavage

1. H_2NNH_2/MeOH, 20°C, 3 h, 90% yield.[1]

2. $Et_2NH/MeOH$, H_2O, 25°C, 24 h or reflux, 2 h, 82% yield.[1]
3. $NaOH/MeOH$, H_2O, 20°C, 45 min, 77% yield.[1]
4. $Zn/HOAc$, 25°C, 12 h, 80% yield.[3]
5. gaseous $HCl/EtOAc$, 20°C, 16 h, 83% yield.[1]
6. $HBr/HOAc$, 20°C, 10–15 min, 80% yield.[1]

1. G. H. L. Nefkens, G. I. Tesser, and R. J. F. Nivard, *Recl. Trav. Chim. Pays-Bas*, **82**, 941 (1963).
2. K. Horiki, *Synth. Commun.*, **8**, 515 (1978).
3. D. L. Turner and E. Baczynski, *Chem. Ind. (London)*, 1204 (1970).

2-Substituted Ethyl Esters

2,2,2-Trichloroethyl Ester: $RCO_2CH_2CCl_3$ (Chart 6)

Upon reaction with $(Me_2N)_3P$ and an amine, the trichloro- and tribromoethyl esters give the amides, and reaction with an alcohol results in conversion to the esters in moderate yields.[1]

Formation

1. CCl_3CH_2OH, DCC, Pyr.[2]
2. CCl_3CH_2OH, TsOH, toluene, reflux.[2,3]
3. CCl_3CH_2OCOCl, THF, Pyr, >60% yield.[4]

Cleavage

1. Zn, AcOH, 0°C, 2.5 h.[2]
2. Zinc, THF buffered at pH 4.2–7.2 (20°C, 10 min, 75–95% yield).[5]
3. Zinc dust, 1 M NH_4OAc, 66% yield.[6]
4. Electrolysis: -1.65 V, $LiClO_4$, MeOH, 87–91% yield.[7] A tribromoethyl ester is cleaved by electrolytic reduction at -0.70 V (85% yield); a dichloroethyl ester is cleaved at -1.85 V (78% yield).[7]
5. Cat. Se, $NaBH_4$, DMF, 40–50°C, 1 h, 77–93% yield.[8]
6. Na_2Te from Te powder and $NaBH_4$, DMF, 74–98% yield.[9]
7. SmI_2, THF, rt, 2 h, quantitative.[10]
8. Cd, DMF, AcOH, 25°C, 15 h, 82% yield.[11]

1. J. J. Hans, R. W. Driver, and S. D. Burke, *J. Org. Chem.*, **65**, 2114 (2000).
2. R. B. Woodward, K. Heusler, J. Gosteli, P. Naegeli, W. Oppolzer, R. Ramage, S. Ranganathan, and H. Vorbrüggen, *J. Am. Chem. Soc.*, **88**, 852 (1966).
3. J. F. Carson, *Synthesis*, 24 (1979).

4. R. R. Chauvette, P. A. Pennington, C.W. Ryan, R. D. G. Cooper, F. L. José, I. G. Wright, E. M. Van Heyningen, and G. W. Huffman, *J. Org. Chem.*, **36**, 1259 (1971).

5. G. Just and K. Grozinger, *Synthesis*, 457 (1976).

6. G. Jou, I. Gonzalez, F. Albericio, P. Lloyd-Williams, and E. Giralt, *J. Org. Chem.*, **62**, 354 (1997).

7. M. F. Semmelhack and G. E. Heinsohn, *J. Am. Chem. Soc.*, **94**, 5139 (1972).

8. Z.-Z. Huang and X.-J. Zhou, *Synthesis*, 693 (1989).

9. G. Blay, L. Cardona, B. Garcia, C. L. Garcia, and J. R. Pedro, *Synth. Commum.*, **28**, 1405 (1998).

10. A. J. Pearson and K. Lee, *J. Org. Chem.*, **59**, 2304 (1994).

11. Y. Génisson, P. C. Tyler, and R. N. Young, *J. Am. Chem. Soc.*, **116**, 759 (1994).

2-Haloethyl Ester: $RCOOCH_2CH_2X$, X=I, Br, Cl (Chart 6)

Formation

1. $ClCH_2CH_2OH$, $Cl_3C_6H_2COCl$, TEA, DMAP, 77% yield.[1]
2. See general methods for ester formation since most of these will apply for this derivative.

Cleavage

2-Haloethyl esters have been cleaved under a variety of conditions many of which proceed by a nucleophilic process.

1. Li^+ or Na^+ Co(I)phthalocyanine/MeOH, 0–20°C, 40 min to 60 h, 60–98% yield.[2]
2. Electrolysis: Co(I)phthalocyanine, $LiClO_4$, EtOH, H_2O, −1.95 V, 95% yield.[3]
3. $NaS(CH_2)_2SNa/CH_3CN$, reflux, 2 h, 80–85% yield.[4]
4. $NaSeH$/EtOH, 25°C, 1 h to reflux, 6 min, 92–99% yield.[5,6]
5. $(NaS)_2CS/CH_3CN$, reflux, 1.5 h, 75–86% yield.[7]
6. Me_3SnLi/THF, 3 h then Bu_4NF, reflux, 15 min, 78–86% yield.[8]
7. $NaHTe$, EtOH, 2–60 min, 80–92% yield.[9]
8. Na_2S, 40–68% yield.[10]
9. Li(cobalt phthalocyanine).[11]
10. Cobalt phthalocyanine, $NaBH_4$[12]
11. SmI_2, THF, rt, 2 h, 88–100% yield.[13] These conditions were found effective when many of the above reagents failed to give clean deprotection.
12. Zn, N-methylimidazole, EtOAc, reflux, 1 h to 5 days, 54–80% yield. The advantage of this method is that azides, nitro groups and conjugated alkenes are not reduced, whereas using the standard Zn/AcOH conditions they are.[14]

1. W. R. Roush and G. C. Lane, *Org. Lett.*, **1**, 95 (1999).

2. H. Eckert and I. Ugi, *Angew Chem., Int. Ed. Engl.*, **15**, 681 (1976).

3. R. Scheffold and E. Amble, *Angew. Chem., Int. Ed. Engl.*, **19**, 629 (1980).

4. T.-L. Ho, *Synthesis*, 510 (1975).

5. T.-L. Ho, *Synth. Commun.*, **8**, 301 (1978).

6. Z.-Z. Huang and X.-J. Zhou, *Synthesis*, 633 (1990).

7. T.-L. Ho, *Synthesis*, 715 (1974).

8. T.-L. Ho, *Synth. Commun.*, **8**, 359 (1978).

9. J. Chen and X. Zhou, *Synth. Commun.*, **17**, 161 (1987).

10. M. Joaquina, S. A. Amaral Trigo, and M. I. A. Oliveira Sartos, in *Peptides 1986*, D. Theodoropoulos, Ed., Walter de Gruyter & Co., Berlin, 1987, p. 61.

11. P. Lemmen, K. M. Buchweitz, and R. Stumpf, *Chem. Phys. Lipids*, **53**, 65 (1990).

12. H. Eckert, Z. Naturforch., *B: Chem. Sci.*, **45**, 1715 (1990).

13. A. J. Pearson and K. Lee, *J. Org. Chem.*, **59**, 2257, 2304 (1994); *idem, ibid.*, **60**, 7153 (1995).

14. L. Somsak, K. Czifrák, and E. Veres, *Tetrahedron Lett.*, **45**, 9095 (2004).

ω-Chloroalkyl Ester: $RCOO(CH_2)_nCl$

ω-Chloroalkyl esters (n = 4, 5) have been cleaved by sodium sulfide (reflux, 4 h, 58–85% yield). The reaction proceeds by sulfide displacement of the chloride ion followed by intramolecular displacement of the carboxylate group by the (now) sulfhydryl group.[1]

1. T.-L. Ho and C. M. Wong, *Synth. Commun.*, **4**, 307 (1974).

2-(Trimethylsilyl)ethyl Ester (TMSE): $RCO_2CH_2CH_2Si(CH_3)_3$

Formation

1. $Me_3SiCH_2CH_2OH$, DCC, Pyr, CH_3CN, 0°C, 5–15 h, 66–97% yield.[1] In the presence of DMAP, this method can be used for the preparation of fairly hindered TMSE derivatives.[2]

2. From an acid chloride: $Me_3SiCH_2CH_2OH$, Pyr, 25°C, 3 h.[3]

3. $Me_3SiCH_2CH_2OH$, Me_3SiCl, THF, reflux, 12–36 h.[4] This method of esterification is also effective for the preparation of other esters.

4. From an anhydride: $Me_2AlOCH_2CH_2SiMe_3$, benzene, heat, >85% yield.[5]

5. $Me_3SiCH_2CH_2OH$, 2-chloro-1-methylpyridinium iodide, Et_3N, 90% yield.[6]

6. From a methyl ester: $Me_3SiCH_2CH_2OH$, $Ti(Oi\text{-}Pr)_4$, 120°C, 4 h, 85% yield.[7]

7. Me₃SiCH₂CH₂OH, EDC, DMAP, Pyr.[8]
8. Me₃SiCH₂CH₂OH, DEAD, Ph₃P, THF, >75% yield. [9]

Cleavage

1. Et₄NF or Bu₄NF, DMF or DMSO, 20–30°C, 5–60 min, quant. yield.[1,10]
2. DMF, Bu₄NCl, KF·2H₂O, 42–62% yield (substrate = polypeptide).[11]
3. DMF, NaH, rt, 82–92% yield. This method most likely proceeds by hydroxide produced by adventitious water, which is consistent with the fact that with the inclusion of molecular sieves the reaction fails to go to completion.[12]
4. TBAF, SiO₂, 100% yield[8] or TBAF, DMF, 20 min.[13] In the following case, TAS-F and other fluoride reagents proved ineffective.[14] It is likely that the more acidic reagents cause N to O migration in the threonine fragment.

5. TBAF, TsOH, THF, 20°C. Other conditions in this sensitive Ivermectin analog led to decomposition.[7]
6. Tris(dimethylamino)sulfonium difluorotrimethylsilicate (TAS-F), DMF, >76% yield.[9] This method was effective where TBAF caused elimination of a β-acetoxyester.[15]

(2-Methyl-2-trimethylsilyl)ethyl (Tms) Ester: TMSCH(Me)CH$_2$O$_2$CR

The ester was prepared from and amino acid and the alcohol using DCC/DMAP. It was developed to prevent diketopiperazine formation during the formation and deprotection at the dipeptide stage of the growing peptide. It is cleaved with TBAF at approximately half the rate of TMSE cleavage.[16]

(2-Phenyl-2-trimethylsilyl)ethyl (PTMSE) Esters: TMSCH(Ph)CH$_2$O$_2$CR

The PTMSE group is introduced via the "Steglich esterification" using DCC and DMAP (57–91% yield). It can be cleaved with TBAF in CH$_2$Cl$_2$, which are milder conditions than when DMF is used as the solvent. In general, its cleavage is significantly faster than the TMSE group. TFA will cleave the PTMSE group, but a BOC group can be cleaved in its presence with either PTSA·H$_2$O (Et$_2$O, EtOH, 65°C, 30 min, 72% yield) or with 1.2 N HCl, CF$_3$CH$_2$OH, rt, 40 min, 83% yield).[17]

1. P. Sieber, *Helv. Chim. Acta*, **60**, 2711 (1977).

2. T. G. Back and J. E. Wulff, *Angew. Chem. Int. Ed.*, **43**, 6493 (2004).

3. H. Gerlach, *Helv. Chim. Acta*, **60**, 3039 (1977).

4. M. A. Brook and T. H. Chan, *Synthesis*, 201 (1983).

5. E. Vedejs and S. D. Larsen, *J. Am. Chem. Soc.*, **106**, 3030 (1984).

6. J. D. White and L. R. Jayasinghe, *Tetrahedron Lett.*, **29**, 2139 (1988).

7. J.-P. Férézou, M. Julia, Y. Li, L. W. Liu, and A. Pancrazi, *Bull. Soc. Chim. Fr.*, **132**, 428 (1995).

8. A. M. Sefler, M. C. Kozlowski, T. Guo, and P. A. Bartlett, *J. Org. Chem.*, **62**, 93 (1997).

9. W. R. Roush, D. S. Coffey, and D. J. Madar, *J. Am. Chem. Soc.*, **119**, 11331 (1997).

10. P. Sieber, R. H. Andreatta, K. Eisler, B. Kamber, B. Riniker, and H. Rink, *Peptides: Proceedings of the Fifth American Peptide Symposium*, M. Goodman and J. Meienhofer, Eds., Halsted Press, New York, 1977, pp. 543–545.

11. R. A. Forsch and A. Rosowsky, *J. Org. Chem.*, **49**, 1305 (1984).

12. M. H. Serrano-Wu, A. Regueiro-Ren, D. R. St.Laurent, T. M. Carroll, and B. N. Balasubramanian, *Tetrahedron Lett.*, **42**, 8593 (2001).

13. C. K. Marlowe, *Bioorg. Med. Chem. Lett.*, **3**, 437 (1993).

14. T. Hu and J. S. Panek, *J. Am. Chem. Soc.*, **124**, 11368 (2002).

15. K. A. Scheidt, H. Chen, B. C. Follows, S. R. Chemler, D. S. Coffey, and W. R. Roush, *J. Org. Chem.*, **63**, 6436 (1998).

16. K. Borsuk, F. L. van Delft, I. F. Eggen, P. B. W. ten Kortenaar, A. Petersen, and F. P. J. T. Rutjes, *Tetrahedron Lett.*, **45**, 3585 (2004).

17. M. Wagner and H. Kunz, *Synlett*, 400 (2000).

2-Methylthioethyl Ester: RCO$_2$CH$_2$CH$_2$SCH$_3$

The 2-methylthioethyl ester is prepared from a carboxylic acid and methylthioethyl alcohol or methylthioethyl chloride (MeSCH$_2$CH$_2$OH, TsOH, benzene, reflux, 55 h,

55% yield; $MeSCH_2CH_2Cl$, Et_3N, 65°C, 12 h, 50–70% yield).[1] It is cleaved by oxidation [H_2O_2, $(NH_4)_6Mo_7O_{24}$, acetone, 25°C, 2 h, 80–95% yield → pH 10–11, 25°C, 12–24 h, 85–95% yield][2,3] and by alkylation followed by hydrolysis (MeI, 70–95% yield → pH 10, 5–10 min, 70–95% yield).[1]

1. M. J. S. A. Amaral, G. C. Barrett, H. N. Rydon, and J. E. Willet, *J. Chem. Soc. C*, 807 (1966).
2. P. M. Hardy, H. N. Rydon, and R. C. Thompson, *Tetrahedron Lett.*, **9**, 2525 (1968).
3. S. Inoue, K. Okada, H. Tanino, K. Hashizume, and H. Kakoi, *Tetrahedron*, **50**, 2729 (1994).

1,3-Dithianyl-2-methyl Ester (Dim Ester):

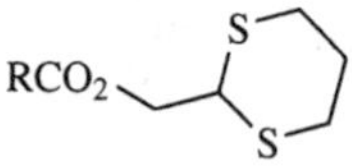

The Dim ester was developed for the protection of the carboxyl function during peptide synthesis. It is prepared by transesterification of amino acid methyl esters with 2-(hydroxymethyl)-1,3-dithiane and $(i\text{-PrO})_3Al$ (reflux, 4 h, 75°C, 12 torr, 75% yield). It is removed by oxidation [H_2O_2, $(NH_4)_2MoO_4$; pH 8, H_2O, 60 min, 83% yield]. Since it must be removed by oxidation, it is not compatible with sulfur-containing amino acids such as cysteine and methionine. It may also be cleaved electrochemically (CH_3CN, aq. AcONa, 65–74% yield).[1] Its suitability for other, easily oxidized amino acids (e.g., tyrosine and tryptophan) must also be questioned. It is stable to CF_3CO_2H and HCl/ether and thus is compatible with the BOC group.[2,3]

1. L. A. Barnhurst, Y. Wan, and A. G. Kutateladze, *Org. Lett.*, **2**, 799 (2000).
2. H. Kunz and H. Waldmann, *Angew. Chem., Int. Ed. Engl.*, **22**, 62 (1983).
3. H. Waldmann and H. Kunz, *J. Org. Chem.*, **53**, 4172 (1988).

2-(*p*-Nitrophenylthio)ethyl Ester: $RCO_2CH_2CH_2SC_6H_4\text{-}p\text{-}NO_2$

This ester is similar to the 2-methylthioethyl ester in that it is prepared from 2-(*p*-nitrophenylthio)ethanol and cleaved by oxidation [H_2O_2, $(NH_4)_6Mo_7O_{24}$].[1] Treatment with base then releases the acid by and E-2 process.

1. M. J. S. A. *Amaral, J. Chem. Soc. C*, 2495 (1969).

2-(*p*-Toluenesulfonyl)ethyl Ester (Tse Ester): $RCO_2CH_2CH_2SO_2C_6H_4\text{-}p\text{-}CH_3$
(Chart 6)

Formation

$TsCH_2CH_2OH$, DCC, Pyr, 0°C, 1 h to 20°C, 16 h, 70–90% yield.[1] Water-soluble carbodiimide can also be used effectively for this esterification.[2]

Cleavage

1. Na_2CO_3, dioxane, H_2O, 20°C, 2 h, 95% yield.[1]
2. 1 N NaOH, dioxane, H_2O, 20°C, 3 min, 60–95% yield.[1]
3. KCN, dioxane, H_2O, 20°C, 2.5 h, 60–85% yield.[1]
4. DBN, benzene, 25°C, quant.[3]
5. DBU, benzene, 11 h, 100% yield.[4]

6. Bu_4NF, THF, 0°C, 1 h, 52–95% yield.[5] A primary alcohol protected as the *t*-butyldimethylsilyl ether is cleaved under these conditions, but a similarly protected secondary alcohol was stable.

1. A. W. Miller and C. J. M. Stirling, *J. Chem. Soc. C*, 2612 (1968).
2. T. Ueda, F. Feng, R. Sadamoto, K. Niikura, K. Monde, and S.-i. Nishimura, *Org. Lett.*, **6**, 1753 (2004).
3. E. W. Colvin, T. A. Purcell, and R. A. Raphael, *J. Chem. Soc., Chem. Commun.*, 1031 (1972). G. V. M. Sharma and C. C. Mouli, *Tetrahedron Lett.*, **44**, 8161 (2003).
4. H. Tsutsui and O. Mitsunobo, *Tetrahedron Lett.*, **25**, 2163 (1984).
5. H. Tsutsui, M. Muto, K. Motoyoshi, and O. Mitsunobo, *Chem. Lett.*, **16**, 1595 (1987).

2-(2′-Pyridyl)ethyl Ester (Pet Ester): $RCO_2CH_2CH_2$-2-C_5H_4N

The Pet ester is stable to (a) the acidic conditions required to remove the BOC and *t*-butyl ester groups, (b) the basic conditions required to remove the Fmoc and Fm groups, and (c) hydrogenolysis. It is not recommended for use in peptides that contain methionine or histidine since these are susceptible to alkylation with methyl iodide.

Formation

1. DCC, HOBt, $HOCH_2CH_2$-2-C_5H_4N, 0°C to rt, CH_2Cl_2 or DMF, overnight, 50–92% yield.[1,2]
2. DCC, DMAP, $HOCH_2CH_2$-2-C_5H_4N, CH_2Cl_2, 61–92% yield.[3]
3. The related 2-(4′-pyridyl)ethyl ester has also been prepared from the acid chloride and the alcohol.[4]

Cleavage

MeI, CH_3CN; morpholine or diethylamine, methanol, 76–95% yield.[1,3] These conditions also cleave the 4′-pyridyl derivative.[4]

1. H. Kessler, G. Becker, H. Kogler, and M. Wolff, *Tetrahedron Lett.*, **25**, 3971 (1984).
2. H. Kessler, G. Becker, H. Kogler, J. Friesse, and R. Kerssebaum, *Int. J. Pept. Protein Res.*, **28**, 342 (1986).
3. H. Kunz and M. Kneip, *Angew. Chem., Int. Ed. Engl.*, **23**, 716 (1984).
4. A. R. Katritsky, G. R. Khan, and O. A. Schwarz, *Tetrahedron Lett.*, **25**, 1223 (1984).

2-(Diphenylphosphino)ethyl Ester (Dppe Ester): $(C_6H_5)_2PCH_2CH_2O_2CR$

The Dppe group was developed for carboxyl protection in peptide synthesis. It is formed from an *N*-protected amino acid and the alcohol (DCC, DMAP, 3–12 h, 0°C, rt). It is most efficiently cleaved by quaternization with MeI followed by treatment with fluoride ion or K_2CO_3. The ester is stable to HBr/AcOH, $BF_3·Et_2O$, and CF_3CO_2H.[1]

1. D. Chantreux, J.-P. Gamet, R. Jacquier, and J. Verducci, *Tetrahedron*, **40**, 3087 (1984).

(*p*-Methoxyphenyl)ethyl Ester: $CH_3OC_6H_4CH_2CH_2O_2CR$

Formation of the ester proceeds under standard DCC coupling conditions (DMAP, THF, 28–93%), and it is cleaved with 1% TFA or dichloroacetic acid in CH_2Cl_2 by DDQ (reflux, CH_2Cl_2, H_2O, 5–15 h, 47–92% yield).[2] Hydrogenolysis (Pd/C, EtOAc, MeOH) cleaves the ester in 23 h, whereas a benzyl ester is cleaved in 10 min under these conditions.

1. M. S. Bernatowicz, H.-G. Chao, and G. R. Matsueda, *Tetrahedron Lett.*, **35**, 1651 (1994).
2. S.-E. Yoo, H. R. Kim, and K. Y. Yi, *Tetrahedron Lett.*, **31**, 5913 (1990).

1-Methyl-1-phenylethyl Ester (Cumyl Ester): $RCO_2C(CH_3)_2C_6H_5$

Formation

$C_6H_5C(CH_3)_2OC(=NH)CCl_3$, CH_2Cl_2, cHex, 78–98% yield.[1,2]

Cleavage

1. TFA/CH_2Cl_2, rt, 15 min, 86% yield. BOC and *t*-BuO groups were stable.[1,3]
2. Note that a cumyl ester can be selectively cleaved in the presence of a *t*-butyl ester and a β-lactam.[4]

1. C. Yue, J. Thierry, and P. Potier, *Tetrahedron Lett.*, **34**, 323 (1993).
2. J. Thierry, C. Yue, and P. Potier, *Tetrahedron Lett.*, **39**, 1557 (1998).
3. I. Hamachi, S. Kiyonaka, and S. Shinkai, *Chem. Commun.*, 1281 (2000).
4. D. M. Brunwin and G. Lowe, *J. Chem. Soc., Perkin Trans. I*, 1321 (1973).

2-(4-Acetyl-2-nitrophenyl)ethyl Ester (Anpe-)

This ester was designed as a base-labile protecting group. Monoprotection of aspartic acid was achieved using the DCC/DMAP protocol. Cleavage is promoted with 0.1 M TBAF. A comparison of other base-labile esters for the β-carboxyl group of aspartic acid to 0.1 M TBAF is provided in the table.[1]

Relative Lability of Aspartic Acid β-Carboxyl Protective Groups

Carboxyl Protective Group	Abbreviation	Deprotection Time
	Npe	1.5–2 h
	Cne	45 min
	Fm	<5 min
	Anpe	<5 min
$O_2NCH_2CH_2-$	Ne	a
	Dnpe	a

aNot prepared because of a lack of stability.

1. J. Robles, E. Pedroso, and A. Grandas, *Synthesis*, 1261 (1993).

1-[2-(2-Hydroxyalkyl)phenyl]ethanone (HAPE)

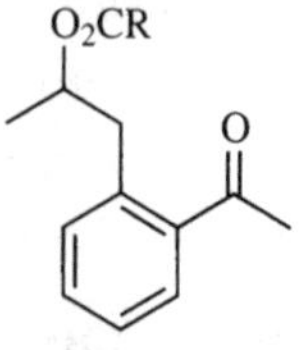

The HAPE group is introduced from the ketal protected alcohol using DCC/DMAP. The ketal is then hydrolyzed with PTSA or wet silica gel/oxalic acid. Cleavage is carried out by irradiation in CH_3CN through a Pyrex filter in the absence of oxygen for 3–6 h to afford the acid in 56–82% yield.[1]

1. W. N. Atemnkeng, L. D. Louisiana, II, P. K. Yong, B. Vottero, and A. Banerjee, *Org. Lett.*, **5**, 4469 (2003).

2-Cyanoethyl Ester: $NCCH_2CH_2O_2CR$

Formation

$HOCH_2CH_2CN$, DCC, DMAP, CH_2Cl_2, 86–97% yield.[1]

Cleavage

1. TBAF, DMF/THF, 64–100% yield. Cleavage occurs in the presence of TMSE and benzyl esters and acetates.[1]
2. K_2CO_3, MeOH, H_2O.[2] Acetates and most other simple esters are cleaved under these conditions.
3. Na_2S, MeOH, 67–91% yield.[3]

1. Y. Kita, H. Maeda, F. Takahashi, S. Fukui, T. Ogawa, and K. Hatayama, *Chem. Pharm. Bull.*, **42**, 147 (1994).
2. P. K. Misra, S. A. N. Hashmi, W. Haq, and S. B. Katti, *Tetrahedron Lett.*, **30**, 3569 (1989).
3. T. Ogawa, K. Hatayama, H. Maeda, and Y. Kita, *Chem. Pharm. Bull.*, **42**, 1579 (1994).

t-Butyl Ester: $RCO_2C(CH_3)_3$ (Chart 6)

Formation

The *t*-butyl ester is a relatively hindered ester and many of the methods reported below should be, and in many cases are, equally effective for the preparation of other hindered esters. The related 1- and 2-adamantyl esters have been used for the protection of aspartic acid[1] and other amino acids (1-AdOH, toluene, dimethyl sulfate, cat.

TsOH, 70–80% yield).[2] The *t*-butyl ester is much less susceptible to nucleophilic additions than is the methyl ester. A fluorous version of this ester [($C_6F_{13}CH_2CH_2)_2$ CH_3C-O_2CR] has been developed for use in fluorous-based synthesis.[3]

1. Isobutylene, concd. H_2SO_4, Et_2O, 25°C, 2–24 h, 50–60% yield.[4] This method works for the preparation of *t*-Bu esters of alkyl acids, amino acids[5,6] and penicillins.[7]

2. Isobutylene, CH_2Cl_2, H_3PO_4 (P_2O_5), BF_3·Et_2O, −78°C, 2 h to 0°C, 24 h.[8]

3. *t*-BuOH, H_2SO_4, $MgSO_4$, CH_2Cl_2, 54–93% yield. These conditions can also be used to prepare *t*-Bu ethers.[9]

4. $(COCl)_2$, benzene, DMF, 7–10°C, 45 min; *t*-BuOH, Et_3N, CH_2Cl_2, 0°C, 3 h, 75% yield.[10]

5. From an aromatic acid chloride: LiO-*t*-Bu, 25°C, 15 h, 79–82% yield.[11]

6. 2,4,6-$Cl_3C_6H_2COCl$, Et_3N, THF; *t*-BuOH, DMAP, benzene, 25°C, 20 min, 90% yield.[12]

7. *t*-BuOH, Pyr, $(Me_2N)(Cl)C=N^+Me_2Cl^-$, 77% yield.[13] This method is also effective for the preparation of other esters.

8. $(Im)_2CO$ (*N,N'*-carbonyldiimidazole), *t*-BuOH, DBU, 54–91% yield.[14]

9. Bu_3PI_2, Et_2O, HMPA; *t*-BuOH, 73% yield.[15]

10. *t*-BuOH, EDCI (EDCI = 1-ethyl-3-[3-(dimethylamino)propyl]carbodiimide hydrochloride, DMAP, CH_2Cl_2, 88% yield.[16] Cbz-Proline was protected without racemization.

11. *i*-PrN=C(O-*t*-Bu)NH-*i*-Pr, toluene, 60°C, 4 h, 90% yield.[17]

12. $Cl_3C(t$-BuO)C=NH, BF_3·Et_2O, CH_2Cl_2, cyclohexane, 70–92% yield.[18] This reagent also forms *t*-butyl ethers from alcohols.

13. $(t$-BuO)_2CHNMe_2$, toluene, 80°C, 30 min, 82% yield.[19,20]

14. From an acid chloride: *t*-BuOH, AgCN, benzene, 20–80°C, 60–100% yield.[21] Alumina also promotes the conversion of an acid chloride to a *t*-Bu ester in 79–96 yield.[22]

15. 2-Cl-3,5-$(NO_2)_2C_5H_2N$, Pyr, rt → 115°C, *t*-BuOH.[23] Other esters are also prepared effectively using this methodology.

16. *t*-BuOCOF, Et_3N, DMAP, CH_2Cl_2, *t*-BuOH, rt, 82–96% yield.[24]

17. $(BOC)_2O$, t-BuOH or THF, DMAP, 99% yield. This methodology is effective for the preparation of allyl, methyl, ethyl, and benzyl esters as well.[25]

18. t-BuBr, K_2CO_3, BTEAC, DMAC, 55°C, 72–100% yield.[26]

19.

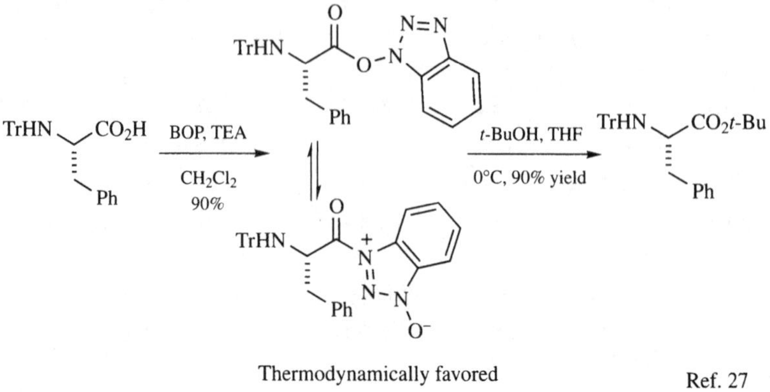

Thermodynamically favored

Ref. 27

20. For acids with α-electron withdrawing groups: t-BuOH, DCC, 60–100% yield. The reaction proceeds through a ketene intermediate. Other sterically hindered alcohols effectively give esters by this method.[28]

21. The section on transesterification should be consulted since this method is applicable to the preparation of t-Bu esters from other esters. For example: by transesterification of a methyl ester with t-BuOH and sulfated SiO_2.[29]

Cleavage

t-Butyl esters are stable to mild basic hydrolysis, to hydrazine and to ammonia. They are cleaved by moderately acidic hydrolysis with the release of isobutylene or the t-Bu cation that often must be scavenged to prevent side reactions.

1. HCO_2H, 20°C, 3 h.[30]

2. CF_3COOH, CH_2Cl_2, 25°C, 1 h.[31] The addition of Et_3SiH to the deprotection step improves the yields over the use of the normal cation scavengers.[32]

3. CF_3COOH, thioanisole, 93% yield. In this case the thioanisole was essential for the cleavage.[33]

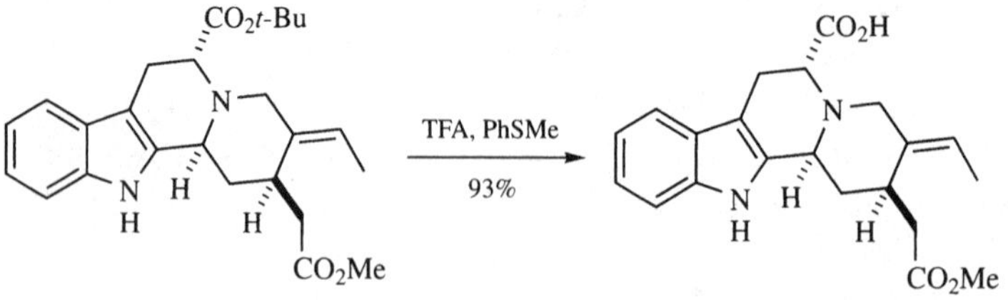

Phenol[34] and 1,3-dimethoxybenzene[35] have also been used as cation scavengers. The use of these cation scavengers is necessary in the presence of very electron-rich aromatics.

4. Montmorillonite KSF clay, reflux, CH₃CN.[36] In this case, an N-BOC group is retained. In other cases, *t*-Bu esters are somewhat more stable to acid than are N-BOC derivatives.[37]

5. AcOH, HBr, 10°C, 10 min, 70% yield.[5] Phthaloyl or trifluoroacetyl groups on amino acids are stable to these conditions; benzyloxycarbonyl (Cbz) or *t*-butoxycarbonyl (BOC) groups are cleaved.

6. HCl, AcOH, CH₂Cl₂, 5°C, 2 h. A *t*-butyl ether and an Fmoc group were not affected.[38]

7. TsOH, benzene, reflux, 30 min, 76% yield.[5] A *t*-butyl ester is stable to the conditions needed to convert an α,β-unsaturated ketone to a dioxolane (HOCH₂CH₂OH, TsOH, benzene, reflux).[39] TsOH with microwave heating has also been used on a few trivial esters.[40]

8. H₂SO₄, CH₂Cl₂, rt, 6 h, 89–98% yield.[41] The method also cleaves BOC and adamantyl groups.

9. HNO₃, CH₂Cl₂, 0°C, 92–99% yield. These conditions were shown to be substantially faster than the use of trifluoroacetic acid which is one of the more commonly used reagents.[42]

10. SiO₂, toluene, reflux, 53–94% yield. Phenolic *t*-Bu ethers are cleaved but more slowly.[43]

11. KOH, 18-crown-6, toluene, 100°C, 5 h, 94% yield.[44] These conditions were used to cleave the *t*-butyl ester from an aromatic ester; they are probably too harsh to be used on more highly functionalized substrates.

12. 50% aq. NaOH, benzyltriethylammonium chloride, CH₂Cl₂, 90–98% yield. This method was selective for (E)-glycinates over (Z)-glycinates.[45]

13. 2 eq. of *t*-BuOK, THF, 0°C, 35–100% yield.[46]

14. NaH, DMF, 2–24 h, rt or 70°C, 60–87% yield. These reagents form Me₂NNa by decomposition of DMF.[47] The liberation of H₂ and CO could be a problem on scale.

15. 190–200°C, 15 min, 100% yield.[48] A thermolysis in quinoline was found advantageous when acid-catalyzed cleavage resulted in partial debenzylation of a phenol.[49] Thermolytic conditions also cleave the BOC group from amines. In the following case the furan was anticipated not to be stable to strong acid.[50]

16. Bromocatecholborane.[51] Ethyl esters are not affected by this reagent, but it does cleave other groups; see the section on methoxymethyl (MOM) ethers.

17. TMSOTf, TEA, 53–90% yield. *t*-Butyl esters are cleaved in preference to *t*-butyl ethers.[52] The somewhat less reactive TESOTf has been used when more moderate conditions are required.[53]

18. TBDMSOTf, 2,6-lutidine, CH$_2$Cl$_2$, rt, 93% yield. In this the *t*-butyl ester is converted to a TBDMS ester.[54]

19. Yb(OTf)$_3$, CH$_3$NO$_2$, 50°C, 80–98% yield. N-BOC groups and phenolic *t*-Bu ethers are also cleaved.[55]

20. MgI$_2$, toluene, 46–111°C, 1–3 days, 41–96% yield.[56]

21. ZnBr$_2$, CH$_2$Cl$_2$, rt, 2–24 h, 62–93% yield. *t*-Bu ethers are also cleaved but more slowly.[57] Allyl esters and PMB groups are unaffected.

22. CeCl$_3$·7H$_2$O, NaI, CH$_3$CN, reflux, 1–6 h, 75–99% yield. N-BOC groups are stable to these conditions.[58]

23. Thermitase, pH 7.5, 45°C, 20% DMF, 70–89% yield.[59]

24. Esterase from *Bacillus subtilis* (BsubpNBE), 16–77% yield.[60]

25. Pig liver esterase.[61]

26. LiI, EtOAc, reflux.[62]

27. TiCl$_4$, CH$_2$Cl$_2$, −10°C to 0°C, 54–91% yield. These conditions were developed for use with cephalosporin *t*-butyl esters.[63]

28. Reduction to the aldehyde by DIBAL, CH$_2$Cl$_2$, −78°C then oxidation with NaClO$_2$, NaH$_2$PO$_4$, 2-methyl-2-butene, THF, H$_2$O, 86% yield.[64] **Do not mix NaClO$_2$ with strong acid because they react violently!**

1. Y. Okada and S. Iguchi, *J. Chem. Soc., Perkin Trans. I*, 2129 (1988).

2. S. M. Iossifidou and C. C. Froussios, *Synthesis*, 1355 (1996).

3. J. Pardo, A. Cobas, E. Guitian, and L. Castedo, *Org. Lett.*, **3**, 3711 (2001).

4. A. L. McCloskey, G. S. Fonken, R. W. Kluiber, and W. S. Johnson, *Org. Synth., Coll. Vol. IV*, 261 (1963).

5. G. W. Anderson and F. M. Callahan, *J. Am. Chem. Soc.*, **82**, 3359 (1960).

6. R. M. Valerio, P. F. Alewood, and R. B. Johns, *Synthesis*, 786 (1988).

7. R. J. Stedman, *J. Med. Chem.*, **9**, 444 (1966).

8. C.-Q. Han, D. DiTullio, Y.-F. Wang, and C. J. Sih, *J. Org. Chem.*, **51**, 1253 (1986).

9. S. W. Wright, D. L. Hageman, A. S. Wright, and L. D. McClure, *Tetrahedron Lett.*, **38**, 7345 (1997).

10. C. F. Murphy and R. E. Koehler, *J. Org. Chem.*, **35**, 2429 (1970).

11. G. P. Crowther, E. M. Kaiser, R. A. Woodruff, and C. R. Hauser, *Org. Synth., Coll. Vol. VI*, 259 (1988).

12. J. Inanaga, K. Hirata, H. Saeki, T. Katsuki, and M. Yamaguchi, *Bull. Chem. Soc. Jpn.*, **52**, 1989 (1979).

13. T. Fujisawa, T. Mori, K. Fukumoto, and T. Sato, *Chem. Lett.*, **11**, 1891 (1982).

14. S. Ohta, A. Shimabayashi, M. Aona, and M. Okamoto, *Synthesis*, 833 (1982).

15. R. K. Haynes and M. Holden, *Aust. J. Chem.*, **35**, 517 (1982).

16. M. K. Dhaon, R. K. Olsen, and K. Ramasamy, *J. Org. Chem.*, **47**, 1962 (1982).

17. R. M. Burk, G. D. Berger, R. L. Bugianesi, N. N. Girotra, W. H. Parsons, and M. M. Ponpipom, *Tetrahedron Lett.*, **34**, 975 (1993); S. C. Bergmeier, A. A. Cobas, and H. Rapoport, *J. Org. Chem.*, **58**, 2369 (1993).

18. A. Armstrong, I. Brackenridge, R.F. W. Jackson, and J. M. Kirk, *Tetrahedron Lett.*, **29**, 2483 (1988). R. N. Atkinson, L. Moore, J. Tobin, and S. B. King, *J. Org. Chem.*, **64**, 3467 (1999).

19. U. Widmer, *Synthesis*, 135 (1983).

20. J. Deng, Y. Hamada and T. Shioiri, *J. Am. Chem. Soc.*, **117**, 7824 (1995).

21. S. Takimoto, J. Inanaga, T. Katsuki, and M. Yamaguchi, *Bull. Chem. Soc. Jpn.*, **49**, 2335 (1976).

22. K. Nagasawa, S. Yoshitake, T. Amiya, and K. Ito, *Synth. Commun.*, **20**, 2033 (1990); K. Nagasawa, K. Ohhashi, A. Yamashita, and K. Ito, *Chem. Lett.*, **23**, 209 (1994).

23. S. Takimoto, N. Abe, Y. Kodera, and H. Ohta, *Bull. Chem. Soc. Jpn.*, **56**, 639 (1983).

24. A. Loffet, N. Galeotti, P. Jouin, and B. Castro, *Tetrahedron Lett.*, **30**, 6859 (1989).

25. K. Takeda, A. Akiyama, H. Nakamura, S.-i. Takizawa, Y. Mizuno, H. Takayanagi, and Y. Harigaya, *Synthesis*, 1063 (1994).

26. P. Chevallet, P. Garrouste, B. Malawska, and J. Martinez, *Tetrahedron Lett.*, **34**, 7409 (1993).

27. K. M. Sliedregt, A. Schouten, J. Kroon, and R. M. J. Liskamp, *Tetrahedron Lett.*, **37**, 4237 (1996).

28. M. Nahmany and A. Melman, *Org. Lett.*, **3**, 3733 (2001).

29. S. P. Chavan, P. K. Zubaidha, S. W. Dantale, A. Keshavaraja, A. V. Ramaswamy, and T. Ravindranathan, *Tetrahedron Lett.*, **37**, 233 (1996).

30. S. Chandrasekaran, A. F. Kluge, and J. A. Edwards, *J. Org. Chem.*, **42**, 3972 (1977).

31. D. B. Bryan, R. F. Hall, K. G. Holden, W. F. Huffman, and J. G. Gleason, *J. Am. Chem. Soc.*, **99**, 2353 (1977).

32. A. Mehta, R. Jaouhari, T. J. Benson, and K. T. Douglas, *Tetrahedron Lett.*, **33**, 5441 (1992).

33. S. F. Martin, K. X. Chen, and C. T. Eary, *Org. Lett.*, **1**, 79 (1999).

34. S. Torii, H. Tanaka, M. Taniguchi, Y. Kameyama, M. Sasaoka, T. Shiroi, R. Kikuchi, I. Kawahara, A. Shimabayashi, and S. Nagao, *J. Org. Chem.*, **56**, 3633 (1991).

35. U. Schmidt, A. Lieberknecht, H. Bökens, and H. Griesser, *J. Org. Chem.*, **48**, 2680 (1983).

36. J. S. Yadav, B. V. S. Reddy, K. S. Rao, and K. Harikishan, *Synlett*, 826 (2002).

37. Y. Zou, N. E. Fahmi, C. Vialas, G. M. Miller, and S. M. Hecht, *J. Am. Chem. Soc.*, **124**, 9476 (2002).

38. G. M. Makara and G. R. Marshall, *Tetrahedron Lett.*, **38**, 5069 (1997).

39. A. Martel, T. W. Doyle, and B.-Y. Luh, *Can. J. Chem.*, **57**, 614 (1979).

40. J. C. Lee, E. S. Yoo, and J. S. Lee, *Synth. Commun.*, **34**, 3017 (2004).

41. P. Strazzolini, N. Misuri, and P. Polese, *Tetrahedron Lett.*, **46**, 2075 (2005).

42. P. Strazzolini, M. G. Dall'Arche, and A. G. Giumanini, Tetrahedron Lett., **39**, 9255 (1998). P. Strazzolini, M. Scuccato, and A. G. Giumanini, *Tetrahedron*, **56**, 3625 (2000).

43. R. W. Jackson, *Tetrahedron Lett.*, **42**, 5163 (2001).

44. C. J. Pedersen, *J. Am. Chem. Soc.*, **89**, 7017 (1967).

45. A. Jonczyk and T. Zomerfeld, *Tetrahedron Lett.*, **44**, 2359 (2003).

46. V. Alezra, C. Bouchet, L. Micouin, M. Bonin, and H.-P. Husson, *Tetrahedron Lett.*, **41**, 655 (2000).

47. S. Paul and R. R. Schmidt, *Synlett*, 1107 (2002).

48. L. H. Klemm, E. P. Antoniades, and D. C. Lind, *J. Org. Chem.*, **27**, 519 (1962).

49. J. W. Lampe, P. F. Hughes, C. K. Biggers, S. H. Smith, and H. Hu, *J. Org. Chem.*, **61**, 4572 (1996).

50. J. A. Marshall, L. M. McNulty, and D. Zou, *J. Org. Chem.*, **64**, 5193 (1999).

51. R. K. Boeckman, Jr., and J. C. Potenza, *Tetrahedron Lett.*, **26**, 1411 (1985).

52. A. Trzeciak and W. Bannwarth, *Synthesis*, 1433 (1996).

53. M. Oikawa, T. Ueno, H. Oikawa, and A. Ichihara, *J. Org. Chem.* **60**, 5048 (1995).

54. D. Meng, P. Bertinato, A. Balog, D.-S. Su, T. Kamenecka, E. J. Sorensen, and S. J. Danishefsky, *J. Am. Chem. Soc.*, **119**, 10073 (1997).

55. P. R. Sridhar, S. Sinha, and S. Chandrasekaran, *Ind. J. Chem.*, **41B**, 157 (2002).

56. A. G. Martinez, J. O. Bardina, G. H. del Veccio, M. Hanack, and L. R. Subramanian, *Tetrahedron Lett.*, **32**, 5931 (1991).

57. Y.-q. Wu, D. C. Limburg, D. E. Wilkinson, M. J. Vaal, and G. S. Hamilton, *Tetrahedron Lett.*, **41**, 2847 (2000). R. Kaul, Y. Brouillette, Z. Sajjadi, K. A. Hansford, and W. D. Lubell, *J. Org. Chem.*, **69**, 6131 (2004).

58. E. Marcantoni, M. Massaccesi, and E. Torregiani, *J. Org. Chem.*, **66**, 4430 (2001).

59. M. Schultz, P. Hermann, and H. Kunz, *Synlett*, 37 (1992).

60. M. Schmidt, E. Barbayianni, I. Fotakopoulou, M. Hohne, V. Constantinou-Kokotou, U. T. Bornscheuer, and G. Kokotos, *J. Org. Chem.*, **70**, 3737 (2005).

61. K. A. Stein and P. L. Toogood, *J. Org. Chem.*, **60**, 8110 (1995).

62. J. W. Fisher and K. L. Trinkle, *Tetrahedron Lett.*, **35**, 2505 (1994).

63. M. Valencic, T. van der Does, and E. de Vroom, *Tetrahedron Lett.*, **39**, 1625 (1998).

64. E. B. Holson and W. R. Roush, *Org. Lett.*, **4**, 3719 (2002).

3-Methyl-3-pentyl Ester (Mpe$-$O$_2$CR): $(C_2H_5)_2CCH_3CO_2CR$

This tertiary ester was developed to reduce aspartimide and piperidide formation during the Fmoc-based peptide synthesis by increasing the steric bulk around the carboxyl carbon. A twofold improvement was achieved over the standard *t*-butyl ester. The ester is prepared from the acid chloride and the alcohol and can be cleaved under conditions similar to those used for the *t*-butyl ester.[1]

1. A. Karlström and A. Undén, *Tetrahedron Lett.*, **37**, 4243 (1996).

Dicyclopropylmethyl Ester (Dcpm—O₂CR)

The Dcpm group can be removed in the presence of *t*-butyl or *N*-trityl group with 1% TFA in CH_2Cl_2.[1]

1. L. A. Carpino, H.-G. Chao, S. Ghassemi, E. M. E. Mansour, C. Riemer, R. Warrass, D. Sadat-Aalaee, G. A. Truran, H. Imazumi, A. El-Faham, D. Ionescu, M. Ismail, T. L. Kowaleski , C. H. Han, H. Wenschuh, M. Beyermann, M. Bienert, H. Shroff, F. Albericio, S. A. Triolo, N. A. Sole, and S. A. Kates, *J. Org. Chem.*, **60**, 7718 (1995).

2,4-Dimethyl-3-pentyl Ester (Dmp—O₂CR): (*i*-Pr)₂CHO₂CR

This group reduces aspartimide formation during Fmoc-based peptide synthesis.

Formation

2,4-Dimethyl-3-pentanol, DCC, DMAP, CH_2Cl_2, 4 h. This group was developed as an improvement over cyclohexanol for aspartic acid protection during peptide synthesis.[1]

Cleavage

Cleavage is affected with acid. The following table compares the acidolysis rates with Bn and cyclohexyl esters in TFA/phenol at 43°C.

Protective Group	$t_{1/2}$ (h)
Bn	6
Dmp	40
cHeX	500

1. A. H. Karlström and A. E. Unden, *Tetrahedron Lett.*, **36**, 3909 (1995).

Cyclopentyl Ester: RCO_2-c-C_5H_9

Cyclohexyl Ester: RCO_2-c-C_6H_{11}

Cycloalkyl esters have been used to protect the β-CO_2H group in aspartyl peptides to minimize aspartimide formation during acidic or basic reactions.[1] Aspartimide formation is limited to 2–3% in TFA (20 h, 25°C), 5–7% with HF at 0°C, and 1.5–4% TfOH (thioanisole in TFA). Cycloalkyl esters are also stable to Et_3N, whereas use of the benzyl ester leads to 25% aspartimide formation during Et_3N treatment. Cycloalkyl esters are stable to CF_3COOH, but are readily cleaved with HF or TfOH.[2–4]

1. For an improved synthesis of cyclohexyl aspartate, see G. K. Toth and B. Penke, *Synthesis*, 361 (1992).

2. J. Blake, *Int. J. Pept. Protein Res.*, **13**, 418 (1979).

3. J. P. Tam, T.-W. Wong, M. W. Riemen, F.-S. Tjoeng, and R. B. Merrifield, *Tetrahedron Lett.*, **20**, 4033 (1979).

4. N. Fujii, M. Nomizu, S. Futaki, A. Otaka, S. Funakoshi, K. Akaji, K. Watanabe, and H. Yajima, *Chem. Pharm. Bull.*, **34**, 864 (1986).

Allyl Ester: $RCO_2CH_2CH=CH_2$

The use of various allyl protective groups in complex molecule synthesis has been reviewed.[1]

Formation

1. Allyl bromide, Aliquat 336, $NaHCO_3$, CH_2Cl_2, 83% yield.[2] The carboxylic acid group of Z-serine (Z = Cbz = benzyloxycarbonyl) is selectively esterified without affecting the alcohol.

2. $R'R''C=CHCH_2OH$, NaH, THF, 1–3 days, 80–95% yield.[3] A methyl ester is exchanged for an allyl ester under these conditions.

3. Allyl bromide, Cs_2CO_3, DMF, 84% yield.[4]

4. Allyl alcohol, TsOH, benzene, $-H_2O$.[5] These conditions were used to prepare esters of amino acids.

5. Allyl alcohol, TsOH, $CHCl_3$, reflux, inverse Dean Stark trap, 72–98% yield. The method was developed for β,γ-unsaturated esters.[6]

6. Allyl alcohol, $[Ir(cod)_2]BF_4$, toluene, 100°C, 5 h, 88–97% yield. This method can also be used to prepare allyl ethers and allyl amines.[7]

7. By transesterification of an ethyl ester: AllylOH, DBU, LiBr, 0°C, 12 h, >54% yield.[8]

8. $AllylOCO_2CO_2allyl$, THF, DMAP.[9]

9. DMAP, 81–100% yield.[10]

10. AllylOC=NH(CCl$_3$), BF$_3$·Et$_2$O, CH$_2$Cl$_2$, cyclohexane, 67–96% yield.[11]

11. Vinyldiazomethane, CH$_2$Cl$_2$, 80–92% yield.[12]

12. From the Oppolzer sultam by exchange: AllylOH, Ti(OR)$_4$, 67–95% yield.[13]

13. Transesterification of an ethyl ester: AllylOH, La(O−i-Pr)$_3$, 60°C, 6h, 67% yield.[14]

Cleavage

1. Pd(OAc)$_2$, sodium 2-methylhexanoate, Ph$_3$P, acetone.[15] Triethyl phosphite could be used as the ligand for palladium.[16]

2. (Ph$_3$P)$_3$RhCl or Pd(Ph$_3$P)$_4$, 70°C, EtOH, H$_2$O, 91% yield.[17]

3. Pd(Ph$_3$P)$_4$, pyrrolidine, 0°C, 5–15 min, CH$_3$CN, 70–90% yield.[18] Morpholine has also been used as an allyl scavenger in this process.[2,4] Allylamines are not affected by these conditions.[19]

4. PdCl$_2$(Ph$_3$P)$_2$, dimedone, THF, 95% yield.[20] This method is also effective for removing the allyloxycarbonyl group from alcohols and amines.

5. Pd(Ph$_3$P)$_4$, 2-ethylhexanoic acid[21] or barbituric acid (THF, 3h, 93% yield)[22] or a polymer supported version (80–100% yield).[23] These conditions are effective for other allyl-based protective groups. Tributylstannane can serve as an allyl scavenger.[24]

6. Me$_2$CuLi, Et$_2$O, 0°C, 1h; H$_3$O$^+$, 75–85% yield.[25]

7. PhSiH$_3$, Pd(Ph$_3$P)$_4$, CH$_2$Cl$_2$, 74–100% yield.[26] CF$_3$CON(SiMe$_3$)CH$_3$ was also used to scavenge the allyl group from the Alloc and allyl ether protected derivatives.

8. Pd(Ph$_3$P)$_4$, BnONH$_2$, CH$_2$Cl$_2$, 80% yield.[27]

9. Pd(OAc)$_2$, Ph$_3$P, TEA, HCO$_2$H, dioxane, 96% yield.[28,29]

10. Papain, dithiothreitol, DMF.[30]

11. TiCl$_4$, Mg-Hg, THF, 40–70% yield.[31] Benzyl esters are also cleaved.

12. Pd(Ph$_3$P)$_4$, RSO$_2$Na, CH$_2$Cl$_2$ or THF/MeOH, 70–99% yield. These conditions were shown to be superior to the use of sodium 2-ethylhexanoate. Methallyl, allyl, crotyl, and cinnamyl ethers, the Alloc group, and allylamines are all efficiently cleaved by this method.[32,33]

13. (Ph$_3$P)CpRu(CH$_3$CN)$_2$PF$_6$, S/C~100–1000, MeOH, 6h, 71–99% yield.[34]

Methallyl Ester: CH$_2$=C(CH$_3$)CH$_2$O$_2$CR

Cleavage of the methallyl ester is achieved in 80–95% yield by solvolysis in refluxing 90% formic acid. Cinnamyl and crotyl alcohols are similarly cleaved.[35] Some of the Pd catalyzed method should also cleave this ester.

2-Methylbut-3-en-2-yl Ester: $CH_2=CHC(CH_3)_2O_2CR$

The advantage of this ester is that it has the resistance to nucleophiles of the *t*-butyl ester, and its deprotection is accomplished under the mild Pd catalysis, thus avoiding strong acids during deprotection.

Formation

1. CuI, KI, Cs_2CO_3, DMF, $HC\equiv C(Me)_2Cl$, 25°C, 72–91%, then H_2, Pd/BaSO$_4$, quinoline, MeOH, 94–98% yield.[36]

2. $(CH_3)_2C=CHCH_2SR_2$, CuBr, RCO_2K, CH_2Cl_2, 80–100% yield.[37]

Cleavage

This ester is cleaved with $Pd(OAc)_2$, Ph_3P, $Et_3NH_2CO_2H$, rt, 30 min.[38]

3-Methylbut-2-enyl (Prenyl) Ester: $(CH_3)_2C=CHCH_2O_2CR$

Cleavage

1. I_2 in cyclohexane, rt, 75–97% yield.[39]
2. TMSOTf, CH_2Cl_2, rt, 2 h, 74–98% yield. Boc groups are not compatible with this method, since they are cleaved with this reagent. Electron-rich aromatics can also be problematic, because the methallyl cation can react to form a chromane.[40] The addition of TESH might possibly prevent this side reaction. The *t*-Bu ester can be cleaved with this method.
3. $NaHSO_4 \cdot SiO_2$, CH_2Cl_2, rt, 4–6 h, 85–96% yield.[41]
4. $CeCl_3 \cdot H_2O$, NaI, CH_3CN, reflux, 1.5–2.5 h, 85–92% yield. Allyl esters are cleaved only after prolonged (~10 h) reaction times. N-Boc, N-Cbz, allyl, THP, and PMB ethers are all stable.[42]
5. K-10 clay, toluene, 1,4-dimethoxybenzene or anisole, heat, 87–98% yield.[43] Microwave heating was also effective. Cinnamyl esters were cleaved similarly.
6. H-β-zeolite, anisole, toluene, reflux, 1.5–8 h, 70–90% yield. Cinnamyl esters are also cleaved in excellent yield, but allyl esters give mixed results with aliphatic allyl esters showing no cleavage.[44]

7. Pd(OAc)$_2$, TPPTS, CH$_3$CN, H$_2$O, Et$_2$NH, 96–100% yield. The allyl carbamate (alloc) group can be cleaved in the presence of the prenyl ester. These conditions will also cleave allyl carbonates, cinnamyl esters, and prenyl carbamates.[45,46]

3-Buten-1-yl Ester: CH$_2$=CHCH$_2$CH$_2$O$_2$CR

This ester, formed from the acid (COCl$_2$, toluene; then CH$_2$=CHCH$_2$CH$_2$OH, acetone, −78°C warm to rt, 70–94% yield), can be cleaved by ozonolysis followed by Et$_3$N or DBU treatment (79–99% yield). The ester is suitable for the protection of enolizable and base-sensitive carboxylic acids.[47]

4-(Trimethylsilyl)-2-buten-1-yl Ester: RCO$_2$CH$_2$CH=CHCH$_2$Si(CH$_3$)$_3$

This ester is formed by standard procedures and is readily cleaved with Pd(Ph$_3$P)$_4$ in CH$_2$Cl$_2$ to form trimethylsilyl esters that readily hydrolyze on treatment with water or alcohol or on chromatography on silica gel (73–98% yield). Amines can be protected using the related carbamate.[48]

Cinnamyl Ester: RCO$_2$CH$_2$CH=CHC$_6$H$_5$ (Chart 6)

The cinnamyl ester, which is somewhat more stable to nucleophiles,[49] can be prepared from an activated carboxylic acid derivative and cinnamyl alcohol or by transesterification with cinnamyl alcohol in the presence of the H-Beta zeolite (toluene, reflux, 8 h, 59–96% yield)[50] or DMAP (CH$_3$CN, heat).[51] It is cleaved under nearly neutral conditions [Hg(OAc)$_2$, MeOH, 23°C, 2–4 h; KSCN, H$_2$O, 23°C, 12–16 h, 90% yield],[52] by treatment with sulfated SnO$_2$, toluene, anisole, reflux[53] or with K-10 clay and microwave heating.[43] The latter conditions will also cleave crotyl and prenyl esters. Pd catalysis may also be used to induce cleavage either with a nucleophile[45] or reductively with TEA/HCO$_2$H.[51]

α-Methylcinnamyl (MEC) Ester: RCO$_2$CH(CH$_3$)CH=CHC$_6$H$_5$

Formation

1. PhCH=CHCH(CH$_3$)OH, DCC, DMAP, THF, 98% yield.[54]
2. From an acid chloride: PhCH=CHCH(CH$_3$)OH, Pyr, DMAP, 75–88% yield.[54]

Cleavage

Me$_2$Sn(SMe)$_2$, BF$_3$·Et$_2$O, PhCH$_3$, 0°C, 3–24 h; AcOH, 75–100% yield.[47,54] An ethyl ester can be hydrolyzed in the presence of an MEC ester with 1 N aqueous NaOH-DMSO (1:1), and MEC esters can be cleaved in the presence of ethyl, benzyl, cinnamyl, and t-butyl esters as well as the acetate, TBDMS, and MEM groups.

Prop-2-ynyl (Propargyl) Ester: $RCO_2CH_2C\equiv CH$

Formation

1. Transesterification from a β-ketoester: toluene, propargyl alcohol, reflux with distillation of low-molecular-weight alcohol, 70–96% yield.[55]
2. Propargyl alcohol, DCC, DMAP.[56]

Cleavage

1. Benzyltriethylammonium tetrathiomolybdate in CH_3CN in 61–97% yield. Deprotection is compatible with esters such as benzyl, allyl, acetate, and t-butyl esters.[56]
2. $Pd(Ph_3P)_2Cl_2$ (Bu_3SnH, benzene)[57] or cobalt carbonyl.[58] The palladium method cleaves allyl esters, propargyl phosphates, and propargyl carbamates as well.
3. SmI_2.[59,60]
4. Hydrogenolysis.[61]
5. Electrolysis, Ni(II), Mg anode, DMF, rt, 77–99% yield. This method is not compatible with halogenated phenols because of competing halogen cleavage.[62]

1. F. Guibé, Tetrahedron, **54**, 2967 (1998); J. Tsuji and T. Mandai, Synthesis, 1, (1996)
2. S. F.-Bochnitschek, H. Waldmann, and H. Kunz, J. Org. Chem., **54**, 751 (1989).
3. N. Engel, B. Kübel, and W. Steglich, Angew. Chem., Int. Ed. Engl., **16**, 394 (1977).
4. H. Kunz, H. Waldmann, and C. Unverzagt, Int. J. Pept. Protein Res., **26**, 493 (1985).
5. H. Waldmann and H. Kunz, Liebigs Ann. Chem., 1712 (1983).
6. P. R. Andreana, J. S. McLellan, Y. Chen, and P. G. Wang, Org. Lett., **4**, 3875 (2002).
7. H. Nakagawa, T. Hirabayashi, S. Sakaguchi, and Y. Ishii, J. Org. Chem., **69**, 3474 (2004).
8. M. J. I. Andrews, and A. B. Tabor, Tetrahedron Lett., **38**, 3063 (1997); D. Seebach, A. Thaler, D. Blaser, and S. Y. Ko, Helv. Chim. Acta, **74**, 1102 (1991).
9. K. Takeda, A. Akiyama, H. Nakamura, S.-i. Takizawa, Y. Mizuno, H. Takayamagi, and Y. Harigaya, Synthesis, 1063 (1994).
10. K. Takeda, A. Akiyama, Y. Konda, H. Takayanagi, and Y. Harigaya, Tetrahedron Lett., **36**, 113 (1995).
11. G. Kokotos and A. Chiou, Synthesis, 168 (1997).
12. S. T. Waddell and G. M. Santorelli, Tetrahedron Lett., **37**, 1971 (1996).
13. W. Oppolzer and P. Lienard, Helv. Chim. Acta, **75**, 2572 (1992).
14. J. R. P. Cetusic, F. R. Green, III, P. R. Graupner, and M. P. Oliver, Org. Lett., **4**, 1307 (2002).
15. L. N. Jungheim, Tetrahedron Lett., **30**, 1889 (1989).
16. M. Seki, K. Kondo, T. Kuroda, T. Yamanaka, and T. Iwasaki, Synlett, 609 (1995).
17. H. Kunz and H. Waldmann, Helv. Chim. Acta, **68**, 618 (1985).

18. R. Deziel, *Tetrahedron Lett.*, **28**, 4371 (1987); C. A. Dvorak, W. D. Schmitz, D. J. Poon, D. C. Pryde, J. P. Lawson, R. A. Amos, and A. I. Meyers, *Angew. Chem. Int. Ed.*, **39**, 1664 (2000).

19. J. E. Bardaji, J. L. Torres, N. Xaus, P. Clapés, X. Jorba, B. G. de la Torre, and G. Valencia, *Synthesis*, 531 (1990).

20. H. X. Zhang, F. Guibé, and G. Balavoine, *Tetrahedron Lett.*, **29**, 623 (1988).

21. P. D. Jeffrey and S. W. McCombie, *J. Org. Chem.*, **47**, 587 (1982).

22. H. Kunz and J. März, *Synlett*, 591 (1992).

23. H. Tsukamoto, T. Suzuki, and Y. Kondo, *Synlett*, 1105 (2003).

24. B. G. de la Torre, J. L. Torres, E. Bardají, P. Clapés, N. Xaus, X. Jorba, S. Calvet, F. Albericio, and G. Valencia, *J. Chem. Soc., Chem. Commun.*, 965 (1990).

25. T.-L. Ho, *Synth. Commun.*, **8**, 15 (1978).

26. M. Dessolin, M.-G. Guillerez, N. Thieriet, F. Guibé, and A. Loffet, *Tetrahedron Lett.*, **36**, 5741 (1995).

27. B. T. Lotz and M. J. Miller, *J. Org. Chem.*, **58**, 618 (1993).

28. G. Casy, A. G. Sutherland, R. J. K. Taylor, and R. G. Urben, *Synthesis*, 767 (1989).

29. E. J. Corey and S. Choi, *Tetrahedron Lett.*, **34**, 6969 (1993).

30. N. Xaus, P. Clapés, E. Bardají, J. L. Torres, X. Jorba, J. Mata, and G. Valencia, *Tetrahedron*, **45**, 7421 (1989).

31. K. Satyanarayana, N. Chidambaram, and S. Chandrasekaran, *Synth. Commun.*, **19**, 2159 (1989).

32. M. Honda, H. Morita, and I. Nagakura, *J. Org. Chem.*, **62**, 8932 (1997).

33. A. Stapon, R. Li, and C. A. Townsend, *J. Am. Chem. Soc.*, **125**, 15746 (2003).

34. M. Kitamura, S. Tanaka, and M. Yoshimura, *J. Org. Chem.*, **67**, 4975 (2002).

35. C. R. Schmid, *Tetrahedron Lett.*, **33**, 757 (1992).

36. M. Sedighi and M. A. Lipton, *Org. Lett.*, **7**, 1473 (2005).

37. B. Badet, M. Julia, M. Ramirez-Munoz, and C. A. Sarrazin, Tetrahedron, **39**, 3111 (1983).

38. M. Yamaguchi, T. Okuma, A. Horiguchi, C. Ikeura, and T. Minami, *J. Org. Chem.*, **57**, 1647 (1992).

39. J. Cossy, A. Albouy, M. Scheloske, and D. G. Pardo, *Tetrahedron Lett.*, **35**, 1539 (1994).

40. M. Nishizawa, H. Yamamoto, K. Seo, H. Imagawa, and T. Sugihara, *Org. Lett.*, **4**, 1947 (2002).

41. G. Mahender, R. Ramu, C. Ramesh, and B. Das, *Chem. Lett.*, **32**, 734 (2003).

42. J. S. Yadav, B. V. S. Reddy, C. V. Rao, P. K. Chand, and A. R. Prasad, *Synlett*, 137 (2002).

43. A. S. Gajare, N. S. Shaikh, B. K. Bonde, and V. H. Deshpande, *Perkin 1*, 639 (2000).

44. R. K. Pandey, V. S. Kadam, R. K. Upadhyay, M. K. Dongare, and P. Kumar, *Synth. Commum.*, **33**, 3017 (2003).

45. S. Lemaire-Audoire, M. Savignac, E. Blart, G. Pourcelot, J. P. Genét, and J-M. Bernard, *Tetrahedron Lett.*, **35**, 8783 (1994).

46. S. Lemaire-Audoire, M. Savignac, G. Pourcelot, J.-P. Genét, and J.-M. Bernard, *J. Mol. Cat. A: Chemical*, **116**, 247 (1997).

47. A. G. M. Barrett, S. A. Lebold, and X.-an Zhang, *Tetrahedron Lett.*, **30**, 7317 (1989).

48. H. Mastalerz, *J. Org. Chem.*, **49**, 4092 (1984).

49. M. A. Ciufolini, D. Valognes, and N. Xi, *Angew. Chem. Int. Ed.*, **39**, 2493 (2000).

50. B. S. Balaji, M. Sasidharan, R. Kumar, and B. Chandra, *J. Chem. Soc., Chem. Commun.*, 707 (1996).

51. Z. D. Aron and L. E. Overman, *J. Am. Chem. Soc.*, **127**, 3380 (2005).

52. E. J. Corey and M. A. Tius, *Tetrahedron Lett.*, **18**, 2081 (1977).

53. S. P. Chavan, P. K. Zubaidha, S. W. Dantale, A. Keshavaraja, A. V. Ramaswamy, and T. Ravindranathan, *Tetrahedron Lett.*, **37**, 237 (1996).

54. T. Sato, J. Otera, and H. Nozaki, *Tetrahedron Lett.*, **30**, 2959 (1989).

55. C. Mottet, O. Hamelin, G. Garavel, J.-P. Depres, and A. E. Greene, *J. Org. Chem.*, **64**, 1380 (1999).

56. P. Ilankumaran, N. Manoj, and S. Chandrasekaran, *Chem. Commun.*, 1957 (1996).

57. H. X. Zhang, F. Guibé, and G. Balavoine, *Tetrahedron Lett.*, **29**, 619, 623 (1988),

58. B. Alcaide, J. Perez-Castels, B. Sanchez-Vigo, and M. A. Sierra, *J. Chem. Soc., Chem. Commun.*, 587 (1994).

59. J. Inanaga, Y. Sugimoto, and T. Hanamoto, *Tetrahedron Lett.*, **33**, 7035 (1992).

60. J. M. Aurrecoechea and R. F.-S. Anton, *J. Org. Chem.*, **59**, 702 (1994).

61. J. Tsuji and T. Mandai, *Synthesis*, 1 (1996); J. Tsuji and T. Mandai, *Angew. Chem., Int. Ed. Engl.*, **34**, 2589 (1995).

62. S. Olivero and E. Duñach, *Tetrahedron Lett.*, **38**, 6193 (1997).

Phenyl Ester: $RCO_2C_6H_5$

$$BOP =$$

Phenyl esters can be prepared from *N*-protected amino acids (PhOH, DCC, CH_2Cl_2, $-20°C$ to $20°C$, 12 h, 86% yield[1]; PhOH, BOP, Et_3N, CH_2Cl_2, $25°C$, 2 h, 73–97% yield).[2] Phenyl esters are readily cleaved under basic conditions (H_2O_2, H_2O, DMF, pH 10.5, $20°C$, 15 min).[3] Phenyl esters are more easily cleaved than an alkyl ester.

1. I. J. Galpin, P. M. Hardy, G. W. Kenner, J. R. McDermott, R. Ramage, J. H. Seely, and R. G. Tyson, *Tetrahedron*, **35**, 2577 (1979).

2. B. Castro, G. Evin, C. Selve, and R. Seyer, *Synthesis*, 413 (1977).

3. G. W. Kenner and J. H Seely, *J. Am. Chem. Soc.*, **94**, 3259 (1972).

2,6-Dialkylphenyl Esters

2,6-Dimethylphenyl Ester

2,6-Diisopropylphenyl Ester

2,6-Di-*t*-butyl-4-methylphenyl (BHT) Ester

2,6-Di-*t*-butyl-4-methoxyphenyl Ester

The esters were prepared from the phenol and the acid chloride plus DMAP (or from the acid plus trifluoroacetic anhydride). In these esters the steric bulk of the ortho substituents protects the carbonyl from nucleophilic reagents, making them difficult hydrolyze. Although the diisopropyl derivative can be cleaved with hot aqueous NaOH, the di-*t*-butyl derivatives could only be cleaved with NaOMe in a mixture of toluene and HMPA.[1] The related 2,6-di-*t*-butyl-4-methoxyphenyl ester can be cleaved oxidatively with ceric ammonium nitrate.[2] These hindered esters have found utility in directing the aldol condensation.[3,4]

1. T. Hattori, T. Suzuki, N. Hayashizaka, N. Koike, and S. Miyano, *Bull. Chem. Soc. Jpn.*, **66**, 3034 (1993).

2. M. P. Cooke, Jr., *J. Org. Chem.*, **51**, 1637 (1986); C. H. Heathcock, M. C. Pirrung, S. H. Montgomery, and J. Lampe, *Tetrahedron*, **37**, 4087 (1981).

3. C. H. Heathcock, in *Asymmetric Synthesis*, Vol. 3, J. D. Morrison, Ed., Academic, New York, 1984, pp. 111–212; D. A. Evans, J. V. Nelson, and T. R. Tabor, in *Topics in Stereochemistry*, Vol. 13, N. L. Allinger, E. L. Eliel, and S. H. Wilen, Eds., Wiley Interscience, New York, 1982, p. 1; C. H. Heathcock, in *Comprehensive Organic Synthesis*; B. M. Trost and I. Fleming, Eds., Pergamon, Oxford, 1991, Vol. 2, pp. 133–238.

4. I. Paterson, O. Delgado, G. J. Florence, I. Lyothier, M. O'Brien, J. P. Scott, and N. Sereinig, *J. Org. Chem.*, **70**, 150 (2005).

p-(Methylthio)phenyl Ester: $RCO_2C_6H_4-p$-SCH_3

The *p*-(methylthio)phenyl ester has been prepared from an *N*-protected amino acid and 4-$CH_3SC_6H_4OH$ (DCC, CH_2Cl_2, 0°C, 1 h to 20°C, 12 h, 60–70% yield). The *p*-(methylthio)phenyl ester serves as an unactivated ester that is activated on oxidation to the sulfone (H_2O_2, AcOH, 20°C, 12 h, 60–80% yield), which then serves as an activated ester in peptide synthesis.[1]

1. B. J. Johnson and T. A. Ruettinger, *J. Org. Chem.*, **35**, 255 (1970).

Pentafluorophenyl Ester (Pfp): $C_6F_5O_2CR$

The active ester was used for carboxyl protection of Fmoc-serine and Fmoc-threonine during glycosylation.[1,2] The esters are then used as an active ester in peptide synthesis.

Formation

1. $C_6F_5O_2CCF_3$, Pyr, DMF, rt, 45 min, 92–95% yield.[3] This reagent converts amines to the trifluoroacetamide.[4]

2. C_6F_5OH, DCC, dioxane or EtOAc and DMF, 0°C, 1 h then rt 1 h, 75–99% yield.[5]

3. From a protected amino acid: $C_6F_5OSO_2C_6H_4NO_2$, HOBt, TEA, DMF, 20–30 min, 61–98% yield.[6] This method can also be used to prepare other electron deficient phenolic esters such as the 4-nitrophenyl, 2,4,5-trichloro-phenyl, and the pentachlorophenyl ester.

1. M. Meldal and K. Bock, *Tetrahedron Lett.*, **31**, 6987 (1990).

2. M. Meldal and K. J. Jensen, *J. Chem. Soc., Chem. Commun.*, 483 (1990).

3. M. Green and J. Berman, *Tetrahedron Lett.*, **31**, 5851 (1990).

4. L. M. Gayo and M. J. Suto, *Tetrahedron Lett.*, **37**, 4915 (1996).

5. L. Kisfaludy and I. Schön, *Synthesis*, 325 (1983); I. Schön and L. Kisfaludy, *ibid.*, 303 (1986).

6. K. Pudhom and T. Vilaivan, *Tetrahedron Lett.*, **40**, 5939 (1999).

2-(Dimethylamino)-5-nitrophenyl (DNAP) Ester

The DNAP group is introduced from the acid and the phenol using DCC/DMAP as a coupling agent. It is cleaved by photolysis at 400 nm in a pH 7 buffer. The group was developed as a caging group for intracellular kinetic investigations.[1]

1. A. Banerjee, C. Grewer, L. Ramakrishnan, J. Jaeger, A. Gameiro, H.-G. A. Breitinger, K. R. Gee, B. K. Carpenter, and G. P. Hess, *J. Org. Chem.*, **68**, 8361 (2003).

Benzyl Ester: $RCO_2CH_2C_6H_5$, RCO_2Bn (Chart 6)

Formation

Benzyl esters are readily prepared by many of the classical methods, (see introduction to this chapter), as well as by many newer methods, since benzyl alcohol is unhindered and relatively acid stable.

1. BnOCOCl, Et$_3$N, 0°C, DMAP, CH$_2$Cl$_2$, 30 min, 97% yield.[1] In the case of very hindered acids the yields are poor, and formation of the symmetrical anhydride is observed. Useful selectivity can be achieved for a less hindered acid in the presence of a more hindered one.[2]

A similar method that uses BOC_2O, BnOH, and DMAP also gives good yields of benzyl esters except for electron poor aromatic acids.[3]

2. A methyl ester can be exchanged for a benzyl ester thermally (185°C, 1.25 h, −MeOH).[4]

3. $BnOC=NH(CCl_3)$, $BF_3·Et_2O$, CH_2Cl_2, cyclohexane, 60–98% yield.[5,6]

4.

5. $(BnO)_2CHNMe_2$.[8]

6. BnBr, DBU, CH_3CN, 75% yield.[9]

7. BnBr, Cs_2CO_3, CH_3CN, reflux, 93–100% yield.[10] Other esters are prepared similarly.

8. For amino acids: DCC, DMAP, BnOH, 92% yield.[11]

9. $cHexN=C(OBn)NHcHex$.[6] A polymer supported version of this reagent has been prepared (97–99% yields).[12] The analogous reagent can be used to prepare allyl and methyl esters in excellent yield.

10. From an anhydride, BnOH, Bu_3P, CH_2Cl_2.[13]

11. KF, ionic liquid, BnCl, 90°C, 76–95% yield.[14]

12. Ph_2POBn, dimethylbenzoquinone, CH_2Cl_2, rt, 0.5 h, 86–98% yield.[15]

13. $BnOC(S)SCH_2C≡CH$, toluene, reflux, 74–98% yield. The method was also successfully tested on a limited set of phenols and heterocyclic amines.[16]

Cleavage

The most useful property of benzyl esters is that they are readily cleaved by hydrogenolysis. It is possible to hydrogenate an olefin and retain the benzyl ester.[17]

1. H_2/Pd–C, 25°C, 45 min to 24 h, high yields.[18] Catalytic transfer hydrogenation (entries 2 and 3 below) can be used to cleave benzyl esters in some compounds that contain sulfur, a poison for hydrogenolysis catalysts.

2. Pd–C, cyclohexene[19] or 1,4-cyclohexadiene,[20] 25°C, 1.5–6 h, good yields. Some alkenes,[6] benzyl ethers, BOM groups, and benzyl amines[21] are compatible with these conditions.

3. Pd–C, 4.4% HCOOH, MeOH, 25°C, 5–10 min in a column, 100% yield.[22]

4. Pd–C(en), H_2, Dabco or DMAP, MeOH.[23] Benzyl esters are cleaved in the presence of N-Cbz groups unless the Cbz is attached to an aromatic amine which gives competitive hydrogenolysis. These conditions also reduce olefins in the presence of benzyl ethers. 2,2′-Dipyridyl also serves as a catalyst poison that will allow the selective hydrogenolysis of a benzyl ester in the presence of a benzyl phenyl ether.[24]

5. t-BuNH$_2$·BH$_3$, 10 Pd/C, MeOH, 90% yield. A 3° benzyl ether was unaffected, but benzyl amines are cleaved.[25]

6. K_2CO_3, H_2O, THF, 0–25°C, 1 h, 75% yield.[26]

7. $AlCl_3$, anisole, CH_2Cl_2, CH_3NO_2, 0–25°C, 5 h, 80–95% yield.[27] These conditions were used to cleave the benzyl ester in a variety of penicillin derivatives.

8. BCl_3, CH_2Cl_2, −10°C to rt, 3 h, 90% yield.[28]

9. $FeCl_3$ or $Re(CO)_5Br$, mesitylene, 50–130°C, 2–72 h, 82–100% yield.[29]

10. Na, ammonia, 50% yield.[30] These conditions were used to cleave the benzyl ester of an amino acid; the Cbz and benzylsulfenamide derivatives were also cleaved. A possible side reaction in this process is reduction of the carbonyl group.

11. Mg, H_2NNH_2, HCO_2H, MeOH, 89–93% yield. These conditions also reduce other benzyl-based protective groups.[31]

12. Aq. $CuSO_4$, EtOH, pH 8, 32°C, 60 min; pH 3; EDTA (ethylenediaminetetraacetic acid), 75% yield.[32]

13. Benzyl esters can be cleaved by electrolytic reduction at −2.7 V.[33]

14. t-BuMe$_2$SiH, Pd(OAc)$_2$, CH_2Cl_2, Et$_3$N, 100% yield.[34] Cbz groups and Alloc groups are also cleaved, but benzyl ethers are stable. PdCl$_2$ and Et$_3$SiH have also been used to cleave a benzyl ester.[35]

15. NaHTe, DMF, t-BuOH, 80–90°C, 5 min, 98% yield.[36] Methyl and propyl esters are also cleaved (13–97% yield).

16. W2 Raney nickel, EtOH, Et$_3$N, rt, 0.5 h, 75–85% yield.[37] A disubstituted olefin was not reduced.

17. NBS, CCl$_4$, Bz$_2$O, reflux, 61–97% yield.[38] Substituted benzyl esters are cleaved similarly. This method proceeds by a free radical induced bromination of the benzyl CH$_2$ group.

18. Bis(tributyltin) oxide, toluene, 70–90°C, 36–96 h, 60–69% yield.[39]

19. Acidic alumina, microwaves, 7 min, 90% yield.[40]

20. Catalyst (HCTf$_3$, Sc(CTf$_3$)$_3$, HNTf$_2$, Bi(NTf$_2$)$_3$, or Yb(NTf$_2$)$_3$), anisole, 100°C, 99% yield. The fastest rate was achieved with Sc(CTf$_3$)$_3$. This method also can be used to cleave benzyl and MPM ethers and MPM amides.[41]

21. Alcatase, t-BuOH, pH 8.2, 35°C, 0.5 h, 91% yield.[42]

22. P. Fluorescens, ROH, MTBE converts a benzyl ester by transesterification to Me, Et, and Bu esters.[43]

23. Pronase, 25°C, pH 7.2, aq. EtOH, 70–73% yield.[44]

24. Esterase from *Bacillus subtilis* (BS2) or lipase from *Candida antarctica*, 39–99% yield.[45] Methyl esters are also cleaved.

25. Alkaline protease from *Bacillus subtilis* DY, pH 8, 37°C, 80–85% yield.[46] Methyl esters are cleaved similarly.

1. S. Kim, Y. C. Kim, and J. I. Lee, *Tetrahedron Lett.*, **24**, 3365 (1983); S. Kim, J. I. Lee, and Y. C. Kim, *J. Org. Chem.*, **50**, 560 (1985).

2. J. E. Baldwin, M. Otsuka, and P. M. Wallace, *Tetrahedron*, **42**, 3097 (1986).

3. L. J. Gooben and A. Dohring, *Synlett*, 263 (2004).

4. W. L. White, P. B. Anzeveno, and F. Johnson, *J. Org. Chem.*, **47**, 2379 (1982).

5. G. Kokotos and A. Chiou, *Synthesis*, 169 (1997).

6. K. C. Nicolaou, E. W. Yue, Y. Naniwa, F. D. Riccardis, A. Nadin, J. E. Leresche, S. La Greca, and Z. Yang, *Angew. Chem., Int. Ed., Engl.*, **33**, 2184 (1994).

7. J. F. Okonya, T. Kolasa, and M. J. Miller, *J. Org. Chem.*, **60**, 1932 (1995).

8. G. Emmer, M. A. Grassberger, J. G. Meingassner, G. Schulz, and M. Schaude, *J. Med. Chem.*, **37**, 1908 (1994).

9. M. J. Smith, D. Kim, B. Horenstein, K. Nakanishi, and K. Kustin, *Acc. Chem. Res.*, **24**, 117 (1991).

10. J. C. Lee, Y. S. Oh, S. H. Cho, and J. I. Lee, *Org. Prep. Proced. Int.*, **28**, 480 (1996).

11. B. Neises, T. Andries, and W. Steglich, *J. Chem. Soc., Chem. Commun.*, 1132 (1982).

12. S. Crosignani, P. D. White, R. Steinauer, and B. Linclau, *Org. Lett.*, **5**, 853 (2003).

13. E. Vedejs, N. S. Bennett, L. M. Conn, S. T. Diver, M. Gingras, S. Lin, P. A. Oliver, and M. J. Peterson, *J. Org. Chem.*, **58**, 7286 (1993).

14. L. Brinchi, R. Germani, and G. Savelli, *Tetrahedron Lett.*, **44**, 6583 (2003).

15. T. Mukaiyama, T. Shintou, and W. Kikuchi, *Chem. Lett.*, **31**, 1126 (2002); T. Mukaiyama, W. Kikuchi, and T. Shintou, *Chem. Lett.*, **32**, 300 (2003).

16. M. Faure-Tromeur and S. Z. Zard, *Tetrahedron Lett.*, **39**, 7301 (1998).

17. D. Misiti, G. Zappia, and G. D. Monache, *Synthesis* 873 (1999).

18. W. H. Hartung and R. Simonoff, *Org. React.*, **VII**, 263 (1953).

19. G. M. Anantharamaiah and K. M. Sivanandaiah, *J. Chem. Soc., Perkin Trans. I*, 490 (1977).

20. A. M. Felix, E. P. Heimer, T. J. Lambros, C. Tzougraki, and J. Meienhofer, *J. Org. Chem.*, **43**, 4194 (1978).

21. J. S. Bajwa, *Tetrahedron Lett.*, **33**, 2299 (1992).

22. B. ElAmin, G. M. Anantharamaiah, G. P. Royer, and G. E. Means, *J. Org. Chem.*, **44**, 3442 (1979).

23. K. Hattori, H. Sajiki, and K. Hirota, *Tetrahedron*, **56**, 8433 (2000); H. Sajiki, K. Hattori, and K. Hirota, *J. Org. Chem.*, **63**, 7990 (1998).

24. H. Sajiki and K. Hirota, *Tetrahedron*, **54**, 13981 (1998).

25. M. Couturier, B. M. Andresen, J. L. Tucker, P. Dube, S. J. Brenek, and J. T. Negri, *Tetrahedron Lett.*, **42**, 2763 (2001).

26. W. F. Huffman, R. F. Hall, J. A. Grant, and K. G. Holden, *J. Med. Chem.*, **21**, 413 (1978).

27. T. Tsuji, T. Kataoka, M. Yoshioka, Y. Sendo, Y. Nishitani, S. Hirai, T. Maeda, and W. Nagata, *Tetrahedron Lett.*, **20**, 2793 (1979).

28. U. Schmidt, M. Kroner, and H. Griesser, *Synthesis*, 294 (1991).

29. T. J. Davies, R. V. H. Jones, W. E. Lindsell, C. Miln, and P. N. Preston, *Tetrahedron Lett.*, **43**, 487 (2002).

30. C. W. Roberts, *J. Am. Chem. Soc.*, **76**, 6203 (1954).

31. D. C. Gowda, *Tetrahedron Lett.*, **43**, 311 (2002).

32. R. L. Prestidge, D. R. K. Harding, J. E. Battersby, and W. S. Hancock, *J. Org. Chem.*, **40**, 3287 (1975).

33. W. G. Mairanovsky, *Angew. Chem., Int. Ed. Engl.*, **15**, 281 (1976).

34. M. Sakaitani, N. Kurokawa, and Y. Ohfune, *Tetrahedron Lett.*, **27**, 3753 (1986).

35. K. M. Rupprecht, R. K. Baker, J. Boger, A. A. Davis, P. J. Hodges, and J. F. Kinneary, *Tetrahedron Lett.*, **39**, 233 (1998).

36. J. Chen and X. J. Zhou, *Synthesis*, 586 (1987).

37. S.-i. Hashimoto, Y. Miyazaki, T. Shinoda, and S. Ikegami, *Tetrahedron Lett.*, **30**, 7195 (1989).

38. M. S. Anson and J. G. Montana, *Synlett*, 219 (1994).

39. C. J. Salomon, E. G. Mata, and O. A. Mascaretti, *J. Chem. Soc., Perkin Trans I*, 995 (1996).

40. R. S. Varma, A. K. Chatterjee, and M. Varma, *Tetrahedron Lett.*, **34**, 4603 (1993).

41. K. Ishihara, Y. Hiraiwa, and H. Yamamoto, *Synlett*, 80 (2000).

42. S. T. Chen, S. C. Hsiao, C. H. Chang, and K. T. Wang, *Synth. Commun.*, **22**, 391 (1992).

43. A. L. Gutman, E. Shkolnik, and M. Shapira, *Tetrahedron*, **48**, 8775 (1992).

44. M. Pugniere, B. Castro, N. Domerque, and A. Previero, *Tetrahedron: Asymmetry*, **3**, 1015 (1992).

45. E. Barbayianni, I. Fotakopoulou, M. Schmidt, V. Constantinou-Kokotou, W. T. Bornscheuer, and G. Kokotos, *J. Org. Chem.*, **70**, 8730 (2005).

46. B. Aleksiev, P. Schamlian, G. Videnov, S. Stoev, S. Zachariev, and E. Golovinskii, *Hoppe-Seylers Z. Physiol. Chem.*, **362**, 1323 (1981).

Substituted Benzyl Esters

Triphenylmethyl (Tr) Ester: $RCO_2C(C_6H_5)_3$ (Chart 6)

Triphenylmethyl esters are not always stable in aqueous solution, but are stable to oxymercuration.[1] The related 4-pyridyldiphenylmethyl and the 9-phenylfluoren-9-yl esters have been prepared of aspartic acid but these were found unsuitable for the prevention of aspartimide formation during peptide synthesis.[2]

Formation

1. TrCl, DBU, THF, reflux.[3]
2. RCO_2M (M = Ag^+, K^+, Na^+), Ph_3CBr, benzene, reflux, 3–5 h, 85–95% yield.[4]
3. RCO_2SiMe_3, Ph_3COTMS, TMSOTf, CH_2Cl_2, 0°C, 0.5 h, 86% yield.[5]
4. Transesterification of a β-ketoester: Ph_3COH, $LiClO_4$, toluene, heat, 8 h, 57% yield.[6]

Cleavage

1. Cleavage of $HCl \cdot H_2NCH_2CO_2CPh_3$: MeOH or H_2O/dioxane, 18°C, 5 h, 72%; 18°C, 24 h, 98%, 100°C, 1 min, 98%.[7]
2. Trityl esters have been cleaved by electrolytic reduction at −2.6 V.[8]
3. 1*H*-tetrazole, CH_3CN. Partial cleavage observed after 15 min. In contrast, the 2-chlorotrityl group was stable up to 1 h under these conditions.[9]

2-Chlorophenyldiphenylmethyl Ester: $RCO_2C(C_6H_5)_2-2-ClC_6H_4$

The 2-chlorotrityl ester is prepared by reaction of the acid with the trityl chloride and TEA in CH_2Cl_2,[9] or from the Cs salt (Cs_2CO_3, DMF, 2-Cl-TrCl).[10] They are cleaved by acid and the following table gives the relative acid stability of the trityl and 2-chlorotrityl esters of 4-hydroxypentanoic acid.[9] As expected, the electron-withdrawing group improves acid stability.

Acid Stability of Trityl and 2-Chlorotrityl Esters of 4-Hydroxypentanoic Acid[a]

Reagent	Trityl Ester	2-Chlorotrityl Ester
0.5 *M* 1-*H*-Tetrazole in MeCN	30 min	>1 h
AcOH, H_2O 4:1 (v/v)	<5 min	15 min
2.5% Cl_2CHCO_2H, CH_2Cl_2	<1 min	<1 min

[a] Time needed for ~50% removal of the protecting group (TLC)

1. W. A. Slusarchyk, H. E. Applegate, C. M. Cimarusti, J. E. Dolfini, P. Funke, and M. Puar, *J. Am. Chem. Soc.*, **100**, 1886 (1978).
2. M. Mergler, F. Dick, B. Sax, P. Weiler, and T. Vorherr, *J. Peptide Sci.*, **9**, 36 (2003).
3. B. Yu, J. Xie, S. Deng, and Y. Hui, *J. Am. Chem. Soc.*, **121**, 12196 (1999).

4. K. D. Berlin, L. H. Gower, J. W. White, D. E. Gibbs, and G. P. Sturm, *J. Org. Chem.*, **27**, 3595 (1962).

5. S. Murata and R. Noyori, *Tetrahedron Lett.*, **22**, 2107 (1981).

6. B. P. Bandgar, V. S. Sadavarte, and L. S. Uppalla, *Synlett*, 1338 (2001).

7. G. C. Stelakatos, A. Paganou, and L. Zervas, *J. Chem. Soc. C*, 1191 (1966).

8. V. G. Mairanovsky, *Angew. Chem., Int. Ed., Engl.*, **15**, 281 (1976).

9. A. V. Kachalova, D. A. Stetsenko, E. A. Romanova, V. N. Tashlitsky, M. J. Gait, and T. S. Oretskaya, *Helv. Chim. Acta*, **85**, 2409 (2002).

10. N. M. A. J. Kriek, D. V. Filippov, H. van den Elst, N. J. Meeuwenoord, G. I. Tesser, J. H. van Boom, and G. A. van der Marel, *Tetrahedron*, **59**, 1589 (2003).

2,3,4,4′,4″,5,6-Heptafluorotriphenylmethyl (TrtF$_7$) Ester:
$(4\text{-}FC_6H_4)_2(C_6F_5)C-O_2CR$

The ester was prepared for glutamic acid protection during peptide synthesis. It is more acid stable than the corresponding trityl ester. It is stable to AcOH/EtOAc, but is cleaved with 1% TFA/CH$_2$Cl$_2$ in 30–60 min. Cleavage is facilitated by the inclusion of triisopropylsilane. The ester is prepared from the trityl chloride (DIPEA, CH$_2$Cl$_2$, rt, 14 h, 49% yield).[1] Cleavage of this trityl group in the presence of the BOC group is not completely selective, but it can be selectively cleaved in the presence of the *t*-butyl ester and ether. The phenylfluorenyl ester was shown to have similar acid stability to the TrtF$_7$ ester.

1. B. Löhr, S. Orlich, and H. Kunz, *Synlett*, 1136 (1999).

Diphenylmethyl Ester (Dpm Ester): $RCO_2CH(C_6H_5)_2$

Diphenylmethyl esters are similar in acid lability to *t*-butyl esters and can be cleaved by acidic hydrolysis from *S*-containing peptides that poison hydrogenolysis catalysts.

Formation

1. Ph$_2$CN$_2$, acetone, 0°C, 30 min to 20°C, 4 h, 70%.[1,2]

2. Ph$_2$C=NNH$_2$, I$_2$, AcOH, >90% yield.[3] Methods based on the hydrazone all proceed by oxidation to the diazo derivative.

3. Ph$_2$C=NNH$_2$, Oxone supported on wet Al$_2$O$_3$, cat. I$_2$, 0°C, 66–95% yield.[4]

4. Ph$_2$C=NNH$_2$, PhI(OAc)$_2$, CH$_2$Cl$_2$, cat. I$_2$, −10°C to 0°C, 1 h, 73–93% yield.[5]

5. Ph$_2$C=NNH$_2$, AcOOH, 91% yield.[6]

6. Ph$_2$CHOH, cat. TsOH, benzene, azeotropic removal of water, 78–83% yield.[7]

7. (Ph$_2$CHO)$_3$PO, CF$_3$COOH, CH$_2$Cl$_2$, reflux, 1–5 h, 70–87% yield.[8] Free alcohols are converted to the corresponding Dpm ethers. This reaction has also

been used for the selective protection of amino acids as their tosylate salts (CCl$_4$, 15 min to 3 h, 63–91% yield).[9]

8. Ph$_2$CHOH, 5 mol% MoO$_2$Cl$_2$, Bz$_2$O, 4°C, 36 h, CH$_2$Cl$_2$, 88–91% yield.[10] Trityl and *t*-butylthio esters may be prepared similarly.

Cleavage

1. H$_2$/Pd black, MeOH, THF, 3 h, 90% yield.[11]

2. CF$_3$COOH, PhOH, 20°C, 30 min, 82% yield.[1]

3. AcOH, reflux, 6 h.[12]

4. BF$_3$·Et$_2$O, AcOH, 40°C, 0.5 h to 10°C, several hours, 65% yield.[13] The sulfur–sulfur bond in cystine is stable to these conditions.

5. H$_2$NNH$_2$, MeOH, reflux, 60 min, 100% yield.[14] In this case the ester is converted to a hydrazide.

6. Diphenylmethyl esters are cleaved by electrolytic reduction at −2.6 V.[15]

7. HF, CH$_3$NO$_2$, AcOH (12:2:1), 91% yield.[16]

8. HCl, CH$_3$NO$_2$, <5 min, 25°C.[17]

9. 98% HCOOH, 40–50°C, 70–97% yield.[2]

10. 1 *N* NaOH, MeOH, rt.[9]

11. AlCl$_3$, CH$_3$NO$_2$, anisole, 3–6 h, 73–95% yield.[18,19] These conditions also cleaved the *p*-MeOC$_6$H$_4$CH$_2$ ester and ether in penam- and cephalosporin-type intermediates.

12. 1 eq. TsOH, benzene, reflux, 78–95% yield.[7]

1. G. C. Stelakatos, A. Paganou, and L. Zervas, *J. Chem. Soc. C*, 1191 (1966).

2. T. Kametani, H. Sekine, and T. Hondo, *Chem. Pharm. Bull.*, **30**, 4545 (1982).

3. R. Bywood, G. Gallagher, G. K. Sharma, and D. Walker, *J. Chem. Soc., Perkin Trans. 1*, 2019 (1975).

4. M. Curini, O. Rosati, E. Pisani, W. Cabri, S. Brusco, and M. Riscazzi, *Tetrahedron Lett.*, **38**, 1239 (1997).

5. L. Lapatsanis, G. Milias, and S. Paraskewas, *Synthesis* 513 (1985); H. Zhou and W. A. v. d. Donk, *Org. Lett.*, **4**, 1335 (2002).

6. R. G. Micetich, S. N. Maiti, P. Spevak, M. Tanaka, T. Yamazaki, and K. Ogawa, *Synthesis*, 292 (1986).

7. R. Paredes, F. Agudelo, and G. Taborda, *Tetrahedron Lett.*, **37**, 1965 (1996).

8. L. Lapatsanis, *Tetrahedron Lett.*, **19**, 4697 (1978).

9. C. Froussios and M. Kolovos, *Synthesis*, 1106 (1987).

10. C.-T. Chen, J.-H. Kuo, C.-H. Ku, S.-S. Weng, and C.-Y. Liu, *J. Org. Chem.*, **70**, 1328 (2005).

11. S. De Bernardo, J. P. Tengi, G. J. Sasso, and M. Weigele, *J. Org. Chem.*, **50**, 3457 (1985).

12. E. Haslam, R. D. Haworth, and G. K. Makinson, *J. Chem. Soc.*, 5153 (1961).

13. R. G. Hiskey and E. L. Smithwick, *J. Am. Chem. Soc.*, **89**, 437 (1967).

14. R. G. Hiskey and J. B. Adams, *J. Am. Chem. Soc.*, **87**, 3969 (1965).

15. V. G. Mairanovsky, *Angew. Chem., Int. Ed. Engl.*, **15**, 281 (1976).

16. L. R. Hillis and R. C. Ronald, *J. Org. Chem.*, **50**, 470 (1985).

17. R. C. Kelly, I. Schletter, S. J. Stein, and W. Wierenga, *J. Am. Chem. Soc.*, **101**, 1054 (1979).

18. T. Tsuji, T. Kataoka, M. Yoshioka, Y. Sendo, Y. Nishitani, S. Hirai, T. Maeda, and W. Nagata, *Tetrahedron Lett.*, **20**, 2793 (1979).

19. M. Ohtani, F. Watanabe, and M. Narisada, *J. Org. Chem.*, **49**, 5271 (1984).

Bis(*o*-nitrophenyl)methyl Ester: $RCOOCH(C_6H_4-o-NO_2)_2$ (Chart 6)

Bis(*o*-nitrophenyl)methyl esters are formed and cleaved by the same methods used for diphenylmethyl esters. They can also be cleaved by irradiation ($h\nu = 320$ nm, dioxane, THF, 1–24 h, quant. yield).[1] Because of the electron withdrawing nitro group, these esters are more acid stable than the unsubstituted Dpm ester.

1. A. Patchornik, B. Amit, and R. B. Woodward, *J. Am. Chem. Soc.*, **92**, 6333 (1970).

9-Anthrylmethyl Ester: $RCOOCH_2-9$-anthryl (Chart 6)

Formation

1. 9-Anthrylmethyl chloride, Et₃N, MeCN, reflux, 4–6 h, 70–90% yield.[1]

2. N₂CH-9-anthryl, hexane, 25°C, 10 min, 80% yield.[2,3]

3. 9-Anthrylmethyl alcohol, DCC, DMAP.[4]

Cleavage

1. 2 *N* HBr/HOAc, 25°C, 10–30 min, 100% yield.[1]

2. 0.1 *N* NaOH/dioxane, 25°C, 15 min, 97% yield.[1]

3. MeSNa, THF-HMPA, $-20°C$, 1 h, 90–100% yield.[5] Cleavage proceeds by addition of thiolate to the 10-position of the anthracene ring followed by release of the acid by elimination.

4. Photolysis at 386 nm in CH_3CN/H_2O, which results in fluorescence emission at 380–480 nm with release of the acid in 43–100% yield.[4]

1. F. H. C. Stewart, *Aust. J. Chem.*, **18**, 1699 (1965).
2. M. G. Krakovyak, T. D. Amanieva, and S. S. Skorokhodov, *Synth. Commun.*, **7**, 397 (1977).
3. K. Hör, O. Gimple, P. Schreier, and H.-U. Humpf, *J. Org. Chem.*, **63**, 322 (1998).
4. A. K. Singh and P. K. Khade, *Tetrahedron Lett.*, **46**, 5563 (2005).
5. N. Kornblum and A. Scott, *J. Am. Chem. Soc.*, **96**, 590 (1974).

2-(9,10-Dioxo)anthrylmethyl Ester (Chart 6)

R' = H, Ph

This derivative is prepared from an *N*-protected amino acid and the anthrylmethyl alcohol in the presence of DCC/hydroxybenzotriazole.[1] It can also be prepared from 2-(bromomethyl)-9,10-anthraquinone (Cs_2CO_3).[2] It is stable to moderately acidic conditions (e.g., CF_3COOH, 20°C, 1 h; HBr/HOAc, $t_{1/2} = 65$ h; HCl/CH_2Cl_2, 20°C, 1 h).[1] Cleavage is effected by reduction of the quinone to the hydroquinone **i**; in the latter, electron release from the $-OH$ group of the hydroquinone results in facile cleavage of the methylene-carboxylate bond.

The related 2-phenyl-2-(9,10-dioxo)anthrylmethyl ester has also been prepared, but is cleaved by electrolysis (-0.9 V, DMF, 0.1 *M* $LiClO_4$, 80% yield).[3]

i

Cleavage

This derivative is cleaved by hydrogenolysis and by the following conditions:[1]

1. $Na_2S_2O_4$, dioxane–H_2O, pH 7–8, 8 h, 100% yield.
2. Irradiation, *i*-PrOH, 4 h, 99% yield.
3. 9-Hydroxyanthrone, Et_3N/DMF, 5 h, 99% yield.
4. 9,10-Dihydroxyanthracene/polystyrene resin, 1.5 h, 100% yield.

1. D. S. Kemp and J. Reczek, *Tetrahedron Lett.*, **18**, 1031 (1977).

2. P. Hoogerhout, C. P. Guis, C. Erkelens, W. Bloemhoff, K. E. T. Kerling, and J. H. Boom, *Recl. Trav. Chim. Pays-Bas*, **104**, 54 (1985).

3. R. L. Blankespoor, A. N. K. Lau, and L. L. Miller, *J. Org. Chem.*, **49**, 4441 (1984).

5-Dibenzosuberyl Ester

The dibenzosuberyl ester is prepared from dibenzosuberyl chloride (which is also used to protect −OH, −NH, and −SH groups) and a carboxylic acid (Et₃N, reflux, 4 h, 45% yield). It can be cleaved by hydrogenolysis and, like *t*-butyl esters, by acidic hydrolysis (aq. HCl/THF, 20°C, 30 min, 98% yield).[1] Because of its doubly benzylic nature, acid promoted cleavage should occur more easily than *t*-Bu ester cleavage.

1. J. Pless, *Helv. Chim. Acta*, **59**, 499 (1976).

1-Pyrenylmethyl Ester (R′ = H, Me, Ph)

These esters are prepared from the diazomethylpyrenes and carboxylic acids in DMF (R′ = H, 60% yield, R′ = Me, 80% yield, R′ = Ph, 20% yield for 4-methylbenzoic acid). They are cleaved by photolysis at 340 nm (80–100% yield, R′ = H).[1,2] The esters are very fluorescent.

1. M. Iwamura, T. Ishikawa, Y. Koyama, K. Sakuma, and H. Iwamura, *Tetrahedron Lett.*, **28**, 679 (1987).

2. M. Iwamura, C. Hodota, and M. Ishibashi, *Synlett*, 35 (1991).

2-(Trifluoromethyl)-6-chromonylmethyl Ester (Tcrom Ester)

The Tcrom ester is prepared from the cesium salt of an *N*-protected amino acid by reaction with 2-(trifluoromethyl)-6-chromylmethyl bromide (DMF, 25°C, 4 h,

53–89% yield). Cleavage of the Tcrom group is affected by brief treatment with
n-propylamine (2 min, 25°C, 96% yield). It is stable to HCl/dioxane, used to cleave
a BOC group.[1]

1. D. S. Kemp and G. Hanson, *J. Org. Chem.*, **46**, 4971 (1981).

2,4,6-Trimethylbenzyl Ester: $RCOOCH_2C_6H_2-2,4,6-(CH_3)_3$

The 2,4,6-trimethylbenzyl ester has been prepared from an amino acid and the benzyl chloride (Et$_3$N, DMF, 25°C, 12 h, 60–80% yield); it is cleaved by acidic hydrolysis (CF$_3$COOH, 25°C, 60 min, 60–90% yield; 2 *N* HBr/HOAc, 25°C, 60 min,
80–95% yield) and by hydrogenolysis. It is stable to methanolic hydrogen chloride
used to remove *N-o*-nitrophenylsulfenyl groups or triphenylmethyl esters.[1]

1. F. H. C. Stewart, *Aust. J. Chem.*, **21**, 2831 (1968).

p-Bromobenzyl Ester: $RCOOCH_2C_6H_4-p$-Br

The *p*-bromobenzyl ester has been used to protect the β-COOH group in aspartic
acid. It is cleaved by strong acidic hydrolysis (HF, 0°C, 10 min, 100% yield), but is
stable to 50% CF$_3$COOH/CH$_2$Cl$_2$ used to cleave *t*-butyl carbamates. It is five to seven
times more stable toward acid than a benzyl ester.[1] It may also be cleaved by hydrogenolysis, but in this case HBr may be liberated do to bromine hydrogenolysis.

1. D. Yamashiro, *J. Org. Chem.*, **42**, 523 (1977).

o-Nitrobenzyl Ester: $RCOOCH_2C_6H_4-o$-NO$_2$

p-Nitrobenzyl Ester: $RCOOCH_2C_6H_4-p$-NO$_2$

The *o*-nitrobenzyl ester, used to protect penicillin precursors, can be cleaved by irradiation (H$_2$O/dioxane, pH 7). Reductive cleavage of benzyl or *p*-nitrobenzyl esters
occurred in lower yields.[1,2]

    p-Nitrobenzyl esters have been prepared from the Hg(I) salt of penicillin precur-
sors and the phenyldiazomethane.[3] They are much more stable to acidic hydrolysis
(e.g., HBr*)* than *p*-chlorobenzyl esters and are recommended for terminal −COOH
protection in solid-phase peptide synthesis.[4] *p*-Nitrobenzyl esters of penicillin and
cephalosporin precursors have been cleaved by alkaline hydrolysis with Na$_2$S (0°C,
aq acetone, 25–30 min, 75–85% yield).[5] They are also cleaved by electrolytic reduction at −1.2 V,[6] by reduction with SnCl$_2$ (DMF, phenol, AcOH),[7] by reduction with
sodium dithionite, by hydrogenolysis,[8] or by transfer hydrogenation with Pd–C (ammonium formate or phosphinic acid).[9]

1. L. D. Cama and B. G. Christensen, *J. Am. Chem. Soc.*, **100**, 8006 (1978).

2. For a reviews covering the photolytic removal of protective groups, see V. N. R. Pillai, *Synthesis*, 1 (1980); C. G. Bochet, *J. Chem. Soc., Perkin Trans. 1*, 125 (2002); P. Pelliccioli Anna and J. Wirz, *Photochemical & Photobiological Sciences : Official Journal of the European Photochemistry Association and the European Society for Photobiology*, **1**, 441 (2002).

3. W. Baker, C. M. Pant, and R. J. Stoodley, *J. Chem. Soc., Perkin Trans. 1*, 668 (1978).

4. R. L. Prestidge, D. R. K. Harding, and W. S. Hancock, *J. Org. Chem.*, **41**, 2579 (1976).

5. S. R. Lammert, A. I. Ellis, R. R. Chauvette, and S. Kukolja, *J. Org. Chem.*, **43**, 1243 (1978).

6. V. G. Mairanovsky, *Angew Chem., Int. Ed. Engl.*, **15**, 281 (1976).

7. M. D. Hocker, C. G. Caldwell, R. W. Macsata, and M. H. Lyttle, *Pept. Res.*, **8**, 310 (1995).

8. J. W. Perich, P. F. Alewood, and R. B. Johns, *Aust. J. Chem.*, **44**, 233 (1991).

9. D. Albanese, M. Leone, M. Penso, M. Seminati, and M. Zenoni, *Tetrahedron Lett.*, **39**, 2405 (1998).

p-Methoxybenzyl Ester (PMB$-$O$_2$CR): RCOOCH$_2$C$_6$H$_4$$-$*p*-OCH$_3$

Formation

1. *p*-Methoxybenzyl esters have been prepared from the Ag(I) salt of amino acids and the benzyl halide (Et$_3$N, CHCl$_3$, 25°C, 24 h, 60% yield).[1]

2. *p*-Methoxybenzyl alcohol, Me$_2$NCH(OCH$_2$-*t*-Bu)$_2$, CH$_2$Cl$_2$, 90% yield.[2]

3. Isopropenyl chloroformate, MeOC$_6$H$_4$CH$_2$OH, DMAP, 0°C, CH$_2$Cl$_2$, 91%.[3]

4. *p*-Methoxyphenyldiazomethane (MeOC$_6$H$_4$CHN$_2$) in CH$_2$Cl$_2$, 80–96% yield.[4]

5. *p*-Methoxybenzyl chloride, NaHCO$_3$, DMF, 45°C, 89% yield.[5]

6. PMBOC(=NH)CCl$_3$, CH$_2$Cl$_2$, 0°C, 85% yield.[6]

Cleavage

1. CF$_3$COOH, PhOMe, 25°C, 3 min, 98% yield.[7,8]

2. HCOOH, 22°C, 1 h, 81% yield.[1]

3. TFA, phenol, 1 h, 45°C, 73–93% yield. [9,10] These conditions were developed for the mild cleavage of acid-sensitive esters of β-lactam-related antibiotics. Diphenylmethyl and *t*-butyl esters were cleaved with similarly high efficiency.

4. TFA, Et₃SiH, CH₂Cl₂, 0°C, 1 h. [11] Conventional hydrolysis and the nearly neutral Me₃SnOH both fail with this substrate.

5. AlCl₃, anisole, CH₂Cl₂ or CH₃NO₂, −50°C; NaHCO₃, −50°C, 73–95% yield. [12,13]
6. CF₃CO₂H, B(OTf)₃. [14]

1. G. C. Stelakatos and N. Argyropoulos, *J. Chem. Soc. C*, 964 (1970).

2. J. A. Webber, E. M. Van Heyningen, and R. T. Vasileff, *J. Am. Chem. Soc.*, **91**, 5674 (1969).

3. P. Jouin, B. Castro, C. Zeggaf, A. Pantaloni, J. P. Senet, S. Lecolier, and G. Sennyey, *Tetrahedron Lett.*, **28**, 1661 (1987).

4. S. T. Waddell and G. M. Santorelli, *Tetrahedron Lett.*, **37**, 1971 (1996).

5. D. L. Boger, M. Hikota, and B. M. Lewis, *J. Org. Chem.*, **62**, 1748 (1997).

6. M. Shoji, T. Uno, H. Kakeya, R. Onose, I. Shiina, H. Osada, and Y. Hayashi, *J. Org. Chem.*, **70**, 9905 (2005).

7. F. H. C. Stewart, *Aust. J. Chem.*, **21**, 2543 (1968).

8. M. Shoji, T. Uno, and Y. Hayashi, *Org. Lett.*, **6**, 4535 (2004).

9. H. Tanaka, M. Taniguchi, Y. Kameyama, S. Torii, M. Sasaoka, T. Shiroi, R. Kikuchi, I. Kawahara, A. Shimabayashi, and S. Nagao, *Tetrahedron Lett.*, **31**, 6661 (1990).

10. S. Torii, H. Tanaka, M. Taniguchi, Y. Kameyama, M. Sasaoka, T. Shiroi, R. Kikuchi, I. Kawahara, A. Shimabayashi, and S. Nagao, *J. Org. Chem.*, **56**, 3633 (1991).

11. T. R. Hoye and J. Wang, *J. Am. Chem. Soc.*, **127**, 6950 (2005).

12. M. Ohtani, F. Watanabe, and M. Narisada, *J. Org. Chem.*, **49**, 5271 (1984).

13. T. Tsuji, T. Kataoka, M. Yoshioka, Y. Sendo, Y. Nishitani, S. Hirai, T. Maeda, and W. Nagata, *Tetrahedron Lett.*, **20**, 2793 (1979).

14. S. D. Young and P. P. Tamburini, *J. Am. Chem. Soc.*, **111**, 1933 (1989).

2,6-Dimethoxybenzyl Ester: 2,6-(CH₃O)₂C₆H₃CH₂O₂CR

2,6-Dimethoxybenzyl esters prepared from the acid chloride and the benzyl alcohol are readily cleaved oxidatively by DDQ (CH₂Cl₂, H₂O, rt, 18 h, 90–95% yield). A

4-methoxybenzyl ester was found not to be cleaved by DDQ. The authors have also explored the oxidative cleavage (ceric ammonium nitrate, CH_3CN, H_2O, 0°C, 4 h, 65–97% yield) of a variety of 4-hydroxy- and 4-amino-substituted phenolic esters.[1]

The dimethoxybenzyl group is cleaved from a hydroxamic acid with TFA, CH_2Cl_2, rt, 2 h.[2]

1. C. U. Kim and P. F. Misco, *Tetrahedron Lett.*, **26**, 2027 (1985).
2. B. Barlaam, A. Hamon, and M. Maudet, Tetrahedron Lett., **39**, 7865 (1998).

4-(Methylsulfinyl)benzyl (Msib) Ester: $4\text{-}CH_3S(O)C_6H_4CH_2O_2CR$

The 4-(methylsulfinyl)benzyl ester was recommended as a selectively cleavable carboxyl protective group for peptide synthesis. It is readily prepared from 4-(methylsulfinyl)benzyl alcohol (EDCI, HOBt, CHCl$_3$, 78–100% yield) or from 4-methylthiobenzyl alcohol followed by oxidation of the derived ester with MCPBA or H_2O_2/AcOH. The Msib ester is exceptionally stable to CF_3COOH (cleavage rate = 0.000038% ester cleaved/min) and only undergoes 10% cleavage in HF (anisole, 0°C, 1 h). Anhydrous HCl/dioxane rapidly reduces the sulfoxide to the sulfide (Mtb ester) that is completely cleaved in 30 min with CF_3CO_2H. A number of reagents readily reduce the Msib ester to the Mtb ester with $(CH_3)_3SiCl$/Ph$_3$P as the reagent of choice.[1]

1. J. M. Samanen and E. Brandeis, *J. Org. Chem.*, **53**, 561 (1988).

4-Sulfobenzyl Ester: Na^+ $^-O_3SC_6H_4CH_2O_2CR$

4-Sulfobenzyl esters were prepared (cesium salt or dicyclohexylammonium salt, $NaO_3SC_6H_4CH_2Br$, DMF, 37–95% yield) from *N*-protected amino acids. They are cleaved by hydrogenolysis (H_2/Pd), or hydrolysis (NaOH, dioxane/water). Treatment with ammonia or hydrazine results in formation of the amide or hydrazide. The ester is stable to 2 *M* HBr/AcOH and to CF_3SO_3H in CF_3CO_2H. The relative rates of hydrolysis and hydrazinolysis for different esters are as follows:

Hydrolysis: $NO_2C_6H_4CH_2O- \gg C_6H_4CH_2O- > {}^-O_3SC_6H_4CH_2O- > MeO-$
Hydrazinolysis: $NO_2C_6H_4CH_2O- > {}^-O_3SC_6H_4CH_2O- > C_6H_5CH_2O- > MeO-$

A benzyl ester can be cleaved in the presence of the 4-sulfobenzyl ester by CF_3SO_3H.[1,2]

1. R. Bindewald, A. Hubbuch, W. Danho, E.E. Büllesbach, J. Föhles, and H. Zahn, *Int. J. Pept. Protein Res.*, **23**, 368 (1984).
2. A. Hubbuch, R. Bindewald, J. Föhles, V. K. Naithani, and H. Zahn, *Angew. Chem., Int. Ed. Engl.*, **19**, 394 (1980).

4-Azidomethoxybenzyl Ester: $N_3CH_2OC_6H_4CH_2O_2CR$

This ester, developed for peptide synthesis, is prepared by the standard DCC coupling protocol, and it is cleaved reductively with $SnCl_2$ (MeOH, 25°C, 5 h) followed by treatment with mild base to effect quinone methide formation with release of the acid in 75–95% yield.[1] Since cleavage is initiated by reduction of the azide group, other reagents that reduce the azide should also cleave this ester.

1. B. Loubinoux and P. Gerardin, *Tetrahedron*, **47**, 239 (1991).

4-{N-[1-(4,4-Dimethyl-2,6-dioxocyclohexylidene)-3-methylbutyl]amino}benzyl Ester (Dmab)

The Dmab group was developed for glutamic acid protection during Fmoc–*t*-Bu-based peptide synthesis. It shows excellent acid stability and stability toward 20% piperidine in DMF. It is formed from the alcohol using the DCC protocol for ester formation and is cleaved with 2% hydrazine in DMF at rt.[1]

1. W. C. Chan, B. W. Bycroft, D. J. Evans, and P. D. White, *J. Chem. Soc., Chem. Commun.*, 2209 (1995); D. H. Live, Z.-G. Wang, U. Iserloh, and S. J. Danishefsky, *Org. Lett.*, **3**, 851 (2001).

Piperonyl Ester: (Chart 6)

The piperonyl ester can be prepared from an amino acid ester and the benzyl alcohol (imidazole/dioxane, 25°C, 12 h, 85% yield) or from an amino acid and the benzyl chloride (Et$_3$N, DMF, 25°C, 57–95% yield). It is cleaved, more readily than a *p*-methoxybenzyl ester, by acidic hydrolysis (CF$_3$COOH, 25°C, 5 min, 91% yield).[1]

1. F. H. C. Stewart, *Aust. J. Chem.*, **24**, 2193 (1971).

4-Picolyl Ester: RCO_2CH_2-4-pyridyl

The picolyl ester has been prepared from amino acids and picolyl alcohol (DCC/ CH$_2$Cl$_2$, 20°C, 16 h, 60% yield) or picolyl chloride (DMF, 90–100°C, 2 h, 50% yield).

It is cleaved by reduction (H_2/Pd–C, aq. EtOH, 10 h, 98% yield; Na/NH_3, 1.5 h, 93% yield) and by basic hydrolysis (1 N NaOH, dioxane, 20°C, 1 h, 93% yield). Photolysis can be used for deprotection of these esters after alkylation of the basic nitrogen.[1] These salts are cleaved at >400 nm by sensitized photolysis in the presence of the radical scavenger cyclohexadiene (76–100% yield). Deprotection of the related phosphates has also been demonstrated in one case.[2] The basic site in a picolyl ester allows its ready separation by extraction into an acidic medium.[3]

1. C. Sundararajan and D. E. Falvey, *J. Org. Chem.*, **69**, 5547 (2004).
2. C. Sundararajan and D. E. Falvey, *J. Am. Chem. Soc.*, **127**, 8000 (2005).
3. R. Camble, R. Garner, and G. T. Young, *J. Chem. Soc. C*, 1911 (1969).

p-Polymer-Benzyl Ester: $RCOOCH_2C_6H_4$-*p-Polymer*

The first,[1] and still widely used, polymer-supported ester is formed from an amino acid and a chloromethylated copolymer of styrene-divinylbenzene. Originally, it was cleaved by basic hydrolysis (2 N NaOH, EtOH, 25°C, 1 h). Subsequently, it has been cleaved by hydrogenolysis (H_2/Pd–C, DMF, 40°C, 60 psi, 24 h, 71% yield)[2] and by HF, which concurrently removes many amine protective groups.[3]

Monoesterification of a symmetrical dicarboxylic acid chloride can be effected by reaction with a hydroxymethyl copolymer of styrene–divinylbenzene to give an ester; a mono salt of a diacid was converted into a dibenzyl polymer.[4]

1. R. B. Merrifield, *J. Am. Chem. Soc.*, **85**, 2149 (1963).
2. J. M. Schlatter and R. H. Mazur, *Tetrahedron Lett.*, **18**, 2851 (1977).
3. J. Lenard and A. B. Robinson, *J. Am. Chem. Soc.*, **89**, 181 (1967).
4. D. D. Leznoff and J. M. Goldwasser, *Tetrahedron Lett.*, **18**, 1875 (1977).

2-Naphthylmethyl Ester (2-NAP-O₂CR)

The 2-naphthylmethyl ester is prepared by conventional means (DCC, DMAP, CH_2Cl_2, NAP-OH). It is cleaved by hydrogenolysis in the presence of a benzyl ester with Pd/C (EtOAc, H_2, 75–240 min, 89–97% yield).[1] Many of the methods used to cleave the benzyl ester should cleave the NAP ester, often more readily.

1. M. J. Gaunt, C. E. Boschetti, J. Yu, and J. B. Spencer, *Tetrahedron Lett.*, **40**, 1803 (1999).

3-Nitro-2-naphthylmethyl (NNM) Ester

This group was developed as a photo-cleavable protective group with improved properties over the parent 2-nitrobenzyl group. It is cleaved by photolysis at 380 nm in aqueous CH_3CN in yields from 88–100%.[1]

1. A. K. Singh and P. K. Khade, *Tetrahedron*, **61**, 10007 (2005).

4-Quinolylmethyl Ester (4-QUI–O₂R)

This ester is readily cleaved with $Pd(dba)_2$, dppe, NH_4O_2CH in DMSO at 50°C, 80–95% yield. This method is also applicable to the cleavage of the 1-NAP ester.[1]

1. A. Boutros, J.-Y. Legros, and J.-C. Fiaud, *Tetrahedron Lett.*, **40**, 7329 (1999); A. Boutros, J.-Y. Legros, and J.-C. Fiaud, *Tetrahedron*, **56**, 2239 (2000).

8-Bromo-7-hydroxyquinoline-2-ylmethyl Ester (BHQ)

The photolytically induced cleavage of the BHQ ester has a greater quantum efficiency than does the 4,5-dimethoxy-4-nitrobenzyl (DMNB) ester and the 6-bromo-7-hydroxycourmarin-4ylmethyl (Bhc) ester. It can be used *in vivo* because it has sufficient sensitivity to multiphoton-induced photolysis. It is also more soluble than the DMNB and the Bhc esters, which is advantageous for *in vivo* applications.[1]

1. O. D. Fedoryak and T. M. Dore, *Org. Lett.*, **4**, 3419 (2002).

2-Nitro-4,5-dimethoxybenzyl (Nitroveratrole) Ester

The nitroveratrole group can be prepared by direct acid-catalyzed esterification with the benzyl alcohol. It is cleaved photochemically by irradiation at 420 nm. It is cleaved in the presence of the 1,2-dihydroxy-2,4,4-trimethyl-3-pentanone, which is cleaved photochemically at 300 nm[1] and the ester of 3',5'-dimethoxybenzoin **ii** at 420 nm.[2]

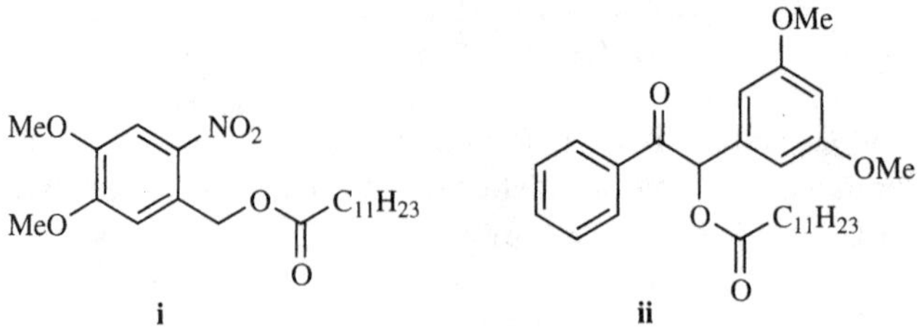

Cleaved at 420 nm in   Cleaved at 254 nm in
the presence of **ii**   the presence of **i**

1. M. Kessler, R. Glatthar, B. Giese, and C. G. Bochet, *Org. Lett.*, **5**, 1179 (2003).
2. A. Blanc and C. G. Bochet, *J. Org. Chem.*, **67**, 5567 (2002).

1,2,3,4-Tetrahydro-1-naphthyl Ester:

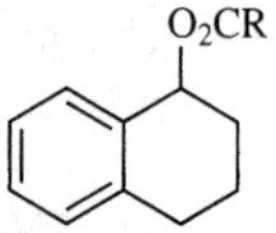

This ester can be prepared using DCC, BOP-Cl or a mixed anhydride method. It is cleaved with TMSCl/NaI in the presence of phenyl, 4-methoxyphenyl and benzhydryl esters (60–82% yield). This ester is also cleaved with TFA and by hydrogenolysis with Pd–C.[1] The chirality of the ester is a liability that may limit its usefulness.

1. C. J. Slade, C. A. Pringle, and I. G. Sumner, *Tetrahedron Lett.*, **40**, 5601 (1999).

Silyl Esters

Silyl esters are stable to nonaqueous reaction conditions, but this is dependent upon the steric environment of the ester and silyl group. A trimethylsilyl ester is cleaved by refluxing in alcohol; the more substituted and therefore more stable silyl esters are cleaved by mildly acidic or basic hydrolysis.

Trimethylsilyl Ester: RCOOSi(CH$_3$)$_3$ (Chart 6)

Some of the more common reagents for the conversion of carboxylic acids to trimethylsilyl esters are listed below. For additional methods that can be used to silylate acids, the section on alcohol protection should be consulted, since many of the methods presented there are also applicable to carboxylic acids. Trimethylsilyl esters are cleaved in aqueous solutions, and thus *in situ* protection is preferred over direct isolation of the ester in most cases.

Formation

1. Me_3SiCl/Pyr, CH_2Cl_2, 30°, 2 h.[1]
2. $MeC(OSiMe_3)=NSiMe_3$, HBr, dioxane, α-picoline, 6 h, 80% yield.[2]
3. $MeCH=C(OMe)OSiMe_3/CH_2Cl_2$, 15–25°C, 5–40 min, quant.[3]
4. $Me_3SiNHSO_2OSiMe_3/CH_2Cl_2$, 30°C, 0.5 h, 92–98% yield.[4]

1. B. Fechti, H. Peter, H. Bickel, and E. Vischer, *Helv. Chim. Acta*, **51**, 1108 (1968).
2. J. J. de Koning, H. J. Kooreman, H. S. Tan, and J. Verweij, *J. Org. Chem.*, **40**, 1346 (1975).
3. Y. Kita, J. Haruta, J. Segawa, and Y. Tamura, *Tetrahedron Lett.*, **20**, 4311 (1979).
4. B. E. Cooper and S. Westall, *J. Organomet. Chem.*, **118**, 135 (1976).

Triethylsilyl Ester (TES): $RCOOSi(C_2H_5)_3$

Formation

1. TESCl, pyridine, 60°C, 0.5 h, 95% yield.[1]

2. TESH, $Pd(OAc)_2$, benzene, reflux, 4 h, 95% yield.[2]

Cleavage

AcOH, THF, H_2O, 20°C, 4 h, 76% yield.[1]

1. T. W. Hart, D. A. Metcalfe, and F. Scheinmann, *J. Chem. Soc., Chem. Commun.*, 156 (1979).
2. M. Chauhan, B. P. S. Chauhan, and P. Boudjouk, *Org. Lett.*, **2**, 1027 (2000).

t-Butyldimethylsilyl Ester (TBDMS): $RCOOSi(CH_3)_2C(CH_3)_3$(Chart 6)

Formation

1. *t*-$BuMe_2SiCl$, imidazole, DMF, 25°C, 48 h, 88%.[1]

2. Morpholine, TBDMSCl, THF, 2 min, 20°C, >80% yield.[2] In this case the ester was formed in the presence of a phenol. The functionally and sterically similar thexyldimethylsilyl ester is also formed under these conditions.[3]

3. t-BuMe$_2$SiH, Pd/C, benzene, 70°C.[4]

Cleavage

1. AcOH, H$_2$O, THF, (3:1:1), 25°C, 20 h.[1]

2. Bu$_4$NF, DMF, 25°C.[1,3]

3. K$_2$CO$_3$, MeOH, H$_2$O, 25°C, 1 h, 88% yield.[5]

4. The TBDMS ester can be converted directly to an acid chloride [DMF, (COCl)$_2$, rt, CH$_2$Cl$_2$] and then converted to another ester, with different properties, by standard means. This procedure avoids the generation of HCl during the acid chloride formation and is thus suitable for acid sensitive substrates.[6]

1. E. J. Corey and A. Venkateswarlu, *J. Am. Chem. Soc.*, **94**, 6190 (1972).
2. J. W. Perich and R. B. Johns, *Synthesis*, 701 (1989).
3. R. C. Claussen, B. M. Rabatic, and S. I. Stupp, *J. Am. Chem. Soc.*, **125**, 12680 (2003).
4. K. Yamamoto and M. Takemae, *Bull. Chem. Soc. Jpn.*, **62**, 2111 (1989).
5. D. R. Morton and J. L. Thompson, *J. Org. Chem.*, **43**, 2102 (1978).
6. A. Wissner and G. V. Grudzinskas, *J. Org. Chem.*, **43**, 3972 (1978).

t-Butyldiphenylsilyl (TBDPS) Ester: t-(CH$_3$)$_3$C(C$_6$H$_5$)$_2$Si$-$O$_2$CR

This ester was used to differentially protect a polyene diacid. It is cleaved with HF (THF, H$_2$O, CH$_3$CN, 1 h, 95% yield) in the presence of a t-butyl ester.[1]

1. U. Schmidt, K. Neumann, A. Schumacher, and S. Weinbrenner, *Angew. Chem. Int. Ed., Engl.*, **36**, 1110 (1997).

i-Propyldimethylsilyl Ester: $RCOOSi(CH_3)_2CH(CH_3)_2$

The *i*-propyldimethylsilyl ester is prepared from a carboxylic acid and the silyl chloride (Et_3N, 0°C). It is cleaved at pH 4.5 by conditions that do not cleave a tetrahydropyranyl ether (HOAc–NaOAc, acetone–H_2O, 0°C, 45 min to 25°C, 30 min, 91% yield).[1]

1. E. J. Corey and C. U. Kim, *J. Org. Chem.*, **38**, 1233 (1973).

Phenyldimethylsilyl Ester: $RCOOSi(CH_3)_2C_6H_5$

The phenyldimethylsilyl ester has been prepared from an amino acid and phenyldimethylsilane (Ni/THF, reflux, 3–5 h, 62–92% yield).[1]

1. M. Abe, K. Adachi, T. Takiguchi, Y. Iwakura, and K. Uno, Tetrahedron Lett., **16**, 3207 (1975).

Di-*t*-butylmethylsilyl Ester (DTBMS Ester): $(t\text{-}Bu)_2CH_3SiO_2CR$

The DTBMS ester was prepared (THF, DTBMSOTf, Et_3N, rt) to protect an ester so that a lactone could be reduced to an aldehyde. The ester is cleaved with aq. HF/THF or Bu_4NF in wet THF. A THP derivative can be deprotected (pyridinium *p*-toluenesulfonate, warm ethanol) in the presence of a DTBMS ester.[1]

1. R. S. Bhide, B. S. Levison, R. B. Sharma, S. Ghosh, and R. G. Salomon, *Tetrahedron Lett.*, **27**, 671 (1986).

Triisopropylsilyl Ester (TIPS)

A TIPS ester, prepared by silylation with TIPSCl, TEA and THF, is cleaved with HF·Pyr (Pyr, THF, 0°C), HF·Pyr (pyridine, THF, 0°C),[1] KF (MeOH, THF, 100% yield)[2], CsF (MeOH, PhH, rt, 10 min, quant.),[3] or irradiation in the presence of CBr_4/MeOH.[4]

1. D. A. Evans, B. W. Trotter, B. Cote, P. J. Coleman, L. C. Dias, and A. N. Tyler, *Angew. Chem. Int. Ed.*, **36**, 2744 (1997).
2. A. B. Smith, III, Q. Lin, V. A. Doughty, L. Zhuang, M. D. McBriar, K. Kerns, C. S. Brook, N. Murase, and K. Nakayama, *Angew. Chem. Int. Ed.*, **40**, 196 (2001).
3. P. Wipf and P. D. G. Coish, *J. Org. Chem.*, **64**, 5053 (1999).
4. A. S.-Y. Lee and F.-Y. Su, *Tetrahedron Lett.*, **46**, 6305 (2005).

Tris(2,6-diphenylbenzyl)silyl (TDS) Ester

The TDS ester is prepared from a carboxylic acid and the silyl bromide by reaction with AgOTf in CH_2Cl_2 (84–93% yield). It is stable to n-BuLi, $LiAlH_4$, AcOH, aqueous NaOH at 50°C, and 1 N HCl at 40°C, but is cleaved with Pyr·HF, THF, 50°C and t-BuOK/DMSO at 25°C. It is not deprotonated at the α-carbon of the ester with n-BuLi and thus this group also serves to protect these hydrogens from enolization.[1]

1. A. Iwasaki, Y. Kondo, and K. Maruoka, *J. Am. Chem. Soc.*, **122**, 10238 (2000).

Activated Esters

Thiol Esters

Thiol esters, more reactive to nucleophiles than the corresponding oxygen esters, have been prepared to activate carboxyl groups, both for lactonization and peptide bond formation. Thioesters also increase the acidity of the hydrogens α to the carbonyl group. For lactonization, S-t-butyl[1] and S-2-pyridyl[2] esters are widely used. Some methods used to prepare thiol esters are shown below. The S-t-butyl ester is included in Reactivity Chart 6.

Formation

1. $$RCOOH + R'SH \xrightarrow[\text{0°C, 5 min} \rightarrow \text{20°C, 3 h}]{\text{DCC, DMAP, } CH_2Cl_2{}^3} RCOSR', \text{ 85–92\%}$$

    $R' = $ Et, t-Bu

    DMAP = 4-dimethylaminopyridine (10^4 times more effective than pyridine)

2.

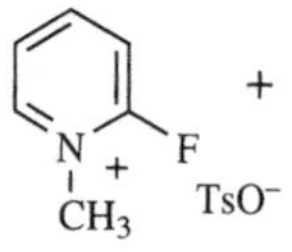

    $$+ \ RCOOH \xrightarrow[\text{−15°C, 1 h}]{Et_3N, CH_2Cl_2} \xrightarrow[\text{2 h, 75–95\%}]{R'SH, Et_3N, CH_2Cl_2} RCOSR'^4$$

    $R' = n$-Bu, s-Bu, t-Bu, Ph, 2-pyridyl

3. $$RCOOH + R'SH \xrightarrow[\text{25°C, 1 h, 70–100\%}]{Me_2NPOCl_2, Et_3N, DME} RCOSR'$$

    $R' = $ Et, i-Pr, t-Bu, c-C_6H_{11}, Ph

These neutral conditions can be used to prepare thiol esters of acid- or base-sensitive compounds including penicillins.[5]

4. $$\underset{\underset{NHPG}{|}}{RCHCOOH} + Ph_2POCl \xrightarrow[\text{0°C, 30min}]{Et_3N, CH_2Cl_2} \xrightarrow[]{R'SH, Et_3N} \xrightarrow[\text{25°C, 1 h}]{\text{or R'STl}} \underset{\underset{NHPG}{|}}{RCHCOSR'}$$

    $R' = t$-Bu, Ph, $PhCH_2$ 70–100%[6]

5.

$$RCOOH + R'SH \xrightarrow[\text{Et}_3\text{N, DMF, 25°C, 3 h, 70–85\%}]{\text{(EtO)}_2\text{POCN or (PhO)}_2\text{PON}_3} RCOSR'^7$$

R = alkyl, aryl, benzyl, amino acids; penicillins

R' = Et, i-Pr, n-Bu, Ph, PhCH$_2$

6. $RCOCl + n\text{-Bu}_3\text{SnSR}' \xrightarrow{\text{CHCl}_3} RCOSR'^8$

R' = t-Bu: 60°C, 0.5 h, 90–95% yield

R' = Ph: 25°C, 12 h, 92–95% yield

R' = PhCH$_2$: 25°C, 0.5–1 h, 87–96% yield

7. $RCOOR' + \text{Me}_2\text{AlS-}t\text{-Bu} \xrightarrow[\text{25°C, 75–100\%}]{\text{CH}_2\text{Cl}_2} RCOS\text{-}t\text{-Bu}^9$

R' = Me, Et

This reaction avoids the use of toxic thallium compounds.

8. $RCOOH + PhSCN \xrightarrow[\text{25°C, 30 min, 80–95\%}]{\text{Bu}_3\text{P, CH}_2\text{Cl}_2} RCOSPh^{10}$

9. $RCOOH + ClCOS\text{-2-pyridyl} \xrightarrow[\text{0.5 h, 95–100\%}]{\text{Et}_3\text{N, 0°C}} RCOS\text{-2-pyridyl} + \text{Et}_3\text{N·HCl}^{11}$

10. $RCO_2H + \text{hydroxybenzotriazole} \xrightarrow{\text{DCC}} \xrightarrow[\text{70–100\%}^6]{R'SH,\ \text{Et}_3\text{N}} \text{or} \xrightarrow{R'STl} RCOSR'$

R' = t-Bu, Ph, PhCH$_2$

Cleavage

1. AgNO$_3$, H$_2$O, dioxane, (1:4), 2 h.[12]
2. ROH, Hg(O$_2$CCF$_3$)$_2$, 90% yield.[1]
3. Electrolysis, Bu$_4$NBr, H$_2$O, CH$_3$CN, NaHCO$_3$.[13] This method is unsatisfactory for substrates containing primary and secondary alcohols, aldehydes, olefins or amines.
4. MeI, ROH (R = t-Bu, PhSH, etc.), 68–97% yield.[14]
5. RCO$_2$H, R'SH, TfOH, toluene, azeotropic reflux, 6 h, 76–97% yield.[15]
6. Hydrolysis of RCOS-t-Bu: KOH, H$_2$O, MeOH, 0–25°C, 99% yield.[16]
7. Treatment of the phenylthio ester with Pd/C and TESH results in reduction to the aldehyde.[17]

1. S. Masamune, S. Kamata, and W. Schilling, *J. Am. Chem. Soc.*, **97**, 3515 (1975).
2. T. Mukaiyama, R. Matsueda, and M. Suzuki, *Tetrahedron Lett.*, **11**, 1901 (1970); E. J. Corey, P. Ulrich, and J. M. Fitzpatrick, *J. Am. Chem. Soc.*, **98**, 222 (1976).
3. B. Neises and W. Steglich, *Angew. Chem., Int. Ed, Engl.*, **17**, 522 (1978).
4. Y. Watanabe, S.-i. Shoda, and T. Mukaiyama, *Chem. Lett.*, **5**, 741 (1976).
5. H. -J. Liu, S. P. Lee, and W. H. Chan, *Synth. Commun.*, **9**, 91 (1979).

6. K. Horiki, *Synth. Commun.*, **7**, 251 (1977).

7. S. Yamada, Y. Yokoyama, and T. Shiori, *J. Org. Chem.*, **39**, 3302 (1974).

8. D. N. Harpp, T. Aida, and T. H. Chan, *Tetrahedron Lett.*, **20**, 2853 (1979).

9. R. P. Hatch and S. M. Weinreb, *J. Org. Chem.*, **42**, 3960 (1977).

10. P. A. Grieco, Y. Yokoyama, and E. Williams, *J. Org. Chem.*, **43**, 1283 (1978).

11. E. J. Corey and D. A. Clark, *Tetrahedron Lett.*, **20**, 2875 (1979).

12. A. B. Shenvi and H. Gerlach, *Helv. Chim. Acta*, **63**, 2426 (1980).

13. M. Kimura, S. Matsubara, and Y. Sawaki, *J. Chem. Soc., Chem. Commun.*, 1619 (1984).

14. D. Ravi and H. B. Mereyala, *Tetrahedron Lett.*, **30**, 6089 (1989).

15. S. Iimura, K. Manabe, and S. Kobayashi, *Chem. Commun.*, 94 (2002).

16. J. P. Vitale, S. A. Wolckenhauer, N. M. Do, and S. D. Rychnovsky, *Org. Lett.*, **7**, 3255 (2005).

17. T. Fukuyama, S.-C. Lin, and L. Li, *J. Am. Chem. Soc.*, **112**, 7050 (1990).

Miscellaneous Derivatives

Oxazoles

Oxazoles, prepared from carboxylic acids (benzoin, DCC; NH_4OAc, AcOH, 80–85% yield), have been used as carboxylic acid protective groups in a variety of synthetic applications. They are readily cleaved by singlet oxygen followed by hydrolysis (ROH, TsOH, benzene[1] or K_2CO_3, MeOH).[2]

2-Alkyl-1,3-oxazoline (Chart 6)

2-Alkyl-1,3-oxazolines are prepared to protect both the carbonyl and hydroxyl groups of an acid. They are stable to Grignard reagents[3] and to lithium aluminum hydride (25°C, 2 h)[4]. The section on amino alcohols should be consulted, since the technology utilized there should be applicable here. They can readily be prepared from a nitrile and the amino alcohol using Bi(III) salts (85–90% yield)[5] or from the acid and an amino alcohol using Deoxo-Fluor as a dehydrating agent (96–99% yield).[6]

Formation

1. $HOCH_2C(CH_3)_2NH_2$, $PhCH_3$, reflux, 70–80% yield.[7]

2. $HOCH_2C(CH_3)_2NH_2$, 2-chloro-4,6-dimethoxy-1,3,5-triazine, morpholine, CH_2Cl_2, 51–89% yield.[8]

3. From an acid chloride: $HOCH_2C(CH_3)_2NH_2$; $SOCl_2$, CH_2Cl_2, 25°C, 30 min, >80% yield.[9]

4. Dimethylaziridine, DCC; 3% H_2SO_4, Et_2O or CH_2Cl_2, rt, 6–16 h, 50–80% yield.[4]

5. $H_2NCH_2CH_2OH$, Ph_3P, Et_3N, CCl_4, CH_3CN, Pyr, rt, 70% yield.[10]

6. From an acid chloride: $BrCH_2CH_2NH_3^+Br^-$; Et_3N, benzene, reflux, 24 h, 46–67% yield.[11]

7. $H_2NCH_2CH_2OH$, Ersorb-4 zeolite, xylene reflux, 5 h, 30–90% yield.[12]

Cleavage

1. 3 N HCl, EtOH, 90% yield.[3]

2. MeI, 25°C, 12 h; 1 N NaOH, 25°C, 15 h, 94% yield.[13]

3. (a)TFA, H_2O, (b) Ac_2O, Pyr, (c) *t*-BuOK, H_2O, THF, quantitative.[14]

4. (a) TFAA, (b) H_2O, (c) diazomethane, (d) KOH, DMSO, 56–88% yield.[15]

1. H. H. Wasserman, K. E. McCarthy, and K. S. Prowse, *Chem. Rev.*, **86**, 845 (1986).

2. M. A. Tius and D. P. Astrab, *Tetrahedron Lett.*, **30**, 2333 (1989).

3. A. I. Meyers and D. L. Temple, *J. Am. Chem. Soc.*, **92**, 6644 (1970).

4. D. Haidukewych and A. I. Meyers, *Tetrahedron Lett.*, **13**, 3031 (1972).

5. I. Mohammadpoor-Baltork, A. R. Khosropour, and S. F. Hojati, *Synlett*, 2747 (2005).

6. C. O. Kangani and D. E. Kelley, *Tetrahedron Lett.*, **46**, 8917 (2005).

7. H. L. Wehrmeister, *J. Org. Chem.*, **26**, 3821 (1961).

8. B. P. Bandgar and S. S. Pandit, *Tetrahedron Lett.*, **44**, 2331 (2003).

9. S. R. Schow, J. D. Bloom, A. S. Thompson, K. N. Winzenberg, and A. B. Smith, III, *J. Am. Chem. Soc.*, **108**, 2662 (1986).

10. H. Vorbrüggen and K. Krolikiewicz, *Tetrahedron Lett.*, **22**, 4471 (1981).

11. C. Kashima and H. Arao, *Synthesis*, 873 (1989).

12. A. Cwik, Z. Hell, A. Hegedus, Z. Finta, and Z. Horvath, *Tetrahedron Lett.*, **43**, 3985 (2002).

13. A. I. Meyers, D. L. Temple, R. L. Nolen, and E. D. Mihelich, *J. Org. Chem.*, **39**, 2778 (1974).

14. T. D. Nelson and A. I. Meyers, *J. Org. Chem.*, **59**, 2577 (1994).

15. D. P. Phillion and J. K. Pratt, *Synth. Commun.*, **22**, 13 (1992).

4-Alkyl-5-oxo-1,3-oxazolidine

1,3-Oxazolidines are prepared to allow selective protection of the α- or ω-CO_2H groups in aspartic and glutamic acids and α-hydroxy acids.

Formation[1,2]

1. CH_2O, Ac_2O, $SOCl_2$, 100°C, 4 h, 80% yield.

$n = 1,2$

The use of paraformaldehyde and acid is equally effective (80–94% yield).[3]

2. CH_2I_2, or CH_2Br_2, K_2CO_3, CH_3CN, reflux, 1 h, 86–94% yield.[4]

3. The related 2-t-butyl derivative has been prepared and used to advantage as a temporary protective group for the stereogenic center of amino acids during alkylations.[5]

4. $PhCH(OMe)_2$, $ZnCl_2$, $SOCl_2$, THF, 0°C, 76% yield.[6]

Cleavage

1. Cleavage with an alcohol and $NaHCO_3$ (reflux, 10 min, 70–89% yield) gives the ester.[7]

2. These derivatives are also cleaved with TMSOK in THF at 60–75°C.[8]

2,2-Bistrifluoromethyl-4-alkyl-5-oxo-1,3-oxazolidine

These derivatives are readily formed by the reaction of hexafluoroacetone with the amino acid.[9,10]

Cleavage is achieved with H_2O, IPA, or MeOH.[10] These derivatives also serve as active esters in peptide bond formation.[11] These derivatives are sufficiently reactive that they will react with amines to form amides and release the HFA group.[12] Reaction of the 5-oxo-1,3-oxazolidine with an alcohol and acid results in cleavage of the HFA group with concomitant ester formation.[13]

2,2-Dimethyl-4-alkyl-2-sila-5-oxo-1,3-oxazolidine

This group was used for transient protection of histidine during its attachment to a trityl-based polymer support. It is introduced by refluxing a mixture of Me_2SiCl_2 and histidine in chloroform. As expected with these unhindered silyl derivatives, they are cleaved simply by stirring in MeOH.[14]

2,2-Difluoro-1,3,2-oxazaborolidin-5-one

This derivative was developed to facilitate side-chain protection of serine and threonine. The oxazaborolidinone is readily prepared from the anhydrous lithium or sodium salt of the amino acid by treatment with $BF_3 \cdot Et_2O$ in THF. These derivatives are sensitive to water, but are sufficiently stable for the introduction of the t-butyl and benzyl groups on the serine and threonine hydroxyl. Cleavage of the oxazaborolidinone is affected with 1 M NaOH.

M = Li or Na

1. M. Itoh, *Chem. Pharm. Bull.*, **17**, 1679 (1969).

2. M. A. Blaskovich and M. Kahn, *Synthesis*, 379 (1998).

3. M. R. Paleo and F. J. Sardina, *Tetrahedron Lett.*, **37**, 3403 (1996); M. W. Walter, R. M. Adlington, J. E. Baldwin, J. Chuhan, and C. J. Schofield, *Tetrahedron Lett.*, **36**, 7761 (1995).

4. S. Karmakar and D. K. Mohapatra, *Synlett*, 1326 (2001).

5. D. Seebach and A. Fadel, *Helv. Chim. Acta*, **68**, 1243 (1985).

6. S. R. Kapadia, D. M. Spero, and M. Eriksson, *J. Org. Chem.*, **66**, 1903 (2001).

7. P. Allevi, G. Cighetti, and M. Anastasia, *Tetrahedron Lett.*, **42**, 5319 (2001).

8. D. M. Coe, R. Perciaccante, and P. A. Procopiou, *Org. Biomol. Chem.*, **1**, 1106 (2003).

9. K. Burger, M. Rudolph, and S. Fehn, *Angew. Chem. Int., Ed.. Engl.*, **32**, 285 (1993).

10. K. Burger, E. Windeisen, and R. Pires, *J. Org. Chem.*, **60**, 7641 (1995).

11. K. Burger, M. Rudolph, E. Windeisen, A. Worku, and S. Fehn, *Monatsh. Chem.*, **124**, 453 (1993).

12. G. Böttcher and H.-U. Reissig, *Synlett*, 725 (2000).

13. J. Spengler and K. Burger, *Synthesis*, 67 (1998); S. N. Osipov, T. Lange, P. Tsouker, J. Spengler, L. Hennig, B. Koksch, S. Berger, S. M. El-Kousy, and K. Burger, *Synthesis*, 1821 (2004).

14. S. Eleftheriou, D. Gatos, A. Panagopoulos, S. Stathopoulos, and K. Barlos, *Tetrahedron Lett.*, **40**, 2825 (1999).

5-Alkyl-4-oxo-1,3-dioxolane

These derivatives are prepared to protect α-hydroxy carboxylic acids; they are cleaved by acidic hydrolysis of the acetal structure (HCl, DMF, 50°C, 7h, 71% yield), or basic hydrolysis of the lactone.[1]

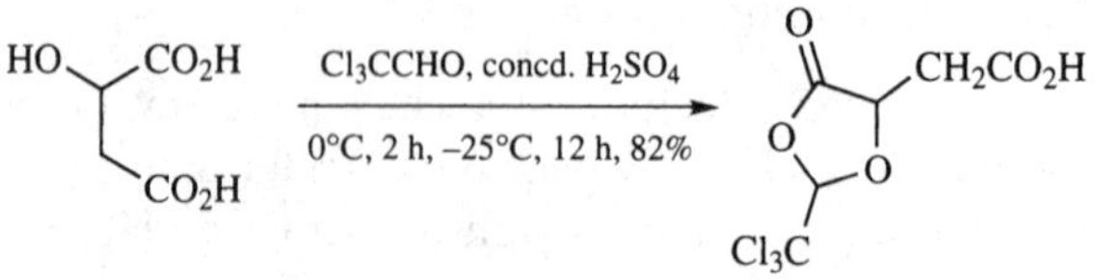

The 2-alkyl derivatives have been prepared to protect the stereogenic center of the α-hydroxy acid during alkylations.[2]

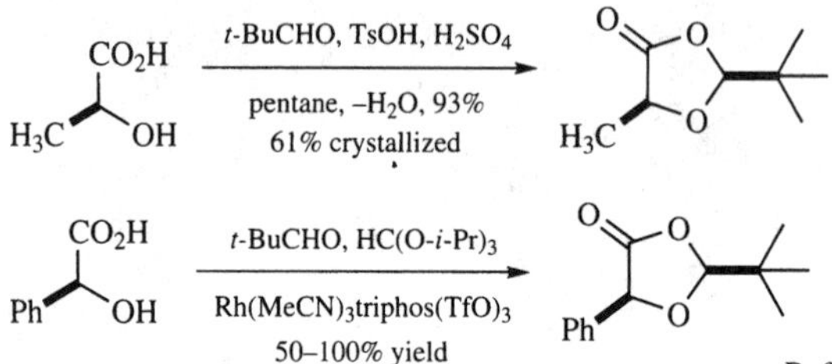

Ref. 3

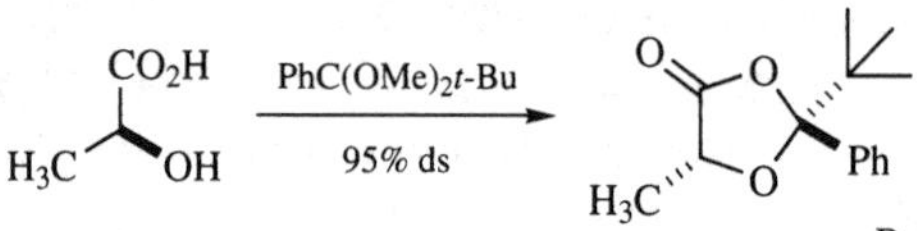

Ref. 4

This methodology is also effective for protection of β-hydroxy acids.[5]

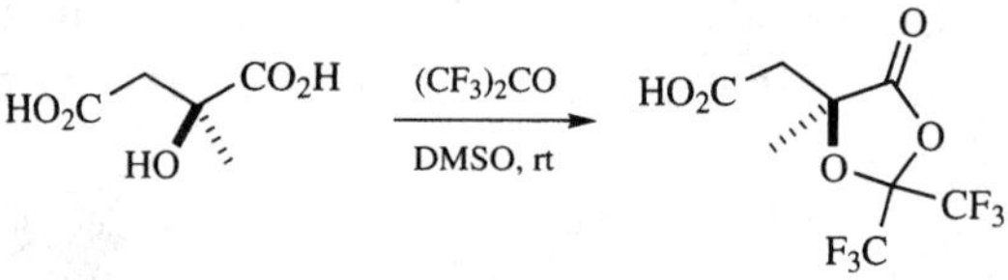

In this case the adduct is sufficiently reactive that amines react to form amides.[6,7]

1. H. Eggerer and C. Grünewälder, *Justus Liebigs Ann. Chem.*, **677**, 200 (1964).
2. D. Seebach, R. Naef, and G. Calderari, *Tetrahedron*, **40**, 1313 (1984).
3. J. Ott, G. M. R. Tombo, B. Schmid, L. M. Venanzi, G. Wang, and T. R. Ward, *Tetrahedron Lett.*, **30**, 6151 (1989).
4. A. Greiner and J.-Y. Ortholand, *Tetrahedron Lett.*, **31**, 2135 (1990).
5. D. Seebach, R. Imwinkelried, and T. Weber, "EPC Synthesis with C,C Bond Formation via Acetals and Enimines," in *Modern Synthetic Methods 1986*, Vol. 4, R. Scheffold, Ed., Springer-Verlag, New York, 1986, p. 125.
6. G. Radics, B. Koksch, S. M. El-Kousy, J. Spengler, and K. Burger, *Synlett*, 1826 (2003); K. Pumpor, E. Windeisen, J. Spengler, F. Albericio, and K. Burger, *Monatsh. Chem.*, **135**, 1427 (2004); F. Albericio, K. Burger, J. Ruiz-Rodriguez, and J. Spengler, *Org. Lett.*, **7**, 597 (2005).

7. C. Böttcher, J. Spengler, S. A. Essawy, and K. Burger, *Monatsh. Chem.*, **135**, 853 (2004); C. Böttcher, J. Spengler, L. Hennig, F. Albericio, and K. Burger, *Monatsh. Chem.*, **136**, 577 (2005).

Dioxanones

$$n = 0, 1$$

Dioxanones have been prepared to protect α- or β-hydroxy acids.

Formation

1. $RR'C=O$, $Sc(NTf_2)_3$ or $Sc(OTf)_3$, CH_2Cl_2, $MgSO_4$ or azeotropic water removal, 54–96% yield. In the case of aldehydes, better stereoselectivity is achieved using $MgSO_4$ as a water scavenger.[1]
2. From a silylated hydroxy acid: RCHO, TMSOTf, 2,6-di-*t*-butylpyridine, 77% yield.[2–4]
3. From a hydroxy acid: pivaldehyde, acid catalyst.[5,6]
4. From a hydroxy acid: pivaldehyde, *i*-PrOTMS, TMSOTf, CH_2Cl_2, −78°C, 4-Å MS, 79% yield.[7]
5. From a hydroxy acid: $RCH(OR)_2$, PPTS, 20–62% yield.[8,9]

1. K. Ishihara, Y. Karumi, M. Kubota, and H. Yamamoto, *Synlett*, 839 (1996).
2. W. H. Pearson and M.-C. Cheng, *J. Org. Chem.*, **51**, 3746 (1986); *idem, ibid.*, **52**, 1353 (1987).
3. S. L. Schreiber and J. Reagan, *Tetrahedron Lett.*, **27**, 2945 (1986).
4. T. R. Hoye, B. H. Peterson, and J. D. Miller, *J. Org. Chem.*, **52**, 1351 (1987).
5. D. Seebach and J. Zimmerman, *Helv. Chim. Acta*, **69**, 1147 (1986).
6. D. Seebach, R. Imwinkelried, and G. Stucky, *Helv. Chim. Acta*, **70**, 448 (1987).
7. A. K. Ghosh and S. Fidanze, *Org. Lett.*, **2**, 2405 (2000).
8. N. Chapel, A. Greiner, and J.-Y. Ortholand, *Tetrahedron Lett.*, **32**, 1441 (1991).
9. J.-Y. Ortholand, N. Vicart, and A. Greiner, *J. Org. Chem.*, **60**, 1880 (1995).

Ortho Esters: $RC(OR')_3$

Ortho esters are one of the few derivatives that can be prepared from acids and esters that protect the carbonyl against nucleophilic attack by hydroxide or other strong nucleophiles such as Grignard reagents. In general, ortho esters are difficult to prepare directly from acids and are therefore more often prepared from the nitrile.[1,2] Simple ortho esters derived from normal alcohols are the least stable in terms of acid stability and stability toward Grignard reagents, but as the ortho ester becomes more constrained its stability increases.

Formation

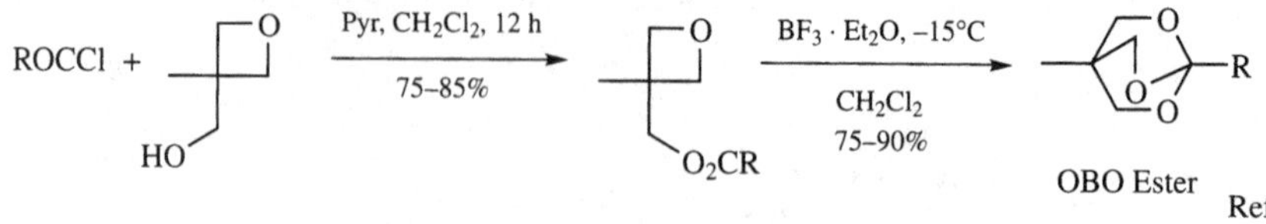

OBO Ester

Ref. 3

This is one of the few methods available for the direct and efficient conversion of an acid, via the acid chloride, to an ortho ester. An alternative esterification using and S_N2 displacement to form the ester is also possible.[4]

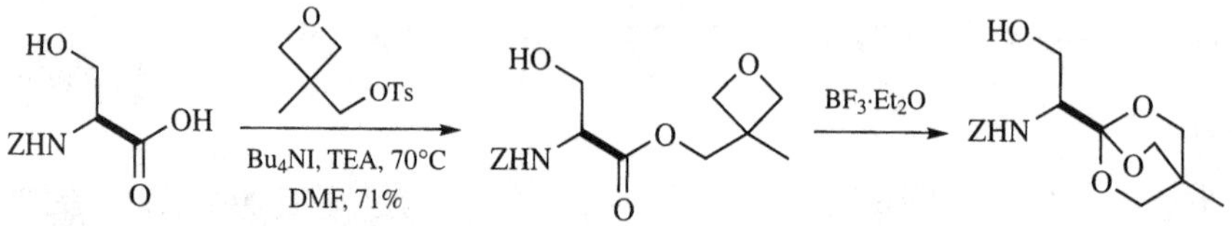

The ester precursor to the OBO group has also been prepared by transesterification using $ClBu_2SnOSnBu_2OH$ as a catalyst.[5] The preparation of the oxetane is straightforward and a large number of them have been prepared [triol, $(EtO)_2CO$, KOH].[6] In addition, the *t*-butyl analog has been used for the protection of acids.[7] During the course of a borane reduction, the ortho ester was reduced to form a ketal. This was attributed to an intramolecular delivery of the hydride.[8]

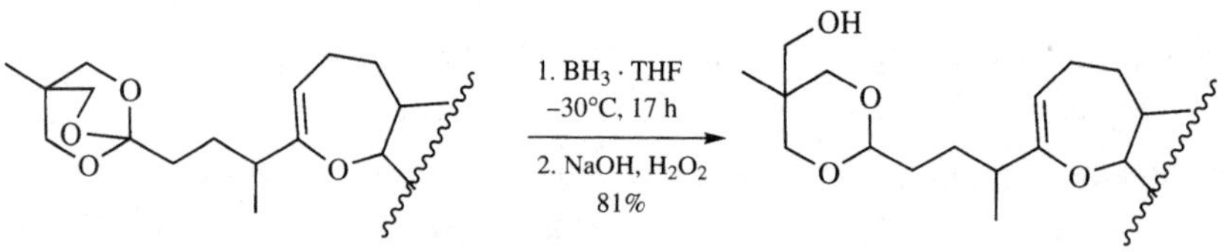

The OBO ester can also be prepared from a secondary or tertiary amide (Tf_2O, CH_2Cl_2, Pyr, then 2,2-bis(hydroxymethyl)-1-propanol, 10–88% yield).[9]

The addition of methyl groups to the oxetane precursor increases the rate of ortho ester formation by a factor of 22,000 over the OBO derivative and decreases its rate of acid catalyzed hydrolysis by a factor of 2.[10]

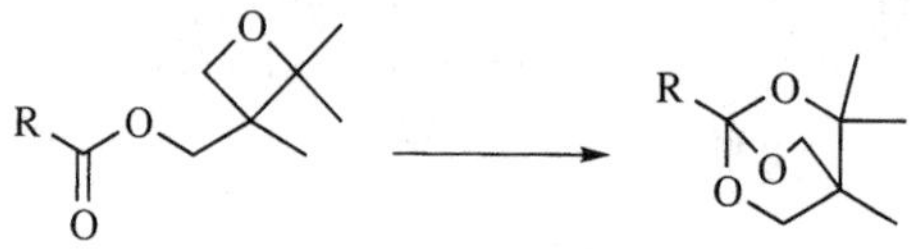

The complementary ABO ester (2,7,8-trioxabicyclo[3.2.1]octane ester) is prepared from the epoxy ester by rearrangement with $Cp_2ZrCl_2/AgClO_4$. The OBO ester is more easily cleaved by Brønsted acids than the ABO ester, but the ABO ester is cleaved more easily by Lewis acids, thus forming an orthogonal set. The ABO ester can be cleaved with PPTS[11] (MeOH, H_2O, 22°C, 2 h; LiOH); the OBO ester is cleaved at 0°C in 2 min.[12]

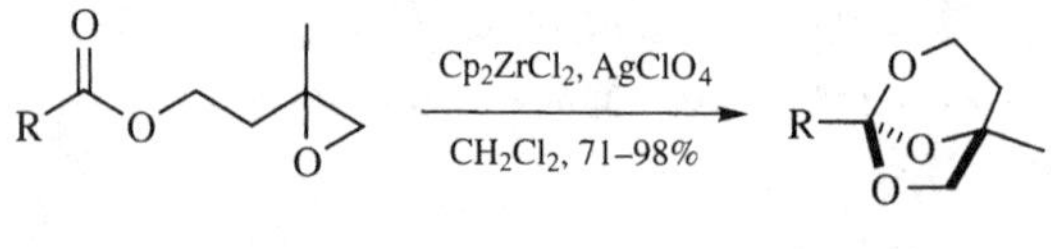

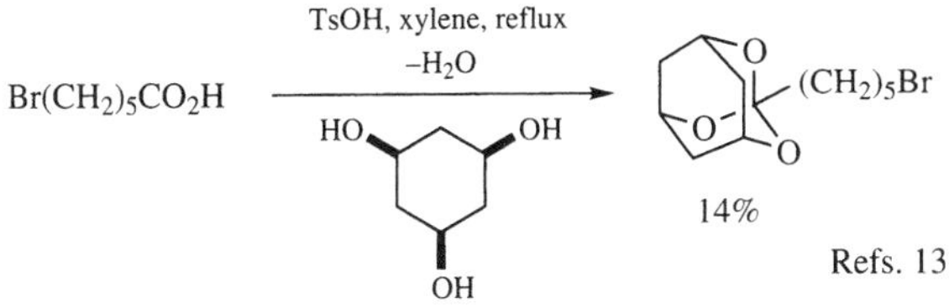

Br(CH$_2$)$_5$CO$_2$H

$\xrightarrow[\text{$-H_2$O}]{\text{TsOH, xylene, reflux}}$

(CH$_2$)$_5$Br

14%

Refs. 13, 14

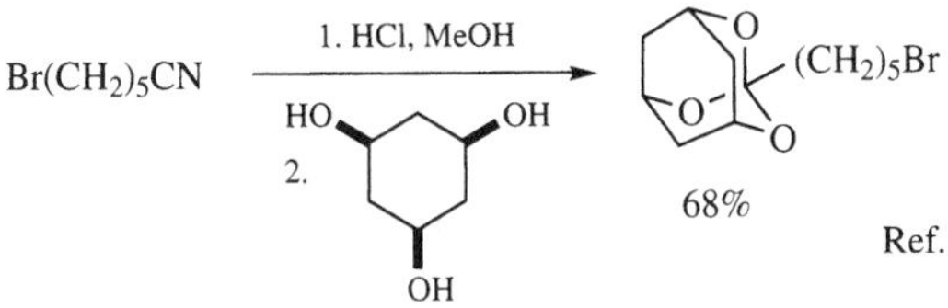

Br(CH$_2$)$_5$CN

$\xrightarrow[\text{2.}]{\text{1. HCl, MeOH}}$

(CH$_2$)$_5$Br

68%

Ref. 13

$\xrightarrow[\text{>88%}]{\substack{\text{HOCH$_2$CH$_2$OH} \\ \text{H$^+$, PhH, reflux}}}$

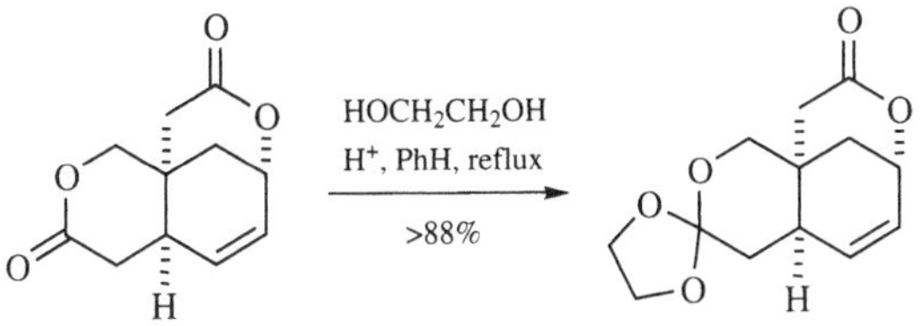

Refs. 15, 16

Note that this method does not work on simple esters. In addition, TMSO-CH$_2$CH$_2$OTMS/TMSOTF has been used to effect this conversion.[17] The same process was used to introduce the cyclohexyl version of this ortho ester in a quassinoid synthesis. Its cleavage was affected with DDQ in aqueous acetone.[18] When (R,R)-2,3-butandiol is used, it can be used to resolve the lactone.[19]

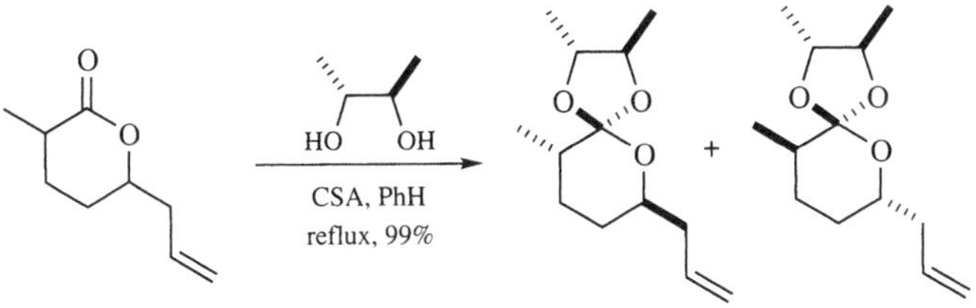

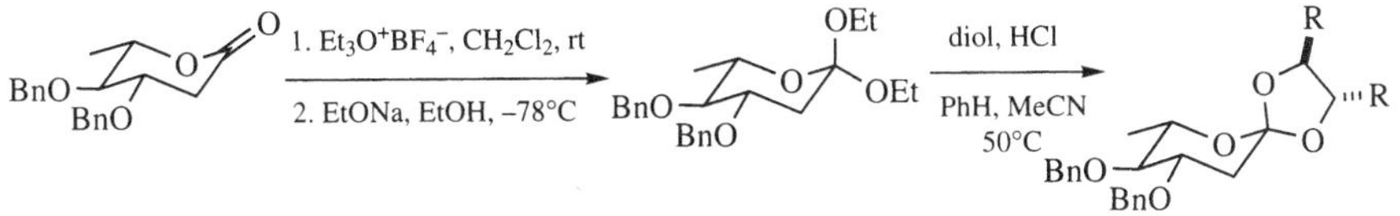

$\xrightarrow[\substack{\text{CSA, PhH} \\ \text{reflux, 99\%}}]{}$

+

$\xrightarrow[\text{2. EtONa, EtOH, $-$78°C}]{\text{1. Et$_3$O$^+$BF$_4$$^-$, CH$_2Cl_2$, rt}}$

$\xrightarrow[\substack{\text{PhH, MeCN} \\ \text{50°C}}]{\text{diol, HCl}}$

2-Substituted gulonolactones failed to react with Meerwein's salt.[20]

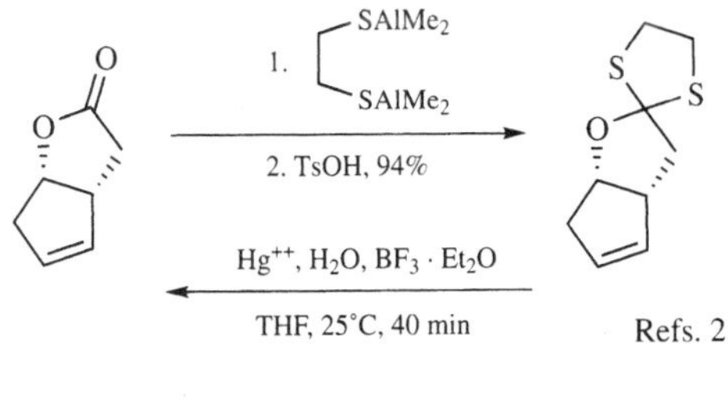

$\xrightarrow[\text{2. TsOH, 94\%}]{\text{1.}}$

$\xleftarrow[\text{THF, 25°C, 40 min}]{\text{Hg^{++}, H$_2$O, BF$_3$ · Et$_2$O}}$

Refs. 21, 22

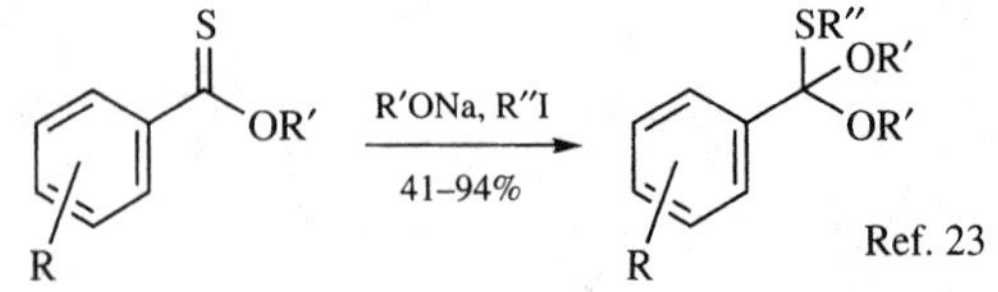

Ref. 23

Cleavage

Oxygen ortho esters are readily cleaved by mild aqueous acid (TsOH·Pyr, H_2O;[24] NaHSO$_4$, 5:1 DME, H_2O, 0°C, 20 min)[25] to form esters that are then hydrolyzed with aqueous base to give the acid. Note that a trimethyl ortho ester is readily hydrolyzed in the presence of an acid-sensitive ethoxyethyl acetal.[24] The order of acid stability is[26]

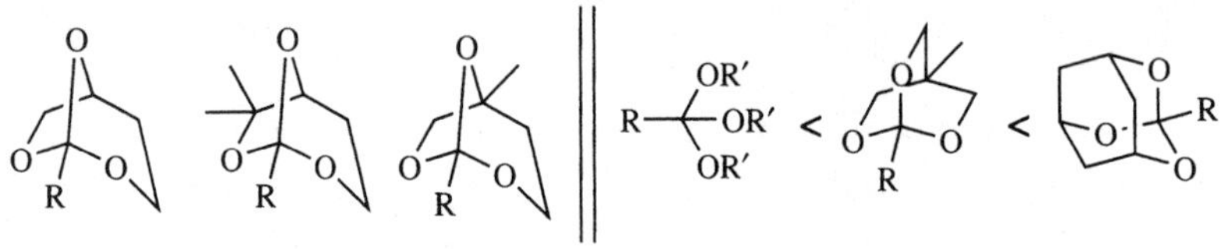

Relative rates of acid-catalyzed rearrangement to the ester = 7:3:1

Relative acid stability

Braun Ortho Ester

Formation/Cleavage[27]

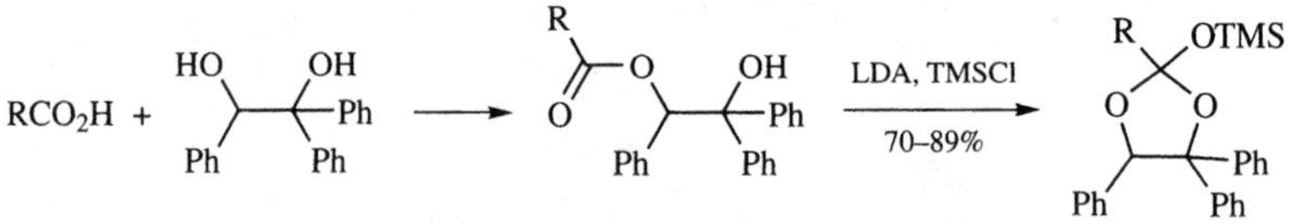

The derivative is stable to *n*-BuLi, *t*-BuLi (−78°C) and pH 6–8. It is cleaved with NaOH, MeOH/H_2O at reflux (96% yield).

1. S. M. McElvain and J. W. Nelson, *J. Am. Chem. Soc.*, **64**, 1825 (1942).
2. The synthesis and interconversion of simple ortho esters has been reviewed: R. H. DeWolfe, *Synthesis*, 153 (1974).
3. E. J. Corey and N. Raju, *Tetrahedron Lett.*, **24**, 5571 (1983).
4. M. A. Blaskovich, G. Evindar, N. G. W. Rose, S. Wilkinson, Y. Luo, and G. A. Lajoie, *J. Org. Chem.*, **63**, 3631 (1998). D. B. Hansen, X. Wan, P. J. Carroll, and M. M. Joullie, *J. Org. Chem.*, **70**, 3120 (2005).
5. A. Oku and M. Numata, *J. Org. Chem.*, **65**, 1899 (2000).
6. C. J. Palmer, L. M. Cole, J. P. Larkin, I. H. Smith, and J. E. Casida, *J. Agric. Food Chem.*, **39**, 1329 (1991).
7. B. R. DeCosta, A. Lewin, J. A. Schoenheimer, P. Skolnick, and K. C. Rice, *Heterocycles*, **32**, 2343 (1991); B. R. DeCosta, A. H. Lewin, K. C. Rice, P. Skolnick, and J. A. Schoenheimer, *J. Med. Chem.*, **34**, 1531 (1991).

8. K. C. Nicolaou, E. A. Theodorakis, F. P. J. T. Rutjes, M. Sato, J. Tiebes, X.-Y. Xiao, C.-K. Hwang, M. E. Duggan, Z. Yang, E. A. Couladouros, F. Sato, J. Shin, H.-M. He, and T. Bleckman, *J. Am. Chem. Soc.*, **117**, 10239 (1995).

9. A. B. Charette and P. Chua, *Tetrahedron Lett.*, **38**, 8499 (1997).

10. J.-L. Giner, *Org. Lett.*, **7**, 499 (2005).

11. P. Wipf and S. R. Spencer, *J. Am. Chem. Soc.*, **127**, 225 (2005).

12. P. Wipf, W. Xu, H. Kim, and H. Takahashi, *Tetrahedron*, **53**, 16575 (1997); P. Wipf and D. C. Aslan, *J. Org. Chem.*, **66**, 337 (2001).

13. G. Voss and H. Gerlach, *Helv. Chim. Acta*, **66**, 2294 (1983).

14. C. Müller, G. Voss, and H. Gerlach, *Liebigs Ann. Chem.*, 673 (1995).

15. T. Wakamatsu, H. Hara, and Y. Ban, *J. Org. Chem.*, **50**, 108 (1985).

16. J. D. White, S.-c. Kuo, and T. R. Vedananda, *Tetrahedron Lett.*, **28**, 3061 (1987).

17. D. J. Collins, L. M. Downes, A. G. Jhingran, S. B. Rutschman, and G. J. Sharp, *Aust. J. Chem.*, **42**, 1235 (1989); H. Ohtake and S. Ikegami, *Org. Lett.*, **2**, 457 (2000).

18. S. Vasudevan and D. S. Watt, *J. Org. Chem.*, **59**, 361 (1994).

19. A. B. Smith, III, J. W. Leahy, I. Noda, S. W. Remiszewski, N. J. Liverton, and R. Zibuck, *J. Am. Chem. Soc.*, **114**, 2995 (1992).

20. C. Zagar and H. D. Scharf, *Liebigs Ann. Chem.*, 447 (1993).

21. E. J. Corey and D. J. Beames, *J. Am. Chem. Soc.*, **95**, 5829 (1973).

22. M. Nakada, S. Kobayashi, S. Iwasaki, and M. Ohno, *Tetrahedron Lett.*, **34**, 1035 (1993).

23. J. Voss and B. Wollny, *Synthesis*, 684 (1989).

24. G. Just, C. Luthe, and M. T. P. Viet, *Can. J. Chem.*, **61**, 712 (1983).

25. E. J. Corey, K. Niimura, Y. Konishi, S. Hashimoto, and Y. Hamada, *Tetrahedron Lett.*, **27**, 2199 (1986).

26. J.-L. Giner, X. Li, and J. J. Mullins, *J. Org. Chem.*, **68**, 10079 (2003). This paper discusses the mechanistic details of the rearrangement.

27. D. Waldmüller, M. Braun, and A. Steigel, *Synlett*, 160 (1991).

Pentaaminecobalt(III) Complex: $[RCO_2Co(NH_3)_5](BF_4)_3$

The pentaaminecobalt(III) complex has been prepared from amino acids to protect the carboxyl group during peptide synthesis [$(H_2O)Co(NH_3)_5(ClO_4)_3$, 70°C, H_2O, 6h; cool to 0°C; filter; HBF_4, 60–80% yield]. It is cleaved by reduction [$NaBH_4$, NaSH, or $(NH_4)_2S$, Fe(II)EDTA]. These complexes do not tend to racemize and are stable to CF_3CO_2H that is used to remove BOC groups.[1-3] The related bisethylenediamine complex of amino acids has been prepared. It is stable to strong acids and is cleaved with ammonium sulfide.[4]

Hydrogens left off for clarity

1. S. Bagger, I. Kristjansson, I. Soetofte, and A. Thorlacius, *Acta Chem. Scand. Ser A*, **A39**, 125 (1985).
2. S. S. Isied, A. Vassilian, and J. M. Lyon, *J. Am. Chem. Soc.*, **104**, 3910 (1982).
3. S. S. Isied, J. Lyon, and A. Vassilian, *Int. J. Pept. Protein Res.*, **19**, 354 (1982).
4. P. M. Angus, B. T. Golding, and A. M. Sargeson, *J. Chem. Soc., Chem. Commun.*, 979 (1993).

Tetraalkylammonium salts: $R'_4N^+ \; {}^-O_2CR$

In a rather nontraditional approach to acid protection, the tetraalkylammonium salts of amino acids allow for coupling of HOBt-activated amino acids in yields of 55–84%.[1]

1. S.-T. Chen and K.-T. Wang, *J. Chem. Soc., Chem. Commun.*, 1045 (1990).

Stannyl Esters

Triethylstannyl Ester: $RCOOSn(C_2H_5)_3$

Tri-*n*-butylstannyl Ester: $RCOOSn(n\text{-}C_4H_9)_3$

Stannyl esters have been prepared to protect a $-COOH$ group in the presence of an $-NH_2$ group [$(n\text{-}Bu_3Sn)_2O$ or $n\text{-}Bu_3SnOH$, C_6H_6, reflux, 88%].[1] An improved method which does not require water removal involves reacting the acid directly with $n\text{-}Bu_3SnH$ at rt or with $n\text{-}Bu_3SnOCH_3$ at rt (50–100% yield).[2] Stannyl esters of *N*-acylamino acids are stable to reaction with anhydrous amines, and to water and alcohols;[3] aqueous amines convert them to ammonium salts.[3] Stannyl esters of amino acids are cleaved in quantitative yield by water or alcohols (PhSK, DMF, 25°C, 15 min, 63% yield, HOAc, EtOH, 25°C, 30 min, 77% yield[3] or KF, H_3O^+, 50–100% yield).[2]

1. P. Bamberg, B. Ekström, and B. Sjöberg, *Acta Chem. Scand.*, **22**, 367 (1968).
2. J. Thibonnet, M. Abarbri, J. L. Parrain, and A. Duchene, *Main Group Met. Chem.*, **20**, 195 (1997).
3. M. Frankel, D. Gertner, D. Wagner, and A. Zilkha, *J. Org. Chem.*, **30**, 1596 (1965).

AMIDES AND HYDRAZIDES

To a limited extent, carboxyl groups have been protected as amides and hydrazides, derivatives that complement esters in the methods used for their cleavage. Amides and hydrazides are stable to the mild alkaline hydrolysis that cleaves esters. Esters are stable to nitrous acid, effective in cleaving amides, and to the oxidizing agents [including $Pb(OAc)_4$, MnO_2, SeO_2, CrO_3, and $NaIO_4$;[1] $Ce(NH_4)_2(NO_3)_6$;[2]

$Ag_2O;^3$ and $Hg(OAc)_2^4$] that have been used to cleave hydrazides. Some amides and hydrazides that have been prepared to protect carboxyl groups are included in Reactivity Chart 6.

Formation

Classically, amides and hydrazides have been prepared from an ester or an acid chloride and an amine or hydrazine, respectively; they can also be prepared directly from the acid. Numerous activating agents have been used for the conversion of carboxylic acids to amides especially with regard to peptide bond formation.[5] It is beyond the scope of this book to give an exhaustive listing, and thus only a few methods are listed here that give a sampling of the types of available methods. The simplest method is generally the reaction of the acid chloride with an amine and an auxillary base such as TEA.

1. $\underset{\underset{NHPG}{|}}{RCHCOOH} + R'NH_2 \xrightarrow[20°C,\ 4\ h,\ 70\text{--}90\%]{DCC,\ THF\ or\ CH_2Cl_2{}^6} \underset{\underset{NHPG}{|}}{RCHCONHR'}$

Polymer supported diimides have also been used which facilitate removal of the coupling agent.[7]

2. $\underset{\underset{NHPG}{|}}{RCHCOOH} \xrightarrow{H_2NNH_2,\ N\text{-hydroxybenzotriazole}^8} \underset{\underset{NHPG}{|}}{RCHCONHNH_2}$

3. $RCOOH + R'R''NH \xrightarrow{Ph_3P^9\ or\ Bu_3P/o\text{-}NO_2\text{-}C_6H_4SCN^{10}} RCONR'R''$

4. $RCOOH + R'NH_2 \xrightarrow{C_6H_3F_3B(OH)_2\ toluene,\ heat^{11}} RCONHR'$

5. $RCO_2R' + NaEt_2Al(NR''_2)_2 \xrightarrow[83\text{--}96\%^{12}]{} RCONR''_2$

6. $RCOOH \xrightarrow[dioxane,\ Pyr,\ 4\text{--}16\ h^{13}]{NH_4OCO_2H,\ (BOC)_2O} RCONH_2$

7. $RCOOH + R'OCOCl + TEA \longrightarrow RCO_2CO_2R' \xrightarrow{R''_2NH} RCONR''_2$

This is a very general and mild method for the preparation of amides, applicable to large structural variations in both the acid and the amine. A variety of chloroformates can be used, but isobutyl chloroformate is used most often. The solvent is not critical, but generally THF is used. Even wet acetone can be used very efficiently.[14] The method has been applied to amino acid derivatives without erosion of chirality.[15]

8. From an aromatic amine: NaHMDS, 0°C to rt, then add an ester to effect aminolysis of the ester (88–99% yield). Even methyl pivalate reacts in high yield.[16]

9. Preparation of a primary amide: RCO_2H, urea, imidazole, microwaves, 47–88% yield.[17]

10. For an amino acid: Me_2SiCl_2, pyridine then RNH_2 (81–98% yield).[18]

11. Transamidation: $Sc(OTf)_3$ or $Ti(NMe_2)_4$ were shown to be effective catalysts for transamidation in toluene at 90°C.[19]

12. *S*-(1-Oxido-2-pyridinyl)1,1,3,3-tetramethylthiouronium tetrafluoroborate, NH_4Cl, DIPEA, DMF, 30 min, rt (46–99% yield of primary amide).[20]

13. From an acid and an amine: toluene, [emim][OTf], azeotropic reflux, 2 h, *N*-alkyl-4-boronopyridinium chloride as catalyst, 74–99% yield.[21]

Cleavage

Examples 1–11 illustrate some mild methods that can be used to cleave amides. Equations 1 and 2 indicate the conditions that were used by Woodward[22] and Eschenmoser,[23] respectively, in their synthesis of vitamin B_{12}. Butyl nitrite,[24] nitrosyl chloride,[25] and nitrosonium tetrafluoroborate ($NO^+BF_4^-$)[26] have also been used to cleave amides. Since only tertiary amides are cleaved by potassium *t*-butoxide (eq. 3), this method can be used to effect selective cleavage of tertiary amides in the presence of primary or secondary amides.[27] (Esters, however, are cleaved by similar conditions.)[28] Photolytic cleavage of nitro amides (eq. 4) is discussed in a review.[29]

1. $RCONH_2 \xrightarrow{N_2O_4/CCl_4^{22,30}} RCOOH$

2. $RCONH_2 \xrightarrow{[ClCH_2CH=N(\rightarrow O)\text{-}c\text{-}C_6H_{11} + AgBF_4]^{23}} \xrightarrow{H_3O^+} RCOOH$

3. $RCONR'R'' \atop R',R'' \neq H$ $\xrightarrow[24°C, 2–48 h, 88–96\%]{KO\text{-}t\text{-}Bu/H_2O \ (6:2), \ Et_2O^{27}} RCOOH$

4. a, b, or c $\xrightarrow[5–10 h, 70–100\%]{350 \ nm^{29}} RCOOH$

a = *o*-nitroanilides[31]
b = *N*-acyl-7-nitroanilides[32,33]
c = *N*-acyl-8-nitrotetrahydroquinolines[34]

5.

Treatment of acyl pyrroles with primary and secondary amines affords amides.[35]

6. The following cleavage proceeds via intramolecular assistance from the alkoxide formed on base treatment.[36,37]

7. For primary and secondary amides: $CuCl_2$, glyoxal, H_2O, pH 3.5, reflux. 92% yield.[38]

8. For primary amides: DMF dimethyl acetal, MeOH, 92–100% yield. The methyl ester is formed, but if MeOH is replaced with another alcohol, other esters can be prepared with similar efficiency.[39,40]

9. $NaNO_2$, AcOH, Ac_2O, 30 min 0°C to rt[41] then hydrolysis with LiOOH. These conditions were developed as a mild method to cleave an amide that was prone to decomposition under the more basic conditions.[42]

10. N_2O_4, −20°C, CH_3CN, 66–100% yield. Additionally, these conditions cleave hydroxamic acids, anilides and sulfonamides.[43,44] The following case illustrates the remarkable selectivity that can be obtained with this method.[45]

11. For a tertiary amide: Me_3OBF_4, Na_2HPO_4, CH_3CN then sat. $NaHCO_3$, 91% yield.[46] This method is only good for aromatic amides.

1. M. J. V. O. Baptista, A. G. M. Barrett, D. H. R. Barton, M. Girijavallabhan, R. C. Jennings, J. Kelly, V. J. Papadimitriou, J. V. Turner, and N. A. Usher, *J. Chem. Soc., Perkin Trans. 1*, 1477 (1977).

2. T.-L. Ho, H. C. Ho, and C. M. Wong, *Synthesis*, 562 (1972).

3. Y. Wolman, P. M. Gallop, A. Patchornik, and A. Berger, *J. Am. Chem. Soc.*, **84**, 1889 (1962).

4. J. B. Aylward and R. O. C. Norman, *J. Chem. Soc. C*, 2399 (1968).

5. C. A. G. N. Montalbetti and V. Falque, *Tetrahedron*, **61**, 10827 (2005).

6. J. C. Sheehan and G. P. Hess, *J. Am. Chem. Soc.*, **77**, 1067 (1955).

7. N. H. Kawahata, J. Brookes, and G. M. Makara, *Tetrahedron Lett.*, **43**, 7221 (2002).

8. For example, see S. S. Wang, I. D. Kulesha, D. P. Winter, R. Makofske, R. Kutny, and J. Meienhofer, *Int. J. Pept. Protein Res.*, **11**, 297 (1978).

9. L. E. Barstow and V. J. Hruby, *J. Org. Chem.*, **36**, 1305 (1971).

10. P. A. Grieco, D. S. Clark, and G. P. Withers, *J. Org. Chem.*, **44**, 2945 (1979).

11. K. Ishihara, S. Ohara, and H. Yamamoto, *J. Org. Chem.*, **61**, 4196 (1996).

12. T. B. Sim and N. M. Yoon, *Synlett*, 827 (1994).

13. V. F. Posdnev, *Tetrahedron Lett.*, **36**, 7115 (1995).

14. P. G. M. Wuts, unpublished results.

15. D. M. Shendage, R. Frohlich, and G. Haufe, *Org. Lett.*, **6**, 3675 (2004).

16. J. Wang, M. Rosingana, R. P. Discordia, N. Soundararajan, and R. Polniaszek, *Synlett*, 1485 (2001).

17. A. Khalafi-Nezhad, B. Mokhtari, and M. N. S. Rad, *Tetrahedron Lett.*, **44**, 7325 (2003).

18. S. H. van Leeuwen, P. J. L. M. Quaedflieg, Q. B. Broxterman, and R. M. J. Liskamp, *Tetrahedron Lett.*, **43**, 9203 (2002).

19. S. E. Eldred, D. A. Stone, S. H. Gellman, and S. S. Stahl, *J. Am. Chem. Soc.*, **125**, 3422 (2003).

20. M. A. Bailén, R. Chinchilla, D. J. Dodsworth, and C. Najera, *Tetrahedron Lett.*, **41**, 9809 (2000).

21. T. Maki, K. Ishihara, and H. Yamamoto, *Org. Lett.*, **7**, 5043 (2005).

22. R. B. Woodward, *Pure Appl. Chem.*, **33**, 145 (1973).

23. U. M. Kempe, T. K. Das Gupta, K. Blatt, P. Gygax, D. Felix, and A. Eschenmoser, *Helv. Chim. Acta*, **55**, 2187 (1972).

24. N. Sperber, D. Papa, and E. Schwenk, *J. Am. Chem. Soc.*, **70**, 3091 (1948).

25. M. E. Kuehne, *J. Am. Chem. Soc.*, **83**, 1492 (1961).

26. G. A. Olah and J. A. Olah, *J. Org. Chem.*, **30**, 2386 (1965).

27. P. G. Gassman, P. K. G. Hodgson, and R. J. Balchunis, *J. Am. Chem. Soc.*, **98**, 1275 (1976).

28. P. G. Gassman and W. N. Schenk, *J. Org. Chem.*, **42**, 918 (1977).

29. B. Amit, U. Zehavi, and A. Patchornik, *Isr. J. Chem.*, **12**, 103 (1974).

30. T. Itoh, K. Nagata, Y. Matsuya, M. Miyazaki, and A. Ohsawa, *Tetrahedron Lett.*, **38**, 5017 (1997).

31. B. Amit and A. Patchornik, *Tetrahedron Lett.*, **14**, 2205 (1973).

32. B. Amit, D. A. Ben-Efraim, and A. Patchornik, *J. Am. Chem. Soc.*, **98**, 843 (1976). G. Papageorgiou, D. C. Ogden, A. Barth, and J. E. T. Corrie, *J. Am. Chem. Soc.*, **121**, 6503 (1999).

33. J. Morrison, P. Wan, E. T. Corrie John, and G. Papageorgiou, *Photochem. Photobiol. Sci.*, **1**, 960 (2002).

34. B. Amit, D. A. Ben-Efraim, and A. Patchornik, *J. Chem. Soc., Perkin Trans I*, 57 (1976).

35. S. D. Lee, M. A. Brook, and T. H. Chan, *Tetrahedron Lett.*, **24**, 1569 (1983).

36. T. Tsunoda, O. Sasaki, and S. Itô, *Tetrahedron Lett.*, **31**, 731 (1990).

37. T. Tsunoda, O. Sasaki, O. Takenchi, and S. Ito, *Tetrahedron*, **47**, 3925 (1991).

38. L. Singh and R. N. Ram, *J. Org. Chem.*, **59**, 710 (1994).

39. P. L. Anelli, M. Brocchetta, D. Palano, and M. Visigalli, *Tetrahedron Lett.*, **38**, 2367 (1997).

40. D. J. Wardrop, A. I. Velter, and R. E. Forslund, *Org. Lett.*, **3**, 2261 (2001).

41. I. R. Vlahov, P. I. Vlahova, and R. R. Schmidt, *Tetrahedron Lett.*, **33**, 7503 (1992).

42. D. A. Evans, P. H. Carter, C. J. Dinsmore, J. C. Barrow, J. L. Katz, and D. W. Kung, *Tetrahedron Lett.*, **38**, 4535 (1997). For aryl amides, see D. T. Glatzhofer, R. R. Roy, and K. N. Cossey, *Org. Lett.*, **4**, 2349 (2002).

43. Y. H. Kim, K. Kim, and Y. J. Park, *Tetrahedron Lett.*, **31**, 3893 (1990).

44. D. S. Karanewsky, M. F. Malley, and J. Z. Gougoutas, *J. Org. Chem.*, **56**, 3744 (1991).

45. D. A. Evans, J. L. Katz, G. S. Peterson, and T. Hintermann, *J. Am. Chem. Soc.*, **123**, 12411 (2001). For another example, see D. A. Evans, J. C. Barrow, P. S. Watson, A. M. Ratz, C. J. Dinsmore, D. A. Evrard, K. M. DeVries, J. A. Ellman, S. D. Rychnovsky, and J. Lacour, *J. Am. Chem. Soc.*, **119**, 3419 (1997).

46. G. E. Keck, M. D. McLaws, and T. T. Wager, *Tetrahedron*, **56**, 9875 (2000).

Amides

N,N-**Dimethylamide:** $RCON(CH_3)_2$ (Chart 6)

Formation/Cleavage[1]

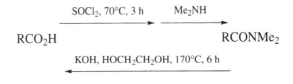

In these papers the carboxylic acid to be protected was a stable, unsubstituted compound. Harsh conditions were acceptable for both formation and cleavage of the amide. Typically, a simple secondary amide is very difficult to cleave. As the pK_a of the conjugate acid of an amide decreases, the rate of hydrolysis of amides derived from these amines increases. The dimethylamide of a cephalosporin was prepared as follows using 2,2'-dipyridyl disulfide.[2]

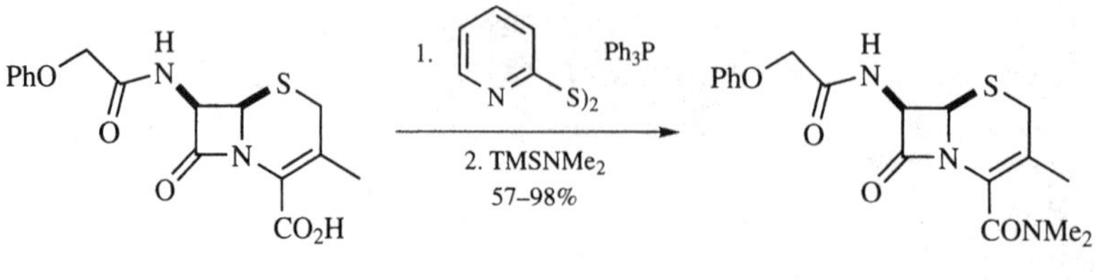

1. D. E. Ames and P. J. Islip, *J. Chem. Soc.*, 351 (1961); *idem, ibid.*, 4363 (1963).
2. R. DiFabio, V. Summa, and T. Rossi, *Tetrahedron*, **49**, 2299 (1993).

Pyrrolidinamide: RCONR'R", [R'R" = (−CH$_2$−)$_4$]

The following example illustrates how difficult it can be to hydrolyze a simple amide.

Formation/Cleavage[1]

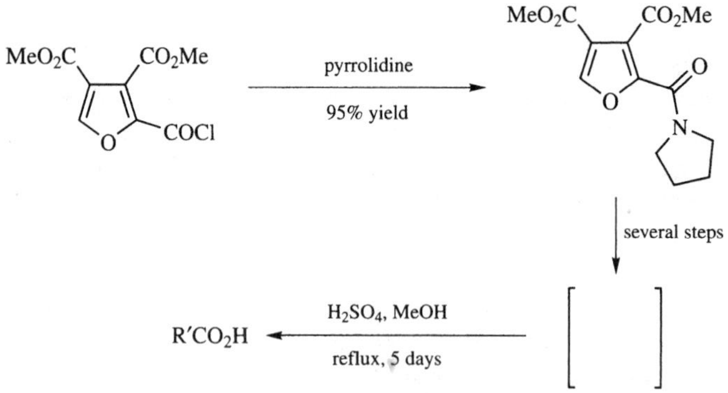

R'CO$_2$H = precursor to DL-camptothecin

1. A. S. Kende, T. J. Bentley, R. W. Draper, J. K. Jenkins, M. Joyeux, and I. Kubo, *Tetrahedron Lett.*, **14**, 1307 (1973).

Piperidinamide: RCONR'R", [R'R" = (−CH$_2$−)$_5$]

Formation/Cleavage[1]

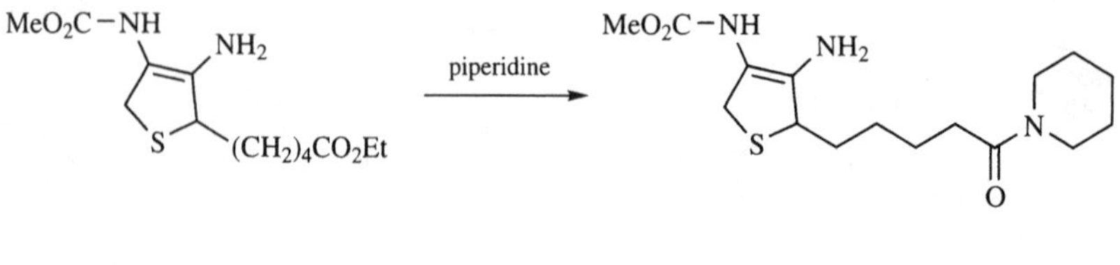

Biotin

1. P. N. Confalone, G. Pizzolato, and M. R. Uskokovic, *J. Org. Chem.*, **42**, 1630 (1977).

5,6-Dihydrophenanthridinamide

Formation/Cleavage

This amide is stable to HCl or KOH (THF, MeOH, H_2O, 70°C, 10 h) and MeMgI, THF, HMPA, −78°C. It can also be formed directly from the acid chloride.[1]

1. T. Uchimaru, K. Narasaka, and T. Mukaiyama, *Chem. Lett.*, **10**, 1551 (1981).

o-**Nitroanilide:** $RCONR'C_6H_4-o-NO_2$, $R' \neq H$

N-**7-Nitroindolylamide:** (Chart 6)

N-8-Nitro-1,2,3,4-tetrahydroquinolylamide

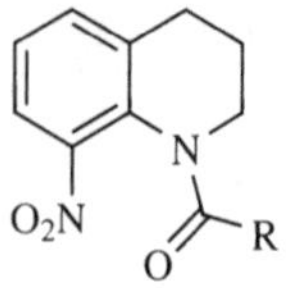

o-Nitroanilides (R' = Me, *n*-Bu, *c*-C$_6$H$_{11}$, Ph, PhCH$_2$; ≠ H)[1], nitroindolylamides,[2] and tetrahydroquinolylamides[3] are cleaved in high yields under mild conditions by irradiation at 350 nm (5–10 h).

1. B. Amit and A. Patchornik, *Tetrahedron Lett.*, **14**, 2205 (1973).
2. B. Amit, D. A. Ben-Efraim, and A. Patchornik, *J. Am. Chem. Soc.*, **98**, 843 (1976).
3. B. Amit, D. A. Ben-Efraim, and A. Patchornik, *J. Chem. Soc., Perkin Trans. I*, 57 (1976).

2-(2-Aminophenyl)acetaldehyde Dimethyl Acetal Amide

Nonaromatic amides are quite stable to hydrolysis, whereas aromatic amides are much more easily hydrolyzed. The amide is readily prepared from the acid chloride (Pyr, rt, 1 h, 77–86% yield) or the acid (DCC, DMAP, CH$_2$Cl$_2$, rt, 1 h, 88% yield). The following method takes advantage of this property in that the stable amide can be converted to the much more labile indole derivative. The acid can be regenerated from the *N*-acylindole by LiOH/H$_2$O$_2$/THF/H$_2$O or NaOH/MeOH. Alternatively, it can be transesterified with MeOH/TEA, converted to an amide by heating with an amine, or converted to an aldehyde by DIBAH (62–85% yield).[1]

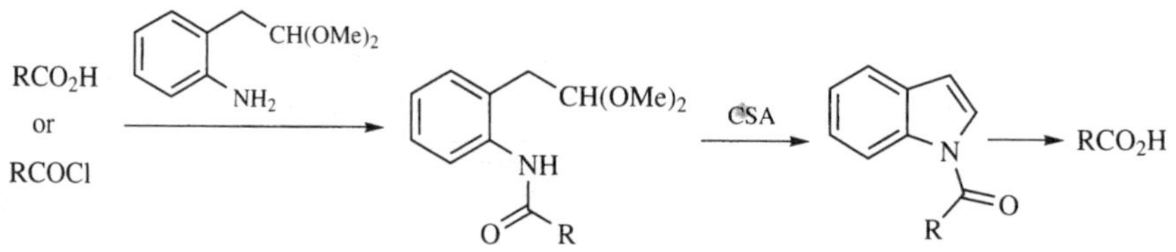

1. E. Arai, H. Tokuyama, M. S. Linsell, and T. Fukuyama, *Tetrahedron Lett.*, **39**, 71 (1998).

p-Polymer-Benzenesulfonamide: RCONHSO$_2$C$_6$H$_4$–*p-Polymer*

A polymer-supported sulfonamide, prepared from an amino acid activated ester and a polystyrene-sulfonamide, is stable to acidic hydrolysis (CF$_3$COOH; HBr/HOAc). It is cleaved by the "safety-catch" method shown below.[1] Prior to methylation basic hydrolysis is inhibited by salt formation at the acidic NH.

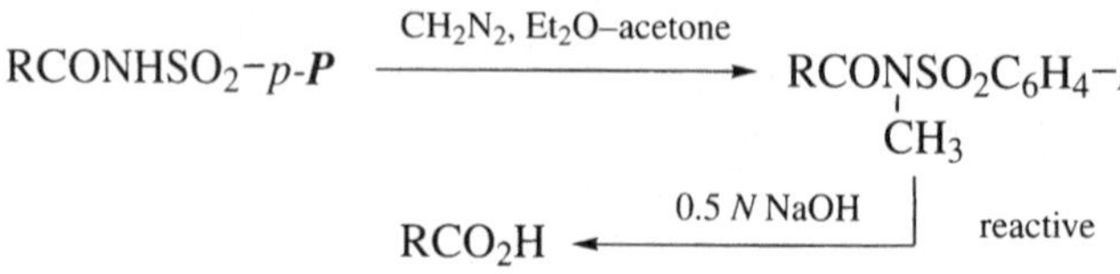

1. G. W. Kenner, J. R. McDermott, and R. C. Sheppard, *J. Chem. Soc., Chem. Commun.*, 636 (1971).

Hydrazides

Hydrazides: $RCONHNH_2$ (Chart 6)

Formation

Hydrazides are formed from an acid chloride, anhydride, or other activated ester and hydrazine.

Cleavage

1. NBS/H_2O, 25°C, 10 min, 74% yield.[1,2]
2. 60% $HClO_4$, 48°C, 24 h, 100% yield.[3]
3. $POCl_3$, H_2O, 94% yield.[3]
4. HBr/HOAc or HCl/HOAc, 94% yield.[3]
5. $CuCl_2$, H_2O, THF.[4] If an alcohol such as ethanol is substituted for H_2O in this reaction, the ester is produced instead of the acid.
6. *t*-BuONO, HOAt, rt. These conditions convert the hydrazide to an acyl–OAt group, which is much more easily hydrolyzed than a hydrazide.[5]

1. H. T. Cheung and E. R. Blout, *J. Org. Chem.*, **30**, 315 (1965).
2. K. J. Hale, L. Lazarides, and J. Cai, *Org. Lett.*, **3**, 2927 (2001).
3. J. Schnyder and M. Rottenberg, *Helv. Chim. Acta*, **58**, 521 (1975).
4. O. Attanasi and F. Serra-Zannetti, *Synthesis*, 314 (1980).
5. P. Wang, R. Layfield, M. Landon, R. J. Mayer, and R. Ramage, *Tetrahedron Lett.*, **39**, 8711 (1998).

***N*-Phenylhydrazide:** $RCONHNHC_6H_5$ (Chart 6)

Formation

Phenylhydrazides have been prepared from amino acid esters and phenylhydrazine in 70% yield.[1] The use of a carbodiimide and HOBt gives the hydrazide of amino acids in 56–99% yield.[2]

Cleavage

1. $Cu(OAc)_2$, 95°C, 10 min, 67% yield.[3,4] A reagent prepared from $CuBr_2$ and *t*-BuOLi in THF will convert a phenylhydrazide to the *t*-butyl ester 49–86% yield.[5]
2. $FeCl_3$/1 *N* HCl, 96°C, 14 min, 85% yield.[6]

3. Dioxane, DMF, 1 M aq. Pyr–AcOH buffer, AcOH, $CuCl_2$, 48 h, air, 86% yield.[7]

4. Horse radish peroxidase, H_2O_2 or Laccase, pH 4, 2% DMSO or DMF. Cleavage occurs by formation of a phenyldiimide, which decomposes to the acid, nitrogen, and benzene. The laccase method is compatible with the readily oxidized tryptophan and methionine because it does not use peroxide.[8]

5. Tyrosinase, rt, pH 7, CH_3CN, O_2, 17–99% yield.[2,9]

1. R. B. Kelly, *J. Org. Chem.*, **28**, 453 (1963).

2. M. Völkert, S. Koul, G. H. Muller, M. Lehnig, and H. Waldmann, *J. Org. Chem.*, **67**, 6902 (2002).

3. E. W.-Leitz and K. Kühn, *Chem. Ber.*, **84**, 381 (1951).

4. F. Stieber, U. Grether, and H. Waldmann, *Angew. Chem. Int. Ed.*, **38**, 1073 (1999).

5. J.-i. Yamaguchi, T. Aoyagi, R. Fujikura, and T. Suyama, *Chem. Lett.*, **30**, 466 (2001).

6. H. B. Milne, J. E. Halver, D. S. Ho, and M. S. Mason, *J. Am. Chem. Soc.*, **79**, 637 (1957).

7. A. N. Semenov and I. V. Lomonosova, *Int. J. Pept. Protein Res.*, **43**, 113 (1994).

8. A. N. Semenov, I. V. Lomonosova, V. I. Berezin, and M. I. Titov, *Biotechnol. Bioengin.*, **42**, 1137 (1993).

9. G. H. Müller and H. Waldmann, *Tetrahedron Lett.*, **40**, 3549 (1999).

N',N'-Dimethylhydrazide: RCONH-NMe₂

The N',N'-dimethylhydrazide is readily prepared from an acid chloride. It is cleaved with PhI(OH)OTs in CH_2Cl_2/H_2O (55–91% yield)[1] or with $MnO_2/AcOH$.[2]

N,N'-Diisopropylhydrazide: RCON(i-C₃H₇)NH−i-C₃H₇ (Chart 6)

The N,N'-diisopropylhydrazide, prepared to protect penicillin derivatives, is cleaved oxidatively by the following methods.[3]

1. Pb(OAc)₄/Pyr, 25°C, 10 min, 90% yield.

2. NaIO₄/H₂O-THF, H_2SO_4, 20°C, 5 min, 89% yield.

3. Aq. NBS/THF-Pyr, 20°C, 10 min, 90% yield.

4. CrO₃/HOAc, 25°C, 10 min, 65% yield.

5. A number of di- and trisubstituted hydrazides of penicillin and cephalosporin derivatives were prepared to study the effect of N-substitution on ease of oxidative cleavage.[4]

1. P. G. M. Wuts and M. P. Goble, *Org. Lett.*, **2**, 2139 (2000).

2. M. Romero and M. D. Pujol, *Synlett*, 173 (2003).

3. D. H. R. Barton, M. Girijavallabhan, and P. G. Sammes, *J. Chem. Soc., Perkin Trans. I*, 929 (1972).

4. D. H. R. Barton and eight co-workers, *J. Chem. Soc., Perkin Trans. I*, 1477 (1977).

Phenyl Group: C_6H_5-

The phenyl group became a practical "protective" group for carboxylic acids when Sharpless published a mild, effective one-step method for its conversion to a carboxylic acid.[1] It has recently been used in a synthesis of the amino acid statine, where it served as a masked or carboxylic acid equivalent.[2]

The furan group also serves as a protected carboxylic acid.[3] It is more readily converted to an acid in most cases.

1. P. H. J. Carlsen, T. Katsuki, V. S. Martin, and K. B. Sharpless, *J. Org. Chem.*, **46**, 3936 (1981).
2. S. Kano, Y. Yuasa, T. Yokomatsu, and S. Shibuya, *J. Org. Chem.*, **53**, 3865 (1988).
3. S. Sasaki, Y. Hamada, and T. Shioiri, *Tetrahedron Lett.*, **38**, 3013 (1997).

PROTECTION OF BORONIC ACIDS

Boronic esters are easily prepared from a diol and the boronic acid with removal of water either chemically or azeotropically (see Protection of Diols). Sterically hindered boronic esters such as those of pinacol can be prepared in the presence of water. Boronic esters of simple unhindered diols are quite water-sensitive and readily hydrolyze. On the other hand, those very hindered esters such as the **pinacol** and **pinanediol** derivatives are very difficult to hydrolyze and often require rather harsh conditions to achieve cleavage.

Pinacol and Pinanediol Esters

Cleavage

1. Ether, water, phenylboronic acid. Cleavage occurs by transesterification.[1,2]
2. (a) $NaIO_4$, NH_4OAc, acetone, water, 24–48 h; (b) pH 3 with HCl, 55–71% yield.[1,3]
3. BCl_3, $-78°C$, CH_2Cl_2, 8 h, 83% yield.[4] BBr_3 has also been used but also results in BOC cleavage.[5]

4. Pinanediolboronate: 3 N HCl, 120°C, 1 h, 55–58% yield.[6]
5. $LiAlH_4$, Et_2O; MeONa, 1,3-propanediol.[7] These conditions reduce the boronate to the hydride.

6. HN(CH$_2$CH$_2$OH)$_2$, ether; 1 N HCl, ~80% yield.[8,9] This method has also been used to cleave cedranediolboronates which are similar to the pinanediol derivative.[10]

7. Polystyrene-boronic acid, TFA, CH$_3$CN, reflux, 78–99% yield.[11]

8. KHF$_2$, CH$_3$CN, or MeOH followed by hydrolysis of the fluroborate salts with LiOH in CH$_3$CN/H$_2$O or TMSCl, H$_2$O, 21–100% yield.[12]

9.

This method was only partially successful with the pinanediol boranate.[13]

1,2-Benzenedimethanol: 1,2-(CH$_2$OH)$_2$C$_6$H$_4$

The ester is formed quantitatively in THF from the diol in the presence of a dehydrating agent such as sodium sulfate. It can be cleaved by hydrogenolysis, but it is also quite susceptible to hydrolytic cleavage.[14]

1,3-Diphenyl-1,3-propanediol: C$_6$H$_5$CH(OH)CH$_2$CH(OH)C$_6$H$_5$

Esterification is readily achieved in THF in the presence of a dehydrating agent.[15] The boronate is stable to chromatography, has good stability to 2M TFA/CH$_2$Cl$_2$, but is not stable to aqueous 1M NaOH. Cleavage is also achieved by hydrogenolysis.[14]

1,1,2,2-Tetraphenyl-1,2-ethanediol: (C$_6$H$_5$)$_2$C(OH)C(OH)(C$_6$H$_5$)$_2$

This group was used as a protecting group that was more stable than the pinacol group for the preparation of cyclopropaneboronic esters. No conditions were described for its removal.[16]

1-(4-Methoxyphenyl)-2-methylpropane-1,2-diol (MPMP-diol)

This group unlike the others is cleaved oxidatively with DDQ in CH$_2$CL$_2$/H$_2$O, rt to 50°C (65–85% yield). Only aromatic boronates were reported.[17]

1. S. J. Coutts, J. Adams, D. Krolikowski, and R. J. Snow, *Tetrahedron Lett.*, **35**, 5109 (1994).

2. J. Wityak, R. A. Earl, M. M. Abelman, Y. B. Bethel, B. N. Fisher, G. S. Kauffman, C. A. Kettner, P. Ma, J. L. McMillan, L. J. Mersinger, J. Pesti, M. E. Pierce, F. W. Rankin, R. J. Chorvat, and P. N. Confalone, *J. Org. Chem.*, **60**, 3717 (1995).

3. J. R. Falck, M. Bondlela, S. K. Venkataraman, and D. Srinivas, *J. Org. Chem.*, **66**, 7148 (2001).

4. D. S. Matteson, T. J. Michnick, R. D. Willett, and C. D. Patterson, *Organometallics*, **8**, 726 (1989).

5. C. Malan and C. Morin, *J. Org. Chem.*, **63**, 8019 (1998).

6. F. Morandi, E. Caselli, S. Morandi, P. J. Focia, J. Blazquez, B. K. Shoichet, and F. Prati, *J. Am. Chem. Soc.*, **125**, 685 (2003).

7. M. V. Rangaishenvi, B. Singaram, and H. C. Brown, *J. Org. Chem.*, **56**, 3286 (1991).

8. D. H. Kinder and M. M. Ames, *J. Org. Chem.*, **52**, 2452 (1987).

9. D. S. Matteson and R. Ray, *J. Am. Chem. Soc.*, **102**, 7590 (1980).

10. Y.-L. Song and C. Morin, *Synlett*, 266 (2001).

11. T. E. Pennington, C. Kardiman, and C. A. Hutton, *Tetrahedron Lett.*, **45**, 6657 (2004).

12. A. K. L. Yuen and C. A. Hutton, *Tetrahedron Lett.*, **46**, 7899 (2005).

13. D. S. Matteson and H.-W. Man, *J. Org. Chem.*, **61**, 6047 (1996).

14. C. Malan, C. Morin, and G. Preckher, *Tetrahedron Lett.*, **37**, 6705 (1996).

15. C. Malan and C. Morin, *Synlett*, 167 (1996).

16. J. Pietruszka and A. Witt, *Synlett*, 91 (2003).

17. J. Yan, S. Jin, and B. Wang, *Tetrahedron Lett.*, **46**, 8503 (2005).

PROTECTION OF SULFONIC ACIDS

Few methods exist for the protection of sulfonic acids. Imidazolides and phenolic esters are too base labile to be useful in most cases. Simple sulfonate esters often cannot be used because these are obviously quite susceptible to nucleophilic reagents.

Neopentyl Ester: $(CH_3)_3CCH_2OSO_2R$

The neopentyl sulfonate, prepared from the sulfonyl chloride (Pyr, 95% yield), is cleaved nucleophilically under rather severe conditions (Me_4NCl, DMF, 160°C, 16 h, 100% yield).[1] These may also be cleaved by acidolysis (CH_3CN, H_2O, 0.1% TFA, 4–5 days), with LiBr, butanone, reflux, 48 h,[2] or liquid HF, *m*-cresol, 100% yield.[3]

N-BOC-4-Amino-2,2-dimethylbutyl Sulfonate:
$BOCNHCH_2CH_2C(CH_3)CH_2OSO_2R$

This sulfonate, prepared from $BOCNHCH_2CH_2C(CH_3)CH_2OH$ and the sulfonyl chloride (Pyr, 100% yield), is cleaved by initial BOC cleavage to release the free amine after pH adjustment to 7–8. Intramolecular displacement occurs to release the sulfonate and a pyrrolidine.[1]

Isobutyl Sulfonate: $(CH_3)_2CHCH_2OSO_2R$

The isobutyl sulfonate was examined as a replacement for the isopropyl sulfonate that had undesirable stability properties. Cleavage occurs with 2 eq. of Bu_4NI and proceeds much more readily than cleavage of the isopropyl sulfonate.[4]

Isopropyl Sulfonate: $(CH_3)_2CHOSO_2R$

This sulfonate is cleaved with Bu_4NI or ammonia.[5] The group has been reported to suffer from stability problems upon storage and use.[4]

2,2,2-Trichloroethyl Sulfonate: $Cl_3CCH_2OSO_2R$

The TCE group is formed from the sulfonyl chloride[2] and can be cleaved by hydrogenolysis (Pd–C, NH_4HCO_2, MeOH, 36 h or Pd/C, AcOH, TFA, 3.5 h, 81–92% yield) or Zn powder and NH_4HCO_2, 95% yield.

2,2,2-Trifluoroethyl Sulfate: $F_3CCH_2OSO_2R$

This ester is prepared by the reaction of the acid with 2,2,2-trifluorodiazoethane in 46–93% yield. These esters are stable to TBAF and NaOMe/MeOH but not to t-BuOK in refluxing t-BuOH, which cleaves the ester to leave the potassium salt.[6,7]

Polymeric Benzyl Sulfonate: $Polymer-OC_6H_4CH_2OSO_2R$

These are introduced onto the polymer through the sulfonyl chloride (CH_3CN, TEA, rt) and are cleaved with TEA, TMG, DBU, or pyridine in either CH_3CN, CH_3OH, CH_2Cl_2, or DMF in 56–91% yield.[8]

2,5-Dimethylphenacyl Sulfonate

The sulfonate is prepared from the silver salt of the sulfonic acid by reaction with the phenacyl bromide in CH_3CN at reflux for 48 h (35–39% yield). It is cleaved by photolysis at >280 nm in benzene (90–94% yield).[9] The corresponding phosphate ester is cleaved similarly.

1. J. C. Roberts, H. Gao, A. Gopalsamy, A. Kongsjahju, and R. J. Patch, *Tetrahedron Lett.*, **38**, 355 (1997).
2. S. Liu, C. Dockendorff, and S. D. Taylor, *Org. Lett.*, **3**, 1571 (2001). G. T. Gunnarsson, M. Riaz, J. Adams, and U. R. Desai, *Bioorg. & Med. Chem.*, **13**, 1783 (2005).
3. L. Bischoff, C. David, B. P. Roques, and M.-C. Fournie-Zaluski, *J. Org. Chem.*, **64**, 1420 (1999); C. C. Kotoris, M.-J. Chen, and S. D. Taylor, *J. Org. Chem.*, **63**, 8052 (1998).
4. M. Xie and T. S. Widlanski, *Tetrahedron Lett.*, **37**, 4443 (1996).
5. B. Musicki and T. S. Widlanski, *Tetrahedron Lett.*, **32**, 1267 (1991); B. Musicki and T. S. Widlanski, *J. Org. Chem.*, **55**, 4231 (1990).
6. A. D. Proud, J. C. Prodger, and S. L. Flitsch, *Tetrahedron Lett.*, **38**, 7243 (1997).
7. N. A. Karst, T. F. Islam, and R. J. Linhardt, *Org. Lett.*, **5**, 4839 (2003); N. A. Karst, T. F. Islam, F. Y. Avci, and R. J. Linhardt, *Tetrahedron Lett.*, **45**, 6433 (2004).
8. A. Hari and B. L. Miller, *Org. Lett.*, **1**, 2109 (1999).
9. P. Klán, A. P. Pelliccioli, T. Pospisil, and J. Wirz, *Photochem. Photobiol. Sci.*, **1**, 920 (2002).

6

PROTECTION FOR THE THIOL GROUP

Greene's Protective Groups in Organic Synthesis, Fourth Edition, by Peter G. M. Wuts and Theodora W. Greene
Copyright © 2007 John Wiley & Sons, Inc.

MISCELLANEOUS DERIVATIVES 687

Protection for the thiol group is important in many areas of organic research, particularly in peptide and protein syntheses, which often involve the amino acid cysteine, $HSCH_2CH(NH_2)CO_2H$, CySH[1]. Protection of the thiol group in β-lactam chemistry has been reviewed.[2] The synthesis[3] of coenzyme A, which converts a carboxylic acid into a thioester, an acyl transfer agent in the biosynthesis or oxidation of fatty acids, also requires the use of thiol protective groups. A free −SH group can be protected as a thioether or a thioester, or oxidized to a symmetrical disulfide, from which it is regenerated by reduction. Thiols are more acidic than normal alcohols: $pK_a \sim 10–11$ for thiols versus $pK_a \sim 15–16$ for alcohols. Thiols are also more nucleophilic than alcohols, especially in basic solution. Thioethers are, in general, formed by reaction of the thiol, in a basic solution, with a halide; they are cleaved by reduction with sodium/ammonia, by acid-catalyzed hydrolysis, or by reaction with a heavy metal ion such as silver(I) or mercury(II), followed by hydrogen sulfide treatment. Some groups, including S-diphenylmethyl and S-triphenylmethyl thioethers and S-2-tetrahydropyranyl and S-isobutoxymethyl hemithioacetals, can be oxidized by thiocyanogen, $(SCN)_2$, iodine, or a sulfenyl chloride to a disulfide that is subsequently reduced to the thiol. Thioesters are formed and cleaved in the same way as oxygen esters; they are more reactive to nucleophilic substitution, as indicated by their use as "activated esters." Several miscellaneous protective groups, including thiazolidines, unsymmetrical disulfides, and S-sulfenyl derivatives, have been used to a more limited extent. This chapter discusses some synthetically useful thiol protective groups.[4,5] Some of the more useful groups are included in Reactivity Chart 7.

1. For a review on cysteine protection, see F. Cavelier, J. Daunis, R. Jacquier, *Bull. Soc. Chim. Fr.*, 210 (1990); see also reference 22 (peptides) in Chapter 1. L. Moroder, H.-J. Musiol, N. Schaschke, L. Chen, B. Hargittai, and G. Barany, "Protection of Functional Groups (Cysteine)," in *Synthesis of Peptides and Peptidomemetics;* M. Goodman, Ed., *Houben-Weyl*, 4th ed., Vol. E22a; Thieme, Stuttgart, 2002, pp 384–423.

2. H. Wild, "Protective Groups in β-Lactam Chemistry," in *The Organic Chemistry of β-Lactams,* G. I. Georg, Ed., 1993, pp. 1–48, VCH.

3. J. G. Moffatt and H. G. Khorana, *J. Am. Chem. Soc.*, **83**, 663 (1961).

4. See also Y. Wolman, "Protection of the Thiol Group," in *The Chemistry of the Thiol Group*, S. Patai, Ed., Wiley-Interscience, New York, 1974, Vol. 15/2, pp. 669–684; R. G. Hiskey, V. R. Rao, and W. G. Rhodes, "Protection of Thiols," in *Protective Groups in Organic Chemistry*, J. F. W. McOmie, Ed., Plenum Press, New York and London, 1973, pp. 235–308.

5. R. G. Hiskey, "Sulfhydryl Group Protection in Peptide Synthesis," in *The Peptides*, E. Gross and J. Meienhofer, Eds., Academic Press, New York, 1981, Vol. 3, pp 137–167.

THIOETHERS

S-Benzyl and substituted S-benzyl derivatives, readily cleaved with sodium/ammonia, are the most frequently used thioethers. n-Alkyl thioethers are difficult to cleave and have not been used extensively as protective groups. Alkoxymethyl or alkylthiomethyl hemithio- or dithioacetals ($RSCH_2OR'$ or $RSCH_2SR'$) can be cleaved by acidic hydrolysis or by reaction with silver or mercury salts, respectively. Mercury(II) salts also cleave dithioacetals, $RS-CH_2SR'$, S-triphenylmethyl thioethers, $RS-CPh_3$, S-diphenylmethyl thioethers, $RS-CHPh_2$, S-acetamidomethyl derivatives, $RS-CH_2NHCOCH_3$, and S-(N-ethylcarbamates), $RS-CONHEt$. S-t-Butyl thioethers, $RS-t$-Bu, are cleaved if refluxed with mercury(II); S-benzyl thioethers, $RS-CH_2Ph$, are cleaved if refluxed with mercury(II)/1 N HCl. Some β-substituted S-ethyl thioethers are cleaved by reactions associated with the β-substituent.

S-Alkyl Thioethers: $C_nH_{2n+1}SR$

Formation

1. *S,S*-Diphenyl-*S*-methoxythiazyne, benzene, 30°C, was used to prepare the methyl thioether.[1]
2. One of the simplest methods for preparation is by reaction of the thiol with KOH and RX in ethanol as solvent.
3. In many cases, a thiol group is introduced into a substrate through the use of a thiol (e.g., monoprotected H_2S) by simple displacement or an addition reaction.[2]
4. By the Mitsunobu reaction: 1,1′-(azodicarbonyl)dipiperidine, Me_3P, 61–85% yield. This reaction was used for the alkylation of thioglycosides. The addition of imidazole improves the process.[3]

Cleavage

1. Na/NH_3, >54% yield. Methyl thioether cleavage of BOC protected methionine.[4]
2. Na/NH_3.[5]

3. Na, naphthalene, THF, rt, 12 h, 97% yield.[6]

4. *t*-BuSNa, DMF, 160°C, 4 h, 95–99% yield. This method was specific for aryl *S*-methyl groups, probably because the cleavage occurs by an S_N2 process. An *S*-ethyl group failed to give clean results.[7]

1. T. Yoshimura, E. Tsukurimichi, Y. Sugiyama, H. Kita, C. Shimasaki, and K. Hasegawa, *Bull. Chem. Soc. Jpn.*, **64**, 3176 (1991).
2. For a review, see J. L. Wardell, in *The Chemistry of the Thiol Group*, part 1, S. Patai, Ed., Wiley, New York, 1974, pp. 179–211.
3. R. A. Falconer, I. Jablonkai, and I. Toth, *Tetrahedron Lett.*, **40**, 8663 (1999); J. R. Falck, J.-Y. Lai, S.-D. Cho, and J. Yu, *Tetrahedron Lett.*, **40**, 2903 (1999).
4. R. Lutgring, K. Sujatha, and J. Chmielewski, *Bioorg. Med. Chem. Lett.*, **3**, 739 (1993).
5. P. W. Ford, M. R. Narbut, J. Belli, and B. S. Davidson, *J. Org. Chem.*, **59**, 5955 (1994).

6. H. V. Huynh, C. Schulze-Isfort, W. W. Seidel, T. Lügger, R. Fröhlich, O. Kataeva, and F. E. Hahn, *Chem. Eur. J.*, **8**, 1327 (2002).

7. A. Pinchart, C. Dallaire, A. V. Bierbeek, and M. Gingras, *Tetrahedron Lett.*, **40**, 5479 (1999).

S-Benzyl Thioether: RSCH$_2$Ph (Chart 7)

For the most part cysteine and its derivatives have been protected by the following reactions.

Formation

1. PhCH$_2$Cl, 2 *N* NaOH or NH$_3$, EtOH, 30 min, 25°C, 90% yield.[1]
2. PhCH$_2$Cl, Cs$_2$CO$_3$, DMF, 20°C.[2]
3. PhCH$_2$Br, *n*-BuLi, THF, 0°C to rt, 30 min, 85% yield.[3]
4. Dibenzyl carbonate, DABCO, DMA, 135°C, 79% yield. Aryl amines, imides, and acids are also benzylated using this method.[4]

Cleavage

1. Na, NH$_3$, 10 min.[5]
2. Sodium in boiling butyl alcohol[6] or in boiling ethyl alcohol[7] can be used if the benzyl thioether is insoluble in ammonia.
3. Li, NH$_3$, THF, −78°C.[8]

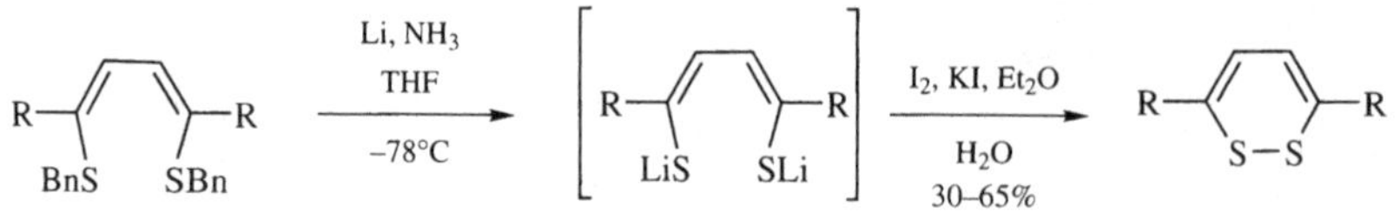

In this case the use of Na/NH$_3$ was slow.

4. Mg, ammonium formate, MeOH, rt, 90% yield. This method also cleaves *O*- and *N*-benzyl groups.[9] Aryl halides and esters are unaffected by this method.
5. Zn, ammonium formate, MeOH or ethylene glycol, rt, 90% yield. *O*- and *N*-benzyl groups are also cleaved.[10]
6. HF, anisole, 25°C, 1 h.[11] The authors list 15 protective groups that are cleaved by this method, including some branched-chain carbonates and esters, benzyl esters and ethers, the nitro-protective group in arginine, and *S*-benzyl and *S*-*t*-butyl thioethers. They report that 12 protective groups are stable under these conditions, including some straight-chain carbonates and esters, *N*-benzyl de-rivatives, and *S*-methyl, *S*-ethyl, and *S*-isopropyl thioethers.[11]
7. 5% Cresol, 5% thiocresol, 90% HF.[12] In the HF deprotection of thioethers and many other protective groups, anisole serves as a scavenger for the liberated cation formed during the deprotection process. If cations liberated during this deprotection are not scavenged, they can react with other amino acid residues,

especially tyrosine. Dimethyl sulfide, thiocresol, cresol, and thioanisole have also been used as scavengers when strong acids are used for deprotection. A mixture of 5% cresol, 5% p-thiocresol, and 90% HF is recommended for benzyl thioether deprotection.[12] These conditions cause cleavage by an S_N1 mechanism. The use of low concentrations of HF in dimethyl sulfide (1:3), which has been recommended for deprotection of other peptide protective groups, does not cleave the S-4-methylbenzyl group. Reactions that use low HF concentrations are considered to proceed via an S_N2 mechanism. The use of low HF concentrations with thioanisole results in some methylation of free thiols. The use of HF in anisole can also result in alkylation of methionine.

8. Electrolysis, NH_3, 90 min.[13]

9. Electrolysis, $-2.8\,V$, DMF, $R_4N^+X^-$, 82% yield.[14,15]

10. Ph_2SO, $MeSiCl_3$, TFA, 4°C, 4 h, 94% yield. The disulfide is formed.[16]

11. Bu_3SnH, AIBN, PhH, 3 h, Δ, >72% yield. The thiol is released as a stannyl sulfide that was used directly in a glycosylation.[17]

1. M. Frankel, D. Gertner, H. Jacobson, and A. Zilkha, *J. Chem. Soc.*, 1390 (1960); M. Dymicky and D. M. Byler, *Org. Prep. Proced. Int.*, **23**, 93 (1991).

2. F. Vögtle and B. Klieser, *Synthesis*, 294 (1982).

3. J. Yin and C. Pidgeon, *Tetrahedron Lett.*, **38**, 5953 (1997).

4. W.-C. Shieh, M. Lozanov, M. Loo, O. Repic, and T. J. Blacklock, *Tetrahedron Lett.*, **44**, 4563 (2003).

5. J. E. T. Corrie, J. R. Hlubucek, and G. Lowe, *J. Chem. Soc., Perkin Trans. 1*, 1421 (1977); K. Akaji, N. Kuriyama, and Y. Kiso, *J. Org. Chem.*, **61**, 3350 (1996).

6. W. I. Patterson and V. du Vigneaud, *J. Biol. Chem.*, **111**, 393 (1935).

7. K. Hofmann, A. Bridgwater, and A. E. Axelrod, *J. Am. Chem. Soc.*, **71**, 1253 (1949).

8. M. Koreeda and W. Yang, *Synlett.*, 201 (1994).

9. S. N. N. Babu, G. R. Srinivasa, D. C. Santhosh, and D. C. Gowda, *J. Chemical Res.*, 66 (2004).

10. G. R. Srinivasa, S. N. N. Babu, C. Lakshmi, and D. C. Gowda, *Synth. Commum.*, **34**, 1831 (2004).

11. S. Sakakibara, Y. Shimonishi, Y. Kishida, M. Okada, and H. Sugihara, *Bull. Chem. Soc. Jpn.*, **40**, 2164 (1967).

12. W. F. Heath, J. P. Tam, and R. B. Merrifield, *Int. J. Pept. Protein Res.*, **28**, 498 (1986).

13. D. A. J. Ives, *Can. J. Chem.*, **47**, 3697 (1969).

14. V. G. Mairanovsky, *Angew. Chem., Inter. Ed., Engl.*, **15**, 281 (1976).

15. C. M. Delerue-Matos, A. M. Freitas, H. L. S. Maia, M. J. Medeiros, M. I. Montenegro, and D. Pletcher, *J. Electroanal. Chem. Interfacial Electrochem.*, **315**, 1 (1991).

16. T. Koide, A. Otaka, H. Suzuki, and N. Fujii, *Synlett*, 345 (1991); K. Akaji, T. Tatsumi, M. Yoshida, T. Kimura, Y. Fujiwara, and Y. Kiso, *J. Chem. Soc., Chem. Comm.*, 167 (1991).

17. W. P. Neumann, *Synthesis*, 665 (1987); H.-S. Byun and R. Bittman, *Tetrahedron Lett.*, **36**, 5143 (1995).

S-p-Methoxybenzyl Thioether: $RSCH_2C_6H_4$-p-OCH_3 (Chart 7)

An S-4-methoxybenzyl thioether is stable to HBr/AcOH and to I_2/MeOH.[1] The latter reagent cleaves S-trityl and S-diphenylmethyl groups.

Formation

1. 4-$MeOC_6H_4CH_2Cl$, NH_3, 78% yield.[2]
2. 4-$MeOC_6H_4CH_2Cl$, Na/NH_3, 87% yield.[3]
3. 4-$MeOC_6H_4CH_2OH$, TFA, CH_2Cl_2, 37–81% yield.[4]
4.

Ar = 4-$MeOC_6H_4$, Ph, 4-MeC_6H_4

Ref. 4

5. 4-$MeOC_6H_4CH_2Cl$, NaH, THF, 60°C, 1 h.[5]

Cleavage

1. $Hg(OAc)_2$, CF_3COOH, 0°C, 10–30 min, or $Hg(OCOCF_3)_2$, aq. AcOH, 20°C, 2–3 h, followed by H_2S or $HSCH_2CH_2OH$, 100% yield.[6–8] An S-t-butyl thioether is cleaved in quantitative yield under these conditions.

$$R = PMB \xrightarrow[\text{10 min, DTT}]{Hg(OAc)_2, \text{ TFA, } 0°C} R = H$$

2. $Hg(OCOCF_3)_2$, CF_3COOH, anisole.[9] The dimethoxybenzyl thioether is also cleaved with this reagent.[10]

3. CF_3COOH, reflux.[2]
4. CF_3COOH, o-cresol, reflux, 24 h, >52% yield.[11]
5. Anhydrous HF, anisole, 25°C, 1 h, quant.[12]
6.

7. During the synthesis of peptides that contain 4-methoxybenzyl-protected cysteine residues, sulfoxide formation may occur. These sulfoxides, when treated with HF/anisole, form thiophenyl ethers that cannot be deprotected; therefore the peptides should be subjected to a reduction step prior to deprotection.[13]

MSA = methanesulfonic acid

8. $AgBF_4$, anisole, TFA, 4°C, 1 h, 87% conversion.[14]
9. $MeSiCl_3$, Ph_2SO, TFA, 4°C, 10 min, 95% conversion to cystine.[15]
10. In the following case, the PMB group was lost because of the nucleophilicity of sulfur reacting with a proximal mesylate.[16] This process was facilitated by the methoxy group, since the benzyl analog resulted in a much slower reaction.

11. BBr_3, CH_2Cl_2, −78°C, then NH_4Cl and warm to rt, 62% yield.[17]
12. Me_2SSMe^+ BF_4^-, MeOH, H_2O, 50–93% yield.[18] Disulfides are readily reduced to thiols.

13. 3-Nitro-2-pyridinesulfenyl chloride, CH_2Cl_2, 0°C, 30 min then Bu_3P, 69% yield.[19]

1. M. Platen and E. Steckhan, *Liebigs Ann. Chem.*, 1563 (1984).

2. S. Akabori, S. Sakakibara, Y. Shimonishi, and Y. Nobuhara, *Bull. Chem. Soc. Jpn.*, **37**, 433 (1964).

3. M. D. Bachi, S. Sasson, and J. Vaya, *J. Chem. Soc., Perkin Trans. I*, 2228 (1980).

4. L. S. Richter, J. C. Marsters, Jr., and T. R. Gadek, *Tetrahedron Lett.*, **35**, 1631 (1994).

5. A. W. Tayor and D. K. Dean, *Tetrahedron Lett.*, **29**, 1845 (1988).

6. O. Nishimura, C. Kitada, and M. Fujino, *Chem. Pharm. Bull.*, **26**, 1576 (1978).

7. E. M. Gordon, J. D. Godfrey, N. G. Delaney, M. M. Asaad, D. Von Langen, and D. W. Cushman, *J. Med. Chem.*, **31**, 2199 (1988).

8. D. Macmillan and D. W. Anderson, *Org. Lett.*, **6**, 4659 (2004); B. Zhou, J. Guo, and S. J. Danishefsky, *Org. Lett.*, **4**, 43 (2002).

9. T. P. Holler, A. Spaltenstein, E. Turner, R. E. Klevit, B. M. Shapiro, and P. B. Hopkins, *J. Org. Chem.*, **52**, 4420 (1987).

10. N. Shibata, J. E. Baldwin, A. Jacobs, and M. E. Wood, *Tetrahedron*, **52**, 12839 (1996).

11. R. Lutgring, K. Sujatha, and J. Chmielewski, *Bioorg. Med. Chem. Lett.* **3**, 739 (1993).

12. S. Sakakibara and Y. Shimonishi, *Bull. Chem. Soc. Jpn.*, **38**, 1412 (1965).

13. S. Funakoshi, N. Fujii, K. Akaji, H. Irie, and H. Yajima, *Chem. Pharm. Bull.*, **27**, 2151 (1979).

14. M. Yoshida, T. Tatsumi, Y. Fujiwara, S. Iinuma, T. Kimura, K. Akaji, and Y. Kiso, *Chem. Pharm. Bull.*, **38**, 1551 (1990).

15. K. Akaji, T. Tatsumi, M. Yoshida, T. Kimura, Y. Fujiwara, and Y. Kiso, *J. Chem. Soc., Chem. Commun.*, 167 (1991).

16. A. S. Grant, K. Chaudhary, L. Stewart, A. Peters, S. Delisle, and A. Decken, *Tetrahedron Lett.*, **45**, 1777 (2004).

17. S. T. Hilton, W. B. Motherwell, and D. L. Selwood, *Synlett*, 2609 (2004).

18. T. Zoller, J.-B. Ducep, C. Tahtaoui, and M. Hibert, *Tetrahedron Lett.*, **41**, 9989 (2000).

19. A. Ino, Y. Hasegawa, and A. Murabayashi, *Tetrahedron Lett.*, **39**, 3509 (1998).

S-o- or *p-*Hydroxy- or Acetoxybenzyl Thioether: $RSCH_2C_6H_4-o$(or *p*-)$-OR'$: $R' = H$ or Ac

Formation/Cleavage[1]

The cleavage process occurs by *p*-quinonemethide formation after acetate hydrolysis.

1. L. D. Taylor, J. M. Grasshoff, and M. Pluhar, *J. Org. Chem.*, **43**, 1197 (1978); J. B. Christensen, *Org. Prep. Proced. Int.*, **26**, 471 (1994); C. Gemmell, G. C. Janairo, J. D. Kilburn, H. Ueck, and A. E. Underhill, *J. Chem Soc., Perkin Trans. 1*, 2715 (1994).

S-p- and *S-o-*Nitrobenzyl Thioether: $RSCH_2C_6H_4$-*p* and *o*-NO_2 (Chart 7)

Formation

$4-NO_2C_6H_4CH_2Cl$, 1 *N* NaOH, 0°C, 1 h to 25°C, 0.5 h[1] or NaH, $PhCH_3$, 68% yield.[2]

Cleavage

1. H_2, Pd–C, HCl or AcOH, 7–8 h, 60–68%; $HgSO_4$, H_2SO_4, 20 h, 60%; H_2S, 15 min., 60% yield[2] or RSSR, 76% yield after air oxidation.[1] Hydrogenation initially produces the *p*-amino derivative, which is then cleaved with Hg(II).
2. Photolysis (366 nm), pH 6 buffer.[3]

1. M. D. Bachi and K. J. Ross-Petersen, *J. Org. Chem.*, **37**, 3550 (1972).
2. M. D. Bachi and K. J. Ross-Petersen, *J. Chem. Soc., Chem. Commun.*, 12 (1974).
3. A. B. Smith, III, S. N. Savinov, U. V. Manjappara, and I. M. Chaiken, *Org. Lett.*, **4**, 4041 (2002).

*S-*2,4,6-Trimethylbenzyl Thioether: $2,4,6-Me_3C_6H_2CH_2SR$

Formation

From cysteine: Na/NH_3, $2,4,6-Me_3C_6H_2CH_2Cl$, 57% yield.[1]

Cleavage

1. HF, anisole, 0°C, 30 min or TfOH, TFA, anisole, 30 min. This group is stable to refluxing TFA whereas the more frequently used 4-methoxybenzyl group is not.[1]
2. Me$_2$Se, HF, *m*-cresol, 0°C, 60 min. These conditions are also excellent for reduction of methionine sulfoxide [Met(O)].[2]
3. AgBF$_4$, anisole, TFA, 4°C, 1 h, 73% conversion.[3]

1. F. Brtnik, M. Krojidlo, T. Barth, and K. Jost, *Collect. Czech. Chem. Commun.*, **46**, 286 (1981).
2. M. Yoshida, M. Shimokura, Y. Fujiwara, T. Fujisaki, K. Akaji, and Y. Kiso, *Chem. Pharm. Bull.*, **38**, 382 (1990).
3. M. Yoshida, T. Tatsumi, Y. Fujiwara, S. Iinuma, T. Kimura, K. Akaji, and Y. Kiso, *Chem. Pharm. Bull.*, **38**, 1551 (1990).

S-2,4,6-Trimethoxybenzyl Thioether (Tmob-SR): 2,4,6-(MeO)$_3$C$_6$H$_2$CH$_2$SR

Formation

2,4,6-(MeO)$_3$PhCH$_2$OH, TFA, CH$_2$Cl$_2$, 84% yield.[1]

Cleavage

1. 5% H$_2$O, 5% phenol, 5% thioanisol in TFA/CH$_2$Cl$_2$ (30% v/v).[1]
2. TFA, CH$_2$Cl$_2$, triisopropylsilane or triethylsilane, 30 min, 25°C.[1]
3. Tl(TFA)$_3$, DMF, anisole, 0°C, 90 min.[1, 2]
4. Formic acid, cysteine, CH$_2$Cl$_2$, rt, 5 h, 40–99% yield. Cysteine is used to scavenge the Tmob cation.[3]

1. M. C. Munson, C. Garcia-Escheverría, F. Albericio, and G. Barany, *J. Org. Chem.*, **57**, 3013 (1992).
2. M. C. Munson, C. Garcia-Echeverría, F. Albericio, and G. Barany, *Pept.: Chem. Biol., Proc. Am. Pept. Symp., 12th*, 605 (1992).
3. C.-E. Lin, S. K. Richardson, and D. S. Garvey, *Tetrahedron Lett.*, **43**, 4531 (2002).

S-4-Picolyl Thioether: RSCH$_2$-4-pyridyl (Chart 7)

Formation

4-Picolyl chloride, 60% yield.[1]

Cleavage

Electrolytic reduction, 0.25 *M* H$_2$SO$_4$, 88% yield. *S*-4-Picolylcysteine is stable to CF$_3$COOH (7 days), to HBr/AcOH, and to 1 *M* NaOH. References for the electrolytic removal of seven other protective groups are included.[1,2]

1. A. Gosden, R. Macrae, and G. T. Young, *J. Chem. Res., Synop.*, 22 (1977).
2. C. M. Delerue-Matos, A. M. Freitas, H. L. S. Maia, M. J. Medeiros, M. I. Montenegro, and D. Pletcher, *J. Electroanal. Chem. Interfacial Electrochem.*, **315**, 1 (1991).

S-2-Picolyl *N*-Oxide Thioether: $RSCH_2$-2-pyridyl *N*-Oxide (Chart 7)

Formation

2-Picolyl chloride *N*-oxide, aq. NaOH, moderate yields.[1]

Cleavage

1. Ac_2O, reflux, 7 min or 25°C, 1.5 h followed by hydrolysis; aq NaOH, 25°C, 3–12 h, 79% yield.[1]
2. Electrolysis on a glassy carbon electrode, DMF, Bu_4NBF_4, 85% yield.[2]

1. Y. Mizuno and K. Ikeda, *Chem. Pharm. Bull.*, **22**, 2889 (1974).
2. M. D. Geraldo and M. J. Medeiros, *Port. Electrochim. Acta*, **9**, 175 (1991).

S-2-Quinolinylmethyl Thioether (Qm–SR)

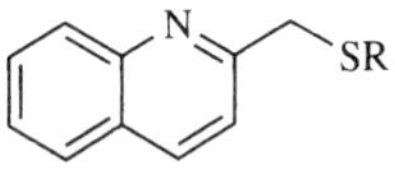

Formation

QmCl, NaH or NaOH or TEA, EtOH, 74% from cysteine.[1]

Cleavage

$FeCl_3$ or $CuCl_2$, DMF, H_2O, 61–99% yield, isolated as the disulfide. The quinoline group is isolated as the aldehyde.[1]

1. H. Yoshizawa, A. Otaka, H. Habashita, and N. Fujii, *Chem. Lett.*, **22**, 803 (1993).

S-9-Anthrylmethyl Thioether: $RSCH_2$-9-anthryl (Chart 7)

Formation

9-Anthrylmethyl chloride, DMF, −20°C, N_2.[1]

Cleavage

CH_3SNa, DMF or HMPA, 0–25°C, 2–5 h, 68–92% yield.[1] Cleavage proceeds by addition to the 10 position which results in expulsion or RS^-.

1. N. Kornblum and A. Scott, *J. Am. Chem. Soc.*, **96**, 590 (1974).

S-9-Fluorenylmethyl Ether (Fm−SR)

Formation

1. Et(i-Pr)$_2$N, DMF, FmCl.[1]
2. FmOTs, DMF, 0°C – 25°C, 71%. This procedure has the advantage that FmOTs is prepared in 83% yield from FmOH whereas the chloride, FmCl, is produced in only 30% yield from the alcohol and SOCl$_2$[2]
3. Conversion of an Fmoc group to an Fm group can be accomplished by treatment with TEA. These conditions do not cleave the Fmoc group from an amine.[3] A direct route uses FmocOSu.

Cleavage

1. 50% Piperidine, DMF or NH$_4$OH, 2 h.[4] The *S*-fluorenylmethyl group is stable to 95% HF/5% anisole for 1 h at 0°C, to trifluoroacetic acid, to 12 *N* HCl, to 0.1 *M* I$_2$ in DMF, and to CF$_3$SO$_3$H in CF$_3$COOH.[2]
2. (Me$_2$N)$_2$C=N−t-Bu, 23°C.[5]

Cleavage is followed by Micheal addition to quinonemethide

3. 1,2,4,6,7,8-Hexahydro-1-methyl-2*H*-pyrimido[1,2-*a*]-pyrimidine (MTBD), CH₃CN, DMF, Teoc-*p*-nitrophenyl.[6] This method results in protective group interchange to the *S*-Teoc derivative.

1. M. Bodanszky and M. A. Bednarek, *Int. J. Pept. Protein Res.*, **20**, 434 (1982).
2. F. Albericio, E. Nicolas, J. Rizo, M. Ruiz-Gayo, E. Pedroso, and E. Giralt, *Synthesis,* 119 (1990).
3. C. W. West, M. A. Estiarte, and D. H. Rich, *Org. Lett.*, **3**, 1205 (2001).
4. M. Ruiz-Gayo, F. Albericio, E. Pedroso, and E. Giralt, *J. Chem. Soc., Chem. Commun.*, 1501 (1986).
5. E. J. Corey, D. Y. Gin, and R. S. Kania, *J. Am. Chem. Soc.*, **118**, 9202 (1996).
6. C. W. West and D. H. Rich, *Org. Lett.*, **1**, 1819 (1999).

S-Xanthenyl Thioether (Xan−SR)

Formation

9*H*-Xanthen-9-ol, TFA, CH₂Cl₂, 25°C, 30 min.[1,2] The 2-methoxy analog can be prepared similarly, and it is cleaved only slightly faster than the unsubstituted derivative.

Cleavage[1]

1. 0.2% TFA, CH₂Cl₂, Et₃SiH. Other scavengers are not nearly as effective, but when the xanthenyl group is used on the solid phase, more acid is required to get efficient cleavage.

2. I_2, MeOH, DMF, or AcOH. AcOH is the most effective solvent, 67–100% yield.

3. $Tl(TFA)_3$, DMF, MeOH, CH_2Cl_2, or acetic acid, 94–100% yield.

1. Y. Han and G. Barany, *J. Org. Chem.*, **62**, 3841 (1997).
2. R. A. Falconer, *Tetrahedron Lett.*, **43**, 8503 (2002).

S-Ferrocenylmethyl Thioether (Fcm−SR):

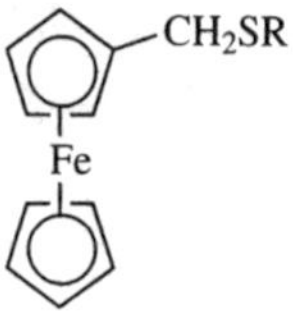

Formation

Cp-Fe-CpCH₂OH, TFA, acetone, H_2O, rt, overnight, 96% yield.[1]

Cleavage

The Fcm group can be removed with TFA, Ag(I) or Hg(II). The use of scavengers such as thiophenol and anisole is recommended. The Fcm group is stable to mild acid, and base, but it is not stable to electrophilic reagents such as $(SCN)_2$, I_2/AcOH, or carboxymethylsulfenyl chloride (CmsCl).[1]

1. A. S. J. Stewart and C. N. C. Drey, *J. Chem. Soc., Perkin Trans. 1*, 1753 (1990).

S-Diphenylmethyl, Substituted S-Diphenylmethyl, and S-Triphenylmethyl Thioethers

S-Diphenylmethyl, substituted S-diphenylmethyl, and S-triphenylmethyl thioethers have often been formed or cleaved by the same conditions, although sometimes in rather different yields. As an effort has been made to avoid repetition in the sections that describe these three protective groups, the reader should glance at all the sections.

S-Diphenylmethyl Thioether (DPM): $RSCH(C_6H_5)_2$ (Chart 7)

Formation

1. Ph_2CHOH, CF_3COOH, 25°C, 15 min or Ph_2CHOH, HBr, AcOH, 50°C, 2 h, >90% yield.[1]

2. Boron trifluoride etherate (in HOAc, 60–80°C, 15 min, high yields)[2] also catalyzes formation of S-diphenylmethyl and S-triphenylmethyl thioethers from aralkyl alcohols.

3. Yields of thioethers, formed under nonacidic conditions (Ph_2CHCl or Ph_3CCl, DMF, 80–90°C, 2 h, N_2), are not as high ($RSCHPh_2$, 50% yield; $RSCPh_3$, 75% yield)[3] as the yields obtained under the acidic conditions described above.

4. $AlPW_{12}O_{40}$, Ph_2CHOH, CH_2Cl_2, rt, 2–24 h, 81–96% yield.[4]

Cleavage

1. CF_3COOH, 2.5% phenol, 30°C, 2 h, 65% yield.[1] Zervas and co-workers tried many conditions for the acid-catalyzed formation and removal of the S-diphenylmethyl, S-4,4′-dimethoxydiphenylmethyl, and S-triphenylmethyl thioethers. The best conditions for the S-diphenylmethyl thioether are shown above. Phenol or anisole act as cation scavengers.

2. Na, NH_3, 97% yield.[3] Sodium/ammonia is an efficient but nonselective reagent ($RS-Ph$, $RS-CH_2Ph$, $RS-CPh_3$, and $RS-SR$ are also cleaved).

3. $2\text{-}NO_2C_6H_4SCl$, AcOH (results in disulfide formation), followed by $NaBH_4$ or $HS(CH_2)_2OH$ or dithioerythritol, quant.[5] S-Triphenylmethyl, S-4,4′-dimethoxydiphenylmethyl, and S-acetamidomethyl groups are also removed by this method.

4. I_2, CH_2Cl_2, reflux. These conditions result in the formation of disulfides, which may then be reduced using conventional methods.[4]

1. I. Photaki, J. T.-Papadimitriou, C. Sakarellos, P. Mazarakis, and L. Zervas, *J. Chem. Soc. C*, 2683 (1970).

2. R. G. Hiskey and J. B. Adams, Jr., *J. Org. Chem.*, **30**, 1340 (1965).

3. L. Zervas and I. Photaki, *J. Am. Chem. Soc.*, **84**, 3887 (1962).

4. H. Firouzabadi, N. Iranpoor, and A. A. Jafari, *Tetrahedron Lett.*, **46**, 2683 (2005).

5. A. Fontana, *J. Chem. Soc., Chem. Commun.*, 976 (1975).

S-Bis(4-methoxyphenyl)methyl Thioether (DMTr): $RSCH(C_6H_4\text{-}4\text{-}OCH_3)_2$ (Chart 7)

Formation

1. DMTrCl (dimethoxytrityl chloride), TEA, 80% aq. AcOH, 91% yield.[1]

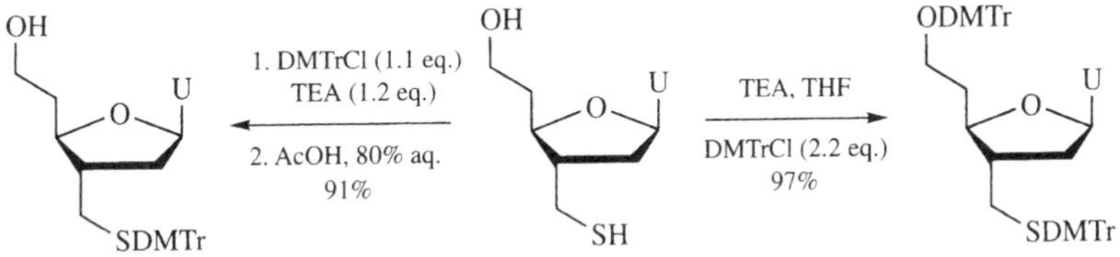

2. $(4\text{-}MeOC_6H_4)_2CHCl$, DMF, 25°C, 2 days, 96% yield.[2]

Cleavage

1. Selective cleavage of the DMT group from oxygen is accomplished with 80% aq. AcOH (rt, 10 min), whereas selective cleavage of the DMT group from the thiol is effected with $AgNO_3$ /NaOAc buffer (rt, 1 min).[1]
2. HBr, AcOH, 50–60°C, 30 min, or CF_3COOH, phenol, reflux, 30 min, quant.[2]

1. Z. Huang and S. A. Benner, *Synlett*, 83 (1993).
2. R. W. Hanson and H. D. Law, *J. Chem. Soc.*, 7285 (1965).

S-5-Dibenzosuberyl Thioether

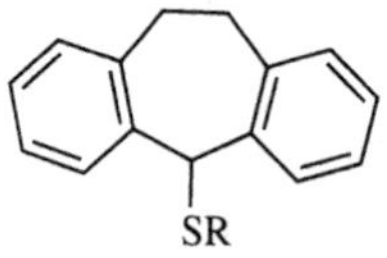

5-Dibenzosuberyl alcohol reacts in 60% yield with cysteine to give a thioether that is cleaved by mercury(II) acetate or oxidized by iodine to cystine. The dibenzosuberyl group has also been used to protect $-OH$, $-NH_2$, and $-CO_2H$ groups.[1]

1. J. Pless, *Helv. Chim. Acta*, **59**, 499 (1976).

S-Triphenylmethyl Thioether (Tr−SR): $RSC(C_6H_5)_3$ (Chart 7)

Formation

S-Triphenylmethyl thioethers have been formed by reaction of the thiol with triphenylmethyl alcohol/anhydrous CF_3COOH (85–90% yield) or with triphenylmethyl chloride (75% yield). Glycosidic triphenylmethyl thioethers are prepared by displacement of the chloride with $TrSN(Bu)_4$ (tetrabutylammonium triphenylmethanethiolate),[1] or simply TrSH in the presence of NaOMe/MeOH (100% yield).[2]

Cleavage

1. HCl, aq. AcOH, 90°C, 1.5 h.[3]
2. Trifluoroacetic acid.[4]
3. $Hg(OAc)_2$, EtOH, reflux, 3 h to 25°C, 12 h; H_2S, 61% yield.[3] Mercury salts will not cleave an N-Tr group except in the presence of $TFA/NaBH_4$.[5]
4. PhHgOAc 1.2 eq., $MeOH-CH_2Cl_2$ (4:1), 96% yield. The Hg salt is liberated with H_2S.[1,6]

5. $AgNO_3$, EtOH, Pyr, 90°C, 1.5 h; H_2S, 47% yield.[3] DTE (dithioerythritol) and NaOAc in MeOH/THF can be used in place of H_2S (97% yield).[7] An S-triphenylmethyl thioether can be selectively cleaved in the presence of an S-diphenylmethyl thioether by acidic hydrolysis or by heavy-metal ions. As a result of the structure of the substrate, the relative yields of cleavage by $AgNO_3$ and $Hg(OAc)_2$ can be reversed.[8]

6. Thiocyanogen [$(SCN)_2$, 5°C, 4 h, 40% yield] selectively oxidizes an S-triphenylmethyl thioether to the disulfide (RSSR) in the presence of an S-diphenylmethyl thioether.[9]

7. S-Triphenylmethylcysteine is readily oxidized by iodine (MeOH, 25°C) to cystine.[10,11]

The S-triphenylmethylcysteine group can be selectively cleaved in the presence of a −Cys(Acm)− group (Acm = acetamidomethyl).[12] S-Benzyl and S-t-butyl thioethers are stable to the action of iodine.

8. Electrolysis, −2.6 V, DMF, R_4NX.[13]

9. Et_3SiH, 50% TFA, CH_2Cl_2, 1 h, rt.[2,14] Et_3SiH is one of the best available scavengers for the trityl cation.

10. $TiCl_4$, CH_2Cl_2, 0°C, 53–96% yield. This deprotection is followed by cyclodehydration to form a thiazoline.[15]

4-Methoxytrityl (Mtt−SR) Thioether

The Mtt thioether is more easily cleaved with acid than the trityl ether because of improved cation stability.

Cleavage

1% TFA, CH_2Cl_2, Et_3SiH, quantitative.[16,17]

1. M. Blanc-Muesser, L. Vigne, and H. Driquez, *Tetrahedron Lett.*, **31**, 3869 (1990).

2. R. L. Harding and T. D. H. Bugg, *Tetrahedron Lett.*, **41**, 2729 (2000).

3. R. G. Hiskey, T. Mizoguchi, and H. Igeta, *J. Org. Chem.*, **31**, 1188 (1966).

4. H. B. Lee, M. Pattarawarapan, S. Roy, and K. Burgess, *Chem. Commun.*, 1674 (2003).

5. M. Maltese, *J. Org. Chem.*, **66**, 7615 (2001).

6. A. J. Pearson and J.-J. Hwang, *Tetrahedron*, **57**, 1489 (2001).

7. Z. Huang and S. A. Benner, *Synlett*, 83 (1993).

8. R. G. Hiskey and J. B. Adams, *J. Org. Chem.*, **31**, 2178 (1966).

9. R. G. Hiskey, T. Mizoguchi, and E. L. Smithwick, *J. Org. Chem.*, **32**, 97 (1967).

10. B. Kamber, *Helv. Chim. Acta*, **54**, 398 (1971).

11. K. W. Li, J. Wu, W. Xing, and J. A. Simon, *J. Am. Chem. Soc.*, **118**, 7237 (1996).

12. B. Kamber, A. Hartmann, K. Eisler, B. Riniker, H. Rink, P. Sieber, and W. Rittel, *Helv. Chim. Acta*, **63**, 899 (1980).

13. V. G. Mairanovsky, *Angew. Chem., Int. Ed. Engl.*, **15**, 281 (1976).

14. D. A. Pearson, M. Blanchette, M. L. Baker, and C. A. Guindon, *Tetrahedron Lett.*, **30**, 2739 (1989).

15. P. Raman, H. Razavi, and J. W. Kelly, *Org. Lett.*, **2**, 3289 (2000).

16. D. Kadereit, P. Deck, I. Heinemann, and H. Waldmann, *Chem. Eur. J.*, **7**, 1184 (2001).

17. B. Denis and E. Trifilieff, *J. Pept. Sci.*, **6**, 372 (2000).

S-Diphenyl-4-pyridylmethyl Thioether: RSC(C$_6$H$_5$)$_2$-4-pyridyl

Formation

Ph$_2$(4-C$_5$H$_4$N)COH, BF$_3$·Et$_2$O, AcOH, 60°C, 48 h.[1]

Cleavage

1. Hg(OAc)$_2$, AcOH, pH 4, 25°C, 15 min.[1]
2. Zn, 80% AcOH, H$_2$O.[2]

   The diphenylpyridylmethyl thioether is stable to acids (e.g., CF$_3$COOH, 21°C, 48 h; 45% HBr/AcOH, 21°C); it is oxidized by iodine to cystine (91%) or reduced by electrolysis at a mercury cathode.[1]

1. S. Coyle and G. T. Young, *J. Chem. Soc., Chem. Commun.*, 980 (1976).

2. S. Coyle, A. Hallett, M. S. Munns, and G. T. Young, *J. Chem. Soc., Perkin Trans. I*, 522 (1981).

S-Phenyl Thioether: RSC$_6$H$_5$

Although a sulfhydryl group generally is not converted to an *S*-phenyl thioether, the conversion can be accomplished through the use of a Pd-catalyzed arylation with an aryl iodide.[1] Thiophenol can be used to introduce sulfur into molecules by simple displacement or by Michael additions, and thus the phenyl group serves as a suitable protective group that can be removed by electrolysis ($-2.7\,V$, DMF, R_4NX).[2] The phenyl thioether is cleaved with $Pd(OAc)_2$ and TBDMS-H in DMA at rt in generally excellent yields. Alkyl thioethers are not effectively cleaved by this method.[3]

1. P. G. Ciattini, E. Morera, and G. Ortar, *Tetrahedron Lett.*, **36**, 4133 (1995).
2. V. G. Mairanovsky, *Angew. Chem., Int. Ed., Engl.*, **15**, 281 (1976).
3. M.-K. Chung and M. Schlaf, *J. Am. Chem. Soc.*, **126**, 7386 (2004).

S-2,4-Dinitrophenyl Thioether: RSC_6H_3-2,4-$(NO_2)_2$ (Chart 7)

Formation

2,4-$(NO_2)_2$-C_6H_3F, base.[1] The sulfhydryl group in cysteine can be selectively protected in the presence of the amino group by reaction with 2,4-dinitrophenol at pH 5–6.[2]

Cleavage

$HSCH_2CH_2OH$, pH 8, 22°C, 1 h, quant.[1]

1. S. Shaltiel, *Biochem. Biophys. Res. Commun.*, **29**, 178 (1967).
2. H. Zahn and K. Traumann, *Z. Naturforsch.*, **9b**, 518 (1954).

S-2-Quinolyl Thioether

2-Quinolinethiol is used to introduce sulfur as the thioether by an S_N2 reaction on a mesylate. The quinoline group is removed by $NaCNBH_3$ reduction in AcOH.[1] In the presence of a 2-amino group, the deprotection process failed.[2]

unsuccessful deprotection

1. J. Zhang and M. D. Matteucci, *Tetrahedron Lett.*, **40**, 1467 (1999).
2. Q. Dai and J. A. Piccirilli, *Org. Lett.*, **6**, 2169 (2004).

S-t-Butyl Thioether: $RSC(CH_3)_3$ (Chart 7)

Formation

1. Isobutylene, H_2SO_4, CH_2Cl_2, 25°C, 12h, 73% yield.[1] The S-t-butyl derivative of cysteine is stable to HBr/AcOH and to CF_3COOH.
2. t-BuOH, 2 N HCl, reflux, 90% yield.[2]
3. t-BuOH, H_2SO_4, H_2O, 0°C, 0.5h and rt, 2h, 98%.[3] A carboxylic acid was left unprotected under these conditions.

Cleavage

1. $Hg(OAc)_2$, CF_3COOH, anisole, 0°C, 15 min; H_2S, quant.[4]
2. $Hg(OCOCF_3)_2$, aq. AcOH, 25°C, 1 h; H_2S, quant.[4]
3. HF, anisole, 20°C, 30 min.[5] No cleavage is observed with HF, m-cresol.[6]
4. $2\text{-}NO_2C_6H_4SCl$; $NaBH_4$.[2,7] Treatment of the thioether with the sulfenyl chloride initially produces a disulfide that is then reduced to afford the free thiol.

5. Tetramethylene sulfoxide, TMSOTf, 4°C, 4 h, 87% yield or Ph_2SO, $MeSiCl_3$ or $SiCl_4$, TFA, 90–96% yield. The latter conditions also cleave the Acm, Bn, MeOBn, and MeBn groups. In all cases, disulfides are isolated.[8]
6. Catalytic Br_2, AcCl, AcOH, rt, 86–97% yield. This method results in the formation of the acetate which can be cleaved with mild base. Substrates containing acetylenes give low yields.[9]
7. BBr_3, CH_2Cl_2, AcCl, toluene, rt, 2h, 89–96% yield. The thioacetate is formed.[10]

1. F. M. Callahan, G. W. Anderson, R. Paul, and J. E. Zimmerman, *J. Am. Chem. Soc.*, **85**, 201 (1963).
2. J. J. Pastuszak and A. Chimiak, *J. Org. Chem.*, **46**, 1868 (1981).
3. R. Breitschuh and D. Seebach, *Synthesis*, 83 (1992).
4. O. Nishimura, C. Kitada, and M. Fujino, *Chem. Pharm. Bull.*, **26**, 1576 (1978).
5. S. Sakakibara, Y. Shimonishi, Y. Kishida, M. Okada, and H. Sugihara, *Bull. Chem. Soc. Jpn.*, **40**, 2164 (1967).
6. K. Akaji, K. Fujino, T. Tatsumi, and Y. Kiso, *J. Am. Chem. Soc.*, **115**, 11384 (1993).
7. J. M. Quintela and C. Peinador, *Tetrahedron*, **52**, 10497 (1996).
8. T. Koide, A. Otaka, H. Suzuki, and N. Fujii, *Synlett*, 345 (1991).

9. A. Blaszczyk, M. Elbing, and M. Mayor, *Org. Biomol. Chem.*, **2**, 2722 (2004).

10. N. Stuhr-Hansen, *Synth. Commun.*, **33**, 641 (2003).

S-1-Adamantyl Thioether: RS-1-Adamantyl

The *S*-adamantyl group is less prone to sulfoxide formation than the *S*-4-methoxy-benzyl group. It is also more stable to CF$_3$COOH.

Formation

1. 1-Adamantyl alcohol, CF$_3$COOH, 25°C, 12 h, 90% yield.[1]
2. From a disulfide: ArI(OCOAd)$_2$, Hg *hv*, CH$_2$Cl$_2$[2]

Cleavage

1. Hg(OAc)$_2$, CF$_3$COOH, 0°C, 15 min, 100% yield.[1]
2. Hg(OCOCF$_3$)$_2$, aq. AcOH, 20°C, 60 min, 100% yield.[1]
3. 1 *M* CF$_3$SO$_3$H, PhSCH$_3$ or Tl(OCOCF$_3$)$_3$.[3]

1. O. Nishimura, C. Kitada, and M. Fujino, *Chem. Pharm. Bull.*, **26**, 1576 (1978).

2. H. Togo, T. Muraki, and M. Yokoyama, *Synthesis*, 155 (1995).

3. N. Fujii, A. Otaka, S. Funakoshi, H. Yajima, O. Nishimura, and M. Fujino, *Chem. Pharm. Bull.*, **34**, 869 (1986); N. Fujii, H. Yajima, A. Otaka, S. Funakoshi, M. Nomizu, K. Akaji, I. Yamamoto, K. Torizuka, K. Kitagawa, T. Akita, K., Ando, T. Kawamoto, Y. Shimonishi, and T. Takao, *J. Chem. Soc., Chem. Commun.*, 602 (1985).

Substituted *S*-Methyl Derivatives: Monothio, Dithio, and Aminothio Acetals

S-Methoxymethyl Monothioacetal: RSCH$_2$OCH$_3$

Formation

1. BrCH$_2$OMe, DBU, CH$_2$Cl$_2$, rt, 10 min, >52% yield.[1]
2. ClCH$_2$OMe, DIPEA, CHCl$_3$, reflux, 8 h, 85% yield.[2]
3. Zn, (CH$_3$O)$_2$CH$_2$, BrCH$_2$CO$_2$Et, 80–82% yield. Formation of the methoxy-methyl thioether with dimethoxymethane[3] avoids the use of the **carcinogen chloromethyl methyl ether**.[4] The reaction forms an intermediate zinc thiolate, which then forms the monothioacetal.
4. ClCH$_2$Br, KOH, BnNEt$_3$Cl, MeOH, 70–90% yield.[5]
5. TEA, CH$_2$Cl$_2$ then MeOH, MeONa.[6]

1. H. Mastalerz, G. Zhang, J. Kadow, C. Fairchild, B. Long, and D. M. Vyas, *Org. Lett.*, **3**, 1613 (2001).

2. J. H. Zaidi, F. Naeem, K. M. Khan, R. Iqbal, and Zia-Ullah, *Syn. Commun.*, **34**, 2641 (2004).

3. F. Dardoize, M. Gaudemar, and N. Goasdoue, *Synthesis,* 567 (1977).

4. T. Fukuyama, S. Nakatsuka, and Y. Kishi, *Tetrahedron Lett.*, **17**, 3393 (1976).

5. F. D. Toste and I. W. J. Still, *Synlett*, 159 (1995).

6. C. Chen and Y.-J. Chen, *Tetrahedron Lett.*, **45**, 113 (2004).

S-Isobutoxymethyl Monothioacetal: $RSCH_2OCH_2CH(CH_3)_2$ (Chart 7)

The S-isobutoxymethyl monothioacetal is stable to 2 N hydrochloric acid and to 50% acetic acid; some decomposition occurs in 2 N sodium hydroxide.[1] The monothioacetal is also stable to 12 N hydrochloric acid in acetone (used to remove an N-triphenylmethyl group), and to hydrazine hydrate in refluxing ethanol (used to cleave an N-phthaloyl group).

Formation

$ClCH_2OCH_2CH(CH_3)_2$, 82% yield.[1]

Cleavage

1. 2 N HBr, AcOH, rapid.[1]
2. The S-isobutoxymethyl monothioacetal is cleaved by boron trifluoride etherate in acetic acid, by silver nitrate in ethanol, and by trifluoroacetic acid. The monothioacetal is oxidized to a disulfide by thiocyanogen, $(SCN)_2$.[2]

1. P. J. E. Brownlee, M. E. Cox, B. O. Handford, J. C. Marsden, and G. T. Young, *J. Chem. Soc.*, 3832 (1964).

2. R. G. Hiskey and J. T. Sparrow, *J. Org. Chem.*, **35**, 215 (1970).

S-Benzyloxymethyl Thioether (BOM−SR): $BnOCH_2SR$

Formation

$BnOCH_2Cl$, 4 N NaOH, 2 h, 0°C, 69% yield.[1]

Cleavage

AgOTf, TFA.[1]

1. A. Otaka, H. Morimoto, N. Fujii, T. Koide, S. Funakoshi, and H. Yajima, *Chem. Pharm. Bull.*, **37**, 526 (1989).

S-1-Ethoxyethyl Thioether (EE−SR): EtOCH(CH₃)SR

Formation/Cleavage

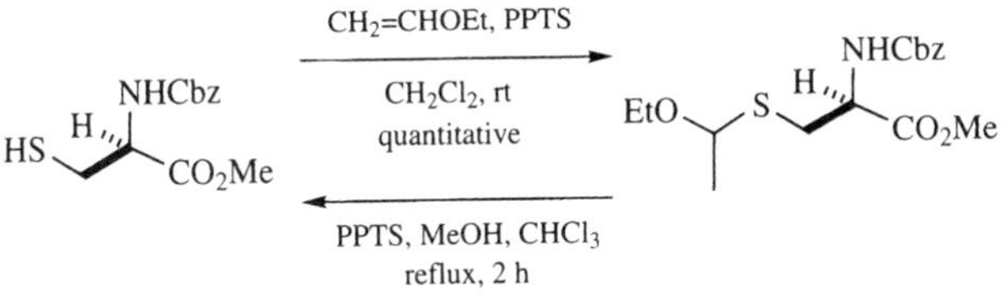

no racemization

Ref. 1

1. J. H. Zaidi, F. Naeem, K. M. Khan, R. I. Zia-Ullah, and S. Perveen, *J. Chem. Soc. Pak.*, **26**, 333 (2004).

S-2-Tetrahydropyranyl Monothioacetal: RS-2-tetrahydropyranyl (Chart 7)

An *S*-tetrahydropyranyl monothioacetal is stable to 4 *N* HCl/CH₃OH, 0°C, and to reduction with Na/NH₃. (An *O*-tetrahydropyranyl acetal is cleaved by 0.1 *N* HCl, 22°C, $t_{1/2} = 4 \, \text{min}$.)[1] An *S*-2-tetrahydropyranyl monothioacetal is oxidized to a disulfide by iodine[5] or thiocyanogen, $(SCN)_2$.[2]

Formation

1. Dihydropyran, BF₃·Et₂O, Et₂O, 0°C, 0.5 h to 25°C, 1 h, satisfactory yields.[3]
2. Dihydropyran, PPTS (pyridinium *p*-toluenesulfonate), 4 hr, 25°C, 92% yield.[4]

Cleavage

The section on monothioacetals in carbonyl protection should be consulted since those methods should be applicable in this case.

1. Aqueous AgNO₃, 0°C, 10 min, quant.[5]
2. HBr, CF₃COOH, 90 min, 100% yield.[6]
3. 37% HCl, rt, 30 min >86% yield.[7]

1. B. E. Griffin, M. Jarman, and C. B. Reese, *Tetrahedron*, **24**, 639 (1968).
2. R. G. Hiskey and W. P. Tucker, *J. Am. Chem. Soc.*, **84**, 4794 (1962).
3. R. G. Hiskey and W. P. Tucker, *J. Am. Chem. Soc.*, **84**, 4789 (1962).
4. E. Block, V. Eswarakrishnan, M. Gernon, G. O.-Okai, C. Saha, K. Tang, and J. Zubieta, *J. Am. Chem. Soc.*, **111**, 658 (1989).
5. G. F. Holland and L. A. Cohen, *J. Am. Chem. Soc.*, **80**, 3765 (1958).
6. K. Hammerström, W. Lunkenheimer, and H. Zahn, *Makromol. Chem.*, **133**, 41 (1970).
7. N. Zhang, M. Tomizawa, and J. E. Casida, *J. Org. Chem.*, **69**, 876 (2004).

S-Benzylthiomethyl Dithioacetal: $RSCH_2SCH_2C_6H_5$

Formation

$ClCH_2SCH_2Ph$, NH_3, 91% yield.[1]

Cleavage

$Hg(OAc)_2$, H_2O, 80% AcOH, $HSCH_2CH_2SH$, 25°C, 5–20 min; H_2S, 2 h, high yield.[1] The removal of an S-benzylthiomethyl protective group from a dithioacetal with mercury(II) acetate avoids certain side reactions that occur when an S-benzyl thioether is cleaved with sodium/ammonia. The dithioacetal is stable to hydrogen bromide/acetic acid used to cleave benzyl carbamates.

S-Phenylthiomethyl Dithioacetal: $RSCH_2SC_6H_5$

S-Phenylthiomethyl dithioacetals ($RSCH_2SC_6H_5$) were prepared and cleaved by methods similar to those used for the S-benzylthiomethyl dithioacetal.[1] The dithioacetal is stable to catalytic reduction (H_2/Pd–C, CH_3OH–HOAc, 12 h, the conditions used to cleave a p-nitrobenzyl carbamate).[2]

1. P. J. E. Brownlee, M. E. Cox, B. O. Handford, J. C. Marsden, and G. T. Young, *J. Chem. Soc.*, 3832 (1964).
2. R. Camble, R. Purkayastha, and G. T. Young, *J. Chem. Soc. C*, 1219 (1968).

Thiazolidine Derivative

Thiazolidines have been prepared from β-aminothiols—for example, cysteine—to protect the −SH and −NH groups during syntheses of peptides, including glutathione.[1] Thiazolidines are oxidized to symmetrical disulfides with iodine[2]; they do not react with thiocyanogen in a neutral solution.[3]

Formation[4]

Cleavage

1. HCl, H_2O, CH_3OH, 25°C, 3 days, high yield.[4]
2. $HgCl_2$, H_2O, 25°C, 2 days or 60–70°C, 15 min; H_2S, 20 min, 30–40% yield.[4]

3. *N*-BOC thiazolidines can be cleaved with ScmCl (methoxycarbonylsulfenyl chloride) (AcOH, DMF, H_2O) to afford the Scm derivative in >90% yield.[5]

4.

Na, NH₃, 95%

Ref. 6

1. F. E. King, J. W. Clark-Lewis, G. R. Smith, and R. Wade, *J. Chem. Soc.*, 2264 (1959).
2. S. Ratner and H. T. Clarke, *J. Am. Chem. Soc.*, **59**, 200 (1937).
3. R. G. Hiskey and W. P. Tucker, *J. Am. Chem. Soc.*, **84**, 4789 (1962).
4. J. C. Sheehan and D.-D. H. Yang, *J. Am. Chem. Soc.*, **80**, 1158 (1958).
5. D. S. Kemp and R. I. Carey, *J. Org. Chem.*, **54**, 3640 (1989).
6. F. D. Deroose and P. J. De Clercq, *J. Org. Chem.*, **60**, 321 (1995).

S-Acetamidomethyl Aminothioacetal (Acm−SR): $RSCH_2NHCOCH_3$ (Chart 7)

Formation

1. $AcNHCH_2OH$, concd. HCl, pH 0.5, 25°C, 1–2 days, 52% yield.[1]
2. $AcNHCH_2OH$, TFA.[2]

Cleavage

1. $Hg(OAc)_2$, pH 4, 25°C, 1 h; H_2S; air, 98% yield of cystine.[1] An *S*-acetamidomethyl group is hydrolyzed by the strongly acidic (6 *N* HCl, 110°C, 6 h) or strongly basic conditions used to cleave amide bonds. It is stable to anhydrous trifluoroacetic acid and to hydrogen fluoride (0°C, 1 h; 18°C, 1 h, 10% cleaved). On the other hand, in the presence of scavengers such as anisole and thioanisole during TFA cleavage of protective groups, the Acm group is susceptible to partial cleavage and to migration to the tyrosine hydroxyl.[3] It is stable to zinc in acetic acid and to hydrazine in acetic acid or methanol.[1] If the Acm group is oxidized, there is no satisfactory method to liberate the cysteine. Cleavage of the sulfoxide with HF/anisole or CH_3SO_3H/anisole affords $Cys(C_6H_4OMe)$.[4]

2. $2\text{-}NO_2C_6H_4SCl$, AcOH; $HO(CH_2)_2SH$ or $NaBH_4$, quant.[5]
3. PhSH. This reagent affords the phenyl disulfide.[4]
4. $ClSCO_2Me$, MeOH, 80% yield.[6]

These conditions convert the Acm group to a methyl *S*-sulfenylthiocarbonate group (Scm group), which can be cleaved with dithiothreitol.[7]

5. ClCOSCl, CHCl$_3$; PhNHMe.[7]

The S-(N'-methyl-N'-phenylcarbamoyl)sulfenyl group (Snm group) produced under these conditions is stable to HF or CF_3SO_3H. Since there are few acid-stable SH protective groups, the Snm group should prove useful where strong acids are encountered in synthesis.

6. MeSiCl$_3$, Ph$_2$SO, TFA, 4°C, 30 min, 93% yield. These conditions also cleave the Tacm, Bam (benzamidomethyl), t-Bu, MeOBn, and MeBn groups in high yield.[8]

7. AgTFA, TFA/anisole (95:5), 3 h, rt; H$_2$S.[9]

8. Tl(TFA)$_3$, TFA, anisole, 1 h, 66% yield.[10]

9. AgBF$_4$, anisole, TFA, 4°C, 1 h, 93% yield. The benzamidomethyl (Bam), 4-methoxybenzyl, and 2,4,6-trimethylbenzyl (Tmb) groups are only partially cleaved under these conditions (87%, 87%, and 73%, respectively).[11]

10. I$_2$. Met, Tyr, His, and Trp are susceptible to over-oxidation with iodine if reaction conditions are not carefully controlled.[12]

11. TFA, triisopropylsilane, 70% yield.[13]

1. D. F. Veber, J. D. Milkowski, S. L. Varga, R. G. Denkewalter, and R. Hirschmann, *J. Am. Chem. Soc.*, **94**, 5456 (1972); J. D. Milkowski, D. F. Veber, and R. Hirschmann, *Org. Synth., Collect. Vol. VI*, 5 (1988).

2. P. Marbach and J. Rudinger, *Helv. Chim. Acta,* **57**, 403 (1974).

3. M. Engebretsen, E. Agner, J. Sandosham, and P. M. Fischer, *J. Peptide Research*, **49**, 341 (1997).

4. H. Yajima, K. Akaji, S. Funakoshi, N. Fujii, and H. Irie, *Chem. Pharm. Bull.*, **28**, 1942 (1980).

5. L. Moroder, F. Marchiori, G. Borin, and E. Schoffone, *Biopolymers*, **12**, 493 (1973); A. Fontana, *J. Chem. Soc., Chem. Commun.*, 976 (1975).

6. R. G. Hiskey, N. Muthukumaraswamy, and R. R. Vunnam, *J. Org. Chem.*, **40**, 950 (1975).

7. A. L. Schroll and G. Barany, *J. Org. Chem.*, **54**, 244 (1989).

8. K. Akaji, T. Tatsumi, M. Yoshida, T. Kimura, Y. Fujiwara, and Y. Kiso, *J. Chem. Soc., Chem. Commun.*, 167 (1991).

9. Z. Chen and B. Hemmasi, *Biol. Chem. Hoppe-Seyler*, **374**, 1057 (1993).

10. N. Fujii, A. Otaka, S. Funakoshi, K. Bessho, and H. Yajima, *J. Chem. Soc., Chem. Commun.*, 163 (1987); C. Garcia-Echeverria, M. A. Molins, F. Alberico, M. Pons, and E. Giralt, *Int. J. Pept. Protein Res.*, **35**, 434 (1990).

11. M. Yoshida, K. Akaji, T. Tatsumi, S. IInuma, Y. Fujiwara, T. Kimura, and Y. Kiso, *Chem. Pharm. Bull.*, **38**, 273 (1990).

12. B. Kamber, *Helv. Chim. Acta*, **54**, 927 (1971); B. Kamber, A. Hartmann, K. Eisler, B. Riniker, H. Rink, P. Sieber, and W. Rittle, *Helv. Chim. Acta*, **63**, 899 (1980).

13. P. R. Singh, M. Rajopadhye, S. L. Clark, and N. E. Williams, *Tetrahedron Lett.*, **37**, 4117 (1996).

S-Trimethylacetamidomethyl Aminothioacetal (Tacm−SR): $(CH_3)_3CCONHCH_2SR$

Formation

$(CH_3)_3CCONHCH_2OH$, TFA, rt, 1 h, >85% yield.[1]

Cleavage

1. I_2, AcOH, EtOH, 25°C, 1 h, 100% yield. These conditions can result in methionine oxidation.[4]

2. $Hg(OAc)_2$, TFA, 0°C, 30 min. The Tacm group is stable to HF (0°C, 1 h); to 1 M CF_3COOH, $PhSCH_3$ (0°C, 1 h); to 0.5 M NaOH/MeOH (0°C, 1 h); to NH_2NH_2, MeOH, and to Zn/AcOH. It is not stable to 25% HBr/AcOH, 2 h, rt.[1] This group was reported to be more useful than the Acm group because it was less susceptible to by-product formation and oxidation.[2] The Pim (phthalimidomethyl) group is stable under these conditions.[3]

3. $AgBF_4$, anisole, 0°C, 1 h, quant. These conditions also cleave the Acm group.[4]

1. Y. Kiso, M. Yoshida, Y. Fujiwara, T. Kimura, M. Shimokura, and K. Akaji, *Chem. Pharm. Bull.*, **38**, 673 (1990).

2. Y. Kiso, M. Yoshida, T. Kimura, Y. Fujiwara, and M. Shimokura, *Tetrahedron Lett.*, **30**, 1979 (1989).

3. Y.-D. Gong and N. Iwasawa, *Chem. Lett.*, **23**, 2139 (1994).

4. M. Yoshida, K. Akaji, T. Tatsumi, S. Iinuma, Y. Fujiwara, T. Kimura, and Y. Kiso, *Chem. Pharm. Bull.*, **38**, 273 (1990).

S-Benzamidomethyl Aminothioacetal (Bam-SR): $RSCH_2NHCOC_6H_5$

S-Benzamidomethyl-*N*-methylcysteine has been prepared as a crystalline derivative $(HOCH_2NHCOC_6H_5$, anhydrous CF_3CO_2H, 25°C, 45 min, 88% yield as the trifluoroacetate salt), and cleaved (100% yield) by treatment with mercury(II) acetate (pH 4, 25°C, 1 h) followed by hydrogen sulfide. Attempted preparation of *S*-acetamidomethyl-*N*-methylcysteine resulted in noncrystalline material, shown by TLC to be a mixture.[1] It is also cleaved with $AgBF_4$/TFA, 4°C, >1 h[2], and $MeSiCl_3$/ Ph_2SO, 4°C, 30 min, 100% cleavage.[3] The latter conditions also cleave the Acm, Tacm, *t*-Bu, 4-methoxybenzyl, and 4-methylbenzyl groups.

1. P. K. Chakravarty and R. K. Olsen, *J. Org. Chem.*, **43**, 1270 (1978).

2. M. Yoshida, T. Tatsumi, Y. Fujiwara, S. Iinuma, T. Kimura, K. Akaji, and Y. Kiso, *Chem. Pharm. Bull.*, **38**, 1551 (1990).

3. K. Akaji, T. Tatsumi, M. Yoshida, T. Kimura, Y. Fujiwara, and Y. Kiso, *J. Chem. Soc., Chem. Commun.*, 167 (1991).

S-Allyloxycarbonylaminomethyl Thioether (Allocam−SR):
CH$_2$=CHCH$_2$OC(O)NHCH$_2$SR

Formation/Cleavage[1]

This group is not totally stable to the conditions for BOC cleavage.

1. A. M. Kimbonguila, A. Merzouk, F. Guibe, and A. Loffet, *Tetrahedron Lett.*, **35**, 9035 (1995).

S-N-[2,3,5,6-Tetrafluoro-4-(*N'*-piperidino)-phenyl]-*N*-allyloxycarbonylaminomethyl Thioether (Fnam-SR)

This group was developed to overcome the acid instability of the allyloxycarbonylaminomethyl group which slowly decomposes during BOC deprotections.[1] The Fnam group is stable to base and to conditions used for BOC cleavage.

Formation[2]

Cleavage

Pd(Ph$_3$P)$_4$, PhSiH$_3$ or *N,N'*-dimethylbarbituric acid, 15–60 min, then HO-CH$_2$CH$_2$SH, AcOH, 77–95% yield.[2]

1. A. M. Kimbonguila, A. Merzouk, F. Guibé, and A. Loffet, *Tetrahedron*, **55**, 6931 (1999).
2. P. Gomez-Martinez, A. M. Kimbonguila, and F. Guibe, *Tetrahedron*, **55**, 6945 (1999).

S-Phthalimidomethyl Thioether (Pim−SR)

Formation[1]

Cleavage[1]

1. NH$_2$NH$_2$, H$_2$O, MeOH, 0°C to rt, 1–2 h; Hg(OAc)$_2$, 2–3 h or Cu(OAc)$_2$, 3–24 h; HSCH$_2$CH$_2$OH, 71–92% yield. These conditions return the free thiol. The use of Hg(OAc)$_2$ cleaves the Acm (acetamidomethyl) group in the presence of the Pim group.
2. NH$_2$NH$_2$, H$_2$O, MeOH, 0°C to rt, 1–2 h; I$_2$, rt, 1–2 h, 79–89% yield. The disulfide is formed.

1. Y.-D. Gong, and N. Iwasawa, *Chem. Lett.*, **23**, 2139 (1994).

S-Phenylacetamidomethyl (Phacm−SR): C$_6$H$_5$CH$_2$C(O)NHCH$_2$SR

The Phacm group is stable to the following conditions: DIEA–CH$_2$Cl$_2$, TFA-CH$_2$Cl$_2$, piperidine–DMF, 0.1 *M* TBAF–DMF, and DBU–DMF for 24 h at rt; to HF-anisole or *p*-cresol (9:1) at 0°C for 1 h; to TFA-scavengers (phenol, HSCH$_2$CH$_2$SH, *p*-cresol, anisole) for 2 h at 25°C. It is partially stable (>80%) to TFMSA-TFA-*p*-cresol for 2 h at 25°C. These stability characteristics make it compatible with BOC or Fmoc-based peptide synthesis.[2]

Formation

The Phacm group is introduced by the same methodology as the Acm group[1] (PhCH$_2$C(O)NHCH$_2$OH, TFMSA).[2]

Cleavage

1. Penicillin G. acylase, pH 7.8 buffer, 35°C, 30 min to 2 h. These conditions result in isolation of the disulfide, but if β-mercaptoethanol is included in the reaction mixture the thiol can be isolated.[2]
2. I₂, 80% aq. AcOH. The disulfide is isolated.[2]

1. F. Albericio, A. Grandas, A. Porta, E. Pedroso, and E. Giralt, *Synthesis*, 271 (1987).
2. M. Royo, J. Alsina, E. Giralt, U. Slomcyznska, and F. Albericio, *J. Chem. Soc., Perkin Trans. 1*, 1095 (1995).

S-Acetyl-, S-Carboxy-, and S-Cyanomethyl Thioethers: ArSCH₂X
X = −COCH₃, −CO₂H, −CN (Chart 7)

In an attempt to protect thiophenols during electrophilic substitution reactions on the aromatic ring, the three substituted thioethers were prepared. After acetylation of the aromatic ring (moderate yields), the protective group was converted to the disulfide in moderate yields, 50–60%, by oxidation with hydrogen peroxide/boiling mineral acid, nitric acid, or acidic potassium permanganate.[1]

1. D. Walker, *J. Org. Chem.*, **31**, 835 (1966).

Substituted *S*-Ethyl Derivatives

A thiol, usually under basic catalysis, can undergo Michael addition to an activated double bond, resulting in protection of the sulfhydryl group as a substituted *S*-ethyl derivative. Displacement of an ethyl tosylate by thiolate also affords an *S*-ethyl derivative.

S-(2-Nitro-1-phenyl)ethyl Thioether: RSCH(C₆H₅)CH₂NO₂ (Chart 7)

Formation

PhCH=CHNO₂, *N*-methylmorpholine, pH 7–8, 10 min, 70% yield.[1]

Cleavage

The protective group is removed by mildly alkaline conditions that do not cleave methyl or benzyl esters. The group is stable to CF₃COOH, HCl–AcOH and HBr–AcOH. A polymer-bound version of this group has also been developed.[2] The generation of a chiral center is a disadvantage when using this group in the presence of chiral substrates.

S-2-(2,4-Dinitrophenyl)ethyl Thioether (Dnpe−SR)

Formation

2-(2,4-Dinitrophenyl)ethyl tosylate, DIPEA, DMF, 63% yield.[3]

Cleavage

Piperidine, DMF (1:1), 30 min, 25°C, 57–90% yield.[3]

1. G. Jung, H. Fouad, and G. Heusel, *Angew. Chem., Int. Ed. Engl.*, **14**, 817 (1975).
2. G. Heusel and G. Jung, *Liebigs Ann. Chem.*, 1173 (1979).
3. M. Royo, C. Garcia-Echeverria, E. Giralt, R. Eritja, and F. Albericio, *Tetrahedron Lett.*, **33**, 2391 (1992).

S-2-(4'-Pyridyl)ethyl Thioether: $C_4H_4NCH_2CH_2SR$

Formation[1]/Cleavage[2]

The intermediate sulfides can be oxidized to the corresponding sulfoxides and sulfones and then liberated to give sulfenic, and sulfinic acids.

1. A. R. Katritzky, I. Takahashi, and C. M. Marson, *J. Org. Chem.*, **51**, 4914 (1986).
2. A. R. Katritzky, G. R. Khan, and O. A. Schwarz, *Tetrahedron Lett.*, **25**, 1223 (1984).

S-2-Cyanoethyl Thioether: $NCCH_2CH_2SR$

Formation

$BrCH_2CH_2CN$, K_2CO_3, DMF.[1]

Cleavage

1. The 2-cyanoethyl group was cleaved from an aromatic sulfide with $K_2CO_3/NaBH_4$ (DMF, 135°C, 70% yield).[2]

2. Concd. NH$_4$OH, rt, quant.[1]
3. t-BuOK, DMF, 50–94% yield.[3]

1. M. S. Christopherson and A. D. Broom, *Nucleic Acids Res.*, **19**, 5719 (1991).
2. Y. Ohtsuka and T. Oishi, *Tetrahedron Lett.*, **27**, 203 (1986).
3. A. Kakehi, S. Ito, N. Yamada, and K. Yamaguchi, *Bull. Chem. Soc. Jpn.*, **63**, 829 (1990).

S-2-(Trimethylsilyl)ethyl Thioether: TMSCH$_2$CH$_2$SR[1]

Formation

1. The *S*-2-(trimethylsilyl)ethyl thioether is typically introduced using 2-(trimethylsilyl)ethanethiol by reaction with an epoxide, halide or sulfonate.
2. TMSCH=CH$_2$, AIBN, 50–70°C, 10 h, 87–92% yield.[2]

Cleavage

1. Bu$_4$NF, 3 Å, THF, rt, >53% yield.[3,4]
2. MeSS$^+$Me$_2$ BF$_4^-$ forms a disulfide in 92% yield that is cleaved to the thiol with Ph$_3$P/MeOH/H$_2$O in 90% yield.[5]
3. AcCl, AgBF$_4$, CH$_2$Cl$_2$, rt, 5 min, <5 to 100% yield.[6]

4. BrCN, CH$_2$Cl$_2$ or MeOH, rt, 34–80% yield of the thiocyanate.[7]

1. S. Chambert, J. Désiré, and J.-L. Décourt, *Synthesis*, 2319 (2002).
2. A. Mahadevan, C. Li, and P. L. Fuchs, *Syn. Commun.*, **24**, 3099 (1994). A. Schwan, D. Brillon, and R. Dufault, *Can. J. Chem.*, **72**, 325 (1994).
3. M. Koreeda and W. Yang, *J. Am. Chem. Soc.*, **116**, 10793 (1994); Y. Wang, M. Koreeda, T. Chatterji, and K. S. Gates, *J. Org. Chem.*, **63**, 8644 (1998).
4. M. L. Hamm, R. Cholera, C. L. Hoey, and T. J. Gill, *Org. Lett.*, **6**, 3817 (2004).
5. M. B., Anderson, M. G. Ranasinghe, J. T. Palmer, and P. L. Fuchs, *J. Org. Chem.*, **53**, 3125 (1988). S. Chambert, I. Gautier-Luneau, M. Fontecave, and J.-L. Décout, *J. Org. Chem.*, **65**, 249 (2000).

6. H. Grundberg, M., Andergran, and U. J. Nilsson, *Tetrahedron Lett.*, **40**, 1811 (1999).

7. S. Chambert, F. Thomasson, and J.-L. DÉcout, *J. Org. Chem.*, **67**, 1898 (2002).

S-2,2-Bis(carboethoxy)ethyl Thioether: $RSCH_2CH(COOC_2H_5)_2$ (Chart 7)

Formation

CH_2=$C(CO_2Et)_2$, EtOH, 1 h, 74% yield.[1]

Cleavage

1 *N* KOH, EtOH, 20°C, 5–10 min, 80% yield. *S*-2,2-Bis(carboethoxy)ethyl thio-ether, stable to acidic reagents such as trifluoroacetic acid and hydrogen bromide/acetic acid, has been used in a synthesis of glutathione.[1]

1. T. Wieland and A. Sieber, *Liebigs Ann. Chem.*, **722**, 222 (1969); *idem, ibid.*, **727**, 121 (1969).

S-(1-*m*-Nitrophenyl-2-benzoyl)ethyl Thioether: $ArSCH(C_6H_4$-*m*-$NO_2)CH_2COC_6H_5$

An *S*-(1-*m*-nitrophenyl-2-benzoyl)ethyl thioether was used to protect thiophenols during electrophilic substitution reactions of the benzene ring.[1]

Formation

$PhCOCH$=CHC_6H_4-*m*-NO_2, piperidine, benzene, 96% yield.[1]

Cleavage

$Pb(OAc)_2$, EtOH, pH 8–10; dil. HCl, 77% yield.[1]

1. A. H. Herz and D. S. Tarbell, *J. Am. Chem. Soc.*, **75**, 4657 (1953).

S-2-Phenylsulfonylethyl Thioether, and *S*-1-(4-Methylphenylsulfonyl)-2-methylprop-2-yl Thioether: $PhSO_2CH_2CH_2SR$, 4-$CH_3C_6H_4SO_2CH_2C(CH_3)_2SR$

Formation/cleavage[1,2]

$$RSH \underset{\text{\textit{t}-BuOK, THF, DME or \textit{t}-BuOH, 80-100\%}}{\overset{PhSO_2CH=CH_2, (Et_3N, THF) \text{ or } (MeONa, MeOH)}{\rightleftarrows}} \begin{array}{l} RSCH_2CH_2SO_2Ph \\ 84-100\% \end{array}$$

1. Y. Kuroki and R. Lett, *Tetrahedron Lett.*, **25**, 197 (1984).

2. L. Horner and H. Lindel, *Phosphorus Sulfur*, **15**, 1 (1983).

S-p-Hydroxyphenacyl Thioether: $4\text{-HOC}_6\text{H}_4\text{C(O)CH}_2\text{-SR}$

p-Hydroxyphenacyl derivative is formed from p-hydroxyphenacyl bromide in an ethanol/pH 7 buffer in 80–92% yield. It is cleaved by photolysis at 312 nm in a Tris-HCl buffer (pH 7.2) containing dithiothreitol in 0–71% yield.[1]

1. A. Specht, S. Loudwig, L. Peng, and M. Goeldner, *Tetrahedron Lett.*, **43**, 8947 (2002).

Silyl Thioethers

Silyl-derived protective groups are also used to mask the thiol function. A complete compilation is not given here, since silyl derivatives are described in the section on alcohol protection. The formation and cleavage of silyl thioethers proceed analogously to simple alcohols. The Si–S bond is weaker than the Si–O bond and therefore sulfur derivatives are more susceptible to hydrolysis. For the most part silyl ethers are rarely used to protect the thiol function because of their instability. Silyl ethers have been used for *in situ* protection of the SH group during amide formation.[1] The use of the sterically demanding and thus more stable triisopropylsilyl thioether may prove useful.[2,3]

1. E. W. Abel, *J. Chem. Soc.*, 4933 (1961); L. Birkofer, W. Konkol, and A. Ritter, *Chem. Ber.*, **94**, 1263 (1961).
2. J. C. Arnould, M. Didelot, C. Cadilhac, and M. J. Pasquet, *Tetrahedron Lett.*, **37**, 4523 (1996).
3. N. Ollivier, J.-B. Behr, Q. El-Mahdi, A. Blanpain, and O. Melnyk, *Org. Lett.*, **7**, 2647 (2005).

THIOESTERS

S-Acetyl Derivative: $RSCOCH_3$

S-Benzoyl Derivative: $RSCOC_6H_5$ (Chart 7)

Two disadvantages are associated with the use of S-acetyl or S-benzoyl derivatives in peptide syntheses: (a) Base-catalyzed hydrolysis of S-acetyl- and S-benzoylcysteine occurs with β-elimination to give olefinic side products, $CH_2=C\text{-(NHPG)CO-}$;[1] (b) the yields of peptides formed by coupling an unprotected amino group in an S-acyl-cysteine are low because of prior S–N acyl migration.[2] An S-acetyl group is stable to oxidation of a double bond by ozone (−20°C, 5.5 h, 73% yield).[3]

Formation

1. Ac_2O, $KHCO_3$, 55% yield.[4]
2. BzCl, NaOH, $KHCO_3$, 0–5°C, 30 min., 50% yield.[5]

3. The base-catalyzed reaction of thiothreitol with methyl dithiobenzoate selectively protects a thiol group as an *S*-thiobenzoyl derivative in the presence of a hydroxyl group.[5]

$$
\begin{array}{c}
\text{CH}_2\text{SH} \\
| \\
\text{CHOH} \\
| \\
\text{CH}_2\text{OH}
\end{array}
\quad
\xrightarrow[\text{MeOH, 25°C, 1.5 h, 54\%}]{\text{PhCSSMe, cat. NaOMe}}
\quad
\begin{array}{c}
\text{CH}_2\text{SCSPh} \\
| \\
\text{CHOH} \\
| \\
\text{CH}_2\text{OH}
\end{array}
$$

4. RC(O)–benzotriazole, TEA, CH_2Cl_2, rt, 76–99% yield.[6] This method is useful for the preparation of a large variety of esters.

Cleavage

1. 0.2 *N* NaOH, N_2, 20°C, 2–15 min, 100% yield.[4]
2. Aqueous NH_3, N_2, 20°C, 95–100% yield.[4]
3. $NH_2NH_2 \cdot H_2O$, CH_3CN, >83% yield. Hydrolysis proceeds in the presence of an ester.[7]
4. 0.1 eq. of Bu_4NCN, MeOH, rt, 41–94% yield.[8]
5. HBr, AcOH, 25°C, 30 min, 5% to a substantial amount.[4]
6. CF_3CO_2H, phenol, reflux, 30 min, 2–5% yield. In this case an *S*-Cbz group is removed.[4]
7. $PS\text{-}SO_3H$, H_2O, reflux, 24 h, 71–100% yield.[9] PS = polystyrene.
8. $Fe(NO_3)_3$–Clayfen.[10]
9. NaSMe, MeOH, 23°C, 81–95% yield.[11] This procedure is chemoselective for removal of a thioacetate in the presence of an acetate.
10. $TiCl_4$, Zn, CH_2Cl_2, 0°C to rt, 82–87% yield.[12] The method was shown to be compatible with esters, aldehydes, ketones, silyl ethers, and urethanes.

S-2-Methoxyisobutyryl Derivative: $MeOC(CH_3)_2CO-SR$

This ester was developed for use as a protecting group for arylthiols that was compatible with Suzuki coupling conditions which typically use some form of base.[13] Its increased stability is the result of steric protection of the carbonyl.

S-Trifluoroacetyl Derivative: $RSCOCF_3$

This group is exceptionally labile to base.

Formation

$CF_3COSC_6F_5$, Pyr, DMF, 75% yield.[14]

1. R. G. Hiskey, R. A. Upham, G. M. Beverly, and W. C. Jones, Jr., *J. Org. Chem.*, **35**, 513 (1970).
2. R. G. Hiskey, T. Mizoguchi, and T. Inui, *J. Org. Chem.*, **31**, 1192 (1966).

3. I. Ernest, J. Gosteli, C. W. Greengrass, W. Holick, D. E. Jackman, H. R. Pfaendler, and R. B. Woodward, *J. Am. Chem. Soc.*, **100**, 8214 (1978).

4. L. Zervas, I. Photaki, and N. Ghelis, *J. Am. Chem. Soc.*, **85**, 1337 (1963).

5. E. J. Hedgley and N. H. Leon, *J. Chem. Soc. C*, 467 (1970).

6. A. R. Katritzky, A. A. Shestopalov, and K. Suzuki, *Synthesis*, 1806 (2004).

7. A. Endo, A. Yanagisawa, M. Abe, S. Tohma, T. Kan, and T. Fukuyama, *J. Am. Chem. Soc.*, **124**, 6552 (2002); A. Dion, P. Dube, and C. Spino, *Org. Lett.*, **7**, 5601 (2005).

8. B. T. Holmes and A. W. Snow, *Tetrahedron*, **61**, 12339 (2005).

9. S. Iimura, K. Manabe, and S. Kobayashi, *J. Org. Chem.*, **68**, 8723 (2003).

10. H. M. Meshram, *Tetrahedron Lett.*, **34**, 2521 (1993).

11. O. B. Wallace and D. M. Springer, *Tetrahedron Lett.*, **39**, 2693 (1998).

12. C. K. Jin, H. J. Jeong, M. K. Kim, J. Y. Kim, Y.-J. Yoon, and S.-G. Lee, *Synlett*, 1956 (2001).

13. B. Zeysing, C. Gosch, and A. Terfort, *Org. Lett.*, **2**, 1843 (2000).

14. L. M. Gayo and M. J. Suto, *Tetrahedron Lett.*, **37**, 4915 (1996).

S-N-[[(*p*-Biphenylyl)isopropoxy]carbonyl]-*N*-methyl-γ-aminothiobutyrate: BpocN(CH₃)CH₂CH₂CH₂COSR, and *S-N-*(*t*-Butoxycarbonyl)- *N*-methyl-γ-aminothiobutyrate: BOCN(CH₃)CH₂CH₂CH₂COSR

Formation/Cleavage[1]

Deprotection is only effected by step 1 (TFA, PhOMe, CH₂Cl₂, 0°C, 2–10 min). Step 3 is for disulfide formation.

1. N. G. Galakatos and D. S. Kemp, *J. Org. Chem.*, **50**, 1302 (1985).

Thiocarbonate Derivatives

When cysteine reacts with an alkyl or aryl chloroformate, both the −SH and −NH groups are protected as a thiocarbonate and as a carbamate, respectively. Selective or simultaneous removal of the protective groups is possible. Thiocarbonates are somewhat more stable than thioesters, but neither of these are as stable as the corresponding ester and carbonate. This is due to the poor overlap of the large sulfur atom $3p$ orbitals with the $2p$ orbitals of the carbonyl group. (See cleavage conditions 3-6 for an S-benzyloxycarbonyl derivative, as shown on p. 000.

S-2,2,2-Trichloroethoxycarbonyl Derivative (Troc−SR): RSCOOCH$_2$CCl$_3$

Cleavage

Electrolysis, -1.5 V, LiClO$_4$, CH$_3$OH, 90% yield. The conditions can be adjusted to form either the sulfide or disulfide.[1] Other sections discussing the Troc group should be consulted for alternative methods of cleavage.

1. M. F. Semmelhack and G. E. Heinsohn, *J. Am. Chem. Soc.*, **94**, 5139 (1972).

S-t-Butoxycarbonyl Derivative (BOC−SR): RSCOOC(CH$_3$)$_3$

t-Butyl chloroformate reacts with cysteine to protect both the amine and thiol groups; as with N,S-bis(benzyloxycarbonyl)cysteine, selective or simultaneous removal of the N- or S-protective groups can be effected.[1] 1-*tert*-Butoxy-*tert*-butoxycarbonyl-1,2-dihydroquinoline can be used to efficiently prepare the BOC derivative of thiophenol (89% yield).[2] Treatment with HCl/EtOAc efficiently cleaves the S-BOC group.[3]

1. M. Muraki and T. Mizoguchi, *Chem Pharm. Bull.*, **19**, 1708 (1971).
2. H. Ouchi, Y. Saito, Y. Yamamoto, and H. Takahata, *Org. Lett.*, **4**, 585 (2002).
3. F. S. Gibson, S. C. Bergmeier, and H. Rapoport, *J. Org. Chem.*, **59**, 3216 (1994).

S-Benzyloxycarbonyl Derivative (RS−Cbz, RS−Z): RSCOOCH$_2$C$_6$H$_5$

Formation[1]

Cleavage

1. Concd. NH$_4$OH, 25°C, 1 h, 90% yield.[1]
2. Na, NH$_3$, 62% yield.[1]
3. 0.1 N NaOCH$_3$, CH$_3$OH, N$_2$, 30 min to 3 h, 100% yield.[2] An S-benzoyl group is removed (95–100% yield) in 5–10 min.

4. CF_3COOH, reflux, 30 min, ca. quant.[2] An *N*-Cbz group is also removed under these conditions.

5. 2 *N* HBr, AcOH, 25°C, 30 min.[2,3] The *S*-Cbz group is removed slowly under these conditions, but the *N*-Cbz group is completely cleaved, thus providing some selectivity in the protection scheme for cysteine.

6. Electrolysis, −2.6 V, R_4NX, DMF.[4] Both an *N*-Cbz group and an *S*-Cbz group are removed under these conditions.

1. A. Berger, J. Noguchi, and E. Katchalski, *J. Am. Chem. Soc.*, **78**, 4483 (1956).
2. L. Zervas, I. Photaki, and N. Ghelis, *J. Am. Chem. Soc.*, **85**, 1337 (1963).
3. M. Sokolovsky, M. Wilchek, and A. Patchornik, *J. Am. Chem. Soc.*, **86**, 1202 (1964).
4. V. G. Mairanovsky, *Angew. Chem., Int. Ed., Engl.*, **15**, 281 (1976).

S-p-Methoxybenzyloxycarbonyl Derivative: $RSCOOCH_2C_6H_4$-*p*-OCH_3

S-p-Methoxybenzyloxycarbonylcysteine has been prepared in low yield (30%). It has been used in peptide syntheses, but is very labile to acids and bases.[1]

1. I. Photaki, *J. Chem. Soc. C*, 2687 (1970).

S-Fluorenylmethylcarbonyl Derivative (Fmoc−SR)

Formation

1. FmocCl, CH_2Cl_2, TEA, 98% yield.[1]
2. FmocCl, dioxane, H_2O, 0°C, pH = 7, 72% yield. These conditions were used to protect both the NH, and SH groups of cysteine simultaneously.[1]
3. Use of FmocOSu with TEA results in the formation of the Fm thioether because the basicity of the medium is greater, resulting in Fmoc cleavage followed by thiol scavenging of the fulvene.

Cleavage

TEA, I_2, MeOH, CH_2Cl_2, 75% yield. These conditions do not cleave an N-Fmoc group, and the selectivity is attributed to the greater leaving group ability of the thiol.[1]

1. C. W. West, M. A. Estiarte, and D. H. Rich, *Org. Lett.*, **3**, 1205 (2001).

Thiocarbamate Derivatives

Thiocarbamates, formed by reaction of a thiol with an isocyanate, are stable in acidic and neutral solutions and are readily cleaved by basic hydrolysis. The β-elimination that can occur when an *S*-acyl group is removed with base from a cysteine derivative does not occur under the conditions needed to cleave a thiocarbamate.[1]

S-(N-Ethylcarbamate): $RSCONHC_2H_5$ (Chart 7)

This protective group is stable to acidic hydrolysis (4.5 *N* HBr/HOAc; 1 *N* HCl; CF_3CO_2H, reflux). There is no evidence of S → N acyl migration in *S*-(*N*-ethylcarbamates) (RS = cysteinyl).[1] Oxidation of *S*-(*N*-ethylcarbamoyl)cysteine with performic acid yields cysteic acid.[2]

Formation

   EtN=C=O, pH 1→ pH 6, 20°C, 70 h, 67% yield.[1]

Cleavage

1. 1 *N* NaOH, 20°C, 20 min, 100% yield.[1]
2. NH_3 or NH_2NH_2, methanol, 20°C, 2 h, 100% yield.[1]
3. Na/NH_3, −30°C, 3 min, 100% yield.[1]
4. $Hg(OAc)_2$, H_2O, CH_3OH, 30 min; H_2S, 4 h, 79% yield.[2]
5. $AgNO_3$, H_2O, CH_3OH; concd HCl, 3 h, 62% yield.[2]

1. St. Guttmann, *Helv. Chim. Acta,* **49**, 83 (1966).
2. H. T. Storey, J. Beacham, S. F. Cernosek, F. M. Finn, C. Yanaihara, and K. Hofmann, *J. Am. Chem. Soc.*, **94**, 6170 (1972).

S-(N-Methoxymethylcarbamate): $RSCONHCH_2OCH_3$

Formation

   $CH_3OCH_2N=C=O$, pH 4–5, 2 min, 100% yield.[1] At pH 4–5 the reaction is selective for protection of thiol groups in the presence of α- or ε-amino groups.

Cleavage

   At pH 9.6, a cysteine derivative is cleaved in 100%, glutathione in 80% yield.[1]

1. H. Tschesche and H. Jering, *Angew. Chem., Inter. Ed. Engl.*, **12**, 756 (1973).

MISCELLANEOUS DERIVATIVES

Unsymmetrical Disulfides

A thiol can be protected by oxidation (with O_2; H_2O_2; I_2...) to the corresponding symmetrical disulfide, which subsequently can be cleaved by reduction: [Sn/HCl;

Na/xylene, Et_2O, or NH_3; $LiAlH_4$; $NaBH_4$; or thiols such as $HO(CH_2)_2SH$]. Unsymmetrical disulfides have also been prepared, and are discussed. A newer method involves the Rh-catalyzed exchange between two symmetrical disulfides.[1]

S-Ethyl Disulfide: $RSSC_2H_5$ (Chart 7)

Formation

EtS(O)SEt, $-70°C$, 1 h, 80–90% yield.[2]

Cleavage

PhSH, $>50°C$ or $HSCH_2CO_2H$, 45°C, 15 h, quant.[3] The *S*-ethyl disulfide is stable to acid-catalyzed hydrolysis (CF_3CO_2H) of carbamates, and to ammonolysis (25% NH_3/CH_3OH).[3]

1. M. Arisawa and M. Yamaguchi, *J. Am. Chem. Soc.*, **125**, 6624 (2003).
2. D. A. Armitage, M. J. Clark, and C. C. Tso, *J. Chem. Soc., Perkin Trans. 1*, 680 (1972).
3. N. Inukai, K. Nakano, and M. Murakami, *Bull. Chem. Soc. Jpn.*, **40**, 2913 (1967).

S-*t*-Butyl Disulfide: $RSSC(CH_3)_3$

Formation

1. $CH_3OC(O)SCl$, 0–5°C, 1.5 h; *t*-BuSH, MeOH, 5 days, 97% crude, 46% pure.[1] The reaction proceeds through an *S*-sulfenyl thiocarbonate.
2. t-$BuO_2CNHN(S$-t-$Bu)CO_2$-t-Bu, H_2O.[2]

Cleavage

1. $NaBH_4$.[3]
2. Bu_3P, trifluoroethanol/water (95/5).[4]
3. β-Mercaptoethanol, DMF, 135°C, 24 h, 77% yield. These conditions were used to cleave an *S*-*t*-Bu group from a peptide when tributylphosphine failed. The failure was attributed to steric factors associated with the peptide sequence.[5]
4. Na (2-sulfanylethansulfonic acid, sodium salt), DMF, DIPEA, rt, 100% yield. These conditions were used to cleave an *S*-*t*-Bu disulfide from a complex glycopeptide when many other conditions all failed to give clean reactions.[6]

1. L. Field and R. Ravichandran, *J. Org. Chem.*, **44**, 2624 (1979).
2. E. Wünsch, L. Moroder, and S. Romani, *Hoppe-Seyler's Z. Physiol. Chem.*, **363**, 1461 (1982).
3. E. Wünsch and R. Spangenberg, in *Peptides, 1969*, E. Schoffone, Ed., North Holland, Amsterdam, 1969, p. 1971.

4. R. Ramage and A. S. J. Stewart, *J. Chem. Soc., Perkin Trans. I*, 1947 (1993).
5. B. Denis and E. Trifilieff, *Journal of Peptide Science*, **6**, 372 (2000).
6. M. Mandal, V. Y. Dudkin, X. Geng, and S. J. Danishefsky, *Angew. Chem. Int. Ed.*, **43**, 2557 (2004).

Substituted *S*-Phenyl Disulfides: $RSSC_6H_4$-Y

Three substituted *S*-phenyl unsymmetrical disulfides—i[1], ii[2], and iii[3]— have been prepared: Compounds **i** and **ii** by reaction of a thiol with a sulfenyl halide, and compound **iii** from a thiol and an aryl thiosulfonate ($ArSO_2SAr$). The disulfides are cleaved by reduction ($NaBH_4$) or by treatment with excess thiol ($HSCH_2CH_2OH$).

$$RSS-C_6H_3-2-NO_2-4-R' \quad RSS-C_6H_4-2-N=N-C_6H_5 \quad RSS-C_6H_4-2-COOH$$

$$\text{i} \quad R' = H, NO_2 \qquad\qquad\qquad \text{ii} \qquad\qquad\qquad\qquad \text{iii}$$

1. A. Fontana, E. Scoffone, and C. A. Benassi, *Biochemistry*, **7**, 980 (1968); A. Fontana, *J. Chem. Soc., Chem. Commun.*, 976 (1975).
2. A. Fontana, F. M. Veronese, and E. Scoffone, *Biochemistry*, **7**, 3901 (1968).
3. L. Field and P. M. Giles, Jr., *J. Org. Chem.*, **36**, 309 (1971).

Sulfenyl Derivatives

S-Sulfonate Derivative: $RSSO_3^-$

Formation

Na_2SO_3, cat. cysteine, O_2, pH 7–8.5, 1 h, quant.[1]

Cleavage

1. $HSCH_2CH_2OH$, pH 7.5, 25°C, 2 h, 100% yield.[1]
2. $NaBH_4$.[1] *S*-Sulfonates are stable at pH 1–9; they are unstable in hot acidic solutions, and in 0.1 *N* sodium hydroxide.

S-Thiosulfonate Derivative: $RS-S_2O_3^-$

The thiosulfonate derivative was prepared from cysteine to protect the thiol during the native chemical ligation method for peptide synthesis. It is prepared from sodium tetrathionate in DMSO with DIPEA. It can be removed with dithiothreitol.[2]

1. W. W.-C. Chan, *Biochemistry*, **7**, 4247 (1968).
2. T. Sato and S. Aimoto, *Tetrahedron Lett.*, **44**, 8085 (2003).

S-Sulfenylthiocarbonate: RSSCOOR′

A number of *S*-sulfenylthiocarbonates have been prepared to protect thiols. A benzyl derivative, R′=CH$_2$Ph, is stable to trifluoroacetic acid (25°C, 1 h), but not to HBr/AcOH, and provides satisfactory protection during peptide syntheses;[1] a *t*-butyl derivative, R′ = *t*-Bu, is too labile in base to provide protection.[1] A methyl derivative, R′=CH$_3$, has been used to protect a cysteine fragment that is subsequently converted to a cystine.[2]

1. K. Nokihara and H. Berndt, *J. Org. Chem.*, **43**, 4893 (1978).
2. R. G. Hiskey, N. Muthukumaraswamy, and R. R. Vunnam, *J. Org. Chem.*, **40**, 950 (1975).

S-3-Nitro-2-pyridinesulfenyl Sulfide (Npys−SR): 3-NO$_2$-C$_5$H$_3$NSSR

These sulfides are prepared from other sulfur-protected cysteine derivatives by reaction with the sulfenyl chloride.[1] The Npys group can also be introduced directly by treatment of the thiol with NpysCl.[2]

Conversion of Conventional *S*-Protective Groups into the NpysSR Derivative[1]

Starting Material	Npys−X, Eq.	Conditions	% Yield
Boc−Cys(Bn)−OH	Cl, 1.2	rt, 24 h, CH$_2$Cl$_2$	No reaction
Boc−Cys(MeOBn)−OH[3]	Cl, 1.2	0°C, 30 min, CH$_2$Cl$_2$	92
Boc−Cys(Me$_2$Bn)−OH	Cl, 1.2	0°C, 30 min, CH$_2$Cl$_2$	90
Z−Cys(MeOBn)−Phe−Phe−Gln−Asn−O−*t*-Bu	Cl, 1.2	rt, 30 min, CH$_2$Cl$_2$, CF$_3$COOH (1:1)	85
Fmoc−Cys(*t*-Bu)−OH	Cl, 1.2	0°C, 30 min, CH$_2$Cl$_2$	80
Boc−Cys(Tr)−OH	Cl, 1.2	−30°C, 3 h, CH$_2$Cl$_2$	91
Boc−Cys(Acm)−OH	Cl, 1.2	0°C, 30 min, AcOH	63
Z-Cys(Bn)−OH	Br, 2.0	rt, 10 h, CH$_2$Cl$_2$	21
Z-Cys(Bn)−OH	Cl, 2.0	rt, 5 h, CF$_3$CH$_2$OH	61
Z-Cys(Bn)−OH	Br, 2.4	rt, 3 h, CF$_3$CH$_2$OH, AcOH (10:1)	73
Z-Cys(Bn)-Pro-Leu−GlyNH$_2$	Br, 2.4	rt, 3 h, CF$_3$CH$_2$OH, AcOH (10:1)	70

The Npys group can be cleaved reductively with Bu$_3$P, H$_2$O or mercaptoethanol. It has also been cleaved with 2-mercaptopyridine, 2-mercaptomethylimidazole, or 2-mercaptoacetic acid in methanol/acetic acid. Selective cleavage of the *O*-Npys bond over the *S*-Npys bond can be achieved with the aromatic thiols.[4] This group is stable to CF_3COOH (24 h), 4 M HCl/dioxane (24 h), and HF (1 h).[2] The related reagent, 2-pyridinesulfenyl chloride, has also been proposed as a useful reagent for the deprotection of the *S*-trityl, *S*-diphenylmethyl, *S*-acetamidomethyl, *S*-*t*-butyl, and *S*-*t*-butylsulfenyl groups, but this reagent is very susceptible to hydrolysis.[5]

1. R. Matsueda, S. Higashida, R. J. Ridge, and G.R. Matsueda, *Chem. Lett.*, **11**, 921 (1982).
2. R. Matsueda, T. Kimura, E. T. Kaiser, and G. R. Matsueda, *Chem. Lett.*, **10**, 737 (1981).
3. O. Ploux, M. Caruso, G. Chassaing, and A. Marquet, *J. Org. Chem.*, **53**, 3154 (1988).
4. O. Rosen, S. Rubinraut, and M. Fridkin, *Int. J. Pept. Protein Res.* **35**, 545 (1990).
5. J. V. Castell and A. Tun-Kyi, *Helv. Chim. Acta*, **62**, 2507 (1979).

S-[Tricarbonyl[1,2,3,4,5-η]-2,4-cyclohexadien-1-yl]-iron(1+) Thioether: $[(\eta\text{-}^5C_6H_7)Fe(CO)_3]SR$

Formation

Cleavage

Treatment with HBF_4 in $CHCl_3$ liberates the thiol, and returns the derivatizing agent, $[(\eta\text{-}^5C_6H_7)Fe(CO)_3]^+$ BF_4^- [tricarbonyl[1,2,3,4,5-η]-2,4-cyclohexadien-1-yl-iron(1+) tetrafluoroborate] as a precipitate.[1]

1. S. Fu, J. A. Carver, and L. A. P. Kane-Maguire, *J. Organomet. Chem.*, **454**, C11 (1993).

Oxathiolones

Oxathiolones are formed by heating a ketone with the mercaptocarboxylic acid in the presence of TsOH. They are cleaved by either acid (TFA, H_2O, THF) or base (NaOH, acetone) hydrolysis.[1]

1. L. M. Gustavson, D. S. Jones, J. S. Nelson, and A. Srinivasan, *Syn. Commun.*, **21**, 249 (1991).

Protection for Dithiols: Dithio Acetals, and Ketals

S,S'-Methylene (i), S,S'-Isopropylidene (ii), and S,S'-Benzylidene (iii) Derivatives

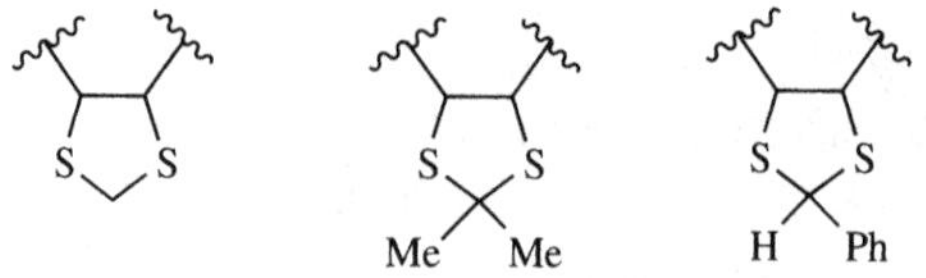

i   ii   iii

Dithiols, like diols, have been protected as S,S'-methylene,[1] S,S'-isopropylidene,[2] and S,S'-benzylidene[3] derivatives, formed by reaction of the dithiol with formaldehyde, acetone, or benzaldehyde, respectively. The methylene and benzylidene derivatives are cleaved by reduction with sodium/ammonia. The isopropylidene[2] and benzylidene[3] derivatives are cleaved by mercury(II) chloride; with sodium/ammonia the isopropylidene derivative is converted to a monothio ether, HSCHRCHRSCHMe$_2$.[1]

1. E. D. Brown, S. M. Igbal, and L. N. Owen, *J. Chem. Soc., C* 415 (1966).
2. E. P. Adams, F. P. Doyle, W. H. Hunter, and J. H. C. Nayler, *J. Chem. Soc.*, 2674 (1960).
3. L. W. C. Miles and L. N. Owen, *J. Chem. Soc.*, 2938 (1950).

S,S'-p-Methoxybenzylidene Derivative: $(RS)_2CHC_6H_4$-4-OCH_3

Formation[1]

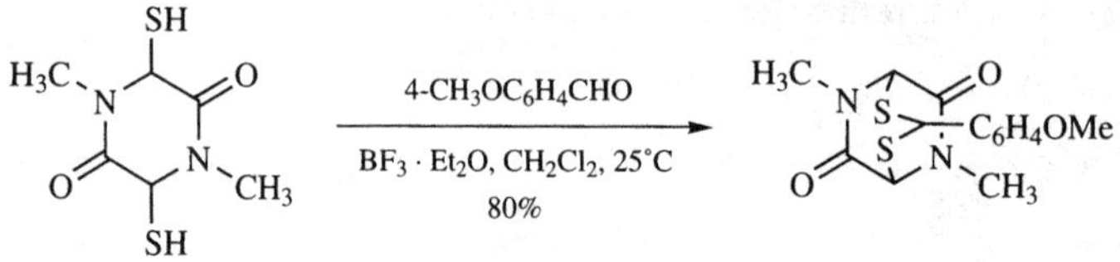

Cleavage[1]

The epidithioketopiperazine shown above is present in natural products including the gliotoxins and sporidesmins.[1]

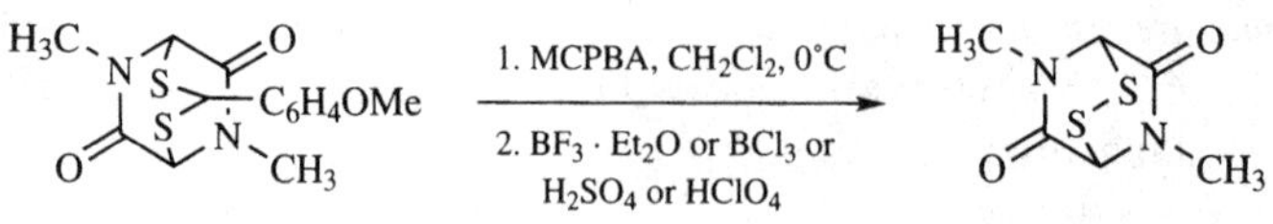

1. Y. Kishi, T. Fukuyama, and S. Nakatusuka, *J. Am. Chem. Soc.*, **95**, 6490 (1973).

Protection for Sulfides

Since sulfides tend to react with electrophiles, a method for protection could be quite useful. Sulfoxides can be used to protect sulfides and are easily formed by a variety of oxidants. Sulfides can be regenerated with thiols,[1] $SiCl_4$ ($0°C$, 15 min, TFA, anisole);[2] $LiBH_4/Me_3SiCl$;[3] $DMF\cdot SO_3/HSCH_2CH_2SH$ (DMF, Pyr, rt, 85% yield);[4] dithiane, NBS ($CHCl_3$, rt, 89–96% yield);[5] Bu_4NBr, TFA, thioanisole, anisole, EDT[6]; Catecholborane, benzene.[7] Sulfides can also be protected as sulfonium salts.

S-Methylsulfonium Salt: $R_2S^+CH_3\ X^-$

A methylsulfonium salt is stable to $NH_3/MeOH$, and to TFA, but not to hydrogenolysis ($H_2/Pd–C$).[9]

Formation

1. $CH_3OSO_2CF_3$, CH_2Cl_2, 99% yield.[8]
2. MeOTs, EtOAc, rt, 4 days, 85% yield.[9]

Cleavage

1. DMF, Et_3N, $HSCH_2CH_2OH$, rt, 78% yield.[9]
2. $LiAlH_4$, THF.[8]

S-Benzyl- and S-4-Methoxybenzylsulfonium Salt: $R_2S^+CH_2Ph\ X^-$

Formation

1. $C_6H_5CH_2OTf$, CH_3CN.[10]
2. $4\text{-}MeOC_6H_4CH_2Cl$, $AgBF_4$, CH_3CN, 97–99% yield.[11]

Cleavage

The benzylsulfonium salt is cleaved by hydrogenolysis ($H_2/Pd–C$, MeOH)[10]; the 4-methoxybenzylsulfonium salt is cleaved by methylamine (100%).[11]

S-1-(4-Phthalimidobutyl)sulfonium Salt

Formation/Cleavage[11]

1. N. Fujii, A. Otaka, S. Funakoshi, H. Yajima, O. Nishimura, and M. Fujino, *Chem. Pharm. Bull.*, **34**, 869 (1986).

2. Y. Kiso, M. Yoshida, T. Fujisaki, T. Mimoto, T. Kimura, and M. Shimokura, *Pept. Chem.*, *1986*, **24**[th], 205 (1987); *Chem. Abstr.*, **108**: 112924j (1988).

3. A. Giannis and K. Sandhoff, *Angew. Chem., Int. Ed. Engl.*, **28**, 218 (1989).

4. S. Futaki, T. Taike, T. Yagami, T. Akita, and K. Kitagawa, *Tetrahedron Lett.*, **30**, 4411 (1989).

5. N. Iranpoor, H. Firouzabadi, and H. R. Shaterian, *J. Org. Chem.*, **67**, 2826 (2002).

6. L. Taboada, E. Nicolas, and E. Giralt, *Tetrahedron Lett.*, **42**, 1891 (2001).

7. D. J. Harrison, N. C. Tam, C. M. Vogels, R. F. Langler, R. T. Baker, A. Decken, and S. A. Westcott, *Tetrahedron Lett.*, **45**, 8493 (2004).

8. V. Cere, A. Guenzi, S. Pollicino, E. Sandri, and A. Fava, *J. Org. Chem.*, **45**, 261 (1980).

9. M. Bodansky and M. A. Bednareck, *Int. J. Pept. Protein Res.*, **20**, 408 (1982).

10. R. C. Roemmele and H. Rapoport, *J. Org. Chem.*, **54**, 1866 (1989).

11. J. T. Doi and G. W. Luehr, *Tetrahedron Lett.*, **26**, 6143 (1985).

S-P Derivatives

***S*-(Dimethylphosphino)thioyl Group (Mpt−SR):** $(CH_3)_2P(S)SR$

***S*-(Diphenylphosphino)thioyl Group (Ppt−SR):** $Ph_2P(S)SR$

Formation

MptCl, $(i\text{-Pr})_2$EtN, CHCl$_3$, 79% yield. The Mpt group on the nitrogen in cysteine can be selectively removed with HCl/Ph$_3$P leaving the S-Mpt group intact.[1]

Cleavage

1. AgNO$_3$, H$_2$O, Pyr, 0°C, 1 h; H$_2$S, 100% yield.[1]

2. KF, 18-crown-6 or Bu$_4$NF, CH$_3$CN, MeOH, 88% yield.[2]

   The related *S*-(diphenylphosphino)thioyl group (Ppt group) has also been cleaved using these conditions.[3] The Mpt derivative of cysteine is not stable to DBU; it forms dehydroalanine. The Mpt group is stable to TFA, and to 1 *M* HCl, but not to HBr/AcOH or 6 *M* HCl.[1]

3. Bu$_4$NF, THF, AcOH, >76% yield.[4]

1. M. Ueki and K. Shinozaki, *Bull. Chem. Soc. Jpn.*, **56**, 1187 (1983).

2. M. Ueki and K. Shinozaki, *Bull. Chem. Soc. Jpn.*, **57**, 2156 (1984).

3. L. Horner, R. Gehring, and H. Lindel, *Phosphorus Sulfur*, **11**, 349 (1981).

4. M. Ueki, H. Takeshita, A. Sacki, H. Komatsu, and T. Katoh, *Pept. Chem. 1994, 32nd*, 173 (1995), *Chem. Abstr.*, **123**: 257332j (1995).

Protection for the Amino Thiol Group

Thiazoline

The phenyl thiazoline is formed from an amino thiol upon reaction with ethyl benzimidate hydrochloride in 87% yield. It is cleaved by heating to reflux with 6 N HCl.[1]

Ninhydrin

Although ninhydrin is typically used as an indicator for terminal amines, it has been used for the protection of N-terminal cysteine peptides. The derivative is readily formed under aqueous conditions at neutral to acidic pH. It is cleaved by reaction with an excess of cysteine using the mass action principle, 3-mercaptopropiosulfonic acid at pH 7.7, or Zn in 10% aqueous TFA.[2]

1. S. Singh, S. J. Rao, and M. W. Pennington, *J. Org. Chem.*, **69**, 4551 (2004).
2. C. T. Pool, J. G. Boyd, and J. P. Tam, *J. Peptide Res.*, **63**, 223 (2004).

7

PROTECTION FOR THE AMINO GROUP

Greene's Protective Groups in Organic Synthesis, Fourth Edition, by Peter G. M. Wuts and
Theodora W. Greene
Copyright © 2007 John Wiley & Sons, Inc.

A great many protective groups have been developed for the amino group, including carbamates ($>NCO_2R$), used for the protection of amino acids in peptide and protein syntheses,[1] and amides ($>NCOR$), used more widely in syntheses of alkaloids and for the protection[2] of the nitrogen bases adenine, cytosine, and guanine in nucleotide syntheses. Carbamates are formed from an amine with a wide variety of reagents, the chloroformate being the most common; amides are formed from the acid chloride. n-Alkyl carbamates are cleaved by acid-catalyzed hydrolysis; N-alkylamides are cleaved under forcing conditions by acidic or basic hydrolysis at reflux, as well as by ammonolysis in cases where the amine is not very basic such as in heterocyclic amine derivatives.

In this chapter, detailed information is provided for the more useful protective groups (some of which are included in Reactivity Charts 8–10); structures and references are given for protective groups that seem to have more limited use.[3] A large variety of alkyl and substituted alkylamines have been developed, each with its own special characteristics for eventual cleavage.

CARBAMATES

Carbamates can be used as protective groups for amino acids to minimize racemization in peptide synthesis. Racemization occurs during the base-catalyzed coupling reaction of an N-protected, carboxyl-activated amino acid, and it takes place in the intermediate oxazolone that forms readily from an N-acyl protected amino acid (R' = alkyl, aryl):

oxazolone

To minimize racemization, the use of nonpolar solvents, a minimum of base, low reaction temperatures, and carbamate protective groups (R' = O-alkyl or O-aryl) is effective.

Many carbamates have been used as protective groups. They are, for the most part, arranged in this chapter in order of increasing complexity of structure. The most useful compounds (not necessarily the simplest structures) are: t-butyl (BOC), readily cleaved by acidic hydrolysis; benzyl (Cbz or Z), cleaved by catalytic hydrogenolysis; 2,4-dichlorobenzyl, stable to the acid-catalyzed hydrolysis of benzyl and t-butyl carbamates; 2-(biphenylyl)isopropyl, cleaved more easily than t-butyl carbamate by dilute acetic acid; 9-fluorenylmethyl, cleaved by β-elimination with base; isonicotinyl, cleaved by reduction with zinc in acetic acid; 1-adamantyl, readily cleaved by trifluoroacetic acid; allyl, readily cleaved by Pd-catalyzed isomerization or by nucleophilic addition to the π-allylpalladium complex; and trimethylsilylethyl, cleaved with fluoride.

1. See reference 22 (peptides) in Chapter 1.
2. See reference 23 (oligonucleotides) in Chapter 1. See also C. B. Reese, *Tetrahedron*, **34**, 3143 (1978); V. Amarnath and A. D. Broom, *Chem. Rev.*, **77**, 183 (1977).
3. See also E. Wünsch, "Blockierung und Schutz der α-Amino-Funktion," in *Methoden der Organischen Chemie (Houben-Weyl)*, Vol. 15/1, Georg Thieme Verlag, Stuttgart, 1974, pp. 46–305; J. W. Barton, "Protection of N-H Bonds and NR$_3$," in *Protective Groups in Organic Chemistry*, J. F. W. McOmie, Ed., Plenum Press, New York and London, 1973, pp. 43–93; L. A. Carpino, *Acc. Chem. Res.*, **6**, 191–198 (1973); Y. Wolman, "Protection of the Amino Group," in *The Chemistry of the Amino Group*, Vol. 4, S. Patai, Ed., Wiley-Interscience, New York, 1968, pp. 669–699; E. Gross and J. Meienhofer, Eds., *The Peptides: Analysis, Synthesis, Biology, Vol. 3: Protection of Functional Groups in Peptide Synthesis*, Academic Press, New York, 1981; P. J. Kocienski, Protecting Groups, 3rd ed., G. Thieme Verlag, New York, 2004, Chapter 8.

Carbamate Derived from an Amine and CO$_2$

CO$_2$ is well known to react with primary and secondary amines. In fact the white solid often found on the mouth of bottles containing these amines is the carbamate salt (R$_2$NCO$_2$HNR$_2$) formed from the CO$_2$ in the air. This type of salt has been used to advantage in a carbapenam synthesis during the hydrogenolysis of a 4-nitrobenzyl ester. Prereduction of the Pd–C was necessary to prevent the formation of colloidal

Pd which carried over in the product.[1] Protection of the amine improves the hydrogenolysis since amines tend to deactivate Pd–C.

1. J. M. Williams, K. M. J. Brands, R. T. Skerlj, R. B. Jobson, G. Marchesini, K. M. Conrad, B. Pipik, K. A. Savary, F.-R. Tsay, P. G. Houghton, D. R. Sidler, U.-H. Dolling, L. M. DiMichele, and T. J. Novak, *J. Org. Chem.*, **70**, 7479 (2005).

Methyl and Ethyl Carbamate: $CH_3OC(O)NR_2$ (Chart 8)

Formation

1. CH_3OCOCl, K_2CO_3, reflux 12 h.[1] Methyl chloroformate is the most common reagent used for the introduction of a methyl carbamates. Pyridine and TEA are the most frequently used bases.
2. *N*-[(Methoxy)carbonyloxy]succinimide, Pyr, rt, >89% yield.[2]
3. $CH_3OCO_2CH_3$, Al_2O_3, reflux, 60–95% yield.[3]
4. $CH_3OCO_2CH_3$, Sc(OTf)$_3$, Yb(OTf)$_3$ or La(OTf)$_3$, rt, 23–86% yield.[4,5]
5. $Ph_2P(O)OC(O)OCH_3$, THF, CO_2.[6]
6.

$$R_2NH \xrightarrow[CH_2Cl_2,\ rt,\ 93\text{--}97\%]{} R_2NCO_2Et \quad \text{Ref. 7}$$

7. CO, O_2, MeOH, HCl, PdCl$_2$, CuCl$_2$.[8]
8. CO, EtOH, O_2, KI, Pd–C[9] or Pd(OAc)$_2$.[10] Electrochemical oxidation has also been used (55–99% yield).[11]
9. CO, O_2, Co(tpp), NaI, EtOH, 68 atm, 3 h, 180 °C.[12]
10. CO_2, HC(OEt)$_3$, 40 h, 120°C 45 atm, 83% yield.[13]
11. CO_2, TEA, RCl, 20–76% yield.[14]

12. DBU·CO$_2$, CH$_3$CN, EtI, 5°C, 77–96% yield.[15] DBU·CO$_2$ is a solid easy to handle complex.

13. From a thiocarbamate: NaOMe, MeOH, reflux, 43 h, 90% yield.[16]

Cleavage

1. *n*-PrSLi, 0°C, 8.5 h, 75–80% yield.[17]

2. Me$_3$SiI, 50°C, 70% yield.[18,19] The most electron rich-carbamate is cleaved preferentially.

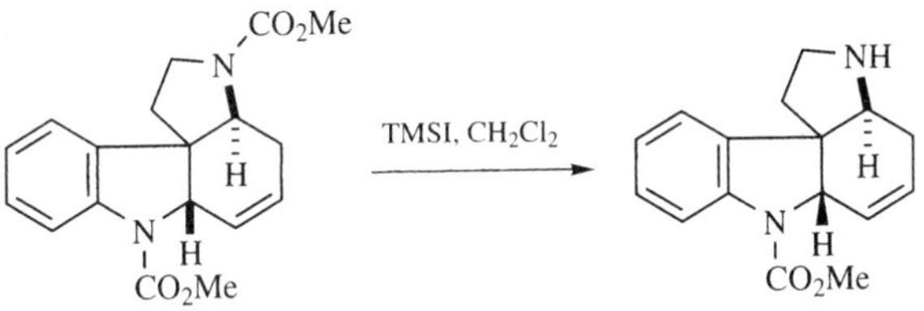

Contains 10% fully
deprotected material   Ref. 20

3. MeSiCl$_3$, TEA, THF, 60°C, 54–93% yield.[21] The method was developed for the cleavage of 2-amino-2-deoxy-D-glucoside methoxycarbonyl derivatives. The acetate, PMB, Bn, Troc, acetonide and azide groups were stable to these conditions.

4. KOH, H$_2$O, ethylene glycol, 100°C, 12 h, 88% yield.[22]

5. HBr, AcOH, 25°C, 18 h.[23, 24]

6. Ba(OH)$_2$, H$_2$O, MeOH, 110°C, 12 h.[25]

7. K$_2$CO$_3$, MeOH, 67% yield. These conditions were used to cleave a methyl carbamate from an aziridine.[26]

8. NH$_2$NH$_2$·H$_2$O, KOH, 98% yield.[27]

9. Dimethyl sulfide, methanesulfonic acid, 5°C, 58–100% yield.[28]

10. NaHTe, 45–83% yield.[29]

11. TMSOK, MeOH, reflux, 48 h, 67% yield.[30]

12. MeLi, THF, 0°C.[31]

13. AcCl, NaI, CH$_3$CN, 16 h, 60°C, 52% yield.[32]

14. NaOH, MeOH, rt, 80% yield.[33] Cleavage occurs under such mild conditions because the N-O nitrogen in this case is a much better leaving group than the typical aliphatic amine.

15. NaAlH$_2$(OCH$_2$CH$_2$OCH$_3$)$_2$, benzene, rt, 80% yield.[34]

16. L-Selectride, THF, rt, 2 days, 51–87% yield. Benzyl carbamates are cleaved sluggishly with this reagent, but BOC derivatives are stable.[35]

17. LiBH$_4$, MeOH, THF, rt, then Pd-catalyzed decomposition of the borane amine complex, 78% yield. This method is not expected to work for normal amides because the leaving group ability of the aziridine is better than that of a simple alkyl amine.[36]

1. E. J. Corey, M. G. Bock, A. P. Kozikowski, A. V. Rama Rao, D. Floyd, and B. Lipshutz, *Tetrahedron Lett.*, **19**, 1051 (1978).

2. S. B. Rollins and R. M. Williams, *Tetrahedron Lett.*, **38**, 4033 (1997).

3. I. Vauthey, F. Valot, C. Gozzi, F. Fache, and M. Lemaire, *Tetrahedron Lett.*, **41**, 6347 (2000).

4. M. Distaso and E. Quaranta, *Tetrahedron*, **60**, 1531 (2004).

5. M. Curini, F. Epifano, F. Maltese, and O. Rosati, *Tetrahedron Lett.*, **43**, 4895 (2002).

6. M. Aresta and A. Dibenedetto, *Chem. Eur. J.*, **8**, 685 (2002).

7. L. C. Chen and S. C. Yang, *J. Chin. Chem. Soc. (Tapei)*, **33**, 347 (1986).

8. H. Alper and F. W. Hartstock, J. Chem. Soc., Chem. Commun., 1141 (1985).

9. S. Fukuoka, M. Chono, and M. Kohno, *J. Org. Chem.*, **49**, 1458 (1984).

10. T. Pri-Bar and J. Schwartz, *J. Org. Chem.*, **60**, 8124 (1995).

11. F. W. Hartstock, D. G. Herrington, and L. B. McMahon, *Tetrahedron Lett.*, **35**, 8761 (1994).

12. T. W. Leung and B. D. Dombek, *J. Chem. Soc., Chem. Commun.*, 205 (1992).

13. S. Ishii, H. Nakayama, Y. Hoshida, and T. Yamashita, *Bull. Chem. Soc. Jpn.*, **62**, 455 (1989).

14. W. McGhee, D. Riley, K. Christ, Y. Pan, and B. Parnas, *J. Org. Chem.*, **60**, 2820 (1995).

15. E. R. Pérez, M. O. D Silva, V. C. Costa, U. P. Rodriques-Filho, and D. W. Franco, *Tetrahedron Lett.*, **43**, 4091 (2002).

16. S. K. Tandel, S. Ragappa, and S. V. Pansare, *Tetrahedron*, **49**, 7479 (1993).

17. E. J. Corey, L. O. Weigel, D. Floyd, and M. G. Bock, *J. Am. Chem. Soc.*, **100**, 2916 (1978).

18. R. S. Lott, V. S. Chauhan, and C. H. Stammer, *J. Chem. Soc., Chem. Commun.*, 495 (1979).

19. S. Raucher, B. L. Bray, and R. F. Lawrence, *J. Am. Chem. Soc.*, **109**, 442 (1987).

20. V. H. Rawal and C. Michoud, *J. Org. Chem.*, **58**, 5583 (1993).

21. B. K. S. Yeung, S. L. Adamski-Werner, J. B. Bernard, G. Poulenat, and P. A. Petillo, *Org. Lett.*, **2**, 3135 (2000).

22. E. Wenkert, T. Hudlicky, and H. D. H. Showalter, *J. Am. Chem. Soc.*, **100**, 4893 (1978).

23. M. C. Wani, H. F. Campbell, G. A. Brine, J. A. Kepler, M. E. Wall, and S. G. Levine, *J. Am. Chem. Soc.*, **94**, 3631 (1972).

24. P. Magnus, J. Rodrigues-Lôpez, K. Mulholland, and I. Matthews, *J. Am. Chem. Soc.*, **114**, 382 (1992).

25. P. M. Wovkulich and M. R. Uskokovic, *Tetrahedron*, **41**, 3455 (1985).

26. K. F. McClure and S. J. Danishefsky, *J. Am. Chem. Soc.*, **115**, 6094 (1993).

27. T. Shono, Y. Matsumura, K. Uchida, K. Tsubata, and A. Makino, *J. Org. Chem.*, **49**, 300 (1984).

28. H. Irie, H. Nakanishi, N. Fujii, Y. Mizuno, T. Fushimi, S. Funakoshi, and H. Yajima, *Chem. Lett.*, **9**, 705 (1980).

29. X.-J. Zhou and Z.-Z. Huang, *Synth. Commun.*, **19**, 1347 (1989).

30. S. J. Hays, P. M. Novak, D. F. Ortwine, C. F. Bigge, N. L. Colbry, G. Johnson, L. J. Lescosky, T. C. Malone, A. Michael, M. D. Reily, L. L. Coughenour, L. J. Brahce, J. L. Shillis, and A. Probert, Jr., *J. Med. Chem.*, **36**, 654 (1993).

31. M. Tius and M. A. Keer, *J. Am. Chem. Soc.*, **114**, 5959 (1992).

32. M. Ihara, A. Hirabayashi, N. Taniguchi, and K. Fukumoto, *Heterocycles*, **33**, 851 (1992).

33. D. Yang, S.-H. Kim, and D. Kahne, *J. Am. Chem. Soc.*, **113**, 4715 (1991).

34. G. R. Lenz, *J. Org. Chem.*, **53**, 4447 (1988).

35. A. Coop and K. C. Rice, *Tetrahedron Lett.*, **39**, 8933 (1998).

36. T. C. Judd and R. M. Williams, *Angew. Chem. Int. Ed.*, **41**, 4683 (2002).

9-Fluorenylmethyl Carbamate (Fmoc−NR$_2$): (Chart 8)

$$CH_2OC(O)NR_2$$

Some advantages of the Fmoc protective group are that it has excellent acid stability; thus, BOC– and benzyl-based groups can be removed in its presence. It is readily cleaved, nonhydrolytically, by simple amines, and the protected amine is liberated as its free base.[1] The Fmoc group is generally considered to be stable to hydrogenation conditions, but it has been shown that under some circumstances it can be cleaved with H$_2$/Pd–C, AcOH, MeOH ($T_{1/2} = 3$–33 h).[2] The use of transfer hydrogenation (Pd–C, HCO$_2$NH$_4$, rt, 4 h) in some cases cleaves an Fmoc group,[3] but this is not universally true.[4] Fmoc cleavage may be the result of the formation of a basic medium as a result of the decomposition of ammonium formate to ammonium carbonate by the Pd catalyst over time.

Hydroxide ion may also cleave the Fmoc group during ester hydrolysis, but the inclusion of $CaCl_2$ in the hydrolysis prevents its cleavage in the presence of hydroxide.[5]

Formation

1. Fmoc$-$Cl, $NaHCO_3$, aq. dioxane, 88–98% yield.[6] Diisopropylethylamine is reported to suppress dipeptide formation during Fmoc introduction with Fmoc$-$Cl.[7]

2. From an amino acid: Silylate the acid with TMSTFA and then treat with Fmoc$-$OSu followed by MeOH to remove the silyl group. This method prevents oligimerization of the amino acid.[8]

3. Fmoc$-$N_3, $NaHCO_3$, aq. dioxane, 88–98% yield.[6,9] This reagent reacts more slowly with amino acids than does the acid chloride. It is not the safest method for Fmoc introduction because of the azide, especially on scale.

4. Fmoc$-$OBt (Bt = benzotriazol-1-yl).[10,11] The method has been used to protect an aziridine.[12] A polymer-supported version of the reagent has been prepared.[13]

5. Fmoc$-$OSu (Su = succinimidyl), H_2O, CH_3CN.[10,11,14,15] The advantage of Fmoc$-$OSu is that little or no oligopeptides are formed when amino acid derivatives are prepared.[16] A polymer-supported version of this reagent has been prepared and used to introduce the Fmoc group onto amino acids (34–96% yield). An indole nitrogen was unreactive.[17]

6. Fmoc$-$OC_6F_5, $NaHCO_3$, H_2O, acetone, rt, 64–99% yield.[18]

7. From a benzyl carbamates: 10% Pd$-$C, 2,2'-dipyridyl, Fmoc$-$OSu, H_2, MeOH, 79–90% yield.[19]

Cleavage

1. The Fmoc group is cleaved under mild conditions with an amine base to afford the free amine and dibenzofulvene. The accompanying table gives the

approximate half-lives for the deprotection of Fmoc–ValOH by a variety of amine bases in DMF.[16] The half-lives shown in the table will vary, depending on the structure of the Fmoc–amine derivative. In the case of solid phase glycopeptide synthesis piperidine was found to be superior to morpholine for Fmoc cleavage.[21] In peptide synthesis a free lysine residue was shown to be sufficiently basic to cause partial Fmoc deprotection.[22]

Amine	$T_{1/2}$
20% Piperidine	6 s
5% Piperidine	20 s
50% Morpholine	1 min
50% Dicyclohexylamine	35 min
10% p-Dimethylaminopyridine	85 min
50% Diisopropylethylamine[20]	10.1 h

2. Bu$_4$NF, DMF, rt, 2 min.[23,24]

3. Bu$_4$NF, n-C$_8$H$_{17}$SH, 92–100% yield.[25] The thiol is used to scavenge the liberated fulvene.

4. Catalytic DBU, n-C$_8$H$_{17}$SH, 70–100% yield. The octanethiol was superior to other thiols in its scavenging ability of dibenzofulvene.[26]

5. DBU, HOBt, DMF. This method was used to remove the Fmoc group on resins containing thioesters.[27]

6. Piperazine attached to a polymer has also been used to cleave the Fmoc group.[28]

7. Tris(2-aminoethyl)amine, CH$_2$Cl$_2$. This amine acts as the deblocking agent and the scavenger for the dibenzofulvene and does not cause the formation of precipitates or emulsions, which sometimes occur.[1b]

8. Direct conversion of an Fmoc group to a Cbz group: KF, TEA, DMF, N-benzyloxycarbonyloxy-5-norbornene-2,3-dicarboximide, 7–12 h, 83–99% yield.[29]

9. AlCl$_3$, toluene, rt, 3 h, 86–95% yield. A limited number of examples were reported.[30]

2,6-Di-t-butyl-9-fluorenylmethyl (Dtb–Fmoc)[31] and 2,7-Bis(trimethylsilyl)fluorenylmethyl (Bts–Fmoc)[32] Carbamate

Both these carbamates were prepared to give derivatives that are more soluble than the conventional Fmoc group. The 2,7-di-t-butyl derivative has similar properties.[33]

Cleavage occurs using the conventional conditions, but the rates vary as a function of the substituents (see accompanying table), showing that what may seem as an innocuous change can have a dramatic effect on the chemistry.

Time for Complete Deblocking of Substituted Urethanes by Various Amines

Base	Deblocking Times		
	PG = Fmoc	PG = Bts-Fmoc	PG = Dtb−Fmoc
Piperidine	<3 min	<3 min	12 min
Ethanolamine	45 min	90 min	4 h
Morpholine	75 min	190 min	10 h
t-Butylamine	5 h	4.5 h	20 h

9-(2-Sulfo)fluorenylmethyl Carbamate

Because of the electron-withdrawing sulfonic acid substituent, cleavage occurs under milder conditions than needed for the Fmoc group (0.1 N NH$_4$OH; 1% Na$_2$CO$_3$, 45 min).[34]

9-(2,7-Dibromo)fluorenylmethyl Carbamate

$$\text{CH}_2\text{OC(O)NR}_2$$

Because of the two electron-withdrawing bromine groups, pyridine can be used to cleave this derivative from its parent amine.[35]

17-Tetrabenzo[a,c,g,i]fluorenylmethyl Carbamate (Tbfmoc−NR$_2$)

$$\text{CH}_2\text{OCONR}_2$$

This Fmoc analog is prepared from the chloroformate, O-succinimide or p-nitrophenyl carbonate and is cleaved with 10% piperidine in 1:1 6M guanidine/IPA.[36] It was designed to interact strongly on a column of porous graphitized carbon so as to aid in the purification of peptides after cleavage from the resin.

2-Chloro-3-indenylmethyl Carbamate (Climoc) and Benz[*f*]inden-3-ylmethyl Carbamate (Bimoc−NR₂)

Climoc   Bimoc

These base-sensitive protective groups were introduced from the chloroformate or azidoformate. They are more sensitive to base than the Fmoc group. Cleavage times with 0.2 mL of piperidine to 0.1 mmole of urethane in 5 mL of CHCl₃ at rt occurs as follows: Climoc, <10 min; Bimoc, <14 h; Fmoc, 18 h.[37]

1,1-Dioxobenzo[*b*]thiophene-2-ylmethyl Carbamate (Bsmoc−NR₂)

During the cleavage of the Fmoc group with base, dibenzofulvene is liberated and must be scavenged to prevent its reaction with the liberated peptides during peptide synthesis. The Bsmoc group was designed so that the cleavage agent [(tris(2-aminoethyl)amine] also serves as the scavenging agent.

The Bsmoc derivative is formed from the chloroformate or the N-hydroxysuccinimide ester.[38] It is cleaved rapidly by a Michael addition with tris(2-aminoethyl)amine at a rate that leaves Fmoc derivatives intact. More hindered bases such as N-methylcyclohexylamine or diisopropylamine do not react with the Bsmoc group, but do cleave the Fmoc group, illustrating the importance of steric effects in additions to Michael acceptors.[39] In the following example, Fmoc protection was unsuccessful because of purification problems associated with removal of the by-products from Fmoc deprotection.[40]

The Bsmoc group is stable to TFA, HCl/EtOAc at rt for 24 h, to tertiary amines, and to hydrogenolysis, but it is not stable to HBr/AcOH. It is readily cleaved by RSH and base (DIPEA).

2-Methylsulfonyl-3-phenyl-1-prop-2-enyloxy Carbamate (Mspoc−NR₂)

The Mspoc group was prepared as an amino protecting group for peptide synthesis. It is introduced with the chloroformate or the succinimide method. It is more stable

Bspoc   Mspoc

to amines than the Bspoc group, which suffered premature cleavage in the presence of amines. It is cleaved with piperidine or with thiolate in the presence of DIPEA.[41]

2,7-Di-*t*-butyl[9-(10,10-dioxo-10,10,10,10-tetrahydrothioxanthyl)]methyl Carbamate (DBD−Tmoc−NR2)

The DBD−Tmoc group is stable to TFA and HBr/AcOH.

Formation

DBD−TmocCl, NaHCO$_3$, H$_2$O, dioxane.[42]

Cleavage[42]

1. 50–75°C in DMSO, 4.5–16 h, 100% yield.
2. Pd–C, HCO$_2$NH$_4$, MeOH.
3. Pyridine. The Fmoc group is stable to pyridine.

1. (a) L. A. Carpino, *Acc. Chem. Res.*, **20**, 401 (1987); (b) L. A. Carpino, D. Sadat-Aalaee, and M. Beyermann, *J. Org. Chem.*, **55**, 1673 (1990).
2. E. Atherton, C. Bury, R. C. Sheppard, and B. J. Williams, *Tetrahedron Lett.*, **20**, 3041 (1979).
3. L. A. Carpino, B. J. Cohen, K. E. Stephens, S. Y. Sadat-Aalaee, J.-H Tien, and D. C. Langridge, *J. Org. Chem.*, **51**, 3732 (1986); D. L. Boger, M. W. Ledeboer, and M. Kume, *J. Am. Chem. Soc.*, **121**, 1098, (1999).
4. R. C. Kelly, I. Gebhard, and N. Wicnienski, *J. Am. Chem. Soc.*, **51**, 4590 (1986).
5. R. Pascal and R. Sola, *Tetrahedron Lett.*, **39**, 5031 (1998).
6. L. A. Carpino and G. Y. Han, *J. Org. Chem.*, **37**, 3404 (1972).
7. F. M. F. Chen and N. L. Benoiton, *Can. J. Chem.*, **65**, 1224 (1987).
8. S. P. Raillard, A. D. Mann, and T. A. Baer, *Org. Prep. Proc. Int.*, **30**, 183 (1998).
9. M. Tessier, F. Albericio, E. Pedroso, A. Grandas, R. Eritja, E. Giralt, C. Granier, and J. Van Rietschoten, *Int. J. Pept. Protein Res.*, **22**, 125 (1983).
10. A. Paquet, *Can. J. Chem.*, **60**, 976 (1982).

11. G. F. Sigler, W. D. Fuller, N. C. Chaturvedi, M. Goodman, and M. Verlander, *Biopolymers*, **22**, 2157 (1983).

12. R. S. Coleman, J.-S. Kong, and T. E. Richardson, *J. Am. Chem. Soc.*, **121**, 9088 (1999).

13. K. G. Dendrinos and A. G. Kalivretenos, *J. Chem. Soc. Perkin Trans. 1*, 1463 (1998).

14. R. C. de L. Milton, E. Becker, S. C. F. Milton, J. E. H. Baxter, and J. F. Elsworth, *Int. J. Pept. Protein Res.*, **30**, 431 (1987).

15. L. Lapatsanis, G. Milias, K. Froussios, and M. Kolovos, *Synthesis*, 671 (1983).

16. For a review of the use of Fmoc protection in peptide synthesis, see E. Atherton and R. C. Sheppard "The Fluorenylmethoxycarbonyl Amino Protecting Group," in *The Peptides*, Vol 9, S. Udenfriend and J. Meienhofer, Eds., Academic Press, New York, 1987, p. 1.

17. R. Chinchilla, D. J. Dodsworth, C. Najera and J. M. Soriano, *Tetrahedron Lett.*, **42**, 7579 (2001); H. Sumiyoshi, T. Shimizu, M. Katoh, Y. Baba, and M. Sodeoka, *Org. Lett.*, **4**, 3923 (2002).

18. I. Schoen and L. Kisfaludy, *Synthesis*, 303 (1986).

19. V. Dzubeck and J. P. Schneider, *Tetrahedron Lett.*, **41**, 9953 (2000).

20. T. Hoeg-Jensen, M. H. Jakobsen, and A. Holm, *Tetrahedron Lett.*, **32**, 6387 (1991).

21. T. Vuljanic, K.-E. Bergquist, H. Clausen, S. Roy, and J. Kihlberg, *Tetrahedron*, **52**, 7983 (1996).

22. J. Farrera-Sinfreu, M. Royo, and F. Albericio, *Tetrahedron Lett.*, **43**, 7813 (2002).

23. M. Ueki and M. Amemiya, *Tetrahedron Lett.*, **28**, 6617 (1987).

24. Y. Rew, D. Shin, I. Hwang, and D. L. Boger, *J. Am. Chem. Soc.*, **126**, 1041 (2004).

25. M. Ueki, N. Nishigaki, H. Aoki, T. Tsurusaki, and T. Katoh, *Chem. Lett.*, **22**, 721 (1993).

26. J. E. Sheppeck II, H. Kar, and H. Hong, *Tetrahedron Lett.*, **41**, 5329 (2000).

27. X. Bu, G. Xie, C. W. Law, and Z. Guo, *Tetrahedron Lett.*, **43**, 2419 (2002).

28. L. A. Carpino, E. M. E. Mansour, and J. Knapczyk, *J. Org. Chem.*, **48**, 666 (1983).

29. W. R. Li, J. Jiang, and M. M. Joullié, *Synlett*, 362 (1993).

30. A. Leggio, A. Liguori, A. Napoli, C. Siciliano, and G. Sindona, *Eur. J. Org. Chem.*, 573 (2000).

31. K. D. Stigers, M. R. Koutroulis, D. M. Chung, and J. S. Nowick, *J. Org. Chem.*, **65**, 3858 (2000).

32. L. A. Carpino and A.-C. Wu, *J. Org. Chem.*, **65**, 9238 (2000).

33. K. D. Stigers, M. R. Koutroulis, D. M. Chung, and J. S. Nowick, *J. Org. Chem.*, **65**, 3858 (2000).

34. R. B. Merrifield and A. E. Bach, *J. Org. Chem.*, **43**, 4808 (1978).

35. L. A. Carpino, *J. Org. Chem.*, **45**, 4250 (1980).

36. A. R. Brown, S. L. Irving, R. Ramage, and G. Raphy, *Tetrahedron*, **57**, 11815 (1995); R. Ramage and G. Raphy, *Tetrahedron Lett.*, **33**, 385 (1992).

37. L. A. Carpino, B. J. Cohen, Y. Z. Lin, K. E. Stephens, Jr., and S A. Triolo, *J. Org. Chem.*, **55**, 251 (1990).

38. L. A. Carpino, M. Ismail, G. A. Truran, E. M. E. Mansour, S. Iguchi, D. Ionescu, A. El-Faham, C. Riemer, and R. Warrass, *J. Org. Chem.*, **64**, 4324 (1999).

49. L. A. Carpino, M. Philbin, M. Ismail, G. A. Truran, E. M. E. Mansour, S. Iguchi, D. Ionescu, A. El-Faham, C. Riemer, R. Warrass, and M. S. Weiss, *J. Am. Chem. Soc.*, **119**, 9915 (1997).

40. R. B. Greenwald, H. Zhao and P. Reddy, *J. Org. Chem.*, **68**, 4894 (2003).

41. L. A. Carpino and E. M. E. Mansour, *J. Org. Chem.*, **64**, 8399 (1999); L. A. Carpino and M. Philbin, *J. Org. Chem.*, **64**, 4315 (1999).
42. L. A. Carpino, H.-S. Gao, G.-S. Ti, and D. Segev, *J. Org. Chem.*, **54**, 5887 (1989).

Substituted Ethyl Carbamates

2,2,2-Trichloroethyl Carbamate (Troc−NR$_2$): Cl$_3$CCH$_2$OC(O)NR$_2$) (Chart 8)

Formation

1. Cl$_3$CCH$_2$OCOCl, Pyr or aq. NaOH, 25°C, 12 h.[1,2]
2. Silylate with Me$_3$SiN=C(OSiMe$_3$)CH$_3$ then treat with Cl$_3$CCH$_2$OCOCl.[3]
3. Cl$_3$CCH$_2$OCO−O-succinimidyl, 1 *N* NaOH or 1 *N* Na$_2$CO$_3$, dioxane, 77–96% yield.[4,5] This method does not result in oligopeptide formation when used to prepare amino acid derivatives.
4. Treatment of a tertiary benzylamine also affords the Troc derivative with cleavage of the benzyl group (Cl$_3$CCH$_2$OCOCl, CH$_3$CN, 93% yield).[6]
5.
   CH$_2$Cl$_2$, rt, 3.5 h, 90–97% yield.[7]

Cleavage

1. Zn, THF, H$_2$O, pH 4.2, 30 min, 86% yield or pH 5.5–7.2, 18 h, 96% yield.[8] Under these conditions the Troc group can be cleaved in the presence of the BOC, benzyl, and trifluoroacetamido groups and these groups can in turn be cleaved individually in the presence of a Troc group.[9] Deprotection in the presence of Ac$_2$O results in acetamide formation.[10]
2. Electrolysis at a Hg cathode, −1.7 V (SCE), DMF, >72% yield.[11]
3. Electrolysis, −1.7 V, 0.1 *M* LiClO$_4$, 85% yield.[12]
4. Zn-Pb couple, 4:1 THF/1*M* NH$_4$OAc.[13]
5. Cd-Pb, AcOH, 89–94% yield.[14] This reagent also cleaves trichloroethyl esters and carbonates.

6. Indium aq. NH$_4$Cl, EtOH, water, reflux, 60–98% yield.[15]
7. Cd, AcOH.[16] These conditions were reported to be superior to the use of Zn/AcOH. The authors also report that the Troc group is not stable to hydrogenation with Pd–C (TsOH, DMF, H$_2$), but is stable to hydrogenation with Ru–C or Pt–C.

8. Cobalt(I) phthalocyanine.[17]

9.

Ref. 18

10. (Bu$_3$Sn)$_2$, DMF, 100°C, 1 day, 99% yield.[19]

1. T. B. Windholz and D. B. R. Johnston, *Tetrahedron Lett.*, **8**, 2555 (1967).

2. J. F. Carson, *Synthesis*, 268 (1981).

3. S. Raucher and D. S. Jones, *Synth. Commun.*, **15**, 1025 (1985).

4. A. Paquet, *Can. J. Chem.*, **60**, 976 (1982).

5. L. Lapatsanis, G. Milias, K. Froussios, and M. Kolovos, *Synthesis*, 671 (1983).

6. V. H. Rawal, R. J. Jones, and M. P. Cava, *J. Org. Chem.*, **52**, 19 (1987).

7. Y. Kita, J.-i. Haruta, H. Yasuda, K. Fukunaga, Y. Shirouchi, and Y. Tamura, *J. Org. Chem.*, **47**, 2697 (1982).

8. G. Just and K. Grozinger, *Synthesis*, 457 (1976).

9. R. J. Bergeron and J. S. McManis, *J. Org. Chem.*, **53**, 3108 (1988).

10. X. Zhu, K. Pachamuthu, and R. R. Schmidt, *Org. Lett.*, **6**, 1083 (2004).

11. L. Van Hijfte and R. D. Little, *J. Org. Chem.*, **50**, 3940 (1985).

12. M. F. Semmelhack and G. E. Heinsohn, *J. Am. Chem. Soc.*, **94**, 5139 (1972).

13. L. E. Overman and R. L. Freerks, *J. Org. Chem.*, **46**, 2833 (1981).

14. Q. Dong, C. E. Anderson, and M. A. Ciufolini, *Tetrahedron Lett.*, **36**, 5681 (1995).

15. T. Mineno, S.-R. Choi, and M. A. Avery, *Synlett*, 883 (2002).

16. G. Hancock, I. J. Galpin, and B. A. Morgan, *Tetrahedron Lett.*, **23**, 249 (1982).

17. H. Eckert and I. Ugi, *Liebigs Ann. Chem.*, 278 (1979).

18. M. V. Lakshmikantham, Y. A. Jackson, R. J. Jones, G. J. O'Malley, K. Ravichandran, and M. P. Cava, *Tetrahedron Lett.*, **27**, 4687 (1986).

19. H. Tokimoto and K. Fukase, *Tetrahedron Lett.*, **46**, 6831 (2005).

2-Trimethylsilylethyl Carbamate (Teoc−NR$_2$): (CH$_3$)$_3$SiCH$_2$CH$_2$OC(O)NR$_2$ (Chart 8)

Formation

1. Teoc−O-succinimidyl, NaHCO$_3$ or TEA, dioxane, H$_2$O, rt, overnight, 43–96% yield.[1,2] The use of Teoc–OSu for the protection of amino acids proceeds without oligopeptide formation. Teoc–*O*-benzotriazolyl was also examined, but was inferior to the succinimide derivative.

2. Teoc−OC$_6$H$_4$-4-NO$_2$, NaOH, *t*-BuOH, 66–89% yield.[3-5]

3. Teoc−Cl or Teoc−N$_3$.[6] The chloroformate is thermally unstable and unstable upon storage and should be freshly prepared. Azides are also hazardous and precautions should be taken with this reagent, especially as the scale increases.

4. The Teoc derivative can be prepared by cleavage of an *N*–Bn bond with Teoc–Cl in THF. This is a general method for removal of benzyl groups from nitrogen.[7] Methyl and ethyl groups are also cleaved, but more slowly (24 h versus 4 h) and in lower yield.

5. From a phenyl carbamate: TMSCH$_2$CH$_2$OH, *t*-BuOK, THF, rt, 12 h, 100% yield. The driving force for this reaction is the leaving group ability of the phenol.[8] The reaction probably proceeds through an isocyanate.

Cleavage

1. Bu$_4$NF, KF·2H$_2$O, CH$_3$CN, 50°C, 8 h, 93% yield or 28°C, 70 h, 93% yield.[9] In some instances, silyl ethers are retained during Teoc cleavage with TBAF.[10]

2. CF$_3$COOH, 0°C, 90% yield[6] or 20% TFA in CH$_2$Cl$_2$.[11,12] TFA cleavage of the Teoc group is competitive with the BOC group. The use of fluoride reagents in this case resulted in partial loss of the boronate. Cleavage of BOC group at this position was unsuccessful.

This method was also used when the typical fluoride reagents resulted in partial elimination of the mesylate below.[13]

3. $ZnCl_2$, CH_3NO_2 or $ZnCl_2$, CF_3CH_2OH.[4] These conditions cause partial BOC cleavage. The BOC group can be removed in the presence of a Teoc group with TsOH.[4]

4. Tris(dimethylamino)sulfonium difluorotrimethylsilicate (TAS–F), DMF, >76% yield.[14]

5. TAS-F, DMF, 0°C, 20 h, 74% yield.[15] The use of TAS-F was superior to TBAF.

6. Bu_4NCl, $KF \cdot 2H_2O$, CH_3CN, 45°C.[16]

7. CsF, DMF, t-BuOH, 90–110°C, 78% yield.[17]

(2-Phenyl-2-trimethylsilyl)ethyl Carbamate (Psoc-NR₂)

The Psoc group was developed for the protection of amino acids. Its stability and orthogonality are similar to that of the Teoc group but it is more susceptible to cleavage with TBAF than is the Teoc group and can be cleaved with trifluoroacetic acid. It is introduced with the 4-nitrophenyl carbonate (72–99% yield) and cleaved with TBAF in CH_2Cl_2 (no yield reported). The liability that this group has is its chirality.[18]

1. R. E. Shute and D. H. Rich, *Synthesis*, 346 (1987).

2. A. Paquet, *Can. J. Chem.*, **60**, 976 (1982).

3. E. Wuensch, L. Moroder, and O. Keller, *Hoppe-Seyler's Z. Physiol. Chem.*, **362**, 1289 (1981).

4. A. Rosowsky and J. E. Wright, *J. Org. Chem.*, **48**, 1539 (1983).

5. J. D. Chisholm and D. L. Van Vranken, *J. Org. Chem.*, **65**, 7541 (2000).

6. L. A. Carpino, J.-H. Tsao, H. Ringsdorf, E. Fell, and G. Hettrich, *J. Chem. Soc., Chem. Commun.*, 358 (1978).

7. A. L. Campbell, D. R. Pilipauskas, I. K. Khanna, and R. A. Rhodes, *Tetrahedron Lett.*, **28**, 2331 (1987).

8. A. I. Meyers, K. A. Babiak, A. L. Campbell, D. L. Comins, M. P. Flemming, R. Henning, M. Heuschmann, J. P. Hudspeth, J. M. Kane, P. J. Reider, D. M. Roland, K. Shimizu, K. Tomioka, and R. D. Walkup, *J. Am. Chem. Soc.*, **105**, 5015 (1983).

9. L. A. Carpino and A. C. Sau, *J. Chem. Soc., Chem. Commun.*, 514 (1979).

10. A. B. Smith III, I. G. Safonov, and R. M. Corbett, *J. Am. Chem. Soc.*, **123**, 12426 (2001).

11. K. K. Hansen, B. Grosch, S. Greiveldinger-Poenaru, and P. A. Bartlett, *J. Org. Chem.*, **68**, 8465 (2003).

12. J. E. Johannes, S. Wenglowsky, and Y. Kishi, *Org. Lett.*, **7**, 3997 (2005).

13. W. H. Pearson, I. Y. Lee, Y. Mi, and P. Stoy, *J. Org. Chem.*, **69**, 9109 (2004).

14. W. R. Roush, D. S. Coffey, and D. J. Madar, *J. Am. Chem. Soc.*, **119**, 11331 (1997).

15. K. C. Nicolaou, S. Vyskocil, T. V. Koftis, Y. M. A. Yamada, T. Ling, D. Y.-K. Chen, W. Tang, G. Petrovic, M. O. Frederick, Y. Li, and M. Sataki, *Angew. Chem. Int. Ed.*, **43**, 4312 (2004).

16. J. H. Van Maarseveen and H. W. Scheeren, *Tetrahedron*, **49**, 2325 (1993).

17. J. Igarashi and Y. Kobayashi, *Tetrahedron Lett.*, **46**, 6381 (2005).

18. M. Wagner, S. Heiner, and H. Kunz, *Synlett*, 1753 (2000).

2-Phenylethyl Carbamate (hZ−NR$_2$): R$_2$NCO$_2$CH$_2$CH$_2$Ph

The 2-phenylethyl carbamate ("homo Z" = homobenzyloxycarbonyl derivative) is prepared from the chloroformate and can be cleaved with H$_2$/Pd–C if the catalyst is freshly prepared [Pd(OAc)$_2$, HCO$_2$NH$_4$]. This derivative is stable to CF$_3$COOH, HBr/AcOH, HCl/Et$_2$O and normal hydrogenation with Pd/C (1 atm). Hydrogenolysis of the hZ group is slower than the Fmoc group, which is slower than the Z group (Cbz).[1]

1. L. A. Carpino and A. Tunga, *J. Org. Chem.*, **51**, 1930 (1986).

2-Chloroethyl Carbamate: ClCH$_2$CH$_2$OC(O)NR$_2$

Cleavage

1. SmI$_2$, THF, 70°C, 7 h, 70% yield.[1]
2. Other reducing agents should also cleave this group.

1,1-Dimethyl-2-haloethyl Carbamate: XCH$_2$C(CH$_3$)$_2$OC(O)NR$_2$, X = Br, Cl
(Chart 8)

Formation[2]

XCH$_2$C(CH$_3$)$_2$OCOCl, THF, Et$_3$N, H$_2$O, CHCl$_3$, 0°C, 1.5 h (X = Br, 41–79% yield; X = Cl, 60–86% yield). These halo-substituted *t*-butyl chloroformates are more stable than an unsubstituted *t*-butyl chloroformate.

Cleavage[2]

1. CH$_3$OH, reflux, 1 h.
2. BF$_3$·Et$_2$O, CF$_3$COOH, 25°C.
3. 4 *N* HBr, AcOH, 25°C, 1 h.
4. Na, NH$_3$.

1,1-Dimethyl-2,2-dibromoethyl Carbamate (DB−*t*-BOC−NR$_2$):

$Br_2CHC(CH_3)_2OC(O)NR_2$

The DB-*t*-BOC group is introduced with the chloroformate and can be cleaved solvolytically in hot ethanol or by HBr/AcOH. It is stable to CF$_3$COOH, 24 h; HCl, MeNO$_2$, 24 h; HCl, AcOH, 24 h; HBr, MeNO$_2$, 5 h.[3]

1,1-Dimethyl-2,2,2-trichloroethyl Carbamate (TCBOC−NR$_2$):

$Cl_3CC(CH_3)_2OC(O)NR_2$

The TCBOC group is stable to the alkaline hydrolysis of methyl esters and to the acidic hydrolysis of *t*-butyl esters. It is rapidly cleaved by the supernucleophile lithium cobalt(I)-phthalocyanine (0.1 eq. NaBH$_4$, EtOH, 77–90% yield),[4] zinc in acetic acid,[5] and Cd/Pb in NH$_4$OAc.[6]

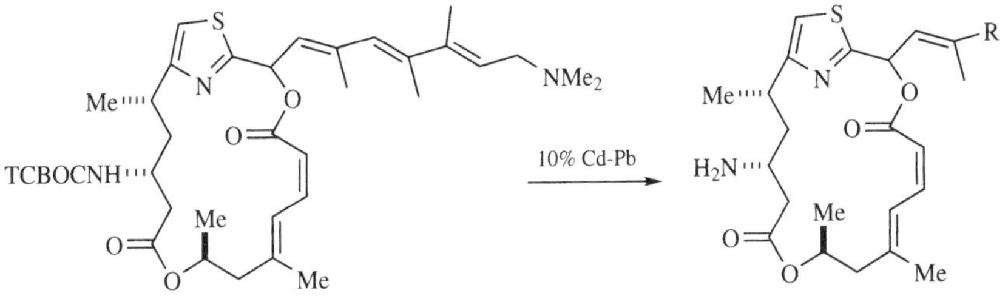

1. T. P. Ananthanarayan, T. Gallagher, and P. Magnus, *J. Chem. Soc., Chem. Commun.*, 709 (1982).
2. T. Ohnishi, H. Sugano, and M. Miyoshi, *Bull. Chem. Soc. Jpn.*, **45**, 2603 (1972).
3. L. A. Carpino, N.W. Rice, E. M. E. Mansour and S. A. Triolo, *J. Org. Chem.*, **49**, 836 (1984).
4. H. Eckert and Y. Kiesel, *Synthesis*, 947 (1980).
5. H. Eckert, M. Listl, and I. Ugi, *Angew. Chem., Int. Ed. Engl.*, **17**, 361 (1978).
6. R. M. Rzasa, H. A. Shea, and D. Romo, *J. Am. Chem. Soc.*, **120**, 591 (1998).
7. D. Romo, R. M. Rzasa, H. A. Shea, K. Park, J. M. Langenhan, L. Sun, A. Akhiezer, and J. O. Liu, *J. Am. Chem. Soc.*, **120**, 12237 (1998).

2-(2′- and 4′-Pyridyl)ethyl Carbamate (Pyoc−NR$_2$)

Formation/Cleavage[1,2]

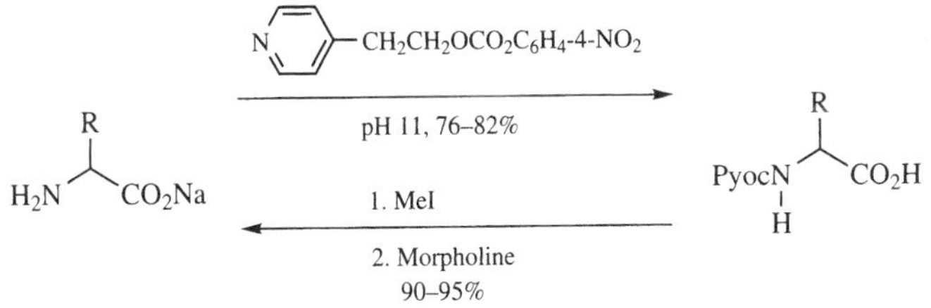

The Pyoc derivative is not affected by H$_2$/Pd–C or TFA.

1. H. Kunz and S. Birnbach, *Tetrahedron Lett.*, **25**, 3567 (1984).

2. H. Kunz and R. Barthels, *Angew. Chem., Int. Ed. Engl.*, **22**, 783 (1983).

2,2-Bis(4'-nitrophenyl)ethyl Carbamate (Bnpeoc−NR₂):
$(4-O_2N-C_6H_4)_2CHCH_2OCONR_2$

The Bnpeoc group was developed as a base-labile protecting group for solid-phase peptide synthesis. The carbamate is formed from the *O*-succinimide (DMF, 10% Na_2CO_3 or 5% $NaHCO_3$), and it is cleaved using DBN, DBU, DBU/AcOH, or piperidine.[1]

1. R. Ramage, A. J. Blake, M. R. Florence, T. Gray, G. Raphy, and P. L. Roach, *Tetrahedron*, **47**, 8001 (1991).

2-[(2-Nitrophenyl)dithio]-1-phenylethyl Carbamate (NpSSPeoc−NR₂)

The protective group was designed as part of the development of an affinity chromatography method for the purification of hydrophobic peptides. S–S cleavage in peptides containing this group, followed by attachment of the resulting S–S moiety to an affinity support (e.g., iodoacetamide resin), was found to be the simplest and highest yielding method. When R′ = H, the protective group is efficiently cleaved with TFA, and when R = NO_2, TfOH in TFA must be used.[1]

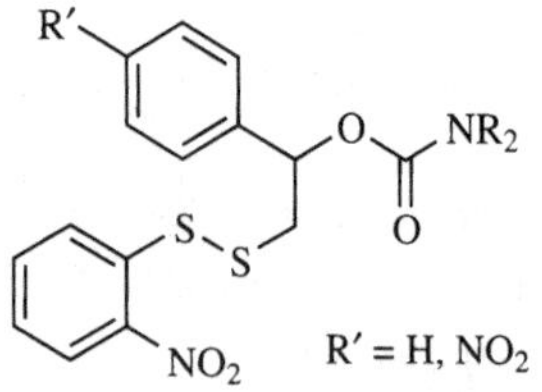

1. I. Sucholeiki and P. T. Lansbury, Jr., *J. Org. Chem.*, **58**, 1318 (1993).

2-(*N,N*-Dicyclohexylcarboxamido)ethyl Carbamate:
$(C_6H_{11})_2NC(O)CH_2CH_2OCONR_2$

This protective group is stable to $LiAlH_4$; 3 *N* NaOH, MeOH, rt; H_2, RaNi, 1500 psi, 100°C, EtOH; and TFA.[1]

Formation

$(C_6H_{11})_2NC(O)CH_2CH_2OCOCl$, diisopropylethylamine, CH_2Cl_2, 0°C, 15 min.[1]

Cleavage

t-BuOK, *t*-BuOH, 18-crown-6, THF, 0°C, 30 min, 100% yield.

1. T. Fukuyama, L. Li, A. A. Laird, and R. K. Frank, *J. Am. Chem. Soc.*, **109**, 1587 (1987).

t-Butyl Carbamate (BOC Group): $(CH_3)_3COC(O)NR_2$ (Chart 8)

The BOC group is used extensively in peptide and heterocyclic synthesis for amine protection.[1] It is not readily hydrolyzed under basic conditions and is inert to many other nucleophilic reagents. It is usually cleaved with strong acid, giving only *t*-BuOH or isobutylene and CO_2 as by-products. As a result, it is one of the most commonly used protective groups for amines. In general, it is considered nonreactive, but there are many cases in which the BOC group participates in reactions—anticipated and unanticipated.[2]

Formation

1. For simple amines, mixing $(BOC)_2O$ and the amine in THF with gentle heating ($\sim40°C$) to drive off CO_2 is often the simplest method for preparing BOC derivatives. If at least 2 equivalents of $(BOC)_2O$ are used, primary amines can be converted to the bis-BOC derivative ($(BOC)_2O$, THF, reflux, 92% yield).[3] Sterically hindered amines often tend to form ureas with $(BOC)_2O$ because of isocyanate formation.[4,5] This can also be a problem with some anilines[6] and when using DMAP as a catalyst, but this side reaction may be avoided by using *N*-methylimidazole as a catalyst.[7,8] Alternatively, isocyanate formation can be avoided by reacting the amine with NaHMDS and then with $(BOC)_2O$.[9] The isocyanates can also be converted to the BOC group by heating with *t*-BuOH. When other alcohols are used, the corresponding carbamate is produced.[10] When DMAP is used as catalyst, excess BOC_2O can be destroyed by adding water and heating to $40–50°C$[11]; in the absence of DMAP, imidazole and water can be used to destroy excess BOC_2O at room temperature.[12]

2. $(BOC)_2O$, NaOH, H_2O, 25°C, 10–30 min, 75–95% yield.[13] This is one of the more common methods for introduction of the BOC group onto amino acids, but does not work efficiently for hindered amines because of reagent destruction. It has the advantage that the by-products are innocuous and are easily removed.

3. $(BOC)_2O$, $Me_4NOH·5H_2O$, CH_3CN, 88–100% yield. These conditions were found to be very good for sterically hindered amino acids.[14]

4. $(BOC)_2O$, TEA, MeOH or DMF, 40–50°C, 87–99% yield. These nonaqueous conditions were used in the protection of ^{17}O-labeled amino acids so that the label would not be lost because of exchange with water.[15]

5. $(BOC)_2O$, EtOH or MeOH, $NaHCO_3$, ultrasound, 84–100% yield.[16]

6. $BOC–ON=C(CN)Ph$, Et_3N, 25°C, several hours, 72–100% yield.[17] This reagent selectively protects primary amines in the presence of secondary amines.[18] This reagent was used to directly convert azides to the BOC derivative in the presence of Me_3P (87–100% yield).[19]

7. $BOC–ONH_2$.[20] This reagent reacts with amines 1.5–2.5 times faster than $(BOC)_2O$. Hydroxylamine can be used catalytically in the presence of $(BOC)_2O$ to generate this reagent *in situ*.

8. BOC−OCH(Cl)CCl$_3$ (1,2,2,2-tetrachloroethyl *tert*-butyl carbonate, BOC-OTCE), THF, K$_2$CO$_3$ or dioxane, H$_2$O, Et$_3$N, 60–91% yield.[21] This reagent is a cheap, distillable solid that has the effectiveness of (BOC)$_2$O.

9.

10. BOC−N$_3$, DMSO, 25°C.[23] Since this is an azide, use of this reagent must be accompanied by the proper safety considerations.

11. BOC−OC$_6$H$_4$S$^+$Me$_2$ MeSO$_4^-$, H$_2$O.[24] This is a water-soluble reagent for the introduction of the BOC group.

12. CF$_3$S(O)$_2$N(BOC)C$_6$H$_4$CF$_3$, THF, DMAP, 0°C, 3 h, 87–97% yield.[25] This method can also be used to introduce the Cbz group. In both cases a primary amine can be protected in the presence of a secondary amine.

13. 1-(*t*-Butoxycarbonyl)benzotriazole, NaOH, dioxane, 20°C, 88–96% yield.[26] Aniline was not reactive with this reagent.[27]

14. Derivatization of urethanes and oxazolidinones with (BOC)$_2$O makes the urethane carbonyl susceptible to hydrolysis under mild conditions and leaves the amine protected as a BOC derivative.[28]

15.

dioxane, H$_2$O, Et$_3$N, 70–94% yield.[29]

16. *t*-BuOCO$_2$Ph, CH$_2$Cl$_2$ or DMF, 49–91% yield. This method selectively protects only the primary amines in polyamines.[30]

17.

50% acetone, H$_2$O, DMAP, TEA, 85–95% yield.[31] This

method was also used to prepare the benzyl, methyl, ethyl, and *p*-methoxybenzyl derivatives. A polymeric version of the reagent was also described.

18.

This reagent is useful for the selective protection of

primary amines.[32] The BOC succinimide derivative reacts similarly (63–80% yield).[33]

19. DME, rt, overnight, 76–98% yield.[34] The reagent also reacts with phenols and thiols to give the corresponding BOC derivatives.

20. Monoprotection of small diamines [$H_2N(CH_2)_xNH_2$, x = 2–6] is achieved by reacting an excess of the amine with $(BOC)_2O$ in dioxane (75–90% yield).[35]

21. t-BuOCOF. [36,37] **The chloroformate is not stable and presents a safety hazard.**

22. $(BOC)_2O$, $Zn(ClO_4)_2 \cdot 6H_2O$, t-BuOH or CH_2Cl_2, 6–168 h, 50–99% yield. This method was developed for the protection of less nucleophilic amines such as anilines and other aryl amines. It avoids the side reactions encountered using methods catalyzed by organic bases.[38] $ZrCl_4$ has also been used as an effective catalyst (CH_3CN, rt, 3–10 h, 85–96% yield).[39]

23. $(BOC)_2O$, $LiClO_4$, CH_2Cl_2, 5 h, rt, 73–90% yield. The reaction is effective for such nonnucleophilic amines as p-nitroaniline (74%).[40]

24. Directly from a carbobenzoxy protected amine: 1,4-cyclohexadiene, Pd/C, $(BOC)_2O$, EtOH, rt, 86–96% yield.[41]

25. Directly from an Fmoc-protected amine: TEA, $(BOC)_2O$, KF, DMF.[42]

26. Directly from an azide: $(BOC)_2O$, Et_3SiH, 20% $Pd(OH)_2-C$, EtOH, 75–99% yield.[43]

27. BOC derivatives can be prepared directly from azides by hydrogenation in the presence of $(BOC)_2O$.[44]

28. From an acetamide or benzamide: $(BOC)_2O$, THF, DMAP; hydrazine, 70–94% yield.[45]

29. From a hydrazine or azo compound: PMHS, Pd–C, $(BOC)_2O$, ethanol, rt, 76–90% yield.[46]

30. From a urea: CuX_2, t-BuOLi, THF, rt or 50°C, 5–30 min, 76–90% yield.[47]

Cleavage

1. 3 M HCl, EtOAc, 25°C, 30 min, 96% yield.[48] With MeOH as the solvent a diphenylmethyl ester is not affected.[49] The combination of HCl/EtOAc leaves TBDMS and TBDPS ethers,[50] and t-butyl esters and nonphenolic ethers[51] intact during BOC cleavage, but an S-BOC and MOM[52] group are cleaved.

2. 4 *M* HCl, dioxane (toxic), 92–100% yield. This method was used to remove BOC groups from peptides in the presence of *t*-Bu esters, ethers, and thioethers, but not phenolic *t*-Bu ethers.[54]

3. Aqueous HCl, toluene, 65°C, 93% yield. This method is a commercially convenient method and has been used on a multikilogram scale.[55]

4. AcCl, MeOH, 95–100% yield. This is a convenient method for generating anhydrous HCl in methanol. These conditions are also used to prepare methyl esters from carboxylic acids and for the formation of amine hydrochlorides.[56]

5. TMSCl, MeOH, 0°C to rt, 5 h, 90–97% yield. This method converts BOC protected amino acids to methyl esters along with cleavage of the BOC group. Again, anhydrous HCl is generated *in situ*.[57]

6. Me$_3$SiCl, PhOH, CH$_2$Cl$_2$, 20 min, 100% yield.[58] Under these conditions benzyl groups are not cleaved and thus provide marked improvement over the conventional 50% TFA/CH$_2$Cl$_2$ used in peptide synthesis.

7. 1 *M* SiCl$_4$, 3 *M* phenol, CH$_2$Cl$_2$, 10 min. The Fmoc, Cbz, Bn ester, and ether groups were not noticeably cleaved even after 18 h of exposure.[59]

8. CF$_3$COOH, CH$_2$Cl$_2$, 5% H$_2$O. Water was added to the mixture to prevent double bond isomerization which occurred under the normal conditions.[60]

9. CF$_3$COOH, PhSH, 20°C, 1 h, 100% yield.[61] Thiophenol is used to scavenge the liberated *t*-butyl cations, thus preventing alkylation of methionine or tryptophan. Other scavengers such as anisole, thioanisole, thiocresol, cresol, and

dimethyl sulfide have also been used.[62] TBDPS[63] and TBDMS[64] groups are stable to TFA during BOC cleavage.

10. TsOH, THF, CH_2Cl_2, 5 min. This method was developed for solid-phase peptide synthesis as a safe large-scale alternative to the use of TFA, which is expensive, corrosive, and a waste problem on a large scale.[65] The reaction is accelerated with microwave irradiation.[66] Polymer-supported sulfonic acids such as Amberlyst 15 effectively cleave the BOC group and leave it loaded on the resin. Washing with NH_4OH releases the free amine from the resin in pure form.[67]

11. 10% H_2SO_4, dioxane.[68] These conditions are similar to the use of 50% TFA/CH_2Cl_2 and are considered safer for large scale use than the use of the volatile, corrosive, and expensive TFA. The authors provide a comparison of many of the acidic methods for BOC cleavage. The only problem with this method is the **toxicity associated with dioxane**, but toluene can be used as a replacement.[55] This method is compatible with high-loading Wang resin for effectively removing BOC groups with minimal cleavage of the substrate from the resin.[69]

12. H_2SO_4, t-BuOAc or CH_3SO_3H, t-BuOAc, CH_2Cl_2, 70–100% yield. This method was used to remove a BOC group in the presence of a t-Bu ester.[70]

13. HNO_3, CH_2Cl_2, 63–92% yield. Cleavage of the BOC group is accomplished in the presence of a t-Bu ester. The by-product from the reaction is t-Bu-ONO_2.[71]

14. Aqueous H_3PO_4, THF, CH_3CN, 86–98% yield. Compatible groups are: Cbz, TBDMS, acetonide, benzyl esters. t-Bu esters are cleaved.[72]

15. Trimethylsilyl triflate (TMSOTf), $PhSCH_3$, CF_3COOH.[73] These conditions also cleave the following protective groups used in peptide synthesis: (MeO)Z−, Bn−, Ts−, $Cl_2C_6H_3CH_2$−, BOM (benzyloxymethyl)−, Mts−, MBS−, t-Bu−SR, Ad−SR, but not a Bn−SR, Acm, or Arg(NO_2) group. The method is partially compatible with acid-cleavable Wang resin.[74] The rate of cleavage is reported to be faster than with TfOH/TFA.

16. Trimethylsilyl triflate (TMSOTf), 2,6-lutidine, CH_2Cl_2, 0°C to rt, 2 h, 100% yield. With TFA the t-Bu ether was also cleaved.[75]

17. The BOC group can be cleaved with TBDMSOTf and the intermediate silyl carbamate converted to other nitrogen protective groups by treatment with fluoride followed by a suitable alkylating agent.[76]

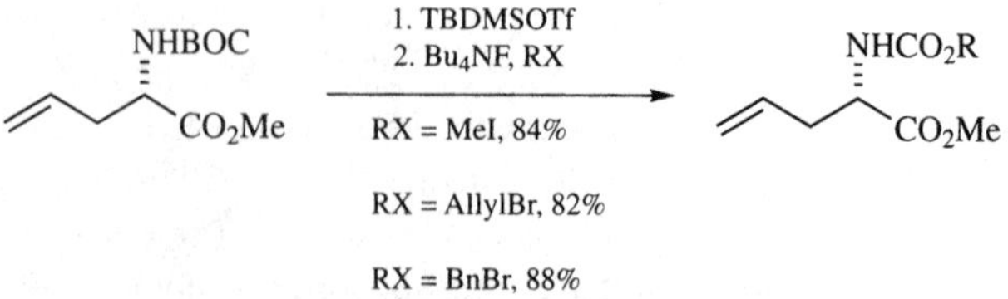

NHBOC / CO₂Me → (1. TBDMSOTf, 2. Bu₄NF, RX) → NHCO₂R / CO₂Me

RX = MeI, 84%

RX = AllylBr, 82%

RX = BnBr, 88%

18. Me₃SiI, CHCl₃ or CH₃CN, 25°C, 6 min, 100% yield.[77,78] Me₃SiI also cleaves carbamates, esters, ethers, and ketals under neutral, nonhydrolytic conditions. Some selectivity can be achieved by control of reaction conditions.

19. AlCl₃, PhOCH₃, CH₂Cl₂, CH₃NO₂, 0–25°C, 2–5 h, 73–88% yield.[79,80] AlCl₃ with microwave heating in the absence of solvent has been used to cleave BOC groups, but the utility of the method is questionable.[81]

20. Bromocatecholborane.[82] A trityl protected amide is preserved under these conditions CH₃CN, 0°C, 3 h).[83]

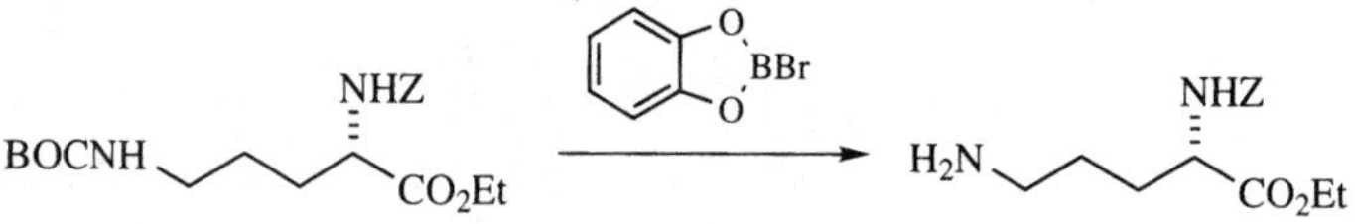

BOCNH—NHZ—CO₂Et → (bromocatecholborane) → H₂N—NHZ—CO₂Et

21. 0.05 M MeSO₃H, dioxane (toxic), CH₂Cl₂ (1:9).[84,85] This reagent also cleaves the Moz (4-methoxybenzyloxycarbonyl) group.

22. Mg(ClO₄)₂, >67% yield.[86] These conditions cleave one of the two BOC groups on a primary amine.

23. BF₃·Et₂O, 4-Å ms, CH₂Cl₂, rt, 20 h, 77–98% yield.[87]

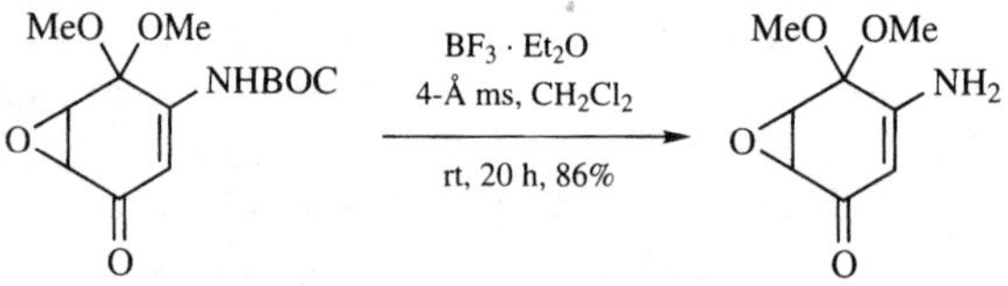

(MeO OMe)...NHBOC → (BF₃·Et₂O, 4-Å ms, CH₂Cl₂, rt, 20 h, 86%) → (MeO OMe)...NH₂

24. SnCl₄, AcOH, THF, CH₂Cl₂, toluene or CH₃CN, 82–98% yield. This method was developed because acid-based methods were incompatible with the presence of a thioamide peptide bond.[88] Guanidines were cleanly deprotected.[89]

25. From RN(Ts)BOC: DMF, 100–120°C, 24 h, >69–75% yield.[90]

26. Silica gel, heat under vacuum, 80–92% yield.[91] These conditions will selective remove only the indole BOC group from a fully *t*-Bu-based, protected tryptophan. An epoxide is compatible with these conditions.[92]

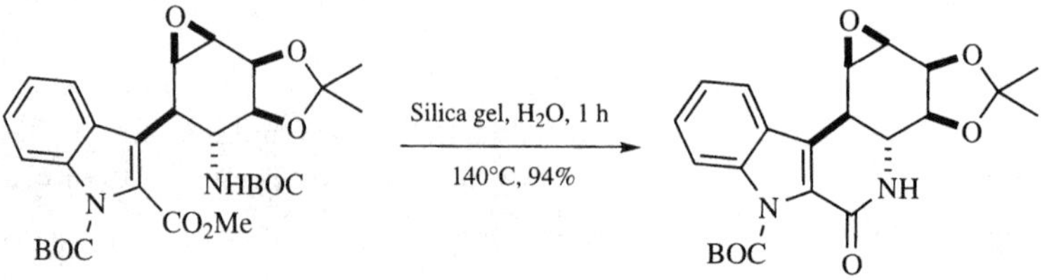

Silica gel, H₂O, 1 h; 140°C, 94%

27. Migration of a BOC is normally not observed, but in the following case a BOC group moved to a hydroxyl. The stabilizing effect of the tosyl group makes this possible.[93]

Note BOC group moved from N to O

28. A BOC-protected primary amine with an adjacent leaving group is slowly converted to an oxazolidone.[94]

29. The conversion of the BOC group to other carbamates is achieved by heating with an alcohol, Ti(O-*i*-Pr)$_4$ in toluene. Teoc-, Cbz- and Alloc-protected primary amines have been prepared in this fashion. The reaction is selective for a primary BOC derivative probably because the reaction proceeds through an isocyanide.[95]

30. CAN, CH$_3$CN, 90–99% yield.[96]

31. ZnBr$_2$, CH$_2$Cl$_2$, 89–94% yield. These conditions selectively cleave the BOC group from secondary amines in the presence of the primary derivatives.[97]

32. Sn(OTf)$_2$, CH$_2$Cl$_2$, 87–95% yield.[98] *t*-Butyl esters are retained under these conditions.

33. Montmorillonite K10 clay, ClCH$_2$CH$_2$Cl, reflux, 64–98% yield. This method selectively cleaves the BOC group from aromatic amines.[99]

34. H-β-zeolite, CH$_2$Cl$_2$, reflux, 77–100% yield. This method is selective for BOC protected aryl amines with BOC protected alkyl amines being stable. BOC protected amides and *t*-butyl esters are also stable.[100]

35. NaI, acetone, 60–100°C, 15–25 min, 88–98% yield.[101] This method was used to remove the BOC from amino acids probably by the formation of catalytic HI.

36. *t*-BuOK, H$_2$O, 2-MeTHF, reflux, 90–100% yield. This method is only good for primary derivatives, since it proceeds through formation of an isocyanate, which is hydrolyzed by water.[102]

37. The BOC group can be removed thermally, either neat (185°C, 20–30 min, 97% yield)[103] or in diphenyl ether.[104,105]

Selective cleavage of a single BOC group from a di-BOC amine

$$(BOC)_2NR \quad \xrightarrow{\text{Reagent}} \quad BOCNHR$$

38. LiBr, CH_3CN, 65°C, 95% yield.[3]

39. $CeCl_3 \cdot 7H_2O$, NaI, CH_3CN, 4–6 h, 83–95% yield. This method can be used to cleave A BOC group from an amine that contains a typical amide.

40. In an amine bearing two BOC groups, 2 eq. of TFA in CH_2Cl_2 will cleave only one BOC leaving a monoprotected primary amine.[106,107] A t-Bu ester was stable.

41. In or Zn, MeOH, reflux, 15–24 h, 80–92% yield.[108]

1. M. Bodanszky, *Principles of Peptide Chemistry*, Springer-Verlag: New York, 1984, p. 99.

2. For a review, see C. Agami and F. Couty, *Tetrahedron*, **58**, 2701 (2002).

3. B. E. Haug and D. H. Rich, *Org. Lett.*, **6**, 4783 (2004).

4. H.-J. Knölker, T. Braxmeier, and G. Schlechtingen, *Angew. Chem., Int. Ed. Engl.* **34**, 2497 (1995).

5. H.-J. Knölker, T. Braxmeier, and G. Schlechtingen, *Synlett*, 502 (1996).

6. A. Kessler, C. M. Coleman, P. Charoenying, and D. F. O'Shea, *J. Org. Chem.*, **69**, 7836 (2004).

7. Y. Basel and A. Hassner, *J. Org. Chem.*, **6**, 6368 (2000).

8. For a review covering some issues of DMAP-catalyzed acylations of amines, see U. Ragnarsson and L. Grehn, *Acc. Chem. Res.*, **31**, 494 (1998).

9. T. A. Kelly and D. W. McNeil, *Tetrahedron Lett.*, **35**, 9003 (1994).

10. H.-J. Knölker and T. Braxmeier, *Tetrahedron Lett.*, **37**, 5861 (1996).

11. P. G. M. Wuts, personal observation.

12. Y. Basel and A. Hassner, *Synthesis*, 550 (2001).

13. D. S. Tarbell, Y. Yamamoto, and B. M. Pope, *Proc. Natl. Acad. Sci., USA*, **69**, 730 (1972).

14. E. M. Khalil, N. L. Subasinghe, and R. L. Johnson, *Tetrahedron Lett.*, **37**, 3441 (1996).

15. E. Ponnusamy, U. Fotadar, A. Spisni, and D. Fiat, *Synthesis*, 48 (1986).

16. J. Einhorn, C. Einhorn, and J. L. Luche, *Synlett*, 37 (1991).

17. M. Itoh, D. Hagiwara, and T. Kamiya, *Bull. Chem. Soc. Jpn.*, **50**, 718 (1977).

18. G. M. Cohen, P. M.Cullis, J. A. Hartley, A. Mather, M. C. R. Symons, and R. T. Wheelhouse, *J. Chem. Soc., Chem. Commun.*, 298 (1992).

19. X. Ariza, F. Urpf, C. Viladomat, and J. Vilarrasa, *Tetrahedron Lett.*, **39**, 9101 (1998).

20. R. B. Harris and I. B. Wilson, *Tetrahedron Lett.*, **24**, 231 (1983).

21. G. Barcelo, J. P. Senet, and G. Sennyey, *J. Org. Chem.*, **50**, 3951 (1985).

22. S. Kim, J. I. Lee, and K. Y. Yi, *Bull. Chem. Soc. Jpn.* **58**, 3570 (1985).

23. J. B. Hansen, M. C. Nielsen, U. Ehrbar, and O. Buchardt, *Synthesis*, 404 (1982).

24. I. Azuse, H. Okai, K. Kouge, Y. Yamamoto, and T. Koizumi, *Chem. Express*, **3**, 45 (1988).

25. T. Yasuhara, Y. Nagaoka, and K. Tomioka, *J. Chem. Soc. Perkin Trans. 1*, 2233 (1999).

26. A. R. Katritzky, C. N. Fali, J. Li, D. J. Ager, and I. Prakash, *Synth. Commun.*, **27**, 1623 (1997).

27. S. Kim and H. Chang, *Bull. Korean. Chem. Soc.*, **7**, 70 (1986).

28. T. Ishizuka and T. Kunieda, *Tetrahedron Lett.*, **28**, 4185 (1987).

29. F. Effenberger and W. Brodt, *Chem. Ber.* **118**, 468 (1985).

30. M. Pittelkow, R. Lewinsky, and J. B. Christensen, *Synthesis*, 2195 (2002).

31. T. Kunieda, T. Higuchi, Y. Abe, and M. Hirobe, *Chem. Pharm. Bull.*, **32**, 2174 (1984).

32. I. Grapsas, Y. J. Cho, and S. Mobashery, *J. Org. Chem.*, **59**, 1918 (1994).

33. M. Adamczyk, J. R. Fishpaugh, and K. J. Heuser, *Org. Prep. Proc. Int.*, **30**, 339 (1998).

34. H. Ouchi, Y. Saito, Y. Yamamoto, and H. Takahata, *Org. Lett.*, **4**, 585 (2002).

35. A. P. Krapcho and C. S.Kuell, *Synth. Commun.*, **20**, 2559 (1990).

36. L. A. Carpino, K. N. Parameswaran, R. K. Kirkley, J. W. Spiewak, and E. Schmitz, *J. Org. Chem.*, **35**, 3291 (1970).

37. For an improved preparation of this reagent, see V. A. Dang, R. A. Olofson, P. R. Wolf, M. D. Piteau, and J.-P. G. Senet, *J. Org. Chem.*, **55**, 1847 (1990).

38. G. Bartoli, M. Bosco, M. Locatelli, E. Marcantoni, M. Massaccesi, P. Melchiorre, and L. Sambri, *Synlett*, 1794 (2004).

39. G. V. M. Sharma, J. J. Reddy, P. S. Lakshmi, and P. R. Krishna, *Tetrahedron Lett.*, **45**, 6963 (2004).

40. A. Heydari and S. E. Hosseini, *Adv. Synth. Catal.*, **347**, 1929 (2005).

41. J. S. Bajwa, *Tetrahedron Lett.*, **33**, 2955 (1992).

42. W.-R. Li, J. Jiang, and M. J. Joullie, *Tetrahedron Lett.*, **34**, 1413 (1993).

43. H. Kotsuki, T. Ohishi, and T. Araki, *Tetrahedron Lett.*, **38**, 2129 (1997).

44. S. Saito, H. Nakajima, M. Inaba, and T. Moriwake, *Tetrahedron Lett.*, **30**, 837 (1989).

45. M. J. Burk and J. G. Allen, *J. Org. Chem.*, **62**, 7054 (1997).

46. S. Chandrasekhar, C. R. Reddy, and R. J. Rao, *Synlett*, 1561 (2001).

47. J.-i. Yamaguchi, Y. Shusa, and T. Suyama, *Tetrahedron Lett.*, **40**, 8251 (1999).

48. G. L. Stahl, R. Walter, and C. W. Smith, *J. Org. Chem.*, **43**, 2285 (1978).

49. Z. Tozuka, H. Takasugi, and T. Takaya, *J. Antibiotics*, **36**, 276 (1983).

50. F. Cavelier and C. Enjalbal, *Tetrahedron Lett.*, **37**, 5131 (1996).

51. F. S. Gibson, S. C. Bergmeier, and H. Rapoport, *J. Org. Chem.*, **59**, 3216 (1994).

52. D. L. Boger, T. V. Hughes, and M. P. Hedrick, *J. Org. Chem.*, **66**, 2207 (2001).

53. P. Wipf and H. Kim, *J. Org. Chem.*, **58**, 5592 (1993).

54. G. Han, M. Tamaki, and V. J. Hruby, *J. Peptide Res.*, **58**, 338 (2001).

55. M. Prashad, D. Har, B. Hu, H.-Y. Kim, M. J. Girgis, A. Chaudhary, O. Repic, and T. J. Blacklock, *Org. Proc. Res. Dev.*, **8**, 330 (2004).

56. A. Nudelman, Y. Bechor, E. Falb, B. Fischer, B. A. Wexler, and A. Nudelman, *Synth. Commun.*, **28**, 471 (1998).

57. B.-C. Chen, A. P. Skoumbourdis, P. Guo, M. S. Bednarz, O. R. Kocy, J. E. Sundeen, and G. D. Vite, *J. Org. Chem.*, **64**, 9294 (1999).

58. E. Kaiser, Sr., T. Kubiak, J. P. Tam, and R. B. Merrifield, *Tetrahedron Lett.*, **29**, 303 (1988); E. Kaiser, Sr., F. Picart, T. Kubiak , J. P. Tam, and R. B. Merrifield, *J. Org. Chem.*, **58**, 5167 (1993).

59. K. M. Sivanandaiah, V. V. S. Babu, and B. P. Gangadhar, *Tetrahedron Lett.*, **37**, 5989 (1996).

60. A. S. Ripka, R. S. Bohacek, and D. H. Rich, *Bioorganic & Med. Chem. Letters*, **8**, 357 (1998).

61. B. F. Lundt, N. L. Johansen, A. Vølund, and J. Markussen, *Int. J. Pept. Protein Res.*, **12**, 258 (1978).

62. See, for example, M. Bodanszky and A. Bodanszky, *Int. J. Pept. Protein Res.*, **23**, 565 (1984); Y. Masui, N. Chino, and S. Sakakibara, *Bull. Chem. Soc. Jpn.*, **53**, 464 (1980).

63. P. A. Jacobi, S. Murphree, F. Rupprecht, and W. Zheng, *J. Org. Chem.*, **61**, 2413 (1996).

64. J. Deng, Y. Hamada, and T. Shioiri, *J. Am. Chem. Soc.*, **117**, 7824 (1995).

65. H. R. Brinkman, J. J. Landi, Jr., J. B. Paterson, Jr., and P. J. Stone, *Synth. Commun.*, **21**, 459 (1991).

66. V. V. S. Babu, B. S. Patil, and G.-R. Vasanthakumar, *Synth. Commum.*, **35**, 1795 (2005).

67. Y.-S. Liu, C. Zhao, D. E. Bergbreiter, and D. Romo, *J. Org. Chem.*, **63**, 3471 (1998).

68. R. A. Houghten, A. Beckman, and J. M. Ostresh, *Int. J. Pept. Protein Res.*, **27**, 653 (1986).

69. H. S. Trivedi, M. Anson, P. G. Steel, and J. Worley, *Synlett*, **12**, 1932 (2001).

70. L. S. Lin, J. T. Lanza, S. E. deLaszlo, Q. Truong, T. Kamenecka, and W. K. Hagmann, *Tetrahedron Lett.*, **41**, 7013 (2000).

71. P. Strazzolini, T. Melloni, and A. G. Giumanini, *Tetrahedron*, **57**, 9033 (2001).

72. B. Li, R. Bemish, R. A. Buzon, C. K.-F. Chiu, S. T. Colgan, W. Kissel, T. Le, K. R. Leeman, L. Newell, and J. Roth, *Tetrahedron Lett.*, **44**, 8113 (2003).

73. N. Fujii, A. Otaka, O. Ikemura, K. Akaji, S. Funakoshi, Y. Hayashi, Y. Kuroda, and H. Yajima, *J. Chem. Soc., Chem. Commun.*, 274 (1987).

74. V. Lejeune, J. Martinez, and F. Cavelier, *Tetrahedron Lett.*, **44**, 4757 (2003).

75. H. M. M. Bastiaans, J. L. van der Baan, and H. C. J. Ottenheijm, *J. Org. Chem.*, **62**, 3880 (1997).

76. M. Sakaitani and Y. Ohfune, *Tetrahedron Lett.*, **26**, 5543 (1985); *idem., J. Org. Chem.*, **55**, 870 (1990).

77. R. S. Lott, V. S. Chauhan, and C. H. Stammer, *J. Chem. Soc., Chem. Commun.*, 495 (1979).

78. For a review on the use of Me₃SiI, see G. A. Olah and S. C. Narang, *Tetrahedron*, **38**, 2225 (1982).

79. T. Tsuji, T. Kataoka, M. Yoshioka, Y. Sendo, Y. Nishitani, S. Hirai, T. Maeda, and W. Nagata, *Tetrahedron Lett.*, **20**, 2793 (1979).

80. K. D. James and A. D. Ellington, *Tetrahedron Lett.*, **39**, 175 (1998); D. S. Bose and V. Lakshminarayana, *Synthesis*, 66 (1999).

81. D. S. Bose and V. Lakshminarayana, *Tetrahedron Lett.*, **39**, 5631 (1998).

82. R. K. Boeckman, Jr., and J. C. Potenza, *Tetrahedron Lett.*, **26**, 1411 (1985).

83. Y. Rew, D. Shin, I. Hwang, and D. L. Boger, *J. Am. Chem. Soc.*, **126**, 1041 (2004).

84. Y. Kiso, A. Nishitani, M. Shimokura, Y. Fujiwara, and T. Kimura, *Pept. Chem., 1987*, 291 (1988), *Chem. Abstr.*: **109**, 190837t (1988).

85. Y. Kiso, Y. Fujiwara, T. Kimura, A. Nishitani, and K. Akaji, *Int. J. Pept. Protein, Res.*, **40**, 308 (1992).

86. F. Burkhart, M. Hoffmann, and H. Kessler, *Angew. Chem., Int. Ed. Engl.*, **36**, 1191 (1997).

87. E. F. Evans, N. J. Lewis, I. Kapfer, G. Macdonald, and R. J. K. Taylor, *Synth. Commun.*, **27**, 1819 (1997).

88. R. Frank and M. Schutkowski, *J. Chem. Soc., Chem. Commun.*, 2509 (1996).

89. H. Miel and S. Rault, *Tetrahedron Lett.*, **38**, 7865 (1997).

90. K. E. Krakowiak and J. S. Bradshaw, *Synth. Commun.*, **26**, 3999 (1996).

91. T. Apelqvist and D. Wensbo, *Tetrahedron Lett.*, **37**, 1471 (1996).

92. T. Hudlicky, U. Rinner, K. J. Finn, and I. Ghiviriga, *J. Org. Chem.*, **70**, 3490 (2005).

93. R. J. Valentekovich and S. L. Schreiber, *J. Am. Chem. Soc.*, **117**, 9069 (1995).

94. T. P. Curran, M. P. Pollastri, S. M. Abelleira, R. J. Messier, T. A. McCollum, and C. G. Rowe, *Tetrahedron Lett.*, **35**, 5409 (1994).

95. G. Shapiro and M. Marzi, *J. Org. Chem.*, **62**, 7096 (1997).

96. J. R. Hwu, M. L. Jain, S.-C. Tsay, and G. H. Hakimelahi, *Tetrahedron Lett.*, **37**, 2035 (1996).

97. S. C. Nigam, A. Mann, M. Taddei, and C.-G. Wermuth, *Synth. Commun.*, **19**, 3139 (1989).

98. D. S. Bose, K. K. Kumar, and A. V. N. Reddy, *Synth. Commun.*, **33**, 445 (2003).

99. N. S. Shaikh, A. S. Gajare, V. H. Deshpande, and A. V. Bedekar, *Tetrahedron Lett.*, **41**, 385 (2000).

100. V. H. Tillu, R. D. Wakharkar, R. K. Pandey, and P. Kumar, Ind. *J. Chem.*, **43B**, 1004 (2004). N. Ravindranath, C. Ramesh, M. R. Reddy, and B. Das, *Adv. Synth. Catal.*, **345**, 1207 (2003).

101. J. Ham, K. Choi, J. Ko, H. Lee, and M. Jung, *Protein and Peptide Letters*, **5**, 257 (1998).

102. N. J. Tom, W. M. Simon, H. N. Frost, and M. Ewing, *Tetrahedron Lett.*, **45**, 905 (2004).

103. V. H. Rawal, R. J. Jones, and M. P. Cava, *J. Org. Chem.*, **52**, 19 (1987).

104. H. H. Wasserman, G. D. Berger, and K. R. Cho, *Tetrahedron Lett.*, **23**, 465 (1982).

105. R. Downham, F. W. Ng, and L. E. Overman, *J. Org. Chem.*, **63**, 8096 (1998).

106. R. A. T. M. van Benthem, H. Hiemstra, and W. N. Speckamp, *J. Org. Chem.*, **57**, 6083 (1992).

107. G. Papageorgiou, D. Ogden, and J. E. T. Corrie, *J. Org. Chem.*, **69**, 7228 (2004).

108. J. S. Yadav, B. V. S. Reddy, K. S. Reddy, and K. B. Reddy, *Tetrahedron Lett.*, **43**, 1549 (2002).

Fluorous BOC (^{F}BOC$-$NR$_2$) Carbamate: $C_8F_{19}CH_2CH_2C(CH_3)_2O_2CNR_2$

This derivative along with four other analogs was prepared for use in the fluorous synthesis technique. It is cleaved with TFA, similarly to the regular BOC derivative.[1]

1. Z. Luo, J. Williams, R. W. Read, and D. P. Curran, *J. Org. Chem.*, **66**, 4261 (2001).

1-Adamantyl Carbamate (Adoc−NR$_2$): 1-Adamantyl−OC(O)NR$_2$ (Chart 8)

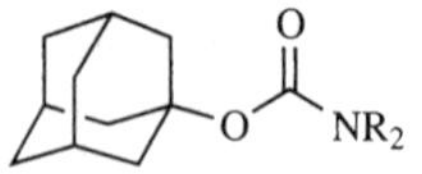

The Adoc group is very similar to the *t*-BOC group in its sensitivity to acid, but often provides more crystalline derivatives of amino acids.

Formation

1. 1-Adoc-Cl, NaOH, 27–98% yield.[1]
2. 1-Adoc-O-2-pyridyl, 70–95% yield.[2]

3. 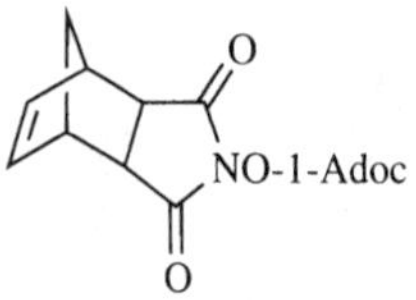 NO-1-Adoc   Dioxane (toxic), H$_2$O, 2–3 h, 82–84% yield.[3]

4. 1-Adoc-F, 84–94% yield.[4,5] The solvolytic decomposition of this reagent has been examined.[6]

Cleavage

CF$_3$CO$_2$H, 25°C, 15 min, 100% yield.[1]

2-Adamantyl Carbamate (2-Adoc-NR$_2$): 2-Adamantyl-OC(O)NR$_2$

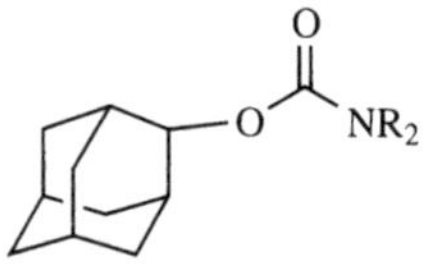

Since this is a derivative of a secondary alcohol, it is much more stable to acid than the 1-Adoc derivative. The 2-Adoc group is stable to HCl/dioxane, TFA, 25% HBr/AcOH, TMSBr/thioanisole/TFA for up to 24 h.[7]

Formation

2-Adamantyl chloroformate.[7]

Cleavage

Trifluoromethanesulfonic acid or anhydrous HF at 0°C.[7]

1. W. L. Haas, E. V. Krumkalns, and K. Gerzon, *J. Am. Chem. Soc.* **88**, 1988 (1966).
2. F. Effenberger and W. Brodt, *Chem. Ber.*, **118**, 468 (1985).

3. P. Henklein, H.-U. Heyne, W.-R. Halatsch, and H. Niedrich, *Synthesis*, 166 (1987).

4. L. Moroder, L. Wackerle, and E. Wuensch, *Hoppe-Seyler's Z. Physiol. Chem.*, **357**, 1647 (1976).

5. R. Presentini and G. Antoni, *Int. J. Pept. Protein Res.*, **27**, 123 (1986).

6. D. N. Kevill and J. B. Kyong, *J. Org. Chem.*, **57**, 258 (1992).

7. Y. Nishiyama, N. Shintomi, Y. Kondo, and Y. Okada, *J. Chem. Soc., Perkin Trans. 1*, 3201 (1994).

1-(1-Adamantyl)-1-methylethyl Carbamate (Adpoc−NR₂):

1-1-(Adamantyl)C(Me)$_2$OC(O)NR$_2$

The Adpoc derivative is cleaved by CF$_3$COOH (0°C, 4–5 min) 10^3 times faster than the *t*-BOC derivative.[1]

1. H. Kalbacher and W. Voelter, *Angew. Chem., Int. Ed. Engl.*, **17**, 944 (1978); *idem., J. Chem. Soc., Chem. Commun.*, 1265 (1980).

1-Methyl-1-(4-biphenylyl)ethyl Carbamate (Bpoc−NR₂):

p-PhC$_6$H$_4$C(Me)$_2$OC(O)NR$_2$ (Chart 8)

Formation

1. Bpoc−N$_3$, 35–80% yield.[1] Large scale detailed experimentals are provided for the reagent and its use.

2. dioxane, H$_2$O, TEA, 79–81% yield.[2]

3. 2-(4-biphenylyl)-prop-2-yl 4′-methoxycarbonylphenyl carbonate, DMF, 50°C, 4 h, >60% yield.[3]

Cleavage

1. This derivative is readily cleaved by acidic hydrolysis (dil. CF$_3$COOH, CH$_2$Cl$_2$, 10 min, quant.). It is cleaved 3000 times faster than the *t*-BOC derivative because of stabilization of the cation by the biphenyl group.[1] BnSH was found to be the most effective scavenger for PhC$_6$H$_4$C$^+$Me$_2$ when deblocking is performed in 0.5% TFA/CH$_2$Cl$_2$.[4]

2. 1% TFA, 5% Et$_3$SiH, CH$_2$Cl$_2$, rt, 30 min.[3]

3. Tetrazole, trifluoroethanol, 24 h, 95% yield.[5] These conditions will also cleave the *N*-trityl group. If deprotection is performed in the presence of an acylating agent, acylation proceeds directly.

4. *N*-Hydroxybenzotriazole, trifluoroethanol, rt.[6] Trityl and Nps (2-nitrophenyl-sulfenyl) groups are also cleaved under these conditions.

5. Mg(ClO)$_4$, CH$_3$CN, 50°C.[7]

1. R. S. Feinberg and R. B. Merrifield, *Tetrahedron*, **28**, 5865 (1972).

2. P. Henklein, H.-U. Heyne, W.-R. Halatsch, and H. Niedrich, *Synthesis*, 166 (1987).

3. A. R. Poreddy, O. F. Schall, G. R. Marshall, C. Ratledge, and U. Slomczynska, *Biorg. Med. Chem. Lett.*, **13**, 2553 (2003).

4. D. S. Kemp, N. Fotouhi, J. G. Boyd, R. I. Carey, C. Ashton, and J. Hoare, *Int. J. Pept. Protein Res.*, **31**, 359 (1988).

5. M. Bodansky, A. Bodansky, M. Casaretto, and H. Zahn, *Int. J. Pept. Protein Res.*, **26**, 550 (1985).

6. M. Bodanzky, M. A. Bednarek, and A. Bodanzky, *Int. J. Pept. Protein Res.*, **20**, 387 (1982).

7. D. Wildemann, M. Drewello, G. Fischer, and M. Schutkowski, *Chem. Commun.*, 1809 (1999).

1-(3,5-Di-*t*-butylphenyl)-1-methylethyl Carbamate (*t*-Bumeoc−NR$_2$)

Formation

The *t*-Bumeoc adduct is prepared from the acid fluoride or the mixed carbonate in dioxane, H$_2$O, NaOH.[1]

Cleavage

Cleavage occurs with acid. The following tables give relative rate data that are useful for comparing other, more commonly employed derivatives of phenylalanine (Phe).[1]

Half-Life of t-Bumeoc-Phe-OH with Different Acids

Acid	Half-Life (min)	Complete Cleavage (min)
3% TFA/CH$_2$Cl$_2$	0.07	0.6
80% AcOH/H$_2$O	2.1	18.8
AcOH/HCO$_2$H/H$_2$O (7:1:2)	22.0	167.0

Comparison of Cleavage Rates for Various Carbamate Protective Groups

Group	$k_{rel}{}^a$	$k_{rel}{}^b$
Boc	1	1
Ppoc[c]	700	750
Adpoc[d]	2400	600
Bpoc[e]	2800	2000
t-Bumeoc	4000	8000

[a]80% AcOH/H_2O.

[b]AcOH/HCO_2H/H_2O (7:1:2).

[c]Ppoc = 1-Methyl-1-(triphenylphosphonio)ethyl.

[d]Adpoc = 1-Methyl-1-(1-adamantyl)ethyl.

[e]Bpoc = 1-Methyl-1-(4-biphenylyl)ethyl.

1. W. Voelter and J. Mueller, Liebigs Ann. Chem., 248 (1983).

Triisopropylsiloxy (Tsoc$-$NR$_2$) Carbamate: $(i\text{-}Pr)_3SiO_2CNR_2$

The reaction of an amine with CO_2 in DMF or CH_2Cl_2 in the presence of TEA gives an adduct that is silylated with TIPSCl or TIPSOTf to form the Tsoc derivative in 56–94% yield.[1,2] The reaction with other silyl chlorides gives similar derivatives but these tend to be more susceptible to hydrolysis. Electron-deficient amines do not react well because the CO_2 adduct does not form efficiently. The Tsoc group is orthogonal to the BOC, Cbz, and Fmoc groups in that it is stable to TFA/CH_2Cl_2, rt, 2 h, to hydrogenolysis over Pd–C, rt, 2 h, and to morpholine/DMF, rt 1 h. Cleavage of the Tsoc group is accomplished by treatment with TBAF (81–96% yield).[1] This group has found utility for the synthesis of peptides using acyl fluorides.[3] Other fluoride sources should also be effective. Tsoc is stable to n-BuLi at $-50°C$.[4]

1. B. H. Lipshutz, P. Papa, and J. M. Keith, *J. Org. Chem.*, **64**, 3792–3793 (1999).
2. L. Pouységu, A.-V. Avellan, and S. Quideau, *J. Org. Chem.*, **67**, 3425 (2002).
3. K. Sakamoto, Y. Nakahara, and Y. Ito, *Tetrahedron Lett.*, **43**, 1515 (2002).
4. P. G. M. Wuts, unpublished results.

Vinyl Carbamate (Voc$-$NR$_2$): $CH_2=CHOC(O)NR_2$ (Chart 8)

The olefin of the Voc group is very susceptible to electrophilic reagents and thus is readily cleaved by reaction with bromine or mercuric acetate.

Formation

1. $CH_2=CHOCOCl$, MgO, H_2O, dioxane, pH 9–10, 90% yield.[1]
2. $CH_2=CHOCOSPh$, Et_3N, dioxane or DMF, H_2O, 25°C, 16 h, 50–80% yield.[2]

Cleavage

1. Anhydrous HCl, dioxane, 25°C, 97% yield.[1]
2. HBr, AcOH, 94% yield.[1]
3. Br_2, CH_2Cl_2 then MeOH, 95% yield.[1]
4. $Hg(OAc)_2$, AcOH, H_2O, 25°C, 97% yield.[1]

1. R. A. Olofson, Y. S. Yamamoto, and D. J. Wancowicz, *Tetrahedron Lett.*, **18**, 1563 (1977).
2. A. J. Duggan and F. E. Roberts, *Tetrahedron Lett.*, **20**, 595 (1979). For an improved preparation of this reagent, see R. A. Olofson and J. Cuomo, *J. Org. Chem.*, **45**, 2538 (1980).

Allyl Carbamate (Alloc−NR₂ or Aloc−NR₂): $CH_2=CHCH_2OC(O)NR_2$ (Chart 8)

The Alloc group has become one of the more frequently used protective groups because of its excellent orthogonality with many other groups and the mild conditions under which it can be removed. The utility of this group has been reviewed.[1] The most commonly used methods for cleavage are based on Pd(0) in the presence of a nucleophile to scavenge the allyl group.

Formation

1. $CH_2=CHCH_2OCOCl$, Pyr.[2]
2. $(CH_2=CHCH_2OCO)_2O$, dioxane, H_2O, reflux or CH_2Cl_2, 1 h, rt, 67–96% yield.[3]
3. $CH_2=CHCH_2OC(O)-O-$benzotriazolyl.[4]
4. Polymer supported Alloc−OSu, TEA, CH_2Cl_2, 72–98% yield.[5]
5. $(CH_2=CHCH_2OCO)_2O$, phosphate buffer pH 8, Subtilisin.[6]

6. $(CH_2=CHCH_2OCO)_2O$, *candida antartica lipase* B, MS4 Å, THF or THF/ pyridine, 64–69% yield.[7]

7. Allyl bromide, CO_2, 18-crown-6, 9–55% yield.[8]
8. (a) $NO_2C_6H_4OCOCl$, (b) allyl alcohol, CH_3CN, 3 h, rt, 88% yield.[9]

9. PhOCO$_2$CH$_2$CH=CH$_2$, 46–98% yield. This reagent allows the protection of primary over secondary amines.[10]

Cleavage

1. I$_2$, CH$_3$CN, H$_2$O, 60°C, 8–16 h, 82–93% yield.[11]

2. Ni(CO)$_4$ (**CAUTION: VERY TOXIC**), DMF, H$_2$O (95:5), 55°, 4 h, 83–95% yield.[2]

3. (Ph$_3$P)$_2$NiCl$_2$, 3% Ph$_3$P, Me$_2$NH·BH$_3$, K$_2$CO$_3$, CH$_3$CN, 40°C, 82–97% yield.[12] Aryl Cbz groups are also cleaved but at a slower rate. The Alloc group can be cleaved in the presence of a Cbz.

4. [Ni(bipy)$_3$](BF$_4$)$_2$, Zn anode, DMF, rt, 70–99% yield.[13]

5. Pd(Ph$_3$P)$_4$, Bu$_3$SnH, AcOH, 70–100% yield.[14]

6. Pd(Ph$_3$P)$_4$, Me$_2$NH·BH$_3$, DMF. This method proved superior for peptide synthesis where the use of PhSiH$_3$ and morpholine proved relatively ineffective.[15]

7. Pd(Ph$_3$P)$_4$, Me$_2$NTMS, 89–100% yield as the TMS carbamate that is easily hydrolyzed. This method was developed to suppress allylamine formation.[16]

8. Pd(Ph$_3$P)$_4$, dimedone, THF, 88–95% yield.[17] The catalyst is not poisoned by the presence of thioethers such as methionine. Diethyl malonate,[18] DABCO and PhSiH$_3$ (15 min, 90–97% yield)[19] and barbituric acid (free and polymer-supported)[20] have also been used as a nucleophiles to trap the π-allylpalladium intermediate and regenerate Pd(0).

9. Pd(Ph$_3$P)$_2$Cl$_2$, Bu$_3$SnH, p-NO$_2$C$_6$H$_4$OH, CH$_2$Cl$_2$, 70–100% yield.[14,21] This reaction works best in the presence of acids. AcOH and pyridinium acetate are also effective.

10. Pd$_2$(dba)$_3$·CHCl$_3$, [tris(dibenzylideneacetone)dipalladium-(chloroform)], HCO$_2$H, 74–100% yield.[22]

11. The Alloc group can be converted to a silyl carbamate that is readily hydrolyzed.[23,24]

12. Pd(Ph$_3$P)$_4$, 2-ethylhexanoic acid.[25]

Ref. 26

13. Pd(OAc)$_2$, TPPTS, CH$_3$CN, H$_2$O, Et$_2$NH, 30 min, 89–99% yield. Deprotection can be achieved in the presence of a prenyl or cinnamyl ester; however, as the reaction times increase, these esters are also cleaved.[27-29] Prenyl carbamates and allyl carbonates are cleaved similarly.

14. Pd(Ph$_3$P)$_4$ and Bu$_3$SnH will convert the Alloc group to other amine derivatives when electrophiles such as (BOC)$_2$O, AcCl, TsCl, or succinic anhydride are added. Hydrolysis of the stannyl carbamate with acetic acid gives the free amine.[30] PhSiH$_3$ also serves as an allyl scavenger in this type of transformation.[31]

15. Pd(Ph$_3$P)$_4$, *N,N'*-dimethylbarbituric acid, 92% yield.[32]

16. Pd(Ph$_3$P)$_4$, HCO$_2$H, TEA[33] or AcOH, NMO.[34]

17. Pd(Ph$_3$P)$_4$, NaBH$_4$, THF, 91% yield. If (BOC)$_2$O or CbzOSu is included in the reaction, transprotection to the BOC or Cbz derivative is achieved.[35]

18. Pd(dba)$_2$, dppb, Et$_2$NH or 2-mercaptobenzoic acid, THF or EtOH, 15–120 min, 90–100% yield.[36] Allyl carbonates and allyl ethers are cleaved similarly.

19. The accompanying table gives the results for the deprotection of the Alloc group with various allyl scavengers. The study was undertaken to determine the best scavenger for solid phase peptide synthesis.[37] From these results the two amine boranes, NH$_3$·BH$_3$ and Me$_2$NH·BH$_3$ were recommended as superior allyl scavenging agents because of there fast kinetics.

Relative Efficiency of Allyl Group Scavengers with Pd(Ph$_3$P)$_4$ as Catalyst

Allyl Group Scavenger	% Yield	
	i	ii
NDMBA	100	0
Thiosalicylic acid	100	0
PhSiH$_3$	95	5
Bu$_4$NBH$_4$	60	40
NH$_3$·BH$_3$	99.6–100	0–0.4
Me$_2$NH·BH$_3$	100	0
t-BuNH$_2$·BH$_3$	96–97	3–4
Me$_3$N·BH$_3$	0	100
Py·BH$_3$	0	100

Prenyl Carbamate (Preoc−NR$_2$): (CH$_3$)$_2$C=CHCH$_2$O$_2$CNR$_2$

A comparison of the Alloc group to the Preoc group shows that the Alloc group is more easily cleaved, but an extensive investigation has not been done.

Formation

1. The use of prenyl chloroformate to introduce the Preoc group is unsatisfactory and the reagent has problems with stability.
2. $Im-CO_2CH_2CH=C(CH_3)_2$, DMF, rt, 0–97% yield. More hindered amines failed to react or were very sluggish.[38]
3. $NO_2C_6H_4OCO_2CH_2CH=C(CH_3)_2$, CH_2Cl_2, DMAP, rt, 39–98% yield.[38]

Cleavage

1. $Pd(OAc)_2$, TPPTS, CH_3CN, H_2O, Et_2NH, 100% yield.[27]
2. I_2, MeOH then Zn, 53–85% yield.[38,39]

1-Isopropylallyl Carbamate (Ipaoc−NR₂)

This group was developed to minimize the problem of nitrogen allylation during the deprotection step, because deprotection proceeds with β-hydride elimination. The derivative is stable to TFA and 6 N HCl.[40]

Formation/Cleavage

Cinnamyl Carbamate (Coc−NR₂): $PhCH=CHCH_2OC(O)NR_2$ (Chart 8)

Formation

$PhCH=CHCH_2OCO-O-$benzotriazolyl, Et_3N, dioxane or DMF, rt, 16 h, 71–100% yield.[41]

Cleavage

1. $Pd(Ph_3P)_4$, THF, Pyr, HCO_2H, heat, 4 min.[41]
2. $Hg(OAc)_2$, CH_3OH, HNO_3, 23°C, 2–4 h, then KSCN, H_2O, 23°C, 12–16 h.[42]
3. Electrolysis: Hg electrode, −2.45 V, 54–80% yield.[43] Cinnamyl ethers are also cleaved, but at more negative potentials.

4-Nitrocinnamyl Carbamate (Noc−NR$_2$): 4-NO$_2$C$_6$H$_4$CH=CHCH$_2$OC(O)NR$_2$

The Noc group, developed for amino acid protection, is introduced with the acid chloride (Et$_3$N, H$_2$O, dioxane, 2 h, 20°C, 61–95% yield). It is cleaved with Pd(Ph$_3$P)$_4$ (THF, N,N-dimethylbarbituric acid, 8 h, 20°C, 80% yield). It is not isomerized by Wilkinson's catalyst, thus allowing selective removal of the allyl ester group.[44]

3-(3′-Pyridyl)prop-2-enyl Carbamate (Paloc−NR$_2$):
3-C$_5$H$_4$N-CH=CHCH$_2$OC(O)NR$_2$

The Paloc group was developed as an amino acid protective group that is introduced with the p-nitrophenyl carbonate (H$_2$O, dioxane, 68–89% yield). It is exceptionally stable to TFA and to rhodium-catalyzed allyl isomerization, but it is conveniently cleaved with Pd(Ph$_3$P)$_4$ (methylaniline, THF, 20°C, 10 h, 74–89% yield).[45]

Hexadienyloxy Carbamate (Hdoc−NR$_2$): CH$_3$CH=CHCH=CHCH$_2$O$_2$CNR$_2$

This group is introduced using E,E-2,4-hexadienyl-(4-nitrophenyl) carbonate. It is cleaved with 1% TFA in CH$_2$Cl$_2$ in 10 min. It is stable to base, milder acids, photolysis, NaBH$_4$ and TBAF, but it is modified with I$_2$ in DMF and unexpectedly it was not cleaved with Pd(0) as are most allyl groups. This group was proposed as a useful alternative to the trityl and Bpoc groups.[46]

Propargyloxy Carbamate (Proc or Poc−NR$_2$): HC≡CCH$_2$O$_2$CNR$_2$

Propargyloxy carbamates are stable to neat TFA.

Formation

1. Poc−Cl, CH$_2$Cl$_2$, TEA, 88–92% yield.[47]
2. Polymer supported Proc−OSu, TEA, CH$_2$Cl$_2$, 62–87% yield.[5]
3. Poc−OC$_6$F$_5$, acetone, H$_2$O, DMF, NaHCO$_3$, −10°C to rt, 76–93% yield.[48]
4. From a 3° amine: Poc−Cl, CHCl$_3$, rt or reflux, 51–92% yield.[49] Methyl, isopropyl, allyl, and benzyl amines were examined in this process.

Cleavage

1. (C$_6$H$_5$CH$_2$Net$_3$)$_2$MoS$_4$, CH$_3$CN, rt, 1–2.5 h, 96–98% yield.[49]
2. Co$_2$(CO)$_8$, TFA, CH$_2$Cl$_2$, rt, 30 min, 88-quant. yield.[50] These conditions also cleave propargyl esters. The cobalt complex helps to stabilize a positive charge, thus facilitating cleavage of the carbamate and the carbonate with acid.

But-2-ynylbisoxycarbamate (Bbc−NR$_2$): R$_2$NCO$_2$CH$_2$C≡CCH$_2$O$_2$CNR$_2$

The Bbc carbamate is formed from the bischloroformate (NaHCO$_3$, CH$_2$Cl$_2$, 3 h, 79–96% yield. It is cleaved with (PhCH$_2$Net$_3$)$_2$MoS$_4$ (CH$_3$CN, 30 min, 28°C, 82–96% yield) as is the propargyloxy carbamate. It is stable to TMSI (28°C, 4 h), formic acid used in BOC removal, piperidine used in Fmoc cleavage, and NaOH used in ester hydrolysis.[51]

1. F. Guibe, *Tetrahedron*, **53**, 13509 (1997); F. Guibe, *Tetrahedron*, **54**, 2967 (1998). A correction: F. Guibe, *Tetrahedron*, **54**, 3645 (1998).

2. E. J. Corey and J. W. Suggs, *J. Org. Chem.*, **38**, 3223 (1973).

3. G. Sennyey, G. Barcelo, and J.-P. Senet, *Tetrahedron Lett.*, **28**, 5809 (1987).

4. Y. Hayakawa, H. Kato, M. Uchiyama, H. Kajino, and R. Noyori, *J. Org. Chem.*, **51**, 2400 (1986); R. G. Bhat, S. Sinha, and S. Chandrasekaran, *Chem. Commun.*, 812 (2002).

5. R. Chinchilla, D. J. Dodsworth, C. Najera, and J. M. Soriano, *Synlett*, 809 (2003).

6. B. Orsat, P. B. Alper, W. Moree, C.-P. Mak, and C.-H. Wong, *J. Am. Chem. Soc.*, **118**, 712 (1996).

7. I. Lavandera, S. Fernandez, M. Ferrero, and V. Gotor, *J. Org. Chem.*, **69**, 1748 (2004).

8. M. Aresta and E. Quaranta, *Tetrahedron*, **48**, 1515 (1992).

9. N. Choy, K. Y. Moon, C. Park, Y. C. Son, W. H. Jung, H. Choi, C. S. Lee, C. R. Kim, S. C. Kim, and H. Yoon, *Org. Prep. Proced. Int.*, **28**, 173 (1996).

10. M. Pittelkow, R. Lewinsky, and J. B. Christensen, *Synthesis*, 2195 (2002).

11. R. H. Szumigala, Jr., E. Onofiok, S. Karady, J. D. Armstrong, III, and R. A. Miller, *Tetrahedron Lett.*, **46**, 4403 (2005).

12. B. H. Lipshutz, S. S. Pfeiffer, and A. B. Reed, *Org. Lett.*, **3**, 4145 (2001).

13. D. Franco and E. Duñach, *Tetrahedron Lett.*, **41**, 7333 (2000).

14. O. Dangles, F. Guibé, G. Balavoine, S. Lavielle, and A. Marquet, *J. Org. Chem.*, **52**, 4984 (1987).

15. D. Fernández-Forner, G. Casals, E. Navarro, H. Ryder, and F. Albericio, *Tetrahedron Lett.*, **42**, 4471 (2001).

16. A. Merzouk and F. Guibe, *Tetrahedron Lett.*, **33**, 477 (1992).

17. H. Kunz and C. Unverzagt, *Angew. Chem., Int. Ed. Engl.*, **23**, 436 (1984).

18. P. Boullanger and G. Descotes, *Tetrahedron Lett.*, **27**, 2599 (1986).

19. C. Zorn, F. Gnad, S. Salmen, T. Herpin, and O. Reiser, *Tetrahedron Lett.*, **42**, 7049 (2001). N. Thieriet, J. Alsina, E. Giralt, F. Guibe, and F. Albericio, *Tetrahedron Lett.*, **38**, 7275 (1997).

20. H. Tsukamoto, T. Suzuki, and Y. Kondo, *Synlett*, **8**, 1105 (2003).

21. P. Four and F. Guibé, *Tetrahedron Lett.*, **23**, 1825 (1982).

22. I. Minami, Y. Ohashi, I. Shimizu, and J. Tsuji, *Tetrahedron Lett.*, **26**, 2449 (1985).

23. M. Sakaitani, N. Kurokawa, and Y. Ohfune, *Tetrahedron Lett.*, **27**, 3753 (1986).

24. M. Sakaitani and Y. Ohfune, *J. Org. Chem.*, **55**, 870 (1990).

25. P. D. Jeffrey and S. W. McCombie, *J. Org. Chem.*, **47**, 587 (1982).

26. S. F. Martin and C. L. Campbell, *J. Org. Chem.*, **53**, 3184 (1988).

27. S. Lemaire-Audoire, M. Savignac, E. Blart, G. Pourcelot, J. P. Genêt, and J.-M. Bernard, *Tetrahedron Lett.*, **35**, 8783 (1994); S. Lemaire-Audoire, M. Savignac, E. Blart, J.-M. Bernard, and J. P. Genêt, *Tetrahedron Lett.*, **38**, 2955 (1997); S. Lemaire-Audoire, M. Savignac, G. Pourcelot, J.-P. Genêt, and J.-M. Bernard, *J. Mol. Catalysis A: Chemical*, **116**, 247 (1997).

28. J. P.Genêt, E. Blart, M. Savignac, S. Lemeune, and J.-M. Paris, *Tetrahedron Lett.*, **34**, 4189 (1993).

29. J. P. Genêt, E. Blart, M. Savignac, S. Lemeune, S. Lemaire-Audoire, J. M. Paris, and J. M. Bernard, *Tetrahedron*, **50**, 497 (1994).

30. E. C. Roos, P. Bernabe, H. Hiemstra, W. N. Speckamp, B. Kaptein, and W. H. J. Boesten, *J. Org. Chem.*, **60**, 1733 (1995).

31. N. Thieriet, P. Gomez-Martinez, and F. Guibe, *Tetrahedron Lett.*, **40**, 2505 (1999).

32. C. Unverzagt and H. Kunz, *Biorg. Med. Chem.*, **2**, 1189 (1994).

33. Y. Kanda, H. Arai, T. Ashizawa, M. Morimoto, and M. Kasai, *J. Med. Chem.*, **35**, 2781 (1992).

34. J. Lee, J. H. Griffin, and T. I. Nicas, *J. Org. Chem.*, **61**, 3983 (1996).

35. R. Beugelmans, L. Neuville, M. Bois-Choussy, J. Chastanet, and J. Zhu, *Tetrahedron Lett.*, **36**, 3129 (1995).

36. J. P. Genêt, E. Blart, M. Savignac, S. Lemeune, S. Lemaire-Audoire, and J. M. Bernard, *Synlett*, 680 (1993).

37. P. Gómez-Martínez, M. Dessolin, F. Guibé, and F. Albericio, *J. Chem. Soc.Perkin I*, 2871 (1999).

38. J.-M. Vatèle, *Tetrahedron*, **58**, 5689 (2002).

39. J.-M. Vatele, *Tetrahedron*, **60**, 4251 (2004).

40. I. Minami, M. Yukara, and J. Tsuji, *Tetrahedron Lett.*, **28**, 2737 (1987).

41. H. Kinoshita, K. Inomata, T. Kameda, and H. Kotake, *Chem. Lett.*, **14**, 515 (1985).

42. E. J. Corey and M. A. Tius, *Tetrahedron Lett.*, **18**, 2081 (1977).

43. J. Hansen, S. Freeman, and T. Hudlicky, *Tetrahedron Lett.*, **44**, 1575 (2003).

44. H. Kunz and J. März, *Angew. Chem., Int. Ed. Engl.*, **27**, 1375 (1988).

45. K. von dem Bruch and H. Kunz, *Angew. Chem., Int., Ed. Engl.*, **29**, 1457 (1990).

46. I. Lingard, G. Bhalay, and M. Bradley, *Synlett*, **12**, 1791 (2003).

47. R. Ramesh, R. G. Bhat, and S. Chandresekaran, *J. Org. Chem.*, **70**, 837 (2005).

48. R. G. Bhat, E. Kerouredan, E. Porhiel, and S. Chandrasekaran, *Tetrahedron Lett.*, **43**, 2467 (2002).

49. R. G. Bhat, Y. Ghosh, and S. Chandrasekaran, *Tetrahedron Lett.*, **45**, 7983 (2004).

50. Y. Fukase, K. Fukase, and S. Kusumoto, *Tetrahedron Lett.*, **40**, 1169 (1999).

51. R. Ramesh and S. Chandrasekaran, *Org. Lett.*, **7**, 4947 (2005).

8-Quinolyl Carbamate: (Chart 8)

Formation/Cleavage

An 8-quinolyl carbamate is cleaved under neutral conditions by Cu(II)- or Ni(II)-catalyzed hydrolysis.[1]

1. E. J. Corey and R. L. Dawson, *J. Am. Chem. Soc.*, **84**, 4899 (1962).

N-Hydroxypiperidinyl Carbamate: (Chart 8)

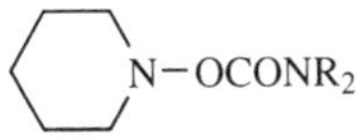

$$\langle\!\!\!\!\!\!\!\bigcirc\!\!N-OCONR_2$$

A piperidinyl carbamate, stable to aqueous alkali and to cold acid (30% HBr, 25°C, several hours) is best cleaved by reduction.[1]

Formation

1-Piperidinyl-OCOX (X = 2,4,5-trichlorophenyl, ...), Et$_3$N, 55–85% yield.

Cleavage

1. H$_2$, Pd–C, AcOH, 20°C, 30 min, 95% yield.
2. Electrolysis, 200 mA, 1 N H$_2$SO$_4$, 20°C, 90 min, 90–93% yield.
3. Na$_2$S$_2$O$_4$, AcOH, 20°C, 5 min, 93% yield.
4. Zn, AcOH, 20°C, 10 min, 94% yield.

1. D. Stevenson and G. T. Young, *J. Chem. Soc. C*, 2389 (1969).

Alkyldithio Carbamate: R$_2$NCOSSR′

Alkyldithio carbamates are prepared from the acid chloride (Et$_3$N, EtOAc, 0°C) and amino acid, either free or as the *O*-silyl derivatives (70–88% yield).[1] They may also be prepared by the addition of carbon disulfide to the amine which can then be alkylated with an alkyl halide using Cs$_2$CO$_2$ as the base.[2] The *N*-(*i*-propyldithio) carbamate has been used in the protection of proline during peptide synthesis.[3] Alkyldithio carbamates can be cleaved with thiols, NaOH, Ph$_3$P/TsOH. They are stable to acid. Cleavage rates are a function of the size of the alkyl group as illustrated in the table below.

Relative Rates of Cleavage of Alkyldithio Carbamates

Alkyl Group (R′)	HSCH$_2$CH$_2$OH	NaOH
CH$_3$	100	100
Et	33	32
i-Pr	1.4	1.3
t-Bu	0.0002	—
Ph	460	500

The rates were determined using the proline derivative as a substrate.[4]

1. E. Wünsch, L. Moroder, R. Nyfeler, and E. Jäger, *Hoppe-Seyler's Z. Physiol. Chem.*, **363**, 197 (1982).

2. A. S. Nagle, R. N. Salvatore, R. M. Cross, E. A. Kapxhiu, S. Sahab, C. H. Yoon, and K. W. Jung, *Tetrahedron Lett.*, **44**, 5695 (2003).

3. F. Albericio and G. Barany, *Int. J. Pept. Protein Res.*, **30**, 177 (1987).

4. G. Barany, *Int. J. Pept. Protein Res.*, **19**, 321 (1982).

Benzyl Carbamate (Cbz− or Z−NR$_2$): PhCH$_2$OC(O)NR$_2$ (Chart 8)

The benzyl carbamate is one of the most popular protective groups that results largely from its facile hydrogenolysis and its orthogonality to numerous other protective groups.

Formation

1. PhCH$_2$OCOCl, Na$_2$CO$_3$, H$_2$O, 0°C, 30 min, 72% yield.[1] Alpha-omega diamines can be protected somewhat selectively with this reagent at a pH between 3.5 and 4.5, but the selectivity decreases as the chain length increases [H$_2$N(CH$_2$)$_n$NH$_2$, n = 2, 71% mono; n = 7, 29% mono].[2] Hindered amino acids are protected in DMSO (DMAP, TEA, heat, 47–82% yield). These conditions also convert a carboxylic acid to the benzyl ester.[3]

2. PhCH$_2$OCOCl, MgO, EtOAc, 3 h, 70°C to reflux, 60% yield.[4] Zinc metal can be used to scavenge the HCl produced in the protection process. ZnCl$_2$ is formed in the reaction.[5]

3. PhCH$_2$OCOCl, DIPEA, CH$_3$CN, Sc(OTf)$_3$, 95% yield. The reaction fails without the Sc(OTf)$_3$.[6]

4. (PhCH$_2$OCO)$_2$O, dioxane, H$_2$O, NaOH or Et$_3$N.[7,8] This reagent was reported to give better yields in preparing amino acid derivatives than when PhCH$_2$OCOCl was used. The reagent decomposes at 50°C.

5. PhCH$_2$OCO$_2$−C(OMe)=CH$_2$, 90–98% yield.[9]

6. PhCH$_2$OCO$_2$−succinimidyl, >70% yield.[10] This reagent avoids the formation of amino acid dimers and is a stable, easily handled solid.

7. PhCH$_2$OCO−benzotriazolyl, NaOH, dioxane, rt.[11] A polymeric version of this reagent has been developed.[12]

8. PhCH$_2$OCOCN, CH$_2$Cl$_2$, CH$_3$CN or 1,2-dimethoxyethane.[13]

9. PhCH$_2$OCO-imidazolyl, 4-dimethylaminopyridine, 16 h, rt, 76% yield.[14] Two primary amines were protected in the presence of a secondary amine.

10.

ROH, TEA
dioxane, 64–91%

ROH = *t*-BuOH, BnOH, FmOH, AdamantylOH, PhC$_6$H$_4$CMe$_2$OH, CH$_3$CH$_2$CH$_2$OH

This method is suitable for the preparation of BOC, Fmoc, Adoc, and Bpoc protected amino acids. The acid chloride is a stable, storable solid.[15]

11. CO$_2$, BnCl, DMF, Cs$_2$CO$_3$, 58–96% yield.[16] Other carbamates can be formed similarly using this methodology.

12.

TEA, 88% yield.[17]

13. 4-NO$_2$PhOCO$_2$Bn, Pyr, DMF, 26°C, 24 h, 74% yield. Primary amines are selectively protected over secondary amines, but anilines are insufficiently nucleophilic to react with this reagent.[18] The less reactive reagent, PhOCO$_2$Bn, will also selectively derivatize primary amines over secondary amines.[19]

14. 1,3-Bis(benzyloxycarbonyl)-3,4,5,6-tetrahydropyrimidine-2-thione,   refluxing dioxane, 7 h.[20]

15. [4-(Benzyloxycarbonyloxy)phenyl]dimethylsulfonium   methyl   sulfate, NaOH, H$_2$O, 51–95% yield.[21] This is a water-soluble reagent for benzyloxy carbamate formation. Analogous reagents for the introduction of BOC and Fmoc were also prepared and give the respective derivatives in similar high yields.

16. 2-Fluoro-*N*-benzyloxycarbonyl-*N*-mesylaniline, pyridine, rt, 1–8 h, 90–93%. This reagent gives good selectivity for primary amines over secondary amines, but α,α-disubstituted primary amines do not react.[22]

17. 4,6-Dimethoxy-1,3,5-triazinylbenzyl carbonate, THF, CH$_3$CN or MeOH, 15–60 min, TEA or NaHCO$_3$, rt, 67–95% yield.[23]

18. 1-(Benzyloxycarbonyl)-3-ethylimidazolium tetrafluoroborate, "Rapoport's reagent," CH$_2$Cl$_2$, 82% yield. More conventional methods failed to give good results.[24]

19. CCl$_3$C(=NH)OBn (TFA, heat, 46–56% yield) will exchange the Teoc and BOC groups for the Z group.[25]

Cleavage

1. H$_2$/Pd–C.[1] If hydrogenation is carried out in the presence of (BOC)$_2$O, the released amine is directly converted to the BOC derivative.[26] The formation of *N*-methylated lysines during the hydrogenolysis of a Z group has been observed with MeOH/DMF as the solvent.[27] Formaldehyde-derived oxidatively from methanol is the source of the methyl carbon.[28] This was not a problem when the reaction was conducted in EtOH.[29] The presence of squaric acid will prevent hydrogenolysis of a Cbz group as well as a benzyl ether, but does not inhibit olefin hydrogenation.[30]

2. H$_2$/Pd–C, NH$_3$, −33°C, 3–8 h, quantitative.[31] When ammonia is used as the solvent, cysteine or methionine units in a peptide do not poison the catalyst. Additionally, amines inhibit the reduction of BnO ethers, thus selectivity can be achieved for the Z group.[32]

3. Pd–C or Pd black, hydrogen donor, solvent, 25°C or reflux in EtOH, 15 min to 2 h, 80–100% yield. Several hydrogen donors, including cyclohexene,[33] 1,4-cyclohexadiene,[34] formic acid,[35,36] *cis*-decalin,[37] and HCO$_2$NH$_4$,[38] have been used for catalytic transfer hydrogenation, in general a more rapid reaction than catalytic hydrogenation. Microwave irradiation accelerates the deprotection process.[39] Use of this technique in the presence of (BOC)$_2$O converts a Z-protected amine to a BOC-protected amine.[40] In the following case, Pd black was the only catalyst that worked to cleave the Cbz group in the presence of a sulfur atom. In most cases, sulfur is a superb catalyst poison; in this case, however, since it is both aromatic and conjugated to the ester, its lone pair of electrons is probably rather unavailable for complexing with the Pd.[41]

4. PdCl$_2$, MeOH, H$_2$, conc. HCl, rt, 100% yield. These conditions also reduce olefins, but a benzylic ether remained in tact.[42] At 80–85°C these conditions will cleave the benzylic amine and ether.

5. Pd–poly(ethylenimine), HCO$_2$H.[43] This catalyst system was reported to be better than Pd–C or Pd black for Z removal.

6. Pd–C, polymethylhydrosiloxane, EtOH, rt, BOC$_2$O, 86–94% yield. This results in exchange of the Z group for a BOC group.[44]

7. Pd supported on hydroxyapatite, H$_2$, MeOH, 40°C, 1 atm, 84–99% yield. This method proved exceptional for the cleavage of a Z group buried in a dendrimer.[45]

8. Pd–C, 2,2'-dipyridyl, MeOH, EtOAc, H$_2$. A phenolic benzyl ether survives these conditions.[46]

9. CaNi$_5$, H$_2$, MeOH, H$_2$O.[47] The catalyst is a hydrogen storage alloy and is partially consumed by the reaction of Ca with water or methanol.

10. Raney Ni (W-2), MeOH, reflux, 65% yield.[48]

11. K$_3$[Co(CN)$_5$], H$_2$, MeOH, 20°C, 3 h.[49] Benzyl ethers are not cleaved under these conditions.

12. Et$_3$SiH, cat. Et$_3$N, cat. PdCl$_2$, reflux, 3 h, 80% yield.[50] If the reaction is performed in the presence of t-BuMe$_2$SiH, the t-butyldimethylsilyl carbamate can be isolated because of its greater stability.[51] S-Benzyl groups are stable to these conditions, but benzyl esters and benzyl ethers are cleaved.[50] A similar procedure has been published, but in this case the benzyl ether was stable to the cleavage conditions.[26] Alkenes are stable to these conditions.[52]

13. t-BuMe$_2$SiH, Pd(OAc)$_2$, TEA, CH$_2$Cl$_2$, rt, 95–100% yield. In this case the relatively stable TBDMS carbamate is isolated.[54]

14. Na/NH$_3$.[55] Lithium is also often used as the reducing metal.[56]

15. Lithium naphthalenide, THF, 0°C, 1–2 h, 71–98% yield. Alloc and Cbz carbonates are also cleaved under these conditions.[57]

16. Me$_3$SiI, CH$_3$CN, 25°C, 6 min, 100% yield.[58,59] Aryl stannanes are stable to this reagent.[60] The cleavage can be performed using in situ-generated TMSI from TMSCl and NaI in CH$_3$CN.[61]

17. TMSBr, PhSMe, TFA, 0°C, 1 h, 70% yield.[62]

18. AlCl$_3$, PhOCH$_3$, 0–25°C, 5 h, 73% yield.[63] These conditions are compatible with β-lactams.

19. BBr$_3$, CH$_2$Cl$_2$, −10°C, 1 h to 25°C, 2 h, 80–100% yield.[64] Benzyl carbamates of larger peptides can be cleaved by boron tribromide in trifluoroacetic acid, since the peptides are more soluble in acid than in methylene chloride.[65]

20. BCl$_3$, CH$_2$Cl$_2$, rt.[66,67]

21. Benzyl carbamates are readily cleaved under strongly acidic conditions: HBr, AcOH[68]; 50% CF$_3$COOH (25°C, 14 days, partially cleaved)[69]; 70% HF, Pyr[70]; CF$_3$SO$_3$H[71]; FSO$_3$H[72]; or CH$_3$SO$_3$H.[72] In cleaving benzyl carbamates from peptides, 0.5 M 4-(methylmercapto)phenol in CF$_3$CO$_2$H has been recommended to suppress Bn$^+$ additions to aromatic amino acids.[73] Thioanisole can also be used as Bn$^+$ scavenger.[74] To achieve deprotection via an S$_N$2 mechanism, which also reduces the problem of Bn$^+$ addition, HF–Me$_2$S–p-cresol (25:65:10, v/v) has been recommended for peptide deprotection.[75]

22. 6 N HCl, reflux, 1 h, 92% yield.[76]

23. AcCl and NaI transform a Z-protected amine into an acetamide (84% yield).[77]

24. Trifluoroacetic anhydride, pyridine, 40°C, 15 h, >70% yield.[78]

25. Catecholborane halides cleave benzyl carbamates in the presence of ethyl and benzyl esters and TBDMS ethers.[79]

26. BF$_3$·Et$_2$O, CH$_3$SCH$_3$, CH$_2$Cl$_2$, 92% yield.[80]

27. BF$_3$·Et$_2$O, EtSH, CH$_2$Cl$_2$, rt. 76–96% yield.[81] It is possible to achieve some selectivity for a secondary derivative over a primary one when the reaction is conducted under more dilute conditions.

Ratio = 1:9

28. 40% KOH, MeOH, H$_2$O, 85–94% yield.[82]

29. 0.15 M Ba(OH)$_2$, heat, 40 h, 3:2 glyme/H$_2$O, 75% yield.[83]

In this case the following reagents failed to afford clean deprotection because of destruction of the acetylene: Me_3SiI, BBr_3, Me_2BBr, $BF_3/EtSH$, $AlCl_3/EtSH$, $MeLi/LiBr$, $KOH/EtOH$.

30. $LiBH_4$ or $NaBH_4$, Me_3SiCl, THF, 24 h, 88–95%.[84] This combination of reagents also reduces all functional groups that can normally be reduced with diborane.

31. $LiEt_3BH$, THF, 0°C to rt, 72–96% yield. Other amides are also cleaved in good yields.[85]

32. Agarose supported penicillin G acylase, 20–192 h, 8–100% yield. The method was used for the deprotection of amino acids and small peptides. The larger peptides tend to give slow and incomplete reaction.[86]

33. Photolysis: 253.7 nm, hv, 55°C, 4 h, CH_3OH, H_2O, 70% yield.[87,88]

34. Electrolysis: -2.9 V, DMF, R_4NX, 70–80% yield[89] or Pd/graphite cathode, MeOH, AcOH, 2.5% $NaClO_4$ (0.5 mole/L), 99% yield.[90] Benzyl ethers and tosylates are stable to these conditions, but benzyl esters are cleaved.

35. Benzyl carbamates of pyrrole-type nitrogens can be cleaved with nucleophilic reagents such as hydrazine; hydrogenation and HF treatment are also effective.[91] See section on protection of aryl amines.

36. *Sphingomonas paucinobilis* SC 16113, 42°C, 18–20 h, 0–100% conversion. This method was only tested on a variety of amino acids and small peptides. Not all protected amino acids were successfully deprotected.[92]

1. M. Bergmann and L. Zervas, *Ber.*, **65**, 1192 (1932).

2. G. J. Atwell and W. A. Denny, *Synthesis*, 1032 (1984).

3. D. B. Berkowitz and M. L. Pedersen, *J. Org. Chem.*, **59**, 5476 (1994).

4. M. Dymicky, *Org. Prep. Proced. Int.*, **21**, 83 (1989).

5. J. S. Yadav, G. S. Reddy, M. M. Reddy, and H. M. Meshram, *Tetrahedron Lett.*, **39**, 3259 (1998).

6. V. K. Aggarwal, P. S. Humphries, and A. Fenwick, *Angew. Chem. Int. Ed.*, **38**, 1985 (1999).

7. W. Graf, O. Keller, W. Keller, G. Wersin, and E. Wuensch, *Peptides 1986: Proc. 19th Eur. Pept. Symp.*, D. Theodoropoulos, Ed., de Gruyter, New York, 1987, p. 73.

8. G. Sennyey, G. Barcelo, and J.-P. Senet, *Tetrahedron Lett.*, **27**, 5375 (1986).

9. Y. Kita, J. Haruta, H. Yasuda, K. Fukunaga, Y. Shirouchi, and Y. Tamura, *J. Org. Chem.*, **47**, 2697 (1982).

10. A. Paquet, *Can. J. Chem.* **60**, 976 (1982).

11. E. Wuensch, W. Graf, O. Keller, W. Keller, and G. Wersin, *Synthesis*, 958 (1986).

12. K. G. Dendrinos and A. G. Kalivretenos, *J. Chem. Soc. Perkin Trans. 1*, 1463 (1998).

13. S. Murahashi, T. Naota, and N. Nakajima, *Chem. Lett.*, 879 (1987).

14. S. K. Sharma, M. J. Miller, and S. M. Payne, *J. Med. Chem.*, **32**, 357 (1989).

15. P. Henklein, H.-U. Heyne, W.-R. Halatsch, and H. Niedrich, *Synthesis*, 166 (1987).

16. K. J. Butcher, *Synlett*, 825 (1994). R. N. Salvatore, F. Chu, A. S. Nagle, E. A. Kapxhiu, R. M. Cross, and K. W. Jung, *Tetrahedron*, **58**, 3329 (2002).

17. M. Allainmat, P. L. Haridon, L. Toupet, and D. Plusquellec, *Synthesis*, 27 (1990).

18. D. R. Kelly and M. Gingell, *Chem. Ind., London*, 888 (1991).

19. M. Pittelkow, R. Lewinsky, and J. B. Christensen, *Synthesis*, 2195 (2002).

20. N. Matsumura, A. Noguchi, A. Kitayoshi, and H. Inoue, *J. Chem. Soc., Perkin Trans. 1*, 2953 (1995).

21. I. Azuse, M. Tamura, K. Kinomura, H. Okai, K. Kouge, F. Hamatsu, and T. Koizumi, *Bull. Chem. Soc. Jpn.*, **62**, 3103 (1989).

22. K. Kondo, E. Sekimoto, K. Miki, and Y. Murakami, *J. Chem. Soc. Perkin Trans. 1*, 2973 (1998).

23. K. Hioki, M. Fujiwara, S. Tani, and M. Kunishima, *Chem. Lett.*, **31**, 66 (2002).

24. N. M. Howarth and L. P. G. Wakelin, *J. Org. Chem.*, **62**, 5441 (1997).

25. A. G. M. Barrett and D. Pilipauskas, *J. Org. Chem.*, **55**, 5170 (1990).

26. M. Sakaitani, K. Hori, and Y. Ohfune, *Tetrahedron Lett.*, **29**, 2983 (1988).

27. J.-P. Mazaleyrat, J. Xie, and M. Wakselman, *Tetrahedron Lett.*, **33**, 4301 (1992).

28. N. L. Benoiton, *Int. J. Pept. Protein Res.*, **41**, 611 (1993).

29. S. Lin, Z.-Q. Yang, B. H. B. Kwok, M. Koldobskiy, C. M. Crews, and S. J. Danishefsky, *J. Am. Chem. Soc.*, **126**, 6347 (2004).

30. T. Shinada, K.-i. Hayashi, Y. Yoshida, and Y. Ohfune, *Synlett*, **10**, 1506 (2000).

31. J. Meienhofer and K. Kuromizu, *Tetrahedron Lett.*, **15**, 3259 (1974).

32. H. Sajiki, *Tetrahedron Lett.*, **36**, 3465 (1995).

33. A. E. Jackson and R. A. W. Johnstone, *Synthesis*, 685 (1976); G. M. Anantharamaiah and K. M. Sivanandaiah, *J. Chem. Soc., Perkin Trans. 1*, 490 (1977).

34. A. M. Felix, E. P. Heimer, T. J. Lambros, C. Tzougraki, and J. Meienhofer, *J. Org. Chem.*, **43**, 4194 (1978).

35. K. M. Sivanandaiah and S. Gurusiddappa, *J. Chem. Res., Synop.*, 108 (1979); B. ElAmin, G. M. Anantharamaiah, G. P. Royer, and G. E. Means, *J. Org. Chem.*, **44**, 3442 (1979).

36. M. J. O. Anteunis, C. Becu, F. Becu, and M. F. Reyniers, *Bull. Soc. Chim. Belg.*, **96**, 775 (1987).

37. Y. Okada and N. Ohta, *Chem. Pharm. Bull.*, **30**, 581 (1982).

38. M. Makowski, B. Rzeszotarska, L. Smelka, and Z. Kubica, *Liebigs Ann. Chem.*, 1457 (1985).

39. M. C. Daga, M. Taddei, and G. Varchi, *Tetrahedron Lett.*, **42**, 5191 (2001).

40. J. S. Bajwa, *Tetrahedron Lett.*, **33**, 2955 (1992).

41. S. V. Downing, E. Aguilar, and A. I. Meyers, *J. Org. Chem.*, **64**, 826 (1999).

42. R. P. Jain, B. K. Albrecht, D. E. DeMong, and R. M. Williams, *Org. Lett.*, **3**, 4287 (2001).

43. D. R. Coleman and G. P. Royer, *J. Org. Chem.*, **45**, 2268 (1980).

44. S. Chandrasekhar, L. Chandraiah, C. R. Reddy, and M. V. Reddy, *Chem. Lett.*, **29**, 780 (2000).

45. M. Murata, T. Hara, K. Mori, M. Ooe, T. Mizugaki, K. Ebitani, and K. Kaneda, *Tetrahedron Lett.*, **44**, 4981 (2003).

46. B. Cao, H. Park, and M. M. Joullie, *J. Am. Chem. Soc.*, **124**, 520 (2002).

47. Y. Kawasaki, H. Konishi, M. Morita, M. Kawanari, S. i. Dosako, and I. Nakajima, *Chem. Pharm. Bull.*, **42**, 1238 (1994).

48. O. Tamura, T. Yanagimachi, T. Kobayashi, and H. Ishibashi, *Org. Lett.*, **3**, 2427 (2001).

49. G. Losse and H. U. Stiehl, *Z. Chem.*, **21**, 188 (1981).

50. L. Birkofer, E. Bierwirth, and A. Ritter, *Chem. Ber.*, **94**, 821 (1961).

51. M. Sakaitani, N. Kurokawa, and Y. Ohfune, *Tetrahedron Lett.*, **27**, 3753 (1986).

52. P. Wipf and Y. Uto, *Tetrahedron Lett.*, **40**, 5165 (1999); P. Wipf and Y. Uto, *J. Org. Chem.*, **65**, 1037 (2000).

53. R. S. Coleman and A. J. Carpenter, *J. Org. Chem.*, **57**, 5813 (1992); R. S. Coleman, *Synlett*, 1031 (1998).

54. M. Sakaitani and Y. Ohfune, *J. Org. Chem.*, **55**, 870 (1990).

55. I. Schon, T. Szirtes, T. Uberhardt, A. Rill, A. Csehi, and B. Hegedus, *Int. J. Pept. Protein Res.*, **22**, 92 (1983).

56. R. M. Williams, P. J. Sinclair, D. Zhai, and D. Chen, *J. Am. Chem. Soc.*, **110**, 1547 (1988).

57. C. Behloul, D. Guijarro, and M. Yus, *Tetrahedron*, **61**, 9319 (2005).

58. R. S. Lott, V. S. Chauhan, and C. H. Stammer, *J. Chem. Soc., Chem. Commun.*, 495 (1979).

59. M. Ihara, N. Taniguchi, K. Nogochi, K. Fujumoto, and T. Kametani, *J. Chem. Soc., Perkin Trans. I*, 1277 (1988).

60. E. Brenner, R. M. Baldwin, and G. Tamagnan, *Tetrahedron Lett.*, **45**, 3607 (2004).

61. S. Tanimori, K. Fukubayashi, and M. Kirihata, *Tetrahedron Lett.*, **42**, 4013 (2001).

62. U. Schmidt, V. Leitenberger, H. Griesser, J. Schmidt, and R. Meyer, *Synthesis*, 1248 (1992).

63. T. Tsuji, T. Kataoka, M. Yoshioka, Y. Sendo, Y. Nishitani, S. Hirai, T. Maeda, and W. Nagata, *Tetrahedron Lett.*, **20**, 2793 (1979).

64. A. M. Felix, *J. Org. Chem.*, **39**, 1427 (1974).

65. J. Pless and W. Bauer, *Angew Chem., Int. Ed. Engl.*, **12**, 147 (1973).

66. T. Fukuyama, R. K. Frank, and C. F. Jewell, Jr., *J. Am. Chem. Soc.*, **102**, 2122 (1980).

67. P. Allevi, R. Cribiu, and M. Anastasia, *Tetrahedron Lett.*, **45**, 5841 (2004).

68. D. Ben-Ishai and A. Berger, *J. Org. Chem.*, **17**, 1564 (1952).

69. A. R. Mitchell and R. B. Merrifield, *J. Org. Chem.*, **41**, 2015 (1976).

70. S. Matsuura, C.-H. Niu, and J. S. Cohen, *J. Chem. Soc., Chem. Commun.*, 451 (1976).

71. H. Yajima, N. Fujii, H. Ogawa, and H. Kawatani, *J. Chem. Soc., Chem. Commun.*, 107 (1974).

72. H. Yajima, H. Ogawa, and H. Sakurai, *J. Chem. Soc., Chem. Commun.*, 909 (1977).

73. M. Bodanszky and A. Bodanszky, *Int. J. Pept. Protein Res.*, **23**, 287 (1984).

74. D. L. Boger, S. Ichikawa, W. C. Tse, M. P. Hedrick, and Q. Jin, *J. Am. Chem. Soc.*, **123**, 561 (2001).

75. J. P. Tam, W. F. Heath, and R. B. Merrifield, *J. Am. Chem. Soc.*, **105**, 6442 (1983).

76. G. Chelucci, M. Falorni, and G. Giacomelli, *Synthesis*, 1121 (1990).

77. M. Ihara, A. Hirabayashi, N. Taniguchi, and K. Fukumoto, *Heterocycles*, **33**, 851 (1992).

78. J. T. Suri, D. D. Steiner, and C. F. Barbas III, *Org. Lett.*, **7**, 3885 (2005).

79. R. K. Boeckman, Jr., and J. C. Potenza, *Tetrahedron Lett.*, **26**, 1411 (1985).

80. I. H. Sanchez, F. J. López, J. J. Soria, M. I. Larraza, and H. J. Flores, *J. Am. Chem. Soc.*, **105**, 7640 (1983).

81. D. S. Bose and D. E. Thurston, *Tetrahedron Lett.*, **31**, 6903 (1990).

82. S. R. Angle and D. O. Arnaiz, *Tetrahedron Lett.*, **30**, 515 (1989).

83. L. E. Overman and M. J. Sharp, *Tetrahedron Lett.*, **29**, 901 (1988).

84. A. Giannis and K. Sandhoff, *Angew. Chem., Int. Ed. Engl.*, **28**, 218 (1989).

85. H. Tanaka and K. Ogasawara, *Tetrahedron Lett.*, **43**, 4417 (2002).

86. G. Alvaro, J. A. Feliu, G. Caminal, J. Lopez-Santin, and P. Clapes, *Biocatalysis and Biotransformation*, **18**, 253 (2000).

87. S. Hanessian and R. Masse, *Carbohydr. Res.*, **54**, 142 (1977).

88. For a review of photochemically labile protective groups, see V. N. R. Pillai, *Synthesis*, 1 (1980).

89. V. G. Mairanovsky, *Angew Chem., Int. Ed., Engl.*, **15**, 281 (1976).

90. M. A. Casadei and D. Pletcher, *Synthesis*, 1118 (1987).

91. M. Chorev and Y. S. Klausner, *J. Chem. Soc., Chem. Commun.*, 596 (1976).

92. R. N. Patel, V. Nanduri, D. Brzozowski, C. McNamee, and A. Banerjee, *Adv. Synth. Catal.*, **345**, 830 (2003).

3,5-Di-*t*-butylbenzyl Carbamate: $3,5\text{-}(t\text{-Bu})_2C_6H_3CH_2O_2CNR_2$

The 3,5-di-*t*-butylbenzyl carbamate group was developed as a more soluble Cbz group for the protection certain aromatic diamines. It is introduced conventionally using the chloroformate method (84% yield). It is cleaved by hydrogenolysis.[1] Most of the methods applicable to benzyl carbamates should be applicable to both the preparation and cleavage of this derivative.

1. G. Festel and C. D. Eisenbach, *J. Prakt. Chem.*, **341**, 29 (1999).

p-Methoxybenzyl Carbamate (Moz−NR₂): $p\text{-MeOC}_6H_4CH_2OC(O)NR_2$

Formation

1. Moz−ON=C(CN)Ph, H_2O, Et_3N, rt, 6 h, 90% yield.[1]
2. MozN₃.[2,3]
3. Moz−OC₆H₄NO₂, CH_2Cl_2, pyridine, 50–60% yield. This method was used for the protection of amidines.[4]

Cleavage

The Moz group is more readily cleaved by acid than the benzyloxycarbonyl or BOC group.[5,6] The section on benzyl carbamates should be consulted since many of the methods for formation and cleavage should be applicable to the Moz group as well.

1. TsOH, CH$_3$CN, acetone, rt.[7,8]

Ref. 5

2. 10% CF$_3$COOH, CH$_2$Cl$_2$, 100% yield.[1,5]
3. CH$_3$SO$_3$H, *m*-cresol, CH$_2$Cl$_2$. The addition of *m*-cresol greatly accelerates the rate of cleavage.[9]

1. S.-T. Chen, S.-H. Wu, and K.-T. Wang, *Synthesis*, 36 (1989).
2. J. M. Kerr, S. C. Banville, and R. N. Zuckermann, *J. Am. Chem. Soc.*, **115**, 2529 (1993).
3. S. S. Wang, S. T. Chen, K. T. Wang, and R. B. Merrifield, *Int. J. Pept. Protein, Res.*, **30**, 662 (1987).
4. C. Bailey, E. Baker, J. Hayler, and P. Kane, *Tetrahedron Lett.*, **40**, 4847 (1999).
5. F. Weygand and K. Hunger, *Chem. Ber.*, **95**, 1 (1962).
6. S. S. Wang, S. T. Chen, K. T. Wang, and R. B. Merrifield, *Int. J. Pept. Protein Res.*, **30**, 662 (1987).
7. H. Yajima, H. Ogawa, N. Fujii, and S. Funakoshi, *Chem. Pharm. Bull.*, **25**, 740 (1977).
8. H. Yamada, H. Tobiki, N. Tanno, H. Suzuki, K. Jimpo, S. Ueda, and T. Nakagome, *Bull. Chem. Soc. Jpn.*, **57**, 3333 (1984).
9. H. Tamamura, J. Nakamura, K. Noguchi, S. Funakoshi, and N. Fujii, *Chem. Pharm. Bull.*, **41**, 954 (1993).

p-Nitrobenzyl Carbamate (PNZ−NR$_2$): *p*-NO$_2$C$_6$H$_4$CH$_2$OC(O)NR$_2$ (Chart 8)

The use of PNZ derivatives in conjunction with Fmoc chemistry results in fewer problems with aspartamide and diketopiperazine formation[1] and it is superior to the use of the Alloc group for ornithine and lysine protection in Fmoc based peptide synthesis.[6]

Formation

p-NO$_2$C$_6$H$_4$CH$_2$OCOCl, base, 0°C, 1.5 h, 78% yield. [2]

Cleavage

1. H_2/Pd–C, 10 h, 87% yield.[2] A nitrobenzyl carbamate is more readily cleaved by hydrogenolysis than a benzyl carbamate; it is more stable to acid-catalyzed hydrolysis than a benzyl carbamate, and therefore selective cleavage is possible.

2. 4 *N* HBr, AcOH, 60°C, 2 h, 68% yield.[1]

3. $Na_2S_2O_4$, NaOH.[3] This method was used for deprotection of a glucosamine.[4] Cleavage occurs by reduction to the amine, which then undergoes a 1,6-elimination.

4. Electrolysis, −1.2 V, DMF, R_4NX.[5]

5. $SnCl_2$, HCl, dioxane, phenol, DMF, rt. This method was very effective at removing the PNZ group from peptides supported on Rink-polystyrene resin.[1,6]

6. Other reagents that reduce nitro groups should be effective at cleaving the PNZ group.

1. A. Isidro-Llobet, J. Guasch-Camell, M. Alvarez, and F. Albericio, *Eur. J. Org. Chem.*, 3031 (2005).

2. J. E. Shields and F. H. Carpenter, *J. Am. Chem. Soc.*, **83**, 3066 (1961); S. Hashiguchi, H. Natsugari, and M. Ochiai, *J. Chem. Soc., Perkin Trans. I*, 2345 (1988).

3. P. J. Romanovskis, P. Henklein, J. A. Benders, I. V. Siskov, and G. I. Chipens, in *5th All-Union Symposium on Protein and Peptide Chemistry and Physics, Abstracts*, Baku, 1980, p. 229 [in Russian]; P. J. Romanovskis, I. V. Siskov, I. K. Liepkaula, E. A. Porunkevich, M. P. Ratkevich, A. A. Skujins, and G. I. Chipens, "Linear and Cyclic Analogs of ACTH Fragments: Synthesis and Biological Activity," in *Peptides: Synthesis, Structure, Function: Proceedings of the Seventh American Peptide Symposium*, University of Wisconsin, Madison, D. H. Rich and E. Gross, Eds., Pierce Chem. Co., Rockford, IL, 1981, pp. 229–232.

4. X. Qian and O. Hindsgaul, *J. Chem. Soc., Chem. Commun.*, 1059 (1997).

5. V. G. Mairanovsky, *Angew. Chem., Int. Ed. Engl.*, **15**, 281 (1976); H. L. S. Maia, M. J. Medeiros, M. I. Montenegro, and D. Pletcher, *Port. Electrochim. Acta*, **5**, 187 (1987), *Chem. Abstr.* **109**: 118114n (1989).

6. P. E. López, A. Isidro-Llobet, C. Gracia, L. J. Cruz, A. Garcia-Granados, A. Parra, M. Alvarez, and F. Albericio, *Tetrahedron Lett.*, **46**, 7737 (2005); A. Isidro-Llobet, M. Álvarez, and F. Albericio, *Tetrahedron Lett.*, **46**, 7733 (2005).

Halobenzyl Carbamates

Benzyl carbamates substituted with one or more halogens are much more stable to acidic hydrolysis than the unsubstituted benzyl carbamates.[1,2] For example, the 2,4-dichlorobenzyl carbamate is 80 times more stable to acid than is the simple benzyl derivative.[3] Halobenzyl carbamates can also be cleaved by hydrogenolysis with Pd-C,[3] but this process is expected to release acid by the simultaneous hydrogenolysis of the halogen group. The following halobenzyl carbamates have been found to be useful when increased acid stability is required: *p*-**Bromobenzyl Carbamate**,[4] *p*-**Chlorobenzyl Carbamate**,[1,2] and **2,4-Dichlorobenzyl Carbamate**[3] (Chart 8).

The 2-BrZ and 2-ClZ derivatives have been cleaved by transfer hydrogenolysis with ammonium formate/Pd–C.[5]

1. K. Noda, S. Terada, and N. Izumiya, *Bull. Chem. Soc. Jpn.*, **43**, 1883 (1970).
2. B. W. Erickson and R. B. Merrifield, *J. Am. Chem. Soc.*, **95**, 3757 (1973).
3. Y. S. Klausner and M. Chorev, *J. Chem. Soc., Perkin Trans. I*, 627 (1977).
4. D. M. Channing, P. B. Turner, and G. T. Young, *Nature*, **167**, 487 (1951).
5. D. C. Gowda, B. Rajesh, and S. Gowda, *Ind. J. Chem., Sect. B*, **39B**, 504 (2000).

4-Methylsulfinylbenzyl Carbamate (Msz$-$NR$_2$): $CH_3S(O)C_6H_4CH_2OCONR_2$

The Msz group is stable to TFA/anisole, NaOH, and hydrazine.

Formation

Msz$-$O$-$succinimidyl, CH_3CN, H_2O, Et_3N, 45% yield.[1]

Cleavage

1. $SiCl_4$, TFA, anisole.[1] $SiCl_4$ serves to reduce the sulfoxide prior to acid-catalyzed cleavage. The reduced form of this group becomes much more sensitive to acidolysis. Other sulfoxide reducing agents can be used.
2. TMSCl, Me$_2$S, THF.[2]

1. Y. Kiso, T. Kimura, M. Yoshida, M. Shimokura, K. Akaji, and T. Mimoto, *J. Chem. Soc., Chem. Commun.*, 1511 (1989).
2. T. Kimura, T. Fukui, S. Tanaka, K. Akaji, and Y. Kiso, *Chem. Pharm. Bull.*, **45**, 18 (1997).

4-Trifluoromethylbenzyl Carbamate (CTFB$-$NR$_2$): $CF_3C_6H_4CH_2OC(O)NR_2$

The CTFB group was developed to be orthogonal to the 2-naphthyl carbamate. Benzyl esters and aromatic benzyl ethers can also be reduced in its presence.[1] It is introduced via the chloroformate. It is cleaved by hydrogenolysis with Pd–C; in sluggish cases, Pearlman's catalyst should be used. It can be cleaved without reduction of an aromatic nitrile group.

1. E. A. Papageorgiou, M. J. Gaunt, J.-q. Yu, and J. B. Spencer, *Org. Lett.*, **2**, 1049 (2000).

Fluorous Benzyloxycarbamate (^{F}Cbz-NR$_2$):
$RC_6H_4CH_2O_2CNR_2$ R $= C_8F_{17}CH_2CH_2-$ and R $= (C_8F_{17}CH_2CH_2)_3Si-$

These reagents were prepared for use in fluorous synthesis methods. With R $=$ $C_8F_{17}CH_2CH_2-$ introduction proceeds using either the chloroformate method or the

N-hydroxysuccinimide method.[1] With R = $(C_8F_{17}CH_2CH_2)_3Si-$ the benzyl alcohol is treated with CDI and then methylated with MeOTf (highly toxic) to form a "Rapoport reagent" which in the presence of the amine and DMAP forms the carbamates.[2] In both derivatives cleavage is affected by hydrogenolysis. The main problem with these reagents is that they require multiple steps to prepare, but they are advantageous in combinatorial synthesis because of the ease by which these are separated by fluorous chromatography.

1. D. P. Curran, M. Amatore, D. Guthrie, M. Campbell, E. Go, and Z. Luo, *J. Org. Chem.*, **68**, 4643 (2003). D. V. Filippov, D. J. van Zoelen, S. P. Oldfield, G. A. van der Marel, H. S. Overkleeft, J. W. Drijfhout, and J. H. van Boom, *Tetrahedron Lett.*, **43**, 7809 (2002).

2. D. Schwinn and W. Bannwarth, *Helv. Chim. Acta*, **85**, 255 (2002).

2-Naphthylmethyl Carbamates (CNAP-NR₂)

The CNAP group was examined as a more easily cleaved group than the Cbz group by hydrogenolysis since the NAP ether could be cleaved in the presence of the Bn ether by hydrogenolysis. Although the desired selectivity was not observed, excellent orthogonality was obtained with the trifluoromethylbenzyl (CTFB) carbamates. The CNAP group is introduced with the chloroformate in excellent yields.[1] An aromatic nitro group was not reduced during the hydrogenolysis of the CNAP group.

1. E. A. Papageorgiou, M. J. Gaunt, J.-q. Yu, and J. B. Spencer, *Org. Lett.*, **2**, 1049 (2000).

9-Anthrylmethyl Carbamate: (Chart 8)

$CH_2OC(O)NR_2$

Formation

9-Anthryl-$CH_2OCO_2C_6H_4$-*p*-NO_2, DMF, 25°C, 86% yield.[1]

Cleavage[1]

1. CH_3SNa, DMF, −20°C, 1–7 h, 77–91% yield or 25°C, 4 min, 86% yield. In this case, cleavage occurs by thiolate addition to the 10-position followed by elimination.

2. CF$_3$COOH, CH$_2$Cl$_2$, 0°C, 5 min, 88–92% yield. The anthrylmethyl carbamate is stable to 0.01 N lithium hydroxide (25°C, 6 h), to 0.1 N sulfuric acid (25°C, 1 h), and to 1 M trifluoroacetic acid (25°C, 1 h, dioxane).

1. N. Kornblum and A. Scott, *J. Org. Chem.*, **42**, 399 (1977).

Diphenylmethyl Carbamate: Ph$_2$CHOC(O)NR$_2$ (Chart 8)

The diphenylmethyl carbamate, prepared from the azidoformate, is readily cleaved by mild acid hydrolysis (1.7 N HCl, THF, 65°C, 10 min, 100% yield).[1]

1. R. G. Hiskey and J. B. Adams, *J. Am. Chem. Soc.*, **87**, 3969 (1965).

Carbamates Cleaved by a 1,6-Elimination

A series of carbamates have been prepared that are cleaved by liberation of a phenol or amine, which when treated with base, cleaves the carbamate by quinone methide formation through a 1,6-elimination.[1]

4-Phenylacetoxybenzyl Carbamate (PhAcOZ-NR$_2$):
4-(C$_6$H$_5$CH$_2$CO$_2$)C$_6$H$_4$CH$_2$OCONR$_2$

Preparation of PhAcOZ amino acids proceeds from the chloroformate and cleavage is accomplished enzymatically with penicillin G acylase (pH 7 phosphate buffer, 25°C, NaHSO$_3$, 40–88% yield).[2,3] In a related approach, the 4-acetoxy derivative is used; in this case, however, deprotection is achieved using the lipase, acetyl esterase from oranges (pH 7, NaCl buffer, 45°C, 57–70% yield).[4]

4-Azidobenzyl Carbamate (ACBZ-NR$_2$): 4-N$_3$C$_6$H$_4$CH$_2$OCONR$_2$

The carbamate, prepared from the 4-nitrophenyl carbonate, is cleaved by reduction with dithiothreitol (DTT) and TEA to give the aniline, which triggers fragmentation releasing the amine.[5]

4-Azidomethoxybenzyl Carbamate: N$_3$CH$_2$OC$_6$H$_4$CH$_2$OC(O)NR$_2$

Amino acids are protected with the 4-nitrophenyl carbonate (H$_2$O, dioxane, 54–85% yield) and cleaved by reduction of the azide with SnCl$_2$. The group is stable to the conditions normally used to cleave a BOC group, but it is not expected to be stable to a large number of strongly reducing conditions.[6]

m-Chloro-p-acyloxybenzyl Carbamate

Cleavage[7,8]

1. $NaHCO_3/Na_2CO_3$ or H_2O_2/NH_3, $NaHSO_3$, 1 h.
2. 0.1 N NaOH, 10 min, 100% yield.
3. H_2/Pd–C.
4. HBr, AcOH.

p-(Dihydroxyboryl)benzyl Carbamate (Dobz–NR₂)

Formation[9]

aqueous base; aqueous acid

Cleavage[10]

1. H_2O_2, pH 9.5, 25°C, 5 min, 90% yield.
2. H_2, Pd–C.
3. HBr, AcOH.

5-Benzisoxazolylmethyl Carbamate (Bic–NR₂) (Chart 8)

Formation

$ClCO_2CH_2$-5-benzisoxazole, pH 8.5–9.0, CH_3CN, 0°C, 1 h, 63% yield.[10]

Cleavage[10]

1. Et_3N, CH_3CN or DMF, 25°C, 30 min; Na_2SO_3, EtOH, H_2O, 40°C, 3 h, pH 7, 92% yield or CF_3COOH, 90 min, 95% yield.

2. H_2, Pd–C.

3. HBr, AcOH. This derivative is stable to trifluoroacetic acid.

2-(Trifluoromethyl)-6-chromonylmethyl Carbamate (Tcroc-NR$_2$)[11,12]

Cleavage

PrNH$_2$ or hydrazine. The Tcroc group resists cleavage by CF$_3$COOH.

1. M. Wakselman, *Nouv. J. Chim.*, **7**, 439 (1983).
2. T Pohl and H. Waldmann, *Angew. Chem., Int. Ed. Engl.*, **35**, 1720 (1996).
3. D. Sebastian, A. Heuser, S. Schulze, and H. Waldemann, *Synthesis*, 1098 (1997); H. Waldmann and E. Nägele, *Angew. Chem., Int. Ed. Engl.*, **34**, 2259 (1995); R. Machauer and H. Waldmann, *Chem. Eur. J.*, **7**, 2940–2956 (2001).
4. H. Waldmann and E. Nägele, *Angew. Chem., Int. Ed. Engl.*, **34**, 2259 (1995); E. Nägele, M. Schelhaas, N. Kuder, and H. Waldmann, *J. Am. Chem. Soc.*, **120**, 6889 (1998).
5. A. Mitchinson, B. T. Golding, R. J. Griffin, and M. C. O'Sullivan, *J. Chem. Soc., Chem. Commun.*, 2613 (1994); B. T. Golding, A. Mitchinson, W. Clegg, M. R. J. Elsegood, and R. J. Griffin, *J. Chem. Soc. Perkin Trans. 1*, 349 (1999); T. Pohl and H. Waldmann, *J. Am. Chem. Soc.*, **119**, 6702 (1997).
6. B. Loubinoux and P. Gerardin, *Tetrahedron Lett.*, **32**, 351 (1991).
7. M. Wakselman and E. G.-Jampel, *J. Chem. Soc., Chem. Commun.*, 593 (1973).
8. G. Le Corre, E. G.-Jampel, and M. Wakselman, *Tetrahedron*, **34**, 3105 (1978).
9. D. S. Kemp and D. C. Roberts, *Tetrahedron Lett.*, **16**, 4629 (1975).
10. D. S. Kemp and C. F. Hoyng, *Tetrahedron Lett.*, **16**, 4625 (1975).
11. D. S. Kemp and G. Hanson, *J. Org. Chem.*, **46**, 4971 (1981).
12. D. S. Kemp, D. R. Bolin, and M. E. Parham, *Tetrahedron Lett.*, **22**, 4575 (1981).

Carbamates Cleaved by β-Elimination

Several protective groups have been prepared that rely on a β-elimination to effect cleavage. Often the protective group must first be activated to increase the acidity of the β-hydrogen. In general, the derivatives are prepared by standard procedures, either from the chloroformate or mixed carbonate.

2-Methylthioethyl Carbamate: MeSCH$_2$CH$_2$OC(O)NR$_2$

A 2-methylthioethyl carbamate is cleaved by 0.01 N NaOH after alkylation to Me$_2$S$^+$CH$_2$CH$_2$OC(O)NR$_2$ or by 0.1 N NaOH after oxidation to the sulfone.[1]

2-Methylsulfonylethyl Carbamate: $MeSO_2CH_2CH_2OC(O)NR_2$

This is the oxidized form of the methylthio derivative above. It is stable to catalytic hydrogenolysis and does not poison the catalyst. It is stable to liq. HF (30 min), but is cleaved in 5 s with 1 N NaOH.[2,3]

2-(p-Toluenesulfonyl)ethyl Carbamate: $4\text{-}CH_3\text{-}C_6H_4SO_2CH_2CH_2OC(O)NR_2$

This derivative is similar to the methylsulfonylethyl derivative. It is cleaved by 1 M NaOH, <1 h.[4] The related 4-chlorobenzenesulfonylethyl carbamate has also been used as a protective group that can be cleaved with DBU or tetramethylguanidine.[5]

2-(4-Nitrophenylsulfonyl)ethoxy (Nse−NR₂) Carbamate:
$4\text{-}NO_2C_6H_4SO_2CH_2CH_2O_2CNR_2$

The Nsc group was explored as an alternative to the Fmoc group in peptide synthesis because of problems encountered with the dibenzofulvene polymers that are often produced during deprotection. The Nsc group is introduced with the chloroformate or the succinimidyl carbonate[6] and is cleaved efficiently with Tris(2-aminoethyl)amine in CH_2Cl_2 or MeOH. Its major advantage over the Fmoc group is that the vinyl sulfone does not polymerize and thus purifications are greatly simplified.[7] Diethylamine or piperdine can also be used for deprotection. The accompanying table compares some of the properties of each group.

Comparison Between Fmoc- and Nsc-Protected Peptides

Properties	Fmoc	Nsc
Cleavage rate ($t_{1/2}$)		
20% Piperidine/DMF	10–15 s	90–110 s
1% DBU/20% piperidine/DMF	—	12–15 s
Decompositon in DMF solution		
1 week	10%	<1%
3 weeks	40%	2%
Olefin–amine adduct formation	Fast and reversible	Very fast and irreversible
Polymerization during removal	Yes	No
UV monitoring range	302 nm	380 nm

2-(2,4-Dinitrophenylsulfonyl)ethoxy (DNse−NR₂) Carbamate:
$2,4\text{-}(NO_2)_2C_6H_3SO_2CH_2CH_2O_2CNR_2$

This derivative was prepared from the chloroformate (88–90% yield) and can be cleaved with piperidine.[8] It is expected to be more labile to base than the Nsc group.

2-(4-Trifluoromethylphenylsulfonyl)ethoxy (Tsc−NR₂) Carbamate:
$4\text{-}CF_3C_6H_4SO_2CH_2CH_2O_2CNR_2$

The Tsc group was developed as a more soluble alternative to the Nsc group for the protection of pyrrole-imidazole polyamides. It is formed from the 4-nitrophenyl carbonate (DIPEA, DMAP, HOBt, CH$_2$Cl$_2$, rt, 66–81% yield) and is cleaved with 20% piperidine/DMF within 5 min. It has better solution stability than the Fmoc group during peptide couplings.[9]

[2-(1,3-Dithianyl)]methyl Carbamate (Dmoc−NR$_2$)

Cleavage occurs by prior activation with peracetic acid to the bissulfone followed by mild base treatment.[10]

2-Phosphonioethyl Carbamate (Peoc−NR$_2$): R$_3$P$^+$CH$_2$CH$_2$OC(O)NR$_2'$ X$^-$

This derivative is stable to trifluoroacetic acid; it is cleaved by mild bases (pH 8.4; 0.1 N NaOH, 1 min, 100% yield).[11]

2-[Phenyl(methyl)sulfonio]ethyl Carbamate (Pms−NR$_2$): Ph(CH$_3$)S$^+$CH$_2$CH$_2$OC(O)NR$_2$ BF$_4^-$

This group was developed as a water soluble carbamate in peptide synthesis. It is prepared by methylating (2-phenylthio)ethylcarbamates of amino acids with methyl iodide and AgBF$_4$. Amines may also be protected using Pms-4-nitrophenyl carbonate as a stable crystalline reagent that can be stored. The Pms group was cleavable with NaHCO$_3$, but Na$_2$CO$_3$ was proved to be more efficient.[12]

1-Methyl-1-(triphenylphosphonio)ethyl (2-Triphenylphosphonioisopropyl) Carbamate (Ppoc−NR$_2$): Ph$_3$P$^+$CH$_2$CH(CH$_3$)OC(O)NR$_2$ X$^-$

This derivative is similar to the Peoc group except that it is four times more stable to base and is not as susceptible to side reactions as is the Peoc group.[13]

1,1-Dimethyl-2-cyanoethyl Carbamate: (CN)CH$_2$C(CH$_3$)$_2$OC(O)NR$_2$ (Chart 8)

This derivative is stable to trifluoroacetic acid and is cleaved by aqueous K$_2$CO$_2$ or Et$_3$N, 25°C, 6 h, 90% yield.[14]

2-Dansylethyl Carbamate (Dnseoc−NR$_2$)

The Dnseoc group was developed as a base labile protecting group for the 5′-hydroxyl in oligonucleotide synthesis. It is cleaved with DBU in aprotic solvents.

The condensation of oligonucleotide synthesis can be determined by UV detection at 350 nm or by fluorescence at 530 nm of the liberated vinylsulfone.[15]

2-(4-Nitrophenyl)ethyl Carbamate (Npeoc—NR₂):
$4\text{-}NO_2C_6H_4CH_2CH_2OCONR_2$

The Npeoc group was introduced for protection of the exocyclic amino functions of nucleic acid bases, but has also been used for simple amines.

Formation

1. $4\text{-}NO_2C_6H_4CH_2CH_2OCOCl.$ [16]

2. DMAP, DMF, 75–97% yield. [17]

Cleavage

1. DBU, CH_3CN or Pyr.[18]
2. Photolysis, for *N-o*-nitrodiphenylmethoxycarbonyl compounds.[19]

1. H. Kunz, *Chem. Ber.*, **109**, 3693 (1976).
2. G. I. Tesser and I. C. Balver-Geers, *Int. J. Pept. Protein Res.*, **7**, 295 (1975).
3. D. Filippov, G.A. van der Marel, E. Kuyl-Yeheskiely, and J. H. van Boom, *Synlett*, 922 (1994).
4. A. T. Kader and C. J. M. Sterling, *J. Chem. Soc.*, 258 (1964).
5. V. V. Samukov, A. N. Sabirov, and M. L. Troshkov, *Zh. Obshch. Khim*, **58**, 1432 (1988); *Chem. Abstr.*, **110**: 76008u (1989).
6. F. Albericio, *Biopolymers (Peptide Science)*, **55**, 123 (2000). C. Carreno, M. E. Mendez, Y.-D. Kim, H.-J. Kim, S. A. Kates, D. Andreu, and F. Albericio, *J. Peptide Res.*, **56**, 63 (2000).
7. X. Zhu, K. Pachamuthu, and R. R. Schmidt, *Org. Lett.*, **6**, 1083 (2004).
8. J. Gurnani, C. K. Narang, and M. R. K. Sherwani, *Hungarian J. Industrial Chem.*, **27**,1 (1999).
9. J. S. Choi, Y. Lee, E. Kim, N. Jeong, H. Yu, and H. Han, *Tetrahedron Lett.*, **44**, 1607 (2003).
10. H. Kunz and R. Barthels, *Chem. Ber.*, **115**, 833 (1982); R. Barthels and H. Kunz, *Angew Chem., Int. Ed. Engl.*, **21**, 292 (1982).
11. H. Kunz, *Angew. Chem., Int. Ed. Engl.*, **17**, 67 (1978).
12. K. Hojo, M. Maeda, and K. Kawasaki, *Tetrahedron*, **60**, 1875 (2004).
13. H. Kunz and G. Schaumloeffel, *Liebigs Ann. Chem.*, 1784 (1985).
14. E. Wünsch and R. Spangenberg, *Chem. Ber.*, **104**, 2427 (1971).
15. F. Bergmann and W. Pfleiderer, *Helv. Chim. Acta*, **77**, 203 (1994).
16. H. Sigmund and W. Pfleiderer, *Helv. Chim. Acta*, **77**, 1267 (1994).
17. G. Walcher and W. Pfleiderer, *Helv. Chim. Acta*, **79**, 1067 (1996).

18. F. Himmelsbach, B. S. Schulz, T. Trichtinger, R. Charubala, and W. Pfleiderer, *Tetrahedron*, **40**, 59 (1984).

19. J. A. Baltrop, P. J. Plant, and P. Schofield, *J. Chem. Soc., Chem. Commun.*, 822 (1966).

4-Methylthiophenyl Carbamate (Mtpc$-$NR$_2$): 4-MeSC$_6$H$_4$OC(O)NR$_2$

2,4-Dimethylthiophenyl Carbamate (Bmpc$-$NR$_2$): 2,4-(MeS)$_2$C$_6$H$_3$OC(O)NR$_2$

After activation with peracetic acid and base treatment, derivatives of primary amines form the isocyanate, which can be trapped with water to effect hydrolysis or with an alcohol to form other carbamates.[1,2]

1. H. Kunz and K. Lorenz, *Angew Chem., Int. Ed. Engl.*, **19**, 932 (1980).

2. H. Kunz and H.-J. Lasowski, *Angew. Chem., Int. Ed. Engl.*, **25**, 170 (1986).

Photolytically Cleaved Carbamates

The following carbamates can be cleaved by photolysis.[1] They can be prepared either from the chloroformate or from the mixed carbonate.

1. *m*-**Nitrophenyl Carbamate.**[2]

2. **3,5-Dimethoxybenzyl Carbamate.**[3]

3. **1-Methyl-1-(3,5-dimethoxyphenyl)ethyl Carbamate (Ddz$-$NR$_2$).** The carbamate, prepared in 80% yield from the azidoformate or pentachlorophenyl carbonate, is cleaved by photolysis and as expected, by acidic hydrolysis (TFA, 20°C, 8 min, 100% yield).[4,5]

4. α-**Methylnitropiperonyl Carbamate (Menpoc$-$NR$_2$).**[6] The half-life for the photochemical cleavage is on the order of 20–30 s for a variety of amino acid derivatives. This rate is substantially faster than the 2-nitrobenzyl carbamate.

5. *o*-**Nitrobenzyl Carbamate.**[7,8]

6. **3,4-Dimethoxy-6-nitrobenzyl Carbamate.**[7,9] (Chart 8) This group was effective for the photochemical deprotection of the guanidine group with a quantum efficiency of 0.023.[10]

7. **3,4-Disubstituted-6-nitrobenzyl Carbamates.** A series of different 3,4-disubstituted 6-nitrobenzyl carbamates were prepared and their cleavage rates examined at 254 nm and 420 nm. These studies showed that the 3-chloro or 3-bromo derivatives cleave faster at 254 nm than does the 2-nitroveratrole-derived carbamates whereas at 420 nm the relative rates are reversed.[11]

8. **Phenyl(*o*-nitrophenyl)methyl Carbamate (Npeoc$-$NR$_2$).**[12]

9. **2-Nitrophenylethyl Carbamate.**[13] The photolytic removal of this group is two-fold faster than the 2-nitrobenzyl carbamate.[14,15] Additionally, substitution at the alpha carbon increases the rate of cleavage even more.

10. **6-Nitroveratryl Carbamate (Nvoc−NR$_2$).**[16] The use of the Nvoc group for protection of an alkoxy amine was demonstrated in a synthesis of a modified tRNA.[17]

11. **4-Methoxyphenacyl Carbamate (Phenoc−NR$_2$).** This group is stable to 50% TFA/CH$_2$Cl$_2$, NaOH, and 20% piperidine/DMF.[18]

12. **3′,5′-Dimethoxybenzoin Carbamate (DMBOCONR$_2$).**[19] The DMB carbamate can also be introduced through the 4-nitrophenyl carbonate.[20] It has been prepared from an isocyanate and 3′,5′-dimethoxybenzoin.[21] The synthesis of a number of other substituted benzoins as possible protective groups has been described.[22]

13. **9-Xanthenylmethyl Carbamate:** The 9-xanthenylmethyl carbamate is introduced using the 4-nitrophenyl carbonate in either DMF, Na$_2$CO$_3$/THF, or TEA/THF in 62–84% yield. It is cleaved photochemically at 300 nm in CH$_3$CN/H$_2$O in 52–90% yield. Liberated xanthone sometimes results in compromised yields.[23]

14. **N-Methyl-N-(o-nitrophenyl) Carbamate.** This carbamate is prepared from the carbamoyl chloride (CH$_2$Cl$_2$, DMAP, TEA or RONa, 88–94% yield). It

is cleaved by photolysis at 248–365 nm in EtOH, H_2O, (91–100% yield) to afford the alcohol and 2-nitrosoaniline.[24]

15. **N-(2-Acetoxyethyl) amine: $AcOCH_2CH_2NR_2$**. N-(2-Acetoxyethyl) derivatives are introduced from bromoethyl acetate in CH_3CN by heating to reflux for 5 h. Cleavage is affected photochemically by irradiation at 350 nm in the presence of 4,4'-dimethoxybenzophenone in CH_3CN/H_2O (60–80% yield). The cleavage fails for secondary amines.[25]

1. For a review of photochemically labile protective groups, see V. N. R. Pillai, *Synthesis*, 1 (1980).
2. Th. Wieland, Ch. Lamperstorfer, and Ch. Birr, *Makromol. Chem.*, **92**, 277 (1966).
3. J. W. Chamberlin, *J. Org. Chem.*, **31**, 1658 (1966).
4. C. Birr, W. Lochinger, G. Stahnke, and P. Lang, *Justus Liebigs Ann. Chem.*, **763**, 162 (1972); C. Birr, *JustusLiebigs Ann. Chem.*, 1652 (1973).
5. J. F. Cameron and J. M. J. Fréchet, *J. Org. Chem.*, **55**, 5919 (1990).
6. C. P. Holmes and B. Kiangsoontra, *Pept.: Chem., Struct. Biol., Proc. Am. Pept. Symp., 13th*, 110 (1994).
7. B. Amit, U. Zehavi, and A. Patchornik, *J. Org. Chem.*, **39**, 192 (1974).
8. A. Misra, S. Tripathi, and K. Misra, *Ind. J. Chem., Sect. B*, **41B**, 1454 (2002).
9. K. Burgess, S. E. Jacutin, D. Lim, and A. Shitangkoon, *J. Org. Chem.*, **62**, 5165 (1997).
10. J. S. Wood, M. Koszelak, J. Liu, and D. S. Lawrence, *J. Am. Chem. Soc.*, **120**, 7145 (1998).
11. C. G. Bochet, *Tetrahedron Lett.*, **41**, 6341 (2000).
12. J. A. Baltrop, P. J. Plant, and P. Schofield, *J. Chem. Soc., Chem. Commun.*, 822 (1966).
13. A. Hasan, K.-P. Stengele, H. Giegrich, P. Cornwell, K. R. Isham, R. A. Sachleben, W. Pfleiderer, and R. S. Foote, *Tetrahedron*, **53**, 4247 (1997).
14. K. R. Bhushan, C. DeLisi, and R. A. Laursen, *Tetrahedron Lett.*, **44**, 8585 (2003).
15. L. Qiao and A. P. Kozikowski, *Tetrahedron Lett.*, **39**, 8959 (1998).
16. S. B. Rollins and R. M. Williams, *Tetrahedron Lett.*, **38**, 4033 (1997).
17. N. Matsubara, K. Oiwa, T. Hohsaka, R. Sadamoto, K. Niikura, N. Fukuhara, A. Takimoto, H. Kondo, and S.-I. Nishimura, *Chem. Eur. J.*, **11**, 6974 (2005).
18. G. Church, J.-M. Ferland, and J. Gauthier, *Tetrahedron Lett.*, **30**, 1901 (1989).
19. M. C. Pirrung and C.-Y. Huang, *Tetrahedron Lett.*, **36**, 5883 (1995).
20. G. Papageorgiou and J. E. T. Corrie, *Tetrahedron*, **53**, 3917 (1997).
21. J. F. Cameron, C. G. Willson, and J. M. J. Fréchet, *J. Chem. Soc., Chem. Commun.*, 923 (1995).
22. J. F. Cameron, C. G. Willson, and J. M. J. Fréchet, *J. Chem. Soc., Perkin Trans.1*, 2429 (1997).

23. H. Du and M. K. Boyd, *Tetrahedron Lett.*, **42**, 6645 (2001).

24. S. Loudwig and M. Goeldner, *Tetrahedron Lett.*, **42**, 7957 (2001).

25. J. Cossy and H. Rakotoarisoa, *Tetrahedron Lett.*, **41**, 2097 (2000).

Miscellaneous Carbamates

The following carbamates have seen little use since the preparation of the first edition of this book; they are listed here for completeness. For the most part they are variations of the BOC and benzyl carbamates, with the exception of the azo derivatives, which are highly colored. The differences between them are largely in the strength of the acid required for their cleavage. Unfortunately, they have not been compared in a single study to more clearly define their relative stability.

1. *t*-Amyl Carbamate[1]
2. 1-Methylcyclobutyl Carbamate[2] (Chart 8)
3. 1-Methylcyclohexyl Carbamate[2] (Chart 8). The half-life for cleavage in neat CF_3CO_2H is 2 min and 180 min in formic acid.
4. 1-Methyl-1-cyclopropylmethyl Carbamate[3]
5. Cyclobutyl Carbamate[2] (Chart 8). The half-life for cleavage in neat CF_3CO_2H is >300 min.
6. Cyclopentyl Carbamate[3]
7. Cyclohexyl Carbamate[3–5] This group was used in BOC-based peptide synthesis and is cleaved with HF.
8. Isobutyl Carbamate[6]
9. Isobornyl Carbamate[7]
10. Cyclopropylmethyl Carbamate[2]
11. *p*-Decyloxybenzyl Carbamate[8]
12. Diisopropylmethyl Carbamate[3]
13. 2,2-Dimethoxycarbonylvinyl Carbamate[9]
14. *o*-(*N*,*N*-Dimethylcarboxamido)benzyl Carbamate[10]
15. 1,1-Dimethyl-3-(*N*,*N*-dimethylcarboxamido)propyl Carbamate[10]
16. Butynyl Carbamate[11]
17. 1,1-Dimethylpropynyl Carbamate[12] (Chart 8)
18. 2-Iodoethyl Carbamate[13]
19. 1-Methyl-1-(4'-pyridyl)ethyl Carbamate[10]
20. 1-Methyl-1-(*p*-phenylazophenyl)ethyl Carbamate[14] Azo derivatives are colored and thus may have certain analytical advantages.
21. *p*-(*p*'-Methoxyphenylazo)benzyl Carbamate[15]
22. *p*-(Phenylazo)benzyl Carbamate[15]
23. 2,4,6-Trimethylbenzyl Carbamate[16]
24. Isonicotinyl Carbamate.[17] (Chart 8)

25. **4-(Trimethylammonium)benzyl Carbamate**[18]
26. ***p*-Cyanobenzyl Carbamate**[19]
27. **Di(2-pyridyl)methyl Carbamate**[10]
28. **2-Furanylmethyl Carbamate**[20]
29. **Phenyl Carbamate**[21]
30. **2,4,6-Tri-*t*-butylphenyl Carbamate**[22]
31. **1-Methyl-1-phenylethyl Carbamate**[23] (Chart 8)
32. ***S*-Benzyl Thiocarbamate**[24] (Chart 8)

1. S. Sakakibara, I. Honda, K. Takada, M. Miyoshi, T. Ohnishi, and K. Okumura, *Bull. Chem. Soc. Jpn.*, **42**, 809 (1969); S. Matsuura, C.-H. Niu, and J. S. Cohen, *J. Chem. Soc., Chem. Commun.*, 451 (1976).

2. S. F. Brady, R. Hirschmann, and D. F. Veber, *J. Org. Chem.*, **42**, 143 (1973).

3. F. C. McKay and N. F. Albertson, *J. Am. Chem. Soc.* **79**, 4686 (1957).

4. G. Mezõ, M. Mihala, G. Kóczán, and F. Hudecz, *Tetrahedron*, **54**, 6757 (1998).

5. H. Nishio, Y. Nishiuchi, T. Inui, K. Yoshizawa-Kumagaye, and T. Kimura, *Tetrahedron Lett.*, **41**, 6839 (2000).

6. R. L. Letsinger and P. S. Miller, *J. Am. Chem. Soc.* **91**, 3356 (1969).

7. M. Fujino, S. Shinagawa, O. Nishimura, and T. Fukuda, *Chem. Pharm. Bull.*, **20**, 1017 (1972).

8. H. Brechbühler, H. Büchi, E. Hatz, J. Schreiber, and A. Eschenmoser, *Helv. Chim. Acta*, **48**, 1746 (1965).

9. A. Gomez-Sanchez, P. B. Moya, and J. Bellanato, *Carbohydr. Res.*, **135**, 101 (1984).

10. S. Coyle, O. Keller, and G. T. Young, *J. Chem. Soc., Perkin Trans. I*, 1459 (1979).

11. S. Miyazawa, K. Okano, T. Kawahara, Y. Machida, and I. Yamatsu, *Chem. Pharm. Bull.*, **40**, 762 (1992).

12. G. L. Southard, B. R. Zaborowsky, and J. M. Pettee, *J. Am. Chem. Soc.* **93**, 3302 (1971).

13. J. Grimshaw, *J. Chem. Soc.*, 7136 (1965).

14. A. T.-Kyi and R. Schwyzer, *Helv. Chim. Acta*, **59**, 1642 (1976).

15. R. Schwyzer, P. Sieber, and K. Zatsko, *Helv. Chim. Acta*, **41**, 491 (1958).

16. Y. Isowa, M. Ohmori, M. Sato, and K. Mori, *Bull. Chem. Soc. Jpn.*, **50**, 2766 (1977).

17. D. F. Veber, W. J. Paleveda, Y. C. Lee, and R. Hirschmann, *J. Org. Chem.*, **42**, 3286 (1977).

18. Y. Zhang, X. Wang, L. Li, and P. Zhang, *Sci. Sin. Ser. B*, **29**, 1009 (1986); *Chem. Abstr.*, **108**: 6354p (1988).

19. K. Noda, S. Terada, and N. Izumiya, *Bull. Chem. Soc. Jpn.*, **43**, 1883 (1970).

20. G. Losse and K. Neubert, *Tetrahedron Lett.*, **11**, 1267 (1970).

21. J. D. Hobson and J. G. McCluskey, *J. Chem. Soc. C*, 2015 (1967); R. W. Adamiak and J. Stawinski, *Tetrahedron Lett.*, **18**, 1935 (1977).

22. D. Seebach and T. Hassel, *Angew. Chem., Int. Ed. Engl.*, **17**, 274 (1978).

23. B. E. B. Sandberg and U. Ragnarsson, *Int. J. Pept. Protein Res.*, **6**, 111 (1974); H. Franzen and U. Ragnarsson, *Acta Chem. Scand. B*, **33**, 690 (1979).

24. H. B. Milne, S. L. Razniak, R. P. Bayer, and D. W. Fish, *J. Am. Chem. Soc.* **82**, 4582 (1960).

Urea-Type Derivatives

Urea: $NH_2C(O)NHR$

Urea derivatives of amino acid derivatives were cleaved using N_2O_4/H_2O.[1]

Phenothiazinyl-(10)-carbonyl Derivative

The derivative is prepared in 51–82% yield and is cleaved with $Ba(OH)_2$ or NaOH in 52–96% yield after oxidation of the sulfur with hydrogen peroxide. It is stable to CF_3COOH and NaOH.[2]

N'-p-Toluenesulfonylaminocarbonyl Derivative: $R_2NCONHSO_2C_6H_4$-p-CH_3

This sulfonyl urea, prepared from an amino acid and p-tosyl isocyanate in 20–80% yield, is cleaved by alcohols (95% aq. EtOH, n-PrOH, or n-BuOH, 100°C, 1 h, 95% yield). It is stable to dilute base, to acids (HBr/AcOH or cold CF_3CO_2H), and to hydrazine.[3]

N'-Phenylaminothiocarbonyl Derivative: $R_2NCSNHC_6H_5$ (Chart 8)

This thiourea, prepared from an amino acid and phenyl isothiocyanate,[4] is cleaved by anhydrous trifluoroacetic acid (an $N-COCF_3$ group is stable)[5] and by oxidation (m-$ClC_6H_4CO_3H$, 0°C, 1.5 h, 73% yield; $H_2O_2/AcOH$, 80°C, 80 min, 44% yield).[6]

4-Hydroxyphenylaminocarbonyl Derivative
and 3-Hydroxytryptaminocarbonyl Derivative

These derivatives are prepared by reacting the amine with triphosgene to form the isocyanate which is then treated with either 4-aminophenol or 3-hydroxytryptamine to give the urea (72–99% yield). These are cleaved enzymatically with mushroom tyrosinase (73–93% yield).[7]

1. H. Collet, L. Boiteau, J. Taillades, and A. Commeyras, *Tetrahedron Lett.*, **40**, 3355 (1999).

2. J. Gante, W. Hechler, and R. Weitzel, in *Peptides 1986: Proc. 19th Eur. Pept. Symp.*, 1986, D. Theodoropoulos, Ed., de Gruyter, Berlin, 1987, pp. 87–90.

3. B. Weinstein, T. N.-S. Ho, R. T. Fukura, and E. C. Angell, *Synth. Commun.*, **6**, 17 (1976).

4. P. Edman, *Acta Chem. Scand.*, **4**, 277 (1950).

5. F. Borrás and R. E. Offord, *Nature*, **227**, 716 (1970).

6. J. Kollonitsch, A. Hajós, and V. Gábor, *Chem. Ber.*, **89**, 2288 (1956).

7. H. M. I. Osborn and N. A. O. Williams, *Org. Lett.*, **6**, 3111 (2004).

N'-Phenylaminothiocarbonyl Derivative: $R_2NCSNHC_6H_5$ (Chart 8)

This thiourea, prepared from an amino acid and phenyl isothiocyanate,[1] is cleaved by anhydrous trifluoroacetic acid (an $N-COCF_3$ group is stable)[2] and by oxidation (m-ClC$_6$H$_4$CO$_3$H, 0°C, 1.5 h, 73% yield; H$_2$O$_2$/AcOH, 80°C, 80 min, 44% yield).[3]

1. P. Edman, *Acta Chem. Scand.*, **4**, 277 (1950).

2. F. Borrás and R. E. Offord, *Nature*, **227**, 716 (1970).

3. J. Kollonitsch, A. Hajós, and V. Gábor, *Chem. Ber.*, **89**, 2288 (1956).

AMIDES

Simple amides are generally prepared from the acid chloride or the anhydride. There are also numerous other coupling agents and methodologies that have been developed for amide formation.[1] Amides are exceptionally stable to acidic or basic hydrolysis, and they are classically hydrolyzed through brute force by heating in strongly acidic or basic solutions. Among simple amides, hydrolytic stability increases from formyl to acetyl to benzoyl. Lability of the haloacetyl derivatives to mild acid hydrolysis increases with substitution: acetyl < chloroacetyl < dichloroacetyl < trichloroacetyl < trifluoroacetyl.[2] It should be noted that amide hydrolysis under acidic or basic[3] conditions is *greatly* facilitated in the presence of a neighboring hydroxyl group that can participate in the hydrolysis.[4] Although a number of imaginative amide-derived protective groups have been developed, most are not commonly used because they contain other reactive functionality or are not commercially available or because other more easily introduced and cleaved groups such as the BOC, Alloc, and Cbz groups serve adequately for amine protection. Amide derivatives of the nucleotides are not discussed in this section since their behavior is atypical of amides. They are generally more easily hydrolyzed than the typical amide because of the reduced basicity of the free amine in these derivatives. Several review articles discuss amides as −NH protective groups.[5–8]

1. C. A. G. N. Montalbetti and V. Falque, *Tetrahedron*, **61**, 10827 (2005).

2. R. S. Goody and R. T. Walker, *Tetrahedron Lett.*, **8**, 289 (1967).

3. B. F. Cain, *J. Org. Chem.*, **41**, 2029 (1976).

4. See, for example, C. K. Lai, R. S. Buckanin, S. J. Chen, D. F. Zimmerman, F. T. Sher, and G. A. Berchtold, *J. Org. Chem.*, **47**, 2364 (1982); E. R. Koft, P. Dorff, and R. Kullnig, *J. Org. Chem.*, **54**, 2936 (1989).

5. E. Wünsch, "Blockierung und Schutz der α-Amino-Function," in *Methoden der Organischen Chemie (Houben-Weyl)*, Georg Thieme Verlag, Stuttgart, 1974, Vol. 15/1, pp. 164–203, 250–264.

6. J. W. Barton, "Protection of N–H Bonds and NR₃," in *Protective Groups in Organic Chemistry*, J. F. W. McOmie, Ed., Plenum Press, New York and London, 1973, pp. 46–56.

7. Y. Wolman, "Protection of the Amino Group," in *The Chemistry of the Amino Group*, S. Patai, Ed., Wiley-Interscience, New York, 1968, Vol. 4, pp. 669–682.

8. *The Peptides; Analysis, Synthesis, Biology*, Vol. 3: *Protection of Functional Groups in Peptide Synthesis*, E. Gross and J. Meienhofer, Eds., Academic Press, New York, 1981.

Formamide: R_2NCHO (Chart 9)

Formation

1. 98% HCO_2H, Ac_2O, 25°C, 1 h, 78–90% yield.[1,2] The use of formic acetic anhydride for esterification and amide formation has been reviewed.[3]

2. HCO_2H, DCC, Pyr, 0°C, 4 h, 87–90% yield.[4] These conditions produce N-formyl derivatives of t-butyl amino acid esters with a minimum of racemization.

3. HCO_2H, $EtN=C=N(CH_2)_3NMe_2$·HCl, 0°C, 15 min; then N-methylmorpholine, 5°C, 20 h, 65–96% yield. This method can be used with amine hydrochlorides.[5]

4. From an aminoester: HCO_2NH_4, CH_3CN, reflux, 63–91% yield.[6]

5. C_6F_5OCHO, $CHCl_3$, rt, 5–30 min, 85–99% yield.[7] The simpler phenyl formate can also be used efficiently (83% yield).[8]

6. This reagent also formylates alcohols in the presence of added base.[9]

7. t-$BuMe_2SiCl$, DMAP, Et_3N, DMF, 35–60°C, 65–85% yield.[10]

8. DMF, silica gel, heat, 5 h, 100% yield,[11] or DMF, ZrO, heat, 5 h, 92% yield.[12]

9. HCO_2Et, heat.[13]

10. Triethyl orthoformate, 50–100% yield.[14]

11. HCO_2CH_2CN, CH_2Cl_2, rt, 12 h, 62–97% yield.[15]

12. Vinyl formates readily react with amines, alcohols and phenols to give the formamide or ester.[16]

13. 2-Chloro-4,6-dimethoxy[1,3,5]triazine, formic acid, N-methylmorpholine, DMAP, CH_2Cl_2, 20°C, 85–99% yield.[17]

Cleavage

1. HCl, H₂O, dioxane, 25°C, 48 h, or reflux, 1 h, 80–95% yield.[1]
2. Hydrazine, EtOH, 60°C, 4 h, 60–80% yield.[18]
3. H₂/Pd-C, THF, HCl, 25°C, 5–7 h, quant.[19]
4. 15% H₂O₂, H₂O, 60°C, 2 h, 80% yield.[20]
5. AcCl, PhCH₂OH, 20°C, 24 h, or 60°C, 3 h, good yields.[21]
6. *h*ν, 254 nm, CH₃CN, 100% yield.[22]
7. NaOH, H₂O, reflux, 18 h, 85% yield.[23]

1. J. C. Sheehan and D.-D. H. Yang, *J. Am. Chem. Soc.*, **80**, 1154 (1958).
2. E. G. E. Jahngen, Jr., and E. F. Rossomando, *Synth. Commun.*, **12**, 601 (1982).
3. P. Strazzolini, A. G. Giumanini, and S. Cauci, *Tetrahedron*, **46**, 1018 (1990).
4. M. Waki and J. Meienhofer, *J. Org. Chem.*, **42**, 2019 (1977).
5. F. M. F. Chen and N. L. Benoiton, *Synthesis*, 709 (1979).
6. S. Kotha, M. Behera, and P. Khedkar, *Tetrahedron Lett.*, **45**, 7589 (2004).
7. L. Kisfaludy and L. Ötvös, Jr., *Synthesis*, 510 (1987).
8. G. Shen and B. S. J. Blagg, *Org. Lett.*, **7**, 2157 (2005).
9. H. Yazawa and S. Goto, *Tetrahedron Lett.*, **26**, 3703 (1985).
10. S. W. Djuric, *J. Org. Chem.*, **49**, 1311 (1984).
11. European Patent to Japan Tobacco Inc., EP. 271093, June 15 (1988).
12. K. Takahashi, M. Shibagaki, and H. Matsushita, *Agric. Biol. Chem.*, **52**, 853 (1988).
13. H. Schmidhammer and A. Brossi, *Can. J. Chem.*, **60**, 3055 (1982).
14. T. Chancellor and C. Morton, *Synthesis*, 1023 (1994).
15. W. Duczek, J. Deutsch, S. Vieth, and H.-J. Niclas, *Synthesis*, 37 (1996).
16. M. Neveux, C. Bruneau, and P. H. Dixneuf, *J. Chem. Soc., Perkin Trans. 1*, 1197 (1991).
17. L. De Luca, G. Giacomelli, A. Porcheddu, and M. Salaris, *Synlett*, 2570 (2004).
18. R. Geiger and W. Siedel, *Chem. Ber.*, **101**, 3386 (1968).
19. G. Losse and D. Nadolski, *J. Prakt. Chem.*, **24**, 118 (1964).
20. G. Losse and W. Zönnchen, *Justus Liebigs Ann. Chem.*, **636**, 140 (1960).
21. J. O. Thomas, *Tetrahedron Lett.*, **8**, 335 (1967).
22. B. K. Barnett and T. D. Roberts, *J. Chem. Soc., Chem. Commun.*, 758 (1972).
23. U. Hengartner, A. D. Batcho, J. F. Blount, W. Leimgruber, M. E. Larscheid, and J. W. Scott, *J. Org. Chem.*, **44**, 3748 (1979).

Acetamide: R₂NAc (Chart 9)

Formation

The simplest method for acetamide preparation involves reaction of the amine with acetic anhydride or acetyl chloride with or without added base. The primary disadvantage of these reagents is that they are quite reactive and thus often are insufficiently selective. Some other methods are listed below tend to be more selective.

1. C_6F_5OAc, DMF, 25°C, 1–12 h, 78–91% yield.[1] These conditions allow selective acylation of amines in the presence of alcohols. If triethylamine is used in place of DMF, alcohols are also acylated (75–85% yield).
2. Ac_2O, 18-crown-6, Et_3N, 98% yield.[2] The crown ether forms a complex with a primary amine, thus allowing selective acylation of a secondary amine.
3. $AcOC_6H_4$-p-NO_2, pH 11.[3]

4. The readily prepared quinazolinone will selectively acyl-

ate a primary amine in the presence of a secondary amine, but more uniquely it will selectively acylate a pyrrolidine over a piperidine with 3:1 selectivity, and dimethylamine over diethylamine with 9:1 selectivity.[4]

5. Vinyl acetate or diethyl carbonate, $Cp_2Sm(THF)_2$, 80–99% yield.[5] Aniline fails to react under these conditions.
6. N,N-Diacetyl-2-trifluoromethylaniline, organic solvents, 3–24 h, rt or reflux, 54–99% yield. Acylation selectivity is a very sensitive function of steric effects; this reagent will selectively acylate pyrrolidine over piperidine (15:1). It is more selective than the diacetylaminoquinazolinones.[6]
7. Ac_2NOMe selectively acylates a primary amine of a spermidine.[7]

8. This reagent acylates amines by photoactivation at 300 nm,

MeCN, 1 h, 89–90% yield. The reaction is general for other amides as well.[8]

9. CH_3SO_2NHAc, heat, 90% yield. This method can also be used to transfer other acyl groups and is selective for primary amines in the presence of secondary amines.[9]

Cleavage

In general, acetamides as well as most other alkyl and aryl amides are quite difficult to hydrolyze and often require rather forcing conditions to achieve hydrolysis.

1. 1.2 N HCl, reflux, 9 h, 61–77% yield.[10]

2. 85% Hydrazine, 70°C, 15 h, 68% yield.[11]

3. $Et_3O^+BF_4^-$, CH_2Cl_2, 25°C, 1–2 h, 90% yield, then aq. $NaHCO_3$, satisfactory yields.[12]

4. Hog kidney acylase, pH 7, H_2O, 36°C, 35 h.[13,14] In this case, deprotection also proceeds with resolution, since only one enantiomer is cleaved.

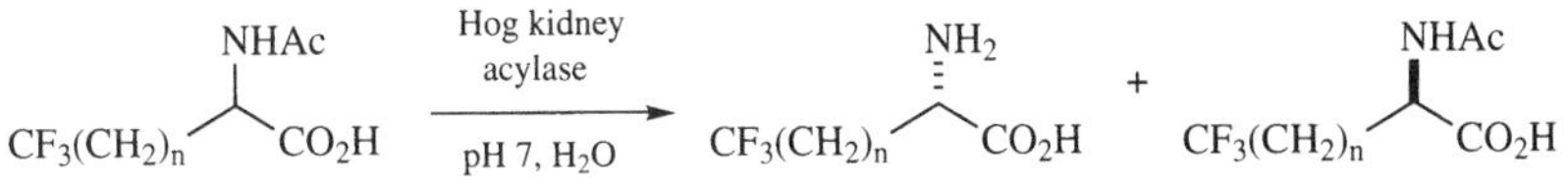

5. Enzymatic hydrolysis with *Aspergillis* Acylase, pH 8.5, 75% yield.[15]

6. Simple amides that are difficult to cleave can first be converted to a BOC derivative by an exchange process that relies on the reduced electrophilicity of the carbamate as well as its increased steric bulk.[16,17]

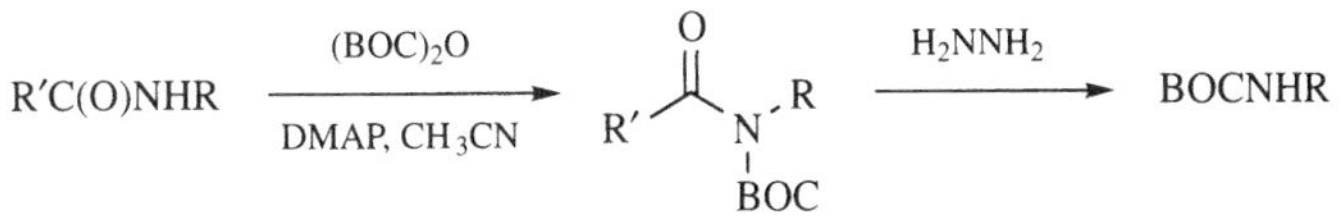

7. Na, BuOH, 120°C, 62% yield.[18]

8. Ca, NH_3, DME, EtOH, 4 h, 96% yield.[19] When using Ca metal, its surface coating must be cleaned before reaction will occur. This can be accomplished mechanically by stirring with sand.

9. For most common amides, cleavage is quite difficult, but in the case of an aziridine which has significantly reduced participation in amide resonance because of the nonplanar amide moiety,[20] hydrolysis is much simpler as shown in the illustration below.[21] As the lone pair and the carbonyl group become more orthogonal (thereby reducing the level of resonance), the rate of amide hydrolysis increases.[22,23] Aziridines are also less basic facilitating hydrolysis.

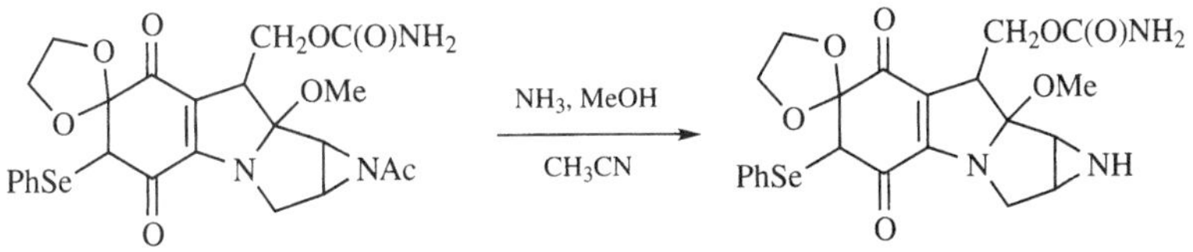

10. In a diacetamide, one acetamide is easily cleaved by hydrolysis with NaOMe and MeOH,[24] which is consistent with the use of N,N-diacetylaminoquinazoline,[4] 2-trifluoromethyl-N, N-diacetylaniline,[6] and N-methoxydiacetamide as amidating agents.[7]

11. The acetamide was shown unexpectedly to be subject to transacylation upon treatment with another acyl chloride.[25]

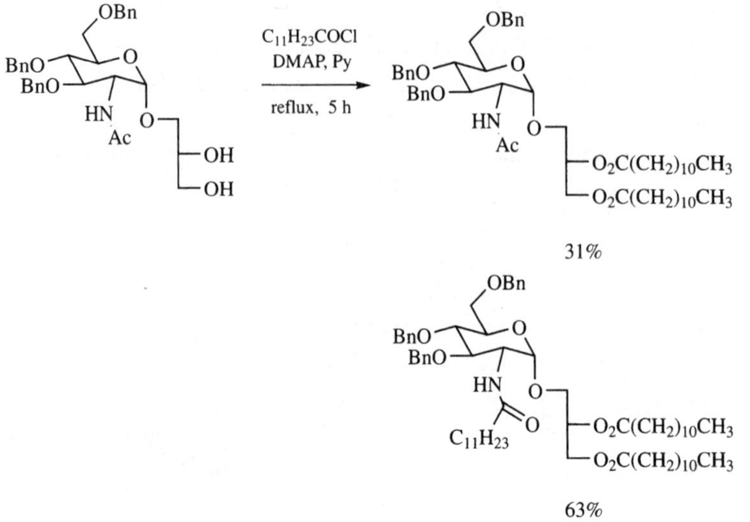

31%

63%

12. For an aniline derived acetamide: $BF_3 \cdot Et_2O$, MeOH, reflux, 5 h, 64–95% yield.[26]

13. Ph_3P, Cl_2, TEA, CH_2Cl_2, −30°C then ethylene glycol, 90% yield. This is a general method applicable to a variety of amides.[27]

14. $NaBH_4$, THF, H_2O, 0°C to rt, 100% yield.[28] This method will not cleave regular acetamides.

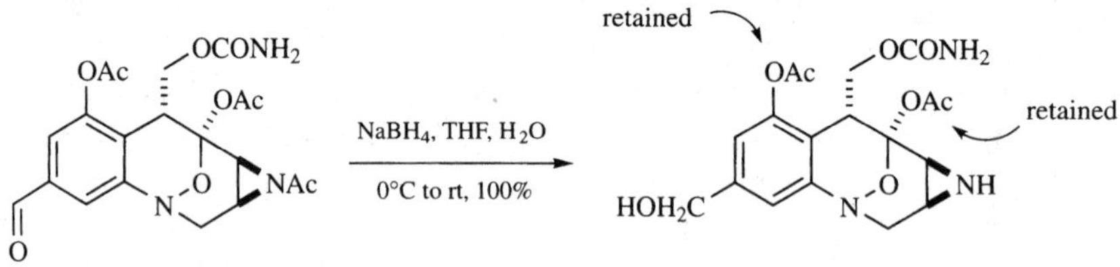

1. l. Kisfaludy, T. Mohacsi, M. Low, and F. Drexler, *J. Org. Chem.*, **44**, 654 (1979).

2. A. G. M. Barrett and J. C. A. Lana, *J. Chem. Soc., Chem. Commun.*, 471 (1978).

3. F. Kanai, T. Kaneko, H. Morishima, K. Isshiki, T. Takita, T. Takeuchi, and H. Umezawa, *J. Antibiotics, (Tokyo)*, **38**, 39 (1985).

4. R. S. Atkinson, E. Barker, and M. J. Sutcliffe, *J. Chem. Soc., Chem. Commun.*, 1051 (1996).

5. Y. Ishii, M. Takeno, Y. Kawasaki, A. Muromachi, Y. Nishiyama, and S. Sakaguchi, *J. Org. Chem.*, **61**, 3088 (1996).

6. Y. Murakami, K. Kondo, K. Miki, Y. Akiyama, T. Watanabe, and Y. Yokoyma, *Tetrahedron Lett.*, **38**, 3751 (1997).

7. Y. Kikugawa, K. Mitsui, and T. Sakamoto, *Tetrahedron Lett.*, **31**, 243 (1990).

8. C. Helgen and C. G. Bochet, *Synlett*, 1968 (2001).

9. S. Coniglio, A. Aramini, M. C. Cesta, S. Colagioia, R. Curti, F. D'Alessandro, G. D'Anniballe, V. D'Elia, G. Nano, V. Orlando, and M. Allegretti, *Tetrahedron Lett.*, **45**, 5375 (2004).

10. G. A. Dilbeck, L. Field, A. A. Gallo, and R. J. Gargiulo, *J. Org. Chem.*, **43**, 4593 (1978).

11. D. D. Keith, J. A. Tortora, and R. Yang, *J. Org. Chem.*, **43**, 3711 (1978).

12. S. Hanessian, *Tetrahedron Lett.*, **8**, 1549 (1967).

13. T. Tsushima, K. Kawada, S. Ishihara, N. Uchida, O. Shiratori, J. Higaki, and M. Hirata, *Tetrahedron*, **44**, 5375 (1988).

14. R. J. Cox, W. A. Sherwin, L. K. P. Lam, and J. C. Vederas, *J. Am. Chem. Soc.*, **118**, 7449 (1996).

15. F. VanMiddlesworth, C. Dufresne, F. E. Wincott, R. T. Mosley, and K. E. Wilson, *Tetrahedron Lett.*, **33**, 297 (1992).

16. L. Grehn, K. Gunnarsson, and U. Ragnarsson, *J. Chem. Soc., Chem. Commun.*, 1317 (1985). L. Grehn, K. Gunnarsson, and U. Ragnarsson, *Acta Chem. Scand. Ser B.*, **B40**, 745 (1986).

17. D. J. Kempf, *Tetrahedron Lett.*, **30**, 2029 (1989).

18. M. Obayashi and M. Schlosser, *Chem. Lett.*, 1715 (1985).

19. A. J. Pearson and D. C. Rees, *J. Am. Chem. Soc.* **104**, 1118 (1982).

20. G. V. Shustov, G. K. Kadorkina, S. V. Varlamov, A. V. Kachanov, R. G. Kostyanovsky, and A. Rauk, *J. Am. Chem. Soc.*, **114**, 1616 (1992).

21. H. Arai and M. Kasai, *J. Org. Chem.*, **58**, 4151 (1993).

22. A. J. Bennet, Q.-P. Wang, H. Stebocka-Tilk, V. Somayaji, and R. C. Brown, *J. Am. Chem. Soc.*, **112**, 6383 (1990).

23. A. J. Kirby, I. V. Konarov, P. D. Wothers, and N. Feeder, *Angew. Chem., Int. Ed. Engl.*, **37**, 785 (1998).

24. J. C. Castro-Palomino and R. R. Schmidt, *Tetrahedron Lett.*, **36**, 6871 (1995).

25. Y. Li, C. Li, P. Wang, S. Chu, H. Guan, and B. Yu, *Tetrahedron Lett.*, **45**, 611 (2004).

26. S. Miltsov, L. Rivera, C. Encinaas, and J. Alonso, *Tetrahedron Lett.*, **44**, 2301 (2003).

27. A. Spaggiari, L. C. Blaszczak, and F. Prati, *Org. Lett.*, **6**, 3885 (2004).

28. T. Katoh, E. Itoh, T. Yoshino, and S. Terashima, *Tetrahedron*, **53**, 10229 (1997).

Chloroacetamide: R_2NCOCH_2Cl (Chart 9)

Monochloroacetamides are cleaved (by "assisted removal") by reagents that contain two nucleophilic groups (e.g., *o*-phenylenediamine,[1] thiourea,[2,3] 1-piperidinethiocarboxamide,[4] 3-nitropyridine-2-thione,[5] 2-aminothiophenol):[6]

The chloroacetamide can also be cleaved by first converting it to the pyridiniumacetamide (Pyr, 90°C, 1 h, 70–90% yield), followed by mild basic hydrolysis (0.1 N

NaOH, 25°C)[7] or by acidic hydrolysis (4N HCl, 60°C, 8 h).[8] In glycosidations it was found to be an effective participating group that directs glycosidations from the β-face in a glucosamine derivative. It can be reduced with Ph_3SnH to give the natively displayed acetamide.[9]

1. R. W. Holley and A. D. Holley, *J. Am. Chem. Soc.*, **74**, 3069 (1952).
2. M. Masaki, T. Kitahara, H. Kurita, and M. Ohta, *J. Am. Chem. Soc.*, **90**, 4508 (1968).
3. J. E. Baldwin, M. Otsuka, and P. M. Wallace, *Tetrahedron*, **42**, 3097 (1986); T. Allmend-inger, G. Rihs, and H. Wetter, *Helv. Chim. Acta*, **71**, 395 (1988).
4. W. Steglich and H.-G. Batz, *Angew Chem., Int. Ed. Engl.*, **10**, 75 (1971).
5. K. Undheim and P. E. Fjeldstad, *J. Chem. Soc., Perkin Trans. I*, 829 (1973).
6. J. D. Glass, M. Pelzig, and C. S. Pande, *Int. J. Pept. Protein Res.*, **13**, 28 (1979).
7. C. H. Gaozza, B. C. Cieri, and S. Lamdan, *J. Heterocycl. Chem.*, **8**, 1079 (1971).
8. B. E. Ledford and E. M. Carreira, *J. Am. Chem. Soc.*, **117**, 11811 (1995).
9. T. Murakami and K. Taguchi, *Tetrahedron*, **55**, 989 (1999).

Trichloroacetamide: $R_2NCOCCl_3$ (Chart 9)

The TCA group has been use in oligosaccharide synthesis and is readily converted to the naturally displayed acetamide by reductive dehalogenation.[1]

Formation

1. $Cl_3CCOCCl_3$, hexane, 65°C, 90 min, 65–97% yield.[2]
2. Cl_3CCOCl, TEA, 81% yield.[1]

Cleavage

1. $NaBH_4$, EtOH, 1 h, 65% yield.[3]
2. Cs_2CO_3, DMF or DMSO, 100°C, 49–86% yield.[4]

1. K. R. Love and P. H. Seeberger, *J. Org. Chem.*, **70**, 3168 (2005).
2. B. Sukornick, *Org. Synth., Collect. Vol. V*, 1074 (1973).
3. F. Weygand and E. Frauendorfer, *Chem. Ber.*, **103**, 2437 (1970).
4. D. Urabe, K. Sugino, T. Nishikawa, and M. Isobe, *Tetrahedron Lett.* **45**, 9405 (2004).

Trifluoroacetamide (TFA): R_2NCOCF_3 (Chart 9)

The trifluoroacetamide group is one of the more useful amides because of the ease in which it may be removed under mildly basic conditions. Otherwise, it is stable to acidic conditions such as TFA and single electron reducing agents such as Na/anthracene, but is reduced with hydride reducing agents.

Formation

1. CF_3CO_2Et, Et_3N, CH_3OH, 25°C, 15–45 h, 75–95% yield.[1] A polymeric version of this approach has also been developed.[2] This reagent selectively protects a primary amine in the presence of a secondary amine.[3] With DMAP catalysis primary anilines are efficiently acylated (75–98% yield).[4]
2. $(CF_3CO)_2O$, 18–crown-6, Et_3N, 95% yield.[5] Complex formation of a primary amine with 18–crown-6 allows selective acylation of a secondary amine.
3. CF_3COO-succinimidyl, CH_2Cl_2, 0°C, 85% yield.[6] These conditions selectively introduced the TFA group onto a primary amine in the presence of a secondary amine.
4. (Trifluoroacetyl)benzotriazole, THF, rt, 85–100% yield.[7,8] The reagent can be used to prepare trifluoroacetate esters.
5. TFA, Ph_3P, NBS, CH_2Cl_2, Pyr, 81–99% yield. This methodology can be used for the preparation of other amides from simple carboxylic acids.[9]
6. $(CF_3CO)_2O$, Pyr, CH_2Cl_2.[10]
7. $CF_3CO_2C_6F_5$, Pyr, DMF, 52–92% yield.[11]
8. CF_3CO_2H, MeOH, rt, 97% yield.[12]
9. 2-Trifluoroacetoxypyridine, ether, 20°C, 30 min, 93% yield.[13]
10. Dodecyltrifluorothioacetate, sat. aq. $NaHCO_3$, CH_3CN, TBAB, 50°C, 71–92% yield. This method was developed for the protection of amino acids.[14]

Cleavage

1. K_2CO_3 or Na_2CO_3, MeOH, H_2O, rt, 55–95% yield.[6,15] Note that the trifluoroacetamide has been cleaved in the presence of a methyl ester, which illustrates the ease of hydrolysis of the trifluoroacetamide group.[16]

2. $LiOH \cdot H_2O$, THF, MeOH, H_2O, rt, 24 h, 100% yield.[17]

also hydrolyzed

3. KOH, triethylbenzylammonium chloride, water, CH_2Cl_2, 75–89% yield.[18]
4. NH_3, MeOH.[19]
5. Lewatit 500, MeOH, 96% yield.[12]
6. By phase transfer hydrolysis: KOH, Et_3BnNBr, H_2O, CH_2Cl_2 or ether, 75–95% yield.[20]
7. 0.2 N $Ba(OH)_2$, CH_3OH, 25°C, 2 h, 79% yield.[21]
8. $NaBH_4$, EtOH, 20°C, or 60°C, 1 h, 60–100% yield.[22,23]
9. $PhCH_2NEt_3OH$, CH_2Cl_2, −40°C, 48 h.[10]
10. HCl, MeOH, 65°C, 24 h.[24]

1. T. J. Curphey, *J. Org. Chem.*, **44**, 2805 (1979).

2. P. I. Svirskaya, C. C. Leznoff, and M. Steinman, *J. Org. Chem.*, **52**, 1362 (1987).

3. D. Xu, K. Prasad, O. Repic, and T. J. Blacklock *Tetrahedron Lett.*, **36**, 7357 (1995); M. C. O'Sullivan and D. M. Dalrymple, *Tetrahedron Lett.*, **36**, 3451 (1995).

4. M. Prashad, B. Hu, D. Har, O. Repic, and T. J. Blacklock, *Tetrahedron Lett.*, **41**, 9957 (2000).

5. A. G. M. Barrett and J. C. A. Lana, *J. Chem. Soc., Chem. Commun.*, 471 (1978).

6. R. J. Bergeron and J. J. McManis, *J. Org. Chem.*, **53**, 3108 (1988); T. S. Rao, S. Nampalli, P. Sekher, and S. Kumar, *Tetrahedron Lett.*, **43**, 7793 (2002); A. R. Katritzky, H.-Y. He, and K. Suzuki, *J. Org. Chem.*, **65**, 8210 (2000).

7. A. R. Katritzky, B. Yang, and D. Semenzin, *J. Org. Chem.*, **62**, 726 (1997).

8. J. M. Ndungu, X. Gu, D. E. Gross, J. Ying, and V. J. Hruby, *Tetrahedron Lett.*, **45**, 3245 (2004).

9. P. Froyen, *Tetrahedron Lett.*, **38**, 5359 (1997).

10. S. G. Pyne, *Tetrahedron Lett.*, **28**, 4737 (1987).

11. L. M. Gayo and M. J. Suto, *Tetrahedron Lett.*, **37**, 4915 (1996).

12. L. F. Tietze, C. Schneider, and A. Grote, *Chem. Eur. J.*, **2**, 139 (1996).

13. T. Keumi, M. Shimada, T. Morita, and H. Kitajima, *Bull. Chem.Soc. Jpn*, **63**, 2252 (1990).

14. M. R. Hickey, T. D. Nelson, E. A. Secord, S. P. Allwein, and M. H. Kress, *Synlett*, 255 (2005).

15. H. Newman, *J. Org. Chem.*, **30**, 1287 (1965); J. Quick and C. Meltz, *J. Org. Chem.*, **44**, 573 (1979); M. A. Schwartz, B. F. Rose, and B. Vishnuvajjala, *J. Am. Chem. Soc.*, **95**, 612 (1973).

16. D. L. Boger and D. Yohannes, *J. Org. Chem.*, **54**, 2498 (1989).

17. R. Baker and J. L. Castro, *J. Chem. Soc., Perkin Trans 1*, 47 (1990).

18. D. Albanese, F. Corcella, D. Landini, A. Maia, and M. Penso, *J. Chem. Soc., Perkin Trans.1* 247 (1997).

19. M. Imazawa and F. Eckstein, *J. Org. Chem.*, **44**, 2039 (1979).

20. D. Albanese, F. Corcella, D. Landini, A. Maia, and M. Penso, *J. Chem. Soc., Perkin Trans. 1*, 247 (1997).

21. F. Weygand and W. Swodenk, *Chem. Ber.*, **90**, 639 (1957).

22. F. Weygand and E. Frauendorfer, *Chem. Ber.*, **103**, 2437 (1970).

23. Z. H. Kudzin, P. Lyzwa, J. Luczak, and G. Andrijewski, *Synthesis*, 44 (1997).

24. S. B. King and B. Ganem, *J. Am. Chem. Soc.*, **116**, 562 (1994).

Phenylacetamide: $R_2NCOCH_2C_6H_5$

This amide, readily formed from an amine and the anhydride[1] or enzymatically using penicillin amidase[2], is readily cleaved by penicillin acylase (pH 8.1, N-methylpyrrolidone, 65–95% yield). This deprotection procedure works on peptides,[3-5] phosphorylated peptides,[6] and oligonucleotides[7] as well as on nonpeptide substrates.[8,9] The deprotection of racemic phenylacetamides with penicillin acylase can result in enantiomer enrichment of the cleaved amine and the remaining amide.[10] An immobilized form of penicillin G acylase has been developed.[11]

1. A. R. Jacobson, A. N. Makris, and L. M. Sayre, *J. Org. Chem.*, **52**, 2592 (1987).

2. C. Ebert, L. Gardossi and P. Linda, *Tetrahedron Lett.*, **37**, 9377 (1996).

3. R. Didziapetris, B. Drabnig, V. Schellenberger, H.-D. Jakubke, and V. Svedas, *FEBS Lett.*, **287**, 31 (1991).

4. For a review on the use of enzymes in protective group manipulation in peptide chemistry, see J. D. Glass in *The Peptides*, Vol. 9, S. Undenfriend and J. Meienhofer, Eds., Academic Press, 1987, p. 167.

5. D. Sebastian, A. Heuser, S. Schulze, and H. Waldmann, *Synthesis*, 1098 (1997).

6. H. Waldmann, A. Heuser, and S. Schulze, *Tetrahedron Lett.*, **37**, 8725 (1996).

7. H. Waldmann and A. Reidel, *Angew. Chem., Int. Ed. Engl.*, **36**, 647 (1997); V. Jungmann and H. Waldmann, *Tetrahedron Lett.*, **39**, 1139 (1998).

8. H. Waldmann, *Tetrahedron Lett.*, **29**, 1131 (1988) and references cited therein.

9. V. M. Vrudhula, H. P. Svensson, and P. D. Senter, *J. Med. Chem.*, **38**, 1380 (1995).

10. A. L. Margolin, *Tetrahedron Lett.*, **34**, 1239 (1993).

11. T. F. Favino, G. Fronza, C. Fuganti, D. Fuganti, P. Grasselli, and A. Mele, *J. Org. Chem.*, **61**, 8975 (1996); N. C. R. van Straten, H. L. Duynstee, E. de Vroom, G. A. van der Marel, and J. H. van Boom, *Liebigs Ann./Recueil*, 1215 (1997).

3-Phenylpropanamide: $R_2NCOCH_2CH_2C_6H_5$ (Chart 9)

A 3-phenylpropanamide, prepared from a nucleoside, is hydrolyzed under mild conditions by α-chymotrypsin (37°C, pH 7, 2–12 h).[1]

1. H. S. Sachdev and N. A. Starkovsky, *Tetrahedron Lett.*, **10**, 733 (1969).

Pent-4-enamide: $CH_2=CHCH_2CH_2C(O)NR_2$

Formation

1. $(CH_2=CHCH_2CH_2CO)_2O$, Pyr, CH_2Cl_2, MeOH, H_2O, 90–99% yield.[1]
2. $CH_2=CHCH_2CH_2CO_2CH_2CN$, 3-methyl-3-pentanol, Subtilisin Carlsberg. These conditions were used to resolve a chiral amine (43% yield, 97% ee).[2]

Cleavage

1. I_2, THF, H_2O, 83–94% yield.[1–3]

2. Dibromantin, CH_3CN, H_2O, rt, 75–80% yield.[4]

1. R. Madsen, C. Roberts, and B. Fraser-Reid, *J. Org. Chem.*, **60**, 7920 (1995).
2. S. Takayama, W. J. Moree, and C.-H. Wong, *Tetrahedron Lett.*, **37**, 6287 (1996).
3. M. Lodder, S. Golovine, A. L. Laikhter, V. A. Karginov, and S. M. Hecht, *J. Org. Chem.*, **63**, 794 (1998).
4. J. Limanto, A. Shafiee, P. N. Devine, V. Upadhyay, R. A. Desmond, B. R. Foster, D. R. Gauthier, Jr., R. A. Reamer, and R. P. Volante, *J. Org. Chem.*, **70**, 2372 (2005).

Picolinamide: R_2NCO-2-pyridyl (Chart 9)

The picolinamide is prepared in 95% yield from picolinic acid/DCC and an amino acid, and hydrolyzed in 75% yield by aqueous $Cu(OAc)_2$[1] or by electrochemical reduction (sulfuric acid, MeOH, 20°C, 20–94% yield).[2]

3-Pyridylcarboxamide: R_2NCO-3-pyridyl

The 3-pyridylcarboxamide, prepared from the anhydride (pyridine, 99% yield), is cleaved (55–86% yield) by basic hydrolysis (0.5 *M* NaOH, rt) after quaternization of the pyridine nitrogen with methyl iodide.[3]

1. A. K. Koul, B. Prashad, J. M. Bachhawat, N. S. Ramegowda, and N. K. Mathur, *Synth. Commun.*, **2**, 383 (1972).
2. N. Auzeil, G. Dutruc-Rosset, and M. Largeron, *Tetrahedron Lett.*, **38**, 2283 (1997).
3. S. Ushida, *Chem. Lett.*, 59, (1989).

N-Benzoylphenylalanyl Derivative: R$_2$NCOCH(NHCOC$_6$H$_5$)CH$_2$C$_6$H$_5$

This derivative, prepared from an amino acid and the acyl azide, is selectively cleaved in 80% yield by chymotrypsin.[1]

1. R. W. Holley, *J. Am. Chem. Soc.*, **77**, 2552 (1955).

Benzamide: R$_2$NCOC$_6$H$_5$ (Chart 9)

Formation

1. PhCOCl, Pyr, 0°C, high yield.[1]
2. PhCOCN, CH$_2$Cl$_2$, −10°C, 92% yield.[2] This reagent readily acylates amines in the presence of alcohols.
3. PhCOCF(CF$_3$)$_2$, Me$_2$NCH$_2$CH$_2$NMe$_2$ (TMEDA), 25°C, 30 min, high yield.[3]

4.
N+ SCOPh aq. NaHCO$_3$ or aq. NaOH, good yields.[4]
CH$_3$ Cl⁻

5. (PhCO)$_2$NOCH$_3$, DMF, H$_2$O, or dioxane, 3–26 h, 66–89%.[5] The reagent is selective for primary amines.
6. *N*-Benzoyltetrazole, CH$_3$CN, DMAP, 65°C, 72–90% yield. This method was used to protect the the exocyclic amino group of nucleic acid bases.[6]
7. 2-Fluoro-*N*-benzoy-*N*-mesylaniline, 0°C, THF, 24 h, 93% yield.[7] Other acyl groups may be introduced similarly. 2-Chloro-*N*,*N*-dibenzoylaniline may also be used in this capacity.[8]

8. 2-Benzoyl-4,5-dichloropyridazin-3-one, CH$_2$Cl$_2$ or THF, 80–99% yield. The use of other 2-acylpyridazin-3-ones are similarly effective acylating agent.[9]
9. The following provides a method for the monoprotection of symmetrical primary amines by using 9-BBN to complex one of the amines.[10]

Ratio = 19.7:1

Cleavage

1. 6 *N* HCl, reflux, 48 h or HBr, AcOH, 25°C, 72 h, 80% yield.[11]
2. (HF)$_n$·Pyr, 25°C, 60 min, 100% yield.[12] Polyhydrogen fluoride/pyridine cleaves most of the protective groups used in peptide synthesis.
3. Electrolysis, −2.3 V, Me$_4$NX, CH$_3$OH, 70 min, 60–90% yield.[13]
4. (Me$_2$CHCH$_2$)$_2$AlH, PhCH$_3$, −78°, 80% yield.[14] Since the *N*-benzoyl group in this substrate could not be removed by hydrolysis, a less selective reductive cleavage with diisobutylaluminum hydride was used.
5. Hydrazine, EtOH, 85% yield.[15] Note that the cleavage of an anilide and a benzoylpyrrole is much more facile than that of a typical aliphatic benzamide.

6. Ph$_3$P, Cl$_2$, TEA, CH$_2$Cl$_2$, −30°C, then ethylene glycol, 90% yield. This is a general method applicable to a variety of amides.[16]
7. LiEt$_3$BH, THF, 0°C to rt, 76–99% yield. The method is good for disubstituted methyl and benzyl carbamates, acetamides and benzamides.[17]

1. E. White, *Org. Synth., Coll. Vol. V*, 336 (1973).
2. S.-I. Murahashi, T. Naota, and N. Nakajima, *Tetrahedron Lett.*, **26**, 925 (1985).
3. N. Ishikawa and S. Shin-ya, *Chem. Lett.*, **5**, 673 (1976).
4. M. Yamada, Y. Watabe, T. Sakakibara, and R. Sudoh, *J. Chem. Soc., Chem. Commun.*, 179 (1979).
5. Y. Kikugawa, K. Mitsui, T. Sakamoto, M. Kawase, and H. Tamiya, *Tetrahedron Lett.*, **31**, 243 (1990).
6. B. Bhat and Y. S. Sanghvi, *Tetrahedron Lett.*, **38**, 8811 (1997).
7. K. Kondo, E. Sekimoto, K. Miki, and Y. Murakami, *J. Chem. Soc. Perkin Trans. 1*, 2973 (1998).
8. K. Kondo and Y. Murakami, *Chem. Pharm. Bull.*, **46**, 1217 (1996).
9. Y.-J. Kang, H.-A. Chung, J.-J. Kim, and Y.-J. Yoon, *Synthesis*, 733 (2002).
10. Z. Zhang, Z. Yin, N. A. Meanwell, J. F. Kadow, and T. Wang, *Org. Lett.*, **5**, 3399 (2003).

11. D. Ben-Ishai, J. Altman, and N. Peled, *Tetrahedron*, **33**, 2715 (1977); P. Hughes and J. Clardy, *J. Org. Chem.*, **53**, 4793 (1988).

12. S. Matsuura, C.-H. Niu, and J. S. Cohen, *J. Chem. Soc., Chem. Commun.*, 451 (1976).

13. L. Horner and H. Neumann, *Chem. Ber.*, **98**, 3462 (1965).

14. J. Gutzwiller and M. Uskokovic, *J. Am. Chem. Soc.*, **92**, 204 (1970).

15. D. L. Boger and K. Machiya, *J. Am. Chem. Soc.*, **114**, 10056 (1992); D. L. Boger, J. A. McKie, T. Nishi, and T. Ogiku, *J. Am. Chem. Soc.*, **119**, 311 (1997).

16. A. Spaggiari, L. C. Blaszczak, and F. Prati, *Org. Lett.*, **6**, 3885 (2004).

17. H. Tanaka and K. Ogasawara, *Tetrahedron Lett.*, **43**, 4417 (2002).

p-Phenylbenzamide: $R_2NCOC_6H_4$-*p*-C_6H_5

The phenylbenzamide is prepared from the acid chloride in the presence of Et_3N (86% yield) and can be cleaved with 3% Na(Hg) (MeOH, 25°C, 4 h, 81% yield).[1] Most amides react only slowly with Na(Hg). Phenylbenzamides are generally crystalline compounds, an aid in purification.[2]

1. R. B. Woodward and 48 co-workers, *J. Am. Chem. Soc.* **103**, 3210 (1981).

2. R. M. Scribner, *Tetrahedron Lett.*, **17**, 3853 (1976).

Assisted Cleavage

A series of amides have been prepared as protective groups that are cleaved by intramolecular cyclization after activation, by reduction of a nitro group, or by activation by other chemical means. These groups have not found much use. A significant consideration when examining the use of any of these amides is that the nature of the amine will have a substantial effect on the rate of deprotection. Amines, such as those of the nucleobases and many aniline derivatives whose basicity is much reduced compared to typical primary and secondary aliphatic amines, tend to be cleaved at a much greater rate. Structural effects such as the "trimethyl lock"[1] effect exert considerable influence on the effectiveness of the deprotection event. Typically amides are quite planer, but when stereochemical constraints force them out of planarity they are also much easier to cleave. The concept of assisted cleavage is generalized in the following scheme:

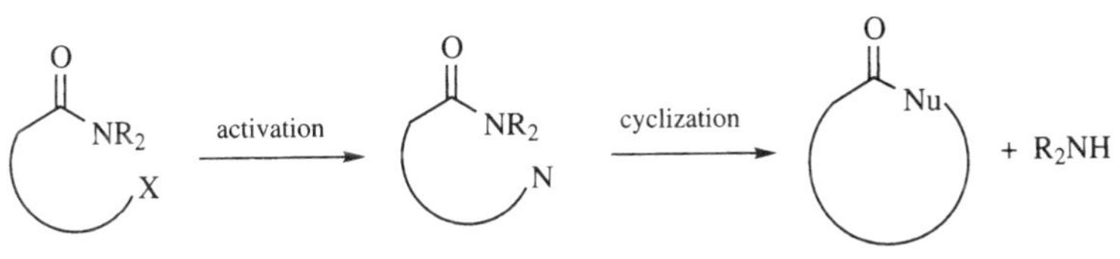

Amide Cleavage Induced by Nitro Group Reduction

In this series of compounds any reagent that is capable of reducing a nitro group should be capable of initiating deprotection.

1. *o*-Nitrophenylacetamide[2] (Chart 9)

2. 2,2-Dimethyl-2-(*o*-nitrophenyl)acetamide.[3] Cleaved by electrolytic reduction to the hydroxylamine.

3. *o*-Nitrophenoxyacetamide[4] (Chart 9)

4. 3-(*o*-Nitrophenyl)propanamide[5]

5. 2-Methyl-2-(*o*-nitrophenoxy)propanamide[2,6] (Chart 9)

6. 3-Methyl-3-nitrobutanamide[7]

7. *o*-Nitrocinnamide[8] (Chart 9)

8. *o*-Nitrobenzamide[9,10]

9. 3-(4-*t*-Butyl-2,6-dinitrophenyl)-2,2-dimethylpropanamide[11]

Amide Cleavage Induced by Release of an Alcohol

In this series of amides, hydrolysis or aminolysis of a simple ester, cleavage of a silyl group, a *cis/trans* isomerization, or reduction of a quinone to a hydroquinone exposes an alcohol that then induces deprotection by intramolecular addition to the amide carbonyl.

1. *o*-(Benzoyloxymethyl)benzamide (BMB).[12] Cleavage is initiated by ester hydrolysis.

2. 2-(Acetoxymethyl)benzamide (AMB).[13,14] Cleavage is initiated by ester hydrolysis.

3. 2-[(*t*-Butyldiphenylsiloxy)methyl]benzoyl (SiOMB).[15,16] Cleavage is induced by silyl ether cleavage.

4. 3-(3',6'-Dioxo-2',4',5'-trimethylcyclohexa-1',4'-diene)-3,3-dimethylpropionamide (Q): The application of this well-known acid [3-(3',6'-dioxo-2',4',5'-trimethylcyclohexa-1',4'-diene)-3,3-dimethylpropionic acid] to protection of the amino function for peptide synthesis has been examined. Reduction of the quinone with sodium dithionite causes rapid "trimethyl lock"[1] -facilitated ring closure with release of the amine.[17,18]

5. *o*-Hydroxy-*trans*-cinnamide. The amide is formed from the acid and an amine using the classical DCC/HOBt coupling protocol (67–98% yield). It is cleaved by photochemical isomerization at 365 nm in MeOH/AcOH to release the amine and coumarin (100% yield).[19] The disadvantage of the method is that an acidic hydrogen is still present.

Amides Cleaved by Other Chemical Reactions

1. 2-Methyl-2-(*o*-phenylazophenoxy)propanamide[20] (Chart 9). Cleaved by reduction.

2. 4-Chlorobutanamide[21] (Chart 9). Cleaved by cyclization induced with silver ion.

3. Acetoacetamide[22] (Chart 9). Cleaved with hydrazine.

4. **3-(*p*-Hydroxyphenyl)propanamide**[23] (Chart 9). Cleaved by oxidation with NBS.

5. **(*N'*-Dithiobenzyloxycarbonylamino)acetamide.**[24] Cleaved by TFA-induced cyclization.

6. **N-Acetylmethionine Derivative**[25] (Chart 9). Cleaved by alkylation of the thioether with iodoacetamide followed by cyclization.

1. B. Wang, M. G. Nicolaou, S. Liu, and R. T. Borchardt, *Biorg. Chem.*, **24**, 39 (1996).

2. F. Cuiban, *Rev. Roum. Chim.*, **18**, 449 (1973).

3. Y. Jiang, J. Zhao, and L. Hu, *Tetrahedron Lett.*, **43**, 4589 (2002).

4. R. W. Holley and A. D. Holley, *J. Am. Chem. Soc.*, **74**, 3069 (1952).

5. I. D. Entwistle, *Tetrahedron Lett.*, **20**, 555 (1979).

6. C. A. Panetta, *J. Org. Chem.*, **34**, 2773 (1969).

7. T.-L. Ho, *Synth. Commun.*, **10**, 469 (1980).

8. G. Just and G. Rosebery, *Synth. Commun.*, **3**, 447 (1973).

9. A. K. Koul, J. M. Bachhawat, B. Rashad, N. S. Ramegowda, A. K. Mathur, and N. K. Mathur, *Tetrahedron*, **29**, 625 (1973).

10. A. Chibani and Y. Bendaoud, *J. Soc. Alger. Chim.*, **11**, 17 (2001).

11. F. Johnson, I. Habus, R. G. Gentles, S. Shibutani, H.-C. Lee, C. R. Iden, and R. Rieger, *J. Am. Chem. Soc.*, **114**, 4923 (1992).

12. B. F. Cain, *J. Org. Chem.*, **41**, 2029 (1976).

13. W. H. A. Kuijpers, J. Huskens, and C. A. A. van Boeckel, *Tetrahedron Lett.*, **31**, 6729 (1990).

14. W. H. A. Kuijpers, E. Kuyl-Yeheskiely, J. H. van Boom, and C. A. A. van Boeckel, *Nucleic Acids Res.*, **21**, 3493 (1993).

15. C. M. Dreef-Tromp, E. M. A. Van Dam, H. Van den Elst, J. E. Van den Boogaart, G. A. Van der Marel, and J. H. Van Boom, *Recl. Trav. Chim. Pays-Bas*, **110**, 378 (1991).

16. C. M. Dreef-Tromp, E. M. A. Van Dam, H. Van den Elst, G. A. Van der Marel, and J. H. Van Boom, *Nucleic Acids Res.*, **18**, 6491 (1990).

17. L. A. Carpino and F. Nowshad, *Tetrahedron Lett.*, **34**, 7009 (1993).

18. B. Wang, S. Liu, and R.T. Borchardt, *J. Org. Chem.*, **60**, 539 (1995).

19. B. Wang and A. Zheng, *Chem. Pharm. Bull.*, **45**, 715 (1997).

20. C. A. Panetta and A.-U. Rahman, *J. Org. Chem.*, **36**, 2250 (1971).

21. H. Peter, M. Brugger, J. Schreiber, and A. Eschenmoser, *Helv. Chim. Acta*, **46**, 577 (1963).

22. C. Di Bello, F. Filira, V. Giormani, and F. D'Angeli, *J. Chem. Soc. C*, 350 (1969).

23. G. L. Schmir and L. A. Cohen, *J. Am. Chem. Soc.*, **83**, 723 (1961); L. Farber and L. A. Cohen, *Biochemistry*, **5**, 1027 (1966).

24. F. E. Roberts, *Tetrahedron Lett.*, **20**, 325 (1979).

25. W. B. Lawson, E. Gross, C. M. Foltz, and B. Witkop, *J. Am. Chem. Soc.*, **84**, 1715 (1962).

Bisprotection of Amines

A number of protective groups have been developed that simultaneously protect both sites of a primary nitrogen. These may prove to be useful for cases where acidic hydrogens on nitrogen cannot be tolerated. The azide group is also used for this purpose.

4,5-Diphenyl-3-oxazolin-2-one (Chart 8)

Formation[1]

DMF, 0.5 h; CF$_3$CO$_2$H, 67–85% yield.

Cleavage

1. H$_2$/Pd–C, aq. HCl, 25°C, 12 h, quantitative. [1,2]
2. Na/NH$_3$, 75–85% yield.[1]
3. m-ClC$_6$H$_4$CO$_3$H, then water, 70% yield.[1]
4. O$_2$, photolysis, −30°C, then Zn, AcOH, quant.[3]

N-Phthalimide: (Chart 9)

The phthalimide group is often used for the bisprotection of primary amines. In most cases, it is readily introduced, but it does have the liability that it is quite sensitive to nucleophilic reagents, which in the case of mild aqueous base results in ring opening. In that case, the ring may be reclosed simply by refluxing the acid in anhydrous alcohols. Propanol is particularly effective, since the water generated may be removed by a very efficient azeotropic distillation.[4] The phthalimide group is photochemically active,[5] which may make it incompatible with some of the photochemically removable protective groups. The phthalimide group has been tested for the protection of adenine, cytosine, and guanine in oligonucleotide synthesis.[6] The phthalimide is normally considered inert toward hydrogenation, but it has been reported to reduce to the lactam by hydrogenation over Pd–C in acetic acid.[7]

Formation

1. Phthalic anhydride, CHCl$_3$, 70°C, 4 h, 85–93% yield.[8]
2. Phthalic anhydride, pyridine then Ac$_2$O, 97% yield.[9]
3. Phthalic anhydride, TaCl$_5$–SiO$_2$, 5 min, 88–92% yield.[10]
4. Phthalic anhydride, HMDS, rt, 1 h, then reflux with ZnBr$_2$ 1 h, 94% yield.[11]
5. Phthalic anhydride, [bmim]PF$_6$ an ionic liquid, 8 h, 90–97% yield. This method was particularly good for anilines.[12]
6. o-(CH$_3$OOC)C$_6$H$_4$COCl, Et$_3$N, THF, 0°C, 2 h, 90–95% yield.[13]
7. Phthalimide−CO$_2$Et, aq. Na$_2$CO$_3$, 25°C, 10–15 min, 85–95% yield.[14] This reagent can be used to protect selectively primary amines in the presence of secondary amines.[15]
8. 3-Chloro-3-(dimethoxyphosphoryl)isobenzofuran-1(3H)-one, DIPEA, CH$_3$CN, H$_2$O, rt, 10 min, 77% yield. The reagent is readily prepared from phthaloyl chloride and (MeO)$_3$P.[16]

9. 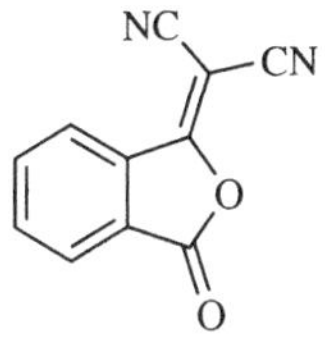 65–91% yield.[17]

10.

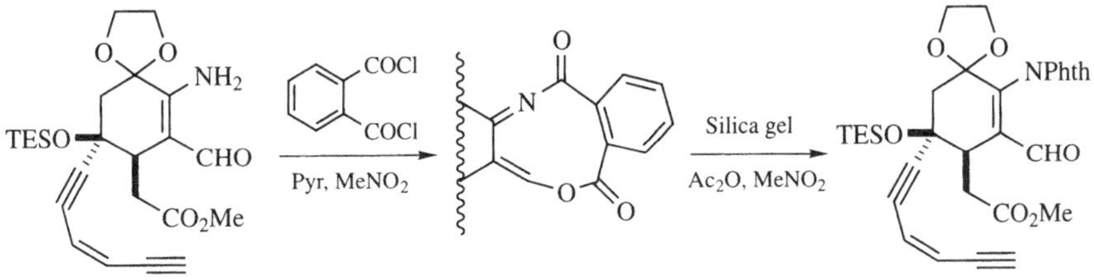

Ref. 18

11. Methyl 2-((succinimidooxy)carbonyl)benzoate (MSB), CH$_3$CN, H$_2$O, rt, 3–6 h, 65–100% yield. This method was developed specifically for the protection of amino acids and peptides without racemization.[19]
12. Monomethyl phthalate, BOP, ZnCl$_2$, DIPEA, CH$_3$CN, sonication, 16 h, 53–95% yield. These conditions result in racemization free protection of amino acid amides and esters.[20] PyBOP can also be used as a dehydrating agent.[21]

Cleavage

1. Hydrazine, EtOH, 25°C, 12 h; H$_3$O$^+$, 76% yield.[8,22] Hydrazine can oxidize to form diimide which will reduce double bonds. This was observed during the deprotection in the following scheme. The problem was solved by including a sacrificial alkene to scavenge any diimide that was formed.[23]

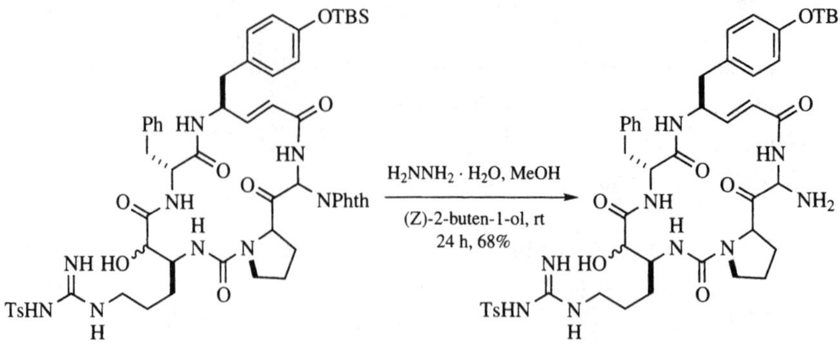

2. MeNHNH$_2$. This reagent was used as a replacement for hydrazine to prevent diimide formation which resulted in acetylene reduction.[24]

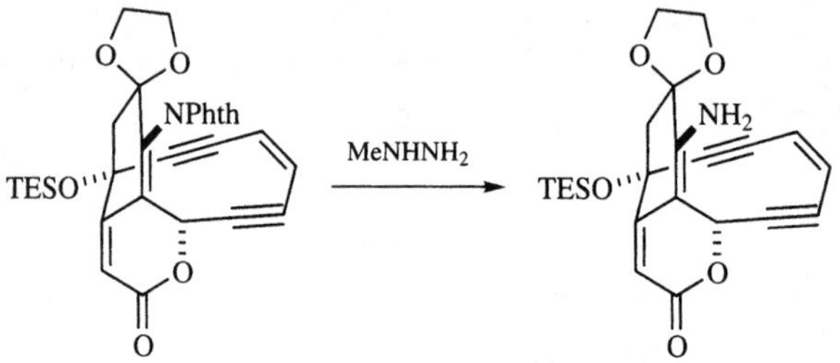

3. PhNHNH$_2$, n-Bu$_3$N, reflux, 2 h, 83% yield.[25]

4. Na$_2$S·H$_2$O, H$_2$O, THF, 68–90% yield; DCC(-H$_2$O), 67–97% yield; hydrazine; dil. HCl, 55–95% yield.[26] This method is used to cleave N-phthalimido penicillins; hydrazine attacks an intermediate phthalisoimide instead of the azetidinone ring. With a β-lactam the typical hydrazinolysis is not always usable because of the reactivity of the azetidinone carbonyl. The following scheme provides an example.[27]

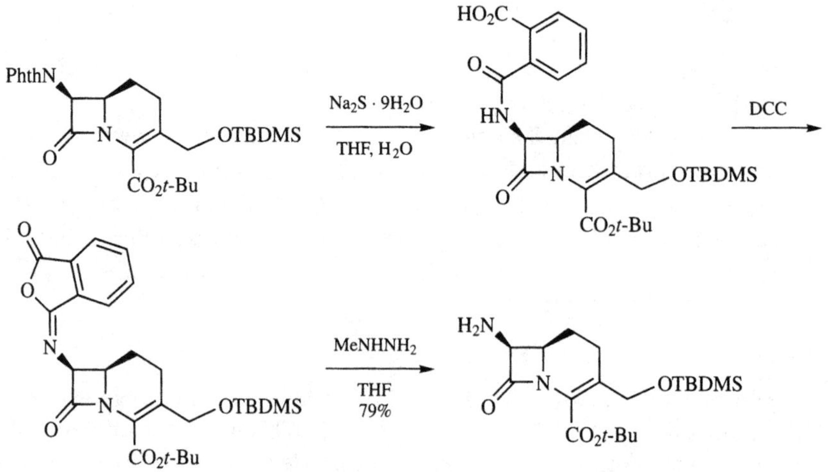

On the other hand, there are cases where hydrazinolysis has been effective.[28]

5. NaBH$_4$, 2-propanol, H$_2$O (6:1); AcOH, pH 5, 80°C, 5–8 h.[29,30] This method was reported to be superior in cases where hydrazine proved to be inefficient.

6. MeNH$_2$, EtOH, rt, 5 min, then heat, 2.5 h, 89% yield.[31] Butylamine has also been used.[32]

7. (a) Base, H$_2$O, CH$_3$CN. (b) 0.2—pH 8 buffer, phthalyl amidase.[33]

8. Me$_2$NCH$_2$CH$_2$CH$_2$NH$_2$, MeOH, TEA, 5°C, 24 h, 60% yield.[34]

9. HONH$_2$, MeONa, MeOH, >72% yield.[35]

10. Hydrazine acetate, MeOH, reflux, >82% yield.[36]

11. The phthalimido group is susceptible to basic reagents and thus must ocassionally be protected. This is accomplished by treatment with pyrrolidine to open the ring (>90%). It can be closed by treatment with HF, B(OH)$_3$, THF, H$_2$O, 73–99% yield.[37]

12. MsOH, HCO$_2$H.[38]

13. Ethylenediamine, butanol, 90°C, 67–96% yield.[39] These conditions were used when heating with butylamine failed to give clean conversions.

14. Diaion WA-20, EtOH, H$_2$O, 80–90°C, 1 h, 87–92% yield.[40]

N-Dichlorophthalimide (DCP or DCPhth)

The dichlorophthalimide group has been examined for 2-amino protection in carbohydrate synthesis. It is intermediate in stability toward base when comparing the Phth, DCP, and TCP groups.[41,42]

Formation

Dichlorophthalic anhydride, TEA, and ClCH$_2$CH$_2$Cl, followed by ring closure with Ac$_2$O, pyridine, 94% yield.[42]

Cleavage

H$_2$NNH$_2$·AcOH, EtOH, 70°C, >82% yield.[43] With these conditions the DCP group can be removed in the presence of acetates.

N-Tetrachlorophthalimide (TCP)

The use of this group was developed to improve the quality and mildness of the cleavage reaction in the synthesis of complex amino sugars.[44] It is possible to remove acetates in the presence of this group with Mg(OMe)$_2$/MeOH.[45] The TCP is stable to piperidine and thus is compatible with Fmoc technology for peptide synthesis.[46]

Formation

1. Tetrachlorophthalic anhydride, microwaves, 90% yield.[47]
2. Tetrachlorophthalic anhydride, TEA; Ac$_2$O, Pyr.[48]

Cleavage

1. Ethylenediamine, CH$_3$CN, THF, EtOH, 60°C.[47,49] The phthalimide group and *O*-acetate are not cleaved with this reagent.[50] These conditions will cause acetate migration in carbohydrates, but this can be avoided if the acetates are replaced with benzoates.[51]

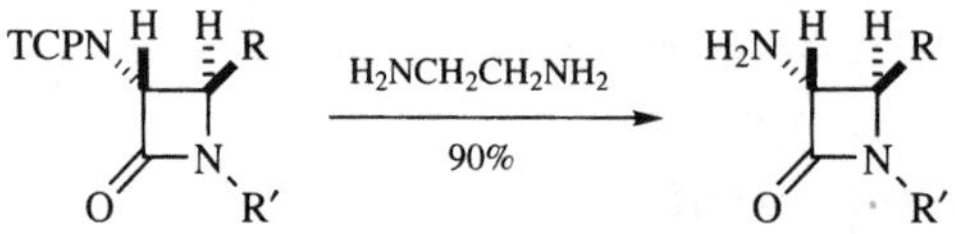

2. Polymer-NH(CH$_2$)$_x$NH$_2$, (x = 2, 4, 6), BuOH, 85°C, 92–96% yield. The polymer supported amine helps in the final purification of oligosaccharides that have used the TCP group for NH$_2$ protection.[52]
3. (a) NaBH$_4$, (b) AcOH, >60–80% yield.[53,54] This method first reduces the imide to an amide alcohol, which, upon acid treatment, releases the amine and a lactone.
4. Hydrazine, DMF, 2 h, 100% yield. This method was used to remove the TCP group from polymer supported peptides.[55]

N-4-Nitrophthalimide

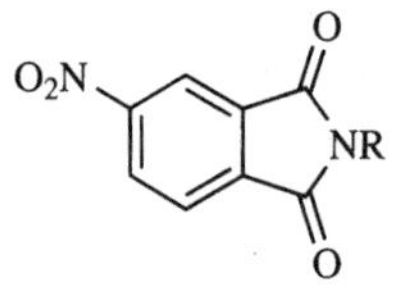

The 4-nitro-*N*-phthalimide, prepared by heating the amine with the anhydride to 130°C for 30 min, is cleaved with MeNHCH$_2$CH$_2$NH$_2$ (71–92% yield). These cleavage conditions were compatible with cephalosporins, where the phthalimide was removed in 92% yield at −50°C in 30 min.[56]

N-Thiodiglycoloyl Amine (TDG−NR):

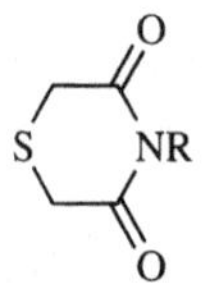

The TDG group was developed for the protection of glucosamine. It is introduced in a 2 step process from the amine and the anhydride followed by ring closure with

Ac$_2$O. It is cleaved by methanolysis with NaOMe/MeOH to open the ring followed by reductive desulfurization with Bu$_3$SnH/AIBN. This leaves the amine protected as an acetamide.[57]

N-Dithiasuccinimide (Dts−NR) (Chart 9)

The Dts group can be used as a participating group in carbohydrate synthesis to direct β-glycosidations of the glucosamine derivative.[58]

Formation

1. EtOCS$_2$CH$_2$CO$_2$H or EtOCS$_2$CSOEt; ClSCOCl, 0–45°C, 70–90% yield.[59-61]
2. PEG(2000)-OCS$_2$CH$_2$CONH$_2$; TMSNH(CO)NHTMS; ClCOSCl.[59]
3. A bis(silyl)amine route to Dts amines.[62]

Cleavage

The Dts group is cleaved by treatment with a thiol and base, e.g., HOCH$_2$CH$_2$SH, Et$_3$N, 25°C, 5 min, HSCH$_2$C(O)NHMe, Pyr, 5 min.[63] Dithiothreitol (DIPEA, CH$_2$Cl$_2$, 87–98% yield) seems to be the most trouble-free method for Dts deprotection.[61b] In the presence of an azide, the Dts group can be removed with NaBH$_4$[64] or with HSCH$_2$CH$_2$CH$_2$SH, (DIPEA, CH$_2$Cl$_2$, 94% yield)[65]; however, when dithiothreitol is used, the azide is reduced. The use of Zn (AcOH, Ac$_2$O, THF, 80–87% yield)[66] cleaves the Dts group in the presence of the extremely sensitive pentafluorophenyl ester.[61a]

The Dts group, stable to acidic cleavage of t-butyl carbamates (12 N HCl, AcOH, reflux; HBr, AcOH), to mild base (NaHCO$_3$), and to photolytic cleavage of o-nitro-benzyl carbamates, can be used in orthogonal schemes for protection of peptides.[63] The treatment of a Dts protected amine with Ph$_3$P in toluene at reflux in the presence of an alcohol such as benzyl alcohol converts it through the isocyanate to the Cbz protected amine (57–92% yield).[67] The Dts amine can also serve as a nitrogen source in the Mitsunobu reaction.[68]

N-2,3-Diphenylmaleimide

The diphenylmaleimide is prepared from the anhydride, 33–87% yield, and cleaved by hydrazinolysis, 65–75% yield.[63] It is stable to acid (HBr, AcOH, 48 h) and to mercuric cyanide. It is colored and easily located during chromatography, and has been prepared to protect steroidal amines and amino sugars.

N-2,3-Dimethylmaleimide (DMN−NR)

The DMN group has been used for the protection of the 2-amino group during carbohydrate synthesis.[69] It is introduced with 2,3-dimethylmaleic anhydride followed by ring closure with Ac$_2$O (55% yield). It is cleaved with NaOH (dioxane, H$_2$O then HCl, pH 3).

N-2,5-Dimethylpyrrole

This group is stable to strong base and LiAlH$_4$. It is also relatively nonnucleophilic, making it unreactive to acid chlorides.[70] It is stable to conditions used to cleave the phthalimide group and was shown to be effective for protection of the 2-amino group in glycoside synthesis.[71] It has also been used to protect anilines during nucleophilic aromatic substitutions when the more typical protective groups failed.[72]

Formation

1. CH$_3$C(O)CH$_2$CH$_2$C(O)CH$_3$, AcOH, 88% yield.[73,74]
2. α-Zr(KPO$_4$)$_2$, CH$_3$C(O)CH$_2$CH$_2$C(O)CH$_3$, neat, rt, 56–95% yield.[75]
3. Montmorillonite KSF or I$_2$, CH$_3$C(O)CH$_2$CH$_2$C(O)CH$_3$, neat, rt, 70–98% yield.[76]
4. CH$_3$C(O)CH$_2$CH$_2$C(O)CH$_3$, Bi(NO$_2$)$_3$·5H$_2$O, CH$_2$Cl$_2$, 70–96% yield.[77]
5. 1,5-hexadyne, Ti(NMe$_2$)$_2$(dpma), 100°C, 34–68% yield.[78]

Cleavage

1. H$_2$NOH·HCl, EtOH, H$_2$O, 73% yield.[73,79]
2. Ozone, −78°C, MeOH; NaBH$_4$; HCl, MeOH, H$_2$O.[80,81]
3. RuCl$_3$, NaIO$_4$, CH$_3$CN, CCl$_4$, H$_2$O, 71% yield.[82]

N-2,5-Bis(triisopropylsiloxy)pyrrole (BIPSOP)

These derivatives are formed from the succinimide by silylation (TIPSOTff, TEA, CH$_2$Cl$_2$, 0°C to rt, 68–87% yield). Deprotection is achieved by hydrolysis of the silyl groups followed by succinimide cleavage with hydrazine (EtOH, H$_2$O, reflux, 72% yield).[83] The succinimides were prepared by heating the amine with succinic anhydride followed by ring closure with AcCl or Ac$_2$O/NaOAc. They may also be prepared by reacting succinic anhydride with the amine and HMDS followed by ring closure with ZnBr$_2$ (reflux, 1 h).[11]

N-1,1,4,4-Tetramethyldisilylazacyclopentane Adduct (STABASE)

Formation/Cleavage[84–87]

1.

2. Me$_2$NSi(Me)$_2$CH$_2$CH$_2$Si(Me)$_2$NMe$_2$, ZnI$_2$, 140°C, 8 h, 72% yield.[88] The amine adducts are stable to the following reagents: *n*-BuLi (THF, −25°C), *s*-BuLi (Et$_2$O, −25°C); lithium diisopropylamide; saturated aqueous ammonium chloride; H$_2$O; MeOH; 2 *N* NaHCO$_3$; pyridinium dichromate, CH$_2$Cl$_2$; KF·2H$_2$O, THF, H$_2$O; saturated aqueous sodium dihydrogen phosphate. The derivative is not stable to strong acid or base; to pyridinium chlorochromate, CH$_2$Cl$_2$; or to NaBH$_4$, EtOH.

N-1,1,3,3-Tetramethyl-1,3-disilaisoindoline (Benzostabase, BSB)

Formation

1. 1,2-Bisdimethylsilylbenzene, Rh(Ph$_3$P)$_3$Cl, toluene, 120°C, 71–92% yield.[89]

2. 1,2-Bisdimethylsilylbenzene, CsF, HMPA, 71–92% yield.[89]

3. 1,2-Bisdimethylsilylbenzene, PdCl$_2$, toluene, rt, 69–87% yield.[90]

4. 1,2-Bis(diethylsilyl)benzene, PdCl$_2$ or CsF, DMPU, 50–86% yield. The tetraethyl analog (TEDI) was found to be more stable to acid than the tetramethyl derivative. Exposure of BnNBSB and BnNTEDI to a phosphate buffer of pH 2.5 resulted in a cleavage half-life of <0.4 min for the BSB derivative and a half-life of ~30 min for the TEDI analog. The TEDI group can also be introduced with the dibromide and TEA. [91]

5. A difluorinated analog was found to be somewhat more stable to acid than the BSB derivative, but overall it showed no major advantage to the original Benzostabase.[92]

Cleavage

Cleavage is achieved by simple acid hydrolysis. The Benzostabase group is reasonably stable to base (KOH, MeOH).[92]

N-Diphenylsilyldiethylene Group (DPSide—NR)

Formation/Cleavage

This group is compatible with BOC, Cbz, and phthalimide cleavage conditions: TFA, hydrogenolysis, and hydrazine, respectively.[93] The DPSide group is introduced by alkylation of the amine with the ditosylate in the presence of TEA in DMF (85–96% yield). Cleavage requires a combination of TBAF and CsF in DMF or THF (80–92% yield).

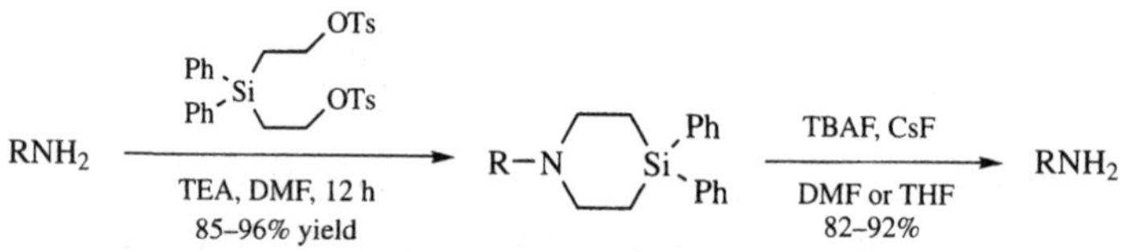

N-5-Substituted 1,3-Dimethyl-1,3,5-triazacyclohexan-2-one and

N-5-Substituted 1,3-Dibenzyl-1,3,5-triazacyclohexan-2-one

The triazone is stable to LiAlH$_4$; PtO$_2$/H$_2$/EtOH, 48 h; Pd-black/H$_2$/THF, 1 h; n-BuLi/THF/−40°C/30 min; PhMgBr/THF/−78°C/30 min; Wittig reagents; DIBAL/THF/rt/3 h; LiBH$_4$/THF/40°C; acylation, silylation, and anhydrous acids (TiCl$_4$, CH$_2$Cl$_2$, −78°C, 30 min; TsOH, toluene, 12 h; neat CF$_3$CO$_2$H, 15 min). Extended exposure (48 h) of a triazone to neat CF$_3$CO$_2$H results in cleavage.[94]

Formation[95]

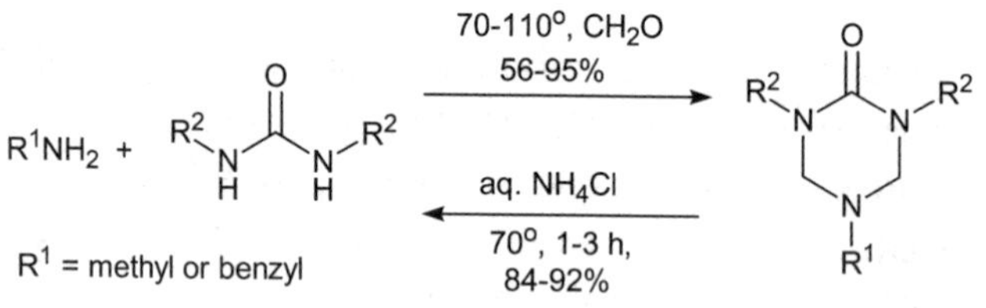

Cleavage

1. Aqueous NH$_4$Cl, 70°C, 1–3 h, 84–92% yield.[95]
2. HN(CH$_2$CH$_2$OH)$_3$.[96]
3. 1 *N* HCl, 23°C, >84% yield.[97,98]

1-Substituted 3,5-Dinitro-4-pyridone

Formation/Cleavage[99]

$$4\text{-}NO_2\text{-}C_6H_4\text{-}N \overset{NO_2}{\underset{NO_2}{\diagdown}} O$$

Pyr, H$_2$O, rt, 2–24 h
72–100%

RNH$_2$

MeNH$_2$, PrNH$_2$ or hexylNH$_2$
Pyr, H$_2$O, 0.5–2 h
83–97%

1,3,5-Dioxazine

The reaction of a cepham primary amine with 20 eq. of 37% formalin produces the dioxazine in 75% yield. The dioxazine is sufficiently stable to allow formation of Wittig reagents and to carry out an olefination with formaldehyde. Treatment of the dioxazine with 6 N HCl in CH$_2$Cl$_2$ releases the amine in excellent yield.[100]

1. J. C. Sheehan and F. S. Guziec, *J. Org. Chem.*, **38**, 3034 (1973).

2. S. V. Pansare and J. C. Vederas, *J. Org. Chem.*, **52**, 4804 (1987); U. Sreenivasan, R. K. Mishra, and R. L. Johnson, *J. Med. Chem.*, **36**, 256 (1993).

3. F. S. Guziec, Jr., and E. T. Tewes, *J. Heterocycl. Chem.*, **17**, 1807 (1980).

4. P. G. M. Wuts, unpublished observations.

5. M. Oelgemöller and A. G. Griesbeck, *Internet Photochemistry & Photobiology [online computer file]*, **3rd**, No pp given (2000).

6. M. Beier and W. Pfleiderer, *Helv. Chim. Acta*, **82**, 633 (1999).

7. X.-B. Meng, H. Li, Q.-H. Lou, M.-S. Cai, and Z.-J. Li, *Carbohydr. Res.*, **339**, 1497 (2004).

8. T. Sasaki, K. Minamoto, and H. Itoh, *J. Org. Chem.*, **43**, 2320 (1978).

9. Z.-G. Wang, X. Zhang, M. Visser, D. Live, A. Zatorski, U. Iserloh, K. O. Lloyd, and S. J. Danishefsky, *Angew. Chem. Int. Ed.*, **40**, 1728 (2001).

10. S. Chandrasekhar, M. Takhi, and G. Uma, *Tetrahedron Lett.*, **38**, 8089 (1997).

11. P. Y. Reddy, S. Kondo, T. Toru, and Y. Ueno, *J. Org. Chem.*, **62**, 2652 (1997).

12. M.-Y. Zhou, Y.-Q. Li, and X.-M. Xu, *Synth. Commun.*, **33**, 3777 (2003).

13. D. A. Hoogwater, D. N. Reinhoudt, T. S. Lie, J. J. Gunneweg, and H. C. Beyerman, *Recl. Trav. Chim. Pays-Bas*, **92**, 819 (1973).

14. G. H. L. Nefkins, G. I. Tesser, and R. J. F. Nivard, *Recl. Trav. Chim. Pays-Bas*, **79**, 688 (1960); C. R. McArthur, P. M. Worster, and A. U. Okon, *Synth. Commun.*, **13**, 311 (1983).

15. G. Sosnovsky and J. Lukszo, *Z. Naturforsch. B*, **41B**, 122 (1986).

16. J. Kehler and E. Breuer, *Synthesis*, 1419 (1998).

17. J. A. Moore and J.-H. Kim, *Tetrahedron Lett.*, **32**, 3449 (1991).

18. K. C. Nicolaou, *Angew. Chem., Int. Ed. Engl.*, **32**, 1377 (1993).

19. J. R. Casimir, G. Guichard, and J.-P. Briand, *J. Org. Chem.*, **67**, 3764 (2002).

20. J. R. Casimir, G. Guichard, and J.-P. Briand, *Synthesis*, 75 (2001).

21. N. Aguilar, A. Moyano, M. A. Pericas, and A. Riera, *Synthesis*, 313 (1998).

22. For a mechanistic study of this reaction, see M. N. Khan, *J. Org. Chem.*, **60**, 4536 (1995).

23. B. E. Maryanoff, M. N. Greco, H.-C. Zhang, P. Andrade-Gordon, J. A. Kauffman, K. C. Nicolaou, A. Liu, and P. H. Brungs, *J. Am. Chem. Soc.*, **117**, 1225 (1995).

24. A. L. Smith, C.-K. Hwang, E. Pitsinos, G. R. Scarlato, and K. C. Nicolaou, *J. Am. Chem. Soc.*, **114**, 3134 (1992).

25. I. Schumann and R. A. Boissonnas, *Helv. Chim. Acta*, **35**, 2235 (1952).

26. S. Kukolja and S. R. Lammert, *J. Am. Chem. Soc.*, **97**, 5582 (1975).

27. M. G. Stockdale, S. Ramurthy, and M. J. Miller, *J. Org. Chem.*, **63**, 1221 (1998).

28. B. Herberich, M. Kinugawa, A. Vazquez, and R. M. Williams, *Tetrahedron Lett.*, **42**, 543 (2001).

29. F. Dasgupta and P. J. Garegg, *J. Carbohydr. Chem.*, **7**, 701 (1988).

30. J. O. Osby, M. G. Martin, and B. Ganem, *Tetrahedron Lett.*, **25**, 2093 (1984).

31. M. S. Motawia, J. Wengel, A. E. S. Abdel-Megid, and E. B. Pedersen, Synthesis, 384 (1989).

32. P. L. Durette, E. P. Meitzner, and T. Y. Shen, *Tetrahedron Lett.*, **20**, 4013 (1979).

33. C. A. Costello, A. J. Kreuzman, and M. J. Zmijewski, *Tetrahedron Lett.*, **37**, 7469 (1996).

34. T. Kamiya, M. Hashimoto, O. Nakaguchi, and T. Oku, *Tetrahedron*, **35**, 323 (1979).

35. D. R. Mootoo and B. Fraser-Reid, *Tetrahedron Lett.*, **30**, 2363 (1989).

36. H. H. Lee, D. A. Schwartz, J. F Harris, J. P. Carver, and J. J. Krepinsky, *Can. J. Chem.*, **64**, 1912 (1986).

37. B. Astleford and L. O. Weigel, *Tetrahedron Lett.*, **32**, 3301 (1991).

38. S. Kotha, D. Anglos and A. Kuki, *Tetrahedron Lett.*, **33**, 1569 (1992).

39. O. Kanie, S. C. Crawley, M. M. Palcic, and O. Hindsgaul, *Carbohydr. Res.*, **243**, 139 (1993).

40. M.Kuriyama, Y. Inoue, and K. Kitagawa, *Synthesis*, 735 (1990).

41. M. Lergenmüller, Y. Ito, and T. Ogawa, *Tetrahedron*, **54**, 1381 (1998).

42. H. Shimizu, Y. Ito, Y. Matsuzaki, H. Iijima and T. Ogawa, *Biosci., Biotech., Biochem.*, **60**, 73 (1996).

43. T. Hashihayata, K. Ikegai, K. Takeuchi, H. Jona, and T. Mukaiyama, *Bull. Chem. Soc. Jpn.*, **76**, 1829 (2003).

44. For a review of the use of TCP in amino sugar synthesis, see J. Debenham, R. Rodebaugh, and B. Fraser-Reid, *Liebigs Ann./Recl.*, 791 (1997).

45. Z. H. Qin, H. Li, M. S. Cai, and Z. J. Li, *Chin. Chem. Lett.*, **11**, 941 (2000).

46. E. Cros, M. Planas, X. Mejias, and E. Bardaji, *Tetrahedron Lett.*, **42**, 6105 (2001).

47. A. K. Bose, M. Jayaraman, A. Okawa, S. S. Bari, E. W. Robb, and M. S. Manhas, *Tetrahedron Lett.*, **37**, 6989 (1996).

48. J. S. Debenham, S. D. Debenham, and B. Fraser-Reid, *Bioorg. Med. Chem.*, **4**, 1909 (1996).

49. J. S. Debenham, R. Rodebaugh, and B. Frasier-Reid, *J. Org. Chem.*, **61**, 6478 (1996).

50. J. S. Debenham, R. Madsen, C. Roberts, and B. Fraser-Reid, *J. Am. Chem. Soc.*, **117**, 3302 (1995); J. S. Debenham, R. Rodebaugh, and B. Fraser-Reid, *J. Org. Chem.*, **62**, 4591 (1997).

51. N. Khiar, I. Fernandez, C. S. Araujo, J.-A. Rodriguez, B. Suarez, and E. Alvarez, *J. Org. Chem.*, **68**, 1433 (2003).

52. P. Stangier and O. Hindsgaul, *Synlett*, 179 (1996).

53. B. A. Roe, C. G. Boojamra, J. L. Griggs, and C. R. Bertozzi, *J. Org. Chem.*, **61**, 6442 (1996).

54. J. C. Castro-Palomino and R. R. Schmidt, *Tetrahedron Lett.*, **36**, 5343 (1995).

55. E. Cros, M. Planas, G. Barany, and E. Bardaji, *Eur. J. Org. Chem.*, 3633 (2004).

56. H. Tsubouchi, K. Tsuji, and H. Ishikawa, *Synlett*, 63 (1994).

57. J. C. Castro-Palomino and R. R. Schmidt, *Tetrahedron Lett.*, **41**, 629 (2000).

58. K. J. Jensen, P. R. Hansen, D. Venugopal, and G. Barany, *J. Am. Chem. Soc.*, **118**, 3148 (1996).

59. S. Zalipsky, F. Albericio, U. Slomczynska, and G. Barany, *Int. J. Pept. Protein Res.*, **30**, 748 (1987).

60. U. Zehavi, *J. Org. Chem.*, **42**, 2819 (1977).

61. For an application in glucosamine chemistry, see (a) K. J. Jensen, P. R. Hansen, D. Venugopal, and G. Barany, *J. Am. Chem. Soc.*, **118**, 3148 (1996); (b) E. Meinjohanns, M. Meldal, H. Paulsen, and K. Bock, *J. Chem. Soc., Perkin Trans. 1*, 405 (1995).

62. M. J. Barany, R. P. Hammer, R. B. Merrifield, and G. Barany, *J. Am. Chem. Soc.*, **127**, 508 (2005).

63. G. Barany and R. B. Merrifield, *J. Am. Chem. Soc.*, **99**, 7363 (1977); *idem*, **102**, 3084 (1980).

64. I. Christiansen-Brams, M. Meldal, and K. Bock, *J. Chem. Soc., Perkin Trans. 1*, 1461 (1993).

65. E. Meinjohanns, M. Meldal, T. Jensen, O. Werdelin, L. Galli-Stampino, S. Mouritsen, and K. Bock, *J. Chem. Soc., Perkin Trans. 1*, 871 (1997).

66. E. Meinjohanns, M. Meldal, T. Jensen, O. Werdelin, L. Galli-Stampino, S. Mouritsen, and K. Bock, *J. Chem. Soc. Perkin Trans. 1*, 871 (1997).

67. D. J. Cane-Honeysett, M. D. Dowle, and M. E. Wood, *Tetrahedron*, **61**, 2141 (2005).

68. M. E. Wood, D. J. Cane-Honeysett, and M. D. Dowle, *J. Chem. Soc. Perkin Trans. 1*, 2046 (2002).

69. M. R. E. Aly, E.-S. I. Ibrahim, E. S. H. El Ashry, and R. R. Schmidt, *Eur. J. Org. Chem.*, 319 (2000). M. R. E. Aly, J. C. Castro-Palomino, E.-S. I. Ibrahim, E.-S. H. El-Ashry, and R. R. Schmidt, *Eur. J. Org. Chem.*, 2305 (1998). D. Hesek, M. Lee, K.-i. Morio, and S. Mobashery, *J. Org. Chem.*, **69**, 2137 (2004); K. R. Love, R. B. Andrade, and P. H. Seeberger, *J. Org. Chem.*, **66**, 8165 (2001).

70. J. E. Macor, B. L. Chenard, and R. J. Post, *J. Org. Chem.*, **59**, 7496 (1994).

71. S. G. Bowers, D. M. Coe, and G.-J. Boons, *J. Org. Chem.*, **63**, 4570 (1998).

72. J. A. Ragan, T. W. Makowski, M. J. Castaldi, and P. D. Hill, *Synthesis*, 1599 (1998).

73. S. P. Bruekelman, S. E. Leach, G. D. Meakins, and M. D. Tirel, *J. Chem. Soc., Perkin Trans. I*, 2801 (1984).

74. J. E. Macor, B. L. Chenard, and R. J. Post, *J. Org. Chem.*, **59**, 7496 (1994).

75. M. Curini, F. Montanari, O. Rosati, E. Lioy, and R. Margarita, *Tetrahedron Lett.*, **44**, 3923 (2003).

76. B. K. Banik, S. Samajdar, and I. Banik, *J. Org. Chem.*, **69**, 213 (2004).

77. B. K. Banik, I. Banik, M. Renteria, and S. K. Dasgupta, *Tetrahedron Lett.*, **46**, 2643 (2005).

78. B. Ramanathan, A. J. Keith, D. Armstrong, and A. L. Odom, *Org. Lett.*, **6**, 2957 (2004).

79. S. P. Breukelman, G. D. Meakins, and M. D. Tirel, *J. Chem. Soc., Chem. Commun.*, 800 (1982).

80. C. Kashima, T. Maruyama, Y. Fujioka, and K. Harada, *J. Chem. Soc., Perkin Trans. I*, 1041 (1989).

81. A. P. Davis and T. J. Egan, *Tetrahedron Lett.*, **33**, 8125 (1992).

82. T. Katagiri, M. Irie, and K. Uneyama, *Org. Lett.*, **2**, 2423 (2000).

83. S. F. Martin and C. Limberakis, *Tetrahedron Lett.*, **38**, 2617 (1997).

84. S. Djuric, J. Venit, and P. Magnus, *Tetrahedron Lett.*, **22**, 1787 (1981).

85. T. Högberg, P. Ström, and U. H. Lindberg, *Acta Chem. Scand., Ser. B.*, **B39**, 414 (1985).

86. T. L. Guggenheim, *Tetrahedron Lett.*, **25**, 1253 (1984).

87. M. J. Sofia, P. K. Chakravarty, and J. A. Katzenellenbogen, *J. Org. Chem.*, **48**, 3318 (1983).

88. K. Deshayes, R. D. Broene, I. Chao, C. B. Knobler, and F. Diederich, *J. Org. Chem.*, **56**, 6787 (1991).

89. R. P. Bonar-Law, A. P. Davis, and B. J. Dorgan, *Tetrahedron Lett.*, **31**, 6721 (1990). *idem, Tetrahedron*, **49**, 9855 (1993).

90. R. P. Boner-Law, A. P. Davis, B. J. Dorgan, M. T. Reetz, and A. Wehrsig, *Tetrahedron Lett.*, **31**, 6725 (1990).

91. A. P. Davis and P. J. Gallagher, *Tetrahedron Lett.*, **36**, 3269 (1995).

92. F. Cavelier-Frontin, R. Jacquier, J. Paladino, and J. Verducci, *Tetrahedron*, **47**, 9807 (1991).

93. B. M. Kim and J. H. Cho, *Tetrahedron Lett.*, **40**, 5333 (1999).

94. S. Knapp, J. J. Hale, M. Bastos, A. Molina, and K. Y. Chen, *J. Org. Chem.*, **57**, 6239 (1992).

95. S. Knapp, J. J. Hale, M. Bastos, and F. S. Gibson, *Tetrahedron Lett.*, **31**, 2109 (1990).

96. S. Knapp and J. J. Hale, *J. Org. Chem.*, **58**, 2650 (1993).

97. S. R. Angle, J. M. Fevig, S.D. Knight, R. W. Marguis, Jr., and L. E. Overman, *J. Am. Chem. Soc.*, **115**, 3966 (1993).

98. W. H. Pearson, I. Y. Lee, Y. Mi, and P. Stoy, *J. Org. Chem.*, **69**, 9109 (2004).

99. E. Matsumura, H. Kobayashi, T. Nishikawa, M. Ariga, Y. Tohda, and T. Kawashima, *Bull. Chem. Soc. Jpn.*, **57**, 1961 (1984); E. Matsumura, M. Ariga, Y. Tohda, and T. Kawashima, *Tetrahedron Lett.*, **22**, 757 (1981).

100. Y. Katsura and M. Aratani, *Tetrahedron Lett.*, **35**, 9601 (1994).

SPECIAL−NH PROTECTIVE GROUPS

N-Alkyl and *N*-Aryl Amines

N-Methylamine: CH_3NR_2

The methyl group, although inert to many chemical transformations, is not often considered a good protective group because of the perceived difficulty in its removal, but as illustrated there are a number of methods that can be used to cleave an *N*-methyl group in highly functionalized substrates.

Formation

1. Methylamines are commonly formed by reacting the amine with a methylating agent such as MeI or dimethyl sulfate.
2. Preparation from an amine and $TMSCHN_2$ (HBF_4, CH_2Cl_2, H_2O) has also been explored.
3. For primary aromatic amines: dimethyl carbonate, Y-zeolite, 130–150°C, 72–93% yields.[1] Y-faujasites have been used as catalysts and require lower temperatures to achieve methylation. CO_2 must be removed with a stream of N_2 to prevent carbamates formation.[2]
4. HCHO, HCO_2H, 5°C then reflux, 12 h, 91% yield.[3,4]
5. For vicinal amino alcohols: CH_2O, PTSA, reflux, benzene, then $NaCNBH_3$, $TMSCl$, CH_3CN, rt, 94–97% yield.[5]

Pg = Ts, Cbz, BOC

Cleavage

1. The cleavage of a methylamine can be accomplished photochemically in the presence of an electron acceptor such as 9,10-dicyanoanthracene.[6]

2. Photolysis with visible light, DAP^{2+}; TMSCN. The photochemical reaction generates an iminium ion that is trapped with cyanide.[7]
3. $CH_2=CHOCOCl$, K_2CO_3, CH_2Cl_2.[8] The *N*-methyl group of a tertiary amine is converted to a vinyl carbamate that is easily hydrolyzed.

4. 1-Chloroethyl chloroformate, EtOAc, 7 eq. 50°C, 5 h followed by treatment with methanol which removes the carbamate by solvolysis. This method was used to cleave the N-methyl from erythromycin B[9] and in the synthesis of a series of *Strychnos* alkaloids.[10]

5. I_2, CaO, THF, MeOH. A dimethylaniline is converted to a monomethylaniline.[11]

6. CS_2, MeI, THF, 6 h, 30°C, 97% yield. N-Methylpiperidine is converted to a dithiocarbamate.

7. t-BuOOH, $RuCl_2(Ph_3P)_2$, benzene, rt, 3 h, 83% yield. The methyl group is converted to t-BuOOCH_2NR_2 that can then be hydrolyzed, releasing the secondary amine.[12] The oxidation of amines has been reviewed.[13]

8. PhSeH, 160°C, 5 days, 68% yield.[14]

9. $RuCl_3$, H_2O_2, MeOH, 55–80% yield.[15] These conditions convert the methyl to a MOM group that can be removed by hydrolysis. In the presence of NaCN, N-cyanomethylamine derivatives are produced,[16] which can be cleaved *vida infra*. The reaction proceeds through an iminium ion.

10. The Polonovski reaction: H_2O_2, MeOH, then 6 M HCl to form the salt of the N-oxide, which is treated with $FeSO_4·7H_2O$, 49–97% yield.[17]

11.

Ref. 18

12. $Na_2CO_3·1.5H_2O_2$ to form amine N-oxide and then Na salt of 4,6-dichloro-2-hydroxy-(1,3,5)-triazine, 89–98% yield. The reactions are carried out in a zoned chromatography column.[19]

13. MCPBA then TEA, TFAA, CH_2Cl_2.[20]

14. For substituted *N,N*-dimethylanilines: $TiCl_4$, CH_2Cl_2, 0–25°C, 8 h, 72–86% yield. Unsubstituted *N,N*-dialkylanilines undergo oxidative dimerization to form *N,N,N,N*-tetraalkylbenzidines.[21]

15. PhIO, $TMSN_3$, CH_2Cl_2, −40°C, 3 h, then workup with aqueous $NaHCO_3$, 92% yield.[22]

16. Diethylazodicarboxylate, acetone then MeOH, NH_4Cl, reflux, 82% yield.[23]

1. M. Selva, A. Bomben, and P. Tundo, *J. Chem. Soc., Perkin Trans. 1*, 1041 (1997).

2. M. Selva and P. Tundo, *Tetrahedron Lett.*, **44**, 8139 (2003).

3. G. Chelucci, M. Falorni, and G. Giacomelli, *Synthesis*, 1121 (1990).

4. For a review of the Leukart reaction, see M. L. Moore, *Org. React.*, **5**, 301 (1949).

5. G. V. Reddy, G. V. Rao, V. Sreevani, and D. S. Iyengar, *Tetrahedron Lett.*, **41**, 949 (2000).

6. J. Santamaria, R. Ouchabane, and J. Rigaudy, *Tetrahedron Lett.*, **30**, 2927 (1989).

7. J. Santamaria, M. T. Kaddachi, and J. Rigaudy, *Tetrahedron Lett.*, **31**, 4735 (1990).

8. J. R. Ferguson, K. W. Lumbard, F. Scheinmann, A. V. Stachulski, P. Stjernlöf, and S. Sundell, *Tetrahedron Lett.*, **36**, 8867 (1995); R. A. Olofson, R. C. Schnur, L. Bunes, and J. P. Pepe, *ibid.*, 1567 (1977).

9. J. E. Hengeveld, A. K. Gupta, A. H. Kemp, and A. V. Thomas, *Tetrahedron Lett.*, **40**, 2497 (1999).

10. J. Bonjoch, D. Solé, S. García-Ribio, and J. Bosch, *J. Am. Chem. Soc.*, **119**, 7230 (1997).

11. K. Acosta, J. W. Cessac, P. N. Rao, and H. K. Kim, *J. Chem. Soc., Chem. Commun.*, 1985 (1994).

12. S.-I. Murahashi, T. Naota, and K. Yonemura, *J. Am. Chem. Soc.*, **110**, 8256 (1988).

13. S.-I. Murahashi, *Angew. Chem,. Int. Ed. Engl.*, **34**, 2443 (1995).

14. R. P. Polniaszek and L. W. Dillard, *J. Org. Chem.*, **57**, 4103 (1992).

15. S.-I. Murahashi, T. Naota, N. Miyaguchi, and T. Nakato, *Tetrahedron Lett.*, **33**, 6991 (1992).

16. S.-I. Murahashi, N. Komiya, H. Terai, and T. Nakae, *J. Am. Chem. Soc.*, **125**, 15312 (2003).

17. K. McCamley, J. A. Ripper, R. D. Singer, and P. J. Scammells, *J. Org. Chem.*, **68**, 9847 (2003).

18. J. P. Gesson, J. C. Jacquesy, and M. Mondon, *Synlett*, 669 (1990).

19. T. Rosenau, A. Hofinger, A. Potthast, and P. Kosma, *Org. Lett.*, **6**, 541 (2004).

20. R. Menchaca, V. Martinez, A. Rodriguez, N. Rodriguez, M. Flores, P. Gallego, I. Manzanares, and C. Cuevas, *J. Org. Chem.*, **68**, 8859 (2003).

21. M. Periasamy, K. N. Jayakumar, and P. Bharathi, *J. Org. Chem.*, **65**, 3548 (2000).

22. S. Saaby, Z. Fang, N. Gathergood, and K. A. Jorgensen, *Angew. Chem. Int. Ed.*, **29**, 4114 (2000).

23. A. Denis and C. Renou, *Tetrahedron Lett.*, **43**, 4171 (2002). A. Zhang, C. Csutoras, R. Zong, and J. L. Neumeyer, *Org. Lett.*, **7**, 3239 (2005).

*N-t-*Butylamine: $(CH_3)_3CNR_2$

The *t*-butyl group can be cleaved from a cyclopropylamine upon prolonged heating in acid (H_3O^+, reflux, 3–5 days).[1] Not all cases require such protracted reaction times as is illustrated in the following case[2]:

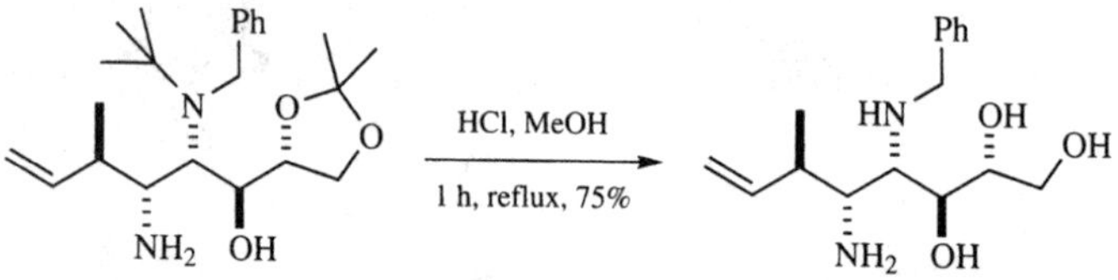

Treatment of a *t*-butylamine (among others with Ac_2O) with a catalytic amount of $BF_3 \cdot Et_2O$ at reflux results in conversion to the acetamide.[3] The acetamides can be removed hydrolytically.

1. N. De Kimpe, P. Sulmon, and P. Brunet, *J. Org. Chem.*, **55**, 5777 (1990).

2. E. Leclerc, E. Vrancken, and P. Mangeney, *J. Org. Chem.*, **67**, 8928 (2002).

3. P. R. Dave, K. A. Kumar, R. Duddu, T. Axenrod, K. Dai, K. K. Das, N. J. Trivedi, and R. D. Gilardi, *J. Org. Chem.*, **65**, 1207 (2000).

*N-*Allylamine: $CH_2=CHCH_2NR_2$ (Chart 10)

Formation

1. Allyl bromide, K_2CO_3, THF, heat, 75% yield.[1] This is a fairly general method that has been used widely for the preparation of allylamines. It is difficult to stop this reaction at the monoallyl stage.

2. Allyl bromide, $CsOH \cdot H_2O$, 4-Å ms, DMF, 85% monoallyl along with 15% of the diallylamine.[2]

3. Allyl bromide, $LiOH \cdot H_2O$, 4-Å ms, DMF, rt, 61–82% yield. This method was developed for the monoalkylation of aminoacid esters.[3]

4. Allyl chloride, Cu(0), $Cu(ClO_4)_2 \cdot 6H_2O$, Et_2O, 97% yield.[4]

5. AllylOAc, Pd(Ph$_3$P)$_4$, diisopropylamine, 80°C, 24 h, 82% yield.[5]

6. Allylbenzotriazole, Pd(OAc)$_2$, PPh$_3$, K$_2$CO$_3$, MeOH, reflux, 85% yield. This method is also good for allylation of sulfonamides.[6]

7. Allyl alcohol, Pd(OAc)$_2$, PPh$_3$, Ti(O-iPr)$_4$, MS4Å, benzene, 50°C, 18–86% yield. Only anilines were examined with this method, but the method could be used to prepare cinnamyl, methallyl, and crotyl derivatives.[7]

8. Ni(cod)$_2$, Bu$_4$NPF$_6$, dppb, THF, 50°C.[8]

9. From a sulfonamide as Li salt: CH$_2$=CHCH$_2$OCO$_2$Me, Rh(Ph$_3$P)$_3$Cl, AgOTf, toluene, rt, >87% yield.[9]

Cleavage

1. Isomerization to the enamine (t-BuOK, DMSO), followed by hydrolysis.[10]

2. Rhodium-catalyzed isomerization.[11] Ru(cod)(cot) has been used to convert an allylamine into an enamine.[12]

In the presence of a nearby hydroxyl, the aminal is formed.[13]

The use of Pd(Ph$_3$P)$_4$, and N,N-dimethylbarbituric acid removed the allyl group in 98% yield.

3. Pd(Ph$_3$P)$_4$, and N,N-dimethylbarbituric acid, 30°C, 1.5–3 h, 91–100% yield.[5]

4. Pd–C, MsOH, H$_2$O, 82% yield.[15] In certain heterocyclic systems this method failed, but was successful when MsOH was replaced with BF$_3$·Et$_2$O.[16]

5. Pd/C, EtOH, H$_2$NCH$_2$CH$_2$OH, reflux, 3 h, then H$_2$SO$_4$, H$_2$O, 77% yield.[17]

6. Pd(Ph$_3$P)$_4$, PMHS, ZnCl$_2$, THF, rt, 89–92% yield.[18] Allyl ethers and esters are cleaved similarly, but a prenyl ether is stable.

7. Pd(Ph$_3$P)$_4$, RSO$_2$Na, CH$_2$Cl$_2$ or THF/MeOH, 70–99% yield. These conditions were shown to be superior to the use of sodium 2-ethylhexanoate. Methallyl, crotyl, allyl, and cinnamyl ethers, the Aloc group, and allyl esters are all efficiently cleaved by this method.[19]

8. Pd(dba)$_2$dppb, 2-thiolbenzoic acid, THF, 70–100% yield.[20] Tertiary allylamines are cleaved efficiently at 20°C, but secondary allylamines require
   heating to 60°C to achieve cleavage. Thus, it is possible to monodeallylate a
   diallylamine.[21,22]

9. DIBAL, Ni(dppp)Cl$_2$, toluene, rt, 69–91% yield.[23]

10. Cl$_2$(Cy$_3$P)$_2$Ru=CHPh (Grubbs' carbene), toluene or CH$_2$Cl$_2$, reflux, 49–78%
    yield. Allyl amines are cleaved in the presence of allyl ethers. An allyl β-lactam
    was converted to its enamide while attempting a ring closing metathesis reac-
    tion.[24] This method was generalized to other amines,[25] but allyl ethers are stable.

11. Ru(η^3:η^2:η^3-C$_{12}$H$_{18}$)Cl$_2$, H$_2$O, 90°C 15 min to 3.5 h, 95–99% yield.[26]

12. Cp$_2$Zr, then water, 66% yield.[27] O-Allyl ethers are cleaved at a faster rate;
    THF, acetonide, Bn ethers and benzoates are stable.

13. CH$_3$CHCl(OCOCl), then methanolysis with MeOH, 74% yield.[28]

14. EtOCOCl, NaI, acetone, reflux, 3 h, 85% yield.[29] The addition of NaI serves
    to generate the more reactive ethyl iodoformate. It also helps preserve the pri-
    mary iodide which could be displaced by released chloride ion to give some
    of the primary chloride.

N-Prenylamine: (CH$_3$)$_2$C=CHCH$_2$NR$_2$

Cleavage

TolSH, benzene, AIBN, reflux, 57–98% yield. This method proceeds by an isomerization of the prenylamine to the enamine which is then readily hydrolyzed.[30]

N-Cinnamylamine: (E)-C_6H_5CH=$CHCH_2NR_2$

Formation

This method failed with the acetamide (R = Ac) and the BOC derivative (R = BOC), but does work with sulfonamides.[31]

$$
\begin{array}{c}
\text{R}\diagdown_{\text{NH}} \\
\text{Ph}
\end{array}
\xrightarrow[\substack{\text{1,4-dioxane, 100°C} \\ \text{52–98\%}}]{\substack{\text{1-phenylpropyne,} \\ \text{PhCO}_2\text{H}}}
\begin{array}{c}
\text{R}\diagdown_{\text{N}}\diagup\diagdown\text{Ph} \\
\text{Ph}
\end{array}
$$

N-2-Phenallylamine: CH_2=$C(Ph)CH_2NR_2$

This group was used as a bulky protective group that could be cleaved in the presence of a propargyl amine using Pd catalyzed cleavage.[32] *t*-BuLi (−78 to 0°C) has also been used to cleave these amines by an addition elimination reaction. The corresponding ethers are similarly cleaved.[33]

N-Propargylamine: HC≡CCH_2NR_2

Cleavage

TiCl$_3$, Li, THF, rt, 0.5–30 h, 35–77% yield. A phenolic propargyl ether is also cleaved.[34]

1. G. A. Molander and P. J. Nichols, *J. Org. Chem.*, **61**, 6040 (1996).
2. R. N. Salvatore, A. S. Nagle, and K. W. Jung, *J. Org. Chem.*, **67**, 674 (2002).
3. J. H. Cho and B. M. Kim, *Tetrahedron Lett.*, **43**, 1273 (2002).
4. J. B. Baruah and A. G. Samuelson, *Tetrahedron*, **47**, 9449 (1991).
5. F. Garro-Helion, A. Merzouk, and F. Guibé, *J. Org. Chem.*, **58**, 6109 (1993).
6. A. R. Katritzky, J. Yao, and O. V. Denisko, *J. Org. Chem.*, **65**, 8063 (2000); A. R. Katritzky, J. Yao, and M. Qi, *J. Org. Chem.*, **63**, 5232 (1998).
7. S.-C. Yang and W.-H. Chung, *Tetrahedron Lett.*, **40**, 953 (1999); S.-C. Yang, C.-L. Yu, and Y.-C. Tsai, *Tetrahedron Lett.*, **41**, 7097 (2000). S.-C. Yang and C.-W. Hung, *J. Org. Chem.*, **64**, 5000 (1999). Y.-J. Shue, S.-C. Yang, and H.-C. Lai, *Tetrahedron Lett.*, **44**, 1481 (2003).

8. H. Bricout, J.-F. Carpentier, and A. Mortreux, *J. Chem. Soc., Chem. Commun.*, 1863 (1995).

9. P. A. Evans and J. E. Robinson, *J. Am. Chem. Soc.*, **123**, 4609 (2001).

10. R. Gigg and R. Conant, *J. Carbohydr. Chem.*, **1**, 331 (1983).

11. B. C. Laguzza and B. Ganem, *Tetrahedron Lett.*, **22**, 1483 (1981).

12. T. Mitsudo, S.-W. Zhang, N. Satake, T. Kondo, and Y. Watanabe, *Tetrahedron Lett.*, **33**, 5533 (1992).

13. S. G. Davies and D. R. Fenwick, *J. Chem. Soc., Chem. Commun.*, 565 (1997).

14. S. D. Bull, S. G. Davies, P. M. Kelly, M. Gianotti, and A. D. Smith, *J. Chem. Soc. Perkin Trans. 1*, 3106 (2001).

15. Q. Liu, A. P. Marchington, N. Boden, and C. M. Rayner, *J. Chem. Soc., Perkin Trans. 1*, 511 (1997).

16. S. Jaime-Figueroa, Y. Liu, J. M. Muchowski, and D. G. Putman, *Tetrahedron Lett.*, **39**, 1313 (1998).

17. M. Karpf and R. Trussardi, *J. Org. Chem.*, **66**, 2044 (2001).

18. S. Chandrasekhar, C. Raji Reddy, and R. Jagadeeshwar Rao, *Tetrahedron*, **57**, 3435 (2001).

19. M. Honda, H. Morita, and I. Nagakura, *J. Org. Chem.*, **62**, 8932 (1997).

20. S. Lemaire-Audoire, M. Savignac, J. P. Genêt, and J.-M. Bernard, *Tetrahedron Lett.*, **36**, 1267 (1995); W. F. Bailey and X.-L. Jiang, *J. Org. Chem.*, **61**, 2596 (1996); I. C. Baldwin, P. Briner, M. D. Eastgate, D. J. Fox, and S. Warren, *Org. Lett.*, **4**, 4381 (2002).

21. S. Lemaire-Audoire, M. Savignac, C. Dupuis, and J. P. Genêt, *Bull. Soc. Chim. Fr.*, **132**, 1157 (1995).

22. C. Koradin, K. Polborn, and P. Knochel, *Angew. Chem. Int. Ed.*, **41**, 2535 (2002).

23. T. Taniguchi and K. Ogasawara, *Tetrahedron Lett.*, **39**, 4679 (1998).

24. B. Alcaide, P. Almendros, J. M. Alonso, and M. F. Aly, *Org. Lett.*, **3**, 3781 (2001). C. Cadot, P. I. Dalko and J. Cossy, *Tetrahedron Lett.*, **43**, 1839 (2002); B. Alcaide and P. Almendros, *Chem. Eur. J.*, **9**, 1259 (2003).

25. B. Alcaide, P. Almendros, and J. M. Alonso, *Chem. Eur. J.*, **9**, 5793 (2003).

26. V. Cadierno, S. E. Garcia-Garrido, J. Gimeno, and N. Nebra, *Chem. Commun.*, 4086 (2005).

27. H. Ito, T. Taguchi and Y. Hanzawa, *J. Org. Chem.*, **58**, 774 (1993).

28. P. Magnus and L. S. Thurston, *J. Org. Chem.*, **56**, 1166 (1991).

29. J. H. Tidwell and S. L. Buchwald, *J. Am. Chem. Soc.*, **116**, 11797 (1994).

30. S. Escoubet, S. Gastaldi, V. I. Timokhin, M. P. Bertrand, and D. Siri, *J. Am. Chem. Soc.*, **126**, 12343 (2004).

31. N. T. Patil, H. Wu, I. Kadota, and Y. Yamamoto, *J. Org. Chem.*, **69**, 8745 (2004).

32. N. Gommermann and P. Knochel, *Chem. Commun.*, 4175 (2005).

33. J. Barluenga, F. J. Fananas, R. Sanz, C. Marcos, and J. M. Ignacio, *Chem. Commun.*, 933 (2005).

34. S. Rele, S. Talukdar, and A. Banerji, *Tetrahedron Lett.*, **40**, 767 (1999).

N-Methoxymethyl amine (MOM−NR₂): $CH_3OCH_2NR_2$

Formation/Cleavage[1]

1. M. A. Zajac and E. Vedejs, *Org. Lett.*, **6**, 237 (2004).

N-[2-(Trimethylsilyl)ethoxy]methylamine (SEM−NR₂): $(CH_3)_3SiCH_2CH_2OCH_2−NR_2$

The SEM derivative of a secondary aromatic amine, prepared from SEMCl (NaH, DMF, 0°C, 100% yield) can be cleaved with HCl (EtOH, >88% yield).[1]

1. Z. Zeng and S. C. Zimmerman, *Tetrahedron Lett.*, **29**, 5123 (1988).

N-3-Acetoxypropylamine: $R_2NCH_2CH_2CH_2OCOCH_3$ (Chart 10)

Formation

1. $CH_2=CHCHO$, CH_2Cl_2, 20°C
2. BH_3, THF, CH_2Cl_2, −78°C
3. Ac_2O, Pyr, 20°C, 78%

Cleavage

1. NaOMe, MeOH, 20°C
2. DMSO, DCC, TFA, Pyr, 20 °C
3. $HClO_4$, $PhNMe_2$, 20°C, 35%

A 3-acetoxypropyl group was used to protect an aziridine −NH group during the synthesis of mitomycins A and C; acetyl, benzoyl, ethoxycarbonyl and methoxymethyl groups were unsatisfactory.[1]

1. T. Fukuyama, F. Nakatsubo, A. J. Cocuzza, and Y. Kishi, *Tetrahedron Lett.*, **18**, 4295 (1977).

N-Cyanomethylamine: NCCH₂NR₂

The cyanomethylamine, formed from the amine and bromoacetonitrile (DMF, TEA, 86–96% yield), is cleaved by reduction of the nitrile followed by hydrolysis (PtO₂, H₂, EtOH, 96–98% yield)[1] or with AgNO₃/EtOH (92% yield).[2] *N*-protected amides and *O*-protected phenols are also cleaved using similar hydrogenation conditions. These are also the products of the Strecker reaction with an amine and formaldehyde.

1. A. Benarab, S. Boyé, L. Savelon, and G. Guillaumet, *Tetrahedron Lett.*, **34**, 7567 (1993).
2. L. E. Overman and J. Shim, *J. Org. Chem.*, **56**, 5005 (1991).

N-2-Azanorbornenes

A primary amine, protected by reaction of the amine with cyclopentadiene and formaldehyde (H₂O, rt 3 h)[1], is cleaved by trapping cyclopentadiene with *N*-methylmaleimide (H₂O, 2.5 h, 23–50°C, 61–97% yield),[2] CuSO₄ (EtOH or MeOH, 70°C, 74–99%) or Bio-Rad AG 50W-X2 acid ion-exchange resin, 82–98% yield.[3]

1. S. D. Larsen and P. A. Grieco, *J. Am. Chem. Soc.*, **107**, 1768 (1985).
2. P. A. Grieco, D. T. Parker, W. F. Forbare, and R. Ruckle, *J. Am. Chem. Soc.*, **109**, 5859 (1987); P. A. Grieco and B. Bahsas, *J. Org. Chem.*, **52**, 5746 (1987).
3. P. A. Grieco and J. D. Clark, *J. Org. Chem.*, **55**, 2271 (1990).

N-2,4-Dinitrophenylamine: 2,4-(NO₂)₂C₆H₃NR₂

The DNP derivative, prepared from 2,4-dinitrofluorobenzene,[1–3] is released from the nitrogen with an anionic ion exchange resin.[4,5] When used for histidine protection the DNP group has been observed to migrate to nearby lysine residues during Fmoc cleavage.[6] The DNP group has been successfully used to protect the glucosamine nitrogen during glycosylation.[7]

1. P. F. Lloyd and M. Stacey, *Tetrahedron*, **9**, 116 (1960).

2. K. Izawa, T. Ineyama, K. Fujii, and T. Suami, *Carbohydr. Res.*, **205**, 415 (1990).

3. Y. Nakamura, A. Ito, and C.-g. Shin, *Bull. Chem. Soc. Jpn.*, **67**, 2151 (1994).

4. H. Tsunoda, J. Inokuchi, K. Yamagishi, and S. Ogawa, *Liebigs Ann.*, 279 (1995).

5. T. E. Nicolas and R. W. Franck, *J. Org. Chem.*, **60**, 6904 (1995).

6. J.-C. Gesquiere, J. Najib, T. Letailleur, P. Maes, and A. Tartar, *Tetrahedron Lett.*, **34**, 1921 (1993).

7. S. Koto, M. Hirooka, K. Yago, M. Komiya, T. Shimizu, K. Kato, T. Takehara, A. Ikefuji, A. Iwasa, S. Hagino, M. Sekiya, Y. Nakase, S. Zen, F. Tomonaga, and S. Shimada, *Bull. Chem. Soc. Jpn.*, **73**, 173 (2000).

N-o- or *p*-Methoxyphenylamine (PMP–NR₂): *o*- or *p*-CH₃OC₆H₄NR₂

o- or *p*-Methoxyphenylamine is often used as a protected ammonia equivalent that must then be removed later in a synthetic sequence, but with the advent of the Buchwald–Hartwig reaction it can now be considered as a protective group that can both be installed and cleaved.

Formation

1. 4-CH₃OC₆H₄Br, *t*-BuONa, Pd(OAc)₂, polymer supported phosphine ligand, toluene, 80°C, 15–20 h, 84% yield.[1]

2. The Buchwald–Hartwig reaction: 4-CH₃OC₆H₄Br, Pd₂(dba)₃, BINAP, *t*-BuONa, 18-C-6, THF, rt, 83% yield. There are number of variants of this reaction that largely involve a change in the phosphine ligand.[2,3] Some of the early work has been reviewed.[4]

3. 4-CH₃OC₆H₄OTf, *t*-BuONa, (NHC)Pd(allyl)Cl, toluene, 70°C, 88–90% yield.[5]

4. (2-CH₃OC₆H₄)₃Bi, TEA, CH₂Cl₂, Cu(OAc)₂, 81% yield.[6]

Cleavage

1. Ceric ammonium nitrate, CH₃CN, H₂O, 78% yield.[7] It has been shown that the addition of NaBH₄ and then Ac₂O after the oxidation improves the yield by reducing the quinone to the hydroquinone. Ac₂O traps the amine and the hydroquinone as the amide and diacetate respectively. The same process was used to cleave the 4-methoxynaphthal group from an amine.[8]

2. PhI(OAc)$_2$, >72% yield.[9] These conditions can also be used to cleave the 4-t-butyldimethylsiloxyphenyl group from an amine.[10]

3. AgNO$_3$, (NH$_4$)$_2$S$_2$O$_8$, THF, H$_2$O, CH$_3$CN, 60°C, 53% yield.[11]

4. Anodic oxidation, 0.85 V vs. SCE, Pt electrode, CH$_3$CN, H$_2$O, HClO$_4$, 68–94% yield. Dithianes and p-methoxybenzylamines are unaffected by this method.[12] Yields were better than when CAN was used.

1. C. A. Parrish and S. L. Buchwald, *J. Org. Chem.*, **66**, 3820 (2001).

2. D. Gerristma, T. Brenstrum, J. McNulty, and A. Capretta, *Tetrahedron Lett.*, **45**, 8319 (2004); Y. Wan, M. Alterman, and A. Hallberg, *Synthesis*, 1597 (2002); S. Urgaonkar and J. G. Verkade, *J. Org. Chem.*, **69**, 9135 (2004).

3. U. Nettekoven, F. Naud, A. Schnyder, and H.-U. Blaser, *Synlett*, 2549 (2004).

4. J. P. Wolfe, S. Wagaw, J.-F. Marcoux, and S. L. Buchwald, *Acc. Chem. Res.*, **31**, 805 (1998).

5. O. Navarro, H. Kaur, P. Mahjoor, and S. P. Nolan, *J. Org. Chem.*, **69**, 3173 (2004).

6. R. J. Sorenson, *J. Org. Chem.*, **65**, 7747 (2000).

7. J. Takaya, H. Kagoshima, and T. Akiyama, *Org. Lett.*, **2**, 1577 (2000); S. Fustero, J. G. Soler, A. Bartolome, and M. S. Rosello, *Org. Lett.*, **5**, 2707 (2003).

8. D. Taniyama, M. Hasegawa, and K. Tomioka, *Tetrahedron Lett.*, **41**, 5533 (2000).

9. I. Ibrahem, J. Casas, and A. Cordova, *Angew. Chem. Int. Ed.*, **43**, 6528 (2004).

10. Y. Hayashi, W. Tsuboi, M. Shoji, and N. Suzuki, *J. Am. Chem. Soc.*, **125**, 11208 (2003).

11. S. Saito, K. Hatanaka, and H. Yamamoto, *Org. Lett.*, **2**, 1891 (2000).

12. S. D. L. Marin, T. Martens, C. Mioskowski, and J. Royer, *J. Org. Chem.*, **70**, 10592 (2005).

N-Benzylamine (R$_2$N–Bn): R$_2$NCH$_2$Ph (Chart 10)

Formation

1. BnCl, aq. K$_2$CO$_3$, reflux, 30 min; H$_2$, Pd–C, 77% yield.[1]

2. BnBr, LiOH·H$_2$O, 4-Å ms, DMF, rt, 12 h, 87% yield of monobenzyl derivative of the methyl ester of phenylalanine.[2] The 4-nitrobenzylamine derivative of other aminoacids could be prepared by this method.

3. BnBr, EtOH, Na$_2$CO$_3$, H$_2$O, CH$_2$Cl$_2$, reflux.[3]

4. BnBr, Et$_3$N, CH$_3$CN.[4] Examples 2 and 3 above produce dibenzyl derivatives from primary amines.

5. CsOH·H$_2$O, DMF, 0°C to rt, 12 h, 4-Å ms, 52–79% yield. Monobenzylamines are prepared from primary amines selectively in the presence of secondary amines.[5]

6. Dibenzyl carbonate, Ph_4PBr, 150–170°C, neat, 76–93% yield. These conditions give dibenzyl amines with only minimal amounts of the carbamates.[6]

7. $PhCHN_2$, HBF_4, −40°C, CH_2Cl_2, 57–68% yield.[7] $SnCl_2$–H_2O has been used to catalyze this transformation.[8]

8. PhCHO, 6 M HCl in MeOH, MeOH, $NaCNBH_3$[9]

9. PhCHO, PhSeSePh, $NaBH_4$, EtOH, 1.5 h, 25°C, 90% yield.[10]

10. PhCHO, $CHCl_3$, 3-Å ms; $NaBH_4$ alcohol solvent, 66% yield. These conditions were used to protect selectively the terminal ends of a polyamine.[11]

Cleavage

Reductive Methods. The following table shows that substituents have a significant effect on the rate of hydrogenolysis of benzyl amines.

Substituent Effect on the Hydrogenolysis of Various Secondary Amines[12]

Entry	R	Conv.	Relative Rate
1	p-H	77%	1
2	p-CH_3	60%	0.78
3	p-C_2H_5	49%	0.64
4	p-CF_3	42%	0.55
5	p-F	9%	0.12
6	m-F	7%	0.09
7	3,5-di-F	0.2%	<0.01

1. Pd–C, 4.4% HCOOH, CH_3OH, 25°C, 10 h, 80–90% yield.[4,13] The cleavage of benzylamines with H_2/Pd–C is often very slow.[14] Note in example 2 below that one of the benzyl groups can be selectively removed from a dibenzyl derivative.

2. Pd–C, ROH, HCO_2NH_4,[15] hydrazine or sodium hypophosphite, 42–91% yield.[16] 2-Benzylaminopyridine and benzyladenine were stable to these reaction conditions. Lower yields occurred because of the water solubility of the product, thus hampering isolation. Cyclohexene can be used as a hydrogen source in the transfer hydrogenation.[17]

*Note that the OBn group is retained
and that the BOC group has migrated*

With cyclohexadiene as the H_2 source tertiary benzylamines are cleaved in the presence of the benzyloxymethyl (BOM) group and benzyl ethers, but alkenes are reduced.[18]

$$\text{BnO} \diagdown \diagup_{\text{NBn}_2} \xrightarrow[\text{cyclohexadiene}]{\text{Pd–C, EtOH}} \text{BnO} \diagdown \diagup_{\text{NHBn}}$$

3. 20% Pd(OH)$_2$, EtOH, H$_2$, 55 psi, 19 h. A benzyl ether was not cleaved.[19] Under typical hydrogenolysis conditions, trifluoromethylbenzylamines are retained while the benzyl group is cleaved.[20]

4. Pd–C, K$_2$CO$_3$, H$_2$, MeOH, 10 min, 94% yield.[21]

5. Polymethylhydrosiloxane, Pd(OH)$_2$, EtOH, BOC$_2$O, rt, 87–92% yield. These conditions cleave the benzyl group with concomitant protection of the amine with a BOC group while maintaining an MPM ether. Trityl and diphenyl-methylamines react similarly.[22]

6. Na, NH$_3$, excellent yields.[23]

7. Li, (CH$_2$NH$_2$)$_2$, TEA, THF, 71% yield. Standard Birch conditions or the chloroformate method failed to cleanly remove the benzyl group from the following piperidine.[24] It may be that allylamine cleavage is competitive under the normal Birch conditions.

Acylative Methods. Benzyl groups, as well as other alkyl groups, can be converted to various carbamates by a variation of the von Braun reaction.[25,26] These can then be cleaved by conditions that are outlined in the section on carbamates.

1. CCl$_3$CH$_2$OCOCl, CH$_3$CN, 93%.[27,28]

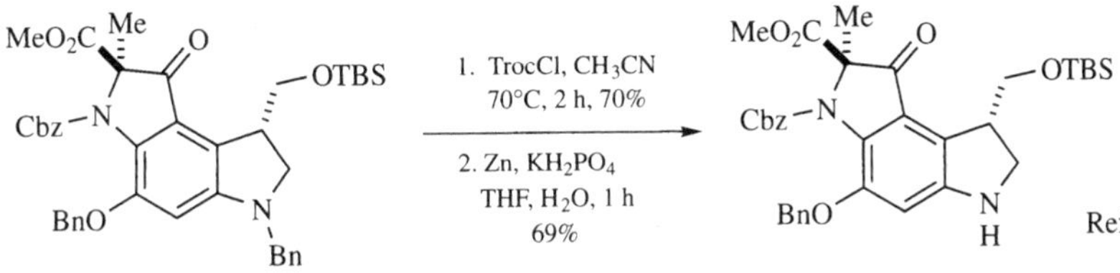

Ref. 29

2. (a) $ClCO_2Et$, CH_2Cl_2, reflux. (b) $PhNEt_2$-BI_3, 25°C, 85–89% yield.[30]
3. $Me_3SiCH_2CH_2OCOCl$, THF, −50°C, then 25°C, overnight, 78–91% yield.[31]
4. α-Chloroethyl chloroformate, NaOH.[32,33] The 4-methoxybenzyl group is selectively cleaved with this reagent, and the benzyl group is cleaved in preference to the 4-nitrobenzyl group.[34] In general, cleavage is expected at the most electron-rich nitrogen.

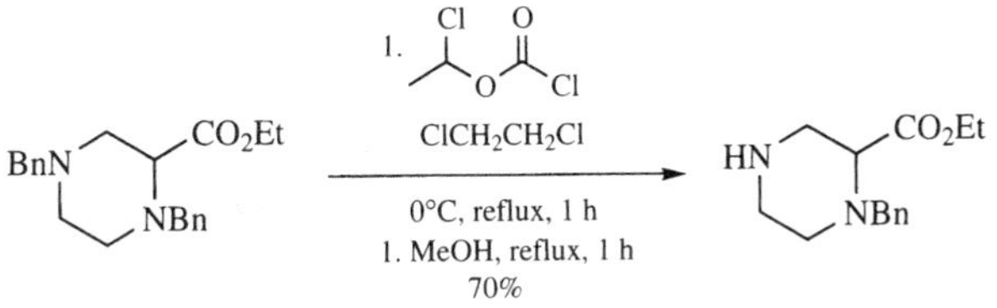

Ref. 35

5. Vinyl chloroformate is reported to be the best reagent for dealkylation of tertiary alkyl amines.[36]
6. Allyl chloroformate, CH_2Cl_2, >80% yield.[37] In this case the benzylamine was converted to an Alloc carbamate.
7. Triphosgene, CH_2Cl_2, 0°C, 77% yield. This method is quite general and in competition experiments the most electron rich amine is converted to the carbamoyl chloride.[38,39] These can be hydrolyzed to the amine or converted to various carbamates if desired.

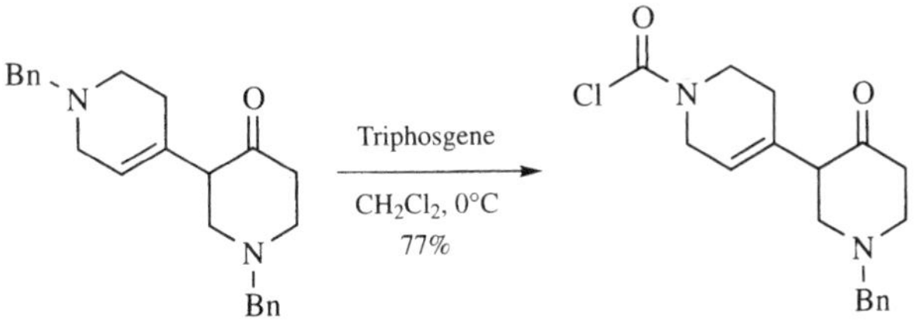

Oxidative Methods

1. RuO_4, NH_3, H_2O, 70% yield.[40]
2. *m*-Chloroperoxybenzoic acid followed by $FeCl_2$, −10°C, 6–80% yield.[41]
3. Co(II)L, *t*-BuOOH, DMSO, 40°C; H_2O, 90–97% yield.[42]
4. *t*-BuOLi, $CuBr_2$, 20 min, THF, rt, 99%.[43]
5. TPAP, NMO, rt, CH_3CN, 89% yield.[44]
6. CAN, CH_3CN, H_2O, rt, 89% yield.[45] A phenylthioether was not oxidized under these conditions.[46] These conditions are selective for acyclic tertiary benzyl

amines. Cyclic and some aromatic amines are inert to these conditions.[47] With dibenzylamines only one benzyl group is removed.

7. *o*-Iodoxybenzoic acid (IBX) in DMSO will oxidize benzylamines and other amines to the imine (49–98% yield) which is easily hydrolyzed with mild aqueous acid.[48,49] The reagent also converts dithianes to ketones in excellent yield.
8. NIS, CH_2Cl_2, rt, 50–98% yield.[50]

9. Diisopropyl azodicarboxylate, THF, then acid hydrolysis.[51] The reaction proceeds through triazane formation which then decomposes to give an imine which is hydrolyzed.

Miscellaneous Methods

1. BBr_3, CH_2Cl_2, rt, 54–88% yield. This method was used for the cleavage or arylbenzylamines.[52] A PMB-protected arylamine can also be cleaved by this method.
2. *h*ν, 405 nm ($CuSO_4$: NH_3 solution filter), CH_3CN, H_2O, 9,10-dicyanoanthracene, 6–10 h, 78–90% yield.[53]

N-4-Methoxybenzylamine (MPM−NR₂): $CH_3OC_6H_4CH_2NR_2$

Formation

1. $MeOC_6H_4CH_2Br$, KI, K_2CO_3, DMF, 92% yield.[54]
2. $MeOC_6H_4CH_2OH$, CH_3CN, cat. PTSA, 90% yield.[55]

Cleavage

1. Pd–C, HCl, MeOH, H_2.[56]
2. $Pd(OH)_2$, H_2. A hydroxamic acid is stable to these conditions.[57]
3. α-Chloroethyl chloroformate, THF, 89–98% yield.[34]
4. DDQ is often used to remove the MPM group from alcohols, and can be used to cleave it from an amine, but in the following case over-oxidation also occurs.[58]

5. Selective removal of the PMB group can be accomplished with DDQ in the presence of the benzyl group but not with the use of CAN.[59,60]

	100	0
DDQ	100	0
CAN	50	50

In the presence of a proximal alcohol the aminal is isolated upon DDQ treatment. This can be cleaved by treatment with NaOH followed by NaBH$_4$.[55]

N-2,4-Dimethoxybenzylamine (Dmb−NR$_2$): 2,4-(CH$_3$O)$_2$C$_6$H$_3$CH$_2$NR$_2$

The dimethoxybenzyl group was used for backbone protection of the pseudopeptides of the form Xaaψ(CH$_2$N)Gly (Xaa = amino acid). It is introduced by reductive alkylation with the aldehyde and NaCNBH$_3$. Acidolysis with TFMSA in TFA/thioanisole is used to remove it from the amine, but the efficiency is dependent upon the peptide sequence.[61] Cleavage of the Dmb group is also achieved by conversion with trifluoroacetic anhydride to the amide, which is then removed with NaBH$_4$/EtOH (93–97% yield).[62] It may also be cleaved with TsOH.[63]

N-2-Hydroxybenzylamine (HBn−NR₂): 2-(HO)C₆H₄CH₂NR₂

Amino acids were protected by reductive alkylation with salicylaldehyde (NaBH₄, KOH, aq. EtOH). The amine is released by treatment with CF_3SO_3H (TFA, EDT, PhSMe, 2 h, >75% yield).[64]

1. L. Velluz, G. Amiard, and R. Heymes, *Bull. Soc. Chim. Fr.*, 1012 (1954).

2. J. H. Cho and B. M. Kim, *Tetrahedron Lett.*, **43**, 1273 (2002).

3. N. Yamazaki and C. Kibayashi, *J. Am. Chem. Soc.*, **111**, 1397 (1989).

4. B. D. Gray and P. W. Jeffs, *J. Chem. Soc., Chem. Commun.*, 1329 (1987).

5. R. N. Salvatore, S. E. Schmidt, S. I. Shin, A. S. Naagle, J. H. Worrell, and K. W. Jung, *Tetrahedron Lett.*, **41**, 9705 (2000).

6. A. Loris, A. Perosa, M. Selva, and P. Tundo, *J. Org. Chem.*, **69**, 3953 (2004).

7. L. J. Liotta and B. Ganem, *Tetrahedron Lett.*, **30**, 4759 (1989).

8. L. J. Liotta and B. Ganem, *Isr. J. Chem.*, **31**, 215 (1991).

9. C. M. Cain, R. P. C. Cousins, G. Coumbarides, and N. S. Simpkins *Tetrahedron*, **46**, 523 (1990).

10. A. Guy and J. F. Barbetti, *Synth. Commun.*, **22**, 853 (1992).

11. J. A. Sclafani, M. T. Maranto, T. M. Sisk, and S. A. Van Arman, *J. Org. Chem.*, **61**, 3221 (1996).

12. M. Kanai, M. Yasumoto, Y. Kuriyama, K. Inomiya, Y. Katsuhara, K. Higashiyama, and A. Ishii, *Chem. Lett.*, **33**, 1424 (2004).

13. B. ElAmin, G. M. Anantharamaiah, G. P. Royer, and G. E. Means, *J. Org. Chem.*, **44**, 3442 (1979).

14. W. H. Hartung and R. Simonoff, *Org. Reactions*, **VII**, 263 (1953).

15. S. Ram and L. D. Spicer, *Tetrahedron Lett.*, **28**, 515 (1987); *idem, Synth. Commun.*, **17**, 415 (1987); O. Germay, N. Kumar, and E. J. Thomas, *Tetrahedron Lett.*, **42**, 4969 (2001).

16. B. M. Adger, C. O'Farrell, N. J. Lewis, and M. B. Mitchell, *Synthesis*, 53 (1987).

17. A. S. Kende, K. Liu, and K. M. J. Brands, *J. Am. Chem. Soc.*, **117**, 10597 (1995).

18. J. S. Bajwa, J. Slade, and O. Repic, *Tetrahedron Lett.*, **41**, 6025 (2000).

19. R. C. Bernotas and R. V. Cube, *Synth. Commun.*, **20**, 1209 (1990).

20. M. Kanai, M. Yasumoto, Y. Kuriyama, K. Inomiya, Y. Katsuhara, K. Higashiyama, and A. Ishii, *Org. Lett.*, **5**, 1007 (2003).

21. A. N. Hulme and E. M. Rosser, *Org. Lett.*, **4**, 265 (2002).

22. S. Chandrasekhar, B. N. Babu, and C. R. Reddy, *Tetrahedron Lett.*, **44**, 2057 (2003).

23. V. du Vigneaud and O. K. Behrens, *J. Biol. Chem.*, **117**, 27 (1937).

24. S. R. Angle and R. M. Henry, *J. Org. Chem.*, **63**, 7490 (1998).

25. H. A. Hageman, *Org. Reactions*, **7**, 198 (1953).

26. For a review, see J. H. Cooley and E. J. Evain, *Synthesis*, 1 (1989).

27. V. H. Rawal, R. J. Jones, and M. P. Cava, *J. Org. Chem.*, **52**, 19 (1987).

28. M. Shirai, S. Okamoto, and F. Sato, *Tetrahedron Lett.*, **40**, 5331 (1999).

29. K. Yamada, T. Kurokawa, H. Tokuyama, and T. Fukuyama, *J. Am. Chem. Soc.*, **125**, 6630 (2003).

30. J. V. B. Kanth, C. K. Reddy, and M. Periasamy, *Synth. Commun.*, **24**, 313 (1994).

31. A. L. Campbell, D. R. Pilipauskas, I. K. Khanna, and R. A. Rhodes, *Tetrahedron Lett.*, **28**, 2331 (1987).

32. R. A. Olofson, J. T. Martz, J.-P. Senet, M. Piteau, and T. Malfroot, *J. Org. Chem.*, **49**, 2081 (1984).

33. P. DeShong and D. A. Kell, *Tetrahedron Lett.*, **27**, 3979 (1986).

34. B. V. Yang, D. O'Rourke, and J. Li, *Synlett*, 195 (1993).

35. S. Gubert, C. Braojos, A. Sacristan, and J. A. Ortiz, *Synthesis*, 318 (1991).

36. R. A. Olofson, R. C. Schnur, L. Bunes, and J. P. Pepe, *Tetrahedron Lett.*, **18**, 1567 (1977).

37. E. Magnier, Y. Langlois, and C. Mérienne, *Tetrahedron Lett.*, **36**, 9475 (1995).

38. M. G. Banwell, M. J. Coster, M. J. Harvey, and J. Moraes, *J. Org. Chem.*, **68**, 613 (2003).

39. L. Lemoucheux, J. Rouden, M. Ibazizene, F. Sobrio, and M.-C. Lasne, *J. Org. Chem.*, **68**, 7289 (2003).

40. X. Gao and R. A. Jones, *J. Am. Chem. Soc.*, **109**, 1275 (1987).

41. T. Monkovic, H. Wong, and C. Bachand, *Synthesis*, 770 (1985).

42. K. Maruyama, T. Kusukawa, Y. Higuchi, and A. Nishinaga, *Chem. Lett.*, **20**, 1093 (1991).

43. J. Yamaguchi and T. Takeda, *Chem. Lett.*, **21**, 1933 (1992).

44. A. Goti and M. Romani, *Tetrahedron Lett.*, **35**, 6567 (1994).

45. S. D. Bull, S. G. Davies, P. M. Kelly, M. Gianotti, and A. D. Smith, *J. Chem. Soc. Perkin Trans. 1*, 3106 (2001).

46. I. C. Baldwin, P. Briner, M. D. Eastgate, D. J. Fox, and S. Warren, *Org. Lett.*, **4**, 4381 (2002).

47. S. D. Bull, S. G. Davies, G. Fenton, A. W. Mulvaney, R. S. Prasad, and A. D. Smith, *J. Chem. Soc. Perkin Trans. 1*, 3765 (2000).

48. K. C. Nicolaou, C. J. N. Mathison, and T. Montagnon, *Angew. Chem. Int. Ed.*, **42**, 4077 (2003).

49. T. Sueda, D. Kajishima, and S. Goto, *J. Org. Chem.*, **68**, 3307 (2003).

50. E. J. Grayson and B. G. Davis, *Org. Lett.*, **7**, 2361 (2005).

51. J. Kroutil, T. Trnka, and M. Cerný, *Synthesis*, 446 (2004).

52. E. Paliakov and L. Strekowski, *Tetrahedron Lett.*, **45**, 4093 (2004).

53. G. Pandey and K. S. Rani, *Tetrahedron Lett.*, **29**, 4157 (1988).

54. M. Yamato, Y. Takeuchi, and Y. Ikeda, *Heterocycles*, **26**, 191 (1987).

55. M. E. Pierce, R. L. Parsons, Jr., L. A. Radesca, Y. S. Lo, S. Silverman, J. R. Moore, Q. Islam, A. Choudhury, J. M. D. Fortunak, D. Nguyen, C. Luo, S. J. Morgan, W. P. Davis, and P. N. Confalone, *J. Org. Chem.*, **63**, 8536 (1998).

56. B. M. Trost, M. J. Krische, R. Radinov, and G. Zanoni, *J. Am. Chem. Soc.*, **118**, 6297 (1996).

57. M. Rowley, P. D. Leeson, B. J. Williams, K. W. Moore, and R. Baker, *Tetrahedron*, **48**, 3557 (1992).

58. S. B. Singh, *Tetrahedron Lett.*, **36**, 2009 (1995).

59. S. D. Bull, S. G. Davies, G. Fenton, A. W. Mulvaney, R. S. Prasad, and A. D. Smith, *J. Chem. Soc. Perkin Trans. 1*, 3765 (2000).

60. B. Hungerhoff, S. S. Samanta, J. Roels, and P. Metz, *Synlett*, 77 (2000).
61. Y. Sasaki and J. Abe, *Chem. Pharm. Bull.*, **45**, 13 (1997).
62. P. Nussbaumer, K. Baumann, T. Dechat, and M. Harasek, *Tetrahedron*, **47**, 4591 (1991).
63. B. M. Trost and D. R. Fandrick, *Org. Lett.*, **7**, 823 (2005).
64. T. Johnson and M. Quibell, *Tetrahedron Lett.*, **35**, 463 (1994).

N-9-Phenylfluorenylamine (Pf−NR$_2$): 9-(C$_6$H$_5$)−(C$_{13}$H$_8$)−NR$_2$

Formation

1. 9-Pf-Br, Pb(NO$_3$)$_2$, CH$_3$CN, rt, 28 h, >80% yield.[1,2]
2. 9-Pf-Br, K$_3$PO$_4$, CH$_3$NO$_2$. This method avoids the use of lead nitrate.[3]

Cleavage

This group was reported to be 6000 times more stable to acid than the trityl group because of destabilization of the cation by the fluorenyl group.[4]

1. CH$_3$CN, H$_2$O, 0°C, 1 h to rt, 1 h.
2. 3% CF$_3$COOH, CH$_2$Cl$_2$, Et$_3$SiH, 0°C, 95% yield. The Et$_3$SiH serves to scavenge the cation.[5]
3. I$_2$, MeOH, 3–5 h, reflux, 72–85% yield. This method only cleaves tertiary Pf groups.[6] TBDMS and isopropylidene groups are also cleaved by this reagent.

4. H$_2$, Pd/C, EtOAc, AcOH.[7,8]

5. H$_2$, Pd(OH)$_2$, THF, MeOH, BOC$_2$O.[9]

6. Li, NH$_3$, THF, 76% yield.[10]

N-Fluorenylamine (Flu−NR₂)

Fluoreneamine was used to introduce a nitrogen through a Schiff base. It was cleaved with DDQ in excellent yield.[11]

1. P. L. Feldman and H. Rapoport, *J. Org. Chem.*, **51**, 3882 (1986).
2. B. D. Christie and H. Rapoport, *J. Org. Chem.*, **50**, 1239 (1985).
3. S. C. Bergmeier, A. A. Cobas, and H. Rapoport, *J. Org. Chem.*, **58**, 2369 (1993).
4. R. Bolton, N. B. Chapman, and J. Shorter, *J. Chem. Soc.*, 1895 (1964).
5. D. Kadereit, P. Deck, I. Heinemann, and H. Waldmann, *Chem. Eur. J.*, **7**, 1184 (2001).
6. J. H. Kim, W. S. Lee, M. S. Yang, S. G. Lee, and K. H. Park, *Synlett* 614 (1999).
7. H.-G. Lombart and W. D. Lubell, *J. Org. Chem.*, **61**, 9437 (1996).
8. J. A. Campbell, W. K. Lee, and H. Rapoport, *J. Org. Chem.*, **60**, 4602 (1995).
9. G. Jeannotte and W. D. Lubell, *J. Org. Chem.*, **69**, 4656 (2004).
10. W. D. Lubbel, T. F. Jamison, and H. Rapoport, *J. Org. Chem.*, **55**, 3511 (1990).
11. M. Takamura, Y. Hamashima, H. Usuda, M. Kanai, and M. Shibasaki, *Angew. Chem. Int. Ed.*, **39**, 1650 (2000).

N-Ferrocenylmethylamine (Fcm−NR₂): $C_{10}H_{10}FeCH_2NR_2$

$$CH_2NR_2$$
$$Fe$$

The Fcm derivative is prepared from amino acids on treatment with formylferrocene and Pd-phthalocyanine by reductive alkylation (60–89% yield). It is cleaved with 2-thionaphthol/CF_3COOH. Its primary advantage is its color, making it easily detected.[1]

1. H. Eckert and C. Seidel, *Angew. Chem., Int. Ed. Engl.*, **25**, 159 (1986).

N-2-Picolylamine *N'*-Oxide: R₂NCH₂−2-pyridyl *N*-Oxide (Chart 10)

N-2-Picolylamine *N'*-oxide, used in oligonucleotide syntheses, is cleaved by acetic anhydride at 22°C, followed by methanolic ammonia (85–95% yield).[1]

1. Y. Mizuno, T. Endo, T. Miyaoka, and K. Ikeda, *J. Org. Chem.*, **39**, 1250 (1974).

N-7-Methoxycoumar-4-ylmethylamine

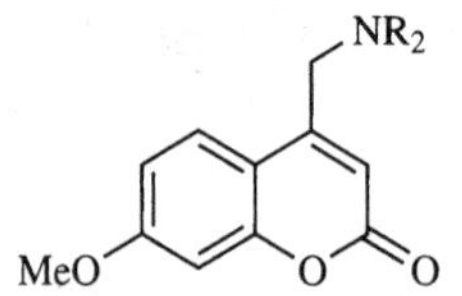

The derivative is formed by reaction of an amine with 4-bromomethyl-7-methoxy-coumarin. Cleavage is affected by irradiation at >360 nm in the presence of an H-donor such as $C_{10}H_{21}SH$ in MeOH, 77–90% yield.[1]

1. R. O. Schoenleber and B. Giese, *Synlett*, **4**, 501 (2003).

N-(Diphenylmethyl)amine (DPM−NR₂): Ph₂CHNR₂

Formation

1. By reduction of a benzophenone imine with $NaCNBH_3$, pH 6, 25°C.[1,2]
2. (Diphenylmethyl)amine is used as a convenient protected source of ammonia.[3]

Cleavage

1. Et_3SiH, TFA, 86% yield.[4]
2. Pd–C, cyclohexene, 1 *M* HCl, EtOH, 83% yield.[5] Ammonium formate[2] and polymethylhydrosiloxane (PMHS)[6] can also be used as a source of hydrogen.
3. $Pd(OH)_2$, H_2, MeOH, 20 bar, 40°C, 8 h, 90% yield.[7]
4. DDQ, benzene, 4-Å ms, 60°C, then 0.1 *N* HCl, Et_2O, 6 h, 70–95% yield.[8]

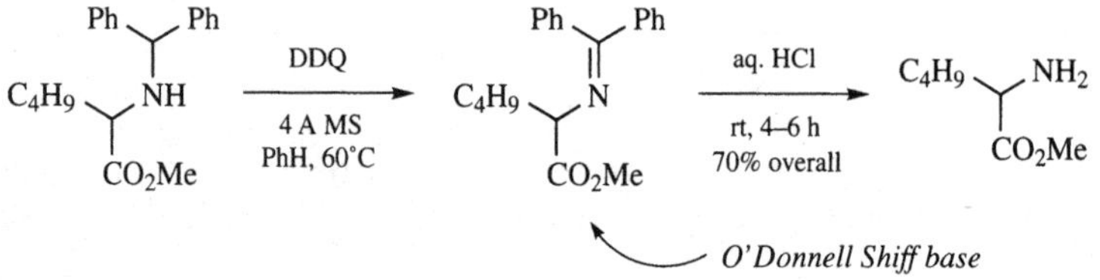

O'Donnell Shiff base

5. Ozonolysis, CH_2Cl_2, −78°C, 3 h, quench with $MeOH/NaBH_4$, 77–81% yield. This method was developed for the cleavage of aziridinyl DPM groups.[9]

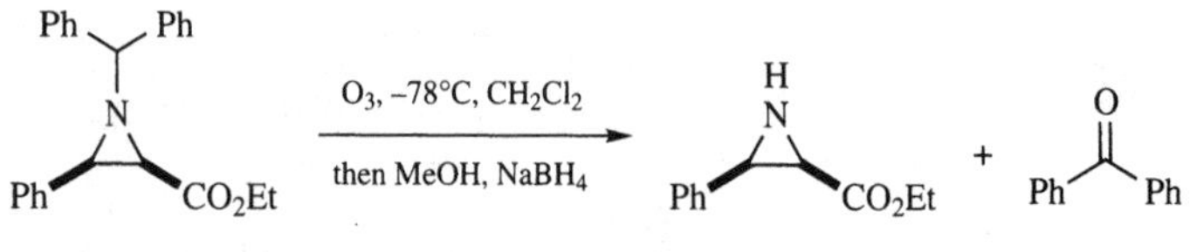

N-Bis(4-methoxyphenyl)methylamine (Dod—NR$_2$): (4-MeOC$_6$H$_4$)$_2$CHNR$_2$ (Chart 10)

This derivative has been used to protect the amines of amino acids [(4-MeOC$_6$H$_4$)$_2$CHCl, Et$_3$N, 0–20°C 20 h, 67% yield]. It is easily cleaved with 80% AcOH (80°C, 5 min, 73% yield).[10] The Dod group can be cleaved in the presence of the Mmd group, which is cleaved with more concentrated TFA/CH$_2$Cl$_2$.[11]

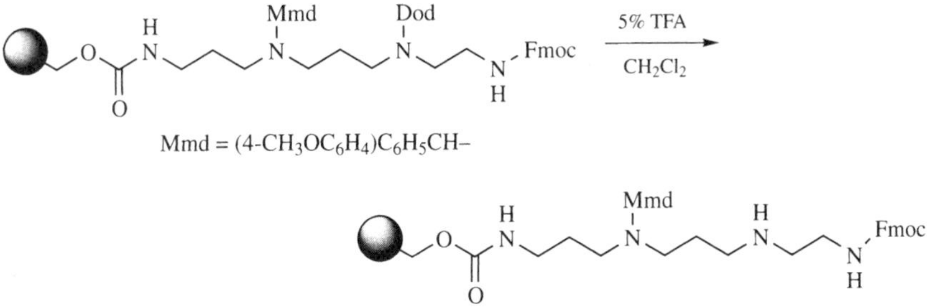

Mmd = (4-CH$_3$OC$_6$H$_4$)C$_6$H$_5$CH–

N-5-Dibenzosuberylamine (DBS—NR$_2$):

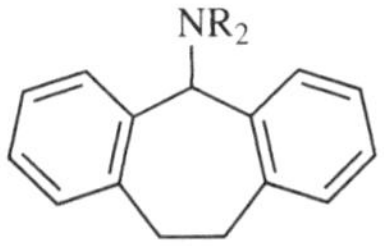

The dibenzosuberylamine is prepared in quantitative yield from an amine or amino acid and suberyl chloride; this chloride has also been used to protect hydroxyl, thiol, and carboxyl groups. This group has been examined for protection of the guanidine group.[12] Although the dibenzosuberylamine is stable to 5 *N* HCl/dioxane (22°C, 16 h) and to refluxing HBr (1 h), it is completely cleaved by some acids (HCOOH, CH$_2$Cl$_2$, 22°C, 2 h; CF$_3$COOH, CH$_2$Cl$_2$, 22°C, 0.5 h; BBr$_3$, CH$_2$Cl$_2$, 22°C, 0.5 h; 4 *N* HBr, AcOH, 22°C, 1 h; 60% AcOH, reflux, 1 h) and by reduction (H$_2$, Pd–C, CH$_3$OH, 22°C, 1 h, 100% cleaved).[13] Hydrogenolysis in the presence of formaldehyde converts the DBS group to a methylamine.[14]

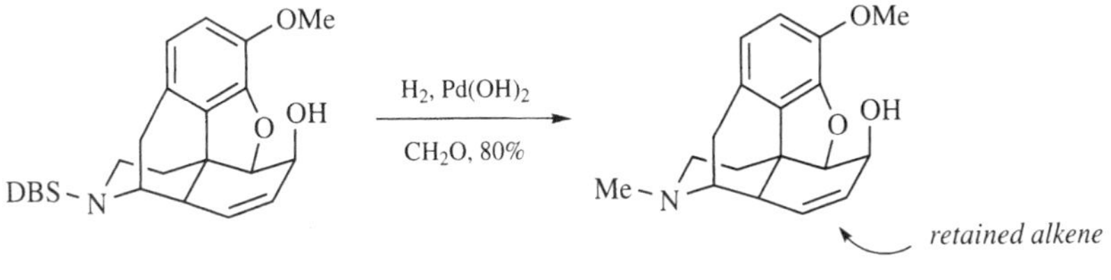

N-Triphenylmethylamine (Tr—NR$_2$): Ph$_3$CNR$_2$ (Chart 10)

The bulky triphenylmethyl group has been used to protect a variety of amines such as amino acids, penicillins, and cephalosporins. Esters of *N*-trityl α-amino acids are shielded from hydrolysis and require forcing conditions for cleavage. The α-proton is also shielded from deprotonation, which means that esters elsewhere in the molecule can be selectively deprotonated.

Formation

1. TrCl, Et$_3$N, 25°C, 4 h.[15]

2. TrBr, CHCl$_3$, DMF, rt, 0.5–1 h; Et$_3$N, rt, 50 min.[16] These conditions also lead to tritylation of carboxyl groups in the amino acids, but they can be selectively hydrolyzed. This method was considered to be an improvement over the standard methods of N-tritylation of amino acids.

3. (i) Silylation of −CO$_2$H with Me$_3$SiCl, Et$_3$N; (ii) TrCl, Et$_3$N; (iii) MeOH, 65–92% yield.[17] To effect N-tritylation of serine, Me$_2$SiCl$_2$ should be used in the silylation step.

Cleavage

1. HCl, acetone, 25°C, 3 h, 80% yield.[15]

2. Yb(OTf)$_3$, THF, 1 eq. H$_2$O, 89–95% yield. Trityl ethers are cleaved similarly.[18]

3. H$_2$, Pd black, EtOH, 45°C, 92% yield.[19] If the hydrogenolysis is performed in the presence of (BOC)$_2$O or Fmoc−OSu, the released amine is converted to the BOC and Fmoc derivatives *in situ*.[20]

4. Pd/C, HCO$_2$NH$_4$, EtOH, AcOH, >82% yield.[21] Polymethylhydrosiloxane (PMHS) can be used as a hydrogen source as well.[6]

5. Na, NH$_3$.[22]

6. Li, naphthalene, THF, 1–6 h, 41–94% yield. A primary tritylamine can be cleaved in the presence of a secondary tritylamine if the reaction is conducted at 0°C and trityl ethers are cleaved in preference to tritylamines.[23]

7. Hydroxybenzotriazole (HOBT), trifluoroethanol, rt.[24]

8. 1-Hydroxy-7-azabenzotriazole, TMSCl, in trifluoroethanol or TMSCl in trifluoroethanol, quant.[25]

9. 0.2% TFA, 1% H$_2$O, CH$_2$Cl$_2$.[25] Under these conditions, an S-Tr group is retained while an N-trityl group is cleaved.[26]

10. (A) TFA, Et$_3$SiH, CH$_2$Cl$_2$, 0°C or (B) MsOH, Et$_3$SiH, CH$_2$Cl$_2$, 0°C or (C) TFA, Me$_3$N·BH$_3$, CH$_2$Cl$_2$, 0°C, 5–88% yield. These conditions were developed for the removal of the trityl group from aziridines. The choice of conditions depends on the substrate and as illustrated in the second example the cleavage process is not always straightforward.[27]

N-[(4-Methylphenyl)diphenylmethyl]amine (Mtt−NR$_2$):
(4-CH$_3$C$_6$H$_4$)(C$_6$H$_5$)$_2$C−NR$_2$

The Mtt group was examined for lysine side-chain protection during peptide synthesis and lipidated peptide synthesis. It is cleaved with 1% TFA in CH$_2$Cl$_2$; however, since this is an equilibrium, it is better to include a cation scavenger such as Et$_3$SiH[28] or (i-Pr)$_3$SiH[29] to drive the equilibrium.

N-[(4-Methoxyphenyl)diphenylmethyl]amine (MMTr−NR$_2$):
(4-CH$_3$O-C$_6$H$_4$)(C$_6$H$_5$)$_2$C−NR$_2$ (Chart 10)

In contrast to the corresponding MMTr ethers, the amine derivatives are substantially more stable and require much stronger acid to cleave them. The MMTr derivative is easily prepared from amino acids (from the silylamine: MMTrCl, rt, 18 h, 91% yield)[30] and is readily cleaved by acid hydrolysis (5% CCl$_3$CO$_2$H, 4°C, 5 min, 100% yield)[31] or (CHCl$_2$CO$_2$H, anisole, CH$_2$Cl$_2$, rt 1 h).[30] MMTBF$_4$ has been recommended as a superior reagent for the introduction of this group because of its ease of purification and good stability.[32] The kinetics of detritylation were shown to be dependent upon the basicity of the amine.[33]

1. K. M. Czerwinski, L. Deng, and J. M. Cook, *Tetrahedron Lett.*, **33**, 4721 (1992).

2. E. D. Cox, L. K. Hamaker, J. Li, P. Yu, K. M. Czerwinski, L. Deng, D. W. Bennett, J. M. Cook, W. H. Watson, and M. Krawiec, *J. Org. Chem.*, **62**, 44 (1997).

3. M. E. Jung and Y. M. Choi, *J. Org. Chem.*, **56**, 6729 (1991).

4. W. L. Neumann, M. M. Rogic, and T. J. Dunn, *Tetrahedron Lett.*, **32**, 5865 (1991); J. R. Porter, W. G. Wirschun, and K. W. Kuntz, *J. Am. Chem. Soc.*, **122**, 2657 (2000).

5. L. E. Overman, L. T. Mendelson, and E. J. Jacobsen, *J. Am. Chem. Soc.*, **105**, 6629 (1983).

6. S. Chandrasekhar, B. N. Babu, and C. R. Reddy, *Tetrahedron Lett.*, **44**, 2057 (2003).

7. E. Bacqué, J.-M. Paris, and S. Le Bitoux, *Synth. Commun.*, **25**, 803 (1995).

8. P. B. Sampson and J. F. Honek, *Org. Lett.*, **1**, 1395 (1999).

9. A. P. Patwardhan, Z. Lu, V. R. Pulgam, and W. D. Wulff, *Org. Lett.*, **7**, 2201 (2005).

10. R. W. Hanson and H. D. Law, *J. Chem. Soc.*, 7285 (1965).

11. D. Jönsson, *Tetrahedron Lett.*, **43**, 4793 (2002); D. Jönsson, A. Uddén *Tetrahedron Lett.*, **43**, 3125 (2002).

12. M. Noda and M. Kiffe, *J. Peptide Res.*, **50**, 329 (1997).

13. J. Pless, *Helv. Chim. Acta*, **59**, 499 (1976).

14. C. Y. Hong, L. E. Overman, and A. Romero, *Tetrahedron Lett.*, **38**, 8439 (1997).

15. H. E. Applegate, C. M. Cimarusti, J. E. Dolfini, P. T. Funke, W. H. Koster, M. S. Puar, W. A. Slusarchyk, and M. G. Young, *J. Org. Chem.*, **44**, 811 (1979).

16. M. Mutter and R. Hersperger, *Synthesis*, 198 (1989).

17. K. Barlos, D. Papaioannou, and D. Theodoropoulos, *J. Org. Chem.*, **47**, 1324 (1982).

18. R. J. Lu, D. Liu, and R. W. Giese, *Tetrahedron Lett.*, **41**, 2817 (2000).

19. L. Zervas and D. M. Theodoropoulos, *J. Am. Chem. Soc.*, **78**, 1359 (1956).

20. C. Dugave and A. Menez, *J. Org. Chem.*, **61**, 6067 (1996).

21. S. K. Sharma, M. F. Songster, T. L. Colpitts, P. Hegyes, G. Barany, and F. J. Castellino, *J. Org. Chem.*, **58**, 4993 (1993). These conditions also cleave benzyl esters.

22. H. Nesvadba and H. Roth, *Monatsh. Chem.*, **98**, 1432 (1967).

23. C. Behloul, D. Guijarro, and M. Yus, *Synthesis*, 1274 (2004).

24. M. Bodansky, M. A. Bednarek, and A. Bodansky, *Int. J. Pept. Protein Res.*, **20**, 387 (1982).

25. J. Alsina, E. Giralt, and F. Albericio, *Tetrahedron Lett.*, **37**, 4195 (1996).

26. M. Maltese, *J. Org. Chem.*, **66**, 7615 (2001).

27. E. Vedejs, A. Klapars, D. L. Warner, and A. H. Weiss, *J. Org. Chem.*, **66**, 7542 (2001); E. Vedejs, B. N. Naidu, A. Klapars, D. L. Warner, V.-s. Li, Y. Na, and H. Kohn, *J. Am. Chem. Soc.*, **125**, 15796 (2003).

28. D. Kadereit, P. Deck, I. Heinemann, and H. Waldmann, *Chem. Eur. J.*, **7**, 1184 (2001); D. Li and D. L. Elbert, *J. Peptide Res.*, **60**, 300 (2002); L. Bourel, O. Carion, H. Gras-Masse, and O. Melnyk, *J. Peptide Sci.*, **6**, 264 (2000).

29. H. B. Lee, M. Pattarawarapan, S. Roy, and K. Burgess, *Chem. Commun.*, 1674 (2003); J. Gariépy, S. Remy, X. Zhang, J. R. Ballinger, E. Bolewska-Pedyczak, M. Rauth, and S. K. Bisland, *Bioconjugate Chem.*, **13**, 679 (2002).

30. G. M. Dubowchik and S. Radia, *Tetrahedron Lett.*, **38**, 5257 (1997).

31. Y. Lapidot, N. de Groot, M. Weiss, R. Peled, and Y. Wolman, *Biochim. Biophys. Acta*, **138**, 241 (1967).

32. A. P. Henderson, J. Riseborough, C. Bleasdale, W. Clegg, M. R. J. Elsegood, and B. T. Golding, *J. Chem. Soc., Perkin Trans* 1, 3407 (1997).

33. M. Canle L., I Demirtas, and H. Maskill, *J. Chem. Soc., Perkin Trans 2*, 1748, (2001).

Imine Derivatives

A number of imine derivatives have been prepared as amine protective groups, but most of these have not seen extensive use. The most widely used are the benzylidene and diphenylmethylene derivatives. The less used derivatives are listed, for completeness, with their references at the end of this section. For the most part, they are prepared from the aldehyde and the amine by water removal; cleavage is effected by acid hydrolysis.

N-1,1-Dimethylthiomethyleneamine: $(MeS)_2C=NR$

This group was used to protect the nitrogen of glycine in a synthesis of amino acids.[1]

Formation

1. CS_2, TEA, $CHCl_3$, 20–40°C, 1 h; MeI, reflux, 1 h, 77% yield.[2]
2. CS_2, NaOH, benzene; MeI, benzene, TEBA, 20°C, 39–86% yield.[3]
3. CS_2, TEA, $BrCH_2CH_2Br$, 70–75% yield.[4]

Cleavage

1. H_2O_2, HCO_2H, TsOH, 0–20°C, 90% yield.[2]
2. HCl, H_2O, THF, rt, 100% yield.[2,5]

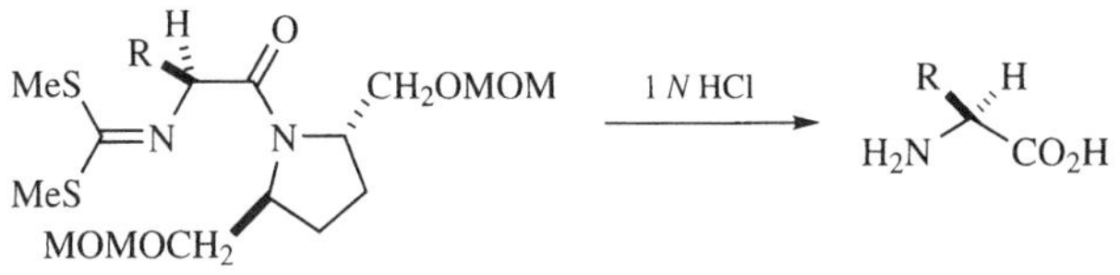

3. Direct conversion to other protective groups is possible.[6]

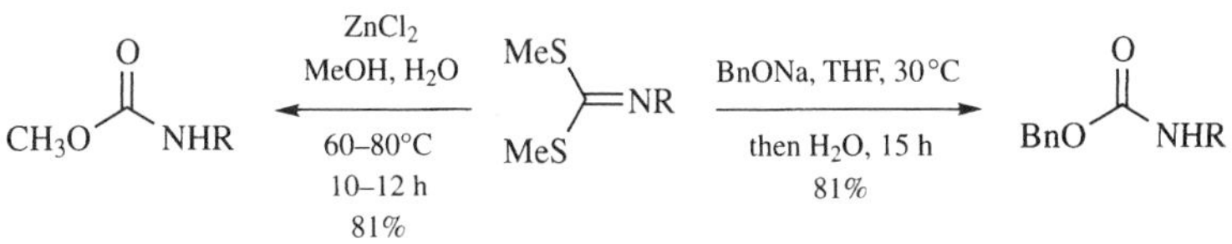

N-**Benzylideneamine:** RN=CHPh (Chart 10)

Most applications of this derivative have been for the preparation and modification of amino acids, although some applications in the area of carbohydrates have been reported. The derivative is stable to *n*-butyllithium, lithium diisopropylamide, and *t*-BuOK.[7] Various substituted benzylidenes have been used for amine protection of amino acids during phase transfer catalyzed alkylations.

Formation

1. PhCHO, Et_3N, 80–90% yield.[8]
2. PhCHO, Na_2SO_4, benzene, rt, 99% yield.[9] A primary amine is protected in the presence of a secondary amine.[10]
3. PhCHO, trimethyl orthoformate, 89–100% yield.[11]

Cleavage

1. 1 *N* HCl, 25°C, 1 h.[1,12]
2. H_2, Pd-C, CH_3OH.[13]
3. Hydrazine, EtOH, reflux, 6 h, 70% yield.[14]
4. Girard-T reagent, >75% yield.[15]

N-*p*-**Methoxybenzylideneamine:** 4-$MeOC_6H_4CH$=NR

The *N*-*p*-methoxybenzylideneamine has been used to protect glucosamines.[16]

Formation

4-$MeOC_6H_4CHO$, benzene, pyridine, heat, >72% yield.[17]

Cleavage

1. MeOH, 10% aq. AcOH, TsNHNH$_2$, >81% yield.[13,18]
2. 5 N HCl.[19]

N-Diphenylmethyleneamine: RN=CPh$_2$

The derivative of glycine, prepared from benzophenone (cat. BF$_3$·Et$_2$O, xylene, reflux, 82% yield), has found considerable use in the preparation of amino acids. It is preferably prepared by an exchange reaction with benzophenonimine (Ph$_2$C=NH, CH$_2$Cl$_2$, rt).[20] It is stable to DIBAH, Grignard reagents, strong base,[21] and osmium oxidations.[22] When used for the protection of serine, it increases the nucleophilicity of the hydroxyl group and improves β-O-glycosylation.[23] Benzophenonimine has been used as a protective group for ammonia in the amination of aromatic rings.[24] The fluorene analog, prepared from fluorenone (TiCl$_4$, toluene, 0°C), has also been used to protect a primary amine.[22]

Cleavage

1. Concd. HCl, reflux, 6 h or aq. citric acid, 12 h.[25]
2. H$_2$, Pd–C, MeOH, rt, 14 h, 90% yield.[26]
3. NH$_2$OH, 3 min, pH 4-6.[27–29]

N-[(2-Pyridyl)mesityl]methyleneamine: (C$_5$H$_4$N)(Me$_3$C$_6$H$_2$)C=NR[30]

The imine, prepared from an amine and (C$_5$H$_4$N)(Me$_3$C$_6$H$_2$)CO (TiCl$_4$, toluene, reflux, 12 h; NaOH, 80% yield), can be cleaved with concd. HCl (reflux). The protective group was used to direct α-alkylation of amines.

N-(N',N'-Dimethylaminomethylene)amine (N,N-Dimethylformamidine): RN=CHN(CH$_3$)$_2$

The formamidine is prepared by heating the primary amine in DMF-dimethylacetal (81–100% yield). Deprotection is effected by heating in EtOH with ZnCl$_2$.[31] LiAlH$_4$ (Et$_2$O, reflux), hydrazine (AcOH, MeOH), KOH (MeOH, reflux),[32] dilute ammonia (high yield)[33] and concd. HCl (reflux, 65–90% yield)[34] are also known to cleave the formamidine group. Treatment of the formamidine in MeOH/H$_2$O with or without TEA results in the formation of a formamide (48–100% yield).[35]

N-(N',N'-Dibenzylaminomethylene)amine (N,N-Dibenzylformamidine): (C$_6$H$_5$CH$_2$)$_2$NCH=NR

Heating a primary amine with dibenzylformamide-dimethyl acetal in CH$_3$CN gives the formamidine in 49–99% yield. N',N'-Dibenzyl chloromethylene iminium chloride is a more reactive reagent that can be used at lower temperatures with excellent yields for amines not bearing unprotected alcohols.[36] It is cleaved by hydrogenolysis (Pd(OH)$_2$, MeOH, H$_2$O, H$_2$, 52–99% yield).[35,37]

N-(*N'*-*t*-Butylaminomethylene)amine (*N*-*t*-Butylformamidine):
$(CH_3)_3CN=CH-NR_2$

The *t*-butylformamidine was used to protect and direct the course of metalation of secondary amines. It is formed from *N,N*-dimethyl-*N'*-*t*-butylformamidine by an acid-catalyzed exchange reaction or from the *N*-*t*-butylimidate tetrafluoroborate salt, and is cleaved with hydrazine.[38]

N,N'-Isopropylidenediamine:[39] (Chart 10)

N-*p*-Nitrobenzylideneamine: $4\text{-}NO_2C_6H_4CH=NR$[40] (Chart 10)

N-Salicylideneamine: $2\text{-}HO\text{-}C_6H_4CH=NR$[41] (Chart 10)

This imine is stabilized by hydrogen bonding of the phenolic hydroxyl with the lone pair on the imine. This group is cleaved with strong acids such as HCl or with Me-ONH$_2$/MeOH/CHCl$_3$, which is preferred over the use of hydroxylamine because it is a poorer nucleophile and thus is compatible with esters.[42]

N-5-Chlorosalicylideneamine: $2\text{-}HO-5\text{-}ClC_6H_3CH=NR$[43]

N-(5-Chloro-2-hydroxyphenyl)phenylmethyleneamine:
$RN=C(Ph)C_6H_3\text{-}2\text{-}OH\text{-}5\text{-}Cl$[44,45]

N-Cyclohexylideneamine: $C_6H_{11}N=CHR$[46]

This imine is stable to the Fe(acac)$_3$-catalyzed Grignard coupling of aryl halides.

N-*t*-Butylideneamine: $(CH_3)_3CCH=NR$[47]

1. S. Ikegami, T. Hayama, T. Katsuki, and M. Yamaguchi, *Tetrahedron Lett.*, **27**, 3403 (1986); S. Ikegama, H. Uchiyama, T. Hayama, T. Katsuki, and M. Yamaguchi, *Tetrahedron*, **44**, 5333 (1988).

2. D. Hoppe and L. Beckmann, *Liebigs Ann. Chem.*, 2066 (1979).

3. C. Alvarez-Ibarra, M. L. Quiroga, E. Martinez-Santos, and E. Toledano, *Org. Prep. Proced. Int.*, **23**, 611 (1991).

4. S. Hanessian and Y. L. Bennani, *Tetrahedron Lett.*, **31**, 6465 (1990).

5. W. Oppolzer, R. Moretti, and S. Thomi, *Tetrahedron Lett.*, **30**, 6009 (1989).

6. M. Anbazhagan, T. I. Reddy, and S. Rajappa, *J. Chem. Soc., Perkin Trans. 1*, 1623 (1997).

7. N. De Kimpe and P. Sulmon, *Synlett*, 161 (1990).

8. P. Bey and J. P. Vevert, *Tetrahedron Lett.*, **18**, 1455 (1977).

9. B.W. Metcalf and P. Casara, *Tetrahedron Lett.*, **16**, 3337 (1975).

10. J. D. Prugh, L. A. Birchenough, and M. S. Egbertson, *Synth. Commun.*, **22**, 2357 (1992).

11. G. C. Look, M. M. Murphy, D. A. Campbell, and M. A. Gallop, *Tetrahedron Lett.*, **36**, 2937 (1995).

12. D. Ferroud, J. P. Genet, and R. Kiolle, *Tetrahedron Lett.*, **27**, 23 (1986).

13. R. A. Lucas, D.F. Dickel, R. L. Dziemian, M. J. Ceglowski, B. L. Hensle, and H. B. MacPhillamy, *J. Am. Chem. Soc.*, **82**, 5688 (1960).

14. G. W. J. Fleet and I. Fleming, *J. Chem. Soc. C*, 1758 (1969).

15. T. Watanabe, S. Sugawara, and T. Miyadera, *Chem. Pharm Bull.*, **30**, 2579 (1982).

16. A. Marra and P. Sinay, *Carbohydr. Res.*, **200**, 319 (1990).

17. D. R. Mootoo and B. Fraser-Reid, *Tetrahedron Lett.*, **30**, 2363 (1989).

18. F. Baumberger, A. Vasella, and R. Schauer, *Helv. Chim. Acta*, **71**, 429 (1988).

19. M. Bergmann and L. Zervas, *Ber.*, **64**, 975 (1931).

20. T. Hvidt, W. A. Szarek, and D. B. Maclean, *Can. J. Chem.*, **66**, 779 (1988); M. A. Peterson and R. Polt, *J. Org. Chem.*, **58**, 4309 (1993).

21. R. Polt and M. A. Peterson, *Tetrahedron Lett.*, **31**, 4985 (1990).

22. E. J. Corey, A. Guzman-Perez, and M. C. Noe, *J. Am. Chem. Soc.*, **117**, 10805 (1995).

23. L. Szabò, Y. Li, and R. Polt, *Tetrahedron Lett.*, **32**, 585 (1991).

24. J. P. Wolfe, J. Ahman, J. P. Sadighi, R. A. Singer, and S. L. Buchwald, *Tetrahedron Lett.*, **38**, 6367 (1997).

25. M. J. O'Donnell, J. M. Boniece, and S. E. Earp, *Tetrahedron Lett.*, **19**, 2641 (1978).

26. L. Wessjohann, G. McGaffin, and A. de Meijere, *Synthesis*, 359 (1989).

27. K.-J. Fasth, G. Antoni, and B. Langström, *J. Chem. Soc., Perkin Trans. I*, 3081 (1988).

28. M. Lögers, L. E. Overman, and G. S. Welmaker, *J. Am. Chem. Soc.*, **117**, 9139 (1995).

29. E. M. Stocking, J. F. Sanz-Cervera, and R. M. Williams, *J. Am. Chem. Soc.*, **122**, 1675 (2000).

30. J. M. Hornback and B. Murugaverl, *Tetrahedron Lett.*, **30**, 5853 (1989).

31. D. Toste, J. McNulty, and I. W. J. Still, *Synth. Commun.*, **24**, 1617 (1994).

32. A. I. Meyers, P. D. Edwards, W. F. Rieker, and T. R. Bailey, *J. Am. Chem. Soc.*, **106**, 3270 (1984); A. I. Meyers, *Aldrichimica Acta*, **18**, 59 (1985).

33. J. Zemlicka, S. Chládek, A. Holy, and J. Smrt, *Collect. Czech. Chem. Commun.*, **31**, 3198 (1966).

34. J. J. Fitt and H. W. Gschwend, *J. Org. Chem.*, **42**, 2639 (1977).

35. S. Vincent, C. Mioskowski, and L. Lebeau, *J. Org. Chem.*, **64**, 991 (1999).

36. S. Vincent, L. Lebeau, and C. Mioskowski, *Synth. Commum.*, **29**, 167 (1999).

37. S. Vincent, S. Mons, L. Lebeau, and C. Mioskowki, *Tetrahedron Lett.*, **38**, 7527 (1997).

38. A. I. Meyers, P. D. Edwards, W. F. Rieker, and T. R. Bailey, *J. Am. Chem. Soc.*, **106**, 3270 (1984).

39. P. M. Hardy and D. J. Samworth, *J. Chem. Soc., Perkin Trans. I*, 1954 (1977).

40. J. L. Douglas, D. E. Horning, and T. T. Conway, *Can. J. Chem.*, **56**, 2879 (1978).

41. J. N. Williams and R. M. Jacobs, *Biochem. Biophys. Res. Commun.*, **22**, 695 (1966).

42. A. R. Khomutov, A. S. Shvetsov, J. J. Vepsalainen, and A. M. Kritzyn, *Tetrahedron Lett.*, **42**, 2887 (2001).

43. J. C. Sheehan and V. J. Grenada, *J. Am. Chem. Soc.*, **84**, 2417 (1962).

44. B. Halpern and A. P. Hope, *Aust. J. Chem.*, **27**, 2047 (1974).

45. A. Abdipranoto, A. P. Hope, and B. Halpern, *Aust. J. Chem.*, **30**, 2711 (1977).

46. L. N. Pridgen, L. Snyder, and J. Prol, Jr., *J. Org. Chem.*, **54**, 1523 (1989).

47. S. Kanemasa, O. Uchida, and E. Wada, *J. Org. Chem.*, **55**, 4411 (1990).

Enamine Derivatives

N-(5,5-Dimethyl-3-oxo-1-cyclohexenyl)amine: (Chart 10)

This vinylogous amide has been prepared in 70% yield to protect amino acid esters. It is cleaved by treatment with either aqueous bromine[1] or nitrous acid (90% yield).[2]

N-2,7-Dichloro-9-fluorenylmethyleneamine

Formation/Cleavage[3]

N-1-(4,4-Dimethyl-2,6-dioxocyclohexylidene)ethylamine (Dde-NR₂)

The Dde group was developed for amine protection in solid-phase peptide synthesis. It is formed from 2-acetyldimedone in DMF and cleaved using 2% hydrazine in DMF[4,5] or ethanolamine.[6] Hydrazinolysis of the Dde group in the presence of the

Aloc group was found to be troublesome because of hydrogenation of the allyl group unless allyl alcohol was included in the deprotection mixture to scavenge diimide that reduces the olefin.[7] This is probably the result of some diimide formation by oxidation of hydrazine. This group can be installed selectively on a primary amine in the presence of a secondary amine.[8] A number of structurally similar analogs employing the concept of stabilization through conjugation and intramolecular hydrogen bonding have been prepared for the same purpose.[9–13] Normally, the Dde and Fmoc groups are not considered orthogonal because hydrazine used to cleave the Dde group will also cleave the Fmoc group. New conditions have been developed that will cleave the Dde group in the presence of an Fmoc group. Treatment $NH_2OH \cdot HCl$ (imidazole, NMP, CH_2Cl_2) quantitatively removes the Dde group in the presence of the Fmoc group.[14]

N-(1,3-Dimethyl-2,4,6-(1*H*,3*H*,5*H*)-trioxopyrimidine-5-ylidene)methylamine (DTPM−NR₂)

This group was developed for the protection of amino sugars that is compatible with the conditions used in typical carbohydrate synthesis.[15] The 5-methyl analog of this group can be used to selectively protect a primary amine in the presence of a secondary amine.[16] The DTPM group is stable to the following conditions: Ac_2O/Py, AcOH/HBr, AcSK/ MeONa/MeOH, DMF/NaH/BnBr/ TsOH/CH_3CN/ PhCH(OMe)₂, NaCNBH₃/HCl/THF, TBDPS/DMAP/ClCH₂CH₂Cl, DDQ/CH_2Cl_2/ H_2O. Cleavage of the DTPM group is affected by treatment with NH_3, hydrazine or primary amines at rt in a few minutes.

N-4,4,4-Trifluoro-3-oxo-1-butenylamine (Tfav−NR₂)

This group was developed for the protection of amino acids. It is formed from 4-ethoxy-1,1,1-trifluoro-3-buten-2-one in aqueous sodium hydroxide (70–94% yield). Primary amino acids form the Z-enamines whereas secondary amines such as proline form the E-enamines. Deprotection is achieved with 1–6 N aqueous HCl in dioxane at rt.[17,18]

N-(1-Isopropyl-4-nitro-2-oxo-3-pyrrolin-3-yl)amine

Formation/Cleavage[19]

1. B. Halpern and L. B. James, *Aust. J. Chem.*, **17**, 1282 (1964).

2. B. Halpern and A. D. Cross, *Chem. Ind. (London)*, 1183 (1965).

3. L. A. Carpino, H. G. Chao, and J.-H. Tien, *J. Org. Chem.*, **54**, 4302 (1989).

4. B. W. Bycroft, W. C. Chan, S. R. Chhabra, and N. D. Hone, *J. Chem. Soc., Chem. Commun.*, 778 (1993).

5. I. A. Nash, B. W. Bycroft, and W. C. Chan, *Tetrahedron Lett.*, **37**, 2625 (1996).

6. J.-C. Truffert, O. Lorthioir, U. Asseline, N. T. Thuong, and A. Brack, *Tetrahedron Lett.*, **35**, 2353 (1994).

7. B. Rohwedder and Y. Mutti, P. Dumy, and M. Mutter, *Tetrahedron Lett.*, **39**, 1175 (1998).

8. F. Wang, S. Manku, and D. G. Hall, *Org. Lett.*, **2**, 1581 (2000); B. Kellam, B. W. Bycroft, W. C. Chan, and S. R. Chhabra, *Tetrahedron*, **54**, 6817 (1998).

9. M. de G. Garcia Martin, C. Gasch, and A. Gomez-Sanchez, *Carbohydr. Res.*, **199**, 139 (1990).

10. J. Svete, M. Aljaz-Rozic, and B. Stanovnik, *J. Heterocycl. Chem.*, **34**, 177 (1997).

11. M. Abarbri, A. Guignard, and M. Lamant, *Helv. Chim. Acta*, **78**, 109 (1995).

12. M. A. Pradera, D. Olano, and J. Fuentes, *Tetrahedron Lett.*, **36**, 8653 (1995).

13. S. R. Chhabra, B. Hothi, D. J. Evans, P. D. White, B. W. Bycroft, and W. C. Chan, *Tetrahedron Lett.*, **39**, 1603 (1998).

14. J. J. Diaz-Mochon, L. Bialy, and M. Bradley, *Org. Lett.*, **6**, 1127 (2004).

15. G. Dekany, L. Bornaghi, J. Papageorgiou, and S. Taylor, *Tetrahedron Lett.*, **42**, 3129 (2001).

16. E. T. d. Silva and E. L. S. Lima, *Tetrahedron Lett.*, **44**, 3621 (2003).

17. M. G. Gorbunova, I. I. Gerus, S. V. Galushko, and V. P. Kukhar, *Synthesis*, 207 (1991).

18. I. I. Gerus, M. G. Gorbunova, and V. P. Kukhar, *J. Fluorine Chem.*, **69**, 195 (1994).

19. P. L. Southwick, G. K. Chin, M. A. Koshute, J. R. Miller, K. E. Niemela, C. A. Siegel, R. T. Nolte, and W. E. Brown, *J. Org. Chem.*, **49**, 1130 (1984).

Quaternary Ammonium Salts: R_3NCH_3I (Chart 10)

Formation

CH_3I, CH_3OH, $KHCO_3$, 20°C, 24 h, 85–95% yield. These salts are generally used to protect tertiary amines during oxidation reactions. The conditions cited above form quaternary salts from primary, secondary, or tertiary amines, including amino acids, in the presence of hydroxyl or phenol groups.[1]

Cleavage

1. PhSNa, 2-butanone, reflux, 24–36 h, 85% yield.[2]
2. From an ammonium iodide: AgCl, then 4-pyridinethiol, NaH, CH_3CN, reflux, 24 h.[3]

1. F. C. M. Chen and N. L. Benoiton, *Can. J. Chem.*, **54**, 3310 (1976).
2. M. Shamma, N. C. Deno, and J. F. Remar, *Tetrahedron Lett.*, **7**, 1375 (1966).
3. W.-M. Chen, H. N. C. Wong, D. T. W. Chu, and X. Lin, *Tetrahedron*, **59**, 7033 (2003).

N-Hetero Atom Derivatives

Six categories of *N*-hetero atom derivatives are considered: N–M (M = boron, copper); N–N (e.g., *N*-nitro, *N*-nitroso); *N*-oxides (used to protect tertiary amines); N–P (e.g., phosphinamides, phosphonamides); N–SiR$_3$ (R = CH$_3$), and N–S (e.g., sulfonamides, sulfenamides).

N-Metal Derivatives

N-Borane Derivatives: $R_3N \cdot BH_3$

Aminoboranes can be prepared from diborane to protect a tertiary amine during oxidation.[1,2]

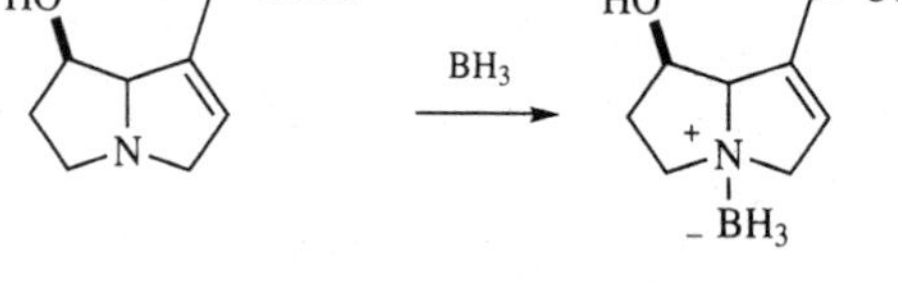

Ref. 3

They are cleaved by refluxing in ethanol,[4] methanolic sodium carbonate,[5] TFA,[6] or ammonium chloride.[7] The aminoborane was found to be stable to LDA and KHMDS.[7] Pd–C was found to be very effective for the cleavage of an intermediate borane complex during the synthesis of the sensitive FR-66979.[8] The hydrogen liberated during this decomposition will cleave benzylamines.[9]

Boranes have been used to protect the basic lone pair on pyridines and phosphines as well.[10]

N-Diphenylborinic Acid Derivative

Formation/Cleavage[11,12]

This derivative is stable to acetic acid and CF_3CO_2H.[12]

N-Diethylborinic Acid Derivative

The diethylborinic acid derivative has been prepared from triethylborane (THF, reflux).[13] After esterification of the remaining carboxyl group the boron was removed with HCl(g) (Et_2O, rt, 15 min, >80% yield).[13,14]

N-9-Borabicyclononane (9-BBN)

This group was developed for the protection and further manipulation of 5-hydroxy-L-Lysine.[15] The group is stable to the formation of carbamates, silyl ethers and azides and a Königs–Knorr glycosidation. It is cleaved by stirring in $MeOH/CHCl_3$, but is stable in the individual solvents. Since $CHCl_3$ often contains some HCl, it is likely that the deprotection is actually acid-catalyzed, and this is consistent with the fact that it may also be cleaved with aqueous HCl. Ethylenediamine in MeOH is used for deprotection by exchange.[16]

These complexes are stable to the conditions of the Sonogashira reaction, silica gel chromatography (EtOAc/Hex), dilute TEA, KF in DMF, POCl₃, PSCl₃, MCPBA, MMPP, Arbuzov conditions (neat (EtO)₃P, 110°C), and NaI/acetone.[16,17] Reagents that release HCl will require an acid scavenger to prevent premature deprotection. The 9-BBN chelate of amino alcohols has been used to selectively monoalkylate primary amines, a process that is often problematic because of bisalkylation.[18]

N-Difluoroborinic Acid

These water sensitive derivatives can be used to cleanly form the *t*-butyl ethers of serine and threonine. They are cleaved with aqueous acid or base.[19]

3,5-Bis(trifluoromethyl)phenylboronic Acid

The free amine can be monoacylated. Without this protection only the bisacylated derivative is obtained.[20]

1. J. L. Brayer, J. P. Alazard, and C. Thal, *Tetrahedron*, **46**, 5187 (1990).

2. C. J. Swain, C. Kneen, R. Herbert, and R. Baker, *J. Chem. Soc., Perkin Trans. 1*, 3183 (1990).

3. J. D. White, J. C. Amedio, Jr., S. Gut, and L. Jayasinghe, *J. Org. Chem.*, **54**, 4268 (1989).

4. A. Picot and X. Lusinchi, *Bull. Soc. Chim. Fr.*, 1227 (1977).

5. M. A. Schwartz, B. F. Rose, and B. Vishnuvajjala, *J. Am. Chem. Soc.*, **95**, 612 (1973).

6. S. Choi, I. Bruce, A. J. Fairbanks, G. W. J. Fleet, A. H. Jones, R. J. Nash, and L. E. Fellows, *Tetrahedron Lett.*, **32**, 5517 (1991).

7. V. Ferey, P. Vedrenne, L. Toupet, T. Le Gall, and C. Mioskowski, *J. Org. Chem.*, **61**, 7244 (1996).

8. T. C. Judd and R. M. Williams, *J. Org. Chem.*, **69**, 2825 (2004).

9. M. Couturier, J. L. Tucker, B. M. Andresen, P. Dube, and J. T. Negri, *Org. Lett.*, **3**, 465 (2001).

10. C. Lutz, C.-D. Graf, and P. Knochel, *Tetrahedron*, **54**, 10317 (1998).

11. I. Staatz, U. H. Granzer, A. Blume, and H. J. Roth, Liebigs *Ann. Chem.*, 127 (1989).

12. G. H. L. Nefkens and B. Zwanenburg, *Tetrahedron*, **39**, 2995 (1983).

13. F. Albericio, E. Nicolás, J. Rizo, M. Ruiz-Gayo, E. Pedroso, and E. Giralt, *Synthesis*, 119 (1990).

14. J. Robles, E. Pedroso and A. Grandas, *Synthesis*, 1261 (1993).

15. B. M. Syed, T. Gustafsson, and J. Kihlberg, *Tetrahedron*, **60**, 5571 (2004).

16. W. H. Dent III, W. R. Erickson, S. C. Fields, M. H. Parker, and E. G. Tromiczak, *Org. Lett.*, **4**, 1249 (2002).

17. W. H. Walker and S. Rokita, *J. Org. Chem.*, **68**, 1563 (2003).

18. G. Bar-Haim and M. Kol, *Org. Lett.*, **6**, 3549 (2004).

19. J. Wang, Y. Okada, W. Li, T. Yokoi, and J. Zhu, *J. Chem. Soc., Perkin Trans. 1*, 621 (1997).

20. K. Ishihara, Y. Kuroki, N. Hanaki, S. Ohara, and H. Yamamoto, *J. Am. Chem. Soc.*, **118**, 1569 (1996).

N-[Phenyl(pentacarbonylchromium- or -tungsten)carbenyl]amine

$$(CO)_5M \mathrm{=\!\!\!<} \begin{array}{c} NR_2 \\ R' \end{array}$$

R' = Ph or CH₃; M = Cr or W

These transition metal carbenes, prepared in 66–97% yield from amino acid esters, are cleaved by acid hydrolysis (CF_3CO_2H, 20°C, 80% yield; 80% AcOH; M = W; BBr_3, −25°C).[1]

1. K. Weiss and E. O. Fischer, *Chem. Ber.*, **109**, 1868 (1976).

N-Copper or *N*-Zinc Chelate: $RNH_2 \cdots M \cdots OH$ M = Cu(II), Zn(II)

Formation/Cleavage

1.

A copper chelate selectively protects the α-NH_2 group in lysine. The chelate is cleaved by $2N$ HCl or by EDTA, $(HO_2CCH_2)_2NCH_2CH_2N(CH_2CO_2H)_2$.[1] This mode of protection is sufficient to allow alkylation of a copper-protected tyrosine at the phenol (75% yield).[2]

2. In an aminoglycoside a vicinal amino hydroxy group can be protected as a Cu(II) chelate. After acylation of other amine groups, the chelate is cleaved by aqueous ammonia.[3] The copper chelate can also be cleaved with $Bu_2NC(S)NHBz$ (EtOH, reflux, 2h).[4]

3. After examination of the complexing ability of Ca(II), Cr(III), Mn(II), Fe(III), Co(II), Ni(II), Cu(II), Zn(II), Ru(III), Ag(I), and Sn(IV), the authors decided that Zn(II) provides the best protection for vicinal amino hydroxy groups during trifluoroacetylation of other amino groups in the course of some syntheses of kanamycin derivatives.[5]

1. R. Ledger and F. H. C. Stewart, *Aust. J. Chem.*, **18**, 933 (1965).

2. K. Nakanishi, R. Goodnow, K. Konno, M. Niwa, R. Bukownik, T. A. Kallimopoulos, P. Usherwood, A. T. Eldefrawi, and M. E. Eldefrawi, *Pure Appl. Chem.*, **62**, 1223 (1990).

3. S. Hanessian, and G. Patil, *Tetrahedron Lett.*, **19**, 1035 (1978).

4. K. H. König, L. Kaul, M. Kuge, and M. Schuster, *Liebigs Ann. Chem.*, 1115 (1987).

5. T. Tsuchiya, Y. Takagi, and S. Umezawa, *Tetrahedron Lett.*, **20**, 4951 (1979).

18-Crown-6 Derivative

The primary amine of an amino acid as its tosylate salt can be protected by coordination with a crown ether. The protection scheme was sufficient to allow the HOBt/DDC coupling of amino acids. The crown is removed by treatment with diisopropylethylamine or KCl solution.[1,2]

1. P. Botti, H. L. Ball, E. Rizzi, P. Lucietto, M. Pinori, and P. Mascagni, *Tetrahedron*, **51**, 5447 (1995).

2. C. B. Hyde and P. Mascagni, *Tetrahedron Lett.*, **31**, 399 (1990).

N−N Derivatives

N-Nitroamine: R_2NNO_2 (Chart 10)

Formation

An *N*-nitro derivative is used primarily to protect the guanidino group in arginine; it is cleaved by reduction: H_2/Pd–C, $AcOH/CH_3OH$, ~80% yield;[1] 10% Pd–C/cyclohexadiene, 25°C, 2 h, good yields;[2] Pd–C/4% HCO_2H-CH_3OH, 5 h, 100% yield;[3] $TiCl_3$/pH 6, 25°C, 45 min, 70–98% yield;[4] $SnCl_2$/60% HCO_2H, 63% yield;[5] electrolysis, $1N$ H_2SO_4, 1–6 h, 85–95% yield,[6] and O_2, H_2O, acid, 79% yield.[7]

1. K. Hofmann, W. D. Peckham, and A. Rheiner, *J. Am. Chem. Soc.*, **78**, 238 (1956).

2. A. M. Felix, E. P. Heimer, T. J. Lambros, C. Tzougraki, and J. Meienhofer, *J. Org. Chem.*, **43**, 4194 (1978).

3. B. ElAmin, G. M. Anantharamaiah, G. P. Royer, and G. E. Means, *J. Org. Chem.*, **44**, 3442 (1979).

4. R. M. Freidinger, R. Hirschmann, and D. F. Veber, *J. Org. Chem.*, **43**, 4800 (1978).

5. T. Hayakawa, Y. Fujiwara, and J. Noguchi, *Bull. Chem. Soc. Jpn.*, **40**, 1205 (1967).

6. P. M. Scopes, K. B. Walshaw, M. Welford, and G. T. Young, *J. Chem. Soc.*, 782 (1965).

7. T. Cupido, J. Spengler, K. Burger, and F. Albericio, *Tetrahedron Lett.*, **46**, 6733 (2005).

N-Nitrosoamine: R_2NNO

N-Nitroso derivatives, prepared from secondary amines and nitrous acid, are cleaved by reduction (H_2/Raney Ni, EtOH, 28°C, 3.5 h[1]; CuCl/concd. HCl[2]). Since many *N*-nitroso compounds are carcinogens, and because some racemization and cyclo-dehydration of *N*-nitroso derivatives of *N*-alkyl amino acids occur during peptide syntheses, [3,4] *N*-nitroso derivatives are of limited value as protective groups.

1. M. Harfenist and E. Magnein, *J. Am. Chem. Soc.*, **79**, 2215 (1957).

2. C. F. Koelsch, *J. Am. Chem. Soc.*, **68**, 146 (1946).

3. P. Quitt, R. O. Studer, and K. Vogler, *Helv. Chim. Acta*, **47**, 166 (1964).

4. F. H. C. Stewart, *Aust. J. Chem.*, **22**, 2451 (1969).

Amine *N*-Oxide: $R_3N{\rightarrow}O$ (Chart 10)

Amine oxides, prepared to protect tertiary amines during methylation[1,2] and to prevent their protonation in diazotized aminopyridines,[3] can be cleaved by reduction (e.g., SO_2/H_2O, 1 h, 22°C, 63% yield[1]; H_2/Pd–C, AcOH, Ac_2O, 7 h, 91% yield;[2] Zn/ HCl, 30% yield,[3] reduction with RaNi).[4] Photolytic reduction of an aromatic amine oxide has been reported [i.e., 4-nitropyridine N-oxide, 300 nm, $(MeO)_3PO/CH_2Cl_2$, 15 min, 85–95% yield].[5] Amine oxides are also substrates for the Cope elimination.

1. F. N. H. Chang, J. F. Oneto, P. P. T. Sah, B. M. Tolbert, and H. Rapoport, *J. Org. Chem.*, **15**, 634 (1950).
2. J. A. Berson and T. Cohen, *J. Org. Chem.*, **20**, 1461 (1955).
3. F. Koniuszy, P. F. Wiley, and K. Folkers, *J. Am. Chem. Soc.*, **71**, 875 (1949).
4. K. Toshima, Y. Nozaki, S. Mukaiyama, T. Tamai, M. Nakata, K. Tatsuta, and M. Kinoshita, *J. Am. Chem. Soc.*, **117**, 3717 (1995).
5. C. Kaneko, A. Yamamoto, and M. Gomi, *Heterocycles*, **12**, 227 (1979).

Azide: RN_3

Azide is often used to introduce nitrogen by nucleophilic displacement on a halide or sulfonate. **Care must be exercised when producing or handling azides, since they can be quite explosive.** In fact, azides are rarely used on an industrial scale. Special facilities are required to work with most azides on scale. The safety factor improves as the carbon-to-nitrogen ratio in the substrate increases. Beyond being a source of nitrogen, they are most commonly used to protect the amine during carbohydrate synthesis.

Formation

1. Tf_2O, NaN_3, 89% yield.[1]

2. TfN_3, $CuSO_4$.[2] **TfN_3 is explosive and should not be distilled. It is best used as a solution.**
3. TfN_3, $ZnCl_2$, CH_2Cl_2, H_2O, 80–99% yield per amine.[3]

Cleavage

Azides are cleaved by reduction. Some methods are provided, but this is not meant to be an exhaustive list.

1. H_2, Pd–C, MeOH.[2a,2b]
2. PMe_3, THF, H_2O, 1 N NaOH, 75% yield.[4]
3. PMe_3, THF, −78°C to rt then CbzCl, 30 min.[3,5] $(BOC)_2O$ can also be used to prepare the BOC derivative.

47%   5%   2%

4. TMSCl, RCOCl, heat, 62–92% yield. This method directly converts an azide to an amide.[6]

5. Et_3NH^+ $[(PhS)_3Sn]^-$, CH_2Cl_2, rt, 4 h., >73% yield.[7] In this case, other more classical methods such as the use of Ph_3P, 1,3-propanethiol and H_2S gave unsatisfactory results.

1. P. H. Seeberger, M. Baumann, G. Zhang, T. Kanemitsu, E. E. Swayze, S. A. Hofstadler, and R. H. Griffey, *Synlett*, 1323 (2003).

2. (a) S.-Y. Luo, S. R. Thopate, C.-Y. Hsu, and S.-C. Hung, *Tetrahedron Lett.*, **43**, 4889 (2002). (b) J. Liu, M. M. D. Numa, H. Liu, S.-J. Huang, P. Sears, A. R. Shikhman, and C.-H. Wong, *J. Org. Chem.*, **69**, 6273 (2004). (c) B. Wu, J. Yang, Y. He, and E. E. Swayze, *Org. Lett.*, **4**, 3455 (2002). (d) J. T. Lundquist and J. C. Pelletier, *Org. Lett.*, **4**, 3219 (2002).

3. P. T. Nyffeler, C.-H. Liang, K. M. Koeller, and C.-H. Wong, *J. Am. Chem. Soc.*, **124**, 10773 (2002).

4. P. B. Alper, M. Hendrix, P. Sears, and C.-H. Wong, *J. Am. Chem. Soc.*, **120**, 1965(1998).

5. X. Ariza, F. Urpi, and J. Vilarrasa, *Tetrahedron Lett.*, **40**, 7515 (1999).

6. A. Barua, G. Bez, and N. C. Barua, *Synlett*, 553 (1999).

7. L. F. Tietze and H. Keim, *Angew. Chem. Int. Ed.*, **36**, 1615 (1997).

Triazene Derivative

This group is stable to metalation of the aromatic ring by metal halogen exchange, Grignard formation, $LiAlH_4$ reduction, NaOH, PDC, hydrogenolysis, $NaBH_4$, and LDA.[1] Reaction of an aromatic triazene with MeI at 120°C gives the aryl iodide.[2]

Formation

1. Protection of primary aryl amines as the triazene is accomplished by diazotization of the amine followed by reaction with a dialkylamine in aq. KOH or other base. *t*-BuONO, $BF_3 \cdot Et_2O$, Et_2NH, K_2CO_3, 99% yield.[3]

2. For secondary amines: PhN_2BF_4, pyridine, 75–90% yield.[4]

Cleavage

1. The amine is recovered by reductive cleavage with Ni–Al alloy (aq. KOH, rt, 37–68% yield).[5]
2. RaNi MeOH.[6]
3. TFA, NaH$_2$PO$_2$, CuCl$_2$. Acids cleave the triazene but the released diazonium salt must be reduced, and it is for this reason that NaH$_2$PO$_2$ is used in the reaction.[4]

1. For a brief review on triazenes and leading references, see D. B. Kimball and M. M. Haley, *Angew. Chem. Int. Ed.*, **41**, 3338 (2002).
2. X. Yang, L. Yuan, K. Yamato, A. L. Brown, W. Feng, M. Furukawa, X. C. Zeng, and B. Gong, *J. Am. Chem. Soc.*, **126**, 3148 (2004).
3. G. Li, X. Wang and F. Wang, *Tetrahedron Lett.*, **46**, 8971 (2005).
4. R. Lazny, M. Sienkiewicz, and S. Brase, *Tetrahedron*, **57**, 5825 (2001). R. Lazny, J. Poplawski, J. Kobberling, D. Enders, and S. Brase, *Synlett*, 1304 (1999).
5. M. L. Gross, D. H. Blank, and W. M. Welch, *J. Org. Chem.*, **58**, 2104 (1993).
6. J. M. Ready, S. E. Reisman, M. Hirata, M. W. Weiss, K. Tamaki, T. V. Ovaska, and J. L. Wood, *Angew. Chem. Int. Ed.*, **43**, 1270 (2004).

N-Trimethylsilylmethyl-N-benzylhydrazine: (CH$_3$)$_3$SiCH$_2$(C$_6$H$_4$CH$_2$)N-NR$_2$

The hydrazine was used to introduce nitrogen during a Diels–Alder reaction. It is readily cleaved with 5% HCl/EtOH at 50°C.[1]

1. B. B. Touré and D. G. Hall, *J. Org. Chem.*, **69**, 8429 (2004).

N–P Derivatives

Diphenylphosphinamide (Dpp−NR$_2$): Ph$_2$P(O)NR$_2$ (Chart 10)

Phosphinamides are stable to catalytic hydrogenation, used to cleave benzyl-derived protective groups, and to hydrazine.[1] The rate of hydrolysis of phosphinamides is a function of the steric and electronic factors around the phosphorus.[2] This derivative has largely been used for the protection of amino acids and has seen little use in the general synthetic literature. It has been used as a protective group that can activate imines (DppN=CR$_2$) for nucleophilic additions to form alkylamines.

Formation

Ph$_2$POCl, *N*-methylmorpholine, 0°C, 60–90% yield.[3]

Cleavage

1. The Dpp group is cleaved by the following acidic conditions: AcOH, HCOOH, H$_2$O, 24 h, 100% yield; 80% CF$_3$COOH, ca. quant; 0.4 *M* HCl, 90% CF$_3$CH$_2$OH, ca. quant.; *p*-TsOH, H$_2$O–CH$_3$OH, ca. quant.; 80% AcOH, 3 days, not completely cleaved.[3] The Dpp group is slightly less stable to acid than the BOC group.[2,3]

2. MeOH, BF$_3$·Et$_2$O, CH$_2$Cl$_2$, 0°C to rt, 81–93% yield.[4] This method cleaves the Dpp group from an aziridine without complications of ring opening.[5]

3. Bu$_2$CuLi, PhLi, or Ph$_2$CuLi cleaved the Dpp group from an aziridine (63–83% yield), but Me$_2$CuLi resulted in ring opening.[4]

Dimethyl- and Diphenylthiophosphinamide (Mpt–NR$_2$ and Ppt-NR$_2$):
(CH$_3$)$_2$P(S)NR$_2$ (Chart 10) and Ph$_2$P(S)NR$_2$

The Mpt and Ppt derivatives can be prepared from an amino acid and the thiophosphinyl chloride (Me$_2$PSCl or Ph$_2$PSCl, respectively, 41–78% yield, lysine gives 16% yield).[6] The Mpt group is cleaved with HCl or Ph$_3$P·HCl[7] and is cleaved 60 times faster than the BOC group. The Ppt group is the more stable of the two groups.

Dialkyl Phosphoramidates: (RO)$_2$P(O)NR$_2$

Formation

1. (EtO)$_2$P(O)H, CCl$_4$, aq. NaOH, PhCH$_2$NEt$_3$Cl, 0°, 1 h to 22°C, 1 h, 75–90% yield.[8,9]
2. (EtO)$_2$P(O)H, NaOCl, pH 9 using NaOH, 80% yield. This procedure was performed on a 200-g scale for the protection of *trans*-4-hydroxy-L-proline.[10]
3. (BuO)$_2$P(O)H, Et$_3$N, CCl$_4$.[11]
4. (*i*-PrO)$_2$P(O)Cl, 73–93% yield.[12]

Cleavage

Phosphoramidates are cleaved with HCl-saturated THF (70–94% yield). Their stability is dependent upon the alkyl group, the methyl derivative being the least stable. They also have good stability to organic acids and Lewis acids.[12,13]

Dibenzyl and Diphenyl Phosphoramidate: $(BnO)_2P(O)NR_2$ and $(PhO)_2P(O)NR_2$

Dibenzyl phosphoramidates have been prepared from amino acids and the phosphoryl chloride, $(BnO)_2P(O)Cl$.[14] A diphenyl phosphoramidate has been prepared from a glucosamine; it was converted by transesterification into a dibenzyl derivative to facilitate cleavage.[15]

Iminotriphenyphosphorane: $Ph_3P=NR$

This derivative is most conveniently prepared by reaction of an azide with triphenylphosphine. It was used because of its stability toward Ph_2PLi. Its aqueous hydrolysis is well-documented.[16,17]

1. G. W. Kenner, G. A. Moore, and R. Ramage, *Tetrahedron Lett.*, **17**, 3623 (1976).

2. R. Ramage, B. Atrash, D. Hopton, and M. J. Parrott, *J. Chem. Soc., Perkin Trans. I*, 1217 (1985).

3. R. Ramage, D. Hopton, M. J. Parrott, G. W. Kenner, and G. A. Moore, *J. Chem. Soc., Perkin Trans. I*, 1357 (1984).

4. H. M. I. Osborn and J. B. Sweeney, *Synlett*, 145 (1994).

5. N. E. Maguire, A. B. McLaren, and J. B. Sweeney, *Synlett*, 1898 (2003).

6. S. Ikeda, F. Tonegawa, E. Shikano, K. Shinozaki, and M. Ueki, *Bull. Chem. Soc. Jpn.*, **52**, 1431 (1979).

7. M. Ueki, T. Inazu, and S. Ikeda, *Bull. Chem. Soc. Jpn.*, **52**, 2424 (1979).

8. A. Zwierzak, *Synthesis*, 507 (1975).

9. A. Zwierzak and K. Osowska, *Synthesis*, 223 (1984). A. Chellini, R. Pagliarin, G. B. Giovenzana, G. Palmisano, and M. Sisti, *Helv. Chim. Acta*, **83**, 793 (2000).

10. K. M. J. Brands, K. Wiedbrauk, J. M. Williams, U.-H. Dolling, and P. J. Reider, *Tetrahedron Lett.*, **39**, 9583 (1998).

11. Y.-F. Zhao, S.-K. Xi, A.-T. Song, and G.-J. Ji, *J. Org. Chem.*, **49**, 4549 (1984).

12. Y. F. Zhao, G. J. Ji, S. K. Xi, H. G. Tang, A. T. Song, and S. Z. Wei, *Phosphorus Sulfur*, **18**, 155 (1983).

13. K. M. J. Brands, R. B. Jobson, K. M. Conrad, J. M. Williams, B. Pipik, M. Cameron, A. J. Davies, P. G. Houghton, M. S. Ashwood, I. F. Cottrell, R. A. Reamer, D. J. Kennedy, U.-H. Dolling, and P. J. Reider, *J. Org. Chem.*, **67**, 4771 (2002).

14. A. Cosmatos, I. Photaki, and L. Zervas, *Chem. Ber.*, **94**, 2644 (1961).

15. M. L. Wolfrom, P. J. Conigliaro, and E. J. Soltes, *J. Org. Chem.*, **32**, 653 (1967).

16. S.-T. Liu and C.-Y. Liu, *J. Org. Chem.*, **57**, 6079 (1992).

17. M. Campbell and M. J. McLeish, *J. Chem. Res., Synop.*, 148 (1993).

N–Si Derivatives

For the most part silyl derivatives such as trimethylsilylamines have not been used extensively for amine protection because of their high reactivity to moisture, although they do provide satisfactory protection when prepared and used under anhydrous conditions.[1,2] They are also reported to increase the nucleophilicity of the nitrogen, thus improving acylations.[3] The more stable and sterically demanding *t*-butyldiphenylsilyl group has been used to protect primary amines in the presence of secondary amines, thus allowing selective acylation or alkylation of the secondary amine.[4] Silylamines are reported not to be stable to oxidative conditions.[4] Silylamines are readily cleaved in the presence of silyl ethers.[5] Primary amines can be bis-silylated and are sufficiently stable during a metalation reaction.[6]

Triphenylsilylamine has been used as a protected ammonia equivalent for displacement of aryl halides to prepare anilines.[7] For a more thorough discussion of silylating reagents the section on alcohol protection should be consulted since many of the reagents described there will also silylate amines.

1. J. R. Pratt, W. D. Massey, F. H. Pinkerton, and S. F. Thames, *J. Org. Chem.*, **40**, 1090 (1975).

2. A. B. Smith, III, M. Visnick, J. N. Haseltine, and P. A. Sprengeler, *Tetrahedron*, **42**, 2957 (1986).

3. V. V. S. Babu, G.-R. Vasanthakumar, and S. J. Tantry, *Tetrahedron Lett.*, **46**, 4099 (2005).

4. L. E. Overman, M. E. Okazaki, and P. Mishra, *Tetrahedron Lett.*, **27**, 4391 (1986).

5. T. P. Mawhinney and M. A. Madson, *J. Org. Chem.*, **47**, 3336 (1982).

6. A. B. Smith, III and H. Cui, *Org. Lett.*, **5**, 587 (2003); S. Das, V. L. Alexeev, A. C. Sharma, S. J. Geib, and S. A. Asher, *Tetrahedron Lett.*, **44**, 7719 (2003).

7. X. Huang and S. L. Buchwald, *Org. Lett.*, **3**, 3417 (2001).

N–S Derivatives

N-Sulfenyl Derivatives

Sulfenamides, R_2NSR', prepared from an amine and a sulfenyl halide,[1,2] are readily cleaved by acid hydrolysis and have been used in syntheses of peptides, penicillins, and nucleosides. They are also cleaved by nucleophiles,[3] and by Raney nickel desulfurization.[4] The synthesis and application of sulfenamides has been reviewed.[5]

Benzenesulfenamide: $R_2NSC_6H_5$, A (Chart 10)

Formation

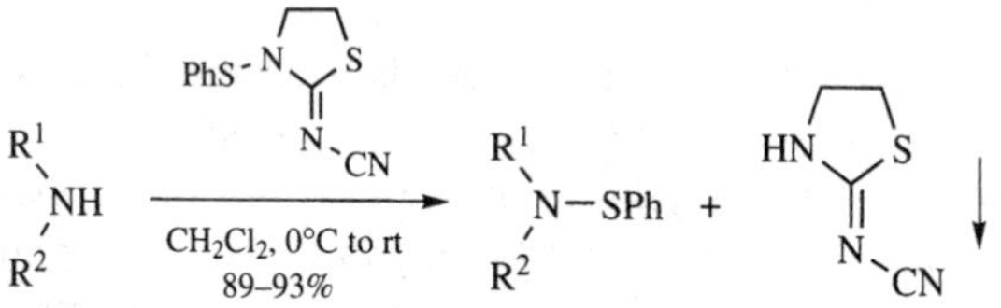

2-Nitrobenzenesulfenamide (Nps–NR$_2$): $R_2NSC_6H_4$–o-NO_2, B (Chart 10)

The 2-nitrobenzenesulfenamide has been used for the protection of amino acids[7,8] or nucleosides.[9]

Formation

1. o-$NO_2C_6H_4SCl$, NaOH, dioxane, 79% yield.[10] The reagent is unstable and often requires recrystallization prior to use.
2. o-$NO_2C_6H_4SSCN$, $AgNO_2$.[11]
3. N-(2-Nitrobenzenesulphenyl)–saccharin, NaOH, dioxane, 75–87% yield.[12]

Cleavage

1. Sodium iodide, CH_3OH, CH_2Cl_2, AcOH, 0°C, 20 min, 53% yield.[13]
2. Acidic hydrolysis: HCl/Et_2O or EtOH, 0°C, 1 h, 95% yield.[14]
3. By nucleophiles: 13 reagents, 5 min to 12 h, 90% cleaved.[3]
4. PhSH or $HSCH_2CO_2H$, 22°C, 1 h.[15]
5. $CH_3C_6H_4SH$, TsOH, CH_2Cl_2, 84% yield.[16,17]

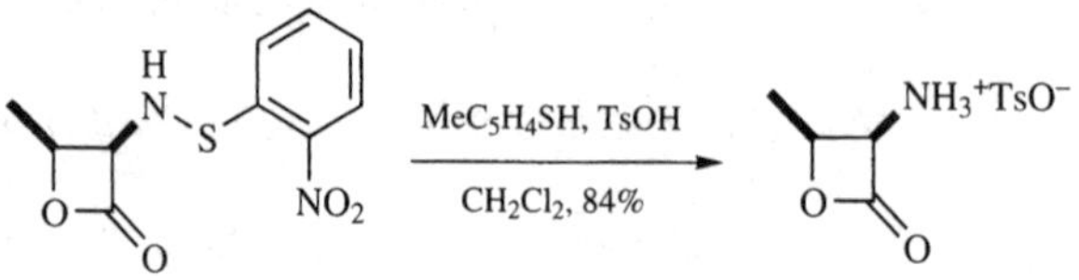

6. 2-Mercaptopyridine/CH_2Cl_2, 1 min, 100% yield.[18]
7. NH_4SCN, 2-methyl-1-indolylacetic acid.[8]

8. HOBt, aniline, DMF. These conditions give the amine as the HOBt salt, which may be acylated without the addition of a tertiary amine.[16]

9. Catalytic desulfurization: Raney Ni/DMF, column, few hours, satisfactory yields.[4]

10. 2-Acylthiomercaptobenzotriazoles, PPTS, 52–80% yield. In this case the amide is formed rather than the free amine.[19]

2,4-Dinitrobenzenesulfenamide: $R_2NSC_6H_3$-2,4-$(NO_2)_2$, C

The 2,4-dinitrobenzenesulfenamide is cleaved with p-thiocresol/TsOH.[20]

Pentachlorobenzenesulfenamide: $R_2NSC_6Cl_5$, D

Benzenesulfenamide, and a number of substituted benzenesulfenamides (compounds B, C, and D) have been prepared to protect the 7-amino group in cephalosporins.

2-Nitro-4-methoxybenzenesulfenamide: $R_2NSC_6H_3$-2-NO_2-4-OCH_3

This sulfenamide, prepared from an amino acid, the sulfenyl chloride and sodium bicarbonate, is cleaved by acid hydrolysis (HOAc/dioxane, 22°C, 30 min, 95% yield).[21]

Triphenylmethylsulfenamide: $R_2NSC(C_6H_5)_3$

The tritylsulfenamide can be prepared from an amine and the sulfenyl chloride (Na_2CO_3, THF, H_2O or Pyr, CH_2Cl_2, 64–96% yield);[22] it is cleaved by hydrogen chloride in ether or ethanol (0°C, 1 h, 90% yield),[14] $CuCl_2$ (THF, EtOH, 58–67% yield), Me_3SiI (77–96% yield),[22] I_2 (0.1 M, THF, collidine, H_2O, 97% yield),[23] Bu_3SnH, 115°C, toluene, 5 min, 82% yield.[24] The tritylsulfenamide is stable to 1 N HCl, base, $NaCNBH_3$, $LiAlH_4$, m-chloroperoxybenzoic acid, pyridinium chlorochromate, Jones reagent, Collins oxidation and Moffat oxidation. The stability of this group is largely due to steric hindrance.

1-(2,2,2-Trifluoro-1,1-diphenyl)ethylsulfenamide (TDE): $CF_3C(Ph)_2S-NR_2$

The sulfenamide is prepared from the sulfenyl chloride (Na_2CO_3, THF, H_2O, rt, 95–100% yield or CH_2Cl_2, TEA, 87–96% yield). It is cleaved with Na/NH_3, (67–94% yield) or with HCl/Et_2O (80–98% yield). In the latter method the sulfenyl chloride can be recovered. The TDE group is stable to strong aqueous HCl, NaOH, $NaBH_4$, $LiAlH_4/Et_2O$ at 0°C, Bu_3SnH (toluene, 90°C), $Pd(OH)_2/H_2$ and Ac_2O/Pyr.[25]

3-Nitro-2-pyridinesulfenamide (Npys−NR$_2$)

This group, which is more stable than the 2-nitrobenzenesulfenamide, has been developed to protect amino acids. It is readily introduced with the sulfenyl chloride[26] (52–74% yield).

Cleavage

1. Triphenylphosphine, pentachlorophenol, or 2-thiopyridine *N*-oxide. It is stable to CF₃COOH, but can be cleaved with 0.1 *M* HCl.[27]
2. 2-Mercaptopyridine and 2-mercapto-1-methylimidazole.[28]
3. 2-Mercaptopyridine *N*-oxide, CH₂Cl₂. The use of a 1000-fold excess of this reagent is required to achieve good yields for cleavage in solid-phase peptide synthesis.[29]

1. For other methods of preparation, see F. A. Davis and U. K. Nadir, *Org. Prep. Proced. Int.*, **11**, 33 (1979).
2. For a review of sulfenamides, see L. Craine and M. Raban, *Chem. Rev.*, **89**, 689 (1989).
3. W. Kessler and B. Iselin, *Helv. Chim. Acta*, **49**, 1330 (1966).
4. J. Meienhofer, *Nature* **205**, 73 (1965).
5. I. V. Koval, *Russ. Chem. Rev.*, **65**, 421 (1996).
6. T. Tanaka, T. Azuma, X. Fang, S. Uchida, C. Iwata, T. Ishida, Y. In, and N. Maezaki, *Synlett*, 32 (2000).
7. S. Romani, G. Bovermann, L. Moroder, and E. Wünsch, *Synthesis*, 512 (1985).
8. I. F. Luescher and C. H. Schneider, *Helv. Chim. Acta*, **66**, 602 (1983).
9. M. Sekine, *J. Org. Chem.*, **54**, 2321 (1989).
10. M. A. Bednarek and M. Bodanzky, *Int. J. Pept. Protein. Res.*, **45**, 64 (1995).
11. J. Savrda and D. H. Veyrat, *J. Chem. Soc.* C, 2180 (1970).
12. S. Romani, G. Bovermann, L. Moroder, and E. Wunsch, *Synthesis*, 512 (1985).
13. T. Kobayashi, K. Iino, and T. Hiraoka, *J. Am. Chem. Soc.*, **99**, 5505 (1977).
14. L. Zervas, D. Borovas, and E. Gazis, *J. Am. Chem. Soc.*, **85**, 3660 (1963).
15. A. Fontana, F. Marchiori, L. Moroder, and E. Schoffone, *Tetrahedron Lett.*, **7**, 2985 (1966).
16. Y. Pu, F. M. Martin, and J. C. Vederas, *J. Org. Chem.*, **56**, 1280 (1991).
17. Y. Pu, C. Lowe, M. Sailer, and J. C. Vederas, *J. Org. Chem.*, **59**, 3643 (1994).
18. M. Stern, A. Warshawsky, and M. Fridkin, *Int. J. Pept. Protein Res.*, **13**, 315 (1979).
19. M. N. Rao, A. G. Holkar, and N. R. Ayyangar, *J. Chem. Soc., Chem. Commun.*, 1007 (1991).
20. E. M. Gordon, M. A. Ondetti, J. Pluscec, C. M. Cimarusti, D. P. Bonner, and R. B. Sykes, *J. Am. Chem. Soc.*, **104**, 6053 (1982).
21. Y. Wolman, *Isr. J. Chem.*, **5**, 231 (1967).
22. B. P. Branchaud, *J. Org. Chem.*, **48**, 3538 (1983).
23. H. Takaku, K. Imai, and M. Nagai, *Chem. Lett.*, **17**, 857 (1988).
24. M. Sekine and K. Seio, *J. Chem. Soc., Perkin Trans.*1, 3087 (1993).
25. T. Netscher and T. Wellar, *Tetrahedron*, **47**, 8145 (1991).
26. For a one-pot preparation of the reagent, see M. Ueki, M. Honda, Y. Kazama, and T. Katoh, *Synthesis*, 21 (1994).
27. R. Matsueda and R. Walter, *Int. J. Pept. Protein Res.*, **16**, 392 (1980).
28. O. Rosen, S. Rubinraut, and M. Fridkin, *Int. J. Pept. Protein Res.*, **35**, 545 (1990).
29. S. Rajagopalan, T. J. Heck, T. Iwamoto, and J. M. Tomich, *Int. J. Pept. Protein Res.*, **45**, 173 (1995).

N-Sulfonyl Derivatives: R₂NSO₂R′

Sulfonamides are prepared from an amine and a sulfonyl chloride in the presence of pyridine or aqueous base.[1] The sulfonamide is one of the most stable nitrogen protective groups. Most arylsulfonamides are stable to alkaline hydrolysis and to catalytic reduction; they are cleaved by Na/NH₃,[2] Na/butanol,[3] sodium naphthalenide,[4] or sodium anthracenide,[5] as well as by refluxing in acid (48% HBr/cat. phenol).[6] Sulfonamides of less basic amines such as pyrroles and indoles are much easier to cleave than those of the more basic alkylamines. In fact, sulfonamides of the less basic amines (pyrroles, indoles, and imidazoles) can be cleaved by basic hydrolysis, which is almost impossible for the alkyl amines. Because of the inherent differences between the aromatic−NH group and simple aliphatic amines, the protection of these compounds (pyrroles, indoles, and imidazoles) will be described in a separate section. One appealing property of sulfonamides is that the derivatives are more crystalline than amides or carbamates.

1. E. Fischer and W. Lipschitz, *Ber.*, **48**, 360 (1915).

2. V. du Vigneaud and O. K. Behrens, *J. Biol. Chem.*, **117**, 27 (1937).

3. G. Wittig, W. Joos, and P. Rathfelder, *Justus Liebigs Ann. Chem.* **610**, 180 (1957).

4. S. Ji, L. B. Gortler, A. Waring, A. Battisti, S. Bank, W.D. Closson, and P. Wriede, *J. Am. Chem. Soc.* **89**, 5311 (1967).

5. K. S. Quaal, S. Ji, Y. M. Kim, W. D. Closson, and J. A. Zubieta, *J. Org. Chem.*, **43**, 1311 (1978).

6. H. R. Synder and R. E. Heckert, *J. Am. Chem. Soc.*, **74**, 2006 (1952).

Methanesulfonamide (Ms−NR₂): CH₃SO₂NR₂

Formation

1. CH₃SO₂Cl, TEA, CH₂Cl₂, 0°C, high yields. This is the most common method for introducing the mesylate.[1]

2. 1*H*-Benzotriazol-1-yl methanesulfonate, 23°C, DMF, 60–87% yield.[2] Primary amines are selectively mesylated.

3. The following method was employed because of the poor nucleophilicity of the amine[3]:

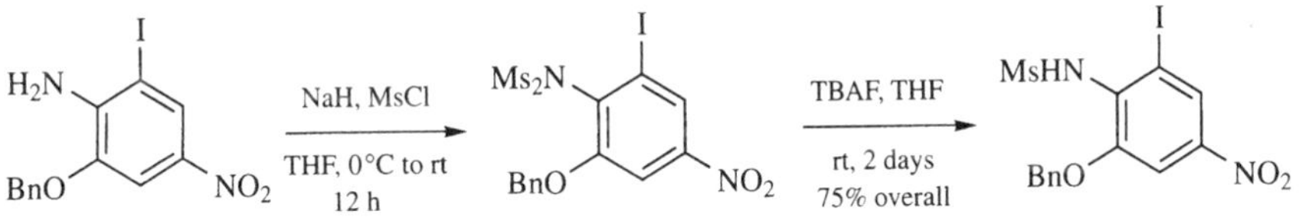

Cleavage

1. LiAlH₄.[1]

2. Na, *t*-BuOH, HMPT, NH₃, 64% yield.[1]

3. Lithium naphthalide, THF, 30–77% yield.

1. P. Merlin, J. C. Braekman, and D. Daloze, *Tetrahedron Lett.*, **29**, 1691 (1988).

2. S. Y. Kim, N.-D. Sung, J.-K. Choi, and S. S. Kim, *Tetrahedron Lett.*, **40**, 117 (1999).

3. K. Hiroya, S. Matsumoto, and T. Sakamoto, *Org. Lett.*, **6**, 2953 (2004).

Trifluoromethanesulfonamide: $R_2NSO_2CF_3$ (Chart 10)

A trifluoromethanesulfonamide can be prepared from a primary amine to allow monoalkylation of that amine.[1] The triflamide is not stable to strong base, which causes elimination to an imine,[2] but when used to protect an indole, it is cleaved with K_2CO_3 in refluxing methanol.[6]

Formation

$(CF_3SO_2)_2O$, CH_2Cl_2, $-78°C$, ~quant.[1]

Cleavage

1. $NaAlH_2(OCH_2CH_2OCH_3)_2$, benzene, reflux, few min, 95% yield[1]
2. 4-Br-$C_6H_4COCH_2Br$, K_2CO_3, acetone, 12 h; H_3O^+, 80% yield.[3]
3. $LiAlH_4$, Et_2O, reflux, 90–95% yield.[1,4]
4. Na (NH_3, *t*-BuOH, THF),[5]
5. $BH_3·THF$, >3h.[6]

1. J. B. Hendrickson and R. Bergeron, *Tetrahedron Lett.*, **14**, 3839 (1973).

2. S. Bozec-Ogor, V. Salou-Guiziou, J. J. Yaouanc, and H. Handel, *Tetrahedron Lett.*, **36**, 6063 (1995).

3. J. B. Hendrickson, R. Bergeron, A. Giga, and D. Sternbach, *J. Am. Chem. Soc.*, **95**, 3412 (1973).

4. K. E. Bell, D. W. Knight, and M. B. Gravestock, *Tetrahedron Lett.*, **36**, 8681 (1995).

5. M. L. Edwards, D. M. Stemerick, and J. R. McCarthy, *Tetrahedron Lett.*, **31**, 3417 (1990); D. F. Taber and Y. Wang, *J. Am. Chem. Soc.*, **119**, 22 (1997).

6. M. Lögers, L. E. Overman, and G. S. Welmaker, *J. Am. Chem. Soc.*, **117**, 9139 (1995).

t-Butylsulfonamide (Bus−NR$_2$): *t*-BuSO$_2$NR$_2$

Since *t*-BuSO$_2$Cl is unstable a two-step procedure was developed for introduction of the Bus group as outlined in the scheme below. The sulfinamide can also be considered a protective group that is acid cleavable[1] but it does impart chirality which may not always be desirable.

The *N*-Bus group is stable to the following conditions: (1) $0.1 N$ HCl, MeOH, (2) $0.1 N$ TFA, CH_2Cl_2, rt, 1 h, or (3) pyrolysis neat at 180°C, 3 h. Primary Bus derivatives are more stable to acid than are secondary derivatives.[2,3] TfOH is the preferred reagent to cleave the Bus group (58–100% yield).

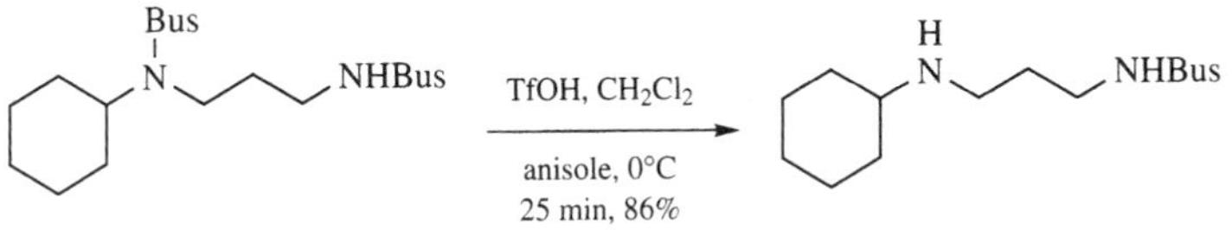

1. T. P. Tang and J. A. Ellman, *J. Org. Chem.*, **67**, 7819 (2002).
2. P. Sun and S. M. Weinreb, *J. Org. Chem.*, **62**, 8604 (1997).
3. G. Borg, M. Chino, and J. A. Ellman, *Tetrahedron Lett.*, **42**, 1433 (2001).

Benzylsulfonamide: $C_6H_5CH_2SO_2NR_2$ (Chart 10)

Benzylsulfonamides, prepared in 40–70% yield, are cleaved by reduction (Na, NH_3, 75% yield; H_2, Raney Ni, 65–85% yield, but not by H_2, PtO_2) and by acid hydrolysis (HBr or HI, slow).[1] They are also cleaved by photolysis (2–4 h, 40–90% yield).[2] The similar *p*-**methylbenzylsulfonamide (PMS–NR₂)** has been prepared to protect the ε-amino group in lysine; it is quantitatively cleaved by anhydrous hydrogen fluoride/anisole ($-20°C$, 60 min).[3] Another example of this seldom-used group is illustrated below.[4]

Formation

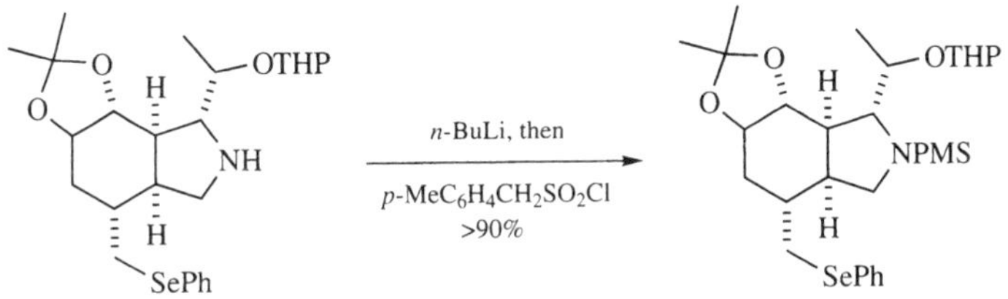

Cleavage

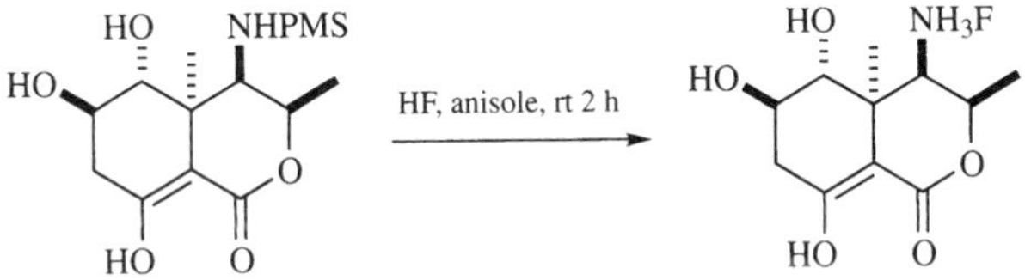

1. H. B. Milne and C.-H. Peng, *J. Am. Chem. Soc.*, **79**, 639, 645 (1957).
2. J. A. Pinock and A. Jurgens, *Tetrahedron Lett.*, **20**, 1029 (1979).
3. T. Fukuda, C. Kitada, and M. Fujino, *J. Chem. Soc., Chem. Commun.*, 220 (1978).
4. M. Yoshioka, H. Nakai, and M. Ohno, *J. Am. Chem. Soc.*, **106**, 1133 (1984).

2-(Trimethylsilyl)ethanesulfonamide (SES−NR$_2$): Me$_3$SiCH$_2$CH$_2$SO$_2$NR$_2$

The SES group is stable to TFA, hot 6 M HCl, THF; LiBH$_4$, CH$_3$CN, BF$_3$·Et$_2$O, 40% HF/EtOH.

Formation

1. SES-Cl, Et$_3$N, DMF, 0°C, 88–95% yield.[1]
2. SES-Cl, AgCN, benzene, 75°C, 22 h, 61% yield. The standard method gave poor yields and more side reactions.[2]

Cleavage

1. DMF, CsF, 95°C, 9–40 h, 80–93% yield.[1] These conditions will cleave 1 SES group from a bis-SES protected amine.[3]
2. Bu$_4$NF, CH$_3$CN, reflux, >85% yield.[1,4]
3. TAS-F, DMF or CH$_3$CN, rt, 60–68% yield for deprotection of aziridines.[5]
4. CsF, DMF, 95°C.[6]
5. CsF, DMF, (BOC)$_2$O, 50°C, 6 h, 0.01M, 96% yield. The amine is converted to a BOC derivative, which prevents diketopiperazine formation.[7]
6. HF, anisole, 0°C, 90 min, 75–85% yield.[8,9]

1. S. M. Weinreb, D. M. Demko, T. A. Lessen, and J. P. Demers, *Tetrahedron Lett.*, **27**, 2099 (1986).
2. K. J. Hale, M. M. Domostoj, D. A. Tocher, E. Irving, and F. Scheinmann, *Org. Lett.*, **5**, 2927 (2003).
3. D. M. Dastrup, M. P. VanBrunt, and S. M. Weinreb, *J. Org. Chem.*, **68**, 4112 (2003).
4. R. S. Garigipati and S. M. Weinreb, *J. Org. Chem.*, **53**, 4143 (1988).
5. P. Dauban and R. H. Dodd, *J. Org. Chem.*, **64**, 5304 (1999).
6. N. Matzanke, R. J. Gregg, and S. M. Weinreb, *J. Org. Chem.*, **62**, 1920 (1997).
7. D. L. Boger, J.-H. Chen, and K. W. Saionz, *J. Am. Chem. Soc.*, **118**, 1629 (1996).

8. Y. Rew, D. Shin, I. Hwang, and D. L. Boger, *J. Am. Chem. Soc.*, **126**, 1041 (2004).

9. D. L. Boger, M. W. Ledeboer, and M. Kume, *J. Am. Chem. Soc.*, **121**, 1098 (1999).

p-Toluenesulfonamide (Ts−NR₂): p-CH$_3$C$_6$H$_4$SO$_2$NR$_2$ (Chart 10)

Benzenesulfonamide: PhSO$_2$NR$_2$

In general, the benzenesulfonyl group is somewhat more reactive than the tosyl group, both in its formation and its ease of cleavage. On the whole, these are extremely robust protective groups and often require very harsh conditions for removal. The exception to this is for aromatic amines *vida infra*. The benzenesulfonyl group also has the advantage that the sulfonyl chloride is a liquid, which is much easier to handle on scale.

Formation

1. Tosylates are generally formed from an amine and tosyl chloride in an inert solvent such as CH$_2$Cl$_2$ with an acid scavenger such as pyridine or triethylamine. They may also be prepared using the Schotten–Baumann reaction.

2. TfO⁻ This reagent is good for the formation of sulfonamides of hindered amines.[1]

Ref. 2

3. 1-Phenylsulfonylbenzotriazole, THF, 1-methylimidazole, reflux, 64–99% yield.[3] The reagent also benzenesulfonates phenols (51–99% yield). A general preparation of these reagents has been published.[4]

4. TsOC$_6$F$_5$, Bu$_4$NCl, CHCl$_3$. The chloride ion accelerates the reaction considerably for the otherwise unreactive PFP sulfonates.[5]

5. TsOH·Pyr(PPTS), Ph$_3$P=O, Tf$_2$O, TEA, CH$_2$Cl$_2$, 96% yield.[6]

Cleavage

1. HBr, AcOH, 70°C, 8 h, 45–50% yield.[32] During the synthesis of L-2-amino-3-oxalylaminopropionic acid, a neurotoxin, cleavage with Na/NH$_3$ or [C$_{10}$H$_8$]⁻ Na⁺ gave a complex mixture of products.

2. HBr, P, reflux, 24 h, 74–88% yield. An *N*-benzyl group survived these brutal conditions.[8]

3. TMSCl, NaI, CH$_3$CN, reflux, 3–4 h, 70–88% yield. Mesylates and besylates are cleaved.[9] This rather harsh method produces TMSI *in situ*, which is known to cleave a large variety of protective groups.

4. HF·Pyr, anisole, rt, >62% yield.[10]

5. NaAlH$_2$(OCH$_2$CH$_2$OCH$_3$)$_2$, benzene or toluene, reflux, 20 h, 65–75% yield.[11] Note that LiAlH$_4$ does not cleave sulfonamides of primary amines; those from secondary amines must be heated to 120°C. In the following case, dissolving metal reduction failed.[12]

6. Electrolysis, Me$_4$NCl, 5°C, 65–98% yield.[13–15] Acylation of a tosylated amine with BOC$_2$O or benzoyl chloride reduces the potential required for electrolytic cleavage so that these aryltosyl groups can be selectively removed in the presence of a simple tosylamide.[16]

7. Electrolysis, ascorbic acid, anthracene, Et$_4$NBF$_4$, DMF.[17]

8. Me$_3$CoLi or Me$_3$FeLi or Me$_3$MnLi, Mg, THF, 83–100% yield.[18] A phenolic allyl ether is cleaved with this reagent.

9. Sodium naphthalenide.[19–21] This reagent has been used to remove the tosyl group from an amide.[22]

Although in this example the Bn and BOM groups were also cleaved,[23] it is possible to retain a Bn group when using this reagent.[24]

10. Sodium anthracenide, DME, 85% yield.[25]

11. Li, catalytic naphthalene, −78°C, THF, 65–99% yield.[26]

12. Li, di-t-butylbiphenyl, −78°C, THF, 1 h, 25–85% yield. The method was used to cleave a toluenesulfonamide.[32]

13. Li, NH$_3$, 75% yield[27] or Na, NH$_3$.[28,29] Note that in the following example enone reduction is slower than benzenesulfonamide cleavage.[30]

14. Na, IPA.[31]

15. Mg, MeOH, 8–75% yield. These conditions were used to cleave a tosyl group from an aziridine, a special case over normal amines.[32] The reaction should work better with a benzenesulfonamide. This method is very good for carbamate and amide protected sulfonamides, but does not work with normal aliphatic amines.[33] Since sulfonamides are readily acylated, this constitutes a relatively mild method for the cleavage of sulfonamides. Lactones and esters are compatible with this methodology.[34]

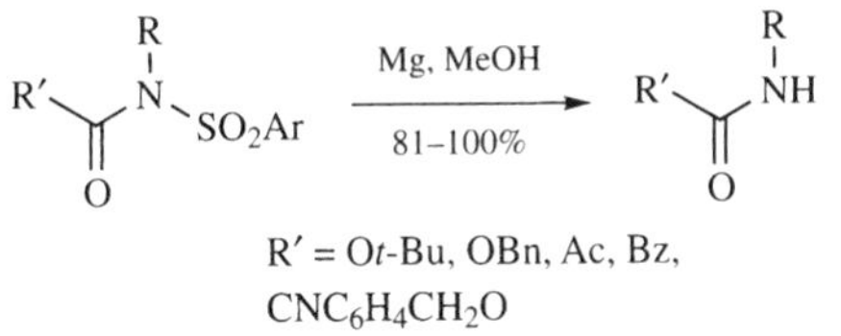

R′ = Ot-Bu, OBn, Ac, Bz,
CNC₆H₄CH₂O

$R' = Ot\text{-}Bu,\ OBn,\ Ac,\ Bz,\ CNC_6H_4CH_2O$

16. SmI₂, DMPU, 50–97% yield.[35,36] The reaction works well for alkyl-substituted aziridines; benzenesulfonamides react faster than tosyl amides. Primary toluenesulfonamides do not give clean reductive cleavage, but benzenesulfonamides do.

17. TiCl₃, Li, THF, 25°C, 18 h, 43–78% yield.[37]

18. 48% HBr, phenol, 30 min, heat, 85% yield.[15,38] 4-Hydroxybenzoic acid has been used in place of phenol to aid in the isolation process. Addition of water to the reaction mixture caused most of the hydroxybenzoic acid derivatives to precipitate, thus greatly simplifying the isolation.[39]

19. HClO₄, AcOH, 100°C, 1 h, 30–75% yield.[40]

20. hν, Et₂O, 6–20 h, 85–90% yield.[41,42]

21. hν, EtOH, H₂O, NaBH₄, 1,2-dimethoxybenzene.[43] This is a photosensitized electron-transfer reaction. Other reductants such as hydrazine and BH₃·NH₃ are also effective.

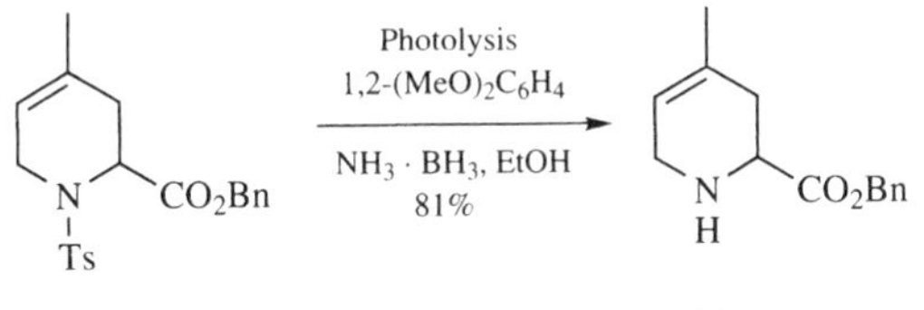

22. hν, β-naphthoxide anion, NaBH₄, quantitative.[44]

23. Na(Hg), Na₂HPO₄.[45,46]

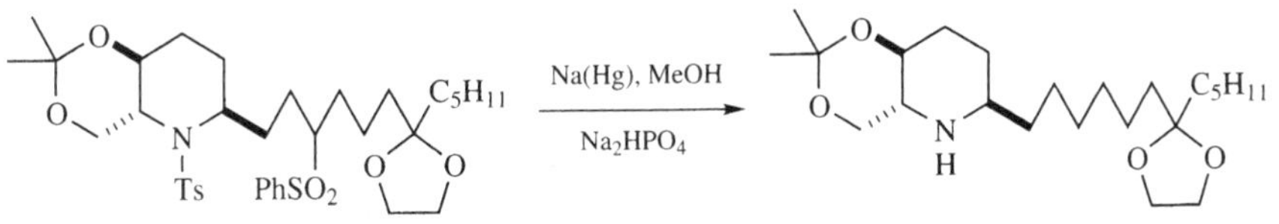

24. In this example the enone was not reduced.[47]

25. SMEAH, *o*-xylene, reflux, 91% yield.[48]

26. PhMe$_2$SiLi, THF, 0°C, 3–6 h, 72–83% yield. Primary tosylates fail to react and tosylaziridines ring open to give trans silyl sulfonamides.[49]

27. During attempted acetonide formation of an amino alcohol derivative, smooth tosyl cleavage was observed. The reaction is general for those cases having a carboxyl group, as in the example below, but fails for simple amino alcohol derivatives that lack this functionality.[50]

1. J. F. O'Connell and H. Rapoport, *J. Org. Chem.*, **57**, 4775 (1992).

2. W. H. Pearson, D. M. Mans, and J. W. Kampf, *Org. Lett.*, **4**, 3099 (2002).

3. A. R. Katritzky, G. Zhang, and J. Wu, *Synth. Commun.*, **24**, 205 (1994).

4. A. R. Katritzky, V. Rodriguez-Garcia, and S. K. Nair, *J. Org. Chem.*, **69**, 1849 (2004).

5. J. D. Wilden, D. B. Judd, and S. Caddick, *Tetrahedron Lett.*, **46**, 7637 (2005).

6. S. Caddick, J. D. Wilden, and D. B. Judd, *J. Am. Chem. Soc.*, **126**, 1024 (2004).

7. B. E. Haskell and S. B. Bowlus, *J. Org. Chem.*, **41**, 159 (1976).

8. U. Jordis, F. Sauter, S. M. Siddiqi, B. Kücnburg, and K. Bhattacharya, *Synthesis*, 925 (1990).

9. G. Sabitha, B. V. S. Reddy, S. Abraham, and J. S. Yadav, *Tetrahedron Lett.*, **40**, 1569 (1999).

10. W. Oppolzer, H. Bienaymé, and A. Genevois-Borella, *J. Am. Chem. Soc.*, **113**, 9660 (1991).

11. E. H. Gold and E. Babad, *J. Org. Chem.*, **37**, 2208 (1972).

12. D. F. Taber, T. D. Neubert, and A. L. Rheingold, *J. Am. Chem. Soc.*, **124**, 12416 (2002).

13. L. Horner and H. Neumann, *Chem. Ber.*, **98**, 3462 (1965).

14. T. Moriwake, S. Saito, H. Tamai, S. Fujita, and M. Inaba, *Heterocycles*, **23**, 2525 (1985).

15. R. C. Roemmele and H. Rapoport, *J. Org. Chem.*, **53**, 2367 (1988).

16. L. Grehn, L. S. Maia, L. S. Monteiro, M. I. Montenegro, and U. Ragnarsson, *J. Chem. Res., Synop.*, 144 (1991).

17. K. Oda, T. Ohnuma, and Y. Ban, *J. Org. Chem.*, **49**, 953 (1984).

18. M. Uchiyama, Y. Matsumoto, S. Nakamura, T. Ohwada, N. Kobayashi, N. Yamashita, A. Matsumiya, and T. Sakamoto, *J. Am. Chem. Soc.*, **126**, 8755 (2004).

19. J. M. McIntosh and L. C. Matassa, *J. Org. Chem.*, **53**, 4452 (1988).

20. C. H. Heathcock, T. A. Blumenkopf, and K. M. Smith, *J. Org. Chem.*, **54**, 1548 (1989).

21. S. C. Bergmeier and P. P. Seth, *Tetrahedron Lett.*, **40**, 6181 (1999).

22. H. Nagashima, N. Ozaki, M. Washiyama, and K. Itoh, *Tetrahedron Lett.*, **26**, 657 (1985); J. R. Henry, L. R. Marcin, M. C. McIntosh, P. M. Scola, G. D. Harris, Jr., and S. M. Weinreb, *Tetrahedron Lett.*, **30**, 5709 (1989).

23. T. Katoh, E. Itoh, T. Yoshino, and S. Terashima, *Tetrahedron Lett.*, **37**, 3471 (1996).

24. W.-S. Zhou, W.-G. Xie, Z.-H. Lu, and X. F. Pan, *Tetrahedron Lett.*, **36**, 1291 (1995).

25. P. Magnus, M. Giles, R. Bonnert, C. S. Kim, L. McQuire, A. Merritt, and N. Vicker, *J. Am. Chem. Soc.*, **114**, 4403 (1992).

26. E. Alonso, D. J. Ramón, and M. Yus, *Tetrahedron*, **53**, 14355 (1997).

27. C. H. Heathcock, K. M. Smith, and T. A. Blumenkopf, *J. Am. Chem. Soc.*, **108**, 5022 (1986).

28. A. G. Schultz, P. J. McCloskey, and J. J. Court, *J. Am. Chem. Soc.*, **109**, 6493 (1987).

29. N. Yamazaki and C. Kibayashi, *J. Am. Chem. Soc.* **111**, 1396 (1989).

30. P. I. Dalko, V. Brun, and Y. Langlois, *Tetrahedron Lett.*, **39**, 8979 (1998).

31. J. S. Bradshaw, K. E. Krakowiak, and R. M. Izatt, *Tetrahedron*, **48**, 4475 (1992).

32. D. A. Alonso and P. G. Andersson, *J. Org. Chem.*, **63**, 9455 (1998).

33. B. Nyasse, L. Grehn and U. Ragnarsson, *Chem. Commun.*, 1017 (1997).

34. K. Juhl, N. Gathergood and K. A. Jørgensen, *Angew. Chem. Int. Ed.*, **40**, 2995 (2001).

35. E. Vedejs and S. Lin, *J. Org. Chem.*, **59**, 1602 (1994).

36. For glucosamines: D. C. Hill, L. A. Flugge and P. A. Petillo, *J. Org. Chem.*, **62**, 4864 (1997).

37. S. K. Nayak, *Synthesis*, 1578 (2000).

38. R. S. Compagnone and H. Rapoport, *J. Org. Chem.*, **51**, 1713 (1986).

39. C. J. Opalka, T. E. D'Ambra, J. J. Faccone, G. Bodson, and E. Cossement, *Synthesis*, 766 (1995).

40. D. P. Kudav, S. P. Samant, and B. D. Hosangadi, *Synth. Commun.*, **17**, 1185 (1987).

41. A. Abad, D. Mellier, J. P. Pète, and C. Portella, *Tetrahedron Lett.*, **12**, 4555 (1971).

42. W. Yuan, K. Fearson, and M. H. Gelb, *J. Org. Chem.*, **54**, 906 (1989).

43. T. Hamada, A. Nishida, and O. Yonemitsu, *J. Am. Chem. Soc.* **108**, 140 (1986); W. Urjasz and L. Celewicz, *J. Phys. Org. Chem.*, **11**, 618 (1998). M. Ayadim, J. L. H. Jiwan, and J. P. Soumillion, *J. Am. Chem. Soc.*, **121**, 10436 (1999).

44. J. F. Art, J. P. Kestemont, and J. P. Soumillion, *Tetrahedron Lett.*, **32**, 1425 (1991).

45. T. N. Birkinshaw and A. B. Holmes, *Tetrahedron Lett.*, **28**, 813 (1987).

46. F. Chavez and A. D. Sherry, *J. Org. Chem.*, **54**, 2990 (1989).

47. P. Somfai and J. Åhman, *Tetrahedron Lett.*, **33**, 3791 (1992).

48. M. Ishizaki, O. Hoshino, and Y. Iitaka, *J. Org. Chem.*, **57**, 7285 (1992).

49. I. Fleming, J. Frackenpohl, and H. Ila, *J. Chem. Soc. Perkin Trans. 1*, 1229 (1998).

50. S. Chandrasekhar and S. Mohapatra, *Tetrahedron Lett.*, **39**, 695 (1998).

o-Anisylsulfonamide (Ans−NR$_2$): 2-CH$_3$OC$_6$H$_4$SO$_2$NR$_2$

Formation

2-CH$_3$OC$_6$H$_4$SO$_2$Cl, TEA, CH$_2$Cl$_2$, 65–97% yield.[1]

Cleavage

i-PrMgCl, Ni(acac)$_2$, Et$_2$O, rt, 2 h, 69–95% yield. This is a fundamentally new approach to sulfonamide cleavage and appears to be quite general. Primary and secondary amines, aryl amines, and aziridines are all smoothly deprotected.[1] These conditions will also cleave the toluenesulfonamide of oxazolidines.

1. R. R. Milburn and V. Snieckus, *Angew. Chem. Int. Ed.*, **43**, 892 (2004).

2- or 4-Nitrobenzenesulfonamide (Nosyl–NR$_2$ or Ns–NR$_2$)

The nosylate has become a popular protective group because of the mild conditions for its cleavage.[1] Its primary liability is in the fact that the nitro group is relatively easy to reduce, which should be remembered in planning a complex synthesis. The nitrobenzenesulfonamide is stable to strong acid and strong base.

Formation

1. NsCl, TEA, CH$_2$Cl$_2$, 97% yield.[2]
2. The Schotten-Baumann protocol can also be used.
3. NsCl, NaHCO$_3$, THF, rt, 56–88% yield. Primary amines are selectively protected.[3]

Cleavage

1. K$_2$CO$_3$ or Cs$_2$CO$_3$, DMF or CH$_3$CN, PhSH, 88–96% yield.[2] This process is not always selective for *p*-nosylate cleavage. Some amines, especially cyclic amines, tend to form 4-phenyl thioethers by nitro displacement as by-products of the cleavage process.[4] This problem has also been observed with the *o*-nosylates.[5] The problem is worse for cyclic amines.

The odorless decanethiol can be substituted effectively for PhSH.[6]

2. K_2CO_3, $MeOC_6H_4SH$, CH_3CN, DMSO, 85% yield. These conditions were developed to cleave the nosylate group from primary amines where isomerization is a concern. The original conditions using PhSH require prolonged heating.[7]

3. LiOH, DMF, $HSCH_2CO_2H$, 93–98% yield. This method has the advantage that the thioether by-products can be washed out by acid/base extraction.[2,8]

4. Electrolysis, DMF.[9] In the case of primary nosylates, −NH deprotonation competes with cleavage.

5. DBU, DMF, $HSCH_2CH_2OH$, >48% yield. These conditions were used to remove the nosyl group from N-methylated peptides.[10]

6. $C_8F_{17}CH_2CH_2SH$, K_2CO_3, CH_3CN, 50°C, 43–96% yield. This reagent was used as part of the "fluorous synthesis" methodology.[11]

7. Nosylaziridines can be opened with a variety of nucleophiles in preference to nucleophilic cleavage of the nosylate.[12]

1. T. Kan and T. Fukuyama, *Chem. Commun.*, 353 (2004).

2. T. Fukuyama, C.-K. Jow, and M. Cheung, *Tetrahedron Lett.*, **36**, 6373 (1995).

3. A. Favre-Réguillon, F. Segat-Dioury, L. Nait-Bouda, C. Cosma, J.-M. Siaugue, J. Foos, and A. Guy, *Synlett*, 868 (2000).

4. P. G. M. Wuts and J. M. Northuis, *Tetrahedron Lett.*, **39**, 3889 (1998).

5. M. De Rosa, N. Stepani, T. Cole, J. Fried, L. Huang-Pang, L. Peacock, and M. Pro, *Tetrahedron Lett.*, **46**, 5715 (2005).

6. T. Hakogi, M. Taichi, and S. Katsumura, *Org. Lett.*, **5**, 2801 (2003).

7. R. S. Narayan and M. S. VanNieuwenhze, *Org. Lett.*, **7**, 2655 (2005).

8. For cleavage of *p*-nosyl group: M. L. Di Gioia, A. Leggio, A. Le Pera, A. Liguori, A. Napoli, C. Siciliano, and G. Sindona, *J. Org. Chem.*, **68**, 7416 (2003).

9. N. R. Stradiotto, M. V. B. Zanoi, O. R. Nascimento, and E. F. Koury, *J. Chim. Phys./ Phys.-Chim. Biol.*, **91**, 75 (1994).

10. S. C. Miller and T. S. Scanlan, *J. Am. Chem. Soc.*, **119**, 2301 (1997).

11. C. Christensen, R. P. Clausen, M. Begtrup, and J. L. Kristensen, *Tetrahedron Lett.*, **45**, 7991 (2004).

12. P. E. Maligres, M. M. See, D. Askin, and P. J. Reider, *Tetrahedron Lett.*, **38**, 5253 (1997).

2,4-Dinitrobenzenesulfonamide ($DNs-NR_2$)

Formation

2,4-Dinitrobenzenesulfonyl chloride, pyridine or lutidine, CH_2Cl_2.[1]

Cleavage

1. Propylamine 20 eq., CH_2Cl_2, 20°C, 10 min, 88–93% yield.[1]

2. $HSCH_2CO_2H$, TEA, CH_2Cl_2, 23°C, 5 min, 91–98% yield. Since the rate of cleavage of the DNs group is much greater than the Ns group, it can be cleaved preferentially. DNs derivatives of primary amines under strongly basic conditions can rearrange to give an aniline with loss of SO_2. A similar process occurs for Ns derivatized primary amines, but much harsher conditions are required.[2]

3. Cleavage with thioacids (RCOSH) results in the formation of amides, $R'_2NC(O)R$.[3] The concept was extended to the formation of ureas, thioureas and thioamides.[4]

4. DMF, PhSH, 91% yield. No base is required.[5]

5. PhOK, CH_3CN, rt, 4 h, >67% yield. The more typical reagents used to cleave the DNs group resulted in Michael addition to the acrylate.[6]

6. An attempt to prepare the DNs derivative of anthranilic acid resulted in an unexpected reaction.

1. T. Fukuyama, M. Cheung, C.-K. Jow, Y. Hidai, and T. Kan, *Tetrahedron Lett.*, **38**, 5831 (1997).

2. P. Müller and N.-T. M. Phuong, *Helv. Chim. Acta*, **62**, 494 (1979).

3. T. Messeri, D. D. Sternbach, and N. C. O. Tomkinson, *Tetrahedron Lett.*, **39**, 1669 (1998).

4. T. Messeri, D. D. Sternbach, and N. C. O. Tomkinson, *Tetrahedron Lett.*, **39**, 1673 (1998).

5. K.-i. Nihei, M. J. Kato, T. Yamane, M. S. Palma, and K. Konno, *Synlett*, 1167 (2001).

6. S. Kobayashi, G. Peng, and T. Fukuyama, *Tetrahedron Lett.*, **40**, 1519 (1999).

2-Naphthalenesulfonamide

The naphthalenesulfonamide is readily prepared from the sulfonyl chloride in the presence of base. Its big advantage over the toluenesulfonamide is that it can be cleaved reductively with the milder Mg/MeOH (~1 h, 96–96% yield).[1,2] These mild cleavage conditions make this a very attractive alternative to the toluenesulfonamide.

1. B. Nyasse, L. Grehn, H. L. S. Maia, L. S. Monteir, and U. Ragnarsson, *J. Org. Chem.*, **64**, 7135 (1999).
2. L. Grehn and U. Ragnarsson, *J. Org. Chem.*, **67**, 6557 (2002).

4-(4',8'-Dimethoxynaphthylmethyl)benzenesulfonamide (DNMBS−NR₂)

The DNMBS derivative, readily prepared from an amine and the sulfonyl chloride, is efficiently ($\phi = 0.65$) cleaved photochemically (*hv* >300 nm, EtOH, $NH_3 \cdot BH_3$, 77–91% yield).[1] A water-soluble version of this group has been prepared and its photolytic cleavage examined.[2]

2-(4-Methylphenyl)-6-methoxy-4-methylsulfonamide

The sulfonamide is prepared from the acid chloride and an amine in IPA at 60° for 1–5 h (~70% yield). Cleavage is affected photochemically at 350 nm in N_2 purged solutions to return the amine in 32–96% yield.[3]

1. T. Hamada, A. Nishida, and O. Yonemitsu, *Tetrahedron Lett.*, **30**, 4241 (1989).
2. J. E. T. Corrie and G. Papageorgiou, *J. Chem. Soc., Perkin Trans. 1*, 1583 (1996).
3. G. A. Epling and M. C. Walker, *Tetrahedron Lett.*, **23**, 3843 (1982).

9-Anthracenesulfonamide

Formation

Anthracenesulfonyl chloride, TEA, THF.[1,2]

Cleavage

1. Hydrogenation: H_2, Pd–C, 24 h[3]
2. SmI_2, THF, t-BuOH.[3]
3. Al(Hg) aqueous NH_4OAc.[4,5]

4. Photolysis with dicyanobenzene sensitizer, 8 h the presence of one of the following hydrogen atom donors: $NaBH_4$, Et_3SiH, $NaCNBH_3$, 9,10-dihydroanthracene.[3]
5. TFA/anisole and thioanisole.[3]
6. $HSCH_2CH_2CH_2SH$, DIPEA.[6] It was reported that the anthracenesulfonamide is cleaved by reduction under these conditions, but treatment with PhSH/DI-PEA/DMF gives cleavage by an addition-elimination mechanism where 9-phenylthioanthracene is isolated as the only by-product.[7]

1. T. M. Kamenecka and S. J. Danishefsky, *Chem. Eur. J.*, **7**, 41 (2001).
2. For an improved preparation of this reagent, see P. G. M. Wuts, *J. Org. Chem.*, **62**, 430 (1997).
3. H. B. Argens and D. S. Kemp, *Synthesis*, 32 (1988).
4. A. J. Robinson and P. B. Wyatt, *Tetrahedron*, **49**, 11329 (1993).
5. J. M. Roe, R. A. B. Webster, and A. Ganesan, *Org. Lett.*, **5**, 2825 (2003).
6. J. Y. Roberge, X. Beebe, and S. J. Danishefsky, *Science*, **269**, 202 (1995).
7. P. G. M. Wuts, R. L. Gu, and J. M. Northuis, *Lett. Org. Chem.*, **1**, 372 (2004).

Pyridine-2-sulfonamide

Formation

Pyridine-2-sulfonyl chloride, aq. K_2CO_3, ether, 64–98% yield.[1]

Cleavage

1. SmI_2, THF or DMPU, rt, 76–94% yield.[1] Deprotection of the pyridinesulfon-amide in the presence of a cinnamoyl group was possible when done without a proton source. BOC, N-benzyl, N-allyl, and trifluoroacetamido groups were all stable to these conditions.[2]
2. Electrolysis, -1.83 V, quantitative.[1,3]

1. C. Goulaouic-Dubois, A. Guggisberg, and M. Hesse, *J. Org. Chem.*, **60**, 5969 (1995).

2. C. Goulaouic-Dubois, A. Guggisberg, and M. Hesse, *Tetrahedron*, **51**, 12573 (1995).

3. J. K. Pak, A. Guggisberg, and M. Hesse, *Tetrahedron*, **54**, 8035 (1998).

Benzothiazole-2-sulfonamide (Betsyl−NR₂ or Bts−NR₂)

Formation

The Bts derivative is formed from the sulfonyl chloride, either using aprotic conditions for simple amines or by the Schotten–Baumann protocol for amino acids (87–97% yield). The primary drawback of this reagent is that its stability depends on its quality. It can, on occasion, rapidly and exothermically lose SO_2 to give 2-chlorobenzothiazole.[1,2]

Cleavage

1. Zn, AcOH, EtOH.[1]
2. Al–Hg, ether, H_2O.[1]
3. Slow addition of excess H_3PO_2 to 1 M DMF solution of substrate at 50°C.[1]
4. PhSH, DIPEA, DMF.[2]
5. $NaBH_4$, EtOH. This method is only good for Bts derivatives of secondary amines. With primary amines the reaction fails to go to completion.[3]
6. $Na_2S_2O_4$ or $NaHSO_3$, EtOH, water, reflux. With peptides these conditions cause racemization.[4]
7. TFA, PhSH, 25% conversion after 2 days.[4]
8. Pd–C, H_2, EtOH. Some cleavage occurs before the catalyst is poisoned.[4]
9. NaOH, rt, 12h. This method can be used for Bts derivatives of secondary amines, but primary amines require 90–100°C and results in racemization of the amino acid.[4]
10. Glutathione S-transferase has also been shown to cleave the Bts group.[5] This has considerable significance when using this group as part of a drug candidate.
11. During the course of a peptide synthesis based on the Bts amine protection the following amine was formed, indicating that amines can react with the benzothiazolesulfonamide.[6]

1. E. Vedejs, S. Lin, A. Klapars, and J. Wang, *J. Am. Chem. Soc.*, **118**, 9796 (1996).

2. P. G. M. Wuts, R. L. Gu, J. M. Northuis, and C. L. Thomas, *Tetrahedron Lett.*, **39**, 9155 (1998). P. G. M. Wuts, R. L. Gu, and J. M. Northuis, *Lett. Org. Chem.*, **1**, 372 (2004).

3. E. Vedejs and C. Kongkittingam, *J. Org. Chem.*, **65**, 2309 (2000).

4. E. Vedejs, S. Lin, and A. Klapars, *J. Am. Chem. Soc.*, **118**, 9796 (1996).

5. Z. Zhao, K. A. Koeplinger, T. Peterson, R. A. Conradi, P. S. Burton, A. S. Suarato, R. L. Heinrikson, and A. G. Tomasselli, *Drug. Met. Disp.*, **27**, 992 (1997).

6. E. Vedejs and C. Kongkittingam, *J. Org. Chem.*, **66**, 7355 (2001).

Phenacylsulfonamide: $R_2NSO_2CH_2COC_6H_5$ (Chart 10)

Like the trifluoromethanesulfonamides, phenacylsulfonamides are used to prevent dialkylation of primary amines. Phenacylsulfonamides are prepared in 91–94% yield from the sulfonyl chloride, and they are cleaved in 66–77% yield by Zn/AcOH/trace HCl.[1]

1. J. B. Hendrickson and R. Bergeron, *Tetrahedron Lett.*, **11**, 345 (1970).

2,3,6-Trimethyl-4-methoxybenzenesulfonamide (Mtr−NR$_2$)[1]

2,4,6-Trimethoxybenzenesulfonamide (Mtb−NR$_2$)[1] (Chart 10)

2,6-Dimethyl-4-methoxybenzenesulfonamide (Mds−NR$_2$)[2]

Pentamethylbenzenesulfonamide (Pme−NR$_2$)[2]

2,3,5,6-Tetramethyl-4-methoxybenzenesulfonamide (Mte−NR$_2$)[2]

4-Methoxybenzenesulfonamide (Mbs−NR$_2$)[2]

2,4,6-Trimethylbenzenesulfonamide (Mts−NR$_2$)[3]

2,6-Dimethoxy-4-methylbenzenesulfonamide (iMds−NR$_2$)[3]

3-Methoxy-4-t-butylbenzenesulfonamide[4]

These sulfonamides have been used to protect the guanidino group of arginine.[5] Their acid stability as determined by TFA cleavage of the N^G-Arg derivative (25°C, 60 min) is as follows: Mtr (52%) > Mds (22%) ≈ Mtb (20%) > Pme (2%) > Mte (1.6%) > Mts ≈ Mbs > iMbs. The Mtr group has been used to protect the ε-nitrogen of lysine. The following table gives the % cleavage of Lys(Mtr) in various acids (MSA = methanesulfonic acid)[6]:

	0.15 M MSA TFA, PhSMe (9:1) 20°C	0.3 M MSA TFA, PhSMe (9:1) 20°C	TFA, PhSMe (9:1) 50°C	HF, PhSMe 0°C	MSA, PhSMe 20°C	TFA 20°C
1 h	80.7	95.1	15.1	3.6	2.3	0
2 h	91.9	99.3	33.6	—	—	0

The rate of cleavage is four to five times faster if dimethyl sulfide is included in the TFA–PhSMe mixture.[7]

The use of $1M$ HBF$_4$ in TFA/thioanisole was found to give significant rate accelerations during cleavage of the Mtr group.[8] Sulfuric acid at 90° has also been used to cleave the Mtr group.[9]

2,2,5,7,8-Pentamethylchroman-6-sulfonamide (Pmc−NR$_2$)

This group was developed for the protection of N^G-Arg. It is effectively an analog of the Mtr group, but has the useful property that it is cleaved in TFA/PhSMe in only 20 min. The enhanced rate of cleavage is attributed to the forced overlap of the oxygen electrons with the incipient cation during cleavage. The Pmc group can also be cleaved with 50% TFA/CH$_2$Cl$_2$, which does not cleave the benzyloxy carbamate.[10,11] It may also be cleaved with HBr/AcOH.[12] One problem associated with the Pmc group is that it tends to migrate to other amino acids, such as tryptophan during acidolysis. This problem, which cannot be completely suppressed with the usual scavenging agents,[13] is also sequence dependent.[14] Another problem observed with both the Mtr and Pmc groups when serine and threonine are present is that of O-sulfonation, which was best suppressed by the addition of 5% water to the cleavage mixture,[15] but the addition of water was not always effective.[16]

Attempts to develop a more acid labile protecting group than the Pmc group[17] has led to the preparation of the related **Pbf** group, which was shown to be 1.2–1.4 times more sensitive to TFA then the Pmc group.[18]

1. E. Atherton, R. C. Sheppard, and J. D. Wade, *J. Chem. Soc., Chem. Commun.*, 1060 (1983).

2. M. Wakimasu, C. Kitada, and M. Fujino, *Chem. Pharm. Bull.*, **29**, 2592 (1981).

3. H. Yajima, K. Akaji, K. Mitani, N. Fujii, S. Funakoshi, H. Adachi, M.Oishi, and Y. Akazawa, *Int. J. Pept. Protein Res.*, **14**, 169 (1979).

4. S. S. Ali, K. M. Khan, H. Echner, W. Voelter, M Hasan, and Atta-ur-Rahman, *J. Prakt Chem./Chem-Ztg.*, **337**, 12 (1995).

5. M. Fujino, M. Wakimasu, and C. Kitada, *Chem. Pharm. Bull.*, **29**, 2825 (1981); M. Fujino, O. Nishimura, M. Wakimasu, and C. Kitada, *J. Chem. Soc., Chem. Commun.*, 668 (1980).

6. M. Wakimasu, C. Kitada, and M. Fujino, *Chem. Pharm. Bull.*, **30**, 2766 (1982).

7. K. Saito, T. Higashijima, T. Miyazawa, M. Wakimasu, and M. Fujino, *Chem. Pharm. Bull.*, **32**, 2187 (1984).

8. K. Akaji, M. Yoshida, T. Tatsumi, T. Kimura, Y. Fujiwara, and Y. Kiso, *J. Chem. Soc., Chem. Commun.*, 288 (1990).

9. T. J. McMurry, M. Brechbiel, C. Wu, and O. A. Gansow, *Bioconjugate Chem.*, **4**, 236 (1993).

10. R. Ramage and J. Green, *Tetrahedron Lett.*, **28**, 2287 (1987).

11. J. Green, O. M. Ogunjobi, R. Ramage, A. S. J. Stewart, S. McCurdy, and R. Noble, *Tetrahedron Lett.*, **29**, 4341 (1988).

12. K. Wisniewski and A. S. Kolodziejczyk, *Tetrahedron Lett.*, **38**, 483 (1997); J. R. Ralbovsky, J. G. Lisko, and W. He, *Synth. Commum.*, **35**, 1613 (2005).

13. C. G. Fields and G. B. Fields, *Tetrahedron Lett.*, **34**, 6661 (1993).

14. A. Stierandova, N. F. Sepetov, G. V. Nikiforovich, and M. Lebl, *Int. J. Pept. Protein Res.*, **43**, 31 (1994).

15. E. Jaeger, H. Remmer, G. Jung, J. Metzger, W. Oberthür, K. P. Rücknagel, W. Schäfer, J. Sonnenbichler, and I. Zetl, *Biol. Chem. Hoppe-Seyler*, **374**, 349 (1993).

16. A. G. Beck-Sickinger, G. Schnorrenberg, J. Metzger, and G. Jung, *Int. J. Pept. Protein Res.*, **38**, 25 (1991).

17. I. M. Eggleston, J. H. Jones, and P. Ward, *J. Chem. Res, Synop.*, 286 (1991).

18. H. N. Shroff, L. A. Carpino, H. Wenschuh, E. M. E. Mansour, S. A. Triolo, G. W. Griffin, and F. Albericio, *Pept.: Chem., Struct., Biol., Proc. Am. Pept. Symp.*, 13th, 121 (1994); L. A. Carpino, H. N. Shroff, S. A. Triolo, E. M. E. Mansour, H. Wenschuh, and F. Albericio, *Tetrahedron Lett.*, **34**, 7829 (1993).

Protection of Amino Alcohols

Oxazolidone

Oxazolidones are cyclic urethanes that are normally very difficult to hydrolyze when compared to esters. Hydrolysis is facilitated if the nitrogen atom bears an electron-withdrawing substituent such as an ester or carbonate. Oxazolidones are stable to a large variety of reagents but terminal oxazolidones in the presence of nucleophilic amines have been shown to react.[1]

Formation

1. Phosgene[2] or triphosgene in the presence of a base such as TEA or pyridine in CH_2Cl_2 is a common method for oxazolidinone formation.[3] Triphosgene has the advantage that it is an easily handled solid.[4]

2. Diethyl carbonate[5]

3. Carbonyldiimidazole. This is a commonly used reagent that is generally effective.

4. $4\text{-}NO_2C_6H_4OCOCl$, Amberlyst IR 120, 76% yield.[6]

5. From an azido alcohol: NaH or BuLi in THF, then CO_2 and Me_3P.[7]

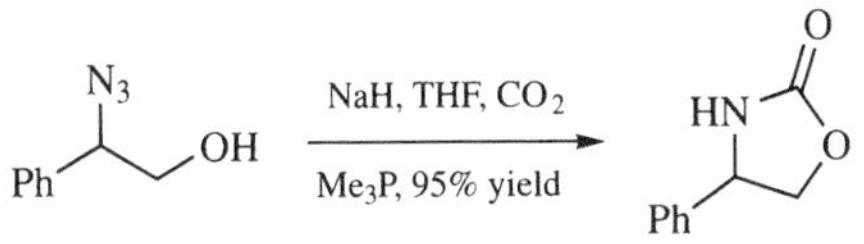

6. PdI_2, CO, O_2, MeOH, KI, 60 atm, 100°C, 86–100% yield.[8]

7. $n\text{-}Bu_2SnO$, CO_2, 5 MPa, 180°C, 16 h, 53–95% yield.[9]

8. Electrogenerated base from 2-pyrrolidone, CO_2, TsCl, CH_3CN, 64–95% yield.[10]

Cleavage

1. *t*-BuOK, THF, 95% yield.[11]

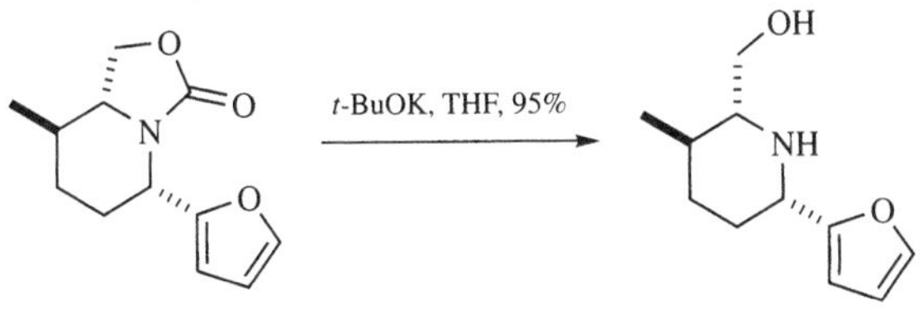

2. $Ba(OH)_2$, EtOH, H_2O; Ac_2O, pyridine, 48–81% yield.[6]

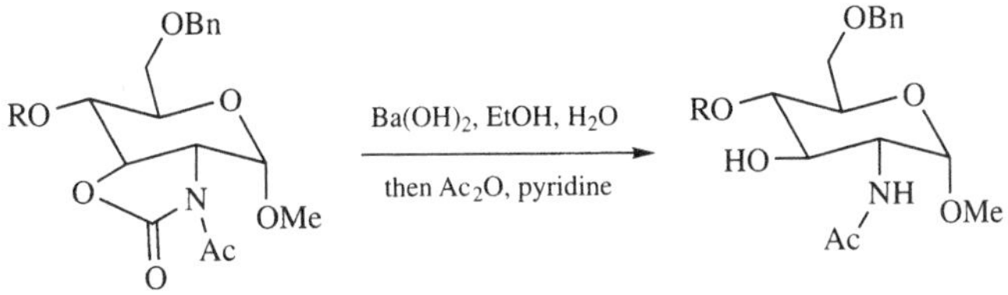

3. Cs_2CO_3, MeOH, 23°C, 3 h, 94% yield.[12]

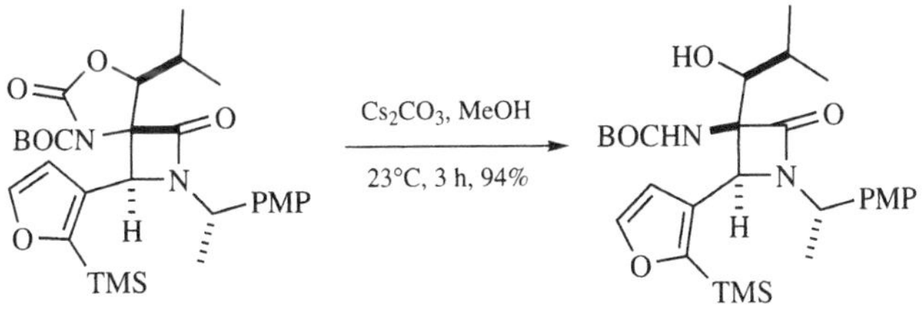

4. LiOH (3000 mol %), EtOH, H_2O, reflux, 76–99% yield.[13]

Oxazolines

One of the main advantages of an oxazoline is that there is no acidic NH as with the oxazolidone.

Formation

Oxazolines are usually formed from an amido alcohol by cyclization with a de-hydrating reagent. There does not seem to be a universal reagent that serves all situations. Some of the reported methods are as follows. The section on the protection of acids as oxazolines should be consulted.

1. Vilsmeier Reagent, pyridine, rt, then DBU.[14]
2. SOCl$_2$ followed by EtOH, KOH, reflux, 100% yield.[4] Thionyl chloride alone is often effective.[15]
3. SOCl$_2$, THF, 4°C, overnight followed by AgOTf, CaCO$_3$, benzene, rt.[16]
4. POCl$_3$, toluene, rt, 92% yield.[17]
5. Ph$_3$P=O or Ph$_2$S=O, Tf$_2$O, CH$_2$Cl$_2$, K$_3$PO$_4$, 46–100% yield.[18]
6. Martin's sulfurane (Ph(CF$_3$)$_2$CO-SPh$_2$, CH$_2$Cl$_2$, rt, 79–94% yield. Oxazoline formation depends on the stereochemistry of the substrate. Threo derivatives give elimination to dehydroamino acids.[19]
7. Ph$_3$P, diisopropylazodicarboxylate, THF, 0°, 56–80% yield.[20–22]
8. Burgess reagent, THF, 70°C, 64–85% yield. A polyethyleneglycol version of this reagent gives improved handling and higher yields (76–98% yield).[23]
9. Ph$_3$P, CCl$_4$, TEA, CH$_3$CN, 20°C, 71% yield.[24]
10. DAST, CH$_2$Cl$_2$, rt.[25]
11. BuSnCl$_2$, xylene reflux, 70% yield. This method proceeds without inversion of the alcohol.[26]
12. MsCl, TEA, CH$_2$Cl$_2$ followed by NaOH, H$_2$O, EtOH, heat, 86% yield.[26] A base treatment is not always required when using MsCl to form oxazo-lines.[27]
13. BF$_3$·Et$_2$O, 120°C, 61–76% yield.[28]
14. *o*-Chlorophenylphosphoro-bis-(1,2,4)-triazolide or phosphoro-tris-triazolide, CH$_3$CN, rt, 47–86% yield.[29]
15. TMSF, reflux.[30]
16. P$_2$O$_5$, refluxing toluene or xylene, 5–90% yield.[31]

Cleavage

TFA, MeOH.[4] Note that in the hydrolysis of these oxazolines the ester is usually produced under relatively mild conditions with the amine protonated. In many cases, if the amine is neutralized after the ring opening, the ester will migrate to the amine to form an amide. In general, to get complete deprotection, much harsher reaction conditions are required: that is, the ester must be hydrolyzed under the acidic conditions.

1. M. K. Sibi and J. W. Christensen, *J. Org. Chem.*, **64**, 6434 (1999).

2. T. Ziegler and C. Jurisch, *Tetrahedron: Asymmetry*, **11**, 3403 (2000).

3. For an extensive compilation of methods, see Y. Wu and X. Shen, *Tetrahedron: Asymmetry*, **11**, 4359 (2000).

4. S. Boisnard, L. Neuville, M. Bois-Choussy, and J. Zhu, *Org. Lett.*, **2**, 2459 (2000).

5. J. R. Gage and D. A. Evans, *Org. Synth.* **68**, 77 (1989).

6. D. Crich and A. U. Vinod, *J. Org. Chem.*, **70**, 1291 (2005).

7. X. Ariza, O. Pineda, F. Urpi, and J. Vilarrasa, *Tetrahedron Lett.*, **42**, 4995 (2001).

8. B. Gabriele, G. Salerno, D. Brindisi, M. Costa, and G. P. Chiusoli, *Org. Lett.*, **2**, 625 (2000). B. Gabriele, R. Mancuso, G. Salerno, and M. Costa, *J. Org. Chem.*, **68**, 601 (2003).

9. K.-i. Tominaga and Y. Sasaki, *Synlett*, 307 (2002).

10. M. A. Casadei, M. Feroci, A. Inesi, L. Rossi, and G. Sotgiu, *J. Org. Chem.*, **65**, 4759 (2000).

11. J. Barluenga, F. Aznar, C. Ribas, and C. Valdes, *J. Org. Chem.*, **64**, 3736 (1999).

12. P. C. Hogan and E. J. Corey, *J. Am. Chem. Soc.*, **127**, 15386 (2005).

13. S. J. Katz and S. C. Bergmeier, *Tetrahedron Lett.*, **43**, 557 (2002).

14. P. G. M. Wuts, J. M. Northuis, and T. A. Kwan, *J. Org. Chem.*, **65**, 9223 (2000).

15. D.-M. Gou, Y.-C. Liu, and C.-S. Chen, *J. Org. Chem.*, **58**, 1287 (1993).

16. F. Yokokawa, Y. Hamada, and T. Shioiri, *SynLett*, 151 (1992); Y. Hamada, M. Shibata, and T. Shioiri, *Tetrahedron Lett.*, **26**, 6501 (1985).

17. N. Langlois and H.-S. Wang, *Synth. Commum.*, **27**, 3133 (1997).

18. F. Yokokawa, Y. Hamada, and T. Shioiri, *SynLett*, 153 (1992).

19. F. Yokokawa, T. Shioiri, *Tetrahedron Lett.*, **43**, 8679 (2002).

20. P. Wipf and C. P. Miller, *Tetrahedron Lett.*, **33**, 907 (1992).

21. N. Galeotti, C. Montagne, J. Poncet, and P. Jouin, *Tetrahedron Lett.*, **33**, 2807 (1992).

22. M. E. Bunnage, S. G. Davies, and C. J. Goodwin, *J. Chem. Soc. Perkin Trans. 1*, 2385 (1994).

23. P. Wipf and S. Venkatraman, *Tetrahedron Lett.*, **37**, 4659 (1996).

24. A. Chesney, M. R. Bryce, R. W. J. Chubb, A. S. Batsanov, and J. A. K. Howard, *Synthesis*, 413 (1998).

25. T. H. Brown, C. A. Campbell, W. N. Chan, J. M. Evans, R. T. Martin, T. O. Stean, G. Stemp, N. C. Stevens, M. Thompson, N. Upton, and A. K. Vong, *Bioorg. Med. Chem. Lett.*, **5**, 2563 (1995).

26. G. Desimoni, G. Faita, and M. Mella, *Tetrahedron*, **52**, 13469 (1996).

27. B. M. Trost and C. B. Lee, *J. Am. Chem. Soc.*, **120**, 6818 (1998); T. Murakami and T. Shimizu, *Synth. Commum.*, **27**, 4255 (1997).

28. I. W. Davies, L. Gerena, N. Lu, R. D. Larsen, and P. J. Reider, *J. Org. Chem.*, **61**, 9629 (1996).

29. C. Sund, J. Ylikoski, and M. Kwiatkowski, *Synthesis*, 853 (1987).

30. D. Choi, J. P. Stables, and H. Kohn, *J. Med. Chem.*, **39**, 1907 (1996).

31. N. Ardabilchi, A. O. Fitton, J. R. Frost, F. K. Oppong-Boachie, A. H. b. A. Hadi, and A. b. M. Sharif, *J. Chem. Soc. Perkin Trans. 1*, 539 (1979).

PROTECTION FOR IMIDAZOLES, PYRROLES, INDOLES, AND OTHER AROMATIC HETEROCYCLES:

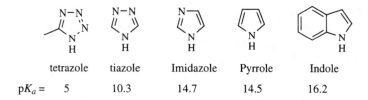

	tetrazole	tiazole	Imidazole	Pyrrole	Indole
pK_a =	5	10.3	14.7	14.5	16.2

Protective group chemistry for these amines has been separated from the simple amines because chemically they behave quite differently with respect to protective group cleavage. The increased acidity of these aromatic amines makes it easier to cleave the various amide, carbamate, and sulfonamide groups that are used to protect this class. A similar situation arises in the deprotection of nucleoside bases (e.g., the isobutanamide is cleaved with methanolic ammonia[1]), again, because of the increased acidity of the NH group.

N-Sulfonyl Derivatives

N,N-**Dimethylsulfonamide:** $R_2N-SO_2NMe_2$

Formation

Imidazole, Me_2NSO_2Cl, Et_3N, PhH, 16 h, 95% yield.[2,3]

Cleavage

1. $2 M$ HCl, reflux, 4 h.[2,4,5]
2. 10% Aqueous TFA.[6]

3. 2% KOH, H$_2$O, reflux, 12 h, 64–92% yield.[4] This group is more stable to *n*-BuLi than is the benzyl group when used to protect imidazoles.
4. TBAF, THF, reflux.[7]
5. From an indole: Electrolysis, DMF, 76–90% yield.
6. SmI$_2$, DMPU, THF, 73% yield.[8] TFA, TfOH, rt was also effective in this case (89% yield).

Methanesulfonamide (Ms−NR$_2$): CH$_3$SO$_2$NR$_2$

Formation

The methanesulfonamide is prepared by reaction of the amine with MsCl and TEA in CH$_2$Cl$_2$.

Cleavage

K$_2$CO$_3$, MeOH, rt, 12 h, 99% yield.[9]

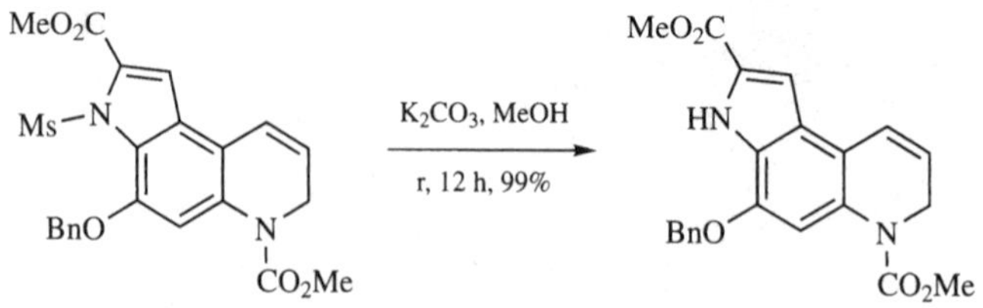

Mesitylenesulfonamide (Mts−NR$_2$): R$_2$N-SO$_2$-C$_6$H$_2$-(2,4,6-CH$_3$)$_3$

Formation/Cleavage[10]

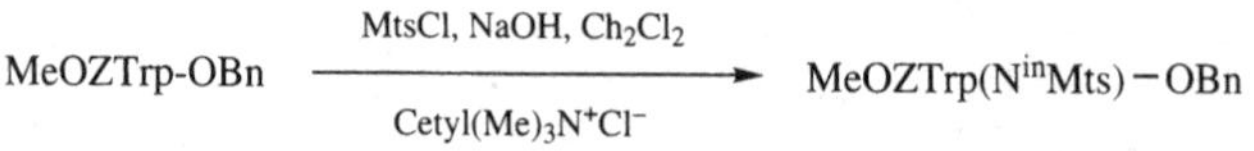

$$\text{MeOZTrp-OBn} \xrightarrow[\text{Cetyl(Me)}_3\text{N}^+\text{Cl}^-]{\text{MtsCl, NaOH, Ch}_2\text{Cl}_2} \text{MeOZTrp(N}^{in}\text{Mts)}-\text{OBn}$$

BuLi and MtsCl (84% yield) can also be used to protect an indole.[11] The Mts group is stable to CF$_3$COOH, 1N NaOH, hydrazine, 4N HCl, 25% HBr–AcOH, and H$_2$–Pd, but is cleaved with 1M CF$_3$SO$_3$H/CF$_3$COOH/thioanisole, CH$_3$SO$_3$H/CF$_3$COOH/thioanisole, HBr/H$_2$O/PhOH/110°C[12] or KOH[13]. Thioanisole is required to obtain clean conversions. The Mts group is not efficiently cleaved by HF.

p-Methoxyphenylsulfonamide (Mps−NR$_2$): R$_2$N-SO$_2$-C$_6$H$_4$-4-OCH$_3$

Formation

p-MeO-C$_6$H$_4$SO$_2$Cl, (imidazole = His).[14,15]

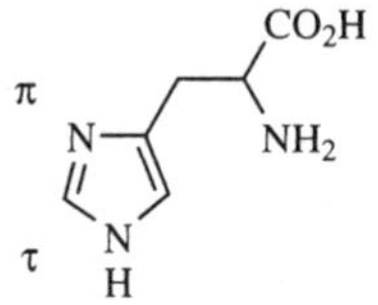

Histidine (His)

Cleavage

1. CF$_3$COOH, Me$_2$S, 40–60 min, 100% [imidazole = His(Mps)].[16]
2. Hydrazine, 1N NaOH, HOBT, and HF.[16] The Mps group on histidine is stable to CF$_3$COOH/anisole and to 25% HBr/AcOH.
3. Mg, MeOH, 60% yield.[17]

Benzenesulfonamide (Bs−NR$_2$): R$_2$N−SO$_2$C$_6$H$_5$ and

p-Toluenesulfonamide (Ts−NR$_2$): R$_2$N−SO$_2$C$_6$H$_4$-4-CH$_3$

Formation

1. For an imidazole, p-toluenesulfonyl chloride, Et$_3$N.[18,19]

2. For a pyrrole, benzenesulfonyl chloride, NaH, DMF, 60% yield.[20]
3. Ts$_2$O, NaH, DMF, >60% yield.[21]

Cleavage

1. Ac$_2$O, Pyr; H$_2$O or trifluoroacetic anhydride, pyridine, 0.5–16 h, 95–100% yield, [imidazole = His(Tos)].[14,19]
2. 1-Hydroxybenzotriazole (HOBT), THF, 1 h, [imidazole = His(Tos)].[15]
3. Pyr/HCl, DMF, [imidazole = His(Tos)].[22]
4. CF$_3$CO$_2$H, Me$_2$S, 40–60 min, 100% yield, [imidazole = His(Tos)].[23] The related benzenesulfonyl group has been used to protect pyrroles and indoles, and is cleaved with NaOH/H$_2$O/dioxane, rt, 2 h.[24,25]
5. KOH, MeOH, 98% yield (indole deprotection).[26,27] Sodium hydroxide can also be used (pyrrole deprotection).[20]
6. Mg, MeOH, sonication 20–40 min, 100% yield.[28] Sulfonamide-protected amides are also efficiently cleaved by this method.[29]
7. Mg, MeOH, NH$_4$Cl, benzene, rt.[30]

Ref. 31

8. PhSH, AIBN, benzene, reflux, 2 h, 90% yield.[32]
9. A benzenesulfonamide is cleaved with TBAF (THF, reflux, 38–100% yield).[33]
10. Electrolysis: CH$_3$CN, Et$_4$NCl, TEA·HCl divided cell, 63–87% yield.[34]
11. LiSCH$_2$CO$_2$Li, DMF, 20°C, 1.5–5 h, 79–95% yield. This method is not compatible with α, β-unsaturated carbonyl compounds or with α-ketoesters.[35]

Carbamates

Benzyl Carbamates (Cbz−NR$_2$ or Z−NR$_2$): C$_6$H$_5$CH$_2$O$_2$CNR$_2$

Formation

1. The section covering benzyl carbamates or normal amines should be consulted since those methods are generally applicable to the formation the heterocyclic derivatives.

2. For nonnucleophilic pyrroles: BnOCOCl, TBAI, K_2CO_3, DMF, 16h, 78% yield.[47]

3. For indoles: Carbonyldiimidazole, DMAP, CH_3CN, reflux then BnOH, 84% yield. Since this process proceeds through an imidazolide, other nucleophiles can be used to prepare a variety of carbamates and ureas.[36]

Cleavage

1. Bu_3SnH, AIBN, benzene, reflux, 1–3.5 h. This method only cleaves Cbz groups from aromatic amines and amides.[37]

2. $(Ph_3P)_2NiCl_2$, 3% Ph_3P, $Me_2NH·BH_3$, K_2CO_3, CH_3CN, 40°C, 82–97% yield.[38] The method is selective for aryl amines and Alloc derivatives.

2,2,2-Trichloroethyl Carbamate (Troc$-NR_2$): $R_2NCO_2CH_2CCl_3$

Formation/Cleavage[39]

$$BOC-TrpOBn \xrightarrow{\text{TrocCl, NaOH, } Bu_4NHSO_4} BOC-Trp(N^{in}-Troc)OBn$$

The Troc group on tryptophan is stable to CF_3COOH, CF_3SO_3H, and H_2–Pd, but can be cleaved with 0.01 M NaOH/MeOH, hydrazine/MeOH/H_2O, Cd/AcOH/DMF. Cleavage with Zn/AcOH is only partially complete. Hydrogenolysis (Pd/C, H_2, 6h) cleaves a Troc group from an imidazole.[150b]

2-(Trimethylsilyl)ethyl Carbamate (Teoc$-NR_2$): $R_2NCO_2CH_2CH_2Si(CH_3)_3$

The Teoc group is introduced onto pyrroles or indoles with 4-nitrophenyl 2-(trimethylsilyl)ethyl carbonate and NaH in 61–64% yield. The Teoc group can be removed with Bu_4NF in CH_3CN.[40]

2-(4-Trifluoromethylphenylsulfonyl)ethoxy Carbamate (Tsc$-NR_2$): $R_2NCO_2CH_2CH_2SO_2C_6H_4$-4-$CF_3$

The Tsc group was examined for the protection of various pyrrole and imidazole nitrogens. It was demonstrated to be orthogonal to the Fmoc group. The use of 1-methylpyrrolidine showed selective deprotection of the Fmoc in the presence of the Tsc group while LiOH will selectively cleave the Tsc group in the presence of the Fmoc group.[41]

t-Butyl Carbamate (BOC-NR_2): R_2N-CO_2-t-C_4H_9

Formation

The BOC group has been introduced onto the imidazole nitrogen of histidine with BOCF, pH 7–8[42]; $BOCN_3$, MgO,[43] and $(BOC)_2O$.[40,44] It can be introduced onto pyrroles and indoles with phenyl *t*-butyl carbonate and NaH, 67–91% yield,[45] or with NaH, BOCN$_3$.[46] Nonnucleophilic pyrroles can be protected with BOC$_2$O (TBAI, K$_2$CO$_3$, DMF, 16 h, 33%).[47]

Cleavage

The section on BOC cleavage for amines should be consulted since most of those methods are applicable for hetercyclic amines as well.

1. The N^{im}-BOC group can be removed under the usual conditions for removing the BOC group: CF$_3$COOH and HF.
2. It can also be removed with hydrazine and NH$_3$/MeOH.
3. NaOMe/MeOH/THF has been used to remove the BOC group from pyrroles in 66–99% yield.[46]
4. Thermolysis at 180°C cleaves the BOC group from indoles and pyrroles in 92–99% yield.[48,49]
5. Bu$_4$NF, THF, rt-reflux, 75–98% yield. This method is specific for electron-deficient amines such as heterocyclic amines and electron poor anilines.[50,51] Because TBAF contains about 4% water and is considered basic, some amides are also cleaved.
6. Sn(OTf)$_2$, CH$_2$Cl$_2$, 89% yield.[52]

1-Adamantyl Carbamate (Adoc−NR$_2$): R$_2$NCO$_2$−1-adamantyl

Formation

AdocCl, histidine, NaOH, Na$_2$CO$_3$, H$_2$O, 86% yield; forms N$^\alpha$,N^{im} (Adoc)$_2$−HisOH.[53]

Cleavage

The Adoc group can be cleaved by the same methods used to cleave the BOC group.[53] The Adoc group is somewhat more stable than the BOC group to acid.

2-Adamantyl Carbamate (2-Adoc−NR$_2$): R$_2$NCO$_2$−2-adamantyl

Formation

2-Adoc-Cl, aq. NaOH, dioxane, 76% yield for His isolated as the cyclohexylamine salt.[54]

Cleavage

The 2-Adoc group is stable to TFA, but cleaved completely within 10 min with 25% HBr/AcOH, HF, and TFMSA/thioanisole/TFA. Under basic conditions, it is slowly cleaved in 10% aq. TEA or 20% piperidine/DMF, but rapidly cleaved in 2 mol dm^{-3} aq. NaOH.[54]

2,4-Dimethylpent-3-yl Carbamate (Doc−NR₂): [(CH₃)₂CH]₂CHOC(O)NR₂

The Doc group, introduced with the chloroformate and either DMAP or *t*-BuOK, is quite acid-stable, but can be cleaved with TFMSA–thioanisole–EDT–TFA (10 min, rt) or with *p*-cresol–HF (1 h, 0°C).[55] The Doc group was found to be suitable for tryptophan protection in *t*-Bu-based peptide synthesis since no alkylation of tryptophan was observed during acid deprotection.

Cyclohexyl Carbamate (Hoc−NR₂): C₆H₁₁OCONR₂

The Hoc group was developed for tryptophan protection to minimize alkylation during BOC-mediated peptide synthesis. It is introduced with the chloroformate (NaOH, CH₂Cl₂, Bu₄N⁺HSO₄⁻) and can be cleaved with HF.[56] The use of HF, 1,4-butanedithiol, cresol reduces the problem of ring alkylation during deprotection with HF alone.[57]

1,1-Dimethyl-2,2,2-trichloroethyl Carbamate (TcBOC−NR₂): R₂NCO₂C(CH₃)₂CCl₃

Formation/Cleavage[58]

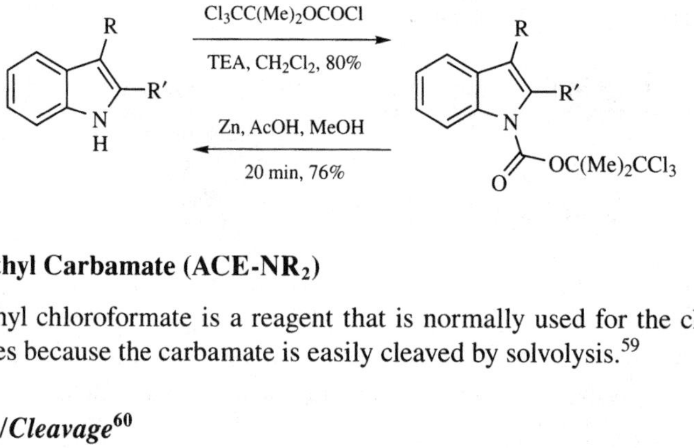

1-Chloroethyl Carbamate (ACE-NR₂)

1-Chloroethyl chloroformate is a reagent that is normally used for the cleavage of alkyl amines because the carbamate is easily cleaved by solvolysis.[59]

Formation/Cleavage[60]

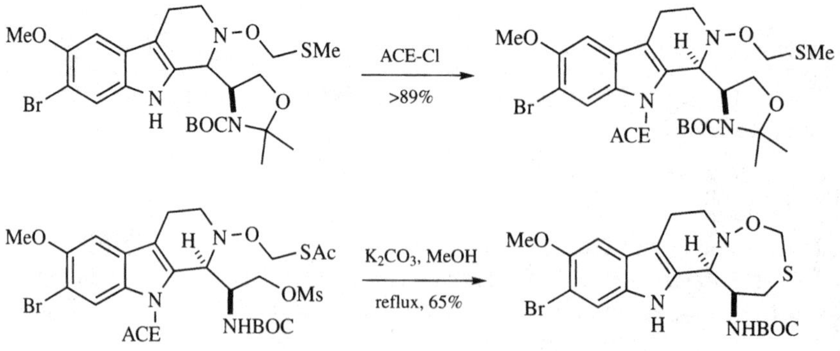

N-Alkyl and *N*-Aryl Derivatives

N-**Vinylamine:** $CH_2=CH-NR_2$

The vinyl group has been used to protect the nitrogen of benzimidazole during meta-lation with lithium diisopropylamide. It is introduced with vinyl acetate [$Hg(OAc)_2$, H_2SO_4, reflux, 24 h] or dibromoethane (TEA, reflux; 10% aq. NaOH reflux)[61] and cleaved by ozonolysis (MeOH, $-78°C$)[62] or $KMnO_4$ (acetone, reflux, 99% yield).[61] Both vinyl silanes and vinyl borates can be used to introduce the vinyl group on to heterocyclic amines.[63,64]

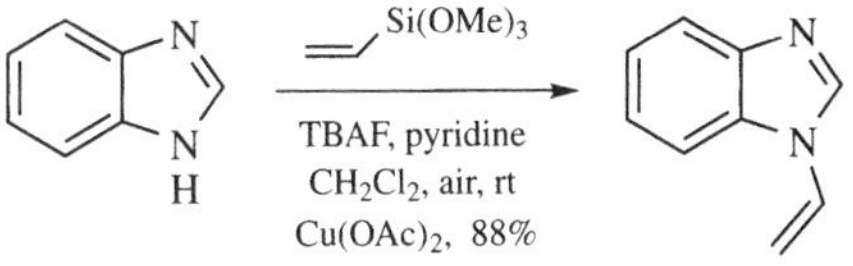

N-**2-Chloroethylamine:** $R_2NCH_2CH_2Cl$

Formation/Cleavage[65]

$ClCH_2CH_2Cl$, 50% NaOH, Bu_4NI, $>84\%$

$$\text{Pyrrole} \xrightleftharpoons[\substack{\text{1. NaH, } CH_3CN \\ \text{2. } Hg(OAc)_2 \\ \text{3. } NaBH_4}]{} C_4H_4NCH_2CH_2Cl$$

N-**(1-Ethoxy)ethylamine (EE−NR₂):** $R_2NCH(OCH_2CH_3)CH_3$

Formation/Cleavage[66]

$$\text{Imidazole} \xrightleftharpoons[\text{1 } N \text{ HCl, } 72°C]{\substack{\text{1. } n\text{-BuLi, } -10°C \\ \text{2. } CH_3CH(Cl)OEt, -20°C \\ 70-86\%}} \text{Imidazole-EE}$$

N-**2-(2′-Pyridyl)ethyl- and *N*-2-(4′-Pyridyl)ethylamine:**
$R_2NCH_2CH_2-2-(C_5H_4N)$ and $R_2NCH_2CH_2-4-(C_5H_4N)$

Formation/Cleavage

$$=\text{NH} \xrightleftharpoons[\substack{AlCl_3, ClCH_2CH_2Cl; \text{ NaOH, } 18-93\% \\ \text{or MeI, acetone, } 25°C; \text{ NaOH}[68]}]{\substack{\text{2- or 4-vinylpyridine, AcOH, } 22-94\%[67] \\ \text{or Na, 4-vinylpyridine}[68]}} =\text{NCH}_2\text{CH}_2\text{Pyr}$$

A series of substituted benzimidazoles and pyrroles were protected and deprot-ected using this methodology.

N-2-(4-Nitrophenyl)ethylamine (PNPE−NR$_2$): NO$_2$C$_6$H$_4$CH$_2$CH$_2$NR$_2$

The PNPE group is cleaved from a pyrrole with DBU (CH$_3$CN, rt, 81% yield).[69,70]

N-2-Phenylsulfonylethylamine: C$_6$H$_4$SO$_2$CH$_2$CH$_2$NR$_2$

Formation

From an indole[71] or pyrrole[72]: PhSO$_2$CH$_2$CH$_2$Cl, NaH, DMF, 67–73% yield.

Cleavage

1. *t*-BuOK, DMF, 34–100% yield.[71,73] The use of amine bases were not as effective. Cleavage occurs by β-elimination.
2. NaH, DMF, >60% yield.[74]

N-Trialkylsilylamines: R$_2$N−SiR$'_3$

Pyrroles and indoles can be protected with the *t*-butyldimethylsilyl group by treatment with TBDMSCl and *n*-BuLi or NaH.[75] **Triisopropylsilyl chloride** (NaH, DMF, 0°C to rt, 73% yield) has been used to protect the pyrrole nitrogen in order to direct electrophilic attack to the 3-position.[76] It has also been used to protect an indole.[77,78] This derivative can be prepared from the silyl chloride and K.[79] The silyl-protective group is cleaved with Bu$_4$NF, THF, rt or with CF$_3$COOH.

N-Allylamine: CH$_2$=CHCH$_2$NR$_2$

Guanine is catalytically protected at the 9-position with allyl acetate [(Pd(Ph$_3$P)$_4$, Cs$_2$CO$_3$, DMSO, 68% yield)].[80] The *N*-τ nitrogen of BOC-protected histidine is protected by bisalkylation with allyl bromide followed by removal of the *N*-π allyl group with Pd(Ph$_3$P)$_4$ (Et$_2$NH, NaHCO$_3$ or PhSiH$_3$, 80–85% yield). Removal of the allyl group is achieved by palladium-catalyzed transfer of the allyl group to *N,N'*-dimethylbarbituric acid.[81] The allyl group is cleaved from various heterocyclic amines as well as other allylamines derivatives with DIBAH (Ni(dppp)Cl$_2$, toluene, rt, 38–86% yield)[82] or *t*-BuMgCl (Ni(dppp)Cl$_2$, toluene, rt).[83] The allyl group was removed from a triazole by isomerization with HRuCl(CO)(Ph$_3$P)$_3$ (toluene, 120°C, 3 h) followed by ozonolysis of the vinyl triazole (88% yield).[84]

N-Benzylamine (Bn-NR$_2$): PhCH$_2$—NR$_2$

Formation

1. BnCl, NH$_3$, Na.[85]

The following benzyl halides were used: PhCH$_2$Br, 82% yield; PhCH(CH$_3$)Br, 33% yield; (Ph)$_2$CHBr, 50% yield; 3,4-(MeO)$_2$C$_6$H$_3$CH$_2$Cl, 52% yield.[86]

2. From an electron-deficient sodium imidazolide: PhCH$_2$OP$^+$(NMe$_2$)$_3$ PF$_6^-$, DMF, 24, heat, 40% yield.[87]

3. From indole: dibenzyl oxalate, *t*-BuOK, DMF, reflux, 86% yield.[88]

4. Dibenzyl carbonate, ionic liquid, DABCO, CH$_3$CN, 85°C, 23 h, 28–93% yield.[89] This method has also been used to methylate indoles in excellent yield by using dimethylcarbonate.[90]

5. MeLi, BnBr, THF, −40°C to rt, 39–74% yield.[91]

6. BnBr, NaH, DMF or DMSO, rt to 50°C, 57–75% yield.[92] This reaction is not regioselective but heating the mixture in the presence of BnBr will drive the reaction to the reaction to the thermodynamically favored product.[3]

The table below shows that this process in general for other alkylating agents.

Entry	RX	Initial Yield and Ratio (A:B)	Post-Heating Yield and Ratio (A:B)
1	BnBr	93 (1:0.3)	93 (1:0)
2	SEMCl	96 (1:0.3)	95 (1:0)
3	MOMCl	86 (1:0.5)	80 (1:0)
4	MeI	93 (1:0.7)	86 (1:0.3)

7. From an indole: Me$_3$P=CHCN, BnOH, 88% yield. These conditions were superior to using either DEAD/PPh$_3$ or TMAD/PBu$_3$.[93]

8. Using phase transfer method: BnBr, Aliquat 336, CH$_2$Cl$_2$, 50% NaOH, 90–96% yield.[94]

Cleavage

1. Cyclohexadiene, Pd-black, 25°C, 100% yield, [imidazole = His(Bn)].[95] With H$_2$/Pd–C, the normal conditions for benzyl group removal, it is difficult to remove the benzyl group on histidine without also causing reduction of other aromatic groups that may be present.[96]

2. AlCl$_3$, benzene or anisole, reflux, 25–91% yield, cleaved from a pyrido[2,3-b]indole[97] and indole.[92]

3. Ca, NH$_3$, >50–88% yield.[98]

4. t-BuOK, DMSO, O$_2$, rt, 20 min, 40–100% yield. This method was good for the cleavage of benzyl group from pyrazoles, indoles, carbazoles, and imidazoles.[99]

N-p-Methoxybenzylamine (PMB−NR$_2$ or MPM−NR$_2$): R$_2$N-CH$_2$C$_6$H$_4$−4-OCH$_3$

The MPM group was used in the preparation of a variety of triazoles,[100] imidazoles,[101] indole,[102] and pyrazoles.[103] This group is typically introduced using the bromide and NaH in DMF. It is readily cleaved with CF$_3$COOH at 65°C (52–100% yield) Anisole is sometimes included during the cleavage to scavenge the PMB cation. It is cleaved from a pyrido[2,3-b]indole (88% yield),[97] carbazole, or indole[104] (79% yield) with DDQ.

N-3,4-Dimethoxybenzylamine: 3,4-(MeO)$_2$C$_6$H$_3$CH$_2$NR$_2$

A 3,4-dimethoxybenzyl derivative, cleaved by acid (concd. H$_2$SO$_4$/anhyd. CF$_3$CO$_2$H, anisole), was used to protect a pyrrole −NH group during the synthesis of a tetrapyrrole pigment precursor. Neither an N-benzyl nor an N-p-methoxybenzyl derivative could be cleaved satisfactorily. Hydrogenolysis of the benzyl derivatives led to cyclohexyl compounds; acidic cleavage resulted in migration of the benzyl groups to the free α-position.[105]

N-3-Methoxybenzylamine and N-3,5-Dimethoxybenzylamine: 3-(MeO)C$_6$H$_4$CH$_2$NR$_2$ and 3,5-(MeO)$_2$C$_6$H$_3$CH$_2$NR$_2$

These benzylamines have been used for the protection of adenine and can be cleaved by photolysis at 254 nm.[106]

N-2-Nitrobenzylamine, (ONB−NR$_2$): R$_2$N−CH$_2$C$_6$H$_4$−2-NO$_2$ (Chart 10)

Formation

BOC−His(N^{im}Ag)OMe, 2-NO$_2$−C$_6$H$_4$CH$_2$Br, PhH, 4 h, reflux.[107]

Cleavage

1. $h\nu$, dioxane, 1 h, 100% yield.[107,108] The ONB group is stable to CF_3COOH, to HCl–AcOH, and to NaOH–MeOH, but is slowly cleaved by hydrogenation.

2. The related **4-nitrobenzyl** group, used to protect a benzimidazole, can be cleaved with H_2O_2 (EtOH, NaOH, 50°C, 72% yield).[109]

N-2,4-Dinitrophenylamine (DNP−NR$_2$): 2,4-$(NO_2)_2$-$C_6H_3NR_2$ (Chart 10)

The dinitrophenyl group has been used to protect the imidazole −NH group in histidines (45% yield) by reaction with 2,4-dinitrofluorobenzene and potassium carbonate[110] or TEA/CH_3CN.[111] Imidazole −NH groups, but not α-amino acid groups, are quantitatively regenerated by reaction with 2-mercaptoethanol (22°C, pH 8, 1 h).[112] The 2,4-dinitrophenyl group on the N^{im} of histidine reduces racemization in peptide synthesis because of its electron-withdrawing character.[113] In Fmoc-based peptide synthesis the DNP group is not stable because it migrates to the ϵ-NH_2 group of lysine[114] and it is also cleaved with 20% piperidine/DMF, conditions used to remove the Fmoc group.[115]

N-Phenacylamine: $R_2NCH_2COC_6H_5$ (Chart 10)

The phenacyl group is stable to HBr−AcOH, CF_3COOH, and CF_3SO_3H.[116] It is used to protect the π-nitrogen in histidine in order to reduce racemization during peptide-bond formation.[117]

N-Triphenylmethylamine, (Tr−NR$_2$): and *N*-Diphenylmethylamine (Dpm−NR$_2$): R_2NCPh_3 and R_2NCHPh_2

Formation

1. BOC−His, TrCl, Pyr.[118]
2. From a tetrazole: TrCl, CH_2Cl_2, TBAB, NaOH, H_2O.[119]

Cleavage

The trityl group can be cleaved with HBr−AcOH, 2 h; CF_3COOH, 30 min; formic acid, 2 min and by hydrogenation.[120] The trityl group in BOC−His(Tr)OH is stable to 1 M HCl/AcOH, rt, 20 h. The **diphenylmethyl** group was introduced in the same manner as the trityl group.[121] It is more stable to acid than the trityl group, but not significantly.[118,120] The trityl group has also been used to protect simple imidazoles.[122] The monomethoxytrityl group has been used to protect a benzotriazole (MMTrCl, pyridine, DMAP, 16 h, 54% yield).[123]

The following table gives the comparative stabilities of the N^α-Tr, N^{Im}-Tr, and N-BOC groups of Tr−His(Tr)−Lys(BOC)−OMe to various acidic conditions.[124]

Cleavage Conditions	% Cleavage		
	N^α-Tr	N^{Im}-Tr	N-BOC
5% HCO$_2$H, ClCH$_2$CH$_2$Cl, 8 min, 20°C	100	1	0
ClCH$_2$CH$_2$Cl, MeOH, TEA, 5 min, 20°C	100	<1	0
2.5 eq. HCl in 90% AcOH, 1 min, 20°C	100	<1	<1
1 N HCl in 90% AcOH, 20 min, 20°C	100	<1	100
90% AcOH, 1.5 h, 60°C	100	100	<1
5% PyrHCl, in MeOH, 2 h, 60°C	100	100	<1
95% TFA, 1 h, 20°C	100	100	100

N-(Diphenyl-4-pyridylmethyl)amine (Dppm−NR$_2$): R$_2$N−C(Ph)$_2$-4-(C$_5$H$_4$N) (Chart 10)

Formation

Ph$_2$-4-(C$_5$H$_4$N)CCl, Et$_3$N, CHCl$_3$, (Z)- or (BOC)-HisOMe.[125,126]

Cleavage

The diphenyl-4-pyridylmethyl group is cleaved by Zn/AcOH, 1.5 h, 91% yield; H$_2$/Pd–C, 91% yield; or by electrolytic reduction, 2.5 h, 0°C, 87% yield. The Dppm group is stable to trifluoroacetic acid.[125,127]

N-(*N'*,*N'*-Dimethyl)hydrazine: R$_2$N−NMe$_2$

The dimethylamine group can be cleaved from a pyrrole in low yield with chromous acetate.[128]

Amino Acetal Derivatives

N-Hydroxymethylamine: HOCH$_2$−NR$_2$

Formation/Cleavage[129]

E$^+$ = Electrophile

N-Methoxymethylamine (MOM$-$NR$_2$): R$_2$NCH$_2$OCH$_3$ (Chart 10)

The MOM group is introduced onto an indole through the sodium salt (NaOH, DMSO, 0°C, 0.5 h; MOMCl, 22°C, 0.5 h, 90% yield). It is removed with BF$_3$·Et$_2$O (Ac$_2$O, LiBr, 20°C, 48 h, 86% yield).[130] Removal of the related **ethoxymethyl** group from an imidazole with 6 N HCl at reflux is slow and low yielding.[131] Small structural effects at a site seemingly remote from the MOM group can have a significant influence on the deprotection process. The MOM group in compound **a** is easily removed with acid, but the cleavage with HCl in compound **b** proved quite difficult.[132]

a   b

N-Diethoxymethylamine (DEM$-$NR$_2$): (EtO)$_2$CH$-$NR$_2$

Formation/Cleavage[133,134]

$$\text{Imidazole} \underset{\text{H}_3\text{O}^+}{\overset{\text{(EtO)}_3\text{CH, TsOH, 130°C}}{\rightleftarrows}} \text{(EtO)}_2\text{CH--Imidazole}$$

DEM protection of an indole is also effective (46–82% yield) and cleavage occurs efficiently with 2 N HCl (EtOH, rt, 0.5 h, 86–93% yield).[135]

N-(2-Chloroethoxy)methylamine: R$_2$NCH$_2$OCH$_2$CH$_2$Cl

This derivative has been prepared from an indole, the chloromethyl ether, and potassium hydride in 50% yield; it is cleaved in 84% yield by potassium cyanide/18-crown-6 in refluxing acetonitrile.[136]

N-[2-(Trimethylsilyl)ethoxy]methylamine, (SEM$-$NR$_2$): R$_2$NCH$_2$OCH$_2$CH$_2$Si(CH$_3$)$_3$

Formation

Imidazole, indole or pyrrole, NaH, SEMCl, 50–85% yield.[137–139]

Cleavage

1. 1 M Bu$_4$NF, THF, reflux, 45 min, 46–90% yield or dil. HCl.[137,138]
2. BF$_3$·Et$_2$O; base.[140,141]
3. Bu$_4$NF, ethylenediamine (ethylenediamine was used as a formaldehyde scavenger), 45–98% yield.[140] Neat TBAF under vacuum has been used (90% yield).[142]

4. 3 *M* HCl, EtOH, reflux, 1 h, 95% yield. [143]
5. PPTS, MeOH, 24 h.[144]

N-t-Butoxymethylamine (Bum−NR₂): $R_2NCH_2O\text{-}t\text{-}C_4H_9$

The Bum derivative has been used to protect the π-nitrogen of histidine to prevent racemization during peptide bond formation.[145] The related 1- and 2-adamantyloxymethylamine has been used similarly for histidine protection.[146,147]

N-t-Butyldimethylsiloxymethylamine: $t\text{-}BuMe_2SiOCH_2NR_2$

The N-9 position of adenine was protected by formylation with basic formalin followed by silylation with TBDMSCl in Pyr, 86% yield. This group is removed with TFA/H₂O, 20°C, 2 h.[148]

N-Pivaloyloxymethylamine (POM−NR₂): $R_2NCH_2OCOC(CH_3)_3$ (Chart 10)

The POM group is introduced onto imidazoles, pyrroles, and indoles by treatment with NaH, $(CH_3)_3CCO_2CH_2Cl$[149] in THF at rt in 65–78% yield.[150] It is removed by hydrolysis with MeOH, NaOH[150] or with NH₃, MeOH (25°C, 4 h, 30–80% yield).[151]

N-Benzyloxymethylamine (BOM−NR₂): $R_2NCH_2OCH_2C_6H_5$ (Chart 10)

The BOM group is introduced onto an indole with the chloromethyl ether and sodium hydride in 80–90% yield. It is cleaved in 92% yield by catalytic reduction followed by basic hydrolysis,[152,153] or by CF₃COOH, HBr or 6 *M* HCl at 110°C.[154] As an alternative to Pd–C for hydrogenolysis, Mg−HCO₂H−NH₂NH₂ has been developed (89% yield). It also cleaves other benzyl-based groups.[155] It has been used to protect the π-nitrogen of histidine, preventing racemization during peptide bond formation. It has also been used to protect the τ-nitrogen of histidine (BnOCH₂Cl, Et₂O; Et₃N, MeOH).[156] During protective group cleavage of BOM-protected histidine, the formaldehyde liberated can react with *N*-terminal cysteine residues to form thiazolidines.[157,158]

N-Dimethylaminomethylamine: $(CH_3)_2NCH_2NR_2$

An indole, protected by a Mannich reaction with formaldehyde and dimethylamine, is stable to lithiation. The protective group is removed with $NaBH_4$ (EtOH, THF, reflux).[159] The related piperidine analog has been used similarly for the protection of a triazole.[160]

N-2-Tetrahydropyranylamine (THP$-$NR$_2$): R_2N-2-Tetrahydropyranyl (Chart 10)

The THP derivative of the imidazole nitrogen in purines has been prepared by treatment with dihydropyran (TsOH, 55°C, 1.5 h, 50–85% yield). It is cleaved by acid hydrolysis.[161] The THP group is useful for the protection of 1,2,4-triazoles.[162] A comparison between the THP and the THF group revealed that the THP is about six times more stable to tartaric acid in methanol.[163]

Amides

Carbon Dioxide: CO_2

The *in situ* generation of the carbon dioxide adduct of an indole provides sufficient protection and activation of an indole for metalation at C-2 with *t*-butyllithium. The lithium reagent can be quenched with an electrophile and quenching of the reaction with water releases the carbon dioxide.[164,165]

Formamide: R_2N-CHO

Formation[166]*/Cleavage*[167]

$$\text{Tryptophan} \xrightarrow{\text{HCO}_2\text{H, HCl}} \text{Tryptophan(N}^{\text{im}}\text{-CHO)}$$

The formyl group is cleaved with HF/anisole/$(CH_2SH)_2$.[167] It is also cleaved at pH 9–10.[166]

N,N-Diethylureide: $(CH_3CH_2)_2NC(O)NR_2$

The ureide, which is stable to BuLi, was used for the protection of indole. It is cleaved with 25% NaOH in EtOH, reflux.[168]

Dichloracetamide: $Cl_2CHCONR_2$

The dichloroacetamide of indole, formed by refluxing a mixture of dichloroacetyl chloride in dichloroethane, is cleaved upon treatment with TEA (CH_2Cl_2, rt).[169]

Pivalamide: $(CH_3)_3CCONR_2$

A pivalamide of an indole, introduced with PvCl (NaH, DMF, 0°C, 1 h, 96% yield) is efficiently cleaved with MeSNa (MeOH, 20°C, 2 h, 96% yield).[170] The use LDA (THF, 45°C, 79–93% yield) cleaves the pivalamide by a Meerwein–Pondorf–Verley reduction.[171]

Diphenylthiophosphinamide: $Ph_2P(S)-NR_2$

This group was used to protect the tryptophan nitrogen.

Formation

Ph$_2$P(S)Cl, NaHSO$_4$, NaOH, CH$_2$Cl$_2$, 0°C, 88% yield.[172]

Cleavage

1. 0.25 M Methanesulfonic acid, thioanisole in CF$_3$COOH, 0°C, 90 min.[172]
2. 0.25 M Trifluoromethanesulfonic acid, 0.25 M thioanisole in CF$_3$COOH, 0°C, 50 min.[172]
3. 0.1 M Bu$_4$NF, DMSO or DMF, 25°C, 10 min.[172,173]
4. 0.5 M KF, 18-crown-6, CH$_3$CN, 25°C, 3 h.[172]

4-Methyl-1,2,4-triazoline-3,5-dione (MTAD)

A special but interesting case is the selective protection of a more reactive indole using an ene reaction with MTAD and then reversing the process after selective functionalization of another indole with singlet oxygen.[174]

1. H. Büchi and H. G. Khorana, *J. Mol. Biol.*, **72**, 251 (1972).
2. D. J. Chadwick and R. I. Ngochindo, *J. Chem. Soc., Perkin Trans. I.*, 481 (1984).

3. Y. He, Y. Chen, H. Du, L. A. Schmid, and C. J. Lovely, *Tetrahedron Lett.*, **45**, 5529 (2004).

4. A. J. Carpenter and D. J. Chadwick, *Tetrahedron*, **42**, 2351 (1986).

5. S. Harusawa, Y. Murai, H. Moriyama, T. Imazu, H. Ohishi, R. Yoneda, and T. Kurihara, *J. Org. Chem.*, **61**, 4405 (1996).

6. D. Guianvarc'h, J.-L. Fourrey, R. Maurisse, J.-S. Sun, and R. Benhida, *Org. Lett.*, **4**, 4209 (2002).

7. J.-H. Liu, H.-W. Chan, and H. N. C. Wong, *J. Org. Chem.*, **65**, 3274 (2000); J.-H. Liu, Q.-C. Yang, T. C. W. Mak, and H. N. C. Wong, *J. Org. Chem.*, **65**, 3587 (2000).

8. A. Batch and R. H. Dodd, *J. Org. Chem.*, **63**, 872 (1998).

9. K. Hiroya, S. Matsumoto, and T. Sakamoto, *Org. Lett.*, **6**, 2953 (2004).

10. N. Fujii, S. Futaki, K. Yasumura, and H. Yajima, *Chem. Pharm. Bull.*, **32**, 2660 (1984).

11. L. W. Boteju, K. Wegner, X. Qian, and V. J. Hruby, *Tetrahedron*, **50**, 2391 (1994).

12. A. Kumar, S. Ghilagaber, J. Knight, and P. B. Wyatt, *Tetrahedron Lett.*, **43**, 6991 (2002).

13. S. Wang, X. Tang, and V. J. Hruby, *Tetrahedron Lett.*, **41**, 1307 (2000).

14. J. M. van der Eijk, R. J. M. Nolte, and J. W. Zwikker, *J. Org. Chem.*, **45**, 547 (1980).

15. T. Fujii and S. Sakakibara, *Bull. Chem. Soc. Jpn.*, **47**, 3146 (1974).

16. K. Kitagawa, K. Kitade, Y. Kiso, T. Akita, S. Funakoshi, N. Fujii, and H. Yajima, *J. Chem. Soc., Chem. Commun.*, 955 (1979).

17. B. Danieli, G. Lesma, M. Martinelli, D. Passarella, and A. Silvani, *J. Org. Chem.*, **62**, 6519 (1997).

18. S. Sakakibara and T. Fujii, *Bull. Chem. Soc. Jpn.*, **42**, 1466 (1969).

19. E. Wuensch in *Methoden der Organischen Chemie (Houben-Weyl)*, Vol. 15/1, E. Mueller, Ed., Georg Thieme Verlag, Stuttgart, 1974, p. 223.

20. C. F. Masaguer, E. Ravina, and J. Fueyo, *Heterocycles*, **34**, 1303 (1992).

21. K. M. Aubart and C. H. Heathcock, *J. Org. Chem.*, **64**, 16 (1999).

22. H. C. Beyerman, J. Hirt, P. Kranenburg, J. L. M. Syrier, and A. Van Zon, *Recl. Trav. Chim. Pays-Bas*, **93**, 256 (1974).

23. K. Kitagawa, K. Kitade, Y. Kiso, T. Akita, S. Funakoshi, N. Fujii, and H. Yajima, *Chem. Pharm. Bull.*, **28**, 926 (1980).

24. J. Rokach, P. Hamel, M. Kakushima, and G. M. Smith, *Tetrahedron Lett.*, **22**, 4901 (1981).

25. W. A. Remers, R. H. Roth, G. J. Gibs, and M. J. Weiss, *J. Org. Chem.*, **36**, 1232 (1971).

26. A. P. Kozikowski and Y.-Y. Chen, *J. Org. Chem.*, **46**, 5248 (1981).

27. M. G. Saulnierand and G. W. Gribble, *J. Org. Chem.*, **47**, 2810 (1982).

28. Y. Yokoyama, T. Matsumoto, and Y. Murakami, *J. Org. Chem.*, **60**, 1486 (1995).

29. B. Nyasse, L. Grehn, and U. Ragnarsson, *J. Chem. Soc., Chem. Commun.*, 1017 (1997).

30. H. Ishibashi, T. Tabata, K. Hanaoka, H. Iriyama, S. Akamatsu, and M. Ikeda, *Tetrahedron Lett.*, **34**, 489 (1993).

31. A. M. Elder and D. H. Rich, *Org. Lett.*, **1**, 1443 (1999).

32. S. H. Kim, I. Figueroa, and P. L. Fuchs, *Tetrahedron Lett.*, **38**, 2601 (1997).

33. A. Yasuhara and T. Sakamoto, *Tetrahedron Lett.*, **39**, 595 (1998).

34. H. L. S. Maia, L. S. Monteiro, and J. Sebastiao, *Eur. J. Org. Chem.*, 1967 (2001).

35. C. M. Haskins and D. W. Knight, *Tetrahedron Lett.*, **45**, 599 (2004).

36. J. E. Macor, A. Cuff, and L. Cornelius, *Tetrahedron Lett.*, **40**, 2733 (1999).

37. M.-L. Bennasar, T. Roca, and A. Padulles, *Org. Lett.*, **5**, 569 (2003).

38. B. H. Lipshutz, S. S. Pfeiffer, and A. B. Reed, *Org. Lett.*, **3**, 4145 (2001).

39. Y. Kiso, M. Inai, K. Kitagawa, and T. Akita, *Chem. Lett.*, **12**, 739 (1983).

40. L. Grehn and U. Ragnarsson, *Angew. Chem., Int. Ed. Engl.*, **23**, 296 (1984).

41. J. S. Choi, H. Kang, N. Jeong, and H. Han, *Tetrahedron*, **61**, 2493 (2005).

42. E. Schnabel, H. Herzog, P. Hoffmann, E. Klauke, and I. Ugi, *Justus Liebigs Ann. Chem.*, **716**, 175 (1968); E. Schnabel, J. Stoltefuss, H. A. Offe, and E. Klauke, *Justus Liebigs Ann. Chem.*, **743**, 57 (1971).

43. M. Fridkin and H. J. Goren, *Can. J. Chem.*, **49**, 1578 (1971).

44. V. F. Pozdnev, *Zh. Obshch. Khim.*, **48**, 476 (1978); *Chem. Abstr.*, **89**: 24739m (1978).

45. D. Dhanak and C. B. Reese, *J. Chem. Soc., Perkin Trans. I*, 2181 (1986).

46. I. Hasan, E. R. Marinelli, L.-C. C. Lin, F. W. Fowler, and A. B. Levy, *J. Org. Chem.*, **46**, 157 (1981).

47. S. T. Handy, J. J. Sabatini, Y. Zhang, and I. Vulfova, *Tetrahedron Lett.*, **45**, 5057 (2004).

48. V. H. Rawal and M. P. Cava, *Tetrahedron Lett.*, **26**, 6141 (1985).

49. P. S. Baran, R. A. Shenvi, and C. A. Mitsos, *Angew. Chem. Int. Ed.*, **44**, 3714 (2005).

50. U. Jacquemard, V. Beneteau, M. Lefoix, S. Routier, J.-Y. Merour, and G. Coudert, *Tetrahedron*, **60**, 10039 (2004).

51. S. Routier, L. Sauge, N. Ayerbe, G. Coudert, and J.-Y. Merour, *Tetrahedron Lett.*, **43**, 589 (2002).

52. D. S. Bose, K. K. Kumar, and A. V. N. Reddy, *Synth. Commum.*, **33**, 445 (2003).

53. W. L. Haas, E. V. Kromkalns, and K. Gerzon, *J. Am. Chem. Soc.*, **88**, 1988 (1966).

54. Y. Nishiyama, N. Shintomi, Y. Kondo, T. Izumi, and Y. Okada, *J. Chem. Soc., Perkin Trans. 1*, 2309 (1995).

55. A. Karström and A. Undén, *J. Chem. Soc., Chem. Commun.*, 1471 (1996).

56. Y. Nishiuchi, H. Nishio, T. Inui, T. Kimura, and S. Sakakibara, *Tetrahedron Lett.*, **37**, 7529 (1996).

57. H. Nishio, Y. Nishiuchi, T. Inui, K. Yoshizawa-Kumagaye, and T. Kimura, *Tetrahedron Lett.*, **41**, 6839 (2000). H. Nishio, Y. Nishiuchi, T. Inui, K. Yoshizawa-Kumagaye, and T. Kimura, *Peptide Science*, **37th**, 9 (2001).

58. S. Raucher, J. E. Macdonald, and R. F. Lawrence, *J. Am. Chem. Soc.*, **103**, 2419 (1981).

59. T. Heinrich, C. Burschka, M. Penka, B. Wagner, and R. Tacke, *J. Organomet.Chem.*, **690**, 33 (2005).

60. T. Yamashita, N. Kawai, H. Tokuyama, and T. Fukuyama, *J. Am. Chem. Soc.*, **127**, 15038 (2005).

61. D. J. Hartley and B. Iddon, *Tetrahedron Lett.*, **38**, 4647 (1997).

62. Y. L. Chen, K. G. Hedberg, and K. J. Guarino, *Tetrahedron Lett.*, **30**, 1067 (1989).

63. P. Y. S. Lam, S. Deudon, K. M. Averill, R. Li, M. Y. He, P. DeShong, and C. G. Clark, *J. Am. Chem. Soc.*, **122**, 7600 (2000).

64. P. Y. S. Lam, G. Vincent, D. Bonne, and C. G. Clark, *Tetrahedron Lett.*, **44**, 4927 (2003).

65. C. Gonzalez, R. Greenhouse, R. Tallabs, and J. M. Muchowski, *Can. J. Chem.*, **61**, 1697 (1983).

66. T. S. Manoharan and R. S. Brown, *J. Org. Chem.*, **53**, 1107 (1988).

67. M. Ichikawa, C. Yamamoto, and T. Hisano, *Chem. Pharm. Bull.*, **29**, 3042 (1981).

68. A. R. Katritzky, G. R. Khan, and C. M. Marson, *J. Heterocycl. Chem.*, **24**, 641 (1987).

69. B. Santiago, C. R. Dalton, E. W. Huber, and J. M. Kane, *J. Org. Chem.*, **60**, 4947 (1995).

70. E. D. Edstrom and Y. Wei, *J. Org. Chem.*, **60**, 5069 (1995).

71. K. E. Bashford, A. L. Cooper, P. D. Kane, and C. J. Moody, *Tetrahedron Lett.*, **43**, 135 (2002).

72. C. Gonzalez, R. Greenhouse, and R. Tallabs, *Can. J. Chem.*, **61**, 1697 (1983).

73. D. M. Dastrup, A. H. Yap, S. M. Weinreb, J. R. Henry, and A. J. Lechleiter, *Tetrahedron*, **60**, 901(2004).

74. E. Vedejs, D. W. Piotrowski, and F. C. Tucci, *J. Org. Chem.*, **65**, 5498 (2000).

75. B. H. Lipshutz, B. Huff, and W. Hagen, *Tetrahedron Lett.*, **29**, 3411 (1988).

76. J. M. Muchowski and D. R. Solas, *Tetrahedron Lett.*, **24**, 3455 (1983).

77. P. J. Beswick, C. S. Greenwood, T. J. Mowlem, G. Nechvatal, and D. A. Widdowson, *Tetrahedron*, **44**, 7325 (1988).

78. M. Iwao, *Heterocycles*, **36**, 29 (1993).

79. K. P. Stefan, W. Schuhmann, H. Parlar, and F. Korte, *Chem. Ber.*, **122**, 169 (1989).

80. L. L. Gundersen, T. Benneche, F. Rise, A. Gogoll, and K. Undheim, *Acta Chem. Scand.*, **46**, 761 (1992).

81. A. M. Kimbonguila, S. Boucida, F. Guibé, and A. Loffet, *Tetrahedron*, **53**, 12525 (1997).

82. T. Taniguchi and K. Ogasawara, *Tetrahedron Lett.*, **39**, 4679 (1998).

83. S. Kamijo, Z. Huo, T. Jin, C. Kanazawa, and Y. Yamamoto, *J. Org. Chem.*, **70**, 6389 (2005).

84. S. Kamijo, T. Jin, Z. Huo, and Y. Yamamoto, *J. Am. Chem. Soc.*, **125**, 7786 (2003).

85. V. du Vigneaud and O. K. Behrens, *J. Biol. Chem.*, **117**, 27 (1937).

86. C. J. Chivikas and J. C. Hodges, *J. Org. Chem.*, **52**, 3591 (1987).

87. M. Searcey, J. B. Lee, and P. L. Pye, *Chem. Ind.(London)*, 569 (1989).

88. J. Bergman, P. Ola Norrby, and P. Sand, *Tetrahedron*, **46**, 6113 (1990).

89. W.-C. Shieh, M. Lozanov, and O. Repic, *Tetrahedron Lett.*, **44**, 6943 (2003); W.-C. Shieh, M. Lozanov, M. Loo, O. Repic, and T. J. Blacklock, *Tetrahedron Lett.*, **44**, 4563 (2003).

90. W.-C. Shieh, S. Dell, A. Bach, O. Repic, and T. J. Blacklock, *J. Org. Chem.*, **68**, 1954 (2003).

91. H. Suzuki, A. Tsukuda, M. Kondo, M. Aizawa, Y. Senoo, M. Nakajima, T. Watanabe, Y. Yokoyama, and Y. Murakami, *Tetrahedron Lett.*, **36**, 1671 (1995).

92. T. Watanabe, A. Kobayashi, M. Nishiura, H. Takahashi, T. Usui, I. Kamiyama, N. Mochizuki, K. Noritake, Y. Yokoyama, and Y. Murkami, *Chem. Pharm. Bull.*, **39**, 1152 (1991).

93. A. Bombrun and G. Casi, *Tetrahedron Lett.*, **43**, 2187 (2002).

94. O. Ottoni, R. Cruz, and R. Alves, *Tetrahedron*, **54**, 13915 (1998).

95. A. M. Felix, E. P. Heimer, T. J. Lambros, C. Tzougraki, and J. Meienhofer, *J. Org. Chem.*, **43**, 4194 (1978).

96. E. C. Jorgensen, G. C. Windridge, and T. C. Lee, *J. Med. Chem.*, **13**, 352 (1970).

97. I. T. Forbes, C. N. Johnson, and M. Thompson, *J. Chem. Soc., Perkin Trans. 1*, 275 (1992).

98. A. Fürstner and H. Weintritt, *J. Am. Chem. Soc.*, **120**, 2817 (1998).

99. A. A. Haddach, A. Kelleman, and M. V. Deaton-Rewolinski, *Tetrahedron Lett.*, **43**, 399 (2002).

100. D. R. Buckle and C. J. M. Rockell, *J. Chem. Soc., Perkin Trans. I*, 627 (1982).

101. T. Kamijo, R. Yamamoto, H. Hirada, and K. Iizuka, *Chem. Pharm. Bull.*, **31**, 1213 (1983).

102. D. L. Boger, B. E. Fink, and M. P. Hedrick, *J. Am. Chem. Soc.*, **122**, 6382 (2000).

103. C. Subramanyam, *Synth. Commun.*, **25**, 761 (1995).

104. Y. Miki, H. Hachiken, Y. Kashima, W. Sugimura, and N. Yanase, *Heterocycles*, **48**, 1 (1998).

105. M. I. Jones, C. Froussios, and D. A. Evans, *J. Chem. Soc., Chem. Commun.*, 472 (1976).

106. A. Er-Rhaimini, N. Mohsinaly, and R. Mornet, *Tetrahedron Lett.*, **31**, 5757 (1990).

107. S. M. Kalbag and R. W. Roeske, *J. Am. Chem. Soc.*, **97**, 440 (1975).

108. T. Voelker, T. Ewell, J. Joo, and E. D. Edstrom, *Tetrahedron Lett.*, **39**, 359 (1998).

109. R. Balasuriya, S. J. Chandler, M. J. Cook, and D. J. Hardstone, *Tetrahedron Lett.*, **24**, 1385 (1983).

110. E. Siepmann and H. Zahn, *Biochim. Biophys. Acta*, **82**, 412 (1964).

111. S. Deechongkit, S.-L. You, and J. W. Kelly, *Org. Lett.*, **6**, 497 (2004).

112. S. Shaltiel, *Biochem. Biophys. Res. Commun.*, **29**, 178 (1967).

113. M. C. Lin, B. Gutte, D. G. Caldi, S. Moore, and R. B. Merrifield, *J. Biol. Chem.*, **247**, 4768 (1972); M. Beltran, E. Pedroso, and A. Grandas, *Tetrahedron Lett.*, **39**, 4115 (1998).

114. J.-C. Gesquière, J. Najib, T. Letailleur, P. Maes, and A. Tartar, *Tetrahedron Lett.*, **34**, 1921 (1993).

115. H. E. Garay, L. J. Gonzalez, L. J. Cruz, R. C. Estrada, and O. Reyes, *Biotecnologia Aplicada*, **14**, 193 (1997).

116. A. R. Fletcher, J. H. Jones, W. I. Ramage, and A. V. Stachulski, in *Peptides 1978*, I. Z. Siemion and G. Kupryszeqski, Eds., Wroclaw University Press, Wroclaw, Poland, 1979, pp. 168–171.

117. A. R. Fletcher, J. H. Jones, W. I. Ramage, and A. V. Stachulski, *J. Chem Soc., Perkin Trans. I*, 2261 (1979).

118. G. Losse and U. Krychowski, *J. Prakt. Chem.*, **312**, 1097 (1970).

119. B. E. Huff, M. E. LeTourneau, M. A. Staszak, and J. A. Ward, *Tetrahedron Lett.*, **37**, 3655 (1996).

120. G. Losse and U. Krychowski, *Tetrahedron Lett.*, **12**, 4121 (1971).

121. V. V. Tolstyakov and I. V. Tselinskii, *Russ. J. Gen. Chem.*, **74**, 399 (2004).

122. N. J. Curtis and R. S. Brown, *J. Org. Chem.*, **45**, 4038 (1980); K. L. Kirk, *J. Org. Chem.*, **43**, 4381 (1978); J. L. Kelley, C. A. Miller, and E. W. McLean, *J. Med. Chem.*, **20**, 721 (1977).

123. R. Brown, W. E. Smith, and D. Graham, *Tetrahedron Lett.*, **42**, 2197 (2001).

124. P. Sieber and B. Riniker, *Tetrahedron Lett.*, **28**, 6031 (1987).

125. S. Coyle and G. T. Young, *J. Chem. Soc., Chem. Commun.*, 980 (1976).

126. S. Coyle, O. Keller, and G. T. Young, *J. Chem. Soc., Chem. Commun.*, 939 (1975).

127. S. Coyle, A. Hallett, M. S. Munns, and G. T. Young, *J. Chem. Soc., Perkin Trans. I*, 522 (1981).

128. G. R. Martinez, P. A. Grieco, E. Williams, K.-i. Kanai, and C. V. Srinivasan, *J. Am. Chem. Soc.*, **104**, 1436 (1982).

129. A. R. Katritzky and K. Akutagawa, *J. Org. Chem.*, **54**, 2949 (1989).

130. R. J. Sundberg and H. F. Russell, *J. Org. Chem.*, **38**, 3324 (1973).

131. T. P. Demuth, Jr., D. C. Lever, L. M. Gorgos, C. M. Hogan, and J. Chu, *J. Org. Chem.*, **57**, 2963 (1992).

132. A. I. Meyers, T. K. Highsmith, and P. T. Bounora, *J. Org. Chem.*, **56**, 2960 (1991).

133. N. J. Curtis and R. S. Brown, *J. Org. Chem.*, **45**, 4038 (1980).

134. S. Ohta, M. Matsukawa, N. Ohashi, and K. Nagayama, *Synthesis*, 78 (1990).

135. P. Gmeiner, J. Kraxner, and B. Bollinger, *Synthesis*, 1196 (1996).

136. A. J. Hutchison and Y. Kishi, *J. Am. Chem. Soc.*, **101**, 6786 (1979).

137. J. P. Whitten, D. P. Matthews, and J. R. McCarthy, *J. Org. Chem.*, **51**, 1891 (1986).

138. B. H. Lipshutz, W. Vaccaro, and B. Huff, *Tetrahedron Lett.*, **27**, 4095 (1986).

139. M. P. Edwards, A. M. Doherty, S. V. Ley, and H. M. Organ, *Tetrahedron*, **42**, 3723 (1986).

140. J. M. Muchowski and D. R. Solas, *J. Org. Chem.*, **49**, 203 (1984).

141. C. R. Dalton, J. M. Kane, and D. Rampe, *Tetrahedron Lett.*, **33**, 5713 (1992).

142. O. A. Moreno and Y. Kishi, *J. Am. Chem. Soc.*, **118**, 8180 (1996).

143. D. P. Matthews, J. P. Whitten, and J. R. McCarthy, *J. Heterocycl. Chem.*, **24**, 689 (1987).

144. J. G. Phillips, L. Fadnis, and D. R. Williams, *Tetrahedron Lett.*, **38**, 7835 (1997).

145. R. Colombo, F. Colombo, and J. H. Jones, *J. Chem. Soc., Chem. Commun.*, 292 (1984).

146. Y. Okada, J. Wang, T. Yamamoto, Y. Mu, and T. Yokoi, *J. Chem. Soc., Perkin Trans.1*, 2139 (1996); Y. Okada, J. Wang, T. Yamamoto, and Y. Mu, *Chem. Pharm. Bull.*, **44**, 871 (1996).

147. Y. Okada, S. Joshi, N. Shintomi, Y. Kondo, Y. Tsuda, K. Ohgi, and M. Irie, *Chem. Pharm. Bull.*, **47**, 1089 (1999).

148. G. C. Magnin, J. Dauvergne, A. Burger, and J.-F. Biellmann, *Tetrahedron Lett.*, **37**, 7833 (1996).

149. For a preparation of the chloride and iodide, see P. P. Iyer, D. Yu, N.-h. Ho, and S. Agrawal, *Synth. Commun.*, **25**, 2739 (1995).

150. (a) D. Dhanak and C. B. Reese, *J. Chem. Soc., Perkin Trans. I*, 2181 (1986). (b) L. Araki, S. Harusawa, M. Yamaguchi, S. Yonezawa, N. Taniguchi, D. M. J. Lilley, Z.-y. Zhao, and T. Kurihara, *Tetrahedron Lett.*, **45**, 2657 (2004).

151. M. Rasmussen and N. J. Leonard, *J. Am. Chem. Soc.* **89**, 5439 (1967).

152. H. J. Anderson and J. K. Groves, *Tetrahedron Lett.*, **12**, 3165 (1971).

153. J. E. Macor, J. T. Forman, R. J. Post, and K. Ryan, *Tetrahedron Lett.*, **38**, 1673 (1997).

154. T. Brown, J. H. Jones, and J. D. Richards, *J. Chem. Soc., Perkin Trans. I*, 1553 (1982).

155. D. C. Gowda, *Tetrahedron Lett.*, **43**, 311 (2002).

156. T. Brown and J. H. Jones, *J. Chem. Soc., Chem. Commun.*, 648 (1981).

157. J.-C. Gesquiere, E. Diesis, and A. Tartar, *J. Chem. Soc., Chem. Commun.*, 1402 (1990).

158. M. A. Mitchell, T. A. Runge, W. R. Mathews, A.K. Ichhpurani, N. K. Harn, P. J. Dobrowolski, and F. M. Eckenrode, *Int. J. Pept. Protein Res.*, **36**, 350 (1990).

159. A. R. Katritzky, P. Lue, and Y.-X. Chen, *J. Org. Chem.*, **55**, 3688 (1990).

160. A. R. Katritzky, P. Lue, and K. Yannakopoulou, *Tetrahedron*, **46**, 641 (1990).

161. R. K. Robins, E. F. Godefroi, E. C Taylor, L. R. Lewis, and A. Jackson, *J. Am. Chem. Soc.* **83**, 2574 (1961).

162. J. S. Bradshaw, C. W. McDaniel, K. E. Krakowiak, and R. M. Izatt, *J. Heterocycl. Chem.*, **27**, 1477 (1990).

163. Z. Song, A. DeMarco, M. Zhao, E. G. Corley, A. S. Thompson, J. McNamara, Y. Li, D. Rieger, P. Sohar, D. J. Mathre, D. M. Tschaen, R. A. Reamer, M. F. Huntington, G.-J. Ho, F.-R. Tsay, K. Emerson, R. Shuman, E. J. J. Grabowski, and P. J. Reider, *J. Org. Chem.*, **64**, 1859 (1999).

164. R. L. Hudkins, J. L. Diebold, and F.D. Marsh, *J. Org. Chem.*, **60**, 6218 (1995).

165. A. R. Katritsky and K. Akutagawa, *Tetrahedron Lett.*, **26**, 5935 (1985).

166. A. Previero, M. A. Coletti-Previero, and J. C. Cavadore, *Biochim. Biophys. Acta*, **147**, 453 (1967).

167. G. R. Matsueda, *Int. J. Pept. Protein Res.*, **20**, 26 (1982).

168. J. Castells, Y. Troin, A. Diez, M. Rubiralta, D. S. Grierson, and H. P. Husson, *Tetrahedron*, **37**, 7911 (1991).

169. W. G. Rajeswaran, and L. A. Cohen, *Tetrahedron Lett.*, **38**, 7813 (1997).

170. K. Teranishi, S.-i. Nakatsuka, and T. Goto, *Synthesis*, 1018 (1994).

171. C. Avedaño, J. D. Sanchez, and J. C. Menendez, *Synlett*, 107 (2005).

172. Y. Kiso, T. Kimura, M. Shimokura, and T. Narukami, *J. Chem. Soc., Chem. Commun.*, 287 (1988).

173. Y. Kiso, T. Kimura, Y. Fujiwara, M. Shimokura, and A. Nishitani, *Chem. Pharm. Bull.*, **36**, 5024 (1988).

174. P. S. Baran, C. A. Guerrero, and E. J. Corey, *J. Am. Chem. Soc.*, **125**, 5628 (2003); P. S. Baran, C. A. Guerrero, and E. J. Corey, *Org. Lett.*, **5**, 1999 (2003).

PROTECTION FOR AMIDES

Protection of the amides −NH, is an area of protective group chemistry that has received little attention, and as a consequence few good methods exist for amide −NH protection. Most of the cases found in the literature do not represent protective groups in the true sense, in that the protective group is often incorporated as a handle to introduce nitrogen into a molecule rather than installed to protect a nitrogen which at some later time is deblocked. For this reason, many of the following examples deal primarily with removal rather than with both formation and cleavage.

Amides

N-Methylamide: $CH_3-NRCO-$

Although a methyl group is usually not considered as a protective group, it is easily introduced with NaH and MeI in THF and amazingly can be cleaved via a free radical process.[1]

N-Allylamide: $CH_2=CHCH_2-NRCO-$

Formation

The allyl group was used to protect the nitrogen in a β-lactam synthesis, but was removed in a four-step sequence.[2]

1. $CH_2=CHCH_2Cl$, CsF, DMF.[3] The use of allyl iodide gives O-alkylation.
2. $CH_2=CHCH_2Br$, P4 base, THF, $-100°C$ to $-78°C$.[4]
3. NaH, LiBr, DME, DMF, allyl bromide, 88% yield.[5]
4. $CH_2=CHCH_2Cl$, 50% aq. NaOH, TBAHSO$_4$, 74–82% yield.[6]
5. $CH_2=CHCH_2Cl$, Pd(Ph$_3$P)$_4$, TEA, 89% yield.[7]

6. $CH_2=CHCH_2OCO_2Et$, (allyl)$_2$PdCl$_2$, 83–99% yield.[8]

Cleavage

Methods that give the enamide are included, since these can be cleaved by ozonolysis and in principle by acid-catalyzed hydrolysis.

1. Rh(Ph$_3$P)$_3$Cl, toluene, reflux, 81% to the enamide; O$_3$, MeOH; DMS; NaHCO$_3$, 87% yield.[9,10]
2. Cleavage of the enamide by the Johnson–Lemieux reaction.[11] The allyl group was the only successfully cleaved group among the many that were examined.

3. Formation of the enamide: $Fe(CO)_5$, 100°C, 44–95% yield.[12] The reaction fails with compounds containing primary bromides.

4. $Pd(Ph_3P)_4$, HCO_2H, TEA, dioxane, reflux, 80% yield. Cleavage is from an imide.[13]

5. Me_3Al, $(dppp)NiCl_2$, toluene reflux, 51–92% yield.[14] Allylsulfonamides are cleaved similarly.

6. For a crotylamide: t-BuOK, DMSO, 80°C, 4 h.[15]

KHMDS[16] and LDA[17] also cause isomerization of allyl amides.

7. $[Ir(COD)Cl]_2$, PCy_3, Cs_2CO_3, toluene, 110°C, 56–96% yield of the enamide.[18]

8. $Cl_2(Cy_3P)_2Ru=CHPh$, CH_2Cl_2, reflux. The enamide is produced.[19] $RuClH(CO)(PPh_3)_3$ is similarly an effective catalyst for this isomerization (87–95% yield).[6] The enamide is cleaved by oxidation with $RuCl_3$-$NaIO_4$ followed by a mildly basic workup (40–78% yield).[20]

9. 4-Methylmorpholine N-oxide, OsO_4, $NaIO_4$, dioxane, water, 60°C, 18 h, 64% yield.[21]

N-t-Butylamide (t-Bu-NRCO-)

The t-butyl group is introduced as a t-butylamine and is cleaved with strong acid (70–97% yield).[22]

N-Dicyclopropylmethylamide (Dcpm−NRCO−): $(C_3H_5)_2CH−NRCO−$

Half-Lives for Cleavage of CH_3CONHR in Neat TFA at rt

R	$t_{1/2}$ (min)
Dicyclopropylmethyl	19
Dimethylcyclopropylmethyl	1–2
$Me_2PhC−$	15
$MePh_2C−$	<1

Cleavage is achieved by acidolysis in neat TFA. N-Cyclopropylmethyl, N-t-butyl, N-t-adamantyl and N-(1-methyl-cyclohexyl)acetamide were not affected by these conditions.[23]

N-Methoxymethylamide (MOM−NRCO−): CH₃OCH₂−NRCO−

The related methoxyethoxymethyl (MEM) group has also been tested but not extensively.[24]

Formation

1. MOMCl, *t*-BuOK, DMSO.[25]
2. MOMCl, CH₂Cl₂, DMAP, DIPEA, 0°C, 1 h, 85% yield.[26]

Cleavage

1. BBr₃, 31% yield.[25]
2. *B*-Bromocatecholborane, CH₂Cl₂, 0°C, 40 min, 78% yield.[27]
3. AlCl₃, toluene, reflux, 48–88% yield.[26]
4. TMSCl, NaI, CH₃CN, 63% yield.[28]
5. Conc. HCl, DME, 55°C, 90% yield.[29] The MOM group on a similar amide was stable to formic acid, conditions used to cleave a *t*-butyl ester.[30]

6. TFA, 4 h, reflux, 92–96% yield. This method will also cleave the MEM group.[31]

N-Methylthiomethylamide (MTM-NRCO-): CH₃SCH₂-NRCO−

Cleavage

SOCl₂; NaHCO₃, H₂O; heat to 120°C under vacuum, 80% yield.[32]

N-*t*-Butylthiomethylamide (BTM−NRCO−): (CH₃)₃CSCH₂−NRCO−

Formation/Cleavage[33]

N-Benzyloxymethylamide (BOM−NRCO−): C₆H₅CH₂OCH₂−NRCO−

Cleavage

1. The BOM group can cleaved with H₂/Pd(OH)₂−C, MeOH, which also removes the BOM group from alcohols.[34]
2. (a) H₂, Pd(OH)₂ EtOAc, MeOH, rt, (b) MeONa, MeOH, 92% yield.[35] Treatment with methoxide was required to remove the formaldehyde from the phthalimide.
3. BBr₃, 25°C, toluene or AlCl₃, toluene, reflux.[36]

N-2-(Trimethylsilyl)ethoxymethylamide: (CH₃)₃SiCH₂CH₂OCH₂−NRCO−

Formation

SEMCl, NaH, 74% yield.[37]

Cleavage

1. Me₂AlCl then DIPEA, MeOH, reflux, 93% yield.[37]

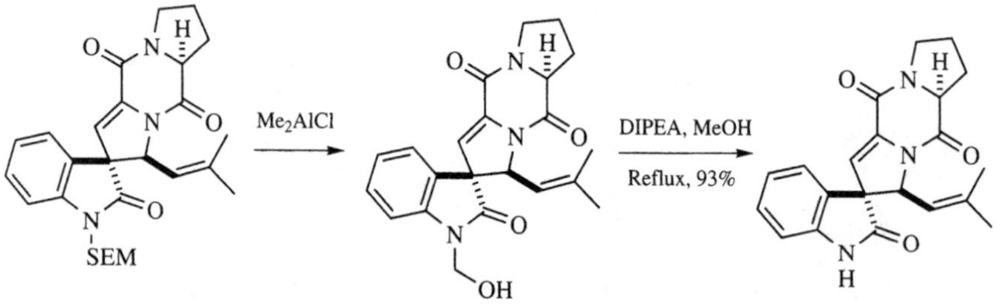

2. TBAF·3H₂O, DMPU, 45°C, 87% yield. Serendipitous ketone reduction was observed which may be due to a Canizzaro like reduction from the released formaldehyde.[38]

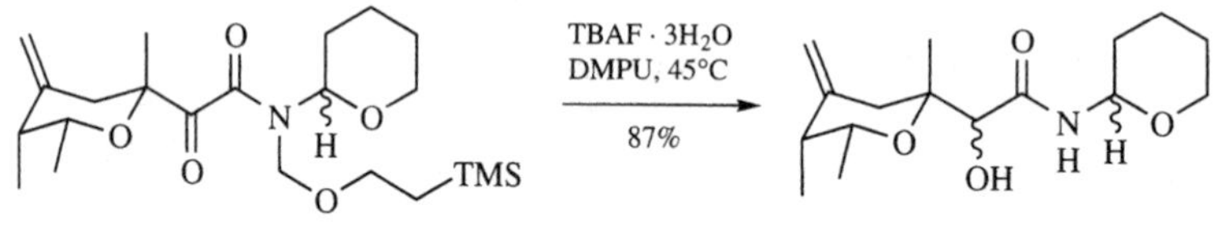

N-2,2,2-Trichloroethoxymethylamide: Cl₃CCH₂OCH₂−NRCO−

Formation

Cl₃CCH₂OCH₂Cl, KH, THF, 0°C to rt, 20 min, 93% yield.[39,40]

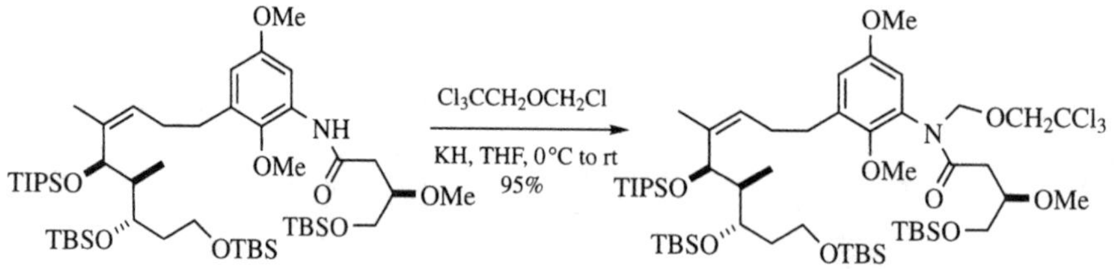

Cleavage

1. 5% Na(Hg), Na$_2$HPO$_4$, MeOH, 67% yield.[39,40]

2. Methods used for the cleavage of the Troc group should also be examined, since these in principle should be effective.

N-2-(*p*-Toluenesulfonyl)ethenylamide (Tsv−NRCO−): *p*-CH$_3$C$_6$H$_4$SO$_2$CH=CH−NRCO

This group was developed as an electron-deficient group that could be converted to an electron-rich group by simple hydrogenation of the double bond. This then affords the tosylethyl group which can be removed by base treatment.

Formation

TsCH=CHTS, NaH, DMF, 20°C, 15 h.[41]

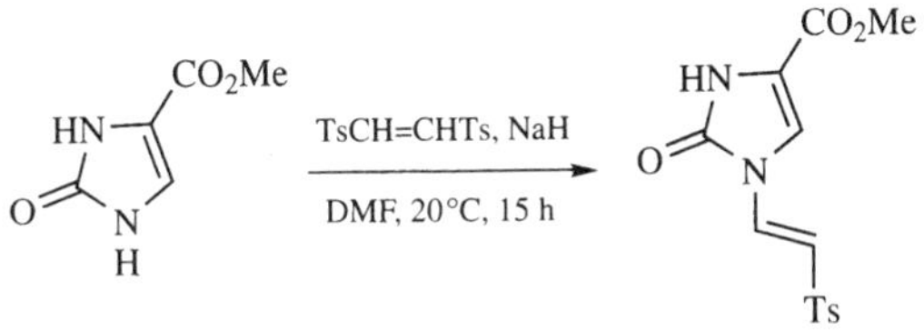

N-t-Butyldimethylsiloxymethylamide: *t*-C$_4$H$_9$(CH$_3$)$_2$SiOCH$_2$−NRCO−

Formation

TBDMSOCH$_2$Cl, TEA, CH$_2$Cl$_2$, −78°C, rt, 24 h, >89% yield.[42,44]

Cleavage

1. Bu$_4$NF, THF, rt, 30 min, 70% yield.[42] Me$_4$NF has also been used to cleave this group.[43]

2. TAS-F, DMF, quantitative.[44]

N-Pivaloyloxymethylamide: (CH$_3$)$_3$CCO$_2$CH$_2$−NRCO−

Formation

NaH, DMF, PvOCH$_2$Cl, rt, 12 h, 80% yield.[45]

Cleavage

NaOH, THF, rt, 4 days, 48% yield.[45]

N-Cyanomethylamide: NCCH$_2$−NRCO−

Formation

BrCH$_2$CN, EtONa, DMF, 82–85% yield.[46] Phenols and amines have also been protected by this method.

Cleavage

H$_2$, PtO$_2$, EtOH, 85–95% yield.[46]

N-Pyrrolidinomethylamide

Formation

HCHO, pyrrolidine, 93% yield.[47,48]

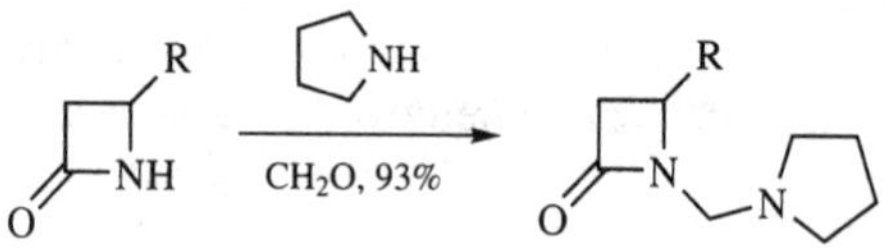

Cleavage

MeOH, 1% HCl, or 1:9 THF, 1% HCl, >52–85% yield.[48] This group was used to protect a β-lactam amide nitrogen during deprotonation of the α-position.

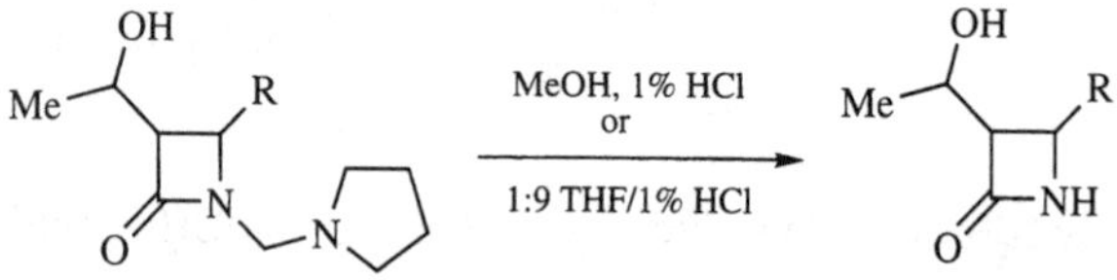

N-Methoxyamide: MeO−NRCO−

The methoxy group on a β-lactam nitrogen was cleaved by reduction with Li (EtNH$_2$, *t*-BuOH, THF, −40°C, 71% yield). A benzyloxy group was stable to these cleavage conditions.[49]

N-Benzyloxyamide (BnO−NRCO-): C$_6$H$_5$CH$_2$O−NRCO−

The benzyloxy group on a β-lactam nitrogen was cleaved by hydrogenolysis (H$_2$, Pd–C) or by TiCl$_3$ [MeOH, H$_2$O, (NH$_4$)$_2$CO$_3$, Na$_2$CO$_3$].[50]

N-Methylthioamide: MeS−NRCO−

Formation

LDA, HMPA, CH$_3$SSO$_2$CH$_3$, −78°C to 0°C, 94% yield.[51]

Cleavage

2-Pyridinethiol, Et$_3$N, CH$_2$Cl$_2$, 95% yield. The methylthioamide group is stable to 2.5 *N* NaOH, THF, H$_2$O and to 10% H$_2$SO$_4$, MeOH, H$_2$O.[51] The section on sulfenamides should be consulted for a related approach to nitrogen protection. Some of the derivatives presented there may also be applicable to amides.

N-Triphenylmethylthioamide: Ph$_3$CS$-$NRCO$-$

Cleavage

1. Bu$_3$P, EtOH, THF, 115°C, 48 h, 75% yield.[52]
2. Me$_3$SiI, CH$_2$Cl$_2$, 25°C, 7 h, 81% yield.[52]
3. Li, NH$_3$.[52]
4. W2 Raney Ni.[52] Li/NH$_3$ and Raney Ni also cleave benzylic C–N bonds.

N-*t*-**Butyldimethylsilylamide (TBDMS$-$NRCO-):** *t*-C$_4$H$_9$(CH$_3$)$_2$Si$-$NRCO$-$

Formation

1. TBDMSCl, Et$_3$N, CH$_2$Cl$_2$, 98% yield.[53-55] This methodology is also used to protect the BOCNH derivatives.[56]
2. TBDMSOTf, collidine.[57]

Silylation of both the primary and secondary hydroxyl groups is followed by selective deprotection to regenerate the primary hydroxyl group.

3. During an attempted esterification of a primary alcohol, a TBDMS group was found to migrate from an amide to the primary alcohol.[58]

4. 10% Pd–C, *t*-BuMe$_2$SiH, hexane, CH$_2$Cl$_2$, rt, 2 h, 80% yield.[59] These conditions also silylate alcohols, amines, and carboxylic acids.

Cleavage

1. 1 *N* HCl, MeOH, rt, 91% yield.[60] The TBDMS derivative of a β-lactam nitrogen is reported to be stable to lithium diisopropylamide, citric acid, Jones oxidation, and BH$_3$–diisopropylamine, but not to Pb(OAc)$_4$ oxidation.
2. Aq. HF, CH$_3$CN, DBU or *t*-BuOK.[61]
3. MeSNa, THF, H$_2$O, >38% yield.[62]
4. KF, MeOH, 90% yield.[63]

N-Triisopropylsilylamide (TIPS−NRCO−): (i-Pr)$_3$Si−NRCO−

Formation

1. TIPSOTf, DBU, CH$_3$CN.[64] Triethylamine is an effective base and is suitable for protection of BOC amines with a variety of silyl groups.[65]
2. TIPSOTf, n-BuLi, >72% yield.[66]

Cleavage

1. HF·Pyr, TBAF or NaOAc in DMSO/H$_2$O at 65°C.[67]
2. AcOH, H$_2$O, DMF, 110°C, 79% yield. In this case the TIPS group was removed from an imide nitrogen.[68] In this case a PMB group could not be cleaved because of the easily oxidized aromatic diamine.

N-4-Methoxyphenylamide (MePh−NRCHO−): 4-CH$_3$O−C$_6$H$_4$−NRCO−

This group has been used extensively in β-lactam syntheses, where it is used to introduce the nitrogen as p-anisidine.

Formation

1. MeOC$_6$H$_4$Si(OMe)$_3$, TBAF, Cu(OAc)$_2$, pyridine, DMF or CH$_2$Cl$_2$, air, rt, 49–98% yield.[69]
2. General arylation of an amide.[70]
3. MeOC$_6$H$_4$I, CuI, glycine, K$_3$PO$_4$, dioxane, 88–98% yield.[71]
4. MeOC$_6$H$_4$I, CuI, KF/Al$_2$O$_3$, toluene, 1,10-phenanthrolene, 90–99% yield.[72]

Cleavage

1. Electrolysis, CH$_3$CN, H$_2$O, LiClO$_4$, 1.5 V, rt, 60–95% yield.[73] The released quinone is removed by forming the bisulfite adduct that can be washed out with water.

2. Ceric ammonium nitrate, CH_3CN, H_2O, 0°C, 95% yield.[74,75] In the presence of chloride ion cleavage fails.[76] The **2-methoxyphenyl** group is cleaved with these conditions as well.[77]

3. Ozonolysis, then reduction with $Na_2S_2O_4$ at 50°C, 57% yield.[78] The **3,4-dimethoxyphenyl derivative** was cleaved in 71% yield using these conditions. Ceric ammonium nitrate was reported not to work in this example.

$R = H, 57\%$
$R = OMe, 71\%$

4. $(NH_4)_2S_2O_8$, $AgNO_3$, CH_3CN, H_2O, 60°C, 57–62% yield.[79]

N-4-(Methoxymethoxy)phenylamide (MOMOC₆H₄-NRCO−):
4-MeOCH₂OC₆H₄-NRCO−

This group was developed for a case where direct oxidation of the methoxyphenyl group with CAN was not very efficient. Prior removal of the MOM group [HCl, $(HC(OMe)_3$, MeOH] followed by oxidation with CAN was reported to be more effective.[80]

N-2-Methoxy-1-naphthylamide: 2-CH₃O-C₁₀H₆-NRCO−

This group was removed from a cyclic urethane with CAN.[81] It more easily oxidized than the *p*-methoxyphenyl group.

N-Benzylamide (Bn-NRCO−): C₆H₅CH₂−NRCO−

Formation

1. BnCl, KH, THF, rt, 100% yield.[82]
2. Et₃BuNBr, toluene, H_2O, BnCl, K_2CO_3, reflux.[83]
3. PhCHO, Pd/C, Na_2SO_4, H_2, 40 bar, 100°C, 93% yield.[84]
4. BnCHO, TFA, Et₃SiH, toluene or CH_3CN, 22–120°C, 87–95% yield.[85]
5. BnBr neat, 120°C.[86] This reaction also works with Ph₂CHBr to give the diphenylmethylamide derivative.

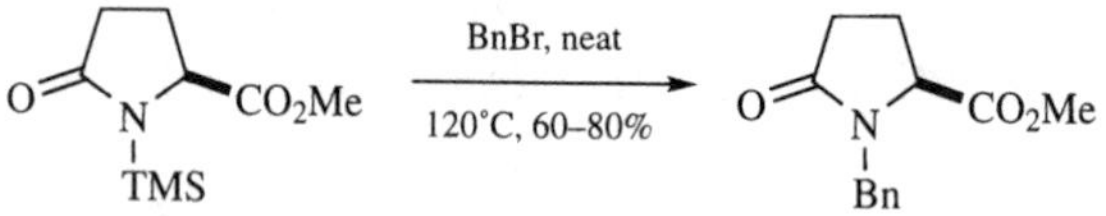

6. BnCl, CsF, DMF, 83% yield.[3]
7. BnBr, KF·alumina, DME, 25°C, 12h, 85% yield.[87]
8. BnCl, Cs$_2$CO$_3$, DMF, TBAI, 90–98% yield.[88]
9. Treatment of an amide with BnOC(=NH)CCl$_3$ (TMSOTf, CH$_2$Cl$_2$, 85–88% yield) protects the amide by O-alkylation.[89]

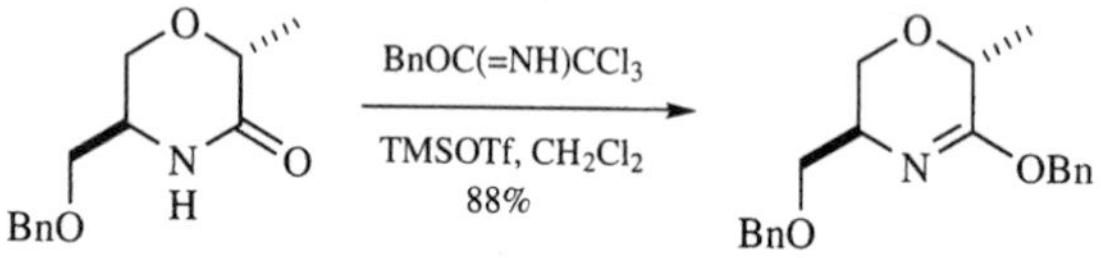

Cleavage

1. H$_2$, Pd–C, AcOH, 2 days.[90] Debenzylation of a benzylacetamide by hydrogenolyis is much slower than hydrogenolysis of a benzyl oxygen bond. Hydroxyl groups protected with benzyl groups or benzylidene groups are readily cleaved without affecting amide benzyl groups. It is often impossible to remove the benzyl group on an amide by hydrogenolysis. On the other hand, a benzyl group can be removed from an imide by transfer hydrogenation.[91]

2. Na or Li and ammonia, excellent yields.[92] This is a very good method to remove a benzyl group from an amide and will usually work when hydrogenolysis does not. A dissolving metal reduction can be effected without cleavage of a sulfur–carbon bond. Note also the unusual selectivity in the cleavage illustrated below. This was attributed to steric compression.[93] Primary benzyl amides are not cleaved under these conditions.[94]

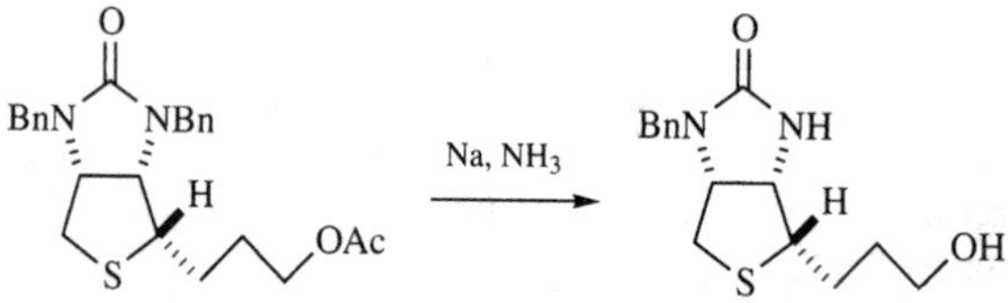

An N-benzyl amide is more easily reduced than a N-benzyl amine.[95] Reactions like this, which must be run for such short periods, are difficult to scale up, since everything on scale takes much longer.

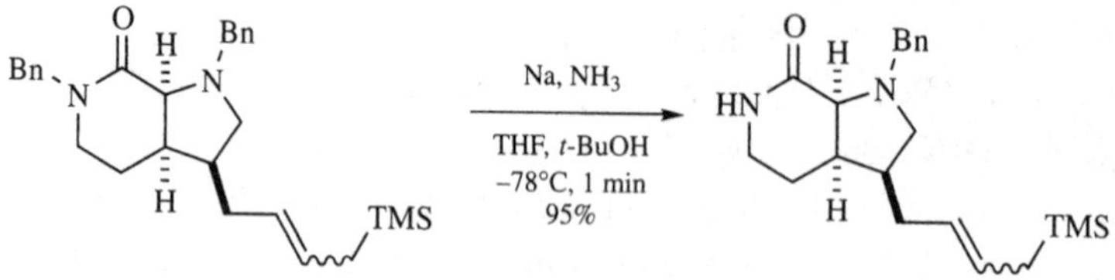

3. Li, catalytic naphthalene, −78°C, THF, 97–99% yield. In addition, tosyl amides and mesyl amides are cleaved with similar efficiency.[96]

4. *t*-BuLi, THF, −78°C; O$_2$ or MoOPH, [oxodiperoxymolybdenum–(hexamethyl phosphorictriamide)(pyridine)], 30–68% yield.[97] This method uses the amide carbonyl to direct benzylic metalation.

5. *t*-BuOK, DMSO, O$_2$, 20°C, 20 min.[98,99]

6. Sunlight, FeCl$_3$, H$_2$O, acetone, 21% yield.[100]

7. 95% HCO$_2$H, 50–60°C, 74–91% yield.[101] This method was used to remove the α-methylbenzyl group from an amide. Methods 7 and 8 were used to remove the benzyl group from a biotin precursor.

8. Aqueous HBr, 85% yield.[102]

9. Orthophosphoric acid, phenol, 53% yield.[103]

N-4-Methoxybenzylamide (PMB−NRCO−): 4-CH$_3$OC$_6$H$_4$CH$_2$-NRCO−

Formation

1. NaH, 4-MeO-C$_6$H$_4$CH$_2$Br, DMF, rt, 12 h, 62% yield.[104]

2. 4-MeO-C$_6$H$_4$CH$_2$Cl, DBU, CH$_3$CN, 45°C, 6 h, 92% yield.[105]

3. 4-MeO-C$_6$H$_4$CH$_2$Cl, Ag$_2$O.[106]

Cleavage

Some of the methods used to cleave the benzyl group should also be effective for cleavage of the PMB group.

1. Ceric ammonium nitrate (CAN), CH$_3$CN, H$_2$O, rt, 12 h, 96% yield.[107,108] Benzylamides are not cleaved under these conditions. This method occasionally results in the formation of imides which must be hydrolyzed with base.[109]

2. *t*-BuLi, THF, −78°C, O$_2$, 60% yield.[110–112]

3. H$_2$, PdCl$_2$, EtOAc, AcOH, rt, 90% yield.[113]
4. AlCl$_3$, anisole, rt, 81–96% yield. An acetonide survived these conditions.[114]
5. TFA, reflux[115] or TFA CHCl$_3$, rt, 1.5 h, 53% yield.[116]

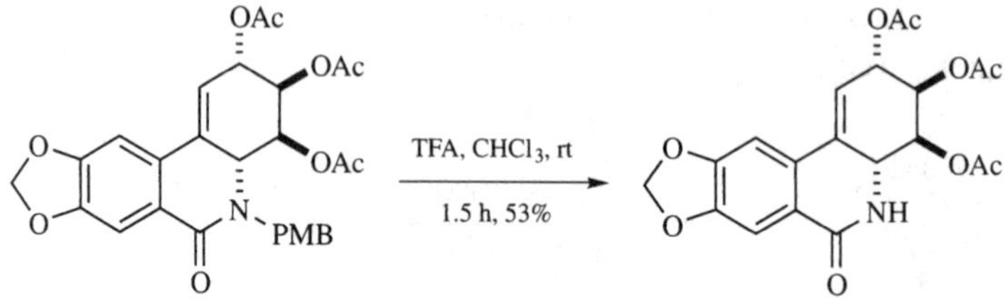

6. Catalyst (HCTf$_3$, Sc(CTf$_3$)$_3$, HNTf$_2$, Bi(NTf$_2$)$_3$, Cu(NTf$_2$)$_2$), anisole, 154°C, 99% yield. The fastest rate was achieved with HCTf$_3$. This method also can be used to cleave benzyl and MPM esters and MPM ethers.[117]

**N-2,4-Dimethoxybenzylamide (DMB−NRCO−) and
N-3,4-Dimethoxybenzylamide:** 2,4- and 3,4-(CH$_3$O)$_2$-C$_6$H$_3$CH$_2$-NRCO−

Cleavage

1. K$_2$S$_2$O$_8$, Na$_2$HPO$_4$, 40% aq. CH$_3$CN, reflux, 1 h, 69% yield.[118]

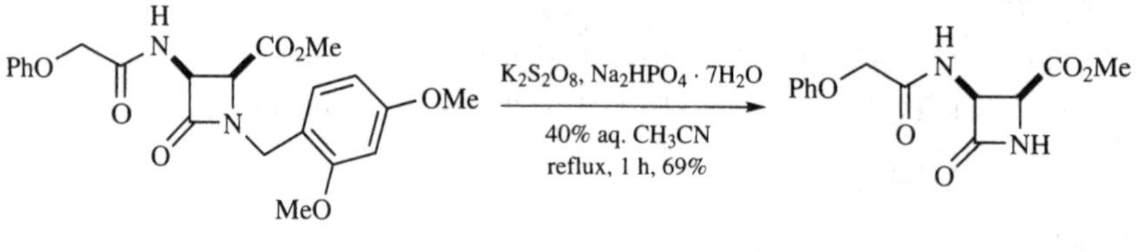

2. TFA, 85% yield.[119,120]

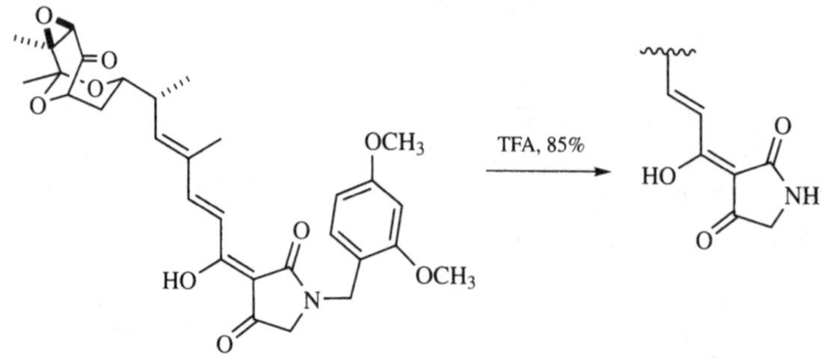

3. TsOH, toluene, reflux, 65–100% yield.[121]
4. TFA, anisole, 75% yield.[122] Thioanisole has been used in this cleavage reaction to scavenge the benzyl cation.[123] Its absence results in considerable alkylation of the indolocarbazole nucleus.[124]

5. DDQ, CHCl$_3$, H$_2$O.[125] The 3,4-dimethoxybenzyl group could be cleaved from a sulfonamide with DDQ (8–50% yield).[126]

6. Ceric ammonium nitrate, CH$_3$CN, H$_2$O, 78% yield.[127]

7. The related 3,4-dimethoxybenzyl group has been cleaved from an amide with Na/NH$_3$, 82% yield.[128]

N-2-Acetoxy-4-methoxybenzylamide (AcHmb-NRCO−): 2-Ac-4-MeOC$_6$H$_4$CH$_2$-NRCO−

This group is used for peptide backbone protection. The acetoxy group makes it stable to TFA that is used to cleave the BOC group during peptide synthesis. When the Ac group is removed (20% piperidine/DMF or 5% hydrazine/DMF) it becomes the Hmb group that is used to improve solubility and prevent as-partamide formation[129–131] and is readily cleaved with TFA.[132] The related 2-Fmoc-4-methoxybenzyl group has also been prepared and used in peptide syn-thesis.[133]

N-o-Nitrobenzylamide (−OCRN-ONB): 2-NO$_2$C$_6$H$_4$CH$_2$-NRCO−

Cleavage[134,135]

N-Cumylamide: (CH$_3$)$_2$C$_6$H$_5$C-NRCO−

This group was used as a bulky protective group to the intramolecular C–H insertion of α-diazo acetamides[136] and in directed orthometalation reactions of aryl amides.[137] The cumyl group is readily cleaved with CF$_3$CO$_2$H. Formic acid has also been used to remove a cumyl group.[138]

N-Bis(4-methoxyphenyl)methylamide (Ddm or Dmbh−NRCO−):
(4-MeOC$_6$H$_4$)$_2$CH-NRCO−

The methoxybenzhydral group was used to protect the −NH group of a β-lactam and a variety of amino acid amides.

Formation

4,4′-Dimethoxybenzhydrol, AcOH, H$_2$SO$_4$, 38–98% yield.[139] Very electron-poor amides give low yields because of there low nucleophilicity.

Cleavage

1. Ceric ammonium nitrate H$_2$O, CH$_3$CN, 0°C, 91% yield.[140,141]
2. TFA, BF$_3$·Et$_2$O, anisole, Et$_3$SiH,[142] TFA, DMS, CH$_2$Cl$_2$,[143] or TFA anisole.[144]
3. HCl (IPA, 60°C, 4 h).[145]
4. AlBr$_3$, BrCH$_2$CH$_2$Br, EtSH, CH$_2$Cl$_2$, rt, >62% yield.[146]

Vancomycin's aglycone

N-Diphenylmethylamide (Dpm-NRCO−): $(C_6H_5)_2CH\text{-}NRCO-$

N-Bis(4-methylphenyl)methylamide (Mbh-NRCO): $(CH_3C_6H_4)_2CH\text{-}NRCO-$

The uracil amide can be protected with the Dpm group by first silylating with BSA in CH_3CN and then reaction with Ph_2CHBr with I_2 or Bu_4NI (93–100% yield). Cleavage is effected with 1% TfOH in TFA (100% yield)[147] or TFA/H_2O at rt.[148] The Mbh derivative prepared by the method of König[149] and is cleaved HBF_4–anisole–TFA.[150]

N-Bis(4-methoxyphenyl)phenylmethylamide (DMTr-NRCO−):
$(4\text{-}MeOC_6H_4)_2PhC\text{-}NRCO-$

Formation

The DMTr group was selectively introduced into a biotin derivative.[151]

R= DMTr, 40%
R = THP, 45%

N-Bis(4-methylsulfinylphenyl)methylamide: $(4\text{-}MeS(O)C_6H_4)_2CH\text{-}NRCO-$

This group was developed for the protection of primary amides of amino acids. It is introduced by amide bond formation with the benzhydryl amine. It is cleaved with 1 M $SiCl_4$/anisole/TFA/0°C or 1 M TMSOTf/thioanisole/TFA, 0°C. Cleavage occurs by initial sulfoxide reduction followed by acidolysis.[152]

N-Triphenylmethylamide (Tr-NRCO−): $(C_6H_5)_3C\text{-}NRCO-$

The trityl group was introduced on a primary amide, $RCONH_2$, in the presence of a secondary amide with TrOH, Ac_2O, H_2SO_4, AcOH, 60°C, 75% yield. Additionally, TsOH acid has been used to catalyze this transformation (72–98% yield)[153] The 4-methyltrityl (Mtt) group has similarly been used for protection of asparagines.[154] The trityl protected amide is stable to BOC removal with 1 N HCl in 50% isopropyl alcohol, 30 min, 50°C, but can be cleaved with TFA.[155] The table below gives the cleavage rates with TFA for a number of protected primary amides.

Compound	$t_{1/2}$ (min)	
Fmoc−Asn(Tr)−OH	8	
Fmoc−Gln(Tr)−OH	2	
Fmoc−Gln(Tmob)−OH	9	Tmob = 2,4,6-trimethoxybenzyl
Fmoc−Gln(Mbh)−OH	27	Mbh = 4,4′-dimethoxybenzyhydryl
Ac−Pro−Asn(Tr)−Gly−Phe−OH	9	

N-9-Phenylfluorenylamide (Pf-NRCO−)

Cleavage

TFA, CH_2Cl_2, 84% yield.[156]

N-Bis(trimethylsilyl)methylamide [(TMS)$_2$CH$_2$-NRCO−]

Cleavage

1. $(NH_4)_2Ce(NO_3)_6$, CH_3CN, H_2O, rt, 3 h, 84–95% yield. These conditions gave a β-lactam formimide that was then hydrolyzed with $NaHCO_3$, Na_2CO_3, H_2O, rt, 2 h, 78–95% yield.[157,158]
2. (i) TBAF, CH_3CHO, (ii) ozonolysis, DMS, (iii) $NaHCO_3$.[158]

N-*t*-Butoxycarbonylamide (BOC-NRCO−): *t*-C$_4$H$_9$OC(O)-NRCO−

Formation

1. $(BOC)_2O$, Et_3N, DMAP, 25°C, 15 h, 78–96% yield.[159,160] The rate of reaction of $(BOC)_2O$ with an amide −NH is a function of its acidity when steric factors are the same. The more acidic, the NH the faster the reaction. For example 4-thiazolidinone, $pK_a = 18.3$, reacts in 2 min whereas pyrrolidinone, $pK_a = 24.2$ requires 2 h to reach completion.[161] If the amide is sufficiently acidic, the same methodology can be used to prepare the methyl and benzyl carbamates.
2. BuLi, $(BOC)_2O$.[162]
3. $(BOC)OCO_2(BOC)$, DMAP.[163]
4. The very similar 1-Adoc derivative of amides can be prepared from $(Adoc)_2O$/ DMAP in CH_3CN. It is a little more reactive than $(BOC)_2O$.[163]

Cleavage

1. It should be noted that when a BOC-protected amide is subjected to nucleophilic reagents such as MeONa, hydrazine, and LiOH the amide bond is cleaved in preference to the BOC group (85–96% yield) because of the difference in steric factors.[164] The BOC group can be removed by the methods used to remove it from simple amines. It is also subject to migration under basic conditions in the presence of a proximal hydroxyl group.[165]

2. $Mg(ClO_4)_2$, CH_3CN, 99% yield.[166,167] These conditions do not cleave a *t*-butyl ester or *t*-butyl carbamate.

3. Yb(OTf)$_3$, SiO$_2$, neat, rt or 40°C, 96–100% yield. Yb(OTf)$_3$ in THF can also be used effectively.[168]

4. TMSOTf, CH$_2$Cl$_2$.[169]

Note migration of the t-Bu group to the hydroxyl

5. Mg(OMe)$_2$, MeOH, 82–90% yield.[170] This method is also effective for the Cbz and MeOCO derivatives, giving 78% and 86% yields, respectively.

6. NaN$_3$, NH$_4$Cl, MeOH, H$_2$O, reflux, 50–98% yield.[171] This method produces hydrazoic acid *in situ* and can present certain safety concerns.

7. Sm, I$_2$, MeOH, reflux 24 h, 95% yield.[172] This reagent also cleaves the Cbz group and other carbamates and esters.

8. Microwave irradiation, silica gel, 56–96% yield.[173] This method was later shown to give variable yields.[168]

N-Benzyloxycarbonylamide (Cbz-NRCO−): C$_6$H$_5$CH$_2$OC(O)-NRCO−

Formation

1. *n*-BuLi, THF, −78°C; CbzCl, −78°C to 0°C, 87–92% yield.[174]
2. (BnO$_2$C)$_2$O, DMAP, CH$_3$CN, 90% yield.[161]

Cleavage

1. Aqueous LiOH, dioxane, 86–92% yield.[174]
2. Et$_2$AlCl, CH$_2$Cl$_2$, −78°C, 10 min then Me$_2$S, 25°C, 4 h, 90–99% yield.[175]

N-Methoxy- and *N*-Ethoxycarbonylamide (MeOC(O)-NRCO−)

Formation

1. (MeO$_2$C)$_2$O, DMAP, CH$_3$CN, 5 min, 71% yield. It appears that only amides having a fairly acidic NH are acylated under these conditions. δ-Valerolactam fails to react.[161]

2. 4-NO$_2$C$_6$H$_4$OCO$_2$Me, DMAP, 92% yield.[176]

3. 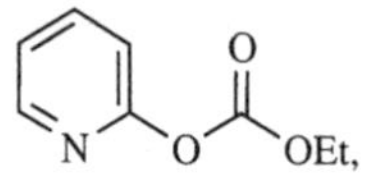 K$_2$CO$_3$, CH$_3$CN, reflux, 94% yield.[177]

Cleavage

NaCN, DMSO, 160°C, 79% yield.[178] This method cleaves the carbonate by nucleophile displacement of the *O*-methyl group.

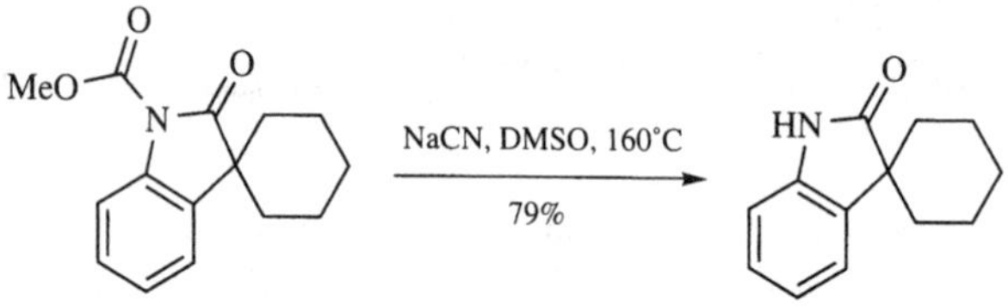

N-*p*-Toluenesulfonylamide: Ts-NRCO−

Cleavage

1. Sodium naphthalenide, DME, 0–20°C, 6 h, 59–94%.[179] A benzyl ether was stable to these reductive conditions.[180]

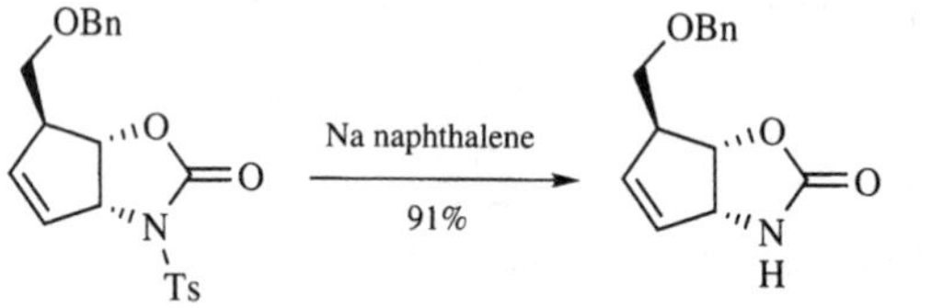

Ref. 180

2. Sodium anthracenide.[181] These conditions will not cleave a normal benzenesulfonamide.[182]

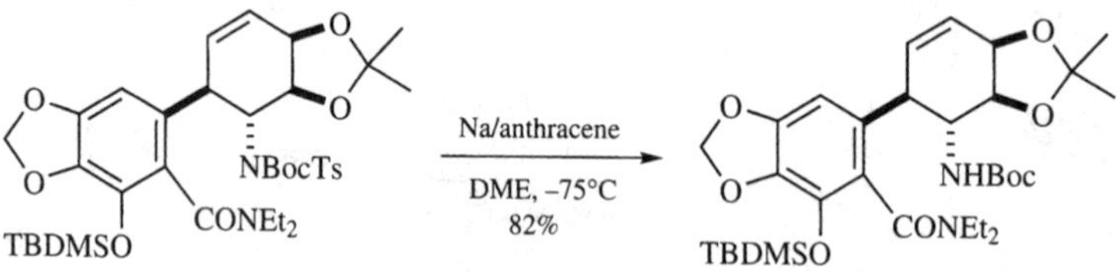

3. Bu₃SnH, AIBN, toluene, 35–94% yield.[183]
4. Electrolysis, TFA, DMF, Hg cathode, 70–98% yield.[184] A number of other sulfonamide are cleaved similarly.[185]
5. Photolysis, CH₃CN, 300 nm, 86% yield.[186]
6. Photolysis, CH₃CN, H₂O, $hv > 300$ nm, 2-phenyl-N,N'-dimethylbenzimidazoline (PDMBI), 82–98% yield.[187] PDMBI serves as a electron and hydrogen donor. Nitrogen bearing both a BOC group and a tosyl group fail to react.

7. Mg, MeOH, sonication, 20–40 min, 93–100% yield. The benzenesulfonyl, cyanophenylsulfonyl, 4-methoxybenzenesulfonyl and the 4-bromosulfonyl groups were all efficiently removed. The reaction is not compatible with the nosyl and Troc groups. The Troc group is converted to a dichloroethoxycarbonyl group.[188]
8. Li, catalytic naphthalene, −78°C, THF, 97–99% yield. In addition, benzylamides and methanesulfonamides are efficiently cleaved.[96]
9. TiCl₄, Zn, THF, 65°C.[189]
10. SmI₂, THF, high yield.[183,190]

N-Trimethylsilylethylsufonylamide (SES-NRCO−):
(CH₃)₃SiCH₂CH₂SO₂-NRCO−

Cleavage

1. Bu₃SnH, toluene, AIBN, reflux, 60% yield. Fluoride-based methods were ineffective in this case.[191]

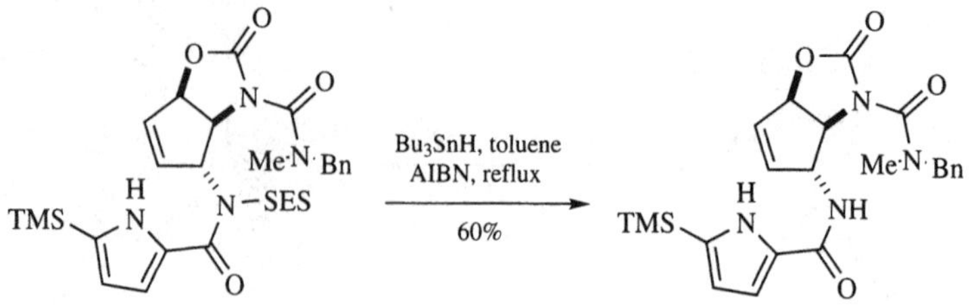

2. TBAF, THF, 99% yield.[192]

N,O-Isopropylidene Acetals

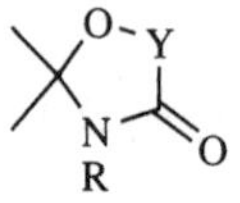

Formation

1. 2-Methoxypropene, BF$_3$·Et$_2$O, CH$_2$Cl$_2$, rt, 0.5 h, 84 % yield.[193]
2. 2,2-Dimethoxypropane, toluene, TsOH, rt, 18 h, >65% yield.[196]
3. (CH$_3$)$_2$C(OCH$_3$)$_2$, acetone, TsOH, rt, 97% yield.[194]
4. For the related cyclohexylidene acetal: cyclohexanone, TsOH, benzene, reflux 40 h with Soxhlet containing 4-Å molecular sieves, 82% yield.[195]

Cleavage

1. Aqueous AcOH, 3 h, >65% yield.[196]
2. Pyridinium chlorochromate. In this case the alcohol cleaved is simultaneously oxidized to give a ketone.[193]
3. BiBr$_3$, MeCN, rt, 85–97% yield. This method is compatible with the BOC and Cbz groups. Terminal acetonides are slowly cleaved.[197]

N,O-Benzylidene Acetals and N,O-4-Methoxybenzylidene Acetals

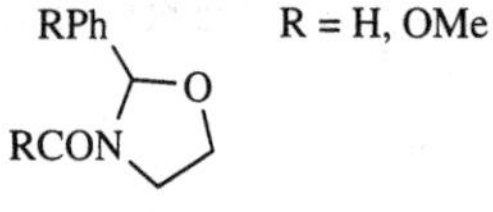

R = H, OMe

Formation

PhCH(OMe)$_2$, BF$_3$·Et$_2$O, 72% yield.[198]

Cleavage

1. Acid hydrolysis.[199]
2. Hydrogenolysis, Pd–C, hydrazine, MeOH, 95% yield.[200]
3. BF$_3$·Et$_2$O, MeOH, rt.[201]

N,O-Formylidene Acetal

These derivatives are often difficult to cleave. The following method relies on the essential irreversibility of dithiolane formation.

Cleavage[202]

N-Butenylamide: $CH_3CH_2CH=CH-NRCO_2-$

Formation

1. Butanal, P_2O_5, toluene, reflux.[203]
2. Butanal, TsOH, toluene, 70% yield.[204]
3. $RCH=CHB(OH)_2$, $Cu(OAc)_2$, TEA or pyridine, O_2, DMF, 61–96% yield.[205]

Cleavage

1. Et_3OBF_4; H_2O; pH 8, 67% yield.[204]
2. $KMnO_4$, acetone, H_2O, 0°C, 10 min, 78–90% yield. These conditions are used for the related ethylidine group.[206]
3. THF, 1% aq HCl, (9:2), reflux, 36 h; THF, H_2O (1:1), Na_2CO_3, reflux, 1 h, 62% yield.[206]
4. $4\text{-}NO_2C_6H_4CO_3H$, THF, H_2O, HCO_2H, (10:10:1), 25°C, 80% yield.[207]

N-[(*E*)-2-(Methoxycarbonyl)vinyl]amide: $MeO_2CC=CH-NRCO-$

Formation

Methyl propiolate, DMAP, rt, <10 min.[208]

Cleavage

1. Pyrrolidine, CH_3CN, rt, <2 h, >98% yield.[208]
2. $CSA\cdot 2H_2O$, MeOH, reflux, 1.5 h, >92% yield.[208]

N-Diethoxymethylamide (DEM−NRCO−): $(EtO)_2CH-NRCO-$

Formation

$CH(OEt)_3$, 160°C, 25–78% yield.[209]

Cleavage

TFA, CH$_2$Cl$_2$, rt, 1 h; 2 N NaOH, rt, 0.5 h, 37–90% yield.[209]

N-(1-Methoxy-2,2-dimethylpropyl)amide

Formation

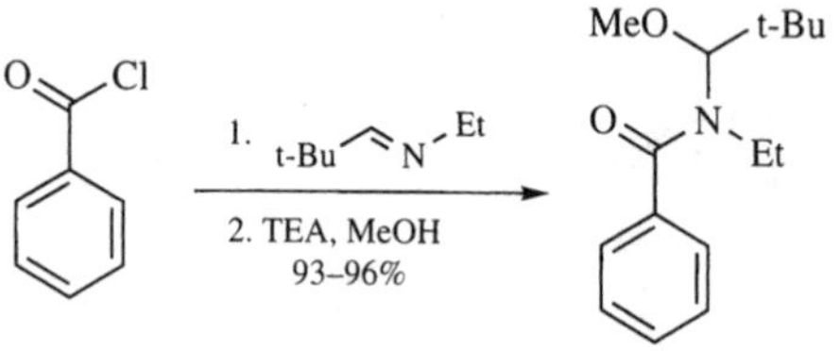

This protective group was used to improve the directed ortho metalation.[210]

Cleavage

HCl, dioxane, >71–82% yield.[210]

N-2-(4-Methylphenylsulfonyl)ethylamide: 4-CH$_3$C$_6$H$_4$SO$_2$CH$_2$CH$_2$-NRCO−

Formation

(4-Methylphenylsulfonyl)ethylamine was used to introduce the nitrogen in a β-lactam synthesis.[211]

Cleavage

By β-elimination with *t*-BuOK, THF, 1,5 h, −35°C to 0°C, 72% yield.[211,212] This group was successfully cleaved from a β-lactam without ring opening.[213]

PROTECTION FOR THE SULFONAMIDE −NH

N-t-Butylsulfonamide: (CH$_3$)$_3$CNRSO$_2$R′

Cleavage

1. BCl$_3$, CH$_2$Cl$_2$, rt, 0.5 h, 74–97% yield.[214]
2. Sc(OTf)$_3$, CH$_3$NO$_2$, 50°C, 4 h, 84–95% yield.[215]

N-Diphenylmethylsulfonamide (DPM-NRSO$_2$R′)

Cleavage

Hydrogenation, H$_2$, 1 atm, Pd(OH)$_2$/C CH$_3$OH, THF, Et$_3$N, 18 h, 87–99% yield.[216]
In this case the use of benzyl, 2,4-dimethoxybenzyl, 3,4-dimethoxybenzyl, and

4-nitrobenzyl protective groups was unsatisfactory because of ring saturation of the benzyl group during the hydrogenolysis. Oxidative cleavage of 2,4- and 3,4-dimethoxybenzyl groups led to complex mixtures.

N-Benzylsulfonamide (BnNRSO$_2$R′)

In the presence of a β-hydroxy group the benzyl group can be removed by hydrogenolysis with Pd(OH)$_2$, but in its absence it is inert unless the nitrogen is acylated.[217]

N-4-Methoxybenzylsulfonamide (PMB–NRSO$_2$R′): 4-CH$_3$OC$_6$H$_4$CH$_2$NRSO$_2$R

Ceric ammonium nitrate is used to cleave the PMB group from a sulfonamide nitrogen.[218]

N-2,4-Dimethoxybenzylsulfonamide (DMB–NRSO$_2$R′)

Cleavage

30% TFA, CH$_2$Cl$_2$, 0°C, 4 h, 81% yield.[219]

N-2,4,6-Trimethoxybenzylsulfonamide (Tmob–NRSO$_2$R′)

Formation

The Tmob group is introduced by reaction of the sulfonyl chloride with 2,4,6-trimethoxybenzylamine.[220]

Cleavage

TFA, CH$_2$Cl$_2$, CH$_3$SCH$_3$, 92% yield.[220]

N-4-Methoxyphenyl sulfonamide (MP–NRSO$_2$R′)

The MP group is introduced on a sulfonamide through a Cu(OAc)$_2$ catalyzed coupling with 4-methoxyphenylboronic acid.[221] It can in principle be cleaved oxidatively with DDQ.

4-Hydroxy-2-methyl-3(2*H*)-isothiazolone 1,1-Dioxide[222]

When the benzylic position was protected, an indole could be prepared without side products.

1. W. G. B. van Henegouwen, R. M. Fieseler, F. P. J. T. Rutjes, and H. Hiemstra, *J. Org. Chem.*, **65**, 8317 (2000).

2. T. Fukuyama, A. A. Laird, and C. A. Schmidt, *Tetrahedron Lett.*, **25**, 4709 (1984).

3. T. Sato, K. Yoshimatsu, and J. Otera, *Synlett*, 845 (1995).

4. T. Pietzonkz and D. Seebach, *Angew. Chem., Int. Ed.*, **31**, 1481 (1992).

5. H. Liu, S.-B. Ko, H. Josien, and D. P. Curran, *Tetrahedron Lett.*, **36**, 8917 (1995).

6. S. Krompiec, M. Pigulla, W. Szczepankiewicz, T. Bieg, N. Kuznik, K. Leszczynska-Sejda, M. Kubicki, and T. Borowiak, *Tetrahedron Lett.*, **42**, 7095 (2001).

7. M. Kimura, K. Fugami, S. Tanaka, and Y. Tamaru, *J. Org. Chem.*, **57**, 6377 (1993).

8. F. L. Zumpe and U. Kazmaier, *Synlett*, 1199 (1998).

9. T. A. Lessen, D. M. Demko, and S. M. Weinreb, *Tetrahedron Lett.*, **31**, 2105 (1990).

10. B. Moreau, S. Lavielle, and A. Marquet, *Tetrahedron Lett.*, **18**, 2591 (1977).

11. P. Wipf and C. R. Hopkins, *J. Org. Chem.*, **66**, 3133 (2001).

12. S. Sergeyev and M. Hesse, *Synlett*, 1313 (2002).

13. T. Koch and M. Hesse, *Synthesis*, 931 (1992); *idem, ibid.*, 251 (1995).

14. T. Taniguchi and K. Ogasawara, *Tetrahedron Lett.*, **39**, 4679 (1998).

15. R. Gigg and R. Conant, *Carbohydrate Res.*, **100**, C5, (1982).

16. S. A. Kozmin, T. Iwama, Y. Huang, and V. H. Rawal, *J. Am. Chem. Soc.*, **124**, 4628 (2002).

17. P. Ribéreau, M. Delamare, S. Celanire, and G. Quéguiner, *Tetrahedron Lett.*, **42**, 3571 (2001).

18. B. Neugnot, J.-C. Cintrat, and B. Rousseau, *Tetrahedron*, **60**, 3575 (2004).

19. B. Alcaide, P. Almendros, J. M. Alonso, and M. F. Aly, *Org. Lett.*, **3**, 3781 (2001).

20. B. Alcaide, P. Almendros, and J. M. Alonso, *Tetrahedron Lett.*, **44**, 8693 (2003).

21. P. I. Kitov and D. R. Bundle, *Org. Lett.*, **3**, 2835 (2001).

22. M. J. Earle, R. A. Fairhurst, H. Heaney, and G. Papageorgiou, *Synlett*, 621 (1990).

23. L. A. Carpino, H.-G. Chao, S. Ghassemi, E. M. E. Mansour, C. Riemer, R. Warrass, D. Sadat-Aalaee, G. A. Truran, H. Imazumi, A. El-Faham, D. Ionescu, M. Ismail, T. L. Kowaleski, C. H. Han, H. Wenschuh, M. Beyermann, M. Bienert, H. Shroff, F. Albericio, S. A. Triolo, N. A. Sole, and S. A. Kates, *J. Org. Chem.*, **60**, 7718 (1995).

24. P. Carato, S. Yous, D. Sellier, J. H. Poupaert, N. Lebegue, and P. Berthelot, *Tetrahedron*, **60**, 10321 (2004).

25. G. W. Kirby, D. J. Robins, and W. M. Stark, *J. Chem. Soc., Chem. Commun.*, 812 (1983).

26. E. Sotelo, A. Coelho, and E. Ravina, *Tetrahedron Lett.*, **42**, 8633 (2001).

27. P. S. Baran and C. A Guerrero, *Angew. Chem. Int. Ed.*, **44**, 3892 (2005).

28. S. Yokoshima, H. Tokuyama, and T. Fukuyama, *Angew. Chem. Int. Ed.*, **39**, 4073 (2000).

29. A. Madin, C. J. O'Donnell, T. Oh, D. W. Old, L. E. Overman, and M. J. Sharp, *Angew. Chem. Int. Ed.*, **38**, 2934 (1999).

30. S. Yokoshima, H. Tokuyama, and T. Fukuyama, *Angew. Chem. Int. Ed.*, **39**, 4073 (2000).

31. P. Carato, S. Yous, D. Sellier, J. H. Poupaert, N. Lebegue, and P. Berthelot, *Tetrahedron*, **60**, 10321 (2004).

32. E. D. Edstrom, X. Feng, and S. Tumkevicius, *Tetrahedron Lett.*, **37**, 759 (1996).

33. A. Ncube, S. B. Park, and J. M. Chong, *J. Org. Chem.*, **67**, 3625 (2002).

34. S. Hanessian, ACS Symp. Ser. **386**, *Trends in Synthetic Carbohydrate Chemistry*, (1989), p. 64.

35. J. T. Link, S. Raghavan, M. Gallant, S. J. Danishefsky, T. C. Chou, and L. M. Ballas, *J. Am. Chem. Soc.*, **118**, 2825 (1996).

36. E. Zara-Kaczian and P. Matyus, *Heterocycles*, **36**, 519 (1993).á

37. L. E. Overman and M. D. Rosen, *Angew. Chem. Int. Ed.*, **39**, 4596 (2000).

38. R. W. Hoffman, S. Breitfelder, and A. Schlapbach, *Helv. Chim. Acta*, **79**, 346 (1996).

39. A. B. Smith, III, J. Barbosa, W. Wong, and J. L. Wood, *J. Am. Chem. Soc.*, **118**, 8316 (1996).

40. G. Evano, J. V. Schaus, and J. S. Panek, *Org. Lett.*, **6**, 525 (2004).

41. P. J. Dransfield, S. Wang, A. Dilley, and D. Romo, *Org. Lett.*, **7**, 1679 (2005).

42. T. Benneche, L. L. Gundersen, and K. Undheim, *Acta Chem. Scand.*, Ser. B, **B42**, 384 (1988).

43. G. Andresen, A. B. Eriksen, B. Dalhus, L.-L. Gundersen, and F. Rise, *J. Chem. Soc. Perkin Trans. 1*, 1662 (2001).

44. M. A. Zajac and E. Vedejs, *Org. Lett.*, **6**, 237 (2004).

45. E. C. Taylor and W. B. Young, *J. Org. Chem.*, **60**, 7947 (1995).

46. A. Benarab, S. Boyé, L. Savelon, and G. Guillaumet, *Tetrahedron Lett.*, **34**, 7567 (1993).

47. G. Cignarella, G. F. Cristiani, and E. Testa, *Justus Liebigs Ann. Chem.*, **661**, 181 (1963).

48. A. B. Hamlet and T. Durst, *Can. J. Chem.*, **61**, 411 (1983).

49. F. Shirai and T. Nakai, *Tetrahedron Lett.*, **29,** 6461 (1988).

50. P. G. Mattingly and M. J. Miller, *J. Org. Chem.*, **46,** 1557 (1981).

51. N. V. Shah and L. D. Cama, *Heterocycles*, **25,** 221 (1987).

52. D. A. Burnett, D. J. Hart, and J. Liu, *J. Org. Chem.*, **51,** 1929 (1986).

53. P. J. Reider and E. J. J. Grabowski, *Tetrahedron Lett.*, **23,** 2293 (1982).

54. H. Hiemstra, W. J. Klaver, and W. N. Speckamp, *Tetrahedron Lett.*, **27,** 1411 (1986).

55. D. J. Hart, C.-S. Lel, W. H. Pirkle, M. H. Hyon, and A. Tsipouras, *J. Am. Chem. Soc.*, **108,** 6054 (1986).

56. J. Roby and N. Voyer, *Tetrahedron Lett.*, **38,** 191 (1997).

57. D. E. Ward and B. F. Kaller, *Tetrahedron Lett.*, **34,** 407 (1993).

58. S. Gérard and J. Marchand-Brynaert, *Tetrahedron Lett.*, **44,** 6339 (2003).

59. K. Yamamoto and M. Takemae, *Bull. Chem. Soc. Jpn.*, **62,** 2111 (1989).

60. R. W. Ratcliffe, T. N. Salzmann, and B. G. Christensen, *Tetrahedron Lett.*, **21,** 31 (1980).

61. S. Knapp, A. T. Levorse, and J. A. Potenza, *J. Org. Chem.*, **53,** 4773 (1988).

62. H.-O. Kim, C. Lum, and M. S. Lee, *Tetrahedron Lett.*, **38,** 4935 (1997).

63. A. Chen, A. Nelson, N. Tanikkul, and E. J. Thomas, *Tetrahedron Lett.*, **42,** 1251 (2001).

64. R. C. F. Jones and A. D. Bates, *Tetrahedron Lett.*, **27,** 5285 (1986).

65. J. Roby and N. Voyer, *Tetrahedron Lett.*, **38,** 191 (1997).

66. Y. Wang, J. Janjic, and S. A. Kozmin, *J. Am. Chem. Soc.*, **124,** 13670 (2002).

67. S. F. Vice, W. R. Bishop, S. W. McCombie, H. Dao, E. Frank, and A. K. Ganguly, *Bioorg. Med. Chem. Lett.*, **4,** 1333 (1994).

68. E. J. Hennessy and S. L. Buchwald, *J. Org. Chem.*, **70,** 7371 (2005).

69. P. Y. S. Lam, S. Deudon, K. M. Averill, R. Li, M. Y. He, P. DeShong, and C. G. Clark, *J. Am. Chem. Soc.*, **122,** 7600 (2000).

70. P. Lopez-Alvardo, C. Avendano, and J. C. Menendez, *Tetrahedron Lett.*, **33,** 6875 (1992), M. S. Akhtar, W. J. Brouillette, and D. V. Waterhous, *J. Org. Chem.*, **55,** 5222 (1990).

71. W. Deng, Y.-F. Wang, Y. Zou, L. Liu, and Q.-X. Guo, *Tetrahedron Lett.*, **45,** 2311 (2004).

72. R. Hosseinzadeh, M. Tajbakhsh, M. Mohadjerani, and H. Mehdinejad, *Synlett*, 1517 (2004).

73. E. G. Corley, S. Karady, N. L. Abramson, D. Ellison, and L. M. Weinstock, *Tetrahedron Lett.*, **29,** 1497 (1988).

74. D. R. Kronenthal, C. Y. Han, and M. K. Taylor, *J. Org. Chem.*, **47,** 2765 (1982).

75. D.-C. Ha and D. J. Hart, *Tetrahedron Lett.*, **28,** 4489 (1987).

76. J. Fetter, L. T. Giang, T. Czuppon, K. Lempert, M. Kajtar-Peredy, and G. Czira, *Tetrahedron*, **50,** 4188 (1994).

77. J. A. Marshall, K. Gill, and B. M. Seletsky, *Angew. Chem. Int. Ed.*, **39,** 953 (2000).

78. H. Yanagisawa, A. Ando, M. Shiozaki, and T. Hiraoka, *Tetrahedron Lett.*, **24,** 1037 (1983).

79. K. Bhattarai, G. Cainelli, and M. Panunzio, *Synlett*, 229 (1990).

80. T. Fukuyama, R. K. Frank, and C. F. Jewell, Jr., *J. Am. Chem. Soc.*, **102,** 2122 (1980).

81. B. M. Trost and A. A. Sudhakar, *J. Am. Chem. Soc.*, **110,** 7933 (1988).

82. Y. Xia and A. P. Kozikowski, *J. Am. Chem. Soc.*, **111,** 4116 (1989).

83. U. R. Kalkote, A. R. Choudhary, and N. R. Ayyangar, *Org. Prep. Proc. Int.*, **24,** 83 (1992).

84. F. Fache, L. Jacquot, and M. Lemaire, *Tetrahedron Lett.*, **35,** 3313 (1994).

85. D. Dubé and A. A. Scholte, *Tetrahedron Lett.*, **40,** 2295 (1999).

86. F. Effenberger, W. Müller, R. Keller, W. Wild, and T. Ziegler, *J. Org. Chem.*, **55,** 3064 (1990).

87. K. C. Nicolaou, J. Hao, M. V. Reddy, P. B. Rao, G. Rassias, S. A. Snyder, H. Huang, D. Y.-K. Chen, W. E. Brenzovich, N. Giuseppone, P. Giannakakou, and A. O'brate, *J. Am. Chem. Soc.*, **126,** 12897 (2004).

88. R. N. Salvatore, S. I. Shin, V. L. Flanders, and K. W. Jung, *Tetrahedron Lett.*, **42,** 1799 (2001).

89. J. Danklmaier and H. Hoenig, *Synth. Commun.*, **20,** 203 (1990).

90. R. Gigg and R. Conant, *Carbohydr. Res.*, **100,** C5 (1982).

91. L. Bérillon, R. Wagner, and P. Knochel, *J. Org. Chem.*, **63,** 9117 (1998).

92. T. Ohgi and S. M. Hecht, *J. Org. Chem.*, **46,** 1232 (1981); M. Y. Kim, J. E. Starrett, Jr., and S. M. Weinreb, *J. Org. Chem.*, **46,** 5383 (1981); S. Sugasawa and T. Fujii, *Chem. Pharm. Bull.*, **6,** 587 (1958); F. X. Webster, J. G. Millar, and R. M. Silverstein, *Tetrahedron Lett.*, **27,** 4941 (1986).

93. G. F. Field, *J. Org. Chem.*, **43,** 1084 (1978).

94. P. A. Jacobi, H. L. Brielmann, and S. I. Hauck, *Tetrahedron Lett.*, **36,** 1193 (1995).

95. J. Sisko, J. R. Henry, and S. M. Weinreb, *J. Org. Chem.*, **58,** 4945 (1993).

96. E. Alonso, D. J. Ramón, and M. Yus, *Tetrahedron,* **53,** 14355 (1997).

97. R. M. Williams and E. Kwast, *Tetrahedron Lett.*, **30,** 451 (1989).

98. R. Gigg and R. Conant, *J. Chem. Soc., Chem. Commun.*, 465 (1983).

99. A. Huang, J. J. Kodanko, and L. E. Overman, *J. Am. Chem. Soc.*, **126,** 14043, (2004).

100. M. Barbier, *Heterocycles*, **23,** 345 (1985).

101. J. E. Semple, P. C. Wang, Z. Lysenko, and M. M. Joullié, *J. Am. Chem. Soc.*, **102,** 7505 (1980).

102. E. G. Baggiolini, H. L. Lee, G. Pizzolato, and M. R. Uskokovic, *J. Am. Chem. Soc.*, **104,** 6460 (1982).

103. G. F. Field, W. J. Zally, L. H. Sternbach, and J. F. Blout, *J. Org. Chem.*, **41,** 3853 (1976).

104. M. Yamaura, T. Suzuki, H. Hashimoto, J. Yoshimura, and C. Shin, *Chem. Lett.*, **13,** 1547 (1984).

105. T. Akiyama, H. Nishimoto, and S. Ozaki, *Bull. Chem. Soc. Jpn.*, **63,** 3356 (1990).

106. Y. Takahashi, H. Yamashita, S. Kobayashi, and M. Ohno, *Chem. Pharm. Bull.*, **34,** 2732 (1986).

107. M. Yamaura, T. Suzuki, H. Hashimoto, J. Yoshimura, T. Okamoto, and C. Shin, *Bull. Chem. Soc. Jpn.*, **58,** 1413 (1985).

108. J. Yoshimura, M. Yamaura, T. Suzuki, and H. Hashimoto, *Chem. Lett.*, **12,** 1001 (1983).

109. A. B. Smith, III, G. K. Friestad, J. Barbosa, E. Bertounesque, K. G. Hull, M. Iwashima, Y. Qiu, B. A. Salvatore, P. G. Spoors, and J. J.-W. Duan, *J. Am. Chem. Soc.*, **121**, 10468 (1999); T. Q. Pham, S. G. Pyne, B. W. Skelton, and A. H. White, *Tetrahedron Lett.*, **43**, 5953 (2002).

110. A. B. Smith, III, I. Noda, S. W. Remiszewski, N. J. Liverton, and R. Zibuck, *J. Org. Chem.*, **55**, 3977 (1990); R. M. Williams, T. Glinka, E. Kwast, H. Coffman, and J. K. Stille, *J. Am. Chem. Soc.*, **112**, 808 (1990).

111. J. H. Rigby and M. E. Mateo, *J. Am. Chem. Soc.*, **119**, 12655 (1997).

112. J. H. Rigby, U. S. M. Maharoof, and M. E. Mateo, *J. Am. Chem. Soc.*, **122**, 6624 (2000).

113. J. H. Rigby and V. Gupta, *Synlett*, 547 (1995).

114. T. Akiyama, Y. Takesue, M. Kumegawa, H. Nishimoto, and S. Ozaki, *Bull. Chem. Soc. Jpn*, **64**, 2266 (1991).

115. G. M. Brooke, S. Mohammed, and M. C. Whiting, *J. Chem. Soc., Chem. Commun.*, 1511 (1997).

116. N. Chida, M. Ohtsuka, and S. Ogawa, *J. Org. Chem.*, **58**, 4441, (1993).

117. K. Ishihara, Y. Hiraiwa, and H. Yamamoto, *Synlett*, 80 (2000).

118. W. F. Huffman, K. G. Holden, T. F. Buckley, J. G. Gleason, and L. Wu, *J. Am. Chem. Soc.*, **99**, 2352 (1977). X. Qian, B. Zheng, B. Burke, M. T. Saindane, and D. R. Kronenthal, *J. Org. Chem.*, **67**, 3595 (2002).

119. R. H. Schlessinger, G. R. Bebernitz, P. Lin, and A. Y. Poss, *J. Am. Chem. Soc.*, **107**, 1777 (1985).

120. P. DeShong, S. Ramesh, V. Elango, and J. J. Perez, *J. Am. Chem. Soc.*, **107**, 5219 (1985); S. S. Shimshock, R. E. Waltermire, and P. DeShong, *J. Am. Chem. Soc.*, **113**, 8791 (1991).

121. C.-Y. Chern, Y.-P. Huang, and W. M. Kan, *Tetrahedron Lett.*, **44**, 1039 (2003).

122. J. L. Wood, B. M. Stoltz, and S. N. Goodman, *J. Am. Chem. Soc.*, **118**, 10656 (1996).

123. J. L. Wood, B. M. Stoltz, and H.-J. Dietrich, *J. Am. Chem. Soc.*, **117**, 10413 (1995).

124. J. L. Wood, B. M. Stoltz, H-J. Dietrich, D. A. Pflum, and D. T. Petsch, *J. Am. Chem. Soc.*, **119**, 9641 (1997). D. J. Watson, E. D. Dowdy, W. S. Li, J. Wang, and R. Polniaszek, *Tetrahedron Lett.*, **42**, 1827 (2001).

125. S. Mori, H. Iwakura, and S. Takechi, *Tetrahedron Lett.*, **29**, 5391 (1988).

126. E. Grunder-Klotz and J. D. Eherhardt, *Tetrahedron Lett.*, **32**, 751 (1991).

127. L. E. Overman and T. Osawa, *J. Am. Chem. Soc.*, **107**, 1698 (1985); F. He, and B. B. Snider, *Synlett*, 483 (1997).

128. T. G. Back, K. Brunner, P. W. Codding, and A. W. Roszak, *Heterocycles*, **28**, 219 (1989).

129. L. C. Packman, *Tetrahedron Lett.*, **36** 7523 (1995); C. Hyde, T. Johnson, D. Owen, M. Quibell, and R. C. Sheppard, *Int. J. Pept. Protein Res.*, **43**, 431 (1994).

130. T. Johnson, L. C. Packman, C. B. Hyde, D. Owen, and M. Quibell, *J. Chem. Soc., Perkin Trans I*, 719 (1996).

131. E. Nicolas, M. Pujades, J. Bacardit, E. Giralt, and F. Albericio, *Tetrahedron Lett.*, **38**, 2317 (1997).

132. M. Quibell, W. G. Turnell, and T. Johnson, *Tetrahedron Lett.*, **35**, 2237 (1994).

133. T. Johnson, M. Quibell, D. Owen, and R. C. Sheppard, *J. Chem. Soc., Chem. Commun.*, 369 (1993).

134. G. F. Miknis and R. M. Williams, *J. Am. Chem. Soc.*, **115**, 537 (1993).

135. B. B. Snider and M. V. Busuyek, *Tetrahedron*, **57**, 3301 (2001).

136. Z. Chen, Z. Chen, Y. Jiang, and W. Hu, *Synlett*, 1763 (2004).

137. C. Metallinos, S. Nerdinger, and V. Snieckus, *Org. Lett.*, **1**, 1183 (1999). C. Metallinos, *Synlett*, 1556 (2002).

138. J. Clayden, F. E. Knowles, and I. R. Baldwin, *J. Am. Chem. Soc.*, **127**, 2412 (2005).

139. K. Ohkawa, K. Ichimiya, A. Nishida, and H. Yamamoto, *Macromol. Biosci.*, **1**, 376 (2001). C. Henneuse, T. Boxus, L. Tesolin, G. Pantano, and J. Marchand-Brynaert, *Synthesis*, 495 (1996).

140. T. Kawabata, Y. Kimura, Y. Ito, and S. Terashima, *Tetrahedron Lett.*, **27**, 6241 (1986).

141. C. Palomo, J. M. Aizpurua, J. M. Garcia, M. Iturburu, and J. M. Odriozola, *J. Org. Chem.*, **59**, 5184 (1994).

142. Y. Kobayashi, Y. Ito, and S. Terashima, *Bull. Chem. Soc., Jpn.*, **62**, 3041 (1989).

143. D. A. Evans, J. C. Barrow, P. S. Watson, A. M. Ratz, C. J. Dinsmore, D. A. Evrard, K. M. DeVries, J. A. Ellman, S. D. Rychnovsky, and J. Lacour, *J. Am. Chem. Soc.*, **119**, 3419 (1997).

144. H. H. Wasserman, J.-H. Chen, and M. Xia, *J. Am. Chem. Soc.*, **121**, 1401 (1999).

145. Y. Kobayashi, Y. Takemoto, Y. Ito, and S. Terashima, *Tetrahedron Lett.*, **31**, 3031 (1990).

146. K. C. Nicolaou, A. E. Koumbis, M. Takayanagi, S. Natarajan, N. F. Jain, T. Bando, H. Li, and R. Hughes, *Chem. Eur. J.*, **5**, 2622 (1999).

147. F. Wu, M. G. Buhendwa, and D. F. Weaver, *J. Org. Chem.*, **69**, 9307 (2004).

148. F. Liu and D. J. Austin, *Org. Lett.*, **3**, 2273 (2001).

149. W. König and R. Geiger, *Chem. Ber.*, **103**, 2041, (1970).

150. K. Akaji, M. Yoshida, T. Tatsumi, T. Kimura, Y. Fujiwara, and Y. Kiso, *J. Chem. Soc., Chem. Commun.*, 288 (1990).

151. A. M. Alves, D. Holland, and M. D. Edge, *Tetrahedron Lett.*, **30**, 3089 (1989).

152. M. Patek and M. Lebl, *Collect. Czech. Chem. Commun.*, **57**, 508 (1992); *idem, Tetrahedron Lett.*, **31**, 5209 (1990).

153. D. R. Reddy, M. A. Iqbal, R. L. Hudkins, P. A. Messina-McLaughlin, and J. P. Mallamo, *Tetrahedron Lett.*, **43**, 8063 (2002).

154. E. Freund, F. Vitali, A. Linden, and J. A. Robinson, *Helv. Chim. Acta*, **83**, 2572 (2000).

155. P. Sieber and B. Riniker, *Tetrahedron Lett.*, **32**, 739 (1991).

156. E. Fernandez-Megía, and F. J. Sardina, *Tetrahedron Lett.*, **38**, 673 (1997).

157. C. Palomo, J. M. Aizpurua, M. Legido, and R. Galarza, *J. Chem. Soc., Chem. Commun.*, 233 (1997); C. Palomo, J. M. Aizpurua, M. Legido, A. Mielgo, and R. Galarza, *Chem. Eur. J.*, **3**, 1432 (1997); C. Palomo, J. M. Aizpurua, A. Benito, R. Galarza, U. K. Khamrai, J. Vazquez, B. d. Pascual-Teresa, P. M. Nieto, and A. Linden, *Angew. Chem. Int. Ed.*, **38**, 3056 (1999).

158. C. Palomo, J. M. Aizpurua, J. M. Garcia, R. Galarza, M. Legido, R. Urchegui, P. Roman, A. Luque, J. Server-Carrio, and A. Linden, *J. Org. Chem.*, **62**, 2070 (1997).

159. D. L. Flynn, R. E. Zelle, and P. A. Grieco, *J. Org. Chem.*, **48**, 2424 (1983).

160. Y. Ohfune and M. Tomita, *J. Am. Chem. Soc.*, **104**, 3511 (1982).

161. M. M. Hansen, A. R. Harkness, D. S. Coffey, F. G. Bordwell, and Y. Zhao, *Tetrahedron Lett.*, **36**, 8949 (1995).

162. A. Giovannini, D. Savoia, and A. Umani-Ronchi, *J. Org. Chem.*, **54**, 228 (1989).

163. A. Könnecke, L. Grehn, and U. Ragnarsson., *Tetrahedron Lett.*, **31**, 2697 (1990).

164. M. J. Burk and J. G. Allen, *J. Org. Chem.*, **62**, 7054 (1997).

165. L. Bunch, P.-O. Norrby, K. Frydenvang, P. Krogsgaard-Larsen, and U. Madsen, *Org. Lett.*, **3**, 433 (2001); S. P. Bew, S. D. Bull, and S. G. Davies, *Tetrahedron Lett.*, **41**, 7577 (2000); S. P. Bew, S. D. Bull, S. G. Davies, E. D. Savory, and D. J. Watkin, *Tetrahedron*, **58**, 9387 (2002).

166. J. A. Stafford, M. F. Brackeen, D. S. Karanewsky, and N. L. Valvano, *Tetrahedron Lett.*, **34**, 7873 (1993).

167. M. Inoue, H. Sakazaki, H. Furuyama, and M. Hirama, *Angew. Chem. Int. Ed.*, **42**, 2654 (2003).

168. S. Calimsiz and M. A. Lipton, *J. Org. Chem.*, **70**, 6218 (2005).

169. G. Casiraghi, F. Ulgheri, P. Spanu, G. Rassu, L. Pinna, G. G. Fava, M. B. Ferrari, and G. Pelosi, *J. Chem. Soc., Perkin Trans. I*, 2991 (1993).

170. Z.-Y. Wei and E. E. Knaus, *Tetrahedron Lett.*, **35**, 847 (1994).

171. J. N. Hernández, M. A. Ramírez, and V. S. Martín, *J. Org. Chem.*, **68**, 743 (2003).

172. R. Yanada, N. Negoro, K. Bessho, and K. Yanada, *Synlett*, 1261 (1995).

173. J. G. Siro, J. Martin, J. L. Garcia-Navio, M. J. Remuinan, and J. J. Vaquero, *Synlett*, 147 (1998).

174. R. W. Hungate, J. L. Chen, K. E. Starbuck, S. A. Macaluso, and R. S. Rubino, *Tetrahedron Lett.*, **37**, 4113 (1996).

175. T. Tsujimoto and A. Murai, *Synlett*, 1283 (2002).

176. M. J. Crossley and R. C. Reid, *J. Chem. Soc., Chem. Commun.*, 2237 (1994).

177. M. A. Neanwell, S. Y. Sit, J. Gao, H. S. Wong, Q. Gao, D. R. St. Laurent, and N. Balasubramanian, *J. Org. Chem.*, **60**, 1565 (1995).

178. Z. Mao and S. W. Baldwin, *Org. Lett.*, **6**, 2425 (2004).

179. J. Martens and M. Scheunemann, *Tetrahedron Lett.*, **32**, 1417 (1991).

180. B. M. Trost and D. L. Van Vranken, *J. Am. Chem. Soc.*, **115**, 444 (1993).

181. T. Hudlicky, X. Tian, K. Königsberger, R. Maurya, J. Rouden, and B. Fan, *J. Am. Chem. Soc.*, **118**, 10752 (1996).

182. H. Uchida, A. Nishida, and M. Nakagawa, *Tetrahedron Lett.*, **40**, 113 (1999).

183. A. F. Parsons and R. M. Pettifer, *Tetrahedron Lett.*, **37**, 1667 (1996); H. S. Knowles, A. F. Parsons, R. M. Pettifer, and S. Rickling, *Tetrahedron*, **56**, 979 (2000).

184. M. A. Casadei, A. Gessner, A. Inesi, W. Jugelt, and F. M. Moracci, *J. Chem. Soc., Perkin Trans. I*, 2001 (1992).

185. B. Nyasse, L. Grehn, U. Ragnarsson, H. L. S. Maia, L. S. Monteiro, I. Leito, I. Koppel, and J. Koppel, *J. Chem. Soc., Perkin Trans. I*, 2025 (1995).

186. C. Li and P. L. Fuchs, *Tetrahedron Lett.*, **34**, 1855 (1993).

187. Q. Liu, Z. Liu, Y.-L. Zhou, W. Zhang, L. Yang, Z.-L. Liu, and W. Yu, *Synlett*, 2510 (2005).

188. B. Nyasse, L. Grehn, and U. Ragnarsson, *J. Chem. Soc., Chem. Commun.*, 1017 (1997).

189. Y.-X. Ding and J. J. Hu, *Chem. Soc., Perkin 1*, 1651 (2000).

190. A. E. Taggi, A. M. Hafez, H. Wack, B. Young, D. Ferraris, and T. Lectka, *J. Am. Chem. Soc.*, **124**, 6626 (2002).

191. M. M. Domostoj, E. Irving, F. Scheinmann, and K. J. Hale, *Org. Lett.*, **6**, 2615 (2004).

192. Y. Gao, P. Lane-Bell, and J. C. Vederas, *J. Org. Chem.*, **63**, 2133 (1998).

193. R. Camerini, M. Panumzio, G. Bonanomi, D. Donati, and A. Perboni, *Tetrahedron Lett.*, **37**, 2467 (1996).

194. T. Yoshino, Y. Nagata, E. Itoh, M. Hashimoto, T. Katoh, and S. Terashima, *Tetrahedron Lett.*, **37**, 3475 (1996); K. Mori, and H. Matsuda, *Liebigs Ann. Chem.*, 131 (1992).

195. L. Williams, Z. Zhang, X. Ding, and M. M. Joullié, *Tetrahedron Lett.*, **36**, 7031 (1995).

196. D. Favara, A. Omodei-Salè, P. Consonni, and A. Depaoli, *Tetrahedron Lett.*, **23**, 3105 (1982).

197. X. Cong, F. Hu, K.-G. Liu, Q.-J. Liao, and Z.-J. Yao, *J. Org. Chem.*, **70**, 4514 (2005).

198. H. Cheng, P. Keitz, and J. B. Jones, *J. Org. Chem.*, **59**, 7671 (1994).

199. E. Didier, E. Fouque, I. Taillepied, and A. Commercon, *Tetrahedron Lett.*, **35**, 2349 (1994).

200. Y. Hamada, A. Kawai, Y. Kohno, O. Hara, and T. Shioiro, *J. Am. Chem. Soc.*, **111**, 1524 (1989).

201. L. C. Dias, A. M. A. P. Fernandes, and J. Zukerman-Schpector, *Synlett*, 100 (2002).

202. E. J. Corey and G. A. Reichard, *Tetrahedron Lett.*, **34**, 6973 (1993).

203. T. W. Kwon, P. F. Keusenkothen, and M. B. Smith, *J. Org. Chem.*, **57**, 6169 (1992).

204. M. B. Smith, C. J. Wang, P. F. Keusenkothen, B. T. Dembofsky, J. G. Fay, C. A. Zezza, T. W. Kwon, J. Sheu, Y. C. Son, and R. F. Menezes, *Chem. Lett.*, **21**, 247 (1992).

205. P. Y. S. Lam, G. Vincent, D. Bonne, and C. G. Clark, *Tetrahedron Lett.*, **44**, 4927 (2003).

206. G. I. Georg, P. He, J. Kant, and J. Mudd, *Tetrahedron Lett.*, **31**, 451 (1990).

207. J. V. Heck and B. G. Christensen, *Tetrahedron Lett.*, **22**, 5027 (1981).

208. M. Faja, X. Ariza, C. Galvez, and J. Vilarrasa, *Tetrahedron Lett.*, **36**, 3261 (1995).

209. P. Gmeiner and B. Bollinger, *Synthesis.*, 168 (1995).

210. D. P. Phillion and D. M. Walker, *J. Org. Chem.*, **60**, 8417 (1995).

211. D. DiPietro, R. M. Borzilleri, and S. M. Weinreb, *J. Org. Chem.*, **59**, 5856 (1994).

212. G. D. Artman, III, J. H. Waldman, and S. M. Weinred, *Synthesis*, 2057 (2002).

213. D. DiPietro, R. M. Borzilleri, and S. M. Weinreb, *J. Org. Chem.*, **59**, 5856 (1994).

214. Y. Wan, X. Wu, M. A. Kannan, and M. Alterman, *Tetrahedron Lett.*, **44**, 4523 (2003).

215. A. K. Mahalingam, X. Wu, Y. Wan, and M. Alterman, *Synth. Commum.*, **35**, 417 (2005).

216. M. A. Poss and J. A. Reid, *Tetrahedron Lett.*, **33**, 7291 (1992).

217. D. C. Johnson, II, and T. S. Widlanski, *Tetrahedron Lett.*, **45**, 8483 (2004).

218. J. Morris, and D. G. Wishka, *J. Org. Chem.*, **56,** 3549 (1991).

219. B. Hill, Y. Liu, and S. D. Taylor, *Org. Lett.*, **6,** 4285 (2004).

220. G. Videnov, B. Aleksiev, M. Stoev, T. Paipanova, and G. Jung, *Liebigs Ann. Chem.* 941 (1993).

221. D. M. T. Chan, K. L. Monaco, R.-P. Wang, and M. P. Winters, *Tetrahedron Lett.*, **39,** 2933 (1998).

222. P. Remuzon, C. Dussy, J. P. Jacquet, M. Soumeillant, and D. Bouzard, *Tetrahedron Lett.*, **36,** 6227 (1995).

8

PROTECTION FOR THE ALKYNE −CH

Protection of an acetylenic hydrogen is often necessary because of its acidity. The bulk of a silane can protect an acetylene against catalytic hydrogenation because of rate differences between an olefin (primary or secondary) vs. the more hindered protected alkyne.[1] Trialkylsilylacetylenes are often used as a convenient method for introduction of an acetylenic unit because they tend to be easily handled liquids or solids as opposed to gaseous acetylene.

Trialkylsilylacetylenes

Formation

1. Trialkylsilanes are usually formed by addition of a lithium or Grignard reagent to the silyl chloride,[2] and thus discussions related to formation of the

Greene's Protective Groups in Organic Synthesis, Fourth Edition, by Peter G. M. Wuts and Theodora W. Greene
Copyright © 2007 John Wiley & Sons, Inc.

silyl acetylene bond will be kept to a minimum. Silyl acetylenes are prepared from the alkynylcopper(I) reagents in the presence of PPh₃, Zn, or TMEDA in CH₃CN at 100°C, 36–98% yield.[3] It is interesting to note that the reaction can be reversed to give the alkynylcopper(I) reagent in the presence of CuCl and 1,3-dimethyl-2-imidazolidinone.[4]

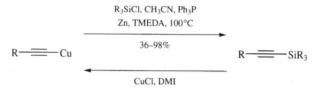

2. Et₂NSiR₃, ZnCl₂, 1,4-dioxane, 100°C, 68–97% yield. This method works for the TMS, TES, and the SiMe₂Ph derivatives but does not work to introduce a TBDMS group.[5]
3. TMSCl, Zn(OTf)₂, TEA, CH₂Cl₂, 75–99% yield. The TES and (*i*-Bu)₃Si derivatives can also be formed using this method but the triphenylsilylalkyne could not be formed.[6]

Trimethylsilylalkyne (TMS−alkyne)

Cleavage

1. KF, MeOH, 50°C, 89% yield.[7,8]
2. AgNO₂, 2,6-lutidine, 90% yield.[9]
3. AgNO₃, MeOH, H₂O, 24°C, cool to 0°C, add KCN, then HCl, 96% yield.[10,11] The reduced electron density of the propargylic alkyne directs the electrophilic silver to the other alkyne and activates it for cleavage. These conditions also resulted in the removal of a primary TBDMS group.[12] AgOTf can also be used, but other inert salts such as AgCl are ineffective.[13] A procedure that does not require the use of cyanide has been developed. The process uses water as a cosolvent with acetone. Since nitric acid is generated in the reaction, TBDMS ethers were also cleaved.[14]

4. AgNO₃, KI, >82% yield. These conditions resulted in partial cleavage of a secondary TES group as well.[15]
5. Bu₄NF, THF, rt, quant.[16]
6. Bu₄NF, 0.4 eq., THF, MeOH, −20°C to −10°C, 98% yield.[17]

7. K_2CO_3, MeOH[16] or KOH, MeOH, 76%, 99% yield.[18-20] Under basic conditions such as these, the more electron-deficient silylalkyne will be cleaved faster.[21]

Very electron-deficient TMS acetylenes such as eynones are unstable and lose the TMS group upon stirring in MeOH.[22]

8. KF, 18-crown-6, aq. THF, 88% yield.[23]

In a similar example, a trimethylsilyl group was cleaved with NaOH, MeOH, H_2O in the presence of a triethylgermyl group.[24] The triethylgermyl group can also be cleaved with methanolic $HClO_4$; the rate increases with increasing electron density.[25]

9. $Na(MeO)_3BH$, THF, H_2O, −20°C, 2.5 h, 60% yield + 20% starting material (SM).[8]

10. MeLi/LiBr.[26]

11. Amberlyst basic resin, MeOH, 80–98% yield.[27] These conditions remove the TMS group in the presence of a secondary TES and TBS.[28]

12. LiOH, THF, H_2O, 1 h, 98% yield. A TIPS alkyne is stable to these conditions.[29]

[(3-Cyanopropyl)dimethylsilyl]-alkyne (CPDMS-alkyne)

This derivative was prepared as a polar analog of the TMS group to facilitate chromatographic purification. It is cleaved using conditions that cleave the TMS group.[30]

Triethylsilylalkyne (TES-alkyne)

The relative rates of cleavage in aqueous, methanolic alkali at 29.4°C for the following silanes are: PhC≡CSiMe₃ / PhC≡CSiEtMe₂ /PhC≡CSiEt₂Me/ PhC≡CSiEt₃/ PhC≡CSiPh₃, 277: 49: 7.4: 1: 11.8.[31] A TES group can be cleaved selectively in the presence of a TBDMS group (t-BuOK, MeOH, 40°C, 65%).[10] A bis TES derivative can be selectively cleaved.[32]

t-Butyldimethylsilylalkyne and Thexyldimethylsilylalkyne (TBDMS- and TDS-alkyne)

Formation

1. For the TBDMS group, KHMDS, THF, TBDMSOTf, −78°C, 98% yield.[10] The TDS group behaves similarly, except that it is slightly more hindered. LHMDS can also be used as a base.[33]

2. TBDMSH, Ir₄(CO)₁₂, Ph₃P, 120°C, 40h, 95% yield. This method works for the introduction of other common silyl ethers such as the TES derivative. The problem with the method is that in some cases, hydrosilylation occurs to form vinylsilanes.[34]

Cleavage

1. Bu₄NF, THF, −23°C, 75% yield.[35,36]

2. Bu₄NF, 2-nitrophenol, THF, 0–23°C, 87% yield. The 2-nitrophenol was added as a weak acid (pK_a = 7.22) to prevent the elimination of a vinyl bromide.[33]

Benzyldimethylsilylalkyne (BDMS−alkyne): $C_6H_5CH_2Si(CH_3)_2$-alkyne

Benzyldimethylsilylacetylene was prepared by the reaction of HC≡CMgBr with the silyl chloride as part of a Fostriecin synthesis.[37]

Dimethyl[1,1-dimethyl-3-(tetrahydro-2*H*-pyran-2-yloxy)propylsilylalkyne)], (DOPS-alkyne)

Cleavage

THF, 0.1 eq. BuLi, −78°C, 2.5 h; −20°C, 2 h.[16]

Protection of the OH with an alcohol protective group gives this approach considerable versatility.

Biphenylyldimethylsilylalkyne (BDMS-alkyne)

Formation

BuLi, BDMSCl, THF, 75–98% yield. The advantage of this group is that many of the derivatives tend to be crystalline and thus provide a safe alternative for purification. Some smaller silylalkynes have been reported to explode upon distillation.[38]

Cleavage

K_2CO_3, MeOH, 72–98% yield. Cleavage occurs selectively in the presence of biphenyldiisopropylalkyne.[38]

Triisopropylsilylalkyne (TIPS-alkyne)

Cleavage

TBAF, THF, H_2O, 20°C, 99% yield.[39,40]

Biphenyldiisopropylsilylalkyne (BDIPS-alkyne)

Formation

BuLi, BDIPSCl, THF, 81% yield.[38]

Cleavage

The cleavage of this group is reported to be similar to the triisopropylsilyl analog.[38]

2-(2-Hydroxypropyl)alkyne: alkyne-CMe₂OH

alkyne-CMe_2OH

Hydroxymethylalkyne: alkyne-CH₂OH

alkyne-CH_2OH

Formation

In this case the low-cost 2-methyl-2-hydroxy-3-butyne is used as a convenient source of acetylene.

Cleavage

1. NaOH, benzene, reflux, >96% yield.[41–43]

Ref. 41

2. For the hydroxylmethyl derivative: MnO_2, KOH, Et_2O, rt, 88% yield.[44]

1. C. J. Palmer and J. E. Casida, *Tetrahedron Lett.*, **31**, 2857 (1990).

2. For a review of the synthesis of silyl and germanyl alkynes, see W. E. Davidsohn and M. C. Henry, *Chem. Rev.*, **67**, 73 (1967).

3. H. Sugita, Y. Hatanaka, and T. Hiyama, *Chem. Lett.*, **25**, 379 (1996).

4. H. Ito, K. Arimoto, H.-o. Senusui, and A. Hosomi, *Tetrahedron Lett.*, **38**, 3977 (1997).

5. A. A. Andreev, V. V. Konshin, N. V. Komarov, M. Rubin, C. Brouwer, and V. Gevorgyan, *Org. Lett.*, **6**, 421 (2004).

6. H. Jiang and S. Zhu, *Tetrahedron Lett.*, **46**, 517 (2005).

7. T. Saito, M. Morimoto, C. Akiyama, T. Matsumoto, and K. Suzuki, *J. Am. Chem. Soc.*, **117**, 10757 (1995).

8. A. G. Myers, P. M. Harrington, and E. Y. Kuo, *J. Am. Chem. Soc.*, **113**, 694 (1991).

9. E. M. Carreira and J. Du Bois, *J. Am. Chem. Soc.*, **117**, 8106 (1995).

10. J. Alzeer and A. Vasella, *Helv. Chim. Acta*, **78**, 177 (1995).

11. E. J. Corey and H. A. Kirst, *Tetrahedron Lett.*, **9**, 5041 (1968).

12. R. A. Pilli, M. M. Victor, and A. deMeijere, *J. Org. Chem.*, **65**, 5910 (2000).

13. A. Orsini, A. Viterisi, A. Bodlenner, J.-M. Weibel, and P. Pale, *Tetrahedron Lett.*, **46**, 2259 (2005).

14. A. Carpita, L. Mannocci, and R. Rossi, *Eur. J. Org. Chem.*, **70**, 1859 (2005).

15. L. K. Geisler, S. Nguyen, and C. J. Forsyth, *Org. Lett.*, **6**, 4159 (2004).

16. C. Cai and A. Vasella, *Helv. Chim. Acta*, **78**, 732 (1995).

17. T. Nishikawa, A Ino, and M. Isobe, *Tetrahedron*, **50**, 1449 (1994).

18. L. T. Scott, M. J. Cooney, and D. Johnels, *J. Am. Chem. Soc.*, **112**, 4054 (1990).

19. Y.-F. Lu and A. G. Fallis, *Tetrahedron Lett.*, **34**, 3367 (1993).

20. M. B. Nielsen and F. Diederich, *Synlett*, 544 (2002).

21. C. Eaborn, R. Eastmond, and D. R. M. Walton, *J. Chem. Soc.* (B) 127 (1971); J. Alzeer and A. Vasella, *Helv. Chem. Acta*, 78, 1219 (1996).

22. T. Nishikawa, D. Urabe, K. Yoshida, T. Iwabuchi, M. Asai, and M. Isobe, *Org. Lett.*, 4, 2679 (2002).

23. A. Ernst, L. Gobbi, and A. Vasella, *Tetrahedron Lett.*, 37, 7959 (1996).

24. R. Eastmond and D. R. M. Walton, *Tetrahedron*, 28, 4591 (1972).

25. C. Eaborn, R. Eastmond, and D. R. M. Walton, *J. Chem. Soc. (B)* 752 (1970).

26. L. Birkoffer, A. Ritter, and H. Dickopp, *Chem. Ber.* 96, 1473 (1963).

27. J. Bach, R. Berenguer, J. Garcia, T. Loscertales, and J. Vilarrasa, *J. Org. Chem.*, 61, 9021 (1996).

28. K. A. Scheidt, T. D. Bannister, A. Tasaka, M. D. Wendt, B. M. Savall, G. J. Fegley, and W. R. Roush, *J. Am. Chem. Soc.*, 124, 6981 (2002).

29. Y. Tobe, N. Utsumi, K. Kawabata, and K. Naemura, *Tetrahedron Lett.*, 37, 9325 (1996).

30. S. Höger and K. Bonrad, *J. Org. Chem.*, 65, 2243 (2000).

31. C. Eaborn and D. R. M. Walton, *J. Organomet. Chem.* 4, 217 (1965).

32. J. M. Wright and G. B. Jones, *Tetrahedron Lett.*, 40, 7605 (1999).

33. A. G. Myers and S. D. Goldberg, *Angew. Chem. Int. Ed.*, 39, 2732 (2000).

34. R. Shimizu and T. Fuchikami, *Tetrahedron Lett.*, 41, 907 (2000).

35. D. Elbaum, T. B. Nguyen, W. L. Jorgensen, and S. L. Schreiber, *Tetrahedron*, 50, 1503 (1994).

36. A. G. Myers, N. J. Tom, M. E. Fraley, S. B. Cohen, and D. J. Madar, *J. Am. Chem. Soc.*, 119, 6072 (1997).

37. B. M. Trost, M. U. Frederiksen, J. P. N. Papillon, P. E. Harrington, S. Shin, and B. T. Shireman, *J. Am. Chem. Soc.*, 127, 3666 (2005).

38. J. Anthony and F. Diederich, *Tetrahedron Lett.*, 32, 3787 (1991).

39. F. Diederich, Y. Rubin, O. L. Chapman, and N. S. Goroff, *Helv. Chim. Acta*, 77, 1441 (1994).

40. P. Wipf and T. H. Graham, *J. Am. Chem. Soc.*, 126, 15346 (2004).

41. C. S. Swindell, W. Fan, and P. G. Klimko, *Tetrahedron Lett.*, 35, 4959 (1994).

42. S. J. Harris and D. R. M. Walton, *Tetrahedron*, 34, 1037 (1978).

43. J. G. Rodriquez, R. Martin-Villamil, F. H. Cano, and I. Fonseca, *J. Chem. Soc., Perkin Trans I*, 709 (1997).

44. H. Kukula, S. Veit, and A. Godt, Eur. *J. Org. Chem.*, 64, 277 (1999).

9

PROTECTION FOR THE PHOSPHATE GROUP

Greene's Protective Groups in Organic Synthesis, Fourth Edition, by Peter G. M. Wuts and Theodora W. Greene
Copyright © 2007 John Wiley & Sons, Inc.

"Phosphate esters and anhydrides dominate the living world."[1] Major areas of synthetic interest include oligonucleotides[2] (polymeric phosphate diesters), phosphorylated peptides, phospholipids, glycosyl phosphates, and inositol phosphates.[2b,3]

a glycosyl phosphate

D-*myo*-inositol 1,4,5-triphosphate

Agrocin 84[4]

The steps involved in automated oligonucleotide synthesis illustrate current use of protective groups in phosphate chemistry (Scheme 1).[5] Oligonucleotide synthesis involves protection and deprotection of the 5′-OH, the amino groups on adenine, guanine, cytosine, and OH groups on phosphorus.

A difference in the problems associated with the protection and deprotection of phosphoric acid species, compared with the other functionalities in this book (alcohols, phenols, aldehydes and ketones, carboxylic acids, amines, and thiols), lies in the fact that phosphoric acid is tribasic ($pK_1 = 2.12$, $pK_2 = 7.21$, $pK_3 = 12.66$). These large differences in pKa's are reflected in large differences in rates of alkaline hydrolysis of the corresponding esters [e.g., $t_{1/2}$ at 1 M NaOH in water, 35°C: $(CH_3O)_3PO$, 30 min; $(CH_3O)_2PO_2^-$, 11 years].[6] Large differences are often found in the rates of successive removal of blocking groups from phosphate derivatives, especially under nonacidic conditions. Phosphate esters are also hydrolyzed by acid[6] but here the relative rates are closer together.

A consequence of the tribasic nature of phosphoric acid (three OH groups attached to phosphorus) is the increased number of options available in the overall process of conversion of alcohol to protected phosphate. This might be carried out by the sequence

$$ROH \longrightarrow ROPO_3H_2 \longrightarrow ROP(O)(OH){-}O{-}PG$$

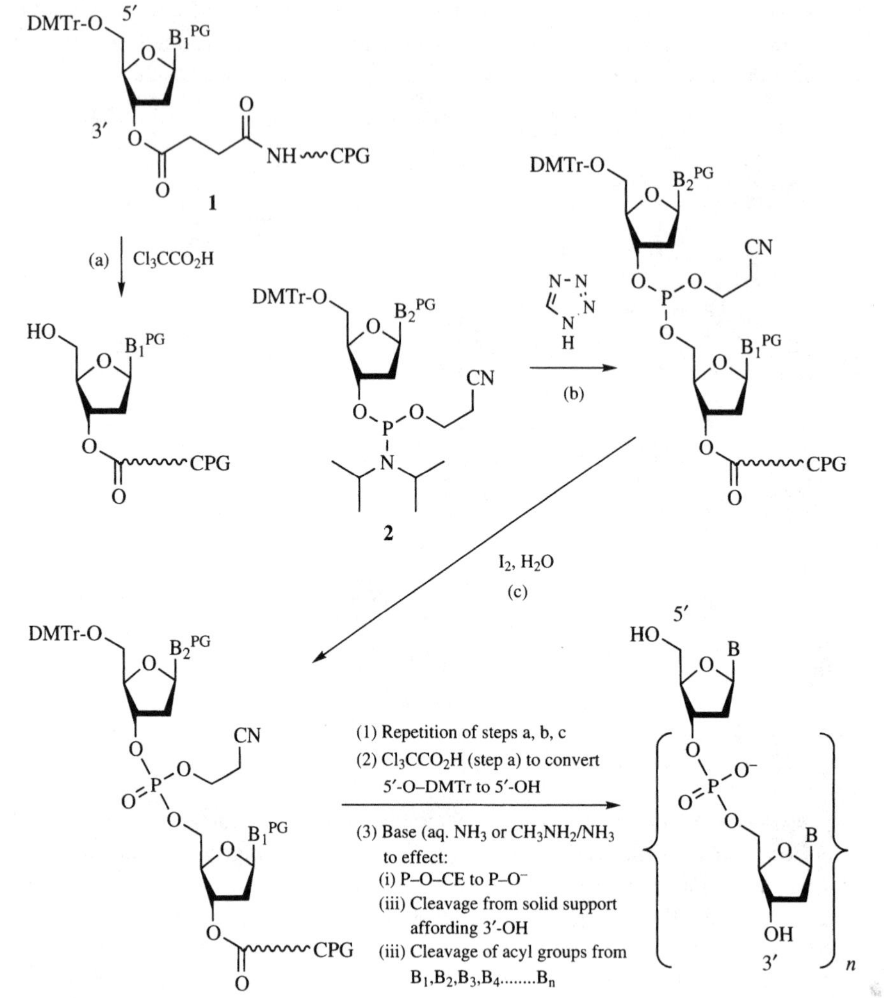

DMTr = 4,4'-dimethoxytrityl

B^{PG} = acetyl, benzoyl, isobutyryl

CPG = "Controlled Pore Glass" (Solid Support)

B_1, B_2, B_3, B_4 = adenyl, cytidyl, guanyl thymidyl

CE = 2-cyanoethyl

1 and 2 (B_1, B_2, B_3, B_4) Commercially available

Scheme 1. Automated Synthesis of Oligonucleotides. Synthetic Cycle for the Phosphoroamidite Method.

or by the formation of the R–O–P attachment *after* the formation of P–O–PG—that is, introduction of the phosphate moiety in a form that is already protected. Another major difference in protection (and deprotection) in the phosphorus area lies in the availability of two major valence states, P(III) and P(V), of this second row element. Both of these aspects [order of formation of the bonds to P and use of P(III) as well as P(V)] are important in current phosphate protection practice.

Phosphate protection may begin at the stage of phosphoryl chloride (phosphorus oxychloride). A protective group may be introduced by reaction of this acid chloride with an alcohol[7] to afford an ester with the desired combination of stability to certain conditions, lability to others.

$$POCl_3 + ROH \longrightarrow ROP(O)Cl_2 \xrightarrow{slower} (RO)_2P(O)Cl \xrightarrow{slower} (RO)_3PO$$

a phosphorodichloridate   a phosphorodichloridate

A disadvantage of phosphoryl chloride reagents is that they are not very reactive but the reactivity can be improved by catalysis with $Ti(O\text{-}t\text{-}Bu)_4$[8]. In the mid 1970s, Letsinger and co-workers introduced a new paradigm that makes use of the more reactive phosphorus(III) reagents.[9] In this approach a monoprotected phosphorodichloridite $(ROPCl_2)$[10,11] is coupled with an alcohol followed by a second condensation with another alcohol to produce a triester. Oxidation with aqueous iodine affords a phosphate.[2,12]

$$ROPCl_2 + R'OH \longrightarrow ROP(OR')Cl \xrightarrow{R''OH} ROP(OR')(OR'') \xrightarrow{I_2, H_2O} ROP(O)(OR')(OR'')$$

The disadvantage of this method is that the dichloridites and monochloridites are sensitive to water and thus cannot be used readily in automated oligonucleotide synthesis. This problem was overcome by Beaucage and Caruthers, who developed the phosphoramidite approach. In this method, derivatives of the form $ROP(NR'_2)_2$ react with one equivalent of an alcohol (catalyzed by species such as $1H$-tetrazole) to form diesters, $R'OP(OR'')NR_2$, which usually are stable, easily handled solids. These phosphoroamidites are easily converted to phosphite triesters by reaction with a second alcohol (catalyzed by $1H$-tetrazole). Certain carboxylic acids have been shown to be good promoters for phosphoramidite couplings.[13] Here, again, oxidation of the phosphite triester with aqueous iodine affords the phosphate triester. Over the years, numerous protective groups and amines have been examined for use in this approach. Much of this work has been reviewed.[2,12] More recent work would indicate that allyl-based protection is superior to some of the older methods that often rely on relatively strong bases for deprotection which can cause side reactions and even internucleotide cleavage to occur. This is especially evident with some of the nonstandard modifications that have been made to the bases and the backbone phosphates. These issues have recently been reviewed.[14]

SOME GENERAL METHODS FOR PHOSPHATE ESTER FORMATION

1. Phosphoric acids may be esterified using an alcohol and an activating agent:
   (a) carbodiimides, e.g., DCC.[15,16]
   (b) arylsulfonyl chloride and a base (TPS, Pyr).[17]
   (c) Various sulfonamido derivatives (ArSO$_2$-Z, Z = 1-imidazolyl, 1-triazolyl, 1-tetrazolyl).[2j,18,19]
   (d) CCl$_3$CN.[20–22]

(e) $SOCl_2$, DMF, $-20°C$, 70–90% yield[23]: $RP(O)(OH)_2 \rightarrow RP(O)(OH)OR$.

(f) $[(Me_2N)_3PBr]^+PF_6^-$, DIEPA, CH_2Cl_2.[24]

2. Nucleophilic (S_N2) reactions for the formation of benzyl, allyl, and certain alkyl phosphates [e.g., $Me_4N^+(RO)_2P(O)O^-$ and an alkyl halide in refluxing DME].[25,26]

3. Reaction of a phosphoric acid with a diazoalkane (CH_2N_2,[21,27] $ArCHN_2$, (N-oxido-α-pyridyl)CHN_2, Ar_2CN_2).[28]

4. Primary alcohols may be phosphorylated by use of the Mitsunobu reaction (Ph_3P, DEAD, HBF_4, Pyr). Of several salts examined, the potassium salt of the phosphate was the best.

5. N-Phosphoryl oxazolidinones are effective phosphorylating agents for a variety of alcohols.[29]

R" = Me or Ph

6. One of the most widely used methods for the formation of phosphate esters involves the conversion of a P-N bond of a phosphorus(III) compound to a P-O bond by ROH, catalyzed by 1H-tetrazole, followed by oxidation to the phosphorus(V) derivative with I_2 or one of several peroxides.[2] The mechanistic aspects of the substitution of phosphoramidites and their congeners have been reviewed.[30]

(a) $R'OH + (R''O)P(NR_2)_2$ $\xrightarrow{\text{1H-tetrazole}}$ $(R'O)P(NR_2)OR''$

      phosphorodiamidite   phosphoramidite

(b) $(R'O)P(NR_2)OR''$ $\xrightarrow[\text{or ROOR}]{\text{I}_2, \text{H}_2\text{O}}$ $(R'O)P(O)(NR_2)OR''$

   phosphoroamidite   phosphoroamidate

7. Preparation of $(MeO)_2P-O-R$: ROH, $(MeO)_3P$, CBr_4, Pyr, 70–98% yield.[31] The alkyl dimethyl phosphite may then be oxidized to the corresponding phosphate by aq. iodine, t-butyl hydroperoxide, or peracid.

REMOVAL OF PROTECTIVE GROUPS FROM PHOSPHORUS

All the approaches for deblocking of protective groups described earlier in this book have found application in the removal of protective groups from phosphorus derivatives. Because phosphate protection and deprotection is commonly associated with compounds that contain acid-sensitive sites (e.g., glycosidic linkages and DMTr−O groups of nucleotides), the most widely used protective groups on phosphorus are those that are deblocked by base.

In the following list, "P^V–O-" stands for phosphorus(V) derivatives—usually $(R^1O)P(O)(OR^2)$–O– in which R^1 and R^2 are not specified:

$$R^1O-\overset{\overset{O}{\|}}{\underset{\underset{R^2O}{|}}{P}}-O-(\text{Protective group}) = \text{``}P^VO-(\text{Protective group})\text{''}$$

1a or **1b**

1a $R^1 = R^2$ = alkyl or aryl

1b R^1 = H, R^2 = alkyl or aryl

1. Groups removed by base (in one step, or the second of two steps).
    (a) One-step removal via β-elimination of various β-substituted ethyl derivatives:
        (i) $P^V-O-CH_2CH_2CN + TEA \rightarrow P^V-O^- + CH_2{=}CHCN$ Ref. 32
        (ii) $P^V-O-CH_2CH_2-SiMe_3 + Bu_4NF, THF \rightarrow P^V\text{-}O^-$ Ref. 4
    (b) Two-step removal:
        (i) oxidation–elimination

$$P^V-O-CH_2CH_2-S-R \xrightarrow{\text{Oxid'n}} P^V-OCH_2CH_2-SO_2R \xrightarrow{\text{base}} P^V-O^- \text{ Ref. 18}$$

        (ii) reduction–elimination

$$P^V-O-CH_2-(\text{2-anthryl-9,10-quinone}) \xrightarrow{\text{Red'n}} \text{corresponding hydroquinone}$$

$$\xrightarrow{\text{base}} P^V-O^- \quad \text{Ref. 33}$$

    (c) Aryl phosphates and strong base. As stated earlier, dialkyl phosphates are quite stable to base. The P^V-O-aryl moiety is more labile to base than the P^V-O-alkyl moiety (hydroxide attack at P and ejection of $Ar-O^-$).

$$P^V\text{-}O\text{-Aryl} + OH^- \longrightarrow P^V-O^- + ArO^- \quad \text{Ref. 34}$$

2. Hydrogenolysis: P^V-O-CH_2Ph, H_2, Pd.[35]
3. Reduction: $P^V-O-CH_2CCl_3$, Zn/Cu, DMF.[36]
4. S_N2 Displacement:
    (a) $P^V-O-CH_2Ph + NaI, CH_3CN \rightarrow P^V\text{-OH}$ (or P^V-O^-).[37]
    (b) $P^V-O-CH_3 + PhS^-$, DMF $\rightarrow P^V-O^- + PhSMe$ Ref. [38]
5. Acid: $P^V-O-t\text{-Bu} + H^+ \rightarrow P^V\text{-OH}$ Ref. 39
6. Photolysis: $P^V-O-R \xrightarrow{h\nu} P^V\text{-OH}$ (or P^V-O^-)   Ref. 40

    R = 3,5-dinitrophenyl, 2-nitrobenzyl, 3,5-dimethoxybenzyl, pyrenylmethyl, desyl, 4-methoxybenzoylmethyl

7. Oxidation: $P^V-O-C_6H_4-p\text{-NHTr}$, I_2, acetone, NH_4OAc.[41]
8. Metal ion catalysis:
    P^V-O-8-quinolinyl, $CuCl_2$, DMSO, $H_2O \rightarrow P^V-O^-$ Ref. 42

9. TMSCl, TMSBr or TMSI: P^v-O-CH_3, TMSI, CH_3CN.[43]

10. Cleavage of P^v-NHR to P^v-OH: $P^v-NH-Ph$, isoamyl nitrite, HOAc.[44]

11. Cleavage of P^v-S-R:

    (a) P^v-S-Et, I_2, Pyr $\rightarrow P^v-O^-$ Ref. 45

    (b) P^v-S-Ph, Zn $\rightarrow P^v-O^-$ Ref. 46

12. Transesterification: conversion of P^v-O-R to P^v-O-R'.

    (a) Transesterification–hydrogenolysis:

$$P^v-O-Ph + Bn-O^- \longrightarrow P^v-O-Bn \xrightarrow{\text{H}_2,\ \text{Pd}} P^v-OH \text{ (or } P^v-O^-)\quad \text{Ref. 47}$$

    (b) Transesterification–elimination:

$$P^v-O-R + R'-CH{=}N-O^- \longrightarrow P^v-O-N{=}CHR \xrightarrow{\text{base}} P^v-O^- + R'CN \quad \text{Ref. 48}$$

13. Electrolysis (has seen little use):

$$P^v-O-CH_2CCl_3 \xrightarrow{\text{electrolytic reduction}} P^v-O^- \quad \text{Ref. 49}$$

The following section primarily describes many of the methods used for the cleavage of some of the more common phosphate protective groups. Since most of these groups are introduced by either the phosphate or phosphite method, little information is included here about their formation. The cited references generally describe the means that were used to introduce the protective group. In some cases, methods of formation are described, but this is done only when alternative methods to the phosphate or phosphite procedure were used.

1. F. W. Westheimer, "Why Nature Chose Phosphates," *Science*, **235**, 1173 (1987).

2. Reviews: (a) S. L. Beaucage and R. P. Iyer, *Tetrahedron*, **48**, 2223 (1992). (b) S. L. Beaucage and R. P. Iyer, *Tetrahedron*, **49**, 10441 (1993). (c) S. L. Beaucage and R. P. Iyer, *Tetrahedron*, **49**, 6123 (1993). (d) R. Cosstick, in Rodd's *Chemistry of Carbon Compounds*, Supplement to the 2nd ed., Suppl. Vol IV, Part L, M. F. Ansell, Ed., Elsevier, New York, 1988, pp. 61–128. (e) H. Kossel and H. Seliger, *Fortschr. Chem. Org. Naturst.*, **32**, 298 (1975). (f) G. C. Crockett, *Aldrichimica Acta*, **16**, 47 (1983). (g) J. W. Engels and E. Uhlmann, *Angew. Chem., Int. Ed. Engl.*, **28**, 716 (1989). (h) V. Amarnath and A. D. Broom, *Chem. Rev.*, **77**, 183 (1977). (i) F. Eckstein, "Protection of Phosphoric and Related Acids," in *Protective Groups in Organic Chemistry*, J. F. W. McOmie, Ed., Plenum Press, New York and London, 1973, p. 217. (j) E. Sonveaux, "The Organic Chemistry Underlying DNA Synthesis," *Bioorg. Chem.*, **14**, 274 (1986). (k) S. L. Beaucage and M. H. Caruthers, "The Chemical Synthesis of DNA/RNA," in *Bioorganic Chemistry: Nucleic Acids*, S. M. Hecht, Ed., Oxford University Press, New York, 1996, Chapter 2, pp. 36–74.

3. B. V. L. Potter and D. Lampe, *Angew. Chem., Int. Ed. Engl.*, **34**, 1933 (1995).

4. T. Moriguchi, T. Wada, and M. Sekine, *J. Org. Chem.*, **61**, 9223 (1996).

5. R. P. Iyer and S. L. Beaucage, *Compr. Nat. Prod. Chem.*, 105 (1999); L. Bellon and F. Wincott, *Solid Phase Synthesis*, 475 (2000); C. B. Reese, *Tetrahedron*, **58**, 8893 (2002).

6. J. R. Cox, Jr., and J. O. B. Ramsay, "Mechanisms of Nucleophilic Substitutions in Phosphate Esters," *Chem. Rev.*, **64**, 317 (1964).

7. A. M. Modro and T. A. Modro, *Org. Prep. Proced. Int.*, **24**, 57 (1992).

8. S. Jones, D. Selitsianos, K. J. Thompson, and S. M. Toms, *J. Org. Chem.*, **68**, 5211 (2003).

9. R. L. Letsinger, J. L. Finnan, G. A. Heavner, and W. B. Lunsford, *J. Am. Chem. Soc.*, **97**, 3278 (1975).

10. C. A. A. Claesen, R. P. A. M. Segers, and G. I. Tesser, *Recl. Trav. Chim. Pays-Bas*, **104**, 119 (1985).

11. K. K. Ogilvie, N. Y. Theriault, J. M. Seifert, R. T. Pon, and M. J. Nemer, *Can. J. Chem.*, **58**, 2686 (1980).

12. C. A. A. Claesen, R. P. A. M. Segers, and G. I. Tesser, *Recl. Trav. Chim. Pays-Bas*, **104**, 209 (1985).

13. Y. Hayakawa, T. Iwase, E. J. Nurminen, M. Tsukamoto, and M. Kataoka, *Tetrahedron*, **61**, 2203 (2005).

14. Y. Hayakawa, *Bull. Chem. Soc. Jpn.*, **74**, 1547 (2001).

15. A. Burger and J. J. Anderson, *J. Am. Chem. Soc.*, **79**, 3575 (1957).

16. W. F. Gilmore and H. A. McBride, *J. Pharm. Sci.*, **63**, 965 (1974).

17. E. Ohtsuka, H. Tsuji, T. Miyake, and M. Ikehara, *Chem. Pharm. Bull.*, **25**, 2844 (1977).

18. C. B. Reese, *Tetrahedron*, **34**, 3143 (1978); R. W. Adamiak, M. Z. Barciszewska, E. Biala, K. Grzeskowiak, R. Kierzek, A. Kraszewski, W. T. Markiewicz, and M. Wiewiorowski, *Nucleic Acids Res.*, **3** 3397 (1976); H. Takaku, M. Kato, and S. Ishikawa, *J. Org. Chem.*, **46**, 4062 (1981).

19. B. L. Gaffney and R. A. Jones, *Tetrahedron Lett.*, **23**, 2257 (1982); M. Sekine, J.-i. Matsuzaki, and T. Hata, *ibid.*, **23**, 5287 (1982).

20. C. Wasielewski, M. Hoffmann, E. Witkowska, and J. Rachon, *Rocz. Chem.*, **50**, 1613 (1976).

21. J. Szewdzyk, J. Rachon, and C. Wasielewski, *Pol. J. Chem.*, **56**, 477 (1982).

22. J. Szewczyk and C. Wasielewski, *Pol. J. Chem.*, **55**, 1985 (1981).

23. M. Hoffmann, *Synthesis*, 557 (1986).

24. N. Galeotti, J. Coste, P. Bedos, and P. Jouin, *Tetrahedron Lett.*, **37**, 3997 (1996).

25. M. Kluba, A. Zwierzak, and R. Gramze, *Rocz. Chem.*, **48**, 227 (1974).

26. A. Zwierzak and M. Kluba, *Tetrahedron*, **27**, 3163 (1971).

27. M. Hoffmann, *Pol. J. Chem.*, **53**, 1153 (1979).

28. G. Lowe and B. S. Sproat, *J. Chem. Soc., Perkin Trans. I*, 1874 (1981).

29. S. Jones and C. Smanmoo, *Tetrahedron Lett.*, **45**, 1585 (2004).

30. E. Nurminen and H. Lonnberg, *J. Phys. Org. Chem.*, **17**, 1 (2004).

31. V. B. Oza and R. C. Corcoran, *J. Org. Chem.*, **60**, 3680 (1995).

32. H. M. Hsiung, *Tetrahedron Lett.*, **23**, 5119 (1982).

33. N. Balgobin, M. Kwiatkowski, and J. Chattopadhyaya, *Chem. Scr.*, **20**, 198 (1982).

34. G. De Nanteuil, A. Benoist, G. Remond, J.-J. Descombes, V. Barou, and T. J. Verbeuren, *Tetrahedron Lett.*, **36**, 1435 (1995).

35. M. M. Sim, H. Kondo, and C.-H. Wong, *J. Am. Chem. Soc.*, **115**, 2260 (1993).

36. J. H. Van Boom, P. M. J. Burgers, R. Crea, G. van der Marel, and G. Wille, *Nucleic Acids Res.*, **4**, 747 (1977).

37. K. H. Scheit, *Tetrahedron Lett.*, **8**, 3243 (1967).

38. B. H. Dahl, K. Bjergaarde, L. Henriksen, and O. Dahl, *Acta Chem. Scand.*, **44**, 639 (1990).

39. J. W. Perich, P. F. Alewood, and R. B. Johns, *Aust. J. Chem.*, **44**, 233 (1991).

40. For a review of phosphate ester photochemistry, see R. S. Givens and L. W. Kueper, III, *Chem. Rev.*, **93**, 55 (1993).

41. E. Ohtsuka, S. Morioka, and M. Ikehara, *J. Am. Chem. Soc.*, **95**, 8437 (1973).

42. H. Takaku, Y. Shimada, and T. Hata, *Chem. Lett.*, **4**, 873 (1975).

43. J. Vepsäläinen, H. Nupponen, and E. Pohjala, *Tetrahedron Lett.*, **34**, 4551 (1993).

44. E. Ohtsuka, T. Ono, and M. Ikehara, *Chem. Pharm. Bull.*, **33**, 3274 (1981).

45. E. Heimer, M. Ahmad, S. Roy, A. Ramel, and A. L. Nussbaum, "Nucleoside *S*-Alkyl Phosphorothioates. VI. Synthesis of Deoxyribonucleotide Oligomers," *J. Am. Chem. Soc.*, **94**, 1707 (1972).

46. M. Sekine, K. Hamaoki, and T. Hata, *Bull. Chem. Soc. Jpn.*, **54**, 3815 (1981).

47. D. C. Billington, R. Baker, J. J. Kulagowski, and I. M. Mawer, *J. Chem. Soc., Chem. Commun.*, 314 (1987).

48. S. S. Jones and C. B. Reese, *J. Am. Chem. Soc.*, **101**, 7399 (1979).

49. J. Engels, *Angew. Chem., Int. Ed. Engl.*, **18**, 148 (1979).

ALKYL PHOSPHATES

Methyl: CH_3-

Formation

1. A phosphonic acid can be esterified with CH_2N_2 in 88–100% yield.[1,2]

2. $(PhO)_2P(O)Cl$, 2 mol % $TiCl_4$, Et_3N, THF, 1 h, 90–98% yield. This is a general method for phosphate formation of a variety of alcohols.[3] $(t-BuO)_4Ti$ is also an effective catalyst.[4]

Cleavage

1. 2-Mercaptobenzothiazole, *N*-methylpyrrolidone, DIPEA. The reagent has the advantage that it is odorless and does not lead to internucleotide cleavage, but the cleavage rate is 10 times slower than when thiophenol is used.[5]

2. Thiophenol, TEA, DMF or dioxane.[6] In the case of dimethyl phosphonates this method can be used to remove selectively only one methyl group.[7] Lithium thiophenoxide is also effective.[8] 2-Methyl-5-*t*-butylthiophenol is an *odorless* replacement for thiophenol.[9]

3. 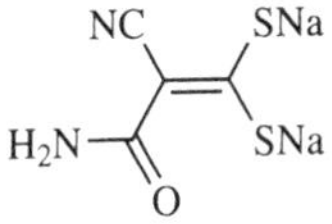 DMF. This odorless and easily prepared reagent is relatively

nonbasic ($pK_B = 8.4$) and cleaves the methyl group about four times faster than thiophenol. It is also used to remove the 2,4-dichlorobenzyl group from phosphates and dithiophosphates.[6]

4. *t*-Butylamine, 46°C, 15 h.[10]

5. Ammonia. Cleavage is not as clean as with thiophenol.[11]

6. Me₃N, toluene, rt, 12 h.[12]

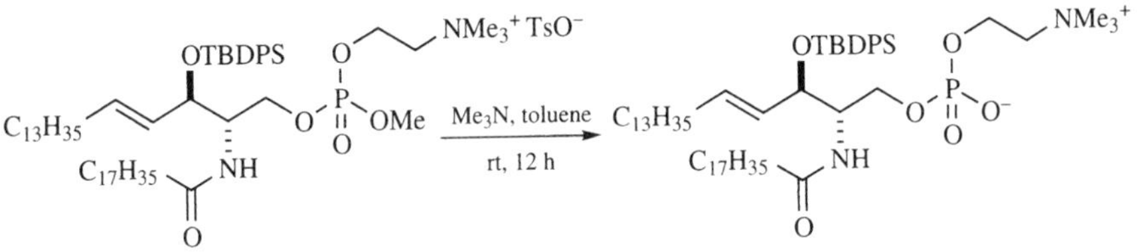

7. 10% Me₃SiBr, CH₃CN, 1–2 h, 25°C, >97% yield.[13,14] This reagent is also useful for the cleavage of ethyl phosphates[15] and phosphonates.[16]

8. BBr₃, toluene, hexane, −30°C to 70°C the MeOH at 20°C, 90% yield. This method will also cleave many other alkyl phosphates with excellent efficiency.[17]

9. 1 *M* Me₃SiBr, thioanisole, TFA.[13,18]

10. 45% HBr, AcOH.[19,20] This method and the use of TMSI were not suitable for the deprotection of phosphorylated serines.[21] Diethyl phosphates are cleaved very slowly.[22]

11. Aqueous pyridine.[23]

12. NaI, acetone.[24,25]

13. LiCN, DMF, rt, 12 h.[26]

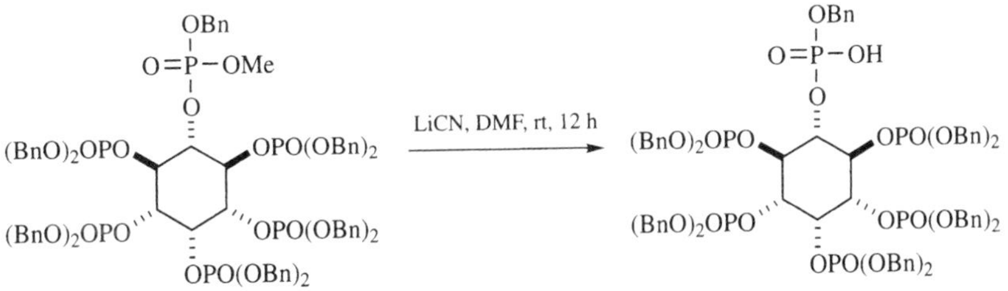

14. The use of TMSOTf and thioanisole results in rapid ($t_{1/2} = 7$ min) cleavage of one methyl in a dimethyl phosphate, whereas the second methyl is cleaved only slowly ($t_{1/2} = 12$ h).[27] The method has been further refined for peptide synthesis.[28]

15. Fmoc chemistry is compatible with methyl phosphates when methanolic K₂CO₃ is used to remove the Fmoc group instead of the usual amines.[29]

16. TMSOK, Et$_2$O, THF or CH$_2$Cl$_2$, 84–98% yield. The reagent also cleaves methyl and ethyl esters.[30] With a mixed ethyl and methyl phosphonate, the methyl ester is cleaved preferencially.

Ethyl: C$_2$H$_5$–

Formation

1. From a phosphinic acid: (EtO)$_4$Si, toluene, reflux, 24 h, 80–100% yield. This method can be used to prepare a variety of phosphinic esters in generally excellent yield.[31]

2. *N,N'*-di-*p*-tolylmethyl pseudourea, benzene, reflux, 2–3 h. The by-product urea is removed by filtration.[32]

Cleavage

1. Ethyl phosphates are usually cleaved by acid hydrolysis.[33]
2. TMSBr, CH$_3$CN.[34]
3. NH$_4$OH, MeOH.[34] These conditions result in cleavage of only one ethyl group of a diethyl phosphonate. Selective monodeprotection of a number of alkyl-protected phosphates is fairly general for cases where cleavage occurs by release of phosphate or phosphonate anions.
4. LiBr has been used to cleave the ethyl group.[35]
5. Et$_3$SiH, 2% (C$_6$F$_5$)$_3$B, toluene, 20°C. This method produces TES phosphates which are readily hydrolyzed.[36]
6. LiN$_3$, DMF, 100°C.[37]

Isopropyl: (CH$_3$)$_2$CH–

A diisopropyl phosphonate is cleaved with TMSBr, TEA, CH$_2$Cl$_2$, rt.[38] Dioxane can also be used as solvent.[39,40]

Cyclohexyl (cHex): C$_6$H$_{11}$–

Cleavage

1. The cyclohexyl phosphate, used in the protection of phosphorylated serine derivatives, is introduced by the phosphoramidite method and cleaved with TFMSA/MTB/*m*-cresol/1,2-ethanedithiol/TFA, 4 h, 0°C to rt.[41]

2. Monocyclohexyl phosphates and phosphonates can be cleaved by a two-step process where the ester is treated with an epoxide such as propylene oxide to form an ester, which upon treatment with base releases the cyclohexyl alcohol.[42]

t-Butyl: $(CH_3)_3C-$

The *t*-butyl phosphate although very stable towards nucleophilic reagents, is extremely susceptible to acidic reagents, which includes chromatography on silica gel.

Cleavage

1. 1 *M* HCl, dioxane, 4 h.[21,43]
2. TFA, water, 7 days, 96% yield.[44]

3. TFA, thiophenol[18] or thioanisole.[45]
4. TMSCl, TEA, CH_3CN, 75°C 2 h.[46]

1-Adamantyl

An adamantyl phosphonate, prepared from adamantyl bromide and Ag_2O, is easily cleaved with TFA in CH_2Cl_2.[47]

Allyl: $CH_2=CHCH_2-$

Typically, the most common method for allyl cleavage is through a Pd-catalyzed process, but in the case of allyl phosphates, nucleophilic reactions are effective and often better because phosphate is such a good leaving group.

Formation[48]

Cleavage

1. Rh(Ph$_3$P)$_3$Cl, acetone, H$_2$O, reflux, 2 h, 86% yield.[49]

2. Pd(Ph$_3$P)$_4$, Ph$_3$P, RCO$_2$K, EtOAc, 25°C, 83% yield.[49,50] Diethylammonium formate,[51] NH$_3$,[52] and BuNH$_2$[53,54] have also been used as allyl scavengers in this process. In a diallyl phosphate, deprotection results in cleavage of only a single allyl group.[55]

3. PdCl$_2$(Ph$_3$P)$_2$, Bu$_3$SnH; ClB(OR)$_2$ then aqueous hydrolysis.[56]

4. Pd$_2$(dba)$_3$-CHCl$_3$, Ph$_3$P, butylamine, formic acid, THF, 50°C, 0.5–1 h.[57]

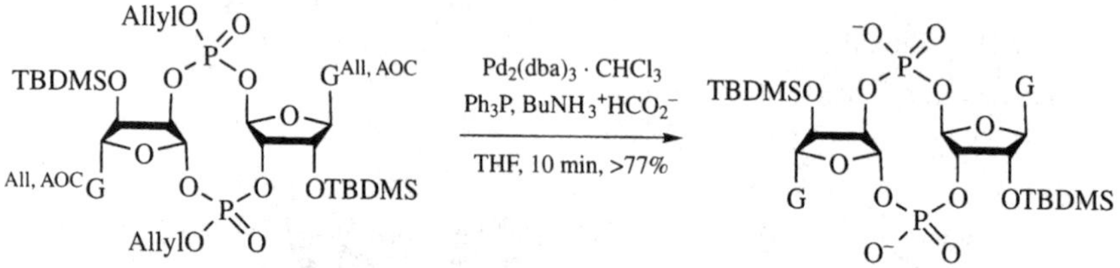

5. Concentrated. ammonia, 70°C.[58]

6. HOCH$_2$CH$_2$SH, NH$_4$OH, 55°C.[59]

7. An allyl phosphate is sufficiently reactive toward nucleophilic reagents that even pyridine can be used to cleave the phosphate, albeit slowly. In this case, stronger bases could not be used because of elimination of phosphate to form a dehydroamino acid.[60]

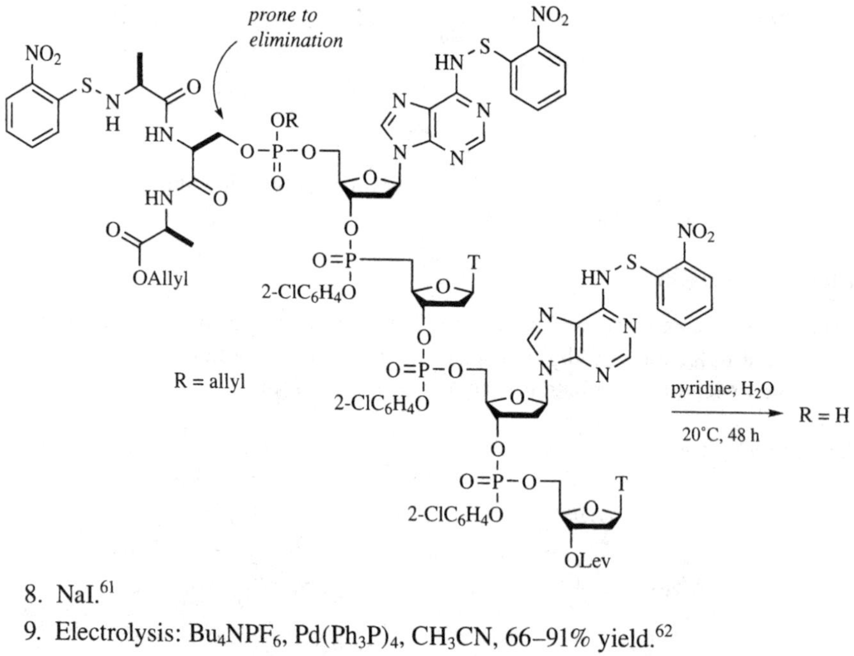

8. NaI.[61]

9. Electrolysis: Bu$_4$NPF$_6$, Pd(Ph$_3$P)$_4$, CH$_3$CN, 66–91% yield.[62]

2-Trimethylsilylprop-2-enyl (TMSP): CH$_2$=C(TMS)CH$_2$−

This derivative is stable to AcOH and methanolic ammonia, but not to 0.5 *N* aq. NaOH.

Cleavage

1. H_2, Pd–C, EtOH.[63]
2. Et_4NF, CH_3CN, 48 h, reflux. TMSF and allene are formed in the cleavage reaction. These conditions are not compatible with phenyl phosphates, which are cleaved preferentially with fluoride.[63] Cleavage of a bis-TMSP phosphate results in cleavage of only one of the TMSP groups.

Hexafluoro-2-butyl (HFB−): $(CF_3)_2CHCH_2-$

Prepared for use in the phosphoramidite approach, the amidite reagent, $(CF_3)_2CHCH_2OP(NiPr)_2$ is stable to distillation unlike the cyanoethyl version which tends to decompose. It is cleaved rapidly with ammonia from the internucleotidic bonds.[64]

Ethylene Glycol Derivative

Cleavage

NaCN, DMSO, rt, 18 h, followed by NaOH, EtOH, rt 2 h.[65]

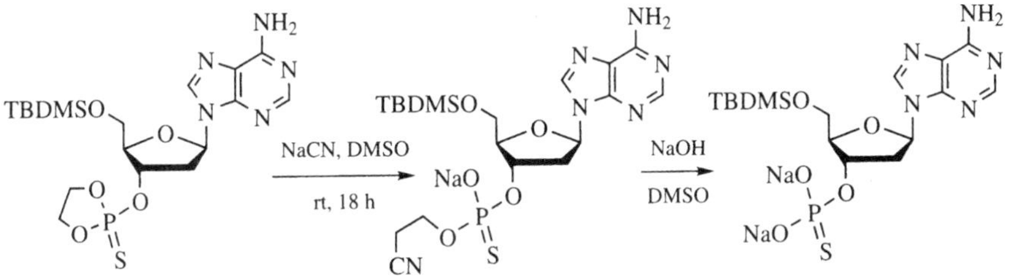

2-Mercaptoethanol Derivative

Cleavage

$HOCH_2CH_2CN$, DBU, CH_3CN, 70–93% yield.[66]

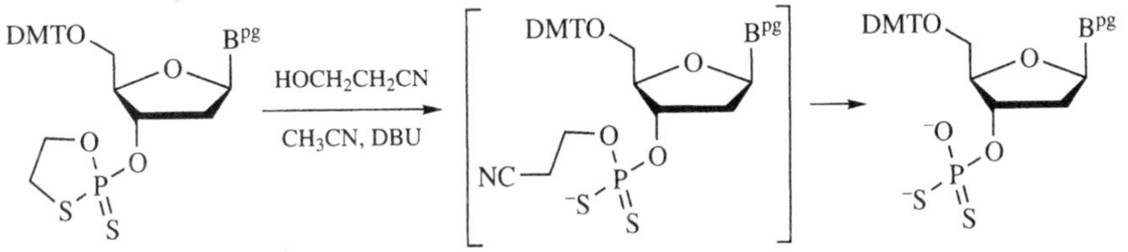

3-Pivaloyloxy-1,3-dihydroxypropyl Derivative

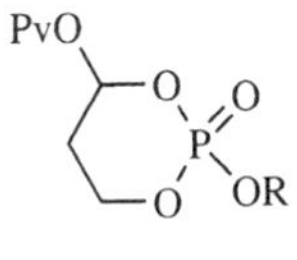

This group was designed as an enzymatically cleavable protective group. Cleavage is achieved using an esterase present in mouse plasma or hog liver carboxylate esterase.[67]

1. M. Hoffmann, *Pol. J. Chem.*, **53**, 1153 (1979).

2. J. Szewdzyk, J. Rachon, and C. Wasielewski, *Pol. J. Chem.*, **56**, 477 (1982).

3. S. Jones and D. Selitsianos, *Org. Lett.*, **4**, 3671 (2002).

4. S. Jones, D. Selitsianos, K. J. Thompson, and S. M. Toms, *J. Org. Chem.*, **68**, 5211 (2003).

5. A. Andrus and S. L. Beaucage, *Tetrahedron Lett.*, **29**, 5479 (1988).

6. B. H. Dahl, K. Bjergaarde, L. Henriksen, and O. Dahl, *Acta Chem. Scand.*, **44**, 639 (1990).

7. B. Müller, T. J. Martin, C. Schaub, and R. R. Schmidt, *Tetrahedron Lett.*, **29**, 509 (1998).

8. G. W. Daud and E. E. van Tamelen, *J. Am. Chem. Soc.*, **99**, 3526 (1977).

9. R. K. Kumar, D. L. Cole, and V. T. Ravikumar, *Nucleosides, Nucleotides & Nucleic Acids*, **22**, 453 (2003).

10. D. J. H. Smith, K. K. Ogilvie, and M. F. Gillen, *Tetrahedron Lett.*, **21**, 861 (1980).

11. T. Tanaka and R. L. Letsinger, *Nucleic Acids Res.*, **10**, 3249 (1982).

12. K. S. Bruzik, *J. Chem. Soc., Perkin Trans 1*, 423 (1988)

13. R. M. Valerio, J. W. Perich, E. A. Kitas, P. F. Alewood, and R. B. Johns, *Aust. J. Chem.*, **42**, 1519 (1989).

14. C. E. McKenna, M. T. Higa, N. H. Cheung, and M.-C. McKenna, *Tetrahedron Lett.*, 155 (1977).

15. A. Holy, *Collect. Czech. Chem. Comm.*, **54**, 446 (1989).

16. L. Qiao and J. C. Vederas, *J. Org. Chem.*, **58**, 3480 (1993).

17. N. Gauvry and J. Mortier, *Synthesis*, 553 (2001).

18. E. A. Kitas, R. Knorr, A. Trzeciak, and W. Bannwarth, *Helv. Chim. Acta*, **74**, 1314 (1991).

19. P. Kafarski, B. Lejczak, P. Mastalerz, J. Szweczyk, and C. Wasielewski, *Can. J. Chem.*, **60**, 3081 (1982).

20. J. Zygmunt, P. Kafarski, and P. Mastalerz, *Synthesis*, 609 (1978).

21. J. W. Perich, P. F. Alewood, and R. B. Johns, *Aust. J. Chem.*, **44**, 233 (1991).

22. R. M. Valerio, P. F. Alewood, R. B. Johns, and B. E. Kemp, *Int. J. Pept. Protein Res.*, **33**, 428 (1989).

23. H. Vecerkova and J. Smrt, *Collect. Czech. Chem. Comm.*, **48**, 1323 (1983).

24. D. V. Patel, E. M. Gordon, R. J. Schmidt, H. N. Weller, M. G. Young, R. Zahler, M. Barbacid, J. M. Carboni, J. L. Gullo-Brown, L. Hunihan, C. Ricca, S. Robinson, B. R. Seizinger, A. V. Tuomari, and V. Manne, *J. Med. Chem.*, **38**, 435 (1995).

25. J. M. Delfino, C. J. Stankovic, S. L. Schreiber, and F. M. Richards, *Tetrahedron Lett.*, **28**, 2323 (1987).

26. K. M. Reddy, K. K. Reddy, and J. R. Falck, *Tetrahedron Lett.*, **38**, 4951, (1997).

27. E. A. Kitas, J. W. Perich, G. W. Tregear, and R. B. Johns, *J. Org. Chem.*, **55**, 4181 (1990).

28. A. Otaka, K. Miyoshi, M. Kaneko, H. Tamamura, N. Fujii, M. Momizu, T. R. Burke, Jr., and P. P. Roller, *J. Org. Chem.*, **60**, 3967 (1995).

29. W. H. A. Kuijpers, J. Huskens, L. H. Koole, and C. A. A. Van Boekel, *Nucleic Acids Res.*, **18**, 5197 (1990).

30. J. Dziemidowicz, D. Witt, M. Sliwka-Kaszynska, and J. Rachon, *Synthesis*, 569 (2005).

31. Y. R. Dumond, R. L. Baker, and J.-L. Montchamp, *Org. Lett.*, **2**, 3341 (2000).

32. H. G. Khorana, *Can. J. Chem.*, **32**, 227 (1954).

33. J. L. Kelley, E. W. McLean, R. C. Crouch, D. R. Averett, and J. V. Tuttle, *J. Med. Chem.*, **38**, 1005 (1995).

34. J. Matulic-Adamic, P. Haeberli, and N. Usman, *J. Org. Chem.*, **60**, 2563 (1995).

35. H. Krawczyk, *Synth. Commun.*, **27**, 3151 (1997).

36. J.-M. Denis, H. Forintos, H. Szelke, and G. Keglevich, *Tetrahedron Lett.*, **43**, 5569 (2002).

37. A. Holý, *Synthesis*, 381 (1998).

38. J.-L. Montchamp, L. T. Piehler, and J. W. Frost, *J. Am. Chem. Soc.*, **114**, 4453 (1992).

39. C. J. Salomon and E. Breuer, *Tetrahedron Lett.*, **36**, 6759 (1995).

40. P. Wainwright, A. Maddaford, R. Bissell, R. Fisher, D. Leese, A. Lund, K. Runcie, P. S. Dragovich, J. Gonzalez, P.-P. Kung, D. S. Middleton, D. C. Pryde, P. T. Stephenson, and S. C. Sutton, *Synlett*, 765 (2005).

41. T. Wakamiya, K. Saruta, J.-i. Yasuoka, and S. Kusumoto, *Bull. Chem. Soc. Jpn.*, **68**, 2699 (1995).

42. M. Sprecher, R. Oppenheimer, and E. Nov, *Synth. Commun.*, **23**, 115 (1993).

43. J. W. Perich and R. B. Johns, *Synthesis*, 142 (1988).

44. A. Burger, D. Tritsch, and J. F. Biellmann, *Carbohydr. Res.*, **332**, 141 (2001).

45. J. M. Lacombe, F. Andriamanampisoa, and A. A. Pavia, *Int. J. Pept. Protein Res.*, **36**, 275 (1990).

46. M. Sekine, S. Iimura, and T. Nakanishi, *Tetrahedron Lett.*, **32**, 395 (1991).

47. A. Yiotakis, S. Vassiliou, J. Jiracek, and V. Dive, *J. Org. Chem.*, **61**, 6601 (1996).

48. K. Miyashita, H. Ikejiri, H. Kawasaki, S. Maemura, and T. Imanishi, *J. Am. Chem. Soc.*, **125**, 8238 (2003).

49. M. Kamber and G. Just, *Can. J. Chem.*, **63**, 823 (1985).

50. D. B. Berkowitz and D. G. Sloss, *J. Org. Chem.*, **60**, 7047 (1995).

51. Y. Hayakawa, H. Kato, T. Nobori, R. Noyori, and J. Imai, *Nucl. Acids Symp. Ser.*, **17**, 97 (1986).

52. W. Bannwarth and E. Küng, *Tetrahedron Lett.*, **30**, 4219 (1989).

53. Y. Hayakawa, M. Uchiyama, H. Kato, and R. Noyori, *Tetrahedron Lett.*, **26**, 6505 (1985).

54. T. Pohl and H. Waldmann, *J. Am. Chem. Soc.*, **119**, 6702 (1997).

55. A. Sawabe, S. A. Filla, and S. Masamune, *Tetrahedron Lett.*, **33**, 7685 (1992).

56. H. X. Zhang, F. Guibé, and G. Balavoine, *Tetrahedron Lett.*, **29**, 623 (1988).

57. Y. Hayakawa, S. Wakabayashi, H. Kato, and R. Noyori, *J. Am. Chem. Soc.*, **112**, 1691 (1990). Y. Hayakawa, R. Nagata, A. Hirata, M. Hyodo, and R. Kawai, *Tetrahedron*, **59**, 6465 (2003).

58. F. Bergmann, E. Kueng, P. Iaiza, and W. Bannwarth, *Tetrahedron*, **51**, 6971 (1995).

59. M. Manoharan, Y. Lu, M. D. Casper, and G. Just, *Org. Lett.*, **2**, 243 (2000).

60. E. M. T. C. Kuyl-Yeheskeily, A. W. M. Lefeber, G. A. van der Marel, and J. H. van Boom, *Tetrahedron*, **44**, 6515 (1988).

61. Y. Hayakawa, M. Hirose, and R. Nyori, *Nucleosides & Nucleotides*, **8**, 867 (1989).

62. Y. Hayakawa, R. Kawai, S. Wakabayashi, M. Uchiyama, and R. Noyori, *Nucleosides & Nucleotides*, **17**, 441 (1998).

63. T.-H. Chan and M. Di Stefano, *J. Chem. Soc., Chem. Commun.*, 761 (1978).

64. S.-G. Kim, K. Eida, and H. Takaku, *Bioorg. Med. Chem. Lett.*, **5**, 1663 (1995).

65. M. B. Szczepanik, L. Desaubry, and R. A. Johnson, *Tetrahedron Lett.*, **39**, 7455 (1998).

66. M. Olesiak, D. Krajewska, E. Wasilewska, D. Korczynski, J. Baraniak, A. Okruszek, and W. J. Stec, *Synlett*, 9671 (2002).

67. D. Farquhar, S. Khan, M. C. Wilkerson, and B. S. Andersson, *Tetrahedron Lett.*, **36**, 655 (1995).

PHOSPHATES CLEAVED BY CYCLODEESTERIFICATION

4-Methylthio-1-butyl: $CH_3SCH_2CH_2CH_2CH_2-$

4-Methylthio-1-butyl group is prepared by the standard phosphoramidite method. Oxidation must be done using I_2 in pyridine rather than hydroperoxides because these will also oxidize the sulfide to the sulfoxide. Cleavage is accomplished by heating the phosphate ester to 55°C for 30 min.[1]

4-[_N_-Methyl-_N_-(2,2,2-trifluoroacetyl)amino]butyl:
$CF_3CONHCH_2CH_2CH_2CH_2-$

This group was developed as an alternative to the cyanoethyl group because of the toxicity associated with the acrylonitrile that is released during deprotection and the problem of nucleobase alkylation with released acrylonitrile. This group is introduced using the phosphoramidite method. It is cleaved by rate limiting aminolysis with concentrated ammonium hydroxide.[2] This group is stable to strong nonnucleophilic bases under anhydrous conditions.

4-(*N*-Trifluoroacetylamino)butyl: $CF_3C(O)NH(CH_2)_4-$

Ammonia treatment removes the TFA group, which then through intramolecular cyclization releases the phosphate and pyrrolidine. The analogous pentyl derivative was also prepared but the cleavage rate was slower.[3]

2-(*S*-Acetylthio)ethyl (SATE): $CH_3C(O)SCH_2CH_2-$

The SATE group is compatible with the fluoride labile trimethylsilylethyl and the [*t*-butyldiphenylsiloxymethyl]benzoyl groups during oligonucleotide synthesis.[4]

Formation

The SATE ester is formed from a phosphite using PvCl activation followed by oxidation to the phosphate with I_2/H_2O.[5,6]

Cleavage

1. Enzymatic hydrolysis exposes the sulfide that undergoes episulfide formation by cyclodeesterification releasing the phosphate.[5] This method was developed for intracellular delivery of a monophosphate. This concept was also extended to the use of an *S*-glucoside (GTE group) that could be activated by a glucosidase to release the thiol.[7]

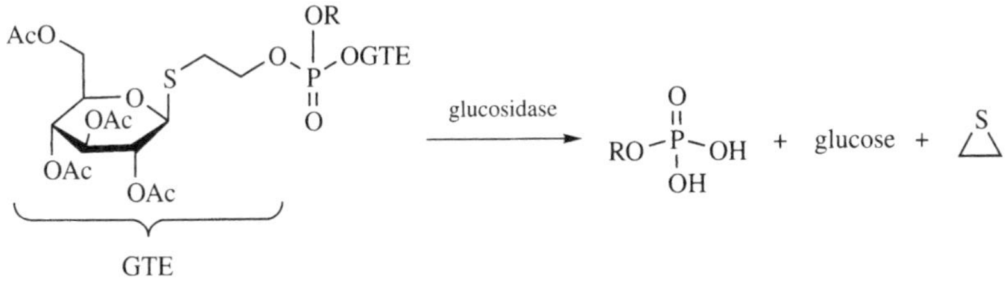

2. Hydrolysis of the thioester of $(EtO)_2P(S)SCH_2CH_2SC(O)R$ (R = Bz was preferred) with ammonia gives $(EtO)_2P(S)S^-$ again, by episulfide formation.[8]

4-Oxopentyl: $CH_3C(O)CH_2CH_2CH_2-$

The 4-oxopentyl group, introduced using the phosphoramidite method, is cleaved using either concentrated ammonia or gaseous ammonia at 10 bar. The ammonia adds to the carbonyl, which initiates the cyclodeesterification process.[9]

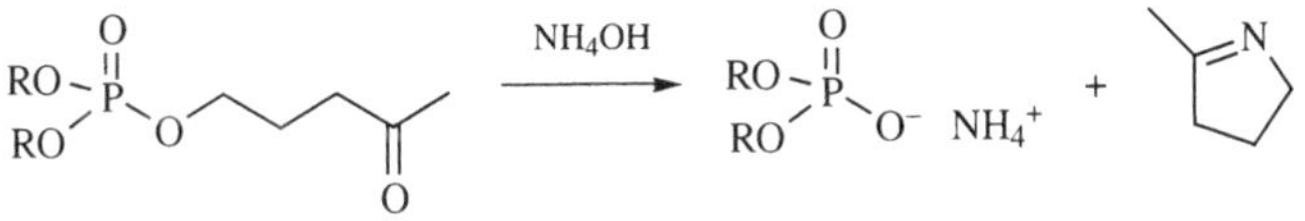

3-(*N*-*t*-Butylcarboxamido)-1-propyl: $(CH_3)_3CNHC(O)CH_2CH_2CH_2-$

Introduced via the phosphoramidite method the 3-(*N*-*t*-butylcarboxamido)-1-propyl group is cleaved thermally by the following process. It was prepared as an alternative to the cyanoethyl group.[10]

3-(Pyridyl)-1-propyl and 2-[N-Methyl-N-(2-pyridyl)]aminoethyl

These groups are introduced using the standard phosphoramidite method. The 3-(pyridyl)-1-propyl group is cleaved from the phosphate within 30 min upon heating at 55°C in concentrated ammonium hydroxide or in an aqueous buffer at pH 7.0, whereas cleavage of the 2-[N-methyl-N-(2-pyridyl)]aminoethyl group occurs spontaneously upon oxidation of the phosphite to phosphate during oligonucleotide synthesis.[11]

2-(2-Pyridyl)-1-propyl   2-N-Methyl-N-(2-pyridyl)]aminoethyl

2-(N-Formyl-N-methyl)aminoethyl

The phosphoramidite method was used to introduce this group. It was developed as a low-cost alternative to the 4-[N-methyl-N-(2,2,2-trifluoroacetyl)amino]butyl group. It is cleaved thermally at 90°C and at pH 7 in 3 h.[12]

2-(N-Isopropyl-N-anisoylamino)ethyl

This group is similar to the 2-(N-formyl-N-methyl)aminoethyl group and is cleaved similarly from a phosphate in CH$_3$CN with a $t_{1/2}$ = 50 min.[13]

2-[(1-Naphthyl)carbamoyloxy]ethyl

Prepared by the phosphoramidite method, this group is cleaved with aqueous ammonium hydroxide at 55°C in 5 h, giving the oxazolidinone and the released phosphate.[14]

2-[*N*-Isopropyl-*N*-(4-methoxybenzoyl)amino]ethyl

This group was one of a family of groups studied to determine if the rates of deprotection could be modified by various substitutions on the backbone. Of the 11 groups studied, the 2-[*N*-isopropyl-*N*-(4-methoxybenzoyl)amino]ethyl group proved to be one of the most easily removed. It was successfully used in the preparation of an oligonucleotide 20-mer. It is rapidly cleaved at 25°C.[15]

1. J. Cieslak, A. Grajkowski, V. Livengood, and S. L. Beaucage, *J. Org. Chem.*, **69**, 2509 (2004).

2. A. Wilk, A. Grajkowski, L. R. Phillips, and S. L. Beaucage, *J. Org. Chem.*, **64**, 7515 (1999).

3. A. Wilk, K. Srinivasachar, and S. L. Beaucage, *J. Org. Chem.*, **62**, 6712 (1997).

4. T. Guerlavais-Dagland, A. Meyer, J.-L. Imbach, and F. Morvan, *Eur. J. Org. Chem.*, 2327 (2003).

5. C. Périgaud, G. Gosselin, I. Lefebvre, J. L. Girardet, S. Benzaria, I. Barber, and J. L. Imbach, *Bioorg. Med. Chem. Lett.*, **3**, 2521 (1993).

6. For a brief review: C. Perigaud, G. Gosselin, and J.-L. Imbach, *Biomed. Chem.*, 115 (2000).

7. N. Schlienger, C. Perigaud, G. Gosselin, and J.-L. Imbach, *J. Org. Chem.*, **62**, 7216 (1997).

8. W. T. Wiesler and M. H. Caruthers, *J. Org. Chem.*, **61**, 4272 (1996).

9. A. Wilk, M. K. Chmielewski, A. Grajkowski, L. R. Phillips, and S. L. Beaucage, *Tetrahedron Lett.*, **42**, 5635 (2001).

10. A. Wilk, M. K. Chmielewski, A. Grajkowski, L. R. Phillips, and S. L. Beaucage, *J. Org. Chem.*, **67**, 6430 (2002).

11. J. Cieslak and S. L. Beaucage, *J. Org. Chem.*, **68**, 10123 (2003).

12. A. Grajkowski, A. Wilk, M. K. Chmielewski, L. R. Phillips, and S. L. Beaucage, *Org. Lett.*, **3**, 1287 (2001).

13. A. P. Guzaev and M. Manoharan, *Nucleosides, Nucleotides & Nucleic Acids*, **20**, 1011 (2001).

14. A. P. Guzaev and M. Manoharan, *Tetrahedron Lett.*, **41**, 5623 (2000).

15. A. P. Guzaev and M. Manoharan, *J. Am. Chem. Soc.*, **123**, 783 (2001); A. P. Guzaev and M. Manoharan, *Org. Lett.*, **3**, 3071 (2001).

2-Substituted Ethyl Phosphates

2-Cyanoethyl: $NCCH_2CH_2-$

This is one of the more commonly used groups for phosphate protection, especially for oligonucleotide synthesis, but its base sensitivity can be a problem in some circumstances. Upon deprotection, acrylonitrile is released, which can result in by-product formation by alkylating nucleophilic substituents.

Formation

1. NCCH$_2$CH$_2$OH, triisopropylbenzenesulfonyl chloride, Pyr, rt, 15 h.[1]
2. NCCH$_2$CH$_2$OH, DCC, Pyr.[2]
3. NCCH$_2$CH$_2$OH, 8-quinolinesulfonyl chloride, 1-methylimidazole, Pyr, rt.[3]
4. For monoprotection of a phosphonic acid: NCCH$_2$CH$_2$OH, Cl$_3$CCN, 74–93% yield.[4]

Cleavage

1. Aqueous ammonia, dioxane.[2,5] The addition of nitromethane in the cleavage reaction will scavenge the released acrylonitrile and prevent it from reacting with the nucleobase during deprotection of oligonucleotides.[6]
2. Alkaline hydrolysis.[2]
3. TMSCl, DBU, CH$_2$Cl$_2$, 25°C. The presence of TMSCl allows for complete deprotection of a biscyanoethyl phosphate. In its absence only one cyanoethyl group was cleaved.[7]
4. Bu$_4$NF, THF, 30 min.[8]
5. In a study of the use of various amines for the deprotection of the cyanoethyl group it was found that primary amines are the most effective in achieving rapid cleavage. The following times for complete cleavage of the cyanoethyl group in phosphate **I** were obtained: TEA, 180 min; DIPA, 60 min; Et$_2$NH, 30 min; s-BuNH$_2$, 20 min, t-BuNH$_2$, 10 min, n-PrNH$_2$, 2 min.[9] Further study showed that t-BuNH$_2$ was most suitable because it did not react with protected nucleobases. Methylamine/ammonia was also a fast (5 min), effective reagent for deprotection.[10]

I

6. Bu$_4$NOH, CH$_2$Cl$_2$, H$_2$O, 100% yield.[11]
7. DBU, Me$_3$SiCl, CH$_2$Cl$_2$, rt, 88% yield.[12] In this case, TMSCl was required to silylate the oxygen after the release of the first cyanoethyl group so as to facilitate the second elimination, which otherwise failed to proceed.

2-Cyano-1,1-dimethylethyl (CDM): $CNCH_2C(CH_3)_2-$

Cleavage

1. Ammonia.[13]
2. DBU, N,O-bis(trimethylsilyl)acetamide.[14] Thiophosphorylated derivatives are cleaved more rapidly than the phosphorylated counterpart.
3. 0.2 N NaOH, dioxane, CH_3OH.[13]
4. Guanidine, tetramethylguanidine, or Bu_4NOH.[15]

4-Cyano-2-butenyl: $NCCH_2CH=CHCH_2-$

This is a vinylogous analog of the cyanoethyl group that is removed by δ-elimination with ammonium hydroxide. It is introduced using the phosphoramidite method.[16]

N-(4-Methoxyphenyl)hydracrylamide, N-Phenylhydracrylamide, and N-Benzylhydracrylamide Derivative: $ArNHC(O)CH_2CH_2-$

These derivatives, used for 5′-phosphate protection, are prepared using the DCC coupling protocol and are cleaved with 2 N NaOH at rt.[17] The protected phosphates can be purified using benzoylated DEAE-Cellulose.

2-(Methyldiphenylsilyl)ethyl (DPSE): $(C_6H_5)_2CH_3SiCH_2CH_2-$

2-(Trimethylsilyl)ethyl (TSE): $(CH_3)_3SiCH_2CH_2-$

These groups along with a number of other trialkylsilylethyl derivatives were examined for protection of phosphorothioates. Only the phenyl-substituted silyl derivative was useful because simple trialkylsilyl derivatives were prone to acid-catalyzed thiono–thiolo rearrangement.[18] Other trialkylsilylethyl derivatives also suffer from inherent instability upon storage,[19] but the trimethylsilylethyl group has been used successfully in the synthesis of the very sensitive agrocin 84[20] and for internucleotide phosphate protection with the phosphoramidite approach.[21]

Formation

The ester is introduced by means of the phosphoramidite method.[18,22]

Cleavage

1. Ammonium hydroxide, rt, 1 h.[18,22,23]
2. Pyr, H_2O.[18,24,25]
3. Bu_4NF THF, AcOH, 62% yield.[26] These conditions prevent the migration of acyl groups in bis(monoacylglycerol)phosphates.[27]

4. Methylamine, H_2O.[18,28]

5. SiF_4, CH_3CN, H_2O, 20 min.[29]

6. NH_4F, methanol, 60°C. One of two DPSE groups is cleaved.[30]

7. HF, CH_3CN, H_2O. In this case, both DPSE groups are removed.[30] This method effectively removes the trimethylsilylethyl group.[31]

8. TFA, CH_2Cl_2 or TFA, phenol, 30 min.[19,32]

9. Catalytic $ZnBr_2$, CH_3NO_2, IPA.[33]

2-(Triphenylsilyl)ethyl: $(C_6H_5)_3SiCH_2CH_2-$

This group, used for 5'-phosphate protection, had hydrophobicity similar to the dimethoxytrityl group and thus was expected to assist in reverse phase HPLC purification of product from failure sequences in oligonucleotide synthesis. It is cleaved with Bu_4NF in DMSO at 70°C.[34]

2-(4-Nitrophenyl)ethyl (Npe): $4-NO_2C_6H_4CH_2CH_2-$

The use of this group in nucleotide and nucleoside synthesis has been reviewed.[35,36]

Cleavage

0.5 M DBU in pyridine or CH_3CN. In this study[37] the cleavage of a series of 2-(pyrazin-2-yl)ethyl phosphates was compared with the NPE group and found to be cleaved with DBU in CH_3CN.[37–39] The related 2-(2-chloro-4-nitrophenyl)ethyl ester is cleaved with the weaker base TEA in CH_3CN.[40] The addition of thymine during DBU deprotection improves the yield because thymine scavenges the released 4-nitrostyrene.[41] The 2-(2-nitrophenyl)ethyl group is cleaved with DBU about 6 times more slowly than the 4-nitrophenyl derivative.[42] A bis-2-(4-nitrophenyl)ethyl phosphate upon DBU treatment releases only a single Npe group.[43]

2-(α-Pyridyl)ethyl (Pyet)

Cleavage

1. NaOMe, MeOH, Pyr or t-BuOK, Pyr, t-BuOH.[44] This group is reasonably stable to aqueous NaOH, ammonia and 80% acetic acid.
2. MeI, CH_3CN.[45]
3. PhOCOCl, CH_3CN, 20°C, 6 h; ammonia, pyridine.[46]

2-(4'-Pyridyl)ethyl

The 4-pyridylethyl group was found to be more effective for internucleotide phosphate protection than the 2-pyridylethyl group because its cleavage proceeded with greater efficiency. It is cleaved in a two-step process: Acylation with PhOCOCl increases the acidity of the benzylic protons, facilitating E-2 elimination by ammonia.[47]

2-(3-Arylpyrimidin-2-yl)ethyl

Cleavage of this ester with DBU is faster than cleavage of the Npes group; it can also be cleaved with the weaker base, TEA/Pyr.[48]

2-(Phenylthio)ethyl: $C_6H_5SCH_2CH_2-$

Formation

1. From $ROP(O)(OH)_2$: $PhSCH_2CH_2OH$, DCC.[49]

2.
   Ref. 50

3. $PhSCH_2CH_2OH$, triisopropylbenzenesulfonyl chloride, DMF, HMPA, rt, 8 h, 65–70% yield.[51]

Cleavage

1. $NaIO_4$, 1 h, rt; 2 N NaOH, 30 min, rt.[49,50]
2. N-Chlorosuccinimide; 1 N NaOH.[52] With this method the sulfide is oxidized completely to the sulfone that is cleaved with hydroxide more readily than the

sulfoxide formed by periodate oxidation. It has been reported that oxidation of the sulfide leads to oxidation of adenine and guanine.[53] However, see the TPTE group, *vida infra.*

2-(4-Nitrophenyl)thioethyl (PTE)

This group is stable to TEA and morpholine in pyridine at 20°C. It is cleaved by oxidation with MCPBA followed by elimination with TEA in Pyr, 10 min, 20°C.[54] The rate of cleavage of a variety of substituted phenylthioethyl derivatives is proportional to the strength of the electron-withdrawing group on the phenyl ring.[55]

2-(4-Tritylphenylthio)ethyl (TPTE): 2-[4-$(C_6H_5)_3CC_6H_4S$]CH_2CH_2-

The TPTE group, an analog of the 2-(phenylthio)ethyl group, was developed to impart lipophilicity to protected oligonucleotides so that they could be isolated by solvent extraction. It is formed from the phosphoric acid and the alcohol using either DCC or TPS as coupling agents. Cleavage is affected by base treatment after oxidation with $NaIO_4$ or NCS.[56]

2-[2-(Monomethoxytrityloxy)ethylthio]ethyl

This easily prepared lipophilic 5′-phosphate protective group is cleaved by NCS oxidation (dioxane, triethylammonium hydrogen carbonate, 2 h, rt) followed by ammonia-induced β-elimination.[3]

Dithiodiethanol Derivatve (DTE): HOCH$_2$CH$_2$SSCH$_2$CH$_2-$

Reduction of the disulfide by a reductase exposes the thiol that then closes to give an episulfide releasing the phosphate.[57]

2-(Methylsulfonyl)ethyl (MSE−): CH$_3$SO$_2$CH$_2$CH$_2-$

The MSE group is introduced using the phosphoramidite method and can be cleaved with 4 M NaOH in dioxane–MeOH.[58]

2-(*tert*-Butylsulfonyl)ethyl (B'SE): $(CH_3)_3CSO_2CH_2CH_2-$

The B'SE group was used for internucleotide protection and is removed with ammonia, also used to remove *N*-acyl protective groups. This group, as compared to the methylsulfonylethyl group,[59] has better solubility properties for solution phase synthesis.[60]

2-(Phenylsulfonyl)ethyl (PSE): $C_6H_5SO_2CH_2CH_2-$

The use of this group avoids the problems associated with the oxidation of the phenylthioethyl group. It is cleaved with TEA in pyridine (20°C, <3 h).[53,61]

2-(Benzylsulfonyl)ethyl: $C_6H_5CH_2SO_2CH_2CH_2-$

This group is cleaved with 2 eq. of TEA in Pyr at a rate somewhat slower than that of the phenylsulfonylethyl group.[62]

1. E. Ohtsuka, H. Tsuji, T. Miyake, and M. Ikehara, *Chem. Pharm. Bull.*, **25**, 2844 (1977).

2. G. M. Tener, *J. Am. Chem. Soc.*, **83**, 159 (1961).

3. K. Kamaike, T. Ogawa, and Y. Ishido, *Nucleosides & Nucleotides*, **12**, 1015 (1993).

4. J. Szewdzyk, J. Rachon, and C. Wasielewski, *Pol. J. Chem.*, **56**, 477 (1982).

5. J. Robles, E. Pedroso, and A. Grandas, *J. Org. Chem.*, **59**, 2482 (1994).

6. T. Umemoto and T. Wada, *Tetrahedron Lett.*, **46**, 4251 (2005).

7. D. A. Evans, J. R. Gage, and J. L. Leighton, *J. Org. Chem.*, **57**, 1964 (1992).

8. K. K. Ogilvie, S. L. Beaucage, and D. W. Entwistle, *Tetrahedron Lett.*, **17**, 1255 (1976).

9. H. M. Hsiung, *Tetrahedron Lett.*, **23**, 5119 (1982).

10. M. P. Reddy, N. B. Hanna, and F. Farooqui, *Tetrahedron Lett.*, **35**, 4311 (1994).

11. D. Crich and V. Dudkin, *Org. Lett.*, **2**, 3941 (2000).

12. D. A. Evans, J. R. Gage, and J. L. Leighton, *J. Org. Chem.* **57**, 1964 (1992).

13. J. E. Marugg, C. E. Dreef, G. A. Van der Marel, and J. H. Van Boom, *Recl., J. R. Neth. Chem. Soc.*, **103**, 97 (1984).

14. M. Sekine, H. Tsuruoka, S. Iimura, and T. Wada, *Nat. Prod. Lett.*, **5**, 41 (1994); M. Kadokura, T. Wada, K. Seio, and M. Sekine, *J. Org. Chem.*, **65**, 5104 (2000).

15. Yu. V. Tumanov, V. V. Gorn, V. K. Potapov, and Z. A. Shabarova, *Dokl. Akad. Nauk SSSR*, **270**, 1130 (1983); *Chem. Abstr.* **99**: 212865e (1983).

16. V. T. Ravikumar, Z. S. Cheruvallath, and D. L. Cole, *Tetrahedron Lett.*, **37**, 6643 (1996). V. T. Ravikumar, Z. S. Cheruvallath, and D. L. Cole, *Nucleosides & Nucleotides*, **16**, 1709 (1997).

17. S. A. Narang, O. S. Bhanot, J. Goodchild, and J. Michniewicz, *J. Chem. Soc., Chem. Commun.*, 516 (1970).

18. A. H. Krotz, P. Wheeler, and V. T. Ravikumar, *Angew. Chem., Int. Ed. Engl.*, **34**, 2406 (1995).

19. H.-G. Chao, M. S. Bernatowicz, P. D. Reiss, and G. R. Matsueda, *J. Org. Chem.*, **59**, 6687 (1994).

20. T. Moriguchi, T. Wada, and M. Sekine, *J. Org. Chem.*, **61**, 9223 (1996).

21. T. Wada, M. Tobe, T. Nagayama, K. Furusawa, and M. Sekine, *Nucleic Acids Symp. Ser.*, **29**, 9 (1993).

22. V. T. Ravikumar, H. Sasmor, and D. L. Cole, *Bioorg. Med. Chem. Lett.*, **3**, 2637 (1993).

23. V. T. Ravikumar, T. K. Wyrzykiewicz, and D. L. Cole, *Tetrahedron*, **50**, 9255 (1994).

24. S. Honda and T. Hata, *Tetrahedron Lett.*, **22**, 2093 (1981).

25. T. Wada and M. Sekine, *Tetrahedron Lett.*, **35**, 757 (1994).

26. T. Moriguchi, T. Yanagi, M. Kunimori, T. Wada, and M. Sekine, *J. Org. Chem.*, **65**, 8229 (2000).

27. J. Chevallier, N. Sakai, F. Robert, T. Kobayashi, J. Gruenberg, and S. Matile, *Org. Lett.*, **2**, 1859 (2000).

28. A. H. Krotz, Z. S. Cheruvallath, D. L. Cole, and V. T. Ravikumar, *Nucleosides & Nucleotides*, **17**, 2335 (1998).

29. V. T. Ravikumar and D. L. Cole, *Gene*, **149**, 157 (1994); V. T. Ravikumar, *Synth. Commun.*, **25**, 2164 (1995).

30. K. C. Ross, D. L. Rathbone, W. Thomson, and S. Freeman, *J. Chem. Soc., Perkin Trans. 1*, 421 (1995).

31. A. Sawabe, S. A. Filla, and S. Masamune, *Tetrahedron Lett.*, **33**, 7685 (1992).

32. S. F. Martin, J. A. Josey, Y.-L. Wong, and D. W. Dean, *J. Org. Chem.*, **59**, 4805 (1994).

33. F. Ferreira, J.-J. Vasseur, and F. Morvan, *Tetrahedron Lett.*, **45**, 6287 (2004).

34. J. E. Celebuski, C. Chan, and R. A. Jones, *J. Org. Chem.*, **57**, 5535 (1992).

35. W. Pfleiderer, F. Himmelsbach, R. Charubala, H. Schirmeister, A. Beiter, B. Schultz, and T. Trichtinger, *Nucleosides & Nucleotides*, **4**, 81 (1985).

36. F. Himmelsbach, B. S. Schulz, T. Trichtinger, R. Charubala, and W. Pfleiderer, *Tetrahedron*, **40**, 59 (1984).

37. W. Pfleiderer, H. Schirmeister, T. Reiner, M. Pfister, and R. Charubala, "Biophosphates, and Their Analogs—Synthesis, Structure, Metabolism, and Activity," *Bioact. Mol.* **3**, 133 (1987).

38. For a brief review, see W. Pfleiderer, M. Schwarz, and H. Schirmeister, *Chem. Scr.*, **26**, 147 (1986).

39. E. Uhlmann and W. Pfleiderer, *Helv. Chim. Acta*, **64**, 1688 (1981).

40. E. Uhlmann and W. Pfleiderer, *Nucl. Acids Res., Spec. Publ.*, **4**, 25 (1978).

41. A. M. Avino and R. Eritja, *Nucleosides & Nucleotides*, **13**, 2059 (1994).

42. E. Uhlmann and W. Pfleiderer, *Tetrahedron Lett.*, **21**, 1181 (1980).

43. E. Uhlmann, R. Charubala, and W. Pfleiderer, *Nucl. Acids Symp. Ser.*, **9**, 131 (1981).

44. W. Freist, R. Helbig, and F. Cramer, *Chem. Ber.*, **103**, 1032 (1970).

45. H. Takaku, S. Hamamoto, and T. Watanabe, *Chem. Lett.*, **15**, 699 (1986).

46. S. Hamamoto, N. Shishido, and H. Takaku, *Nucl. Acids Symp. Ser.*, **17**, 93 (1986).

47. S. Hamamoto, Y. Shishido, M. Furuta, H. Takaku, M. Kawashima, and M. Takaki, *Nucleosides & Nucleotides*, **8**, 317 (1989).

48. T. Reiner and W. Pfleiderer, *Nucleosides & Nucleotides*, **6**, 533 (1987).

49. R. H. Wightman, S. A. Narang, and K. Itakura, *Can. J. Chem.*, **50**, 456 (1972).

50. N. T. Thuong, M. Chassignol, U. Asseline, and P. Chabrier, *Bull. Soc. Chim. Fr.*, II-51 (1981).

51. S. A. Narang, K. Itakura, C. P. Bahl, and N. Katagiri, *J. Am. Chem. Soc.*, **96**, 7074 (1974).

52. K. L. Agarwal, M. Fridkin, E. Jay, and H. G. Khorana, *J. Am. Chem. Soc.*, **95**, 2020 (1973).

53. N. Balgobin, S. Josephson, and J. B. Chattopadhyaya, *Tetrahedron Lett.*, **22**, 1915 (1981).

54. N. Balgobin and J. Chattopadhyaya, *Chem. Scr.*, **20**, 144 (1982).

55. N. Balgobin, C. Welch, and J. B. Chattopadhyaya, *Chem. Scr.*, **20**, 196 (1982).

56. K. L. Agarwal, Y. A. Berlin, H.-J. Fritz, M. J. Gait, D. G. Kleid, R. G. Lees, K. E. Norris, B. Ramamoorthy, and H. G. Khorana, *J. Am. Chem. Soc.*, **98**, 1065 (1976).

57. C. Périgaud, G. Gosselin, I Lefebvre, J. L. Girardet, S. Benzaria, I. Barber, and J. L. Imbach, *Bioorg. Med. Chem. Lett.*, **3**, 2521 (1993).

58. N. C. R. van Straten, G. A. van der Marel, and J. H. van Boom, *Tetrahedron Lett.*, **37**, 3599 (1996).

59. C. Claesen, G. I. Tesser, C. E. Dreef, J. E. Marugg, G. A. van der Marel, and J. H. van Boom, *Tetrahedron Lett.*, **25**, 1307 (1984).

60. C. A. A. Claesen, C. J. M. Daemean, and G. I. Tesser, *Recl. Trav. Chim. Pays-Bas*, **105**, 116 (1986).

61. S. Josephson and J. B. Chattopadhyaya, *Chem. Scr.,*, **18**, 184 (1981).

62. E. Felder, R. Schwyzer, R. Charubala, W. Pfleiderer, and B. Schulz, *Tetrahedron Lett.*, **25**, 3967 (1984).

Haloethyl Phosphates

2,2,2-Trichloroethyl: Cl_3CCH_2O-

Myoinositol bis(trichloroethyl)phosphates were not as stable to pyridine at 20°C as were the related benzyl analogs.[1] This group is not compatible with Fmoc chemistry because of its instability to piperidine. The trichloroethyl phosphates are compatible with TFA, and with hydrogenolysis under acidic conditions. Neutral conditions result in cleavage.[2]

Formation

1. Trichloroethanol, DCC, Pyr, rt, 15 h.[3]

2. A phosphonic acid was monoesterified with trichloroethanol, CCl_3CN in Pyr at 100°C.[4]

3. Bis-(2,2-trichloro)ethyl phosphochloride can be used to introduce the protected phosphate on tyrosine in excellent yield.[5]

Cleavage

1. Electrolysis at a Hg cathode, $-1.2\,V$ (Ag wire), CH_3CN, DMF, $Bu_4N^+BF_4^-$, 2,6-lutidine[6] LiCl or $LiClO_4$ have been used as electrolytes in the electrochemical removal of haloethyl phosphates.[7]

2. Zn, acetylacetone, DMF, Pyr.[8,9] Chelex resin can be used to remove the zinc from these deprotections.[10]

3. Na, ammonia.[11] These conditions also remove cyanoethyl and benzyl protecting groups. Phosphorothioates are similarly deprotected.

4. Zn(Cu), DMF.[12,13]

5. NaOH, aqueous dioxane.[14]

6. The trichloroethyl group is stable to Pd-catalyzed hydrogenolysis in AcOH/TFA, but when hydrogenolysis was attempted using EtOAc/MeOH as solvent, partial removal of the trichloroethyl group occurred along with Fmoc cleavage. Clean cleavage was observed in aqueous ethanol as solvent.[15,16]

7. Hydrogenolysis: Pd, Pyr.[17]

8. Bu₄NF, THF.[18]

9. Zn, anthranilic acid. Anthranilic acid was used to prevent complexation of the zinc with the oligonucleotides.[19]

2,2,2-Trichloro-1,1-dimethylethyl (TCB): Cl₃CC(CH₃)₂O−

Formation

The ester is introduced as the bis-TCB monochlorophosphate.[20]

Cleavage

1. Cobalt(I)-phthalocyanine, CH₃CN, 48 h. In a phosphate with two TCB groups the first is cleaved considerably faster than the second.[20,21]

2. Bu₃P, DMF, TEA, 80°C, quant.[22,23] Trichloroethyl phosphates are also cleaved.

3. Zn, AcAc, TEA, CH₃CN.[24]

4. Zn-Cu, 2,4-pentanedione, pyridine, rt, 1 h, 60% yield.[25] The 2,4-pentanedione is used to maintain a clean surface on the zinc.

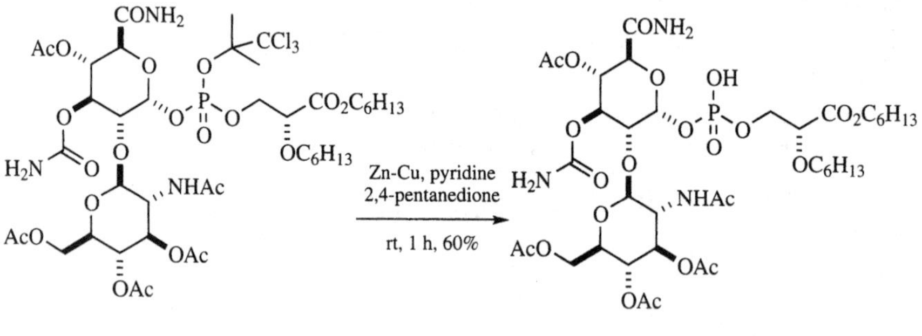

2,2,2-Tribromoethyl: Br₃CCH₂−

Formation

(RO)(Cl₃CCH₂O)P(O)Cl, Br₃CCH₂OH.

Cleavage

1. Electrolysis at a Hg cathode, -0.5 to -0.6 V, LiClO$_4$, CH$_3$CN, Pyr. The trichloroethyl ester, which requires a greater reduction potential for cleavage, is retained under these conditions.[6]
2. Zn(Cu), DMF, 20°C.[26]
3. Zn(Cu), Bu$_3$N, H$_3$PO$_4$, Pyr, rt.[27]

2,3-Dibromopropyl: BrCH$_2$CHBrCH$_2$—

Treatment of this protective group with KI/DMF for 24 h results in complete cleavage. This group is stable to Pyr/TEA/H$_2$O but not to 7 M NH$_4$OH/MeOH.[28]

2,2,2-Trifluoroethyl: CF$_3$CH$_2$—

The trifluoroethyl group was used as an activating group in the phosphotriester approach to oligonucleotide synthesis as well as a protective group that could be removed with 4-nitrobenzaldoxime (tetramethylguanidine, dioxane, H$_2$O).[29]

1,1,1,3,3,3-Hexafluoro-2-propyl: (CF$_3$)$_2$CH—

Cleavage of this group is achieved with tetramethylguanidinium *syn*-2-pyridinecarboxaldoxime.[30,31] Tris(hexafluoro-2-propyl) phosphites are sufficiently reactive to undergo transesterification with alcohols in a stepwise fashion.[32]

1. T. Desai, A. Fernandez-Mayoralas, J. Gigg, R. Gigg, and S. Payne, *Carbohydr. Res.*, **234**, 157 (1992).
2. A. Paquet, B. Blackwell, M. Johns, and J. Nikiforuk, *J. Peptide Res.*, **50**, 262 (1997).
3. E. Ohtsuka, H. Tsuji, T. Miyake, and M. Ikehara, *Chem. Pharm. Bull.*, **25**, 2844 (1977).
4. J. Szewczyk and C. Wasielewski, *Pol. J. Chem.*, **55**, 1985 (1981).
5. A. Paquet, B. Blackwell, M. Johns, and J. Nikiforuk, *J. Peptide Res.*, **50**, 262, (1997).
6. J. Engels, *Angew, Chem,. Int. Ed. Engl.*, **18**, 148 (1979).
7. J. Engels, *Liebigs Ann. Chem.*, 557 (1980).
8. M. Sekine, K. Hamaoki, and T. Hata, *Bull. Chem. Soc. Jpn.*, **54**, 3815 (1981).
9. R. W. Adamiak, E. Biala, K. Grzeskowiak, R. Kierzek, A. Kraszewski, W. T. Markiewicz, J. Stawinski, and M. Wiewiorowski, *Nucleic Acids Res.*, **4**, 2321 (1977).
10. Y. Ichikawa and Y. C. Lee, *Carbohydr. Res.*, **198**, 235 (1990).
11. N. J. Noble, A. M. Cooke, and B. V. L. Potter, *Carbohydr. Res.*, **234**, 177 (1992).
12. F. Eckstein, *Chem. Ber.*, **100**, 2236 (1967).
13. M. Heuer, K. Hohgardt, F. Heinemann, H. Kühne, W. Dietrich, D. Grzelak, D. Müller, P. Welzel, A. Markus, Y. van Heijenoort, and J. van Heijenoort, *Tetrahedron*, **50**, 2029 (1994).
14. T. Neilson and E. S. Werstiuk, *Can. J. Chem.*, **49**, 3004 (1971).
15. A. Paquet, *Int. J. Pept. Protein Res.*, **39**, 82 (1992).
16. N. Mora, J. M. Lacombe, and A. A. Pavia, *Int. J. Pept. Protein Res.*, **45**, 53 (1995).

17. K. Grzeskowiak, R. W. Adamiak, and M. Wiewiorowski, *Nucleic Acids Res.*, **8**, 1097 (1980).

18. K. K. Ogilvie, S. L. Beaucage, and D. W. Entwistle, *Tetrahedron Lett.*, **17**, 1255 (1976).

19. A. Wolter and H. Köster, *Tetrahedron Lett.*, **24**, 873 (1983).

20. H. A. Kellner, R. G. K. Schneiderwind, H. Eckert, and I. K Ugi, *Angew. Chem., Int. Ed. Engl.*, **20**, 577 (1981).

21. P. Lemmen, K. M. Buchweitz, and R. Stumpf, *Chem. Phys. Lipids*, **53**, 65 (1990).

22. R. L. Letsinger, E. P. Groody, and T. Tanaka, *J. Am. Chem. Soc.*, **104**, 6805 (1982).

23. R. L. Letsinger, E. P. Groody, N. Lander, and T. Tanaka, *Tetrahedron*, **40**, 137 (1984).

24. A. B. Kazi and J. Hajdu, *Tetrahedron Lett.*, **33**, 2291 (1992).

25. N. El-Abaddla, M. Lampilas, L. Hennig, M. Findeisen, P. Welzel, D. Muller, A. Markus, and J. van Heijenoort, *Tetrahedron*, **55**, 699 (1999).

26. J. H. Van Boom, P. M. J. Burgers, R. Crea, G. van der Marel, and G. Wille, *Nucleic Acids Res.*, **4**, 747 (1977).

27. L. Desaubry, I. Shoshani, and R. A. Johnson, *Tetrahedron Lett.*, **36**, 995 (1995).

28. A. Kraszewski and J. Strawinski, *Nucleic Acids Symp. Ser.*, **9**, 135 (1981).

29. H. Takaku, H. Tsuchiya, K. Imai, and D. E. Gibbs, *Chem. Lett.*, **13**, 1267 (1984).

30. S. Yamakage, M. Fujii, H. Takaku, and M. Uemura, *Tetrahedron*, **45**, 5459 (1989).

31. H. Takaku, T. Watanabe, and S. Hamamoto, *Tetrahedron Lett.*, **29**, 81 (1988).

32. T. Watanabe, H. Sato, and H. Takaku, *J. Am. Chem. Soc.*, **111**, 3437 (1989).

BENZYL PHOSPHATES

Benzyl (Bn): $C_6H_5CH_2-$

Formation

1. From a tributylstannyl phosphate: BnBr, Et_4NBr, CH_3CN, reflux. Phenacyl, 4-nitrobenzyl and simple alkyl derivatives were similarly prepared. Yields are substrate and alkylating-agent dependent.[1]

2. Diphenyl phosphates are converted by transesterification to dibenzyl phosphates upon treatment with BnONa in THF at 25°C in 83% yield.[2]

Cleavage

1. Pd-C, H_2, formic acid.[3]

2. Pd-C, EtOH, $NaHCO_3$, H_2.[4] Hydrogenolysis in the presence of NH_4OAc cleaves only one benzyl group of a dibenzyl phosphate.[5]

3. Na, ammonia.[6,7] Cyanoethyl and trichloroethyl phosphates are also deprotected.

4. 1 M TFMSA in TFA, thioanisole.[8] Dibenzyl phosphates are only partially labile to TFA alone.[9]

5. TFA, thiophenol.[10]

6. A dibenzyl phosphate is monodeprotected with TFA, CH_2Cl_2.[11]

7. LiSPh, THF, HMPA, 30 min, >95% yield.[12]

8. NaI, CH_3CN,[13] DMF[14] or 2-butanone.[15]

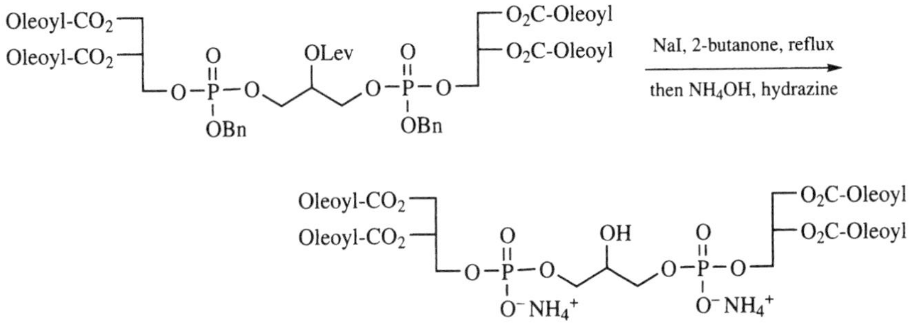

9. TMSBr, Pyr, CH_2Cl_2, rt, 1.5 h.[16] Phenolic phosphates were stable to this reagent.[17]

10. In dibenzyl phosphates or phosphonates treatment with refluxing N-methylmorpholine results in monodebenzylation (60–100% yield).[18]

11. Quinuclidine, toluene, reflux.[19] In dibenzyl phosphates, only one benzyl group is removed.

4-Methoxybenzyl: $CH_3OC_6H_4CH_2-$

Cleavage

HF, CH_3CN, H_2O, rt, 15 min, then add pyridine.[20]

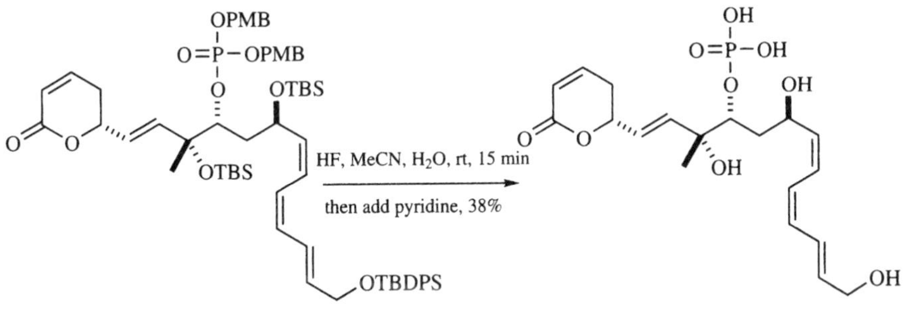

4-Nitrobenzyl: $4-NO_2C_6H_4CH_2-$

The 4-nitrobenzyl group, used in the synthesis of phosphorylated serine, is introduced by the phosphoramidite method and can be cleaved with TFMSA/MTB/m-cresol/1,2-ethanedithiol/TFA, 4 h, 0°C to rt.[21] N-Methylmorpholine at 80°C also cleaves a 4-nitrobenzyl phosphate triester.[22] This ester is more acid stable than the benzyl ester.[9]

2,4-Dinitrobenzyl: $2,4\text{-}(NO_2)_2\text{-}C_6H_3CH_2-$

This group has been used for protection of a phosphorodithioate and is cleaved with 4-methylthiophenol and TEA.[23]

4-Chlorobenzyl: $4\text{-}ClC_6H_4CH_2-$

Cleavage

1. Hydrogenolysis: Pd-C, *t*-BuOH, NaOAc, H_2O.[24-26]
2. From a phosphorothioate: TFMSA, *m*-cresol, thiophenol, TFA. These conditions minimized the migration of the benzyl group to the thione.[27]
3. TFA, EDT, TIS, H_2O. These conditions readily cleave the benzyl phosphate but also result in some methyl ester hydrolysis of a cyclic peptide.[28] The problem was avoided by using hydrogenolysis to affect cleavage, but this also reduced an olefin in the molecule.

4-Chloro-2-nitrobenzyl: $4\text{-}Cl\text{-}2\text{-}NO_2C_6H_3CH_2-$

The 4-chloro-2-nitrobenzyl group was useful in the synthesis of dithymidine phosphorothioates. It could be cleaved with a minimum of side reactions with PhSH, TEA, Pyr.[29]

4-Acyloxybenzyl: $4\text{-}RCO_2C_6H_4CH_2-$

4-Acyloxybenzyl esters were designed to be released under physiological conditions. Porcine liver carboxyesterase efficiently releases the phosphate by acetate hydrolysis and quinone methide formation. In a diester the first ester is cleaved faster than the second.[30]

1-Oxido-4-methoxy-2-picolyl

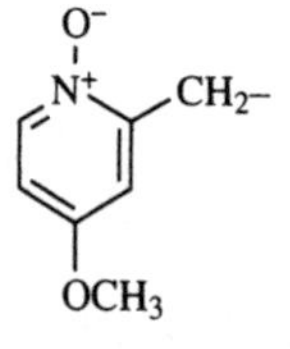

The oxidopicolyl group increases the rate and efficiency of internucleotide phosphodiester synthesis.[31] It is cleaved with piperidine.[32]

Fluorenyl-9-methyl (Fm):

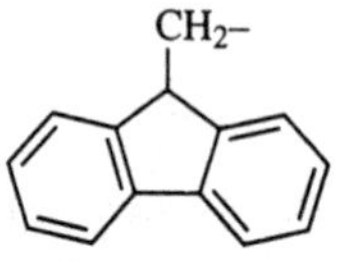

The fluorenyl-9-methyl group has been shown to be of particular value in studies of deoxynucleoside dithiophosphates.[33]

Formation

1. 5′-Nucleoside phosphates are protected using triisopropylbenzenesulfonyl chloride in Pyr.[34]
2. The Atherton–Todd reaction[35]:

$$PhO-\overset{\overset{\displaystyle O}{\|}}{\underset{\underset{\displaystyle H}{|}}{P}}-OPh \xrightarrow{\text{FmOH, pyridine}} PhO-\overset{\overset{\displaystyle O}{\|}}{\underset{\underset{\displaystyle H}{|}}{P}}-OFm$$

Cleavage

1. TEA, Pyr, 20°C, 2 h.[36] These conditions were developed for use with 2-chloro-phenyl protection at the internucleotide junctions.
2. TEA, CH$_3$CN, 14 h, rt.[37] In the case of a bis Fm phosphate, the first Fm group is easily cleaved at rt, but the second is cleaved upon heating to reflux.[38]
3. 0.1 *M* NaOH, 0°C, 10 min.[34]
4. Concd. NH$_4$OH, 50°C, 2 h.[34]
5. *t*-BuNH$_2$, pyridine, 70–80%.[39]

2-(9,10-Anthraquinonyl)methyl or 2-Methyleneanthraquinone (MAQ)

This group is stable to TEA/Pyr and to 80% acetic acid. It is cleaved by reduction with sodium dithionite at pH 7.3.[40]

5-Benzisoxazolylmethylene (Bim)

This group was effective in the synthesis of oligonucleotides using the phosphotriester approach. It is cleaved with TEA, pyridine in < 2 h.[41]

Cleavage Rates of Various Arylmethyl Phosphates

The accompanying table compares the cleavage rates for a variety of benzyl phosphates using thiols or pyridine for the following reaction[42,43]:

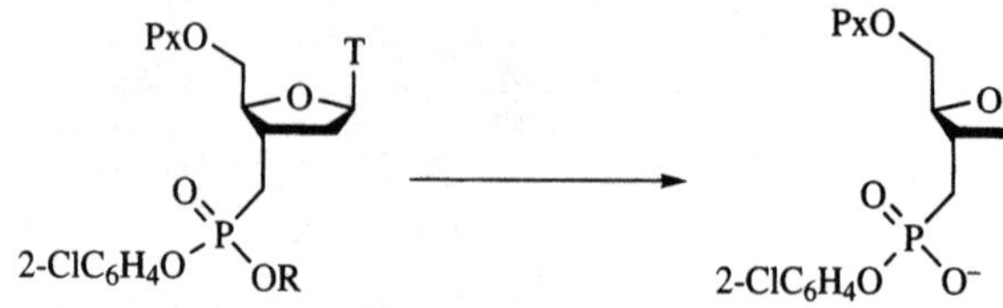

Px = 9-phenylxanthen-9-yl (pixyl)

Substrate R =	p-Thiocresol/TEA/ACN		Pyridine	Ration of Half–Lives
	$t_{1/2}$ (min)	t_∞ (min)	$t_{1/2}$ (h)	(Pyr/RSH)
CH$_3$–	45	—	12	16
Bn–	30	—	12	24
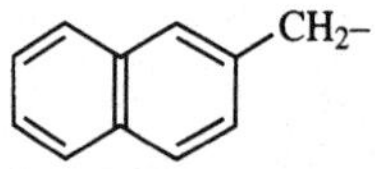 naphthyl-CH$_2$–	5	60	5	60
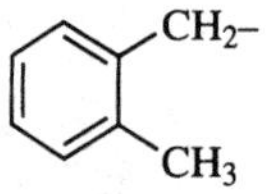 2-CH$_3$-C$_6$H$_4$-CH$_2$–	7	90	3	26
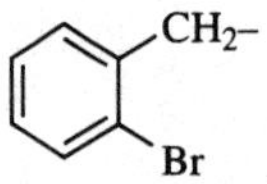 2-Br-C$_6$H$_4$-CH$_2$–	4	45	10	150
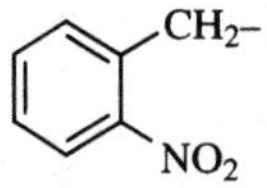 2-NO$_2$-C$_6$H$_4$-CH$_2$–	5	60	68	820
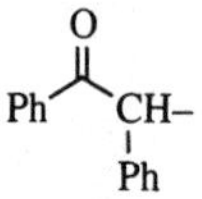 Ph-CO-CH(Ph)–	2	20	40	1200
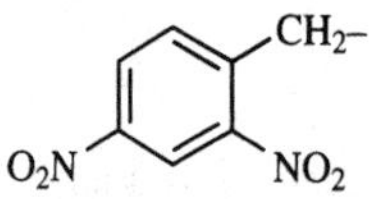 2,4-dinitro-benzyl-CH$_2$–	~10 s	~1	120	~43,000
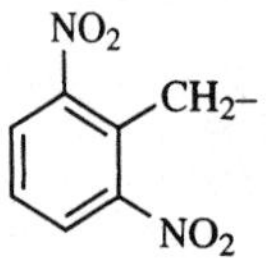 2,6-dinitro-benzyl-CH$_2$–	~10 s	~1	45	~16,000

Diphenylmethyl (Dpm): $(C_6H_5)_2CH$–

The reaction of phosphoric acid with diphenyldiazomethane in dioxane gives the triphosphate.[44,45]

Cleavage

1. (DpmO)$_3$PO upon reaction with NaI, Pyr at 100°C gives (DpmO)$_2$P(O)ONa quantitatively. Bu$_3$NHI can also be used to remove a single Dpm group.[44]
2. H$_2$, Pd–C, aqueous methanol.[44]
3. Trifluoroacetic acid.[45]

o-Xylene Derivative

This group is introduced using the phosphoramidite method and is cleaved by hydrogenolysis (H$_2$, Pd-C, rt, 17 h).[46–48]

1. H. Ayukawa, S. Ohuchi, M. Ishikawa, and T. Hata, *Chem. Lett.*, 81, **24** (1995).

2. D. C. Billington, R. Baker, J. J. Kulagowski, and I. M. Mawer, *J. Chem. Soc., Chem. Commun.*, 314 (1987).

3. J. W. Perich, P. F. Alewood, and R. B. Johns, *Aust. J. Chem.*, **44**, 233 (1991).

4. M. M. Sim, H. Kondo, and C.-H. Wong, *J. Am. Chem. Soc.*, **115**, 2260 (1993).

5. J. Scheigetz, M. Gilbert, and R. Zamboni, *Org. Prep. Proced. Int.*, **29**, 561 (1997).

6. N. J. Noble, A. M. Cooke, and B. V. L. Potter, *Carbohydr. Res.*, **234**, 177 (1992).

7. A. M. Riley, P. Guedat, G. Schlewer, B. Spiess, and B. V. L. Potter, *J. Org. Chem.*, **63**, 295 (1998).

8. T. Wakamiya, K. Saruta, S. Kusumoto, K. Nakajima, K. Yoshizawa-Kumagaye, S. Imajoh-Ohmi, and S. Kanegasaki, *Chem. Lett.*, **22**, 1401 (1993).

9. T. Wakamiya, *Chemistry Express*, **7**, 577 (1992).

10. E. A. Kitas, R. Knorr, A. Trzeciak, and W. Bannwarth, *Helv. Chim. Acta*, **74**, 1314 (1991).

11. Z. Tian, C. Gu, R. W. Roeske, M. Zhou, and R. L. Van Etten, *Int. J. Pept. Protein Res.*, **42**, 155 (1993).

12. G. W. Daud and E. E. van Tamelen, *J. Am. Chem. Soc.*, **99**, 3526 (1977).

13. K. H. Scheit, *Tetrahedron Lett.*, **8**, 3243 (1967).

14. D. Majumdar, G. A. Elsayed, T. Buskas, and G.-J. Boons, *J. Org. Chem.*, **70**, 1691 (2005).

15. U. M. Krishna, M. U. Ahmad, and I. Ahmad, *Tetrahedron Lett.*, **45**, 2077 (2004).

16. P. M. Chouinard and P. A. Bartlett, *J. Org. Chem.*, **51**, 75 (1986).

17. S. Lazar and G Guillaumet, *Synth. Commun.*, **22**, 923 (1992); H.-G. Chao, M. S. Bernatowicz, C. E. Klimas, and G. R. Matsueda, *Tetrahedron Lett.*, **34**, 3377 (1993).

18. M. Saady, L. Lebeau, and C. Mioskowski, *J. Org. Chem.*, **60**, 2946 (1995).

19. M. Saady, L Lebeau, and C. Mioskowski, *Tetrahedron Lett.*, **36**, 4785 (1995).

20. D. L. Boger, S. Ichikawa, and W. Zhong, *J. Am. Chem. Soc.*, **123**, 4161 (2001); K. Miyashita, M. Ikejiri, H. Kawasaki, S. Maemura, and T. Imanishi, *J. Am. Chem. Soc.*, **125**, 8238 (2003).

21. T. Wakamiya, K. Saruta, J.-i. Yasuoka, and S. Kusumoto, *Bull. Chem. Soc. Jpn.*, **68**, 2699 (1995).

22. J. Smrt, *Collect. Czech. Chem. Commun.* **37**, 1870 (1972).

23. G. M. Porritt and C. B. Reese, *Tetrahedron Lett.*, **31**, 1319 (1990).

24. A. H. van Oijen, C. Erkelens, J. H. Van Boom, and R. M. J. Liskamp, *J. Am. Chem. Soc.*, **111**, 9103 (1989).

25. H. B. A. de Bont, J. H. Van Boom, and R. M. J. Liskamp, *Recl. Trav. Chim. Pays-Bas*, **109**, 27 (1990).

26. A. H. van Oijen, H. B. A. de Bont, J. H. van Boom, and R. M. J. Liskamp, *Tetrahedron Lett.*, **32**, 7723 (1991).

27. D. B. A. de Bont, W. J. Moree, J. H. van Boom, and R. M. J. Liskamp, *J. Org. Chem.*, **58**, 1309 (1993).

28. F. J. Dekker, N. J. de Mol, M. J. E. Fischer, J. Kemmink, and R. M. J. Liskamp, *Org. Biomol. Chem.*, **1**, 3297 (2003).

29. A. Püschl, J. Kehler, and O. Dahl, *Nucleosides Nucleotides*, **16**, 145 (1997).

30. A. G. Mitchell, W. Thomson, D. Nicholls, W. J. Irwin, and S. Freeman, *J. Chem. Soc., Perkin Trans. 1*, 2345 (1992).

31. T. Szabo, A. Kers and J. Stawinski, *Nucleic Acids Res.*, **23**, 893 (1995).

32. N. N. Polushin, I. P. Smirnov, A. N. Verentchikov, and J. M. Coull, *Tetrahedron Lett.*, **37**, 3227 (1996).

33. P. H. Seeberger, E. Yau, and M. H. Caruthers, *J. Am. Chem. Soc.*, **117**, 1472 (1995).

34. N. Katagiri, C. P. Bahl, K. Itakura, J. Michniewicz, and S. A. Narang, *J. Chem. Soc., Chem. Commun.*, 803 (1973).

35. J. Zhu, H. Fu, Y. Jiang, and Y. Zhao, *Synlett*, 1927 (2005).

36. C. Gioeli and J. Chattopadhyaya, *Chem. Scr.*, **19**, 235 (1982).

37. Y. Watanabe and M. Nakatomi, *Tetrahedron Lett.*, **39**, 1583 (1998).

38. Y. Watanabe, M. Ishimura, and S. Ozaki, *Chem. Lett.*, **23**, 2163 (1994). Y. Watanabe, T. Nakamura, and H. Mitsumoto, *Tetrahedron Lett.*, **38**, 7407 (1997).

39. H. Almer, T. Szabo, and J. Stawinski, *Chem. Commun.*, 290 (2004).

40. N. Balgobin, M. Kwiatkowski, and J. Chattopadhyaya, *Chem. Scr.*, **20**, 198 (1982).

41. N. Balgobin and J. Chattopadhyaya, *Chem. Scr.*, **20**, 142 (1982).

42. C. Christodoulou and C. B. Reese, *Nucl. Acids Symp. Ser.*, **11**, 33 (1982).

43. C. Christodoulou and C. B. Reese, *Tetrahedron Lett.*, **24**, 951 (1983).

44. G. Lowe and B. S. Sproat, *J. Chem. Soc., Perkin Trans. 1*, 1874 (1981).

45. M. Hoffmann, *Pol. J. Chem.*, **59**, 395 (1985).

46. Y. Watanabe, T. Shinohara, T. Fujimoto, and S. Ozaki, *Chem. Pharm. Bull.*, **38**, 562 (1990).

47. S. Ozaki, Y. Kondo, N. Shiotani, T. Ogasawara, and Y. Watanabe, *J. Chem. Soc., Perkin Trans. 1*, 729 (1992).

48. Y. Watanabe, Y. Komoda, K. Ebisuya, and S. Ozaki, *Tetrahedron Lett.*, **31**, 255 (1990).

PHENYL PHOSPHATES

Phenyl: C_6H_5-

Cleavage

1. PtO_2 (stoichiometric), TFA, AcOH, H_2, 91% yield.[1,2] This method cannot be used in substrates that contain a tyrosine because tyrosine is easily reduced

in the acidic medium. Neutral conditions do not always cleave phenyl phosphates.[3] Trichloroethyl esters are stable.[4]

2. Aqueous HCl, reflux.[5]
3. Bu$_4$NF, THF, Pyr, H$_2$O, rt, 30 min.[6] These conditions result in the formation of a mixture of fluorophosphate, and phosphate. In the case of oligonucleotides some internucleotide bond cleavage is observed with this reagent.
4. NaOH, THF[7] or LiOH, dioxane.[8]
5. Li, NH$_3$, 99% yield.[9]

6. See cleavage of 2-chlorophenyl for oximate rate comparisons.

2-Methylphenyl: 2-CH$_3$C$_6$H$_4$– **and 2,6-Dimethylphenyl:** 2,6-(CH$_3$)C$_6$H$_3$–

These groups were more effective than the phenyl group for protection of phosphoserine during peptide synthesis. They are cleaved by hydrogenolysis with stoichiometric PtO$_2$ in AcOH.[10]

2-Chlorophenyl: 2-Cl-C$_6$H$_4$–

Cleavage

1. Tetramethylguanidinium 4-nitrobenzaldoxime, dioxane, H$_2$O, 20°C, 22 h.[11] This reagent cleaves the 2-chlorophenyl ester 2.5 times faster than the 4-chlorophenyl ester and 25 times faster than the phenyl ester. The use of *syn*-2-nitrobenzaldoxime increases the rate an additional 2.5 to 4 times.[12] Oximate cleavage proceeds by nucleophilic addition–elimination to give an oxime ester that, with base, undergoes another elimination to give a nitrile and phosphate anion.[13]
2. NaOH, Pyr, H$_2$O, 0°C.[14]
3. *syn*-Pyridine-2-aldoxime, tetramethylguanidine, dioxane, Pyr, H$_2$O.[15] This method involves the addition of the oximate to the phosphate with release of the phenol. Dehydration then leads to a nitrile and the unprotected phosphate.

4-Chlorophenyl: 4-Cl−C₆H₄−

Halogen-substituted phenols were originally introduced for phosphate protection to minimize internucleotide bond cleavage during deprotection.[16]

Cleavage

1. NH₄OH, 55°C, 3 h.[17]
2. Treatment of an internucleotide 4-chlorophenyl ester with CsF and an alcohol (MeOH, EtOH, neopentylOH) results in transesterification.[18]

2,4-Dichlorophenyl: 2,4-Cl₂C₆H₃−

Cleavage

1. 4-Nitrobenzaldoxime, tetramethylguanidine, THF.[19]
2. Aqueous ammonia, dioxane, 12 h, 60°C.[20]

2,5-Dichlorophenyl: 2,5-Cl₂C₆H₃−

Cleavage

1. 4-Nitrobenzaldoxime, TEA, dioxane, H₂O.[21] Cleavage occurs in the presence of 4-nitrophenylethyl phosphate.
2. Pyridine-2-carbaldoxime, TEA, H₂O, dioxane. The 2-(1-methyl-2-imidazolyl)-phenyl group is not removed under these conditions.[22]

2,6-Dichlorophenyl: 2,6-Cl₂C₆H₃−

Cleavage of the 2,6-dichlorophenyl group is accomplished with 4-nitrobenzaldoxime, TEA, dioxane, H₂O.[23]

2-Bromophenyl: 2-BrC₆H₄−

Cleavage of the bromophenyl group is achieved with Cu(OAc)₂ in Pyr, H₂O. The 2-chlorophenyl group is stable to these conditions.[24]

4-Nitrophenyl (PNP): 4-NO₂C₆H₄−

Cleavage

1. p-Thiocresol, TEA, CH₃CN.[11] The 4-nitrophenyl group is removed in the presence of a 2-chlorophenyl group.

2.

(organoiodinane) aqueous micellar cetyltrimethylammonium

chloride, pH 8.[25]

3. Tetrabutylammonium acetate, 20 h, 20°C. For comparison, the 2,4-dichloro-phenyl group was removed in 100 h.[26]

4. syn-4-Nitrobenzaldoxime, tetramethylguanidine, dioxane, CH_3CN, 16 h.[26]

5. 0.125 N NaOH, dioxane.[26]

6. 4-Nitrophenyl phosphonates are transesterified in the presence of DBU and an alcohol.[27]

7. Zr^{4+}, H_2O, pH 3.5, 37°C.[28]

8. La(OTf)$_3$, MeOH converts the 4-nitrophenol derivative to a methyl derivative with a billion-fold rate acceleration and was used as a method to destroy the pesticide paraoxon.[29]

4-Chloro-2-nitrophenyl: 4-Cl-2-$NO_2C_6H_3$−

Cleavage is achieved with refluxing NaOH (15 min), but some deamination occurs with deoxyriboadenosine-5′-phosphate.[30] The ester is formed using the DCC protocol for phosphate ester formation.

2-Chloro-4-tritylphenyl

The lipophilicity of this phosphate protective group helps in the chromatographic purification of oligonucleotides. It is removed by the oximate method.[31]

2-Methoxy-5-nitrophenyl

This ester is cleaved by photolysis at >300 nm in basic aqueous acetonitrile.[32]

1,2-Phenylene

The phenylene group is removed oxidatively with Pb(OAc)$_4$ in dioxane.[33]

4-Tritylaminophenyl: $4-[(C_6H_5)_3CNH]C_6H_4-$

Formation

TrNHC$_6$H$_4$OH, DCC, Pyr.

Cleavage

Iodine, acetone or DMF, ammonium acetate, rt, 2 h. The tritylaminophenyl group is stable to isoamyl nitrite/acetic acid.[34]

4-Benzylaminophenyl: $4-[C_6H_5CH_2NH]C_6H_4-$

Cleavage

Electrolysis: 0.6–1.0 V, 3 h, DMF, H$_2$O, NaClO$_4$.[35] The related 4-tritylaminophenyl and 4-methoxyphenyl groups were not cleanly cleaved.

1-Methyl-2-(2-hydroxyphenyl)imidazole Derivative

The rate of oligonucleotide synthesis by the triester method using mesitylenesulfonyl chloride was increased 5- to 10-fold when this group was used as a protective group during internucleotide bond formation. It was removed with concd. NH$_4$OH at 60°C for 12 h[20] or by the oximate method.[22]

8-Quinolyl

This group is stable to acid and alkali. It has been used as a copper-activated leaving group for triphosphate protection.[36]

Formation

1. 8-Hydroxyquinoline, Ph$_3$P, 2,2′-dipyridyl disulfide, Pyr, rt, 6 h.[37]
2. 8-Hydroxyquinoline, (PhO)$_3$P, 2,2′-dipyridyl diselenide, Pyr, rt, 12 h.[38]

Cleavage

CuCl$_2$, DMSO, H$_2$O, 40–45°C, 5 h.[37]

5-Chloro-8-quinolyl

Formation

1. 5-Chloro-8-hydroxyquinoline, POCl$_3$, Pyr, 92% yield.[39]
2. 5-Chloro-8-hydroxyquinoline, 2,2'-Dipyridyl diselenide, (PhO)$_3$P, Pyr, rt, 12 h, 80–85% yield.[40]

Cleavage

1. Aqueous ammonia, 2 days, 27°C.[41]
2. Zn(OAc)$_2$, Pyr, H$_2$O, 28 h, 98% yield.[14]
3. 2-Pyridinecarboxaldoxime, tetramethylguanidine, dioxane, H$_2$O, 90% yield.[14]
4. ZnCl$_2$, aq. Pyr, rt, 12 h.[40,42]
5. Pyridine, t-BuNH$_2$, H$_2$O. Cleavage occurs in the presence of the 2,6-dichlorophenyl phosphate.[43]
6. The 5-chloro-8-quinolyl group can also be activated with CuCl$_2$ under anhydrous conditions and used in triphosphate formation.[44,45]

Thiophenyl: C$_6$H$_5$S−

The phosphorodithioate is stable to heating at 100°C, 80% acetic acid (1 h), dry or aqueous pyridine (days), and refluxing methanol, ethanol, or isopropyl alcohol for 1 h.

Formation

(ArS)$_2$P(O)O$^-$ C$_6$H$_{11}$NH$_3^+$ is prepared from the phosphinic acid with TMSCl, TEA, PhSSPh in THF at rt, 20 h in 83% yield.[46]

Cleavage

1. Treatment of ROP(O)(SPh)$_2$ (1) with 0.2 N NaOH (dioxane, rt, 15 min)[46] or pyridinium phosphinate (Pyr, TEA)[47] quantitatively gives ROP(O)(SPh)O$^-$ (2).

Ref. 47

2. AgOAc (Pyr, H$_2$O) cleaves both thioates of **1** to give a phosphate.[46]
3. Treatment of **2** with I$_2$ or AgOAc also gives the phosphate.[46]
4. Treatment of **1** with Zn (acetylacetone, Pyr, DMF) gives the phosphate.[46]
5. Treatment of **1** with phosphinic acid and triazole gives **2**.[46]
6. Treatment of (RO)$_2$P(O)SPh with Bu$_3$SnOMe converts it to (RO)$_2$P(O)OMe.[48,49]
7. (Bu$_3$Sn)$_2$O; TMSCl; H$_2$O.[50,51]
8. Treatment of ROP(O)(SPh)$_2$ with H$_3$PO$_3$/Pyr gives ROP(O)(SPh)OH.[52]
9. Phosphorothioates, when activated with AgNO$_3$ under anhydrous conditions in the presence of monophosphates, are converted into diphosphates.[53]
10. Tributylstannyl 2-pyridine-*syn*-carboxaldoxime, Pyr.[50]

Salicylic Acid Derivative

Salicylic acid was used for phosphite protection in the synthesis of glycosyl phosphites and phosphates. This derivative is very reactive and readily forms a phosphite upon treatment with an alcohol or a phosphonic acid upon aqueous hydrolysis.[54]

1. W. H. A. Kuijpers, J. Huskens, L. H. Koole, and C. A. A. Van Boekel, *Nucleic Acids Res.*, **18**, 5197 (1990).
2. J. W. Perich, P. F. Alewood, and R. B. Johns, *Aust. J. Chem.*, **44**, 233 (1991).
3. Y. Ichikawa and Y. C. Lee, *Carbohydr. Res.*, **198**, 235 (1990).
4. H. K. Chenault and R. F. Mandes, *Tetrahedron*, **53**, 11033 (1997).
5. C. C. Tam, K. L. Mattocks, and M. Tishler, *Synthesis*, 188 (1982).
6. K. K. Ogilvie and S. L. Beaucage, *Nucleic Acids Res.*, **7**, 805 (1979).
7. G. De Nanteuil, A. Benoist, G. Remond, J.-J. Descombes, V. Barou, and T. J. Verbeuren, *Tetrahedron Lett.*, **36**, 1435 (1995).
8. R. Plourde and M. d'Alarcao, *Tetrahedron Lett.*, **31**, 2693 (1990).
9. A. J. Morgan, Y. K. Wang, M. F. Roberts, and S. J. Miller, *J. Am. Chem. Soc.*, **126**, 15370 (2004).
10. M. Tsukamoto, R. Kato, K. Ishiguro, T. Uchida, and K. Sato, *Tetrahedron Lett.*, **32**, 7083 (1991).
11. S. S. Jones and C. B. Reese, *J. Am. Chem. Soc.*, **101**, 7399 (1979).
12. C. B. Reese and L. Zard, *Nucleic Acids Res.*, **9**, 4611 (1981).
13. C. B. Reese and L. Yau, *Tetrahedron Lett.*, **19**, 4443 (1978).
14. K. Kamaike, S. Ueda, H. Tsuchiya, and H. Takaku, *Chem. Pharm. Bull.*, **31**, 2928 (1983).
15. T. Tanaka, T. Sakata, K. Fujimoto, and M. Ikehara, *Nucleic Acids Res.*, **15**, 6209 (1987).
16. J. H. van Boom, P. M. J. Burgers, P. H. van Deursen, R. Arentzen, and C. B. Reese, *Tetrahedron Lett.*, **16**, 3785 (1974).
17. E. Ohtsuka, T. Tanaka, T. Wakabayashi, Y. Taniyama, and M. Ikehara, *J. Chem. Soc., Chem. Commun.*, 824 (1978).
18. U. Asseline, C. Barbier, and N. T. Thuong, *Phosphorus Sulfur*, **26**, 63 (1986).

19. B. Mlotkowska, *Liebigs Ann. Chem.*, 1361 (1991).

20. B. C. Froehler and M. D. Matteucci, *J. Am. Chem. Soc.*, **107**, 278 (1985).

21. E. Uhlmann and W. Pfleiderer, *Helv. Chim. Acta*, **64**, 1688 (1981).

22. B. S. Sproat, P. Rider, and B. Beijer, *Nucleic Acids Res.*, **14**, 1811 (1986).

23. H. Takaku, S. Hamamoto, and T. Watanabe, *Chem. Lett.*, **15**, 699 (1986).

24. Y. Stabinsky, R. T. Sakata, and M. H. Caruthers, *Tetrahedron Lett.*, **23**, 275 (1982).

25. R. A. Moss, B. Wilk, K. Krogh-Jespersen, J. T. Blair, and J. D. Westbrook, *J. Am. Chem. Soc.*, **111**, 250 (1989).

26. J. A. J. Den Hartog and J. H. Van Boom, *Recl.: J. R. Neth. Chem. Soc.*, **100**, 285 (1981).

27. D. S. Tawfik, Z. Eshhar, A. Bentolila, and B. S. Green, *Synthesis*, 968 (1993).

28. R. A. Moss, J. Zhang, and K. G. Ragunathan, *Tetrahedron Lett.*, **39**, 1529 (1998).

29. J. S. Tsang, A. A. Neverov, and R. S. Brown, *J. Am. Chem. Soc.*, **125**, 7602 (2003).

30. S. A. Narang, O. S. Bhanot, J. Goodchild, and R. Wightman, *J. Chem. Soc., Chem. Commun.*, 91 (1970).

31. J. J. Vasseur, B. Rayner, and J. L. Imbach, *Tetrahedron Lett.*, **24**, 2753 (1983).

32. N. R. Graciani, D. S. Swanson, and J. W. Kelly, *Tetrahedron*, **51**, 1077 (1995).

33. L.-d. Liu and H.-w. Liu, *Tetrahedron Lett.*, **30** 35 (1989).

34. E. Ohtsuka, S. Morioka, and M. Ikehara, *J. Am. Chem. Soc.*, **95**, 8437 (1973).

35. E. Ohtsuka, T. Miyake, M. Ikehara, A. Matsumoto, and H. Ohmori, *Chem. Pharm. Bull.*, **27**, 2242 (1979).

36. K. Fukuoka, F. Suda, M. Ishikawa, and T. Hata, *Nucl. Acids Symp. Ser.*, **29**, 35 (1993).

37. H. Takaku, Y. Shimada, and T. Hata, *Chem. Lett.*, **4**, 873 (1975).

38. H. Takaku, R. Yamaguchi, and T. Hata, *J. Chem. Soc., Perkin Trans. 1*, 519 (1978).

39. H. Takaku, K. Kamaike, and M. Suetake, *Chem. Lett.*, **12**, 111 (1983).

40. H. Takaku, R. Yamaguchi, and T. Hata, *Chem. Lett.*, **8**, 5 (1979).

41. S. C. Srivastava and A. L. Nussbaum, *J. Carbohydr. Nucleosides, Nucleotides*, **8**, 495 (1981).

42. H. Takaku, R. Yamaguchi, T. Nomoto, and T. Hata, *Tetrahedron Lett.*, **20**, 3857 (1979).

43. H. Takaku, K. Imai, and M. Nagai, *Chem. Lett.*, **17**, 857 (1988).

44. K. Fukuoka, F. Suda, R. Suzuki, M. Ishikawa, H. Takaku, and T. Hata, *Nucleosides Nucleotides*, **13**, 1557 (1994).

45. K. Fukuoka, F. Suda, R. Suzuki, H. Takaku, M. Ishikawa, and T. Hata, *Tetrahedron Lett.*, **35**, 1063 (1994).

46. M. Sekine, K. Hamaoki, and T. Hata, *Bull. Chem. Soc. Jpn.*, **54**, 3815 (1981).

47. T. Hata, T. Kamimura, K. Urakami, K. Kohno, M. Sekine, I. Kumagai, K. Shinozaki, and K. Miura, *Chem. Lett.*, **16**, 117 (1987).

48. S. Ohuchi, H. Ayukawa, and T. Hata, *Chem. Lett.*, **21**, 1501 (1992).

49. Y. Watanabe and T. Mukaiyama, *Chem. Lett.*, **8**, 389 (1979).

50. M. Sekine, H. Tanimura, and T. Hata, *Tetrahedron Lett.*, **26**, 4621 (1985).

51. H. Tanimura, M. Sekine, and T. Hata, *Tetrahedron*, **42**, 4179 (1986); H. Tanaka, H. Hayakawa, K. Obi, and T. Miyasaka, *idem*, 4187 (1986).

52. M. Sekine, K. Hamaoki, and T. Hata, *J. Org. Chem.*, **44**, 2325 (1979).

53. K. Fukuoka, F. Suda, R. Suzuki, M. Ishikawa, H. Takaku, and T. Hata, *Nucleosides & Nucleotides*, **13**, 1557 (1994).

54. J. P. G. Hermans, E. De Vroom, C. J. J. Elie, G. A. Van der Marel, and J. H. Van Boom, *Recl. Trav. Chim. Pays-Bas*, **105**, 510 (1986).

PHOTOCHEMICALLY CLEAVED PHOSPHATE PROTECTIVE GROUPS

The use of these for phosphate protection has been reviewed.[1-3] The following examples are representative.

Pyrenylmethyl Ester

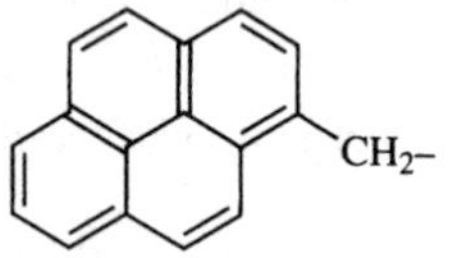

This derivative, synthesized by a silver oxide-promoted condensation of pyrenyl-methyl chloride and a dialkyl phosphate (92% yield), is quantitatively cleaved by photolysis at >300 nm in 60 min.[4]

Benzoin Ester

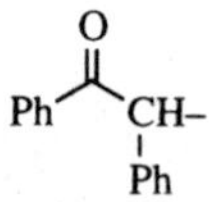

Formation

1. From $(EtO)_2P(O)Cl$: benzoin, Ag_2O.[4]
2. Bu_3NH-cAMP, desyl bromide.[5]

Cleavage

Photolysis, >300 nm.[4,6,7]

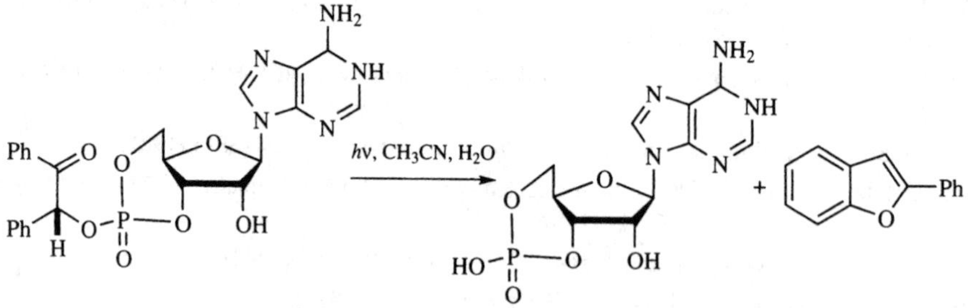

3′,5′-Dimethoxybenzoin Ester (3′,5′-DMB)

The phosphate ester, prepared through either phosphoramidite or phosphoryl chloride protocols, is cleavable by photolysis (350 nm, benzene, 83–87% yield).[8,9]

4-Hydroxyphenacyl Ester: $4\text{-}HOC_6H_4C(O)CH_2-$

The 4-hydroxyphenacyl group is removed by photolysis (300 nm, CH_3CN, tris buffer).[10,11]

The 4-hydroxyphenacyl group is also effectively cleaved from a thiophosphate derivative by photolysis.[12]

4-Methoxyphenacyl Ester: $4\text{-}CH_3OC_6H_4C(O)CH_2-$

Introduced with α-diazo-4-methoxyacetophenone, the phenacyl group is cleaved by photolysis with Pyrex-filtered mercury light in 74–86% yield.[13]

1-(2-Nitrophenyl)ethyl Ester

o-Nitrobenzyl Ester: $2\text{-}NO_2\text{-}C_6H_4CH_2-$

Formation

o-Nitrobenzyl alcohol, DCC, rt, 2 days. Pyridine slowly reacts to displace the nitrobenzyl ester, forming a 2-nitrobenzylpyridinium salt.[14]

Cleavage

1. Photolysis.[15-17]

2. Cleavage of an S-2-nitrobenzyl phosphorothioate is achieved with thiophen-oxide in 5 min.[18]

3,5-Dinitrophenyl Ester: $3,5-(NO_2)_2C_6H_3-$

Photolysis through a Pyrex filter in Pyr, EtOH, H_2O cleaves this phosphate ester.[19] The rate increases with increasing pH.

1. For reviews on photochemically cleaved phosphates, see C. G. Bochet, *J. Chem. Soc., Perkin Trans.* 1, 125 (2002); P. Pelliccioli Anna and J. Wirz, *Photochemical & Photobiological Sciences: Official Journal of the European Photochemistry Association and the European Society for Photobiology,* **1**, 441 (2002).

2. For a review of phosphate ester photochemistry, see R. S. Givens, and L. W. Kueper, III, *Chem. Rev.,* **93**, 55 (1993).

3. R. S. Givens, J. F. W. Weber, A. H. Jung, and C.-H. Park, *Methods in Enzymology,* **291**, 1 (1998).

4. T. Furuta, H. Torigai, T. Osawa, and M. Iwamura, *Chem. Lett.,* **22**, 1179 (1993).

5. R. S. Givens, P. S. Athey, L. W. Kueper, III, B. Matuszewski, and J.-y. Xue, *J. Am. Chem. Soc.,* **114**, 8708 (1992).

6. R. S. Givens and B. Matuszewski, *J. Am. Chem. Soc.,* **106**, 6860 (1984).

7. C.-H. Park and R. S. Givens, *J. Am. Chem. Soc.,* **119**, 2453 (1997).

8. M. C. Pirrung and S. W. Shuey, *J. Org. Chem.,* **59**, 3890 (1994).

9. J. E. Baldwin, A. W. McConnaughie, M. G. Moloney, A. J. Pratt, and S. B. Shim, *Tetrahedron,* **46**, 6879 (1990).

10. R. S. Givens and C.-H. Park, *Tetrahedron Lett.,* **37**, 6259 (1996).

11. C.-H. Park and R. S. Givens, *J. Am. Chem. Soc.,* **119**, 2453 (1997).

12. K. Zou, W. T. Miller, R. S. Givens, and H. Bayley, *Angew. Chem. Int. Ed.,* **40**, 3049 (2001).

13. W. W. Epstein and M. Garrossian, *J. Chem. Soc., Chem. Commun.,* 532 (1987).

14. E. Ohtsuka, H. Tsuji, T. Miyake, and M. Ikehara, *Chem. Pharm. Bull.,* **25**, 2844 (1977).

15. M. Rubenstein, B. Amit, and A. Patchornik, *Tetrahedron Lett.,* **16**, 1445 (1975).

16. J. W. Walker, G. P. Reid, J. A. McCray, and D. R. Trentham, *J. Am. Chem. Soc.,* **110**, 7170 (1988).

17. E. Ohtsuka, T. Tanaka, S. Tanaka, and M. Ikehara, *J. Am. Chem. Soc.,* **100**, 4580 (1978).

18. Z. J. Lesnikowski and M. M. Jaworska, *Tetrahedron Lett.,* **30**, 3821 (1989).

19. A. J. Kirby and A. G. Varvoglis, *J. Chem. Soc., Chem. Commun.,* 406 (1967).

AMIDATES

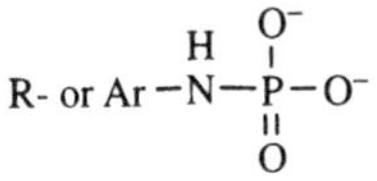

$$\text{R- or Ar} - \overset{H}{\underset{}{N}} - \overset{O^-}{\underset{\overset{\|}{O}}{P}} - O^-$$

Anilidate: C_6H_5NH-

A polymeric version of this group has been developed for terminal phosphate protection in ribooligonucleotide synthesis.[1]

Formation

Ph$_3$P, 2,2'-dipyridyl disulfide, aniline, 60% yield.[2]

Cleavage

Isoamyl nitrite, Pyr, acetic acid.[3,4]

4-Triphenylmethylanilidate: $4\text{-}(C_6H_5)_3CC_6H_4NH-$

This highly lipophilic group is cleaved with isoamyl nitrite in Pyr/AcOH.[5] The use of a lipophilic 5'-phosphate protective group aids in reverse phase HPLC purification of oligonucleotides.

[N-(2-Trityloxy)ethyl]anilidate: $(C_6H_5)_3COCH_2CH_2\text{-}C_6H_4\text{-}NH-$

This lipophilic group, developed for 5'-phosphate protection in oligonucleotide synthesis, is removed with 80% AcOH in 1 h.[6,7] The related trityloxyethylamino group has been used in a similar capacity for phosphate protection and is also cleaved with 80% AcOH.[8]

p-(N,N-Dimethylamino)anilidate: $p\text{-}(CH_3)_2NC_6H_4NH-$

This group was developed to aid in the purification of polynucleotides by adsorbing the phosphoroanilidates on an acidic ion-exchange resin.[9] Derivatives containing this as a terminal phosphate protective group could be adsorbed on an acid ion-exchange resin for purification. The group is removed with 80% acetic acid at 80°C for 3 h.[10]

Formation

DCC, N,N-dimethyl-p-phenylenediamine.

Cleavage

1. 80% acetic acid, 80°C, 3 h.
2. Isoamyl nitrite, Pyr, AcOH.[11]

3-(N,N-Diethylaminomethyl)anilidate: 3-[(C$_2$H$_5$)$_2$NCH$_2$]C$_6$H$_4$NH−

Cleavage is affected with isoamyl nitrite in Pyr/AcOH.[12,13]

p-Anisidate: p-CH$_3$OC$_6$H$_4$NH−

Cleavage

1. Pyr, AcOH, isoamyl nitrite.[14,15]
2. Bu$_4$NNO$_2$, Ac$_2$O, Pyr, rt, 10 min.[16]

2,2′-Diaminobiphenyl Derivative

Formation

2,2′-Diaminobiphenyl, Ph$_3$P, (PyS)$_2$.[17]

Cleavage

Isoamyl nitrite, Pyr, AcOH, AgOAc, benzoic anhydride.[17]

n-Propylamine and i-Propylamine Derivatives

These derivatives provide effective protection for phosphotyrosine in Fmoc-based peptide synthesis. They are cleaved with 95% TFA.[18]

N,N-Dimethyl-(R,R)-1,2-diaminocyclohexyl

This group was used as a protective group and chiral directing group for the asymmetric synthesis of α-aminophosphonic acids. It is cleaved by acid hydrolysis.[19]

Morpholino

Morpholine has been used for 5′-phosphate protection in oligonucleotide synthesis and can be cleaved with 0.01 N HCl without significant depurination of bases having free exocyclic amino functions.[20,21]

1. E. Ohtsuka, S. Morioka, and M. Ikehara, *J. Am. Chem. Soc.*, **94**, 3229 (1972).
2. E. Ohtsuka, H. Tsuji, T. Miyake, and M. Ikehara, *Chem. Pharm. Bull.*, **25**, 2844 (1977).
3. M. Sekine and T. Hata, *Tetrahedron Lett.*, **24**, 5741 (1983).
4. E. Ohtsuka, T. Ono, and M. Ikehara, *Chem. Pharm. Bull.*, **29**, 3274 (1981).
5. K. L. Agarwal, A. Yamazaki, and H. G. Khorana, *J. Am. Chem. Soc.*, **93**, 2754 (1971).
6. T. Tanaka, Y. Yamada, S. Tamatsukuri, T. Sakata, and M. Ikehara, *Nucl. Acids Symp. Ser.*, **17**, 85 (1986).
7. T. Tanaka, Y. Yamada, and M. Ikehara, *Tetrahedron Lett.*, **27**, 3267 (1986).
8. T. Tanaka, Y. Yamada, and M. Ikehara, *Tetrahedron Lett.*, **27**, 5641 (1986).
9. K. Tajima and T. Hata, *Bull. Chem. Soc. Jpn.*, **45**, 2608 (1972).

10. T. Hata, K. Tajima, and T. Mukaiyama, *J. Am. Chem. Soc.*, **93**, 4928 (1971).

11. K. Tajima and T. Hata, "Simple Protecting Group Protection-Purification Handle for Polynucleotide Synthesis," II. *Bull. Chem. Soc. Jpn.* **45**, 2608 (1972).

12. T. Hata, I. Nakagawa, and N. Takebayashi, *Tetrahedron Lett.*, **13**, 2931 (1972).

13. T. Hata, I. Nakagawa, and Y. Nakada, *Tetrahedron Lett.*, **16**, 467 (1975).

14. S. Iwai, M. Asaka, H. Inoue, and E. Ohtsuka, *Chem. Pharm. Bull.*, **33**, 4618 (1985).

15. E. Ohtsuka, M. Shin, Z. Tozuka, A. Ohta, K. Kitano, Y. Taniyama, and M. Ikehara, *Nucl. Acids Symp. Ser.*, **11**, 193 (1982).

16. S. Nishino, Y. Nagato, Y. Hasegawa, K. Kamaike, and Y. Ishido, *Nucl. Acids Symp. Ser.*, **20**, 73 (1988).

17. M. Nishizawa, T. Kurihara, and T. Hata, *Chem. Lett.*, **13**, 175 (1984).

18. M. Ueki, J. Tachibana, Y. Ishii, J. Okumura, and M. Goto, *Tetrahedron Lett.*, **37**, 4953 (1996).

19. S. Hanessian and Y. L. Bennani, *Synthesis*, 1272 (1994).

20. C. van der Marel, G. Veeneman, and J. H. van Boom, *Tetrahedron Lett.*, **22**, 1463 (1981).

21. A. Kondo, Y. Uchimura, F. Kimizuka, and A. Obayashi, *Nucl. Acids Symp. Ser.*, **16**, 161 (1985).

MISCELLANEOUS DERIVATIVES

Ethoxycarbonyl: EtO_2C-

The ethoxycarbonyl group was developed for the protection of phosphonates. The derivative is prepared by reaction of tris(trimethylsilyl) phosphite with ethyl chloroformate and can be cleaved by hydrolysis of the ester followed by silylation with bistrimethylsilylacetamide.[1]

(Dimethylthiocarbamoyl)thio: $(CH_3)_2NC(S)S-$

This group, used for internucleotide protection, is introduced with 8-quinolinesulfonyl chloride, $[(CH_3)_2NC(S)S]_2$, and Ph_3P and is cleaved with $BF_3 \cdot Et_2O$ (dioxane, H_2O, rt).[2]

1. M. Sekine, H. Mori, and T. Hata, *Bull. Chem. Soc. Jpn.*, **55**, 239 (1982).

2. H. Takaku, M. Kato, and S. Ishikawa, *J. Org. Chem.*, **46**, 4062 (1981).

10

REACTIVITIES, REAGENTS, AND REACTIVITY CHARTS

REACTIVITIES

In the selection of a protecting group, it is of paramount importance to know the reactivity of the resulting protected functionality toward various reagents and reaction conditions. The number of reagents available to the organic chemist is large: Approximately >8000 reagents are reviewed in the excellent series of books by the Fiesers.[1] In an effort to assess the effect of a wide variety of standard types of reagents and reaction conditions on the different possible protected functionalities, 108 prototype reagents have been selected and grouped into 16 categories:[2]

1. Aqueous
2. Nonaqueous Bases
3. Nonaqueous Nucleophiles
4. Organometallic
5. Catalytic Reduction
6. Acidic Reduction
7. Basic or Neutral Reduction
8. Hydride Reduction
9. Lewis Acids
10. Soft Acids

Greene's Protective Groups in Organic Synthesis, Fourth Edition, by Peter G. M. Wuts and Theodora W. Greene
Copyright © 2007 John Wiley & Sons, Inc.

11. Radical Addition
12. Oxidizing Agents
13. Thermal Reactions
14. Carbenoids
15. Miscellaneous
16. Electrophiles

These 108 reagents are used in the Reactivity Charts that have been prepared for each class of protective groups. The reagents and some of their properties are described on the following pages.

REAGENTS

1. AQUEOUS

1. pH < 1, 100°C	Refluxing HBr
2. pH < 1	1 N HCl
3. pH 1	0.1 N HCl
4. pH 2–4	0.01N HCl; 1-0.01 N AcOH
5. pH 4–6	0.1 N H$_3$BO$_3$; phosphate buffer; AcOH–NaOAc
6. pH 6–8.5	H$_2$O
7. pH 8.5–10	0.1 N HCO$_3^-$; 0.1 N OAc$^-$; satd. CaCO$_3$
8. pH 10–12	0.1 N CO$_3^{2-}$; 1–0.01 N NH$_4$OH; 0.01 N NaOH; satd. Ca(OH)$_2$
9. pH > 12	1–0.1 N NaOH
10. pH > 12, 150°C	

2. NONAQUEOUS BASES

11. NaH	
12. (C$_6$H$_5$)$_3$CNa	pK_a = 32
13. [C$_{10}$H$_8$]$^{-\cdot}$ Na$^+$	pK_a ≅ 37
14. CH$_3$SOCH$_2^-$Na$^+$	pK_a = 35
15. KO-t-C$_4$H$_9$	pK_a = 19
16. LiN(i-C$_3$H$_7$)$_2$ (LDA)	pK_a = 36
17. Pyridine; Et$_3$N	pK_a = 5; 10
18. NaNH$_2$; NaNHR	pK_a = 36

3. NONAQUEOUS NUCLEOPHILES

19. NaOCH$_3$/CH$_3$OH, 25°C	pK_a = 16
20. Enolate anion	pK_a = 20
21. NH$_3$; RNH$_2$; RNHOH	pK_a = 10
22. RS$^-$; N$_3^-$; SCN$^-$	
23. OAc$^-$; X$^-$	pK_a = 4.5

24. NaCH, pH 12
25. HCN, cat. CN$^-$, pH 6 $pK_a = 9$. For cyanohydrin formation

4. ORGANOMETALLIC

26. RLi
27. RMgX
28. Organozinc   Reformatsky reaction. Similar: R_2Cu; R_2Cd
29. Organocopper R_2CuLi
30. Wittig; ylide   Includes sulfur ylides

5. CATALYTIC REDUCTION

31. H_2/Raney Ni
32. H_2/Pt, pH 2–4
33. H_2/Pd–C
34. H_2/Lindlar
35. H_2/Rh–C or H_2/Rh–Al_2O_3 Avoids hydrogenolysis of benzyl ethers

6. ACIDIC REDUCTION

36. Zn/HCl
37. Zn/HOAc; $SnCl_2$/HCl
38. Cr(II), pH 5

7. BASIC OR NEUTRAL REDUCTION

39. Na/l NH$_3$
40. Al(Hg)
41. $SnCl_2$/Py
42. H_2S or HSO_3^-

8. HYDRIDE REDUCTION

43. LiAlH$_4$
44. Li–s-Bu$_3$BH, −50°C   Li-Selectride
45. [(CH$_3$)$_2$CHCH(CH$_3$)]$_2$BH   Disiamylborane
46. B$_2$H$_6$, 0°C
47. NaBH$_4$
48. Zn(BH$_4$)$_2$ Neutral reduction
49. NaBH$_3$CN, pH 4–6
50. (i-C$_4$H$_9$)$_2$AlH, −60°C   Dibal
51. Li(O−t-C$_4$H$_9$)$_3$AlH, 0°C

9. LEWIS ACIDS (ANHYDROUS CONDITIONS)

52. AlCl$_3$, 80°C
53. AlCl$_3$, 25°C

54. $SnCl_4$, 25°C; $BF_3 \cdot Et_2O$
55. $LiClO_4$; $MgBr_2$ — For epoxide rearrangement
56. TsOH, 80°C — Catalytic amount
57. TsOH, 0°C — Catalytic amount

10. SOFT ACIDS

58. Hg(II)
59. Ag(I)
60. Cu(II)/Py — For example, for Glaser coupling

11. RADICAL ADDITION

61. HBr/initiator — "Acidic" HX addition; acidity $\cong$ TsOH, 0°C
62. HX/initiator — Neutral HX addition; X = P, S, Se, Si
63. NBS/CCl_4, $h\nu$ or heat — Allylic bromination
64. $CHBr_3$; $BrCCl_3$; $CCl_4/In \cdot$ — Carbon–halogen addition

12. OXIDIZING AGENTS

65. OsO_4
66. $KMnO_4$, 0°C, pH 7
67. O_3, −50°C
68. RCO_3H, 0°C — Epoxidation of olefins; prototype for H_2O_2/H^+
69. RCO_3H, 50°C — Baeyer–Villiger oxidation of hindered ketones
70. CrO_3/Py — Collins oxidation
71. CrO_3, pH 1 — Jones oxidation
72. H_2O_2/OH^-, pH 10–12
73. Quinone — Dehydrogenation
74. $^1O_2$ — Singlet oxygen
75. CH_3SOCH_3, 100°C — (DMSO); HCO_3^- may be added to maintain neutrality
76. NaOCl, pH 10
77. Aq. NBS — Nonradical conditions
78. I_2
79. C_6H_5SCl; C_6H_5SeX
80. Cl_2; Br_2
81. MnO_2/CH_2Cl_2
82. $NaIO_4$, pH 5–8
83. SeO_2, pH 2–4
84. SeO_2/Pyridine — In EtOH/cat. Pyridine
85. $K_3Fe(CN)_6$, pH 7–10 — Phenol coupling
86. Pb(IV), 25°C — Glycol and α-hydroxy acid cleavage
87. Pb(IV), 80°C — Oxidative decarboxylation
88. $Tl(NO_3)_3$, pH 2 — Oxidative rearrangement of olefins

13. THERMAL REACTIONS

89. 150°C   Some Cope rearrangements and Cope eliminations

90. 250°C   Claisen or Cope rearrangement

91. 350°C   Ester cracking; Conia "ene" reaction

14. CARBENOIDS

92. $:CCl_2$

93. $N_2CHCO_2C_2H_5$/Cu, 80°C

94. CH_2I_2/Zn–Cu   Simmons–Smith addition

15. MISCELLANEOUS

95. n-Bu$_3$SnH/initiator

96. $Ni(CO)_4$

97. CH_2N_2

98. $SOCl_2$

99. Ac_2O, 25°C   Acetylation

100. Ac_2O, 80°C   Dehydration

101. DCC   Dicyclohexylcarbodiimide, $C_6H_{11}N=C=NC_6H_{11}$

102. CH_3I

103. $(CH_3)O^+BF_4^-$ Or CH_3OSO_2F = Magic Methyl: **SEVERE POISON**

104. 1. LiN$-i$-Pr$_2$; 2. MeI   For C-alkylation

105. 1. K_2CO_3; 2. MeI   For O-alkylation

16. ELECTROPHILES

106. RCHO

107. RCOCl

108. C^+ ion/olefin   For cation–olefin cyclization

REACTIVITY CHARTS

One requirement of a protective group is stability to a given reaction. The charts that follow were prepared as a guide to relative reactivities and thereby as an aid in the choice of a protective group. The reactivities in the charts were estimated by the individual and collective efforts of a group of synthetic chemists. *It is important to realize that not all the reactivities in the charts have been determined experimentally and that considerable conjecture has been exercised.* For those cases in which a literature reference was available concerning the use of a protective group and one of the 108 prototype reagents, the reactivity is printed in italic type. However, an exhaustive search for such references has not been made;

therefore, the absence of italic type does not imply an experimentally unknown reactivity.

There are four levels of reactivity in the charts:

"H" (high) indicated that under the conditions of the prototype reagent, the protective group is readily removed to regenerate the original functional group.

"M" (marginal) indicates that the stability of the protected functionality is marginal and depends on the exact parameters of the reaction. The protective group may be stable, may be cleaved slowly, or may be unstable to the conditions. Relative rates are always important, as illustrated in the following example[5] (in which monothioacetal is cleaved in the presence of a dithiane), and may have to be determined experimentally.

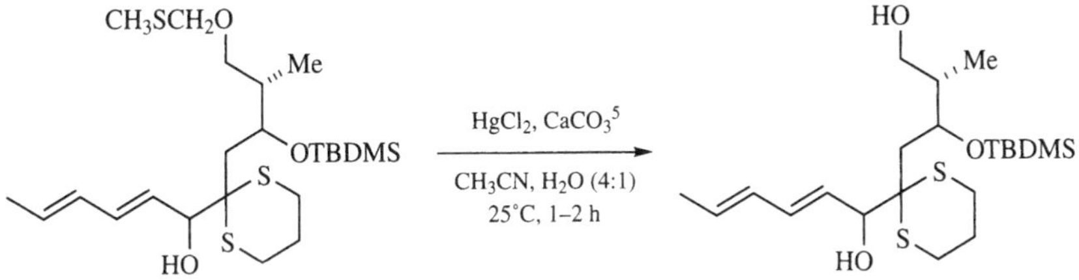

"L" (low) indicates that the protected functionality is stable under the reaction conditions.

"R" (reacts) indicates that the protected compound reacts readily, but that the original functional group is not restored. The protective group may be changed to a new protective group (eq. 1) or to a reactive intermediate (eq. 2), or the protective group may be unstable to the reaction conditions and react further (eq. 3).

(1) $ROCOC_6H_4\text{-}p\text{-}NO_2 \xrightarrow{H_2,\ Pd\text{-}C} ROCOC_6H_4\text{-}p\text{-}NH_2$

(2) $RCONR'_2 \xrightarrow{Me_3O^+\ BF_4^-}$

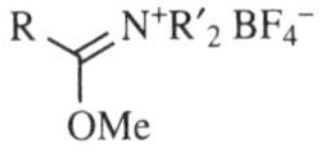

(3) $RCH(OR')_2 \xrightarrow{pH,\ <1,\ 100°C} [RCHO] \longrightarrow$ condensation products

The reactivities in the charts refer *only* to the protected functionality, not to atoms adjacent to the functional group; for example, RCOOEt $\xrightarrow{LDA}$: "L" (low) reactivity of PG(Et).

However, if the protected functionality is $R_2CHCOOEt$, this substrate obviously *will* react with LDA. Reactivity of the entire substrate must be evaluated by the chemist.

Five reagents [#25: HCN, pH 6; #88: $Tl(NO_3)_3$; #103: $Me_3O^+BF_4^-$; #104: LiN−*i*-Pr_2/MeI; and #105: K_2CO_3/MeI] were added after some of the charts had been completed; Reactivities to these reagents are not included for all charts.

Protective group numbers in the Reactivity Charts correspond to the list at the beginning of each chart. The protective groups that are included in the Reactivity

Charts are, in general, those that have been used most widely; consequently, considerable experimental information is available for them.

The Reactivity Charts were prepared in collaboration with the following chemists, to whom we are most grateful: John O. Albright, Dale L. Boger, Dr. Daniel J. Brunelle, Dr. David A. Clark, Dr. Jagabandhu Das, Herbert Estreicher, Anthony L. Feliu, Dr. Frank W. Hobbs, Jr., Paul B. Hopkins, Dr. Spencer Knapp, Dr. Pierre Lavallee, John Munroe, Jay W. Ponder, Marcus A. Tiu, Dr. David R. Williams, and Robert E. Wolf, Jr.

1. L. F. Fieser and M. Fieser, *Reagents for Organic Synthesis*, Wiley-Interscience, New York, 1967, Vol 1; M. Fieser and L. F. Fieser, Vols. 2–7, 1969–1979; M. Fieser, Vols. 8–17, 1980–1994, T. -L. Ho, Vols. 18–19, 1999.

2. The categories and prototype reagents used in this study are an expansion of an earlier set of 11 categories and 60 prototype reagents[3] originally compiled for use in LHASA (Logic and Heuristics Applied to Synthetic Analysis),[4] a long-term research program at Harvard University for Computer-Assisted Synthetic Analysis.

3. E. J. Corey, H. W. Orf, and D. A. Pensak, *J. Am.Chem. Soc.*, **98**, 210 (1976).

4. Selected references include E. J. Corey, *Q. Rev., Chem. Soc.*, **25**, 455 (1971); H. W. Orf, Ph.D. Thesis, Harvard University, 1976.

5. E. J. Corey and M. G. Bock, *Tetrahedron Lett.*, 2643 (1975).

Reactivity Chart 1.   Protection for Hydroxyl Group: Ethers

  1. Methyl Ether
  2. Methoxymethyl Ether (MOM)
  3. Methylthiomethyl Ether (MTM)
  4. 2-Methoxyethoxymethyl Ether (MEM)
  5. Bis(2-chloroethoxy)methyl Ether (BOM)
  6. Tetrahydropyranyl Ether (THP)
  7. Tetrahydrothiopyranyl Ether
  8. 4-Methoxytetrahydropyranyl Ether
  9. 4-Methoxytetrahydrothiopyranyl Ether
 10. Tetrahydrofuranyl Ether
 11. Tetrahydrothiofuranyl Ether
 12. 1-Ethoxyethyl Ether
 13. 1-Methyl-1-methoxyethyl Ether
 14. 2-(Phenylselenyl)ethyl Ether
 15. *t*-Butyl Ether
 16. Allyl Ether
 17. Benzyl Ether (PMB)
 18. *o*-Nitrobenzyl Ether

19. Triphenylmethyl Ether
20. alpha-Naphthyldiphenylmethyl Ether
21. *p*-Methoxyphenyldiphenylmethyl Ether
22. 9-(9-Phenyl-10-oxo)anthryl Ether (Tritylone)
23. Trimethylsilyl Ether (TMS)
24. Isopropyldimethylsilyl Ether
25. *t*-Butyldimethylsilyl Ether (TBDMS)
26. *t*-Butyldiphenylsilyl Ether
27. Tribenzylsilyl Ether
28. Triisopropylsilyl Ether

(See chart, pp. 994–996.)

Reactivity Chart 1. Protection for the Hydroxyl Group: Ethers

Reagent	Group	1	2	3	4	5	6	7	8	9	10	11	12	13	14	15	16	17	18	19	20	21	22	23	24	25	26	27	28
pH < 1, 100°C	Aqueous	H	H	H	H	H	H	H	H	H	H	H	H	H	H	H	H	H	H	H	H	H	H	H	H	H	H	H	H
pH < 1	Aqueous	M	H	H	H	H	H	H	H	H	H	H	H	H	H	H	H	H	H	H	H	H	H	H	M	H	H	H	H
pH 1	Aqueous	L	H	M	L	H	H	H	H	H	H	M	L	L	L	H	H	L	H	H	H	M	L	H	H	H	M	H	H
pH 2–4	Aqueous	L	L	M	L	H	H	H	H	H	H	M	H	H	L	L	L	L	H	H	L	H	H	H	L	L	H	H	H
pH 4–6	Aqueous	L	L	L	M	M	L	M	M	H	L	L	M	L	L	L	L	L	M	L	L	M	L	L	L	L	L	L	L
pH 6–8.5	Aqueous	L	L	L	L	L	L	L	L	L	L	L	L	L	L	L	L	L	L	L	L	L	L	L	L	L	L	L	L
pH 8.5–10	Aqueous	L	L	L	L	L	L	L	L	L	L	L	L	L	L	L	L	L	L	L	L	L	L	L	L	L	L	L	L
pH 10–12	Aqueous	L	L	L	L	L	L	L	L	L	L	L	L	L	L	L	L	L	L	L	L	L	L	H	L	L	L	L	L
pH 12	Aqueous	L	L	L	M	L	L	L	L	L	L	L	L	M	L	L	L	L	L	L	L	L	H	H	H	H	H	L	H
pH > 12, 150°C	Aqueous	L	M	M	L	H	L	M	L	M	L	M	L	L	M	L	R	L	H	H	L	M	L	H	H	H	H	H	H
NaH	Basic	L	L	L	L	L	L	L	L	L	L	L	L	L	L	L	L	L	L	L	L	R	L	L	L	L	L	L	L
Ph₃CNa	Basic	L	L	M	R	L	M	L	L	M	L	L	R	L	L	L	L	M	L	L	L	L	L	L	L	L	L	L	L
$C_{10}H_8^- Na^+$	Basic	L	L	R	L	R	L	M	L	L	R	L	L	R	L	L	R	H	H	H	R	L	L	L	L	H	L	L	L
$CH_3S(O)CH_2^- Na^+$	Basic	L	L	M	L	R	L	L	L	R	L	L	L	R	L	L	L	L	L	L	L	L	H	L	L	L	L	L	L
t-BuOK	Basic	L	L	L	R	L	L	L	L	L	L	M	L	R	L	L	L	L	L	L	L	L	L	L	L	L	L	L	L
LDA	Basic	L	L	M	L	R	L	M	L	L	L	M	L	L	R	L	L	L	R	L	L	L	L	L	L	L	L	L	L
Py, R_3N	Basic	L	L	L	L	L	L	L	L	L	L	L	L	L	L	L	L	L	L	L	L	L	L	L	L	L	L	L	L
Na/NH₃	Basic	L	L	L	R	L	L	L	L	R	L	L	L	R	L	L	L	L	L	L	L	H	L	L	L	L	L	L	L
NaOMe	Nucleophilic	L	L	L	R	L	L	L	L	L	L	L	L	M	L	R	L	L	L	L	L	L	H	L	L	L	L	L	L
Enolate	Nucleophilic	L	L	R	L	L	L	L	L	L	L	L	L	L	L	L	L	L	L	M	L	L	L	L	L	L	L	L	L
NH₃, RNH_2	Nucleophilic	L	L	L	M	L	L	L	L	L	L	L	L	L	L	L	L	L	L	L	L	R	L	L	L	L	L	L	L
RS^-, N_3^-, SCN^-	Nucleophilic	L	L	L	R	L	L	L	L	L	L	L	L	L	L	L	L	L	L	L	L	L	L	L	L	L	L	L	L
AcO^-, X^-	Nucleophilic	L	L	L	L	L	L	L	L	L	L	L	L	L	L	L	L	L	L	L	L	L	L	L	L	L	L	L	L
NaCN, pH 12	Nucleophilic	L	L	L	R	L	L	L	L	L	L	L	L	L	L	L	L	L	L	L	L	H	L	L	L	L	L	L	L
HCN, pH 6	Nucleophilic	L	L	L	L	M									L	L	L	L					H	L					
RLi	Organometallic	L	L	L	L	R	L	L	M	L	L	L	L	R	L	L	M	L	L	L	L	R	H	H	L	L	L	L	L
RMgX	Organometallic	L	L	L	L	R	L	L	L	L	L	L	L	R	L	L	L	L	L	L	L	R	H	H	L	L	L	L	L
Organozinc	Organometallic	L	L	L	L	M	L	L	L	L	L	L	L	L	L	L	L	L	L	L	L	R	L	L	L	L	L	L	L
Organocopper	Organometallic	L	L	L	R	L	L	L	L	L	L	L	L	L	L	L	L	R	L	L	L	L	L	L	L	L	L	L	L
Wittig, ylide	Organometallic	L	L	L	L	L	L	L	L	L	L	L	L	L	L	L	L	L	L	L	L	R	L	H	L	L	L	L	L
H₂, Raney Ni	Cat. Reduction	L	L	R	L	R	L	R	L	R	L	L	L	R	L	R	L	R	L	H	H	H	R	L	L	M	L	L	L
H₂/Pt pH 2–4	Cat. Reduction	L	M	R	L	R	H	R	H	R	H	R	L	R	L	R	L	R	H	R	H	H	R	H	H	H	L	L	L
H₂/Pd	Cat. Reduction	L	L	M	L	R	L	R	L	R	L	R	L	L	R	L	R	H	H	H	H	R	H	M	L	L	L	L	L
H₂/Lindlar	Cat. Reduction	L	L	L	L	L	L	L	L	L	L	L	L	L	L	L	L	L	L	L	L	L	L	L	L	L	L	L	L
H₂/Rh	Cat. Reduction	L	L	R	L	L	R	L	R	L	R	L	L	R	L	R	L	R	L	L	L	R	L	L	L	R	L	L	L
Zn/HCl	Misc.	L	H	R	M	H	H	H	H	H	H	H	H	L	L	L	R	H	H	H	R	H	H	H	R	H	H	M	H
Zn/AcOH	Misc.	L	M	R	L	H	H	H	H	H	H	H	H	L	L	L	R	H	H	H	R	H	H	L	L	H	H	L	H
Ch(II), pH 5	Misc.	L	L	M	L	M	M	M	M	H	M	L	M	L	L	L	R	L	L	M	R	H	M	L	L	M	R	H	M

994

Reactivity Chart 1. Protection for the Hydroxyl Group: Ethers (Continued)

PG	Na/NH₃	Al(Hg)	SnCl₂/py	HSO₃⁻, H₂S	LiAlH₄	Li-s-Bu₃BH	Chex₂BH	B₂H₆, 0°C	NaBH₄	Zn(BH₄)₂	NaBH₃CN, pH 4–6	i-Bu₂AlH	Li(t-BuO)₃AlH	AlCl₃, 80°C	AlCl₃, 25°C	SnCl₄, BF₃	LiClO₄, MgBr₂	TsOH, 80°C	TsOH, 0°C	Hg(II)	Ag(I)	Cu(II)/Py	HBr, In•	HX/In•	NBS/CCl₄	Br₃CCl/In•	OsO₄	KMnO₄, pH 7, 0°C	O₃, −50°C	RCO₃H, 0°C	RCO₃H, 50°C	CrO₃/Py	CrO₃, pH 1	H₂O₂, pH 10–12	Quinone	¹O₂	DMSO, 100°C	NaOCl, pH 10	Aq. NBS
	Single Elec. Red.				**Hydride Reductions**									**Acid and Lewis Acid**						**Soft Acids**			**Free Rad. Rxn**				**Oxidants**												
1	L	L	L	L	L	L	L	L	L	L	L	L	L	H	H	H	L	L	L	L	L	L	L	L	M	L	L	L	L	L	L	L	L	L	L	L	L	L	L
2	R	L	L	L	L	L	L	L	L	L	L	L	L	H	H	H	L	L	L	L	L	L	H	L	R	L	L	L	R	L	M	L	L	L	L	L	L	L	R
3	R	L	M	L	L	L	L	L	L	L	L	L	L	H	H	M	M	M	L	H	H	L	M	L	R	M	L	R	R	R	R	L	H	L	L	R	M	R	R
4	M	M	L	L	M	L	L	L	L	L	L	M	L	H	H	H	H	M	L	L	L	L	H	L	R	L	L	L	R	L	L	L	H	L	L	L	L	L	M
5	L	L	M	L	L	L	L	L	L	L	L	L	L	H	H	H	L	L	M	L	R	L	H	L	M	R	L	L	H	L	M	L	H	L	L	R	L	L	L
6	R	L	L	L	L	L	L	L	L	L	L	L	L	H	H	H	M	H	L	L	L	L	H	L	M	L	L	R	R	L	H	L	H	L	L	R	L	L	R
7	R	L	L	L	L	L	L	L	L	L	M	L	L	H	M	M	L	H	L	H	H	L	M	L	M	L	L	L	R	R	H	L	H	L	L	L	M	R	R
8	R	L	L	L	L	L	L	L	L	L	H	L	L	H	H	H	L	L	L	L	L	L	H	L	M	L	L	R	R	L	R	L	H	L	L	R	L	L	R
9	M	L	L	L	L	L	L	L	L	L	L	L	L	H	H	H	M	M	L	L	L	L	H	L	R	L	L	L	R	R	H	L	H	L	L	L	M	R	M
10	M	M	L	L	L	L	L	L	L	L	L	L	L	H	M	H	L	H	M	L	L	L	H	L	M	L	L	R	R	L	R	L	H	L	L	R	L	L	R
11	R	L	L	L	L	L	L	L	L	L	M	L	L	H	H	M	M	H	L	H	H	L	M	L	L	L	R	L	L	R	H	L	H	L	L	L	M	R	M
12	L	L	L	L	L	L	L	L	L	L	L	R	L	H	H	H	L	M	L	L	L	L	H	L	R	L	L	L	R	L	R	L	R	L	L	R	L	L	M
13	H	L	L	L	L	L	L	L	L	L	L	L	L	H	H	H	L	H	L	L	L	L	L	L	R	R	L	R	L	H	L	R	R	L	L	L	M	L	R
14	H	R	H	L	L	L	L	L	L	L	L	L	L	H	M	M	L	H	L	L	L	L	R	R	L	L	L	L	R	R	R	L	L	L	L	R	L	R	L
15	R	M	H	L	L	L	L	L	L	L	L	L	L	H	H	H	L	M	H	L	L	L	M	L	L	L	R	R	H	L	L	L	M	L	L	L	L	L	R
16	H	M	L	R	R	R	R	R	R	R	M	R	M	H	H	H	M	M	L	R	L	L	R	L	L	L	L	L	R	R	L	L	L	L	L	L	L	L	L
17	H	M	L	H	L	H	L	L	H	H	M	H	H	H	H	H	L	H	H	L	L	L	R	L	L	L	L	L	R	L	L	L	L	L	L	L	L	L	L
18	R	R	M	L	L	L	L	L	L	L	R	L	L	H	M	H	L	H	L	L	L	L	R	L	L	L	L	L	L	L	H	L	H	L	L	L	L	L	L
19	H	L	L	L	L	L	L	L	L	L	L	L	L	H	H	L	L	H	L	L	L	L	H	M	L	L	L	L	L	L	H	R	H	L	L	L	L	L	L
20	H	L	L	L	R	L	L	L	L	L	M	R	L	H	H	H	M	H	L	L	L	L	H	L	R	M	L	L	L	M	R	L	L	L	L	L	L	L	L
21	R	L	L	L	H	L	R	L	L	L	L	H	L	H	H	H	L	H	M	L	L	L	H	L	M	M	L	L	L	L	H	L	H	L	L	L	L	L	L
22	H	R	L	L	L	L	L	L	L	L	L	L	L	H	M	H	L	H	L	L	L	L	H	L	M	L	L	H	L	L	L	L	L	H	L	L	H	H	H
23	L	L	L	R	L	L	L	L	L	L	L	L	L	H	L	L	L	H	L	L	L	L	H	L	R	L	L	L	L	L	L	L	M	L	L	L	L	L	L
24	L	L	L	H	L	L	L	L	L	L	L	L	L	H	M	L	L	H	L	L	L	L	H	L	M	L	L	L	L	L	L	L	H	L	L	L	L	L	L
25	L	L	L	L	L	L	L	L	L	L	L	L	L	H	L	L	L	M	L	L	L	L	H	M	M	L	L	L	L	L	L	L	H	L	L	L	L	L	L
26	R	R	L	L	L	L	L	L	L	L	L	L	L	H	L	L	L	H	L	L	L	L	L	L	R	L	L	L	L	L	L	L	H	H	L	L	L	L	L
27	L	L	L	L	L	L	L	L	L	L	L	L	L	H	H	L	L	H	L	L	L	L	R	L	M	M	L	L	L	L	L	L	L	L	L	L	L	L	L
28	L	L	L	L	L	L	L	L	L	L	L	L	L	H	M	M	L	H	L	L	L	L	H	L	M	L	L	L	L	L	L	L	H	L	L	L	L	L	L

995

Reactivity Chart 1. Protection for the Hydroxyl Group: Ethers (*Continued*)

PG	Oxidants											Thermal			Carbenes			Miscellaneous											Electrophiles		
	I_2	PhSeX, PhSCl	Br_2, Cl_2	MnO_2/CH_2Cl_2	$NaIO_4$, pH 5–6	SeO_2, pH 2–4	SeO_2, Py	$K_3Fe(CN)_6$, pH 8	Pb(IV), 25°C	Pb(IV), 80°C	$Tl(NO_3)_3$	150°C	250°C	350°C	:CCl_2	N_2CHCO_2R, Cu or Rh	CH_2I_2, Zn(Cu)	R_3SnH, In·	$Ni(CO)_4$	CH_2N_2	$SOCl_2$	Ac_2O, 25°C	Ac_2O, 80°C	DCC	MeI	Me_3O^+ BF_4^-	1. LDA, 2. MeI	1. K_2CO_3, 2. MeI	RCHO	RCOCl	C^+
1	L	L	M	L	L	L	L	L	L	L	L	L	L	L	L	L	L	L	L	L	L	L	L	L	L	L	L	L	L	L	L
2	L	L	R	L	L	H	L	L	L	H	H	L	L	R	L	L	L	L	L	L	L	L	L	L	L	M	L	L	L	L	H
3	L	L	R	L	R	H	M	L	R	R	R	L	M	R	M	R	L	M	L	L	L	L	L	L	R	R	M	R	L	L	M
4	L	L	M	L	L	L	L	L	L	M	M	L	L	R	L	L	R	R	L	L	L	M	M	L	L	M	L	L	L	L	H
5	L	L	M	L	L	H	L	L	L	H		L	L	H	L	M	L	L	L	L	L	L	L	L	L	M	L	L	L	L	
6	L	L	R	L	L	H	L	M	M	H	H	M	M	H	L	L	L	R	L	L	L	L	L	L	L	R	R	R	L	L	R
7	L	L	R	L	R	H	M	L	H	H		L	M	H	M	M	R	L	L	L	L	L	M	L	R	M	R	R	L	L	
8	L	L	R	L	L	H	L	M	R	H		L	M	H	L	M	L	M	L	L	L	L	H	L	L	R	R	R	L	L	
9	L	L	R	R	R	H	M	L	H	H		R	M	R	M	L	R	L	L	L	L	L	L	L	R	M	R	R	L	L	
10	L	L	M	L	L	H	L	M	R	H		L	R	R	L	M	L	M	L	L	L	L	M	L	L	R	L	M	L	L	
11	R	R	M	L	M	H	M	L	M	H	M	L	L	R	M	L	L	L	L	L	L	L	L	L	L	L	L	L	L	L	H
12	L	L	R	L	R	R	L	L	R	R	R	L	L	R	L	M	L	M	L	L	L	L	L	L	R	M	L	L	L	L	H
13	M	L	L	L	L	L	L	L	L	L	L	R	R	R	M	L	M	L	M	L	L	L	M	L	L	M	M	L	L	L	L
14	L	L	R	R	M	R	R	L	L	R	M	L	R	R	L	L	L	M	L	L	L	L	M	L	L	L	L	L	L	L	L
15	L	L	M	L	R	L	L	L	L	L	R	L	R	R	M	L	M	L	L	L	L	L	M	L	L	L	L	L	L	L	R
16	M	L	L	L	L	H	L	L	L	L	L	L	M	R	L	L	L	M	M	L	L	L	H	L	L	H	L	L	L	L	L
17	L	L	L	L	L	H	L	L	L	H	M	L	M	R	L	L	L	L	L	L	L	L	H	L	L	M	L	L	L	L	L
18	L	R	L	L	L	L	L	L	L	L	R	L	M	R	L	L	L	L	L	L	L	L	M	L	L	L	L	L	L	L	L
19	L	L	L	L	L	H	L	L	L	H	L	R	R	R	L	L	L	L	L	L	L	M	H	L	L	L	M	L	L	L	L
20	L	L	L	L	L	H	L	H	L	H		L	L	R	L	L	L	L	L	L	L	M	H	L	L	L	L	L	L	L	H
21	L	L	L	L	L	L	L	L	H	L		L	L	R	L	L	L	L	L	L	L	L	L	L	L	L	L	L	L	L	L
22	L	L	L	L	L	H	L	L	H	L	H	L	L	R	L	L	L	M	L	L	L	L	M	L	L	H	L	M	L	L	L
23	M	L	L	L	L	L	L	L	M	L		L	L	R	L	L	L	L	L	L	L	L	H	L	L	M	M	L	L	L	L
24	L	L	M	L	M	H	L	L	L	H		L	L	R	L	L	L	L	L	L	L	L	L	L	L	M	L	L	L	L	L
25	L	L	L	L	L	H	L	L	L	H	M	L	L	R	L	L	L	L	L	L	L	L	L	L	L	L	L	L	L	L	L
26	L	L	L	L	L	L	L	L	M	L		L	L	R	L	L	L	L	L	L	L	L	L	L	L	L	L	L	L	L	L
27	L	L	M	L	L	H	L	L	L	M		L	L	R	L	L	L	L	L	L	L	L	L	L	L	M	M	L	L	L	L
28	L	L	L	L	L	H	L	L	M	H		L	L	R	L	L	L	L	L	L	L	M	M	L	L	M	L	L	L	L	L

996

Reactivity Chart 2.   Protection for Hydroxyl Group: Esters

1. Formate Ester
2. Acetate Ester
3. Trichloroacetate Ester
4. Phenoxyacetate Ester
5. Isobutyrate Ester
6. Pivaloate Ester
7. Adamantoate Ester
8. Benzoate Ester
9. 2,4,6-Trimethylbenzoate (Mesitoate) Ester
10. Methyl Carbonate
11. 2,2,2-ATrichloroethyl Carbonate
12. Allyl Carbonate
13. *p*-Nitrophenyl Carbonate
14. Benzyl Carbonate
15. *p*-Nitrobenzyl Carabonate
16. *S*-Benzyl Thiocarbonate
17. *N*-Phenylcarbamate
18. Nitrate Ester
19. 2,4-Dinitrophenylsulfenate Ester

(See chart, pp. 998–1000.)

Reactivity Chart 2. Protection for the Hydroxyl Group: Esters

PG	Aqueous										Basic								Nucleophilic							Organometallic					Catalytic Reduction					1 elec. Red.		
	pH < 1, 100°C	pH < 1	pH 1	pH 2-4	pH 4-6	pH 6-8.5	pH 8.5-10	pH 10-12	pH > 12	pH > 12, 150°C	NaH	Ph₃CNa	$(C_{10}H_8)^-$·Na⁺	CH₃S(O)CH₂⁻Na⁺	t-BuOK	LDA	Py; R₃N	NaNH₂	MeONa	Enolate	NH₃; RNH₂	RS⁻; N₃⁻; SCN⁻	AcO⁻; X⁻	NaCN, pH 12	HCN, pH 6	RLi	RMgX	Organozinc	Organocopper	Wittig; ylide	H₂/Raney (Ni)	H₂/Pt pH 2-4	H₂/Pd	H₂/Lindlar	H₂/Rh	Zn/HCl	Zn/AcOH	Cr(II), pH 5
1	H	H	H	M	L	L	H	H	H	H	H	L	H	H	H	H	L	H	H	H	H	L	L	M		H	H	M	L	H	M	M	M	M	L	M	M	L
2	H	M	L	L	L	L	M	H	H	H	R	H	H	H	H	H	L	R	R	R	M	H	L	R	L	H	H	L	L	L	L	M	M	L	L	L	M	L
3	H	M	M	L	L	L	H	H	H	H	L	M	H	H	H	H	L	L	H	R	M	H	L	L		H	H	L	L	L	R	R	R	L	L	R	R	H
4	H	M	L	L	L	L	M	H	H	H	M	M	H	H	H	L	L	R	H	R	M	H	L	L		H	H	L	L	L	L	L	L	L	L	L	L	L
5	H	M	M	L	L	L	L	M	M	H	H	H	H	H	H	H	L	H	M	M	M	H	L	L	L	M	M	L	L	L	L	L	L	L	L	L	L	L
6	H	M	L	L	L	L	L	L	H	H	L	L	H	M	R	M	L	M	M	L	H	H	L	L		L	L	L	L	L	L	L	L	L	L	L	L	L
7	H	H	L	L	L	L	L	L	L	H	L	L	L	L	L	L	L	L	L	M	M	H	L	L		H	H	L	L	L	L	L	L	L	L	L	L	L
8	H	M	L	L	L	L	M	M	H	H	L	L	H	H	L	L	L	H	M	M	M	H	L	L	M	L	L	L	L	L	L	L	L	L	L	L	L	L
9	H	H	L	L	L	L	L	L	H	H	L	L	H	H	H	L	L	H	L	L	L	L	L	L		R	R	R	L	M	R	R	R	L	L	R	R	R
10	H	M	L	L	L	L	L	H	H	H	L	L	H	H	R	H	L	R	R	R	M	L	L	R	M	R	R	M	H	M	R	R	R	L	R	R	R	H
11	H	M	L	L	L	L	M	H	H	H	L	L	H	H	R	H	L	H	M	R	M	H	L	L		R	R	R	H	L	R	R	R	L	R	H	H	H
12	H	M	L	L	L	L	L	H	H	H	L	L	H	H	M	H	L	H	R	R	M	L	L	L	M	R	R	M	M	M	R	H	H	L	M	L	L	L
13	H	M	L	L	L	L	L	H	H	H	L	L	H	H	L	L	L	H	M	R	M	L	L	M		R	R	L	L	L	H	H	H	L	H	H	H	R
14	H	M	L	L	L	L	L	H	H	H	L	L	H	H	R	H	L	H	R	R	M	L	L	L		R	R	L	L	M	R	H	R	L	H	R	R	R
15	H	L	L	L	L	L	L	L	M	H	L	L	H	H	H	R	L	H	R	M	H	M	M	L		H	H	M	L	H	H	H	R	L	H	R	R	L
16	H	L	L	L	L	L	L	L	H	H	R	H	H	H	H	M	L	H	M	L	M	L	L	L		R	R	L	L	M	R	R	R	L	L	R	R	R
17	H	H	M	L	L	L	L	L	H	H	L	H	H	H	R	R	L	R	R	L	H	M	L	L		R	R	M	M	M	H	H	H	R	H	R	R	R
18	H	M	M	L	L	L	L	L	H	H	L	L	H	H	H	M	L	H	H	M	H	M	L	L		H	H	M	L	M	R	R	R	L	H	L	L	L
19	H	M	L	L	L	L	H	H	H	H	L	L	H	H	H	H	L	H	H	H	H	H	H	H		M	M	H	H	H	H	H	H	L	H	H	H	H

PG	Na/NH₃	Al(Hg)	SnCl₂/py	HSO₃⁻, H₂S	LiAlH₄	Li-s-Bu₃BH	Chex₂BH	B₂H₆, 0°C	NaBH₄	Zn(BH₄)₂	NaBH₃CN, pH 4–6	i-Bu₂AlH	Li(t-BuO)₃AlH	AlCl₃, 80°C	AlCl₃, 25°C	SnCl₄, BF₃	LiClO₄, MgBr₂	TsOH, 80°C	TsOH, 0°C	Hg(II)	Ag(I)	Cu(II)/Py	HBr, In•	HX/In•	NBS/CCl₄	Br₂CCl/In•	OsO₄	KMnO₄, pH 7, 0°C	O₃, −50°C	RCO₃H, 0°C	RCO₃H, 50°C	CrO₃/Py	CrO₃, pH 1	H₂O₂, pH 10–12	Quinone	¹O₂	DMSO, 100°C	NaOCl, pH 10	Aq. NBS
1	H	L	L	L	H	H	M	M	M	M	M	H	M	H	H	L	L	H	L	L	L	L	M	L	L	L	L	L	L	L	L	L	L	H	L	L	M	H	L
2	H	L	L	L	H	M	L	M	L	L	L	H	L	L	L	L	L	M	L	L	L	L	L	L	L	L	L	L	L	L	L	L	L	L	L	L	L	L	L
3	H	H	L	L	H	H	L	L	L	M	L	H	L	R	R	L	L	M	L	L	R	L	R	L	L	L	L	L	M	L	L	L	L	H	L	L	H	H	L
4	H	H	L	L	H	M	L	L	L	L	L	H	L	R	R	L	L	M	L	L	L	L	M	L	L	L	L	L	L	L	L	L	L	L	L	L	L	M	L
5	H	H	L	L	H	L	L	L	L	L	L	H	L	L	L	L	L	L	L	L	L	L	L	L	L	L	L	L	L	L	L	L	L	L	L	L	L	L	L
6	H	L	L	L	H	L	L	L	L	L	L	L	L	L	L	L	L	L	L	L	L	L	L	L	L	L	L	L	L	L	L	L	L	L	L	L	L	L	L
7	H	L	L	L	H	L	L	L	L	L	L	L	L	L	L	L	L	M	L	L	L	L	L	L	L	L	L	L	L	L	L	L	L	L	L	L	L	L	L
8	H	L	L	L	H	M	L	L	L	L	L	H	L	L	L	L	L	L	L	L	L	L	L	L	L	L	L	L	L	L	L	L	L	L	L	L	L	L	L
9	H	L	L	L	H	L	L	L	L	L	L	L	L	L	L	L	L	M	L	L	L	L	L	L	R	L	L	L	L	L	L	L	L	L	L	L	L	L	L
10	H	R	L	L	H	M	H	R	L	M	L	H	L	R	R	H	L	M	L	L	R	L	M	R	L	R	L	L	H	L	L	L	L	H	L	R	M	M	R
11	H	L	L	L	H	L	L	L	L	L	L	H	R	R	R	L	L	M	L	M	L	L	R	L	R	L	H	H	L	H	H	L	L	H	L	L	M	H	L
12	H	H	L	L	H	H	L	L	L	L	L	H	L	R	M	L	L	M	L	L	L	L	L	L	L	L	L	L	L	L	L	L	L	R	L	L	L	H	L
13	H	L	L	L	H	L	L	L	L	M	L	H	L	R	R	L	L	M	L	L	L	L	L	L	R	L	L	L	L	L	L	L	L	H	L	L	M	L	L
14	H	R	L	L	H	M	M	H	M	L	L	H	M	R	M	L	L	M	L	L	L	L	L	L	L	L	L	L	H	L	L	L	L	L	L	L	L	L	R
15	H	L	L	L	H	H	H	H	L	H	L	H	H	R	L	R	L	M	L	R	M	L	L	L	L	L	L	R	L	M	R	L	R	L	L	L	L	L	L
16	H	L	L	L	H	H	H	H	R	H	L	H	M	R	R	R	L	M	L	L	L	L	L	L	L	L	L	L	L	L	L	L	L	H	L	L	L	H	L
17	H	H	H	M	H	H	H	H	L	H	H	H	H	R	R	H	L	M	L	L	L	L	H	H	H	H	L	L	L	L	L	L	M	L	L	L	L	L	L
18	H	H	H	H	H	H	H	H	H	H	H	H	H	R	H	H	L	M	L	L	L	L	H	H	H	H	L	L	R	R	R	L	L	L	L	L	L	L	L
19	H	H	H	H	H	H	H	H	H	H	H	H	H	H	H	H	L	M	L	M	M	L	R	R	R	R	L	R	R	R	R	L	R	H	L	M	H	H	H

999

Reactivity Chart 2. Protection for the Hydroxyl Group: Esters (*Continued*)

PG	I₂	PhSeX, PhSCl	Br₂, Cl₂	MnO₂/CH₂Cl₂	NaIO₄, pH 5–6	SeO₂, pH 2–4	SeO₂, Py	K₃Fe(CN)₆, pH 8	Pb(IV), 25°C	Pb(IV), 80°C	Tl(NO₃)₃	150°C	250°C	350°C	:CCl₂	N₂CHCO₂R, Cu or Rh	CH₂I₂, Zn(Cu)	R₃SnH, In•	Ni(CO)₄	CH₂N₂	SOCl₂	Ac₂O, 25°C	Ac₂O, 80°C	DCC	MeI	Me₃O⁺ BF₄⁻	1. LDA, 2. MeI	1. K₂CO₃, 2. MeI	RCHO	RCOCl	C⁺/olefin
1	L	L	L	L	L	M	M	M	L	L		L	M	H	L	L	L	L	L	L	M	L	L	L	L			M	L	L	L
2	L	L	L	L	L	L	M	M	L	L	L	L	M	H	L	L	R	L	L	L	L	L	L	L	L	M	R	M			
3	L	L	R	L	L	L	L	H	L	L		M	H	H	L	L	L	R	M	L	L	L	L	L	L						
4	L	L	L	L	L	L	L	L	L	L	L	L	L	H	L	L	L	L	L	L	L	L	L	L	L		L	L	L	L	L
5	L	L	L	L	L	L	L	L	L	L		L	M	H	L	L	L	L	L	L	L	L	L	L	L		L	L			
6	L	L	L	L	L	L	L	L	L	L	L	L	L	H	R	L	L	L	L	L	L	L	L	L	L	L	L	L	L	L	L
7	L	L	L	L	L	L	L	L	L	L	L	L	M	H	L	L	L	L	L	L	L	L	L	L	L	L	L	L	L	L	L
8	L	L	L	L	L	L	L	L	L	L		L	H	H	L	L	L	L	L	L	L	L	L	L	L	R	R	M	L	L	L
9	L	L	L	L	L	L	L	L	L	L	L	M	H	H	L	H	H	R	M	L	L	L	L	L	L	R	R	H			
10	L	L	R	L	L	L	L	L	L	R		M	H	H	L	R	R	H	H	L	L	L	L	L	L				L	L	L
11	M	R	L	L	L	H	H	L	L	R		M	H	H	R	H	H	R	L	L	L	L	L	L	L	R	L	M	L	L	L
12	L	L	L	L	L	L	M	L	L	L		M	H	H	L	R	L	R	L	L	L	L	L	L	L		M				
13	L	L	L	L	L	M	M	L	L	L		M	H	H	L	L	L	L	L	L	L	L	L	L	L				L	L	L
14	M	L	H	L	L	H	H	L	L	R		M	M	H	L	L	L	H	L	L	H	R	L	L	L	R		M	L	L	L
15	L	L	L	L	L	L	H	L	L	L		M	H	H	L	L	L	R	L	L	L	L	L	L	L		R				
16	L	L	L	L	M	L	L	L	L	R		M	H	H	L	L	L	L	L	L	L	R	L	L	L			L	L	L	L
17	L	L	L	L	L	L	L	L	L	L		H	M	H	L	L	L	L	L	L	H	L	L	L	L						
18	M	L	R	L	L	L	L	L	L	L		H	H	H	L	L	L	H	L	L	H	L	L	L	L						
19	M	L	R	L	M	M	M	L	R	R		H	H	H	L	L	M	H	M	L	M	L	L	L	H			L	L	L	L

1000

Reactivity Chart 3. Protection for 1,2- and 1,3-Diols

1. Methylenedioxy Derivative
2. Ethylidene Acetal
3. Acetonide Derivative
4. Benzylidene Acetal
5. *p*-Methoxybenzylidene Acetal
6. Methoxymethylene Acetal
7. Dimethoxymethylenedioxy Derivative
8. Cyclic Carbonates
9. Cyclic Boronates

(See chart, pp. 1002–1004.)

Reactivity Chart 3. Protection for 1,2 and 1,3-Diols

PG	Cr(II), pH 5	Zn/AcOH	Zn/HCl	H₂/Rh	H₂/Lindlar	H₂/Pd	H₂/Pt pH 2–4	H₂/Raney (Ni)	Wittig ylide	Organocopper	Organozinc	RMgX	RLi	HCN, pH 6	NaCN, pH 12	AcO⁻; X⁻	RS⁻; N₃⁻; SCN⁻	NH₃; RNH₂	Enolate	MeONa	NaNH₂	Py; R₃N	LDA	t-BuOK	CH₃S(O)CH₂⁻Na⁺	(C₁₀H₈)⁻˙Na⁺	Ph₃CNa	NaH	pH >12, 150°C	pH >12	pH 10–12	pH 8.5–10	pH 6–8.5	pH 4–6	pH 2–4	pH 1	pH <1	pH <1, 100°C
		1 elec. Red.			**Catalytic Reduction**					**Organometallic**					**Nucleophilic**							**Basic**								**Aqueous**								
1	L	L	L	L	L	L	L	L	L	L	L	L	L		L	L	L	L	L	L	L	L	L	L	L	L	L	L	L	L	L	L	L	L	L	L	H	H
2	L	M	H	L	L	L	L	L	L	L	L	L	L		L	L	L	L	L	L	L	L	L	L	L	L	L	L	L	L	L	L	L	L	M	H	H	H
3	L	M	H	L	L	L	M	L	L	L	L	L	L		L	L	L	L	L	L	L	L	L	L	L	L	L	L	L	L	L	L	L	M	M	H	H	H
4	L	H	H	L	L	H	M	H	L	L	L	L	L		L	L	L	L	L	L	L	L	L	L	L	R	L	L	M	L	L	L	L	L	H	H	H	H
5	M	H	H	L	L	H	H	H	L	L	H	L	L	L	L	L	L	L	L	L	L	L	L	L	M	R	L	L	M	L	L	L	L	M	H	H	H	H
6	H	H	H	L	L	L	H	L	L	L	H	H	M		L	L	L	L	L	L	L	L	L	L	M	L	L	L	M	L	L	L	L	H	H	H	H	H
7	H	H	H	L	L	L	H	L	L	L	L	H	M		L	M	M	L	M	L	L	L	L	L	H	L	L	L	M	L	L	L	L	H	H	H	H	H
8	L	L	L	L	L	L	L	L	L	L	L	H	H	L	L	M	M	M	M	M	H	L	L	L	H	H	L	L	H	H	H	L	L	L	L	L	L	H
9	M	H	H	L	L	L	H	L	H	H	H	H	H	M	H	L	L	L	H	H	H	L	L	H	H	H	L	L	H	H	H	H	H	H	H	H	H	H

1002

Reactivity Chart 3. Protection for 1,2- and 1,3-Diols (*Continued*)

PG	Single Elec. Red.				Hydride Reductions									Acid and Lewis Acid						Soft Acids			Free Rad. Rxn				Oxidants											
	Na/NH₃	Al(Hg)	SnCl₂/py	HSO₃⁻, H₂S	LiAlH₄	Li-*s*-Bu₃BH	Chex₂BH	B₂H₆, 0°C	NaBH₄	Zn(BH₄)₂	NaBH₃CN, pH 4–6	*i*-Bu₂AlH	Li(*t*-BuO)₃AlH	AlCl₃, 80°C	AlCl₃, 25°C	SnCl₄, BF₃	LiClO₄, MgBr₂	TsOH, 80°C	TsOH, 0°C	Hg(II)	Ag(I)	Cu(II)/Py	HBr, In•	HX/In•	NBS/CCl₄	Br₃CCl/In•	OsO₄	KMnO₄, pH 7, 0°C	O₃, −50°C	RCO₃H, 0°C	RCO₃H, 50°C	CrO₃/Py	CrO₃, pH 1	H₂O₂, pH 10–12	Quinone	¹O₂	DMSO, 100°C	NaOCl, pH 10
1	L	L	L	L	L	L	L	L	L	L	L	L	L	H	H	*H*	L	M	L	L	L	L	R	L	L	L	L	L	R	L	L	L	L	L	L	L	L	L
2	L	L	L	L	L	L	L	L	L	L	L	L	L	H	H	*H*	L	M	L	L	L	L	M	L	L	L	L	L	L	L	M	L	H	L	L	L	L	L
3	L	L	L	L	L	L	L	L	L	L	L	L	L	H	H	*H*	L	M	L	L	L	L	M	L	L	L	L	L	R	L	L	L	H	L	L	L	L	L
4	H	L	L	L	L	L	L	L	L	L	L	L	L	H	H	*H*	L	H	L	L	L	L	H	L	R	L	L	L	R	L	H	L	H	L	L	L	L	L
5	H	L	L	L	L	L	L	L	L	L	M	L	L	H	H	H	L	R	H	L	L	L	H	L	R	L	L	L	L	M	H	L	H	L	L	L	L	L
6	L	L	L	L	R	L	L	L	L	L	M	L	L	H	H	H	M	R	H	L	L	L	H	L	H	L	L	M	L	H	H	L	H	L	L	L	L	L
7	L	L	L	L	R	L	L	L	L	L	L	L	L	H	H	H	M	M	L	L	L	L	L	L	L	L	L	M	L	H	H	L	H	L	L	L	L	L
8	H	L	L	L	H	H	H	L	L	H	H	H	L	H	H	L	L	M	L	L	L	L	L	L	L	L	L	L	L	L	L	L	L	H	L	L	L	M
9	H	M	L	L	H	H	H	H	H	H	H	H	H	H	H	L	L	L	L	H	H	L	L	L	L	L	L	H	H	H	*H*	L	H	H	L	L	L	H

Reactivity Chart 3. Protection for 1,2- and 1,3-Diols (*Continued*)

PG	I₂	PhSeX, PhSCl	Br₂, Cl₂	MnO₂/CH₂Cl₂	NaIO₄, pH 5-6	SeO₂, pH 2-4	SeO₂, Py	K₃Fe(CN)₆, pH 8	Pb(IV), 25°C	Pb(IV), 80°C	Tl(NO₃)₃	150°C	250°C	350°C	:CCl₂	N₂CHCO₂R, Cu or Rh	CH₂I₂, Zn(Cu)	R₃SnH, In•	Ni(CO)₄	CH₂N₂	SOCl₂	Ac₂O, 25°C	Ac₂O, 80°C	DCC	MeI	Me₃O⁺ BF₄⁻	1. LDA, 2. MeI	1. K₂CO₃, 2. MeI	RCHO	RCOCl	C⁺/olefin
Oxidants											Thermal			Carbenes			Miscellaneous											Electrophilic			
1	L	L	L	L	L	L	L	L	L	L	L	L	L	M	L	L	L	L	L	L	L	L	M	L	L	M	L	L	L	L	H
2	L	L	L	L	H	M	M	L	L	M	H	L	L	M	L	L	L	L	L	L	L	L	M	L	L	M	L	L	*L*	*L*	*H*
3	*L*	L	L	L	L	M	L	L	L	*M*		L	*L*	L	L	L	L	L	L	L	L	L	*M*	L	L	M	L	L	*L*	*L*	
4	L	L	R	L	L	H	L	L	L	M		L	M	H	L	L	L	L	L	L	L	L	M	L	L	M	L	L			
5	L	L	R	L	L	H	L	L	*L*	*H*		L	M	H	*L*	*M*	L	L	L	L	L	L	*M*	L	L	R	L	L			
6	L	L	L	L	M	H	L	L	*L*	*H*		L	M	H	L	L	L	L	L	L	L	M	*R*	L	L	R	L	L			
7	L	L	L	L	M	R	L	L	*L*	*L*		M	H	H	L	L	L	L	L	L	L	M	*R*	L	L	R	L	L			
8	L	L	L	L	L	L	L	L	L	L	L	L	M	H	L	L	L	L	L	L	L	L	L	L	L	R	L	H	L	L	L
9	L	L	L	L	*H*	H	L	H	H	H	H	L	H	H	L	L	L	M	L	L	M	M	H	L	L	R	L	H	L	M	H

Reactivity Chart 4.   Protection for Phenols and Catechols

Phenols

1. Methyl Ether
2. Methoxymethyl Ether
3. 2-Methoxyethoxymethyl Ether
4. Methylthiomethyl Ether
5. Phenacyl Ether
6. Allyl Ether
7. Cyclohexyl Ether
8. *t*-Butyl Ether
9. Benzyl Ether
10. *o*-Nitrobenzyl Ether
11. 9-Anthrylmethyl Ether
12. 4-Picolyl Ether
13. *t*-Butyldimethylsilyl Ether
14. Aryl Acetate
15. Aryl Pivaloate
16. Aryl Benzoate
17. Aryl 9-Fluorenecarboxylate
18. Aryl Methyl Carbonate
19. Aryl 2,2,2-Trichloroethyl Carbonate
20. Aryl Vinyl Carbonate
21. Aryl Benzyl Carbonate
22. Aryl Methanesulfonate

Catechols

23. Methylenedioxy Derivative
24. Acetonide Derivative
25. Diphenylmethylenedioxy Derivative
26. Cyclic Borates
27. Cyclic Carbonates

(See chart, pp. 1006–1008.)

Reactivity Chart 4. Protection for Phenols and Catechols

PG	Cr(II), pH 5	Zn/AcOH	Zn/HCl	H₂/Rh	H₂/Lindlar	H₂/Pd	H₂/Pt pH 2–4	H₂/Raney (Ni)	Wittig; ylide	Organocopper	Organozinc	RMgX	RLi	HCN, pH 6	NaCN, pH 12	AcO⁻; X⁻	RS⁻; N₃⁻; SCN⁻	NH₃; RNH₂	Enolate	MeONa	NaNH₂	Py; R₃N	LDA	t-BuOK	CH₃S(O)CH₂⁻Na⁺	(C₁₀H₈⁺)⁻Na⁺	Ph₃CNa	NaH	pH > 12, 150°C	pH > 12	pH 10–12	pH 8.5–10	pH 6–8.5	pH 4–6	pH 2–4	pH 1	pH < 1	pH < 1, 100°C
1	L	L	L	L	L	L	L	L	L	L	L	L	L	L	H	L	H	L	L	L	L	L	L	L	L	L	L	L	L	L	L	L	L	L	L	L	M	H
2	L	M	H	L	L	L	M	L	L	L	L	L	L	L	L	L	M	L	L	L	L	L	L	L	L	L	L	L	L	L	L	L	L	L	M	H	H	H
3	L	M	L	L	L	R	M	L	L	L	L	L	L	L	L	L	M	L	L	L	L	L	L	L	L	R	L	L	L	L	L	L	L	L	L	L	H	H
4	M	H	H	R	L	R	R	R	R	L	M	L	M	L	L	L	M	L	R	R	L	L	R	M	R	R	L	R	L	L	L	L	L	L	M	M	H	H
5	H	H	H	L	L	R	R	M	L	M	L	R	R	L	M	L	M	L	L	R	R	L	L	R	L	L	R	L	H	M	L	L	L	L	L	L	L	H
6	L	L	M	R	L	R	R	R	L	M	L	M	R	L	M	L	L	L	L	L	L	L	L	L	L	L	L	L	L	L	L	L	L	L	L	L	L	H
7	L	L	M	L	L	L	L	L	L	L	L	L	L	L	L	L	L	L	L	L	L	L	L	L	L	R	L	L	L	L	L	L	L	L	L	H	H	H
8	L	L	H	L	L	L	L	L	L	L	M	L	L	L	L	L	M	L	L	L	L	L	R	L	R	R	L	L	L	L	L	H	L	L	L	L	L	H
9	L	M	L	H	L	H	H	H	L	L	L	R	M	L	L	L	H	L	L	L	R	L	L	L	L	R	R	R	R	L	H	L	L	M	H	L	M	H
10	R	L	R	H	L	H	H	H	L	R	L	M	R	L	L	L	M	L	L	L	L	L	L	L	L	L	L	L	L	M	L	L	L	L	M	L	L	H
11	L	R	L	H	L	H	H	H	L	L	L	M	R	L	M	L	L	L	L	L	L	L	L	R	H	R	L	L	L	H	M	L	L	L	L	H	H	H
12	L	L	M	H	L	H	H	H	L	L	L	L	L	L	L	L	M	L	L	H	L	L	L	L	M	R	L	L	H	M	H	L	L	L	L	H	M	H
13	M	L	L	L	L	L	M	L	L	L	L	R	R	L	L	L	M	L	H	M	H	L	R	R	H	R	R	R	H	H	H	L	L	L	H	L	H	H
14	L	M	L	L	L	L	L	L	L	L	L	L	L	L	H	L	M	H	M	M	M	L	L	M	R	R	L	L	H	H	M	L	L	L	L	H	M	H
15	L	L	L	L	L	L	L	L	L	L	L	R	R	L	M	L	M	L	M	M	H	L	L	L	H	R	L	R	H	H	L	M	L	M	L	L	M	H
16	L	L	L	L	L	L	L	L	L	L	L	H	H	L	H	L	M	M	R	R	H	L	R	M	R	R	R	L	H	H	H	L	L	L	L	M	H	H
17	L	L	L	L	L	L	L	L	L	L	L	H	H	L	H	L	M	M	R	M	H	L	L	L	H	R	L	R	H	H	H	L	L	L	L	L	L	H
18	M	L	H	L	L	L	H	L	L	R	M	H	H	L	H	L	M	M	R	M	H	L	R	L	H	L	R	L	H	H	L	L	L	L	H	H	M	L
19	L	H	H	R	L	R	L	R	M	L	L	H	H	M	H	L	L	M	M	L	R	L	L	M	L	L	L	L	L	L	L	L	L	L	L	H	H	H
20	L	H	H	R	L	H	R	H	L	L	L	M	H	L	H	L	L	M	L	L	L	L	L	L	L	R	L	R	L	L	L	L	L	L	L	L	H	H
21	L	L	M	L	L	L	H	R	L	R	M	M	L	L	M	L	L	L	L	L	L	L	L	L	H	R	R	L	L	L	L	L	L	L	L	L	L	H
22	L	L	L	L	L	L	L	L	M	L	L	L	L	L	L	L	L	L	L	L	L	L	L	L	H	L	L	L	L	L	L	L	L	L	L	L	L	H
23	L	L	L	L	L	L	L	R	L	L	L	L	L	L	L	L	L	L	H	M	R	L	R	L	L	L	L	L	H	H	H	M	L	M	M	H	M	H
24	M	M	H	L	L	H	M	L	L	L	L	L	H	M	L	L	L	H	M	L	H	L	L	R	L	L	L	R	H	H	L	L	L	L	H	H	H	H
25	M	M	H	L	L	L	H	L	L	L	L	R	R	L	H	L	L	M	L	L	L	L	L	L	L	R	R	L	L	L	L	L	L	L	L	L	H	H
26	L	H	H	L	L	L	L	L	L	L	L	R	L	L	H	L	L	L	L	L	L	L	R	L	L	R	L	L	L	L	L	L	L	L	L	L	H	H
27	L	L	M	L	L	L	L	L	H	H	H	R	R	M	H	L	L	H	M	M	L	L	L	L	H	L	L	L	H	H	H	M	L	L	L	H	H	H

1006

PG	Aq. NBS	NaOCl, pH 10	DMSO, 100°C	$^1O_2$	Quinone	H_2O_2, pH 10–12	CrO_3, pH 1	CrO_3/Py	RCO_3H, 50°C	RCO_3H, 0°C	O_3, –50°C	$KMnO_4$, pH 7, 0°C	OsO_4	Br_3CCl/In·	NBS/CCl_4	HX/In·	HBr, In·	Cu(II)/Py	Ag(I)	Hg(II)	TsOH, 0°C	TsOH, 80°C	$LiClO_4$, $MgBr_2$	$SnCl_4$, BF_3	$AlCl_3$, 25°C	$AlCl_3$, 80°C	Li(t-BuO)$_3$AlH	i-Bu$_2$AlH	NaBH$_3$CN, pH 4–6	Zn(BH$_4$)$_2$	NaBH$_4$	B_2H_6, 0°C	Chex$_2$BH	Li-s-Bu$_3$BH	LiAlH$_4$	HSO_3^-, H_2S	SnCl$_2$/py	Al(Hg)	Na/NH$_3$
	Oxidants													**Free Rad. Rxn**				**Soft Acids**			**Acid and Lewis Acid**						**Hydride Reductions**									**Single elec. Red.**			
1	L	L	L	L	L	L	L	L	L	L	L	L	L	L	R	L	L	L	L	L	L	L	L	L	M	H	L	L	L	L	L	L	L	L	L	L	L	L	R
2	L	L	L	L	L	L	R	L	M	L	L	L	L	L	R	L	H	L	L	L	L	H	L	H	H	H	L	L	L	L	L	L	L	L	L	L	L	L	R
3	L	L	L	L	L	L	M	L	L	L	L	L	L	L	R	L	M	L	M	L	H	M	M	M	H	H	L	L	L	L	L	L	L	L	L	L	L	L	R
4	R	R	L	R	L	L	L	L	R	R	R	R	R	L	L	L	L	L	L	H	M	M	L	L	H	H	R	R	R	R	R	R	R	R	R	L	L	R	R
5	L	L	L	L	L	L	L	L	R	R	L	L	L	L	R	L	L	L	L	L	L	L	L	L	H	H	L	L	L	L	L	L	L	L	L	L	L	L	R
6	R	L	L	R	L	L	L	L	L	L	R	L	R	R	R	R	R	L	L	R	L	L	L	L	L	M	L	L	L	L	L	L	L	L	L	L	L	R	R
7	L	L	L	L	L	L	R	L	L	L	L	L	L	L	L	L	M	L	L	L	L	H	L	L	M	H	L	L	L	L	L	L	L	L	L	L	L	L	R
8	L	L	L	L	L	L	L	L	M	L	L	L	L	R	R	R	R	L	L	L	L	L	L	L	M	M	L	L	L	L	R	L	L	L	L	L	L	L	R
9	L	L	L	L	L	L	L	L	M	L	L	L	L	M	R	M	M	L	L	L	L	L	L	L	H	H	L	L	L	L	L	L	L	L	L	L	L	L	R
10	L	L	L	L	L	L	L	L	M	R	L	L	L	R	R	R	R	L	L	L	L	M	L	L	R	H	L	L	L	L	L	L	L	L	L	L	L	L	R
11	L	L	L	L	L	L	M	L	R	L	L	L	L	R	R	R	R	L	L	R	L	H	L	M	L	R	M	L	L	L	L	L	L	L	H	L	L	L	R
12	L	L	L	L	L	L	R	L	M	L	L	L	L	L	L	L	H	L	L	L	L	H	L	L	R	R	L	L	L	L	L	L	L	L	H	L	L	L	R
13	L	L	L	L	L	L	L	L	L	L	R	L	L	L	L	L	L	L	L	L	L	H	L	L	H	H	M	M	L	L	L	L	L	L	H	L	L	L	R
14	L	H	L	L	L	H	L	L	L	L	L	L	L	L	L	L	L	L	L	H	M	H	L	L	H	H	L	H	M	L	L	L	L	L	H	L	L	M	R
15	L	L	L	L	L	L	L	L	L	L	L	L	L	L	L	L	R	R	R	L	L	H	L	L	R	R	M	H	L	L	L	L	L	L	H	L	L	L	R
16	L	M	L	L	L	H	L	L	R	L	L	L	L	R	R	R	H	L	L	L	L	H	L	L	L	H	M	H	L	L	M	L	L	L	H	L	L	M	R
17	R	M	L	L	L	M	H	L	R	R	L	R	L	R	R	R	M	R	L	H	H	H	L	M	H	H	S	H	M	L	L	R	L	L	H	L	L	L	R
18	L	M	L	L	L	R	L	L	M	L	L	L	L	L	L	L	R	L	L	L	M	H	L	L	L	R	M	H	L	L	L	L	L	L	L	L	L	L	R
19	L	R	L	L	L	R	M	L	L	L	L	L	L	L	L	L	M	L	L	L	M	H	M	M	H	R	M	H	L	L	M	R	L	L	L	L	M	L	R
20	M	R	L	L	L	R	R	L	L	L	L	L	L	R	R	R	L	L	R	L	M	H	L	L	H	H	M	H	L	L	L	L	L	L	L	L	L	L	R
21	R	L	L	R	L	M	L	L	R	R	R	R	R	L	L	L	M	L	L	H	H	H	L	L	H	H	M	M	M	L	M	L	L	L	L	L	L	M	R
22	L	L	L	R	L	L	M	L	R	R	L	L	L	L	R	L	R	L	L	L	H	M	L	L	L	H	L	L	L	L	L	L	L	L	L	L	L	L	R
23	R	L	L	L	L	L	R	L	L	L	L	L	L	L	R	L	M	L	L	L	H	M	L	L	H	H	L	L	L	L	L	L	L	L	L	L	L	L	R
24	L	L	L	L	L	L	R	L	L	L	L	L	L	L	L	L	H	L	L	L	M	H	L	M	H	H	H	H	L	L	L	L	L	L	H	L	L	M	R
25	L	L	L	L	L	L	R	L	L	L	L	L	L	L	R	L	H	L	L	L	H	H	L	L	H	H	M	H	M	H	H	H	H	H	H	H	L	M	R
26	L	L	L	L	L	L	R	L	R	R	L	L	L	L	L	L	H	L	L	L	M	H	L	M	H	H	L	L	L	L	L	L	L	L	H	L	L	M	R
27	L	M	L	L	L	R	R	L	L	L	L	L	L	L	L	L	H	L	L	L	H	H	L	L	H	H	M	H	L	H	H	H	H	L	H	L	L	L	R

Reactivity Chart 4. Protection for Phenols and Catechols (*Continued*)

PG	I₂	PhSeX, PhSCl	Br₂, Cl₂	MnO₂/CH₂Cl₂	NaIO₄, pH 5-6	SeO₂, pH 2-4	SeO₂, Py	K₃Fe(CN)₆, pH 8	Pb(IV), 25°C	Pb(IV), 80°C	Tl(NO₃)₃	150°C	250°C	350°C	:CCl₂	N₂CHCO₂R, Cu or Rh	CH₂I₂, Zn(Cu)	R₃SnH, In•	Ni(CO)₄	CH₂N₂	SOCl₂	Ac₂O, 25°C	Ac₂O, 80°C	DCC	MeI	Me₃O⁺ BF₄⁻	1. LDA, 2. MeI	1. K₂CO₃, 2. MeI	RCHO	RCOCl	C⁺/olefin
1	L	L	L	L	L	L	L	L	L	L	L	L	L	L	L	L	L	L	L	L	L	L	L	L	L	L	L	L	L	L	L
2	L	L	L	L	L	M	L	L	M	R	R	L	M	H	L	L	L	L	L	L	L	L	L	L	L	L	L	L	L	L	H
3	L	L	L	L	L	L	L	L	L	L	L	L	M	H	L	L	L	M	L	L	L	L	L	L	L	L	L	L	L	L	L
4	L	M	R	L	R	M	L	L	R	R	R	M	L	H	M	R	L	M	L	L	L	L	L	L	R	R	R	M	L	L	M
5	L	R	M	L	L	R	M	L	M	R	R	L	R	R	R	R	R	L	M	L	L	L	L	L	L	M	L	L	L	L	R
6	L	L	L	L	L	H	L	L	L	L	R	L	M	L	R	L	L	M	L	L	L	L	L	L	L	M	L	L	L	L	L
7	L	L	L	L	L	L	L	L	L	M	L	L	L	H	L	L	L	L	L	L	L	L	L	L	L	L	L	L	L	L	H
8	L	L	L	L	L	L	L	L	M	L	L	L	L	L	L	L	L	L	L	L	L	L	L	L	L	L	L	L	L	L	L
9	L	L	L	L	L	L	L	L	L	L	L	L	L	L	L	L	L	L	L	L	L	L	L	L	L	L	L	L	L	L	L
10	L	L	M	L	L	L	L	L	L	L	L	L	L	L	L	L	L	L	L	L	L	L	L	L	L	L	L	L	L	L	L
11	L	L	L	L	L	L	L	L	L	L	L	L	L	L	L	L	L	L	L	L	L	L	L	L	R	R	R	L	L	L	L
12	L	L	L	L	L	L	L	R	L	R	R	L	L	L	L	L	L	L	L	L	L	L	L	L	L	L	L	H	L	L	L
13	L	L	L	L	L	M	L	L	L	L	M	L	L	L	L	L	L	L	L	L	L	L	M	L	L	L	L	L	L	L	L
14	L	L	L	L	L	L	L	L	L	L	L	L	M	H	L	L	L	L	L	L	L	L	M	L	L	L	R	M	L	L	L
15	L	L	L	L	L	L	L	L	L	L	L	L	M	M	L	L	L	L	L	L	L	L	L	L	L	L	L	L	L	L	L
16	L	L	L	L	L	H	L	M	M	M	R	L	M	H	L	L	R	L	L	L	L	L	L	L	L	R	R	M	L	L	L
17	L	R	L	L	L	L	L	L	L	L	L	L	L	H	L	L	R	L	L	L	L	L	L	L	L	M	L	H	L	L	M
18	L	L	L	L	L	L	L	L	L	R	R	L	L	H	L	R	L	L	M	L	L	L	L	L	L	R	L	M	L	L	L
19	M	L	R	L	L	L	L	L	L	L	L	L	L	M	L	L	L	R	R	L	L	L	L	L	L	L	R	L	L	L	R
20	L	L	L	L	L	L	L	L	M	R	R	L	M	L	L	L	L	R	L	L	L	L	L	L	L	L	L	L	L	L	L
21	L	L	L	L	L	L	L	L	L	L	L	L	M	H	L	L	L	L	L	L	L	L	L	L	L	L	L	L	L	L	L
22	L	L	L	L	L	M	L	L	L	L	L	L	L	H	L	L	L	L	L	L	L	L	L	L	L	L	L	L	L	L	L
23	L	L	L	L	L	R	L	L	L	M	M	L	L	H	L	L	L	L	L	L	M	L	L	L	L	L	L	L	L	L	M
24	L	L	L	L	L	L	L	L	L	R	R	L	M	M	L	L	L	L	L	L	L	R	R	L	R	L	R	L	R	R	H
25	L	L	L	L	L	L	L	L	L	R	R	L	M	H	L	L	L	L	L	L	L	L	L	L	L	L	L	R	L	R	H
26	L	L	L	L	L	L	L	L	L	L	R	L	L	H	L	L	L	L	L	L	R	R	R	L	R	R	L	L	R	R	L
27	L	L	L	L	L	L	L	M	L	L	M	L	L	M	L	L	L	L	L	L	L	L	L	L	L	R	L	L	L	L	L

Reactivity Chart 5. Protection for the Carbonyl Group

1. Dimethyl Acetals and Ketals
2. Bis(2,2,2-trichloroethyl) Acetals and Ketals
3. 1,3-Dioxanes
4. 5-Methylene-1,3-dioxanes
5. 5,5-Dibromo-1,3-dioxanes
6. 1,3-Dioxolanes
7. 4-Bromomethyl-1,3-dioxolanes
8. 4-*o*-Nitrophenyl-1,3-dioxolanes
9. *S,S'*-Dimethyl Acetals and Ketals
10. 1,3-Dithianes
11. 1,3-Dithiolanes
12. 1,3-Oxathiolanes
13. *O*-Trimethylsilyl Cyanohydrins
14. *N,N*-Dimethylhydrazones
15. 2,4-Dinitrophenylhydrazones
16. *O*-Phenylthiomethyl Oximes
17. Substituted Methylene Derivatives
18. Bismethylenedioxy Derivatives

(See chart, pp. 1010–1012.)

Reactivity Chart 5. Protection for the Carbonyl Group

PG	pH < 1, 100°C	pH < 1	pH 1	pH 2–4	pH 4–6	pH 6–8.5	pH 8.5–10	pH 10–12	pH < 12	pH < 12, 150°C	NaH	Ph₃CNa	(C₁₀H₈)⁻ Na⁺	CH₃S(O)CH₂⁻Na⁺	t-BuOK	LDA	Py; R₃N	NaNH₂	MeONa	Enolate	NH₃; RNH₂	RS⁻; N₃⁻; SCN⁻	AcO⁻; X⁻	NaCN, pH 12	HCN, pH 6	RLi	RMgX	Organozinc	Organocopper	Wittig; ylide	H₂/Raney (Ni)	H₂/Pt pH 2–4	H₂/Pd	H₂/Lindlar	H₂/Rh	Zn/HCl	Zn/AcOH	Cr(II), pH 5
											Basic								*Nucleophilic*							*Organometallic*					*Catalytic Reduction*					*1 Elec. Red.*		
1	H	H	H	L	L	L	L	L	L	L	L	L	L	L	L	L	L	L	L	L	L	L	L	L		L	L	L	L	L	L	L	L	L	L	H	H	L
2	H	H	H	L	L	L	L	L	M	R	R	R	H	R	R	R	L	R	R	R	M	R	R	R		R	R	R	R	R	R	L	R	L	M	H	R	M
3	H	H	H	L	L	L	L	L	L	L	L	L	L	L	R	L	L	L	L	L	L	L	L	L		L	L	L	L	L	L	M	L	L	L	H	H	M
4	H	H	H	L	L	L	L	L	L	R	L	L	H	R	R	R	L	R	R	L	L	L	R	R		R	R	R	R	R	R	R	R	L	R	H	R	M
5	H	H	H	M	L	L	L	M	R	R	R	R	H	R	L	L	L	L	L	R	M	R	L	L		L	L	L	L	L	L	R	L	L	M	H	L	M
6	H	H	H	M	L	L	L	L	L	L	L	L	L	L	R	R	L	R	R	L	L	L	R	R		L	R	R	M	L	R	R	R	L	L	H	H	L
7	H	H	H	M	L	L	L	L	R	R	R	R	H	R	L	L	L	M	L	R	L	R	L	L		R	R	M	R	L	R	R	R	L	L	H	R	M
8	H	H	H	M	L	L	L	L	L	H	L	M	H	H	L	R	L	L	L	M	L	L	L	L		L	L	L	L	L	R	R	R	M	R	L	L	R
9	R	R	L	L	L	L	L	L	L	L	L	L	L	L	L	L	L	L	L	L	L	L	L	L		L	L	L	L	L	R	R	R	H	R	L	L	L
10	R	M	L	L	L	L	L	L	L	L	L	L	L	L	L	L	L	L	L	L	L	L	L	L		L	L	L	L	L	R	R	L	H	R	L	L	L
11	H	R	H	H	L	L	L	L	L	L	L	L	L	L	L	L	L	L	L	L	L	H	L	L		L	L	H	L	R	R	R	R	H	R	H	M	L
12	H	H	H	H	H	H	H	H	H	R	L	L	L	L	M	L	L	R	H	L	R	L	L	L		M	M	L	L	L	R	R	R	L	R	H	H	L
13	L	H	L	L	L	L	L	L	L	H	L	L	R	R	L	L	L	L	L	L	L	L	L	L		L	L	L	L	R	R	R	R	L	R	H	H	R
14	H	H	L	L	L	L	L	L	L	H	L	L	M	L	R	R	L	R	M	M	L	L	L	L		R	R	R	L	L	R	R	R	L	R	R	R	M
15	R	L	L	L	L	L	L	L	L	R	R	R	R	R	L	R	L	L	M	L	R	H	L	L		R	L	R	L	R	R	R	R	L	R	R	R	R
16	R	L	M	L	L	L	L	L	L	H	L	R	M	L	R	M	L	R	L	R	L	L	L	L		R	R	R	L	L	R	R	R	L	R	M	M	R
17	R	M	L	L	L	L	L	L	H	H	H	H	R	R	L	R	L	R	H	R	M	L	L	L		R	L	L	R	R	R	R	L	L	R	M	M	M
18	H	H	M	M	L	L	L	L	L	L	L	L	L	L	L	L	L	L	L	L	L	L	L	L		L	L	L	L	L	L	L	L	L	L	H	H	L

Reactivity Chart 5. Protection for the Carbonyl Group (continued)

PG	Single elec. Red.				Hydride Reductions									Acid and Lewis Acid						Soft Acids			Free Rad. Rxn				Oxidants												
	Na/NH₃	Al(Hg)	SnCl₂/py	HSO₃⁻, H₂S	LiAlH₄	Li-s-Bu₃BH	Chex₂BH	B₂H₆, 0°C	NaBH₄	Zn(BH₄)₂	NaBH₃CN, pH 4-6	i-Bu₂AlH	Li(t-BuO)₃AlH	AlCl₃, 80°C	AlCl₃, 25°C	SnCl₄, BF₃	LiClO₄, MgBr₂	TsOH, 80°C	TsOH, 0°C	Hg(II)	Ag(I)	Cu(II)/Py	HBr, In•	HX/In•	NBS/CCl₄	Br₂CCl/In•	OsO₄	KMnO₄, pH 7, 0°C	O₃, -50°C	RCO₃H, 0°C	RCO₃H, 50°C	CrO₃/Py	CrO₃, pH 1	H₂O₂, pH 10-12	Quinone	¹O₂	DMSO, 100°C	NaOCl, pH 10	Aq. NBS
1	L	L	L	L	L	L	L	M	L	L	L	L	L	H	R	H	L	H	M	L	L	L	H	L	L	R	L	L	L	L	L	L	H	H	L	L	L	L	L
2	M	M	L	L	R	L	L	M	L	L	L	L	L	H	H	H	L	H	M	L	R	L	R	L	R	R	L	L	L	L	L	L	H	L	L	L	H	L	L
3	L	L	L	L	L	L	R	M	L	L	L	L	L	H	H	H	L	L	L	L	L	L	H	R	L	L	L	L	L	L	L	L	M	L	L	L	H	L	L
4	R	L	M	L	R	L	L	R	L	L	L	M	M	H	H	H	L	H	M	R	L	L	H	L	H	R	R	R	R	R	R	L	H	L	L	M	M	L	R
5	R	H	L	L	L	L	L	L	L	L	L	L	L	H	H	H	L	L	L	L	R	L	H	R	H	H	L	L	L	L	L	L	H	L	L	L	L	L	L
6	L	L	L	L	M	L	L	M	L	L	L	L	L	H	H	H	L	L	L	L	L	L	H	L	L	L	L	L	L	L	L	L	H	L	L	L	H	L	L
7	L	H	L	L	R	L	L	M	L	L	L	L	L	H	H	H	L	L	L	L	R	L	H	H	L	L	L	L	L	L	L	L	L	L	L	L	L	L	L
8	R	R	L	L	L	R	L	M	L	L	L	L	L	H	H	H	L	L	L	L	L	H	H	L	L	L	L	L	L	L	L	L	H	L	L	L	R	R	L
9	R	R	L	L	L	L	L	L	L	L	L	L	L	H	L	L	L	L	L	H	H	H	H	L	L	L	L	R	R	R	R	L	M	M	L	R	L	R	R
10	R	L	L	L	L	L	L	L	L	L	L	L	L	H	L	L	L	L	L	H	H	H	H	L	L	L	L	R	R	R	R	L	M	L	L	R	M	M	R
11	R	L	L	L	L	L	R	L	M	M	L	L	L	H	H	M	M	H	H	H	H	H	H	L	L	L	L	R	R	R	R	L	L	L	L	R	M	R	R
12	R	L	M	L	R	R	L	R	M	M	R	R	R	H	L	H	L	H	L	L	L	L	R	M	H	M	L	R	R	R	R	L	H	L	L	L	M	R	R
13	R	R	L	L	R	L	L	L	L	L	R	R	L	H	H	L	L	L	L	L	L	R	R	L	L	L	L	R	R	R	R	L	R	R	L	L	R	R	R
14	L	L	M	L	R	R	R	R	L	M	R	R	M	L	L	L	L	L	L	L	L	H	R	L	L	L	L	H	H	H	H	H	H	H	L	H	L	H	H
15	R	R	M	L	L	L	L	R	M	L	R	R	M	H	L	L	L	L	L	R	R	M	R	L	R	L	L	L	H	R	H	L	L	L	L	M	L	L	R
16	R	R	M	L	R	R	L	L	L	M	L	R	M	H	M	L	L	L	L	L	L	L	R	R	R	R	L	R	H	R	R	L	R	L	L	R	M	H	R
17	R	R	L	L	L	L	R	L	L	L	L	L	L	M	L	M	M	L	L	L	L	L	L	L	L	L	R	R	H	R	R	L	M	L	L	M	M	M	M
18	L	L	L	L	L	L	L	L	L	L	L	L	L	R	R	L	M	L	L	L	L	L	M	L	L	L	L	L	L	L	L	L	L	L	L	L	L	L	M

Reactivity Chart 5. Protection for the Carbonyl Group (*Continued*)

PG	I₂	PhSeX, PhSCl	Br₂, Cl₂	MnO₂/CH₂Cl₂	NaIO₄, pH 5–6	SeO₂, pH 2–4	SeO₂, Py	K₃Fe(CN)₆, pH 8	Pb(IV), 25°C	Pb(IV), 80°C	Tl(NO₃)₃	150°C	250°C	350°C	:CCl₂	N₂CHCO₂R, Cu or Rh	CH₂I₂, Zn(Cu)	R₃SnH, In•	Ni(CO)₄	CH₂N₂	SOCl₂	Ac₂O, 25°C	Ac₂O, 80°C	DCC	MeI	Me₃O⁺ BF₄⁻	1. LDA, 2. MeI	1. K₂CO₃, 2. MeI	RCHO	RCOCl	C⁺/olefin
1	L	L	L	L	L	M	L	L	L	L	L	L	H	H	L	L	L	L	L	L	L	L	L	L	L	M			L	L	M
2	L	L	L	L	L	M	L	L	L	L		L	M	R	L	M	H	R	M	L	L	L	L	L	L	M			L	L	M
3	L	L	R	L	L	M	L	L	L	L		L	L	R	L	L	L	L	L	L	L	L	L	L	L	L			L	L	L
4	R	R	R	L	R	M	R	L	R	R		L	L	R	R	R	R	L	R	L	L	L	R	L	L	M			L	L	R
5	L	L	R	L	L	L	L	L	L	L		L	L	R	L	M	H	R	M	L	L	L	L	L	L	M			L	L	M
6	L	L	R	L	L	H	L	L	L	L		L	L	R	L	L	L	L	L	L	L	L	L	L	L	L			L	L	R
7	L	L	R	L	M	H	L	L	L	L		L	L	R	L	L	H	R	L	L	L	L	L	L	L	M			L	L	M
8	H	L	R	M	M	M	L	L	L	M		L	L	R	M	L	L	R	L	L	L	L	L	L	R	R			L	L	M
9	L	L	R	M	R	M	L	L	R	R		L	L	R	M	M	L	M	L	L	L	L	L	L	H	L			L	L	L
10	H	L	R	M	R	M	L	L	L	M		L	L	R	M	M	L	M	L	L	L	L	L	L	H	R			L	L	M
11	L	L	R	L	R	M	L	L	R	R		L	L	R	M	M	L	M	L	L	L	L	M	L	R	M			L	L	R
12	L	L	H	R	R	R	L	R	R	R		L	L	R	L	M	L	M	L	L	L	R	R	L	L	R			L	L	M
13	L	L	L	M	H	L	M	L	R	R		L	M	R	R	R	L	L	L	L	L	L	L	L	R	R			L	L	L
14	L	L	R	M	H	L	L	L	R	R		L	R	R	R	R	R	L	L	L	L	L	L	L	L	R			L	L	R
15	L	L	R	L	R	M	L	L	R	R		L	M	R	R	R	R	L	L	L	L	L	L	L	M	M			L	L	L
16	L	L	L	L	L	L	L	L	R	R		L	R	R	R	R	R	M	L	L	L	L	L	L	R	R			L	L	R
17	L	L	R	R	L	L	L	L	L	L		L	M	R	R	R	R	R	L	L	L	L	M	L	L	M			L	L	M
18	L	L	L	L	L	L	L	L	L	M		L	L	L	L	L	L	L	L	L	L	L	L	L	L	M			L	L	L

Reactivity Chart 6.　Protection for the Carboxyl Group

Esters

　　1. Methyl Ester
　　2. Methoxymethyl Ester
　　3. Methylthiomethyl Ester
　　4. Tetrahydropyranyl Ester
　　5. Benzyloxymethyl Ester
　　6. Phenacyl Ester
　　7. *N*-Phthalimidomethyl Ester
　　8. 2,2,2-Trichloroethyl Ester
　　9. 2-Haloethyl Ester
　10. 2-(*p*-Toluenesulfonyl)ethyl Ester
　11. *t*-Butyl Ester
　12. Cinnamyl Ester
　13. Benzyl Ester
　14. Triphenylmethyl Ester
　15. Bis(*o*-nitrophenyl)methyl Ester
　16. 9-Anthrylmethyl Ester
　17. 2-(9,10-Dioxo)anthrylmethyl Ester
　18. Piperonyl Ester
　19. Trimethylsilyl Ester
　20. *t*-Butyldimethylsilyl Ester
　21. *S*-*t*-Butyl Ester
　22. 2-Alkyl-1,3-oxazolines

Amides and Hydrazides

　23. *N,N*-Dimethylamide
　24. *N*-7-Nitroindoylamide
　25. Hydrazides
　26. *N*-Phenylhydrazide
　27. *N,N'*-Diisopropylhydrazide

　　(See chart, pp. 1014–1016.)

Reactivity Chart 6. Protection for the Carboxyl Group

PG	Cr(II), pH 5	Zn/AcOH	Zn/HCl	H₂/Rh	H₂/Lindlar	H₂/Pd	H₂/Pt, pH 2–4	H₂/Raney (Ni)	Wittig ylide	Organocopper	Organozinc	RMgX	RLi	HCN, pH 6	NaCN, pH 12	AcO⁻; X⁻	RS⁻; N₃⁻; SCN⁻	NH₃; RNH₂	Enolate	MeONa	NaNH₂	Py; R₃N	LDA	t-BuOK	CH₃S(O)CH₂⁻Na⁺	(C₁₀H₈)⁻·Na⁺	Ph₃CNa	NaH	pH>12, 150°C	pH>12	pH 10–12	pH 8.5–10	pH 6–8.5	pH 4–6	pH 2–4	pH 1	pH<1	pH<1, 100°C
1	L	L	L	L	L	L	L	L	L	L	L	R	R		L	L	H	M	R	L	L	L	L	L	R	R	L	L	H	H	M	L	L	L	L	L	H	H
2	L	M	H	L	L	L	L	L	L	L	L	R	R		L	L	L	M	R	R	L	L	L	L	R	R	L	L	H	H	L	L	L	L	M	H	H	H
3	M	M	H	R	L	R	R	R	L	L	L	R	R		L	L	M	M	R	R	L	L	L	L	R	R	L	L	H	M	L	L	L	L	H	L	M	H
4	L	H	H	L	L	L	H	L	L	L	L	R	R		L	L	M	M	R	R	L	L	L	L	R	R	L	L	H	H	M	L	L	L	M	H	H	H
5	L	H	H	M	L	H	H	H	L	L	R	R	R		L	L	L	M	R	R	R	L	R	L	R	R	R	R	H	H	L	L	L	L	L	H	H	H
6	H	H	H	M	L	H	H	L	R	L	L	R	R		H	L	L	H	R	R	L	L	L	M	R	R	L	L	H	H	H	H	L	L	H	H	L	R
7	L	R	H	L	L	L	M	R	L	L	M	R	R		L	L	H	M	R	L	R	L	R	L	R	R	R	R	H	H	H	M	L	L	L	L	L	H
8	M	H	H	L	L	L	R	R	L	R	L	R	R		M	L	L	M	R	R	R	L	L	R	R	R	R	R	H	H	H	L	L	L	L	L	H	H
9	M	M	M	L	L	R	R	R	R	R	L	R	R		H	L	M	L	R	R	R	R	R	R	R	R	R	R	H	H	H	M	L	L	H	M	H	H
10	L	M	M	L	L	L	L	L	H	L	L	R	R		H	M	R	M	L	L	L	R	R	L	R	R	L	L	H	H	L	L	L	L	L	L	H	H
11	L	L	H	L	L	L	L	L	L	L	L	R	R		L	L	R	M	R	R	L	L	L	L	R	R	L	L	H	H	H	L	L	M	H	H	H	H
12	L	L	M	R	L	R	R	R	L	H	L	R	R		L	L	L	L	R	R	L	L	L	L	R	R	L	L	H	H	H	L	L	L	L	H	M	H
13	M	L	H	L	L	H	H	H	L	L	M	R	R		L	L	L	M	R	R	L	L	L	L	R	R	L	L	H	H	M	L	L	L	M	L	M	H
14	R	H	L	M	L	R	R	H	L	L	L	R	R		L	L	M	M	R	R	L	L	L	L	R	R	L	L	H	H	M	L	L	L	L	M	H	H
15	L	R	H	R	L	H	H	R	L	L	H	R	R		L	L	L	H	R	R	L	L	L	L	R	R	L	L	H	H	L	L	L	L	H	L	H	H
16	L	L	R	L	L	L	H	H	L	L	L	R	R		L	L	L	M	R	R	L	L	L	L	R	R	L	L	H	M	M	L	H	H	H	H	H	H
17	L	R	M	M	L	H	H	H	M	L	R	R	R		L	L	H	H	R	R	L	L	L	L	R	R	L	L	H	M	H	H	M	H	M	H	H	H
18	H	M	R	L	L	H	H	H	L	H	L	R	R		H	H	M	M	R	R	L	L	L	L	L	R	L	L	H	L	H	M	L	L	L	M	L	H
19	L	H	H	M	L	L	H	L	H	L	L	R	R		L	L	M	L	L	L	L	L	L	L	M	M	L	L	H	H	L	L	L	L	M	L	L	H
20	H	H	H	H	L	L	R	R	L	R	L	R	R		H	H	L	L	L	L	L	L	L	H	H	R	L	R	H	H	L	L	H	H	L	L	M	H
21	L	L	H	L	L	R	R	R	L	L	L	R	R		L	L	R	L	M	L	L	L	L	R	H	R	R	R	H	M	L	L	M	H	L	L	H	H
22	R	R	H	R	L	L	L	M	L	L	L	L	L		L	L	R	L	L	L	R	L	R	L	R	R	R	R	H	M	M	L	L	L	L	M	H	H
23	L	L	L	L	R	M	M	H	L	L	L	R	R		L	L	L	L	L	R	R	L	R	L	R	R	L	L	H	L	L	L	L	L	L	L	L	H
24	L	L	R	L	L	M	H	H	L	L	L	R	R		L	L	L	L	L	R	R	L	R	H	L	M	R	R	H	M	L	L	L	L	L	L	M	H
25	L	L	L	L	L	M	H	H	L	L	L	R	L		L	L	L	L	M	L	R	L	R	R	H	R	R	R	H	M	M	L	L	L	L	M	H	H
26	R	L	M	L	R	M	H	H	L	L	L	R	R		L	L	L	L	L	R	R	L	R	L	R	R	L	L	H	L	L	L	L	L	L	L	H	H
27	L	L	M	L	L	M	H	H	L	L	L	R	R		L	L	L	L	L	L	R	L	R	L	R	H	L	L	H	L	L	L	L	L	L	L	H	H

1014

Reactivity Chart 6. Protection for the Carboxyl Group (*Continued*)

PG	Na/NH₃	Al(Hg)	SnCl₂/py	HSO₃⁻,H₂S	LiAlH₄	Li-s-Bu₃BH	Chex₂BH	B₂H₆,0°C	NaBH₄	Zn(BH₄)₂	NaBH₃CN,pH4–6	i-Bu₂AlH	Li(t-BuO)₃AlH	AlCl₃,80°C	AlCl₃,25°C	SnCl₄,BF₃	LiClO₄,MgBr₂	TsOH,80°C	TsOH,0°C	Hg(II)	Ag(I)	Cu(II)/Py	HBr,In·	HX/In·	NBS/CCl₄	Br₃CCl/In·	OsO₄	KMnO₄,pH7,0°C	O₃,–50°C	RCO₃H,0°C	RCO₃H,50°C	CrO₃/Py	CrO₃,pH1	H₂O₂,pH10–12	Quinone	¹O₂	DMSO,100°C	NaOCl,pH10	Aq.NBS
1	R	L	L	L	R	M	M	L	L	L	L	R	M	R	M	L	L	M	L	L	L	L	L	L	L	L	L	L	L	L	L	L	L	M	L	L	L	M	L
2	R	L	L	L	R	M	L	L	L	L	L	R	M	H	H	L	L	H	L	L	L	L	L	L	L	L	L	L	M	L	H	L	H	L	L	L	L	L	L
3	R	L	L	L	R	M	L	L	L	L	L	R	M	H	H	M	L	R	L	H	H	L	L	L	H	L	L	R	R	R	R	L	R	L	L	R	M	R	R
4	R	R	L	L	R	M	L	L	L	L	L	R	M	H	R	M	L	H	M	L	L	L	M	L	M	L	L	L	M	L	H	L	H	M	L	L	L	H	L
5	R	M	L	L	R	R	L	L	M	L	L	L	M	R	R	H	L	R	L	L	L	L	L	L	R	L	L	L	L	L	R	L	H	L	L	L	L	L	L
6	R	M	M	L	R	M	L	R	L	L	L	R	L	R	L	H	L	L	L	L	L	L	L	L	L	R	L	L	L	L	L	L	R	R	L	L	L	M	R
7	R	M	M	L	R	M	R	L	L	M	L	R	M	R	R	M	L	L	L	L	R	L	L	R	R	L	L	L	L	L	L	L	L	H	L	L	H	H	L
8	R	L	L	L	R	M	R	L	L	L	L	R	M	R	L	L	L	L	L	L	L	L	R	L	L	L	L	L	L	L	L	L	L	L	L	L	M	H	L
9	R	L	L	L	R	L	L	L	L	L	R	R	L	R	H	L	L	L	L	L	M	L	L	L	R	R	R	L	L	L	L	L	M	M	L	L	M	H	L
10	R	L	L	L	R	M	L	L	L	L	L	L	M	R	R	L	L	M	H	L	L	L	L	L	R	L	R	L	R	L	H	L	L	H	L	L	L	L	L
11	H	R	M	L	R	M	L	L	L	L	L	R	M	R	R	L	L	H	L	R	L	L	R	L	R	L	L	R	L	R	R	L	H	L	L	L	L	H	L
12	R	L	L	L	R	L	L	L	L	L	L	R	L	R	R	H	L	L	L	L	L	L	L	L	R	L	L	L	L	L	H	L	L	R	L	R	L	M	R
13	H	H	L	L	R	M	R	L	M	L	M	R	M	R	R	R	L	H	L	L	L	L	M	L	R	L	L	L	L	M	L	L	H	M	L	L	H	M	L
14	H	L	L	L	R	M	L	M	L	L	L	R	M	R	R	M	L	L	M	L	L	L	L	L	R	L	L	L	L	L	L	L	L	M	L	L	H	L	L
15	H	L	L	L	R	R	L	L	H	M	L	R	M	R	R	L	H	L	L	L	L	L	L	L	R	L	L	L	L	L	L	L	M	L	L	L	M	M	L
16	R	L	L	L	R	M	L	L	L	L	L	R	M	R	H	L	L	R	L	H	L	L	L	L	L	L	L	L	L	L	L	L	H	M	L	L	L	H	L
17	R	R	L	L	R	H	L	L	L	H	L	R	H	H	H	M	L	H	L	L	L	L	H	L	L	L	L	L	L	L	H	L	H	L	L	L	L	H	L
18	R	L	L	L	R	M	M	R	L	L	H	M	M	H	H	M	L	H	L	R	L	L	L	L	L	L	L	L	L	L	H	L	H	R	L	L	L	L	L
19	R	H	L	H	L	L	L	L	L	L	L	R	L	R	M	H	L	H	M	L	L	R	L	L	L	L	R	M	R	H	M	L	L	M	L	R	H	L	M
20	R	L	L	L	R	L	H	L	L	L	L	R	L	R	L	L	L	R	L	L	L	R	L	L	L	L	L	R	L	H	R	L	M	H	L	L	H	L	M
21	L	M	L	L	R	L	L	L	L	L	M	R	L	L	L	L	L	L	L	R	L	L	L	L	R	R	L	L	L	L	L	L	R	H	L	L	M	R	H
22	R	R	L	L	R	L	L	R	L	L	L	R	R	M	L	L	L	L	L	R	R	R	R	L	R	L	L	L	L	R	L	R	L	L	R	R	L	H	L
23	R	L	L	L	R	L	M	R	L	L	L	R	R	L	L	L	L	L	L	H	H	L	L	L	L	L	L	L	L	L	L	R	L	L	R	R	R	L	L
24	R	L	L	L	R	L	R	R	L	L	L	R	R	M	L	L	L	L	L	L	L	H	L	R	R	L	R	M	R	L	R	R	L	L	R	R	R	L	L
25	L	R	L	L	L	L	R	R	L	L	L	R	R	M	L	L	L	L	L	L	L	L	R	L	R	R	R	R	R	R	R	R	M	R	R	R	R	R	M
26	L	L	L	L	R	L	R	R	L	L	L	R	R	M	L	L	L	L	L	R	L	H	L	L	R	L	R	R	R	R	R	R	H	R	R	R	R	H	H
27	R	L	L	L	R	L	R	R	L	L	L	R	R	M	L	L	L	L	L	H	L	H	L	L	R	L	R	R	R	R	R	R	H	R	R	R	R	H	H

Reagent groups: **Single Elec. Red.** (Na/NH₃, Al(Hg), SnCl₂/py, HSO₃⁻ H₂S); **Hydride Reductions** (LiAlH₄, Li-s-Bu₃BH, Chex₂BH, B₂H₆, NaBH₄, Zn(BH₄)₂, NaBH₃CN, i-Bu₂AlH, Li(t-BuO)₃AlH); **Acid and Lewis Acid** (AlCl₃ 80°C, AlCl₃ 25°C, SnCl₄ BF₃, LiClO₄ MgBr₂, TsOH 80°C, TsOH 0°C); **Soft Acids** (Hg(II), Ag(I), Cu(II)/Py); **Free Rad. Rxn** (HBr In·, HX/In·, NBS/CCl₄, Br₃CCl/In·); **Oxidants** (OsO₄, KMnO₄, O₃, RCO₃H 0°C, RCO₃H 50°C, CrO₃/Py, CrO₃ pH 1, H₂O₂, Quinone, ¹O₂, DMSO 100°C, NaOCl pH 10, Aq. NBS).

Reactivity Chart 6. Protection for the Carboxyl Group (continued)

PG	I₂	PhSeX, PhSCl	Br₂, Cl₂	MnO₂/CH₂Cl₂	NaIO₄, pH 5-6	SeO₂, pH 2-4	SeO₂, Py	K₃Fe(CN)₆, pH 8	Pb(IV), 25°C	Pb(IV), 80°C	Tl(NO₃)₃	150°C	250°C	350°C	:CCl₂	N₂CHCO₂R, Cu or Rh	CH₂I₂, Zn(Cu)	R₃SnH, In·	Ni(CO)₄	CH₂N₂	SOCl₂	Ac₂O, 25°C	Ac₂O, 80°C	DCC	MeI	Me₃O⁺ BF₄⁻	1. LDA, 2. MeI	1. K₂CO₃, 2. MeI	RCHO	RCOCl	C⁺/olefin
																													Electrophilic		
1	L	L	L	L	L	L	L	L	L	L		L	L	L	L	L	L	L	L	L	L	L	L	L	L	L			L	L	L
2	L	L	R	L	L	L	L	L	L	L		L	L	R	L	L	L	L	L	L	L	L	L	L	L	L			L	L	M
3	L	L	R	L	R	M	L	L	R	H		L	L	R	M	M	L	M	L	L	L	L	L	L	R	R			L	L	L
4	L	L	M	L	L	H	L	L	L	R		M	H	R	L	L	L	L	L	L	M	M	M	L	L	L			L	L	M
5	L	R	R	L	L	M	L	L	L	H		L	L	R	L	L	L	L	L	L	L	L	M	L	L	L			L	L	L
6	L	L	R	L	L	L	L	L	L	H		L	L	R	L	L	L	L	L	L	L	L	L	L	L	R			L	L	L
7	L	L	L	L	L	L	L	H	L	L		L	M	R	L	L	L	L	M	L	L	L	L	L	L	L			L	L	L
8	L	L	L	L	L	L	L	M	L	L		M	H	R	L	M	R	R	L	L	L	L	L	L	L	L			L	L	L
9	L	L	L	L	L	L	L	L	L	L		L	M	H	L	L	R	R	L	L	L	L	L	L	L	L			L	L	L
10	L	L	L	L	L	L	L	L	L	L		M	H	R	L	L	L	L	L	L	L	L	L	L	L	L			L	L	L
11	R	R	R	L	L	H	L	L	L	M		M	R	R	L	L	L	L	R	L	L	M	M	L	L	L			L	L	H
12	L	L	L	L	L	M	L	L	L	H		R	L	R	R	R	R	L	L	L	L	L	L	L	L	L			L	L	R
13	L	L	L	L	L	H	L	L	L	M		L	H	R	L	L	L	L	L	L	L	L	L	L	L	L			L	L	L
14	L	L	L	L	L	L	L	L	L	L		M	L	R	L	L	L	L	L	L	L	L	L	L	L	L			L	L	H
15	L	L	M	L	L	M	L	L	L	H		L	L	R	L	L	R	L	L	L	L	H	M	L	L	L			L	L	L
16	L	L	L	L	L	L	L	L	L	L		L	L	R	L	L	L	L	L	L	L	L	L	L	L	L			L	L	L
17	H	L	M	L	L	M	L	L	L	L		L	L	R	L	M	L	L	L	L	L	L	L	L	L	L			L	L	L
18	H	L	L	L	H	M	L	H	H	L		L	L	R	L	L	L	L	L	L	L	L	L	L	L	H			L	M	R
19	L	L	L	L	M	M	L	H	H	R		L	L	R	M	M	L	L	L	L	L	H	H	L	R	L			L	M	L
20	L	L	R	L	L	H	L	L	L	H		L	H	L	L	L	L	L	L	L	L	L	L	L	L	R			L	L	L
21	L	L	L	L	L	L	M	L	L	H		M	L	M	R	L	L	L	L	L	L	L	L	L	L	R			L	R	L
22	L	L	L	L	L	M	L	L	L	R		L	L	R	L	L	L	L	L	L	L	L	L	L	L	R			L	L	R
23	R	R	M	R	L	L	L	L	L	R		L	L	R	L	L	L	L	L	L	L	L	L	L	R	R			L	L	L
24	R	R	R	L	R	L	M	R	H	L		L	R	R	R	R	R	L	L	L	L	R	L	L	R	R			R	R	M
25	H	R	R	R	H	R	M	H	H	M		M	R	R	L	R	L	L	L	M	R	R	R	L	R	R			R	R	R
26	H	R	R	H	H	R	M	H	H	R		M	R	R	L	M	L	L	L	L	R	R	R	L	R	R			R	R	R
27	H	R	R	H	H	R	M	H	H	R		M	R	R	L	M	L	L	L	L	R	R	R	L	R	R			R	R	R

1016

Reactivity Chart 7.   Protection for the Thiol Group

1. *S*-Benzyl Thioether
2. *S*-*p*-Methoxybenzyl Thioether
3. *S*-*p*-Nitrobenzyl Thioether
4. *S*-4-Picolyl Thioether
5. *S*-2-Picolyl *N*-Oxide Thioether
6. *S*-9-Anthrylmethyl Thioether
7. *S*-Diphenylmethyl Thioether
8. *S*-Di(*p*-methoxyphenyl)methyl Thioether
9. *S*-Triphenylmethyl Thioether
10. *S*-2,4-Dinitrophenyl Thioether
11. *S*-*t*-Butyl Thioether
12. *S*-Isobutoxymethyl Monothioacetal
13. *S*-2-Tetrahydropyranyl Monothioacetal
14. *S*-Acetamidomethyl Aminothioacetal
15. *S*-Cyanomethyl Thioether
16. *S*-2-Nitro-1-phenylethyl Thioether
17. *S*-2,2-Bis(carboethoxy)ethyl Thioether
18. *S*-Benzoyl Derivative
19. *S*-(*N*-Ethylcarbamate)
20. *S*-Ethyl Disulfide

(See chart, pp. 1018–1020.)

Reactivity Chart 7. Protection for the Thiol Group

Reagent	1	2	3	4	5	6	7	8	9	10	11	12	13	14	15	16	17	18	19	20
1 Elec. Red.																				
Cr(II), pH 5	L	L	R	L	R	L	L	L	L	R	L	L	L	L	R	L	L	L	L	H
Zn/AcOH	L	L	R	L	R	L	L	L	L	R	L	L	L	L	R	R	L	L	L	H
Zn/HCl	L	M	R	L	R	L	L	L	M	R	L	H	H	L	R	R	L	H	H	H
Catalytic Reduction																				
H₂/Rh	R	R	R	R	R	R	R	R	R	M	R	R	R	R	R	R	R	R	R	R
H₂/Lindlar	M	M	M	M	R	M	M	M	R	L	L	L	M	R	L	L	L	L	L	H
H₂/Pd	R	R	R	R	R	R	R	R	R	M	R	R	R	R	R	R	R	R	R	R
H₂/Pt pH 2–4	R	R	R	R	R	R	R	R	R	R	R	R	R	R	R	R	R	R	R	R
H₂/Raney (Ni)	R	R	R	R	R	R	R	R	R	R	R	R	R	R	R	R	R	R	R	R
Wittig: ylide	L	L	L	L	L	L	L	L	L	L	L	L	L	L	L	L	L	L	H	H
Organometallic																				
Organocopper	L	L	R	L	L	L	L	L	L	R	L	L	L	L	R	R	L	L	L	M
Organozinc	L	L	L	L	L	L	L	L	L	L	L	L	L	L	L	R	H	L	L	H
RMgX	R	R	R	R	R	R	R	R	L	R	L	L	L	R	R	R	H	H	H	H
RLi	R	R	R	R	R	R	R	L	L	L	L	R	R	R	R	H	H	H	H	H
Nucleophilic																				
HCN, pH 6																				
NaCN, pH 12	L	L	M	L	L	L	L	L	L	H	L	L	L	L	M	M	H	L	L	H
AcO⁻; X⁻	L	L	L	L	L	L	L	L	L	L	L	L	L	L	L	L	L	L	L	L
RS⁻; N₃⁻; SCN⁻	L	L	M	L	L	H	M	M	R	H	L	H	R	L	L	M	M	H	H	M
NH₃; RNH₂	L	L	L	L	L	L	L	L	M	M	L	L	L	L	L	H	R	H	H	L
Enolate	L	L	L	L	L	L	L	L	M	L	L	L	L	L	R	H	H	R	H	R
MeONa	L	L	M	L	L	L	L	L	M	L	L	L	L	R	R	H	R	H	H	R
Basic																				
NaNH₂	L	L	R	L	L	L	M	M	L	R	L	L	L	R	H	H	H	H	H	R
Py; R₃N	L	L	L	L	L	L	L	L	L	L	L	L	L	L	L	H	L	H	H	L
LDA	R	R	R	R	R	R	R	L	R	L	L	L	R	R	H	H	H	H	R	R
t-BuOK	R	L	L	L	L	L	L	L	L	R	L	L	L	L	M	H	H	H	H	R
CH₃S(O)CH₂⁻ Na⁺	R	R	R	R	R	R	R	L	R	L	H	R	R	H	H	H	H	H	H	H
(C₁₀H₈)⁻ Na⁺	R	R	R	R	R	R	R	R	R	L	R	R	R	R	R	H	R	R	R	H
Ph₃CNa	R	R	R	R	R	R	R	L	R	L	L	L	R	R	H	H	L	R	R	R
NaH	L	L	M	L	L	L	L	L	L	R	L	L	L	R	R	H	H	L	R	R
Aqueous																				
pH > 12, 150°C	M	M	H	M	M	M	M	M	H	M	H	H	H	H	R	R	H	H	H	H
pH > 12	L	L	M	L	L	L	L	L	L	H	L	M	M	L	M	H	H	H	H	H
pH 10–12	L	L	L	L	L	L	L	L	L	H	L	L	L	M	H	H	H	H	L	L
pH 8.5–10	L	L	L	L	L	L	L	L	L	L	L	L	L	L	H	L	M	L	L	L
pH 6–8.5	L	L	L	L	L	L	L	L	M	L	L	L	L	L	L	L	L	M	L	L
pH 4–6	L	L	L	L	L	L	L	L	L	L	L	L	L	L	L	L	L	L	L	L
pH 2–4	L	L	L	L	L	L	L	L	L	L	L	L	L	L	L	L	L	L	L	L
pH 1	L	M	L	L	M	L	L	L	M	L	L	L	L	L	L	L	L	L	L	L
pH < 1	L	H	L	H	M	L	H	H	H	L	M	H	H	M	R	L	L	L	L	L
pH < 1, 100°C	H	H	H	H	H	H	H	H	H	H	H	H	H	H	H	H	H	H	H	H

Reactivity Chart 7. Protection for the Thiol Group (*Continued*)

PG	Single Elec. Red.				Hydride Reductions									Acid and Lewis Acid						Soft Acids			Free Rad. Rxn				Oxidants												
	Na/NH₃	Al(Hg)	SnCl₂/py	HSO₃⁻, H₂S	LiAlH₄	Li-s-Bu₃BH	Chex₂BH	B₂H₆, 0°C	NaBH₄	Zn(BH₄)₂	NaBH₃CN, pH 4–6	i-Bu₂AlH	Li(t-BuO)₃AlH	AlCl₃, 80°C	AlCl₃, 25°C	SnCl₄, BF₃	LiClO₄, MgBr₂	TsOH, 80°C	TsOH, 0°C	Hg(II)	Ag(I)	Cu(II)/Py	HBr, In•	HX/In•	NBS/CCl₄	Br₂CCl/In•	OsO₄	KMnO₄, pH 7, 0°C	O₃, −50°C	RCO₃H, 0°C	RCO₃H, 50°C	CrO₃/Py	CrO₃, pH 1	H₂O₂, pH 10–12	Quinone	¹O₂	DMSO, 100°C	NaOCl, pH 10	Aq. NBS
1	*H*	L	L	L	L	L	L	L	L	L	L	L	L	L	L	L	L	M	L	L	M	L	R	L	R	R	L	R	R	R	R	L	R	L	L	R	L	R	R
2	*H*	L	L	L	L	L	L	L	L	L	L	L	L	M	L	L	L	M	L	R	R	L	R	L	R	R	L	R	R	R	R	L	R	L	L	R	L	R	R
3	H	R	L	L	R	M	L	L	L	L	L	L	L	L	L	L	L	M	L	L	L	L	R	L	R	R	L	R	M	R	R	L	R	L	L	R	M	R	R
4	H	L	L	R	L	R	L	L	L	L	L	L	L	M	M	L	L	L	L	L	L	L	R	L	R	R	L	R	R	R	R	L	R	L	L	M	M	R	R
5	H	R	R	L	R	L	L	R	M	L	L	R	R	H	M	M	L	M	L	L	L	L	R	L	R	R	L	R	R	R	R	L	R	L	L	M	L	R	R
6	H	L	M	L	L	L	L	L	L	L	L	L	L	M	H	M	L	M	L	M	M	L	R	L	R	R	L	R	R	R	R	L	R	L	L	R	L	R	R
7	*H*	L	L	L	L	M	L	L	L	L	L	L	L	H	M	L	L	M	L	*M*	*M*	L	R	L	R	R	L	R	R	R	R	L	R	L	L	R	L	R	R
8	*H*	L	L	L	L	M	M	*H*	L	L	L	L	L	H	H	M	L	M	L	*R*	L	L	L	L	L	L	L	R	R	M	R	L	R	L	L	M	L	R	R
9	R	R	L	L	L	L	M	*M*	L	L	L	L	L	H	H	M	L	M	L	R	R	L	L	L	L	L	L	R	L	R	R	L	R	R	L	*M*	L	R	R
10	L	L	L	L	R	L	L	M	L	L	L	R	L	L	L	L	L	M	L	R	R	L	R	L	R	L	L	R	R	R	R	M	R	L	L	M	L	R	R
11	M	L	L	L	L	L	L	L	L	L	L	L	L	H	H	L	L	M	L	R	R	L	R	L	L	L	L	R	R	R	R	M	R	L	L	M	M	R	R
12	L	L	L	L	L	L	L	L	L	L	L	L	L	H	H	L	L	M	L	R	R	L	L	L	R	L	L	R	R	R	R	L	R	L	L	R	L	R	R
13	R	L	L	L	L	M	R	L	L	L	L	R	L	L	L	L	L	M	L	R	R	L	L	L	R	L	L	R	R	R	R	L	R	L	L	R	L	R	R
14	R	R	L	L	R	M	L	R	L	L	L	R	L	M	L	L	L	M	L	L	L	L	L	L	R	L	L	R	R	R	R	L	R	R	L	R	L	R	R
15	R	L	L	L	R	R	L	R	H	L	L	L	R	L	L	L	L	M	L	L	L	L	L	L	R	R	L	R	R	R	R	L	R	R	L	M	L	R	R
16	H	L	L	L	R	R	L	L	L	M	L	R	L	L	M	L	L	M	L	L	L	L	R	L	M	L	L	R	M	R	R	L	R	R	L	R	M	R	R
17	R	H	L	L	R	H	M	L	H	L	L	H	L	M	M	L	L	L	L	R	R	L	L	L	L	L	L	R	R	R	R	L	R	R	L	R	L	R	R
18	H	L	L	L	H	M	M	M	H	H	L	H	L	R	H	L	L	M	L	R	R	L	L	R	M	L	L	R	R	R	R	L	R	H	L	R	L	R	R
19	H	L	L	H	H	H	H	H	L	L	L	H	H	R	H	L	L	M	L	R	R	L	R	R	L	L	L	R	R	R	R	L	R	H	L	R	M	R	R
20	H	H	H	H	H	H	H	H	H	H	M	H	H	R	H	L	L	M	L	R	R	M	R	R	R	R	M	R	R	R	R	R	R	H	R	R	L	R	R

1019

Reactivity Chart 7. Protection for the Thiol Group (Continued)

PG	Electrophilic			Miscellaneous											Carbenes			Thermal			Oxidants										
	C^+/olefin	RCOCl	RCHO	1. K_2CO_3, 2. MeI	1. LDA, 2. MeI	Me_3O^+ BF_4^-	MeI	DCC	Ac_2O, 80°C	Ac_2O, 25°C	$SOCl_2$	CH_2N_2	$Ni(CO)_4$	R_3SnH, In·	CH_2I_2, Zn(Cu)	N_2CHCO_2R, Cu or Rh	:CCl_2	350°C	250°C	150°C	$Tl(NO_3)_3$	Pb(IV), 80°C	Pb(IV), 25°C	$K_3Fe(CN)_6$, pH 8	SeO_2, Py	SeO_2, pH 2–4	$NaIO_4$, pH 5–6	MnO_2/CH_2Cl_2	Br_2, Cl_2	PhSeX, PhSCl	I_2
1	M	L	L	R	R	R	R	L	L	L	L	L	L	R	L	R	M	M	L	L		R	R	L	M	R	R	M	R	L	L
2	M	L	L	R	R	R	R	L	L	L	L	L	L	R	L	R	M	M	L	L		R	R	L	M	R	R	M	R	L	M
3	L	L	L	M	R	M	M	L	L	L	L	L	L	R	L	R	M	M	L	L		R	R	L	L	R	R	L	R	L	R
4	L	H	L	R	R	R	R	L	R	R	R	L	L	R	L	R	M	M	L	L		R	R	L	M	R	R	L	R	L	R
5	L	L	L	R	R	R	R	L	L	L	L	L	L	R	L	R	R	M	L	L		R	R	L	M	R	R	L	R	L	R
6	M	L	L	R	R	R	R	L	L	L	L	L	L	R	L	R	M	M	L	L		R	R	L	R	R	R	M	R	L	M
7	M	L	L	R	R	R	R	L	L	L	L	L	L	R	L	R	M	M	L	L		R	R	L	R	R	R	M	R	L	M
8	M	L	L	R	L	R	R	L	L	L	L	L	L	L	L	R	M	R	L	L		R	R	L	R	L	R	M	M	L	R
9	L	L	L	L	R	L	L	L	L	L	L	L	L	R	L	R	M	M	L	L		R	R	L	L	M	L	L	M	L	R
10	M	L	L	L	M	L	L	L	L	L	L	L	L	L	L	R	M	R	M	L		R	R	R	L	L	R	L	M	L	R
11	M	L	L	M	M	M	M	L	L	L	L	L	L	M	L	R	M	R	M	M		R	R	L	L	M	R	L	R	L	L
12	M	L	L	R	M	R	R	L	M	L	L	L	L	M	L	R	M	M	H	M		R	R	L	L	M	R	L	R	L	R
13	L	L	L	R	R	R	R	L	L	L	L	L	L	M	L	R	M	H	L	L		R	R	L	L	M	R	L	R	L	R
14	L	L	L	R	R	R	R	L	L	L	L	L	L	M	L	R	M	H	L	L		R	R	L	L	M	R	L	R	L	R
15	L	L	L	R	R	R	R	L	L	L	L	L	L	L	L	R	M	R	H	M		R	R	M	L	M	R	L	R	L	R
16	L	L	L	R	R	R	R	L	L	L	L	L	L	L	L	M	M	R	H	M		M	R	H	L	M	R	L	R	L	R
17	L	L	L	R	L	R	L	L	M	L	L	L	L	L	L	M	M	R	R	M		M	L	H	L	L	L	L	R	L	R
18	L	L	L	L	L	R	L	L	L	L	L	L	L	L	L	M	M	H	R	R		R	L	H	L	L	L	L	R	L	L
19	L	L	L	L	L	R	L	L	L	L	L	L	L	R	L	M	M	R	R	R		R	R	H	R	R	R	L	R	L	L
20	R	L	L	M	R	R	M	L	L	L	L	L	R	R	R	R	M	R	R	H		R	R	R	R	R	R	R	R	L	R

Reactivity Chart 8. Protection for the Amino Group: Carbamates

1. Methyl Carbamate
2. 9-Fluorenylmethyl Carbamate
3. 2,2,2-Trichloroethyl Carbamate
4. 2-Trimethylsilylethyl Carbamate
5. 1,1-Dimethylpropynyl Carbamate
6. 1-Methyl-1-phenylethyl Carbamatae
7. 1-Methyl-1-(4-biphenylyl)ethyl Carbamate
8. 1,1-Dimethyl-2-haloethyl Carbamate
9. 1,1-Dimethyl-2-cyanoethyl Carbamate
10. *t*-Butyl Carbamate
11. Cyclobutyl Carbamate
12. 1-Methylcyclobutyl Carbamate
13. 1-Adamantyl Carbamate
14. Vinyl Carbamate
15. Allyl Carbamate
16. Cinnamyl Carbamate
17. 8-Quinolyl Carbamate
18. *N*-Hydroxypiperidinyl Carbamate
19. 4,5-Diphenyl-3-oxazolin-2-one
20. Benzyl Carbamate
21. *p*-Nitrobenzyl Carbamate
22. 3,4-Dimethoxy-6-nitrobenzyl Carbamate
23. 2,4-Dichlorobenzyl Carbamate
24. 5-Benzisoxazolylmethyl Carbamate
25. 9-Anthrylmethyl Carbamate
26. Diphenylmethyl Carbamate
27. Isonicotinyl Carbamate
28. *S*-Benzyl Carbamate
29. *N*-(*N'*-Phenylaminothiocarbonyl) Derivative

(See chart, pp. 1022–1024.)

Reactivity Chart 8. Protection for the Amino Group: Carbamates

Reaction	Reagent	1	2	3	4	5	6	7	8	9	10	11	12	13	14	15	16	17	18	19	20	21	22	23	24	25	26	27	28	29
1 Elec. Red.	Cr(II), pH 5	L	L	H	L	L	L	H	H	L	M	L	L	L	M	L	L	L	L	L	L	R	R	L	L	L	L	L	L	R
	Zn/AcOH	L	L	H	H	H	H	H	H	R	H	M	H	H	H	H	H	H	L	L	R	R	L	M	H	H	H	L	L	H
	Zn/HCl	L	L	H	M	H	H	H	H	R	H	H	H	H	H	H	H	H	H	L	M	R	R	L	H	M	H	H	L	H
Catalytic Reduction	H₂/Rh	L	R	L	L	R	R	L	R	L	L	L	L	R	R	L	H	R	R	R	R	R	R	R	R	R	R	R	R	R
	H₂/Lindlar	L	L	L	L	H	L	L	L	L	L	L	L	L	L	L	L	L	L	L	L	L	L	L	L	L	L	L	L	L
	H₂/Pd	L	M	L	L	H	L	L	R	L	L	L	L	H	H	L	H	H	H	H	H	H	H	H	H	H	H	H	R	R
	H₂/Pt pH 2-4	L	L	L	M	H	H	H	R	H	R	M	H	H	H	H	H	H	H	H	H	H	H	H	H	H	H	H	R	R
	H₂/Raney (Ni)	L	L	L	L	H	M	M	R	L	L	L	L	H	H	R	L	L	R	H	H	H	H	H	H	H	H	H	R	H
Organometallic	Wittig; ylide	L	R	L	R	L	L	L	M	M	L	L	L	L	L	L	L	L	L	L	L	L	L	L	L	L	L	L	L	L
	Organocopper	L	L	R	L	M	L	L	R	L	L	L	L	L	L	L	L	R	L	L	L	L	L	R	L	L	L	L	L	L
	Organozinc	L	M	R	L	R	L	L	M	M	H	L	L	L	L	L	L	L	L	L	L	L	M	L	L	M	L	L	L	M
	RMgX	L	H	H	H	H	H	H	H	H	H	H	H	H	H	H	H	H	H	H	H	H	H	H	H	H	H	H	H	H
	RLi	H	H	H	H	H	H	H	H	H	H	H	H	H	H	H	H	H	H	H	H	H	H	H	H	H	H	H	H	H
Nucleophilic	HCN, pH 6	L	L	L	L	L	L	L	L	L	L	L	L	L	L	L	L	L	L	L	L	L	L	L	L	L	L	L	L	L
	NaCN, pH 12	L	M	L	L	L	L	L	H	L	L	L	L	H	L	L	L	L	L	L	L	L	M	L	L	L	M	L	L	L
	AcO⁻; X⁻	L	L	L	L	L	L	L	L	L	L	L	L	L	L	L	L	L	L	L	L	L	L	L	L	L	L	L	L	L
	RS⁻; N₃⁻; SCN⁻	H	L	M	L	L	L	L	H	L	L	H	L	L	L	L	L	L	L	H	H	L	L	L	L	H	H	H	L	L
	NH₃; RNH₂	L	H	M	L	L	L	M	M	L	L	L	L	H	L	L	L	L	H	H	M	M	H	L	L	H	M	H	L	L
	Enolate	L	L	R	L	L	L	L	R	H	L	L	L	L	L	L	L	L	M	L	M	M	M	H	L	L	L	L	M	L
	MeONa	L	M	L	L	L	L	L	R	H	L	L	L	L	L	L	L	M	L	M	L	M	H	H	L	L	L	L	H	L
Basic	NaNH₂	L	H	R	L	R	L	L	L	M	L	L	L	L	L	L	H	L	L	M	L	M	H	L	L	L	L	L	L	L
	Py; R₃N	L	M	L	L	L	L	L	L	H	L	L	L	L	L	L	L	L	L	L	L	H	L	L	H	R	L	L	L	L
	LDA	L	L	R	L	R	L	L	L	H	L	L	L	L	L	L	L	L	L	L	L	H	L	L	H	L	L	L	L	L
	t-BuOK	L	L	R	L	R	L	L	L	H	L	L	L	L	L	L	L	L	L	L	L	H	L	L	H	L	L	L	L	L
	CH₃S(O)CH₂⁻Na⁺	L	L	R	L	R	L	L	L	H	L	L	L	L	L	L	L	L	L	L	L	H	L	L	H	L	L	L	L	L
	(C₁₀H₈)⁻̇Na⁺	L	L	R	L	R	L	L	H	M	L	L	L	L	L	L	M	H	L	R	L	H	L	L	H	L	L	L	L	L
	Ph₃CNa	L	H	R	L	R	L	L	L	R	L	L	L	L	L	L	L	R	L	L	R	R	L	L	R	R	R	R	L	L
	NaH	L	M	R	L	L	L	L	L	M	L	L	L	L	L	L	L	L	L	L	L	L	L	L	H	L	L	L	L	L
Aqueous	pH >12, 150°C	H	H	H	M	M	M	M	H	H	M	H	H	M	H	M	H	M	M	H	M	H	M	H	H	M	H	M	H	H
	pH >12	M	M	M	L	L	L	H	L	L	L	L	L	L	L	L	L	L	H	L	L	L	L	L	M	L	L	L	H	H
	pH 10-12	L	L	L	L	L	L	H	L	L	L	L	L	L	L	L	L	L	L	L	L	L	L	L	L	L	L	L	L	L
	pH 8.5-10	L	L	L	L	L	M	L	L	L	L	L	L	L	L	L	L	L	L	L	L	L	L	L	L	L	L	L	L	L
	pH 6-8.5	L	L	L	L	L	L	L	L	L	L	L	L	L	L	L	L	L	L	L	L	L	L	L	L	L	L	L	L	L
	pH 4-6	L	L	L	L	L	L	L	L	L	L	M	L	L	L	L	L	L	L	L	L	L	L	L	L	L	L	L	L	L
	pH 2-4	L	L	M	L	L	M	L	L	M	L	M	H	M	M	L	L	L	L	L	L	L	L	L	L	M	L	L	L	L
	pH 1	L	L	L	H	M	H	M	M	H	H	H	H	M	M	L	L	L	L	L	L	M	M	H	L	L	M	H	L	M
	pH <1	H	M	H	H	H	H	H	L	H	H	H	H	H	H	H	L	L	H	H	H	M	H	H	L	M	H	H	L	H
	pH <1, 100°C	H	H	H	H	H	H	H	H	H	H	H	H	H	H	H	H	H	H	H	H	H	H	H	H	H	H	H	H	H

Reactivity Chart 8. Protection for the Amino Group: Carbamates (Continued)

PG	Single Elec. Red.				Hydride Reductions									Acid and Lewis Acid						Soft Acids			Free Rad. Rxn				Oxidants												
	Na/NH₃	Al(Hg)	SnCl₂/py	HSO₃⁻, H₂S	LiAlH₄	Li-s-Bu₃BH	Chex₂BH	B₂H₆, 0°C	NaBH₄	Zn(BH₄)₂	NaBH₃CN, pH 4–6	i-Bu₂AlH	Li(t-BuO)₃AlH	AlCl₃, 80°C	AlCl₃, 25°C	SnCl₂, BF₃	LiClO₄, MgBr₂	TsOH, 80°C	TsOH, 0°C	Hg(II)	Ag(I)	Cu(II)/Py	HBr, In•	HX/In•	NBS/CCl₄	Br₃CCl/In•	OsO₄	KMnO₄, pH 7, 0°C	O₃, −50°C	RCO₃H, 0°C	RCO₃H, 50°C	CrO₃/Py	CrO₃, pH 1	H₂O₂, pH 10–12	Quinone	¹O₂	DMSO, 100°C	NaOCl, pH 10	Aq. NBS
1	L	L	L	L	R	L	L	L	L	L	L	R	L	R	L	L	L	M	L	L	L	L	L	L	L	L	L	L	L	L	L	L	H	L	L	L	L	L	L
2	H	L	L	L	M	L	L	L	L	L	L	M	L	R	L	L	L	L	L	L	L	H	R	R	R	L	L	L	L	L	L	H	L	L	L	L	L	L	L
3	R	M	H	L	R	L	L	L	L	L	L	M	L	R	R	R	L	L	L	L	L	L	M	M	M	M	L	L	L	L	L	L	L	L	L	L	R	L	L
4	L	L	L	L	R	L	R	L	L	L	L	M	L	R	R	L	L	H	H	L	L	L	L	L	L	R	R	R	R	L	R	L	L	L	L	L	L	L	L
5	R	L	L	L	R	L	L	R	L	L	L	M	L	R	L	L	L	R	H	R	L	L	R	L	L	L	L	L	L	M	L	L	H	L	L	L	L	L	R
6	R	L	L	L	R	M	L	L	L	L	L	M	L	R	H	L	L	R	H	L	L	R	L	L	L	L	L	L	L	L	M	L	L	L	L	L	L	L	L
7	R	L	L	L	R	L	L	L	L	L	M	M	L	R	H	M	L	R	L	L	L	L	H	L	L	L	L	L	L	L	L	H	H	L	L	L	L	L	L
8	H	L	H	L	R	L	L	L	L	L	L	L	L	R	H	M	L	M	L	L	L	L	M	L	L	L	L	L	L	L	L	L	H	R	L	L	M	M	L
9	H	L	L	L	M	L	L	M	L	L	L	M	L	R	H	M	L	R	L	R	L	L	L	L	L	L	R	L	L	L	M	L	M	L	L	L	L	L	L
10	L	L	L	L	M	L	L	L	L	L	M	M	L	R	H	L	L	M	H	L	M	H	H	L	L	L	L	L	L	L	L	L	L	L	L	R	L	L	L
11	L	L	L	L	M	L	L	L	L	L	L	M	L	R	M	M	L	R	L	L	L	L	M	R	R	R	L	R	R	L	L	L	H	L	L	R	L	L	L
12	L	L	L	L	R	L	L	R	L	L	L	M	L	R	L	L	L	L	L	R	L	L	R	R	R	R	R	R	R	L	R	L	H	L	L	M	L	L	L
13	L	L	L	L	R	L	R	R	L	L	M	M	L	R	L	L	L	R	L	L	L	L	R	R	R	R	L	L	L	R	R	L	H	L	L	L	L	L	L
14	L	L	M	L	M	L	R	L	L	L	L	L	L	R	L	L	L	R	L	L	L	L	L	L	L	L	L	L	L	R	R	L	L	L	L	L	L	L	R
15	H	L	L	L	M	L	R	R	L	L	L	M	L	R	H	L	L	R	L	R	M	L	R	L	L	L	M	M	M	R	R	L	M	L	L	L	L	L	R
16	H	L	L	L	R	L	L	L	L	L	L	M	L	R	L	H	L	L	L	L	L	H	L	R	R	R	L	L	L	R	R	L	L	L	L	R	L	L	R
17	H	L	L	L	R	L	L	L	L	L	L	M	L	R	L	H	L	L	L	L	L	L	R	R	R	L	L	L	L	H	H	L	M	L	L	L	L	L	L
18	H	L	L	L	R	L	M	R	L	L	L	M	L	R	L	L	L	L	L	L	L	L	L	L	R	L	L	L	L	L	L	H	L	L	L	L	L	L	L
19	H	R	M	L	M	L	L	L	L	L	L	M	L	R	L	L	L	M	L	L	L	H	H	L	L	R	L	L	L	L	L	L	L	L	L	L	L	R	R
20	H	R	L	L	M	L	L	L	L	L	L	M	L	R	L	L	L	R	L	L	L	L	M	L	R	L	R	R	R	L	L	H	M	L	L	M	L	M	L
21	H	L	L	L	M	L	L	R	L	L	L	M	L	R	L	L	L	L	L	L	L	H	R	L	R	L	L	L	L	L	L	R	L	L	L	R	L	L	L
22	H	L	L	L	M	L	L	L	L	L	L	M	L	R	R	L	L	R	L	L	L	L	H	R	R	L	L	L	L	R	R	L	L	L	L	L	L	L	L
23	H	L	M	L	M	L	L	L	L	L	L	M	L	R	H	L	L	M	L	L	L	H	R	R	R	L	L	R	R	L	L	L	L	L	L	L	L	L	L
24	H	L	L	L	M	L	L	L	L	L	L	M	L	R	H	L	L	R	L	L	L	L	H	M	R	R	L	L	R	L	L	H	M	L	L	L	L	L	R
25	H	L	L	L	M	L	L	L	L	L	L	M	L	R	H	L	L	R	L	L	L	L	R	R	R	L	L	L	L	R	R	L	L	L	L	L	L	L	L
26	H	R	L	L	M	L	L	L	L	L	L	M	L	R	M	L	L	R	L	L	L	L	H	R	R	L	L	L	L	H	H	L	H	L	L	L	L	L	L
27	H	L	M	L	M	L	L	L	L	L	L	M	L	R	L	L	L	R	L	L	L	L	R	R	R	L	L	L	L	H	R	L	L	L	L	L	L	L	L
28	H	L	L	L	M	L	L	L	L	L	L	M	L	R	R	L	L	R	L	H	H	L	R	R	R	L	L	R	R	H	H	L	L	M	L	L	L	R	R
29	H	R	L	L	M	L	L	L	L	L	L	M	L	R	M	M	L	L	L	R	M	M	R	R	R	L	L	R	R	H	H	L	M	H	L	R	R	R	R

Reactivity Chart 8. Protection for the Amino Group: Carbamates (*Continued*)

PG	Electrophilic RCHO	Electrophilic RCOCl	Electrophilic C+/olefin	1. K2CO3, 2. MeI	1. LDA, 2. MeI	Me3O+ BF4-	MeI	DCC	Ac2O, 80°C	Ac2O, 25°C	SOCl2	CH2N2	Ni(CO)4	R3SnH, In•	CH2I2, Zn(Cu)	N2CHCO2R, Cu or Rh	:CCl2	350°C	250°C	150°C	Tl(NO3)3	Pb(IV), 80°C	Pb(IV), 25°C	K3Fe(CN)6, pH 8	SeO2, Py	SeO2, pH 2-4	NaIO4, pH 5-6	MnO2/CH2Cl2	Br2, Cl2	PhSeX, PhSCl	I2
1	L	L	L	L	L	R	L	L	L	L	L	L	L	L	L	L	L	M	L	L	L	L	L	L	L	L	L	L	L	L	L
2	L	L	L	L	L	R	L	L	L	L	L	L	L	L	L	L	L	H	M	L	L	L	L	L	L	M	L	L	L	L	L
3	L	L	R	L	L	R	L	L	L	L	L	L	M	R	R	L	L	H	M	L	L	L	L	L	L	L	L	L	L	L	L
4	L	L	L	L	R	R	L	L	L	L	L	L	L	L	L	R	R	H	M	L	R	R	R	L	R	M	L	L	R	R	L
5	L	L	R	L	L	R	L	L	L	L	L	L	R	R	R	L	L	H	M	L	L	L	L	L	L	R	L	L	L	L	L
6	L	L	M	L	L	R	L	L	L	L	L	L	L	L	L	L	L	H	H	L	L	L	L	L	L	H	L	L	L	L	L
7	L	L	M	L	L	R	L	L	L	L	L	L	L	L	M	L	L	H	H	H	L	L	L	L	L	H	L	L	L	L	L
8	L	L	H	H	R	R	L	L	L	L	L	L	L	R	L	L	L	H	M	M	L	L	L	M	H	L	L	L	L	L	L
9	L	L	H	L	L	R	L	L	L	L	L	L	L	L	L	L	L	H	L	H	L	L	L	L	L	L	L	L	L	L	L
10	L	L	M	L	L	R	L	L	L	L	L	L	L	L	L	L	L	H	M	L	R	L	L	L	L	H	L	L	L	L	L
11	L	L	H	L	L	R	L	L	L	L	L	L	L	L	L	L	R	M	M	L	R	R	M	L	L	M	L	L	L	R	L
12	L	L	R	L	L	R	L	L	L	L	L	L	L	L	L	R	R	H	L	L	R	R	M	L	R	M	L	L	L	R	L
13	L	L	R	L	L	R	L	L	L	L	L	L	H	R	R	R	R	R	M	L	L	L	M	L	M	M	L	L	R	R	L
14	L	L	R	L	L	R	R	L	L	L	L	L	H	R	R	R	L	M	L	L	L	L	L	L	L	H	L	L	R	L	L
15	L	L	M	R	L	R	L	L	L	L	L	L	H	L	R	L	L	H	L	L	M	R	R	L	L	M	L	L	R	L	L
16	L	L	L	L	L	R	L	L	L	L	L	L	L	L	R	L	R	M	L	L	L	L	L	L	R	H	L	L	L	L	L
17	L	L	L	L	L	R	L	L	L	L	L	L	L	L	R	R	L	M	L	L	L	L	L	L	M	M	L	L	R	R	L
18	L	L	M	R	L	R	L	L	L	L	L	L	R	R	R	R	R	M	L	L	L	R	R	L	R	M	L	L	R	R	L
19	L	L	M	R	L	R	L	L	L	L	L	L	L	L	L	L	L	M	L	L	M	L	L	L	M	L	L	L	L	L	L
20	L	L	L	L	L	R	R	L	L	L	L	L	L	R	M	M	M	M	L	L	L	R	M	L	R	M	L	L	L	L	L
21	L	L	L	L	L	R	L	L	L	L	L	L	L	R	M	M	M	M	L	L	L	L	L	L	M	M	L	L	L	L	L
22	L	L	M	L	L	R	L	L	L	L	L	L	L	M	L	L	L	M	L	L	L	L	L	L	R	H	L	L	L	L	L
23	L	L	M	L	L	R	L	L	L	L	L	L	L	L	M	M	M	M	L	L	L	L	L	L	L	M	L	L	L	L	L
24	L	L	L	L	L	R	L	L	L	L	L	L	L	L	L	L	L	M	L	L	L	L	L	L	L	L	L	L	L	L	L
25	L	L	H	L	L	R	L	L	L	L	L	L	L	L	M	L	L	M	L	L	M	L	L	L	L	L	L	L	L	L	L
26	L	L	H	L	L	R	L	L	L	L	L	L	L	R	L	L	L	M	L	L	L	L	L	L	L	L	R	R	R	L	L
27	L	L	L	L	L	R	L	L	L	L	L	L	L	R	L	L	L	M	L	L	L	L	L	L	L	L	R	L	L	L	L
28	L	L	L	R	L	R	R	L	M	L	L	L	R	R	L	R	L	M	L	L	M	M	R	L	L	L	L	L	R	L	L
29	L	L	M	R	R	R	R	L	L	L	R	R	R	R	R	R	R	H	M	L	M	R	M	L	M	M	R	L	R	R	R

1024

Reactivity Chart 9.   Protection for the Amino Group: Amides

1. *N*-Formyl
2. *N*-Acetyl
3. *N*-Chloroacetyl
4. *N*-Trichloroacetyl
5. *N*-Trifluoroacetyl
6. *N*-*o*-Nitrophenylacetyl
7. *N*-*o*-Nitrophenoxyacetyl
8. *N*-Acetoacetyl
9. *N*-3-Phenylpropionyl
10. *N*-3-(*p*-Hydroxyphenyl)propionyl
11. *N*-2-Methyl-2-(*o*-nitrophenoxy)propionyl
12. *N*-2-Methyl-2-(*o*-phenylazophenoxy)propionyl
13. *N*-4-Chlorobutyryl
14. *N*-*o*-Nitrocinnamoyl
15. *N*-Picolinoyl
16. *N*-(*N'*-Acetylmethionyl)
17. *N*-Benzoyl
18. *N*-Phthaloyl
19. *N*-Dithiasuccinoyl

(See chart, pp. 1026–1028.)

Reactivity Chart 9. Protection for the Amino Group: Amides

PG	Aqueous										Basic								Nucleophilic							Organometallic					Catalytic Reduction					1 Elec. Red.		
	pH < 1, 100°C	pH < 1	pH 1	pH 2–4	pH 4–6	pH 6–8.5	pH 8.5–10	pH 10–12	pH > 12	pH > 12, 150°C	NaH	Ph₃CNa	(C₁₀H₈)⁻·Na⁺	CH₃S(O)CH₂⁻Na⁺	t-BuOK	LDA	Py; R₃N	NaNH₂	MeONa	Enolate	NH₃; RNH₂	RS⁻; N₃⁻; SCN⁻	AcO⁻; X⁻	NaCN, pH 12	HCN, pH 6	RLi	RMgX	Organozinc	Organocopper	Wittig; ylide	H₂/Raney (Ni)	H₂/Pt pH 2–4	H₂/Pd	H₂/Lindlar	H₂/Rh	Zn/HCl	Zn/AcOH	Cr(II), pH 5
1	H	*H*	*H*	*L*	L	L	L	M	H	H	*H*	L	R	L	L	L	L	L	L	L	*H*	L	L	M	L	H	H	L	L	L	L	*H*	L	L	L	H	L	L
2	*H*	M	L	L	L	L	L	L	M	H	L	R	R	R	L	R	L	R	L	L	H	L	L	L	L	H	M	L	L	L	L	L	L	L	L	M	L	L
3	*H*	M	L	L	L	L	M	H	*H*	H	M	R	R	R	L	R	L	R	R	L	H	R	L	R	L	H	M	M	R	L	M	R	R	L	L	H	H	R
4	H	L	L	L	L	L	M	*H*	H	H	L	L	R	L	L	R	L	L	R	R	H	M	L	R	L	H	M	M	R	M	R	R	R	L	R	H	H	M
5	H	L	L	L	L	L	L	L	*H*	H	L	L	R	L	L	L	L	L	R	R	*H*	M	L	R	L	H	M	M	M	M	R	R	R	L	L	H	H	M
6	H	L	L	L	L	L	L	L	M	H	M	R	R	R	L	R	L	R	L	L	H	M	L	L	L	H	R	L	L	L	R	*H*	L	L	R	H	H	R
7	H	L	L	L	L	L	L	L	H	H	R	R	R	R	R	L	L	R	M	M	M	L	L	M	L	H	R	M	L	L	R	R	R	L	R	H	H	R
8	H	L	L	L	L	L	L	L	M	H	M	R	R	R	R	R	L	R	R	R	*H*	L	L	L	R	H	M	L	R	R	M	R	R	L	M	L	L	L
9	H	L	L	L	L	L	L	L	M	H	R	R	R	R	L	R	L	R	L	L	M	L	L	L	L	H	R	R	L	L	L	L	L	L	R	L	L	L
10	H	L	L	L	L	L	L	L	M	H	L	L	R	R	R	R	L	R	R	L	M	L	L	L	L	H	R	L	L	L	L	L	L	L	R	L	L	L
11	H	L	L	L	L	L	L	L	H	H	L	L	R	L	L	L	L	L	L	L	M	L	L	L	L	H	R	L	L	L	R	*H*	R	R	R	H	H	R
12	H	L	L	L	L	L	L	L	L	H	M	R	R	M	L	H	L	L	L	L	M	R	L	R	L	H	M	R	L	L	R	R	R	L	R	H	H	H
13	H	L	L	L	L	L	L	L	M	H	L	L	R	R	R	R	L	R	R	R	M	M	L	M	L	H	R	L	R	L	M	R	R	L	L	L	L	L
14	H	L	L	L	L	L	L	L	M	H	L	R	R	R	L	L	L	L	M	L	H	L	L	L	L	H	M	R	L	L	L	*H*	L	L	R	H	H	R
15	H	L	L	L	L	L	L	L	*H*	H	R	L	R	R	R	L	L	L	L	L	H	L	L	L	L	H	R	L	L	L	R	L	R	R	R	L	L	L
16	H	L	L	L	L	L	L	L	R	H	L	L	R	L	L	R	L	R	L	L	*H*	L	L	M	L	H	M	L	L	L	L	L	L	L	R	L	L	L
17	*H*	*H*	L	L	L	L	L	L	H	H	L	L	R	L	L	L	L	L	L	L	H	L	L	L	L	H	M	L	L	L	L	L	L	L	R	L	L	L
18	*H*	L	L	L	L	L	L	L	R	H	L	L	R	L	L	L	L	L	L	L	*H*	L	L	L	L	H	M	R	L	L	L	L	L	L	R	L	L	L
19	M	M	L	L	L	L	L	M	H	H	L	L	R	R	L	L	L	L	H	H	H	*H*	L	H	L	H	R	R	M	H	R	R	R	R	R	R	R	R

1026

Reactivity Chart 9. Protection for the Amino Group: Amides (*Continued*)

PG	Na/NH₃	Al(Hg)	SnCl₂/py	HSO₃⁻, H₂S	LiAlH₄	Li-s-Bu₃BH	Chex₂BH	B₂H₆, 0°C	NaBH₄	Zn(BH₄)₂	NaBH₃CN pH 4–6	i-Bu₂AlH	Li(t-BuO)₃AlH	AlCl₃ 80°C	AlCl₃ 25°C	SnCl₄, BF₃	LiClO₄, MgBr₂	TsOH 80°C	TsOH 0°C	Hg(II)	Ag(I)	Cu(II)/Py	HBr, In•	HX/In•	NBS/CCl₄	Br₃CCl/In•	OsO₄	KMnO₄ pH 7, 0°C	O₃, −50°C	RCO₃H 0°C	RCO₃H 50°C	CrO₃/Py	CrO₃, pH 1	H₂O₂ pH 10–12	Quinone	¹O₂	DMSO 100°C	NaOCl pH 10	Aq. NBS
	Single Elec. Red.					Hydride Reductions								Acid and Lewis Acid						Soft Acids			Free Rad. Rxn				Oxidants												
1	R	L	L	L	R	L	H	H	R	L	L	H	L	L	L	L	L	L	L	*L*	L	L	L	L	L	L	L	L	L	L	M	L	H	M	L	L	L	L	L
2	R	L	L	L	R	L	H	R	L	L	L	H	L	L	L	L	L	L	L	L	L	L	L	L	L	L	L	L	L	L	L	L	L	M	L	L	L	L	L
3	R	M	L	L	R	L	H	R	L	L	L	H	L	M	M	L	L	L	L	L	H	L	L	L	L	L	L	L	L	L	L	L	L	L	L	L	R	L	L
4	R	M	L	L	*H*	M	H	R	*H*	M	M	H	M	M	L	L	L	L	L	L	H	L	R	R	R	R	L	L	L	L	L	L	L	L	L	L	L	R	L
5	R	L	L	L	R	M	H	R	*H*	M	M	H	M	L	L	L	L	L	L	L	M	L	L	L	L	L	L	L	L	L	R	L	L	L	L	L	L	R	L
6	R	R	L	L	H	M	H	R	L	L	L	H	L	L	L	L	L	L	L	L	L	L	L	L	M	L	L	L	L	L	L	L	L	L	L	L	L	L	L
7	R	L	L	L	R	M	H	R	L	L	L	H	L	L	R	M	L	L	L	L	L	L	L	L	L	L	L	L	L	L	L	L	L	L	L	L	L	L	L
8	R	L	L	L	R	M	H	R	L	R	R	H	R	R	L	L	L	L	L	L	L	L	L	L	L	L	L	*H*	L	L	M	M	*H*	M	M	L	L	R	R
9	R	L	L	L	R	R	H	R	L	L	L	H	L	L	L	L	L	L	L	L	L	L	R	R	R	R	L	L	L	L	L	L	L	L	L	L	L	L	*H*
10	R	L	L	L	R	L	H	R	L	L	L	H	L	L	L	L	L	L	L	L	L	L	L	L	L	L	L	R	L	L	L	L	R	R	L	L	L	R	L
11	R	R	L	L	R	L	H	R	R	L	L	H	M	L	L	L	L	L	L	L	L	L	L	L	R	R	L	L	R	M	R	L	L	L	L	L	L	L	L
12	R	*H*	R	L	R	M	H	R	L	L	L	H	L	R	M	L	L	L	L	L	R	L	L	L	L	L	L	L	L	L	L	M	M	R	L	R	L	L	L
13	R	L	L	L	R	L	H	R	L	L	L	H	L	M	M	L	L	L	L	H	*H*	L	R	R	R	R	R	L	R	L	M	L	R	L	L	L	R	L	R
14	R	R	L	L	R	M	H	R	L	L	L	H	L	M	L	L	L	L	L	L	L	*H*	L	L	L	L	L	R	L	M	R	L	L	L	L	L	L	L	L
15	R	L	L	L	R	M	H	R	L	L	L	H	L	L	L	L	L	L	L	L	L	L	R	L	R	L	L	L	R	R	L	L	L	L	L	L	L	R	R
16	R	L	L	L	R	L	H	R	L	L	L	H	L	L	L	L	L	L	L	L	L	L	L	L	L	L	L	L	L	L	L	L	L	L	L	L	L	L	L
17	*R*	L	L	L	*R*	L	H	R	*L*	L	L	*H*	L	L	L	L	L	L	L	L	L	*H*	L	L	L	L	L	L	L	L	L	L	L	L	L	L	L	L	L
18	R	L	L	L	R	L	H	R	L	L	L	H	L	L	L	L	L	R	L	L	L	L	L	L	L	L	L	L	L	L	L	L	L	L	L	L	L	L	L
19	R	R	R	L	R	R	H	R	R	L	L	H	R	R	R	L	L	L	L	R	M	L	R	R	R	R	M	R	R	R	R	L	R	R	R	R	R	R	R

1027

Reactivity Chart 9. Protection for the Amino Group: Amides (Continued)

PG	I₂	PhSeX, PhSCl	Br₂, Cl₂	MnO₂/CH₂Cl₂	NaIO₄, pH 5–6	SeO₂, pH 2–4	SeO₂, Py	K₃Fe(CN)₆, pH 8	Pb(IV), 25°C	Pb(IV), 80°C	Tl(NO₃)₃	150°C	250°C	350°C	:CCl₂	N₂CHCO₂R, Cu or Rh	CH₂I₂, Zn(Cu)	R₃SnH, In•	Ni(CO)₄	CH₂N₂	SOCl₂	Ac₂O, 25°C	Ac₂O, 80°C	DCC	MeI	Me₃O⁺ BF₄⁻	1. LDA, 2. MeI	1. K₂CO₃, 2. MeI	RCHO	RCOCl	C⁺/olefin
							Oxidants						Thermal			Carbenes			Miscellaneous											Electrophilic	
1	L	L	M	L	L	L	L	L	L	L	L	L	L	L	L	L	L	L	L	L	L	L	L	L	L	R	R	L	L	H	L
2	L	L	L	L	L	L	L	L	L	L	L	L	L	L	L	L	L	L	L	L	L	L	L	L	L	R	R	L	L	L	L
3	L	L	L	L	L	L	L	L	L	L	L	L	M	L	L	L	M	R	L	L	L	L	L	L	L	R	R	L	L	L	L
4	L	L	L	L	M	L	L	M	L	L	L	L	M	R	L	L	H	R	M	L	L	L	L	L	L	R	L	M	L	L	L
5	L	L	L	L	L	L	L	M	L	L	L	L	L	R	L	L	L	R	L	L	L	L	L	L	L	R	L	L	L	L	L
6	L	L	L	L	L	L	L	L	L	L	L	L	L	L	L	L	L	L	L	L	L	L	L	L	L	R	R	L	L	L	L
7	L	L	L	L	M	H	M	L	M	R	L	L	L	L	L	M	L	L	L	L	L	L	M	L	L	R	R	L	L	L	M
8	M	L	R	L	L	L	L	R	L	L	M	L	L	M	L	L	L	R	L	M	L	M	M	L	L	R	R	R	L	M	L
9	L	L	L	L	M	M	M	L	M	R	L	L	L	L	L	L	L	L	L	L	L	L	R	R	L	R	R	L	L	L	L
10	L	L	H	M	L	L	L	R	L	L	L	R	L	L	L	L	L	L	L	R	L	R	R	L	L	R	L	R	L	R	L
11	L	L	L	L	L	L	L	L	L	M	L	R	R	L	L	R	R	L	L	L	L	L	L	L	L	R	L	L	L	L	R
12	L	L	L	L	L	L	L	L	L	L	R	L	R	R	M	L	R	R	L	L	L	L	L	L	L	R	R	M	L	L	L
13	L	L	L	L	L	L	L	L	L	R	L	M	L	R	L	R	R	R	L	L	L	L	L	L	L	R	L	L	L	L	M
14	L	L	M	L	L	L	L	L	L	L	M	L	H	L	R	M	M	L	L	L	L	L	L	L	L	R	L	L	L	L	L
15	L	L	L	L	R	L	L	L	R	R	L	L	L	L	L	M	L	L	L	L	L	L	L	L	M	R	R	M	L	L	L
16	L	L	R	L	L	M	L	L	L	L	L	L	L	H	M	L	L	L	L	L	L	L	L	L	R	R	L	R	L	L	L
17	L	L	L	L	L	L	L	L	L	L	L	L	L	L	L	L	L	L	L	L	L	L	L	L	L	R	L	L	L	L	L
18	M	L	L	L	L	L	L	L	L	L	L	L	L	L	L	M	L	L	L	L	L	L	L	L	L	R	L	L	L	L	L
19	M	M	R	R	M	R	M	R	R	R	L	L	M	H	M	M	H	R	R	L	L	L	L	L	M	R	M	L	L	L	L

1028

Reactivity Chart 10.   Protection for the Amino Group: Special —NH Protective Groups

1. *N*-Allyl
2. *N*-Phenacyl
3. *N*-3-Acetoxypropyl
4. Quaternary Ammonium Salts
5. *N*-Methoxymethyl
6. *N*-Benzyloxymethyl
7. *N*-Pivaloyloxymethyl
8. *N*-Tetrahydropyranyl
9. *N*-2,4-Dinitrophenyl
10. *N*-Benzyl
11. *N*-*o*-Nitrobenzyl
12. *N*-Di(*p*-methoxyphenyl)methyl
13. *N*-Triphenylmethyl
14. *N*-(*p*-Methoxyphenyl)diphenylmethyl
15. *N*-Diphenyl-4-pyridylmethyl
16. *N*-2-Picolyl *N'*-Oxide
17. *N,N'*-Isopropylidene
18. *N*-Benzylidene
19. *N*-*p*-Nitrobenzylidene
20. *N*-Salicylidene
21. *N*-(5,5-Dimethyl-3-oxo-1-cyclohexenyl)
22. *N*-Nitro
23. *N*-Oxide
24. *N*-Diphenylphosphinyl
25. *N*-Dimethylthiophosphinyl
26. *N*-Benzenesulfenyl
27. *N*-*o*-Nitrobenzenesulfenyl
28. *N*-2,4,6-Trimethylbenzenesulfonyl
29. *N*-Toluenesulfonyl
30. *N*-Benzylsulfonyl
31. *N*-Trifluoromethylsulfonyl
32. *N*-Phenacylsulfonyl

(See chart, pp. 1030–1032.)

Reactivity Chart 10. Protection for the Amino Group: Special –NH Groups

PG	Cr(II), pH 5	Zn/AcOH	Zn/HCl	H₂/Rh	H₂/Lindlar	H₂/Pd	H₂/Pt pH 2–4	H₂/Raney (Ni)	Wittig; ylide	Organocopper	Organozinc	RMgX	RLi	HCN, pH 6	NaCN, pH 12	AcO⁻; X⁻	RS⁻; N₃⁻; SCN⁻	NH₃⁺, RNH₂	Enolate	MeONa	NaNH₂	Py; R₃N	LDA	t-BuOK	CH₃S(O)CH₂⁻Na⁺	(C₁₀H₈⁻) Na⁺	Ph₃CNa	NaH	pH > 12, 150°C	pH > 12	pH 10-12	pH 8.5-10	pH 6-8.5	pH 4-6	pH 2-4	pH 1	pH > 1	pH < 1, 100°C
1	L	L	L	R	L	H	R	R	L	L	L	L	L	L	L	L	L	R	L	L	R	L	R	R	R	R	R	L	R	L	L	L	L	L	L	L	L	H
2	H	H	H	R	L	R	L	L	R	L	L	R	R	R	R	L	L	M	M	M	R	L	R	R	R	R	R	R	R	L	L	L	L	L	L	L	L	H
3	L	L	L	L	L	L	L	L	L	L	L	R	R	L	M	L	L	M	R	R	R	L	R	L	M	R	R	L	R	R	R	M	L	L	L	L	R	H
4	L	L	H	L	L	L	L	L	R	L	M	R	R	L	L	L	L	R	L	L	R	L	R	L	M	L	R	L	L	L	L	L	L	L	L	L	L	H
5	L	L	H	L	L	L	L	L	L	L	L	L	L	L	L	L	L	L	L	L	L	L	L	L	L	L	L	L	L	L	L	L	L	L	L	L	H	H
6	L	L	L	L	L	R	H	H	L	L	L	L	L	L	L	L	L	L	L	H	L	L	L	L	L	L	L	L	L	L	H	L	L	L	L	M	H	H
7	L	L	L	L	L	L	L	L	L	L	L	L	L	L	L	L	L	L	L	M	H	L	L	L	L	L	L	L	H	H	L	M	L	H	H	H	H	H
8	H	H	M	L	L	R	L	H	L	L	L	R	R	L	R	L	H	L	L	L	H	L	L	L	L	R	M	L	H	L	L	L	L	L	H	L	H	H
9	R	R	R	R	L	L	L	R	L	L	L	L	L	L	L	L	L	L	L	L	H	L	L	L	L	L	L	L	H	L	L	L	L	L	L	L	H	H
10	L	L	R	H	L	R	R	R	L	L	L	H	H	L	R	L	L	L	L	H	H	L	L	L	L	R	M	L	R	H	L	L	L	L	L	L	H	H
11	R	R	L	R	L	M	M	R	L	L	L	L	L	L	L	L	L	L	L	L	L	L	L	L	L	L	L	L	L	L	L	L	L	L	L	K	K	M
12	L	L	H	L	L	H	H	H	L	L	M	R	R	R	R	H	H	L	M	M	H	L	L	L	L	L	L	L	R	M	H	K	K	M	H	H	H	H
13	L	L	H	L	L	H	H	H	L	L	L	L	L	L	L	L	L	L	L	L	L	L	L	L	L	L	M	L	H	L	H	L	L	L	H	H	H	H
14	L	L	R	L	L	H	R	R	L	L	L	H	H	L	L	L	L	L	L	L	H	L	L	L	L	R	L	L	L	L	L	K	L	L	L	H	L	M
15	L	L	H	L	L	M	M	R	L	L	M	R	R	L	L	L	L	L	L	L	H	L	L	L	L	L	M	L	R	K	K	K	K	L	L	L	L	H
16	R	R	R	L	L	H	H	R	L	L	L	R	R	L	R	L	H	H	L	L	L	L	L	L	L	R	R	L	H	L	L	L	L	H	M	H	H	H
17	R	R	R	L	L	H	H	R	L	L	L	R	R	L	R	L	L	H	L	L	L	L	L	L	L	R	R	L	H	L	L	L	L	L	M	L	H	M
18	R	R	R	R	L	H	H	R	L	L	R	R	R	R	R	H	R	H	L	R	L	L	L	L	L	R	L	L	H	M	L	L	L	L	M	H	H	M
19	R	R	R	R	L	H	R	R	R	L	R	R	R	M	R	H	R	H	H	H	H	L	R	R	R	R	M	R	H	L	L	L	L	L	M	L	H	H
20	R	R	R	R	L	H	H	R	M	R	R	R	R	L	R	L	L	H	L	H	L	L	L	L	L	R	R	H	H	M	L	L	L	L	L	H	H	H
21	R	R	R	R	L	R	R	R	M	R	H	R	R	L	R	L	R	H	R	H	R	L	R	R	R	R	R	L	H	L	L	L	L	L	L	H	H	H
22	R	R	R	R	L	R	R	R	M	M	H	R	R	R	R	L	L	H	H	H	H	L	R	R	R	R	R	R	H	M	L	L	L	L	L	L	M	H
23	R	R	R	R	L	H	R	R	M	M	L	R	R	M	R	H	L	R	H	H	L	L	L	L	L	R	L	L	H	L	L	L	L	L	L	H	H	H
24	H	R	R	H	L	R	H	H	M	M	H	H	R	H	R	H	R	L	H	H	L	L	M	M	L	L	M	L	H	M	L	L	L	L	M	L	M	H
25	L	L	L	L	L	H	R	R	L	L	L	L	L	L	L	L	L	L	L	L	L	L	L	L	L	R	L	L	H	M	L	L	L	L	L	L	L	H
26	L	L	L	R	L	R	R	R	L	L	M	M	M	L	L	L	L	L	L	L	H	L	L	L	L	R	M	L	H	L	L	L	L	L	M	H	M	H
27	L	L	R	L	L	L	L	R	L	L	L	R	R	H	H	H	L	L	L	L	L	L	L	L	L	R	R	L	H	L	L	L	L	L	L	L	M	H
28	L	L	L	R	L	H	H	H	L	L	H	H	H	H	H	L	L	L	L	L	H	L	L	L	L	H	H	L	H	L	L	L	L	L	L	L	L	H
29	R	H	H	L	L	L	L	L	L	L	H	H	H	H	H	H	L	L	L	L	L	L	L	L	L	H	L	L	H	L	L	L	L	L	L	L	L	H
30	L	L	L	L	L	L	L	L	L	L	L	L	L	H	H	L	L	L	L	L	L	L	L	L	L	H	L	L	L	M	L	L	L	L	L	L	M	H
31	M	M	H	L	L	L	H	L	L	L	H	L	L	L	L	L	R	L	R	R	R	L	R	R	R	R	R	R	H	L	L	L	L	L	L	L	L	H
32	M	L	H	R	L	H	L	H	R	R	H	H	H	R	R	L	L	R	R	R	R	L	R	R	R	R	R	R	H	R	L	L	L	L	L	L	L	H

1030

PG	Single Elec. Red.				Hydride Reductions									Acid and Lewis Acid						Soft Acids			Free Rad. Rxn				Oxidants												
	Na/NH₃	Al(Hg)	SnCl₂/py	HSO₃⁻, H₂S	LiAlH₄	Li-s-Bu₃BH	Chex₂BH	B₂H₆, 0°C	NaBH₄	Zn(BH₄)₂	NaBH₃CN, pH 4-6	i-Bu₂AlH	Li(t-BuO)₃AlH	AlCl₃, 80°C	AlCl₃, 25°C	SnCl₄, BF₃	LiClO₄, MgBr₂	TsOH, 80°C	TsOH, 0°C	Hg(II)	Ag(I)	Cu(II)/Py	HBr, In·	HX/In·	NBS/CCl₄	Br₃CCl/In·	OsO₄	KMnO₄, pH 7, 0°C	O₃, -50°C	RCO₃H, 0°C	RCO₃H, 50°C	CrO₃/Py	CrO₃, pH 1	H₂O₂, pH 10-12	Quinone	¹O₂	DMSO, 100°C	NaOCl, pH 10	Aq. NBS
1	R	L	L	L	L	L	R	R	L	L	L	L	L	L	L	L	L	L	L	R	L	L	R	R	R	R	R	R	R	R	R	R	L	R	R	R	M	M	R
2	R	R	L	R	R	R	R	R	R	M	L	M	R	L	L	L	L	L	L	L	L	L	L	L	L	L	R	R	R	R	R	R	L	R	R	R	M	R	R
3	R	L	L	L	R	L	L	L	L	L	L	L	M	R	L	L	L	R	L	L	M	L	L	L	L	L	L	L	L	L	L	L	L	L	L	L	M	L	L
4	L	L	L	L	L	L	L	L	L	L	L	L	L	R	H	H	M	M	L	L	L	L	L	L	R	L	R	R	R	R	R	R	L	R	R	L	H	M	L
5	H	L	L	L	R	L	L	L	L	L	L	L	L	H	H	M	L	M	L	H	L	L	L	L	R	L	R	R	R	R	R	R	L	R	R	R	M	M	R
6	R	L	L	L	L	L	L	L	L	L	L	L	M	H	H	H	L	R	L	R	L	L	L	L	R	L	R	R	R	R	R	R	L	R	R	R	M	M	R
7	L	L	L	L	L	L	L	L	L	L	L	L	L	H	H	H	L	M	L	L	L	L	L	L	R	L	R	R	R	R	R	L	H	L	L	L	L	L	L
8	H	M	R	L	L	L	L	L	M	M	M	L	M	H	H	H	L	R	L	L	M	R	L	L	R	R	R	M	M	M	H	R	H	L	L	M	L	M	R
9	H	L	L	L	R	L	L	L	L	L	L	L	L	L	L	L	L	L	L	L	M	H	L	L	R	R	R	R	R	R	R	R	L	R	R	R	L	M	R
10	H	L	L	M	L	L	L	L	M	M	L	L	L	R	R	L	M	L	L	L	M	H	L	L	R	R	R	R	R	R	R	R	L	R	R	R	L	M	R
11	H	L	L	R	L	L	L	R	M	L	R	M	L	L	L	L	L	L	L	L	M	H	L	L	R	R	R	R	R	R	R	R	L	R	R	L	L	M	R
12	H	R	L	M	L	L	L	R	M	L	L	L	M	R	R	L	M	R	L	L	M	H	L	L	R	R	R	R	R	R	R	R	H	R	R	L	L	M	R
13	R	M	L	R	R	L	L	R	M	L	L	L	L	L	L	L	L	L	L	L	L	R	L	L	R	R	R	R	R	R	R	R	H	R	R	L	R	R	R
14	R	R	L	L	R	M	L	R	L	L	L	L	M	L	L	L	L	L	L	L	L	R	L	L	R	R	R	R	R	R	R	R	H	R	R	L	R	R	R
15	R	M	L	H	R	L	L	R	M	L	L	R	R	L	R	L	L	L	L	L	L	R	L	L	R	R	R	R	R	R	R	R	L	R	R	R	R	R	R
16	R	L	L	L	R	L	L	R	M	R	R	M	R	R	L	L	L	L	L	L	L	R	L	R	R	R	R	R	R	R	R	R	L	M	R	L	R	R	R
17	R	R	R	L	R	L	L	R	L	L	M	R	R	L	R	R	R	L	L	L	L	R	L	R	R	R	R	R	R	R	R	R	L	M	R	L	R	R	R
18	H	H	H	L	M	L	L	R	L	L	R	L	H	L	L	L	L	L	L	M	M	R	L	R	R	R	R	R	R	R	R	R	H	L	R	M	L	R	R
19	H	L	L	M	M	L	L	R	L	L	M	L	L	R	R	L	L	L	L	M	M	R	R	R	R	R	R	R	R	R	R	R	L	M	R	R	R	R	R
20	H	L	L	H	H	R	L	R	L	L	R	M	H	L	L	L	L	L	L	M	M	R	R	R	R	R	R	R	R	R	R	R	L	L	R	R	R	R	L
21	H	H	L	M	R	L	L	R	M	R	M	L	H	R	R	R	R	R	R	H	L	L	R	L	R	R	R	R	R	R	R	R	R	L	R	R	R	R	L
22	H	H	M	H	H	R	R	R	M	R	M	L	H	L	L	L	L	L	L	H	H	L	R	L	R	L	R	R	R	R	R	R	L	L	R	L	R	R	L
23	H	H	L	L	R	L	L	L	M	M	L	L	H	L	L	R	L	M	L	H	H	L	R	L	R	L	R	R	R	R	R	R	H	L	R	L	L	L	L
24	H	H	L	L	R	L	L	L	L	M	L	L	H	L	L	L	L	M	L	H	H	L	R	L	R	L	R	R	R	R	R	L	L	L	L	L	L	L	L
25	H	H	L	M	H	H	L	L	L	M	L	L	H	R	L	L	L	M	L	H	H	L	R	L	R	L	R	R	R	R	R	R	L	L	L	R	L	L	L
26	H	H	L	M	M	M	R	R	L	R	L	L	H	L	L	L	L	M	L	H	H	L	R	L	R	R	R	R	R	R	R	R	H	L	L	L	L	R	L
27	H	L	L	H	H	L	L	R	R	R	L	L	L	L	L	L	L	L	L	H	H	L	R	L	R	L	R	R	R	R	R	R	L	L	L	L	L	L	L
28	H	H	L	L	L	L	L	R	L	L	L	H	L	L	L	L	L	L	L	H	L	H	L	L	L	L	L	L	L	L	L	L	L	L	L	M	L	L	L
29	H	H	M	M	H	M	L	L	L	L	L	H	H	H	L	L	L	M	L	H	H	H	L	L	R	L	L	L	L	L	L	L	L	L	L	L	L	M	L
30	M	H	L	H	H	L	L	R	M	L	M	H	H	L	L	L	L	M	L	H	H	L	L	L	R	L	L	L	L	L	L	L	L	L	L	L	L	M	L
31	H	H	L	L	L	L	L	R	L	L	M	R	R	L	L	L	L	M	L	H	H	H	L	L	M	L	L	L	L	L	L	L	L	L	L	L	L	M	L
32	H	H	L	L	R	R	R	R	M	L	M	R	R	L	L	L	L	L	L	H	M	H	L	L	L	L	L	L	L	L	L	L	L	L	L	R	L	M	R

Reactivity Chart 10. Protection for the Amino Group: Special −NH Groups (*Continued*)

PG	I_2	PhSeX, PhSCl	Br_2, Cl_2	MnO_2/CH_2Cl_2	$NaIO_4$, pH 5–6	SeO_2, pH 2–4	SeO_2, Py	$K_3Fe(CN)_6$, pH 8	Pb(IV), 25°C	Pb(IV), 80°C	$Tl(NO_3)_3$
	Oxidants										
1	L	R	R	R	R	R	R	R	R	R	R
2	L	M	R	R	R	R	R	R	R	R	R
3	L	L	R	R	R	M	L	R	R	R	R
4	L	L	L	L	L	L	L	L	L	L	L
5	L	L	R	R	R	M	L	R	R	R	R
6	L	L	R	R	R	R	M	R	R	R	R
7	L	L	R	R	R	M	L	R	R	R	R
8	L	L	R	R	R	M	L	R	R	R	R
9	L	L	*M*	*L*	*L*	*L*	L	*L*	*L*	*L*	*L*
10	L	L	R	R	R	R	M	R	R	R	R
1	L	L	R	R	R	M	L	R	R	R	R
12	L	L	R	R	R	R	M	R	R	R	R
13	L	L	R	R	R	M	L	R	R	R	R
14	L	L	R	R	R	M	L	R	R	R	R
15	L	L	R	R	R	M	L	R	R	R	R
16	L	L	R	R	R	R	M	R	R	R	R
17	R	R	R	R	R	R	R	R	R	R	R
18	R	R	R	R	R	R	M	R	R	R	R
19	R	R	R	R	R	M	M	R	R	R	R
20	R	R	R	R	R	R	M	R	R	R	R
21	L	*R*	R	L	R	R	R	M	R	R	R
22	R	R	R	R	R	R	R	R	R	R	R
23	L	L	L	L	L	L	L	L	L	L	L
24	L	L	L	L	L	M	L	L	L	L	L
25	L	L	L	L	R	M	L	L	L	L	L
26	R	L	R	*R*	R	R	L	R	R	R	R
27	R	L	R	R	R	M	L	M	R	R	R
28	L	L	L	L	L	M	L	L	L	L	L
29	L	L	L	L	L	M	L	L	L	L	L
30	L	R	*M*	L	L	R	M	L	L	L	L
31	L	L	L	L	L	L	L	L	L	L	L
32	L	R	R	L	L	R	R	L	L	L	L

PG	150°C	250°C	350°C	:CCl_2	N_2CHCO_2R, Cu or Rh	CH_2I_2, Zn(Cu)
	Thermal			Carbenes		
1	L	L	M	R	R	R
2	L	L	M	L	L	L
3	L	L	R	L	L	L
4	R	R	R	L	L	L
5	L	L	M	L	L	L
6	L	L	M	L	L	L
7	L	L	H	L	L	L
8	L	L	H	L	L	L
9	L	M	H	L	L	L
10	L	L	M	L	L	L
1	L	L	R	L	L	L
12	L	L	M	L	L	L
13	L	L	M	L	L	L
14	L	L	M	L	L	L
15	L	L	M	L	L	L
16	R	R	R	L	L	L
17	L	M	H	R	R	R
18	L	M	H	R	R	R
19	L	M	H	R	R	R
20	L	M	H	R	R	R
21	L	L	M	R	R	R
22	L	R	R	R	R	R
23	H	R	R	L	L	L
24	L	L	M	L	L	L
25	L	L	M	L	L	L
26	L	M	H	*H*	L	L
27	L	M	R	L	L	L
28	L	M	H	L	L	L
29	L	M	H	L	L	L
30	L	M	H	L	L	L
31	L	M	R	L	L	L
32	L	M	H	L	L	L

PG	R_3SnH, In•	$Ni(CO)_4$	CH_2N_2	$SOCl_2$	Ac_2O, 25°C	Ac_2O, 80°C	DCC	MeI	$Me_3O^+ BF_4^-$	1. LDA, 2. MeI	1. K_2CO_3, 2. MeI	RCHO	RCOCl	C^+/olefin
	Miscellaneous											Electrophilic		
1	L	R	L	L	L	L	L	R	R	L	R	L	*L*	R
2	R	L	R	L	L	L	*L*	R	R	R	R	L	L	R
3	L	L	L	L	L	L	L	R	R	M	R	L	L	R
4	L	L	L	L	L	L	L	L	L	L	L	L	*R*	L
5	L	L	L	L	L	L	L	R	R	L	R	L	L	R
6	L	L	L	L	L	*L*	L	R	R	L	R	L	L	R
7	L	L	L	L	L	L	L	R	R	L	R	L	L	R
8	L	L	L	L	L	L	L	R	R	L	R	L	L	R
9	L	L	L	L	L	L	L	M	R	L	M	L	L	R
10	L	L	*L*	L	L	L	L	R	R	*L*	R	L	L	R
1	L	L	L	L	L	L	L	M	R	L	M	L	L	R
12	L	L	L	L	L	L	L	R	R	L	R	L	L	R
13	L	L	L	L	L	L	L	R	R	L	R	L	L	R
14	L	L	L	L	L	L	L	R	R	L	R	L	L	R
15	L	M	L	L	R	R	L	R	R	L	R	L	L	R
16	M	M	M	H	*H*	H	L	R	R	L	R	L	R	R
17	L	L	L	R	R	R	*L*	M	R	L	M	L	L	R
18	R	L	L	L	*L*	H	L	M	R	*L*	M	L	L	R
19	R	L	L	L	M	R	L	M	R	L	M	L	L	R
20	R	L	M	R	R	R	R	M	R	R	R	L	R	R
21	L	L	L	L	M	H	*L*	L	R	R	L	L	L	R
22	R	R	R	L	M	H	*L*	M	R	R	M	R	R	R
23	L	L	L	H	H	H	L	L	R	L	L	L	R	R
24	L	L	L	L	L	L	L	R	R	L	R	L	L	M
25	L	L	L	L	L	L	L	R	R	L	R	L	L	M
26	*H*	L	L	*L*	M	R	L	R	R	R	R	L	L	R
27	L	L	L	*L*	R	R	L	R	R	R	R	L	L	R
28	L	L	L	*L*	L	L	L	L	R	R	R	L	L	M
29	L	L	L	L	L	L	*L*	L	R	R	R	L	L	M
30	L	L	L	*L*	L	L	L	L	R	R	M	L	L	M
31	L	L	L	*L*	L	L	L	L	R	R	R	L	L	M
32	R	L	R	*R*	R	R	L	M	R	R	R	L	L	M

Reactivity Chart 11. Selective Deprotection of Silyl Ethers

Deprotection of:	1 1° TMS	2 2° TMS	3 3° TMS	4 1° TES	5 2° TES	6 3° TES	7 1° TBS	8 2° TBS	9 3° TBS	10 1° TIPS	11 2° TIPS	12 3° TIPS	13 1° TBDPS	14 2° TBDPS	15 3° TBDPS
													In the Presence of:		
A 1° TMS		1[a]		3	4		5	6		7			8		
B 2° TMS	9	10	11		12	13	14	15		16	17		18	19	
C 3° TMS							20	21		22	23	24	25	26	
D 1° TES		27			28	29	30	31	32	33	34		35	36	37
E 2° TES					38	39	40	41	42	43	44		45	46	
F 3° TES							47	48		49	50		51		
G 1° TBS			52		53	54	55	56	57	58	59		60	61	
H 2° TBS					62		63	64	65	66	67	68	69	70	
I 3° TBS								71	72		73		74		
J 1° TIPS								75	76		77		78	79	
K 2° TIPS								80			81			82	
L 3° TIPS															
M 1° TBDPS					83	84	85	86	87	88	89		90	91	
N 2° TBDPS							92								
O 3° TBDPS															

[a] Numbers refer to references and reagents on following pages.

1033

Reactivity Chart 11. Selective Deprotection of Silyl Ethers

1. DMSO, $(COCl)_2$ TL, **1999**, *40*, 5161
   AcOH   JACS, **2003**, *125*, 6697
   Rexyn 101   JOC, **1986**, *51*, 3451
   K_2CO_3, MeOH   CJC, **1965**, *43*, 2004
   Alumina   T, **1994**, *50*, 8539
2. PPTS   CEJ, **1995**, *1*, 467
3. $NaHCO_3$ JACS, **2000**, *122*, 10033
4. $NaHCO_3$ JACS, **2000**, *122*, 10033
   Swern   TL, **1999**, *40*, 5161
5. NaOH, EtOH   JCSPT1, **1992**, 3043
   $Cu(NO_3)_2$ SC, **1990**, *20*, 757
   $Ce(NO_3)_3$ SC, **1990**, *20*, 757
   $[Bu_2(NCS)Sn]_2O$ TL, **1986** *27*, 5743
   $BiCl_3$ SC, **2001**, *31*, 905
   $Bi(OTf)_3$ SC, **2001**, *31*, 905
   K_2CO_3 OL, **2002**, *4*, 3655
   $NaHCO_3$ JACS, **2000**, *122*, 10033
   MCM-41   SL, **1999**, 357
6. $BF_3 \cdot Et_2O$ JCSCC, **1993**, 1823
7. NaOH/EtOH   JCSPT1, **1992**, 3043
   MCM-1   SL, **1999**, 357
8. NaOH, EtOH   JCSPT1, **1992**, 3043
   $Cu(NO_3)_2$ SC, **1990**, *20*, 757
   $Ce(NO_3)_3$ SC, **1990**, *20*, 757
   HCl   JCSPT1, **1992**, 3043, SL, **1994**, 40
9. Amberlyst 15   JOC, **1986**, *51*, 3451
10. SiO_2Cl, NaI   TL, **2002**, *43*, 7139
    TBAF   S, **1992**, 1112
11. TsOH   JOC, **1993**, *58*, 3201, TL, **1990**, *31*, 4965
12. HF-Pyr   ACIEE, **1997**, *36*, 2744, JACS, **2002**, *124*, 5661
    TBAF   JACS, **2002**, *124*, 4552
    KF   ACIEE, **2001**, *40*, 196
13. CSA   JACS, **1998**, *120*, 3518
    TsOH   TL, **1995**, *36*, 4927
    AcOH   JMC, **1994**, *37*, 3730
    HF-Pyr   JACS, **2002**, *124*, 5661, ACIEE, **1997**, *36*, 2744
    KF   OL, **2003**, *5*, 761, ACIEE, **2001**, *40*, 196

14.	TsOH	ACIEE, **1999**, *38*, 2258
	HCl	TL, **1995**, *36*, 819
	Citric Acid	JOC, **2003,** *68*, 4215
	H$_2$SiF$_6$	JOC, **2003**, *68*, 4215
	NaOH	JACS, **2003**, *125*, 11514
	K$_2$CO$_3$, MeOH	JACS, **1986**, *108*, 3112
15.	PPTS	JOC, **2002**, *67*, 2751, JACS, **1998**, *120*, 9084, JOC, **1994**, *59*, 3113
	AcOH	OL, **2001**, *3*, 1685, BMCL, **2003**, *13*, 809, JMC, **1994**, *37*, 3730,
	Citric Acid/MeOH	JACS, **1995**, *117*, 12013, JACS, **1991**, *113*, 5378
	TsOH	OL, **1999**, *1*, 451, JACS, **1992**, *114*, 9414, JOC, **1993**, *58*, 3201,
	CSA	JACS, **2003**, *125*, 15443
	HCl	JOC, **1987**, *52*, 622, JOC, **1985**, *50*, 5005
	HF-Pyr	ACIEE, **1997**, *36*, 2744, JACS, **2002**, *124*, 5661
	HF·TEA	JACS, **1997**, *119*, 2404, JACS, **1986**, *108*, 5549
	BF$_3$·Et$_2$O	JOC, **1999**, *54*, 5511, JACS, **2003**, *125*, 15433
	K$_2$CO$_3$	LA, **1996**, 1717, S, **2003**, 1827, ACIEE, **2003**, *42*, 4685
	TBAF	OL, **2002**, *4*, 2953
	KF	ACIEE, **2001**, *40*, 196, OL, **2003**, *5*, 761
16.	TBAF	TL, **2003**, *44*, 8935
17.	TBAF	TL, **2003**, *44*, 8935
	TBAF/AcOH	TL, **1992**, *33*, 7469, JACS, **1993**, *115*, 9345
	KF	OL, **2003**, *5*, 761
	NaOH	JACS, **2003**, *125*, 11514
	H$_2$SiF$_6$	JOC, **2003**, *68*, 4215
	Citric Acid/MeOH	TA, **1995**, *6*, 2127
	FeCl$_3$	TL, **1994**, *35*, 5069
18.	AcOH	TL, **2003**, *44*, 5547
	CSA	TL, **1997**, *38*, 3879
	Acetone, Me$_2$C(OMe)$_2$,	
	CSA	TL, **1994**, *35*, 7601
	TsOH	JACS, **1997**, *119*, 8381, JACS, **1998**, *120*, 2534, OL, **1999**, *1*, 451, JOC, **1995**, *60*, 7343
	HCl	TL, **1992**, *33*, 1813
	PPTS	TA, **1993**, *4*, 399, TL, **1991**, *32*, 1073
	Alumina	T, **1994**, *50*, 8539

PhSeCl, K$_2$CO$_3$	TL, **1991**, *32*, 4015
TBAF	JACS, **1993**, *113*, 10400
BF$_3$·Et$_2$O	TL, **2003**, *44*, 5547
TMSOTf	JACS, **2002**, *124*, 11102
19. K$_2$CO$_3$	S, **2003**, 1827, ACIEE, **2003**, *42*, 4685
NaIO$_4$	TL, **2002**, *43*, 8727
TBAF	TL, **1992**, *33*, 671
Cu(NO$_3$)$_2$	SC. **1990**, *20*, 757
Ce(NO$_3$)$_3$	SC, **1990**, *20*, 757
20. HCl	JACS, **1997**, *119*, 2784
ClCH$_2$CO$_2$H, MeOH	JACS, **1995**, *117*, 8106
BH$_3$·Me$_2$S	SL, **2003**, 353
21. HCl	JACS, **1997**, *119*, 2784, SL, **2000**, 1733, JACS, **1987**, *109*, 7063
HF	ACIEE, **1997**, *36*, 1524
TBAF	TL, **1987**, *28*, 2491
TBAF, AcOH	JACS, **1997**, *119*, 962
BH$_3$·THF	JOC, **2003**, *68*, 1367
K$_2$CO$_3$	SL, **2000**, 1733
LiAlH$_4$	TL, **1980**, *21*, 445
FeCl$_3$	SL, **1992**, 969
22. HCl	JACS, **2003**, *125*, 8228
PPTS	JACS, **1997**, *119*, 11353
23. TBAF, AcOH	JACS, **1997**, *119*, 962, JACS, **1993**, *115*, 9345
24. 1 M HCl, THF	JACS, **1993**, *115*, 8871
25. HCl	SL, **2000**, 1733
AcOH	JACS, **2002**, *124*, 2137
BF$_3$·Et$_2$O	JOC, **2003**, *68*, 9050, TL, **2003**, *44*, 2319
K$_2$CO$_3$	SL, **2000**, 1733
26. H$_2$SiF$_6$	TL, **1995**, *36*, 2427, JACS, **1999**, *121*, 2056
HCl	CEJ, **1995**, *1*, 467
K$_2$CO$_3$	JACS, **1996**, *118*, 7513
27. DMSO, (COCl)$_2$	TL, **1999**, *40*, 5161
28. DMSO, (COCl)$_2$	TL, **1999**, *40*, 5161
Ph$_3$P·HBr, MeOH	JACS, **2003**, *125*, 12844
CSA	JOC,**1999**, *64*, 8267
PPTS	ACIEE, **1997**, *36*, 2520
AcOH	JOC, **2001**, *66*, 6410, OL, **2000**, *2*, 2897, JOC, **1990**, *55*, 5451

	HF-Pyr	JACS, **2000**, *122*, 10033
	TBAF/AcOH	ACIEE, **2000**, *39*, 2290, T, **2002**, *58*, 10353
	TBAF	JACS, **1998**, *120*, 2523, TL, **1998**, *39*, 1865, OL, **2001**, *3*, 4307, JACS, **2003**, *125*, 5393, TL, **2002**, *43*, 3381, JACS, **2002**, *124*, 4552
	KF	OL, **2003**, *5*, 761
	LiOH	JACS, **2000**, *122*, 10033
	Swern	JOC, **2003**, *68*, 3023, TL, **1999**, *40*, 5161, BMC, **2002**, *10*, 2031
	CrO$_3$-Pyr	TL, **2003**, *44*, 7411
29.	CSA	OL, **2000**, *2*, 2905, JOC, **2003**, *68*, 1693
	PPTS	ACIEE, **1997**, *36*, 2520
	HF-Pyr	ACIEE, **2000**, *39*, 2536, JACS, **2002**, *124*, 5661, SL, **1994**, 417
30.	HF-Pyr	JACS, **1990**, *112*, 7079
	TCNQ, MeCN, H$_2$O	BCSJ, **1994**, *67*, 290
	DDQ, MeCN, H$_2$O	BCSJ, **1994**, *67*, 290, JCSPT1, **1992**, 2997
	CSA	TL, **1999**, *40*, 7135
	IBX, DMSO	OL **2002**, *4*, 2141
	MCM-41	SL **1999**, 357
	H$_2$ Pd/C	TL, **2004**, *45*, 1973
31.	H$_2$SiF$_6$, IPA	TL, **1999**, *40*, 4145
	Ph$_3$P·HBr, MeOH	JACS, **2003**, *125*, 12844
	HCl	BMCL, **1999** 9, 3047
	CSA	JOC, **1999**, *64*, 8267, ACIEE, **2001**, *40*, 2063, TL, **1999**, *40*, 7135
	PPTS	OL, **1999**, *1*, 941
	AcOH, THF,H$_2$O	JOC, **1990**, *55*, 5451, JACS, **1992**, *114*, 5427, TL, **1993**, *34*, 3993
	TFA	T, **2003**, *59*, 6819, JACS, **2001**, *123*, 12432
	HF-Pyr	JACS, **1996**, *118*, 11054, JOC, **1998**, *63*, 7885
	HF	OL, **2002**, *4*, 897
	TMSOTf, DIPEA	OL, **2003**, *5*, 3159
	TBAF	JOC, **1999**, *64*, 8267
	TBAF, AcOH	ACIEE, **2000**, *39,* 2290
	KF	OL, **2003**, *5*, 761
	Swern	SL, **2003**, 1698
32.	Amberlyst-15	TL, **1998**, *39*, 6373
	TBAF, AcOH	T, **2002**, *58*, 10353

33. TFA   T, **1998**, *54*, 4591
    H₂, Pd/C   CC, **2003**, 654
    MCM-41   SL, **1999**, 357
34. H₂SiF₆, IPA   TL **1999**, *40*, 4145
    TMSOTf-DIPEA   OL, **2003**, *5*, 3159
    HF-Pyr   JACS, **2000**, *122*, 10033, ACIEE, **2000**, *39*, 2536
    LiOH   JACS, **2000**, *122*, 10033
    AcOH   JOC, **1990**, *55*, 5451
35. SiF₄, CH₂Cl₂   TL, **1992**, 33, 2289
    DDQ, MeCN, H₂O   BCSJ, **1994**, *67*, 290, JCSPT1, **1992**, 2997
    CSA   OL, **2002**, *4*, 2181, JOC, **2003**, *68*, 1693
    H₂, Pd/C   CC, **2003**, 654, TL, **2004**, *45*, 1973
    TMSOTf, HCO₂DPM,   SC, **2001**, *31*, 2761
    Silica
    ZnBr₂, H₂O,   TL **2002**, *43*, 7151
36. Citric Acid   JACS, **1997**, *119*, 10935
    TsOH   JOC, **2003**, *68*, 3026
    TfOH   TL, **1994**, *35*, 7801
    TMSOTf, DIPEA   OL, **2003**, *5*, 3159
    TMSOTf, TEA, MeOH   TL, **1999**, *40*, 3643
37. AcOH   JACS, **2002**, *124*, 2137
38. TsOH   BMCL, **2002**, *12*, 2815
    TFA   JACS, **2002**, *124*, 6981, OL, **2003**, *5*, 377
    AcOH   JCSCC, **1979**, 156
    HF-Pyr   ACIEE, **1997**, *36*, 2744, JACS, **2002**, *124*, 5661
    TBAF, AcOH   ACIEE, **2000**, *39*, 2290
    (NH₄)HF₂   OL, **2002**, *4*, 3979
    TBAF   OL, **2002**, *4*, 3549, JACS, **2003**, *125*, 12844, TL,
                                 1996, *37*, 447
    KF   OL, **2003**, *5*, 761
    NaOH, DMPU   ACIEE, **2001**, *40*, 196
39. HCl   CC, **2002**, 742, JACS, **2003**, *125*, 8238
    PPTS   TL, **2003**, *44*, 7741
    HF-Pyr   JACS, **2002**, *124*, 5661, ACIEE, **1997**, *36*, 2744,
                                 JOC, **1999**, *64*, 8267
    TBAF   JACS, **2003**, *125*, 12844
40. Pd/C, MeOH   OL, **2002**, *4*, 4701, TL, **2004**, *45*, 1973
    HF-Pyr   OL, **2002**, *4*, 3655
    HCl·Pyr   JOC, **1988**, *53*, 706

	DDQ, MeCN, H$_2$O	JCSPT1, **1992**, 2997
	I$_2$, Ag$_2$CO$_3$	ACIEE, **2002**, *41*, 1392, SL, **2003**, 393
	TBAF, NH$_4$Cl	JACS, **1999**, *121*, 5589
41.	PPTS, MeOH, TMOF	OL, **2003**, *5*, 4477
	HCl	CC, **2002**, 742, JACS, **2003**, *125*, 8238
	AcOH	JACS, **2003**, *125*, 6042, OL, **1999**, *1*, 909, JACS, **2001**, *123*, 10942, JACS, **1982**, *104*, 5523, LAC, **1986**, 1281, JACS, **1994**, *116*, 1753, SL, **1994**, 601, JOC, **1992**, *57*, 4793
	CSA	TL, **1999**, *40*, 7135, TL, **1997**, *38*, 8241
	TsOH	JOC, **1998**, *63*, 7885
	PPTS	TL, **2001**, *42*, 5505, TL, **1999**, *40*, 3351, TL, **1996**, *37*, 8581, OL, **2001**, *3*, 1385, OL, **2001**, *3*, 949, ACIEE, **2001**, *40*, 3854, JACS, **2003**, *125*, 15443, ACIEE, **2001**, *40*, 603, OL, **2003**, *5*, 4477, JOC, **1990**, *55*, 5451, T, **1995**, *51*, 8771, TL, **1995**, *36*, 273
	TFA	ACIEE, **1999**, *38*, 1652, JACS, **2002**, *124*, 6981, OL, **2003**, *5*, 377, JACS, **1990**, *112*, 2998, JACS, **1990**, *112*, 5583, JACS, **1989**, *111*, 1157
	HF-Pyr	JOC, **1999**, *64*, 8267, BMCL, **1999**, *9*, 3047, JOC, **2001**, *66*, 6410, ACIEE, **1997**, *36*, 2744, TL, **1999**, *40*, 4955, JACS, **2002**, *124*, 5661, ACIEE, **2000**, *39*, 581, BMCL, **2001**, *11*, 1683, JACS, **1994**, *116*, 1599, JACS, **1994**, *116*, 7443, TL, **1985**, *26*, 5239, JACS, **1993**, *115*, 11446
	HF, TEA	JACS, **1998**, *120*, 8674, ACIEE, **1997**, *36*, 2520
	HF	OL, **2002**, *4*, 4615, TL, **1985**, *26*, 5239
	Zn(OTf)$_2$, EtSH	JACS, **2003**, *125*, 14294
	TiCl$_3$(O-*i*Pr)	OL, **1999**, *1*, 1459
	TBAF, AcOH	ACIEE, **2000**, *39*, 2290, JACS, **1998**, *120*, 3935
	TBAF	JACS, **2003**, *125*, 12844, TL, **2003**, *44*, 3175, OL, **2002**, *4*, 995, OL, **2002**, *4*, 3549, JOC, **2003**, *68*, 8162
	KF	OL, **2003**, *5*, 761
	NaOH, DMPU	ACIEE, **2001**, *40*, 196
	MCM-41	SL, **1999**, 357
	PdCl$_2$, CuCl, H$_2$O	OL, **2003**, *5*, 3535
	DDQ, MeCN, H$_2$O	JCSPT1, **1992**, 2997
	MoO$_5$·HMPA	TL, **1987**, *28*, 6191, BCSJ, **1990**, *63*, 1039
	WO$_5$·HMPA	TL, **1987**, *28*, 6191, BCSJ, **1990**, *63*, 1039

42.	HCl	T, **2002**, *58*, 10353
	CSA	JACS, **1995**, *117*, 1171
	TfOH	JACS, **1997**, *119*, 6739
	AcOH	TL, **1990**, *31*, 431
43.	HCl	JACS, **2003**, *125*, 8228
	AcOH	ACIEE, **2003**, *42*, 1258
	CSA	JOC, **2000**, *65*, 4145, TL, **1999**, *40*, 7135
	PPTS	TL, **1997**, *38*, 8241
	H₂SO₄	JOC, **2000**, *65*, 4145
	TFA	JACS, **1990**, *112*, 2998
	Ph₃P-HBr	JOC, **2000**, *65*, 4145
	HF-Pyr	JOC, **1999**, *64*, 8267, JACS, **2000**, *122*, 10033, ACIEE, **2000**, *39*, 2536
	HF, TEA	JACS, **1998**, *120*, 8661
	H₂SiF₆, HF, H₂O	CC, **1996**, 21
44.	PPTS	OBC, **2003**, *1*, 4173, TL, **2000**, *41*, 983, JOC, **1990**, *55*, 5451
	H₂SO₄	JOC, **2000**, *65*, 4145
	TFA	SL, **1999**, 49, JACS, **1990**, *112*, 2998, JACS, **1990**, *112*, 5583, JACS, **1989**, *111*, 1157
	AcOH	JOC, **1994**, *59*, 715, JACS, **1993**, *115*, 4497, JACS, **1992**, *114*, 2260
	HF-Pyr	JACS, **2003**, *125*, 7822, JACS, **1998**, *120*, 5921, JACS, **2000**, *121*, 10033
	NH₄F	JACS, **1997**, *119*, 2757
	Zn(OTf)₂, EtSH	JACS, **2003**, *125*, 14294
	Amberlyst-15	JACS, **1999**, *121*, 6944
	MoO₅·HMPA	TL, **1987**, *28*, 6191, BCSJ, **1990**, *63*, 1039
	WO₅·HMPA	TL, **1987**, *28*, 6191, BCSJ, **1990**, *63*, 1039
45.	AcOH	TL, **2003**, *44*, 3175, CPB, **1990**, *38*, 2890
	CSA	TL, **2004**, *45*, 351
	PPTS	TL, **2001**, *42*, 6035
	TsOH	JOC, **1998**, *63*, 6200, JACS, **1999**, *121*, 4542, JACS, **2000**, *121*, 619, TL, **2001**, *42*, 6035, JOC, **1989**, *54*, 3009
	HCl	JACS, **2001**, *123*, 12426, JACS, **2002**, *124*, 11102, TL, **1997**, *38*, 3651, CEJ, **1999**, *5*, 2241, SL, **1994**, 40, JACS, **1990**, *112*, 6348
	H₂SO₄	TL, **1996**, *37*, 9073
	Cl₃CCO₂H	JACS, **1990**, *112*, 8997

	BF$_3$·Et$_2$O	JOC, **1998**, *63*, 6200, TL, **2001**, *42*, 6035
	HF, TEA	OL, **2000**, *2*, 3913
	HF-Pyr	TL, **1999**, *40*, 4955
	HF, MeCN	JACS, **1988**, *110*, 6914, JACS, **1990**, *112*, 3018
	TBAF	JACS, **2002**, *124*, 9726, TL, **2001**, *42*, 6035, T, **1990**, *46*, 4517
	SiF$_4$, CH$_2$Cl$_2$	TL, **1992**, *33*, 2289
	2,4,4,6-Br$_4$-2,	
	5-Cyclohexadienone, PPh$_3$	TL, **1997**, *38*, 7223
	Pd/C, H$_2$	TL, **2003**, *44*, 2105
	ZnBr$_2$, H$_2$O	TL, **2002**, *43*, 7151
	PhSeCl, K$_2$CO$_3$	TL, **1991**, *32*, 4015
	DDQ, MeCN, H$_2$O	JCSPT1, **1992**, 2997
46.	HCl	OL, **2003**, *5*, 515
	AcOH	ACIEE, **2003**, *42*, 1258, TL, **1997**, *38*, 5119, TL, **1993**, *34*, 8439
	CSA	JACS, **1995**, *117*, 12013
	TBAF	ACIEE, **1991**, *30*, 299
	SiF$_4$, CH$_2$Cl$_2$	TL, **1992**, *33*, 2289
	HF, MeCN	JACS, **1995**, *117*, 12013
	HF·Pyr	JACS, **1993**, *115*, 4419, JACS, **1986**, *108*, 5549
	DDQ, MeCN, H$_2$O	BCJ, **1994**, *67*, 290
47.	TBAF, NH$_4$Cl	JACS, **1999**, *121*, 5589
	SiO$_2$	TL, **1995**, *36*, 8799
48.	HF·TEA	ACIEE, **1997**, *36*, 2520, JACS, **1998**, *120*, 8674, JOC, **1984**, *49*, 5279, JOC, **1987**, *52*, 4898
	TBAF, AcOH	JACS, **1997**, *119*, 962
49.	HF·TEA	JACS, **1998**, *120*, 8661
	TBAF	TL, **1996**, *37*, 7695
50.	TBAF, AcOH	JACS, **1997**, *119*, 962
51.	TBAF	TL, **1996**, *37*, 7695
	HF	JACS, **1997**, *119*, 7897, JACS, **1997**, *119*, 12976, JACS, **1996**, *118*, 7502
	TBAF	JOC, **1997**, *62*, 5672, OL, **2002**, *4*, 2953
	K$_2$CO$_3$	ACIEE, **2003**, *42*, 5996
	NaIO$_4$	TL, **2002**, *43*, 8727
53.	HF-Pyr	TL, **1999**, *40*, 7135
54.	HF·Pyr, Pyr, THF	OL, **2003**, *5*, 4819, OBC, **2003**, *1*, 4173, ACIEE, **2001**, *40*, 191, ACIEE, **1997**, *36*, 2744
	CSA	JOC, **2003**, *68*, 1693, JACS, **1999**, *121*, 890

55. TsOH, THF, H$_2$O	JCSCC, **1987**, 992
Cl$_2$CHCO$_2$H	JACS, **1995**, *117*, 8106, JACS, **1994**, *116*, 10825
TBAF	CL, **1986**, 1185, JCM, **2001**, *13*, 15
TBSOTf	TL, **1987**, *28*, 3189
AcOH	TL, **1997**, *38*, 4429
PPTS	ACIEE, **2003**, *42*, 4779
DDQ	OL, **2001**, *3*, 2661
MnO$_2$, AlCl$_3$	SC, **1999**, *29*, 4333
DMSO, H$_2$O	TL, **1997**, *38*, 495
H$_2$, Pd/C	TL, **2004**, *45*, 1973
56. H$_2$SiF$_6$	JOC, **2002**, *67*, 2751, JOC, **2003**, *68*, 4215, JOC, **1993**, *58*, 5130
BCl$_3$, THF	SL, **2000**, 1634
CBr$_4$, MeOH, hv	TL, **2002**, *43*, 2777, TL, **2004**, *45*, 635
InCl$_3$, Aq. ACN	NJC, **2000** *24*, 853
PPTS, EtOH	OL, **2003**, *5*, 1729, JOC, **2002**, *67*, 733, ACIEE, **2003**, *42*, 4779, H, **2003**, *59*, 347, ACIEE, **1999**, *38*, 3662, JACS, **2001**, *123*, 765, OL, **2002**, *5*, 1729, TL, **1996**, *37*, 2253, JACS, **2002**, *124*, 5958, TL, **2003**, *44*, 7949, TL, **1987**, *28*, 5865, TL, **1988**, *29*, 4591, CL, **1992**, 1851, TL, **1993**, *34*, 4981, CPB, **1989**, *37*, 586
CAN, IPA	JACS, **2002**, *124*, 4956
HCl	OL, **2003**, *5*, 4405, NAR, **1989**, *17*, 7663
AcOH	TL, **1997**, *38*, 1703, JOC, **2000**, *65*, 7792, JOC, **2003**, *68*, 187, JOC, **1990**, *55*, 5451, JOC, **1994**, *59*, 5192, TL, **1993**, *49*, 785, TL, **1990**, *35*, 5041, JMC, **1992**, *35*, 56, TL, **1998**, *29*, 6331, JOC, **1991**, *56*, 5493
CSA	JOC, **2000**, *65*, 7456, JACS, **2003**, *125*, 46, JACS, **1997**, *119*, 4557, JOC, **1998**, *63*, 6200, OBC, **2003**, *1*, 4173, ACIEE, **1998**, *37*, 81, ACIEE, **2003**, *42*, 2521, H, **2003**, *59*, 347, ACIEE, **1999**, *38*, 3662, TL, **2003**, *44*, 7949, JACS, **1998**, *120*, 4123, OL, **2002**, *4*, 3549, TL, **1998**, *39*, 8633, JACS, **2002**, *124*, 384, TL, **2000**, *41*, 7635, ACIEE, **1999**, *38*, 1263, TL, **2001**, *42*, 3649, T, **2003**, *59*, 6851, JACS, **1995**, *117*, 1171, T, **1990**, *46*, 4517, JACS, **1992**, *114*, 7935, TL, **1992**, *33*, 1557
TsOH	TL, **2002**, *43*, 6377, JOC, **2003**, *68*, 7967, JOC, **2002**, *67*, 4316
Amberlite(H$^+$)	JOC, **1989**, *54*, 5841

TsOH, Bu$_4$NHSO$_3$	JACS, **2003**, *125*, 13531
TFA	SL, **2000**, 1733, JOC, **1997**, *62*, 1368, JMC, **1998**, *41*, 5094, JACS, **1999**, *121*, 5661, JOC, **2002**, *67*, 9331, JCSPT1, **1999**, 839, JOC, **1990**, *55*, 410, T, **1995**, *51*, 7131, JOC, **1992**, *57*, 1070
Acidic CHCl$_3$	JOC, **2001**, *66*, 1885
Cu(OTf)$_2$, Ac$_2$O	TL, **2001**, *42*, 5309
NH$_4$F	ACIEE, **1999**, *38*, 3542, TL, **1993**, *34*, 3385, SL, **1993**, 535
HF	OL, **2000**, *2*, 2983, S, **2000** 399, JACS, **1989**, *111*, 2967, JACS, **1987**, *109*, 8117
HF·Pyr	EJOC, **2001**, 1701, ACIEE, **2003**, *42*, 3515, JACS, **1998**, *120*, 4113, JACS, **2001**, *123*, 12426, OBC, **2003**, *1*, 4173, TL, **1998**, *39*, 4421, JACS, **1998**, *120*, 4123, JACS, **1999**, *121*, 9229, ACIEE, **2001**, *40*, 191, OL, **2000**, *2*, 2575, JOC, **1997**, *62*, 8290, TL, **1998**, *39*, 3567, TL, **2000**, *41*, 8569, JACS, **2002**, *124*, 12806, JOC, **2003**, *68*, 1780, JACS, **2002**, *124*, 11102, JOC, **2002**, *67*, 7158, ACIEE, **1997**, *36*, 2744, TL, **1999**, *40*, 2279, TL, **2002**, *43*, 8507, T, **1998**, *54*, 7127, CC, **1999**, 519, JOC, **2003**, *68*, 6646, TL, **1999**, *40*, 4267, JACS, **2000**, *122*, 5473, OL, **2003**, *5*, 181, JOC, **2003**, *68*, 5320, OL, **2002**, *4*, 4443, JACS, **2004**, *126*, 36, JACS, **1998**, *120*, 7647, JACS, **1998**, *120*, 13287, TL, **1996**, *37*, 5049, JACS, **1990**, *112*, 7079, JCSCC, **1989**, 378, JACS, **1992**, *114*, 9434, JOC, **1992**, *57*, 1964, ACIEE, **1994**, *33*, 673, TL, **1993**, *34*, 6559, TL, **1993**, *34*, 8403, JOC, **1992**, *57*, 5058, TL, **1992**, *33*, 2641, JACS, **1995**, *117*, 7289, TL, **1995**, *36*, 1003, JCSCC, **1993**, 619
TBAF, AcOH	CC, **2002** 1624, OBC, **2003**, *1*, 1664
TBAF	JCSPT1, **2001**, 3338, JCSPT1, **2002**, 1581, TL, **2001**, *42*, 1187, TL, **2002**, *43*, 6609, OL, **2002**, *4*, 2921, JACS, **1995**, *117*, 1173
TAS-F	JACS, **1998**, *120*, 6627
Jones Reagent	TL, **1998**, *39*, 4421
LiBr, 18-C-6	SC, **1997**, *27*, 2953
POCl$_3$, DMF	TL, **1999**, *40*, 7043, JOC, **2001**, *66*, 693, SL, **2004**, 564
Tf$_2$O, DMF	JOC, **2001**, *66*, 693, SL, **2004**, 564
SnCl$_2$·H$_2$O	CCL, **2004**, 15, 1430
AgOAc	TL, **1986**, *27*, 291

NBS, DMSO	JOC, **1995**, *60*, 143
MeOH, CCl$_4$, MW	TL, **1995**, *36*, 6891
NaOH	JOC, **1980**, *45*, 4797
Alumina	T, **1994**, *50*, 8539, JCSCC, **1992**, 1451
57. CSA	OL, **2002**, *4*, 2981, OL, **2000**, *2*, 207, BMCL, **2003**, *13*, 2519, JACS, **1992**, *114*, 7935, JACS, **1993**, *115*, 3558
HF-Pyr	JACS, **1998**, *120*, 13287
TBAF, AcOH	JACS, **1997**, *119*, 3193
TBAF	JACS, **1998**, *120*, 7647
NH$_4$F	JOC, **1992**, *57*, 2270
H$_2$SiF$_6$	JOC, **1993**, *58*, 5130
SiF$_4$, CH$_2$Cl$_2$	TL, **1992**, *33*, 2289
Oxone	OL, **1999**, *1*, 1701
MeOH, CCl$_4$, MW	TL, **1995**, *36*, 6891
58. HCl, EtOH	JCSPT1, **1992**, 3043
H$_2$SiF$_6$, *t*-BuOH	JOC, **1992**, *57*, 2492, JOC, **1993** *58*, 5130
NaOH, EtOH	JOC, **1980**, *45*, 4797
Cyclohexene, PdO	TL, **1993**, *34*, 243
Alumina	T, **1994**, *50*, 8539
H$_2$SO$_4$	JOC, **2000**, *65*, 4145
CSA	JACS, **1997**, *119*, 4557, JOC, **1998**, *63*, 6200
PPTS	CEJ, **2000**, *6*, 3116
H$_2$SiF$_6$	JOC, **1999**, *64*, 8267
TMSOTf, TEA, MeOH	TL, **1999**, *40*, 3643
Decaborane	JCSPT1, **2002**, 1223
CeCl$_3$·7H$_2$O, NaI	SL, **1998**, 209
H$_2$, Pd/C	TL, **2004**, *45*, 1973, CC, **2003**, 654
59. HCl	T, **2003**, *59*, 6833, JACS, **2003**, *125*, 11514, JOC, **1980**, *45*, 4797
H$_2$SO$_4$	JOC, **1999**, *64*, 23
AcOH	JACS, **2000**, *122*, 10482, JACS, **2001**, *123*, 9974, JOC, **1997**, *62*, 4961, OL, **2002**, *4*, 3463, **TL**, **2003**, *44*, 2557, TL, **1974**, 2865, JOC, **1994**, *59*, 5192, TL, **1993**, *34*, 7107, TL, **1989**, *30*, 3757, JACS, **1990**, *112*, 7659
CSA	ACIEE, **2003**, *42*, 343, JACS, **2003**, *125*, 46, JACS, **1997**, *119*, 4557, JOC, **1998**, *63*, 6200, OBC, **2003**, *1*, 4173, T, **2003**, *59*, 6833, JACS, **1998**, *120*, 4123, ACIEE, **2002**, *41*, 4686

PPTS	JCSPT1, **2002**, 1693
TsOH	JOC, **2000**, *65*, 7070, TL, **1993**, *49*, 7385, JOC, **1995**, *60*, 7870
TsOH/PPTS	JACS, **1993**, *115*, 7906
NH$_4$Cl, MeOH	JACS, **1997**, *119*, 2755
HF-Pyr	OBC, **2003**, *1*, 4173, JACS, **1998**, *120*, 4123, JOC, **1997**, *62*, 8290, TL, **1999**, *40*, 2279, TL, **2002**, *43*, 8507, TL, **1995**, *36*, 1003, JCSCC, **1993**, 619, CEJ, **1995**, *1*, 318
H$_2$SiF$_6$	JOC, **2003**, *68*, 4215
TBAF	JACS, **2002**, *124*, 1664, JOC, **1999**, *64*, 8267
Polymeric DCKA	SL, **1999**, 1960
NaOH	JOC, **1980**, *45*, 4797
Cyclohexene/PdO	TL, **1993**, *34*, 243
60. HCl	JCSPT1, **1992**, 3043, HCA, **1986**, *69*, 1273, OL, **2003**, *5*, 749, TL, **2003**, *44*, 3175
AcCl, MeOH	SL, **2003**, 694
H$_2$SO$_4$	TL, **2001**, *42*, 2701
HOAc, THF, H$_2$O	CJC, **1975**, *53*, 2975, JOC, **1991**, *56*, 5496, JOC, **1981**, *46*, 1506, JOC, **1985**, *50*, 1440, T, **1988**, *44*, 619, JACS, **1998**, *120*, 1337, JCSPT1, **2002**, 949, JOC, **2000**, *65*, 5785
TFA	JACS, **2003**, *125*, 13132, SL, **2000**, 1733
CSA, MeOH	ACIEE, **1991**, *30*, 299, T, **1990**, *46*, 4517, JACS, **2003**, *125*, 8112, JACS, **2003**, *125*, 8798, OL, **2002**, *4*, 2981, BMCL, **2003**, *13*, 2519
PPTS	JOC, **1994**, *59*, 1457, TL, **1989**, *30*, 19, JACS, **1997**, *119*, 12425, T, **2002**, *58*, 6433, ACIEE, **1999**, *38*, 3662, T, **2000**, *56*, 7123, SL, **1999**, 780
TsOH	JCSCC, **1993**, 125, TL, **2002**, *43*, 6377, JOC, **2003**, *62*, 7967,
LL-ALPS-SO$_3$H	JOC, **2003**, *68*, 8723
MeOH, CCl$_4$	TL, **1995**, *36*, 6891, TL, **2003**, *44*, 2713
Cu(NO$_3$)$_2$	SC, **1990**, *20*, 757
Ce(NO$_3$)$_3$	SC, **1990**, *20*, 757
Cyclohexene, PdO	TL, **193**, *34*, 243
Alumina	JCSCC, **1987**, 992
SiF$_4$, CH$_2$Cl$_2$	TL, **1992**, *33*, 2289
DDQ, MeCN, H$_2$O	BCSJ, **1994**, 67, 290, JCSPT1, **1992**, 2997
AcBr, CH$_2$Cl$_2$	TL, **1994**, *35*, 2027
TMSOTf	TL, **1990**, *31*, 567

	Cu(OTf)₂, Ac₂O	TL, **2001**, *42*, 5309
	Decaborane	JCSPT1, **2002**, 1223
	CeCl₃·7H₂O, NaI	SL, **1998**, 209
	Ce(OTf)₃, THF, H₂O	TL, **2002**, *43*, 5945
	ZrCl₄, IPA	LOC, **2005**, *2*, 57
	PdCl₂(CH₃CN)₂	TL, **1998**, *39*, 6369
	CeCl₃·7H₂O	OL, **2001**, *3*, 1149
	InCl₃	NJC, **2000**, *24*, 853
	Zn(BF₄)₂	TL, **1999**, *40*, 1985
	ZnBr₂, H₂O	TL, **2002**, *43*, 7151
	ZrCl₄, Ac₂O	TL, **2003**, *44*, 4693
	TBAF	JACS, **1996**, *118*, 6096
	HF-Pyridine	OL, **2001**, *3*, 979
	H₂, Pd/C	TL, **2004**, *45*, 1973, CC, **2003**, 654
	I₂, KOH	T, **2002**, *58*, 6433
	I₂, MeOH	T, **2000**, *56*, 6511
	Br₂, MeOH	SL, **2001**, 1146
	IBr	SL, **1999**, 311
	Bu₄NBr₃, MeOH	OL, **2000**, *2*, 4177
	LiCl, DMF	TL, **1998**, *39*, 327
	TMSOTf, HCO₂DPM, Silica	SC, **2001**, *31*, 2761
	Oxone, MeOH	OL, **1999**, *1*, 1701
61.	TMSOTf, MeOH	TL, **1999**, *40*, 3643
	AcOH	JOC, **2000**, *65*, 5785, JCSCC, **1986**, 497, JOC, **1986**, *51*, 4840
	CSA	OL, **2002**, *4*, 2981, JCSPT1, **2002**, 1693, JCSPT1, **2002**, 1701, BMCL, **2003**, *13*, 2519
	PPTS	JOC, **2001**, *66*, 5875, OL, **2002**, *4*, 4301, MC, **2002**, *133*, 1147, OL, **2003**, *5*, 2335, TL, **1989**, *30*, 19, JACS, **1994**, *116*, 549, JMC, **1992**, *35*, 3280
	TsOH	OL, **2003**, *5*, 2335, ACIEE, **2002**, *41*, 4763, JACS, **2003**, *125*, 15512, JMC, **1992**, *35*, 3388, T, **1995**, *51*, 5193, TA, **1995**, *6*, 559
	HF-Pyr	ACIEE, **2003**, *42*, 3934, ACIEE, **1994**, *33*, 2320
	HF, TEA	TL, **1994**, *35*, 6417
	HF	ACIEE, **2001**, *40*, 901, TL, **2000**, *41*, 3755, OBC, **2003**, *1*, 2348, ACIEE, **2003**, *42*, 1258
	H₂SiF₆	JACS, **1999**, *121*, 205
	TBAF	JACS, **1987**, *109*, 2208

	TMSOTf, TEA, MeOH	TL, **1999**, *40*, 3643
	Cu(OTf)$_2$, Ac$_2$O	TL, **2001**, *42*, 5309
	Zn(OTf)$_2$	TL, **1999**, *40*, 1985
	K$_2$CO$_3$	JACS, **1996**, *118*, 7513
	NaOH	TL, **2000**, *41*, 10013
	TBTU	SL, **1999**, 709
	QFC	JOC, **1997**, *62*, 2628
	Polymeric DCKA	SL, **1999**, 1960
	InCl$_3$, Aq. ACN	NJC, **2000**, *24*, 853
	Cu(OTf)$_2$, Ac$_2$O	TL, **2001**, *42*, 5309
62.	TBAF	BMCL, **2001**, *11*, 1683
63.	MnO$_2$, AlCl$_3$	SC, **1999**, *29*, 4333
	DIBALH	TL, **1992**, *33*, 6259
64.	TBAF, THF	OL, **1999**, *1*, 1431, JOC, **2000**, *65*, 7456, ACIEE, **1998**, *37*, 81, ACIEE, **2003**, *42*, 2521, T, **1998**, *54*, 7127, CC, **1999**, 519, ACIEE, **2001**, *40*, 603, OL, **2003**, *5*, 4477, JACS, **1997**, *119*, 7974, TL, **1998**, *39*, 8633, OL, **1999**, *1*, 1431, OL, **2001**, *3*, 2221, JOC, **2000**, *65*, 4070, JACS, **2002**, *124*, 5654, TL, **2000**, *41*, 775, TL, **1996**, *37*, 9361, TL, **1995**, *36*, 5761, JACS, **1986**, *108*, 8105, TL, **1987**, *28*, 5457, TL, **1992**, *33*, 671
	KF·H$_2$O	JACS, **1994**, *116*, 4697
	H$_2$SO$_4$	ACIEE, **1998**, *37*, 1880
	HCl	JOC, **1980**, *45*, 4797, CPB, **1978**, *26*, 2209
	CSA	ACIEE, **2000**, *39*, 2290, JACS, **2001**, *123*, 9535, CC, **1986**, 874
	TsOH	TL, **1986**, *27*, 5281
	HF-Pyr	SL, **2002**, 2007, ACIEE, **1999**, *38*, 1485
	HF	ACIEE, **1997**, *36*, 1524, JOC, **1992**, *57*, 1070, JCSPT1, **1981**, 2055, JCSPT1, **1981**, 1729, BMCL, **1994**, *4*, 921
	BF$_3$·Et$_2$O	ACIEE, **2000**, *39*, 3656
	P$_2$O$_5$, (MeO)$_2$CH$_2$	JOC, **2003**, *68*, 7967
	MnO$_2$, AlCl$_3$	SC, **1999**, *29*, 4333
	LiAlH$_4$	JOC, **1994**, *59*, 7133
65.	TfOH	JACS, **1997**, *119*, 6739
	HF	TL, **1997**, *38*, 5583
	AcOH	JOC, **1989**, *54*, 3354
	CSA	N, **1994**, *367*, 630
	TBAF	JOC, **1988**, *53*, 5885, SL, **1993**, 20

66. CSA   TL, **1997**, *38*, 8241

    H$_2$SiF$_6$, HF, H$_2$O   JOC, **1993**, *58*, 5130, JACS, **1991**, *113*, 8791

67. HCl   TL, **2000**, *41*, 7667, TL, **2003**, *44*, 7829, JOC, **1980**, *45*, 4797

    CSA   OBC, **2003**, *1*, 4173, OL, **2002**, *4*, 2043

    AcOH   JACS, **1993**, *115*, 4497, JACS, **1992**, *114*, 2260

    HF   JOC, **1997**, *62*, 6098, JOC, **2003**, *68*, 6646, JOC, **1992**, *57*, 1070

    Et$_3$N-3HF   ACIEE, **2002**, *41*, 1062

    H$_2$SiF$_6$, Et$_3$N   TL, **1997**, *38*, 1117, JOC, **1993**, *58*, 5130

    TBAF   CEJ, **2000**, *6*, 3116, SL, **1994**, 967, JACS, **1993**, *115*, 9345

68. H$_2$SiF$_6$, *t*-BuOH, H$_2$O   JOC, **1993**, *58*, 5130

69. Sc(OTf)$_3$, H$_2$O, ACN   SL, **1998**, 1047

    H$_2$SiF$_6$ JACS, **2001**, *123*, 10942

    HCl   JOC, **1980**, *45*, 4797, JACS, **1995**, *117*, 8258

    TsOH   JACS, **1987**, *109*, 7553, JACS, **1991**, *113*, 5337

    PPTS   JCSPT1, **2000**, 2429, TL, **2001**, *42*, 5505, TL, **1995**, *36*, 5271, TL, **1989**, *30*, 19

    AcOH, H$_2$O, THF   T, **1995**, *51*, 3691, JACS, **1992**, *114*, 8464

    HF, MeCN   JOC, **1992**, *57*, 1070

    TBAF   ACIEE, **1991**, *30*, 299

    SiF$_4$, CH$_2$Cl$_2$ TL, **1992**, *33*, 2289

    TMSOTf, TEA   TL, **1999**, *40*, 3643

    TMSOTf   JCSPT1, **2000,** 2429, TL, **1998**, *39*, 6095, TL, **1990**, *31*, 567

    Cu(OTf)$_2$, Ac$_2$O   TL, **2001**, *42*, 5309

    InCl$_3$ NJC, **2000**, *24*, 853

    LiAlH$_4$ TL, **1998**, *39*, 6525, TL, **2000**, *41*, 941

    IBr   TL, **2002**, *43*, 6771

    P$_2$O$_5$, (MeO)$_2$CH$_2$ JOC, **2003**, 68, 7967

    LiCl, DMF   TL, **1998**, *39*, 327

    Polymeric DCKA   SL, **1999**, 1960

    ZnBr$_2$, H$_2$O   TL, **2002**, *43*, 7151

    Zn(BF$_4$)$_2$ TL, **1999**, *40*, 1985

    BF$_3$·Et$_2$O   JCSCC, **1994**, 293

70. HCl   TL, **2003**, *44*, 251, JOC, **2003**, *68*, 2183

    TsOH   JOC, **1987**, *52*, 3541, JOC, **1993**, *58*, 7185, JOC, **1994**, *59*, 2910

    CSA   JCSPT1, **1991**, 667

	AcOH	TL, **1997**, *38*, 1271, JACS, **1998**, *120*, 2553, JACS, **1998**, *120*, 2543
	HCO$_2$H, THF,H$_2$O	TL, **1995**, *36*, 4741
	PPTS	OL, **2000**, *2*, 3023, JACS, **2000**, *122*, 1235, CL, **1989**, 1063, TL, **1995**, *36*, 1981
	TBAF	JACS, **1991**, *113*, 1830, JOC, **1992**, *57*, 5071
	SiF$_4$, CH$_2$Cl$_2$	TL, **1992**, *33*, 2289
	HF·Pyr	TL, **1999**, *40*, 2287, ACIEE, **2003**, *42*, 3934
	TMSOTf	JCSPT1, **2000**, 2429, TL, **1998**, *39*, 6095
	BF$_3$·EtO	TL, **2002**, *43*, 8195
	Sc(OTf)$_3$	SL, **1998**, 1047
	NaIO$_4$	TL, **2002**, *43*, 8727
	Cu(NO$_3$)$_2$	SC, **1990**, *120*, 757
	Ce(NO$_3$)$_3$	SC, **1990**, *120*, 757
	DDQ, MeCN, H$_2$O	JCSPT1, **1992**, 2997
71.	TBAF, AcOH	JACS, **1997**, *119*, 962
	TBAF	TL, **1990**, *31*, 431
72.	CSA	JACS, **1995**, *117*, 634 & 8690
73.	TBAF, AcOH	JACS, **1997**, *119*, 962
74.	LiAlH$_4$	N, **1994**, *367*, 630, JACS, **1995**, *117*, 645
75.	TBAF, AcOH	OL, **2001**, *3*, 929
	TFA,H$_2$O,THF	JACS, **1990**, *112*, 2998
76.	TBAF	OL, **2000**, *2*, 2695, TL, **1995**, *36*, 5777
	SiF$_4$	JOC, **1998**, *63*, 6597
77.	CBr$_4$, IPA, reflux	
	CSA	TL, **2003**, *44*, 8935
	TBAF	ACIEE, **2000**, *39*, 2536, TL, **1995**, *36*, 5777
	SiF$_4$	JOC, **1998**, *63*, 6597
	POCl$_3$-DMF	SL, **2004**, 564, JOC, **2001**, *66*, 693
	(CF$_3$SO$_2$)$_2$O, DMF	SL, **2004**, 564, JOC, **2001**, *66*, 693
	CBr$_4$, MEOH	TL, **1998**, *39*, 5249
	CAN, SiO$_2$	JOC, **2000**, *65*, 5077
78.	TMSOTf, HCO$_2$DPM,	
	Silica	SC, **2001**, *31*, 2761
79.	CSA	ACIEE, **2003**, *42*, 1258
	TFA, H$_2$O, THF	TL, **1994**, *35*, 5849
80.	TBAF	CEJ, **2000**, *6*, 3116, TL, **1996**, *37*, 8069, ACIEE, **1999**, *38*, 3340, JACS, **1982**, *104*, 6818
	LiAlH$_4$	JACS, **2003**, *125*, 2374

81. HF   JOC, **1997**, *62*, 6098, JOC, **2003**, *68*, 6646
82. NaIO₄   TL, **2002**, *43*, 8727
83. NaOH, DMPU, H₂O   ACIEE, **2001**, *40*, 196
    TBAF, AcOH   OL, **2003**, *5*, 3583, TL, **2003**, *44*, 7747
84. TBAF, AcOH   TL, **2003**, *44*, 7747
85. TBAF, AcOH   SL, **2000**, 1306
    NaOH   OL, **1999** *1*, 1491, JOC, **2000**, *65*, 3738, TL, **2001**, *42*, 3223, ACIEE, **2001**, *40*, 196
    n-Bu₄NOH   SL, **2000** 1306
86. TBAF, AcOH   ACIEE, **2002**, *41*, 1787, EJOC, **2001**, 1701, OL, **2001**, *3*, 3149, SL, **2000**, 1306, JACS, **2003**, *125*, 13531, OL, **2000**, *2*, 2575, OL, **2003**, *5*, 3583, ACIEE, **2002**, *41*, 1787, OBC, **2003**, *1*, 3343, TL, **2002**, *43*, 493, TL, **2003**, *44*, 7747, JACS, **1995**, *117*, 1173, JACS, **1993**, *115*, 3558, JOC, **1987**, *52*, 1372
    TBAF   JACS, **1999**, *121*, 9229, JACS, **2002**, *124*, 12806, SL, **2003**, 1500
    NH₄F   JACS, **2003**, *125*, 14722, OL, **2003**, *5*, 3029 TL, **1992**, *33*, 1177
    TAS-F   JOC, **1998**, *63*, 6436
    NaH, Propargyl-OH   TL, **2003**, *44*, 3175
    NaOH, DMPU   ACIEE, **2001**, *40*, 196, TL, **1990**, *31*, 1669
    NaOH   T, **1994**, *50*, 13369
    KOH, 18-C-6   JACS, **2001**, *123*, 10942
    KOH, MeOH   TL, **1992**, *33*, 7701
87. TAS-F   JACS, **1999**, *121*, 9873
88. KOH   JACS, **2003**, *125*, 46
89. HF-Pyr   TL, **2003**, *44*, 5229
    KOH, 18-C-6   JACS, **2001**, *123*, 10942
    NaH/HMPA   TL, **1994**, *35*, 4907
90. LiAlH₄   JOC, **1994**, *59*, 7133
    HF-Pyr   JCSPT1, **2000**, 2429
    TBAF   JACS, **1996**, *118*, 6096
    Br₂, MeOH   SL, **2001**, 1146
91. CSA   JOC, **2003**, *68*, 5754
    HF-Pyr   JCSPT1, **2000**, 2429, TL, **2002**, *43*, 3825, JACS, **1993**, *115*, 5815
    NH₄F   OL, **2003**, *5*, 3029
    TBAF   JACS, **2000**, *122*, 11090, ACIEE, **2003**, *42*, 1258

POCl$_3$, DMF	JOC, **2001**, *66*, 693, SL, **2004**, 564
(CF$_3$SO$_2$)$_2$O, DMF	JOC, **2001**, *66*, 693, SL, **2004**, 564
Alumina	T, **1994**, *50*, 8539
92. NaH, HMPA	TL, **1990**, *31*, 1669

List of Journal Abbreviations

AEICC	*Angew. Chem. Int. Ed., Eng.*
BCSJ	*Bull. Chem. Soc. Jpn.*
BMCL	*Bioorg. Med. Chem. Lett.*
CC	*Chem. Commun.*
CEJ	*Chem. Eur. J.*
CJC	*Can. J. Chem.*
CL	*Chem. Lett.*
CPB	*Chem. Pharm. Bull.*
H	*Heterocycles*
HCA	*Helv. Chim. Acta*
JACS	*J. Am. Chem. Soc.*
JCSCC	*J. Chem. Soc. Chem. Commun.*
JCSPT1	*J. Chem. Soc. Perkins Trans. 1*
JMC	*J. Med. Chem.*
JOC	*J. Org. Chem.*
LA	*Liebigs Ann.*
N	*Nature*
NJC	*New J. Chem.*
OBC	*Org. Biomol. Chem.*
OL	*Org. Lett.*
S	*Synthesis*
SC	*Synth. Commun.*
SL	*Synlett*
T	*Tetrahedron*
TA	*Tetrahedron: Asymm.*
TL	*Tetrahedron Lett.*

INDEX

Greene's Protective Groups in Organic Synthesis, Fourth Edition, by Peter G. M. Wuts and
Theodora W. Greene
Copyright © 2007 John Wiley & Sons, Inc.